Microscopy of Semiconducting Materials 1999

Other Titles in the Series

The Institute of Physics Conference Series regularly features papers presented at important conferences and symposia highlighting new developments in physics and related fields.

Microscopy of Semiconducting Materials 1999

Proceedings of the Institute of Physics Conference held at Oxford University, 22–25 March 1999

Edited by A G Cullis and R Beanland

S
M   M
XI

Institute of Physics Conference Series Number 164
Institute of Physics Publishing, Bristol and Philadelphia

British Library Cataloguing in Publication Data

A catalogue record for this book is available from the British Library.

ISBN 0 7503 0650 5

Library of Congress Cataloging-in-Publication Data are available

Conference Chairmen
    A G Cullis and R Beanland

Honorary Editors
    A G Cullis and R Beanland

Scientific Sponsors
    The Institute of Physics
    The Royal Microscopical Society
    The Materials Research Society

Production: Clare McGurrell
Production Control: Sarah Plenty and Jenny Troyano
Commissioning Editors: Kathryn Cantley and Ann Berne
Editorial Assistant: Victoria Le Billon
Cover Design: Jeremy Stephens
Marketing Executive: Colin Fenton

Published by Institute of Physics Publishing, wholly owned by The Institute of Physics, London
Institute of Physics Publishing, Dirac House, Temple Back, Bristol BS1 6BE, UK
US Office: Institute of Physics Publishing, The Public Ledger Building, Suite 1035, 150 South Independence Mall West, Philadelphia, PA 19106, USA

Printed in the UK by J W Arrowmith Ltd, Bristol

Contents

Self-organized and quantum domain structures

Epitaxy: growth phenomena

Epitaxy: defect formation

Device studies and specimen preparation

Scanning probe microscopy

Advanced scanning electron and optical microscopy

Preface

This volume contains the invited and contributed papers presented at the conference on 'Microscopy of Semiconducting Materials' held at Oxford University from 22–25 March 1999. The event was organized with scientific sponsorship by the Electron Microscopy and Analysis Group of the Institute of Physics, the Royal Microscopical Society and the Materials Research Society. This conference was the eleventh in the series which focuses on the most recent international advances in semiconductor studies carried out by all forms of microscopy.

The advance of modern IC technology continues apace and never has it been more important to have the capability of analysing the very small structures produced. Transmission and scanning electron microscopy remain key elements in the analytical armoury although, especially for the former where cross-sectioning is required, location of individual circuit elements has become a particular problem. In this regard, the introduction of focused ion beam milling for local sectioning has assumed special importance. Furthermore, the wider availability of ultrafine electron probes provided by field emission instruments has broadened the application of high spatial resolution microanalysis, especially in combination with analytical methods such as advanced electron energy loss spectroscopy. The conference highlighted many experimental approaches of these types, together with much detailed work involving many more conventional microscopy applications. Also, work demonstrating advances in the exploitation of scanning probe microscopy (AFM, STM, SCM) was much in evidence. Materials subjected to study ranged from finished electronic devices through to partly processed materials, epitaxial layer structures and nanostructural elements including quantum wires and dots. Overall, the vigorous activity taking place in all semiconductor analysis areas was strongly evident in the presentations given by conference delegates from more than 20 countries.

Each camera-ready manuscript submitted for publication in this volume has been reviewed by at least two referees and modified accordingly. The editors are extremely grateful to the following scientific referees for their rapid and careful work:

M A Al-Khafaji, A Bright, R M Anderson, U Bangert, H Bender, G R Booker, P D Brown, J Bruley, C R Brundle, A Cavallini, H Cerva, D Cherns, A Cornet, R Davis, K Durose, P F Fewster, E A Fitzgerald, R Garcia, P J Goodhew, A Gustafsson, C Hetherington, P B Hirsch, D B Holt, Y Homma, C J Humphreys, J A Mardinly, R T Murray, A G Norman, C E Norman, B Pecz, P Pirouz, T Plamann, V R Raineri, F Ross, J Rouviere, P Ruterana, M Shiojiri, B Sieber, J W Steeds, H P Strunk, K G Tillmann, T Walther, P Werner, G M Williams and C Zanotti Fregonara.

The conference organizers are most grateful for the financial support provided by Hitachi Scientific Instruments Ltd, JEOL (UK) Ltd and FEI Ltd. The conference organization over the whole two-year cycle was underpinned by the meticulous work carried out by J Watts and C Pantlin (Institute of Physics), who deserve very special thanks. Assistance in correcting the proof copies of many manuscripts was ably provided by S Gledhill (University of Oxford).

A G Cullis
R Beanland
September 1999

Inst. Phys. Conf. Ser. No 164
Paper presented at Microsc. Semicond. Mater. Conf., Oxford, 22–25 March 1999

Advances in high resolution imaging and microanalysis of Si, GaAs and GaN

C J Humphreys, A N Bright and S L Elliott

Department of Materials Science and Metallurgy, University of Cambridge, Pembroke Street, Cambridge, CB2 3QZ, UK

ABSTRACT: Some recent advances in instrumentation for electron microscopy and microanalysis are reviewed. Results are presented for dopant imaging using secondary electron contrast in a scanning electron microscopy for Si, GaAs and GaN. It is shown that dopant levels as low as 5×10^{14} cm^{-3} can be revealed using this technique. A variety of analytical TEM methods have been used to study a Ti/Al contact on GaN and the structural nature of this ohmic contact is revealed, in particular the presence of a TiN layer only 1.5 nm thick at the metal-semiconductor interface. EELS has been used to study individual defects in GaN, and it has been shown that a stacking fault known as DB1 has a state in the bandgap and is charged, whereas another fault, DB2, does not have a state in the bandgap and is uncharged.

1. INTRODUCTION

This paper will review some recent advances in techniques and instrumentation in electron microscopy and analysis relevant to semiconductor materials, and it will then focus on recent applications to Si, GaAs and GaN. What is remarkable about the microscopy of semiconductor materials is the pace of advance, clearly revealed in the Microscopy of Semiconductor Materials conference series, known as the Cullis Conferences, which started in 1979. Twenty years later this advance continues even more rapidly, fuelled by the shrinking dimensions of existing semiconductor devices, the discovery of new materials like GaN, and unexpected new microscopy techniques such as doping contrast in the scanning electron microscope (SEM).

2. ADVANCES IN INSTRUMENTATION AND TECHNIQUES

About fifteen years ago the semiconductor device industry moved away from transmission electron microscopes (TEMs) and towards SEMs because of more rapid specimen preparation and because SEMs (particularly FEGSEMs) gave the resolution required. However, the last few years have seen commercial silicon devices with gate oxide thicknesses of only 1 nm, strained layer superlattices with layer thicknesses of about 1 nm and metal-semiconductor contacts with interdiffusion distances of about 1 nm. Hence the semiconductor device industry now routinely needs microscopy with 1 nm resolution in order to characterise the devices it manufactures and occasionally much better resolution (~2Å) is needed. The semiconductor device industry has therefore moved back to TEMs because of their higher resolution and contrast, so that today both SEMs and TEMs are important for the device industry. The return to TEMs has been accompanied by a need for the fast preparation of cross-sectional specimens for TEM, and here the focussed ion beam (FIB) instrument is proving invaluable. Specimen preparation in a FIB can take only a few hours. In addition, the secondary electrons generated by the ion beam in a FIB can be used to produce images of the precise region being machined by the FIB so that cross-sections from particular locations can easily be produced. The FIB is therefore a major advance in TEM and SEM specimen preparation which has been developed mainly because of the market pull of the semiconductor industry for rapid TEM specimen preparation (and also for mask repair, etc).

In addition to the semiconductor market providing motivation and funding for FIB development, the science base has been pushing through the development of a number of important techniques in the last few years. An interesting example is that of the high-angle annular dark-field (HAADF) method of high resolution imaging which was developed a number of years ago, first by Crewe et al (1970) for single atom imaging and then by Pennycook and Jesson (1990) for the imaging of single atomic columns in crystals. The method has never become widespread because it requires a field-emission gun STEM (FEGSTEM), and the main manufacturer, VG, went out of business a few years ago. However demand from scientists for this technique has led to HAADF detectors and high resolution scanning attachments being fitted to the latest generation 200 kV and 300 kV FEGTEMs manufactured by mainstream producers such as Philips and JEOL. These instruments have small electron probes and high efficiency annular dark field detectors so that the HAADF technique, after a long gestation period, is now likely to develop rapidly in popularity in the next few years.

The sensitivity of the secondary electron intensity to the work function of a material has been known for some time (for example Sawyer et al, 1970). However, this sensitivity has only recently been exploited to map the dopant distributions in semiconductors (Perovic et al 1995) using a field-emission SEM. Some recent work in this area will be presented in Section 3.

Another recent development in microanalysis is the use of energy-filtered TEM (EFTEM) to produce high spatial resolution maps of the elemental distributions in a sample. An application of this technique to metal-semiconductor contacts will be discussed in Section 4. A further recent development is electron energy loss spectroscopy (EELS) of individual defects in materials, and some recent results will be given in Section 5.

A most remarkable feature of electron microscopy and analysis is that new techniques continue to appear. One might expect that electron microscopy, like x-ray diffraction, would be a mature subject with the main developments being brighter and brighter sources. But electron microscopy continues to advance on many fronts, and it is the ability to employ a wide range of techniques in solving a particular problem which makes EM so powerful. In addition, it is now possible to have most of these techniques available in the same instrument, rather than having, for example, separate machines dedicated to HREM or Analytical EM. This greatly facilitates studying the same area of the specimen using a variety of methods. Only in the last few years has it become possible to have a high resolution TEM with a field emission gun, STEM attachment, HAADF, electron biprism for holography, Lorentz lens for magnetic specimens, imaging filter, energy dispersive x-ray analysis and more all in the same microscope. Recent advances in instrumentation and techniques have truly been remarkable, and will continue with monochromators and aberration correctors becoming commercially available in the next few years.

3. RAPID 2-D MAPPING OF CARRIER CONCENTRATIONS IN SEMICONDUCTORS USING AN SEM

The semiconductor device industry has a major need for a new technique for the rapid two-dimensional mapping of carrier concentrations. Existing techniques, such as secondary ion mass spectroscopy (SIMS), capacitance-voltage (CV) profiling and spreading resistance (SR) profiling all have disadvantages, for example being slow or having low spatial resolution.

Recently it has been found (Perovic et al, 1995) that the SEM can be used to form dopant profiles and two-dimensional dopant maps from semiconductor materials and devices because the secondary electron intensity is a sensitive function of the dopant concentration. The method is very rapid since the semiconductor structure is simply cleaved, inserted into the SEM and imaged using secondary electrons.

We used a Philips XL 30 SFEG at an accelerating voltage of 1 kV, a working distance of 5-6 mm, and an upper secondary electron detector for the results shown in Figs. 1, 2 and 3. The high brightness from a field emission gun and the low accelerating voltage of 1 kV maximise the dopant contrast. In addition, the secondary electrons which strike the upper secondary electron detector have had to spiral up through the objective lens, and this inherently provides a degree of energy filtering which significantly increases the dopant contrast, relative to the contrast on a standard in-chamber secondary electron detector.

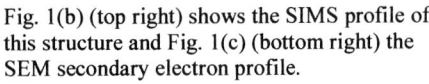

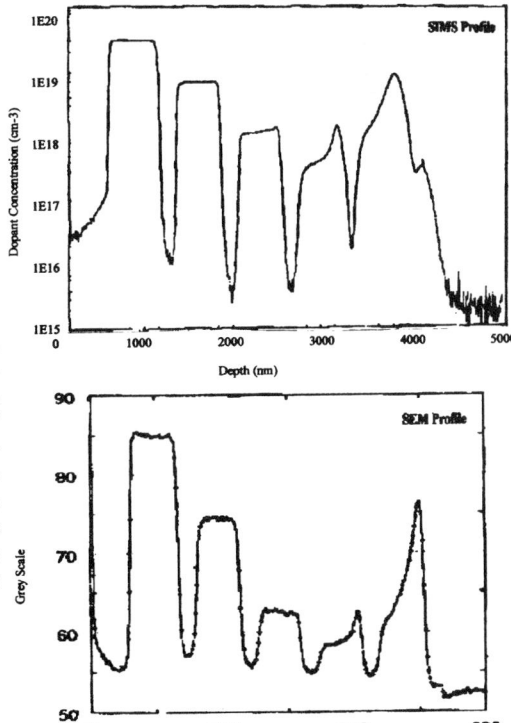

Fig. 1(a) (above) shows a secondary electron image of a silicon structure which contains five B-doped Si layers of concentrations 3×10^{19}, 5×10^{18}, 1×10^{18}, 4×10^{17} and 8×10^{18} cm^{-3} (from the left-hand edge), on an n+ substrate (6-10 mΩcm Sb). The grey region to the right of the image is the substrate. The bright stripes running along the left edge of the sample are the p-type doped layers. The highest doped layer appears brightest in the image.

Fig. 1(b) (top right) shows the SIMS profile of this structure and Fig. 1(c) (bottom right) the SEM secondary electron profile.

Fig. 1(a) shows a secondary electron image from a Si structure p-type doped with five different concentrations of boron. The layers are clearly visible with high contrast. Fig. 1(b) shows the SIMS profile for the same sample, and Fig. 1(c) the SEM secondary electron profile taken from Fig. 1(a). The SIMS profile gives the total concentration of B atoms on an absolute scale whereas the SEM profile gives the concentration of carriers. The SIMS and SEM results are therefore complementary, although it should be noted that for most device applications it is the carrier concentration which is required rather than the total dopant concentration so the SEM technique may be more useful than SIMS, in addition to being much more rapid. Fig. 1(c) is similar to Fig. 1(b) since the number of carriers is proportional to the total dopant concentration.

Fig. 2 shows dopant contrast from GaAs containing nine Si-doped layers of various concentrations. On the original secondary electron image, dopant contrast is visible for the first seven Si-doped layers, i.e. down to a concentration of 5.3 x 10^{14} Si atoms cm^{-3}. This demonstrates the very high sensitivity of the doping contrast method, and it is the highest sensitivity yet published using this technique.

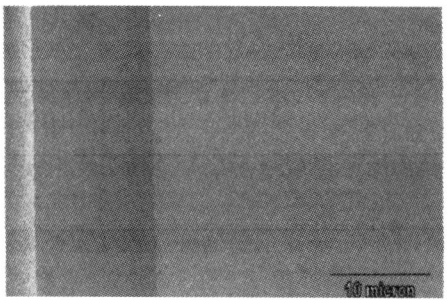

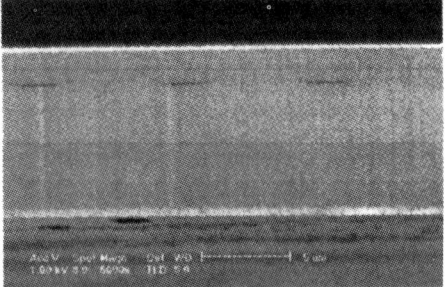

Fig. 2 shows a secondary electron image of a GaAs structure which contains nine Si-doped GaAs layers. The grey region to the right of the image is the undoped substrate. Starting from the substrate, the electron concentration (cm^{-3}) in the layers is 1.8×10^{18}, 3.9×10^{17}, 9.2×10^{16}, 2.3×10^{16}, 1.8×10^{15}, 5.3×10^{14}, 1.7×10^{14} and 5.6×10^{13}. The dark stripes running along the left edge of the sample are the n-type doped layers. The highest doped layer appears darkest.

Fig. 3 shows a LED on an ELOG-GaN sample. From the sapphire substrate (top, dark region) there is ~7.2 μm of undoped GaN, followed by 3.2 μm of Si-doped GaN, a 50 nm GaN/LED MQW and 0.5 μm of Mg-doped GaN. The middle greyish region is the undoped and n-type GaN (these cannot be distinguished from one another) and the bright stripe running along the bottom edge is the p-type doped GaN. The three dark lines that lie horizontally in the image are ELOG stripes. The vertical bright lines are due to surface topography resulting from the cleave.

Fig. 3 shows a secondary electron image of a GaN LED. The image not only reveals details of the device structure, but also maps out the doping within this structure. For example, the image reveals the three SiO_2 stripes (horizontal dark lines in the image) deposited as part of the ELOG (epitaxial lateral overgrowth) process and the p-n junction lying horizontally near the bottom of the figure.

The mechanism for the doping contrast is not yet clear. Perovic et al (1995) emphasised the effects of band-bending at the semiconductor surface. Sealy (1997) produced a model in which the built-in potential from a p-n junction is responsible for the doping contrast. Clearly the different secondary electron signals from p and n regions arise because these electrons have to surmount different energy barriers, and Howie (1999 and private communication) argues that taking into account band-bending and surface patch fields, the energy barrier for secondary electrons from n-type regions is raised by about half the bandgap energy relative to p-type regions, hence secondary electron emission is greater from p-type regions.

A major drawback of the doping contrast method described above is that it is not yet fully quantitative. However a detailed understanding of the contrast mechanisms may enable the technique to be made quantitative. If so, then we will have a rapid technique for characterising semiconductor doping with a sensitivity range from 10^{15} to 10^{21} carriers cm^{-3} (corresponding to 1 atom in 10^7 to 1 atom in 10^2).

4. CHARACTERISATION OF METAL-SEMICONDUCTOR CONTACTS FOR GaN

The analysis of metal contacts on semiconductors is a good example of a situation where a variety of microscopy techniques which provide complementary information are often required. Understanding the process of stable, low resistance ohmic contact formation or rectifying Schottky barrier formation requires correlation of the contact microstructure with the electrical behaviour. In the case of an alloy contact, a polycrystalline structure containing several phases of different composition will generally develop after annealing, together with thin interfacial layers formed at the metal/semiconductor interface. Grain size, film thickness, and the extent of interdiffusion can be determined from bright or dark field conventional TEM, and phase identification can be made using selected area diffraction patterns. To determine the distribution of the different metals deposited after the anneal, energy-filtered TEM is extremely useful, and helps provide chemical analysis of grains already viewed in bright field. Interfacial phases may also be visible in such elemental maps, and comparing dark field images with EFTEM, High Resolution Imaging (perhaps down several zones) and diffraction patterns from the interface area should allow interface phase identification and distribution to be established.

It is known empirically that Ti/Al contacts on n-GaN are ohmic (Fan et al 1996, Lee et al 1998), but the reason for this good ohmic behaviour and the variations observed with different processing conditions are not well understood. We have therefore studied a test contact structure, a 200nm Al / 20nm Ti / GaN ohmic contact annealed in Argon for 60s at 650°C. The specific contact resistance was measured as 1.6×10^{-6} Ωcm^2, which would be good enough for many types of device. The bright field (BF) image in Fig. 4 shows the resulting large-grained polycrystalline structure, but in this Al/Ti alloy system we also need information about the location of the elements in the annealed structure. Energy-filtered TEM is an efficient method for observing the elemental distribution over the same area, as we can see from Ti (Fig. 5) and Al (Fig. 6) jump ratio EFTEM images taken from the same area. These show alternating light and dark regions corresponding to the grain structure observed in BF. It is clear that the large grains contain both Al and Ti, and in the intermediate regions are Al only. The identity of the two regions as Al metal and Al$_3$Ti is confirmed by selected area diffraction patterns taken from individual grains (Fig. 7). This combination of techniques can be used to effectively evaluate such metallisation microstructures.

The electrical properties of metal contacts are largely determined by the behaviour of the metal/semiconductor interface, and so characterisation of interfaces is of crucial importance. In this contact, dark field imaging of the interface area (Fig. 8) shows the presence of a polycrystalline interfacial layer about 1.5 nm thick, with larger grains interspersed along it. To establish the identity of this phase, which is present at the interface of GaN with both the Al and Al$_3$Ti grains, energy filtered imaging was performed, and a Ti jump-ratio EFTEM image (Fig. 9) shows the presence of Ti in the interfacial layer between an Al grain and the GaN. To identify the phase more conclusively, high resolution TEM was performed, from which layer thicknesses and plane spacings and angles can be measured. HREM down different zones provides additional information. In this case the interface phase is TiN, shown here down <110> (Fig. 10). SAD patterns down the same zone using a very small selected area aperture show diffraction spots from the TiN phase. Thus the use of multiple techniques provides complementary or reinforcing information which allows a much greater degree of confidence in phase identification in such contact structures.

TiN is a metallic phase whose presence is thought to be very important in the formation of effective Ti/Al ohmic contacts to n-GaN (Ruvimov et al 1996).

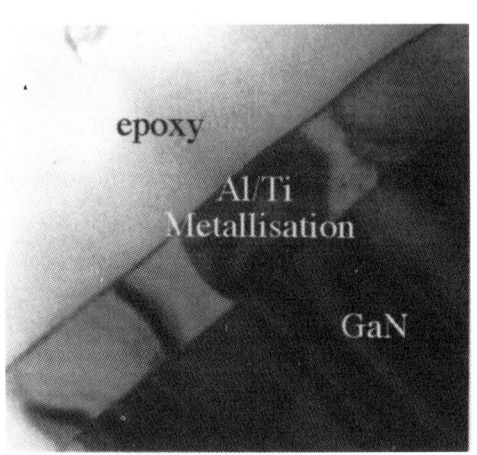

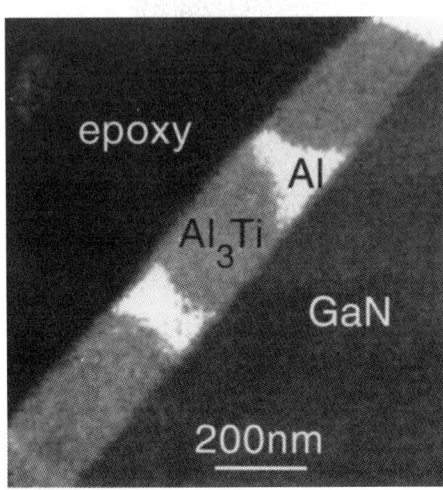

Fig. 4 Bright field image of the 200 nm Al/ 20 nm Ti/GaN ohmic contact after annealing in Ar at 650°C for 60 s. Dark grains are Al₃Ti and light grains Al. The top surface is quite flat.

Fig. 5 Energy Filtered TEM Ti jump-ratio image showing the Ti distribution in the contact (same area as Fig. 4).

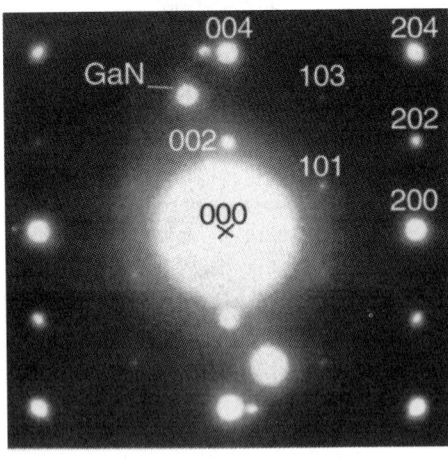

Fig. 6 Energy Filtered TEM Al jump-ratio image showing Al grains (bright), and the presence of Al in the Ti-rich regions of Fig. 5.

Fig. 7 <010> diffraction pattern from an Al₃Ti grain. The 0002 GaN reflection is also present, but shows no obvious Al₃Ti/GaN epitaxial relation.

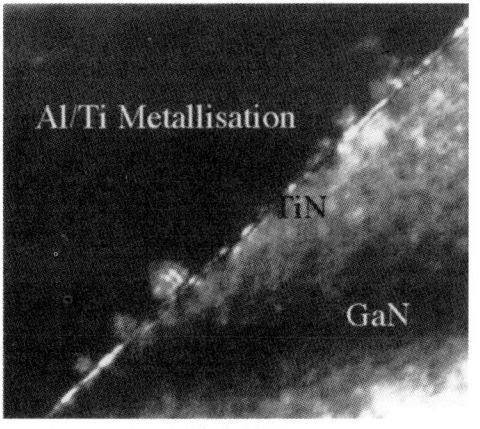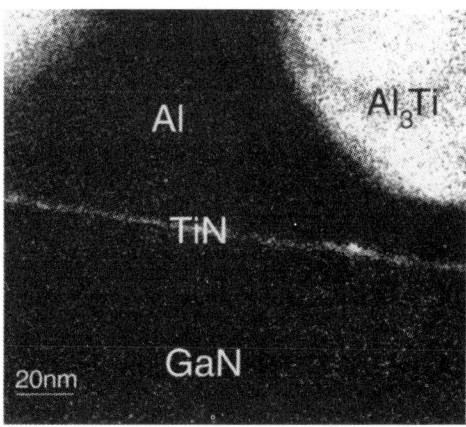

Fig. 8 Dark field image from the interface region showing a patchy interfacial layer (about 1.5 nm thick) with some larger grains spaced along it.

Fig. 9 EFTEM Ti map from the interface area. A patchy interfacial layer rich in Ti is visible. Here Ti has diffused to form Al₃Ti away from the interface, and so the interfacial layer can be easily seen.

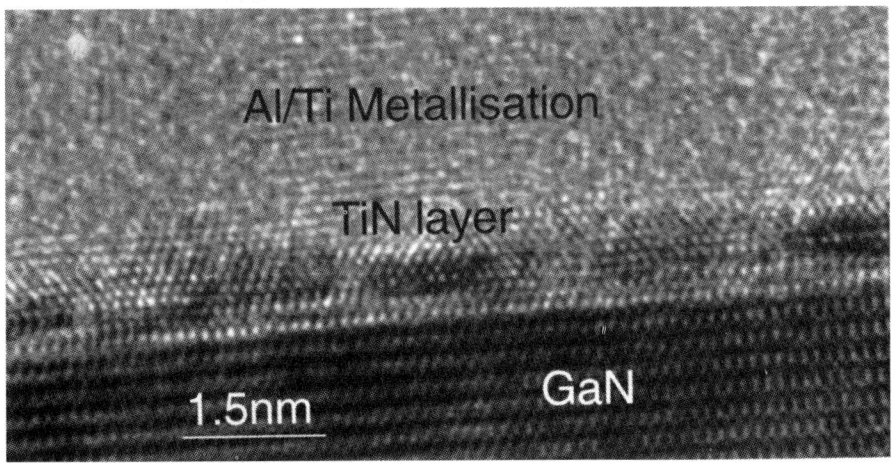

Fig. 10 High Resolution lattice image of the interface layer, from which cubic-TiN can be identified. Orientation is <110> TiN//<11-20>GaN

5. EELS OF INDIVIDUAL DEFECTS IN GaN

As is well known, LEDs fabricated from GaN can be highly defective yet still emit intense light. Even with a dislocation density of 10^{10} cm^{-2}, such GaN devices are very bright, whereas for light emission from other optoelectronic materials (e.g. InP, GaAs, etc.) the dislocation density has to be less than 10^3 cm^{-2} or the dislocations quench the light emission. The reason for the tolerance of GaN-based LEDs to dislocations and stacking faults is not understood. For efficient GaN-based lasers the dislocation and stacking fault densities must be low, but the relative importance of different types of dislocations and stacking faults on laser efficiency is not yet clear. It is therefore important to understand the effects of defects on the local electronic structure of GaN. Electron energy loss spectroscopy (EELS) using the focused electron probe from a field-emission gun scanning transmission electron microscope (FEGSTEM) has the potential to provide information on the modification of the electronic structure by individual defects at high spatial resolution.

There are two equivalent approaches to the interpretation of EELS spectra (see Egerton, 1996). First, the intensity of the energy loss spectrum is proportional to the joint density of states, i.e.

$$I \propto |\mathbf{M}|^2 \rho_f \cdot \rho_i \tag{1}$$

where M is the matrix element for the transition from the initial state i to the final state f, and ρ_i and ρ_f are the density of the initial states and of the unoccupied final states respectively. Assuming parabolic bands for the conduction and valence bands then for interband transitions to the conduction band,

$$\rho_f \propto (E - E_g)^{1/2} \tag{2}$$

Fitting the low loss EELS spectrum to the expected band shape therefore yields the energy gap E_g, and for bulk gallium nitride this was shown to be 3.4 ± 0.3 eV (Natusch et al 1997). Similarly, by recording spectra of individual stacking faults in GaN we have shown that the fault known as DBI gives rise to a state in the band gap whereas the fault called DB2 does not. DB1 has a state about 0.6 eV above the valence band edge or about 0.6 eV below the conduction band edge, but without further information, EELS cannot distinguish between these (in the first case the incident beam is exciting the 2.8 eV transition from a filled acceptor state up to the conduction band; in the second case the beam is exciting the 2.8 eV transition from the top of the valence band to an empty donor state. In both cases EELS reveals a loss process corresponding to 2.8 eV).

The second approach to EELS is the dielectric formulation, which relates the intensity of the electrons to the dielectric constant ($\varepsilon = \varepsilon_1 + i\varepsilon_2$),

$$I \propto -\mathrm{Im}(1/\varepsilon) = \varepsilon_2/|\varepsilon|^2 \tag{3}$$

Kramers-Kronig Analysis (KKA) then yields both the real (ε_1) and the imaginary (ε_2) parts of the dielectric constant as a function of the energy loss $\Delta\varepsilon$.

For wurtzite GaN both the real and imaginary parts of the dielectric function measured as outlined above using EELS (Natusch et al 1997) agree well with optical measurements using synchrotron x-ray sources. However, the major advantage of EELS over synchrotron sources is that, using a focussed probe, the real and imaginary parts of the dielectric function can be measured from a single defect.

A series of EELS spectra were recorded from DB1 and DB2 type stacking faults in MBE grown GaN. The stacking faults were oriented parallel to the incident beam so that they were viewed end-on. Spectra were recorded along an imaginary line of length 12 nm perpendicular to the fault with the electron probe being positioned every 0.4 nm along this line. For comparison a similar series of EELS spectra were recorded every 0.4 nm along an imaginary line of length 12 nm in a region of perfect crystal. Kramers-Kronig analysis was performed on the spectra.

The stacking fault DB1 exhibited large changes in both the real and the imaginary parts of the dielectric constant compared with the perfect crystal (Natusch et al 1998a), and a difference in bandgap of 0.6 ± 0.2 eV was observed at the fault. The effective number of valence electrons per atom on the fault is significantly different to that off the fault, indicating that fault DB1 is negatively charged (Natusch et al, 1998b). The results indicate that this charged fault is highly screened by layers of opposite charge within 1 nm of the fault on each side (Natusch et al 1998c). The fault may therefore act like a p-n-p transistor on a nanometre scale.

On the other hand, our similar studies of fault DB2 show that it is uncharged, and does not give rise to a state in the bandgap. It therefore appears that fault DB1 will have a much greater effect on the electrical and optical properties of GaN than DB2. It may also be significant that MBE grown GaN has a much higher density of DB1 faults than MOCVD grown material, which correlates with MBE grown GaN yielding much poorer quality devices that MOCVD grown material.

It is clear that EELS is a very powerful technique for the study of the electrical properties of individual defects in materials.

ACKNOWLEDGEMENTS

The authors with to thank DA Ritchie, Cavendish Laboratory, University of Cambridge; WB de Boer, Philips Research Laboratories, Eindhoven; and T Foxon, Department of Physics, University of Nottingham for the provision of GaAs, Si and GaN samples respectively. They are also grateful to R Beanland, R Broom and GA Botton for discussions. Financial support from Philips, Marconi Materials Technology and the EPSRC are gratefully acknowledged.

REFERENCES

Crewe A V, Wall J and Langmore J 1970 Science **168**, 1338

Egerton R F 1996 Electron Energy-Loss Spectroscopy (Plenum Press: New York and London)

Fan Z, Mohammad S N, Kim W, Aktas O, Botchkarev A E and Morkoç H 1996 Appl. Phys. Lett. **68**, 1672

Howie A 1999 Proc. Microscopy and Microanalysis '99 **5**(2), Portland, Oregon, 662

Lee H J, Yu S J, Asahi H, Gonda S-I, Kim Y H, Ree J K and Noh S J 1998 J. Electronic Mat. **27**, 829

Natusch M K H, Botton G A and Humphreys C J 1997 Inst. Phys. Conf. Ser. **157**, 213

Natusch M K H, Botton G A and Humphreys C J 1998a Proc. Int. Conf. Electron Microsc. (ICEM14) Cancum Mexico **3**, 647

Natusch M K H, Botton G A and Humphreys C J 1998b Proc. Int. Conf. Electron Microsc. (ICEM14) Cancun Mexico **3**, 391

Natusch M K H, Botton G A, Broom R F, Brown P D, Tricker D M and Humphreys C J 1998c Mat. Res. Soc. Symp. Proc. **482**, 763

Pennycook S J and Jesson D E 1990 Phys. Rev. Lett. **64**, 938

Perovic D D, Castell M R, Howie A, Lavoie C, Tiedje T and Cole J S W 1995 Ultramicroscopy **58**, 104

Ruvimov S, Liliental-Weber Z, Washburn J, Duxstad K J, Haller E E, Fan Z-F, Mohammad S N, Kim W, Botchkarev A E and Morkoç H 1996 Appl. Phys. Lett. **69**, 1556

Sawyer G R and Page T F 1978 J. Mat. Sci. **13**, 885

Sealy C P, 1997 DPhil Thesis, University of Oxford

Tricker D M, Natusch M K H, Boothroyd C B, Xin Y, Brown P D, Cheng T S, Roxon C T and Humphreys C J 1997 Inst. Phys. Conf. Ser. **157**, 217

Inst. Phys. Conf. Ser. No 164
Paper presented at Microsc. Semicond. Mater. Conf., Oxford, 22–25 March 1999
© *1999 IOP Publishing Ltd*

Atomic resolution microscopy of semiconductor defects and interfaces

E M James[1], N D Browning[1], Y Xin[1], J L Reno[2] and A G Baca[2]

[1]Department of Physics, University of Illinois at Chicago, 845 West Taylor Street, Chicago, IL 60607-7059, USA.

[2]Sandia National Laboratories Albuquerque, NM 87185, USA.

ABSTRACT: The optical arrangement of the scanning transmission electron microscope (STEM) allows formation of incoherent images by use of a large annular detector. Here we show this capability in the imaging of defects in GaN and the interfacial region of an Au/GaAs ohmic contact. A resolution of around 0.15 nm is attained. Such "Z-contrast" images show strong atomic number contrast and allow the probe to be positioned accurately at the defect or interface for the purpose of performing high spatial resolution electron energy-loss spectroscopy (EELS).

1. INTRODUCTION

The ability to routinely determine atomistic and electronic structures of defects and interfaces is one of the goals of contemporary electron microscopy. In semiconductor devices the overall performance is often strongly affected by these regions of the specimen. For instance, the propensity for undesirable electron hole recombination at defects, and the height of Schottky barriers at interfaces are correlated to their electronic structures. As a first step towards understanding these effects at the atomic level, it is desirable to be able to form images of the relevant structure. One method to obtain such atomic resolution images is Z-contrast, high-angle annular dark field (HAADF) imaging (Pennycook *et al* 1997). This requires the microscope to operate in small probe-forming mode: resolution is given by the probe size. For crystalline materials oriented close to a zone axis, the images are generally easy to interpret qualitatively (intensity peaks corresponding to the atomic column locations): strong atomic number (Z) contrast is also shown. The limited information on chemical composition obtained from the image can be greatly augmented by then performing electron energy-loss spectroscopy (EELS). At high spatial resolutions, the optimal probe-forming microscope settings for EELS are identical to those for Z-contrast imaging. Switching between imaging and spectroscopy is therefore easy.

In the case of most semiconductor defects and interfaces, a resolution of between 0.1 nm and 0.2 nm is desirable for clear imaging of the structure. To be able to generate enough current in such a small probe, a high brightness, field emission source is used. Here we demonstrate the formation of Z-contrast images at around 0.14 nm resolution on the 200 kV JEOL JEM-2010F: a widely available instrument with a Schottky field emission source. As illustration, an image of the $\{1\bar{1}20\}$ prismatic stacking fault in GaN is presented. Also, shown are atomic resolution images from an ohmic contact between Au and GaAs showing faceting of the interface on the atomic scale.

2. SMALL PROBE FORMATION

Figure 1 shows the experimental arrangement for HAADF imaging. A small probe is incident on the specimen and the high angle scattering is collected by a large annular detector. The image is formed serially by displaying collected intensity for each position of the probe as it is scanned in a raster over the specimen. The smallest probe size is determined by four factors: spherical aberration of the probe-forming optics; probe convergence angle; incoherent broadening due to finite brightness of the electron source; and incoherent broadening due to electrical and mechanical instabilities. The microscope we used has a spherical aberration coefficient (when the lenses are set to form a probe) experimentally determined to be 0.57 mm (James *et al* 1998). From wave optical calculations, the optimum aperture size at 200 kV is then 12 mrad and a probe size at Scherzer focus of around 0.13 nm is expected. In general though, it is the incoherent probe broadening effects that dominate the final probe size. However, if a large demagnification of the electron source is set, using the condenser lenses and electrostatic gun lens, we find it is possible to largely avoid this limitation on an unmodified JEM 2010F and approach the Scherzer limit of 0.13 nm, still with enough probe current for imaging (James and Browning 1999).

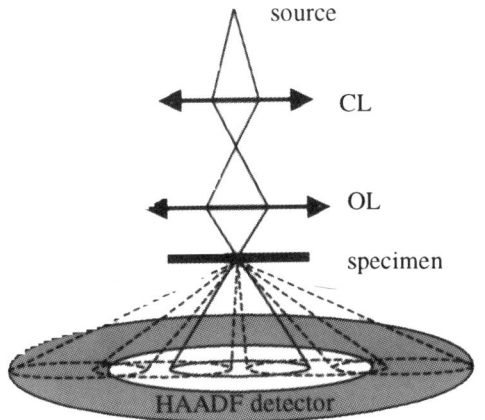

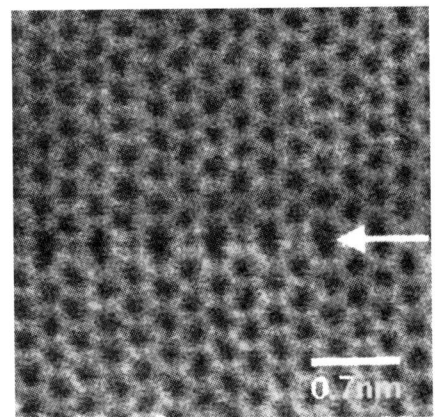

Fig. 1: Schematic illustration of the optical arrangement for HAADF imaging. CL=condenser lens system OL=objective lens pre-field

Fig. 2: Prismatic stacking fault in GaN viewed down <0001>

3. IMAGES OF SEMICONDUCTOR DEFECTS AND INTERFACES

Figure 2 shows the image of a prismatic stacking fault in GaN, with the beam oriented parallel to the <0001> zone axis. The specimen was grown by MBE on a GaP (111) substrate. The electronic structures of these faults, which lie on $\{1\bar{1}20\}$ planes, and also of threading dislocations, which are present with a high density, are currently under investigation since it is desirable to know if they act as electron hole recombination sites and impede the light emitting efficiency of GaN devices (Natusch *et al* 1998). Figure 2 is a raw image with a 4.3 s acquisition time and shows the structure as previously directly observed in the 300 kV STEM (Nellist *et al* 1997) and as predicted by the analysis of Cherns *et al* (1998).

The second structure imaged was the interface in a sample of Au grown on n-type GaAs (001) by MBE. This ohmic contact was alloyed at 420°C for 15 seconds after growth. The interface region exhibits spiking with Au protruding approximately 50 nm into the GaAs. The Au grains are aligned with the $(1\bar{1}0)$ plane approximately parallel to the interface. Figure 3a shows a Z-contrast image at relatively low magnification showing one of the Au spikes, and the atomic resolution structure of the interface 50 nm away from the spike is shown in Fig. 3b. The 256×256 image was acquired in 4.3 s. The image shows Au terminating on the $(1\bar{1}0)$ plane which results in a lattice mismatch of 2.5% with respect to the (001) terminated GaAs. An atomic step in the interface is present near the image center and the resolution here is approximately 0.15 nm. In Fig. 3c an enlarged section of one of the protruding spikes is shown. The Au forms a (111) tilt boundary with GaAs (the angle is

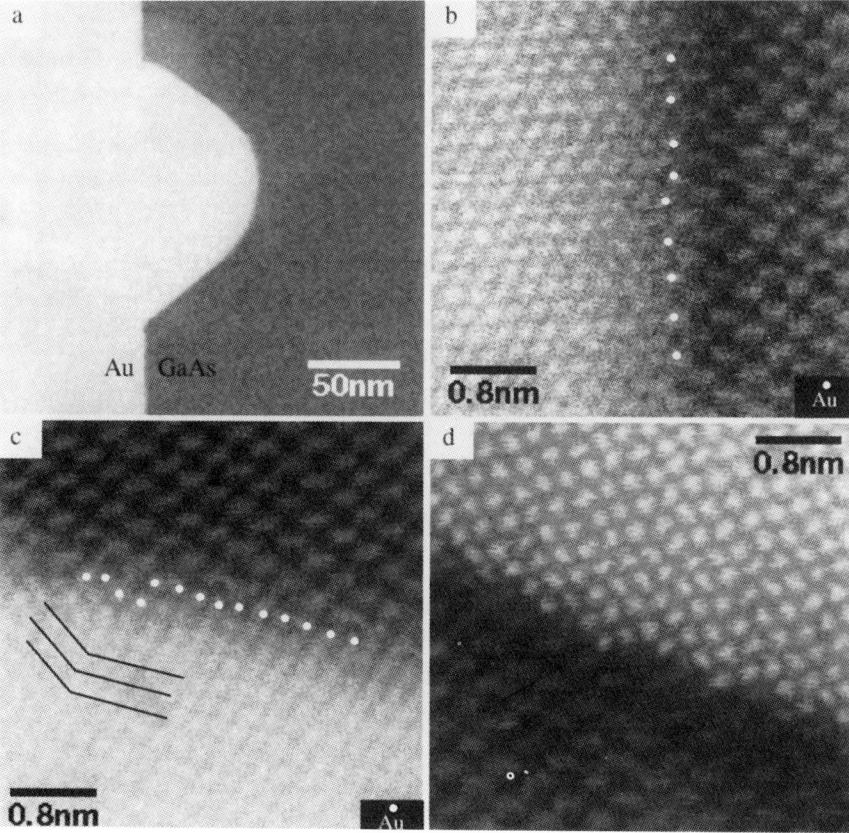

Fig. 3: (a) Raw Z-contrast image at low magnification showing Au metal deposited on the GaAs substrate and the occurrence of spiking. (b) Raw Z-contrast image of the Au/GaAs interface at approx. 0.15 nm resolution (c) Raw atomic resolution Z-contrast image from the Au spike upper region. A (111) tilt boundary is formed with the GaAs with regularly spaced dislocation cores along the interface. (d) Raw Z-contrast image of the lower spike region. Here the Au interface is faceted along the close packed {111} planes.

approximately 18°) and a twin boundary is seen in the bulk Au where the angle of the interface changes. Dislocation cores are visible in the raw image, lying at the interface approximately every 0.5 nm along the interface. Finally, Fig. 3d shows a third example of the interface structure along one side of a spike. Au is faceted along the close packed {111} plane with steps every one to two atomic columns.

4. DISCUSSION AND CONCLUSIONS

Atomic resolution HAADF imaging is possible on a 200 kV transmission electron microscope fitted with a standard Schottky field emission gun and an annular dark field detector. Defect structures in GaN and the atomistic nature of an Au contact to GaAs, in the vicinity of an Au spike, have been imaged at a resolution of around 0.15 nm.

The inelastic scattering signal sampled by EELS is mainly confined to low angles: therefore, it can be collected simultaneously with or immediately following Z-contrast image acquisition. Using the image as a map, the probe can then be stopped at a particular position (such as at a dislocation core or site at an interface) and a spectrum acquired with the same microscope settings. Typically, most of the scattering not collected by the dark field detector is used, in order to approach incoherent EELS conditions.

Our JEM-2010F instrument is equipped with a Gatan imaging filter. In a 0.2 nm probe we observe a current of approximately 15 pA: an energy resolution of around 1 eV is obtainable (a measurement of the full-width-half-maximum of the zero-loss peak). This performance is sufficient to begin examining fine structure details for a number of core-loss edges at atomic spatial resolution. Work is under way to investigate the electronic structure of GaN threading dislocations and the role of Ni in Ni/Au/Ge contacts to GaAs.

This research is funded by NSF grants DMR-9601792 and DMR-9733895.

REFERENCES

Cherns D, Young W T, Saunders M, Steeds J W, Ponce F A and Nakamura S 1998 Phil. Mag. **A77**, 273

James E M, Browning N D, Nicholls A W, Kawasaki M, Xin Y and Stemmer S 1998 J. Electron Microsc. **47**, 561

James E M and Browning N D 1999 Ultramicroscopy in press

Natusch M K H, Botton G A and Humphreys C J, ICEM14, **III**, 39

Nellist P D, Xin Y and Pennycook S J 1997 Inst. Phys. Conf. Ser. **153**, 109

Pennycook S J, Jesson D E, McGibbon A J and Nellist P D 1997 J. Electron Microsc. **45**, 36

Inst. Phys. Conf. Ser. No 164
Paper presented at Microsc. Semicond. Mater. Conf., Oxford, 22–25 March 1999
© 1999 IOP Publishing Ltd

Impact of strain relaxation induced local crystal tilts on the quantitative evaluation of microstructure by high-resolution transmission electron microscopy

K Tillmann, M Lentzen* and R Rosenfeld*

Centre for Microanalysis, University of Kiel, Kaiserstraße 2, D-24143 Kiel, Germany
*Institute for Solid State Research, Research Centre Jülich, D-52425 Jülich, Germany

ABSTRACT: The reliability of strain profiles obtained by a quantitative analysis of lattice fringe spacings from high-resolution micrographs is discussed. Focusing on highly lattice mismatched GaAs/InAs/GaAs heterostructures the layer local strain distribution is calculated by finite element method simulations to determine atom positions in elastically relaxed transmission electron microscopy specimens. By analysing simulated images a significant decoupling between the contrast pattern motifs and the layer structure is found for relevant imaging conditions, which may result in an incorrect determination of strain profiles and layer compositions when examining experimental micrographs.

1. INTRODUCTION

High-resolution transmission electron microscopy (HRTEM) offers a unique possibility of locally determining the geometry, the composition and the elastic strain of pseudomorphic thin layers on an almost atomic scale. In order to calculate strain profiles across heterointerfaces several approaches based upon locating contrast peaks in HRTEM images have been evolved in recent years. These peak finding methods (Bierwolf et al 1993, Bayle et al 1994, Jouneau et al 1994, Robertson et al 1995) rely on fitting a two-dimensional reference lattice to a subset of peaks associated with a non-distorted region of the specimen and measuring local deviations from reference lattice positions. The lattice displacements are then used to determine the strain distribution of the crystal and to calculate compositional maps by use of the continuum theory of elasticity. The application of corresponding algorithms to highly lattice-mismatched heterostructures has given evidence for strain and compositional gradients of $In_xGa_{1-x}As$ layers epitaxially grown on GaAs substrates (Tillmann et al 1996, Rosenauer et al 1997, Kret et al 1998), which have been attributed to an indium segregation during the growth process.

On the other hand, high-resolution imaging requires specimen thicknesses $t \le 30$ nm along the direction of the electron beam and micrographs must not be interpreted as a direct image of the layer structure, i.e. fringe spacings may be altered by contrast pattern changes across heterointerfaces, by Fresnel diffraction due to the discontinuity of the mean crystal potential, by local crystal tilts and by surface relaxation of samples. For instance, it is well known that the elastic relaxation of strained layers at the free surfaces of specimens will change the strain state of mismatched super lattices (Gibson and Treacy 1984) and can lead to an anomalous contrast in conventional micrographs taken under two-beam diffraction conditions (Perovic, Weatherly and Houghton 1991). In the present study the influence of thin-foil relaxation on the image contrast of high-resolution micrographs is discussed. The accuracy of strain profiles gained by measuring fringe spacings of simulated images is judged by a comparison with the real strain state of the specimen used as input for image simulations. It is demonstrated that strain profiles may contain severe artificial strain gradients along the direction of compositional modulation for a variety of relevant imaging conditions.

2. PROCEDURE

Firstly, a model system consisting of an InAs layer with a thickness of $d = 6.0583$ nm equivalent to ten unit cells and embedded in a GaAs matrix assuming atomically sharp interfaces is chosen. The lattice-parameters of both materials result in a lattice mismatch of $f = 0.06686$. The actual wedge-shaped sample geometry is approximated by a plate since standard preparation procedures usually result in acute wedge angles which assure negligible variations in the specimen thickness over a sufficiently small field of view. Unit cell edge coordinates of elastically relaxed samples are determined by finite element simulations taking the anisotropic elastic material properties into account. A sample characterized by a limited thickness t along the [110] direction and with a compositional modulation along the [001] direction is presumed. The finite element mesh is scaled in unit cell dimensions to ensure the calculation of local strains on an almost atomic scale. Details are given elsewhere (Tillmann, Lentzen and Rosenfeld 1999). The contour representation of the elastic strain component ζ_{001} in fig. 1 (a) shows that the lattice planes show a banana-like bending close to the interface with local tilt angles ranging between $\alpha \approx 15$ mrad for $t/d \rightarrow 0$ and $\alpha \rightarrow 0$ mrad in the case of a specimen-to-layer thickness ratio of $t/d \geq 6$.

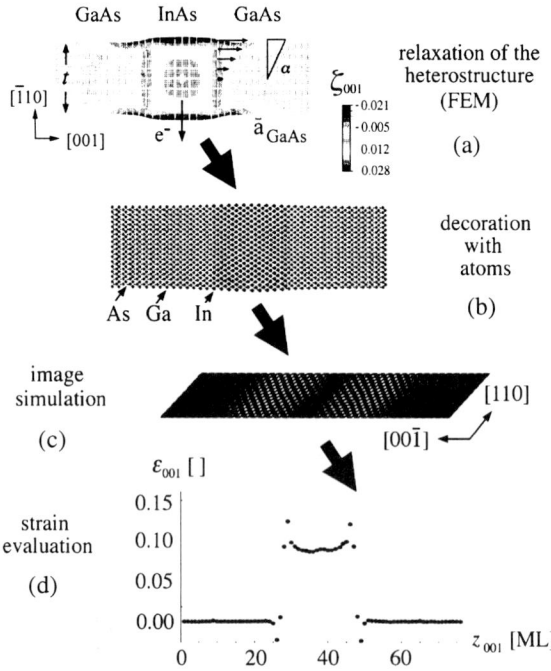

Figure 1: Schematic representation of the four basic steps of the analysis. (a) Finite element simulations are applied to calculate the strain distribution inside an elastically relaxed HRTEM specimen assuming an electron wave e^- propagating along the [110] direction. The heterostructure is meshed in unit cell dimensions. Numerical values of the elastic strain component ζ_{001} along the [001] direction can be taken from the contour legend for a specimen thickness of $t = 6.4$ nm. (b) The FEM data are used to calculate actual atom positions and to create a three-dimensional supercell. Finally, high-resolution images are simulated (c) and subjected to a digital image processing routine (d) calculating lattice strain profiles $\varepsilon_{001}(z_{001}) = f(z_{001}) + \zeta_{001}(z_{001})$ along the direction of the compositional modulation.

Secondly, a supercell (fig 1 (b)) is created for sequencing image simulations. This is performed by a finite element to supercell converter that decorates the finite element mesh with atoms. In detail unit cell edge coordinates inside the front (110) plane follow directly from the finite element data and are decorated with arsenic atoms while gallium, indium and arsenic atom positions inside this plane are interpolated from nearest unit cell edge coordinates with regard to the crystal structure. Atom positions along the [110] direction are calculated from the periodic extrapolation of the front plane coordinates with a period length of the GaAs lattice parameter therefore preserving an InAs layer fully strained along the [110] direction.

Subsequently, high-resolution images (fig. 1 (c)) are calculated using electron-optical parameters typical for a JEOL 4000EX electron microscope operating at an acceleration voltage of 400 kV, i.e. 1.0 mm for the coefficient of spherical aberration, 0.9 mrad for the semi-angle of beam convergence, 10 nm for the halfwidth of a Gaussian spread of focus and 5.2 nm^{-1} for the radius of the objective aperture. Absorption constants of 0.04 and Debye-Waller factors of 0.006 nm^{-2} are chosen for all atoms. Multi-slice calculations with a slice thickness of 0.1 nm are used to determine the electron wavefield at the specimen exit surface. Since the difference in lattice-parameters of InAs and GaAs is in the order of the electron wavelength of 1.6 pm we have to ensure a sufficiently high image discretization in order to measure strains accurately (Walther and Humphreys 1995) and we choose a sampling of 320 picture elements per nanometre.

Finally, the simulated images are analysed by an image processing routine (fig. 1 (d)) to measure the local distribution of the image strain component ε_{001} along the direction of the compositional modulation. The procedure (Tillmann et al 1996) involves the detection of peak positions by a centre-of-mass analysis within contrast dots which is followed by fitting a two-dimensional reference lattice to the subset of peaks associated with unstrained GaAs unit cells far away from the interfaces. Displacement vectors $u = r_{pos} - r_{ref}$ between the measured peak positions r_{pos} and extrapolated reference lattice coordinates r_{ref} are then used to calculate strains $\varepsilon_{001} = f + \zeta_{001}$ $= \partial u_{001} / \partial z_{001}$ as a function of monolayer positions z_{001} along the [001] direction.

3. SIMULATED IMAGES AND RESULTING STRAIN PROFILES

Although simulations have been carried out for a wide range of imaging parameters and compositions we restrict to some representative results. Fig. 2 shows the projected potential map and a defocus series of images as well as corresponding strain profiles for a specimen thickness of $t = 6.4$ nm. In case of defoci $\Delta f = -10$ nm, -30 nm, -90 nm the images show only minor differing fringe motifs in regions of different composition while for other focusing conditions ($\Delta f = -50$ nm, -70 nm) the different scattering properties of GaAs and InAs result in contrast pattern differences which are, however, not directly linked to the local layer stoichiometry since the inhomogeneous elastic relaxation of the specimen generates contrast changes not only at the abrupt interfaces but within each layer of the heterostructure.

The strain profile obtained by measuring fringe spacings between the arsenic positions of the projected potential map is mirror-symmetric with respect to the centre (001) plane thus reflecting the real strain state of the specimen. While large parts of both GaAs layers are unstrained by definition ($\varepsilon_{001} = 0$) since they form the reference lattice, the average strain of the InAs layer calculated as $<\varepsilon_{001,InAs}> = \Sigma \, \varepsilon_{001} \, (z_{001}) \, / \, \Delta z_{001}$ with $\Delta z_{001} = 20$ denoting the layer thickness amounts to $<\varepsilon_{001,InAs}> = 0.105$. Due to thin-foil relaxation this value is distinctly smaller compared with that expected for a completely strained layer predicting $<\varepsilon_{001,InAs}> = 0.1397$. When strains are determined from the HRTEM images defocusing, however, has a strong impact on the

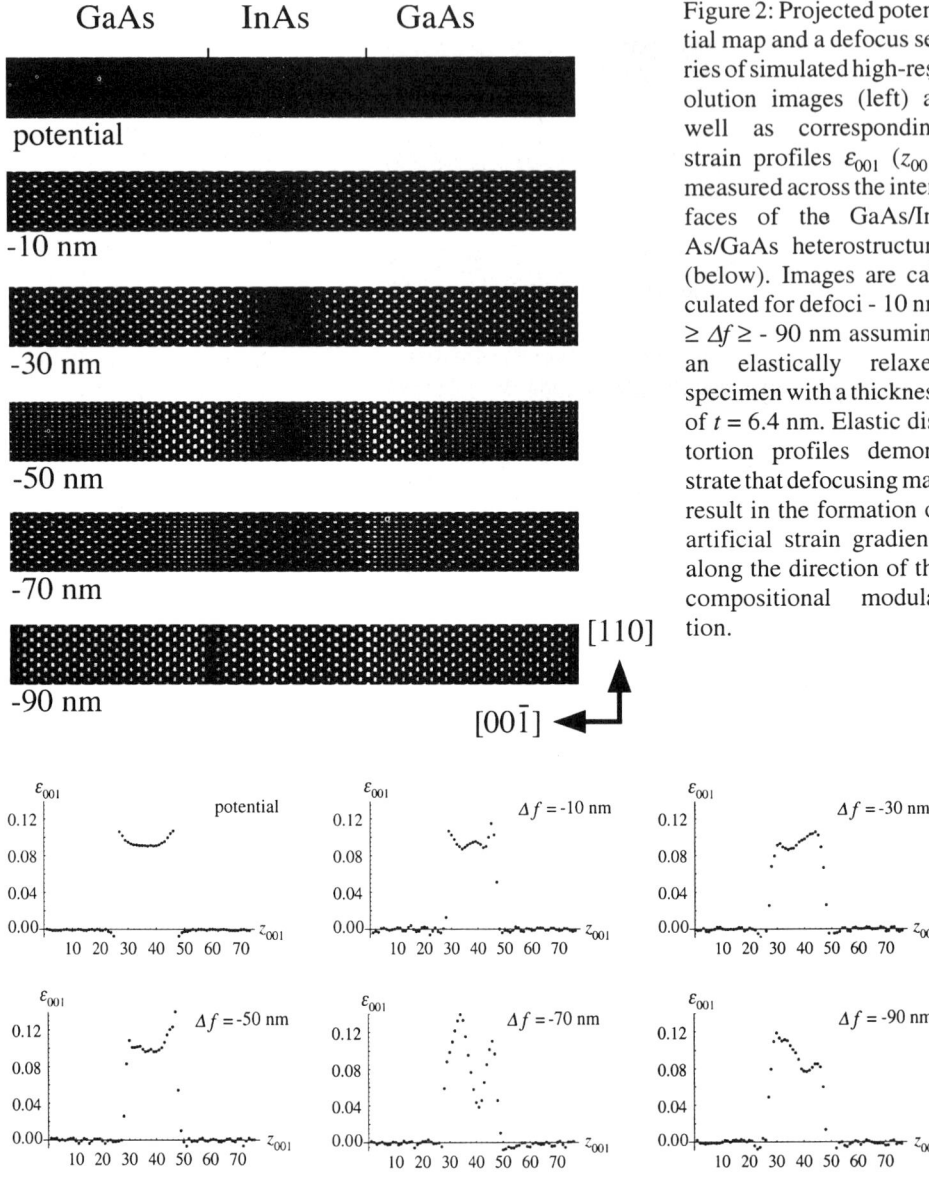

Figure 2: Projected potential map and a defocus series of simulated high-resolution images (left) as well as corresponding strain profiles ε_{001} (z_{001}) measured across the interfaces of the GaAs/InAs/GaAs heterostructure (below). Images are calculated for defoci - 10 nm $\geq \Delta f \geq$ - 90 nm assuming an elastically relaxed specimen with a thickness of $t = 6.4$ nm. Elastic distortion profiles demonstrate that defocusing may result in the formation of artificial strain gradients along the direction of the compositional modulation.

shape of the obtained profiles. While a defocus of $\Delta f = -10$ nm results in a profile which is rather in agreement with the projected potential a defocus of $\Delta f = -70$ nm indicates alternating strains inside the InAs layer resulting from contrast pattern alterations. In case of images showing only minor contrast variations across the interfaces the strain profiles give the impression of an increasing ($\Delta f = -70$ nm) or a decreasing ($\Delta f = -90$ nm) strain of the InAs layer along the [001] direction. Despite these misleading strain gradients the average strain values of the InAs layers extracted from the images are in agreement with that value derived from the projected potential map

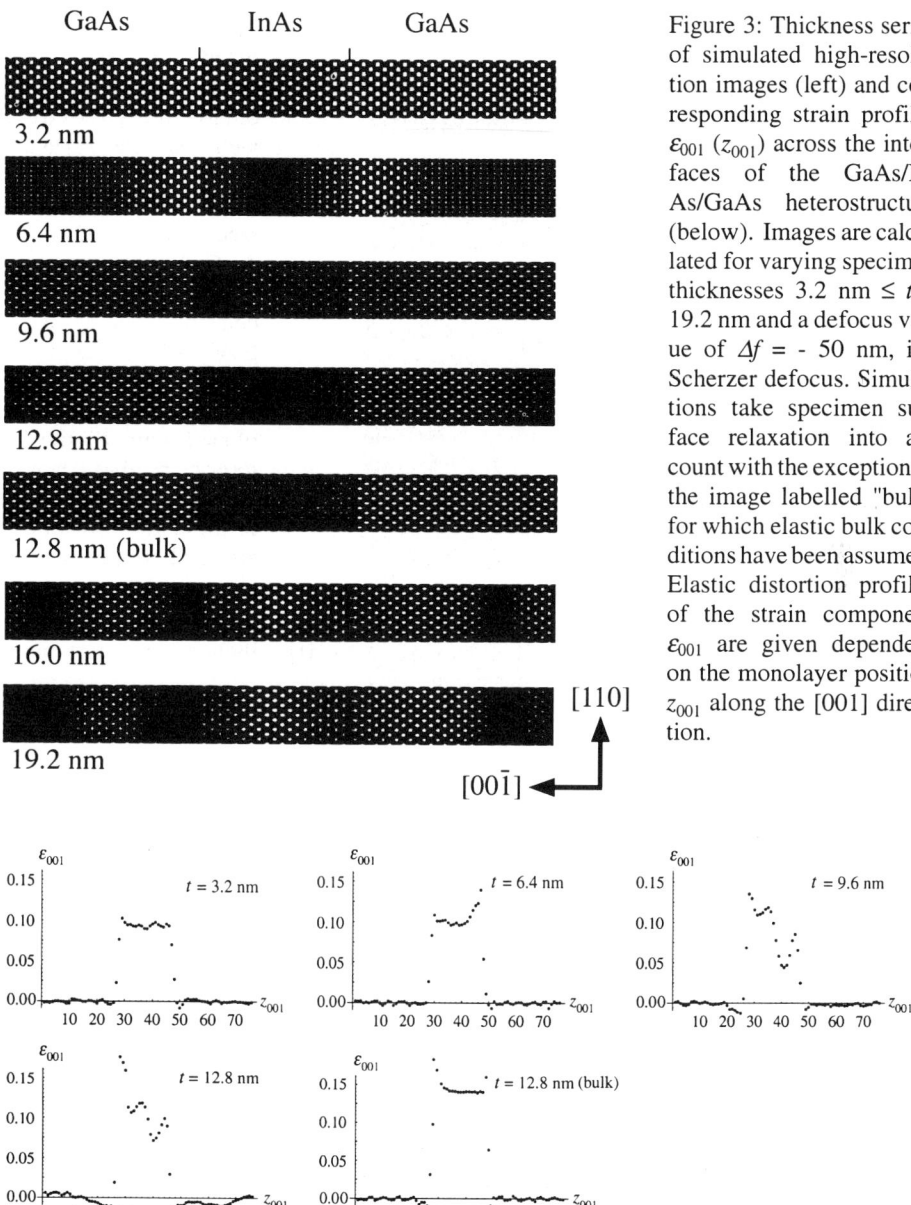

Figure 3: Thickness series of simulated high-resolution images (left) and corresponding strain profiles ε_{001} (z_{001}) across the interfaces of the GaAs/InAs/GaAs heterostructure (below). Images are calculated for varying specimen thicknesses 3.2 nm $\leq t \leq$ 19.2 nm and a defocus value of $\Delta f = -50$ nm, i.e. Scherzer defocus. Simulations take specimen surface relaxation into account with the exception of the image labelled "bulk" for which elastic bulk conditions have been assumed. Elastic distortion profiles of the strain component ε_{001} are given dependent on the monolayer position z_{001} along the [001] direction.

with a margin of error of 0.008, at most. If all ε_{001}-values larger than the maximum value of the fluctuation inside the GaAs area of the profiles are associated to belong to the InAs layer, we gain a total thickness of 22 monolayers along the [001] direction for all defoci except of $\Delta f = -10$ nm. This value represents a slight overestimate of the layer thickness actually amounting to 20 monolayers.

In order to discuss the influence of the specimen thickness on the image contrast fig. 3

shows a thickness series of simulated images and corresponding strain profiles for a defocus value $\Delta f = -50$ nm, i.e. Scherzer defocus. Simulations take specimen surface relaxation into account except of the image labelled "bulk" for which a pure tetragonal distortion of the InAs unit cells has been assumed. For a very thin sample ($t = 3.2$ nm) no motif variations across the interfaces are observed. With increasing sample thickness ($t = 6.4$ nm, 9.6 nm) the variation of the projected potential results in contrast pattern differences between both materials. These continuous changes of the fringe motifs extend approximately eight monolayers from the interfaces and are only absent in case of a pure tetragonal lattice deformation, as can be seen by the direct comparison between images assuming a relaxed and a completely strained specimen ($t = 12.8$ nm). This effect is much more pronounced for defocusing conditions far away from Scherzer focus but is not shown here in detail. The comparison demonstrates that the alteration in contrast motifs is primarily due to the inhomogeneous elastic relaxation of the sample. Fresnel diffraction at the abrupt interfaces will only play a minor role since otherwise comparable motif variations would also be observed in case of the completely strained sample. In case of rather thick specimens ($t = 16.0$ nm, 19.2 nm) the bending of lattice planes even causes a full knock-over of contrast patterns in the extreme left- and right-handed regions of the GaAs layers. It is worth mentioning that independently of the specimen thickness the strain state of the sample has mirror symmetry with respect to the centre (001) plane of the heterostructures but not with respect to the image contrast patterns (best viewed in case of $t = 16.0$ nm, 19.2 nm) since atom columns at the left and the right part of the heterostructure are tilted in opposite directions on both sides of the sample.

In case of $t = 3.2$ nm the strain profile obtained from the Scherzer image is almost mirror-symmetric thus reflecting the shape of the strain state of the sample. Due to thin-foil relaxation absolute strain values are smaller compared with those at elastic bulk conditions. With increasing sample thickness (6.4 nm $\leq t \leq 12.8$ nm) the average strain of the InAs layer increases due to a reduced elastic relaxation, but the profiles are strongly distorted on account of contrast pattern shifts in the images. Especially at the interfaces the changes of fringe motifs result in extremely high or low measured image strains. In case of $t \geq 16.0$ nm the micrographs are not appraisable at all since the knock-over of contrast patterns prevents the definition of a reference lattice within the GaAs layers.

In conclusion, the entirety of measured strain profiles can be divided into four basic types dependent on the imaging conditions. For very thin samples where contrast motif variations across the interfaces are absent the shapes of strain profiles reflect the real strain state of the specimen ($t = 3.2$ nm, $\Delta f = -50$ nm). Compared to elastic bulk conditions absolute strain values are reduced due to thin-foil relaxation. For intermediate thickness values slight contrast pattern shifts

Δf [nm] / t [nm]	-90	-70	-50	-30	-10
3.2	①	②	①	①	①
6.4	②	③	②	②	③
9.6	③	②	③	②	②
12.8	③	③	③	④	③
16.0	③	④	④	④	④
19.2	④	④	④	④	④

Table 1: Overview on the basic type of strain profiles dependent on the specimen thickness t and the defocus Δf values. A good agreement between measured profiles and the real strain state of the specimens is found for conditions labelled ①, while ② and ③ symbolize conditions resulting in the formation of artificial strain gradients and alternating profiles, respectively. Parameters labelled ④ are not suitable at all since a knock-over of the image contrast prevents the calculation of a reference lattice.

indicate strain gradients within the heterostructure, different from the true strain ($t = 6.4$ nm, $\Delta f = -30$ nm). With regard to the determination of strains from experimental micrographs this case is at most problematic since the absence of clearly visible motif variations may lead to the assumption that the measured profiles correspond to the real strain distribution of the sample. For larger thickness values the alternating strain profiles are easily detected as nonphysical since variations in the image contrast pattern motifs are clearly visible ($t = 9.6$ nm, $\Delta f = -50$ nm). Complete contrast pattern knock-overs in rather thick specimens prevent the determination of image strains at all since no reference lattice within the GaAs layers may be established ($t = 16.0$ nm, $\Delta f = -50$ nm). Table 1 gives an overview on the obtained type of strain profiles dependent on the imaging conditions.

4. DISCUSSION

For the chosen model structure elastic relaxation of thin samples causes lattice plane bending with local tilt angles α in the order of low-indexed Bragg reflections in high energy electron diffraction. Local crystal misorientations decrease with increasing t/d-ratios but will have an impact on electron diffraction in case of $t/d \leq 6$. Although local misorientation angles are at maximum for very thin specimens severe contrast motif variations are only found for medium sample thicknesses. This results from the fact that the total elongation of projected atom columns is given by αt thus vanishing in both limiting cases, $t \to 0$ and $t/d \geq 6$, and that it is exerting the strongest influence on electron diffraction for intermediate specimen thickness values.

If heterostructures characterized by a smaller indium content or by thinner buried layers are investigated, local crystal tilts are decreased yielding a spread of the band width of imaging parameters under which strain profiles represent the true strain of the specimen. In those cases the onset of contrast motif variations resulting in artifacts of the strain profile will be shifted to specimen thicknesses larger than those summarized in table 1. On the other hand actual layer thicknesses may be slightly overestimated by one or two monolayers dependent on the imaging conditions, as was shown in the last section. This effect, most probably caused by contrast delocalisation (Coene and Jansen 1992) at the interfaces, will become increasingly important in the microstructure evaluation of extremely thin buried layers.

In many experimental situations interdiffusion decreases local strains at the interfaces thus reducing crystal misorientations and decreasing the amount of non tilt-related contrast variations within different layers of the heterostructures. On the other hand specimen preparation by ion milling techniques usually causes the formation of an amorphous overlayer with typical extensions in the order of 3 nm on both sides of the crystalline sample. Since ion milling at liquid-nitrogen cooling conditions and microscopic image recording are performed at different temperatures the different thermal expansion coefficients of the amorphous and the crystalline structure may further alter the strain state of the specimen yielding modified strain values with regard to the presented results.

The calculations in this study demonstrate that experimental strain profiles may contain severe artifacts dependent on the imaging conditions. It is therefore recommended to determine lattice strains not by measurements from single micrographs but to take up the common idea of measuring strains from experimental exit plane wave functions which provide structural information independent from the optical parameters of the microscope. In this analysis simulations have been carried out for an electron microscope equipped with a conventional LaB$_6$ emitter. For microscopes with a field emission gun image contrast variations across the heterointerfaces will be much more pronounced because the higher spatial coherence introduces larger delocalisation effects (Stobbs, Sato and Stobbs 1995) and the band width of imaging parameters under which experimental strain profiles represent the true strain distribution of the specimen will be further reduced when performing measurements with those kind of instruments.

5. CONCLUSIONS

In this study we have investigated the impact of elastic thin-foil relaxation on the image contrast of high-resolution micrographs of GaAs/InAs/GaAs heterostructures taken along the [110] direction. Lattice strains have been evaluated by the measurement of fringe spacings along the direction of compositional modulations in simulated images. Dependent on the imaging conditions the shapes of measured strain profiles may contain severe artifacts which result from continuous contrast variations extending several monolayers from interface positions and which are caused by local crystal tilts in elastically relaxed specimens. Nevertheless the average strain of thicker layers can be determined with sufficient accuracy and it may be used to get a rough estimate of layer compositions when focusing on experimental micrographs.

ACKNOWLEDGEMENTS

Thanks to A Thust for useful discussions. One of the authors (ML) gratefully acknowledges financial support by the Volkswagen-Foundation.

REFERENCES

Bayle P, Deutsch T, Gilles B, Lancon F, Marty A and Thibault J 1994 Ultramicroscopy **56**, 94
Bierwolf R, Hohenstein M, Phillipp F, Brandt O, Crook, GE and Ploog K 1993 Ultramicroscopy **49**, 273
Coene W and Jansen AJEM 1992 Scanning Microsc. Suppl. **6**, 379
Gibson JM and Treacy MMJ 1984 Ultramicroscopy **14**, 345
Jouneau PH, Tardot A, Feuillet B, Mariette H and Cibert J 1994 J. appl. Phys. **75**, 7310
Kret S, Delamarre C, Laval JY and Dubon A 1998 Phil. Mag. Lett. **77**, 249
Perovic DD, Weatherley GC and Houghton DC 1991 Phil. Mag. A **64**, 1
Robertson MD, Corbett, JM, Webb JB, Jagger J and Currie JE 1995 Micron **26**, 521
Rosenauer A, Remmele T, Gerthsen D and Förster A 1997 Optik **105**, 99
Stobbs SH, Sato K and Stobbs WM 1995 Ultramicroscopy **58**, 275
Tillmann K, Thust A, Lentzen M, Swiatek P, Förster A and Urban K 1996 Phil. Mag. Lett. **74**, 309
Tillmann K, Lentzen M and Rosenfeld R 1999 submitted Phil. Mag. A
Walther T and Humphreys CJ 1995 Inst. Phys. Conf. Ser. **147**, 103

Inst. Phys. Conf. Ser. No 164
Paper presented at Microsc. Semicond. Mater. Conf., Oxford, 22–25 March 1999
© *1999 IOP Publishing Ltd*

Strain mapping in semiconductor heterostructures using HREM

T Plamann, M J Hÿtch, S Kret[1], J Y Laval[1] and C Delamarre[1]

Centre d'Etudes de Chimie Métallurgique, CNRS, 15 rue G. Urbain, 94407 Vitry-sur-Seine, France
[1]Laboratoire de Physique du Solide, CNRS-ESPCI, 10 rue Vauquelin, 75231 Paris, France

ABSTRACT: We present experimental high-resolution electron microscope (HREM) images of $In_xGa_{1-x}As$ islands grown on a GaAs substrate in the [110] orientation and discuss a technique of determining local crystal deformations from the spatially resolved phase of the Fourier components in the image. We investigate the limits of the technique using lens transfer theory and show that, for centrosymmetric structures, astigmatism and thickness variations have no effect on the measurement of displacements. In order to assess errors in the less favourable case of non-centrosymmetric structures, we examine simulated images of crystal wedges of $In_xGa_{1-x}As$ and conclude that in the absence of misalignments, the error in the measurement of the local lattice parameter along (002) is about 1% for a 90° wedge.

1. INTRODUCTION

Strain mapping is being increasingly used in high resolution electron microscopy (HREM) to characterise semiconductor heterostructures (Bierwolf et al 1993, Jouneau et al 1994, Rosenauer et al 1997). The positions of the unit cells are located in the image by comparison of the image features with a reference lattice, using such techniques as « peak-finding », cross-correlation or geometric phase analysis (Hÿtch et al 1998). The problem in all these techniques is that lattice fringes in an image do not necessarily coincide with the atomic plane positions. Errors are introduced by the transfer characteristics of the microscope and, for example, local crystal tilts. In this contribution we examine the accuracy with which local lattice distortions can be measured using the geometric phase technique, though the results can be generalised to the other methods of analysis. Systematic errors will be examined in image simulations. It will be shown that for projected structures with a centre of symmetry and slowly varying strain fields, the phase image predicts the displacement of the atomic planes correctly even in the presence of strong dynamical scattering and of lens imperfections such as spherical aberration and astigmatism. We discuss to what extent thickness changes influence the reconstruction of a displacement map for structures without a centre of symmetry.

2. MEASUREMENT OF DISPLACEMENT FIELDS USING GEOMETRIC PHASE

Fig. 1 shows an experimental HREM image of an $In_xGa_{1-x}As$ island grown on a GaAs substrate in the [110] orientation. Images of this type have recently been analysed using a modified peak-finding technique, and the measured displacement field compared with predictions based on finite element analysis (Kret et al 1998). Most of the noise in the image is introduced by the amorphous film whose thickness may change locally. In order to establish a map of deformations inside the structure using the geometric phase approach, we Fourier transform the HREM image and centre a mask around a strong Bragg spot, in our case corresponding either to the $(1\bar{1}1)$ or the $(\bar{1}11)$ fringes. The size of the mask determines the amount of averaging in real space so there is a compromise to be found between the noise reduction and lateral resolution. In our case the mask was Lorentzian with an averaging in

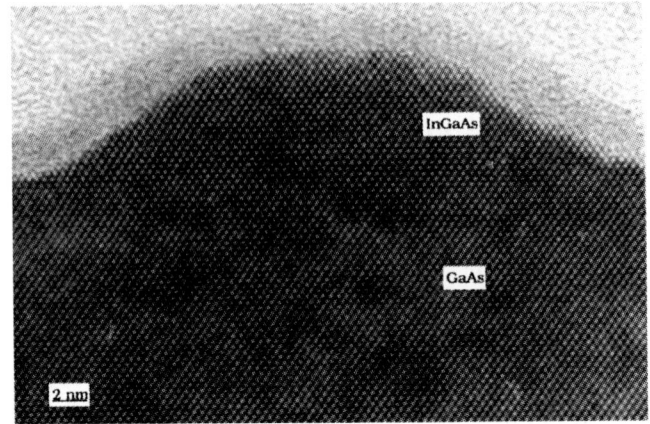

Figure 1: Experimental HREM image of an In$_x$Ga$_{1-x}$As island grown on a GaAs substrate along [$\bar{1}$10].

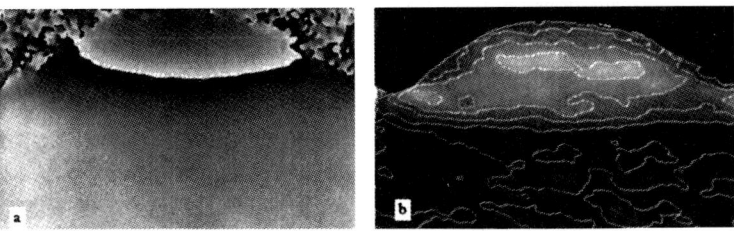

Figure 2: (a) Geometric phase image for **g** = 002; (b): strain along (002) with contours every 1%.

real space of about 1.2 nm. The inverse Fourier transform results in a complex image of the relevant Fourier component mapped as a function of position from which we obtain the geometric phase image, which is linearly related to the displacement field **u** by:

$$P_g(\mathbf{r}) = -2\pi i\mathbf{g}.\mathbf{u} .$$ (1)

Fig. 2a shows the sum of the geometric phase images for the ($1\bar{1}1$) and ($\bar{1}11$) fringes which is equivalent to an image of the component of the displacement field along (002) since:

$$P_g(\mathbf{r}) = P_{g-g'}(\mathbf{r}) + P_{g'}(\mathbf{r}) .$$ (2)

The strain corresponding to the same direction is obtained by calculating the component of the gradient of the phase image along (002) and is shown in Fig. 2b. The contour map shows that there is a maximum of the deformation in the middle of the island which exceeds the misfit parameter (2.7%) and has been interpreted as a local increase in the indium composition (Kret et al 1998).

3. IMAGING THEORY

In order to describe the local thickness or orientation changes of the specimen, we allow the scattered beam amplitudes to be a function of the position in the image. In the presence of a slowly-varying displacement field **u**, the exit surface function is then given by:

$$\psi(\mathbf{r}) = \sum_g \tilde{\psi}_g(\mathbf{r}) exp\{2\pi i\mathbf{g}.(\mathbf{r} - \mathbf{u})\}.$$ (3)

The image intensity is given by the modulus squared of the image wave function and reads

$$I(\mathbf{r}) = \sum_{gg'} \tilde{\psi}_{g+g'}(\mathbf{r})\tilde{\psi}^*_{g'}(\mathbf{r})T(\mathbf{g}',\mathbf{g}+\mathbf{g}')exp\{2\pi i\mathbf{g}.(\mathbf{r}-\mathbf{u})\}, \qquad (4)$$

where $T(\mathbf{g}',\mathbf{g}+\mathbf{g}')$ is the complex lens transfer for each pair of beams and includes the effects of spatial and temporal coherence. By regrouping the terms of periodicity $\mathbf{g}$, we find that the image Fourier components are given by:

$$\tilde{I}_g(\mathbf{r}) = exp\{-2\pi i\mathbf{g}.\mathbf{u}\}\sum_{gg'} \tilde{\psi}_{g+g'}(\mathbf{r})\tilde{\psi}^*_{g'}(\mathbf{r})T(\mathbf{g}',\mathbf{g}+\mathbf{g}'). \qquad (5)$$

If the phase of the sum in Eq. (5) is constant, the phase image will predict the displacement field along $\mathbf{g}$ (cf Eq. (1)). This is always true if specimen thickness and orientation of the crystal foil do not change, even in the case of strong dynamical scattering. For projected structures with a centre of symmetry (for instance in the case of the projections [100] in the sphalerite structure) and a centrosymmetric lens transfer, Eq. (5) can be recast in the form:

$$\tilde{I}_g(\mathbf{r}) = exp\{-2\pi i\mathbf{g}.\mathbf{u}\}\Re e\left[\sum_{g'} \tilde{\psi}_{g+g'}(\mathbf{r})\tilde{\psi}^*_{g'}(\mathbf{r})T(\mathbf{g}',\mathbf{g}+\mathbf{g}').\right] \qquad (6)$$

In this case the phase image predicts the displacement of the atomic planes correctly even in the presence of thickness variations and astigmatism. Fig. 3a shows a simulated image of a crystal wedge of InP [100] with a large amount of astigmatism (18 nm at 40°) and using the parameters of the Philips CM-20UT used to record Fig. 1 (accelerating voltage 200 kV, spherical aberration 0.5 mm, defocus – 40 nm). The diagrams in Figs 3b and c show the magnitude and phase of the (002) fringes, illustrating the insensitivity of the phase to lens imperfections (apart from phase jumps of π, which are caused by contrast reversal). For projected structures with a centre of symmetry the most dangerous factor is beam tilt in the presence of large variations in thickness.

4. SIMULATIONS FOR NON-CENTROSYMMETRIC STRUCTURES

For non-centrosymmetric structures the requirement of a slowly varying specimen thickness becomes crucial. In the more general case, simplifications like in Eq. (6) can only be made for certain reflections and if the lens transfer has at least radial symmetry, i.e. in the absence of misalignments, which will be assumed in the following example. Fig. 4a shows a simulated image of a crystal wedge of $In_{0.5}Ga_{0.5}As$ [110]. In Figs 4b and 4c we show the magnitude and phase of the $(1\bar{1}1)$ fringes. The variation of the phase is caused by a shift of the image features along (002) and cancels if we subtract the phase images of the $(1\bar{1}1)$ and $(\bar{1}11)$ fringes. This results in a map of the displacement field parallel to $(2\bar{2}0)$, as can be verified from Eq. (2). The error introduced by thickness variations restricts itself to the measurement of the displacement along (002), where a phase change of the $(1\bar{1}1)$ fringes of 2π translates into a real space shift of 0.585 nm parallel to (002). It can be seen clearly that the phase in Fig. 4c changes most rapidly at thickness values where the magnitude of the Fourier component is small. The displacement field should therefore be measured only in image regions where the relevant Fourier component is maximal. In order to quantify the error in the measurement of the strain field, it is useful to analyse the derivative of the apparent displacement with respect to the thickness (Fig. 4d), which is equivalent to the apparent (erroneous) strain field for a 90° wedge of perfect crystal. In the regions where the Fourier component is weak the apparent strain can be as great as about 30%. Where the fringe contrast is maximal, we observe a plateau (indicated) that does not exceed 1.2%. Realistically, the wedge angle will be much smaller than 90°, more likely 10 to 15° and therefore the error in the measurement of the local lattice parameter will be of the order of 0.2%. If a strong thickness variation causes a variation of the phase one can often observe a change of the magnitude as well. Additionally, a merit of the technique is that it can be applied separately for different reflections. In general, a discrepancy will occur between the reconstructed displacement field for the different Fourier components if thickness or orientation change, which provides a simple check on the fidelity of the data.

26

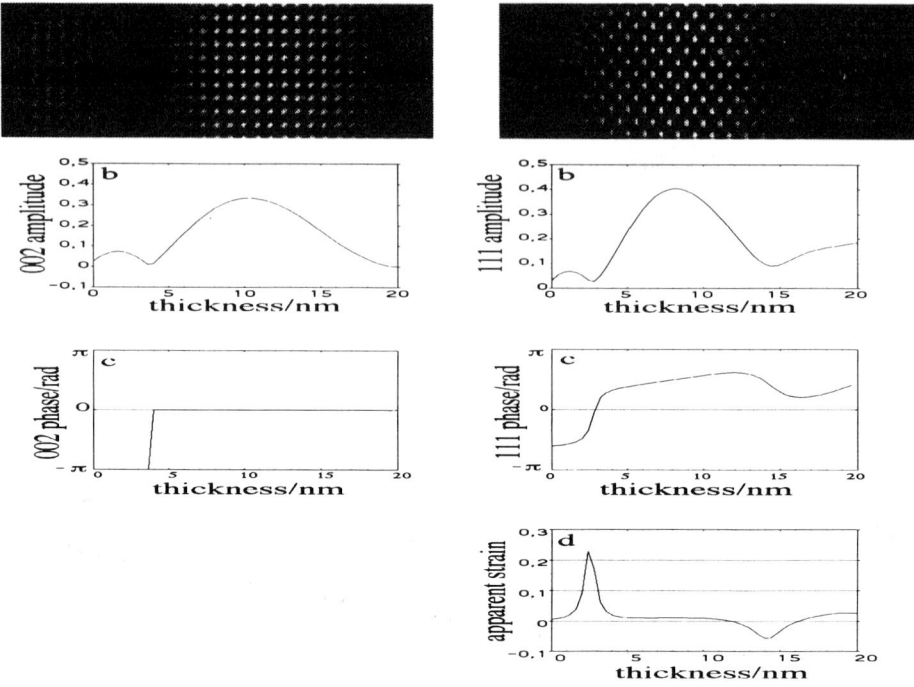

Figure 3: (a) Simulated HREM image of a
crystal wedge of InP [100] (strongly astigmatic,
see text); (b) and (c): amplitude and phase of the
(002) fringes.

Figure 4: (a) Simulated HREM image of a
crystal wedge of In$_{0.5}$Ga$_{0.5}$As [110]; (b) and (c)
amplitude and phase of the (1$\bar{1}$1) fringes; (d)
apparent strain field for a 90° crystal wedge.

5. CONCLUSIONS

In the absence of variations in thickness or orientation, slowly varying displacement fields can be
measured even in the presence of strong dynamical scattering. For centrosymmetric projections,
thickness variations and astigmatism have no effect on the fringe positions. For non-centrosymmetric
structures, errors can be minimised by choosing conditions where the fringe contrast is maximal, and
verifications can be carried out using other reflections or by analysing the contrast of the relevant
fringes. In the case studied the error was estimated as ±0.2% strain.

REFERENCES

Bierwolf B, Hohenstein H, Philipp F, Brandt O, Crook G E and Ploog K 1993 Ultramic. **49**, 273
Hÿtch M J, Snoeck E and Kilaas R 1998 Ultramic. **74**, 131
Jouneau P H, Tardot A, Feuillet G, Mariette H and Cilbert J 1994 J. Appl. Phys. **75**, 731
Kret S, Delamarre C, Laval J Y and Dubon A 1998 Phil. Mag. Lett. **77**, 249
Rosenauer A, Remmele T, Gerthsen D, Tillmann K and Förster A 1997 Optik **105**, 99

Inst. Phys. Conf. Ser. No 164
Paper presented at Microsc. Semicond. Mater. Conf., Oxford, 22–25 March 1999
© 1999 IOP Publishing Ltd

PEELS imaging and linescan study of concentration anisotropies in $Al_xGa_{1-x}As$ and $In_yGa_{1-y}As$ heterostructures grown on non-planar substrates

K Leifer, A Rudra, G Biasiol, P A Buffat[1], H Michler[2], E Blank[2] and E Kapon

Department of Physics, Swiss Federal Institute of Technology (EPFL), Switzerland
[1] Centre Interdepartemental de Microscopie Electronique, EPFL, Lausanne, Switzerland
[2] Departement de Matériaux, EPFL, Lausanne, Switzerland

ABSTRACT: In the particular case of III/V $Al_xGa_{1-x}As$ and $In_yGa_{1-y}As$ semiconductor structures electron energy loss spectroscopy (EELS) is more sensitive than energy dispersive X ray spectroscopy(EDS) This makes EELS useful for measuring compositional variations in these material systems with nanometer resolution in both linescan and imaging mode. An absolute precision of ±0.02-0.03 and a relative precision of ±0.01–0.02 in x can be reached at acquisition times of 1.5s/spectrum. Using optimised acquisition and evaluation methods, concentration profiles and maps with a resolution in the nanometer range were obtained. When $Al_xGa_{1-x}As$ layers are grown by MOCVD on non planar GaAs substrates containing V-grooves and ridges, both self ordered Ga rich vertical layers (VQW) and Al rich vertical quantum barrier (VQB) regions were observed. For $In_yGa_{1-y}As$ layers deposited on such V–grooved structures, growth is furthermore influenced by strain. For the first time we could demonstrate that a vertical In rich region is formed in the centre of the groove with a local In excess concentration of $\Delta y=0.1$ for a nominal $y=0.15$.

1. INTRODUCTION

The high signal intensity in the elemental edges of electron energy loss spectra (EELS) results in a high statistical precision of the concentrations even for acquisition times of the order of one second for many elements. Nevertheless, in the case of the trace element analysis Leapman and Hunt (1991) have shown that EELS is more sensitive than energy dispersive X ray spectroscopy (EDS) only when elements with $Z\leq26$ are analysed. We will show that for non–trace concentrations of selected III/V semiconductor elements (Z>26) EELS is more sensitive than EDS. Consequently EELS was used to acquire concentration maps using a nanoprobe in a field emission microscope, a technique proposed by Jeanguillaume and Colliex (1989). To obtain a high precision of the absolute concentrations, some precautions have to be taken: 1) the specimen orientation has to be chosen such that electron channeling is minimized, 2) the specimen has to be thick enough such that non stoichiometric surface layers are negligible and 3) the evaluation of the spectra has to be optimised (Leifer and Buffat 1997).

In $Al_xGa_{1-x}As$ layers grown by metal organic chemical vapour deposition (MOCVD) on non–planar surfaces the Al concentration depends on the local growth facets (Biasiol et al 1997). In AlGaAs layers grown on V grooves this manifests itself as a vertical quantum well (VQW) in the centre of the grooves (Walther et al 1992, Vermeire et al 1992). The carrier capture mechanisms through this VQW play an important role in diodes realized with such structures (Weman et al 1999). In the $In_yGa_{1-y}As$ system growth is influenced by the presence of strain in these structures. The thickness of the InGaAs layers is limited by the critical thickness that depends on the In concentration. In the present work, we describe a first growth study that was carried out to analyse and understand concentration anisotropies in such InGaAs layers.

2. COMPARISON OF PEELS AND EDS SENSITIVITY

Parallel EELS spectra (PEELS) were compared with spectra taken with a Tracor Voyager EDS instrument on the same microscope. For acquisition times of 1.5s and 20s per spectrum for PEELS and EDS standard deviations σ of the concentrations of 0.7% and 5% were obtained. The relative sensitivity $r=(I^{PEELS}/\sigma^{PEELS})/(I^{EDS}/\sigma^{EDS})$ is reported in figure 1a for the following edge and line intensities I^{PEELS} and I^{EDS}: Ga–L edge/Ga K–line, In M_{45}–edge/In L–line and P K–edge/P K–line. In the case of PEELS I/σ is calculated from the error of the background extrapolation model and the statistical error of the edge intensity. For EDS the error of the multiple least square fit of a standardless analysis was taken as σ. The sensitivity of PEELS is 20-35 times higher than that of EDS for Ga and In. Therefore, when short acquisition times are needed, e.g., in linescans or maps, PEELS will give more precise concentration data. The difference in the relative sensitivity of the P edge/line between Leapman and this work can probably be explained by the use of the second difference technique by that author. In this technique typical fit intervals are of the order of several 10eV. Using the simple power law background for edges beyond 1keV, the fit intervals can exceed 100eV. These wide intervals will contribute to a reduction of the statistical error as compared with the second difference method.

3. EXPERIMENTAL: EELS ANALYSIS AND MOCVD GROWTH

The parallel EELS (PEELS) spectra were acquired in a Hitachi HF-2000 field emission microscope (acceleration voltage: 200keV, probe current: 1nA, semi collection angle: 50mrad) using a GATAN PEELS spectrometer model 666. For the acquisition of the linescans and maps the electron probe was focused down to 1.5nm (FWHM) and the acquisition time was 1.5s/spectrum for the linescans and 0.3-1s/spectrum for the maps. The influence of non–stoichiometric surface layers on the analysis is negligible when the specimen thickness interval is t=60-120nm as described by Leifer and Buffat (1997). Absolute concentrations can be obtained with a precision of $\Delta x \approx \pm 0.02$. The Al and In cross sections are neglected assuming that the ratio of III/V element equals 1.

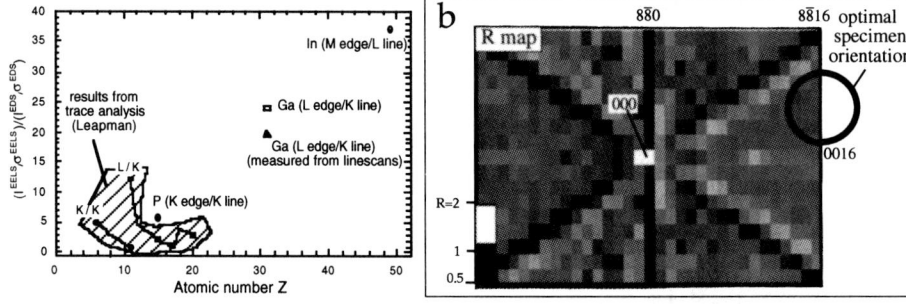

Figure 1: a) Comparison of the sensitivities of EELS/EDS. The region in the left lower corner shows the results from Leapman and Hunt 1991. b) Map of the relative R ratio close to the [110] zone axis. The tilt angle is expressed by the centre of the Laue circle. The grey scale of the R values is given in the lower right corner.

The cross sectional samples are analysed close to a [110] zone axis. On such high–symmetry orientations electron channeling modifies the electron intensities on element III and V sites. We calculated maps of the relative edge intensity R in an AlAs crystal as a function of the tilt angle as described by Leifer and Buffat (1997). The influence of electron channeling on the edge intensities can be quantified and optimal acquisition conditions can be found (fig. 1b). R variations of a factor 2 are observed in the vicinity of the [110] zone axis. When interfaces parallel to ($1\bar{1}0$) planes are analysed, these interfaces stay parallel to the electron beam direction for tilts along the (000)-(0016) trajectory. The circle in figure 1b indicates a tilt range of the specimen where 1) the interfaces are nearly parallel to the electron probe, 2) electron channeling is small and 3) R changes due to small orientation variations are negligible. The latter is important when a large number of spectra is acquired and the specimen orientation can not be verified at each point.

Wet chemical etching techniques are used to pattern the surface of the GaAs (001) substrates with V-grooves along the [110] directions. On this non-planar substrate with a pitch ranging from 0.25 to 3µm $Al_xGa_{1-x}As$ and $In_yGa_{1-y}As$ layers were deposited using MOCVD at temperatures of 680°C and 550°C respectively. Details on the growth process can be found in Biasiol et al (1997).

4. CONCENTRATION ANISOTROPY IN AlGaAs and InGaAs

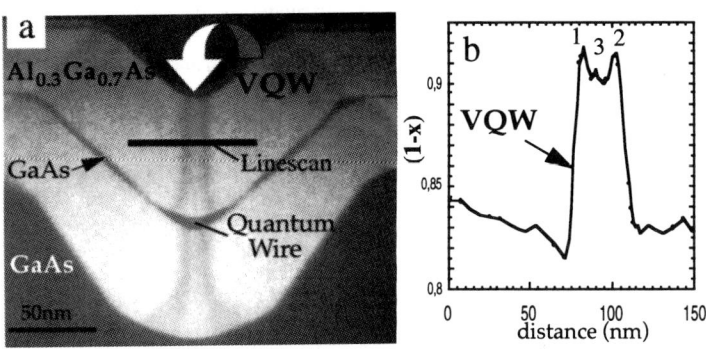

Figure 2: a) (200) Dark field image of an AlGaAs/GaAs heterostructure grown on V-groove substrate. b) Ga concentration profile from PEELS linescan indicated in a.

A typical AlGaAs V-groove heterostructure is shown in figure 2a. It contains a vertical quantum well (VQW) in the centre of the groove, visible as dark contrast. The higher growth rate on the sidewall facets as compared to the (100) facet at the bottom of the groove leads to a shrinking of the bottom facet. For sufficiently narrow groove widths, however, capillarity induced adatom fluxes towards the bottom of the groove stabilize its width. Since in the AlGaAs system the Ga adatoms are the more rapidly diffusing species, the interplay between the capillarity, the growth rate anisotropy and an additional effect related to the entropy of mixing leads to a characteristic Ga enrichment in the centre of the groove. The Ga concentration of this VQW and its width depend on the Al concentration and the growth temperature (Biasiol and Kapon 1998). Furthermore, when a thin GaAs well is grown on top of a self-limiting AlGaAs V-groove, its thickness increases at the groove's bottom where a quantum wire (QWR) is formed.

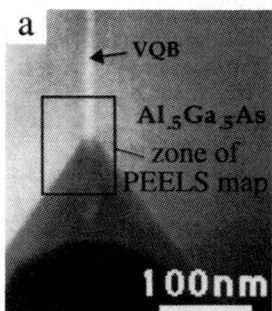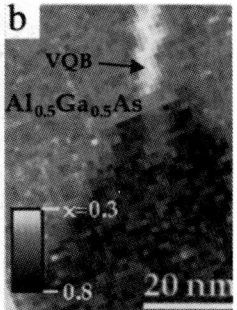

Figure 3: a) Dark field image showing a vertical quantum barrier (VQB) at the ridge. b) The PEELS map of the Al concentration x was acquired in the zone shown in a.

The lateral variation in Ga content (1-x) across the VQW as measured with EELS is shown in figure 2b. Standard deviations of the concentrations of $\sigma(x)=0.01$ or less are obtained in EELS profiles through the VQWs as demonstrated in figure 2b. The linescan was acquired in a specimen containing a VQW similar to the specimen in figure 2a. Since the data presented here were averaged over 3 neighbouring spectra, the effective acquisition time per spectrum was 4.5s and we obtain a standard deviation x in the region of constant x of $\sigma(x)=0.0025$. Clearly two branches of the VQW can be distinguished with a difference between the maxima 1 and 2 and the local minima of $\Delta x=0.015$ (see fig. 2b). This concentration difference is real as it is more than five times higher than the standard deviation $\sigma(x)$. Although a third branch (3) can be observed in the profile, it differs only by $\Delta x=0.005=2\cdot\sigma(x)$ from the local minima which is not statistically significant. It has been shown that electrons and holes are quantum confined in such multi-branch structures (Martinet et al 1997).

For (100) ridge facets neighboured by exact (111)A facets, we also observe an Al rich vertical quantum barrier (VQB) on the ridge (fig. 3a). In this particular case the growth rate on the (111)A sidewalls is small as compared to the (100) ridge facet. Therefore, during growth the ridge facet shrinks and the capillarity fluxes come into play. The equilibrium between capillarity and growth rate anisotropy is established when the mobile adatoms, here mainly the Ga atoms, diffuse away from the ridge to the (111) facets. Consequently a narrow region of Al rich AlGaAs forms, as evidenced in the EELS maps with an Al excess concentration of $\Delta x = +0.22$ (fig. 3b). It should be noted that this VQB is only 7nm wide as compared to widths of typically 15-30nm for the Ga–rich VQW.

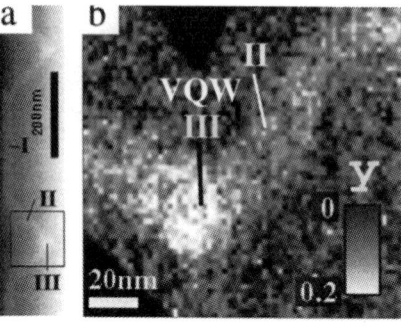

Figure 4: a) HAADF image of $In_{0.15}Ga_{0.85}As$ layer. b) PEELS map of In concentration y acquired in region indicated by the square in a.

When $In_{0.15}Ga_{0.85}As$ layers are grown on the V grooved substrates, not only the above mentioned effects, but also the stress that increases with the layer thickness may influence the growth. In these InGaAs layers a QWR is formed at the bottom of the groove and in this ternary system it is furthermore possible that an additional VQW might develop in this QWR. In the high angle annular dark field image (HAADF) in fig. 4a the ridge, the sidewall and the wire at the bottom of the groove of a 25 nm thick $In_{0.15}Ga_{0.85}As$ layer can be clearly distinguished. In the PEELS image in figure 4b we measure an In concentration of $y=0.11$ on the sidewalls and of $y=0.21$ in the centre of the groove. To our knowledge this is the first experimental evidence for an In rich vertical region in the centre of V–groove InGaAs structures. Its lateral width is 25nm, which corresponds to the width of the same feature observed in (200) dark field images. The In concentration at the bottom of the V–groove is the same as those on the ridge facet (I in fig. 4a). To understand the influence of strain on the growth, first calculations of the strain distribution in this specimen have been carried out using the Abaqus finite element program assuming a homogenous In concentration of $y=0.15$. The result is that the elastic energy density is 1) high on the sidewalls and on the bottom of the groove and 2) low on the ridges. Point 1 is surprising since in the centre of the groove the InGaAs layer thickness is by a factor 1.5 higher than the one on the sidewalls. The reason for 1) is that the elastic constants are higher close to (111) oriented surfaces than for (100) facets. Since the elastic energy at these two parts of the systems is the same, it is not supposed to be an additional driving force in the initial stage of the formation of the vertical InGaAs layer. Nevertheless as soon as there is an In excess concentration at the bottom of the groove, the elastic energy will become higher than on the sidewall and the energy difference might become an additional driving force for In diffusion away from the centre of the groove. The increase of stress between ridges and sidewalls coincides with a decrease of the In concentration. This observation suggests that In diffuses to regions of lower stress. In addition we find that the $In_{0.15}Ga_{0.85}As$ layer is thicker on the ridges than on the side walls. In unstrained GaAs layers the sidewalls are thicker than the ridges. This is another indication for In diffusion from the sidewalls to the ridges.

In conclusion we have characterized and discussed the formation of Ga and Al rich vertical quantum wells and barriers forming in V–grooved AlGaAs layers using PEELS linescans and maps. In InGaAs layers we find In rich regions in the centre of the V–groove. The profile of the elastic energy density of the InGaAs layer correlates with the local In concentration.

We would like to thank C. Constantin, F. Lelarge, E. Martinet and P. Stadelmann for many interesting discussions and the Swiss Priority Program OPTIQUE II for its support.

REFERENCES

Biasiol G, Martinet E, Reinhardt F, Gustafsson A and Kapon E 1997 J. of Cryst. Growth **170**, 600.

Biasiol G and Kapon E 1998 Phys. Rev. Lett. **81(14)**, 2962.

Jeanguillaume C and Colliex C 1989 Ultramicroscopy **28**, 252.

Leapman R D and Hunt J A 1991 Microsc.Microanal.Microstruct. **2**, 231.

Leifer K and Buffat B A 1997 Inst. Phys. Conf. Ser. **157**, 381.

Martinet E, Gustafsson A, Biasiol G, Reinhardt F and Kapon E 1997 Phys. Rev. Rap. Com. **56**, 7096.

Vermeire G, Yu Z, Vermaerke F, Buyden L, Van Daele P and Demeester P 1992 J. of Cryst. Growth **124**, 513.

Walther M, Kapon E, Christen J, Hwang D M and Bhat R 1992 Appl. Phys. Lett. **60(2)**, 521.

Weman H, Martinet E, Rudra A and Kapon E 1999 to be published in Appl. Phys. Lett.

Inst. Phys. Conf. Ser. No 164
Paper presented at Microsc. Semicond. Mater. Conf., Oxford, 22–25 March 1999
© 1999 IOP Publishing Ltd

Analytical electron microscopy of III-V quantum dot structures

R Schneider, H Kirmse, W Neumann, F Heinrichsdorff*, A Krost* and D Bimberg*

Humboldt University of Berlin, Institute of Physics, Chair of Crystallography, Invaliden-strasse 110, D-10115 Berlin, Germany
*Technical University of Berlin, Institute of Solid State Physics, Hardenbergstrasse 36, D-10623 Berlin, Germany

ABSTRACT: The nanochemistry of interfacial regions in III-V compound semiconductor quantum dot (QD) structures grown by metal-organic chemical vapour deposition (MOCVD) was investigated by analytical transmission electron microscopy. Besides structural characterization of the QDs by high-resolution transmission electron microscopy (HRTEM), energy-dispersive X-ray spectroscopy (EDXS) and energy-filtered TEM (EFTEM) were particularly applied to image the element distribution. Solely approximate information on the distribution was gained by X-ray line-profile analysis, whereas energy-filtered images of the QDs show atomic resolution.

1. INTRODUCTION

The optoelectronic properties of quantum dot devices essentially depend on the perfection of the active elements, i.e. the size, shape, arrangement, and chemical composition of the QDs. There are many investigations dealing with the refinement of structural data of QDs (see e.g. Strunk et al (1997) and Tillmann et al (1995)), and sophisticated means have been developed for digital analysis of HRTEM images (cf. e.g. Ourmazd et al (1993) and Rosenauer et al (1996)) to probe the microchemistry due to displacement of atom columns caused by the lattice mismatch and the attributed strain field.

In the present work, EDXS and EFTEM were used as analytical techniques for direct elucidation of the chemical composition in the interfacial region between QD, wetting layer, substrate and cap layer. The potential applicability of these methods is demonstrated for self-organized (In,Ga)As QDs on GaAs.

2. EXPERIMENTAL

Several (In,Ga)As QD samples were produced by depositing (In,Ga)As layers of different thicknesses on (001) GaAs wafers by MOCVD (for details, see Heinrichsdorff et al (1998)). TEM cross-section specimens were prepared in the conventional manner (face-to-face gluing, drilling of cylinders, cutting of thin slices, mechanically polishing down to ca. 100 μm, and finally dimpling and ion milling). TEM studies were performed at 200 kV on a HITACHI H-8110 and a PHILIPS CM20 FEG, both equipped with a GATAN imaging filter (GIF). EDXS line profiles were taken in the scanning TEM (STEM) mode with an electron probe of about 0.7 nm in diameter. During X-ray analyses the specimen was kept at liquid nitrogen temperature via a double-tilt cooling holder.

3. STRUCTURAL CHARACTERIZATION BY HIGH-RESOLUTION IMAGING

The HRTEM image (Fig. 1) of an $In_{0.5}Ga_{0.5}As$/GaAs specimen with a nominally 1.07 nm thick active layer shows the (In,Ga)As wetting layer (WL) and two individual QDs hinting to the Stranski-Krastanov growth of the self-organized dots. Both the underlying GaAs layer and the capping layer are almost free of defects. The QDs are assumed to be truncated pyramids with a width of about 15 - 20 nm and a height of ca. 5 nm in [110] projection. Owing to the lattice mismatch in the (In,Ga)As QD region the phase contrast is influenced by the strain field. Thus, it is difficult to separate this contribution from that of the specific chemical composition by means of digital image analysis.

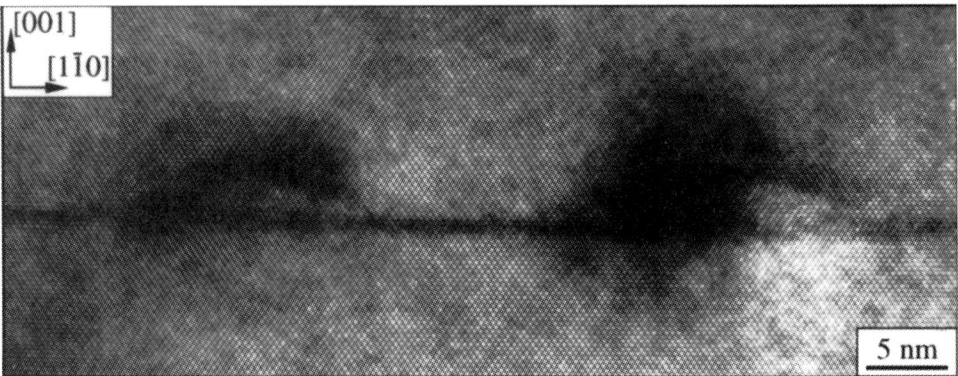

Fig. 1. HRTEM image of two individual (In,Ga)As quantum dots.

4. ENERGY-DISPERSIVE X-RAY SPECTROSCOPY

An STEM dark-field image of a QD region and the corresponding In line profiles are given in Fig. 2. To minimize the effect of specimen drift the In-L_α signal was acquired for only 5 s at 20 points with a step width of about 0.8 nm, which yields data of very low signal-to-noise ratios. Nevertheless, the In profile across the wetting layer clearly identifies its position. However, owing to beam broadening it is smeared out to about 3 nm.

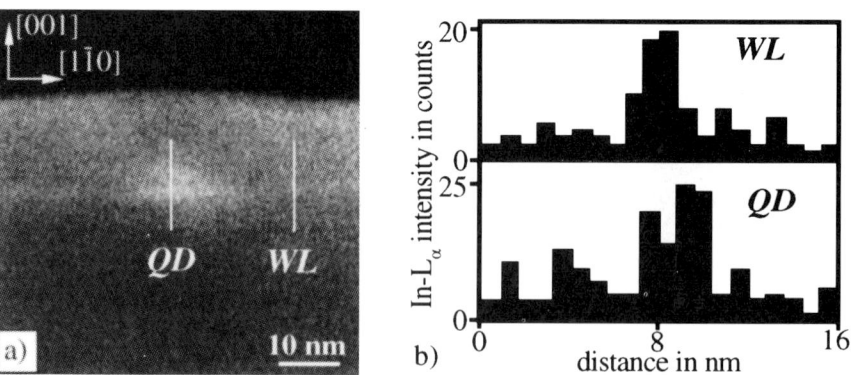

Fig. 2. An (In,Ga)As dot region, a) STEM dark-field image and b) EDXS line profiles.

From the profile across the QD an extension of about 7 nm can be estimated. Moreover, there are hints to two local maxima in the corresponding In distribution, which have also been observed in some other profiles across QD regions. Hence, an inhomogeneous In distribution is assumed in the dot region. Ga should behave vice versa, and this is better resolved by EFTEM Ga maps presented in the following.

5. ENERGY-FILTERED TRANSMISSION ELECTRON MICROSCOPY

The Ga distribution of the QD region was imaged at high lateral resolution using the Ga-M_{23} edge at 103 eV. Fig. 3a shows a zero-loss filtered HRTEM image of a single (In,Ga)As QD. The Ga maps were recorded under identical imaging conditions, where three single images were taken at 89 eV (pre-edge 1), 97 eV (pre-edge 2), and at 113 eV (post-edge), resp., each for 5 s with 8 eV slit width. Subsequently, the Ga map (cf. Fig. 3b) was attained by applying a constant R power-law fit of the background. In the resulting image some phase contrast is conserved and lattice fringes are still observed (see region marked by an arrow), albeit at low contrast. Both the position of the (In,Ga)As QD and that of the wetting layer are clearly revealed by the depletion in Ga.

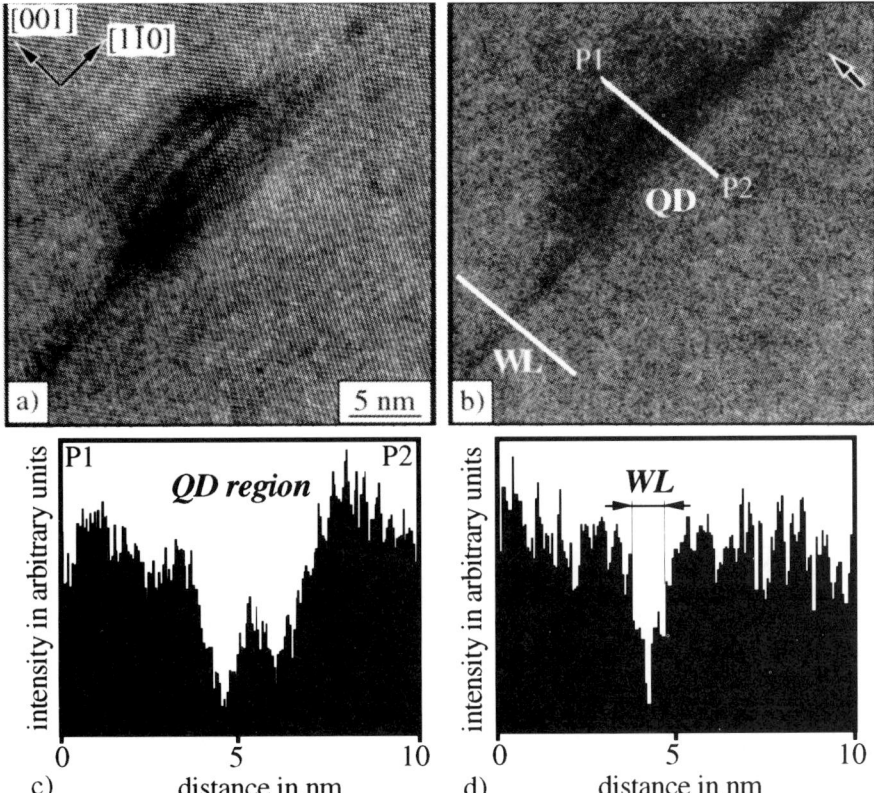

Fig. 3. EFTEM imaging of a single (In,Ga)As QD, a) zero-loss filtered HRTEM image, b) corresponding Ga-M_{23} element map, c) Ga profiles across the QD region, and d) across the wetting layer.

The findings were checked for artefacts by processing of the corresponding jump-ratio images. The pre-edge1/pre-edge 2 ratio image shows almost no inner structure. Contrary to that the post-edge/pre-edge 1 image shows similar contrast to the Ga map. Thus, it can be concluded that the contrast visible in the map is mainly caused by the presence of Ga.

Line profiles were gained from the Ga element map by integrating the counts in a rectangular frame (40 pixels wide) in the growth direction. The results are drawn in Figs. 3c,d yielding an approximately 1 nm thick wetting layer and a QD height of ca. 5 nm. In addition, crossing the QD region reveals two minima of the Ga content with a local maximum in-between. This particular Ga profile indicates a locally varying amount of Ga, which indirectly hints to an In segregation in the top region of the (In,Ga)As dot.

6. CONCLUSIONS

(In,Ga)As QD structures grown by MOCVD were characterized by HRTEM imaging, EDXS, and EFTEM. The existence of (In,Ga)As QDs was proved by EDXS line profiles of the In distribution. In comparison to the wetting layer a wider In-rich zone in the growth direction was detected in the dot region. Moreover, the Ga distribution was visualized by EFTEM, where the individual partial images were taken under high-resolution conditions. Profiles across the QD in growth direction attained from these maps exhibit two zones depleted in Ga, where one zone is in the position of the wetting layer and the other one is separated by a Ga enriched intermediate region.

Since elastic electron scattering obviously contributes to these elemental maps attempts have to be undertaken to differentiate between the individual contrast contributions, i.e. thickness variations, strain-field, and element-specific inelastic scattering. Therefore, simulations of HRTEM images of QD structures have to be performed with the ionization losses taken into account.

ACKNOWLEDGEMENTS

The authors are grateful to Prof U Gösele for providing the CM20 FEG microscope at the Max-Planck-Institut für Mikrostrukturphysik in Halle. Thanks are also due to the Deutsche Forschungsgemeinschaft for financial support of this work within the Sonderforschungsbereich 296.

REFERENCES

Heinrichsdorff F, Krost A, Schatke K, Bimberg D, Kosogov A O and Werner P 1998 Appl. Surf. Science **123/124**, 725

Ourmazd A, Schwander P, Kisielowski C, Seibt M, Baumann H and Kim Y O 1993 Inst. Phys. Conf. Ser. **134**, 1

Rosenauer A, Kaiser S, Reisinger T, Zweck J, Gebhardt W and Gerthsen D 1996 Optik **102**, 63

Strunk H P, Albrecht M, Christiansen S, and Dorsch W 1997 Inst. Phys. Conf. Ser. **157**, 323

Tillmann K, Thust A, Lentzen M, Swiatek P, Förster A, Urban K, Laufs W, Gerthsen D, Remmele T and Rosenauer A 1996 Phil. Mag. Lett. **74**, 309

Inst. Phys. Conf. Ser. No 164
Paper presented at Microsc. Semicond. Mater. Conf., Oxford, 22–25 March 1999
© 1999 IOP Publishing Ltd

Elemental mapping of semiconductor devices using energy-filtering transmission electron microscopy

W Grogger, F Hofer, P Warbichler and O Leitner*

Research Institute for Electron Microscopy, Graz University of Technology, Steyrergasse 17, A-8010 Graz, Austria; * Austria Mikro Systeme International AG, A-8141 Unterpremstätten, Austria

ABSTRACT: During the last couple of years transmission electron microscopy has gained more and more importance in semiconductor applications since device dimensions have been continuously decreasing. Simultaneously, energy-filtering transmission electron microscopy (EFTEM) has established itself as a powerful analytical technique to explore the chemical composition of a thin specimen on a nanometer range. Concentrating on application examples we will highlight the possibilities of EFTEM using a post-column imaging filter on a 200 kV microscope and point out the advantages and limitations of the method.

1. INTRODUCTION

Over the last decade there has been a large increase in the application of transmission electron microscopy (TEM) to semiconductor devices to evaluate and characterize new process modules and technologies and perform analyses for quality control. The increasing use of TEM is caused by several factors: first, cost reduction in terms of price per electrical function drives chip manufacturers to move to smaller critical dimensions and towards higher packing densities in integrated circuits. Continuous shrinking leads to the introduction of new technologies and materials with structural components, which are frequently in the nanometer range. Electron microscopes are ideally suited for imaging such devices making them invaluable for routine constructional analyses.

Much work, however, has been only devoted to morphological investigations and to the analysis of the crystalline structure and crystal defects in devices (e g. dislocations, stacking faults), which can be major yield detractors. Another reason for wider acceptance of TEM analyses is the refinement of the preparation of thin electron transparent samples, which enables the production of plan-views and cross-sections routinely and rapidly. The third important factor contributing to the increased usage of transmission electron microscopy is the possibility to combine it with advanced analytical techniques, such as energy dispersive x-ray spectrometry (EDXS) and the main topic here, electron energy loss spectrometry (EELS) which can provide chemical information at nanometer resolution. A recent and major improvement in this respect is the availability of energy-filtering TEM's (EFTEM) providing exceptional advantages for semiconductor device analyses. EFTEM is a very rapid method for both overview and nanoscale characterization of thin samples, applicable to most chemical elements and especially sensitive to lightweight elements ranging from lithium to zinc.

In this article we briefly outline the principles of EFTEM providing examples in order to show the types of applications where EFTEM generates unique information, which cannot be provided by any other method.

2. EXPERIMENTAL

These investigations were performed on a Philips CM20/STEM (200 kV, LaB$_6$-cathode) equipped with a Gatan imaging filter (GIF) (Krivanek et al 1992). For reasons of sensitivity, all

EFTEM images were recorded using a binning of 2 giving 512 x 512 images. For the acquisition of bright field images 1024 x 1024 pixels were used. Exposure times were between 10 and 60 s. Elemental maps (three window technique) were acquired as well as jump ratio images (two window technique) (Hofer et al 1995). Principally, these two types of images yield the same information; however, they are susceptible to different artifacts. In this paper only jump ratio images are shown, as they mostly exhibit a better signal-to-noise ratio.

In order to compute the quantitative line profile, we quantified elemental maps (not shown here) using k-factors experimentally determined from EEL spectra (Hofer et al 1997). The specimen was prepared using standard cross-sectional TEM preparation techniques with final low angle ion milling.

3. RESULTS

Energy-filtering TEM's are ideally suited for characterizing semiconductor devices, making them invaluable for routine constructional analysis. As shown in fig. 1 EFTEM elemental maps provide an overview of a large specimen area, which can be very important for rapid defect identification.

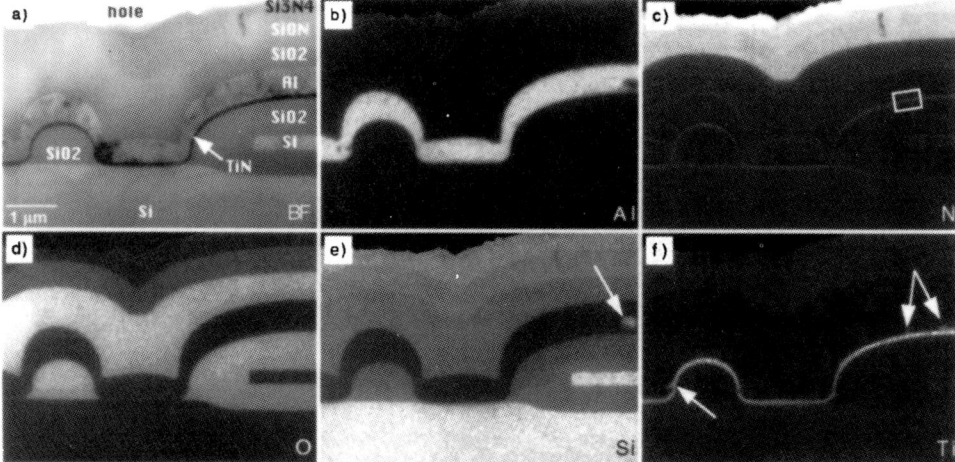

Fig. 1: TEM bright field image and elemental distribution images (jump ratio images of Al K, N K, O K, Si K and Ti L_{23}) of a cross-section through a semiconductor device.

In the TEM image (fig. 1a) only particular structures can be visualized, but all structures and phases are visualized in the elemental distribution images. The silicon map (fig. 1e) was recorded with the Si K-edge thus revealing bright regions which correspond to the silicon wafer and to a thin poly-silicon layer. The grey regions depict the silicon dioxide, silicon oxynitride and silicon nitride layers, respectively. It is even possible to detect the small gradient in silicon concentration in the silicon dioxide layer, which has been introduced by doping with boron and phosphorus. Additionally, a silicon-rich inclusion in the conductive layer (Al-Si-Cu alloy) can be detected (right margin of figs. 1b and 1e); this inclusion would have been overlooked by conventional methods.

The titanium map (fig. 1f) was recorded with the Ti L_{23}-edge and clearly shows the bright TiN layer, which exhibits spikes into the conductive layer (see arrows). The aluminium map (fig. 1b) was recorded with the Al K-edge and clearly shows the conductive layer; the silicon-rich defect and the titanium-rich spikes are visible as dark regions. In the nitrogen map (fig. 1c) we can see the thin titanium nitride layer, the silicon oxynitride and on the top of the structure the silicon nitride. Similarly, the oxygen map (fig. 1d) reveals the silicon oxide and the silicon oxynitride phases.

To study the titanium nitride layer and its interfaces in more detail, we acquired elemental distribution maps at higher magnification. Both, elemental maps and jump ratio images were recorded. Jump ratio images are less affected by diffraction contrast, whereas elemental maps are needed for quantification. In fig. 2 jump ratio images are shown, but the line profile (fig. 3) was calculated using quantified elemental maps.

The titanium nitride layer visible in fig. 1 contains some visible defects, so the rectangular area drawn in fig. 1c was chosen for fig. 2. Fig. 2a shows a bright field image of the titanium nitride layer, which exhibits a nominal thickness of about 90 nm. The aluminium map (fig. 2b) shows the conductive layer (Al-Si-Cu alloy), but highlights also some defective regions with a lower aluminium concentration. The titanium map (fig. 2d) shows a 90 nm thin layer and bright spikes penetrating into the conductive layer. However, the nitrogen map in fig. 2c clearly reveals that the effective titanium nitride layer is essentially thinner (approximately 70 nm) than the titanium layer. A comparison of the silicon and the oxygen map (not shown here) shows that the oxygen rich zone extends further into the titanium layer suggesting that a titanium oxynitride layer developed in the interface region.

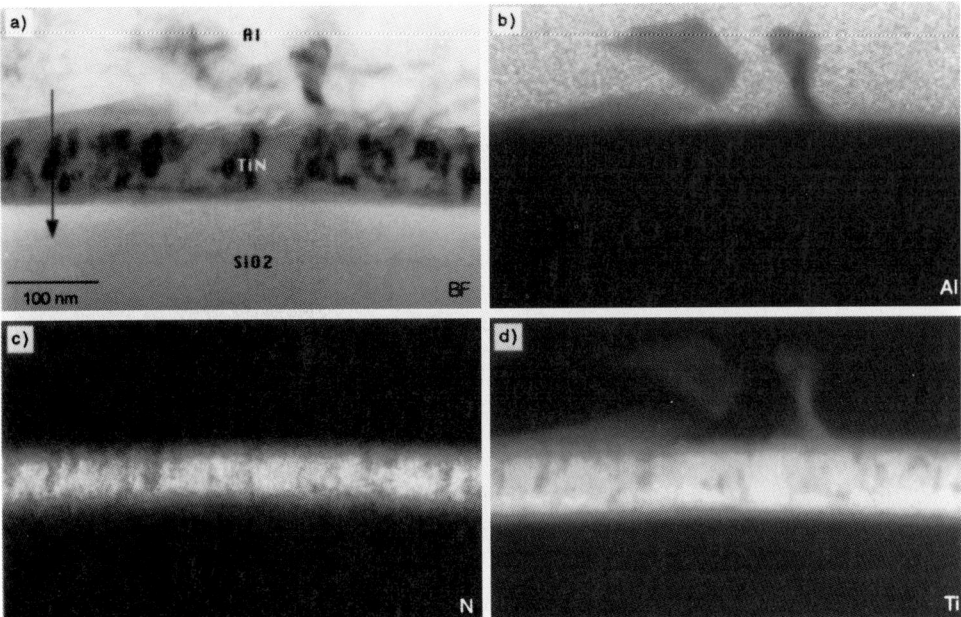

Fig. 2: TEM bright field image and jump ratio images (Al K, N K and Ti L_{23}) of a titanium nitride layer and its interfaces (enlarged section of fig. 1). The line indicates the position and length of a line profile, which is shown in fig. 3.

A powerful feature of EFTEM lies in the possibility of quantification. As previously shown by Hofer et al 1997 and Grogger et al 1998 elemental maps can be quantified by calculating atomic ratio images and multiplying them by the appropriate ionization cross section ratios (k-factors). Using EEL spectra acquired from distinct specimen areas of the titanium nitride layer and the interface regions concentration maps can be computed showing concentration values in atomic percent for each pixel. These concentration maps look similar to the jump ratio images shown in fig. 2 and are therefore not shown in this paper. However, these concentration maps make it possible to draw a quantitative line profile across the titanium nitride layer, thus revealing the concentration profiles for Al, N, O, Si and Ti (fig. 3).

From the line profile one can easily see the formation of an interface layer between Al and TiN consisting of Ti and Al. On the other hand the interface towards the SiO_2 also exhibits an intermediate layer containing O. The composition of the SiO_2 layer is very close to the nominal values (~ 33 at% Si, ~ 66 at% O), whereas the concentrations within the titanium nitride layer are about 60 at% Ti and about 40 at% N. These results were confirmed by the acquisition of an EEL

38

linescan, where individual EEL spectra were recorded along a line using the STEM probe of the microscope.

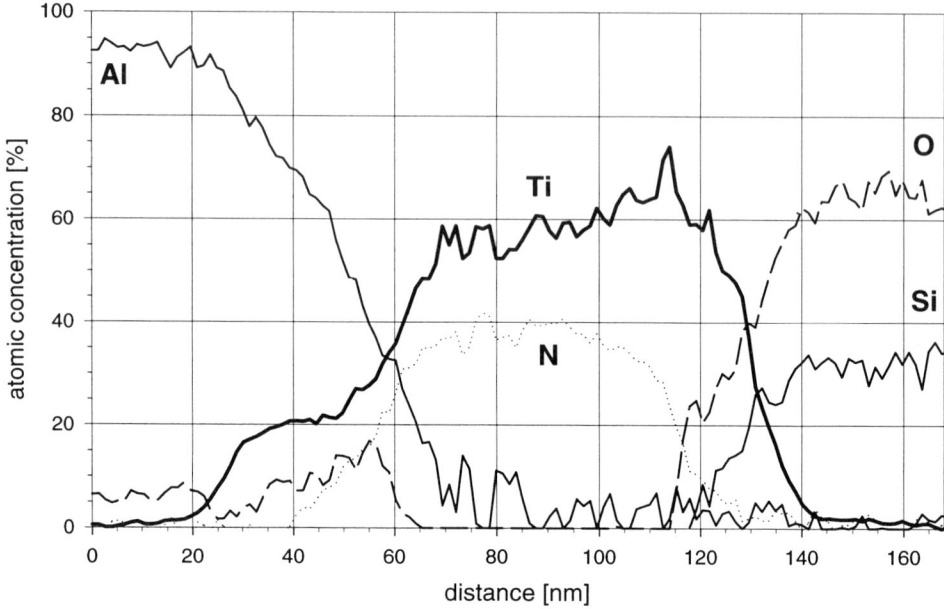

Fig. 3: Quantitative line profile drawn across quantified elemental maps of the individual elements (for the position of this line profile see fig. 2a).

4. CONCLUSION

We have demonstrated that EFTEM elemental distribution maps are a very powerful method for characterizing the chemical composition of semiconductor devices. To illustrate the potential of EFTEM, we have presented investigations of a semiconductor device. First the two dimensional distribution of the elements present in the specimen could be shown in a low magnification overview. In this state, inhomogeneities and defects especially in and near the titanium nitride layer could be detected. Subsequently one of the defective areas was chosen for investigation at higher magnification. There we found interfacial layers on both sides of the titanium nitride layer. The concentration profiles across the layer could be determined quantitatively.

Additionally, it should be mentioned, that further image processing could lead to even more information about and a better visualization of the chemical situation within the specimen; such as RGB images and scatter diagram analysis (Grogger et al 1998).

EFTEM is just now becoming used more widely in semiconductor research and will soon be an essential part of the equipment in advanced research laboratories.

The Bundesministerium für wirtschaftliche Angelegenheiten, Wien, Austria, is gratefully acknowledged for financial support.

REFERENCES

Grogger W, Hofer F and Kothleitner G 1998 Micron **29**, 43
Hofer F, Grogger W, Kothleitner G and Warbichler P 1997 Ultramicroscopy **67**, 83
Hofer F, Warbichler P and Grogger W 1995 Ultramicroscopy **59**, 15
Krivanek O L, Gubbens A J, Dellby N and Mayer C E 1992 Microsc. Microanal. Microstruct. **3**, 187

Inst. Phys. Conf. Ser. No 164
Paper presented at Microsc. Semicond. Mater. Conf., Oxford, 22–25 March 1999
© 1999 IOP Publishing Ltd

Atomic reorganisation studies from HRTEM images

J C Ferrer[1,2]**, F Peiró**[2]**, A Cornet**[2] **and G Armelles**[3]

[1]Serveis Cientifico-Tècnics, Univ. de Barcelona, Lluís Solé i Sabarís 1-3, Barcelona, Spain.
[2]EME Universitat de Barcelona, Av. Diagonal 645-647, 08028 Barcelona, Spain.
[3]Instituto de Microelectrónica de Madrid, CNM-CSIC, Isaac Newton 8, Parque Tecnológico de Madrid, 28760 Tres Cantos, Madrid, Spain.

ABSTRACT: A program based on the analysis of the local atom displacements measured from high resolution electron microscopy images has been applied to the analysis of the quality of GaAs/GaP superlattices. Features such as the homogeneity of the layers or the strain distribution are analysed by means of the grey map representation of the interatomic distances and bond angles. The results show that the GaAs layers are not homogeneous, tending to reorganise and produce accumulations in some regions.

1. INTRODUCTION

Increasing computer processing speed and storage capability have allowed the development of a broad set of applications for the analysis and simulation of high resolution transmission electron microscopy (HRTEM) images that are used frequently as an aid for the interpretation of experimental results. The classical simulation methods are based on the proposition of initial structure or supercell and observation conditions, i.e. specimen thickness and lens defocus, which must be changed for the experimental and calculated images to match. The disadvantage of this method is the need to trial and discard structures and observation parameters until the desired image is well fitted.

Another set of applications is based on the extraction of information directly from the experimental image without any prior assumption of the observation parameters. Kisielowsky et al (1995) propose a method to measure variations in the projected potential in crystalline solids with no knowledge of imaging conditions (QUANTITEM). Robertson et al (1995) measure deviations from averaged lattice fringe images to determine elastic strains in the InAlSb/InSb system. This is also achieved by Bierwolf et al (1993) by measuring local atomic deviations from a reference lattice for InAs/GaAs and Si/GaAs systems. Finally, Rosenauer et al (1997) use a combination of the QUANTITEM approach and the measurement of interatomic distances to detect the presence of InGaAs quantum dots buried in GaAs in high resolution TEM images.

In this work we present the application of the method of measuring atomic distances and bond angles to the characterisation of GaAs strained films, with a thickness ranging between 3 and 6 atomic layers, inserted in a GaP matrix. We discuss the homogeneity of the layers, as well as the strain distribution and the tendency of the GaAs to produce accumulations. The program is also tested on a relaxed InSb layer grown on an InP substrate.

2. EXPERIMENTAL

The epitaxially grown samples were prepared for TEM observation by conventional mechanical polishing and subsequent ion thinning in a cooled stage until perforation. Observations were performed in a Philips CM30 microscope. Following Bierwolf et al (1993), after the images are scanned, they are filtered in Fourier space in order to cut off high frequencies to avoid fine intensity fluctuations, and low frequencies to discard long term intensity fluctuations arising from thickness variations and sample bending. Instead of using an annular aperture in the frequency domain, a signal-to-noise ratio filter is employed to enhance the images. See the paper of Möbus et al (1993) for details on Fourier space filters to improve HRTEM images.

After the image filtering, the local intensity maxima of the high resolution image are located and associated to the basis of the zinc blende structure. The next step involves the calculation of the distances and angles between nearest basis neighbours (d and θ in Fig. 1) in the HRTEM image. These magnitudes are translated into grey maps that give us information about the chemical composition, using Vegard's law, and strain distribution, associating deviations from the equilibrium bond angles with tetragonal distortions. Future work involves the calculation of the effects due to the finite size of the samples.

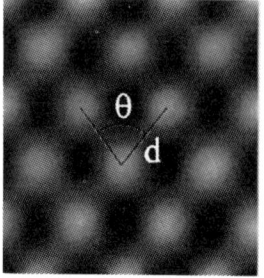

Figure 1. Parameters measured in HRTEM images.

3. RESULTS

In order to test the method in a well-known system, the program was applied to the image of a fully relaxed InSb layer grown on an InP substrate. In this case we also point out the complementary information given by the bond length and angle grey maps. Figure 2a shows the original HRTEM image of the region near the interface between both materials. We can observe the presence of a uniformly distributed array of Lomer dislocations which relax the system. As can be expected, associating brighter tones to larger bond lengths, the distance map (Fig. 2b) shows a brighter region corresponding to InSb (larger lattice parameter) and a darker region below corresponding to InP. We can also find a periodic contrast change located at the interface region associated with the lattice

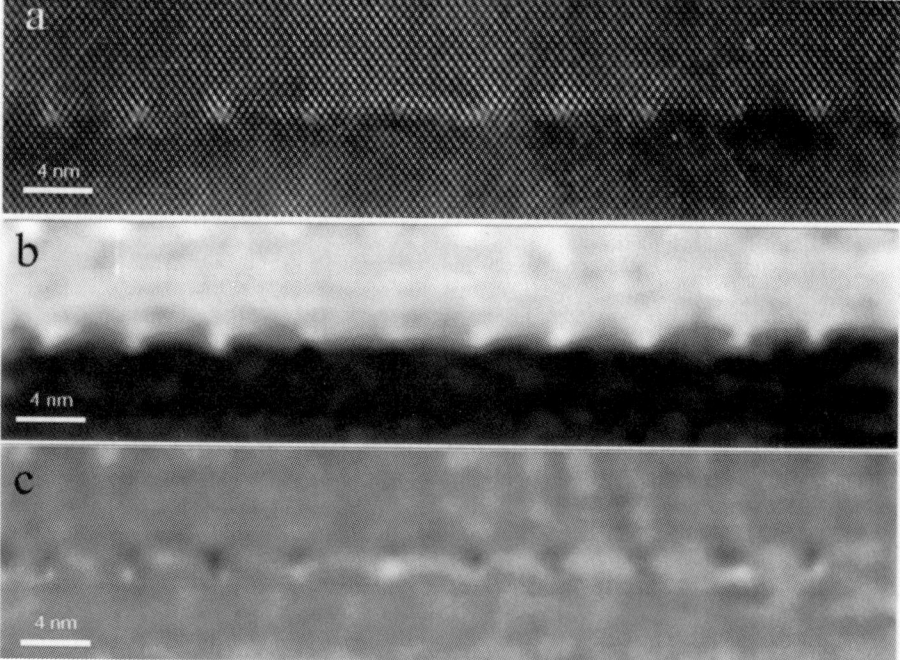

Figure 2. (a) HRTEM image of a fully relaxed InSb layer on InP; notice the array of dislocations at the interface. (b) and (c) Grey maps of bond length and angle. Brighter tones correspond to larger bonds in (b) and narrower angles in (c). The former clearly shows the difference between the lattice parameters of both materials, and the uniform contrast far from the dislocations in the latter confirms the layer relaxation.

distortions near the dislocations. On the other hand, excluding the same locations where the dislocations would be placed, the angle map shows a uniform contrast image because of the relaxed state of the structure and, hence, the similar bond angles in the InSb and InP bulk.

The results of the analysis of a 3 atomic layer (ML) GaAs film in a GaP matrix are presented in Fig. 3. It can be seen from the original high resolution image (Fig. 3a) taken along the [011] zone axis that the chemical contrast is very poor, and that the contrast fluctuations are an obstacle in evaluating the exact position and boundary of the layer. However, the local interatomic distance map (Fig. 3b) shows a layer with a lattice parameter higher than the rest of the area. Notice that this layer is not completely homogeneous probably due to a tendency of the GaAs to reorganise and to form quantum dots. In this case, as we found no defect which relaxes the structure in the HRTEM image, we expect to detect a tetragonal distortion and some strain variations near the GaAs layer. This is illustrated by the contrast in the angle map (Fig. 3c) with a variation at the same place we previously found the GaAs layer. The brighter intensity indicates narrower bond angles, i.e. tetragonal distortion for a material with a lattice parameter higher than that of the matrix. Notice the irregular distribution of the strain that reasonably fits the distribution of larger bonds in the distance map.

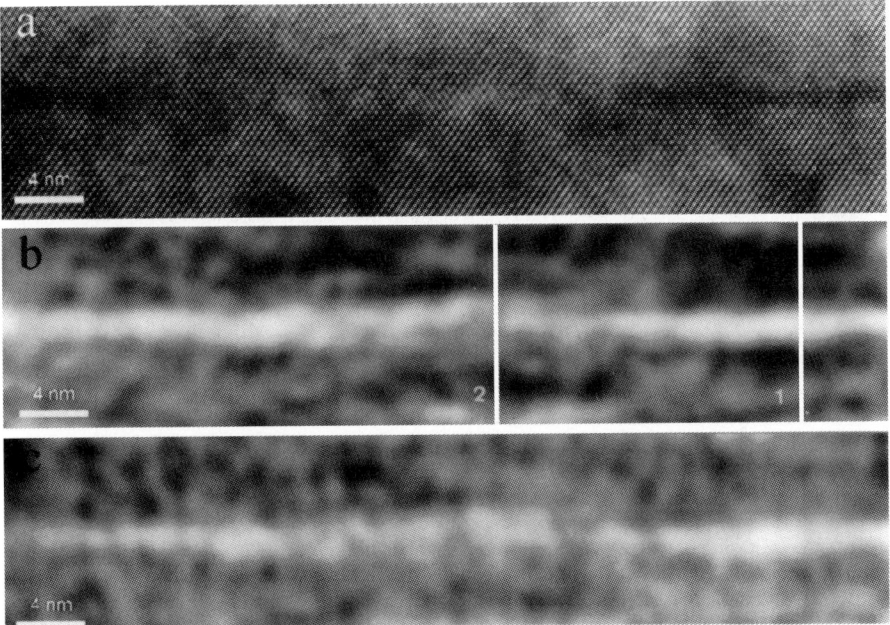

Figure 3. (a) HRTEM image of a 3 monolayer strained GaAs film in a GaP matrix. (b) and (c) Grey maps of bond length and angle, showing the layer location and homogeneity (brighter horizontal band). The angle narrowing (brighter band) confirms the strain distribution around the GaAs.

Figure 4 shows the profiles of the (200) interplanar distances vs. depth, calculated by averaging the values inside a GaAs rich region (section 1 in Fig. 3b) and outside it (section 2). The lattice parameter in the [100] direction is twice the value of this distance. The profile for 3 ML shows that inside the GaAs rich region the (200) plane distance is between half the lattice parameter of pure GaAs and that of GaAs strained on a GaP substrate (0.282 nm and 0.291 nm respectively). This indicates an enlargement of the unit cell in the [100] direction. The thickness of the full strained layer agrees rather well with the nominal thickness of the GaAs film. There are three planes with a separation of almost 0.291nm and the FWHM of the profile is 5 ML. The profile for the same sample in section 2 shows a wider layer with a FWHM of about 8 ML. Moreover, the (200) plane

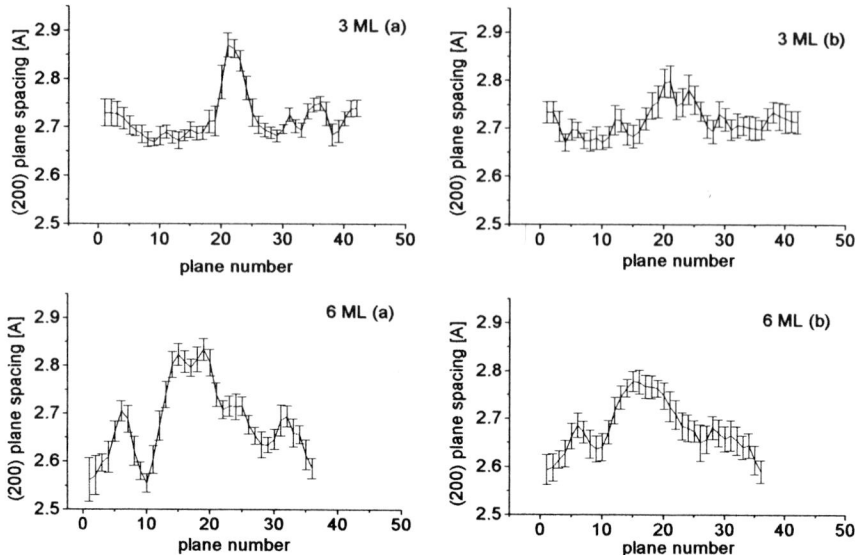

Figure 4. (200) interplanar spacing of 3 ML (upper) and 6 ML (lower) samples in: (a) an As rich region (section 1 in Fig. 3b) and (b) outside it (section 2 in Fig. 3b). In the first case the As concentration is enough to obtain a (200) planar spacing larger than that of GaAs due to the tetragonal distortion. In the second case, layers are wider and the plane spacing increase is not as obvious.

spacing is narrower than that of region 1 and, therefore, the concentration of As atoms in the layer should he lower, suggesting As interdiffusion with P.

The 6 ML sample shows a similar behaviour, with As accumulations at some locations. Hence, in As rich regions we obtain narrower and higher profiles than in regions with indications of interdiffusion. However, in this case the (200) plane spacing does not reach the value of 0.291 nm, corresponding to a completely strained pure GaAs layer; indicating an As loss. The profiles (Fig. 4 lower) suggest the presence of a secondary thinner GaAs layer on the left of the main peak (upper side), that may account for this As deficiency and that could be related to the interchange of As and P atoms during the epitaxial growth.

4. CONCLUSIONS

We have applied a program based on the measurement of the local lattice distortions in HRTEM images to the determination of the atomic reorganisation of a strained GaAs layer buried in a GaP matrix. Results are presented as grey scale maps of the bond lengths and angles. The complementary information of both types of maps has been shown by applying the program to relaxed InSb/InP as well as the strained layers. The application locates the situation of the GaAs layer as well as its inhomogeneities in the bond length map, and compares them to the stain distribution by means of the bond angle map.

REFERENCES

Bierwolf R, Hohenstein M, Phillipp F, Brandt O, Crook G E and Ploog K 1993 Ultramicroscopy **49**, 273
Kisielowski C, Schwander P, Baumaun F H, Seibt M, Kim Y and Ourmazd A 1995 Ultramicroscopy **58**, 131
Möbus G, Necker G and Rühle M 1993 Ultramicroscopy **49**, 46
Robertson M D, Currie J E, Corbett J M and Webb J B 1995 Ultramicroscopy **58**, 175
Rosenauer A, Remmele T, Fischer U, Förster A and Gerthsen D 1997 Inst. Phys. Conf Ser. No **157**, 39

Inst. Phys. Conf. Ser. No 164
Paper presented at Microsc. Semicond. Mater. Conf., Oxford, 22–25 March 1999
© 1999 IOP Publishing Ltd

The modified structure of amorphous Ge near Si(111) substrates

B Plikat, N I Borgardt*, M Seibt, T Wagner** and W Schröter

IV.Physikalisches Institut der Universität Göttingen, Bunsenstr.13-15, D-37073 Göttingen, Germany
* permant address: Moscow Institute of Electronic Technology, Moscow 103498, Russia
** Max-Planck-Institut für Metallforschung, Seestr.29, D-70174 Stuttgart, Germany

ABSTRACT: Using high-resolution transmission electron microscopy (HRTEM) we have investigated the interfaces between (111)- oriented crystalline silicon and amorphous germanium. Periodically averaged images reveal the existence of an interfacial layer which mediates between the crystalline substrate and the amorphous bulk. The width of this layer is estimated   1nm.

1. INTRODUCTION

Unlike crystalline materials where atomic positions are given in terms of periodic arrangements of a unit cell, amorphous solids are described by distribution functions of bond angles and lengths and atomic positions are uncorrelated for distances exceeding the spatial extension of the short-range order (SRO). In the vicinity of a crystalline material, however, long range correlations of atomic positions imposed by the translational symmetry of the substrate are expected. As a result, an interfacial layer exists which structurally mediates between the crystalline substrate and the bulk of the amorphous solid. How this structural transition is realized and of what extension the interfacial layer is for a given system, are open questions which are important for various problems in semiconductor science and technology like e.g. dielectric properties of thin silicon oxide and oxynitride layers or the crystallization of amorphous films on crystalline substrates.

Using high resolution transmission electron microscopy (HREM), we have investigated interfaces between amorphous germanium and crystalline silicon produced by molecular beam epitaxy of Ge on Si(111) at room temperature. We show that periodically averaged images are suitable to extract long range correlations from experimental images. From detailed comparison of simulated and experimental focus series it is concluded that the interfacial layer width is about 1nm.

2. EXPERIMENTAL

In our experiment Ge was evaporated on a Si(111) surface with a misorientation of less than $0.1°$ in $[11\bar{2}]$ direction as determined with atomic force microscopy by Suhren et al (1996).Such misorientations result in terrace widths of 180nm or more. They kindly provided pieces of their starting material which was covered with 100 nm thermal oxide grown in dry oxygen at 1000 °C. This substrate was cleaned in SC1 solution at room temperature for 10 minutes, rinsed in de-ionized (DI) water and etched in pH 5 buffered HF for 2.5 minutes to remove the oxide. After another rinse in DI water the specimen was smoothed by etching in 40%-NH₄F for 30 s and finally rinsed in DI water. This results in an hydrophobic surface with atomically flat hydrogen terminated unreconstructed terraces (Suhren et al 1996, Neuwald et al 1992). TEM cross-sections along [1$\bar{1}$0] were prepared by standard techniques involving mechanical polishing followed by ion beam thinning using Ar ions. Electron micrographs were obtained at 200kV in a Philips CM200-UT-FEG with a spherical aberration coefficient of $C_s = 0.48$mm resulting in a point resolution of 0.187nm. All HREM images shown in this paper were recorded with an objective aperture with a diameter corresponding to 9.6nm^{-1} which includes seven beams for imaging.

3.RESULTS

Due to the statistical nature of atomic positions in amorphous solids, a direct comparison of experimental and simulated images is not appropriate. Instead, we use periodically averaged image intensities $I'(x, y)$

$$I'(x,y) = \frac{1}{N} \sum_{n=-\frac{N-1}{2}}^{\frac{N-1}{2}} I(x, y + nd) \qquad \text{where} \qquad -\frac{d}{2} \leq y < \frac{d}{2} \qquad (1)$$

(d: periodicity of the lattice image along the interface, x: direction along the interface normal, y: direction perpendicular to the interface normal) which extract that part of image intensities with a periodicity of the crystal image along the interface (Borgardt et al 1997).

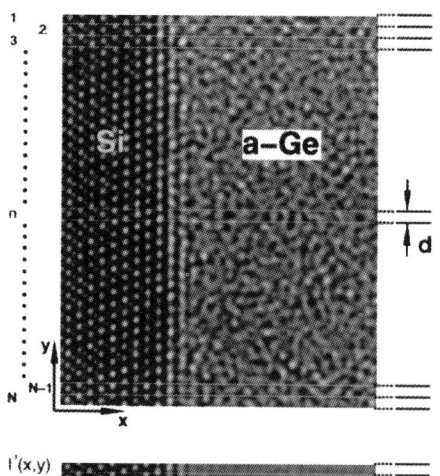

Fig.1: Illustration of periodic averaging described by Eq.1: the experimental image (top) is divided into N stripes of width d which are averaged to obtain the average image repeated five times (bottom). The latter reveals intensity modulations along the x- and y-directions.

Fig.1 shows an experimental image obtained at a c-Si(111)/a-Ge interface and its periodic average. Periodic averaging is only possible when changes of imaging conditions as e.g. sample thickness along the interface can be neglected. This can be checked by calculating the difference of the experimental and periodically averaged image (not shown). Difference images should be free from periodic signals in the crystalline part of the image.

As is expected from the random nature of positions of atoms at distances larger than the extension of the SRO, a constant intensity is observed for the bulk region of the a-Ge. Near the interface, however, the averaged image intensity is modulated along the x- and y-direction. Such intensity modulations may originate from different sources:

1. Scattering of electrons from the crystalline into the amorphous part of the sample.

2. Fresnel diffraction (Wilson et al 1978/79), also termed delocalization in recent work (Coene et al 1992).

3. Long range correlations of atomic positions in the interfacial layer mediating between the crystalline substrate and the amorphous bulk.

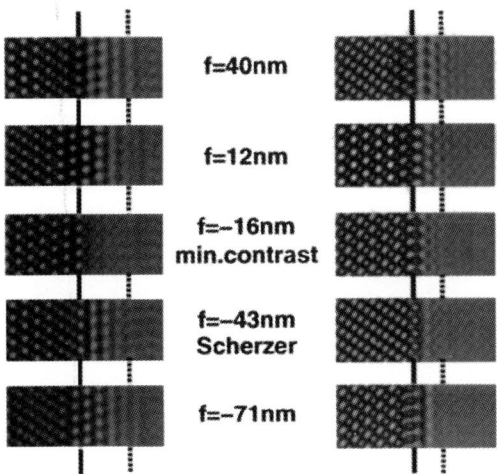

f=40nm

f=12nm

f=-16nm
min.contrast

f=-43nm
Scherzer

f=-71nm

Fig.2: Part of experimental (left column) and simulated (right column) focus series. In order to estimate the effects of Fresnel diffraction and scattering of electrons from the crystal into the amorphous material, amorphous Ge was simulated using its average potential. Solid lines indicate the position of the minimum extension of intensity modulations into the amorphous region within the focus series, dotted lines their maximum extension.

In order to estimate the effect of 1 and 2, we compare experimental focus series with simulations modeling the amorphous part of the sample by a homogeneous potential. The left hand column in Fig.2 shows averaged experimental images obtained from a focus series. The extension of intensity modulations into the a-Ge has a minimum at f=-16nm which corresponds to the focus of minimum contrast in the amorphous bulk. Statistically relevant intensity modulations can be observed for distances of about four (111) layers for large focus values, i.e. f=40nm and f=-90nm. Simulated images in the right-hand column of Fig.2 show a minimum extension of intensity modulations into the a-Ge at about Scherzer defocus (f=-43nm). Furthermore, these modulations reach far less into the amorphous region which is visualized by the solid and dotted lines which indicate the position of minimum and maximum extension of intensity modulations within the focus series, respectively. Two additional features should be noted:

- The contrast inversion in the amorphous region near the interface at minimum contrast is not observed in simulated images and hence can be attributed to the modified structure of the a-Ge in the interfacial layer.

- The distance between successive intensity maxima in simulated images is about 75% of that in experimental images.

4. CONCLUSIONS

In summary, the above simulations show that scattering of electrons from the crystal into the amorphous region as well as Fresnel diffraction do not account for the experimentally observed features of averaged images obtained from focus series. Hence, the above comparison of experimental and simulated images reveals the existence of an interfacial layer where atomic positions show long range correlations at a distance given by the periodicity of the underlying substrate along the interface.

Preliminary image simulations show that the distribution functions of bulk a-Ge as measured from X-ray diffraction (Etherington et al 1982) are not suitable to describe atomic positions in the interfacial layer. Furthermore, we can estimate a width of 1nm corresponding to three (111) double layers for the c-Si/a-Ge system from these simulations. This number is considerably larger than recent theoretical results obtained by Tu et al (1998). On the basis of energy calculations, these authors estimate an interfacial layer width of roughly one (111) double layer (Tu et al 1998). More detailed image simulations are currently carried out to confirm our estimated width of the interfacial layer and determine its atomic structure.

ACKNOWLEDGEMENTS

The authors are grateful to the VW Stiftung and the Sonderforschungsbereich 345 for financial support. We further thank M.Suhren (Wacker Siltronic, Burghausen) for the provision of low-miscut silicon substrates.

REFERENCES

Borgardt N I, Plikat B, Seibt M and Schröter W 1997 Bull. Russ. Acad. of Sciences, Physics **61**, 1544

Coene W and Jansen A J E M 1992 Scanning Microscopy Supplement **6**, 379

Etherington G, Wright A C, Wenzel J T, Dore J C, Clarke J H and Sinclair R N 1982 J.Non-Cryst. Solids **48**, 265

Neuwald V , Hessel H E, Feltz, Memmert U and Behm R J 1992 Appl. Phys. Lett. **60**, 1307

Suhren M, Gräf D, Schmolke R, Piontek H and Wagner P 1996 Inst.Phys.Conf.Ser. **No. 149**, 301

Tu Y, Tersoff J, Grinstein G and Vanderbilt D 1998 Phys.Rev.Lett. **81**, 4899

Wilson A R, Bursill L A and Spargo A E 1978/79, Optik **52**, 313

Inst. Phys. Conf. Ser. No 164
Paper presented at Microsc. Semicond. Mater. Conf., Oxford, 22–25 March 1999
© 1999 IOP Publishing Ltd

Measurements of local lattice strains in self-assembled InAs quantum dots: A high resolution transmission electron microscopy study

Chih-Hang Tung, G Muralidharan, George T T Sheng, Yaohui Zhang, John L F Wang, W J Fan[§], C H Wang[§] and J Jiang[§]

Institute of Microelectronics, Singapore 117685, Rep. of Singapore
[§]MBE Technology Inc., Singapore Science Park, Singapore 118226, Rep. of Singapore

ABSTRACT: Local lattice strains in self-assembled InAs quantum dots grown on GaAs were measured using high resolution transmission electron microscopy. The high resolution lattice images were processed using the cumulative sum of deviations (CUSUM) of lattice fringe spacing from a target value to measure the local lattice strains. The strain distribution along the [001] direction in the (110) plane was measured and the results are shown to be in good agreement with theoretical predictions as well as the results from the scanning tunneling microscopy measurements.

1. INTRODUCTION

Self-assembled quantum dot (SAQDs) structures, either the Si-Ge based system (Peng et al 1998) or the InAs-GaAs based system (Solomon et al 1996) have demonstrated great potential to exhibit novel electronic and optical properties. The Stranski-Krastanov grown self-assembled islands are formed, and strain is actually the driving force for the creation of dots. But the size and size distribution of the dots with respect to the growth conditions need to be understood. Local crystal lattice stress/strain is the key to understand the growth and formation mechanisms of the quantum dot islands (Solomon et al 1996, Tersoff et al 1996, Xie et al 1995). For direct measurement of strains in epitaxial structures on the nanometer scale, high resolution transmission electron microscopy (HRTEM) images recorded using slow scan charge-coupled device (CCD) cameras and analyzed using Fourier filtering techniques can provide reliable, quantitative results that are independent of image contrast (Robertson et al 1995a). The method is based on the analysis of the cumulative sum of deviations (CUSUM) of lattice fringe spacing from a target value (Robertson et al 1995b). This method has been shown to be very sensitive to lattice strain, and independent of the surface relaxation within a reasonably large sample thickness range. A preliminary study of the local lattice strain of Si-Ge SAQDs system using this method has demonstrated the capability of the technique (Tung et al 1998).

2. EXPERIMENTS

The sample was grown using a Riber 49 MBE system. A 4-inch semi-insulating GaAs (100) substrate was slightly etched before loading into the MBE system. The growth rates were 0.70μm/h for GaAs, 0.84 μm/h for $In_{0.17}GaAs$ and 0.11 μm/h for InAs. After oxide desorption, a 300 nm GaAs buffer was grown at the oxide desorption temperature of 580°C. The temperatures were measured by a pyrometer. Then a 3 nm $In_{0.17}GaAs$ layer was grown at 500°C, followed by a 100 nm GaAs at 580°C. Then a 10 period 0.6nm/10nm InAs/GaAs quantum dot superlattice was deposited at 500°C. Finally, a 20 nm GaAs cap layer was grown at 580°C. Reflection high energy electron diffraction (RHEED) patterns were used to monitor the growth. InAs was deposited at a rate of 0.1 monolayer

(ML) per second with 8-second growth interruptions after every 0.4 ML InAs. The RHEED pattern changed from streaks into spots after 1.6 ML InAs was grown. A total of 2 ML InAs was grown. The RHEED pattern changed rapidly from spots to streaks and resumed the 2X4 pattern when a GaAs layer was grown over the InAs quantum dots. The substrate was rotated at a speed of about 20 rpm during the growth.

Cross-sectional transmission electron microscopy samples were prepared using grinding and polishing techniques, and subsequent ion milling. To reduce possible artifacts, the ion milling time of the sample was kept below 15 minutes. High resolution transmission electron microscopy (HRTEM) was done on a Philips CM200 FEG TEM with a CCD camera. High resolution digital TEM images were first acquired and then selected and Fast Fourier Transform (FFT) performed on the selected areas. The (002) reflections were selected and filtered in for inverse FFT to obtain the (002) lattice fringes. After obtaining the filtered (002) lattice fringe images, the fringe intensities as a function of pixel position were digitized, fitted with a polynomial and the peak positions were used for CUSUM computation.

3. RESULTS AND DISCUSSIONS

The CUSUM technique calculates the sum of deviations of the individual lattice fringe spacing from a target value. The average fringe spacing is a useful choice for the target value since regions of the CUSUM plot where the lattice fringe spacing is less than the average value will have negative slope while regions where the lattice fringe spacing is greater than the average will have a positive slope. From the slopes in different areas in the same CUSUM plot, the misfit perpendicular to the interface, measured with respect to a reference region (assumed to be a strain-free lattice region in this study), can be calculated using a simple relationship:

$$e = \frac{(m_2 - m_1)}{(1 - m_2)},$$

where m_1 is the slope of the CUSUM plot in the strain-free matrix lattice area while m_2 is the slope in the area where the strain is to be calculated.

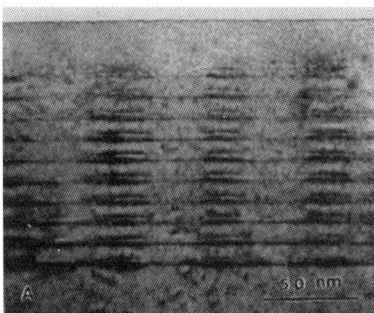 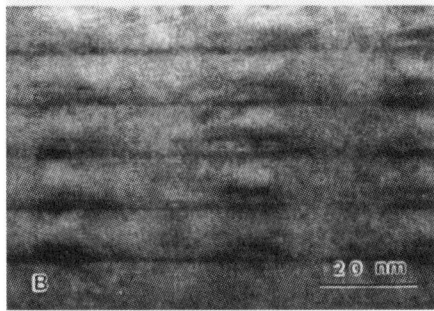

Fig. 1 Cross-sectional TEM images of the vertically aligned InAs dots. The island height is approximately 6 nm and the in-plane dimension is approximately 20 nm.

Cross sectional TEM images of the sample showed distinctive vertically aligned InAs dot columns, as shown in Fig. 1. The island height, measured visually using image contrast was found to be approximately 6 nm and the in-plane dimension was approximately 20 nm. The observed shape of the InAs dots is more like a liquid wetting a substrate with a wetting angle of approximately 30° than like a pyramidal model used for theoretical calculations (Grundmann et al 1995). It should be noted that it is difficult to accurately determine the island height since at the island peaks only a small number of InAs atoms remain in cross section to contribute to the contrast, and the observed cross section may not necessarily intersect the island centre. Exactly the same problem contributes to the strain calculations as well and will be discussed later.

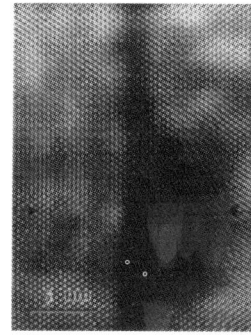

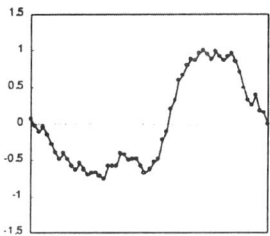

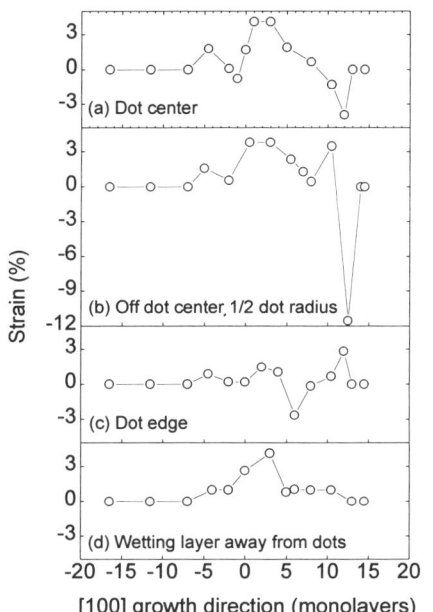

[100] growth direction (monolayers)

Fig. 2 High resolution TEM micrograph of an InAs SAQD rotated 90° clockwise and the corresponding CUSUM plot from the dot centre, as indicated.

Fig. 3 Lattice strain (%) calculated from the CUSUM plots. (a) dot centre, (b) off dot centre at about ½ dot radius, (c) dot edge, and (d) wetting layer away from any dot. The horizontal axis is chosen so that the InAs wetting layer is from 0 to about 2.5 monolayers and the InAs dot is from 0 to about 11 monolayers of GaAs lattice.

Figure 2 shows a typical single InAs dot (rotated 90°) along with the CUSUM plot extracted from the dot centre area perpendicular to the interface. The strain as calculated from the CUSUM plots from the dot centre, dot edge, one-half radius off the centre, and the wetting areas are shown in Fig. 3. Table I shows a comparison of the strains obtained from the theoretical calculations (Grundmann et al 1995), and the current measurement based on the high resolution TEM and CUSUM calculations. Also shown in the table, for comparison, are the results from the STM measurements (Legrand et al 1998). The highest measured tensile strain in the present study is approximately 4% and is located at the lower interface across the dot centre, in reasonable agreement with the theoretical calculations (Grundmann et al 1995). The highest compressive strain of -11.5% is observed at the dot top surface, about a half radius away from the dot centre. Although the theoretical calculations also predicted a similar compressive strain at the top surface (-9%), it was predicted to occur at the apex of the pyramidal dot. It should be noted that in the theoretical model the dot was assumed to be pyramidal in shape while a "half-lens" shape was actually observed in the images. Further, inaccurate determination of the dot centre in the images may also contribute to this discrepancy. Note that the STM measurements show significantly larger magnitudes of both the compressive and the tensile strain. It is also possible to determine, from the CUSUM plots, the actual height of the InAs dots and the wetting layer thickness. The present study showed that the InAs dot height is about 6.2 nm and the wetting layer thickness is about 1.5 nm. This is very close to some of the previously published data (Soloman et al 1996, Grundmann et al 1995). Several important features have been experimentally verified in this study. These include the high tensile strain at the dot centre lower interface, the change in the sign of the strain (from tensile to compressive) at or near the centre of the dot cross section, and the high compressive strain at or near the top surface of the dot.

There are, however, several differences between the theoretical predictions and the current measurements. One noticeable difference is the consistent presence of a small region with a tensile strain of about 1.5 % near the InAs/GaAs interface at or near the dot centre bottom area that was not predicted by the theory. Although the origin of this strain is not clear, such an expansion strain hump may coincide with the large compressive strain on top of the dot peak and induce the alignment mechanism for the dots in the following layers. Another difference between the theoretical calculations and the measurements is the more gradual change in strains over the dot centre area and the wetting areas. Since the high resolution images are taken with the averaged lattice spacings across the sample thickness direction, the much-smoothed strain measured is less than surprising. The dot diameter measured about 20 nm at the bottom area, the full TEM sample thickness, estimated to be about 50 nm in the present analysis, covers not only the InAs dot area, but also the GaAs matrix lattice. This inevitably will average out the lattice strain and results in a less drastic strain profile. The influence needs to be further studied quantitatively through high resolution image simulation.

Table I A comparison among the theoretical prediction, the STM measurement, and the current HRTEM measurement on the local lattice strain within and around an InAs self-assembled quantum dots.

Strain in %		Theory	STM	HRTEM
Dot Center	max.	3	7	4
	min.	-9	-15	-4
1/2 Off Center	max.	3.5	---	3.8
	min.	-2.5	---	-11
Dot Edge	max.	---	---	3
	min.	---	---	-3
Wetting	max.	7	---	4
	min.	0	---	0

In conclusion, it is demonstrated, in this study, that the local lattice strain within and around an InAs self-assembled quantum dot can be measured by using high resolution TEM images. The strains derived from the slopes of CUSUM plots of lattice fringe spacings were in good agreement with the theoretical predictions as well as the results from STM measurements.

The authors would like to thank the Institute of Microelectronics (IME) and the National Science and Technology Board (NSTB) of Singapore for supporting this study.

REFERENCES

Peng C S, Huang Q, Cheng W Q, Zhou J Z, Zhang Y H, Sheng T T and Tung C H 1998 Appl. Phys. Lett. **72**, 2541

Solomon G S, Trezza J A, Marshell A F and Harris Jr. J S 1996 Phys. Rev. Lett. **76**, 952

Tersoff J, Teichert C, and Lagally M G 1996 Phys. Rev. Lett. **76**, 1675

Xie Q, Madhukar A, Chen P and Kobayashi N P 1995 Phys. Rev. Lett. **75**, 2542

Robertson M D, Corbett J M, Currie J E and Webb J B 1995a Proc. Microscopy of Semiconductor Materials 1995, eds A G Cullis and A E Staton-Bevan (Bristol:IOP Publishing) pp 48-52

Robertson M D, Corbett J M, Currie J E and Webb J B 1995b Ultramicroscopy **58**, 175

Tung C-H, Muralidharan G, Sheng T T, Zhang Y H, Wang J L-F and Li M F 1998 Proc. 4th Nat. Symp. Progress in Materials Research (Singapore) 165

Grundmann M, Stier O and Bimberg D 1995 Phys. Rev. B **52**, 11969

Legrand B, Grandidier B, Nys J P, Stievenard D, Gerard J M and Thierry-Mieg V 1998 Appl. Phys. Lett. **73**, 96

Inst. Phys. Conf. Ser. No 164
Paper presented at Microsc. Semicond. Mater. Conf., Oxford, 22–25 March 1999
© *1999 IOP Publishing Ltd*

The atomic structure of dislocations and planar boundaries in GaN

P Ruterana, V Potin, P Vermaut[1], G Nouet, B Barbaray, A Botchkarev[2] and H Morkoç[2]

Laboratoire d'Etudes et de Recherches sur les Matériaux , UPRESA 6004 CNRS, Institut des Sciences de la Matière et du Rayonnement, 6 Bd Maréchal Juin, 14050 Caen Cedex, France.
1 Now at the Laboratoire de Métallurgie Structurale, Ecole Nationale de Chimie de Paris, 11, rue Pierre et Marie Curie, 75231 Paris Cedex 05
2 Department of Electrical Engineering, Virginia Commonwealth University, P.O. Box 843072, Richmond, VA23284-3072, USA
email: ruterana@lermat8.ismra.fr

ABSTRACT: GaN layers contain large densities (10^{10} cm^{-2}) of threading dislocations, nanopipes, (0001) and $\{11\bar{2}0\}$ stacking faults, and $\{10\bar{1}0\}$ inversion domains. Three configurations have been found for pure edge dislocations, mainly inside high angle grain boundaries where the 4 atom ring cores can be stabilized. Two atomic configurations, related by a $1/6 <10\bar{1}0>$ stair rod dislocation, have been observed for the $\{11\bar{2}0\}$ stacking fault in (Ga-Al)N layers. Such defects form on steps in epitaxial layers on the top of (0001) 6H-SiC; on (0001) sapphire, they appear on coalescence of two adjacent islands related by the I1 stacking fault. For the $\{10\bar{1}0\}$ inversion domain boundaries, a configuration corresponding to the Holt model was observed, as well as another with no N-N or Ga-Ga bonds. The two configurations are related by a c/2 translation. These boundaries are quite efficient in minimizing the shift along **c** which is introduced by substrate surface steps.

1. INTRODUCTION

Due to their direct bandgap, III-V nitrides are excellent candidates for optoelectronic application from red to ultra-violet (1.89 eV : InN, 6.2 eV : AlN). However the fabrication of devices is encountering intrinsic problems which are now motivating a world wide research effort. This has been exponentially increasing for the last ten years since the discovery of Mg for p doping (Amano et al 1989). Light emitting diodes were available from 1993 and laser emission was demonstrated at the end of 1995, all in layers containing up to 10^{10} cm^{-2} extended defects (Ponce et al 1994, Nakamura et al 1996). For lack of bulk GaN, active layers are grown either on sapphire or SiC substrates where the misfit in the basal plane is 13 % and 3.5 %, respectively. The large majority of the extended defects are threading dislocations which originate from the interface with the substrate and mostly cross the whole epitaxial layer (Lester et al 1995, Vermaut et al 1995, Rouvière et al 1997). Planar boundaries on the $\{11\bar{2}0\}$ prismatic planes have also been investigated and have led to some controversial reports. They have been characterized as double positioning boundaries (Tanaka et al 1995) stacking mismatch boundaries (Sverdlov et al 1996) and inversion domain boundaries (Rouvière et al 1995), as well as stacking faults (Vermaut et al 1997 a,b,c). In this report, we discuss our results on the atomic structure of the extended defects obtained in

(Ga, Al) N layers grown by molecular been epitaxy on SiC and sapphire in the light of recent literature and we propose some possible mechanisms for their formation.

2. EXPERIMENTAL DETAILS

The investigated GaN layers were grown on the (0001) sapphire or 6H-SiC surfaces by electron cyclotron resonance (ECR) or NH_3 gas source MBE. The ECR-MBE GaN layers were directly grown at 800°C at a rate of 40 nm/h, up to a 2 μm thickness. The NH_3-MBE GaN layers were also deposited at 800° C but on a low temperature (550°C) GaN buffer layer (40 nm thickness).

The TEM samples were thinned down to 100 μm by mechanical grinding and dimpled down to 10 μm. Electron transparency was obtained by ion milling at 5 kV with a liquid N_2 cold stage. A final step at 3 kV was used to decrease ion-beam damage. HREM experiments were carried out along the $[11\bar{2}0]$ and $[0001]$ GaN directions on a Topcon 002B electron microscope operating at 200 kV with a point to point resolution of 0.18 nm (Cs = 0.4 mm). Convergent beam electron diffraction (CBED) experiments were carried out on a JEOL EM 2010 microscope operating at 200 kV with a probe diameter of 25 nm. For displacement fringe analysis, it was necessary to tilt specimens to angles larger than 40°, so we had to use a JEOL 200 CX analytical microscope. HREM images and CBED patterns were simulated using the electron microscopy software (Stadelmann 1987).

3. CRYSTALLOGRAPHIC CONSIDERATIONS

Due to the lack of bulk GaN, the growth of active layers is carried out by hetero-epitaxy on a variety of substrates among which sapphire and SiC clearly dominate. Most of the growth has been carried out on the (0001) surfaces where one has to consider the misfits in the basal plane and the shifts which can be introduced at steps (table 1).

	a (nm)	$\Delta a/a_{SiC}$ (%)	$\Delta a/a_{sap}$ (%)	c (nm)	$\Delta c/c_{SiC}$ (%)	$\Delta c/c_{Sap}$ (%)	Polytype	S G
GaN	0.3189	3.54	16.09	0.5185	2.9	19.7	2H	
AlN	0.3112	1.04	13.29	0.4982	1.15	15.06		$P6_3mc$
SiC	0.308			1.512			6H	
Sapphire	0.476			1.2991				$R\bar{3}c$

Table1: crystallographic data on wurtzite GaN, AlN and the substrates: sapphire and 6H-SiC

Along the **c** axis, this misfit is quite small in the case of 6H-SiC, whereas it is close to 20 % on sapphire.Quite important is the difference in stacking sequences. Whereas wurtzite is a stacking of two interpenetrating hcp lattices, the 6H packing of SiC is more complex as it is deduced from fcc by twinning every three sequences. Recent work has shown that steps at the (0001) 6H-SiC connect terraces by a displacement vector which is either zero, or equal to one of the three stacking fault vectors of the hexagonal lattice (Vermaut et al 1997a). When this vector has a component along [0001], an equivalent translation will exist between islands grown on the adjacent terraces.

For growth on (0001) sapphire, the situation is more complex. A simple geometrical analysis shows that even on a flat surface, there are eight possibilities for the growth of a GaN layer. They are either related by displacement vectors and/or an inversion operation (Barbaray et al 1999). Using 5c GaN ~ 2c sapphire, we were able to identify four types of steps which lead to residual translations upon growth on adjacent terraces (table 2).

Steps	$h/c_{Al_2O_3}$ unit	T_R/c_{GaN} unit	T_R/nm
A-A or B-B	nc (n: integer)	0	0
A-B	1/6, 5/6, 7/6 ,11/6	~ 1/12	~ 0.0432
A-A or B-B	1/3, 2/3	~ 1/6	~0.0863
A-B	1/2, 3/2	~ 1/4	~ 0.1295

Table 2: The 4 values of the residual translation T_R due to steps h between oxygen terraces of the (0001) surface of sapphire

4. THE DISLOCATIONS

A lot of work has been devoted to the study of these defects and it has been demonstrated that most of the strain is released at the interface and inside the buffer layers.

4.1 The misfit dislocations

The interfacial area of GaN/sapphire can exhibit zones without extended defects, but even then, it is not easy to locate the misfit dislocations at the interface. A Fourier filtering shows that they are regularly spaced 60° dislocations. And, if we take the core to be located at the interface, the measured average distance of 1.7 nm shows that they define a stepless relaxed area (Fig. 1).

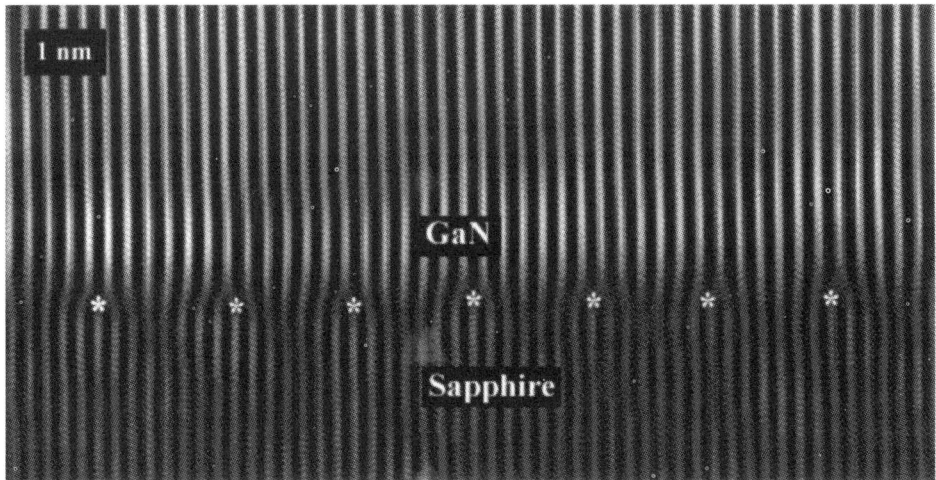

Fig.1 Regularly spaced misfit dislocations at the GaN/sapphire interface, defining a relaxed area.

So in agreement with other published work, the threading dislocations do not contribute to the relaxation of the strain (Rouvière et al 1997, Kaiser et al 1998).

4.2 The threading dislocations

Conventional TEM has shown that these dislocations originate at the interface and the large majority propagate to the sample surface. Their line is roughly parallel to the **c** growth axis and the large majority are edge type. It has been shown that **c** and **a**+**c** dislocations tend to bend and annihilate ; the density of those that reach the layer surface is only a few percent of that near the interface (Rouvière et al 1997). Calculations have shown that the low optical activity of the edge threading dislocations may be explained by the formation of dimer bonds inside an eight atom ring core (Elsner et al 1997). The most accurate observations published so far have shown that this is the case in MOCVD grown samples (Xin et al 1998).

However, our HREM observations suggest that this may not be the case, at least in MBE grown GaN layers. In addition to the above atomic configuration, we have noticed typical contrast (Fig. 2) which is not explained by an eight atom core. And from our observations, this configuration has more or less an equal frequency with the 8-atom ring core. Image simulations show that this core is compatible with a 5/7 atom-ring configuration (Ruterana et al 1998).

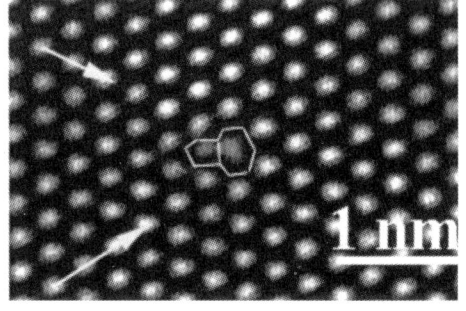

Fig. 2 HREM along [0001] of an edge disloc-ation in GaN with 5/7 atom-ring core inset, arrows show the additional lattice fringes.

Inside the layers we have observed high angle grain boundaries rotated about the [0001] axis, in coincidence orientation relationships. Although these may not be present in good quality epitaxial layers, they were found to exhibit 5/7, and 4/6 as well as the 8-atom ring dislocation cores   (Potin et al 1999a).

4.3 Nanopipes

Such defects are empty or filled holes which exhibit a dislocation character, they mostly have a regular shape. They can extend up to a few tens of nanometres and are usually limited by $\{10\overline{1}0\}$ planes. They have been called nanopipes in analogy to SiC in which such features can measure more than 10 μm and are called macropipes. From our experience they can be classified into two classes, those which have an edge component and those which are pure screw dislocations. As shown in Fig. 3, as large as 2a edge component may be exhibited by such defects. In our specimens, it was noticed that the former class of nanopipes are confined inside the first 200 nm of the epitaxial layer (Vermaut et al 1996); they contain amorphous material in layers grown on 6H-SiC, whereas they are empty on top of sapphire (Ruterana et al 1997). Other authors have shown that the pure screw nanopipes cross the entire epitaxial layer (Cherns et al 1998). They may either keep the same section or close and open up a few times on their way to the surface; in this case they can terminate into surface pinholes (Liliental-Weber et al 1997).

They originate from the coalescence of the mosaic growth islands when dislocations are pinned and they form in order to minimize the elastic energy as already explained by Frank (1951).

5. THE {11$\bar{2}$0} STACKING FAULT

These defects were first reported in epitaxial ultrathin (10 nm) layers from HREM observations carried out in a AlN thick layer, along the [11$\bar{2}$0] zone axis. As the habit plane could not be determined, the planar boundary was then called a DPB (Tanaka et al 1995).

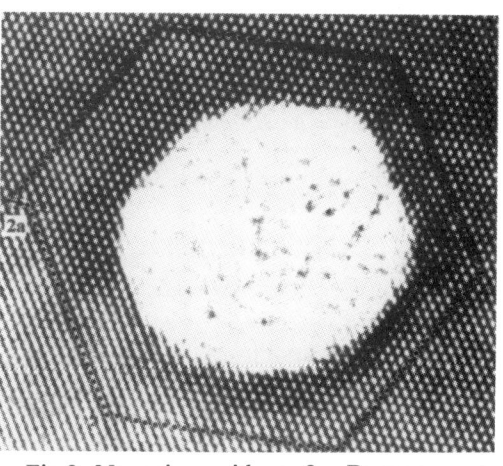

Fig.3 Nanopipe with a 2a Burgers vector component, limited by {10$\bar{1}$0} facets.

A subsequent report, which concentrated on the analysis of the stacking difference between SiC and GaN, named it a stacking mismatch boundary (Sverdlov at al 1995). In the next observations in GaN carried out along [0001], the habit plane was determined. Then several

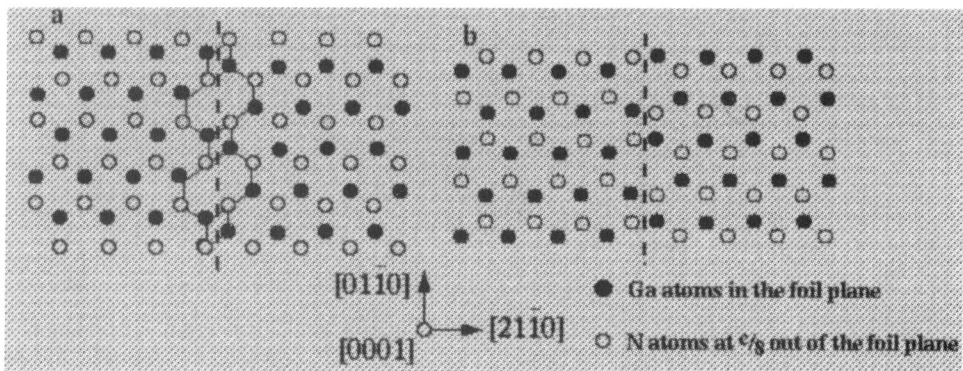

Fig.4 The two observed atomic models for the {11$\bar{2}$0} stacking faults: a) Blank et al. 1964, b) Drum 1965.

possible atomic configurations for the boundary were considered with a particular support for an inversion domain boundary model (Rouvière et al 1995). In HREM investigations carried out in (Ga, Al) N layers grown on 6H-SiC (0001), Vermaut et al 1997a showed that similar boundaries originated systematically from I1 =1/6 [20$\bar{2}$3] types of steps at the substrate surface. These boundaries could either form closed domains or be limited by partial dislocations. For the closed domains convergent beam electron diffraction was used along the [10$\bar{1}$1] zone axis and it allowed the inversion character to be ruled out (Vermaut et al 1997b). Then two atomic configurations (fig.4),which have already been reported in wurtzite materials were compared to the experimental images recorded along the [0001] and [11$\bar{2}$0] zone axes (Blank et al 1964, Drum 1965). Along [0001], where the main results were obtained inside the GaN layers, the Drum model ($\frac{1}{2}$ $[10\bar{1}1]$) was found to agree with all our observations (Vermaut et al 1997a). In cross section along the $[11\bar{2}0]$ zone axis, on an image recorded inside the AlN buffer layers, it was possible to identify the two atomic configurations which were connected by a basal I1 stacking fault (fig. 5). Therefore, as this boundary can be limited by partial dislocations and   it continuously folds in the {0001} and {11$\bar{2}$0} planes as for example in 2 and 3 on fig.5, one can use the name stacking fault to designate it.

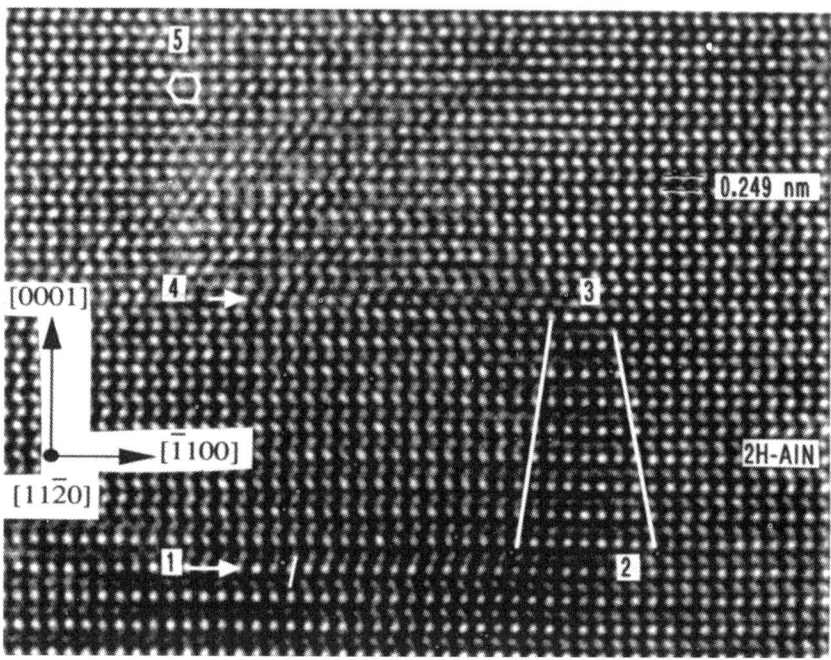

Fig. 5 Simultaneous observation of the Amelinckx(2-3) and Drum (4-5) models of the {11$\bar{2}$0} stacking faults, they are a mere continuation of the basal I1, as seen from 1 to 2 and 3 to 4.

6. INVERSION DOMAIN BOUNDARIES

Inversion domains are naturally possible in the wurtzite structure due to the two tetrahedral sites of the anion sublattice which cannot be simultaneously occupied by cation atoms (Vermaut 1997a). However their formation was found to depend on the growth parameters such as substrates used, buffer layer thickness and growth temperature (Vermaut et al 1997, Rouvière et al 1997). In epitaxial GaN, it is now admitted that they tend to develop in N polar matrix in which they grow at a higher rate and usually reach the layer surface in the center of pyramidal features. In Ga polar layers, the inversion domains are confined to the interfacial area (Rouviere et al 1997). The domains that reach the epitaxial layer surface are limited by {10$\bar{1}$0} boundary planes and, although they originate from the sapphire surface, the presence of oxygen has not yet been associated with them. Ab initio calculations have shown that, due to the ionic character of the compound, the IDB can have low energy if it only contains Ga-N bonds (Northrup et al 1996). We have investigated many GaN layers grown simultaneously on SiC and sapphire substrates. In the GaN films grown on

SiC, we never found inversion domains which supports a strong substrate effect underlying their formation (Ponce et al 1996). On top of sapphire, two types of samples containing inversion domains were identified by their layer surface morphologies (Potin et al 1999b). Some have flat surfaces, whereas others exhibit small pyramids with a surface roughness of about 100-200 nm. In this case, inversion domains up to 50 nm in width were located in the centers of the pyramids. In the former, a high density of smaller inversion domains could be found inside the layer, all less than 20 nm in size. In the model proposed by Holt (1969), the Ga and N (fig.6a,b) atoms are only interchanged across the boundary, whereas in the "V" configuration, there is an additional $c/2$ translation of one crystal in order to avoid the formation of Ga-Ga or N-N bonds (Fig. 6c,d). Our observations did not indicate any switching from one configuration to the other inside the same sample. In contrast, as shown in figure 7, it was possible to notice that in the two configurations the boundary plane could be located in the glide or shuffle positions, thus breaking one or two bonds per atom.

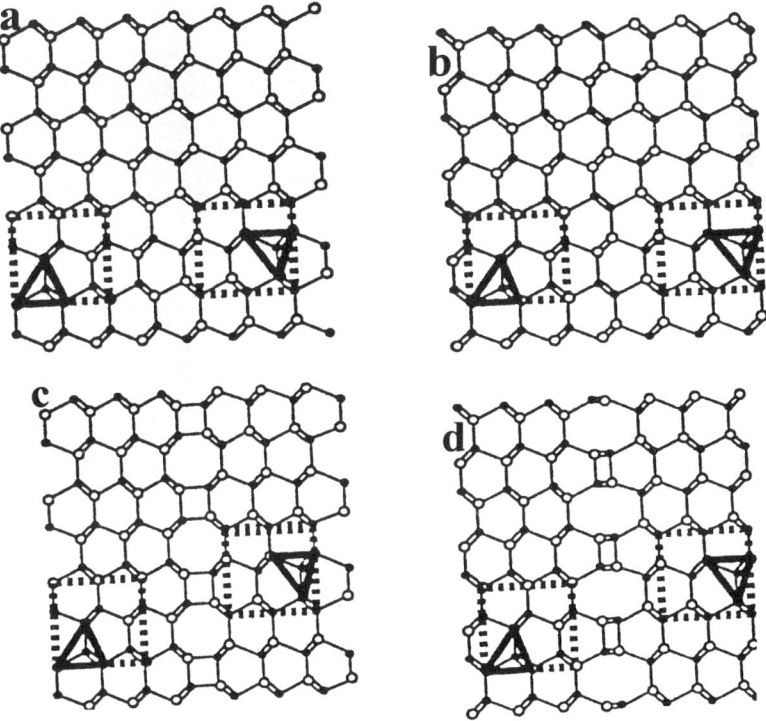

Fig. 6 The two types of observed $\{10\overline{1}0\}$ inversion domain boundary models, a,b): Holt type, c,d): V or IDB*, in a and c one bond per atom is in the boundary, whereas there are two in b and d.

HREM analysis of IDB boundaries in GaN have allowed us to determine that two atomic configurations exist. CBED analysis showed that the two types of layer matrices exhibited an N polarity (Potin et al 1999b). The two atomic configurations are of Holt type (Potin et al 1997), which contains Ga-Ga and N-N and IDB* or V type(Potin et al 1999b) which agrees with theoretical ab initio results, respectively (Northrup 1996).

Fig. 7 An HREM image of an area where an IDB of "V" atomic configuration switches: A-B, two Ga-N bonds are in the boundary plane (as in fig. 6d), B-C only one Ga-N bond is in the boundary plane (as in fig. 6c).

7. FORMATION MECHANISMS

The large density of threading dislocations has been tentatively explained as due to growth errors which lead to slight rotations of islands about the c growth axis (Ning et al 1996). This mechanism agrees quite well with the microstructure of the layers (Rouvière et al 1997) and observation of nanopipes, which may form at the coalescence of islands when a number of dislocations are pinned (Cherns et al 1998).

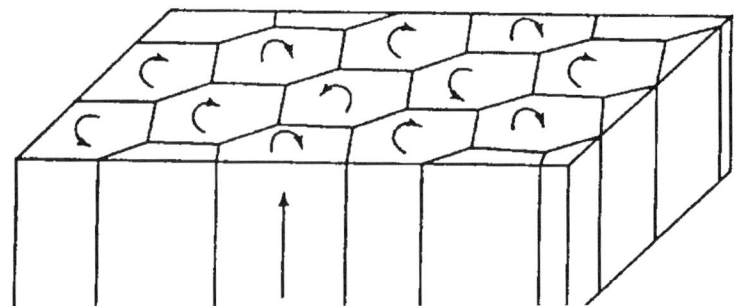

Fig. 8 Diagram showing islands rotated about [0001].

As illustrated in figure 8, such dislocations often appear at the junctions of adjacent islands. The pure screw nanopipes should cross the whole epitaxial layer (Cherns et al 1998), whereas, as seen our samples, those that are buried in the interfacial area are probably of pure edge type. On the way to the surface of the epitaxial layer some have been seen to open and close a few times (Rouvière et al 1997). Liliental et al (1998) have related this to the presence of pinholes and impurities. Although we have not observed such nanopipes in our samples, another way of explaining this behaviour may be the mixed character of these dislocations.

The $\{11\overline{2}0\}$ planar defects which occur in the epitaxial layers of GaN are identical to those identified earlier by Blank et al 1964 in AlN which exhibited a $1/6[20\overline{2}3]$ displacement vector. They were analyzed as growth defects which could not have been generated under deformation.

In GaN, they predominantly exhibit a $\frac{1}{2}[10\overline{1}1]$ displacement vector. In the case of epitaxial layers on SiC, they were related to the steps on the substrate surface which were identified to have $1/6[20\overline{2}3]$ character. They form on coalescence of islands which have nucleated on adjacent terraces related by such a step (fig.9).

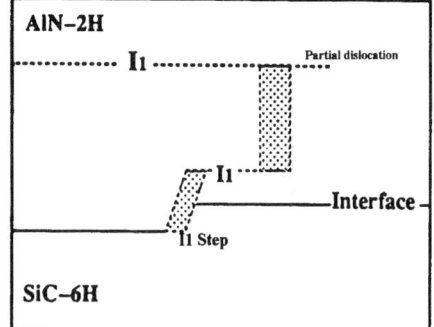

Fig. 9 Diagram for the formation and evolution of the $\{11\overline{2}0\}$ stacking fault during growth of the epitaxial layer.

On sapphire such displacement vectors can even relate islands which nucleate on a flat surface due to the many possible stacking possibilities which face the first Ga and N atoms impinging on that surface. At steps, the residual translation is more efficiently minimized by the formation of IDB's, as it is reduced to 1/24**c**. as shown in column 4 of Table. 3.

Step notation in c_{Al2O3} unit	$W_{ce} = T_{step}$	$W_{ce}' = W_{ce} \pm m$	$W_{ce}' = W_{IDB} + \alpha[0001]$
(A–B, c/6)	$R_E - 1/12\ [0001]$	$R_E - 1/12\ [0001]+m$	$W_{Holt} + 1/24\ [0001]$
(A–A, c/3)	$-1/6\ [0001]$	$-1/6\ [0001]+m$	$W_V - 1/24\ [0001]$
(A–B, c/2)	$1/4\ [0001]$	$\frac{1}{4}\ [0001]+m$	W_B
(A–A, 2c/3)	$R_E + 1/6\ [0001]$	$R_E + 1/6\ [0001]-m$	$-W_{Holt} + 1/24\ [0001]$
(A–B, 5c/6)	$1/12\ [0001]$	$1/12\ [0001]-m$	$-W_V - 1/24\ [0001]$
(A–B, c+c/6)	$-1/12\ [0001]$	$-1/12\ [0001]+m$	$W_V + 1/24\ [0001]$
(A–A, c+c/3)	$R_E - 1/6\ [0001]$	$R_E - 1/6\ [0001]+m$	$W_{Holt} - 1/24\ [0001]$
(A–B, c+c/2)	$-1/4\ [0001]$	$-1/4\ [0001]-m$	W_B
(A–A, c+2c/3)	$1/6\ [0001]$	$1/6\ [0001]-m$	$-W_V + 1/24\ [0001]$
(A–B, c+5c/6)	$R_E + 1/12\ [0001]$	$R_E + 1/12\ [0001]-m$	$-W_{Holt} - 1/24\ [0001]$

Table 3: Steps expressed in terms of the IDB's operation, α is the remaining shift along the **c** axis, after the formation of an IDB, WB is an Austerman type of IDB(Austerman et al 1966)

60

This analysis shows that the two observed configurations of the boundary plane are equally efficient and their occurrence may possibly be explained by the surface structure of the substrate (Barbaray et al 1999a). The Holt model IDBs appear to form on small hexagonal domains limited by steps, whereas the V configuration will occur on slightly larger ones. When extended steps are present, as on misoriented surfaces, both configurations are found to disappear (Barbaray et al 1999b).

REFERENCES

Amano H, Kito M, Hiramatsu K and Aksaki I 1989 Jpn. J. Appl. Phys. **28**, L2112
Barbaray B, Potin V, Ruterana P and Nouet G 1999a Diamond and Related Materials **8**, 314
Barbaray B, Ruterana P, Nouet G, di Forte Poisson M A, Huet F and Tordjman M 1999b these proceedings
Austerman S B and Gehman W G 1966 J. Mat. Sci. **1**, 249
Blank H, Delavignette P, Gevers R, and Amelinckx S 1964 Phys. Status Solidi **7**, 747
Cherns D, Young W T, Sawders M, Steeds J W, Ponce F A and Nakamura S 1998 Philos. Mag. **A77**, 273
Drum C M 1965 Philos. Mag. **A11**, 313
Elsner J, Jones R, Sitch P K, Porezag V D, Elstner M, Fraunheim T, Heggie M I, Oberg S, and Briddon P R 1997 Appl. Phys. Lett. **79**, 3672
Frank F C 1951 Acta Cryst. **4**, 497
Holt D B 1969 J. Phys. Chem. Solids **30**, 1297
Kaiser S, Pries H, Gebhardt W, Ambacher O, Angerer H, Stutzmann M, Rosenauer A and Gerthsen D 1998 Jpn. J. Appl. Phys. **37**, 84
Lester S D, Ponce F A, Crawford M G and Steigewald D A 1995 Appl. Phys. Lett. **66**, 1249
Liliental-Weber Z, Chen Y, Ruvimov S and Washburn J 1997 Phys. Rev. Lett. **79**, 2835
Nakamura S, Senoh M, Nagahama S, Iwasa N, Yamada T, Mitsushita T, Kiyoku H and Sugimoto Y 1996 Jpn. J. Appl. Phys. **35**, L74
Ning K J, Chien F R, Pirouz P, Yang J W and Khan A M 1996 J. Mater. Res. **11**, 580
Northrup J E, Neugebauer J and Romano L T 1996 Phys. Rev. Lett. **77**, 103
Northrup J E 1998 Appl. Phys. Lett. **72**, 2316
Ponce F A, Major S J, Plano W E and Welch D F 1994 Appl. Phys. Lett. **65**, 2302
Ponce F A, Bour D P, Götz W, Johnson N M, Helava H I, Grzegory I, Jun J and Porowski S 1996 Appl. Phys. Lett. **68**, 917
Potin V, Ruterana P and Nouet G 1997 J. Appl. Phys. **82**, 1276
Potin V, Ruterana P, Nouet G and Pond R C 1999a these proceedings
Potin V, Nouet G and Ruterana P 1999b Appl. Phys. Lett. **74**, 947
Rouvière J-L, Arlery M, Bourret A, Niebuhr R and Bachem K 1995 Inst. Phys. Conf. Ser. **146**, 285
Rouvière J-L, Arlery M and Bourret A 1997 Inst. Phys. Conf. Ser. **157**, 173
Ruterana P, Vermaut P, Nouet G, Botchkarev A, Salvador A and Morkoç H 1997 Mater. Sci. Eng. **B50**, 72
Ruterana P, Potin V and Nouet G 1998 Mat. Res. Symp. Proc. **482**, 72
Sverdlov B N, Martin G A, Morkoç H and Smith D J 1995 Appl. Phys. Lett. **67**, 2063
Stadelmann P 1987 Ultramicroscopy **21**, 131
Tanaka S, Kern R S and Davis R F 1995 Appl. Phys. Lett. **66**, 37
Vermaut P, Ruterana P, Nouet G, Salvador A, Botchkarev A and Morkoç H 1995 Inst. Phys. Conf. Ser. **146**, 289
Vermaut P, Ruterana P, Nouet G, Salvador A and Morkoç H 1996 MRS Int. J. Nitride Res. **1**, 42
Vermaut P, Ruterana P, Nouet G, and Morkoç H 1997a Philos. Mag. **A75**, 239
Vermaut P, Ruterana P and Nouet G 1997b Philos. Mag. **A76**, 1215
Vermaut P, Ruterana P, Nouet G, Salvador A and Morkoç H 1997c Inst. Phys. Conf. Ser. **157** 183
Vermaut P, Nouet G and Ruterana P 1999 Appl. Phys. Lett. **74**, 694
Xin Y, Pennycook S J, Browning N D, Nellist P D, Sivanathan S, Beaumont B, Fourié J P and Gibart P 1998 Mat. Res. Symp. Proc. **482**, 781

Inst. Phys. Conf. Ser. No 164
Paper presented at Microsc. Semicond. Mater. Conf., Oxford, 22–25 March 1999
© *1999 IOP Publishing Ltd*

On deformation and fracture of semiconductors

P Pirouz[*]**, A V Samant**[*]**, M H Hong**[*]**, A Moulin**[**]** and L P Kubin**[**]

[*]Department of Materials Science and Engineering, CWRU, Cleveland, Ohio 44106, U.S.A.
[**]LEM, CNRS-ONERA, B.P. 72, Av. de la Division Leclerc, 9322 Châtillon Cedex, France.

ABSTRACT: Recent experiments on deformation of semiconductors show an abrupt change in the variation of the critical resolved shear stress with temperature. This implies a change in the deformation mechanism at a critical temperature T_c. In the cases examined so far in our laboratory and elsewhere, this critical temperature appears to coincide approximately with the brittle-ductile transition temperature of the material. In this paper, we describe TEM investigations that have been performed on the wide bandgap semiconductor, SiC, deformed at temperatures below and above T_c in order to understand the change of mechanism. Based on these deformation and TEM experiments, suggestions are made as to the different mechanisms operating at low and high temperatures in semiconductors, and a possible relation between T_c and the brittle-ductile transition temperature in these materials.

1. INTRODUCTION

Recently, Suzuki *et al.* (1998, 1999a,b) have re-measured the variation of yield stress, τ_Y, of a few semiconductors (InP, GaAs, and InSe) as a function of temperature, T, and find a break at a critical point (τ_c, T_c) in the $\tau_Y(T)$ plot. Instead of a single curve with a smooth increase in the slope of the $\tau_Y(T)$ plot with decreasing temperature, the new $\tau_Y(T)$ plots appear to consist of two segments with different slopes. This transition appears perhaps more clearly in a plot of $\ln(\tau_Y)$ versus $1/T$, usually linear with a slope proportional to the activation energy for dislocation glide in the material. In the new experiments of Suzuki *et al.* (1999a), the $\ln[\tau_Y(1/T)]$ plot is separated into two straight lines with different slopes joined at T_c. Similar results were found by Samant (1999) in two wide bandgap semiconductors, 4H and 6H SiC. In all cases mentioned, the slope of the $\ln[\tau_Y(1/T)]$ linear plot is smaller in the low-temperature regime $(T<T_c)$ than that in the high-temperature regime $(T>T_c)$. Also, intriguingly, in every case, this critical transition temperature in $\tau_Y(T)$ or $\ln[\tau_Y(1/T)]$ is close to the brittle-ductile transition (BDT) temperature, T_{BDT}, of the semiconductor (Pirouz *et al.* 1999).

2. EXPERIMENTAL RESULTS

2.2. Temperature-Dependence of the Yield Stress

The yield stress, τ_Y, of two SiC polytypes, 4H-SiC and 6H-SiC, was measured in the temperature range ~550-1300°C under different strain rates. The most relevant results to the present work occur at the slowest strain rate, 3.1×10^{-5} s^{-1}, for both polytypes. The results are shown in Fig. 1(a) and 1(b) in terms of a plot of $\ln(\tau_Y)$ versus $1/T$. It can be seen that each plot is divided into two linear regimes with different slopes separated at a critical temperature, $T_c \approx 1100 \pm 100$°C. The fact that the plots are linear indicates that the kink pair mechanism can

62

be applied to both regimes, albeit with different activation energies for dislocation glide. In the low-temperature regime, $550°C<T<T_c$, the slope of the plot for both 4H-SiC and 6H-SiC corresponds to an activation energy for dislocation glide of 2.1±0.7 eV. In contrast, in the high-temperature regime, $T>T_c$, the slope of the plots for both materials corresponds to an activation energy of 4.8±1.8 eV. It is noteworthy that an activation energy of ~2 eV corresponds to that for dislocation glide in elemental silicon while an activation energy of ~4 eV corresponds to that for dislocation glide in diamond. Interestingly, the cores of the two 30° partials of a dissociated screw dislocation in SiC are Si(g) and C(g).

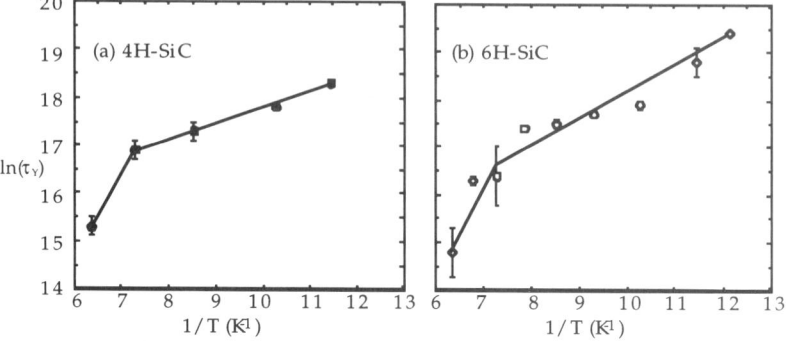

Fig. 1. $\ln(\tau_Y)$ versus $1/T$ for (a) 4H-SiC, (b) 6H-SiC.

2.2. TEM of the Deformed Samples

The dislocation configurations of the deformed 4H-SiC and 6H-SiC crystals were investigated by TEM. From samples deformed above T_c (1300°C for 4H and 1350°C for 6H-SiC) and below T_c (750°C for 4H-SiC and 900°C for 6H-SiC), slices were cut parallel to the primary slip plane, (0001). From these slices, three TEM specimens each were prepared and investigated in a Philips CM20 at 200 kV using bright-field (BF), weak-beam dark-field (WB) and LACBED. The results of these experiments were as follows. The dislocations in all the samples deformed above T_c (three each for 4H-SiC and 6H-SiC) were $\frac{1}{3}\langle 11\bar{2}0 \rangle$ total dislocations, albeit dissociated into two $\frac{1}{3}\langle 1\bar{1}00 \rangle$ partials according to the following reaction:

$$\frac{1}{3}\langle 11\bar{2}0 \rangle \rightarrow \frac{1}{3}\langle 10\bar{1}0 \rangle + \frac{1}{3}\langle 01\bar{1}0 \rangle$$

This is shown in Fig. 2 for both 4H- and 6H-SiC.

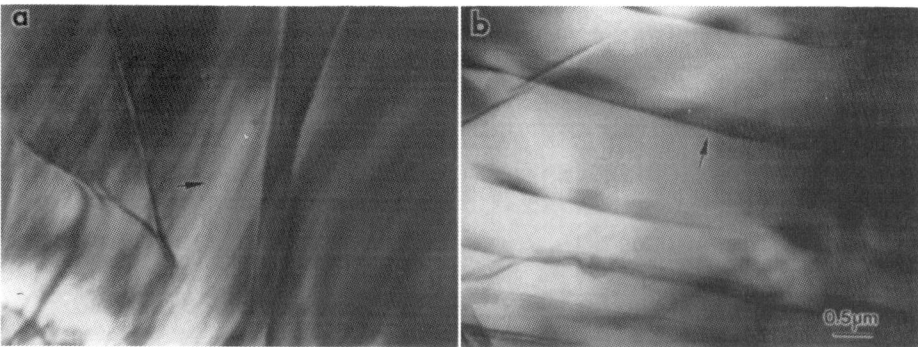

Fig. 2. BF TEM micrographs of dislocations in (a) 4H-SiC, (b) 6H-SiC induced by deformation at $T>T_c$.

By contrast, in all the TEM specimens of 4H and 6H-SiC deformed at temperatures below T_c that have been investigated, the dislocations were predominantly, but not exclusively, $\frac{1}{3}\langle 10\overline{1}0\rangle$ single leading partials, bordered on one side by a staking fault (Fig. 3).

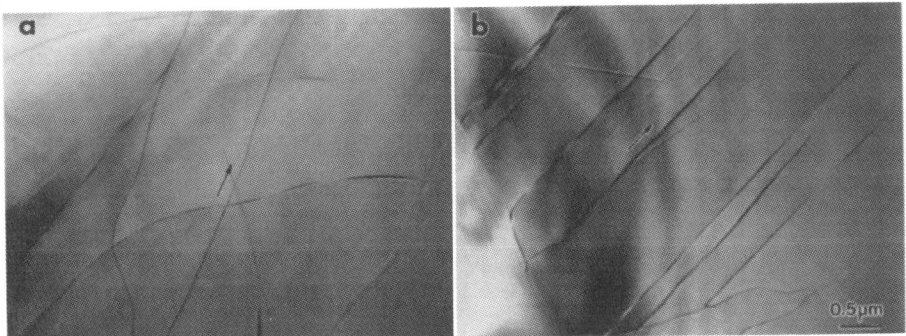

Fig. 3. BF TEM micrographs of single leading dislocations in (a) 4H-SiC, (b) 6H-SiC induced by deformation at $T<T_c$.

In no case was it possible to observe the trailing partial, i.e. the stacking fault either disappeared in the hole within the foil, or extended to the thicker parts of the specimen where the contrast became too weak to be observed. The nature of the single partial in the case of deformation at $T<T_c$ was determined by the LACBED technique (Ning and Pirouz 1996). In the three cases examined, the partial was found to have a silicon core in 4H-SiC (Fig. 4).

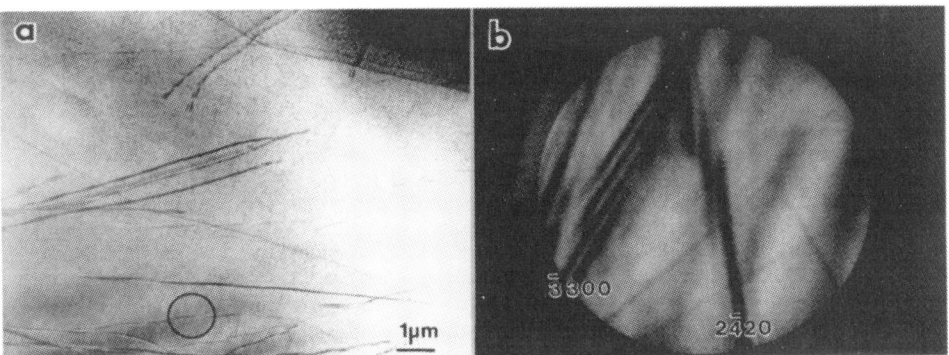

Fig. 4. Partial dislocations in 4H-SiC (a) BF (b) LACBED of the circled segment.

3. DISCUSSION

3.1. The transition in the yield stress versus temperature curves

It may be assumed that to nucleate a dislocation half-loop at a surface (or at a crack tip), the applied stress must overcome primarily the attractive image force and that this is easier at an edge heterogeneity, such as a surface step (or a ledge at a crack tip). Recently, Brochard *et al.* have compared the nucleation of the leading and trailing Shockley partial dislocations from a surface step in tetrahedrally coordinated cubic materials (Brochard, Junqua and Grilhé 1998a) taking into account the possibly different mobilities of the two partials of a dissociated dislocation (Brochard, Rabier and Grilhé 1998b). They found that both the mobility of the partial as well as the average friction force (=velocity/mobility) opposing its motion are important in the nucleation of that partial, and that over a wide range of the τ-γ plane (where γ

is the stacking fault energy), the stress necessary to nucleate a partial dislocation of higher mobility is lower. Although the model of Brochard *et al.* (1998b) does not take into account any activation energy, it is expected that thermal activation will decrease the stresses required for dislocation nucleation, and that the value of the activation energy will be different for the two partials. Conversely, it can be said that at a fixed value of stress, the two partials will be nucleated at different temperatures, with the more mobile dislocation having a lower activation energy and nucleating at a lower temperature.

We therefore propose that, at temperatures below T_c, only the more mobile dislocations, assumed to have a lower activation energy, are nucleated (as the leading partials), and deformation occurs primarily by their glide on different planes. This is despite the fact that creation of a stacking fault (of energy γ) by the leading partial creates an additional force ($=\gamma/b$) for the creation of the trailing partial. In fact, we propose that the trailing partials only start to nucleate at the critical temperature T_c when the thermal energy becomes sufficient. Thus, T_c in Fig. 1, and in the $\tau_Y(T)$ plots of Suzuki (1999), is identified with the temperature at which a transition in the deformation mode of the material takes place: deformation by glide of single leading partials at $T<T_c$ versus glide of total dislocations at $T\geq T_c$. This may be valid for all the tetrahedrally coordinated materials with a deep Peierls valley.

3.2. Brittle-Ductile Transition (BDT)

Since in the five semiconductors for which a transition in the plot of $\tau_Y(T)$ has been observed, the transition temperature, T_c, roughly coincides with the BDT, it is possible that T_c and T_{BDT} are also related. To expand on this point, we shall start with a brief description of the Hirsch and Roberts (HR) model for the BDT (Hirsch, Roberts and Samuels 1989, Hirsch and Roberts 1991) that is able to explain most of the experimental features observed in the fracture of silicon (Samuels and Roberts 1989). Two important features of the BDT in semiconductors are that, for a given geometry: (i) the value of the transition temperature in dislocation-free and non-prestrained crystals is very sharp, and (ii) the strain-rate dependence of T_{BDT} has an activation energy close to that of the dislocation velocity in the material.

The HR model is based on the shielding of the crack front by the dislocations nucleated there (Lin and Thomson 1986). It can be explained by means of Fig. 5(a) that shows a crack front under Mode III (Hirsch *et al.* 1989), or Mode I loading (Hirsch and Roberts 1991). Under the applied stress intensity factor K, dislocations can be nucleated only at special sites, e.g. X and Y, along the crack front. Dislocation loops of Burgers vector, **b**, are emitted when K reaches a critical value K_N, for the first loop, and K_o for the subsequent loops. The value of K_o is assumed to be very close to K_{Ic}, the critical stress intensity factor, and much larger than that of K_N; typically $K_N\approx0.2K_{Ic}$ and $K_o\approx 0.9K_{Ic}$. The dislocation loops so formed shield the crack tip and reduce the effective stress intensity factor on it to a lower value

$$K_{eff} = K - \sum K_d = K - \sum_j \frac{\mu b}{\left(2\pi x_j\right)^{1/2}}$$ (for the simpler case of Mode III), where K_d is the

shielding effect of a dislocation loop (summed over all the loops), μ is the shear modulus, and x_j is the distance of the *j*th dislocation from the crack tip.

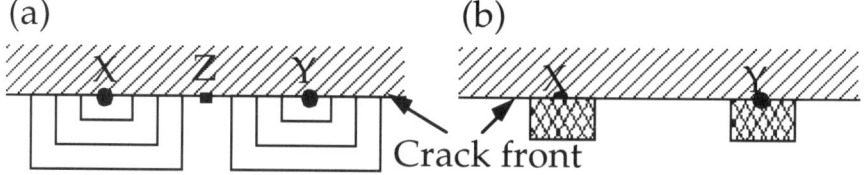

Fig. 5. Nucleation of (a) total, (b) partial dislocations at special sites (shown by solid circles) along a crack front; the solid square is a point in the initially dislocation-free zone.

The main feature of the HR model is that the crack front will propagate catastrophically if the increasing (under constant $\dot{K}$) applied stress intensity factor K increases K_{eff} to the critical value, K_{Ic}, unless the emitted dislocations shield the whole of the crack front, i.e., points X and Y (that are shielded from the start of dislocation nucleation), as well as points such as Z (that are not initially shielded). It is important that sites, such as Z, that are initially in a dislocation-free zone, become shielded by the expanding dislocation loops around points X and Y. In this case, the effective stress intensity factor, K_{eff}, becomes less than K_{Ic} everywhere along the crack tip, for all values of the applied stress intensity factor up to the value at which macroscopic yield takes place. According to the HR model, the sharp yield point in silicon arises because of nucleation events, whereas the strain-rate dependence of T_{BDT} is explained by the time taken for the nucleation events to occur along the crack front, and the time taken for the expanding dislocation loops to shield all the points along the crack front from reaching K_{Ic}.

Using the same equations and terminology as the HR model, let us now consider a crack tip under an increasing K at temperatures lower than T_c. In this case, it is suggested that only leading partial dislocation loops, with a Burgers vector, $\mathbf{b}_l$, can be nucleated at heterogeneities (e.g., ledges) along the crack front, e.g., at sites such as X and Y in Fig. 5(b). At each such site, a partial dislocation loop is able to nucleate only once after which it will expand until the resolved shear stress on it due to the applied stress intensity factor, K, is balanced by the line tension/stacking fault/image stresses. Inevitably, the leading partial will drag a stacking fault behind it (shown shaded inside the loops in Fig. 5(b)) and, more importantly, shut off the source leaving a single partial loop that will not be able to effectively shield the crack tip. As a result, the shielding of the crack tip is very limited and the effective stress intensity factor, K_{eff}, on the crack tip (given by $K_{eff} = K - K_d = K - \mu b_l / \sqrt{2\pi x}$ for Mode III) is nearly the same as the applied K. Because of the lack of effective shielding, the crack front will propagate catastrophically as soon as K_{eff} reaches K_{Ic} under the increasing K (at a constant $\dot{K}$).

At a temperature higher than T_{BDT} [Fig. 5(a)], a trailing partial immediately follows the nucleation of the leading partial to form a total dislocation loop, with a Burgers vector $\mathbf{b}$. Now, however, since the stacking fault has been removed by the trailing partial, the source can remain active and further dislocation loops can form and expand as long as the back stress of the initial dislocation loops does not reduce the stress to a value below that required for nucleation. The situation is now somehow similar to that described by Hirsch and Roberts (1991). Under these conditions, an avalanche of total dislocation loops form that can shield the crack tip effectively, maintaining K_{eff} at a value less than K_{Ic} up to the value at which macroscopic yield takes place (i.e. to the value of K corresponding to the shear stress τ_y).

However, if, as we suggest, T_c does in fact coincide with T_{DBT}, then nucleation, rather than mobility, of dislocations probably controls the BDT, in some ways, similar to the original Rice-Thomson model (1974). To clarify this point, consider Model B in Hirsch and Roberts (1991), i.e. the case where the starting crystal is basically dislocation-free (corresponding to the experiments of Michot and George (1989), and Brede and Haasen (1988)).

In the HR model, it is assumed that dislocations are emitted at a few special sites along the crack tip (e.g. at ledges) spaced d_{crit}^i apart at values of $K = K_N < K_{Ic}$. Subsequently, loops emitted from these sites help to activate nucleation from other, more difficult, sites along the crack front because of the anti-shielding stresses ahead of these dislocation loops. The separation between these secondary nucleation sites is assumed to be d_{crit}^f $\left(<< d_{crit}^i \right)$. In this model, the condition for the BDT is that the loops from the original sources must have traversed a distance $\left(d_{crit}^i - d_{crit}^f \right)$. This is when the applied K reaches K_o (with K_o just below K_{Ic}) such that the dislocations nucleated at the secondary sources have time to traverse d_{crit}^f and shield the whole crack front before the applied K reaches K_{Ic}. Since the effective shielding of the crack tip depends on how quickly a sufficient number of dislocation loops can form and expand to allow the shielding of the crack tip, dislocation velocity, $v = A\tau^m exp(-U/kT)$, plays a predominant role in this model. As Hirsch and Roberts (1991) show, the slope of the linear plot of $\ln(\dot{K})$ versus $1/T_{BDT}$ is the same as the activation energy for dislocation glide, U, and T_{BDT} will depend on the strain-rate because the dislocation velocity is stress-dependent.

Within the present speculation, the sudden formation of an avalanche of full dislocations at T_c $(=T_{BDT})$ releases the stored elastic energy required for further propagation of the crack. The $\dot{K}$-dependence of T_c comes from the assumption that nucleation of a (partial) dislocation depends on the rate of change of shear stress with time, $\dot{\tau}$. This is not unreasonable because the upper yield point, τ_y^u, in semiconductors, which is probably also controlled by the start of dislocation nucleation events in the crystal, is a sensitive function of the strain rate, $\dot{\varepsilon}$, and thus of $\dot{\tau}$. A simple starting point is to assume that the time, t, for a partial dislocation half-loop to nucleate from the crystal surface under an applied shear stress, τ, is given by $t = (1/\nu_o)exp(\Delta E - \tau\Omega/kT)$, where Ω and ν_o are the atomic volume and vibration frequency, respectively. Equating t with the time $(K_{Ic}/\dot{K})$ to reach K_{Ic} gives the $\dot{K}$-dependence of T_{BDT}: a linear variation of $\ln(\dot{K})$ with $1/T_{BDT}$ with a slope $(\Delta E - \alpha K_{Ic}\Omega)$ where $\tau=\alpha K$ (α is a geometrical factor). Thus, the present model suggests an approximate equivalence of the stress-dependent activation energy for dislocation nucleation, $(\Delta E - \tau\Omega)$, with U.

4. CONCLUSION

It is proposed that deformation of semiconductors at low temperatures $(T<T_c)$ takes place by the nucleation and glide of single leading partials dragging a stacking fault behind them. This, together with the low mobility of the partial dislocations at low temperatures, gives rise to very limited plasticity. Also, the very limited shielding of crack tips results in a brittle behaviour at temperatures below T_c. Above the critical temperature, T_c, both partials can be nucleated and deformation of the semiconductor takes place by the glide of total dislocations. In this case, plasticity is extensive and, provided that the dislocations are sufficiently mobile, crack tips can be shielded resulting in a ductile behaviour.

ACKNOWLEDGEMENTS

This work was supported by grant number FG02-93ER45496 from the DoE and subcontract number 95-SPI-420757-CWRU from the SiC Consortium. Thanks are due to Dr. Calvin Carter, Jr. (of Cree Research, Inc.) and Dr. Don Hobgood (previously of Northrop-Grumman) for providing single-crystal samples of 6H-SiC and 4H-SiC, respectively.

REFERENCES

Brede, M and Haasen, P, 1988, *Acta Metall.*, **36**, 2003-2018.
Brochard, S, Junqua, N and Grilhé, J, 1998a, *Phil. Mag. A*, **77**, 911-922.
Brochard, S, Rabier, J and Grilhé, J, 1998b, *Eur. Phys. J. - AP*, **2**, 99-105.
Hirsch, P B and Roberts, S G, 1991, *Phil. Mag. A*, **64**, 55-80.
Hirsch, P B, Roberts, S G and Samuels, J, 1989, *Proc. R. Soc. Lond. A*, **421**, 25-53.
Lin, I-H and Thomson, R, 1986, *Acta Metall.*, **34**, 187-206.
Michot, G, 1988, *Crystal Properties and Preparation*, **17-18**, 55.
Michot, G and George, A, 1989, Inst. Phys. Conf. Ser., **104**, 385-396.
Ning, X J and Pirouz, P, 1996, *J. Mater. Res.*, **11**, 884-894.
Pirouz, P, Samant, A, Hong, M, Moulin, A and Kubin, L, 1999, *J. Mater. Res.* In press.
Rice, J R and Thomson, R, 1974, *Phil. Mag. 29*, 73-97.
Samant, A V, Ph.D. thesis, Case Western Reserve University (1999).
Samuels, J and Roberts, S G, 1989, *Proc. R. Soc. Lond. A*, **421**, 1-23.
Suzuki, T, Nishisako, T, Taru, T and Yasutomi, T, 1998, *Phil. Mag. Lett.*, **77**, 173-180.
Suzuki, T, Yasutomi, T, Tokuoka, T and Yonenaga, I, 1999b, *Phil. Mag. A.* In press.
Suzuki, T, Yasutomi, T, Tokuoka, T and Yonenaga, I, 1999a, *phys. stat. sol. (a)*, **171**, 47.

Inst. Phys. Conf. Ser. No 164
Paper presented at Microsc. Semicond. Mater. Conf., Oxford, 22–25 March 1999
© *1999 IOP Publishing Ltd*

Dislocation bundles in GaAs substrates: assessed by X-ray and Makyoh topography, X-ray diffraction, TEM, scanning infrared polariscopy, light interferometry, and Nomarski microscopy

P Möck[1], M Fukuzawa[2], Z Laczik[3], G W Smith[4], G R Booker[1], M Yamada[2], M Herms[2*] and B K Tanner[5]

[1]Department of Materials, University of Oxford, Parks Road, Oxford OX1 3PH, UK
[2]Kyoto Institute of Technology, Department of Electronics and Information Science, Matsugasaki, Sakyoku, Kyoto 606-8585, Japan; *now at Fraunhofer Institute of Non-destructive Testing, Krügerstraße 22, D-01326 Dresden, Germany
[3]Department of Engineering Science, University of Oxford, Parks Road, Oxford OX1 3PJ, UK
[4]Defence Evaluation and Research Agency, St Andrews Road, Malvern, Worcestershire WR14 3PS, UK
[5]Department of Physics, University of Durham, Science Laboratories, South Road, Durham DH1 3LE, UK

ABSTRACT: Different types of dislocation bundles are identified in (001) GaAs substrates of III-V heterostructures. Comparisons of X-ray transmission topograms with scanning infrared polariscopy images show a one to one correlation of stripes of reduced residual shear stain and dislocation bundles of the majority type. Makyoh topography, visible light interferometry and standard Nomarski microscopy, on the other hand, show a slip-line distribution that is in correspondence to the distribution of dislocation bundles of a minority type. Utilising several complementary techniques, the number and density of heat treatment induced dislocations in two-inch diameter substrates are estimated.

1. INTRODUCTION

It is well known that heat treatment induced plastic deformation of GaAs substrates is a key factor that reduces the yield of opto- and microelectronic devices in manufacturing processes on an industrial scale (Tatsumi et al 1994, Sawada et al 1996, Kiyama et al 1997). A classification for dislocation bundles in such GaAs substrates has been introduced on the basis of the crystallographic parameters of the constituting dislocations (Möck 2000). We distinguished between a majority type and (at least) two minority types of dislocation bundles, the latter accounting only for a few percent of the whole plastic deformation. A complete Burgers vector analysis has been performed for dislocation bundles of the majority type. There are dislocations which possess two different Burgers vectors in each majority type dislocation bundle and the extended segments of all of these dislocations are glissile 60° segments. We suggested that these two Burgers vectors can in effect "pair up" in combinations that cancel surface steps and slip lines to some extent and provide effective shear strain relief perpendicular to the wafer normal (Möck 2000).

Corroborating observations for this hypothesis will be presented in this paper. On the basis of Möck's (2000) classification, we will elucidate the spatial distribution of the different types of dislocation bundles on a variety of samples. In order to help in quantifying the possible adverse effects of dislocation bundles on devices, we will estimate the number and density of those dislocations which are introduced in two-inch diameter GaAs substrates by thermal treatments that are typical for molecular beam epitaxy (MBE) processes.

We observed severe plastic deformation in the GaAs substrates of epitaxial III-V compound semiconductor samples that were grown in MBE machines of four different makes, but finally, we overcame the technical problem of unwanted plastic deformation by means of modifications to the sample holder of a user built MBE machine (Möck and Smith 2000). It should be noted that the observed plastic deformation was entirely caused by the heat treatments used for the epitaxial growth, but bore no other relation to the epitaxial growth processes (Yamada et al 1996a).

2. EXPERIMENTAL DETAILS

Structural and thermal processing details of the samples and experimental details of the X-ray topography studies in transmission and reflection geometry are given elsewhere (Möck 2000, Möck and Smith 2000). The substrates were mainly 0.45 mm thick, (001) oriented, undoped, two-inch diameter GaAs wafers that were grown by the vertical gradient freeze Bridgman (VGFB) technique. The thermal treatments that were used for the epitaxial growth typically involved temperatures of up to about 650 °C. We used epitaxial III-V compound semiconductor samples for most of our study because they were, on the one hand, readily available and, as stated above, completely adequate for this purpose.

While the user built scanning infrared polariscopy microscope that has been developed by Yamada (1993) and Yamada et al (1996b) was used, the Makyoh topography was performed in an experimental set-up by Laczik et al (1994). Visible light (632.8 nm wavelength) interferometry was performed using a Specac FOTI-100 Fizeau interferometer. The instrument was modified to enable phase-shifting interferometry and the observed interference fringes were automatically analysed and converted to surface height maps.

The transmission electron microscopy (TEM) investigation was performed on a Cl_2 / methanol thinned plan-view sample using a JEOL JEM 200 CX microscope operating at an acceleration voltage of 200 kV. High-resolution X-ray diffraction reciprocal space maps of the (004) reflection were taken at different sample positions, employing a Philips MRD diffractometer equipped with a Bartels 220 Ge monochromator, a three-bounce 220 Ge analyser crystal, and a sealed Cu tube.

3. RESULTS AND DISCUSSION

3.1. Observations on dislocation bundles and residual strains

A comparison of Figs. 1a and b shows how the strains which are caused by the heat treatment that was used for the epitaxial growth are distributed spatially. There is an annular region of higher tensile strain ($4.5 - 5 \ 10^{-5}$ immediately on the wafer edge, $\approx 2.5 \ 10^{-5}$ at about a fifth of the wafer radius) that falls off towards the centre of the sample, which is at a strain level that is comparable to that of the untreated substrate ($1 - 2 \ 10^{-5}$). In addition, the strains around <100> peripheral areas are about 10 % higher than the strains around <110> peripheral areas. The distribution of the strain as a whole is modified by dislocation bundles and the plastic deformation that is realised by them.

Figures 2a-c show a scanning infrared polariscopy image, an X-ray transmission topogram, and a Makyoh topogram of the same sample. Due to a corresponding spatial distribution between dislocation bundles with "effective" Burgers vectors $\pm \frac{1}{2}$ <110> (glide directions <110> on inclined {111} planes), and stripes of significantly reduced residual shear strain, Figs. 2a and 2b show predominantly a pseudo-symmetric, four-fold dislocation bundle set, that originated around the four <100> peripheral areas and which we call the majority type set. Conversely, Makyoh topography (Fig. 2c), visible light interferometry and standard Nomarski microscopy (images not presented) show a slip-line distribution that is in correspondence to the distribution of dislocation bundles of a minority type, which originated around <110> peripheral areas. The height of these slip lines ranges from about 30 to 50 nm for the sample shown in Fig. 1, but could in other samples reach up to about 150 nm (at which height they would be visible to the unaided human eye). Usually, dislocation bundles of the minority types are more pronounced in the area of the major flat.

These observations indicate that only dislocation bundles of some minority types do possess a significant Burgers vector component parallel to the wafer normal, for, as marked by arrows in Fig. 2c, only they are visible by the employed surface sensitive techniques.

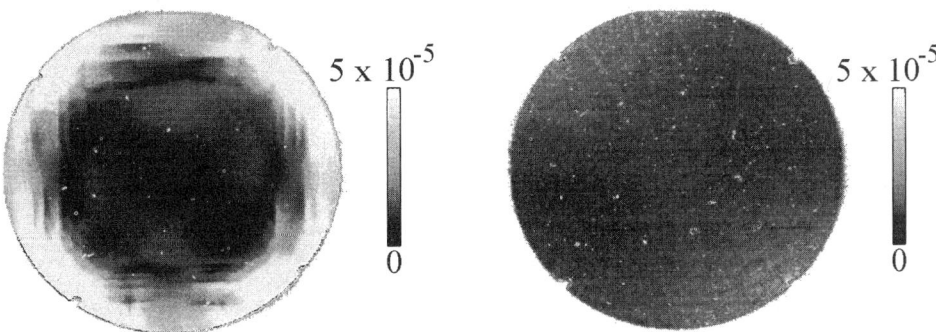

Fig. 1: Scanning infrared polariscopy images **a) left side,** of the sample that is analysed in detail elsewhere (Möck) and shown as well in Figs. 2a-c and 4; **b) right side,** of an unused substrate of the same type "as delivered". The differences in tensile strains parallel to the radial and tangential directions in a polar co-ordinate system are shown. The strain quantifying markers are in both cases linear and the images are in the same orientation. The major flats, i.e. [-1-10] directions, point up towards the top of the page and whole two-inch diameter wafers are depicted except for small mounting artefacts.

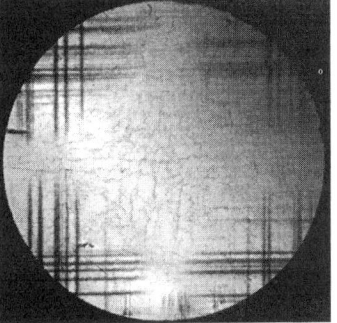

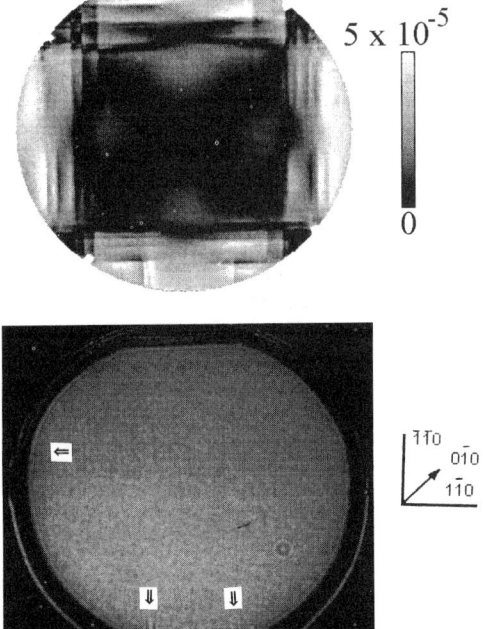

Fig. 2: Comparison of an **a) upper left,** scanning infrared polariscopy image, **(b) upper right,** (0-22) X-ray transmission topogram, and **c) lower left,** Makyoh topogram of the sample that is shown in Figs. 1a and 4. Fig. 2a shows the distribution of shear strains between the <100> directions that are perpendicular to the wafer normal. The quantifying marker of the strain is linear. All three images are in the same orientation and the major flats (i.e. the [-1-10] directions) point up towards the top of the page. Dislocation bundles of a minority type are marked by arrows in Fig. 2c.

This supports the "pairing up" hypothesis for dislocation bundles of the majority type (Möck 2000), for e.g. ± ½ [101] and ± ½ [01-1] could in effect "pair up" to form ± ½ [110], which possesses no such component. The related hypothesis that the surface steps and slip lines of this type of dislocation bundle are partially cancelled is supported by TEM observations.

Figure 3, a two-beam electron diffraction contrast image, shows surface ripples on the free epilayer side of a plan-view sample in a very high magnification. We interpret these surface ripples as to be caused by a majority type dislocation bundle and note that similar contrasts have been observed from surface steps earlier on plan-view samples of similar epitaxial structures (Goodhew et al 1989).

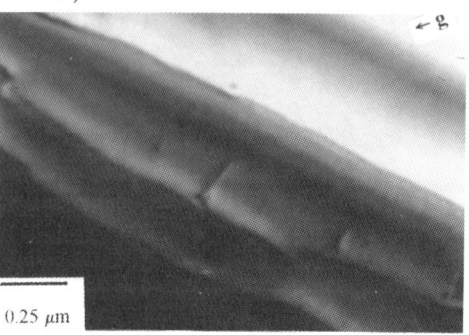

0.25 μm

Fig. 3: (400) two-beam electron diffraction contrast image from the free epilayer surface of a partially relaxed, low misfit (In,Ga)As double-heterostructure above a majority type dislocation bundle. Since the spacing of the dislocations that are threading up from the substrate is smaller than 10 μm (see section 3.2.), the dislocations in the substrate must be clustered into bundles.

Employing the well known extinction distance (ξ) formula, one can estimate the height differences in Fig. 3. With ξ ≈ 91.5 nm we obtain height differences in the order of magnitude of 50 nm for these surface ripples. Thus, the crucial point is that sufficiently high slip-line bunches with a lateral extension of several mm that would have been visible by means of Makyoh topography, visible light interferometry, and standard Nomarski microscopy do not exist above dislocation bundles of the majority type. We interpret the existing surface ripples as being caused by some dynamic imbalance in the number of dislocations with different Burgers vectors in a bundle. With dislocations threading through the front and backside of the wafer, i.e. ending at different points on a dislocation bundle, the number of dislocations with different Burgers vector in a bundle is bound to fluctuate.

Comparing Figs. 1a, 2a, and 4, it can be seen that the pseudo-symmetric set of dislocation bundles relieves shear strains very effectively but not the difference of tensile strains parallel to the <100> directions that are perpendicular to the wafer normal. This is compatible with the determined "effective" Burgers vector, i.e. the "pairing up" hypothesis in other words, and suggests that there are about equal numbers of dislocations with both Burgers vectors in each bundle of the majority type.

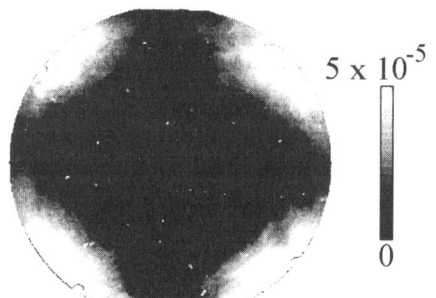

5×10^{-5}

0

Fig. 4: Scanning infrared polariscopy image of the sample that is shown in Figs. 1a and 2a-c. The difference in the tensile strains parallel to the <100> directions that are perpendicular to the wafer normal is shown. The quantifying marker is linear and the whole two-inch diameter wafer is depicted. The orientation is the same as in Figs. 1a and 2a-c, i.e. the major flat points up towards the top of the page. The bright dots are artefacts due to dust particles on the wafers.

Based on the classification that is founded on the crystallographic parameters of their constituting dislocations (Möck 2000), we are able to clarify the spatial distribution of majority and minority type dislocation bundles conclusively. By means of a combination of the five imaging techniques mentioned above and double-crystal X-ray reflection topography (Möck and Smith 2000) we observed in a variety of samples that while dislocation bundles of the first minority type tend to be located at or rather close (< ± 10°) to a <110> pole, other minority type dislocation bundles typically deviate from about ± 10° up to about ± 35° from a <110> pole, partially overlapping the areas that are occupied by the pseudo-symmetric set of majority type dislocation bundles, which typically extend up to about ± 25° around <100> poles.

We note that the models by Sawada et al (1996), Kiyama et al (1997), and Yamada et al (1997) describe the predominance of minority type dislocation bundles in the area of the major flat well, but are not applicable to majority type dislocation bundles since they predict a spatial distribution with an eight-fold symmetry. Applying Curie's (1894) well known symmetry principle, we found, in agreement with the Burgers vector determination on the whole set of majority type dislocation bundles, that the symmetry of this set is the point group $\bar{4}$2m (Möck 2000), which does not contain an eight-fold symmetry axis. The models mentioned above must therefore be oversimplifications. A new model has been developed and will be presented elsewhere.

3.2. Estimation of the number and density of additional dislocation

Figure 2a shows a reduction in the residual shear strain of about $4 \cdot 10^{-5}$ due to the presence of the majority type dislocation bundles and, thus, allows for a first estimation of the number of dislocations in such a bundle. If we assume for the sake of the estimation that the dislocations are evenly distributed both across the width of a dislocation bundle and throughout the wafer thickness, we can deduce an "average" dislocation distance of about 10 µm. For a typical bundle widths of 0.8 mm and a wafer thickness of 0.45 mm, we obtain $3.6 \cdot 10^3$ dislocations in a bundle. With 40 bundles in a typical two-inch diameter sample, this gives $1.4 \cdot 10^5$ dislocations in a wafer, in addition to the grown-in dislocations (which can be seen as cellular structure in Fig. 2b). Taking the area that is affected by the dislocation bundles as to be typically about one third of the whole wafer, a dislocation density of about $2 \cdot 10^4$ cm^{-2} is obtained. About half of the additional dislocations may thread up through the epitaxial structure (Möck 2000).

Figure 5 shows the diffuse scatter around the substrate and epilayer peaks in a reciprocal space map that was taken above a dislocation bundle of the majority type. This diffuse scatter is slightly enhanced in comparison to its counterpart in a reciprocal space map that was taken at the centre of the same sample. The latter showed only the background diffuse scatter which is due to the grown-in cellular dislocation structure that is typical for undoped VGFB GaAs substrates (Fig. 2b). The enhanced diffuse scatter in Fig. 5 allows for an independent estimation of the dislocation density on the basis of the resolution limit for this particular X-ray technique (i.e. about 10^3 cm^{-1}) for the common case of dislocations in epitaxial structures (Tanner et al 1998, Goorsky et al 1995).

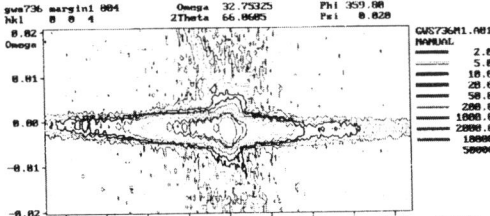

Fig. 5: (004) reciprocal space map around the substrate and epilayer peaks of a fully strained (Al,Ga)As/GaAs multiquantum-well structure. The orientation of the dislocation bundle was such that it was perpendicular to the plane that is defined by the incident and scattered wavevectors.

Supported by X-ray transmission topograms that were taken with the incoming X-ray beam directed towards either the front or back side of one sample (Möck 2000), we assume that the dislocation bundles are located on average in the middle of the substrate (i.e. at a depth of 225 µm). Assuming further that the distortion field of the dislocations diminishes inversely proportional with a characteristic distance to the free epilayer surface that equals the extinction distance (≈ 7.2 µm) of the employed reflection, we estimate that the dislocation density in a bundle of the majority type may be about 30 (i.e. $\approx 225 / 7.2$) times higher than the detection limit, given above. For an average dislocation bundle width of 0.8 mm, we obtain in good agreement to the first estimation 2.4 10^3 dislocations. Note that the reciprocal space map indicates in agreement with the TEM and X-ray topography (Möck and Smith 2000, Möck 2000) results that the "pairing up" of dislocations in bundles of the majority type is not physical, since it is mainly the Burgers vector component parallel to the wafer normal that causes the diffuse scatter (Tanner et al 1998).

3.3. Eradication of dislocation bundles

As described in more detail elsewhere (Möck and Smith 2000), modifications to the sample holder of a user built MBE machine, which are supported by later published theoretically considerations (Sawada et al 1996), led to the reduction of the number of additional dislocations to zero. The crucial point of these modifications was to replace the annular ledge on which the wafer rests while it is heated radiatively by a three point mount. We could eradicate all types of dislocation bundles by our modifications, indicating that if one observes for example dislocation bundles of minority types on a particular sample by a surface sensitive technique, there are probably dislocation bundles of the majority type present as well in the bulk of the substrate.

4. CONCLUSIONS

The number of additional dislocations in most of our two-inch diameter GaAs wafers is about 10^5, with threading dislocation densities of about 10^4 cm^{-2} in the affected one third of the wafer area on the epilayer side. Indirect evidence for the "effective" but not physical "pairing up" of dislocations with different Burgers vectors in dislocation bundles of the majority type has been presented. Different imaging techniques have been applied and the spatial distribution of different types of dislocation bundles has been determined on the basis of a classification scheme that is founded on the crystallographic parameters of the constituting dislocations. Fortunately, the additional dislocations can be eradicated by modifications to the sample holder of molecular beam epitaxy machines.

ACKNOWLEDGEMENT

Peter Möck is grateful to the Engineering and Physical Science Research Council for sponsoring his "Direct Access" projects (ref. No. 30047, 31098) at the U.K. Synchrotron Radiation Source, Daresbury Laboratory.

REFERENCES

Curie P 1894, J. Physique **3**, 393
Goodhew PJ, Dixon R, Colclough A, Homewood KP, Emeny M and Whitehouse CR 1989, Inst. Phys. Conf. Ser. No. **100**, 325
Goorsky MS, Meshkinpour M, Streit DC and Block TR 1995, J. Phys. D: Appl. Phys. **28**, A92
Kiyama M, Takebe T and Fujita K 1997, Inst. Phys. Conf. Ser. No. **155**, 945
Laczik Z, Booker GR and Mowbray A 1994, J. Cryst. Growth **153** 1
Möck P, submitted to J. Appl. Cryst.
Möck P and Smith GW, submitted to J. Cryst. Growth
Sawada S, Yoshida H, Kiyama M, Mukai H, Nakai R, Takebe T, Tatsumi M, Kaji M and Fujita K 1996, Technical Digest - IEEE GaAs Integrated Circuit Symp. **74**, 50
Tanner BK, Möck P and Mizuno K 1998, Inst. Phys. Conf. Ser. No. **160**, 177
Tatsumi M, Kawase T, Iguchi Y, Fujita K and Yamada M 1994, Proc. 8th Conf. on Semi-insulating III-V Materials, Warsaw, Poland, 6th -10th June, World Scientific, p. 11-8
Yamada M 1993, Rev. Sci. Instrum. **64**, 1815
Yamada M, Fukuzawa M, Kawase T, Tatsumi M and Fujita K 1996a, Inst. Phys. Conf. Ser. No. **145**, 447
Yamada M, Ito K and Fukuzawa M 1996b, Proc. 9th IEEE Semiconducting and Semi-Insulating Materials Conf. (IEEE SIMC-9), Toulouse, France, 29th April - 3rd May, p. 177-80
Yamada M, Fukuzawa M and Ito K 1997, Inst. Phys. Conf. Ser. No. **155**, 901

Inst. Phys. Conf. Ser. No 164
Paper presented at Microsc. Semicond. Mater. Conf., Oxford, 22–25 March 1999
© 1999 IOP Publishing Ltd

Transmission electron microscopy investigations of the defect formation during Zn-diffusion in GaP and GaSb

C Jäger, W Jäger, G Bösker*, J Pöpping* and N Stolwijk*

Mikrostrukturanalytik, Technische Fakultät, Christian-Albrechts-Universität, D-24143 Kiel, Germany
* Institut für Metallforschung, Universität Münster, Wilhelm-Klemm-Straße 10, D-48149 Münster

ABSTRACT: Defect formation during Zn diffusion into intrinsic and Te-doped GaP (001) and into intrinsic GaSb (001) single crystals is characterized for different diffusion times and temperatures by analytical transmission electron microscopy of cross-sectional samples. The observations are compared with Zn concentration profiles obtained by secondary ion mass spectrometry and electron microprobe measurements. Independent on the diffusion conditions dislocation loops of interstitial type and Ga precipitates in voids are formed in the diffusion front regions of the crystals. Furthermore, network dislocations and crystalline Zn-rich precipitates are present in the near-surface regions. The formation of these defects in GaP and GaSb can be understood as similar to the diffusion-induced defect formation during Zn diffusion in GaAs.

1. INTRODUCTION

The compound semiconductors GaP and GaSb are used as substrate materials in electronic and optoelectronic devices, with zinc as one of the most commonly used p-dopants. From earlier analyses of diffusion profiles it was concluded that Zn diffusion in GaP and GaSb proceeds via an interstitial-substitutional exchange mechanism (Tuck 1988, Conibeer et al 1996). Earlier transmission electron microscopy (TEM) investigations of GaAs have demonstrated that the formation and temporal evolution of defects during Zn diffusion can yield important informations about the role of point defects and the influence of the defect structure on the diffusion profiles (Luysberg et al 1992, Jäger et al 1993, Bösker et al 1995, Rucki et al 1997). Early experiments of Zn diffusion in GaP used crystals which contained large numbers of defects before the diffusion anneals, and Zn diffusion was reported to have no significant effects on the microstructure (Ball et al 1980). This paper summarizes results of a first transmission electron microscopy investigation of the defect formation by Zn-diffusion into single crystals of GaP and GaSb. A more detailed report of the results will be published elsewhere (Jäger et al 1999).

2. EXPERIMENTS

The diffusion anneals are performed in argon flushed, evacuated and sealed quartz ampoules with elemental Zn as diffusion source. Substrates are intrinsic GaP(001) and GaP(111) predoped with Te (C_{Te} = (2-5)$\cdot 10^{17}$cm^{-3}), with an etch pit density of < 10^5 cm^{-2}. For GaP the diffusion anneals are performed at 900°C for 30 min, at 1100°C for 90 min and at 1100°C for 30 min. Zn diffusion into GaSb is performed for 90 min at 600°C. Zn concentration profiles are measured by secondary-ion mass spectroscopy (SIMS) and by electron microprobe (EMP) on identically treated wafer cross-sections. The defect structures are investigated using a Philips CM30 transmission electron microscope (TEM) at 300 kV. Precipitate compositions are determined by spatially resolved energy-dispersive X-ray microanalyses (XEDS) using a small beam probe (diameter ≈ 30 nm).

3. EXPERIMENTAL RESULTS

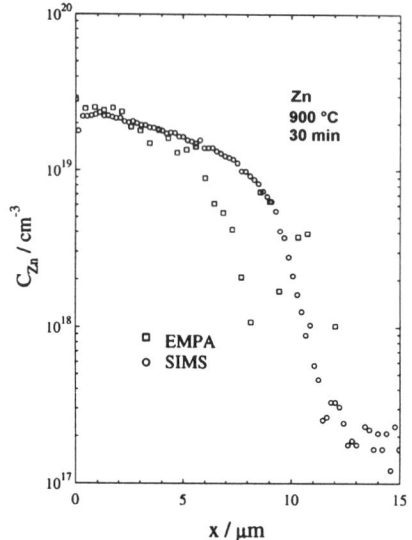

Fig 1 Zinc concentration depth profiles in GaP obtained by SIMS (O) and EMP (□).

Fig. 1 shows a typical box-like concentration profile measured by EMP and by SIMS for Zn diffusion in GaP at 900°C for 30 min. Characteristic are the high surface concentration of $2.5 \cdot 10^{19}$ cm^{-3} and the abrupt decrease at the maximum penetration depth to a value of about $1 \cdot 10^{17}$ cm^{-3}. The surface concentration values measured by SIMS and by EMP are in fair agreement. The shapes of Zn concentration depth profiles in GaSb are similar to those of Zn in GaP but show local fluctuations of the maximum penetration depth (between 16 µm and 26 µm for the diffusion condition investigated, not shown).

Our TEM investigations show that diffusion-induced defects are observed only within the Zn-diffused crystal regions, independent of the diffusion conditions and substrate materials used. Fig. 2 shows examples of the defect arrangements in the diffusion regions of GaP and of GaSb. The diffusion-induced defects consist of dislocations, dislocation loops and smaller precipitates. These precipitates are found to be always spatially corre-lated with dislocations or dislocation segments. Such defects are absent in the diffusion regions for short-term anneals. The nature of the dislocation loops in GaP and in GaSb has been determined by

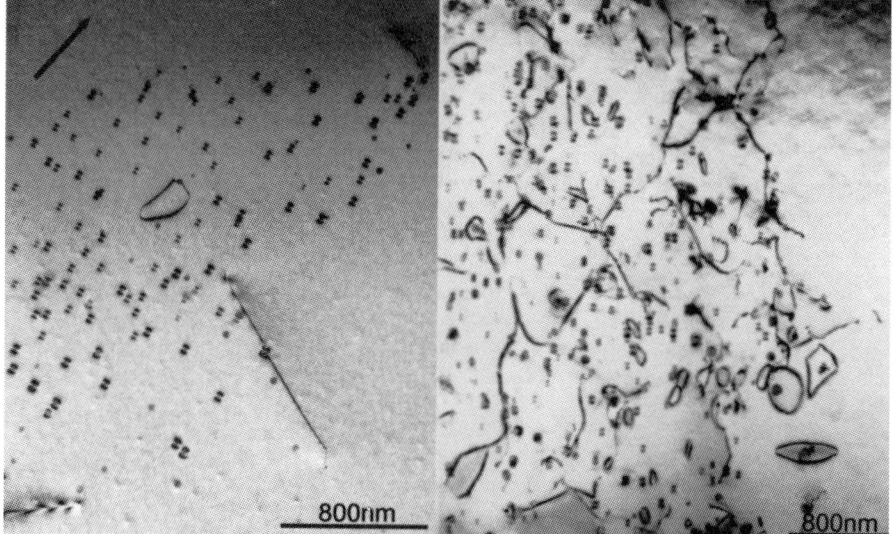

Fig. 2 (a) Defect arrangement in the central part of the Zn-diffusion region of GaP with a dislocation loop, dislocations and precipitates. Imaging vector $\vec{g} = (\bar{2}20)$ (arrow). (b) Diffusion-induced dislocation loops, dislocations and precipitates near the diffusion front in GaSb.

means of the inside-outside contrast method (Föll et al. 1975). For all cases analysed, the nature of the loops was identified as interstitial type, irrespective of their depth location. Fig. 3 shows an example of the inside-outside contrast reversal for a dislocation loop in GaP. Most of these loops were found to be perfect with {110} habit planes and Burgers vectors of the type a/2 <110>.

Analyses of the defocus contrast behaviour of those precipitates which are spatially correlated with the dislocation structure show that most of these precipitates in GaP and in GaSb are connected with voids and are facetted on {111} and on {001} planes (Fig. 5). Analyses of the precipitate composition have been performed by comparing XEDS spectra of precipitate regions and of adjacent matrix regions (Fig. 5). The intensity ratios, $I(Ga-K_\alpha)/I(P-K_\alpha)$, measured for the precipitate region are significantly increased with respect to those of the matrix region in all cases. This result indicates that the precipitates consist predominantly or completely of gallium. While Ga with a melting temperature of about 30°C is liquid at the diffusion temperatures it appears to exist in a solid amorphous state within a void during the TEM observation at ambient temperature.

In near-surface regions of high Zn concentration ($x \geq 10$ μm) Ga precipitates are found to be connected with crystalline precipitates exhibiting moiré contrast (Fig. 4). XEDS analyses indicate that such precipitates contain also a significant amount of Zn. Lattice parameters have been deduced both from fringe distances and from electron diffraction patterns. The experimental values obtained are in excellent agreement with the lattice parameters of the cubic Zn_3P_2 phase. Furthermore, a polycrystalline Zn_3P_2 layer forms for long term anneals at the surface of GaP.

4. DISCUSSION

In the following we will briefly discuss the basic aspects of the formation of defects by Zn diffusion into GaP and GaSb. The types of defects produced in the diffusion front regions of the crystals are similar to those observed during Zn diffusion in GaAs, and the formation of defects in GaP and GaSb under these conditions can be explained on the basis of the model suggested earlier for defect formation in GaAs (Luysberg et al. 1992). The formation of interstitial-type dislocation loops is clear evidence that the incorporation of Zn on Ga sublattice sites leads to a local supersaturation of Ga interstitial atoms via a kick-out reaction. The diffusion of Ga interstitial atoms either to the crystal surface or into deeper crystal regions is apparently not fast enough at the diffusion temperature so that the local Ga interstitial supersaturation persists. Since Ga interstitial atoms are mobile at the diffusion temperature they may condense and form dislocation loops. In order to

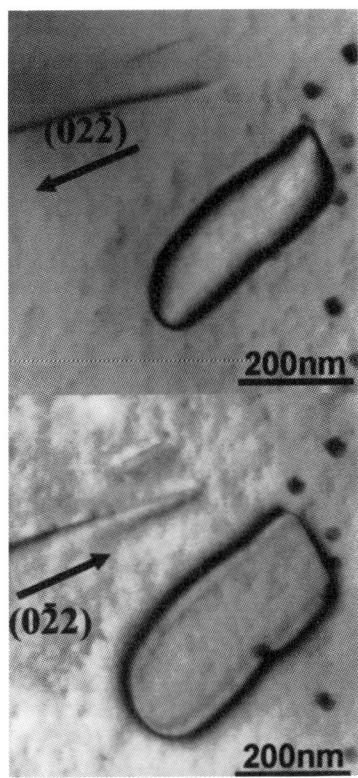

Fig. 3 Inside-contrast (top) and outside-contrast (bottom) of an interstitial-type dislocation loop in GaP.

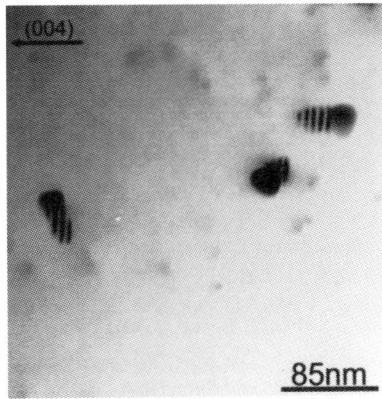

Fig. 4 Precipitates with moiré contrast in regions with high Zn-concentration near the crystal surface. Imaging vector $\vec{g} = (004)$.

76

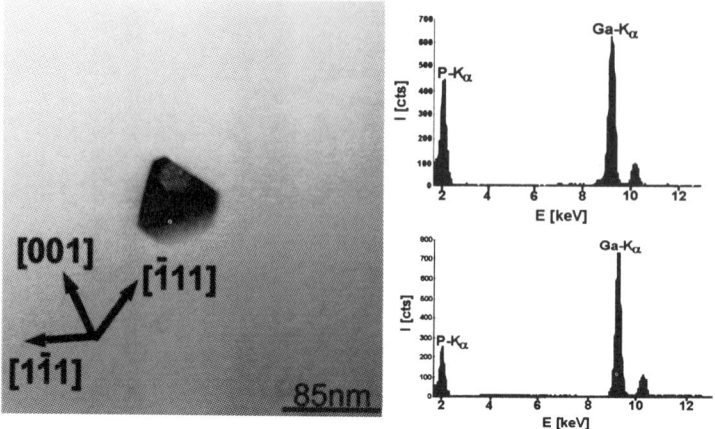

Fig. 5 (left) TEM bright field micrograph of a precipitate with a small void in GaP. (right) XEDS spectra, I (E), of the matrix region (top) and the precipitate region (bottom). The $I(Ga-K_\alpha)/I(P-K_\alpha)$-ratio of the precipitate region is significantly increased.

avoid an energetically unfavourable stacking fault and to maintain stoichiometry during loop formation phosphorus interstitials have to be provided simultaneously. In initially defect-free crystals phosphorus interstitials may be provided by emission of phophorus vacancies. The mobile phosphorus vacancies agglomerate and form voids which act as sinks for Ga interstitials created during further Zn diffusion. The observation of the Ga-rich precipitates with only a small void fraction is indicative of these processes. Similar to GaAs (Luysberg et al 1992, Rucki et al 1997) and to InP (Wittorf et al 1995), our observations of dislocation networks can be understood as the result of entanglement of loops during growth, and the formation of arrays of precipitates as a result of dislocation climb induced by the supersaturation and absorption of Ga interstitials. The role of dislocation climb during supersaturation of only one type of point defect has been recognized previously also in the analyses of GaAs-based lasers (Petroff and Kimerling 1976). Supporting our interpretation, the observation that larger loops and dislocation networks are absent for short-term anneals confirms that wafer surfaces may act as effective sinks for Ga interstitial atoms.

ACKNOWLEDGEMENTS

We acknowledge experimental support by Dr U Södervall (Chalmers University, Göteborg) for the SIMS measurements and funding support by the Deutsche Forschungsgemeimschaft (contracts DFG JA 908/2-1, STO 210/6-1).

REFERENCES

Ball R K and Hutchinson P W 1980 J. Mater. Sci, 2376
Conibeer G J, Willoughby A F W, Hardingham C M and Sharma V K M 1996 Optical Mat. 6, 21
Föll H and Wilkens M 1975 Phys. Stat. Sol. 31,519
Jäger W, Rucki A, Urban K, Hettwer H G, Stolwijk N A, Mehrer H, Tan T Y 1993 J. Appl. Phys. 74, 4409
Jäger C, Jäger W, Bösker G, Pöpping J and Stolwijk N A 1999 Phil. Mag., accepted
Luysberg M, Jäger W, Urban K, Schänzer M, Stolwijk N A and Mehrer H 1992 Mat. Sci. Eng. B13, 137
Petroff P M and Kimerling L C 1976 Appl. Phys. Lett. 29, 461
Rucki A and Jäger W 1997 Defect and Diffusion Forum Vols. 143, 1095
Tuck B 1988 Atomic Diffusion in III-V Semiconductors (Bristol Adam Hilger)
Wittorf D, Rucki A, Jäger W, Dixon R H, Urban K, Stolwijk N A, Hettwer H G, Mehrer H 1995 J. Appl. Phys. 77, 2843

Inst. Phys. Conf. Ser. No 164
Paper presented at Microsc. Semicond. Mater. Conf., Oxford, 22–25 March 1999
© 1999 IOP Publishing Ltd

A study of defects in LEC GaAs after copper diffusion

C Frigeri[1], J L Weyher[2], S Müller[3] and P Hiesinger[3]

[1] CNR-MASPEC Institute, Parco Area delle Scienze 37/A, 43100 Parma (I)
[2] High Pressure Research Center, Polish Academy of Sciences, ul. Sokolowska 29/37, 01-142 Warszawa (PL) and University of Nijmegen, RIM, Exp. Solid State Physics III, Toernoiveld 1, 6525 Nijmegen (NL)
[3] Fraunhofer-IAF, Tullastrasse 72, 79108 Freiburg (D)

ABSTRACT: The defects present after the diffusion of copper at 770 °C in semi-insulating LEC GaAs have been studied by TEM, photoetching and atomic force microscopy. Clusters of microloops in the matrix and around the dislocations have been observed. The enhanced etching velocity in the surroundings of the dislocations suggests that they have gettered Cu. The relationship between such defects and gettering and the generation of non-equilibrium point defects associated with Cu diffusion and incorporation in the GaAs lattice is discussed.

1. INTRODUCTION

The study of the behaviour of copper in GaAs and of its interaction with dislocations and the host lattice by means of diffusion experiments is of interest due to the fact that Cu can be introduced into GaAs either deliberately as an acceptor dopant or as an unwanted contaminant during the growth and subsequent processing of the crystal. In the former case diffused Cu is used as a compensating deep acceptor in n-type GaAs to produce semi-insulating crystals (Roush et al 1993, Yang et al 1994). Previous works have shown that the interaction of Cu with the GaAs crystal lattice is quite complex, several different types of crystal defects being formed (Leon et al 1995, Griehl et al 1996, Leipner et al 1997). Very likely the great variety of the observed crystal defects associated with the presence of Cu atoms depends on the difference in parameters of the diffusion and cooling down processes. This has been shown in the detailed work by Leipner et al (1997) who compared samples treated in different ways. The substrate type can also play a role. In particular, significant discrepancies exist as regards the interaction of Cu with dislocations since some authors did not detect any preferential diffusion or concentration of Cu at dislocations (MacQuistan and Weinberg 1991), whereas other authors did (Leipner et al 1997, Griehl et al 1996). As to the electrical behaviour, it has been reported that Cu gives rise to a deep acceptor with two ionisation levels whose identities are not fully clarified (Wang et al 1985, Leon et al 1995).

2. EXPERIMENTAL

The (100) semi-insulating GaAs samples investigated were grown by the LEC method. Cu diffusion was achieved by depositing on the substrate a 300 nm thick Cu layer and by annealing it at 770 °C for 5 min in closed ampoules with As counterpressure. Cooling down took place at a rate of 20 °C/s down to 700 °C and then at 10 °C/s to room temperature. Analysis of the defect structures after diffusion has been carried out by TEM and by DSL (Diluted Sirtl-like with Light) etching (Weyher and van de Ven 1986) in combination with differential interference contrast optical microscopy (DIC-OM) and atomic force microscopy (AFM). Prior to thinning, the TEM specimens were photoetched in order to localize the defects for the TEM observations. Due to its sensitivity to the hole concentration DSL etching is able to give qualitative information on the electrically active impurities (Weyher and van de Ven 1986, Kelly et al 1985, Frigeri and Weyher 1990). Such information is contained in the morphological profile of the etch features whose assessment was performed by AFM used in the tapping mode as well as by a step profilometer.

3. RESULTS

Figs. 1-2 are the TEM results. As shown in Fig. 1 dislocations are often surrounded by microloops ~20-80 nm in diameter. By diffraction contrast analysis with the inside/outside contrast method of Dahmen (1989) they turned out to be intrinsic. The majority of the microloops are faulted of the Frank type with b = a/3<111> as they vanish by changing from one to another of the two orthogonal <022> diffraction vectors in the (100) plane and do not vanish with g = <004>. Faulted dislocation loops in high density have also been detected in the matrix far from the dislocations (Fig. 2). They are extrinsic and of the Frank type, suggesting that they are due to condensation of point defects. They have sizes of ~80-150 nm and are very often gathered into clusters that are uniformly distributed in the dislocation cells.

Analysis of DSL etched samples has shown that the dislocations are surrounded by surface depressions indicating that in such areas the etching velocity has increased with respect to the matrix (Figs. 3-4). Inside such depressions and close to the dislocations a fine etch structure of tiny hillocks could be seen by both AFM (Fig. 4) and step profilometer. They are ascribed to the intrinsic dislocation microloops detected by TEM. Small etch hillocks are also detected everywhere in the sample far from the dislocations (Fig. 4), corresponding to the matrix extrinsic loops seen by TEM. The formation of etch hillocks at loops and dislocations is due to the recombination of photogenerated holes at the loops, as discussed elsewhere (Weyher and van de Ven 1986, Frigeri and Weyher 1990, Frigeri et al 1991).

4. DISCUSSION

The interstitial type dislocation loops far from the dislocations have been produced during and due to the in-diffusion of Cu. Similarly to other p-type dopants in III-V compounds, like Zn, Be and Cr, very likely the in-diffusion of copper in GaAs occurs via the substitutional-interstitial kick-out mechanism, whereby an interstitial p-type impurity occupies

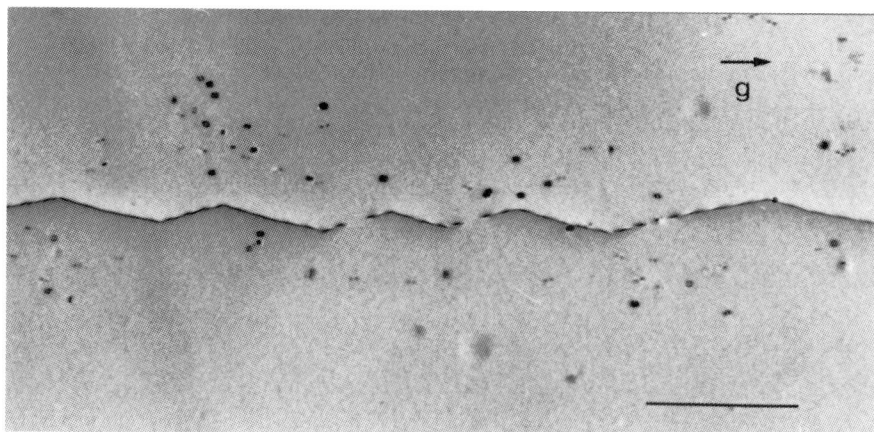

Fig. 1 - Bright field TEM plan-view image of a dislocation surrounded by microloops.
g = [022]. Bar = 1 μm.

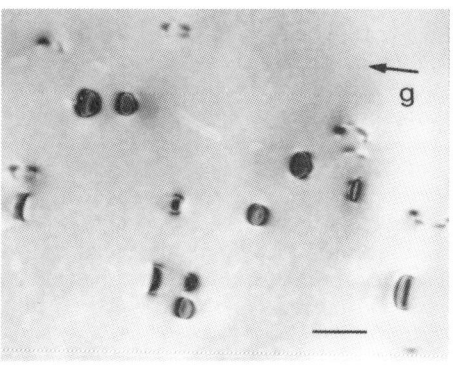 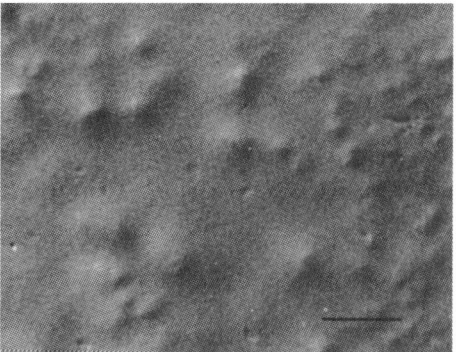

Fig. 2   Fig. 3

Fig. 2 - Bright field TEM plan-view image of a cluster of dislocation loops in the GaAs matrix. $\mathbf{g} = [022]$. Bar = 0.2 μm.

Fig. 3 - DIC-OM image after DSL etching of the (100) surface of the Cu diffused GaAs sample. Bar = 10 μm.

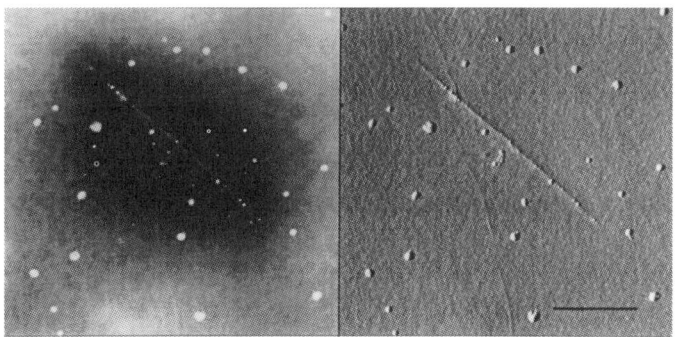

Fig. 4 - AFM images after DSL etching of the (100) surface of the Cu diffused GaAs sample. The image on the left is taken in the tapping mode whereby surface depressions appear dark. Bar = 2 μm.

substitutionally a Ga position by pushing the host Ga atom into an interstitial position, $A_i \Rightarrow A_s + I$ where A_i and A_s indicate the impurity on interstitial and substitutional sites, respectively, and I the intrinsic interstitial (Gösele and Morehead 1981, Gösele and Tan 1988). By considering the charge state of the point defects involved, for Cu in GaAs it is

$$Cu_i^+ + Ga_{Ga} \Rightarrow Cu_{Ga}^{2-} + Ga_i^{2+} + h^+$$

The diffusion of Cu can thus create a large number of Ga interstitials in supersaturation with respect to the equilibrium concentration of Ga interstitials, which is known to be very low in LEC GaAs (Bublik et al 1980, Giling et al 1986). Such Ga interstitial supersaturation is the driving force for the condensation of interstitials of both the As and Ga sublattices into nuclei of extrinsic dislocation loops and for their growth by climb. The As interstitials necessary for dislocation loop formation are readily available in the host matrix as in the LEC semi-insulating GaAs crystals a supersaturation of As interstitials always exists (Hurle et al 1990). Additional As interstitials should come from the matrix arsenic precipitates of the original semi-insulating substrate which are expected to have dissolved at 770 °C as their formation temperature is around 600 °C (Weyher et al 1992). They could be created also at cores of dislocation loops during climb by the Petroff-Kimerling (1976) mechanism.

The DSL etching velocity is proportional to the density of photogenerated holes available for the dissolution of the GaAs molecules at the sample surface in contact with the etching solution (Weyher and van de Ven 1986, Kelly et al 1985, Frigeri and Weyher 1990). The increase of the etching velocity in the surroundings of the dislocations thus indicates an increased density of holes with respect to the matrix. Such extra holes can be accounted for by assuming that in the surroundings of the dislocations there is an increase of the density of electrically active acceptor-like impurities. The most likely acceptor is the Cu_{Ga} double acceptor substitutionally sitting on the Ga site, which is reported to have two ionisation levels at 0.15 and 0.45 eV above the valence band edge (Leon et al 1995). This clearly indicates that Cu has been gettered by the dislocations, contrary to the findings of MacQuistan and Weinberg (1991) but in agreement with the results of Griehl et al (1996) and of Leipner et al. (1997) who found by cathodoluminescence an enrichment of Cu in the strain field of fresh dislocations that were introduced by scratching. A model for the formation of the vacancy type microloops surrounding the dislocations can be envisaged by assuming that those Cu atoms that are able to reach the core of the dislocation during cooling down leave behind Ga vacancies in a high density, very likely in supersaturation, which can later collapse into intrinsic microloops around the dislocation. The increase of the hole concentration in the surroundings of the dislocation implies a higher equilibrium concentration of Ga_i^{2+} (Gösele and Tan 1988, Gösele and Tan 1991, Leipner et al 1997) so that the probability of supersaturation decreases which may explain the absence of extrinsic dislocation loops.

REFERENCES

Bublik V T, Mil'vidsky M G and Osvensky V B 1980 Sov. Phys. Cryst. **23**, 1
Dahmen U 1989 Ultramicrosc. **30**, 102
Frigeri C and Weyher J L 1990 J. Crystal Growth **103**, 268
Frigeri C, Weyher J L and De Potter M 1991 Appl. Surface Sci. **50**, 115
Giling L J, Montree A, Weyher J L, Fornari R and Zanotti L 1986 J. Crystal Growth **79**, 271
Gösele U M and Morehead F 1981 J. Appl. Phys. **52**, 4617
Gösele U M and Tan T Y 1988 Ann. Rev. Mater. Sci. **18**, 257
Gösele U M and Tan T Y 1991 MRS Bulletin, vol XVI, p. 42
Griehl St, Herms M, Klöber J, Niklas J R and Siegel W 1996 Appl. Phys. Lett. **69**, 1767
Hurle D T J, Wenzl H and Henkel D 1990 in *Semi-insulating III-V Materials*, Toronto, (Adam Hilger, Bristol) p. 143
Kelly J J, van de Ven J and van der Meerakker J E A M 1985 J. Electrochem. Soc. **132**, 3026
Leipner H S, Scholz R, Syrowatka F, Uniewski H and Schreiber J 1997 J. Phys. III France **3**, 1495
Leon R, Werner P, Yu K M, Kaminska M and Weber E R 1995 Appl. Phys. A **61**, 7
MacQuistan D A and Weinberg F 1991 J. Crystal Growth **110**, 745
Petroff P M and Kimerling L C 1976 Appl. Phys. Lett. **29**, 461
Roush R A, Stoudt D C and Mazzola M S 1993 Appl. Phys. Lett. **62**, 2670
Yang B, Egilsson T and Gislason H P 1994 in *Semi-insulating III-V Materials*, Warsaw, (World Scientific, Singapore) p. 263
Wang Z G, Gislason H P and Monemar B 1985 J. Appl. Phys. **58**, 230
Weyher J L and van de Ven P 1986 J. Crystal Growth **78**, 191
Weyher J L, Gall P, Le Si Dang, Fillard J P, Bonnafé J, Rüfer H, Baumgartner M and Löhnert K 1992 Semicond. Sci. Technol. **7**, A45

Inst. Phys. Conf. Ser. No 164
Paper presented at Microsc. Semicond. Mater. Conf., Oxford, 22–25 March 1999
© 1999 IOP Publishing Ltd

Micropipes in SiC represent Frank's hollow dislocations

J Heindl and H P Strunk

Institut für Werkstoffwissenschaften - Mikrocharakterisierung, Universität Erlangen - Nürnberg, Cauerstraße 6; 91058 Erlangen; Germany; e-mail: heindl@ww.uni-erlangen.de

ABSTRACT: Micropipes are hollow tubes that extend along the growth direction of sublimation grown SiC single crystals. The Burgers vector b associated with such a micropipe can have a length between one and more than 100 lengths of an elementary Burgers vector. We quantify Frank's r-b-squared relation by scanning force microscope measurements of radii r and Burgers vectors at emerging points of micropipes. All data can consistently be described with the mixed Burgers vector that was analyzed by transmission electron microscopy.

1. INTRODUCTION

SiC is one of the most promising materials for high power and high frequency devices. The most significant problem in growing high grade material is the formation of so called micropipes: hollow tubular defects penetrating the crystal along the growth direction and intersecting the growth surface in the center of step spirals (for review see eg Heindl et al 1997a). The existence of these spirals on the growth surface indicates a screw dislocation to be associated with the micropipe (Heindl et al 1997b,c). The presence of a possible additional Burgers vector component within the surface, i.e. one of edge type, which would turn the dislocation into one of mixed type, cannot be discriminated this way (Strunk 1996). An explanation for the existence of micropipes was proposed by Frank (1951) who considered the model of the hollow core dislocation: the strain field around a dislocation whose Burgers vector exceeds a critical dimension (approximately 1 nm in typical semiconductor materials) contains such a high energy that it is energetically more favorable to remove the crystalline material around the dislocation line and to create an additional surface in the form of a hollow tube. The dependence of the equilibrium radius r_0 of such a hollow dislocation and its Burgers vector b is given by Frank's formula:

$$r_0 = \frac{m y}{8 \pi^2 \gamma} b^2 \ .$$

(1)

μ: shear modulus (for SiC: $\mu = 1.9 \cdot 10^{11}$ J/m^3); γ: specific surface energy of the inner surface of the micropipe. This surface energy will be used as a fit parameter in the following discussion.

In general, Frank's model is applicable to any kind of dislocations. In the past, obviously due to its easy accessibility, only screw components of the Burgers vectors were considered and the results were discussed as if micropipes were formed by pure screw dislocations.

Within Frank's model however, the origin of the extremely large Burgers vectors is not explained. In earlier papers (Heindl et al 1997a,d,1998) we discussed a microscopic model of micropipe formation based on our observations by transmission electron microscopy (TEM) of the nucleation sites of micropipes. According to this model, the micropipes contain multiples n of partial dislocations with mixed type Burgers vector 1/6<2,-1,1> which is lying obliquely to the growth surface. (To avoid problems in indexing we use the cubic notation with [1,1,1] parallel to the c-axis.) The total Burgers vector of the micropipe is thus n/6<2,-1,1>. A simple geometrical consideration then leads to a relation between the screw component and the total Burgers vector:

$$|B_{total}| \ : \ |B_\odot| \ = 2.13 \ : \ 1. \tag{2}$$

For measurements of the heights of spiral steps at the surface, only the screw component can be evaluated. Frank´s relation becomes then:

$$r_0 \ = (2.13)^2 \ \frac{\mu}{8\,\pi^2\gamma K} B_\odot^{\,2} \tag{3}$$

The mixed character of the Burgers vector requires an appropriate energy factor K = 0.88.

2. EXPERIMENTAL

We investigate three 6H-SiC single crystals that have been grown by the modified Lely method at temperatures of about 2300 °C and at an Argon pressure of about 20 mbar (Heydemann 1996). The Si-terminated (1,1,1)-facet of a 6H-SiC single crystal was used as the seed. Radii and step heights of growth steps adjacent to micropipes are measured on the as-grown surface by atomic force microscopy (AFM). A Park Scientific Instruments universal system with high aspect ratio cantilever was used. Round and slightly oval micropipes (typically less then 10% difference between large and small half-axes) were evaluated (Fig.1). As diameter value we used the distance of the falling and the rising edges in the line scan over the center of the micropipe. The total step height of the growth steps associated with the

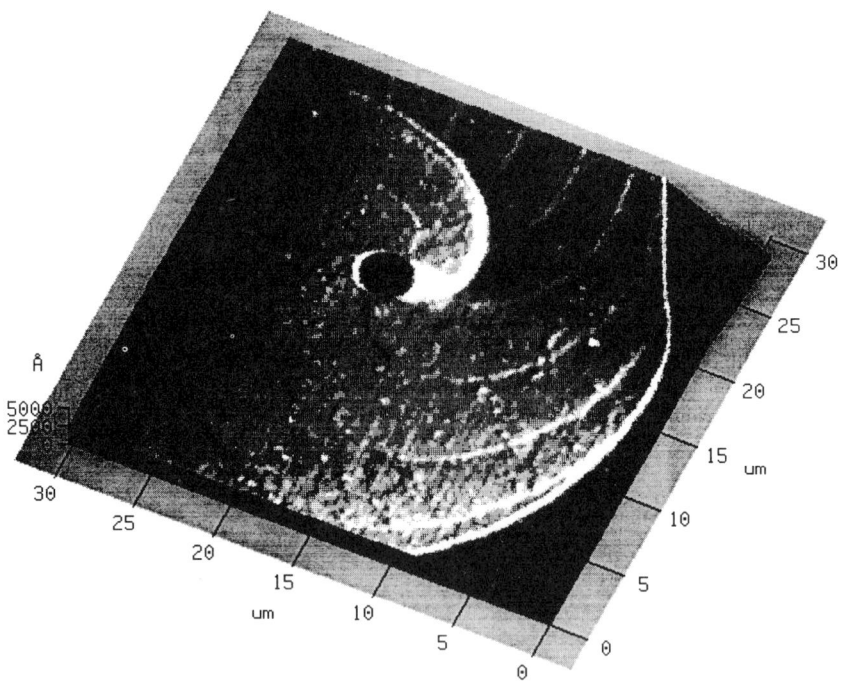

Fig. 1: Typical micropipe on the as-grown surface of a SiC single crystal. The edge of the micropipe as well as several growth steps are clearly visible. Atomic force microscopy.

micropipe was obtained by adding upward steps with positive and downward steps with negative sign during one revolution around the hole. We take this total step height as the screw component of the Burgers vector of the micropipe. On might object that a systematic error in the radii of the micropipes may result from a possible crater-type widening of the micropipe at the emergence point (as deduced by Frank (1951) by continuum mechanics arguments). However, in terms of spiral growth such a crater is descernible from a reversal of the winding sense at the spiral center, which is synonymous with the local formation of a dissolution spiral. This dissolution part of spiral is clearly visible in our AFM data and we took as the value for the diameter the distance between the points where the spiral steps end tangentially in the tube.

3. RESULTS

The measured micropipe parameters, total step heights and radii, range from 1.5 nm to 37.5 nm and from 25 nm to 6000 nm, respectively. We added to our data values obtained by Dudley et al (1997) with a combination of synchrotron white beam x-ray topography (Burgers vector) and scanning electron microscopy (diameter). All radii are plotted in Fig. 2. versus the square of the screw Burgers vector according to Frank´s relation eq.1. The curves fitting four data subsets were obtained by least square fits. A fit curve with a steep slope (especially at small Burgers vectors; in the following called type I-micropipes) and three curves with a smaller slope at larger Burgers vectors (type II-micropipes) are obtained. The type I-micropipes obey Frank´s relation, since the fit curve intersects the origin. In contrast type II-micropipe curves show an offset, which is different for each investigated crystal.

4. DISCUSSION

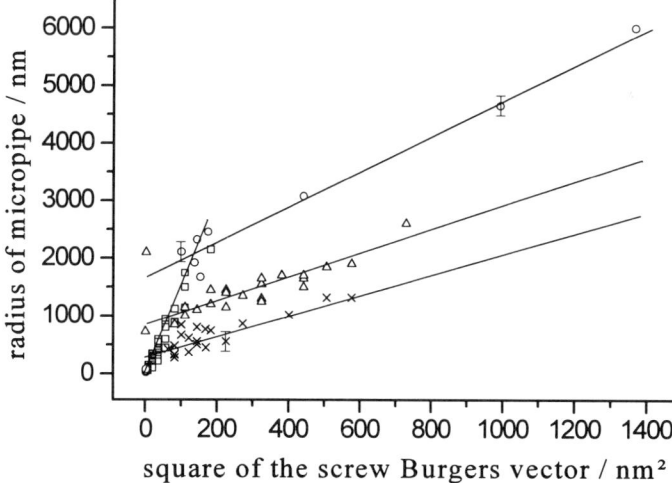

Fig. 2: Micropipe radii as dependent on Burgers vector length squared.
X,○,Δ: Data obtained by AFM.
□: Data from Dudley et al (1997) obtained by synchrotron white beam x-ray topography.
The error bars indicate the typical measurement errors. The fit curves are obtained by least square fits to the respective data subsets. Two different slopes of the fit curves can be distinguished although both can be fitted with one surface energy (see text).

The presented data, though appearing complicated, can in fact be interpreted on the basis of Frank's model of a hollow dislocation, however, with a mixed Burgers vector $B = n/6<2,-1,1>$. The intersection of the type I-curve (which is followed also by Dudley et al's (1997) data) with the origin (Fig. 2) corresponds to Frank's model (eq. 2). Application of the mixed Burgers vector yields a surface energy of $\gamma = (0.93 \pm 0.08)$ J/m², which is very reasonable. Recently, Dudley's group (Huang et al 1999) interpreted micropipes as pure screw dislocations. In this case a direct application of Frank's formula to Dudley's data (see Fig. 2) would lead to a surface energy of approximately 0.2 J/m², which seems too low a value to be physically significant.

The interpretation of type II-curves is a little bit more complicated. The average slopes of the fit curves of type II-micropipes lead to a surface energy of $\gamma = (1.08 \pm 0.04)$ J/m² if we use the measured screw Burger vector as the total Burgers vector. This value agrees well with the one obtained from the type I-micropipes. This agreement is the basis of the interpretation: we assume therefore that in the range of the smaller slopes only the screw Burgers component changes. The edge component stays essentially constant at a value that is reached at the intersection points with the type I-curve. In consequence in the type II-micropipe range each of the investigated crystals exhibits a particular and constant maximum edge Burgers vector component which fact is the topic of current investigations.

5. SUMMARY

We analyze the emergence points of micropipes on the as-grown surface of 6H-SiC single crystals grown with the modified-Lely-method by AFM. All data points can consistently be explained within Frank's model of hollow dislocations under the light of our TEM-results: the Burgers vector in the micropipes is a multiple of $n/6[2,-1,1]$. However the edge component is limited to a maximum value which is characteristic for each crystal. This maximum will be the topic of further work.

ACKNOWLEDGEMENT

This work was supported by the Deutsche Forschungsgemeinschaft under contract number SFB 292 B3.

REFERENCES

Dudley M, Si W, Wang S, Carter C Jr, Glass R and Tsvetkov V 1997 Nuovo Cimento D **19D**, 153

Frank F C 1951 Acta. Cryst. **4**, 497

Heindl J, Strunk H P, Heydemann V and Pensl G 1997a in 'Fundamental Questions and Applications of SiC' eds M Stutzmann, W J Choyke, H Matsunami, G Pensl, phys. stat. sol. (a) **162**, 251

Heindl J, Dorsch W, Eckstein R, Hofmann D, Marek T, Müller S G, Strunk H P and Winnacker A 1997b Proc. 1st European Conference on Silicon Carbide and Related Materials 1996, Diam. Relat. Mater. **6**, 1269

Heindl J, Dorsch W, Eckstein R, Hofmann D, Marek T, Müller S G, Strunk H P and Winnacker A 1997c J. Crystal Growth **179**, 510

Heindl J, Heydemann V D, Pensl G and Strunk H P 1997d Proc. 7th Int. Conf. on Defect Recognition and Image Processing, eds J Donecker and I Rechenberg (IOP) Int Phys Conf Ser **160**, 331

Heindl J, Dorsch W, Strunk H P, Eckstein R, Hofmann D, Müller S G and Winnacker A 1998 Phys. Rev. Lett. **80**, 740

Heydemann V D 1996 PhD Thesis, Friedrich-Alexander-Universität Erlangen-Nürnberg

Huang X R, Dudley M, Vetter W M, Huang W, Wang S and Carter C H Jr. 1999 Appl. Phys. Lett. **74**, 353

Strunk H P 1996 J. Crystal Growth **160**, 184

Inst. Phys. Conf. Ser. No 164
Paper presented at Microsc. Semicond. Mater. Conf., Oxford, 22–25 March 1999
© 1999 IOP Publishing Ltd

Planar, linear and point defects in high purity CVD diamond

J W Steeds[1], S J Charles[1] and J E Butler[2]

[1]Physics Department, University of Bristol, Bristol BS8 1TL, UK
[2]Chemistry Division, Naval Research Laboratory, Washington DC 20375, USA

ABSTRACT: Planar and linear defects in thick CVD diamond specimens have been studied by laser cutting cross-sections prior to reactive ion etching to electron transparency. Undoped samples reveal nano-twins and microscopic precipitates in bands located near the columnar grain boundaries. Boron doped samples show relatively defect-free grain boundary regions and on a macroscopic scale, grain boundaries that are rather accurately parallel to the growth direction. In detail, many of these boundaries reveal localised stress and undulatory character depending on the relative grain misorientations and the grain boundary planes. The boron doped samples have also been investigated by micro-Raman spectroscopy and high spatial resolution imaging SIMS to reveal gross inhomogeneities of the boron distribution from one grain to another.

1. INTRODUCTION

High quality polycrystalline CVD diamond layers are generally grown at the present time at or near the <110> orientation. Previous efforts at <100> growth produced near-epitaxy on <100> Si but the films were highly stressed and contained regions of very high defect density. Reasons for this behaviour have now been advanced (Steeds et al 1998). In contrast, the near <110> oriented layers can be fabricated with large grain sizes, some twins parallel to the growth direction and it is only in the immediate vicinity of the columnar grain boundaries that considerable defect densities have been observed (Steeds et al 1999). Although still containing residual stresses these are at a much lower level than in the <100> oriented layers. B doped layers are of particular interest because B is the chief dopant, at present, that is used to control the conductivity of diamond for various practical applications. In this paper we will report the grain structure and its relationship to the B distribution of some B doped diamond films. Particular attention will be given to the local determination of B concentration.

2. EXPERIMENTAL TECHNIQUES

Several samples from different sources have been studied. Most were in the form of thick layers >100 μm in thickness; one was only 5 μm in thickness. The boron doping range varied from 10^{16} cm^{-3} to 10^{21} cm^{-3} (normal values). Cross sections were cut from the thick layers using a scanned laser beam and they were thinned for TEM investigation using argon/oxygen ion mixtures. Plan view sections were prepared by mechanical polishing and ion thinning. Optical characterization was performed using an Oxford Instruments Microstat attached to a Renishaw micro-Raman system operated with a 488 nm argon ion laser line. This set-up was used to perform both micro-Raman and low temperature (6K) microscopic photoluminescence spectroscopy. Electron irradiation experiments and TEM investigations were carried out using a Philips EM30. SIMS images were generated using an in-house developed Ga FIB gun interfaced to a magnetic sector analyzer.

3. RESULTS AND DISCUSSION

Examination of the thin specimen in plan view revealed a number of different nuclei. These included five-fold twins, rectangular crystals with two opposite quadrants that were defect free and two opposite quadrants that were highly defective (Fig. 1) and very irregular and highly defective clusters of grains. Plan view sections near the top surface of thick layers revealed a mixture of five-fold twins and larger irregular grains, usually containing one or two twins, in a <112> orientation. The large <112> oriented grains had very low dislocation densities and internal flat coherent $\Sigma 3$ twin boundaries without additional dislocations. The five-fold twins were rather common, sometimes occurring in clusters of twos or threes. The individual five-fold units were relatively small, not exceeding 10 μm in diameter and appeared in a wide variety of forms. Some were almost defect-free, others had large densities of dislocations. None were perfectly symmetric. Those that had very low dislocation densities were less symmetric and contained narrow supplementary twin bands. Those that were more symmetric, at least near the centre, tended to have some sectors with high dislocation densities. Some had high dislocation densities in all sectors. It was evident that the dislocations present, sometimes in the forms of dislocation pile-ups, had mostly entered from the periphery. Most of the main radial twin boundaries were flat and close to {111} orientation but had high densities of supplementary screw $\frac{a}{2}$ <100> dislocations parallel to the growth direction. The twin plane was the {111} plane common to the two adjoining grains. A few had one wavy grain boundary originating at the centre and eventually reaching the periphery. In this case the two adjoining planes lacked a common {111} plane parallel to the growth direction. A reasonable proportion, say 10%, had cracks along the radial twin boundary that opened up along the outer half of the boundary but did not penetrate to the centre. Some of the five-fold twins had unusual splays of twins that started at about half way towards the periphery and then grew in number with distance from the centre. These were associated with wavy grain boundaries. The grain boundaries at the periphery of the five-fold twins were generally highly irregular. Many of these observations may be explained by the known stress fields for these star disclinations (de Wit, 1972). The stress at the centre is purely compressive but at a distance from the centre of about 60% of the radius of the star disclinations, the tangential stress becomes tensile while the radial stress remains compressive throughout. In the outer part, with perpendicular tension and compression, shear stresses are generated that may be relaxed by plastic flow as observed. The energy of the star disclination is proportional to the square of its radius which prevents their growth to large dimensions since it is energetically favourable to have a number of smaller stars. The extra dislocations observed along the straight radial boundaries are required to accommodate small deviations from ideal coherency at the $\Sigma 3$ boundaries because 5 x 70°32' ≠ 360°.

Cross-sections through these thick samples showed features that are compatible with the observations reported above. In particular, the wavy grain boundaries were evident and found to occur in a number of different cases (Fig. 2). Some were close to coincidence orientations $\Sigma 3$, $\Sigma 9$, $\Sigma 27$, others were apparently purely random. None of them gave any evidence of residual stress in diffraction contrast or large angle convergent beam electron diffraction micrographs. By tilting them to near-edge-on orientation (this could only be done locally because the average orientation varied greatly along the boundary) the undulations appeared more regular in some cases. Individual facets came close to low index planes, mainly {111}, some {100} oriented. However, the situation observed was far from regular and it may be that in addition to energy reduction by forming facets of lower energy there is some form of stress release involving Asaro-Tiller-Grinfeld instability. At the temperature of growth, 700 to 900°C. both vacancies and self-interstitials are mobile. The vacancy diffusion freezes out below about 650°C but self-interstitials may remain mobile down to room temperature.

At high doping levels ($\geq 10^{18}$ cm^{-3}) there are several techniques to reveal the B distribution. The most dramatic results have been obtained using our Ga SIMS machine (Fig. 3). It is clear that the B distribution is grossly inhomogeneous and, in fact, the variations from grain to grain in this figure are as large as a factor of 8 or so, consistent with recent results showing the relative uptake on

{111} and {100} facets. Similar results have also been demonstrated by changes in the diamond LO phonon Raman profile (Steeds et al, 1999).

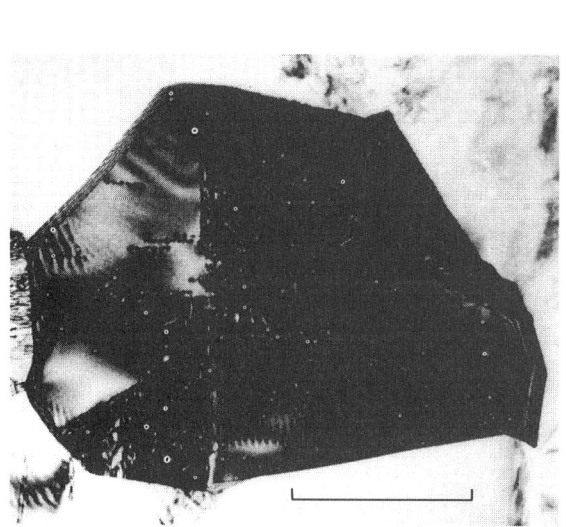

Fig. 1. <110> oriented grain in highly B doped
CVD diamond. The line is of length 1 μm
(plan view)

Fig. 2. A wavy Σ3 boundary taken under two beam bright field conditions in the common 220 reflection (vertical, also the growth direction). The line of length 1 μm.

In fact the lattice parameter changes for boron concentrations above 10^{19} cm^{-3} are large enough to measure by CBED as we have demonstrated recently. At B concentrations below 10^{18} cm^{-3} the only technique reported in the literature with sufficient sensitivity is that of infra-red absorption (Collins & Williams, 1971). However, this is not a local technique and it cannot be applied without the danger of serious ambiguities in the case of polycrystalline material. We have now discovered a microscopic technique which might be effective at low concentrations. Normally B incorporation is associated with the appearance of a broad green band in the luminescence spectrum, centred at about 540 nm. We have found that electron irradiation of diamond in a TEM with 300 keV electrons is sufficient to displace the carbon atoms leading to vacancy and self-interstitial formation with characteristic sharp, optical signatures. The self-interstitials are mobile at room temperature and form complexes with impurities in the diamond lattice. Some of the resulting sharp luminescence peaks only appear in B doped samples and we suggest that they involve B-C complexes. In particular, by reducing the microscope accelerating voltage below the C displacement value we still see one of these PL peaks. This result supports our suggestion since it is likely B can be displaced (but will not diffuse) creating a B interstitial related centre. More work remains to explore the use of these newly observed zero-phonon peaks in investigating the B content of B doped diamond. At higher levels of B the Fermi level is pinned close to the valence band and this leads to quenching of luminescence.

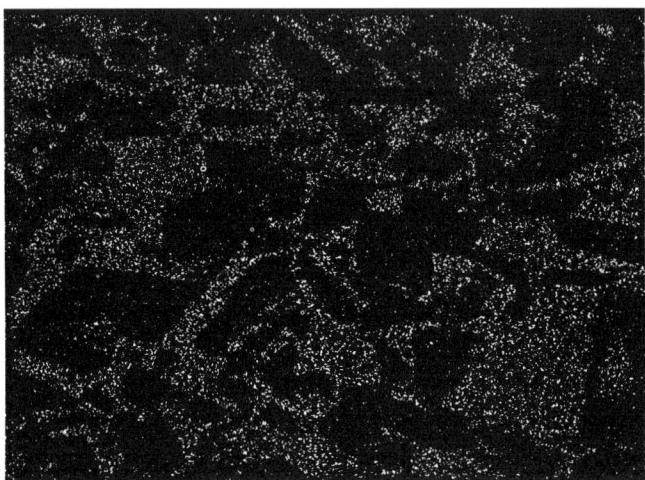

Fig. 3. B SIMS image of B doped diamond layer. The width of the image is 430 μm.

4.   CONCLUSIONS

The microstructure of B-doped polycrystalline CVD diamond layers has been investigated. Five-fold twins are found to be common and the source of a certain amount of internal stress in the layers. However stress leads to wavy boundary morphology. The B uptake has been found to be extremely inhomogeneous varying by up to a factor of 8 from one grain to the next. A number of methods have been found suitable for investigating this inhomogeneity in highly doped films. At low doping levels a novel technique involving near-threshold electron irradiation followed by low temperature photoluminescence microscopy offers interesting prospects for B analysis.

REFERENCES

Collins A T and Williams A W S 1971 J. Phys. C: Sol. State Phys. **4**, 1789
De Wit R 1992 J. Phys. C: Sol. State Phys. **5**, 529
Steeds J W, Gilmore A, Bussmann K M, Butler J E and Koidl P 1999 Diamond and Relat. Mat. (in press)
Steeds J W, Gilmore A, Wilson J A and Butler J E 1998 Diamond and Relat. Mat. **7**, 1437
Steeds J W, Gilmore A, Charles S, Heard P, Howarth B and Butler J E 1999 Acta Materialia (in press)

Inst. Phys. Conf. Ser. No 164
Paper presented at Microsc. Semicond. Mater. Conf., Oxford, 22–25 March 1999
© *1999 IOP Publishing Ltd*

Electrically active grain boundaries in GaP

D B Holt and E Napchan

Department of Materials, Imperial College of Science, Technology and Medicine, London
SW7 2BP

ABSTRACT: SEM REBIC was used to study electrically active grain boundaries in
polycrystalline GaP. The electrically active boundaries were found to run parallel,
independently of the direction from one contact to the other. The REBIC signal profiles
were suppressed at low scan speeds due, it is thought, to the resistance and capacitance
of the specimens themselves acting as high-pass filters. Boundary contrast due to high
resistivity boundary layers, to trapped charge or to enhanced or reduced recombination
was found.

1. INTRODUCTION

Grain boundaries in GaP produce a distinctive peak-and-trough contrast in scanning
electron microscope (SEM) electron beam induced current (EBIC) micrographs (Ziegler et al
1982) which is the characteristic signature of charged boundaries (Holt 1994). Recently
remote EBIC (REBIC) has been applied to characterize grain boundaries (GBs) in many
wide-bandgap materials (Holt 1994, Holt et al 1996, 1997, Russell and Leach 1998). This
paper reports an examination of polycrystalline GaP.

2. EXPERIMENTAL METHODS

Slices of polycrystalline GaP, produced for use as source material for liquid
encapsulated Czochralski (LEC) growth of single crystals, were kindly supplied by Dr. I.R.
Grant of Wafer Technology plc. Specimens were cut and fixed on insulating mounts with
two top surface contacts made with silver epoxy. They were examined in a JEOL JSM 840A
SEM using a Matelect EBIC system and a Stanford Research Systems SR570 amplifier
operated by TestPoint software.

3. RESULTS AND DISCUSSION

Electrically active GBs in the first GaP sample examined (Fig. 1) all ran roughly
in the direction joining one contact to the other. This is at right angles to the direction of
alignment seen previously (e.g. Holt et al 1997). It is not optimum for the observation of
terrace contrast (step changes of brightness across the boundary) arising from high-
resistivity GB layers. Nevertheless, terrace contrast appears i.e. the grain to the right is
brighter than that to the left, for each boundary in Fig. 1. This contrast did not change
with a 90° change in scan direction.

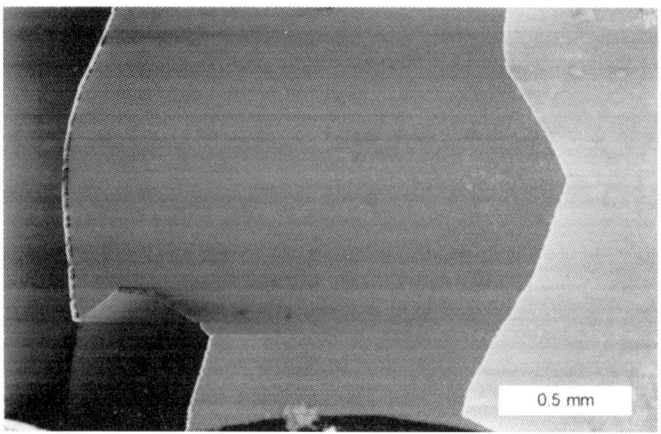

Figure 1. REBIC image of two electrically active GBs in poly-crystalline GaP. They are aligned vertically, i.e. in the direction running from the one contact visible at the bottom of the field of view to the other, above the top of the field of view.

In Fig. 2, however, the electrically active boundaries ran normal to the contact-to-contact direction so the preferred direction for active GBs was real and independent of the arbitrary positioning of the contacts.

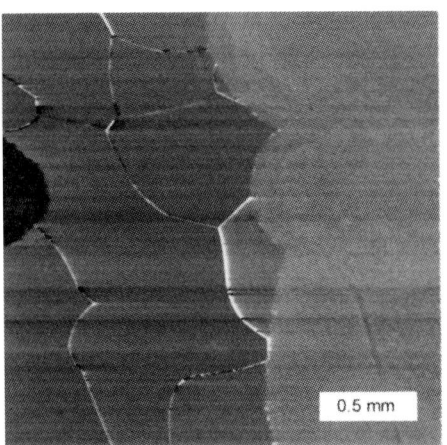

Figure 2. REBIC image of GBs in another specimen. They are predominantly aligned vertically. This is at right angles to the direction from one contact, dark, at the left, to the other, light, at the right.

GB REBIC linescan profiles could only be seen for fast scans. At the SEM's slower scan rates, the GB contrast profile disappeared. This is believed to be due to signal autoprocessing. The specimens were large (square mms in area and a mm thick). The contacts were mms apart. Consequently the capacitance and resistance of the specimens are large and they act as RC high-pass filters. For fast scans, the signal varies rapidly and the profile is rich in high-frequency Fourier components that pass the sample. For slow scans the variation consists more of low-frequency components that are stopped. That is, the capacitance weakens the low frequency components of the waveform. Some evidence for this was seen in other materials but the effect was not as marked.

Numbers of successive scans were recorded and signal averaged to obtain low-noise REBIC signal profiles like the peak and trough (PAT) profile in Fig. 3. Ziegler et al (1982) explained such contrast in GaP as due to charge on the boundary.

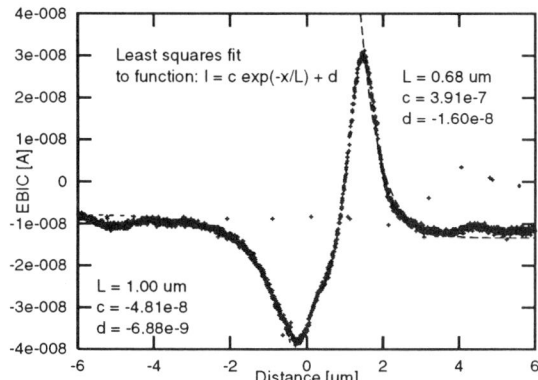

Figure 3. Plot of the signal averaged REBIC linescan recorded across GB A in Fig. 1 at the "TV" scan speed of the JSM 840A SEM. The linescan profile was also visible at the other fast scan speed "SR" but not at e.g. the "slow 2" speed. The dotted curves are exponential. That on the left fits well but that on the right shows evidence of surface recombination.

Trapped boundary charge bends the energy bands which can be modelled as two Schottky barriers back-to-back (Palm 1993). Ziegler et al forward and reverse biased such boundaries and showed that the result was to increase the peak and decrease the trough for one bias and vice versa for the opposite one. For sufficiently large bias it is possible to suppress one so only a peak or only a trough is seen (Holt 1994). The tails of the signal to the left (in grain 1) and to the right (in grain 2) are often of accurately exponential form, exp(-x/L), so values for L can be obtained as in Fig. 3. Often the two values were unequal indicating a difference in trap densities. The occurrence of tails of simple exponential form is indicative of negligible surface recombinaion effects. As is well known, surface recombination makes the curves concave upward when plotted semilogarithmically (e.g. Palm 1993).

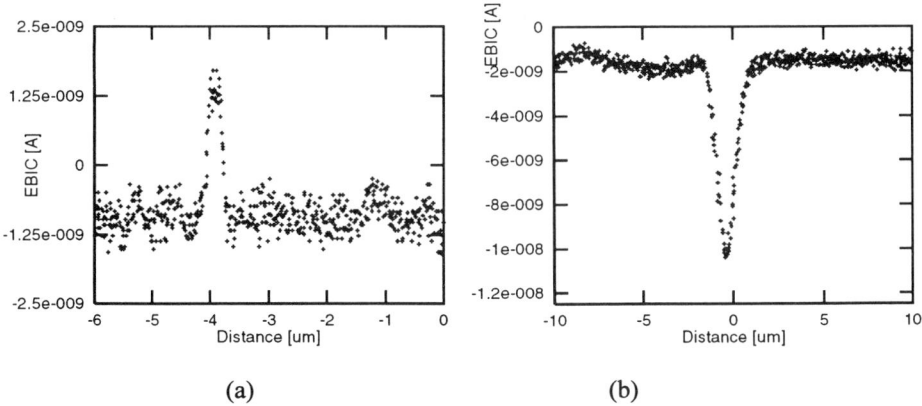

(a)   (b)

Figure 4. REBIC linescan profiles recorded across GBs showing (a) a peak only and consequently giving bright line contrast and (b) a trough only and so exhibiting dark line contrast.

The PAT profiles were always assymmetric. For instance, the Fig. 3 peak is of greater amplitude and narrower than the trough.

Contrast linescans showing either a peak alone (bright contrast – Fig. 4a - due to reduced recombination) or a trough alone (dark contrast – Fig. 4b - due to enhanced GB recombination) were seen. These observations are similar to the results of previous studies of other materials (e.g. Russell and Leach 1998, Holt et al 1997).

4. DISCUSSION

The wide occurrence of peak and trough contrast showed most GBs were charged. The assymmetry of the PAT profiles can be due to three causes. (i) Boundaries steeply inclined to the surface give assymmetric profiles (Palm 1993, Russell and Leach 1998). These grain boundaries ran approximately normally through the slices, however. (ii) Differences in doping on either side of the boundary can give different depletion region widths and hence different peak and trough amplitudes. The differences in the left and right decays in Fig. 3 indicate unequal doping. (iii) REBIC autobiasing often occurs since currents flowing through the high-resistivity boundary layer with the beam incident on either side are generally unequal. Hence the autobiased potential drops across the boundary will generally be different on the two sides. The consequent unequal depletion region widths produce different amplitudes for the peak and trough (Holt et al 1996). Separating the three contributions to the assymmetry would not be easy.

The occurrence of terrace contrast (Fig. 1) showed that some boundaries were of a high local resistivity. The parallel alignment of most of the electrically active boundaries and, conversely, the inactivity of most boundaries joining them has not been reported previously. At present we have no explanation for this.

REFERENCES

Holt D B 1994 in Sol. State Phenomena **37-38**, eds H P Strunk, J H Werner, B Fortin and O Bonnaud (Scitec Publishers: Zurich) pp 171 - 182
Holt D B, Raza B and Wojcik A 1996 Mat. Sci. & Eng. **B42**, 14
Holt D B, Raza B and Wojcik A 1997 in Microscopy of Semiconducting Materials, Inst. Phys. Conf. Ser. **157** (Inst. Phys.: Bristol) pp 639 - 642
Palm J 1993 J. Appl. Phys. **74**, 1169
Russell J D and Leach C 1998 Acta Materiallia **46**, 6227
Ziegler W, Siegel W, Blumtritt H and Breitenstein O 1982 Phys. Stat. Sol. **a72**, 593

Inst. Phys. Conf. Ser. No 164
Paper presented at Microsc. Semicond. Mater. Conf., Oxford, 22–25 March 1999
© 1999 IOP Publishing Ltd

Formation of microcracks in an annealed cubic boron nitride

M Aki, Y Ohno and S Takeda

Department of Physics, Graduate School of Science, Osaka University,
1-16, Machikane-yama, Toyonaka, Osaka 560-0043, Japan

ABSTRACT: We have found that microcracks are formed in cubic boron nitride, grown by a hot-press method, by annealing at temperatures above 673 K. Microcracks originate from segments of 60° dislocations and they extend to the dilation side of the dislocations. We have shown that the microcracks are formed by the condensation of excess nitrogen atoms onto the {111}-type atomic planes.

1. INTRODUCTION

Cubic boron nitride (c-BN), a kind of III-V nitride, has many outstanding properties. It has been widely used as cutting tools and protective coatings due to its large bulk moduli and high thermal conductivity. Since c-BN has a high electrical breakdown strength (Davis 1991), it is also employed in opto- and microelectronic devices working under high-temperature conditions (Mishima 1988). Among the properties, the mechanical characteristics have especially attracted both physical and technological interest.

So far, single crystals of c-BN, whose size is in the sub-millimeter range, have been produced usually by a hot-press method. We have found that small microcracks of ~100 nm size are formed in the crystals by annealing at temperatures above 673 K under a pressure of about 10^{-4} Pa, and we have shown that the microcracks are introduced due to the condensation of nitrogen atoms. Formation of the microcracks results in a decrease of the degree of crystallization and may vary the mechanical properties. Moreover, similar microcracks may be introduced in other nitrides such as GaN. Therefore the structural and dynamical properties of the microcracks are of interest. In the present paper we have systematically characterized the microcracks in c-BN. The first aim of this article is a detailed study of the structural properties of the microcracks by means of transmission electron microscopy and diffraction (TEM and TED) and electron energy-loss spectroscopy (EELS). We will then discuss the mechanism of the formation of the microcracks.

2. EXPERIMENTS

Samples were single crystal c-BN (Showa Denko Co., SBN-B). They were nominally undoped, but some impurity atoms were introduced during crystal growth. The spatial distributions of the impurity concentrations in the crystal were roughly estimated by EELS and no impurities were detectable near the crystal surface. We used these 'impurity-free' parts for the present experiments.

The [110] thin foils for TEM and TED were prepared by conventional techniques. They were annealed up to 1000 K in a JEM-2000EX transmission electron microscope. Extended defects were introduced by annealing, and the growth of the defects during the annealing was systematically observed *in-situ* in the microscope. These defects were characterized by TEM, TED, EELS, and convergent-beam electron diffraction (CBED), using electron microscopes, JEM-2000EX and JEM-2010.

3. RESULTS

A number of dislocations were introduced in specimens during crystal growth. We found that extended defects are formed from the segments of the dislocations by annealing at temperatures T_{an} above 673 K. Figure 1 shows dark field (DF) TEM images $\{\mathbf{g}(3\mathbf{g}), \mathbf{g} = [2\bar{2}0]\}$ of an annealed (T_{an} = 835 K) specimen. Two extended defects along the dislocation (indicated by arrows in Fig. 1b) are clearly observable. A geometrical analysis shows that these extended defects are roundish on the {111}-type planes.

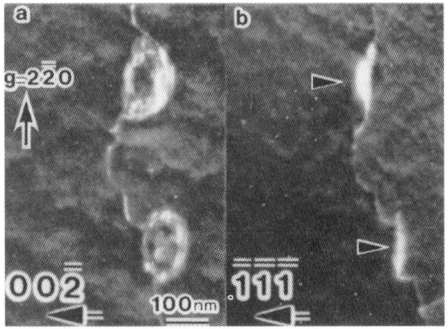

Fig. 1 DF TEM images of an annealed c-BN crystal (incident-beam direction; (a) [$\bar{1}$10] and (b) [$\bar{1}1\bar{2}$]). Extended defects along the dislocation indicated by arrows in (b) show edge-on contrasts.

Figure 2a shows a plane-view image of a (111) platelet parallel to the foil surfaces and nearly in the centre of the foil (the foil thickness is 60 nm). The platelet showed a roundish contrast. Both sides of the foil were etched and we found that the platelet exhibits the characteristic bend contour contrast when the foil is thinned by 40 nm (figure 2b). We confirmed that the (111) atomic plane on the platelet are buckled outwards by the inside-outside method. In a TED pattern from a (111) platelet with plane-view, we observed extra spots attributed to the interior surfaces on the platelet (Aki *et al.* 1999). We have concluded that the platelet is filled with fluid of high pressure, i.e., fluid-filled microcrack (figure 3), and the (111) atomic planes on the microcrack are buckled outwards due to the pressure when the planes are close to the foil surfaces. Such microcracks are introduced in hydrogen (Ponce *et al.* 1987, Neethling *et al.* 1988) or helium (Nomachi *et al.* 1997) implanted semiconductors.

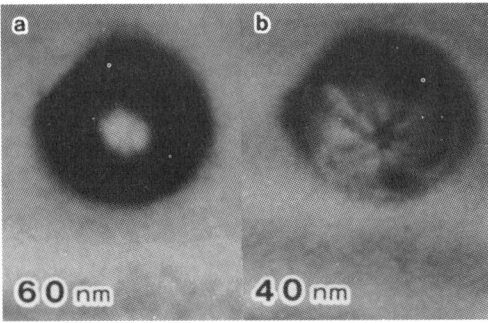

Fig. 2 Variation of the plane-view image of a (111) microcrack with thinning the foil.

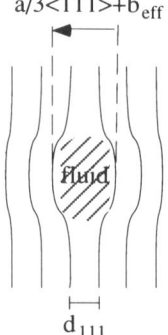

Fig. 3 A schematic representation of a fluid-filled {111} microcrack.

The origin of the fluid in the microcracks was determined by means of EELS. Electron probes for the spectroscopy (8 nm in diameter on a specimen surface) were focused on a microcrack with edge-on view and on a microcrack-free area, respectively. The peak intensities of the B-K spectra from the areas were almost the same. In contrast, the peak intensity of the N-K spectrum from a microcrack was obviously stronger than that from a microcrack-free area. This result clearly shows that excess nitrogen atoms concentrate into a proximity of the microcrack.

The nature of fluid-filled {111} microcrack may be characterized by an effective Burgers vector $\mathbf{b}_{eff}$ (Neethling and Snyman 1986). $\mathbf{b}_{eff}$ is defined as $\alpha <111>$ in which α is a positive constant value (figure 3). The value of α for a microcrack can be determined from the observation of an edge-on image of the microcrack, as discussed in detail in a previous paper (Aki *et al.* 1999). A microcrack showed typical strain contrast when viewed edge-on (Figs. 4a-4c), and the image profile along the line through the centre and perpendicular to the cleaved atomic plane of the microcrack can be calculated with the two-beam dynamical theory of electron diffraction. The calculated profiles (the broken lines in Figs. 4d-4f) fitted well with the experimental (the solid lines in Figs. 4d-4f), and we found that α increases as the microcrack radius r increases. Suppose the microcracks are filled with nitrogen atoms. The pressure in a microcrack can be approximately estimated with α and r (Neethling *et al.* 1988), and we found that the estimated pressure (6±4 GPa) is almost independent of r. This result is consistent with the thermodynamical requirement that the chemical potentials of the gaseous species inside and outside the microcracks are equal, even though the estimated pressure depends on a set of elastic parameters.

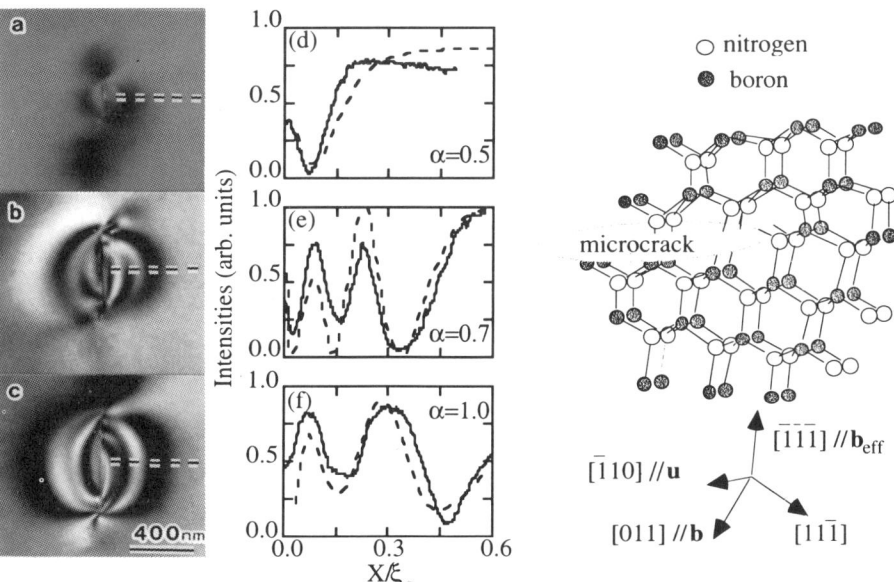

Fig. 4 Edge-on images of microcracks with the radii of (a) 95, (b) 240, and (c) 385 nm, respectively ($\mathbf{g} = [2\bar{2}0]$ and with $\mathbf{s} = 0$). The solid and broken lines in (d), (e), and (f) are the experimental and theoretical image profiles along the broken lines shown in (a), (b), and (c), respectively.

Fig. 5 A schematic representation of the geometry of a microcrack and the dislocation shown in Fig. 1.
$\mathbf{u}$: line direction of the dislocation
$\mathbf{b}$: Burgers vector of the dislocation
$\mathbf{b}_{eff}$: effective Burgers vector of the microcrack

As mentioned above, microcracks are formed along dislocations. Analyses of the data obtained by conventional TEM techniques and a CBED method have shown that;
1) the cleaved atomic planes of the microcracks are the {111}-type planes:
2) a microcrack is formed from the segment of a 60° dislocation line, and
3) it extends to the dilation side of the dislocation.
Figure 5 shows a schematic representation of the geometry of a microcrack and the dislocation shown in Fig. 1. The dislocation was assumed to be a perfect dislocation since a small separation between the two extended partial dislocations.

4. DISCUSSION

Platelets in an annealed c-BN have been well recognized as microcracks filled with nitrogen atoms. Similar nitrogen-agglomerates, i.e., voidites, are observed in Ia-type diamond (Hirsch *et al.* 1986, Luyten *et al.* 1994), and these materials are believed to be crystal-phase N_2 that exist at pressures above 2 GPa at 300 K. We could not obtain any kinds of extra diffraction spots due to crystal-phase N_2 from the platelets at the present moment. Figures 2 and 4 strongly suggest that the platelets are filled with fluid of high pressure. It is unlikely that microcracks are filled with crystal-phase N_2.

Formation of fluid-filled microcracks is generally attributed to the condensation of fluid into the easily cleaved atomic planes. We have found that microcracks are formed along both α- and β-type 60° dislocations and extend to the dilation sides of the dislocations. This result may be easily explained that excess nitrogen atoms condense into the proximity of 60° dislocations to reduce the strain around the dislocation core. Similar to the case of hydrogen in metals (Oriani and Josephic 1974), excess atoms dissolved in the lattice, i.e., nitrogen interstitials, may reduce the interatomic cohesive forces and lower the fracture stress. The condensation of nitrogen atoms into dislocation cores may result in the reduction of the cleavage energy for the {111}-type planes.

The most easily cleaved plane in c-BN is the {110}-type plane, similar to other III-V compounds such as GaAs. The cleaved planes of the microcracks are, however, the {111}-type planes with the (1x1) surface structure. The most stable structures of the clean (111) surfaces having nitrogen and boron termination, (111)N and (111)B surfaces, are theoretically expected to be the (1x1) and (2x2) structures, respectively; the (1x1) structure will change into the (2x1) structure on the (111)N surface with adsorbed nitrogen atoms (Widany *et al.* 1996). We could not find any evidence of these reconstructed structures on the microcracks. The {111}-(1x1) surface structure may be the most stable structure for nitrogen fluid of high pressure. This surface structure has been also observed on a hydrogenated c-BN (Loh *et al.* 1997). It is proposed that a small decrease of the surface energy of the lattice can induce a large decrease of the plastic deformation work (McMahon and Vitek 1979). The condensation of nitrogen atoms on {111}-type planes may decrease the cleavage energy for the {111}-type planes, and the condensation can promote the extension of microcracks.

5. CONCLUSION

We have found the introduction of microcracks in c-BN by annealing at temperatures above 673 K. We have concluded that the formation of the microcracks is due to the condensation of excess nitrogen atoms into the proximity of 60° dislocations.

REFERENCES

Aki M, Ohno Y and Takeda S, submitted to Philos. Mag.
Davis R F 1991 Proc. IEEE **79**, 702.
Hirsch P B, Hutchison J L and Titchmarsh J 1986 Philos. Mag. A **54**, L49.
Loh K P, Sakaguchi I, Gamo M -N, Taniguchi T and Ando T 1997 Phys. Rev. B **56**, R12791.
Luyten W, Van Tendeloo G, Fallon P J and Woods G S 1994 Philos. Mag. A **69**, 767.
McMahon C J Jr, and Vitek V 1979 Acta Metall. **27**, 507.
Mishima O 1988 Appl. Phys. Lett. **53**, 932.
Neethling J H and Snyman H C 1986 J. Appl. Phys. **60**, 941.
Neethling J H, Snyman H C and Ball A B 1988 J. Appl. Phys. **63**, 704.
Nomachi T, Muto S, Hirata M, Kohno H, Yamasaki Jun and Takeda S 1997 Appl. Phys. Lett. **71**, 255.
Oriani R A and Josephic P H 1974 Acta Metall. **22**, 1065.
Ponce F A, Johnson N M, Tramontana J C and Walker J 1987 Proc. Inst. Phys. Conf. **87**, eds A G Cullis and P D Augustus (Oxford: IOP) pp 49.
Widany J, Weich F, Kohler Th, Porezag D and Frauenheim Th 1996 Diam. Relat. Mater. **5**, 1031.

Inst. Phys. Conf. Ser. No 164
Paper presented at Microsc. Semicond. Mater. Conf., Oxford, 22–25 March 1999

97

Growth phenomena of quantum dot structures in the InGaAs system

P Werner

Max-Planck-Institut für Mikrostrukturphysik, D-06120 Halle / Saale, Germany

ABSTRACT: Semiconductors having structures of reduced dimensions as, e.g., quantum dots (QDs) are expected to exhibit special optical and electronic properties leading to new kind of opto-electronic devices. In the past 15 years, the generation of QDs has been attempted using different techniques. However, there was a breakthrough recently initiated by employing of self-ordering mechanisms during the epitaxial growth of lattice-mismatched materials. As an introduction, electronic properties of small islands (Ø< 30nm) are presented. Principles of the formation of such coherent islands applying the "Stranski-Krastanow" growth mode and further developments in the scope of TEM are briefly demonstrated. The applicability of such quantum structures further can be improved using not single layers of QDs, but, e.g., stacked layers. This has been successfully demonstrated for systems such as GaInAs, SiGe, InAsN. The paper is mainly focused on the formation and properties of InGaAs islands ("dots") in a GaAs matrix, a system showing similar properties as found in other material systems. The goal of TEM investigation of these systems, especially QD layers, is a correlation between morphology/structure and their optical behaviour. TEM techniques applied for such analysis will be discussed.

1. INTRODUCTION

The current research of semiconducting clusters is especially focused on the properties of so-called quantum dots - small particles consisting of hundreds to many thousands of atoms - which are embedded in a semiconducting bulk or other material. Quantum dots (QD) are characterised by their strong size dependent electrical and optical properties. An overview of fundamentals and applications of thin semiconducting heterostructures has been given, e.g., by Weisbuch and Vinter (1991). Fig.1 compares, as an example, the different electronic structures between the InAs bulk material with a small InAs inclusion in a GaAs matrix. The scheme on the left represents the concept of the band structure with continuous energy levels in the valence (v) and conduction (c) band. For a perfect crystal no energy levels exist in the band gap. For thin layers or islands (< 30 nm) of InAs in GaAs continuous energy bands no longer exist and we can observe a quantization of the energy levels (Fig.1b), which depends on the size, the shape and the lattice strain. Generated electron-hole pairs (excitons) are normally at the ground state (GS). If they are excited, recombination normally emits light in the infrared region (0.8 - 1.5 µm).

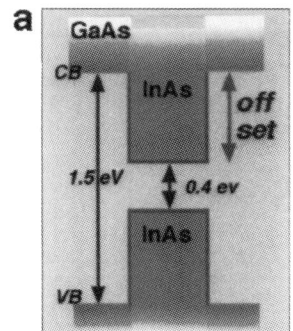
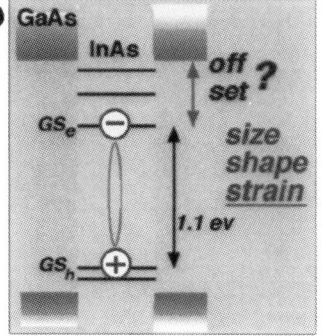

Fig.1 Band scheme of a "thick" InAs layer in GaAs (a) and of the energy levels of a small InAs inclusion in GaAs (b). The quantization of the energy also depends on size, shape and strain of the island.

The ability to arrange such particles or "dots" into complex assemblies creates many opportunities for scientific discoveries. In the field of basic physics, QDs are an exciting subject to study effects in quantum physics, in 'single electron' and laser physics. In the last 10 years different ways have been successfully developed to create such semiconductor nanoparticles, which show the special properties - different from bulk behaviour (MRS Bulletin (1998) gives an overview).

Concerning their technological applications, QD structures are very attractive for active optoelectronic devices (lasers such as 'vertical cavity surface emitting lasers (VCSELs), detectors). A breakthrough occurred recently when it was possible to create high-density arrays of QDs by epitaxial growth of lattice mis-matched heterostructures ("Stranski-Krastanow" growth mode), such as InAs on a GaAs substrate. Depending on the growth techniques applied (mainly MBE and MOCVD) the islands differ in size, shape, chemical composition and lattice strain. To correlate the optical and electrical properties of QD arrays with their morphology and structure (see, e.g., Bimberg, Grundmann, Ledentsov 1999) following investigation methods have been successfully applied: photoluminescence (PL), cathodoluminescence (CL), x-ray diffraction, atomic force microscopy (AFM), techniques of transmission electron microscopy (conventional TEM, HREM)), including energy-filtered TEM.

2. PRINCIPLES OF GROWTH

Most of the semiconductor QD structures are grown on the basis of the "Stranski-Krastanow" mode, an intermediate form between van-der Merve and Vollmer-Weber layer growth. The scheme of Fig.2 explains the basic features of the formation of InAs dots/islands on a GaAs substrate. The first 1-2 deposited monolayers (ML) create a pseudomorphic closed layer, which is strained due to the lattice misfit (7% in the case of InAs/GaAs). During a further deposition the formation of small islands can be observed (b). This transition from the layer growth to island formation is the result of the minimisation of the system energy. This is realised by a complex process, which includes lattice strain and kinetic processes like surface diffusion and incorporation of ad-atoms. The shape and size of the islands depend, on the one hand, on the amount of lattice misfit. On the other hand, they depend on the amount of deposited material and growth parameters, e.g., substrate orientation, growth velocity and temperature. For semiconducting materials, this correlation has been intensively investigated during recent years. The interaction between surface strain fields and growth energetic is predicted to lead to improved lateral ordering (see, e.g., Ch. Teichert et al. 1997). Owing to growth parameters different shapes of islands are observed varying between flat pyramids and spherical islands. During further growth the islands usually are covered by a GaAs capping layer. As a result of a complex diffusion processes the island shape is transformed, e.g., to a truncated pyramid. This important fact has to be taken into account in the study of grown structures. Whereas AFM investigations show only a specific situation of islands at the surface at a given time, TEM is necessary to correlate optical measurements with structural features.

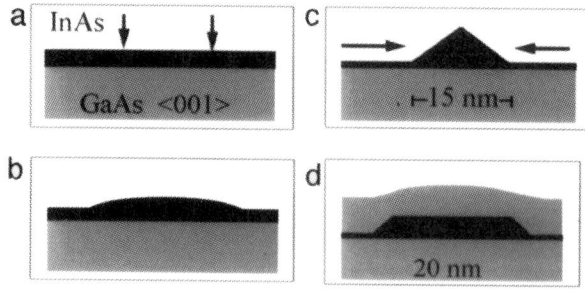

Fig.2 Scheme of different growth steps of the island formation in the lattice strained InAs/GaAs system. a) pseudomorphic layer growth up to about 1.7 ML's, b) island formation for >1.7 ML deposition, c) pyramid formation due to surface diffusion, d) embedded InAs island.

3. SIZE, SHAPE AND STRAIN OF QDs

Concerning the InGaAs system, the scientific interest and technological goal is directed to QD arrays emitting light in the range 1...1,5 μm. This can be controlled through the following parameters of the QDs: size, shape, lattice strain and chemical composition. In the following some examples of TEM investigations are given.

If the islands are not relaxed the mismatched lattice region is 3-dimensionally distorted. HREM images yield only a rough estimate of the shape and crystallography of the interface (if sharp interfaces exist at all). Fig. 3 shows a typical plan-view TEM image of InAs dots on a GaAs substrate. Due to the conventional diffraction contrast technique applied the dots are mainly visible by their strain fields. From contrast analysis the dots seem to have a quadratic base face and edges along <100> directions. From such images one can not only determine the size distribution (see upper right) and dot density (here about 10^{11} cm^{-2}), but also the relation between neighbouring islands. The islands have spacings of about 250 Å, often arranged in <110> directions.

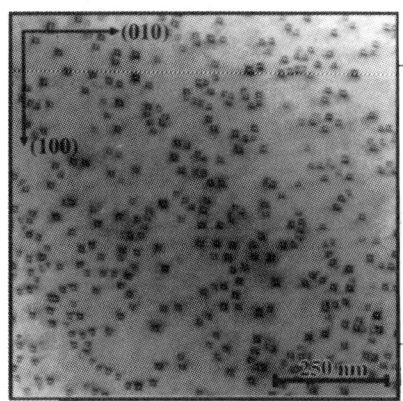
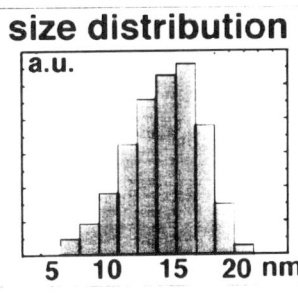

Fig.3 Bright-field TEM image of a single InAs dots layer grown on a <001> GaAs substrate by MBE. Insert right - size distribution.

Lattice strain and local chemical composition of a QD are directly correlated. The question arises if both components can be deconvoluted by TEM techniques. Numerous attempts have been made during recent years based on the comparison between experimental images and the simulated contrast of model systems. One attempt concerns the calculation of strain fields of a given QD model (shape, chemical composition) using, e.g., "finite elements" approach (Christiansen et al. 1994, Grundmann et al 1995) or "molecular dynamics" structure calculations (MD) (Ruvimov, Scheerschmidt 1995). Fig.4 shows bright-field images of InGaAs QDs taken in in-zone <001> orientation. They are taken from samples characterised by different growth condition. It seems that, in the case of single QD-layers (upper row), the islands

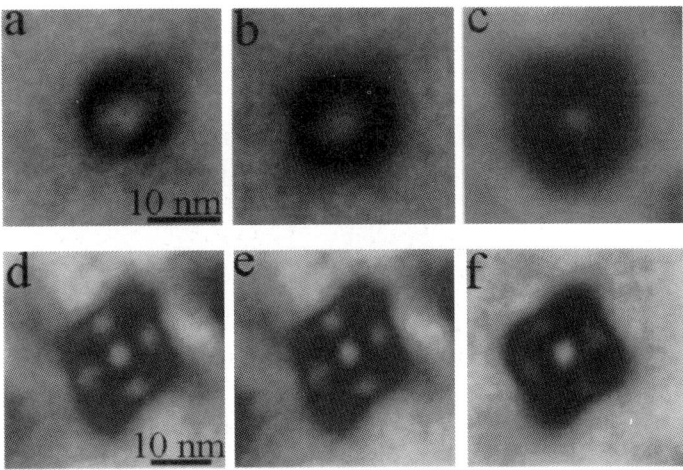

Fig.4 Series of bright-field, plan-view images of InGaAs QDs from different samples. Whereas the first row shows examples from a single QD layer, the lower row relates to samples with multiple stacked QD layers.

are smaller, whereas in the case of multiple stacked layers (see also section 4) the QDs are getting larger and facets and strain are responsible for the typical 'cross' contrast features. Corresponding cross-section images show that in most cases islands (from single row) are flat and pyramids mostly have a truncated shape.

Due to the growth process and post-growth annealing of the samples, an interdiffusion of elements between matrix and island has to be considered, which influences energy states of the exitons. The change in stoichiometry, e.g. the In/Ga ratio, can be observed by PL spectroscopy as an integral measurement. Several approaches have been developed to determine an element distribution in the nm-range by image processing of HREM micrographs (Thoma, 1991, Ourmazd 1993, Stenkamp 1993, Rosenauer 1998). The application of these techniques to the analysis of QD structures is partly restricted, e.g., one has to separate at first lattice distortions, which would disturb the analysing process. As an example, Fig.5 demonstrates a simple procedure of image processing in 'real space'. It bases on the fact that in a special concentration ratio of an In_xGa_{1-x} the amplitude of the (200) beam directly correlates to x. Extracting the '(200) content' from HREM micrographs taken at specific imaging conditions by

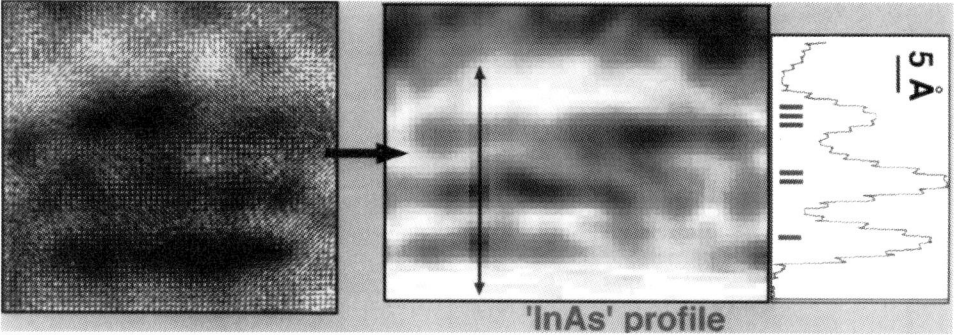

Fig.5 Determination of the In distribution in an arrangement of 3 stacked InAs islands in a GaAs matrix. Starting from HREM cross-section image, Fourier transforms allow to extract the content of the chemical sensitive (200) beam (middle). InAs line profile on the right.

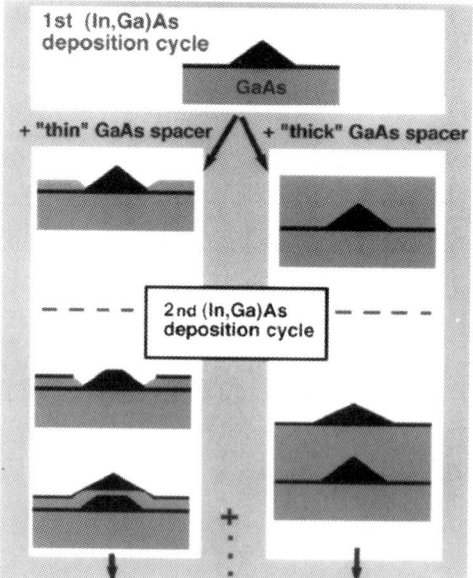

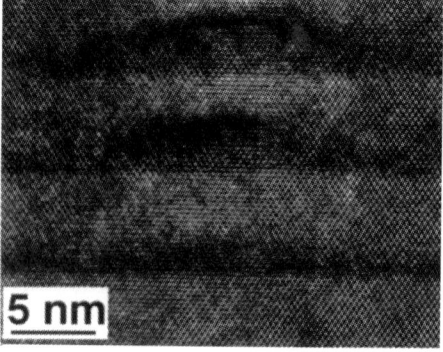

Fig.6. left - growth scheme of the formation of multiple-stacked QD layers given for the <001> InAs/GaAs system. Independent whether a thin GaAs spacer ($\approx$ 20Å, left) or a thick spacer(> 50Å, right) is grown, QDs are formed just above each other due to local lattice strain. below- cross-section image of 3 stacked QD layers.

filtering yields the information on the In/Ga ratio. The InAs content is received as grey levels. A line profile across the 'processed' islands demonstrates not an abrupt In/Ga interface but a smooth transition due to interdiffusion. It has to be mentioned that a general problem of HREM image analysis concerns the deconvolution of imaging parameters to obtain an independent knowledge of them. This includes the determination of the local specimen thickness, imaging parameters as well as lattice strain fields.

Beside image analysis of HREM micrographs, the energy -filtered TEM technique has been established during the recent years allowing a chemical mapping with a lateral resolution down to the 5Å range. In such element specific images even single lattice planes are resolved, however, also in this technique lattice strain fields have to be taken into account for a qualitative/quantitative interpretation of the chemical distribution (Schneider et al. 1999).

4. 3-DIMENSIONAL QD ARRAYS

Light emission from a single QD layer is characterised by a broad peak, which relates to the size distribution of QD (s. also Fig.3). To improve this situation (higher dot density, sharper size distribution) promising solutions were developed during recent years. An attractive growth concept is the multiple-stacking of QD layers to get a 3-dimensional array of islands. It could be successfully applied to several systems of semiconductors (Tersoff 1996, Heinrichsdorff 1997). The basic idea is represented in the scheme of Fig.6. One starts with a first QD layer, which is then covered by a GaAs capping layer. As a next step, again about 2 ML of InAs are deposited. Due to surface diffusion the first QD is covered by GaAs and a second QD layer is formed. In the case of <001> oriented substrates the interaction between adjacent layers due to lattice strain variations generates a vertical alignment of islands. Therefore, this concept is often referred to a form of „self-organisation". Fig.7 shows a cross-section image of such a assembly of InAs islands in GaAs. It demonstrates i) an increase of the island size with increasing layer number, ii) an improving of narrow size-distribution, iii) a flattening of the growth surface by the GaAs spacer. The strong periodicity of the layers in the vertical direction is demonstrated by the occurrence of satellite reflections near the main reflections in the SAD pattern (left); the horizontal periodicity of QD is presented by sub-reflections in the corresponding diffractogram (fast-Fourier transformation, FFT, right). The first layer of deposited InAs is characterised by small, flat islands (as also seen in the upper row of plane-view images of Fig.4). Their homogeneously distributed strain fields initiate an increasing size of InAs islands.

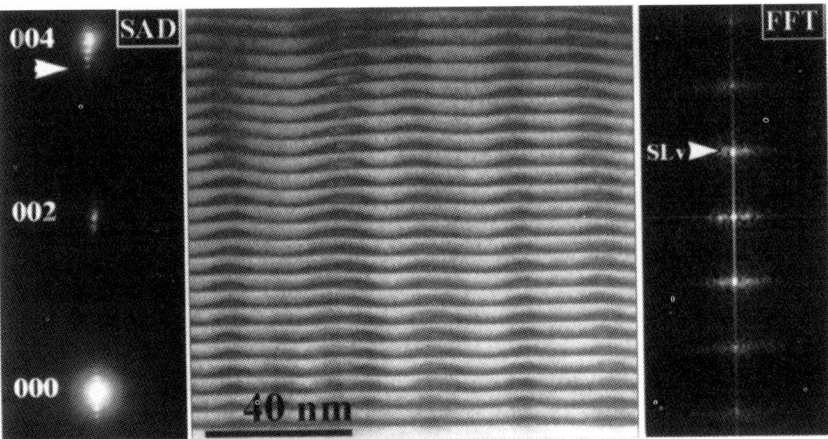

Fig.7. Cross-section image of a multiple-stacked array of 25 InGaAs QD layers (dark islands) in a GaAs matrix (grey regions). The strong periodicity of the QD assembly is represented by sub-reflections in the SAD pattern (left) and FFT diffractogram (right).

PL and CL measurements have demonstrated that such strained coupled arrays are also characterised by a strong electronic coupling. It could be shown that in such stacked QD layers excitons are localised not only at a single, but at several stacked islands. This behaviour allows not only an emission (lasing) at room temperatures, but due to a quantization in 'larger volumes' a tailoring of the emitted wave length. To improve QD size homogeneity and QD density further growth concepts were developed, e.g., island growth on vicinal surfaces, or using sub-MLs of InAs as seeding layers for the further conventional island growth.

With the expected complexity of 3-dimensional QD structures advanced TEM techniques in combination with above mentioned analytical methods will play an essential role in the correlation of the morphology and structure with the optoelectronic properties of semiconducting QD assemblies.

ACKNOWLEDGEMENT

The author would like to thank very much the following colleagues participating in a cooperation. From the Technical University Berlin: F. Heinrichsdorff, M. Grundmann, D. Bimberg; from the Ioffe Physical Technical Institute St.Petersburg/Russia: V. Ustinov, I. Soshnikov, N.N. Ledentsov; from the MPI Halle: N.D. Zakharov, K. Scheerschmidt, R. Hillebrand; from CNR Lecce/Italy: A. Taurino; from Warsaw University: J. Jasinsky. Additionally , the financial support by the European INTAS program and by the DFG/Germany is gratefully acknowledged.

REFERENCES

The investigation on the formation and structural characterisation of QDs have generated a large number of publications during the recent years. Here, the author can only refer to some publications which should allow the reader to enter this subject.

Bimberg D, Grundmann M, Ledentsov N N 1999, „Quantum Dot Heterostructures", John Wiley & Sons, New York

Christiansen S, Albrecht M, Strunk H P, and Maier H J 1994, Appl.Phys.Lett. **64**, p 3617, Computational Mat. Sci. **7** (1996), 213

Grundmann M, Stier O, and Bimberg D 1995, Phys.Rev. **B52**, 11969

Heinrichsdorff F, Mao M H, Kirstaedter N, Krost A, Bimberg D, Kosogov A O, and Werner P 1997, Appl. Phys. Lett. **71**, 22

MRS Bulletin **23** 1998, „Semiconductor Quantum Dots", 15-53

Ourmazd A., Schwander P, Kiselowski C, Seibt M, Baumann F H, Kim Y O 1993, Inst. Phys. Conf. Ser. **134**, 1

Rosenauer A, Fischer U, Gerthsen D, Förster A 1998, Ultramicroscopy **72**, 121

Ruvimov S, Scheerschmidt K 1995, phys.stat.sol **a150**, 471

Schneider R, Kirmse H, Neumann W, Heinrichsdorff F, and Bimberg D 1999, Inst. Phys. Conf. Ser., this volume

Stenkamp D, Jäger W 1993, Inst. Phys. Conf. Ser. **134**, 15

Teichert C, Phang Y H, Peticolas L J, J.C. Bean, and M.G. Lagally 1997, Surface diffusion: atomistic and collective processes, Ed. M.C. Tringides, NATO-ASI Series, Plenum Press, New York

Tersoff J, Teichert C, Lagalli M G 1996, Phys.Rev.Lett **76**, 1675

Thoma S, Cerva H 1991, Ultramicroscopy **38**, 265

Weisbuch C, Vinter B 1991, „Quantum Semiconductor Structures", Academic Press, New York

Inst. Phys. Conf. Ser. No 164
Paper presented at Microsc. Semicond. Mater. Conf., Oxford, 22–25 March 1999

103

Real time observations of the growth and development of self-assembled GeSi islands on Si(001)

F M Ross, R M Tromp, J Tersoff and M C Reuter

IBM Research Division, TJ Watson Research Center, PO Box 218, Yorktown Heights NY 10598, USA

ABSTRACT: We have examined the growth and evolution of Ge and GeSi islands on Si(001) using a UHV transmission electron microscope and a low energy electron microscope, both with *in situ* growth capabilities. Small Ge islands are known to be pyramidal in shape while larger islands are dome shaped. We find that the transition from pyramids to domes occurs through a series of asymmetric transition states. Cooling below the growth temperature transforms these transition shapes to domes. We explain these results with an anomalous coarsening model for island growth.

1. INTRODUCTION

Island formation during strained layer epitaxial growth of semiconductors is an important phenomenon now used in fabricating novel optoelectronic devices. These devices make use of the unusual electronic properties of small islands, or "quantum dots", and strained layer epitaxy is a convenient way of creating arrays of such islands. When fabricating devices, the question of whether the islands are thermodynamically stable or only transient is of great importance, and island growth has therefore been studied extensively. In the Ge/Si system, island growth turns out to be a complex phenomenon characterised by several different island shapes and unusual size distributions.

When Ge is deposited on Si(001), the first 3 monolayers (ML) form a flat "wetting layer", and islands form on subsequent deposition. At low growth temperatures the islands are shaped like rectangular "huts" with {105} facets (Mo et al 1990) while at higher temperatures two different island shapes exist, depending on size (Kamins et al 1997). Smaller islands are *pyramids*, which are similar to huts, although square, while larger islands are multifaceted *domes*, which have a higher aspect ratio and include facets such as {113} (Tomitori et al 1994; Kamins et al 1997; Medeiros-Ribeiro et al 1998). The same island morphology is observed, although with larger lateral dimensions, for lower- strained Ge_xSi_{1-x} alloys on Si(001) (Floro et al 1999). Over a wide range of growth conditions a bimodal distribution of pyramids and domes is seen (Kamins et al 1997). Several models have been proposed to explain how the lower volume pyramids develop into the higher volume domes (Medeiros-Ribeiro et al 1998; Ross et al 1998; Kamins et al 1999) and debate continues about the factors determining the island size distribution.

In this paper we report a study of island growth using low energy electron microscopy (LEEM). This technique is capable of distinguishing between island shapes and determining sizes in real time during growth. Our observations show that the shape change from pyramid to dome is surprisingly complex. We will show that pyramidal islands pass through a series of asymmetric "transition shapes" until they reach the final dome shape. We will discuss the implications of these results for the understanding of island growth.

2. EXPERIMENTAL DETAILS AND RESULTS

The LEEM used in this study has a base pressure of 2×10^{-10} Torr and a point resolution of 5nm, and has been designed with *in situ* deposition capabilities (Tromp et al 1998). Polished Si(001) substrates were flash cleaned then heated to the desired growth temperature of ~650-700°C. Ge_xSi_{1-x} was then grown by chemical vapour deposition using a mixture of disilane and digermane gases introduced through capillary tubes into the specimen area. Typical growth rates were 1-5 ML per minute and images were recorded at video rate during growth at electron energies of 5-10eV.

The characteristic LEEM contrast of pyramids and domes is shown in Fig. 1a. Pyramids display a cross while domes show bright areas separated by a narrow dark line. These bright "facet beams" are caused by diffraction from higher angle facets and indicate that additional flat facets have developed on the domes, although the intensity on the image may not correspond exactly to their spatial position due to local focusing effects in the LEEM. An examination of domes by *ex situ* scanning electron microscopy (Fig. 1b) confirmed the presence of extra pairs of facets and indicated that these pairs were close to {518}. Consideration of the stability of Ge surfaces vicinal to {518} (Gai et al 1998) actually suggests that these facets are probably {15 3 23} (Ross et al 1999). Small areas of {113} and {501} facets are also present on the domes.

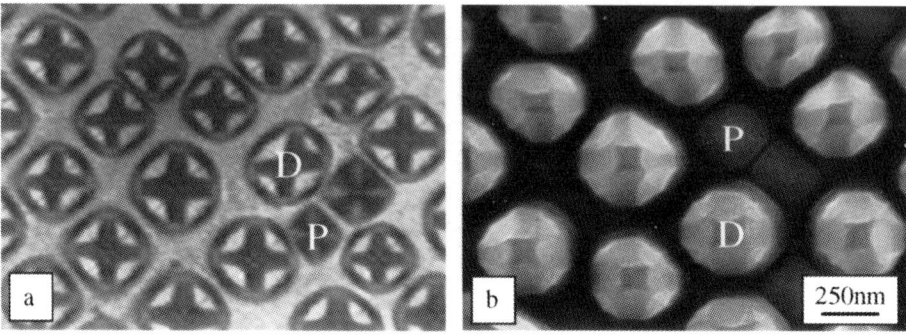

Fig. 1 (a) Densely packed pyramids and domes (P and D respectively) formed after 56 minutes of growth of $Ge_{0.3}Si_{0.7}$ at 710°C. $Si_2H_6/He/Ge_2H_6$ gases in the ratio 100/10/1 were used at a total pressure of $1.0x10^{-5}$ Torr. The image was recorded at room temperature using an electron energy of 4eV. The {110} directions are parallel to the edges of the image. (b) Scanning electron micrograph of a similar specimen grown for 39 minutes at 690°C. The islands are ~70nm high, and the facet edges have been enhanced by filtering.

Video-rate observation shows that the process leading up to the array of pyramids and domes in Fig. 1 is complex. The first stage of growth is a roughening process whose characteristic length scale depends on strain. The amplitude of the roughness increases until {501} facets appear and pyramids eventually form from these facets (Tromp et al 1999). The pyramids coarsen, decreasing in density by at least a factor of 100 and increasing in size. They eventually transform into domes, but pass through a series of transition states on the way.

These states are shown in Fig. 2, where we see that the projected shape of an island first changes from square to more circular and facet beams appear near the rounded corners. These facet beams persist up to the final dome shape and are presumably from {113} facets. The overall shape at this stage thus appears to be a truncated pyramid (TP in Fig. 2). The next stage is the appearance of contrast from one or more of the {15 3 23} facets. This contrast appears gradually, brightening and increasing in extent. The facets always appear in pairs, although one facet may be larger than the other particularly if the original pyramid was not perfectly square or had an asymmetric neighbourhood. Each dome develops these facet pairs in its own order, with some domes passing through all the transition states (D1-D3 in Fig. 2) while others develop 2, 3 or even all 4 facet pairs simultaneously. We suppose that slight asymmetries in the local environment or the original island shape account for the exact behaviour of any island. For example, if the original pyramid were rectangular a variant of D2 with opposite pairs is likely to form. The transition shapes may persist for many minutes at the growth temperature and growth must be continued for several hours before most islands have attained the D4 shape. Small populations of pyramids and intermediate shapes are always present.

Figure 3 shows the evolution of the diameter and shape of a group of islands and gives a clearer picture of the development of an ensemble of islands. Islands have one of two fates: they either reach the TP state and eventually become domes, or remain as pyramids and eventually shrink and disappear. In other words, in these experiments we do not see stable pyramids, and we do not observe the disappearance of islands once they have reached the TP state.

The length of time spent in transition between pyramid and dome can be surprisingly long, and it could be argued that island development is being limited by the flux of Ge and Si. However, coarsening experiments in which a population of pyramids was prepared and then annealed with no additional flux displayed similar kinetics to growth experiments with flux on: the pyramids were not stable, and coarsening and formation of domes via transition states still occurred, demonstrating that the flux does not limit island development.

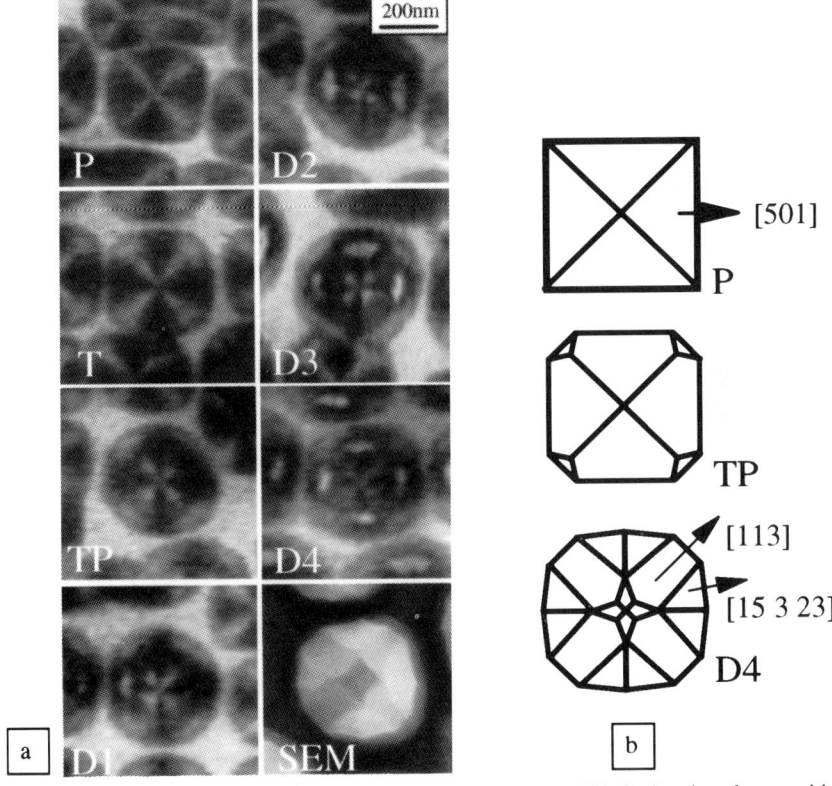

Figure 2 (a) Images recorded during growth of $Ge_{0.25}Si_{0.75}$ at 730°C showing the transition states (TP, D1, D2, D3) between the pyramid (P) and dome (D4). The (100) direction is horizontal. (b) Schematic diagrams indicating the facets present on the P, TP and D4 shapes.

Island growth has been the subject of intense theoretical interest and several models have been proposed to explain island development. The bimodal distribution, and particularly the lack of many islands at intermediate sizes, led Medeiros-Ribiero et al (1998) to propose that the size distribution reflects an equilibrium state. In this model the pyramid and dome shapes are supposed to have energy minima at two discrete volumes, and the width of the bimodal distribution is due to thermal broadening. As the islands grow they make the transition from one energy minimum to the other by a thermally excited process which overcomes the energy barrier and involves rapid accumulation of 10^5 atoms. A very short time scale was suggested for this transition to account for the few islands seen with intermediate sizes. However, our previous real time observations of island growth in the transmission electron microscope (Ross et al 1998) showed that islands actually increase in size very smoothly. Although island shapes could not be distinguished in the TEM experiments, the data was more consistent with an "anomalous coarsening" model. This is a kinetic rather than a thermodynamic model: islands grow by a process similar to Ostwald ripening, but with kinetics modified by an abrupt drop in chemical potential which occurs as islands grow past a critical volume and change their shape. The bimodal distribution is a transient phenomenon which forms as the drop in chemical potential accelerates the coarsening of the largest islands in the distribution, separating them from the smaller islands.

The anomalous coarsening model did not include transition states, but we expect that the addition of intermediate configurations will not change the qualitative pattern of the kinetics. The population of pyramids will coarsen until some islands reach the critical volume for changing to the TP shape; the chemical potential of these islands will drop, accelerating their growth, and the process will repeat at each transition volume until the islands become domes. With their high chemical potential, any remaining pyramids will rapidly shrink and disappear. This simple model is in good agreement with our experimental data, while a slow transformation occurring via transition shapes is not consistent with the sudden change suggested by the equilibrium model.

We finally discuss why the transition shapes have not previously been observed, especially given their long lifetime. Preliminary experiments suggest that rapid cooling causes dome facets to appear on TP islands (Ross et al 1999) so that at room temperature the specimen appears to consist almost entirely of pyramids and domes. It is reasonable to assume that the energies of all facets are temperature dependent and we speculate that the $\{15\ 3\ 23\}$ facet becomes slightly more favourable at lower temperatures compared to other TP facets. This result emphasises the importance of observations made under growth conditions for understanding growth mechanisms.

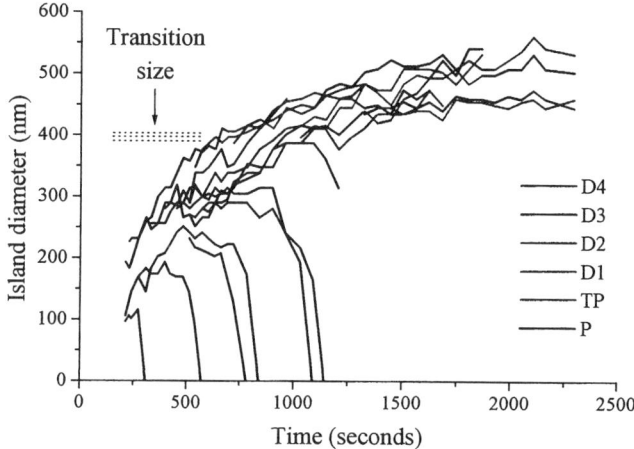

Figure 3 The size and shape of representative islands measured during growth of $Ge_{0.3}Si_{0.7}$ at 690°C. Some pyramids shrink to zero diameter due to coarsening (although trajectories ending above zero indicate islands which drifted out of the field of view).

3. CONCLUSIONS

Observations of island growth *in situ* have revealed that self-assembled GeSi islands on Si(001) change from pyramids to domes through a series of intermediate, asymmetric transition shapes. The shape change is quite slow, requiring several minutes at temperatures around 650-700 °C. The process we observe is consistent with an anomalous coarsening model for island growth, suggesting that island configurations and shapes are controlled more by kinetics than thermodynamics. This is interesting in terms of our attempts to understand and control the growth of self-assembled islands using strained layer epitaxy.

REFERENCES

Floro JA, Chason E, Freund LB, Twesten RD, Hwang RQ and Lucadamo GA 1999 Phys. Rev. B **59**, 1990
Gai Z, Li X, Zhao RG and Yang WS 1998 Phys. Rev. B **57**, 15060
Kamins TI, Carr EC, Williams RS and Rosner SJ 1997 J. Appl. Phys. **81**, 211
Kamins TI, Medeiros-Ribeiro G, Ohlberg DAA and Williams RS 1999 J. Appl. Phys. **85**, 1159
Medeiros-Ribiero G, Bratkovsky AM, Kamins TI, Ohlberg DAA and Williams RS 1998 Science **279**, 353
Mo Y-W, Savage DE, Swartzentruber BS and Lagally MG 1990 Phys. Rev. Lett. **65**, 1020
Ross FM, Tersoff J and Tromp RM 1998 Phys. Rev. Lett. **80**, 984
Ross FM, Tromp RM and Reuter MC 1999 Science, submitted
Tomitori M, Watanabe K, Kobayashi M and Nishikawa O 1994 Appl. Surf. Sci. **76/77**, 322
Tromp RM, Mankos M, Reuter MC, Ellis AW and Copel M 1998 Surf. Rev. Lett. **5**, 1189
Tromp RM, Ross FM and Reuter MC 1999 Phys. Rev. Lett, submitted

Inst. Phys. Conf. Ser. No 164
Paper presented at Microsc. Semicond. Mater. Conf., Oxford, 22–25 March 1999
© *1999 IOP Publishing Ltd*

Size and shape engineering of vertically-stacked InAs quantum dots

J P McCaffrey[1,2]**, M D Robertson**[3]**, Z R Wasilewski**[1]**, S Fafard**[1] **and L D Madsen**[2]

[1] Institute for Microstructural Sciences, National Research Council of Canada, Ottawa, CANADA, K1A 0R6.
[2] Department of Physics (IFM), Linköping University, S-581 83 Linköping, SWEDEN
[3] JDS Fitel, Inc., Nepean, CANADA, K2G 5W8

ABSTRACT: Spontaneous formation of nanostructures during epitaxial growth is one of the most successful techniques for *in situ* fabrication of quantum dots (QDs). We describe new size and shape engineering procedures for the growth of layers of stacked QDs, resulting in a significant narrowing of the size and shape distribution. Transmission electron microscopy (TEM) and photoluminescence (PL) indicate that the main effects of this size and shape engineering are to convert the quantum dots into a population of quantum disks with highly uniform height and diameter.

1. INTRODUCTION

Quantum dot laser diodes have been successfully produced using InAs self-assembled quantum dots (QDs) in the active region of separate confinement heterostructures. The lasers are grown by Molecular Beam Epitaxy (MBE) with either single layers or self-assembled stacks of InAs QDs grown on GaAs substrates. These structures display up to five well-defined electronic shells, as measured by photoluminescence (PL). These confined levels, with an inter-level energy spacing between 25 and 90 meV, are obtained by adjusting the growth temperature or with post-growth annealings. The uniformity and reproducibility of InAs/GaAs QDs are optimized by adjusting growth parameters affecting the evolution and the equilibrium shape of the QDs – primarily the amount of strained material deposited and the annealing time following the InAs deposition (Fafard et al 1999). Well-defined excited-states are also obtained with stacked layers of vertically self-assembled quantum dots with a process referred to as an "indium flush" (Wasilewski et al 1999).

The motivation for the growing interest in stacked QDs is that the much higher dot density leads to a significant reduction of the threshold current. Since the considerable scatter of the dot size and shape in isolated QD layers or uncontrolled stacked layers is one of the reasons for the observed broad PL lines, the most apparent benefit of QD size and shape engineering is a strong narrowing of the PL emission. Proposed models of the observed TEM contrast are discussed herein.

2. QD SIZE ENGINEERING THROUGH ANNEALING

Control of post-deposition annealing times is a growth procedure that can be used to significantly improve the uniformity of QDs. This can be clearly observed in plan-view TEM in conjunction with PL measurements. In Fig. 1, plan-view TEM images of the same region of the sample are shown as acquired using the preferred (200) two-beam conditions [1a], as well as (220) conditions [1b]. The measured diameter of these QDs (below) has been supported using atomic force microscopy (AFM). The dark/light lobes [1a] indicate a radially-symmetric strain field typical of coherent precipitates.

Fig. 2 illustrates the relation between PL and plan-view TEM images of the material. Histograms are also shown illustrating the variations in diameter of the

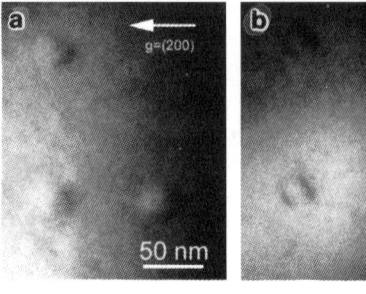

Fig. 1: Two-beam plan view TEM images of QDs, taken with $g=(200)$ in a) and $g=(220)$ in b).

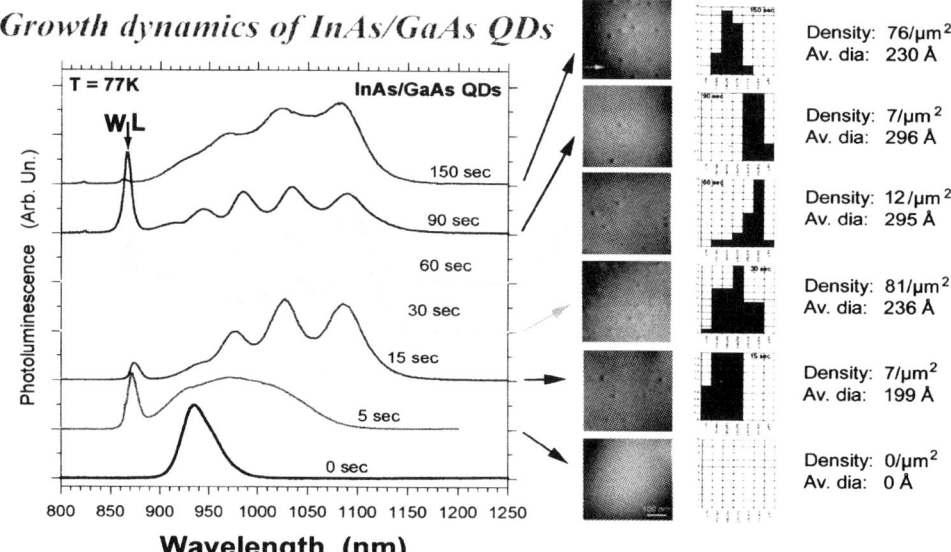

Growth dynamics of InAs/GaAs QDs

Density: 76/µm² Av. dia: 230 Å

Density: 7/µm² Av. dia: 296 Å

Density: 12/µm² Av. dia: 295 Å

Density: 81/µm² Av. dia: 236 Å

Density: 7/µm² Av. dia: 199 Å

Density: 0/µm² Av. dia: 0 Å

Fig. 2: Size engineering of QDs and tuning of the intersublevel energy spacing with anneal time between 0 and 150 sec., following the deposition of 1.9 monolayers (ML) of InAs grown in 27 sec. at 515°C. Each plan-view TEM image displays an area of approximately 1 µm². The histograms show the evolution of QD diameter, and QD densities and average diameters for each sample are given.

QDs with different annealing times. Note that the QDs initially increase in diameter with time, then decrease as some material is desorbed. The best uniformity is obtained for an anneal time of 60 seconds, with the PL displaying well-defined excited states. For longer anneal times, a slight degradation in the uniformity is observed. For the 5 sec. sample, no individual dots were observed, but shifts in the position of the wetting layer (WL) from the original wavelength in the 0 sec. sample and the formation of a bulge in the PL spectrum suggest that the initial stages of dot formation are underway.

3. SIZE AND SHAPE ENGINEERING OF VERTICALLY-STACKED QDs

For certain devices such as QD laser diodes, it is desirable to grow multiple layers of QDs. This can best be observed by cross-sectional TEM. Fig. 3 illustrates the variation in contrast that can be observed when viewing QDs in cross-section. Note the very bright dot in the thinnest region of the sample, with a mid-contrast dot above in a thin region, then dark dots as the sample thickens. This distinctive contrast can be modeled with the dynamical theory of electron diffraction. However, accurate thickness measurements are required if the technique is to be quantitative. These measurements can be acquired with samples prepared by the small-angle cleavage technique (McCaffrey 1993) when applied to III-V materials to produce a series of cleavage steps (McCaffrey 1997). Fig. 4 shows how a cleaved wedge can be mounted in such a way as to provide a cross-sectional view (Fig. 4b). By bending the mounting grid, a "plan-view" image (Fig. 4a) can be observed. In such a way, an accurate measure of sample thickness in the exact region of interest can be obtained.

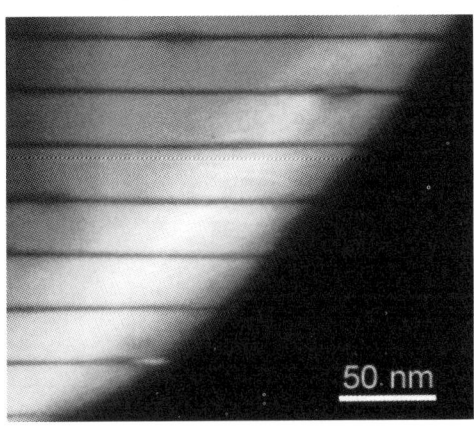

Fig. 3: (200) DF image of InAs QDs in GaAs. Note the contrast variation with thickness.

Fig. 5 presents the results of a 12-beam (200) systematic row Bloch wave calculation of the intensity ratio (contrast) of the center

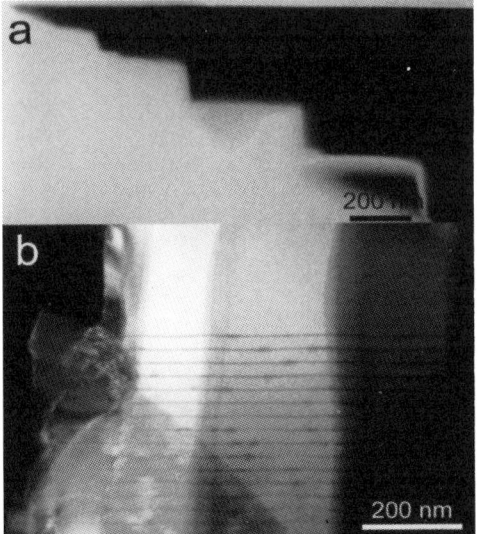

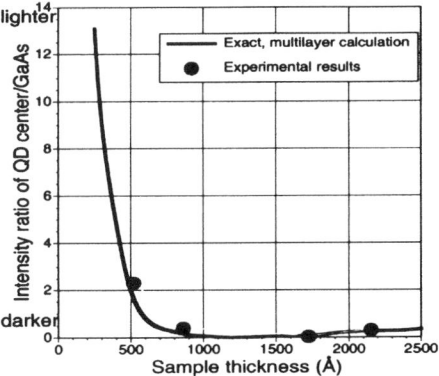

Fig. 5: Comparison between experimental and theoretical intensity ratios.
< Fig. 4: a)"Plan view" and b) cross-section geometry of same TEM sample.

of an InAs QD to the GaAs matrix. The graph shows very good agreement between the theory and data, revealing new evidence that the QD centers are composed of pure InAs.

Stacked QDs displaying vertical self-alignment can be produced by growing thin GaAs cap layers between the QD layers. The strain field produced by the original QD appears to produce conditions favorable to the formation of vertical stacks of QDs in subsequent layers. The QD size uniformity can be increased by providing an "indium flush" step. In this procedure, a thin 2.5 to 11 nm layer of GaAs is grown on top of the QD layer, after which the temperature is ramped up ~100°C to remove surface resident indium. The temperature is then lowered and the remainder of the GaAs cap is deposited to give a total cap thickness of 11 nm. A new QD layer is then grown, and the procedure repeated. Fig. 6 provides examples of size and shape engineering using this "indium flush" technique in conjunction with varying initial cap thicknesses. From observations based on TEM micrographs and PL measurements, a 5 to 5.5 nm cap followed by an "indium flush" produces the optimum stack of QDs, displaying up to five well-defined electronic shells.

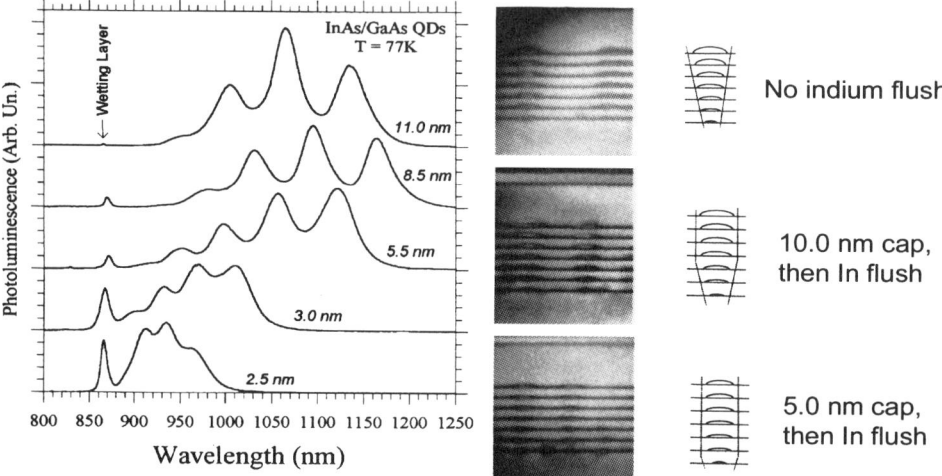

Fig. 6: Variations in PL with an "indium flush" executed after the deposition of thin GaAs caps between 2.5 and 11.0 nm. The sets of TEM micrographs and drawings at the center and right of the figure reveal the morphological behavior of the quantum dots with no indium flush (top set), a 10.0 nm cap and flush (middle set; corresponds ~ to the top PL plot), and a 5.0 nm cap and flush (bottom set; corresponds ~ to the middle PL plot).

4. CONCLUSION

Control of the anneal time and the "indium flush" procedure allows size and shape engineering of single layers as well as vertically stacked arrays of QDs. TEM is an ideal technique for monitoring these changes in size and shape. These procedures permit the manipulation of the energy levels of theses structures and will allow such studies as the physics of tunneling between coupled QDs (artificial molecules), of recombination in charged QDs, and the carrier injection and lasing in QDs with well-defined excited states.

REFERENCES

Fafard S, Wasilewski Z R, Allen C , Picard D, Spanner M, McCaffrey J P 1999 PRB in press
McCaffrey J P 1993 Microsc. Res. and Tech. **24**, 180
McCaffrey J P 1997 Microsc. Res. and Tech. **36**, 362
Wasilewski Z R, Fafard S and McCaffrey J P, 1999 J. Cryst. Growth in press

Inst. Phys. Conf. Ser. No 164
Paper presented at Microsc. Semicond. Mater. Conf., Oxford, 22–25 March 1999

Effect of growth rate on the morphology and composition of InAs quantum dots grown on GaAs by MBE

M A Al-Khafaji, A G Cullis, M Hopkinson, D J Mowbray[*] and M S Skolnick[*]

Electrical and Electronic Engineering department, University of Sheffield, Sheffield S1 3JD, UK
[*]Physics and Astronomy Department, University of Sheffield, Sheffield S3 7RH, UK

ABSTRACT: We report a study of the morphology and composition of InAs quantum dots using high resolution transmission electron microscopy (HRTEM) and electron energy loss spectroscopy (EELS). Three uncapped samples of InAs on GaAs, grown at three different growth rates; 0.01ML/sec, 0.04 ML/sec and 0.4ML/sec, were studied. A total thickness of 2.4 monolayers of InAs was used to grow all samples, and all other growth conditions were kept constant. The dot size for the highest growth rate sample was observed to be about three times smaller than that for the lowest growth rate, while a substantial increase in the dot density was found. Composition mapping using EELS showed an extended segregation of In into the GaAs buffer layer for the case of the low growth rate sample. Effects of GaAs capping and multi stacking of quantum dots on the observed strain contrast in plan view samples were also discussed.

1. INTRODUCTION

InAs self-organised nanoscale 3D quantum dots grown on GaAs(001) have gained considerable interest due to their potential optoelectronics applications (Grundmann et al 1995). These dots are known to grow in the Stranski-Krastanov growth mode, where the first deposited 2ML completely wets the GaAs (i.e. a 2D growth mode) whilst subsequent growth results in a transition to 3D growth (Eaglesham et al 1990). The resultant quantum dots are bonded by {110} and {111} facets and their final equilibrium shape is determined by the competition between the surface and elastic energies (Pehlke et al 1997). Different growth conditions and nucleation processes contribute to the final size, density, composition and distribution of the quantum dots. In this work the effects of growth rate on dot morphology, size, density and composition are investigated using high resolution electron microscopy (HREM) and electron energy loss spectroscopy (EELS). TEM zone axis imaging were used to compare strain contrast from plan view and cross sectional specimens for capped and uncapped samples.

2. EXPERIMENTAL

The growth rates used for the samples investigated in this study were 0.01ML/sec, 0.04ML/sec and 0.4ML/sec. Since both the degree of In segregation and the diffusion distance are affected by the growth temperature (Dosanjh et al 1993), all samples were grown at same temperature of 510°C. All other growth conditions, except growth rates, were kept constant for all samples. The growth rate is defined here as the V/III ratio at fixed As composition; this ratio was monitored continuously by RHEED. TEM cross sectional and plan view specimens were prepared using conventional sample preparation techniques, involving mechanical thinning followed by ion milling using Ar^+ at 6keV (reduced to 4keV at final stages of thinning). Prepared specimens were examined using a field emission JEOL2010F TEM equipped with a Gatan image filter (GIF) for elemental mapping and electron energy loss spectroscopy (EELS).

3. RESULTS AND DISCUSSION

3.1 Dot density and strain contrast analysis

Plan view TEM bright field images of the three samples are shown in Fig. 1. All images were acquired close to the [001] zone axis and are displayed under the same conditions, allowing the direct comparison of dot density and size. Measurements of dot densities in the three samples revealed that, as the growth rate is increased, the dot density also increased, (Fig. 1d). An exact determination of the relationship between the dot density and the growth rate requires the examination of additional samples at intermediate growth rates. However, this plot indicates a rapid increase in the dot density at low growth rates (<1.5ML/sec), followed by a slower increase at intermediate growth rates (between ~0.2-0.4ML/sec) and saturation at high growth rates (>0.4ML/sec).

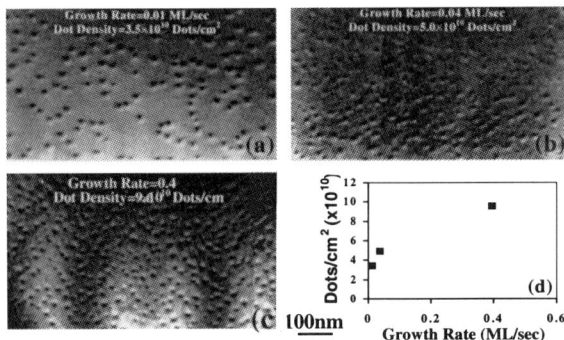

Figure 1. Plan view [001] zone axis TEM bright field images of uncapped InAs/GaAs(001) samples grown at different growth rates: (a) 0.01ML/sec, (b) 0.04ML/sec and (c) 0.4ML/sec.

A general increase in the dot density with growth rate is expected. But for large dots (~few nanometers in diameter), this will eventually lead to dot coalescence due to inter-dot transfer of material and hence a reduction in dot density. This behaviour of dot density as a function of growth rate (particularly in the low growth rate regime) indicates the possibility of differences in dot formation and growth at different growth rates. Since dot formation depends on the surface strain state, it appears that a fast growth rate results in a uniformly roughened surface and hence an increase in the dot density. This may be because the formation of new islands is induced by strain field interaction which gives rise to preferential In adatom migration (Xie et al 1995).

Strain contrast was found to have distinct effect on the observed TEM images for both plan view and cross sectional samples and also for both capped and uncapped samples. This effect can be utilised to determine the size, shape and composition of the quantum dots. In the case of uncapped samples, the strain contrast in a [001] zone axis TEM image take the shape of 'cross contrast' passing through the centre of the dot (see Figs. 2a and b from slow and high growth rate samples, respectively). This contrast shows that the dots in both cases have pyramidal shape with the long side along the [100] directions and they are close in lateral dimensions. The shape and dimensions of the dots were found to be uniform within each sample. Although the strain contrast in both samples seems to be similar, close examination reveals that there are differences particularly in the central region of the dot. This can be attributed to the area of the (001) top surface of the dot, which is larger for the case of the high growth rate sample (see next section), i.e. the pyramids of the low growth samples have larger aspect ratio.

Comparing these contrasts with those from capped samples reveals distinct differences as shown in Figs. 2c and d for single and ten self-assembled, vertically-aligned capped dots. Note that although the dots in both capped samples were grown using the same growth conditions, there is a major transfer in the strain contrast from simple square shape (Fig. 2c) to a 'butterfly' shape with complex detailed contrast in the centre (Fig. 2d). These differences in the strain contrast may reflect differences in strain relief mechanism and the extent of strain penetration into the surrounding GaAs matrix, which can be clearly seen in the cross sectional images shown in Fig. 3.

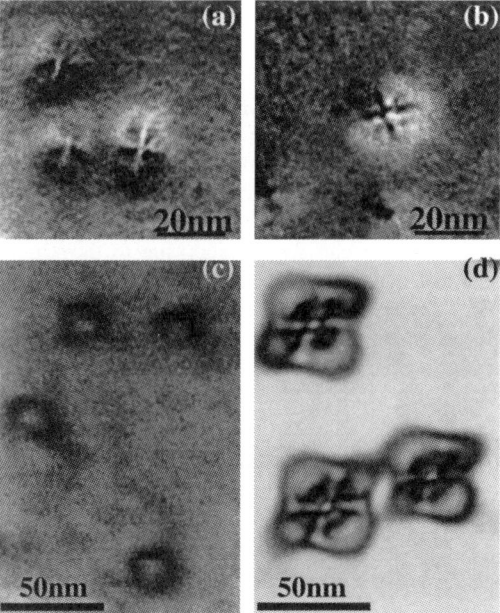

Figure 2. Plan view images close to [001] showing strain contrast from uncapped samples at (a) low growth rate and (b) high growth rate. Strain contrast from capped samples for single and multiple layers are shown in (c) and (d), respectively.

Consideration of the symmetry of the strain contrast shows that uncapped samples exhibit a four fold symmetry, while that from the multi-stacked samples shows a two fold symmetry with two perpendicular mirrors. This suggests that, as the number of stacked quantum dots is increased, the strain distribution on the capping layer could, in some cases, become non-uniform. Such behaviour may force an overlaying dot to initially grow as two small components which then coalesce as the growth proceeds, giving a large dot with a depression in the centre. Some evidence of this effect can be seen from the cross sectional image of the two capped sample shown in Fig. 3.

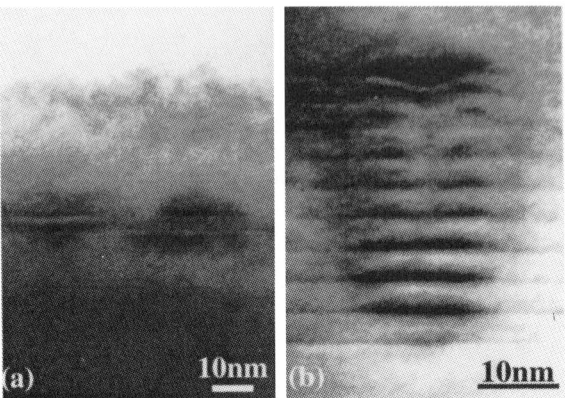

Figure 3. [110] cross sectional bright field images from (a) single capped InAs dot and (b) ten self-assembled InAs quantum dots.

3.2 Dot shape analysis from cross sectional samples

High resolution, cross sectional images of the three samples are shown in Fig. 4. Note that the 2D wetting layer associated with the Stranski-Krastanov growth mode usually observed in capped samples is not visible here. This results because the typical thickness of the wetting layer is only ~1.7ML and hence it can easily relax due to the absence of a capping layer. Measurements of dot size and density for the three samples are listed in table 1. For a fixed growth rate, previous work suggests that the dot density is virtually established at the growth stage when the 2D to 3D transition occurs and that further growth results in little increase in dot density or change in dot shape, although there is an increase in the dot size (Moison et al 1994).

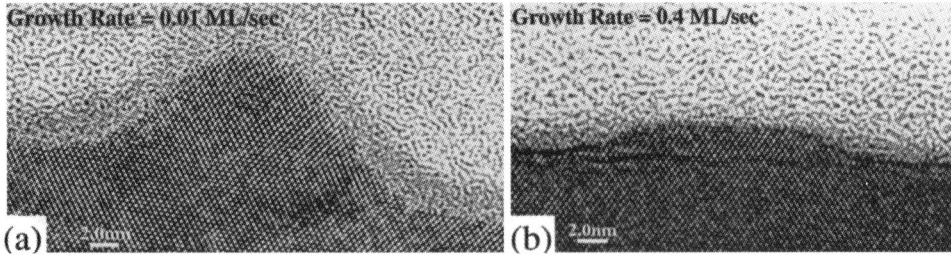

Figure 4. High resolution [110] zone axis TEM bright field images of InAs/GaAs(001) samples for (a) slow growth rate and (b) high growth rate.

Assuming a similar behaviour here, with the majority of dots formed in the early stages of the 2D to 3D transition, then for the low growth rate sample the In adatoms have to migrate large distances to become attached to existing dots rather than forming new dots. This condition is supported by observing that the inter-dot spacing in the low growth rate sample is about three times larger than that for the high growth rate sample (see Fig. 1). This behaviour of the In adatoms in the low growth rate samples probably encourages In migration into the GaAs substrate.

Table 1. Comparison of dot measurements from samples grown at different rates.

Growth rate (ML/sec)	0.01	0.04	0.4
Dot density (dots/cm^2)	3.5×10^{10}	5.0×10^{10}	9.6×10^{10}
Dot height (nm)	8.3±0.5	3.2±0.8	3.2±0.5
Dot diameter (nm)	18±1	16±1	16±1

3.3 EELS analysis

Electron energy loss spectroscopy can have subnanometer spatial resolution (Egerton 1996). This resolution makes this technique suitable for the analysis and mapping of the In composition in the quantum dots and the underlying GaAs substrate. In Fig. 5a an In elemental map obtained using the M4,5 edge at 443eV, of the low growth rate sample, is shown. The corresponding image obtained using the zero loss peak is shown in Fig. 5b. In order to examine the In intensity profile across the dot-substrate interface, line scans across the wetting layer and across the dot are shown in Fig. 5c. Since one of the limitations of EELS in compositional analysis of mismatched interfaces results from the involvement of strain contrast associated with the TEM images (section 3.1), the sample was oriented such that the effect of strain was substantially reduced. This can be readily seen from a comparison of the In profile across the dot (using an elemental map image) and the intensity profile along the same line using the zero loss image, Fig. 5c.

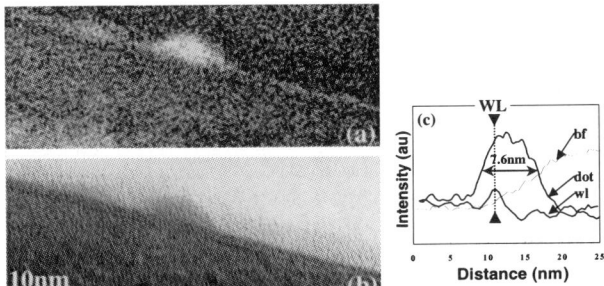

Figure 5. EELS images from the slow growth rate sample; (a) In elemental map, (b) zero loss image and (c) In intensity profiles from the wetting layer and the dot (using (a)) and zero loss image (using (b)).

This comparison shows clearly that the In composition profile data has no associated strain component. In addition, the position of the wetting layer with respect to the quantum dot shows that more In segregation occurs underneath the dot than underneath the wetting layer alone. It has been calculated previously that strain accumulates at the edges of the islands whilst there is strain relaxation at the crest of the island (Benbbas et al 1996). This finding indicates that atoms at the central interface with the GaAs are more relaxed, and this could be associated with atom exchange with the GaAs and hence increased In segregation. In this region, the In can be traced into the substrate to about one third of the thickness of the wetting layer

Using wetting layer measurements and the information listed in table 1, we can calculate the total volume of dots as 19.5 $\times 10^{13}$ nm^3/cm^2 for the low growth sample. This volume is larger than the total volume of deposited InAs (7.3$\times 10^{13}$ cm^3/cm^2) corresponding to a deposition of 2.4ML. There are two main factors contributing to this effect. Firstly, the alloying of the InAs with the GaAs results in Ga segregation into the dot and the wetting layer, resulting in a InGaAs alloy rather than pure InAs. Secondly, the electron beam broadening and limitation of spatial resolution results in an overestimation of the dot wetting layer dimensions. A full comparison of these observations with those observed from the high growth rate samples will be published elsewhere.

4. CONCLUSIONS

Three uncapped InAs quantum dot samples, grown at different rates, and capped samples (single and ten layer repeats) have been studied. For the uncapped samples, TEM measurements using plan view and cross sectional samples reveal that as the growth rate is increased the dot density increases, while the dot height is reduced. Tall pyramids were observed at low growth rate and truncated pyramids at high growth rates. Strain contrast from uncapped samples are cross-shaped indicating pyramidal shaped islands while capped shows four-fold square contrast for single layer and two-fold butterfly contrast for ten layers of multi-stacked InAs dots. These changes are due to morphological changes in dot shape and composition due to capping with GaAs and due to dot-dot strain interaction.

Segregation measurements using EELS reveal greater In segregation into the GaAs buffer layer for the case of a low growth rate, reflecting the large inter-dot distance observed in plan view samples. This behaviour results in a thicker wetting layer and rougher InAs/GaAs interface.

ACKNOWLEDGEMENT

MA acknowledges financial support from the EPSRC.

REFERENCES

Benabbas T, Francois P, Androussi Y and Lefebvre A 1996 J. Appl. Phys. **80**, 2763

Dosanjh S S, Zhang X M, Sansom D, Harris J J, Fahy M R, Joyce B A and Clegg J B 1993 J. Appl. Phys. **74**, 2481

Eaglesham D J and Cerullo M 1990 Phys. Rev. Lett. **64**, 1943

Egerton R 1996 Electron Energy Loss Spectroscopy in the Electron Microscope (New York: Plenum Press)

Grundmann N, Christen J, Ledentsov M N, Böhrer J and Bimberg D 1995 Phys. Rev. Lett. **74**, 4043

Moison J M, Houzay F, Barthe F and Leprince L 1994 Appl. Phys. Lett. **64**, 196

Pehlke E, Moll N, Kley A and Scheffler M 1997 Appl. Phys. **65**, 525

Xie Q, Madhukar A, Chen P and Kobayashi N P 1995 Phys. Rev. Lett. **75**, 2542

Inst. Phys. Conf. Ser. No 164
Paper presented at Microsc. Semicond. Mater. Conf., Oxford, 22–25 March 1999
© 1999 IOP Publishing Ltd

Investigation of CdSe/ZnSe quantum dot structures by composition evaluation by lattice fringe analysis

A Rosenauer, N Peranio and D Gerthsen

Laboratorium für Elektronenmikroskopie, Universität Karlsruhe, Kaiserstraße 12, 76128 Karlsruhe, FRG

ABSTRACT: A new analysis technique for the composition evaluation of ternary layers in sphalerite-type semiconductor nanostructures is presented. The method is based on lattice fringe images obtained with the (000), the chemically sensitive (002), and the (004) reflection. The investigation of the structural and chemical morphology of monolayer CdSe-insertions in a ZnSe matrix reveals quantum dots with a diameter between 10 and 100 nm embedded in a CdZnSe quantum well. An anomalous broadening of the quantum well from nominally < 3 monolayers to a FWHM of 10 monolayers is observed.

1. INTRODUCTION

The compositional analysis of semiconductor nanostructures constitutes a highly interesting challenge to high-resolution transmission electron microscopy (HRTEM). Low-dimensional semiconductor heterostructures are used for the fabrication of optoelectronic devices. Especially quantum dot (QD) structures are of increasing interest due to their enhanced radiative recombination probability and possible application in laser devices (Grundmann et al 1995). A detailed knowledge of structure and composition is essential for the correct interpretation of optical data gained by e.g. photoluminescence spectroscopy (PL) or photoluminescence excitation spectroscopy (PLE). We present here a composition analysis based on the processing of defocus series of images taken under off-axis conditions.

2. COMPOSITION EVALUATION BY LATTICE FRINGE ANALYSIS

Our approach to evaluate the composition c of ternary layers $A_c B_{1-c} C$ in sphalerite-type semiconductor heterostructures is based on the exploitation of the chemically sensitive (002) reflection. To avoid its excitation by multiple scattering an electron beam direction close to a $\langle 100 \rangle$ zone axis is used. The sample is tilted about 3° towards the $\langle 100 \rangle$ zone axis. The (000) and (004) reflections are strongly excited and the (002) reflection is set on the optic axis. The procedure described here is similar to that suggested by Rosenauer et al (1998, 1999). The application of off-axis conditions was previously suggested by Jia et al (1993), where they showed qualitatively that the resulting fringe pattern reveals chemical information in the case of high-temperature superconducting layers, as well as a decreased sensitivity to sample thickness fluctuations due to the larger extinction distance compared to the exact zone-axis orientation. In the following, we give a brief description of the implementation of the composition evaluation by the lattice fringe analysis (CELFA) technique that is used here.

The procedure is based on the acquisition of a defocus series of 10 images with an on-line CCD camera with a (nominal) defocus stepsize of 10 nm. To avoid windowing effects, the images are multiplied with an appropriate filter prior to the application of the noise reduction procedure that was suggested by Rosenauer et al (1996). The filter is 1 inside an area of interest (AOI) that is defined by a closed polygon selected by the user. The filter values continuously drop from 1 to 0 from the rim of the AOI towards the edges of the images. Then, the images are Fourier transformed. The amplitude $|J_n^{002}|$ of the (002) reflection in the diffractogram is given by

$$|J_n^{002}| = T_n^{002} |a_{002}| \sqrt{a_{000}^2 + a_{004}^2 + 2a_{000}a_{004}\cos(\varphi_n)}, \tag{1}$$

where the a_{00j} are the amplitudes of the $(00j)$ beams and n is the image number relating to the defocus series ($n=1..10$). The real number T_n^{002} is connected with the transmission cross coefficients defined by Ishizuka et al (1980) by $T(\mathbf{g}_{002},0;\Delta f_n) =: T_n^{002} \exp(i(\chi_{\Delta f_n} + \chi_S))$ where $\chi_{\Delta f_n}[= \pi\Delta f_n \lambda \mathbf{g}_{002}^2$ (λ: electron-wave length)] and χ_S are the phase shifts introduced by the objective lens defocus Δf_n and the spherical aberration, respectively. The phase φ_n is linked with the phases p_{00j} of the $(00j)$ beams by

$$\varphi_n = -2(\chi_{\Delta f_n} + \chi_S) + (p_{002} - p_{000}) - (p_{004} - p_{002}) = -2\chi_n + 2p_{002} - p_{000} - p_{004}. \tag{2}$$

To obtain spatially resolved information on local values $J_n^{002} = A_n^{002} \exp(i P_n^{002})$ we use the following procedure:

In the Fourier transformed image, a circular area is marked around the (002) reflection. The information outside the circle is deleted. The information inside the circle is centred in such a way that the pixel with largest intensity lies on the zero-frequency position in Fourier space. The result is inverse-Fourier transformed and the subsequently calculated local phase $P_n^{002}(x,y)$ and the amplitude $A_n^{002}(x,y)$ are stored with x and y being spatial image coordinates.

In the next step, the local sample thickness in electron beam direction is evaluated for one position (x_0, y_0) in the image where the composition is known (e.g. the ZnSe buffer or cap layer) by exploiting the dependence of $A_n^{002}(x_0, y_0)$ on n which is an oscillation with a period $\delta(\Delta f_n) = 1/(\lambda \mathbf{g}_{002}^2)$ according to Eq. (1). For this purpose, the data $A_n^{002}(x_0, y_0)$ ($n=1..10$) are fitted by an appropriate fit curve

$$A_n^{002}(x_0, y_0) = (1 + An)\sqrt{B^2 + C^2 + 2BC\cos(D(n-E))}, \quad n = 1...10, \tag{3}$$

with A, B, C, D and E being fit parameters. The resulting values yield the phase angles φ_n of Eq. (1) by $\varphi_n(x_0, y_0) = D(n-E)$. For the evaluation of the sample thickness $t(x_0, y_0)$ two adjacent maximum and minimum values of the measured oscillation of $A_n^{002}(x_0, y_0)$ are used. They are linked with the beam amplitudes $a_{000}(t(x_0, y_0))$ and $a_{004}(t(x_0, y_0))$ by the relation:

$$\left(\left[A_n^{002}(x_0, y_0)\right]_{max} + \left[A_n^{002}(x_0, y_0)\right]_{min}\right) \Big/ \left(\left[A_n^{002}(x_0, y_0)\right]_{max} - \left[A_n^{002}(x_0, y_0)\right]_{min}\right) = \frac{a_{000}(t(x_0, y_0))}{a_{004}(t(x_0, y_0))} \tag{4}$$

Eq. (4) yields the sample thickness $t(x_0, y_0)$ by using values for the right-hand side of the equation that are obtained from Bloch-wave calculations performed with the EMS-program package of Stadelmann (1987). With the known sample thickness $t(x_0, y_0)$ as well as calculated beam phases $p_{00j}(t(x_0, y_0))$ Eq. (2) is used to calculate $\chi_n(x_0, y_0)$. It is important to note that $t(x_0, y_0)$ and $\chi_n(x_0, y_0)$ are considered constant throughout the image n. However, in order to take into account spatial variations of sample thickness and defocus, Rosenauer et al (1999) showed that it is a good approximation to regard T_n^{002} of Eq. (1) as fit parameter $T_n^{002}(x,y)$ that varies locally. In regions with known composition (e.g. inside the buffer and cap layers), local values $T_n^{002}(x,y)$ are calculated from Eqs. (1,2), again using the beam amplitudes $a_{00j}(t(x_0, y_0))$ computed with EMS. A completed map of $T_n^{002}(x,y)$ is obtained by filling regions with unknown values by extrapolation from areas with known composition where $T_n^{002}(x,y)$ was calculated. Finally, the composition $c(x,y)$ is obtained as follows: $\varphi_n(x,y)$ is computed from Eq. 2 using $\chi_n(x_0,y_0)$. For each pixel (x,y), a table is calculated that lists the right-hand side of Eq. 1 for $c = 0..1$ with a stepsize $\Delta c = 0.01$ using values of $a_{00j}(t(x_0,y_0),c)$ and $p_{00j}(t(x_0,y_0),c)$ obtained with EMS as well as local values $T_n^{002}(x,y)$. $A_n^{002}(x,y)$ is compared with each element of the table and the entrance with the best fit yields the composition $c(x,y)$.

It is appropriate to note that the procedure described in the preceding is ambiguous for several materials. In a good approximation, the relation $a_{002}(t,c) \propto a_{002}(t,0)(c-c_0)$ holds for most ternary sphalerite type semiconductor materials. For $c = c_0$ (e.g. $c_0 = 0.41$ for $Cd_cZn_{1-c}Se$), a_{002} changes its sign. In this case, an ambiguity occurs because the measurement of $|J_n^{002}|$ that is proportional to $|a_{002}|$ cannot distinguish between the compositions $c = c_0 + \delta c$ and $c = c_0 - \delta c$. However, a change of the sign of a_{002} is identical with a phase shift of π of J_n^{002}. Therefore, a phase shift of π is registered in the measured phase $P_n^{002}(x,y)$ between regions with $c < c_0$ and $c > c_0$ that is used to overcome the ambiguity.

3. EXPERIMENTAL PROCEDURES

The epitaxial layers were grown pseudomorphically on a GaAs(001) substrate at a temperature of 280°C. The samples contain a 50 nm thick ZnSe buffer grown by molecular beam epitaxy (MBE), a CdSe epilayer deposited by migration enhanced epitaxy (MEE) with a nominal thickness of 0.5, 1, 2 or 3 monolayers (ML) and a 10 nm thick ZnSe cap layer grown by MBE. The TEM cross-sectional samples were conventionally prepared with a final stage of Ar^+ or Xe^+ ion milling at an energy of 3 keV in a liquid nitrogen cooled specimen holder. We used a Philips CM200 FEG/ST electron microscope with a spherical aberration constant of C_S=1.2 mm. The images were recorded with an on-line CCD camera with 1024×1024 picture elements.

4. RESULTS

Figure 1a shows a grayscale map of local amplitudes $A_{10}^{020}(x,y)$ obtained from the evaluation of the nominally 2 ML thick CdSe layer buried in ZnSe. It is appropriate to note that the sample was oriented in such a way that the lattice fringes were running perpendicular to the surface plane. This orientation is advantageous in pseudomorphically grown strained heterostructures because of the constant lattice parameter of the corresponding $(0j0)$ lattice planes. This is not the case for the $(00j)$ planes parallel to the surface that have a larger lattice parameter inside the CdZnSe layer. According to Eq. (2), the contrast pattern is modified by lattice parameter fluctuations.

Figure 1b depicts the local variations of $T_{10}^{020}(x,y)$ that is used to approximately take into account variations of local thickness and defocus. Finally, Fig. 1c gives a grayscale map of the resulting Cd concentration. Obviously, the CdZnSe layer thickness is broadened significantly compared to the nominal thickness of 2 ML. As a consequence, the local Cd concentration is below 24 %. Furthermore, the Cd concentration is not homogeneous along the layer. In Fig 1c, an inclusion with enlarged Cd concentration with a lateral size of approximately 8 nm is visible which will be considered as a quantum dot (QD) in the following. The QDs are embedded in a CdZnSe quantum well (QW) with a maximum Cd concentration of approximately 15%. Fig. 2 shows Cd concentration profiles from QW regions of all four investigated samples. Taking into account the sample tilt, all profiles have a FWHM of approximately 10 ML. The broadening is most likely due to interdiffusion of Cd and Zn. The contribution of segregation is assumed to be small because the concentration profiles exhibit only a small asymmetry with a slightly wider decay towards the ZnSe cap layer. The computation of the total amount of Cd in the QW by integration of the Cd concentration profiles reveals that 20-30% of the Cd must be concentrated in the QDs.

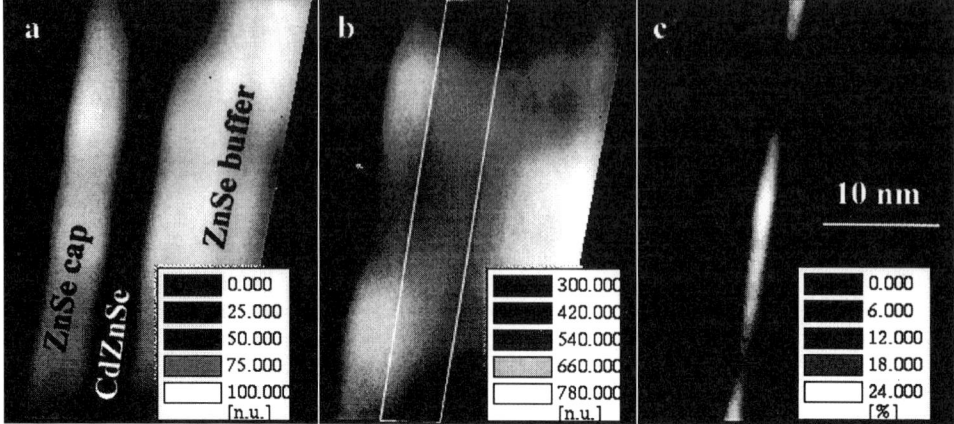

Fig. 1: *Grayscale maps of a) $A_{10}^{020}(x, y)$ evaluated from a fringe image of a CdZnSe interlayer buried in ZnSe. b) $T_{10}^{020}(x, y)$ and c) the evaluated local Cd concentration. The values of $T_{10}^{002}(x, y)$ inside the white frame in b) are calculated by extrapolation from regions outside the frame.*

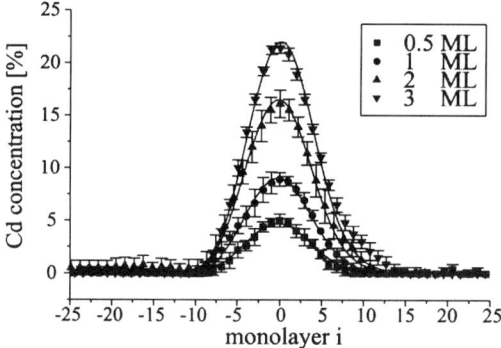

Fig. 2: *Cd concentration profiles in growth direction of all investigated samples with a nominal thickness of 0.5, 1, 2 and 3 ML. The cap layer corresponds to the right side of the diagram.*

5. DISCUSSION

Here we briefly outline possible sources of error of the CELFA technique. Rosenauer et al (1999) showed that the errors due to variations of thickness and defocus are rather small. This is also the case for errors due to deviations of the sample orientation. Other sources of error that should be addressed are delocalization and the occurrence of Fresnel fringes due to the potential step introduced by the layer. Both effects cause increasing errors for increasing sharpness of the interfacial transitions. It is not possible to minimise both effects in microscopes without C_S-corrector. Fresnel effects are minimised for vanishing defocus whereas delocalization of the (000) and (004) beams is minimised by adjusting the maxima of the wave aberration function at the positions of the (000) and (004) reflections. However, a delocalization of the (000) and (004) beams mainly affects the sample thickness determination. It does not cause significant errors for the composition evaluation because the chemical information is mainly carried by the (002) beam. The impact of the point-spread function causes a blurring of the transition regions of the concentration profiles. However, image simulations showed that this effect cannot be responsible for the observed anomalous broadening of the CdSe layer.

ACKNOWLEDGMENT

This work is supported by the Deutsche Forschungsgesellschaft under contract number Ge 841/7. We thank S.V. Ivanov (Ioffe Physico-Technical Institute, St. Petersburg) for supplying the samples.

REFERENCES

Jia C L, Thust A, Jakob G and Urban K 1993 Ultramicroscopy **49**, 330
Grundmann M, Stier O and Bimberg D 1995 Phys. Rev. B **52**, 11969 and references therein
Ihizuka K 1980 Ultramicrocopy **51**, 55
Rosenauer A, Fischer U, Gerthsen D and Förster A 1998 Ultramicroscopy **72**, 121
Rosenauer A and Gerthsen D 1999 Ultramicroscopy **76**, 46
Stadelmann P A 1987 Ultramicroscopy **51**, 131

Inst. Phys. Conf. Ser. No 164
Paper presented at Microsc. Semicond. Mater. Conf., Oxford, 22–25 March 1999

Application of spatially resolved electron energy-loss spectroscopy to the quantitative analysis of semiconducting layer structures

T Walther and W Mader

Institut für Anorganische Chemie, Universität Bonn, Römerstr. 164, D-53117 Bonn, Germany

ABSTRACT: We investigate the potential of spatially resolved electron energy-loss spectroscopy for obtaining quantitative information on the profile of semiconducting layer structures. Operating a transmission electron microscope with an attached imaging filter in the line focus mode and orienting the layer structure accurately parallel to the energy-dispersive direction, we record spectrum images with a space co-ordinate (the growth direction) as a function of the energy loss. We present two practical examples: 1. Using Ge and Si core losses, quantitative chemical profiles across a SiGe/Si quantum well structure are obtained. 2. Recording spectrum images with high dispersions, we study the near-surface region of an oxidised and carbon-capped SiGe layer in cross-section. The fine structure of the Si_L edge allows us to distinguish the Si bonds of diamond-like Si(Ge) from those of amorphous SiO_2. From the experimental data we estimate the accuracy of elemental mapping of SiGe by energy-selected imaging and the detectability of thin layers of C and SiO_2 in Si(Ge) by spectrum imaging.

1. INTRODUCTION

For many applications in the semiconductor industry the structural and chemical quality of heterointerfaces determine the device performance (e.g. Kelly 1995). While transmission electron microscopy (TEM) has traditionally been a tool for microstructural studies, the availability of energy filters with low-aberration imaging optics now also allows highly localised chemical characterisation. In this study we apply the rather new method of spatially resolved electron energy-loss spectroscopy (SR-EELS) (sometimes also referred to as *spectrum imaging* to distinguish it from the serial acquisition of energy-loss spectra with a focussed electron probe scanned across the sample). Our aim is to investigate the possibilities of obtaining reliable values of the local chemical composition and the limits of spatial and chemical resolution achievable. Reimer (1995) and Kimoto et al. (1997) have already used this technique to study the chemical shift from the energy-loss near-edge structure (ELNES) of somewhat thicker layers.

When the electron microscope is operated in a line focus and the attached imaging filter in spectroscopy mode, the energy-loss signal stems from a specimen region determined by the entrance aperture (or, if available, entrance slit) of the spectrometer. Hence, structures oriented exactly parallel to the energy-dispersive direction can be directly imaged with their perpendicular direction displayed as a function of the energy-loss. This corresponds to a simultaneous acquisition of spectra across the structure, made possible by the two-dimensional electron detector (a charge-coupled device camera, CCD). This has specific advantages over the conventional method of recording a series of energy-loss spectra in the scanning mode in that even small drifts of either the specimen or the energy are readily detected during spectrum image acquisition, at the cost of longer recording times. In the scanning mode the sample may drift unnoticeably such that spectra are not recorded from the positions intended. Likewise, when subtracting spectra acquired from a matrix off similar spectra taken near or from a defect structure, as employed in the *spatial difference method* (Müllejans and Bruley 1994), then even tiny drifts of the energy are sufficient to produce a difference signal which can be misinterpreted as being due to the defect. In the case of sharp edges on a slowly-varying background, short-term energy instabilities of 0.1-0.3eV are sufficient to cause artefacts which will only be noted if another distinct spectrum feature is available for reference.

2. EXPERIMENTAL

2.1 Specimen preparation and transmission electron microscopy

The SiGe layers were grown by low-pressure chemical vapour deposition. A thin carbon film was sputtered onto the single SiGe layer sample prior to sample preparation for TEM. The cross-sectional samples were prepared by standard techniques of glueing stacks, cutting, grinding, dimpling, polishing and Ar^+ ion milling until perforation.

We use a Philips CM300UT electron microscope equipped with a field-emitter source (ΔU_{FWHM}=0.8eV), a high-resolution objective lens (Cs=0.6mm, Cc=1.9mm), a corrector for three-fold astigmatism and a Gatan imaging filter (GIF). The high voltage is set to E_0=297keV for which the alignment has been performed such that for energy losses up to 3keV the high voltage can be increased accordingly while the objective lens is still well aligned for focussing 297keV electrons. For our experiments presented here we used an objective aperture of 14mrad radius and an indicated magnification of 5200 on the screen which gives an effective magnification in imaging mode of about 109000 on the CCD of the GIF.

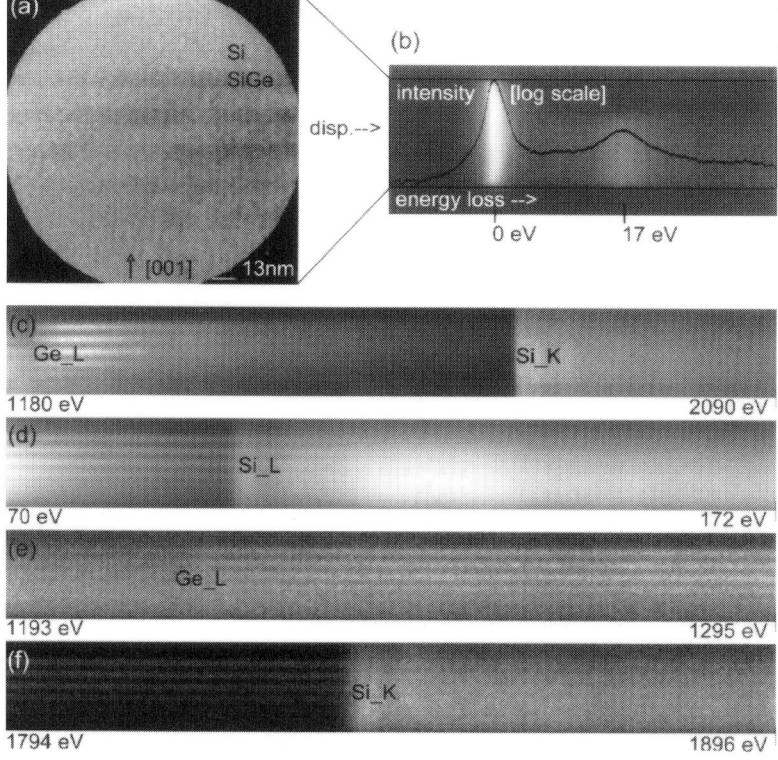

Fig.1: (a) Bright-field TEM image of the multi-quantum well structure 42F84; (b) part of a spectrum image acquired without energy offset at a dispersion of 0.1eV/pixel in 0.002s; (c) spectrum image with 1180eV offset, 1eV/pixel dispersion, 60s exposure time; (d) spectrum image with 70eV offset, 0.1eV/pixel dispersion, 5s exposure; (e) 1193eV offset, 0.1eV/pixel dispersion, 60s exposure; (f) 1794eV offset, 0.1eV/pixel dispersion, 60s exposure.

2.2 Quantification of the germanium composition

Fig.1a shows a bright-field image of a SiGe/Si multi-quantum well structure consisting of five SiGe layers of nominal widths of 10nm and Ge concentrations of $x(Ge)$=0.05, 0.1, 0.15, 0.2 and 0.3. The [001] growth direction points upwards, and the 2mm entrance aperture used for our study limits the field of view. Fig.1b displays a spectrum image (on a logarithmic scale for better visibility as the zero loss peak is very intense) as recorded on the CCD. In the spectroscopy mode the image is compressed in the vertical direction in order to use only the central parts of the imaging lenses where the aberrations are negligibly small. Thus, the spectrum images are only about 1024 x 180 pixels in size for the 3mm and 1024 x 120 pixels for the 2mm entrance aperture. Adjusting the spectrum offset to higher values, core losses of various elements can be recorded. In Fig.1c we have chosen a dispersion of 1eV/pixel and an offset of 1180eV, which gives access to the Ge_L and the Si_K edge. Because of the chromatic aberration of the objective lens, only electrons of a certain energy are well focussed, while electrons with energy losses differing from the high-voltage *offset* by an amount of ΔE will suffer an additional defocus of $\Delta f = Cc \cdot \Delta E/E_0$ which leads to a blurring in the image by a chromatic point spread function $\sim (r^2 + \Delta r^2)^{-1}$ with $\Delta r \approx 0.5 \cdot Cc \cdot \Delta E^2/E_0^2$ (Egerton 1996). This means that in Fig.1c the Ge_L edge is well in focus, but the Si_K edge appears blurred in the vertical direction (point resolution $2\Delta r = 8nm$ for $\Delta E = 1839eV - 1217eV = 622eV$). Figs.1d-f display spectrum images with a higher dispersion of 0.1eV/pixel such that the energy range covered is about 100eV and all features are in focus ($2\Delta r = 0.1nm \ll 1$ pixel for ΔE up to 100eV).

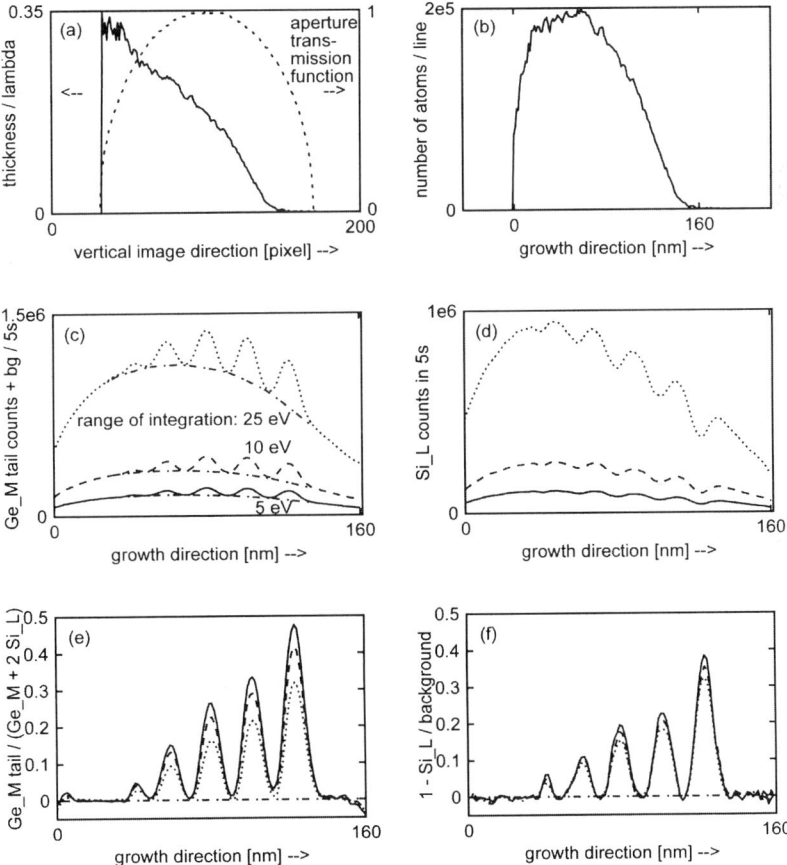

Fig.2: Line-by-line analysis of the low-loss spectrum images: (a) relative thickness plot from Fig.1b along [001] (solid) and projected integration of the area of the circular aperture (dotted); (b) number of atoms per CCD line, given by the product of the two functions shown in (a); (c) intensity plot of Ge_M and background (dot-dashed) before the Si_L edge for three different integration windows ending at 94eV; (d) plot of Si_L net along [001] for three integration windows of 5, 10, 25eV; (e) plot of ratio Ge_M/ (Ge_M+2Si_L) (net signal of Ge_M obtained by background extrapolation from Si); (f) plot of the relative drop of the Si_L signal. Both, (e) and (f) should give x(Ge) if the scattering cross-sections were correct and Si and Ge the only atomic species in the specimen (no surface oxides).

124

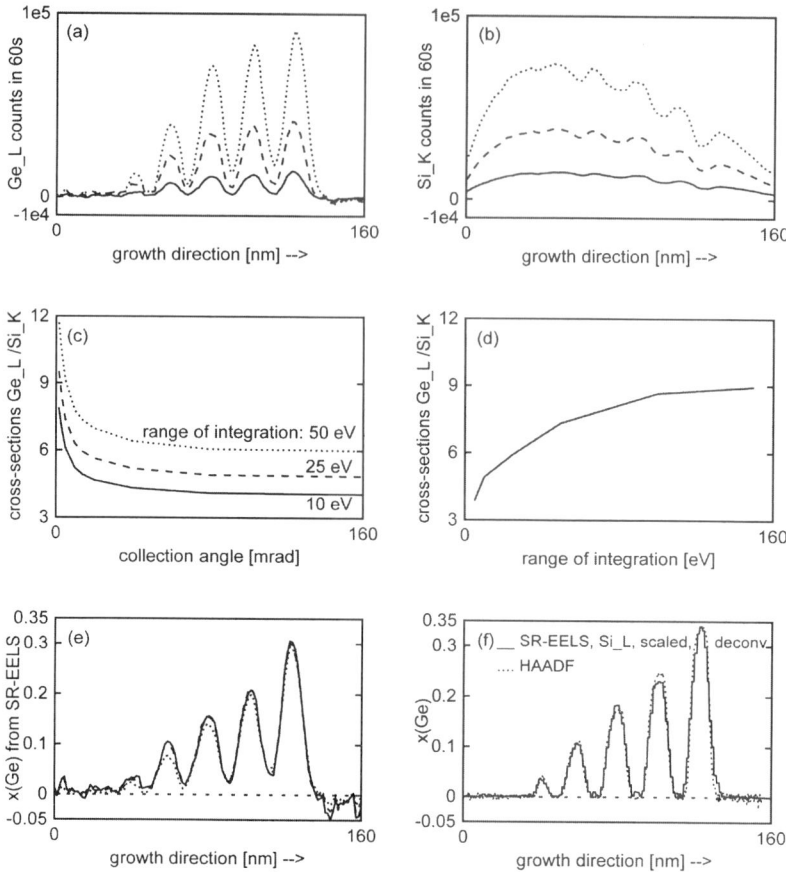

Fig.3: Line-by-line analysis of spectrum images with high energy losses: (a) plot of the Ge_L net signal along [001] from Fig.1e for three integration windows of 5 (solid), 10 (dashed) and 25eV (dotted); (b) plot of the Si_K net signal for the same integration windows; (c,d) plots of the dependence of the ratio of inelastic scattering cross-sections of Ge_L and Si_K on the collection angle (c) and integration range for 14mrad collection angle (d); (e) plot of the SR-EELS signal $x(Ge)=I(Ge\_L)/[I(Ge\_L) + I(Si\_K)\cdot\sigma(Ge\_L)/\sigma(Si\_K)]$ along [001]; (f) comparison of the results from SR-EELS (solid, after deconvolution by the line spread function) and high-angle annular dark-field imaging (HAADF, dotted).

Fig.2 analyses features of Figs.1b,d. From Fig.1b we calculate the relative thickness, expressed in terms of the inelastic mean free path, λ, which is plotted as a function of the vertical image direction in Fig.2a. The additional dotted line describes the transmission function of the aperture as given by a horizontal projection of the almost circular area. The product of both, the thickness and the lateral width of the aperture, yields the plot in Fig.2b, which we have scaled to the total number of atoms contributing to the signal in each line (with λ_{Si}=170±2nm, 1.33nm/pixel, a_{Si}=0.5431nm and 4 atoms per unit cell). The maximum thickness of t=0.35λ at the bottom edge of the spectrum images allows us to refrain from a Fourier-log deconvolution of the spectra. Fig.2c depicts plots of the counts *before* the Si_L edge, as calculated for each of the 120 lines in the spectrum image of Fig.1d by extracting the spectrum from each line and integrating the intensities over a window of 5,10 and 25eV up to an energy of 94eV (i.e. 5eV in front of the Si_L edge at 99eV). This signal contains, on a secondary electron background which resembles the thickness contour, contributions from the Ge_M edge at 29eV which itself is hardly distinguishable from the second plasmon loss at about 33eV but is the principal reason for the SiGe layers appearing light in front of the Si_L edge (Fig.1d). Fig.2d shows the corresponding Si_L net counts after subtracting from each spectrum the background (using a least-square inverse power-law fit). This signal is comparable to the intensities recorded using energy-selected imaging (ESI) with the 3-window method (Jäger and Mayer 1995). Assuming that the specimen thickness of the SiGe layers does not differ much from that of the surrounding Si and that all atoms that are not Si must be Ge (in pure SiGe), one can extrapolate the Si_L signal from the regions of presumably pure Si into the SiGe and calculate a relative drop of the Si_L signal and from this the Ge profile depicted in Fig.2f. The values produced using the different integration windows as parameters, however, are not self-consistent. If, instead, we calculate the ratio of Ge_M to Si_L intensities as can be done from Figs.2c,d we need to take into account for an absolute quantification the relative cross-sections of the ionisation edges. While data for Si_L are available (Rez 1982 and EL/P3.0 software by Gatan, USA), no literature values for Ge_M could be found. Extrapolating cross-sections and their dependence on the position of the integration window relative to the edge onset for elements from $Z=36$ to 42 down to $Z=32$ (Ge) allows an estimate for σ_{Ge_M} of 1.2, 2.4 and 5.8×10^4 barns for energy windows extending from 90-95, 85-95 and 70-95eV, respectively, which is exactly twice the corresponding values of the Si_L edge over 5,10 and 25eV after the edge onset. Using this ratio to calculate the Ge concentration yields Fig.2e which is even further off the nominal values and less consistent than Fig.2f. Hence, we conclude that there is no reliable method at hand for calibrating Ge profiles of SiGe layers from the Si_L edge without using a standard.

In Fig.3 we deal with the higher energetic core losses of Si and Ge (>1keV) which have much smaller cross-sections and thus need much longer exposure times. Figs.3a,b depict the net counts of the Ge_L and Si_K edges, respectively, as determined for each line of the spectrum images by background extrapolation. Again, the width of the integration window is used as a parameter, as the relative cross-sections of Ge_L and Si_K depend strongly on the integration range as well as on the collection angle used (Figs.3c,d), The Ge_L / Si_K intensity ratio scaled with the appropriate inelastic cross-sections, σ, now produces self-consistent values of $x(Ge)=I(Ge\_L) / [I(Ge\_L)+I(Si\_K) \cdot \sigma(Ge\_L)/\sigma(Si\_K)]$ as can be seen in Fig.3e (Walther et al. 1995). The difference between the form of the profiles of Figs.2e and 3e is a result of some drift during the 60s exposures needed for the latter data set (as compared to 5s for the other) and possibly also a relative drift between the Ge_L and Si_K edge spectrum images which had to be recorded separately because of the large energy difference and the chromatic aberration. On the other hand, the difference between the profiles of Figs.2e,3e and those in Fig.3f is due to the influence of the detector point spread function. Fig.3f compares the Ge profiles obtained by the SR-EELS method after deconvolution of the profile by the line spread function of the CCD (FWHM of 1.8pixels, i.e. 2.4nm with our sampling) with a previous study by high-angle annular dark-field (HAADF) imaging (Walther and Humphreys 1999). For the deconvolution which enhances noise we have chosen the profiles of Fig.2e (the Si_L signal yields a much higher signal-to-noise ratio than the Si_K edge!), scaled to the absolute concentrations given by Fig.3e (the Ge_L / Si_K ratio method yields reproducibly $x_{max}(Ge)=0.3$) and set the outermost pixels to zero on both edges of the scan to eliminate wrap-around effects during Fourier transformation. The results thus obtained by SR-EELS and HAADF agree with respect to the chemical compositions within ±1at%Ge and yield identical FWHM-values of the layers within 0.5nm.

2.3 Energy-selected imaging

In Fig.4 we calculate what energy-selected imaging (ESI) with power-law background subtraction based on just two pre-edge images instead of a complete spectrum would yield under identical imaging conditions. Si_L spectra of Si and $Si_{0.7}Ge_{0.3}$ differ in two important points (Fig.4a): the Si_L signal is of course lower in SiGe than in Si of similar thickness, but the background just before the Si_L edge is much higher due to the preceding Ge_M edge. Hence, images recorded with an energy slit *just* behind the Si_L edge will show almost no contrast as both effects cancel. This is corroborated by Figs.4c-e which are patchworks of line scans of energy-selected images that would be recorded with a slit width of 8eV for energies of 74-86eV (pre1), 82-94eV (pre2) and 106-118eV (post-edge). These scans have all been calculated from Fig.1d. Extrapolation of the background based on the intensities in pre1 and pre2 and subtraction from the corresponding post-edge image would yield chemical maps as displayed in Fig.4g (still, however, the vertical direction is an energy co-ordinate!). Fig.4f is simply a two-dimensional representation of Fig.2d for 8eV slit width which gives the true Si_L net signal for proper background subtraction from each spectrum of the spectrum image in Fig.1d.

Although Figs.4f and g are quite similar, there are small differences (e.g. the better visibility of the first $Si_{0.95}Ge_{0.05}$ layer in Fig.4f). The relative difference between the two, calculated line-by-line (after scaling to the same mean intensities as the absolute intensity in the Si_L net signal depends on the position of the post-edge slit) can be up to 17%. Interestingly, the

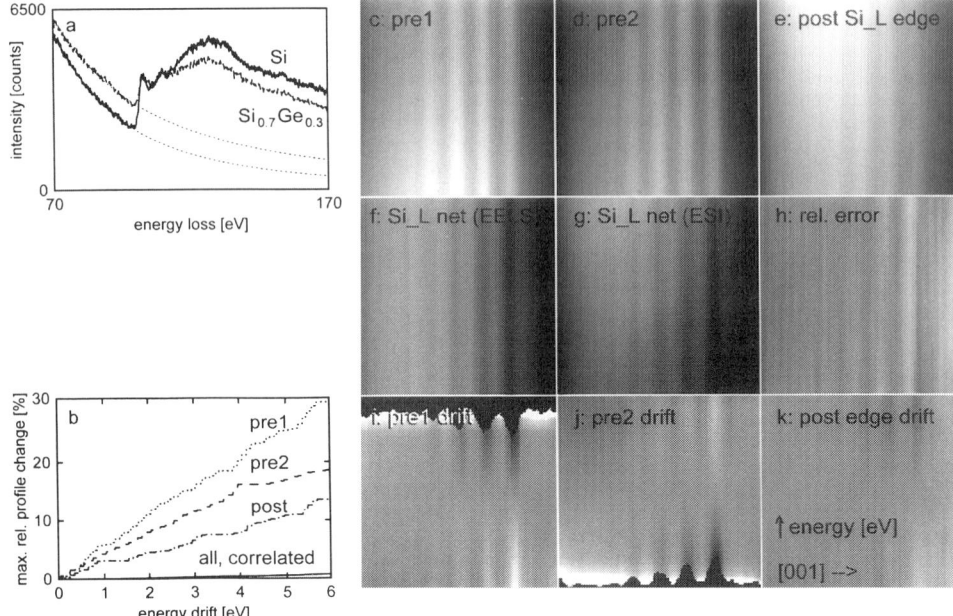

Fig.4: (a) Energy-loss spectra of Si (solid), $Si_{0.7}Ge_{0.3}$ (dashed) and background extrapolation (dotted); (b) plot of maximum relative change in the compositional profile when all images of an ESI series drift in energy simultaneously by a certain amount (solid line, corresponds to a wrong energy calibration) or only one of the series drifts in energy (data from (i-k)); (c-e) patchworks of line scans along [001] of ESI images calculated for 8eV slit widths at energies of 74-86eV (pre1), 82-94eV (pre2) and 106-118eV (post-edge); (g) resulting Si_L net signal from ESI using inverse power-law background subtraction from (e) based on (c) and (d); (f) 'true' Si_L net signal obtained from proper line-by-line background subtraction using spectra; (h) relative difference between (g) and (f); (i-k) relative changes of (g) if the images pre1, pre2 and post-edge deviate from their assumed central positions at 80, 88 and 112eV by −6eV (bottom) to +6eV (top). (h-k): black=−0.3, white=+0.3.

maxima and minima in Fig.4h correlate with the positions of the interfaces, i.e. the error from an ESI map is largest where it is most relevant to get precise values! In Figs.4i-k we have calculated the relative changes expected if the pre1-, pre2- and post-edge images, respectively, were recorded not at the energies assumed in the background fit but at energies shifted by up to ±6eV. For the pre1-images being not at 80eV but at 84eV or higher (or, conversely, the pre2-images not at 88eV but 84eV or lower), both pre-edge images would be so close in energy that the background fit would yield negative values for the background of the post-edge image and thus fails (black areas in Figs.4i,j). The relative changes again are highest near the interfaces, with the maximum change displayed as a function of the deviation in energy loss in Fig.4b. If all images had been acquired with a wrong energy offset (mimicking a long-term energy drift or a wrong energy calibration), however, the result would be hardly changed at all as the background fit routine would just use different fit parameters but still be fairly stable. For short-term energy drifts one may assume a statistical independence of the fluctuations of the energies of the various images around their mean energies and hence add their squares in quadrature. This yields a maximum systematic error due to energy instabilities of about 6%/eV. For our instrument with $\Delta E \leq 0.3$eV within ≈ 1min this amounts to $\leq 2\%$ relative error which in the view of the bad signal-to-noise ratios in typical elemental maps is negligible (Golla and Kohl 1997).

2.4 Detection of carbon and oxygen in silicon

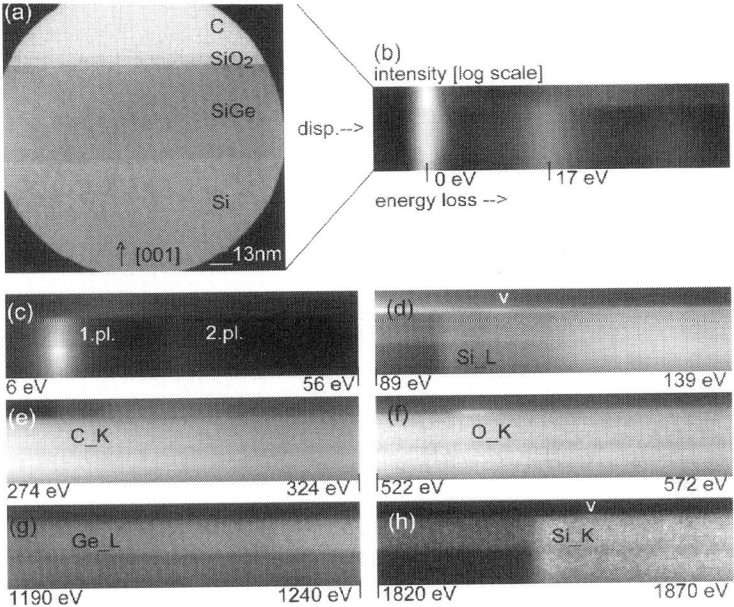

Fig.5: (a) Bright-field image of single SiGe layer on Si (wafer 22C86), oxidised and carbon-capped; (b-h) spectrum images with various energy offsets at a dispersion of 1eV/pixel. Arrows in (d) and (h) indicate shifts of the Si_L and Si_K edges in SiO$_2$ (pl.=plasmon loss).

Fig.5a shows a bright-field image of the single SiGe layer sample (nom. 66nm Si$_{0.83}$Ge$_{0.17}$) and reveals the oxide (now 6nm thickness) as well as the sputtered carbon on top. Figs.5b-h are spectrum images of this structure with a dispersion of 0.1eV/pixel. In Figs.5d,h the Si_L and K edges, respectively, of the SiO$_2$ layer are both shifted to higher energies, compared to Si(Ge). This is best seen when looking at the spectrum images under an oblique angle from the side. Fig.6a is a relative thickness plot calculated from Fig.5b and shows that the specimen thickness decreases along [001] from ≈0.45 to ≈0.20. Fig.6b depicts line scans of the intensities of the Si_L, C_K, O_K, Ge_L and Si_K edges calculated line-by-line from the spectrum images after proper cross-correlation to compensate drift. The difference between the plots of the Si_L and the Si_K signals near the SiGe surface is explained by the different integration windows used as only the latter signal also contains the contribution of the SiO$_2$. The chemical shift is analysed in Fig.6c which compares the Si_L spectra of Si, SiGe (integrated over 50nm along the layers) and SiO$_2$ (integrated over 8nm). While the Si and SiGe spectra are indistinguishable, SiO$_2$ produces a shifted double-peak structure which is only weakly discernible due to the poor signal-to-noise ratio. The chemical shift from Si(Ge) to SiO$_2$ is measured as a function of the growth direction in Fig.6d and amounts to +7.0±0.3eV at the Si_L and +7.1±0.6eV at the Si_K edge. The weaker pre-peak of the Si_L signal of SiO$_2$ lies about 5.2±0.3eV after the Si_L edge of Si(Ge). These values agree with measurements from thicker layers by Kimoto et al. (1997).

Extrapolating mean and standard deviation of the C_K and O_K signals from Si(Ge) to a thickness of 0.25λ as appropriate for the carbon and SiO$_2$ region and considering this as "noise", we calculate from the maxima of the C_K and O_K counts the signal-to-noise ratios (SNR) for detecting a single pixel (=1.3nm) thin layer of C and SiO$_2$, respectively, in a Si(Ge) matrix of ≈40nm thickness. The values of SNR_{C_K}=8.9 and SNR_{O_K}=15.6 mean that 0.45nm C and 0.26nm SiO$_2$ would give signals in our spectrum images at the C_K and O_K positions, respectively, of 3 times the corresponding standard deviation above the background noise. As the CCD was not saturated, increasing the illumination or exposure time by factors of 4 and 8, respectively, should accordingly reduce the noise by factors of 2 and √8 and even make the detection of the equivalent of 0.23nm C or 0.09nm SiO$_2$ possible! The worse detection limit of C is mainly due to carbon contamination on the specimen: the fluctuations of the dotted curve (O_K net signal) in Fig.6b indicate surface contamination of preferentially SiGe and cannot be reduced by improving the statistics (e.g. a wider integration window).

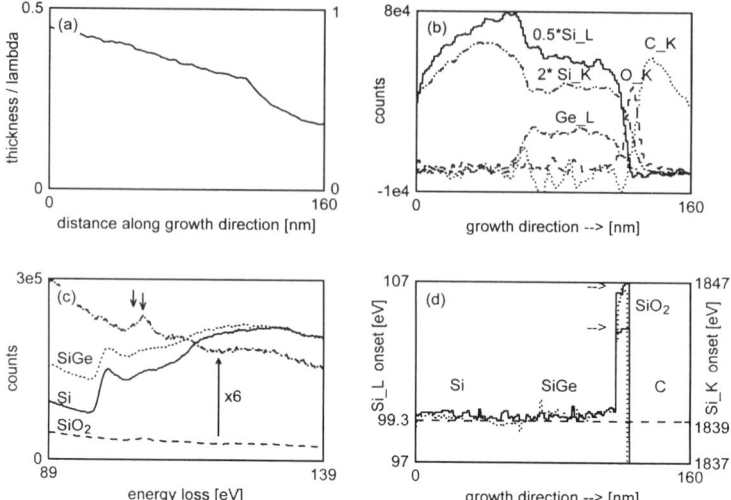

Fig.6: (a) Plot of rel. thickness along [001]; (b) plots of net counts after background subtraction of Si_L (Δ=8eV integration window, t=4s exposure time, solid line), C_K (Δ=8eV, t=10s, dotted), O_K (Δ=8eV, t=30s, dashed), Ge_L (Δ=25eV, t=60s, dot-dashed), Si_K (Δ=25eV, t=60s, double dot-dashed); (c) Si_L ELNES of Si (solid), SiGe (dotted) and SiO$_2$ (dashed and magnified dot-dashed); (d) shift of Si_L and Si_K onset (inflection point) along [001].

3. CONCLUSION

We have investigated SiGe/Si layers using spatially resolved electron energy-loss spectroscopy (SR-EELS). The Si_L signal can be used to record compositional profiles but does not allow calibration of the chemical signal without a standard specimen. The Ge_L / Si_K ratio is better suited for this but suffers from the smaller cross-sections (so that longer exposures are necessary) and the need for cross-correlating two spectrum images recorded separately.

As the spectrum image is squeezed within 120-180 lines on the CCD, it is crucial for the calculation of absolute values to deconvolute either the images by the detector point spread function or the profiles by the corresponding line spread function.

Si_L ESI maps can produce relative errors of up to 17% at interfaces of Si$_{1-x}$Ge$_x$/Si for $x\leq0.3$ (i.e. ±5at% Ge) due to deviations of the background from the inverse power-law behaviour in the SiGe alloy compared to Si. Typical energy drifts are irrelevant for a quantification.

SR-EELS can also be applied to distinguish locally (with nm-resolution) between the Si bonds of Si(Ge) and SiO$_2$, and the detection limits for the elements C and O in this operation mode have been determined to about 0.45nm C and 0.26nm SiO$_2$, while an improvement by a factor of 2-3 should in principal be possible.

ACKNOWLEDGEMENT

We thank DJ Robbins of the DERA Malvern, UK for growing the SiGe/Si structures used for illustrating the technique.

REFERENCES

Egerton RF 1996 EELS in the Electron Microscope (New York: Plenum, 2nd edition)
Golla U and Kohl H 1997 Micron **28**, 397
Jäger W and Mayer J 1995 Ultramicroscopy **59**, 33
Kelly MJ 1995 Low-dimensional semiconductors (Oxford: University Press)
Kimoto K, Sekiguchi T and Aoyama T 1997 J. Electr. Microsc. **46**, 369
Müllejans H and Bruley J 1994 Ultramicroscopy **53**, 351
Reimer L 1995 Energy-filtering Transmission Electron Microscopy (Berlin: Springer), p 30
Rez P 1982 Ultramicroscopy **9**, 283
Walther T, Humphreys CJ, Cullis AG and Robbins DJ 1995 Mater. Sci. For. **196-201**, 505
Walther T and Humphreys CJ 1999 J. Cryst. Growth **197**, 113

Inst. Phys. Conf. Ser. No 164
Paper presented at Microsc. Semicond. Mater. Conf., Oxford, 22–25 March 1999
© 1999 IOP Publishing Ltd

Morphological instability of strained-layer superlattices grown on vicinal substrates

G Patriarche, A Ougazzaden* and F Glas

France Télécom, Centre National d'Etudes des Télécommunications, Laboratoire Concepts et Dispositifs pour la Photonique (CNRS URA 250) and Laboratoire Composants pour l'Optoélectronique*, 196 avenue Henri Ravéra, BP 107, 92225 BAGNEUX Cedex, France

ABSTRACT: We studied by TEM how the morphological instability of misfitting epitaxial layers depends on the misorientation of the substrate with respect to (001) and on the sign of the misfit. For layers under compression, the surface non-planarity develops by bunching of a large fraction of the original substrate steps, whereas for layers under tension new steps are created.

1. INTRODUCTION

It has now been widely observed that the surface of a growing misfitting semiconducting layer tends to develop a non-planarity, which, in complex structures involving such layers, manifests itself by an interface non-planarity (Ponchet et al 1993). This is usually interpreted in terms of the morphological instability of half spaces under non-hydrostatic stress (Asaro and Tiller 1972, Grinfel'd 1986). In the continuum description of this effect, the driving force for surface undulation is the reduction of elastic energy which, beyond a certain critical undulation wavelength depending on the misfit, more than overcomes the increase of surface energy. Experimentally, these undulations seem frequent for misfits larger than about 1% and are very difficult to suppress for misfits of the order of 2% (Patriarche et al 1996), thereby confirming the paramount importance of stress in their generation. Several aspects of this phenomenon are however not well understood. In this paper, we address two of these open questions.

Any continuous surface undulation (as opposed to facetting) requires, at the atomic level, first the presence of steps and, second, a modulation of the spacing of these steps along the undulation direction (normal to the crests and troughs). Tersoff (1995) has however pointed out that there is a large energy barrier to the nucleation of new steps and that surface undulation should only develop by bunching of the steps existing before growth, namely those provided by a vicinal substrate.

It has also been observed that layers under tension and compression often behave differently with respect to the morphological instability (Xie et al 1994, Ponchet et al 1995, Okada et al 1997). Explanations based on strain-dependent step energies or surface atomic mobilities have been proposed, but the question remains controversial.

2. EXPERIMENTS

In order to investigate simultaneously these two related questions, we grew two series of epitaxial structures on vicinal InP substrates. Each sample consists of five layers L_i (i=1 to 5 starting near the substrate) of a misfitting $In_xGa_{1-x}As_yP_{1-y}$ alloy, separated by InP layers. These layers are all nominally 10 nm thick (thicker buffer and capping layers are also present). The alloy is under 1% compression (x=0.82, y= 0.7) in the first series and under 1% tension (x=0.53, y= 0.7) in the second. Within each series, the samples differ by the misorientation θ of the substrate with respect to (001); we studied misorientations of 0, 0.2, 0.5, 1.2 and 2° around a <100> direction (arbitrarily called [100] below). Each series was grown in a single metalorganic vapour phase epitaxy run at 600°C.

We studied mainly (110) and ($\bar{1}$10) cross-sectional specimens in a 200 keV Transmission Electron Microscope (TEM). We analysed diffraction contrast images taken in the symmetric **g,-g** condition for **g**=400. We concentrated on the interfaces I_i between the five misfitting layers L_i and the overlying InP layers. We assume as usual that each interface is geometrically identical to the corresponding free surface just before it was covered by InP. Strictly speaking, these experiments give access to the intersections of I_i with the (110) and ($\bar{1}$10) planes. Due to the overlap of the undulations in the various interfaces, plane-view images are difficult to interpret.

3. LAYERS UNDER COMPRESSION

3.1 Results

Except in the sample with θ=0°, the interfaces I_i are not planar. On the other hand, the upper surfaces of most intermediate InP layers are planar or only slightly undulating. This confirms that InP tends to fill the troughs of an underlying undulating misfitting layer (Patriarche et al 1996). These InP top surfaces provide a reference plane and the profile of I_i can be assessed by measuring the local thickness t_i of L_i. Along a <110> direction, the non-planarity of the layers manifests itself mainly by sudden drops in t_i. Between these drops, t_i varies slowly and linearly with lateral position (zones of uniform thickness also possibly exist). The phenomenon repeats itself roughly periodically at each interface of each sample and is thus characterised by a lateral wavelength Λ and by an amplitude h, the crest to trough height difference. This characteristic pattern of thickness modulation is of course best seen when its amplitude is large (Fig. 1).

The measurements performed in various areas of each specimen are somewhat dispersed, but the non-planarity shows definite properties (Fig. 2):

(i)   for all misorientations except 1.2°, the wavelengths cluster around a single value. For θ=1.2°, we find two wavelengths, depending on the area; the larger is not a multiple of the smaller.

(ii)   the wavelength does not depend on layer index i in any systematic fashion.

(iii) wavelengths and amplitudes are the same along [110] and [$\bar{1}$10].

(iv)   the wavelength decreases when θ increases (for θ=1.2°, we retain the smaller wavelength).

(v)   for a given θ, the amplitude increases with layer index i.

(vi)   for layers with a given index i, the amplitude increases with θ.

(vii) the interface becomes fuzzy in the region of the drops when the specimen thickness increases.

3.2 Discussion

Recall three basic facts. First, any globally planar interface must be parallel to the original misoriented InP substrate and contain the same type of steps (here arbitrarily termed 'down') with the same linear density $d'_0 = \theta/b$ along [010], where b is the step height. Second, any locally planar section of an interface must contain a uniform step density. Third, in any interface, the number of additional down and up steps introduced over an undulation wavelength must be equal.

The sudden drops in t_i must contain a large density of steps. If these were additional up steps, they would have to be compensated by additional down steps in the preceding region of slowly varying thickness. The latter would thus decrease towards the drop, which is not the case. Hence, the drops in t_i are bunches of down steps. (iii) and (vii) show that even if these steps are constituted of <110> segments, these are sufficiently short to ensure that, over distances of the order of 100 nm, the bunch is oriented along the original <100> misorientation axis. Assuming this holds for all steps, the observed step densities d and wavelengths Λ transform into densities d' and wavelengths Λ' along [010]: $d = d'\sqrt{2}$, $\Lambda = \Lambda'/\sqrt{2}$. The density of steps participating in bunches of height h' separated by Λ' is $d' = h'/b\Lambda'$ per unit length of interface. Comparison of this density at the top of the most non-planar layer (i=5) with the original density on the substrate shows that for any θ the proportion of bunched steps is higher than 70% (Table). These observations are easily interpreted in the following way: a portion of the original steps accumulates in bunches whose heights increase with layer index. No new step is created. The remaining steps (only about 10% for θ=2° and i=5) are evenly distributed

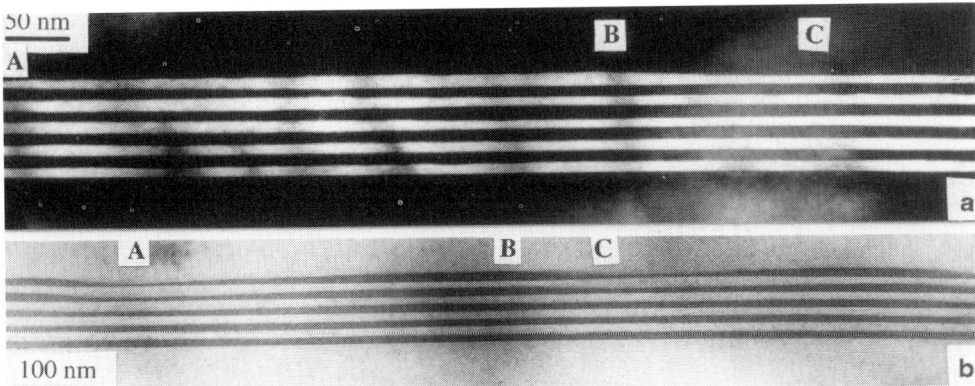

Fig. 1: TEM images of specimens with five layers under compression grown on misoriented InP substrates. (a) $\theta = 0.5°$, bright field image, layers brighter than InP; (b) $\theta = 2°$, dark field image, layers darker than InP. Note the linearly varying thickness of the top layer between A and B and the drop between B and C.

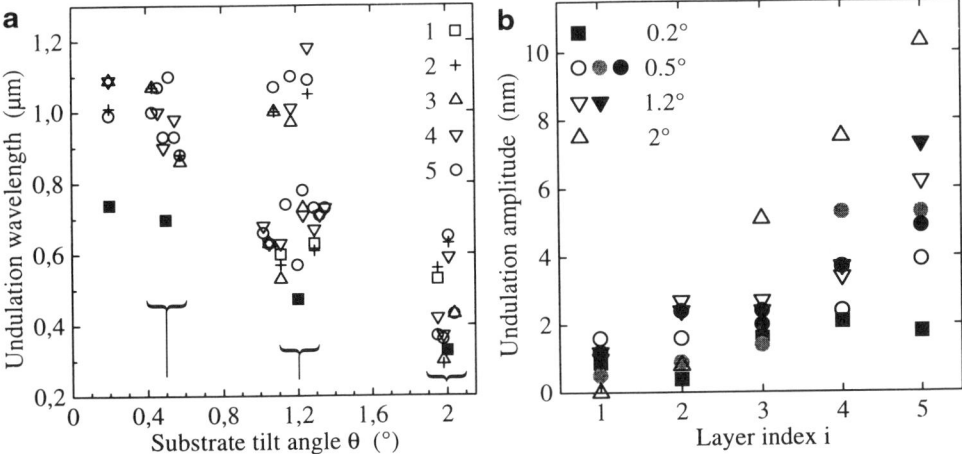

Fig. 2: (a) Variation with θ of the undulation wavelength measured at the top of each layer (key gives layer index i). For clarity, measurements in different areas of the same specimen are slightly offset along the θ axis; ■: average wavelength Λ' projected along 010. (b) Variation with layer index of the undulation amplitude measured in differently misoriented samples (as given by key, where different shadings of the same symbol correspond to different areas of the same specimen).

Fig. 3: TEM dark field image of a specimen with layers under tension. $\theta = 2°$, layers darker than InP.

in the rest of the interface (regions of linearly varying thickness).

θ (°)	d'$_0$ (μm^{-1})	h' (nm) for i=5	d'= h'/b Λ' (μm^{-1})	fraction of steps in bunches
0.2	11.9	1.8	8.3	0.70
0.5	29.8	5.3	25.7	0.86
1.2	71.4	7.3	52	0.73
2	119	10.3	106	0.89

Table: Variation with misorientation of the densities of original and bunched steps. Data from Fig. 2.

4. LAYERS UNDER TENSION

These layers develop a very different kind of non-planarity. These are narrow troughs separated by regions where the thickness of the layers varies only slowly, with a slight excess in the zones bording the troughs. The whole pattern is symmetrical around the centre axes of the troughs and of the intermediate plateaux (Fig. 3). This is very akin to the valley/mesa thickness variations observed by Ponchet et al (1995) in similar specimens. Here again, the amplitude increases with layer index and with θ. The most severely affected layers may locally be reduced to near zero thickness. The repeat distances are fairly dispersed, but for all layers we observe a shorter wavelength between 60 and 100 nm independently of θ, although in some areas only multiples of this distance can be observed. Calculations similar to those of section 3 show that here there must be many more steps involved in the troughs than originally present on the substrate (about 10 times for complete pinching of a layer every 60 nm). This and the symmetry of the thickness variations imply that both up and down steps are present along a <110> direction. The plane-view images suggests that the troughs are finite in all directions so that the steps are very likely closed circuits. Finally, shallow troughs are detected even in the θ=0° specimen.

This type of non-planarity is prominent in the images of the thin areas of the specimens. Because of their narrowness, the troughs tend to disappear in thicker areas. There, at least for $\theta \geq 0.5°$, we observe in addition undulations similar to those present in the samples under compression, with similar wavelengths in the μm range. We did not study this extra non-planarity in detail.

5. SUMMARY AND CONCLUSIONS

The main original points of the present work are the systematic study of the effect of the misorientation, the choice of specimens with misfitting layers all either under compression or tension, and the misorientation axis. In the layers under compression, the surface non-planarity of the misfitting layers develops only through bunching of the steps existing on the original substrate. Even for small misorientations, a large fraction of the latter may gather in widely separated bunches. In layers under tension, the surface non-planarity develops mainly by creation of an equal number of new up and down steps. These steps form closely spaced symmetrical U-shaped bunches. Independently, bunching of the initial steps seems to happen in some of these specimens as well.

REFERENCES

Asaro R J and Tiller W A 1972 Metall. Trans. **3**, 1789

Grinfel'd M A 1986 Sov. Phys. Dokl. **31**, 831

Okada T, Weatherly G C and McComb D W 1997 J. Appl. Phys. **81**, 2185

Patriarche G, Ougazzaden A and Glas F 1996 Appl. Phys. Lett. **69**, 2279

Ponchet A, Le Corre A, Godefroy A, Salaün S and Poudoulec A 1995 J. Cryst. Growth **153**, 71

Ponchet A, Rocher A, Emery J-Y, Starck C and Goldstein L 1993 J. Appl. Phys. **74**, 3778

Tersoff J 1995 Phys. Rev. Lett. **75**, 2730

Xie Y H, Gilmer G H, Roland C, Silverman P J, Buratto S K, Cheng J Y, Fitzgerald E A, Kortan A R, Schuppler S, Marcus M A and Citrin P H 1994 Phys. Rev. Lett. **73**, 3006

Inst. Phys. Conf. Ser. No 164
Paper presented at Microsc. Semicond. Mater. Conf., Oxford, 22–25 March 1999
© 1999 IOP Publishing Ltd

Self-assembled InSb quantum dots in InAs and GaSb matrices assessed by means of TEM, AFM and PL

P Möck[1], G R Booker[1], E Alphandéry[2], R J Nicholas[2] and N J Mason[2]

[1]Department of Materials, University of Oxford, Parks Road, Oxford OX1 3PH
[2]Department of Physics, Clarendon Laboratory, University of Oxford, Parks Road, Oxford OX1 3PU

ABSTRACT: Transmission electron microscopy (TEM), atomic force microscopy (AFM) and photoluminescence (PL) with and without additional magnetic fields (magneto-PL) were employed to study metal organic vapour phase epitaxy (MOVPE) grown, self-assembled InSb rich quantum dots (QDs) embedded in InAs and GaSb matrices. Depending on the growth conditions, coherently strained, partly relaxed, and completely relaxed InSb rich agglomerates were observed by means of TEM. AFM delivered more reliable QD number densities and accurate heights of InSb islands on GaSb. For the case of InSb embedded in GaSb, PL showed QD emission at 1.69 µm and wetting layer (WL) emission at 1.61 µm.

1. INTRODUCTION

Over recent years, there has been increasing interest in the growth and characterisation of semiconductor quantum dots (QDs) for potential use in opto-electronic devices. One of the most promising routes towards the fabrication of opto-electronic devices with low threshold current densities and improved temperature stability is heteroepitaxy in the Stranski-Krastanow growth mode. This process has been shown to produce in (In,Ga)As on GaAs (e.g. Seifert et al 1996) and (Si,Ge) on Si (e.g. Schmidbauer et al 1998) not only an array of QDs that is self-assembled, but also self-organised, if the QD number density is sufficiently high.

Little work has been done so far to extend the range of wavelength at which potential QD based opto-electronic devices might operate into the middle infrared region of the electromagnetic spectrum (Norman et al 1997, Tsatsul'nikov et al 1998, Alphandéry et al 1999). Embedding the III-V semiconductor with the smallest bandgap (i.e. InSb, 0.23 eV at 4K) and largest lattice constant (0.679 nm) into III-V semiconductor matrices with rather small bandgaps (i.e. InAs, 0.42 eV, and GaSb, 0.8 eV) with lattice constants that could produce up to either 6.9% (InSb in InAs) or 6.3% (InSb in GaSb) strain is thought to be a promising route to achieve this goal. The aim of this paper is to expand on our previous reports on such systems (Norman et al 1997, Alphandéry et al 1999) and present new experimental observations concerning the formation of self-assembled InSb QDs in GaSb and InAs matrices.

2. EXPERIMENTAL DETAILS

All samples of this study were grown by MOVPE at the Clarendon Laboratory in an atmospheric pressure reactor that is described by Booker et al (1997). An overview of structural and growth parameters of two series of samples is given in Table 1. Five seconds growth pauses under flowing H_2 were inserted in the growth sequence before and after deposition of InSb. One second InSb deposition time is equivalent to approximately an epilayer thickness of about 1 to 2 monolayers, assuming a uniform coverage. A JEOL JEM 200CX operating at 200 kV was used for

134

the TEM investigations and plan-view specimens were prepared by conventional Cl$_2$/methanol thinning. AFM was performed in the contact mode using a Parks Scientific Instrument SFM-BD2 scanning probe force microscope. PL and magneto-PL modulation spectroscopy was performed on samples of InSb in GaSb at the Clarendon Laboratory, employing user built equipment and magnetic fields of up to 15 T. The PL on samples of InSb in InAs was performed in the Advanced Materials and Photonics Groups at Lancaster University (A Krier, unpublished work).

sample names	substrate	buffer layer and growth temperature (°C)*	InSb growth temperature (°C) and deposition time	capping layer and growth temperature (°C)*
A	InAs	0.3 μm InAs	480 for 1 second	0.3 μm InAs
B	InAs	0.3 μm InAs	520 for 2 seconds	0.3 μm InAs
C	InAs	0.3 μm InAs	500 for 5 seconds	0.3 μm InAs
D	GaAs	1.95 μm GaSb	545 for 2 seconds	either 3 μm GaSb, or none
E	GaAs	1.8 μm GaSb at 545 + 0.15 μm GaSb at 480	480 for 3 seconds	either 25 nm GaSb at 480 + 3 μm GaSb at 545, or none

Table 1: Structural and growth parameters. *Growth temperatures were constant for all layers of a sample if not otherwise specified. Uncapped samples were grown for AFM and TEM only.

3. RESULTS AND DISCUSSION

3.1. Identification of InSb rich particles and islands in plan-view TEM images

The strain fields around the smallest strained InSb particles and islands (in cases of uncapped samples) appeared in 220 two-beam diffraction contrast images close to the 001 zone axis in all samples as typical black-white contrasts (marker BW in Fig.1, as observed previously for other systems by Ashby and Brown 1963, Glas et al 1987). Following McIntyre's and co-workers formula and estimating the constrained strain as to be two thirds of the lattice misfit (Ashby and Brown 1963), we can derive from the presence of this type of contrast upper limits of the particle diameters in the interface plane. If we assume that no alloying has taken place, we obtain an upper limit of 14.5 nm for the size of InSb particles and an upper limit of 18.1 nm for the size of In(Sb$_{0.5}$As$_{0.5}$) particles. For InSb particles in GaSb matrix an analogous estimation yields 14.8 and 18.7 nm. This estimation corresponds well with our observations of the smallest particle sizes (5 - 10 nm) under kinematical conditions at the [001] zone axis.

Fig. 1: Typical contrasts of InSb rich particles in InAs matrix, 220 dark field, sample B: black-white contrast for smallest particles (maker BW); black-lobe contrast (marker DL) and structure factor contrast (marker SF) of small particles.

Fig. 2: Structure factor and orientation contrast of medium size InSb rich islands on GaSb buffer layer, 220 dark field, sample E.

Since the direction vector of the black-white contrasts was found to be not uniformly oriented with respect to the diffraction vectors (although we can assume the same interface location for all particles), we conclude that the particle shapes are not uniform. Changing the direction of the diffraction vector would, as expected from Ashby's and Brown's theory, change the sense of the direction vector of the black-white contrast for embedded particles, but not for free standing islands on GaSb buffer layers. Applying Ashby's and Brown's rule, it can be concluded that the particles possess a bigger lattice constant than the matrix, i.e are indeed InSb rich.

Small InSb rich particles appeared in a wedge shaped foil under 220 two-beam diffraction conditions (Fig. 1) in a mixture of typical structure-factor (marker SF) and faint "dark lobe" contrast (marker DL), which is due to the remaining strain fields in the matrix. Note that "dark" refers to an excess of electrons on the recording film. We observed these types of contrast only for the latter two samples (B and C) of the InSb/InAs series and conclude that the higher growth temperatures and longer deposition times led to significant alloying of both the WLs and the QDs, leading in the latter case probably to increased particle sizes with lower InSb content. A comparison of extinction distances of the 220 and 400 reflections of the (In,Ga)(Sb,As) alloy system shows that these quantities are for 200 kV electrons only slightly (i.e. order of magnitude 1%) dependent on the composition. Thus, a possible alloying effect cannot be quantified by measurements of the extinction distances of the particles and the matrix. The structure factors (F), on the other hand, vary more strongly (i.e order of magnitude 10%) with alloy composition, but without densitometric quantification of the contrasts, we can only conclude that the small particles in Fig. 1 possess a higher InSb content than the matrix, since $|F_{220}|^2$ is about 37.4 nm^2 for InSb and 25.1 nm^2 for InAs.

Figure 2 shows a mixture of structure factor contrast and orientation contrast around medium size InSb islands (diameters $\approx$ 50 - 75 nm) on a GaSb buffer layer (sample E). This image has been taken under 220 dark-field conditions and shows strain fields (i.e. orientation variations) in the substrate around these sorts of islands. The general trend was that the bigger the medium size islands were, the smaller the orientation contrast halo around them would be. Another 220 dark-field image (not shown) from the same sample revealed that a much bigger island of about 0.15 μm^2 base area possessed a uniform array of dislocations with a spacing of about 3.4 nm, indicating complete relaxation via 60° dislocations. We infer from the latter observation that large InSb islands are not alloyed to a significant amount before a capping layer is grown. Since it is not precisely known how much possible alloying may affect medium size islands when the capping layer is grown, estimations of the amount of relaxation via the analysis of Moiré fringes have to be undertaken with caution and no further attempts to quantify relaxation effects have been made. In addition, assuming the bulk lattice constant of the substrate is problematic for such estimations, Glas et al (1987) have shown that the area underneath a strained island is elastically dilated.

3.2. PL analysis of the sample series InSb/InAs

PL from the samples of this series showed up to five peaks at 3.02 μm , 3.1 μm, 3.27 μm, 3.5 μm, and 4 μm. While the first three of these peaks were previously attributed to bulk InAs, the latter two peaks possessed full widths at half maximum (FWHM) of 26 - 30 meV and were previously thought to be caused by InSb QDs (Norman et al 1997). However, only a very low density of QDs ($\approx 10^7$ - 10^8 cm^{-2}) was observed by TEM for the samples of this series. This density is up to about two orders of magnitude smaller than the one for which we previously obtained qualitative similar PL results (Norman et al 1997). From this we assume that the alloyed In(Sb,As) WLs, effectively forming quantum wells, could be responsible for the PL peaks at 3.5 μm and 4 μm.

3.3. AFM and PL analyses of the sample series InSb/GaSb

AFM performed on uncapped specimens, revealed the co-existence of three different types of three-dimensional InSb islands, and enabled accurate measurements of the islands' heights as well as estimations of their base area (Alphandéry et al 1999). A brief summary of AFM results is given in table 2. PL was performed on capped structures at 4 K and showed typically WL emission at 1.61μm and QD emission at 1.69 μm for sample E only. Magneto-PL was used to assign the QD and WL peaks and to deduce the height and base widths of the luminescent QDs. The values obtained were in reasonable agreement with TEM and AFM measurements on the smallest InSb rich islands (Alphandéry et al 1999).

sample names	size of smallest strained islands, base area (nm^2) / height (nm)	number density of small (strained) islands (cm^{-2})	size of smallest relaxed islands, base area (μm^2) / height (nm)	number density of relaxed islands (cm^{-2})
D	1000 / 3	$2 - 6 \bullet 10^7$	0.1 / 17	$0.6 - 1 \bullet 10^9$
E	400 / 3.5	$3 - 5 \bullet 10^9$	0.006 / 8	$0.8 - 1.2 \bullet 10^{10}$

Table 2: Summary of AFM results

As a working hypothesis for the interpretation of AFM images such as the one presented by Alphandéry et al (1999), we followed Seifert's and co-workers' speculation that a barrier of the chemical potential is surrounding every strained island and this allowed us to distinguish between fully strained and partly relaxed islands. According to this hypothesis, the substrate around the vicinity of a strained island should be under compression, and we indeed observed orientation contrast halos in the vicinity of medium size islands (see Fig. 2). AFM and TEM measurements of the base area of strained medium size islands are comparable and we consider this working hypothesis now as being confirmed by indirect means.

As the samples of the InSb/GaSb series were grown on pseudo-substrates, we observed dislocations and stacking fault tetrahedra in the buffer layers by means of TEM and pits of up to 200 nm diameter and 30 nm depth by means of AFM for sample E. Reasonably good QD PL properties were however obtained (Alphandéry et al 1999) and this suggests, in agreement with results by Pan et al 1999, that the formation of QDs is not too critically dependent on the surface morphology.

4. CONCLUSIONS

In conclusion, self-assembled InSb rich QDs were grown by means of MOVPE in InAs and GaSb matrices. There is a wide variety of InSb particle and island sizes and shapes. This is thought to be the result of the low number densities in most of the samples, which may hinder self-organisation processes. The QDs in the InAs matrix were observed to be somewhat smaller than their counterparts on GaSb buffer layers. This might be the result of both size reduction due to the capping process and the higher maximal strain in the case of the InAs matrix.

REFERENCES

Alphandéry E, Nicholas R J, Mason N J, Zhang B, Möck P and Booker G R 1999, Appl. Phys. Lett. **74**, 2041
Ashby M F and Brown L M 1963, Phil. Mag. **8**, 1083, 1649
Booker G R, Daly M, Klipstein P C, Lakrimi M, Kuech T F, Jiang Li, Lyapin S G, Mason N J, Murgatroyd I J, Portal J C, Nicholas R J, Symons S G, Vicente P and Walker P J 1997 J, Cryst. Growth **170**, 777
Glas F, Guille C, Hénoc P and Houzay F 1987, Inst. Phys. Conf. Ser. No. **87**, 71
McIntyre K G and Brown L M 1966, J. Phys. Radium **27**, C3
Norman A G, Mason N J, Fisher M J, Richardson J, Kier A, Walker P J and Booker G R 1997, Inst Phys. Conf. Ser. No. **157**, 353
Pan D, Xu J and Towe E 1999, J. Cryst. Growth **196**, 23
Schmidbauer M, Wiebach Th, Raidt H, Hanke M, Köhler R, Wawra H 1998, Phys. Rev. **B 58**, 10523
Seifert W, Carlsson N, Miller M, Pistol, M-E, Samuelson L and Wallenberg L R 1996, Progr. Cryst. Growth and Charcter. **33**, 423
Tsatsul'nikov A F, Ivanov S V, Kop'ev PS, Kryganovskii A K, Ledentsov N N, Maximov M V, Mel'tser B YA, Nekludov P V, Suvorova A A, Titkov A N, Volovik B V, Grundmann M, Bimberg D, Alferov Zh I. 1998, J. Electron. Mat. **27**, 414

Inst. Phys. Conf. Ser. No 164
Paper presented at Microsc. Semicond. Mater. Conf., Oxford, 22–25 March 1999
© *1999 IOP Publishing Ltd*

TEM characterisation of free standing GaInP Stranski-Krastanow islands grown by MOVPE on GaP substrates

C Svensson, J-O Malm, A Gustafsson[1], L K L Falk[2], J Johansson[1], C Thelander[1] and W Seifert[1]

National Centre for HREM/Inorganic Chemistry 2, Lund University, S-221 00 Lund, Sweden
[1]Solid State Physics/nm-Structure Consortium, Lund University, S-221 00 Lund, Sweden
[2]School of Physics and Engineering Physics, Chalmers University of Technology, S-412 96 Göteborg, Sweden

ABSTRACT: GaInP Stranski-Krastanow islands have been grown on GaP substrates. The fabricated islands can be divided in three classes: flat rounded islands, small faceted islands and larger faceted islands. The predominant class is the small faceted islands, which are on average 10 nm high and have a base width of approximately 30 nm. The Ga/In ratio in these islands is 40/60.

1. INTRODUCTION

The growing interest in nanometer-sized semiconductor structures has resulted in the development of techniques and procedures for preparing Stranski-Krastanow (SK) islands from various materials (see e.g. Seifert et al 1996). In the SK growth mode, the lattice mismatch (generally >3%) between the substrate and the island material results in a self-assembly of the islands. Nevertheless, there are only a few reports of GaInP islands grown on GaP (Lee et al 1997). This system is of special interest due its potential for integration of optoelectronic components into Si based structures (the lattice mismatch between Si and GaP is only 0.4%). This study focuses on the fabrication and characterisation of GaInP islands on GaP. The GaInP islands have been prepared by metal organic vapour phase epitaxy (MOVPE) and characterised by transmission electron microscopy (TEM), and atomic force microscopy (AFM).

2. EXPERIMENTAL

The fabrication was carried out in a low-pressure (100 mbar) MOVPE reactor. After growing a 160 nm thick GaP barrier layer on GaP (001) at 850°C and 0.5 monolayers/s, the GaInP was deposited with a nominal Ga/In ratio of 1 ($Ga_{0.5}In_{0.5}P$). A series of samples with GaInP deposited to different thicknesses (2, 4, 6, 8 and 15 monolayers) at two temperatures (600 and 650°C) were prepared using a deposition rate of 1.0 monolayer/s.

The samples intended for TEM and AFM studies were annealed for 12 s at the growth temperature and then cooled down. Some samples were overgrown with a 90 nm thick GaP layer and studied by photoluminescence (PL). These results will be reported elsewhere.

The TEM studies were performed on uncapped samples grown at both temperatures with a GaInP layer of 6 monolayers. The size, shape, and structure of the 3D islands were studied by high resolution TEM (HRTEM) in cross-section. To evaluate the shape of the islands the samples were investigated in both the [110] and the [1-10] direction. The orientation was determined from the flats of the substrate wafer, and two sets of samples with different orientations were prepared from each growth run. The instrumentation used was a TEM operated at 400kV with a structural resolution of better than 0.17 nm. The local composition of the islands was studied by energy dispersive x-ray spectroscopy (EDS) using a field emission gun TEM operated at 200kV and with a nominal beam diameter of less than 1 nm at FWHM. For the preparation of cross-sectional samples for EDS analyses it was found that the most suitable glue was Araldite (as compared to Gatan G1 and M-Bond 610). Using Araldite, parts of the glueline covering the uncapped SK-island could be dissolved by dipping the ion-milled specimen in acetone for a few minutes, thus leaving uncovered island for analysis. This procedure minimises contamination induced by the interaction of the electron beam with the glue.

The AFM images were recorded in high resolution tapping mode using a super-tip.

3. RESULTS

The islands can be divided into three classes: flat rounded islands, small faceted islands and larger faceted islands. All three kinds of island are present in both samples. The flat rounded islands have the shape of a watch glass slightly elongated along the [110] direction. The small faceted islands are the dominant features in the samples. They are about 10 nm high, being predominantly terminated by {111} and {110} surfaces and the top (001) surface. The third class of islands is less well-defined, containing all larger faceted islands. Here we find islands with heights up to 60-70 nm. Still these islands are terminated by the same surfaces as the smaller ones. However, the ratio between the area of the exposed surfaces can differ from the larger to the smaller islands.

The overall geometry of the islands is exemplified in fig. 1. This figure shows an AFM- and a TEM-image of two of the larger islands. The surface labelled A is the top (001) surface, the surfaces labelled B are {111}B surfaces, the surfaces labelled C are {110} surfaces and the small surfaces labelled D are {111}A surfaces. The small {111}A surfaces labelled D are more pronounced in the larger islands (cf. the smaller island below the labelled larger one). This geometry shows similarities with the geometry previously found for InP islands grown on GaInP (Georgsson et al 1995).

From the AFM image in fig 1a the faceted islands appear to be elongated along the [110] direction. This seems to be the case for some of the larger islands, but measurements from cross-sectional TEM along [110] and [1-10] respectively show no apparent elongation when studying the smaller faceted islands. These islands have

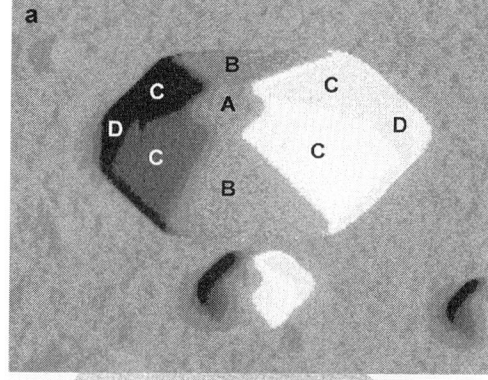
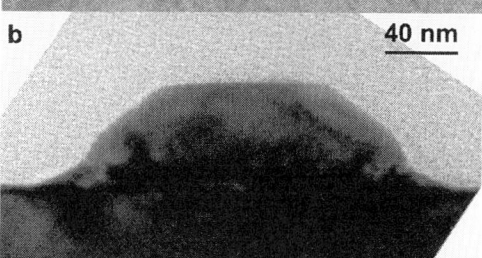

Figure 1. Figure a shows an AFM image of one of the larger faceted GaInP islands. The facets can be identified as: A=(001); B={111}B; C={110}; D={111}A. Figure b is a cross-sectional TEM image of a similar island.

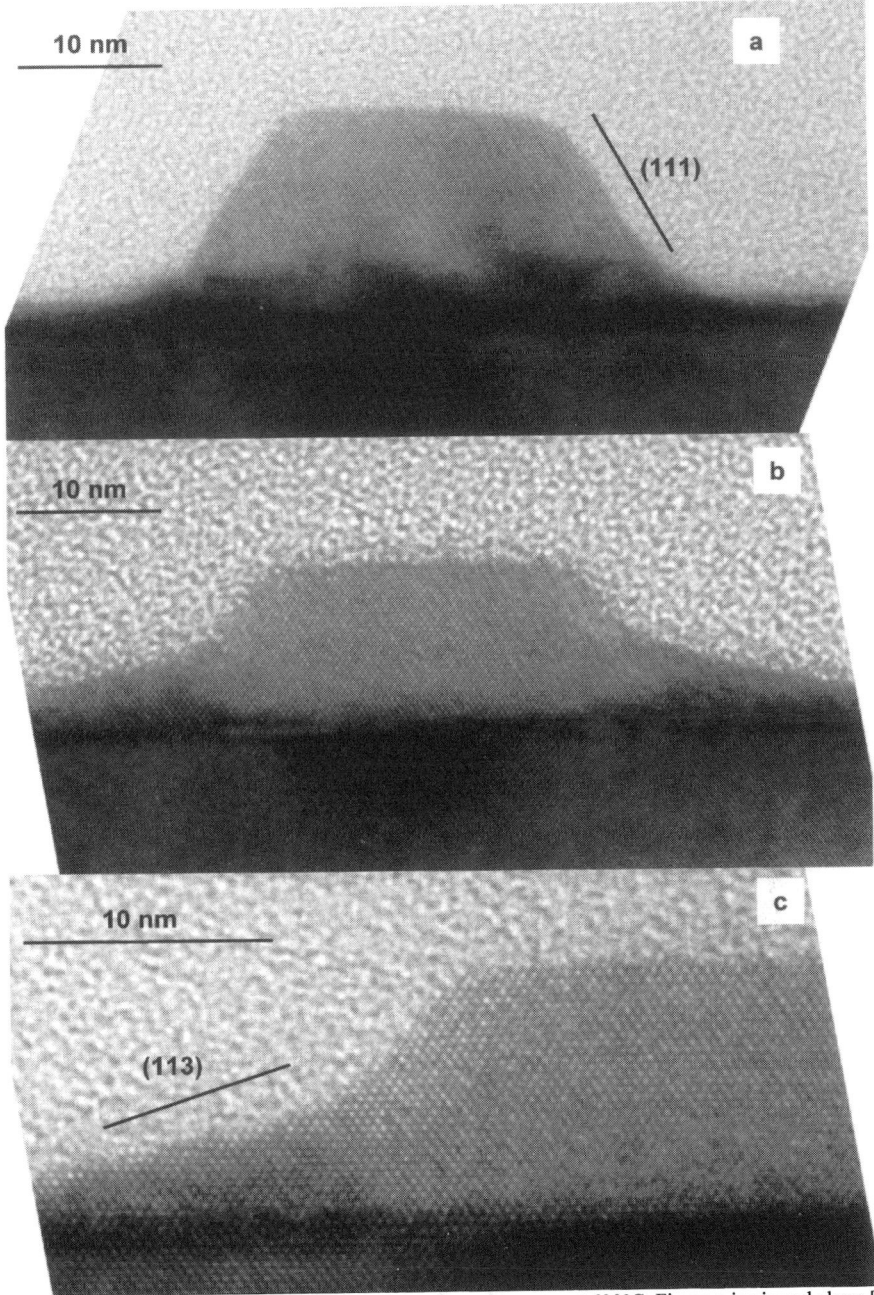

Figure 2. Three cross-sectional TEM images from a sample grown at 600°C. Figure a is viewed along [110] showing no or very small tails (see text). In this direction the terminating surfaces are {111}B (sides) and (001) (top). Figure b is viewed along [1-10]. In this direction the tails are more accentuated indicating an elongation in the perpendicular direction. The surfaces terminating the top of the tails have indices close to {113} (see fig c).

an average height of 10 nm, and the projected base width is around 30 nm in both directions.

In many cases faceted islands appear to be situated on top of the flat rounded islands. This configuration will, when observed in cross-sectional TEM, generate an image of a faceted island with tails extending from the base of the island (see fig. 2b). Since the flat rounded islands are preferably elongated in the [110] direction, these tails will be most accentuated when imaged perpendicular to this direction (i.e. along the [1-10] direction). The geometry of these tails differs depending on preparation temperature. In the case of the samples prepared at 600 °C the tails have a more well-defined geometry and the upper terminating surface can in most cases be identified as a {113} surface. The samples prepared at 650 °C have tails that are smaller and have a less well-defined terminating upper surface.

The TEM investigation did not show any evidence of long-range ordering in the islands, which has been observed for bulk material.

To study the composition of single islands, an EDS investigation was performed. By placing the electron beam a few nanometers below the top surface of the islands, keeping a safe distance from the substrate, and using only thin regions, reliable analyses of the composition of single islands could be made. To further ensure the absence of substrate contribution in the EDS spectra, a glue less prone to contamination by electron beam interaction (Araldite) was used. The EDS analyses show an average Ga/In atomic ratio of 40/60 in the small faceted islands, while EDS analyses of the tails suggest a gallium enrichment. For the larger faceted islands several analyses could be performed within single islands. Some compositional variations were found within the islands, but there is no evidence for any consistent compositional trends.

4. CONCLUSIONS

It is shown that the GaInP islands have a size ranging from a few nm to 60-70 nm measured as the height of the islands. Nevertheless, most islands have a height of approximately 10 nm and are clearly faceted. The base width of such islands is generally close to 30 nm as measured by cross-sectional TEM. The terminating surfaces are (001) (top surface), {111} (sides) and {110} (sides). The geometry is given in fig 1a.

Elemental analyses of the smaller faceted islands show an approximate composition of $Ga_{0.4}In_{0.6}P$ although the input ratio between Ga and In would nominally yield a composition of $Ga_{0.5}In_{0.5}P$. These islands are in most cases situated on top of a flat rounded island. The flat islands are elongated along the [110] direction resulting in cross-sectional TEM images like fig. 2b and c, when imaged perpendicular to the elongation. The tails extending from the base of the faceted islands show the flat islands in projection. These tails appear to be enriched in Ga and the top terminating surface is close to {113}.

No evidence for long-range chemical ordering in the islands has been seen during this investigation.

This work was supported by grants from the Swedish research councils SSF, TFR, and NFR.

REFERENCES

Georgsson K, Carlsson, N, Samuelson L, Seifert W and Wallenberg LR 1995 Appl. Phys. Lett. **67**, 2981

Lee JW, Schremer AT, Fekete D, Shealy JR and Ballantyne JM 1997 J. Electronic Materials **26** 1199

Seifert W, Carlsson N, Miller M, Pistol M-E, Samuelson L, Wallenberg LR 1996 Prog. Crystal Growth and Charact. **33**, 423

Inst. Phys. Conf. Ser. No 164
Paper presented at Microsc. Semicond. Mater. Conf., Oxford, 22–25 March 1999
© *1999 IOP Publishing Ltd*

Impact of surface morphology on InAs(P) island formation on InP

C F Carlström, S Anand*, E Niemi and G Landgren

Royal Institute of Technology, Dept. of Electronics, Laboratory of Semiconductor
Materials, Electrum 229, S-164 40 Kista, Sweden
* Phone : +46 8 752 14 70, Fax : +46 8 752 12 40, E-mail : anand@ele.kth.se

ABSTRACT: The influence of the surface morphology on island formation by the As/P exchange reaction on InP surfaces is investigated. It is demonstrated that island formation is extremely sensitive to the physical nature of the surface. Variation in the surface RMS roughness, caused by controlled ion beam etching (shallow) under different conditions, is shown to drastically affect island formation. The determined total island volume increases with the surface roughness, which is consistent with increased surface area available for the As/P exchange reaction. Modification of surface morphology is suggested as an alternative route to vary the island size distribution.

1. INTRODUCTION

InAs on InP are attractive for opto-electronic applications. These dots (islands) can be formed either by conventional deposition of InAs on InP (Taskinen et al 1997, Carlsson et al 1998), or by using the As/P exchange reaction (Wang et al 1998). For device applications, good control over the size and density of the dots is desirable. The effects of parameters such as deposited layer thickness, deposition rate, etc. on the InAs/InP island distribution has been studied (Taskinen et al 1997, Carlsson et al 1998). However, investigations of the dependence of island formation on substrate morphology are limited (Varma et al 1997). The role of substrate morphology on island formation could be different for the As/P exchange reaction.

Controlled modification of the surface roughness is essential for such an investigation. Ion beam etching is one possible method. Recently, we reported on a chemically assisted ion beam etching (CAIBE) process based on $N_2/CH_4/H_2$ chemistry for etching of InP (Carlström et al 1998). In this process, polymer formation is negligible and depending on ion energy and etching duration, the surface morphology can be modified from being extremely smooth (rms. roughness < 1 nm) to being reasonably rough (RMS roughness ~10 nm). It is therefore possible to study systematically InAs(P) island formation on differently prepared InP surfaces. In this work the dependence of InAs(P) island formation on the substrate morphology is investigated.

2. EXPERIMENTAL

The samples used in this work, are epitaxial InP grown by metal organic vapour phase epitaxy (MOVPE) on epi-ready InP substrates. $N_2/CH_4/H_2$ based CAIBE is used for controlled modification of the substrates. The ion energies varied from 75 eV to 500 eV, and the etch depth was restricted to about 100 nm to minimise stoichiometric damage. The morphologies of the etched surfaces were quantified by roughness measurements using tapping mode atomic force microscopy (AFM). The islands were formed using the As/P exchange reaction by annealing the InP samples in an AsH_3/PH_3 ambient. The annealing was performed in a MOVPE reactor at 650°C, and during the entire annealing cycle, the AsH_3 and PH_3 gas flows were fixed at 5 sccm and 105 sccm, respectively. For studying the effect of As/P exchange reaction on as-grown surfaces an InP layer was first grown at 650°C, cooled to 250°C, and then annealed at 500°C or 650°C under AsH_3/PH_3 flow for 10 min. The island formation on the samples was characterised by tapping mode AFM.

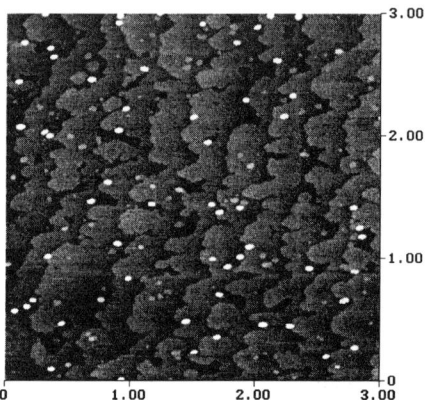

Fig. 1.   3µm x 3µm AFM scan of as-grown epi-InP surface annealed in PH_3 + AsH_3 flow at 650 C for 10 min.

3. RESULTS AND DISCUSSION

Figure 1 shows the typical AFM image of the 650°C annealed epitaxial InP surface. The islands are homogeneous in height (4-8 nm), and were about 70 nm in diameter, which represents an upper bound due to tip-induced widening. The island density varied over the wafer; $4\text{-}6\times10^8$ cm^{-2} at the centre and about 2×10^9 cm^{-2} near the periphery. in addition to some wide 'flat' islands which are about 1-2 ML thick, irregular step-like roughening is seen. In contrast, on samples annealed at 500°C islands were absent. For gas flows used here, the As/P exchange reaction is clearly more effective at 650°C and results in island formation. However, a systematic investigation of gas flows, annealing time, and temperature is not attempted in this study. Here, these parameters were fixed to isolate the effect of surface morphology on island formation. Below we discuss our results on island formation on InP surface etched by CAIBE.

For low ion energy (75 eV), the as-etched surfaces are very smooth with rms. roughness < 1 nm. In striking contrast, the island formation on this sample (Fig. 2 (a)) differs dramatically from that on the as-grown sample (Fig. 1). Here, small islands (4-8 nm high), as well as larger islands (30-40 nm high), are present; their densities are $6\text{-}8\times10^9$ cm^{-2} and $1\text{-}3\times10^9$ cm^{-2} respectively. In addition to these islands, much larger islands (~100 nm high) appear in the sample etched at 150 eV, but they occur at a very low density (~10^7 cm^{-2}). Increasing the ion energy to 300 eV makes the surfaces rougher (RMS. 2-3 nm) and remarkably, after annealing only very large islands are present (Fig. 2(b)), some of them as high as 300 nm. Similar results were obtained on surfaces etched at 500 eV. The above results clearly demonstrate impact of surface morphology on the InAs(P) island formation. The rougher the surface, higher is the incorporation of As in the material. The observed differences in island size and distribution for the different surfaces is qualitatively consistent with available surface area for As/P exchange reaction.

Fig. 3. shows the determined total volume of the islands and rms. roughness of the pre-annealed surfaces for different ion energies. The total island volume was determined by computing the bearing volume using nanoscope software. As seen in Fig. 3, both island volume and RMS roughness increase with ion energy. The data in Fig. 3 represent averaged values obtained over 4μm x 4μm scans at 9 different spots on each sample. The dependence of the island volume plotted on the RMS roughness is shown in the insert. The data clearly shows that the total island volume increases with the rms. roughness and is qualitatively consistent with increased surface area available for reaction. From the island volume determined for surfaces etched at 300 eV and 500 eV and the total surface area of the etched surface, the equivalent 2D film thickness is estimated as 7 and 14 nm, for 300 eV and 500 eV cases, respectively. It must be mentioned that the bearing volume obtained for the other samples represent only an upper bound since tip induced widening causes overestimation of the island lateral dimensions.

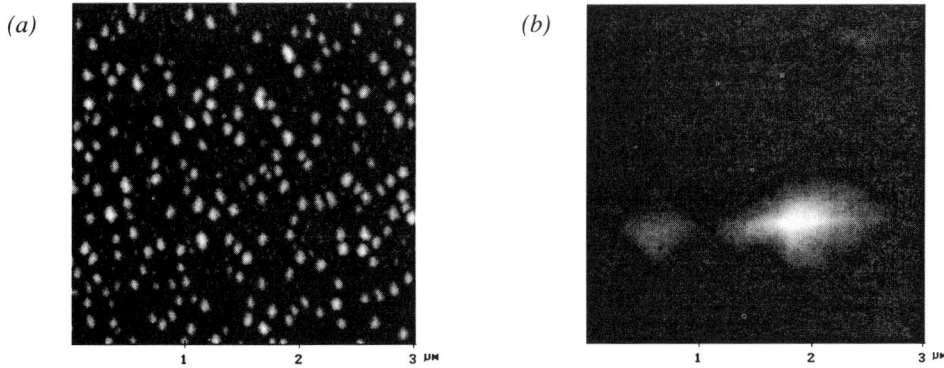

(a) *(b)*

Fig. 2.   3μm x 3μm AFM scans of differently processed InP surfaces annealed in PH₃ + AsH₃ flow. (a) etched at an ion energy of 75 eV (b) etched at an ion energy of 500 eV.

However, dry etching is known to induce stoichiometric damage due to physical sputtering component increasing with energy. In the case of InP preferential phosphorus removal is expected. Auger electron spectroscopy analysis of the as-etched surfaces suggests that the stoichiometry of the differently etched surfaces is not significantly different. One possible reason for this could be due to higher etch rates at higher energies results in rapid removal of the damaged layer as it is formed. That is, the residual damage may not be significantly different. This agrees with our damage evaluation studies (Anand et al 1998). Although our results show that island formation is related to microscopic roughness, more experiments are needed to fully understand the exact mechanism and chemical composition of the islands.

144

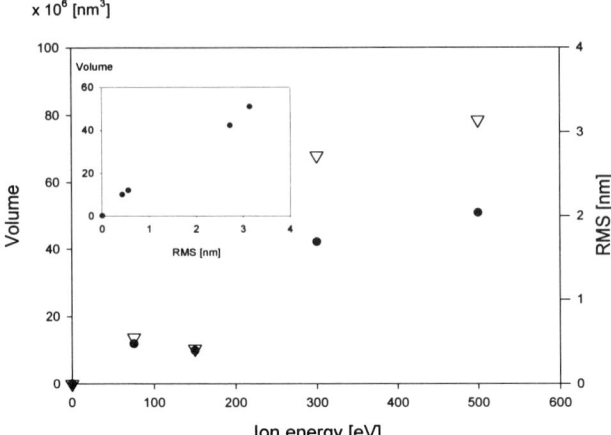

Fig. 3. Bearing volume of the islands and as-etched RMS roughness versus ion energy. The volume is normalised to a 1μm x 1 μm scan area. Insert: The normalised bearing volume as a function of the RMS roughness of the pre-annealed surface. The results for the as-grown sample is shown at 0 eV(main Figure) and at 0 RMS (on the insert).

(●) Bearing volume; (∇) *RMS roughness*

4. CONCLUSION

We have investigated the influence of InP substrate morphology on the InAs(P) island formation by the As/P exchange reaction. The island formation was shown to be extremely sensitive to the physical nature of the surface. The total island volume was found to scale with surface RMS roughness, due to increased available area for the reaction. Finally, our results suggest that modifying the substrate morphology could be an alternative route to vary the island size distribution.

ACKNOWLEDGEMENT

The authors gratefully acknowledge B. Stålnacke and A Patel for MOVPE operation. This work was financially supported by grants from the Foundation for Strategic Research (SSF) and Tekniska Forsknings Rådet (TFR).

REFERENCES

Anand S, Carlström C.F, Landgren G, Söderström D and Lourdudoss S 1998 Proc. InP and Related Materials, Tsukuba, Japan, pp 175-8

Carlsson N, Junno T, Montelius L, Pistol M-E, Samuelson L and Seifert W 1998 J. Crystal Growth **191**, 347

Carlström C.F, Landgren G and Anand S 1998 J. Vac. Sci. Technol. **B16(3)**, 1018

Taskinen M, Sopanen M, Lispanen H, Tulkki J, Tuomi T and Ahopelto J 1997 Surf. Sci. **376**, 60

Varma S, Reaves C M, BresslerHall V, DenBaars S P and Weinberg W H 1997 Surf. Sci. **393**, 24

Wang B, Zhao F, Peng Y, Jin Y, Li Y and Liu S 1998 Appl. Phys. Lett. **72**, 2433

Inst. Phys. Conf. Ser. No 164
Paper presented at Microsc. Semicond. Mater. Conf., Oxford, 22–25 March 1999
© *1999 IOP Publishing Ltd*

Computer-aided analysis of TEM micrographs of CdSe quantum dots on ZnSe

W Neumann, H Kirmse, R Schneider, K Scheerschmidt*, D Conrad*, T Wiebach and R Köhler

Humboldt University of Berlin, Institute of Physics, Invalidenstraße 110, D-10115 Berlin, Germany
* Max Planck Institute of Microstructure Physics, Weinberg 2, D-06120 Halle, Germany

ABSTRACT: Plan-view TEM diffraction contrast images of CdSe/ZnSe quantum dots generated by a modified MBE growth regime are interpreted by corresponding computer simulations. The relationship between experimental diffraction contrast features and possible shapes and orientations of quantum dots is verified. Additionally, finite-element method calculations of the strain field surrounding the CdSe quantum dots were carried out to consider its contribution to the image contrast.

1. INTRODUCTION

Self-organized growth of quantum dots (QDs) has been described for a number of II-VI compounds. The state of the art is reviewed by Bimberg et al (1998). In order to have a thorough understanding of the electrical and optical properties of real QDs the knowledge of their geometrical structure and their chemical composition is necessary.

Using a specific molecular beam epitaxy (MBE) growth regime, CdSe QDs can be formed on ZnSe as verified in-situ by reflection high-energy electron diffraction (RHEED) (Rabe et al 1998). The CdSe/ZnSe QDs were investigated in detail by transmission electron microscopy (TEM), especially by plan-view and cross-section diffraction contrast imaging. In addition, their existence was proved by analyzing of digitized high-resolution TEM images and by energy dispersive X-ray spectroscopy (EDXS) (Kirmse et al 1998).

Reliable interpretation of the diffraction contrast features as well as of the high-resolution contrast images of the QDs requires respective contrast simulations. For the CdSe/ZnSe QDs under investigation the experimental TEM findings and first simulations hint to the presence of truncated pyramids with the edges of the basal plane orientated parallel to $\langle 100 \rangle$ (Kirmse et al 1999). Here, the influence of the particular type of facet on the image contrast is studied. Moreover, since the image contrast observable in the QD regions is strongly affected by the strain field, accompanying finite-element method (FEM) calculations were performed.

2. EXPERIMENTAL

CdSe/ZnSe QDs were generated via thermal activation during growth interruption after CdSe deposition. Details of this modified MBE growth regime are described elsewhere (Rabe et al 1998). Plan-view TEM samples were prepared by conventional mechanical thinning (lapping, dimpling) from the back to preserve the QD layer. Electron transparency was attained by Xe$^+$ ion milling where the acceleration voltage was finally decreased to 0.7 kV in order to minimize the formation of preparation artifacts.

TEM investigations were performed on a HITACHI H-8110 operating at 200 kV. Diffraction contrast simulations were carried out by means of the EMS software package (Stadelmann 1987) while the computer program MARC was applied to the FEM calculations.

3. TEM DIFFRACTION CONTRAST INVESTIGATIONS

The plan-view bright-field image of Fig. 1 represents an arrangement of contrast features arising from CdSe/ZnSe QDs. The size of the features varies from 5 to 50 nm and the average area density was found to be about 100 μm^{-2}. Cross-section diffraction contrast images of these QDs yielded an average size of about 15 nm in width and about 5 nm in height (Kirmse et al 1998).

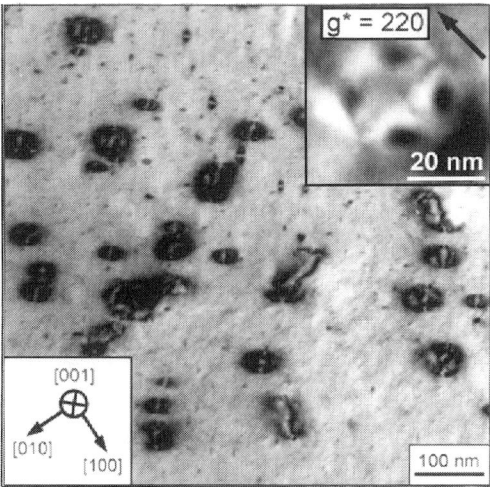

The inset of Fig. 1 shows the contrast feature of a single QD exhibiting a four-fold symmetry suggesting a pyramidal shape. Moreover, there seems to be an orientation of the edges of the basal plane parallel to the ⟨100⟩ directions.

Contrary to this assumption is the observed fine structure (diffuse scattering) along ⟨110⟩ in the diffraction pattern of the [001] zone axis. From that point of view one can conclude that the shortest dimensions of the pyramids, viz. the edges of the basal plane, are orientated parallel ⟨110⟩.

To find out which orientation is present (either parallel ⟨100⟩ or ⟨110⟩) plan-view diffraction contrast images were simulated.

Fig. 1. Plan-view TEM bright-field image of CdSe/ZnSe quantum dots, inset: bright-field image of a single dot.

4. DIFFRACTION CONTRAST SIMULATIONS

Structure models of the QDs consisting of a truncated pyramid with different facets were constructed for the simulations. The structures were relaxed by molecular dynamics calculations where the interatomic forces were described by Stillinger-Weber potentials (Stillinger and Weber 1985).

Diffraction contrast simulations of [001] zone axis images of the supercells having about 7 nm thickness were performed by multi-slice calculations. A small contrast aperture of about 2 nm^{-1} was chosen permitting that only the 000 beam contributes to the contrast feature. According to the HITACHI microscope the following parameters were chosen: U = 200 kV, spherical aberration c_s = 1 mm, beam semi-convergence α = 0. 5 mrad, defocus Δ = -25 nm, and defocus spread δ = 5 nm.

The atomic arrangements of three different truncated CdSe pyramids on a 2 MLs thick CdSe layer are shown in Fig. 2 where pyramids having {101}, {113}, and {112} facets, resp., are illustrated. The results of the corresponding diffraction contrast simulations of the ZnSe supercells clearly exhibit a strong dependence on the particular facetting. The orientation and

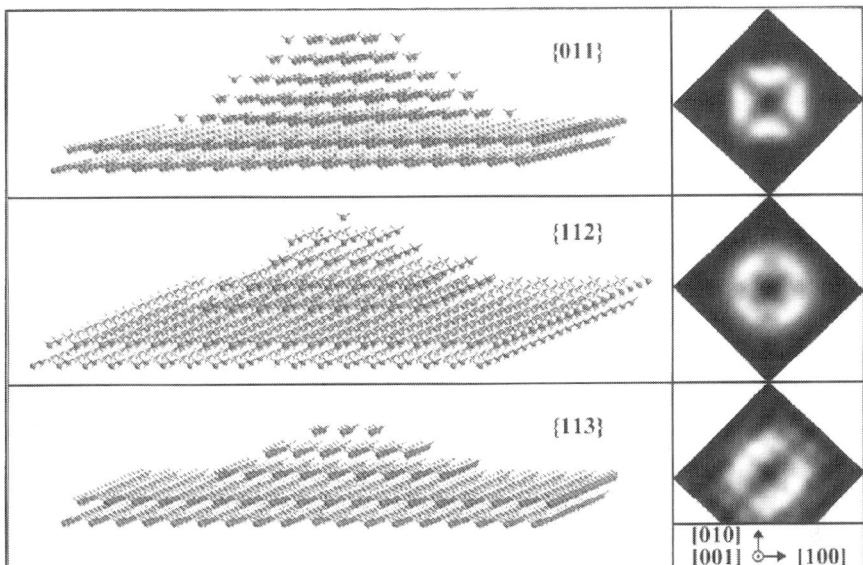

Fig. 2. Correlation between the shape of the CdSe island and the resulting calculated diffraction contrast feature.

the four-fold symmetry of the contrast feature of a pyramid with {101} facets agree very well with the experimentally observed image contrast of the QDs. The visible contrast reversal of the simulated image is caused by the chosen thickness value of about 7 nm.

5. FEM CALCULATIONS

For FEM calculations of the QD structure a configuration of a 15 nm x 15 nm wide and 3 nm thick ZnSe substrate was applied. According to the findings of X-ray diffraction measurements subsequently a 0.5 nm thick (about 2 MLs) CdSe wetting layer was placed on the substrate. The wetting layer is followed by a complete CdSe pyramid having {101} facets. The edges of the basal plane (5 nm x 5 nm in dimensions) of the pyramid were oriented parallel ⟨100⟩.

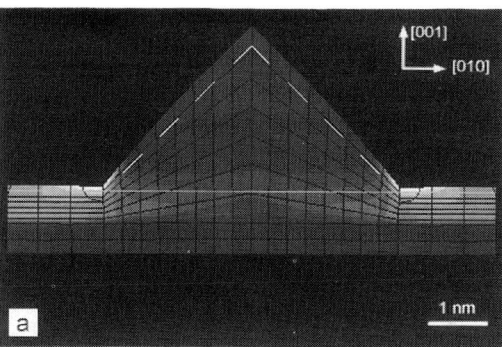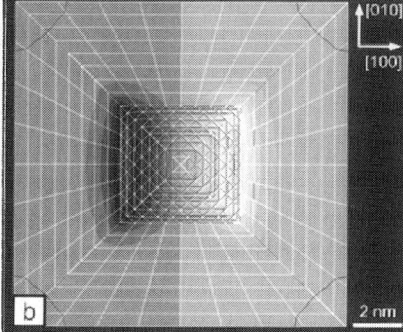

Fig. 3. Strain-field calculations by finite-element method:
a) density of the elastic-strain energy of a CdSe pyramid cross sectioned, b) [100] component of the displacement field.

In Fig. 3a the density of the elastic-strain energy is given in cross section of the uncapped CdSe pyramid where the dashed line denotes the initial shape. The highest density indicated by high brightness is found near the edges of the basal plane. This is also confirmed by the displacement-field analysis in plan view of Fig. 3b. The largest displacement (bright) in the [100] direction is found on the right hand side of the pyramid cut at the horizontal line brightened in Fig 3a. The dark area at the left hand side of the pyramid (Fig. 3b) represents the maximum value of the displacement in the [-100] direction. Taking into consideration also the displacement parallel to the [010] direction lying perpendicular to [100] a contrast feature exhibiting a four-fold symmetry can be realized.

6. CONCLUSIONS

The diffraction contrast features visible in plan-view TEM images of CdSe/ZnSe QDs are in good agreement with the results of corresponding simulations, based on truncated pyramids having {101} facets and edges of the basal plane orientated parallel to ⟨100⟩. The orientation relationships of the CdSe QDs differ from those found for III-V QDs (Ruvimov and Scheerschmidt 1995). This can be explained by the higher ionicity of the II-VI materials making the neutral {101} facets more favorable than the polar {111} facets.

The FEM calculations reveal the maximum of the ⟨100⟩ component of both the elastic-strain energy density as well as of the displacement field at the edges of the basal plane of the pyramid. For a better fit to the sample setup, FEM strain-field computations of capped pyramids are currently under investigation.

ACKNOWLEDGEMENTS

Special thanks are due to Dr M Rabe and Prof F Henneberger for providing the QD structures. Financial support by the Deutsche Forschungsgemeinschaft in the framework of the Sonderforschungsbereich 296 is gratefully acknowledged.

REFERENCES

Bimberg D, Grundmann M and Ledentsov N N 1998 Quantum dot heterostructures (Chichester: John Wiley & Sons)

Kirmse H, Schneider R, Rabe M, Neumann W and Henneberger F 1998 Appl. Phys. Lett. **72**, 1329

Kirmse H, Schneider R, Scheerschmidt K, Conrad D and Neumann W 1999 J. Microscopy, in press

Lowisch M, Rabe M, Stegemann B, Henneberger F, Grundmann M, Turck V and Bimberg D 1996 Phys. Rev. **B54**, 11074

Rabe M, Lowisch M and Henneberger F 1998 J. Crystal Growth **185**, 248

Ruvimow S and Scheerschmidt K 1995 phys. stat. sol. (a) **150**, 471

Stadelmann P A 1987 Ultramicroscopy **21**, 131

Stillinger F H and Weber T A 1985 Phys. Rev. B**31**, 5262

Inst. Phys. Conf. Ser. No 164
Paper presented at Microsc. Semicond. Mater. Conf., Oxford, 22–25 March 1999
© 1999 IOP Publishing Ltd

TEM investigation of MBE grown self-assembled CdSe/ZnSe islands

H Preis, S Kaiser, S Blümel, S Miethaner, S Bauer, K Fuchs and W Gebhardt

Institut für Experimentelle und Angewandte Physik, Universität Regensburg, D-93040 Regensburg, Germany

ABSTRACT: We present a TEM analysis of MBE grown self-assembled CdSe/ZnSe island structures. Homogeneous sized CdSe islands on a ZnSe(001) surface with densities between 5×10^7 cm^{-2} and 8×10^{10} cm^{-2} appear at a coverage of 2.7 - 3.1 monolayers (ML) CdSe. The islands have a lateral dimension of 18 nm × 33 nm, a height of 8 - 9 ML and contain at least 4 ML CdSe. They are mainly oriented in [1$\bar{1}$0]-direction and embedded in a 7 - 8 ML thick ternary $Zn_{1-x}Cd_xSe$ layer with a varying Cd content x.

1. INTRODUCTION

Self-assembled quantum dots provide novel physical properties for optoelectronic devices (Ledentsov 1998). In II-VI heterostructures a high quantum efficiency due to zero dimensional carrier localization, the breaking of the phonon bottleneck (Efros 1995) and the realization of a refractive index modulation (Ledentsov 1996) is expected. Whereas stable island sizes in the III-V system InAs/GaAs have been achieved with good reproducibility and a homogeneous size distribution there is still a controversial discussion concerning the II-VI system CdSe/ZnSe. Atomic force microscopic (AFM) investigations of MBE grown uncovered CdSe islands on ZnSe revealed island diameters ranging from 35 nm to 100 nm (Suemune 1997 and refs.). However, Lee et al (1998) observed Ostwald ripening of CdSe/ZnSe islands at room temperature and ripened islands are also present on pure ZnSe surfaces which are correlated with SeO$_2$ clusters (Smathers 1998). Therefore, AFM investigations of these uncovered islands have to be carried out carefully and correlated with covered structures. In TEM studies of covered MBE grown CdSe/ZnSe islands lateral sizes of 10-20 nm are reported by Flack (1996). Kirmse (1998) found similar results in thermally activated reorganized CdSe layers which indicate the existence of two classes (< 10 nm; 10 - 50 nm) of island sizes. Surprisingly, the optical transitions of these covered CdSe/ZnSe islands are in the range of 2.2 - 2.4 eV (Merz 1998 and refs.) with the exception of the submonolayer systems (Ivanov 1998). Consequently, a stable island size seems to exist in the CdSe/ZnSe system. The strong blue-shift compared to the 2K gap energy E_g = 1.752 eV of bulk zincblende CdSe (Fujita 1992) may be due to interdiffusion, strain or confinement effects. These three effects are actually discussed by Leonardi et al (1998) and Merz et al (1998) but have not been separated in these experiments. We report on the MBE growth and TEM investigations of stable homogeneous sized CdSe/ZnSe island structures. Plan-view TEM (PVTEM) is applied to evaluate the lateral size and shape of the islands. Their height and composition is determined with cross sectional high resolution TEM (HRTEM).

2. EXPERIMENTAL SET-UP

The investigated structures are produced in an elemental source MBE equipment on GaAs(001) substrates, beginning with the growth of a 20 nm thick pseudomorphic ZnSe buffer layer at a substrate temperature of T_S = 300°C. The flux ratio is J_{Se}/J_{Zn} = 4 and the growth rate 0.15 ML/s. After ZnSe deposition T_S is ramped up to 350°C under Se pressure. This is done to enhance the CdSe island formation according to Merz (1998). The CdSe growth is carried out at a flux ratio of J_{Se}/J_{Cd} = 4 and a growth rate of 0.15 ML/s. When this growth step is finished the surface is kept under Se flux for 5s. This enables a reevaporation of excess Cd and enhances surface diffusion necessary for island formation or island ripening. Finally, the island structures are covered with a 20 nm thick ZnSe layer at $T_S = 350°C$. We use reflection high energy electron diffraction (RHEED) as an in-situ probe of the growth mode transition from Frank-van der Merwe (FM) to island growth. The capped island structures are characterized with PVTEM and HRTEM in a Philips CM30 equipped with a TWIN lens ($C_S = 2.0$ mm) and a point resolution of 0.23 nm. PVTEM samples are prepared by a combination of mechanical thinning and a subsequent selective chemical etch procedure of GaAs proposed by Rosenauer (1996a). The HRTEM samples are first mechanically thinned and then ion milled in <110>-orientation using a Balzers RES010.

3. RESULTS

During growth of the first 1.5 ML CdSe on ZnSe at $T_S = 350°C$ a distinct RHEED specular spot oscillation is observed which characterizes the initial FM mode. A precise measurement of the RHEED oscillations allows a determination of the CdSe coverage within an error of ±0.1 ML. All values of CdSe coverages are extrapolated from this initially evaluated RHEED oscillation. A streaky to spotty transition of the RHEED pattern between 2.4 ML and 2.7 ML clearly marks the onset of island formation and indicates a Stranski-Krastanow growth mode. PVTEM images reveal an island density of 5×10^7 cm^{-2} in a sample with 2.7 ML CdSe. When the CdSe coverage increases the island density rises dramatically and reaches a maximum of 8×10^{10} cm^{-2} at 3.1 ML. Additional CdSe supply leads to island coalescence and the formation of dislocations. Figure 1 shows the PVTEM image of a

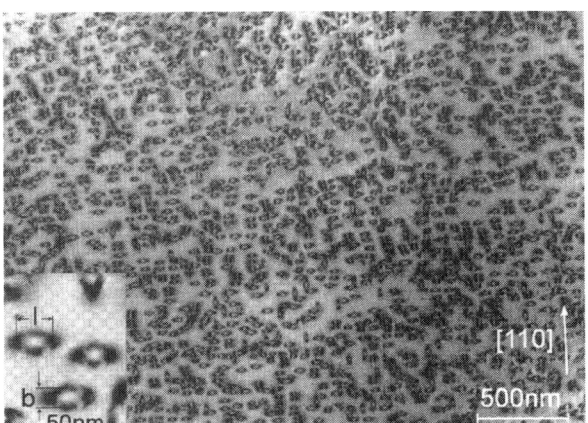

sample containing 3.0 ML CdSe and an island density of 5×10^{10} cm^{-2}. We note that a variation of the CdSe coverage between 2.7 ML and 3.1 ML affects only the island density but not the lateral size and shape. Figure 1 is a bright field PVTEM image with an exact sample orientation in (001)-incidence. Matsamura (1990) proposed this imaging method for a determination of the lateral size and shape of coherently strained nanoparticles. In fig. 1 the contrast pattern of single islands vary from the upper left to the lower right side. This effect is due to sample bending. Furthermore we observe a strong dependence of the diffraction contrast on the electron beam convergence and the exact sample orientation. Optimum imaging conditions are reached when

Fig. 1: Plan-view TEM bright field image of 3.0 ML CdSe sample with an island density of 5×10^{10} cm^{-2}. The lateral size is l = 33 nm and b =18 nm (insert).

parallel illumination is applied and the sample orientation is adjusted using selected area electron diffraction. The lateral island size is determined from the symmetric diffraction contrast pattern of single islands (s. insert of fig. 1). However neighboring islands have to be excluded since strain field interaction falsifies the measured island sizes. With these precautions a lateral island size of 33 nm × 18 nm is found with a preferential [1$\bar{1}$0]-orientation. Similar sized but isotropic islands observed Lee (1998) in AFM investigations of uncapped CdSe/ZnSe. Obviously, there exists a stable lateral island size in this system under the above mentioned growth conditions. Although Lee et al (1998) reported Ostwald ripening the authors discussed the possibility that a metastable minimum of the free energy as a function of the island volume exists and may be responsible for the size homogeneity.

The high island density which is achieved in the present work allows the structural investigation of single islands with cross sectional HRTEM. In fig. 2 an inhomogeneous contrast pattern dominates the HRTEM image. Three areas are marked which belong to different positions in the sample: position A represents the wetting layer region, position B and C are cross sections of single islands. The insert of fig. 2 shows a contrast enhanced image around position C and emphasizes the strain contrast caused by an island with a lateral extension of about 33nm. This size is in exact agreement with the PVTEM results in fig.1.

In order to obtain quantitative information about the local Cd content as well as the layer thickness we used the digital analysis of lattice images (DALI) method previously published by Rosenauer et al (1996b). The evaluation procedure starts with the determination of intensity maxima in a noise reduced HRTEM image which represent the positions of the atomic columns. Subsequently a reference lattice is created within the unstrained region of the lower ZnSe barrier and extended over the whole image. Each experimentally obtained lattice position is compared with the corresponding position in the reference lattice. The difference quantifies the local lattice distortion and allows a calculation of the local lattice parameter (LLP).

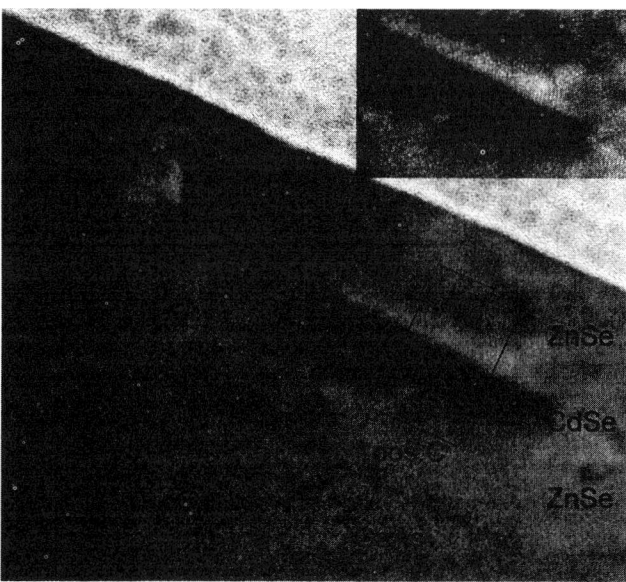

Fig. 2: HRTEM image in <110>-orientation: three marked positions represent the wetting layer region (A) and cross sections of islands (B,C) respectively. The contrast enhanced region around position C depicts a 33 nm extended strain contrast in the lower ZnSe barrier, corresponding exactly to the island length found in fig. 1.

Figure 3 depicts the grey scale coded map of the resulting [001]-component of the LLPs for each sample position in fig. 2. Black lattice points represent no displacement relative to pseudomorphic ZnSe/GaAs. White points correspond to the displacement of binary CdSe, biaxially distorted on ZnSe/GaAs. We should note that effects of elastic lattice relaxation caused by thinning of the TEM samples must be taken into account using the theory of Treacy and Gibson (1986).

Position A: The LLPs in fig. 3a reveal an inhomogeneous ternary $Zn_{1-x}Cd_xSe$ layer. The Cd content x varies in growth direction and extends over 7 - 8 ML, also lateral inhomogeneities are visible. There seems to be no continuous binary CdSe wetting layer.

Position B: Figure 3b contains the evaluation of the cross section of a binary CdSe island with a maximum CdSe-height of 3 ML. The composition in growth direction changes within 2 ML from pure ZnSe in the lower barrier to a Cd content of x = 1 in the island and back to ZnSe in the upper barrier. The influence of interdiffusion is calculated with the Cd diffusion coefficient in ZnSe previously obtained by Rosenauer et al (1995). An intermixing region of ±1 ML around the initially existing island interfaces is calculated as a result of the used growth conditions. This confirms the DALI evaluation in fig. 3b. Consequently, interdiffusion may change the composition at the interface region, but it does not influence the Cd concentration within the island. The LLPs in fig. 3b above the right and the left island edge are slightly increased (grey coded). This effect is due to an inhomoge-

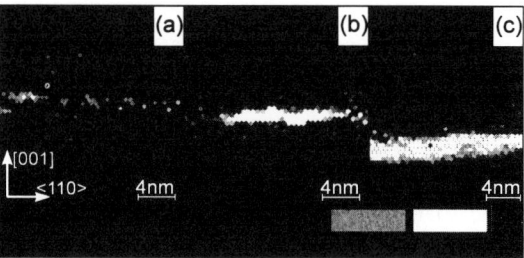

Fig. 3: Local lattice parameters of the three positions (fig. 2) in grey scale coded representation evaluated with the DALI strain state analysis. Black coded LLPs correspond to the ZnSe lattice parameter and white coded to biaxially distorted CdSe pseudomorphically grown on ZnSe/GaAs. Position A shows a ternary $Zn_{1-x}Cd_xSe$ wetting layer region (a), position B and C cross sections of CdSe islands (b,c).

neous strain field which modulates the HRTEM contrast. This is better seen in fig. 4. The left one (a) is the noise reduced HRTEM image of position B the right one (b) the corresponding arrow diagram with local lattice distortion vectors. Obviously, the local lattice parameter in the upper part of fig. 3b reflects an inplane strain component. The problems which arise from these inhomogeneous strain fields should be avoided

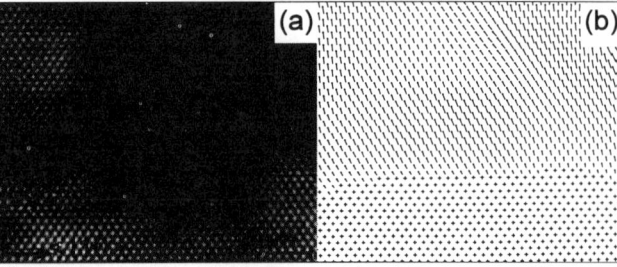

Fig. 4: Noise reduced HRTEM image (a) and the corresponding arrow diagram (b) of fig. 3b. The influence of inhomogeneous strain on the DALI evaluation is shown (s. text).

by a careful selection of the evaluated sample position. Especially, in the present example of fig. 4b the ZnSe/CdSe/ZnSe region of interest shows only negligibly small inplane strain.

Position C: The LLPs in fig. 3c correspond to a maximum island height of 8 - 9 ML. The island consists of at least 4 ML CdSe and contains a 2 - 3 ML mixed ZnSe/CdSe region. There is an excellent agreement between the extent of the island's strain field found in the PVTEM image (fig. 1) and the result of the HRTEM analysis (fig. 2). Therefore we may conclude that the cross section of an island center has been reliably evaluated and that it contains at least 4 ML binary CdSe.

4. CONCLUSION

We report on TEM-investigations of MBE grown self-assembled homogeneous sized CdSe islands on ZnSe(001) of high density. PVTEM investigations reveal a lateral dimension of 18 nm × 33 nm with a predominant $[1\bar{1}0]$-direction of the capped islands and a maximum density of $8 \times 10^{10}\,cm^{-2}$. The DALI method is used to obtain a maximum island height of 8 - 9 ML, including at least 4 ML binary CdSe and 2 - 3 ML ternary $Zn_{1-x}Cd_xSe$. The islands are embedded in a ternary $Zn_{1-x}Cd_xSe$ layer with an inhomogeneous distribution of Cd.

REFERENCES

Al L Efros, V A Kharchenko and M Rosen 1995 Solid State Commun. **93** (4), 281

F Flack and N Samarth, V Nikitin, P A Crowell, J Shi, J Levi and D D Awschalom 1996 Phys. Rev. B **54** (24), R17312

S Fujita, Y Wu, Y Kawakami, S Fujita 1992 J. Appl. Phys. **72** (11), 5233

S V Ivanov, A A Toropov, T V Shubina, S V Sorokin, A V Lebedev, I V Sedova and P S Kop'ev, G R Pozina, J P Bergmann and B Monemar 1998 J. Appl. Phys. **83** (6), 3168

H Kirmse, R Schneider, M Rabe, W Neumann and F Henneberger 1998 Appl. Phys. Lett. **72** (11), 1329

S Lee, I Daruka, C S Kim, A-L Barabási, J L Merz and J K Furdyna 1998 Phys. Rev. Lett. **81** (16), 3479

N N Ledentsov, I L Krestnikov, M V Maximov, S V Ivanov, S L Sorokin, P S Kop'ev and Zh I Alferov, D Bimberg, C M Sotomayor Torres 1996 Appl. Phys. Lett. **69** (10), 1343

N N Ledentsov, V M Ustinov, V A Shchukin, P S Kop'ev and Zh I Alferov, D Bimberg 1998 Semiconductors **32**(4), 343

K Leonardi, H Heinke, K Ohkawa and D Hommel 1997 Appl. Phys. Lett. **71** (11), 1510

S Matsamura, M Toyohara and Y Tomokiyo 1990 Phil. Mag. A, **62** (6), 653

J L Merz, S Lee, J K Furdyna 1998 J. Cryst. Growth **184/185**, 228 3479

A Rosenauer, T Reisinger, F Franzen, G Schütz, B Hahn, K Wolf, J Zweck and W Gebhardt 1996 J. Appl. Phys. **79** (8), 4124

A Rosenauer, S Kaiser, T Reisinger, J Zweck, W Gebhardt 1996 Optik **102**, No. 2, 63

M M J Treacy, J M Gibson 1986 J. Vac. Sci. Technol. B **4**, 1458

A Rosenauer, T Reisinger, E Steinkirchner, J Zweck and W Gebhardt 1995 J. Cryst. Growth **152**, 42

J B Smathers, E Kneedler, B R Bennett and B T Jonker 1998 Appl. Phys. Lett. **72** (10), 1238

I Suemune, T Tawara, T Saitoh and K Uesugi 1997 Appl. Phys. Lett. **71** (26), 3886

Inst. Phys. Conf. Ser. No 164
Paper presented at Microsc. Semicond. Mater. Conf., Oxford, 22–25 March 1999
© *1999 IOP Publishing Ltd*

Characterisation of SnO_2 nanopowders obtained by liquid pyrolysis for gas monitoring

A Vilà, A Diéguez, A Cirera, A Cornet and J R Morante

Department of Electronics, University of Barcelona, Martí i Franqués 1, E-08028-Barcelona, Spain

ABSTRACT: Liquid pyrolysis is tested as an alternative technological process able to give SnO_2 nanopowders suitable for gas sensors. The structure and morphology of the powders obtained by this technique are analysed by conventional and high-resolution transmission electron microscopy as a function of technological parameters such as concentration of the initial bath or processing temperature. Our results show that nanometer-sized crystallites with good crystalline quality are obtained by this technique.

1. INTRODUCTION

Tin dioxide is an n-type semiconductor well known for its applications in gas sensors, especially to monitor explosive and toxic gases such as CO or NO_x. As such, much research has been undertaken in the recent years in order to improve their selectivity and sensitivity. These objectives require fine powders of nanometric scale, in order to enlarge the sensing surface without introducing defects able to disturb the electronic behaviour of the semiconductor. Moreover, good crystallinity is necessary for stability. As a consequence, synthesis of the SnO_2 powder is of major importance, as during the process a variety of technological parameters can influence the properties of the powder and, in consequence, of the sensor (Göpel and Reinhardt 1996).

Up to now, several methods have been used to prepare SnO_2 nanopowders: sol-gel, spray pyrolysis, laser ablation, sputtering, etc. However, none of them has shown to be definitive in producing powders of optimal characteristics. Some methods (like laser ablation and sputtering) are expensive and have problems with material stoichiometry and crystallinity. Other methods (like sol-gel or spray pyrolysis) have shown their usefulness, but the influence of some parameters is still not well understood and controlled (Göpel and Schierbaum 1995). Therefore, new techniques are nowadays proposed with the idea of optimising the SnO_2 production. In this work liquid pyrolysis is tested, as an alternative technological process able to give the desired crystalline characteristics together with industrial mass production. TEM techniques are used to study grain dimensions, dispersions and crystallinity as the parameters important in obtaining materials with good sensor characteristics.

2. EXPERIMENTAL FEATURES

Liquid pyrolysis is based on the application of a thermal treatment to microdrops of an organic solution deposited onto a substrate (Cirera et al 1998). Following this scheme, $SnCl_4·5H_2O$ was first dissolved in methanol in concentrations ranging from 0.1 to 5 M. The pastes obtained this way were then spread in droplets onto a silicon substrate and submitted to a thermal treatment at a temperature up to 900°C for 24 minutes. This procedure allows producing stabilised SnO_2 with small grain size in a simple, repetitive and economic way. For structural characterisation of the powders, TEM observations were carried out in a Philips CM30-ST microscope acting at 300 kV, having a fringe resolution of 1.9 Å. Complementary measurements were made in an X-ray diffractometer Siemens D-500 with radiation Cu Kα ($\lambda = 1.5418$ Å) in the Debye-Scherrer geometry.

154

3. RESULTS

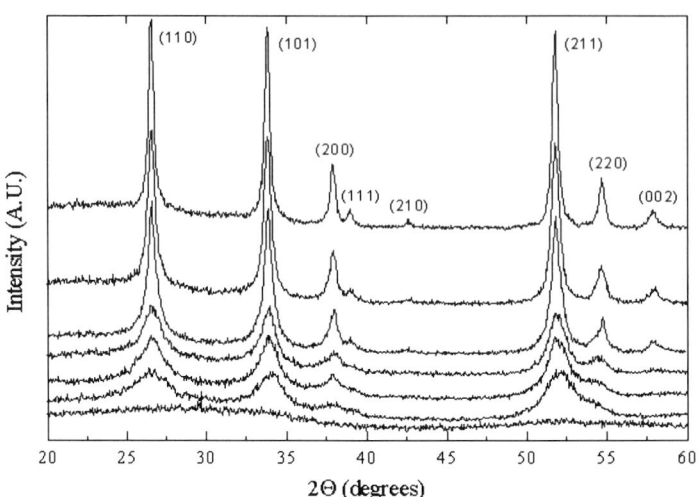

Fig. 1. X-Ray spectra of the powders obtained by liquid pyrolysis at 200, 400, 500, 600, 700, 800 and 900°C (from bottom to top).

A preliminary characterisation of the powders obtained by liquid pyrolysis was made by x-ray diffraction (XRD), in order to obtain a first approximation to the material characteristic. Spectra corresponding to the powders 5M are presented in Fig. 1. It can be seen that thermal treatments up to 300°C leave amorphous material, agreeing with previous reports (Cao et al 1996). Above this temperature, powders become more and more crystalline as process temperature increases. Comparison of these spectra with the cassiterite one (crystalline phase of SnO_2 with rutile structure) indicates that the samples contain only SnO_2 polycrystals in this phase.

Electron diffraction (TED) also indicates an important difference in crystallinity of samples processed at extreme temperatures (Fig. 2). However, contrarily to XRD results, even in the 200°C one there is evidence of some crystalline phase, as indicated by spots in the diffraction pattern. High-resolution images corroborate this result, showing the presence of small crystalline clusters in a textured matrix strongly tending to crystallisation (Fig. 3). Both techniques indicate that high-temperature samples are completely polycrystalline.

Fig. 2. Electron diffraction patterns corresponding to the samples processed at 200 (left) and 800°C (right).

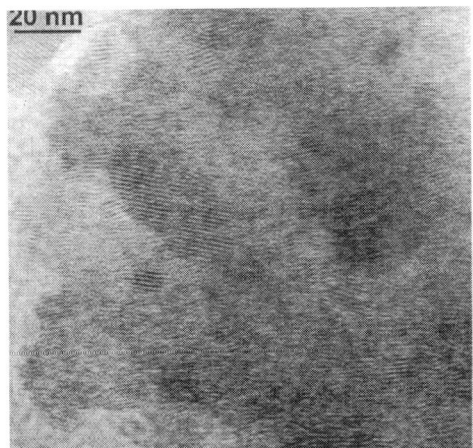

Fig. 3. High-resolution images of the samples obtained at 200°C. There is evidence of small crystalline clusters.

In addition, TEM images at lower magnification reveal that different thermal processes give rise to several effects on the size and shape of the powders. Firstly, as shown in Fig. 4 (upper part), the more the deposition temperature increases, the larger are the nanocrystals generated. The histograms of grain sizes corresponding to these samples (Fig. 4, lower part) clearly corroborate this trend.

Moreover, the histograms indicate a quite uniform distribution for the grain sizes, even at high temperatures. The average grain dimension is calculated to be of 16 and 27 nm for 600 and 800°C, respectively. These values are significantly smaller than the reported ones for other technologies, such as sol-gel (see for example Diéguez et al 1996). Even the standard deviations from the mean values (5.1 and 8.2 nm respectively) are much less important for liquid pyrolysis than for these other techniques. All these observations support the suitability and repeatability of this technological method.

On the other hand, it can be observed that after deposition at higher temperatures flat surfaces are more evidently developed and less defects remain inside the grains. Probably, when the processing temperature is large enough, the atoms at the surface have more mobility. Then they can be distributed in such a manner that the total energy of the system is decreased, generally by faceting of the crystallite through the formation of planes of minimum energy at their surface.

Finally, it seems that thermal treatments also diminish the defect density in the powders, as at a higher temperature fewer defects are visible, especially less dislocations and local distortions observed after low-temperature processing.

4. CONCLUSIONS

TEM has proved to be very useful to analyse the structural properties of SnO_2 nanocrystals for their improvement in order to develop gas sensors. Electron diffraction and high-resolution imaging give even better information than other diffraction techniques such as XRD. They show the presence of small isolated nanocrystals inside a textured matrix in the samples grown at 200°C, implying a progressive transition from amorphous SnO_2 to crystalline as the processing temperature increases. As expected, the higher the processing temperature, the larger are the observed grain dimensions. Their distribution is quite uniform even at high temperature, giving an average grain size (16 and 27 nm for 600 and 800°C, respectively) which is significantly smaller than the obtained for other technological methods. The size distribution is also less extended. All these results demonstrate the suitability of liquid pyrolysis for SnO_2 production for gas sensing.

REFERENCES

Cao X, Cao L, Yoo W and Ye X 1996 Surf. Interf. Analysis **24**, 662

Cirera A, Diéguez A, Diaz R, Cornet A and Morante J R 1998 Proc. Eurosensors XII, 673

Diéguez A, Romano-Rodríguez A, Morante J R, Weimar U, Schweizer-Berberich M and Göpel W 1996 Sensors and Actuators B **31**, 1

Göpel W and Reinhardt G 1996 Sensors Update Vol. 1, eds H Baltes, W Göpel and J Hesse (Weinheim)

Göpel W and Schierbaum K D 1995 Sensors and Activators B **26-27**, 1

156

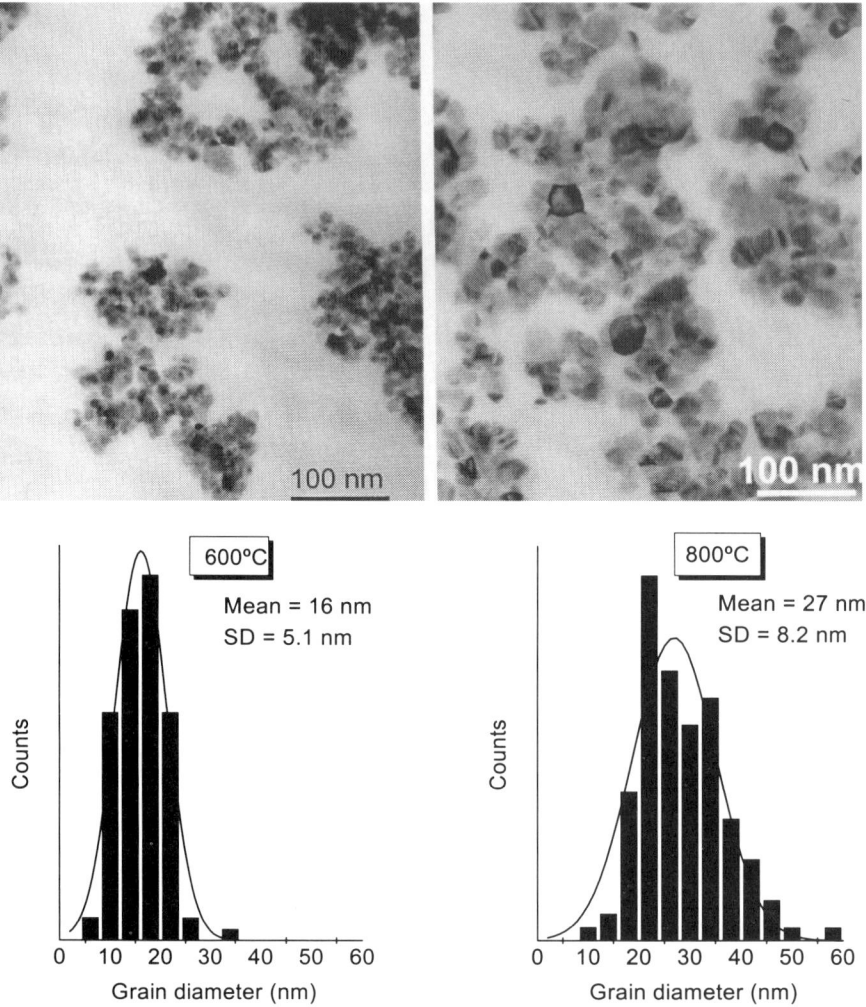

Fig. 4. Low-magnification TEM images of samples obtained at 600 and 800°C and the corresponding histograms for the grain size.

Inst. Phys. Conf. Ser. No 164
Paper presented at Microsc. Semicond. Mater. Conf., Oxford, 22–25 March 1999
© 1999 IOP Publishing Ltd

A kinetics study of the growth of Ge precipitates in SiO$_2$ by coupling TEM and EELS

C Bonafos, M Lopez, B Garrido, A Perez-Rodriguez, J R Morante, J Montserrat[1], M Toufella[2], Y Kihn[2], G Ben Assayag[2], A Claverie[2], A Nejim[3] and P L F Hemment[3]

Departament d'Electronica, unitat associada CNM/CSIC, Universitat de Barcelona, carrer Marti i Franques 1, Barcelona 08028, Spain
[1]CNM/CSIC, campus Universitat Autonoma de Barcelona, Bellaterra 08193, Spain
[2]CEMES/CNRS, BP 4347, 31055 Toulouse Cedex, France
[3]Ion beam Facility for Microelectronics, University of Surrey, Guildford, GU2 5XH, UK

ABSTRACT: We present a new method, to be used as a routine, to access the size-distribution and the density of precipitates embedded in an amorphous matrix. This method, which combines TEM and EELS, is used to follow the size and density evolution upon annealing of a population of Ge precipitates in a SiO$_2$ layer. It is clearly shown that the precipitates undergo a conservative competitive growth in which they exchange Ge atoms. This behaviour is also supported by SIMS measurements.

1. INTRODUCTION

Nanosized semiconductor particles have been investigated during the past years because their optical properties are different from those of the corresponding bulk crystals. When their dimensions are smaller than the corresponding Bohr radius, these particles exhibit a number of striking effects due to exciton quantum confinement (Brus 1990). This allows Si-based materials, such as group IV nanoparticles embedded in a SiO$_2$ matrix, to emit light in the visible range even at room temperature. While the physical origin of the light emission is still under debate, it is admitted that the particular properties of the nanosized materials are critically linked to their densities and size distributions. Thus, a systematic study of the kinetics evolution of the growth of such nanoparticles is a basic prerequisite step towards the understanding of their properties and would also permit their controlled synthesis.

This paper is dedicated to the rigorous study of the kinetics behaviour of Ge nanoparticles prepared by ion implantation and annealing in thermally grown amorphous SiO$_2$. While the evolution of the size distribution of the precipitates can be directly obtained by transmission electron microscopy (TEM), the evolution of the density of precipitates has never been evaluated because of the amorphous nature of the matrix. We present here a method, to be used as a routine, that has been developed and tested on a large number of samples, aimed at the measurement of the density of precipitates embedded in an amorphous matrix. The method involves conventional TEM and Electron Energy Loss Spectroscopy (EELS). The method has allowed us to obtain the time-evolution during annealing of the mean radius, the density per unit volume and the number of atoms within the precipitates. We conclude that Ge precipitates undergo a conservative Ostwald ripening process.

2. EXPERIMENTAL CONDITIONS

The matrix is a thermally grown 500 nm-thick amorphous SiO$_2$ film. A 100 nm-thick region with a constant Ge concentration (5%) has been processed in order to avoid the effect of a concentration gradient on the precipitation kinetics. For this, four different energies and doses have been implanted (20 keV, 3.6×10^{15} ions/cm^2 + 50 keV, 7.2×10^{15} ions/cm^2 + 100 keV, 9×10^{15} ions/cm^2 + 150 keV, 1.5×10^{16} ions/cm^2). The samples were subsequently annealed at temperatures ranging from 900°C to 1200 °C, for times in a range from 1min to 4 hours. Samples for TEM were prepared in cross-section by mechanical thinning followed by ion milling in a Gatan Precision Ion Polishing System using 4 keV Ar$^+$ ions incident at an angle of 6° to the surface of the sample. TEM was performed in a Philips 300 kV electron microscope. EELS analysis was performed in a Philips 200 kV microscope equipped with a Gatan 666 spectrometer.

3. KINETICS STUDY

The kinetics study consists in determining the evolution of the structural characteristics of the population, i e the mean radius, the density of precipitates and the number of atoms stored within the precipitates as a function of the annealing time.

3.1 Size measurement

We observe the formation of Ge precipitates only for annealing at T>900°C. The size histograms have been extracted from images taken under bright field and small out-of-focus conditions, as shown in Fig. 1. Numerous tests have shown that this method is more reliable than high resolution TEM. Indeed, under out-of-focus conditions, precipitates of all sizes (down to 2 nm in diameter) show a high contrast Fresnel fringe (due to a phase difference) at the matrix/precipitate interface provided the sample is reasonably thin. When imaging the same precipitates by HREM, the probability of « seeing » a precipitate depends on different factors. On the one hand, for very thin regions, the probability of forming « fringes » when imaging along a random direction increases as the size of the particle decreases. On the other hand, the contrast arising from small particles is more easily lost in the noise when the thickness of the overall sample increases. These two antagonistic criteria explain the non-reproducibility of size measurements by HREM.

The evolution of the mean radius as a function of the annealing time, at a temperature of 1000°C, is summarised in Fig. 2. When annealing at 1000°C, the mean radius of the precipitates increases from 3 nm after 1 min to 4 nm for 1h (uncertainty 5%).

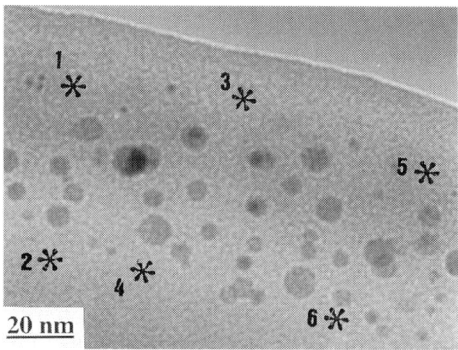
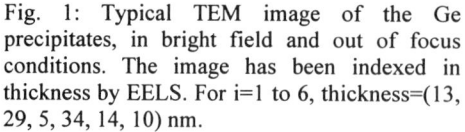

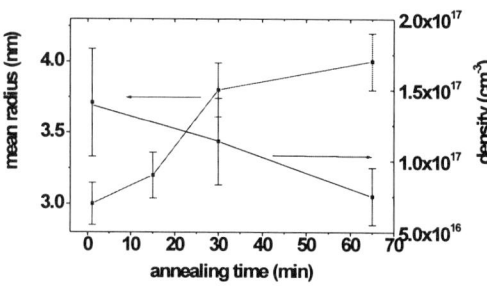

Fig. 1: Typical TEM image of the Ge precipitates, in bright field and out of focus conditions. The image has been indexed in thickness by EELS. For i=1 to 6, thickness=(13, 29, 5, 34, 14, 10) nm.

Fig. 2: Evolution with annealing time of the mean radius and volume density of precipitates, for annealing at 1000°C.

3.2 Volume density measurement

In the case of a crystalline matrix, the thickness of the region under statistical analysis can be estimated through the observation of thickness fringes. This is obviously not possible in amorphous

SiO_2. For this reason, we have developed and tested a method, to be used as a routine, to measure the volume density of precipitates by combining conventional TEM and electron energy loss spectroscopy (EELS).

The amount of inelastic scattering the incident beam suffers increases with the specimen thickness. In principle, this information could be extracted from the ratio between the integral intensity in the EELS spectrum I_T and the intensity in the zero-loss peak (I_0). In practice, the intensity in the EELS spectrum falls off so rapidly as the energy loss increases that we can reasonably approximate I_T by the intensity in the low-loss region in the spectrum up to about 50 eV. For thickness not exceeding 200 nm, this ratio only depends on the bulk plasmon mean free path (MFP) λ_p, and thus, the thickness of the analysed region t is given by (Egerton 1996),

$$t = \lambda_p \frac{I_L}{I_O} \quad (1).$$

We have used a probe diameter of about 5 nm while working at a magnification of x150,000. In principle, the ratio t/λ_p can be directly extracted from the spectrum using the Gatan software, but we have also tested a 'spectrum re-building' method (Aitouchen et al 1997). In the case of silica, the two results do not differ by more than 4%. The accuracy of the thickness measurement has been estimated to be of about 20%. This uncertainty is partially related to the 'quality' of the sample. The thickness of the probed specimen area must be as uniform as possible. The overall uncertainty also depends on the inelastic MFP. Indeed, λ_p depends on a variety of experimental parameters, such as the incident beam energy, the specimen composition and the collection angle. For pure silica, λ_p has been estimated by using a calculation method based on the Ashley-Ritchie theory (Ashley and Ritchie 1970) coupled with the experimental determination of the number of free electrons per unit volume in the matrix. At 200 kV in silica we have found λ_p to be about 130 nm (+/-10%).

On each region of interest, the local thickness is measured by EELS on 8 points of the precipitate region so that a 'window' is built. Figure 3 shows two typical EELS spectra obtained at different locations in the image, one near the free surface of the sample and the other one deeper. This allows us to build up a thickness map of the area as shown in Fig. 1 and estimate the volume of the region under analysis. Then, the number of precipitates found into this volume can be measured and the density of precipitates by unit volume, d_v, can be deduced with an uncertainty of 35%.

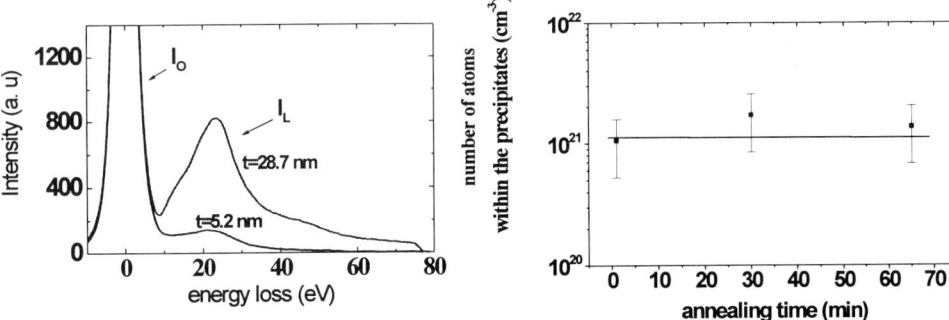

Fig. 3: EELS spectra obtained in two different points on the image Fig. 1 and their associated evaluated thicknesses.

Fig. 4: Evolution with the number of atoms within the precipitates per cm^3, for annealing at 1000°C.

By applying this method to the Ge/SiO_2 system, we have found that, when annealing at 1000°C, the density of precipitates decreases from 1.4×10^{17} cm^{-3} to 7.5×10^{16} cm^{-3} when increasing the annealing time from 1 min to 1h. This evolution is plotted on Fig. 2 together with the growth law of the precipitates.

3.3 Validation of the method

The number of Ge atoms per unit volume contained within the precipitates can be deduced from the size histograms and the density of precipitates per unit volume measurements and is given by,

$$N_v = d_{Ge} d_v \frac{4}{3} \pi \frac{\sum_i n_i \, r_i^{-3}}{\sum_i n_i} \quad (2)$$

where $d_{Ge} = 4.42 \times 10^{22}$ at/cm^3 is the atomic density of bulk Ge and n_i is the number of precipitates having a radius r_i.

The evolution of N_v with the annealing time at T=1000°C is given in Fig. 4. When increasing the annealing time, N_v remains constant within the error bars (45%), and is about 1.2×10^{21} at/cm^3. In order to compare with results obtained by secondary ion mass spectroscopy (SIMS), this number has to be multiplied by L, the thickness of the layer within which the precipitates are found (90 nm). We find a value for N_s, the number of Ge atoms per unit surface area contained within the precipitates, of about 1.2×10^{16} at/cm^2 (+/-50%) in excellent accordance with the SIMS results at 1.2×10^{16} at/cm^2. This comparison gives us good confidence of the accuracy of the method we are proposing.

Therefore, and as shown in Figs. 2 and 4, when increasing the annealing time, the mean radius of the precipitates increases while the density of precipitates decreases. In the mean time, the number of atoms within the nanocrystals remains constant. This behaviour constitutes proof that the precipitates undergo an Ostwald ripening process and that this competitive growth is conservative. During this process, the precipitates exchange Ge atoms with the net result that small ones dissolve while larger ones grow (Lifshitz and Slyosov 1961).

4. CONCLUSION

In this paper, a rigorous kinetics study of Ge nanocrystals embedded in SiO$_2$ has been carried out. An original method has been proposed and tested in order to measure the density of precipitates in an amorphous matrix, This method combines conventional TEM and EELS. The estimated number of Ge atoms contained within the population of precipitates, as calculated from the size and density measurements, is of the same order as the value obtained by SIMS, and this confirms the validity of the method. This method, which is not that time consuming, can be used as a routine and applied to any system made of a population of precipitates embedded in an amorphous matrix, provided the image contrast is sufficiently high, i e the difference of atomic numbers between the precipitates and the matrix is large enough and the incident beam is highly coherent.

REFERENCES

Aitouchen A, Kihn Y and Zanchi G 1997 Microsc. Microanal. Microstruct. **8**, 369
Ashley J C and Ritchie R H 1970 Phys. Stat. Sol. **38**, 425 and **40**, 623
Brus L E 1990 J. Phys. Chem. **90**, 2555
Egerton R F 1996 EELS in Electron Microscope - 2nd edition (New York: Plenum press)
Lifshitz I M and Slyosov V V 1961 J. Phys. Chem. Solids **19**, 35

Inst. Phys. Conf. Ser. No 164
Paper presented at Microsc. Semicond. Mater. Conf., Oxford, 22–25 March 1999
© 1999 IOP Publishing Ltd

In situ scanning electron microscopy of epitaxial processes

Y Homma, H Yamaguchi and P Finnie

NTT Basic Research Laboratories, Atsugi-shi, Kanagawa 243-0198, Japan

ABSTRACT: Using scanning electron microscopy we have observed the nucleation and growth of 2-D islands in the initial stages of GaAs growth by molecular beam epitaxy and migration enhanced epitaxy. We have made *in situ* observations of site selective growth on an atomic-step-controlled Si substrate, and demonstrate the formation of a regular network pattern of GaAs and a regular array of Au dots on the Si substrate.

1. INTRODUCTION

In situ imaging of growing surfaces is useful to achieve control of growth as well as to reveal the elemental processes occurring during it. Precise growth control is receiving particular attention in the fabrication of nanostructures. The progress in surface observation techniques has enabled direct imaging of growing surfaces. Such techniques involve reflection electron microscopy (Shima1991), low energy electron microscopy (´Swiech 1991), and scanning tunneling microscopy (Voigtländer 1996). These have successfully been applied to observe molecular beam epitaxy (MBE) of Si growth at elevated temperatures. For III-V compound materials, however, the situation is rather complicated. They are two component systems, and the vapor pressure of the V-element is high (10^{-3} Pa) under the growth conditions actually employed in the crystal growth technology. Easy access to the sample surface both for surface probing and material supply is necessary for *in situ* imaging of III-V compound growth. Thus, we used scanning electron microscopy (SEM), which has a large working space, to observe GaAs MBE growth. It has been demonstrated that SEM can provide atomic step and 2D-island images on growing GaAs surfaces (Homma 1995, 1996a), so no growth interruption is necessary.

In this paper, we present observations of both fundamental processes and growth site control. Surface morphology evolution during the initial stage of GaAs growth is shown. For the growth site control, a step-structure-arranged Si(111) substrate was used and the growth properties were investigated by *in situ* imaging. GaAs wire growth and Au island growth in a self-assembled fashion has been achieved.

2. INSTRUMENT

In situ observations were performed using an ultrahigh vacuum SEM/MBE system. The instrument was equipped with a cold-cathode field-emission electron gun mounted on top of the sample chamber, in which GaAs MBE growth was performed. The electron beam energy used was 25 keV, with a beam current of 0.1-0.3 nA. For *in situ* SEM imaging, the sample stage was tilted so that both the electron beam and depositing flux impinged on the sample surface. The configuration for *in situ* imaging is illustrated in Fig. 1. The electron beam was directed onto the sample surface from above at a glancing angle θ of 10-15°. The Ga and As fluxes were directed upward onto the surface at 40°- θ. The GaAs samples were mounted on a silicon substrate using indium soldering, and heated resistively by passing a direct current through the silicon substrate. The silicon samples were directly resistive-heated. The growth rate was typically 12-15 s per GaAs monolayer. The scanning rate for SEM imaging was 20-80 s per frame. Thus, the growth of several GaAs layers was imaged in one micrograph frame during growth. It should be noted that time-dependent morphology variations were superimposed on a normal SEM micrograph. The aspect ratio of the SEM micrographs was adjusted to compensate for image foreshortening due to oblique incidence.

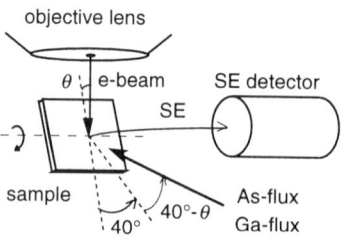

Fig. 1. Configuration for *in situ* SEM imaging.

3. OBSERVATION OF NUCLEATION AND GROWTH

3.1 GaAs(001) MBE

The initial stage of layer-by-layer growth in the 2-D-island-nucleation mode was directly observed for GaAs (001) (Homma 1995, 1996a). Figure 2 shows the initial GaAs(001) surface and the morphology evolution during MBE growth at 580°C with a Ga deposition rate of about 4 monolayers/min. Monolayer steps are seen on the initial surface (a) with a step band on the right side. The central terrace is the lowest in the atomic step staircase. The step image appears bright or dark with the primary electron beam traveling down or up the steps, respectively. After growth started [top of image (b)], the morphology evolution due to the growth of 4 GaAs layers was recorded in one micrograph frame: the growth time increases from the top to the bottom of the micrograph. The surface morphology oscillates clearly in the first three cycles as a result of the nucleation and coalescence of 2-D islands. Monolayer steps become hard to discern when the surface roughens and then become visible again when a monolayer is completed [compare images (a) and (b)]. The surface is fairly smooth after completion of the first monolayer, but the smoothness of the surface degrades as the growth progresses. In images (c) and (d), which show successive morphology changes, it is difficult to determine the monolayer completion period. This is because nucleation and coalescence are out of phase, resulting in the accumulation of islands and holes on the terraces.

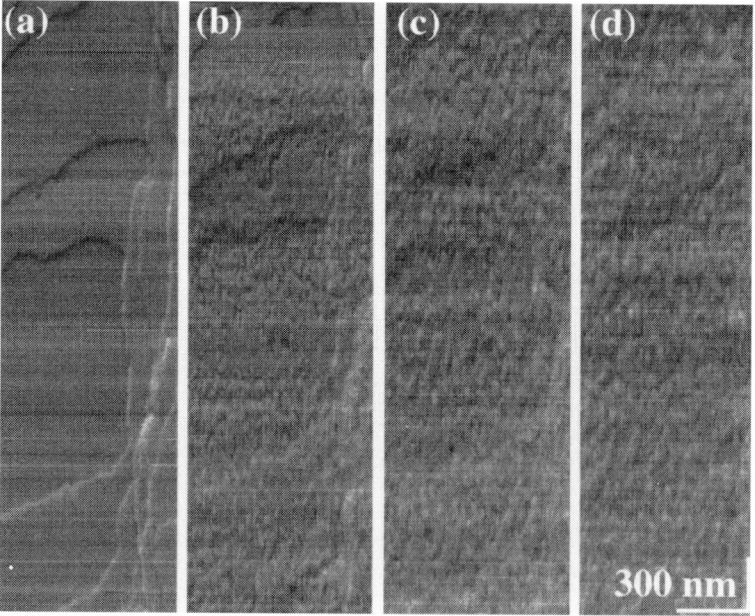

Fig. 2. SEM images of GaAs(001) surface during MBE at 580°C. Image (a) shows the initial surface and (b-d) show the morphology evolution during growth. The scanning time for each image is about 70 s.

3.2 GaAs(111) MBE

On the GaAs(001) surface, individual island sizes were 10 nm or less, which was too small to be resolved in detail with the SEM resolution. This situation is greatly improved on the GaAs (111)A surface, where the adatom diffusion length is much larger than that on (001). Figure 3 shows successive SEM images of a GaAs (111)A surface during MBE growth at 577°C with a Ga deposition rate of about 0.7 monolayers/min (Yamaguchi 1998). These five images trace just one monolayer of growth. After the first nucleation of 2-D islands [middle of image (a)], the islands become larger while retaining their triangular shape. The shape reflects the threefold symmetry of the (111) surface, and the island edges consist of < 1$\bar{1}$0 > steps. Growth by step propagation also takes place. Most islands are two dimensional but some show multilayer growth; i.e., the second layer nucleates on top of the initial islands.

On the (111)A surface, the islands are large enough that it is possible to analyze the details of the growth processes. The step propagation velocity and the adatom diffusion length were directly derived from images like those in Fig. 3 (Yamaguchi 1998). It turns out that desorption of Ga is significant on such a large terrace, and thus growth occurs under near-equilibrium conditions, unlike normal GaAs MBE. The step propagation velocity and the adatom diffusion length were confirmed to fit well to the standard theory of crystal growth by Burton, Cabrera, and Frank (1951).

3.3 GaAs(001) MEE

An alternate supply of group-III and group-V elements to the substrate has been reported to greatly increase the surface diffusion of group-III elements (Horikoshi 1986). This growth mode is called migration enhanced epitaxy (MEE), and is useful for growing high quality GaAs and AlGaAs layers, even at low substrate temperatures. We examined the smoothness of an MEE-grown surface by *in situ* SEM (Homma 1996b).

Figure 4 shows a comparison of the surface morphologies of the initial surface (a), during growth (b, c) and immediately after the growth of 10 layers (c). The initial surface shows large terraces separated by monolayer steps. During MEE growth, Ga and As were supplied alternately in amounts corresponding to 1 monolayer. For one cycle, the Ga shutter was opened for 11.5 s and the As shutter

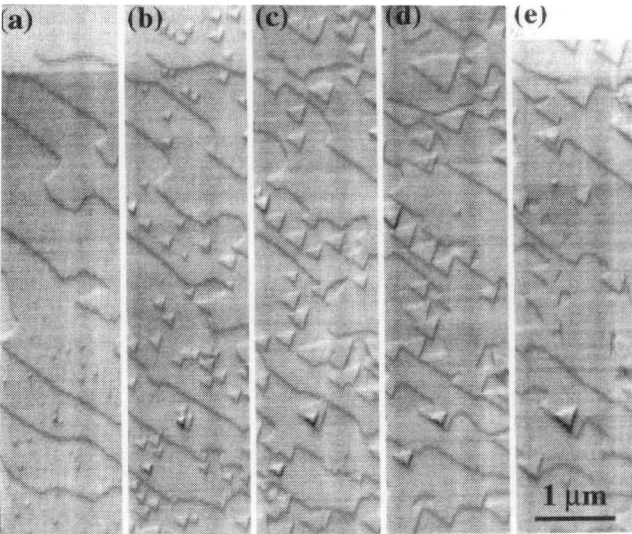

Fig. 3. SEM images of GaAs(111) surface during MBE at 577°C. Growth started at the top of (a). Each image is trimmed to show the scanning time of about 15 s. The interval between each image scan is about 12 s and the total growth time is 123 s.

was opened for 3.5 s. The increase in Ga coverage causes the increase in secondary electron emission, while As desorption decreases the secondary electron emission. Thus, the surface brightness oscillates with the growth cycle.

The surface morphology in image (b) is almost the same as the initial surface. During the third cycle, small islands can be seen, and the size of the islands increases for successive cycles. Growth was terminated at the top of image (d). Throughout the growth process and immediately after growth, monolayer steps almost identical to the initial surface are visible, indicating that the surface roughness due to island nucleation is smaller than that for MBE. This is because these islands are only a monolayer high, and their coverage is much smaller than unity. These islands are almost annealed out at 100 s after growth termination. These results indicate that the surface diffusion is much greater in MEE than in MBE, so only monolayer-high roughness develops during growth.

4. CONTROL OF GROWTH SITE

4.1 Selective growth using atomic step network

Selective growth of wire or dot structures was explored using atomic step bands in combination with *in situ* observation. Ordered structures of atomic-step bands (Ogino 1997) were created by annealing a Si substrate containing an array of small holes, as schematically shown in Fig. 5. An array of holes is formed using a lithographic method on a vicinal Si substrate (a). Then, the substrate is heated up to 1200-1300°C in ultrahigh vacuum. Due to sublimation and surface diffusion, the hole shape changes to cone-like and the bottom terrace expands to form a wide (111) plane. The steps around the hole gather at the higher end of holes (b). Finally, the holes are completely filled in, and the surface contains regularly-spaced step-bands separated by (111) terraces (c). Because of the misorientation between the hole array and step directions, a step band branches to the step band nearest it, forming a 2-D network. Major bands are perpendicular to the misorientation direction of the vicinal surface.

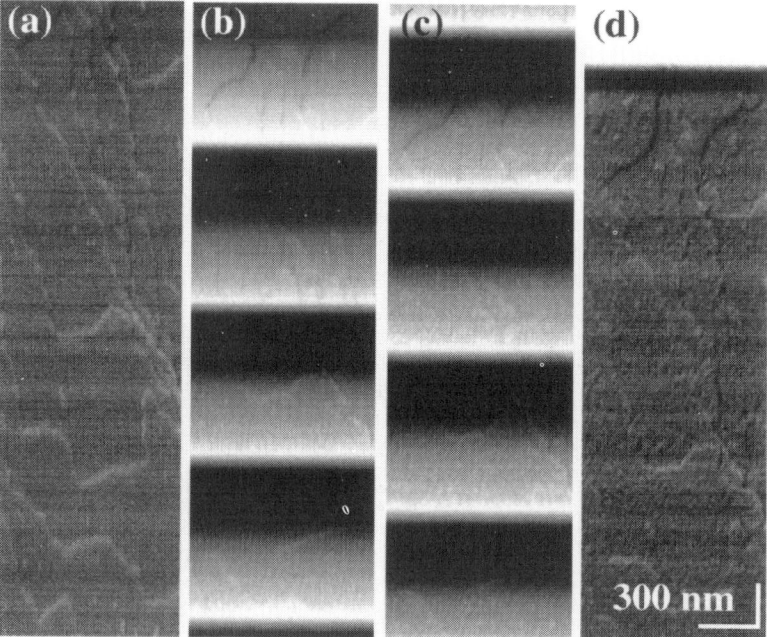

Fig. 4. SEM images of GaAs(001) surface during MEE at 500°C. Those are (a) the initial surface, (b, c) during growth, and (d) after growth. The scanning time of each image is about 70 s.

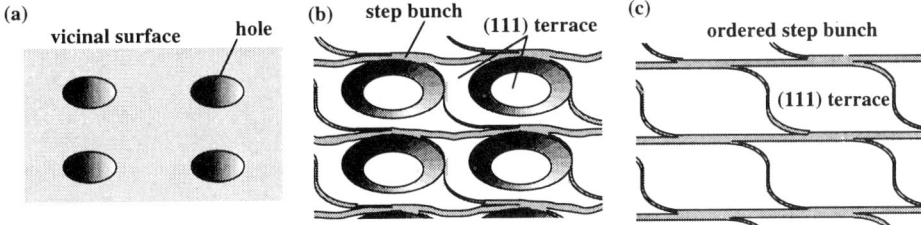

Fig. 5. Schematic illustration of the step-bunching process on a surface with a regular hole array. (a) A hole array is formed on a vicinal Si(111) suface. (b) Heating in ultrahigh vacuum causes the hole diameter to expand and the depth to become shallow due to release of adatoms from steps and filling-in of the bottom terrace. It also makes steps gather at the higher end of each hole. (c) After the holes are filled in completely, a regular array of step bands and (111) terraces remain.

Using the regularly step bunched substrate, GaAs was deposited by MBE. By choosing the appropriate growth temperature and deposition rate, selectivity of island growth (comparing between step bands and flat terraces) increases, and growth on terraces can be eliminated (Finnie 1998). Figure 6 demonstrates GaAs selective growth at step bands. At lower temperatures, selectivity is enhanced by the increasing diffusion length of deposited atoms. When the effective diffusion length, which is large for lower deposition rates, is larger than the half width of a terrace, all the deposited atoms form nuclei at the step bands. Image (a) shows this situation for 400°C growth. Three dimensional GaAs islands have nucleated along the step bands. At a higher temperature, deposited atoms are desorbed more quickly on terraces than on step bands. Image (b) shows selective growth of GaAs at 550°C. In this case, desorption inhibits island nucleation on the terraces. After a long deposition, 3-D GaAs islands merge with each other and form a rather continuous layer. Complete selectivity was obtained.

4.2 Selective island formation

Another type of selective growth is dot-like island formation at a fixed location in each pattern unit. When the diffusion length of deposited material is comparable to the step network size and the mobility of the material along the step band is high, only one island nucleates in every network unit. Such a situation can be realized for liquid metals (Homma 1999). Figure 7 shows the ordering process of Au islands. Here, the holes were only partly filled in, and regularly arranged step bands with shallow holes remained, like Fig. 5(b). A high density of islands were nucleated on the Si surface by depositing Au at 400°C [image (a)]. They are liquid-phase Au-silicide islands, which are made more mobile by annealing at higher temperatures. Image (b) shows the same surface annealed at 560 °C for 4 min. The islands have merged together and grown larger. A regular alignment of islands can already be discerned between the holes at this stage. Further annealing made the aligned islands larger and the other islands smaller. Annealing for another 20 min eliminated extra islands, leaving almost perfectly aligned islands at the same positions in the periodic structure [image (c)]. These islands are located at a convex corner of a

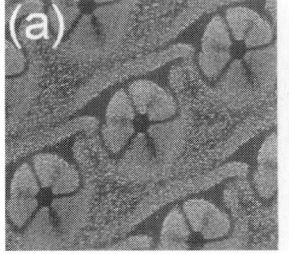
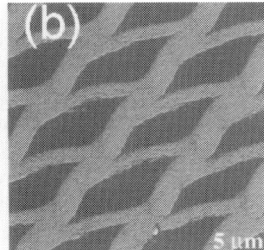

Fig. 6. SEM images of selectively grown GaAs on step bands. GaAs was deposited on (a) a Si(111) substrate with half-filled-in holes at 400°C, and (b) one with completely-filled-in holes at 550°C. The deposited thickness ıs larger for (b) than for (a). Brighter regions are GaAs and darker ones are the Si substrate.

step band. The island size is controllable, depending on the pattern period, the total amount of deposition, and the total amount of desorption. The variation in island size is small.

5. SUMMARY

Surface morphology evolution due to 2D island nucleation and coalescence is clearly imaged in GaAs MBE. The larger diffusion length of adatoms on GaAs(111)A surface makes the islands larger and enables the derivation of step velocity and other growth parameters, which are confirmed to fit well to the standard theory of near-equilibrium crystal growth. For MEE growth the surface diffusion is shown to be so large that only monolayer-high roughness develops during growth.

Site selective growth is achieved on a patterned Si(111) surface that was heated in ultrahigh vacuum to obtain a regularly-arranged step-band structure. GaAs grows selectively at step bands, eliminating any growth on (111) terraces. Au forms silicide islands at fixed locations on the array of step bands. The size of grown structures can be precisely designed by controlling the size of the pattern. Large-scale complex networks of nanostructures can be fabricated using such a step-structure-controlled substrate.

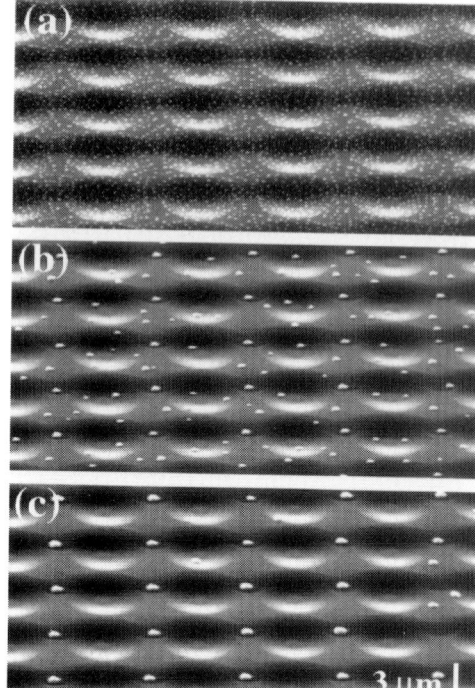

Fig. 7. Successive SEM images of the Au-island alignment process. Images were taken after (a) deposition of Au at 400°C, (b) annealing for 4 min at 560°C, and (c) annealing for an additional 22 min at 560°C.

ACKNOWELGEMENTS

The authors would like to thank Jiro Osaka and Naohisa Inoue for their collaboration on the development of the MBE observation technique. They also would like to thank Yoshiji Horikoshi and Toshio Ogino for their encouragement of this work.

REFERENCES

Burton W K, Cabrera N and Frank F C 1951 Philos. Trans. R. Soc. London, Ser A **243**, 299
Finnie P and Homma Y 1998 Appl. Phys. Lett. **72**, 827
Homma Y, Osaka J and Inoue N 1995 Jpn. J. Appl. Phys. **34**, L1187
Homma Y, Osaka J and Inoue N 1996a Surf. Sci. 357, 441
Homma Y, Yamaguchi H and Horikoshi Y 1996b Appl. Phys. Lett. **68**, 63
Homma Y, Finnie P and Ogino T 1999 Appl. Phys. Lett. **74**, 815
Horikoshi Y, Kawashima M and Yamaguchi H 1986 Jpn. J. Appl. Phys. **25**, L868
Ogino T, Hibino H and Homma Y 1997 Appl. Surf. Sci., **117/118**, 642
Shima M, Kobayashi K, Tanishiro Y and Yagi K 1991 J. Cryst. Growth **115**, 359
'Swiech W and Bauer E 1991 Surf. Sci. **255**, 219
Voigtländer B and Weber T 1996 Phys. Rev., B **54**, 7709
Yamaguchi H and Homma Y 1998 Appl. Phys. Lett. **73**, 3079

Inst. Phys. Conf. Ser. No 164
Paper presented at Microsc. Semicond. Mater. Conf., Oxford, 22–25 March 1999
© *1999 IOP Publishing Ltd*

Bismuth and antimony nanolines in a Si epitaxial layer

K Miki, H Matsuhata, K Sakamoto, G A D Briggs,[a] J H G Owen [a,*] and D R Bowler [a,]**

Electrotechnical Laboratory, 1-1-4 Umezono, Tsukuba, 305-8568, Japan
[a] Department of Materials, University of Oxford, Parks Rd., Oxford, OX1 3PH, UK
[*] present address: Department of Mathematics, University of California, Los Angeles, CA90095, USA
[**] present address: Department of Physics and Astronomy, University College, London WC1E 6BT

ABSTRACT: We succeeded in fabrication of bismuth and antimony nanolines in a Si epitaxial layer. Perfect Bi nanolines form in the terrace of Si(001) in the case of Bi adsorption around its desorption temperature. Although simple epitaxy on this surface allows the surface segregation of Bi from the lines, additional Bi desorption on the line-formed surface before Si overgrowth forbids Bi surface segregation. Additional Sb adsorption instead of Bi led to Sb nanolines in the Si epitaxial layer. Those mechanisms can be explained by use of the analogy of a surfactant.

1. INTRODUCTION

Many Group III and V elements tend to form atomic lines at the initial stage of epitaxial growth on Si(001) (e.g. Baski 1990). We have found that Bi is able to form lines in the terraces of a Si(001) surface. These lines lie within the surface, rather than epitaxially on it, and they are atomically perfect, without any kinks or other defects. These lines form on Si(001) by a selective desorption process around the temperature at which most of the bismuth desorbs (T_D = 500 – 600 °C) from bismuth epitaxial layers. In the present paper we would also suggest a new idea to bury the bismuth line in a Si epitaxial layer on the basis of an atomic exchange model of the surface segregation mechanism. The obtained buried Bi line is very useful for either electrical or optical measurements.

2. EXPERIMENTAL

We used an electron-beam evaporator for Si and resistively heated effusion cells for Bi. Typical Bi flux rate was 3×10^{12} atoms cm^{-2} s^{-1}. A Si(001) substrate was subjected to a standard cleaning process and was loaded into the growth chamber (Miki 1998). The growth rate of Si was 0.3 ML s^{-1} (1 ML=6.8×10^{14} sites cm^{-2}) A 30 keV RHEED system was used for surface analysis. SIMS measurements were made on a CAMECA IMS-4f instrument with O_2^+ primary ions of 3kV for detecting ^{209}Bi and Physical Electronics 6650 with Cs$^+$ primary ions of 3kV for detecting ^{121}Sb. Specimens for cross-sectional HREM observation were prepared using a standard Ar-ion etching technique. A 400 keV HRTEM was operated at 200 keV to reduce the interdiffusion induced by electron beam irradiation.

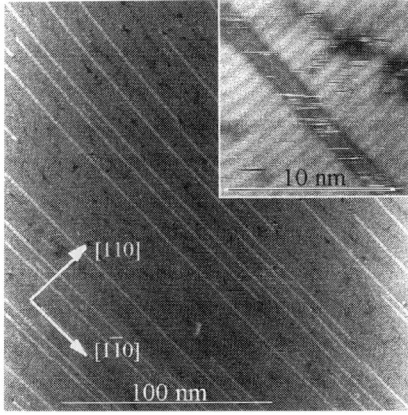

Fig. 1. 200 nm wide STM image of the Bi line at sample bias +2 V. Inset shows a close up view of a Bi line (scan width 12.9 nm and sample bias 0.2 V). The width of the Bi line is 1 nm, equivalent to three Si dimers.

3. Bi PERFECT LINES ON Si(001)

The Bi lines can be formed either by first covering a Si(001) surface with a monolayer or so of Bi and then heating it until most of the bismuth desorbs, or by depositing Bi on a Si(001) surface at a temperature at which the desorption rate is close to the incident flux (Miki 1999). When bismuth adsorbs onto silicon surface well below T_D, an epitaxial Bi monolayer with a $(2 \times n)$ reconstruction is formed (cf. Park 1993). Subsequent annealing close to T_D causes the surface to undergo a radical change. Fig. 1 shows a scanning tunneling microscope (STM) image of the surface after desorption of most of the bismuth epitaxial layer. The remaining Bi forms lines, perpendicular to the silicon dimer rows, which extend beyond the maximum field of view (200 nm × 200 nm). The lines are shown at higher magnification in the inset of Fig. 1: they take up the equivalent width of three dimers in the Si(001) surface, and have an internal structure which shows two dimer-shaped features, aligned with the surrounding silicon dimer rows, darker than the surrounding terraces at bias voltages in the range -0.6 V to + 0.6 V, and brighter than them at greater bias voltage. This suggests that they are at the same height as the silicon dimers, but with a larger band gap.

We have investigated the structure of the lines by atomistic modeling using tight binding total energy calculations (Goringe 1997), using the density matrix method (Nunes 1993). The most energetically favorable structure is shown in Fig. 2 (a) and (b). It involves rebonding in the trench; we know that this rebonded structure is favored in pure silicon with a missing dimer defect (Owen 1995), and this is likely to be even more so with Bi because of the greater length of Bi bonds. For this structure the calculated excess surface energy plus bismuth adsorption energy per pair of Bi atoms is -14.5 eV. Local density of states (LDOS) calculations indicate that this structure is indeed a quantum antiwire, with a lower density of states close to the Fermi level than a pure Si(001)-2×1 surface, in agreement with the bias-dependent imaging in STM.

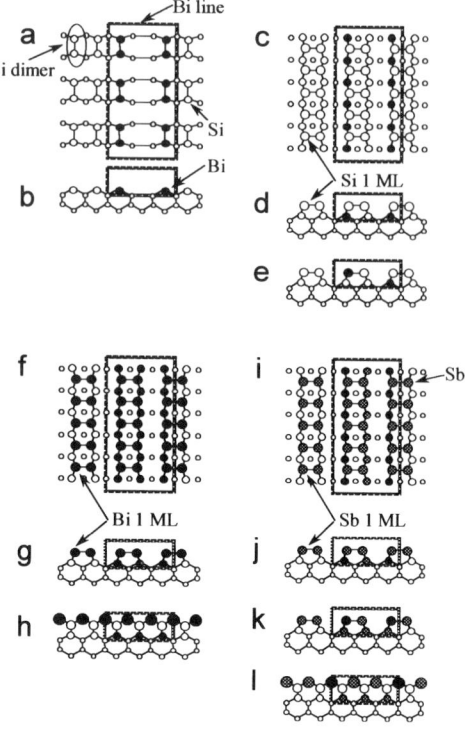

Fig. 2. Ball and Stick models for surface segregation and suppression mechanisms of Bi from the Bi line. (a) and (b) are plan and cross sectional views of our Bi line model. (c) and (d) are the case with Si 1 ML overgrowth and an exchange of Si atom on top and Bi in the Bi line causes surface segregation (e). (f) and (g) are the case with Bi 1 ML overgrowth and Bi atoms in the Bi line remain. After additional 1 ML Si growth, Bi atoms tend to go to the top layer and Bi in the Bi line could be free from the surface segregation mechanism (h). Additional Sb adsorption (i and j) instead of Bi leads the Sb nanoline in a Si epitaxial layer (k) and further 1 ML Si growth leads to surface segregation of Bi and Sb and Sb could be fixed at the place where originally the Bi line existed (l).

4. Bi LINE IN Si EPITAXIAL LAYER

Since the surface structure of the Bi line differs from the $(2 \times n)$ structure, Bi segregation from the line structure may differ from the complete segregation upon subsequent growth on a Bi($2 \times n$) saturated surface (Sakamoto 1993). Fig. 3 shows a secondary ion mass spectroscopy (SIMS) profile after Si epitaxial growth on the Bi lines formed on a Si(001) surface at 400°C. The existence of the peak at the place where the Bi line initially existed demonstrates this expectation. The residual amount of 3.5×10^{11} Bi atoms cm^{-2} is two to three orders of magnitude smaller than the initial value. The maximum value is estimated at around 1.6×10^{14} Bi atoms cm^{-2}. Here we assumed our Bi line model where four Bi atoms are in a (2×3) unit and the distances between nearest neighbors normal to the Bi line is 8×. The final value depends on the initial density of Bi lines, so it may be less than 1.6×10^{14} Bi atoms cm^{-2}. We measured this value by growing an amorphous Si layer on the surface in order to prevent further Bi segregation. The SIMS results show that actual amount of Bi of the Bi line in the terrace is 6×10^{13} atoms cm^{-2}. Therefore our results show that although surface segregation from the Bi line is less than from the (2×n) structure, the remaining Bi is greatly reduced.

We consider the surface segregation mechanism with use of Fig. 2 (c)-(e). (c) and (d) are plan and cross sectional views after 1 ML Si overgrowth on the Bi-line-formed surface. Exchange between Si atoms on top and Bi atoms in the Bi lines could account for the surface segregation. Another Bi layer as indicated in Fig. 2 (f) - (g), could inhibit surface segregation. This resembles the 'self-limiting process' observed in initial stages of Ge surface segregation during Si epitaxial growth (Fukatsu 1991). At normal growth temperature it is impossible to make the second layer since the (2×n) Bi layer is saturated at around 1 ML and the Bi line formation stabilizes at 8× spacing. However we can combine their structures by cooling below the Bi desorption temperature and dosing Bi onto the surface (Fig. 2g) leaving the Bi line in the third layer so that segregation does not occur. Surface segregation of Bi in the top layer continues throughout the Si epitaxial growth.

We demonstrated this idea as indicated in Fig. 5. We formed the Bi lines by dosing Bi at 600°C and monitoring the surface structure with reflection high-energy electron diffraction, then cooled down the substrate to 400 °C and covered the surface with (2×n) Bi layer. This surface corresponds to I_1 in the structure model shown in Fig. 4(b). On this

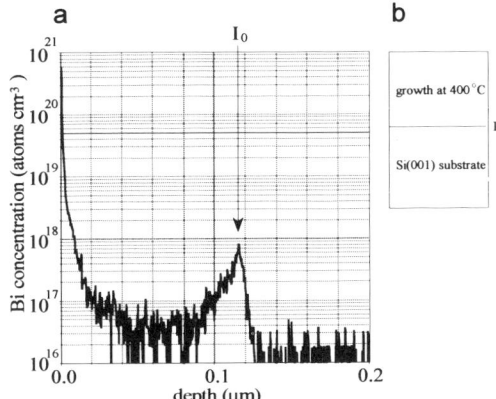

Fig. 3. SIMS profile of Bi after simple Si overgrowth on the Bi line formed Si(001) surface. Unlike (2×n) Bi structured surface some Bi remains, however its concentration is less than the initial value.

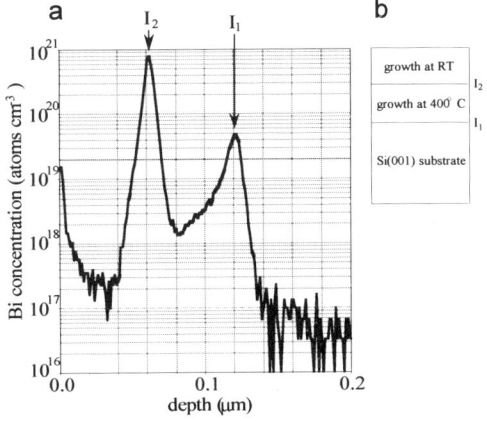

Fig. 4. SIMS profile of Bi after 1 ML Bi overgrowth on the Bi line formed Si(001) surface and following overgrowth of Si on the surface. There are two peaks. I_1 corresponds to surface segregation of Bi during the Si overgrowth and I_2 locates around the place where the Bi line exists and the Bi concentration is very close to the initial value.

surface we grew 60 nm Si, then stopped the growth once at the place that corresponds to I_2 in the structure model which is shown in Fig. 4(b). Then we reduced the growth temperature to room temperature in order to freeze the surface segregated Bi at I_2. We see peaks at both I_1 and I_2 in the SIMS profile (Fig. 4), with 7×10^{13} and 6×10^{14} Bi atoms cm^{-2}. The Bi amount around I_1 is very close to the initial amount, suggesting that Bi lines followed the overgrowth of 1 ML (2×n) Bi. Final evidence lies in cross sectional HREM along [110] of a sample prepared as above, except that the Bi line density was less (Fig. 5). The [001] separation between dots corresponds to the thickness of 2 ML.

We identify two Bi lines; the right line (l_d) has misfit dislocations along d_1 and d_2, while the left one (l_p) has none. The dislocations may form because of the larger size of Bi atoms.

Can we extend this exchange mechanism to make another kind of nanoline, using another metal instead of Bi? A schematic model is indicated in Fig 2 (i)-(l). After putting 1 ML Sb layer on the Bi line surface, if an exchange of Sb with Bi from the line is favorable, then the exchanged Bi and any remaining Sb will segregate during Si epitaxial growth (Fig. 6). Except for the growth temperature of 360 °C, the fabrication conditions were the same as before. We again see peaks of either Bi or Sb concentrations at I_3 and I_4 in the SIMS profile. At I_3 the Sb concentration is thirty times higher than Bi, which, together with the Bi peak at I_4, confirms our expectation. Although we did not take a cross sectional TEM image of I_4, we believe there is a high possibility of making Sb nanolines in Si epitaxial layers.

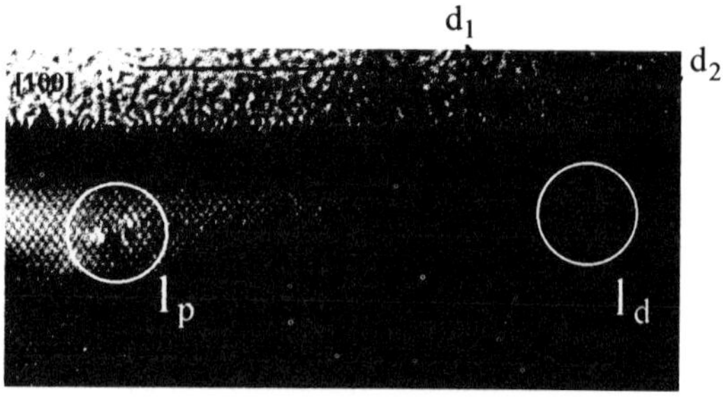

5. CONCLUSION

We discovered how to make a new type of perfect nanoline of Bi in a terrace of Si(001) and extended this to a structure more suitable for either electronic or optical applications. We postulate an atom exchange model based on surface segregation and surfactant phenomena. The obtained buried metal line could be useful for future nanoelectronics.

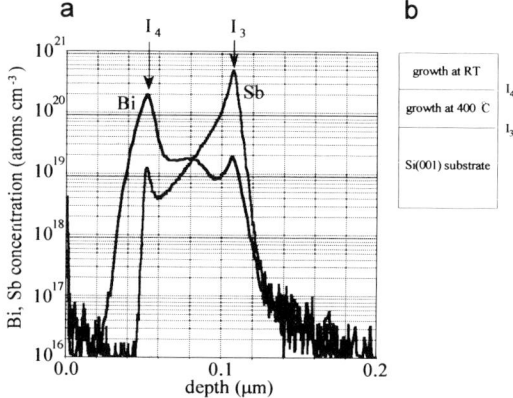

Fig. 5. A cross sectional TEM image of Bi lines buried in Si epitaxial layer. The left one (l_p) is defect free while the right one (l_d) caused misfit dislocation along d_1 and d_2.

Fig. 6. SIMS profiles of Bi and Sb after 1 ML Sb overgrowth on the Bi-line-formed Si(001) surface and following overgrowth of Si on the surface. Both profiles have two peaks. Most of the Bi in the Bi line was exchanged into Sb.

ACKNOWLEDGEMENTS

High-resolution microscopy was carried out at Kyushu University. The authors would like to thank Professor Tomokiyo and Mr. Tanabe for use of the electron microscopy. SIMS measurements were carried out in Foundation for Promotion of Materials Science and Technology of Japan. We thank the British Council for support for our collaboration.

REFERENCES

Baski A A, Nogami J and Quate C F 1990 J. Vac. Sci. & Technol. **A8**, 245
Fukatsu S, Fujita K, Yaguchi H, Shiraki Y and Ito R 1991 Appl. Phys. Lett. **59**, 2103
Goringe C M, Bowler D R and Hernandez E H 1997 Rep. Prog. Phys. **60**, 1447
Li X P, Nunes W and Vanderbilt D 1993 Phys. Rev. **B47**, 10891
Miki K, Owen J H G, Bowler D R, Briggs G A D, Sakamoto K 1999 Surf. Sci. **421**, 397
Miki K, Sakamoto K and Sakamoto T 1998 Surf. Sci. **406**, 312
Owen J H G, Bowler D R, Goringe C M , Miki K, and Briggs G A D 1995 Surf. Sci. **341**, L1042.
Park C, Bakhtizin R Z, Hashizume T and Sakurai T 1993 Jpn. J. Appl. Phys. **32**, L528
Sakamoto K, Matsuhata H, Kyoya K, Miki K and Sakamoto T 1993 Jpn. J. Appl. Phys. **32**, L204

Inst. Phys. Conf. Ser. No 164
Paper presented at Microsc. Semicond. Mater. Conf., Oxford, 22–25 March 1999
© 1999 IOP Publishing Ltd

Phase separation and facet formation during the growth of (GaAs)$_{1-x}$(Ge$_2$)$_x$ alloy layers by metal organic vapour phase epitaxy

A G Norman, J M Olson, J F Geisz, H R Moutinho, A Mason, M M Al-Jassim and S M Vernon[1]

National Renewable Energy Laboratory, 1617 Cole Boulevard, Golden, CO 80401, USA
[1]Spire Corporation, One Patriots Park, Bedford MA 01730, USA

ABSTRACT: Metal organic vapour phase epitaxy (GaAs)$_{1-x}$(Ge$_2$)$_x$ alloy layers, $0<x<0.22$, were grown at temperatures between 640° and 690°C, on vicinal (001) GaAs substrates. Phase separation occurred in all the layers. The phase-separated microstructure changed with alloy composition, growth temperature, and substrate orientation. In $x \sim 0.1$ layers grown at 640°C, Ge segregation occurred on {115}B planes associated with a {115}B surface faceting. Increase in growth temperature led to the formation of large, (001)-oriented, irregular-shaped platelets of Ge-rich material. Growth on {115}B substrates resulted in a "natural superlattice" of GaAs/Ge along the growth direction.

1. INTRODUCTION

Two-junction Ga$_{0.52}$In$_{0.48}$P/GaAs solar cells have demonstrated record-breaking efficiencies (Bertness et al 1994, Takamoto et al 1997). They are in production for space photovoltaic applications and are also leading candidates for concentrator cells in terrestrial applications. More efficient solar cells may be achieved by adding extra junctions in layers with lower band gaps. An ideal material for such an extra junction would be lattice matched to GaAs and have a 1 eV band gap (Kurtz et al 1997). Possible materials fulfilling these requirements are (GaAs)$_{1-x}$(Ge$_2$)$_x$ metastable alloys, the subject of this work, and GaInAsN alloys (Friedman et al 1998). GaAs and Ge, despite being size matched, are mutually insoluble in the equilibrium bulk solid state resulting in almost complete phase separation into GaAs-rich and Ge-rich regions at all temperatures below the melting point (Takeda et al 1965, Osório et al 1991). The reason for this phase separation is the high energy required to form Ga-Ge and As-Ge bonds, which do not satisfy the octet rule for valence electrons, observed in the pure components, and the even higher energies predicted for As-As and Ga-Ga antisite bonds (Osório et al 1991). Despite the strong tendency of this alloy toward phase separation, there have been several reports of the growth of relatively homogeneous epitaxial layers of metastable (GaAs)$_{1-x}$(Ge$_2$)$_x$ alloys across the composition range using non-equilibrium techniques such as metal organic vapour phase epitaxy (MOVPE), ion-assisted sputter deposition, and molecular beam epitaxy (MBE). Growth of single-phase, metastable alloys was reported by MOVPE in the temperature range 700°–750°C (Alferov et al 1982) and by sputter deposition in the temperature range 450°–550°C (Barnett et al 1982, Romano et al 1987). GaAs-rich sputter-deposited layers, however, contained a network of Ge-mediated antiphase boundaries that percolated between zinc-blende phase/antiphase domains without causing significant antisite formation (Romano et al 1987). Banerjee et al (1984, 1985) reported phase separation in MBE (GaAs)$_{1-x}$(Ge$_2$)$_x$ layers grown between 550° and 620°C on (001), (110) and (211) GaAs substrates, resulting in the formation of 10–30 nm, {110}-oriented Ge-rich regions in the surrounding GaAs-rich material. Growths at 430°C on (001) substrates appeared to be single phase. Baird et al (1991) did not find any evidence of phase separation in MBE (GaAs)$_{1-x}$(Ge$_2$)$_x$ layers grown on (001) GaAs substrates at temperatures up to 580°C. This system has also attracted considerable theoretical interest (e.g., Osório et al 1991). This is because a transition from the GaAs, zinc-blende structure to the Ge diamond cubic structure has been reported to occur in single-phase metastable alloys at some critical composition x (due to the different crystal structures of the two end-point constituents). In this work we report evidence of phase separation in MOVPE-grown (GaAs)$_{1-x}$(Ge$_2$)$_x$ layers (see also Norman et al 1999). The observed segregation exhibits a microstructure completely different from that reported before (to the best of our knowledge) in $(GaAs)_{1-x}(Ge_2)_x$ layers. The phase-separated microstructure depends on alloy composition, growth temperature, and substrate orientation and, in some cases, is associated with a surface faceting that occurs during growth.

2. EXPERIMENTAL DETAILS

$(GaAs)_{1-x}(Ge_2)_x$ layers, $0<x<0.22$, were grown by low-pressure ($\approx$ 50–70 Torr) MOVPE in two different reactors, at growth temperatures between 640–690°C, on vicinal (001) GaAs substrates. The source chemicals used for growth were trimethylgallium, arsine, and germane. Substrate rotation was used in one reactor but not in the other. The average Ge content of the layers was measured from a 20 μm diameter area using wavelength-dispersive electron-probe X-ray microanalysis at 10 kV, using Lα lines and GaAs and Ge as standards, and to an accuracy of $\approx$0.5 at. %. Transmission electron microscopy (TEM) cross-section samples were prepared by conventional mechanical and ion-milling techniques and examined in a Philips CM30. (110) and $(\bar{1}10)$ cross sections were distinguished using convergent-beam electron diffraction (Taftø and Spence 1982). Atomic Force Microscopy (AFM) was performed in air on the growth surface topography using a Park Scientific Instruments Autoprobe LS in the noncontact mode.

3. RESULTS

Fig. 1 shows 002 dark-field (DF) TEM micrographs of (110) and $(\bar{1}10)$ cross sections of a $(GaAs)_{0.78}(Ge_2)_{0.22}$ layer, grown at 675°C at Spire, which exhibits pronounced phase separation. This layer was grown on a (001) GaAs substrate, miscut 2° toward (010), at a rate of $\approx$ 2.4 μm per hour. The substrate was rotated at $\approx$ 15 revolutions per minute during growth. In this picture, the Ge-rich regions appear dark because the 002 reflection is forbidden for the diamond cubic structure of Ge. In the (110) cross section, the Ge-rich regions in the layer form an interconnected network of ribbons forming a cell-like structure embedded in GaAs-rich zinc-blende material. The Ge-rich regions are not antiphase boundaries in these layers, and so are different from the Ge-mediated antiphase boundaries previously reported in sputter-deposited layers (Romano et al 1987). Thicker Ge-rich plates, oriented close to (001), occur in some areas, and are connected by Ge-rich ribbons having a tendency to lie on {115}B planes. These {115}B Ge-rich ribbons in many cases are not continuous and show spot-like contrast, indicating that they are composed of closely spaced clusters or rods of Ge-rich material. In the orthogonal $(\bar{1}10)$ cross section, Fig. 1(b), the Ge-rich regions show a completely different morphology, and form a series of dark contrast bands, $\approx$5-10 nm thick, inclined by $\approx$2° to the $(GaAs)_{1-x}(Ge_2)_x$ layer/GaAs buffer layer interface. This inclination we believe is associated with the offcut of the substrate from (001). The bands are not continuous and gradually appear and disappear as you move along them. A low density of small antiphase domains was observed in some regions associated with the growth of zinc-blende GaAs-rich material on thick, Ge-rich, diamond cubic plates. No extra diffraction spots were observed in transmission electron diffraction (TED) patterns, which rules out the existence of GeAs or GeAs$_2$ phases because their crystal structures are different from GaAs and Ge (Pearson 1967). A $(GaAs)_{0.78}(Ge_2)_{0.22}$ layer, grown at the National Renewable Energy Laboratory (NREL), at 640°C without substrate rotation, on an (001) GaAs substrate offcut 2° toward $(\bar{1}10)$, showed a similar phase-separated microstructure (Norman et al 1999), indicating that substrate rotation was not responsible for the phase-

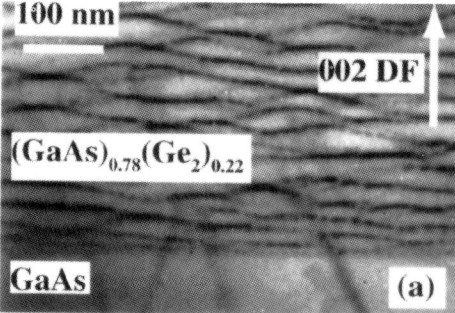

Fig. 1. 002 DF TEM micrographs of $(GaAs)_{0.78}(Ge_2)_{0.22}$ layer, grown at Spire at 675°C, showing pronounced phase separation: (a) (110) cross section; (b) $(\bar{1}10)$ cross section.

separated microstructure in the sample of Fig. 1.

Fig. 2 shows (110) cross-section, 002 DF micrographs of $(GaAs)_{0.90}(Ge_2)_{0.10}$ layers grown at NREL at 643°, 666°, and 689°C, separated by thin InGaP spacer layers, on a (001) GaAs substrate offcut 2° toward ($\bar{1}$10). In the layer grown at 643°C, Fig. 2(a), the phase-separated microstructure is remarkably regular, with thin sheets of Ge-rich material lying on both sets of {115}B planes forming a diamond pattern as they intersect (Norman et al 1999). The Ge-rich sheets are not continuous in some areas and are formed of closely spaced clusters or rods of Ge-rich material lying on the {115}B planes. AFM of the growth surface of a similar $(GaAs)_{0.90}(Ge_2)_{0.10}$ layer, Fig. 3, clearly shows {115}B surface facets, which are identical to the planes observed for the Ge segregation in this sample (Norman et al 1999), suggesting that the Ge segregation and the growth surface morphology are related. In the layer grown at 666°C, Fig. 2(b), it can be seen that thicker, (001)-oriented, Ge-rich plates are starting to form and are connected by thin Ge-rich sheets on {115}B planes. Antiphase domains, e.g., marked APD in Fig. 2 (b), are sometimes formed in this layer during overgrowth of the Ge-rich, diamond cubic plates by the zinc-blende GaAs-rich material. In the layer grown at 689°C, Fig. 2(c), only thick, irregular cross-section, (001)-oriented plates of Ge-rich material are present. Convergent beam electron diffraction indicates that the Ge-rich plates have the diamond cubic structure, whilst the GaAs-rich material is zinc-blende. In the orthogonal ($\bar{1}$10) cross section, the Ge-rich regions again appeared as discontinuous bands, inclined at a slight angle to (001), whose thickness and length increased with growth temperature. As far as we know, the only previous report of similar phase-separated microstructures in a semiconductor alloy was that of Seong et al (1993) for MBE $InAs_ySb_{1-y}$ alloys grown at low temperatures. Growth of a $(GaAs)_{0.90}(Ge_2)_{0.10}$ layer at 640°C, on a {115}B GaAs substrate, resulted in the phase separation only occurring on the {115}B planes parallel to the growth surface, Fig. 4, forming a "natural" GaAs/Ge superlattice along the growth direction.

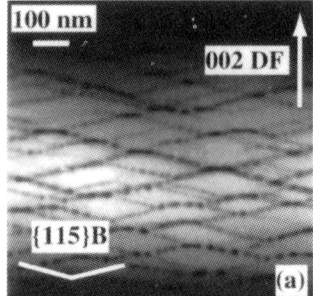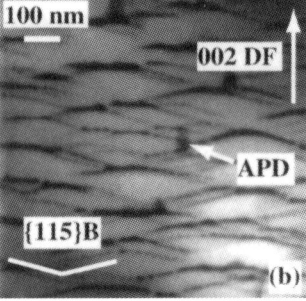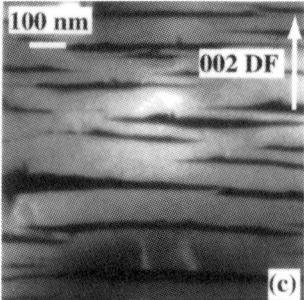

Fig. 2. (110) cross-section, 002 DF TEM images showing phase-separated microstructure of $(GaAs)_{0.90}(Ge_2)_{0.10}$ layers, grown at NREL at: (a) 643°C; (b) 666°C; and (c) 689°C.

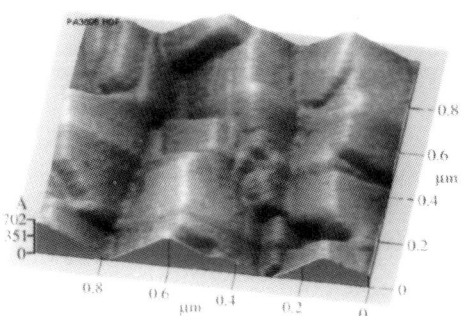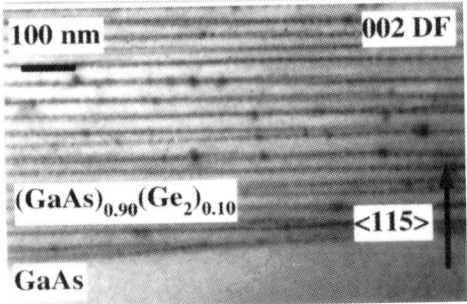

Fig. 3. AFM image of growth surface of $(GaAs)_{0.90}(Ge_2)_{0.10}$ layer, grown at 640°C, showing {115}B surface facets.

Fig. 4. (110) cross-section, 002 DF, TEM image of {115}B $(GaAs)_{0.90}(Ge_2)_{0.10}$ layer grown at 640°C containing "natural" superlattice along [115]B growth direction.

174

4. DISCUSSION AND CONCLUSIONS

The characteristic phase-separated microstructure found in the $(GaAs)_{0.90}(Ge_2)_{0.10}$ layers grown at 640°C, we suggest, may develop as follows. As the $(GaAs)_{0.90}(Ge_2)_{0.10}$ layer starts growing, the GaAs-rich phase deposits first, with the excess Ge segregating to the growing layer surface because the formation of the high-energy As-Ge and Ga-Ge bonds is unfavourable. The accumulation of excess Ge at the surface triggers the spontaneous formation of {115}B surface facets to lower the surface energy. After the surface Ge concentration reaches a critical value, nucleation of Ge-rich material occurs on the {115}B facets. The excess surface Ge then precipitates out, conformal to the growth surface, because it can now form low-energy Ge-Ge bonds at the edges of the Ge-rich nuclei. The GaAs-rich phase continues to grow and repetition of the above growth behaviour results in the observed microstructure. The repeated surface segregation of Ge, followed by nucleation and growth of Ge-rich material once a critical surface Ge concentration is reached, could explain the quasi-periodic nature of the GaAs/Ge "natural" superlattice along the growth direction of sample grown on a {115}B substrate, Fig. 4. The growth process is really a simple eutectic solidification, but from the vapour phase rather than from the more normal liquid phase. A low density of antiphase domains is observed in the GaAs-rich phase, despite the growth of the polar, zinc-blende GaAs-rich material on top of the non-polar diamond cubic Ge-rich material. This may be a consequence of epitaxial lateral overgrowth of the Ge-rich phase by GaAs-rich material emanating from holes in the Ge rich sheets or gaps between Ge-rich plates. The polarity of this GaAs-rich material is determined by the underlying GaAs-rich phase, thus reducing the formation of antiphase domains. Photoluminescence measurements on a series of phase-separated $(GaAs)_{1-x}(Ge_2)_x$ layers, grown across the composition range, by Spire (Vernon et al 1994) revealed pronounced band-gap narrowing which we believe may be a consequence of the phase separation.

In conclusion, we have observed pronounced phase separation in $(GaAs)_{1-x}(Ge_2)_x$ alloy layers, grown by low-pressure MOVPE, that may cause substantial band-gap narrowing in these samples. The phase-separated microstructure depended on alloy composition, growth temperature, and substrate orientation.

ACKNOWLEDGEMENTS

The work performed at the National Renewable Energy Laboratory was supported by the Office of Energy Research, Basic Energy Sciences, and the work at Spire by the U.S. Air Force and Ballistic Missile Defence Organisation.

REFERENCES

Alferov Zh I, Zhingarev M Z, Konnikov S G, Mokan I I, Ulin V P, Umanskii V E and Yavich B S 1982 Sov. Phys. Semicond. **16**, 532
Baird R J, Holloway H, Tamor M A, Hurley M D and Vassell W C 1991 J. Appl. Phys. **69**, 226
Banerjee I, Kroemer H, and Chung D W 1984 Mater. Lett. **2**, 189
Banerjee I, Chung D W and Kroemer H 1985 Appl. Phys. Lett. **46**, 494
Barnett S A, Ray M A, Lastras A, Kramer B, Greene J E, Raccah P M and Abels L L 1982 Electron. Lett. **18**, 891
Bertness K A, Kurtz S R, Friedman D J, Kibbler, A E, Kramer C and Olson J M 1994 Appl. Phys. Lett. **65**, 989
Friedman D J, Geisz J F, Kurtz S R and Olson J M 1998 J. Crystal Growth **195**, 409
Kurtz S R, Myers D and Olson J M 1997 Proc. 26th IEEE Photovoltaic Specialists Conf., (New York: IEEE) pp875–878
Norman A G, Olson J M, Geisz J F, Moutinho H R, Mason A, Al-Jassim M M and Vernon S M 1999 Appl. Phys. Lett. **74**, 1382
Osório R, Froyen S and Zunger A 1991 Phys. Rev. B **43**, 14055
Pearson W B 1967 A Handbook of Lattice Spacings and Structures of Metals and Alloys (Oxford: Pergamon) p 141
Romano L T, Robertson I M, Greene J E, and Sundgren J E 1987 Phys. Rev. B **36**, 7523
Seong T-Y, Norman A G, Ferguson I T and Booker G R 1993 J. Appl. Phys. **73**, 8227
Taftø J and Spence J C H 1982 J. Appl. Crystallogr. **15**, 60
Takamoto T, Ikeda E, Hurita H and Ohmori M 1997 Appl. Phys. Lett. **70**, 381
Takeda Y, Hirai T and Hirao M 1965 J. Electrochem. Soc. **112**, 363
Vernon S M, Sanfacon M M and Ahrenkiel R K 1994 J. Electron. Mater. **23**, 147

Inst. Phys. Conf. Ser. No 164
Paper presented at Microsc. Semicond. Mater. Conf., Oxford, 22–25 March 1999
© *1999 IOP Publishing Ltd*

Optical properties of anti-phase boundaries and Frenkel-type defects in CuPt-ordered GaInP studied by optical spectroscopy in a transmission electron microscope

Y Ohno and S Takeda

Department of Physics, Graduate School of Science, Osaka University,
1-16, Machikane-yama, Toyonaka, Osaka 560-0043, JAPAN

ABSTRACT: Optical properties of anti-phase boundaries (APBs) and Frenkel-pairs (FPs) in CuPt-ordered GaInP has been examined by *in-situ* photoluminescence and cathodoluminescence spectroscopy in a transmission electron microscope. We have found: 1) the decrease of the band gap energy E_g with decreasing the APB density and 2) three APB luminescence bands peaking at the photon energy of about E_g - 8, E_g - 18, and E_g - 30 meV, respectively. We have shown that the FPs on the Ga and In sublattices, generated by electron-irradiation, act as nonradiative recombination centers.

1. INTRODUCTION

The ternary semiconductor GaInP has been widely studied due to its potential for optoelectronic devices. It is well known that GaInP alloys grown on a GaAs substrate by organo-metallic vapor-phase epitaxy (OMVPE) exhibit ordering of cation atoms on the group-III sublattice (Gomyo *et al* 1988). Transmission electron microscope (TEM) studies have shown that the microstructure of ordered GaInP often consists of domains of ordered crystals bounded by anti-phase boundaries (APBs) (Su *et al* 1994). The sample shows typical luminescence bands at low temperatures, and their properties depend on the microstructures of the domains (Ernst *et al* 1996). Recently, effects of APBs on the luminescence bands have been intensively studied by near-field photoluminescence spectroscopy (Cheong *et al* 1998, Gregor *et al* 1995) and cathodoluminescence (CL) spectroscopy in a scanning electron microscope (Nasi *et al* 1996) combined with TEM studies. Nevertheless, the optical properties have not yet been fully clarified, since the same microscopic area has not been studied.

In the present study, we have investigated the optical properties of APBs by means of *in-situ* CL spectroscopy in a transmission electron microscope. This method enables us to obtain simultaneously structural data in higher spatial resolution by TEM studies and luminescence spectra by CL of the same microscopic area (Ohno and Takeda 1995). We have shown that APBs strongly influence the optical properties of ordered GaInP. We have also studied the optical properties of Frenkel-type defects (Ohno *et al* 1999a).

2. EXPERIMENTS

Ordered GaInP samples were grown on a GaAs substrate by OMVPE at 700 °C; the substrate was 2° off from (001) towards [110]. Cross sections of the sample (thickness of about 500 nm) were prepared with a conventional etching method with Ar$^+$ ions.

CL spectroscopy with the spectral resolution of a few meV was carried out at the temperature of 20 K. An 100 keV electron-beam (probe size of about 100 nm on a specimen surface) was used for CL spectroscopy to avoid irradiation effects (Ohno *et al.* 1999a). The spatial resolution of CL, affected by the generation volume and minority carrier diffusion as well as the probe size, was roughly estimated to be in the range from 200 to 300 nm.

Frenkel-type defects were intentionally introduced by electron irradiation in a TEM; the direction of the electron beam was kept parallel to the [110] zone axis. Incident-electron energy E ranged from 100 to 170 keV, and electron dose D, i.e. electron flux f (7.8×10^{16} cm^{-2} s^{-1}) multiplied by irradiation time, was up to 1.0×10^{23} cm^{-2}.

3. RESULTS AND DISCUSSION

A transmission electron diffraction (TED) pattern of a sample showed intense superlattice reflections (Fig. 1a), indicating that the ordering occurs on the ($\bar{1}11$) plane. This ordering has been attributed to step-flow growth from particular step-edge structures (Suzuki and Gomyo 1991). Figure 1b shows the DF-TEM image created by using the superlattice spot at 1/2, -1/2, 3/2. A number of anti-phase boundaries (APBs), nucleated at the substrate surface, appear to be dark lines in the figure. We found that the density of APBs n_{APB}, the length of APB lines in a unit area in a TEM image, decreases as the distance from the substrate increases. The decrease is presumably due to the mutual annihilations of APBs.

Specimens showed strong CL emission, peaking at the photon energy of about 1.9 eV, due to a band-to-band ("excitonic") recombination (large arrows in Fig. 2). We found that the photon energy of the excitonic peak E_{ex} increased with increasing n_{APB} (circles in Fig. 3). Since the increase was independent of specimen thickness, it may not be caused by surface effects. The increase of E_{ex} corresponds to the increase of the band gap energy E_g, and E_g increases as the degree of atomic ordering, i.e., an order parameter S, decreases. The order parameter can be also estimated by means of TED (Noda and Takeda 1996), and we confirmed the decrease of S with increasing n_{APB} (Ohno *et al* 1999b). The decrease may be explained by an APB propagation model that APBs on a growth surface act as an origin of crystal growth and heterogeneous nucleation occurs at the APBs along with the step-flow growth (Takeda *et al* 1999). According to the model, heterogeneous nucleation may result in the decrease of the degree of atomic ordering around APBs, and therefore S decreases with increasing n_{APB}.

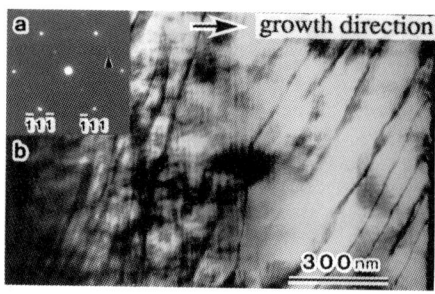

Fig. 1 (a) A TED pattern and (b) DF-TEM image of an ordered GaInP. The superlattice spot that was used to create the DF image is marked by the arrow.

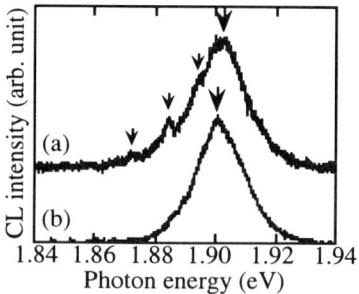

Fig. 2 CL spectra obtained from areas with the APB densiities of (b) 1.8×10^{-2} and (c) 1.3×10^{-2} nm^{-1} respectively.

In addition to the excitonic peak, we found three satellite peaks at the photon energies of E_{ex} - 8, E_{ex} - 18, and E_{ex} - 30 meV, respectively (indicated by small arrows in Fig. 2). Figure 3 shows the satellite-peak energies E_{sat} vs. n_{APB}, indicating that E_{ex} - E_{sats} is independent of n_{APB}. The normalized peak intensities of the satellite peaks I_{nom}, defined as I_{sat}/I_{ex} where I_{sat} and I_{ex} denote the observed intensities of the satellite- and excitonic-peaks, increase as n_{APB} increases (Fig. 3). Similar satellite peaks were also observed in the ordered samples grown at 650 °C (Ohno et al 1999b). These satellite bands are presumably due to recombination via localized energy levels introduced by APBs or impurities segregated to the APBs.

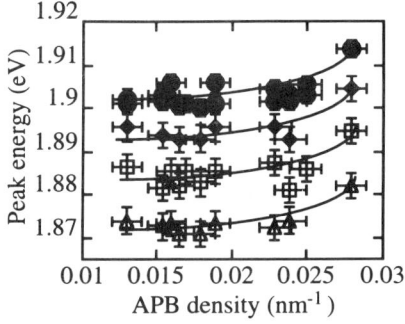

Fig. 3 Peak energies of exitonic (circles) and satellite (open marks) peaks vs. APB density.

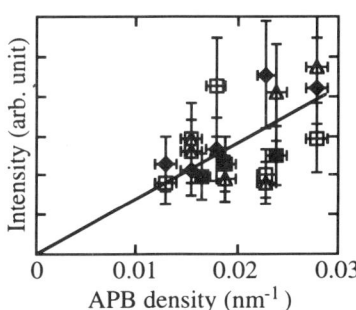

Fig. 4 CL intensities of satellite peaks vs. APB density.

I_{ex} of a specimen irradiated with electrons at the temperature of 110 K shifted to a lower value compared to the as-grown one. We found that the dose-dependent excitonic peak-intensity $I_{ex}(D)$ is well expressed as

$$I_{ex}(D) = I_{ex}(0)/(1+\sigma_l D), \tag{1}$$

where σ_l represents a fit parameter; Fig. 5 shows experimental (marks) and simulated (solid lines) normalized peak intensity, $I_{ex}(D)/I_{ex}(0)$ vs. D for several electron energies. The profiles of the CL spectra of irradiated specimens were the same as the as-grown one. We have shown that the decrease is quite well explained by a recombination-center model that some localized energy levels of irradiation-induced defects act as non-radiative recombination centers (Ohno et al 1996).

It is generally believed that Frenkel-type defects in semiconductors form localized energy levels in the band gap and some of the levels act as recombination centers. Since the threshold electron energy for the displacement of P atoms, $E_{(P)}$, is suggested to be 100 keV (Noda and Takeda 1996), we deduce that the centers are not related to the Frenkel-type defect on the P sublattices. We hence consider that Frenkel-type defects on the Ga and In sublattices are related to the non-radiative centers. The concentration of each defect is theoretically expected, and σ_l was expressed as

$$\sigma_l = v(1-f) \Sigma_\xi \sigma_{D(\xi)} \sigma_{(\xi)}/\{\tau_r^{-1} + \tau_{nr}^{-1}\}, \tag{2}$$

where f represents the correlated recombination factor, $\sigma_{(\xi)}$ the cross sections for displacement damage of ξ-atoms (ξ = Ga and In); $\sigma_{D(\xi)}$ is the capture cross section for carriers (electrons and holes) becoming trapped in the Frenkel-type defect on the ξ-sublattice, and v the velocity of carriers. τ_r and τ_{nr} denote the carrier-lifetimes for the radiative recombination and for the non-radiative recombination due to electron-phonon interactions, respectively. The electron-energy-dependence of σ_l (closed circles in Fig. 6) was well expressed by the theoretical equation when $E_{(Ga)}$ = 143 keV and $E_{(In)}$ = 120 keV, respectively (the solid line in Fig. 6). We thus conclude that I_{ex} decreased owing to the Frenkel-type defects on the Ga and In sublattices generated by electron-irradiation.

178

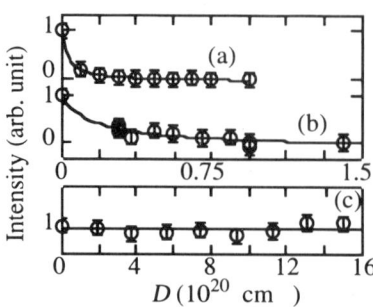
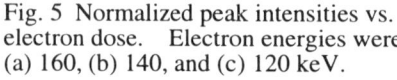
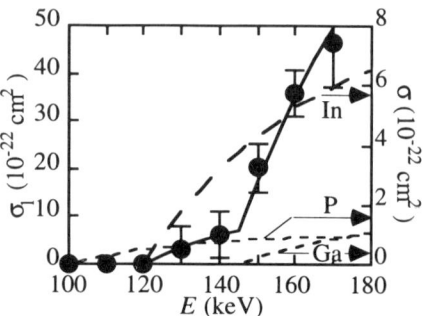

Fig. 5 Normalized peak intensities vs. electron dose. Electron energies were (a) 160, (b) 140, and (c) 120 keV.

Fig. 6 Fitting parameter σ_f and calculated cross sections for atomic displacement σ vs. incident-electron energy E.

We have shown that group-III vacancies in GaInP can migrate even at the temperature of 110 K under electron-irradiation (Ohno *et al* 1999a). As discussed above, CL peak intensity decreases owing to non-radiative electron-hole recombination, and the energy of the recombination may enhance the motion of the vacancies. Such recombination-enhanced effect has widely been investigated in semiconductors, since the electronic and optical properties may vary drastically owing to the effect; as an example, it is considered that dislocation climb (Kimerling *et al* 1986) and rapid-migration of impurities (Uematsu and Wada 1992) due to the effect cause the degradation of some GaAs-based laser diodes. The method of *in-situ* optical spectroscopy in a TEM is useful for studying point defect reactions under electron irradiation.

4. CONCLUSION

We have investigated the optical properties of ordered GaInP by *in-situ* optical spectroscopy in a TEM, and shown that APB's and Frenkel-type defects strongly influence the property. We have found: 1) the increase of the band gap energy with decreasing APB density and 2) new APB luminescence bands. We have also found that 3) Frenkel-type defects on the group-III sublattice act as nonradiative recombination centers.

REFERENCES

Cheong H M, Mascarenhas A, Geisz J F, Olson J M, Keller M W and Wendt J R 1998 Phys. Rev. B **57**, R9400
Ernst P, Geng G, Hahn G, Scholz F, Schweizer H, Phillipp F and Mascarenhas A 1996 J. Appl. Phys. **79**, 2633
Gomyo A, Suzuki T and Iijima S 1988 Phys. Rev. Lett. **60**, 2645
Gregor M J, Blome P G, Ulbrich R G, Grossmann P, Grosse S, Feldmann J, Stolz W, Gobel E O, Arent D J, Bode M, Bertness K A and Olson J M 1995 Appl. Phys. Lett. **67**, 3572
Kimerling L C, Petroff P and Leamy J 1986 Appl. Phys. Lett. **28**, 297
Nasi L, Salviati G, Mazzer M and Zanotti-Fregonara C 1996 Appl. Phys. Lett. **68**, 3263
Noda N and Takeda S 1996 Phys. Rev. B **53**, 7197
Ohno Y and Takeda S 1995 Rev. Sci. Instr. **66**, 4866
Ohno Y, Kawai Y and Takeda S 1999a Phys. Rev. B **59**, 2694
Ohno Y and Takeda S 1999b unpublished
Ohno Y and Takeda S 1996 J. Electron Microsc. **45**, 73
Su L C, Ho I H and Stringfellow G B 1994 J. Appl. Phys. **75**, 5135 and **76**, 3520
Suzuki T and Gomyo A 1991 J. Cryst. Growth **111**, 353
Takeda S, Kuno Y, Hosoi N and Shimoyama , submitted to J. Cryst. Growth
Uematsu M and Wada K 1992 Appl. Phys. Lett. **38**, 9913

Inst. Phys. Conf. Ser. No 164
Paper presented at Microsc. Semicond. Mater. Conf., Oxford, 22–25 March 1999
© 1999 IOP Publishing Ltd

Formation and evolution of antiphase boundaries during epitaxial growth of partially ordered Ga$_{0.5}$In$_{0.5}$P

E Spiecker, M Seibt, W Schröter, M Wenderoth, R Winterhoff[1], C Geng[1] and F Scholz[1]

IV. Physikalisches Institut, Universität Göttingen, D-37073 Göttingen Germany
[1]4. Physikalisches Institut, Universität Stuttgart, D-70550 Stuttgart, Germany

ABSTRACT: The antiphase boundaries in CuPt$_B$-type ordered Ga$_{0.5}$In$_{0.5}$P layers grown by metal-organic vapor phase epitaxy on GaAs(001) substrates misoriented towards [1$\bar{1}$0] have been investigated at the interface to the buffer layer by plan-view transmission electron microscopy. The patterns of antiphase boundaries indicate that the growth of Ga$_{0.5}$In$_{0.5}$P starts by the formation of ordered islands rather than by step-flow as believed so far. Furthermore, the connection between the antiphase boundaries and the surface undulations has been studied by transmission electron microscopy and atomic force microscopy. No direct spatial coincidence of the antiphase boundaries and surface features has been found, especially by plan-view transmission electron microscopy which allows imaging of both at the same time.

1. INTRODUCTION

Ga$_{0.5}$In$_{0.5}$P grown by metal-organic vapor phase epitaxy on GaAs(001) is well known to show spontaneous atomic ordering of the CuPt$_B$-type in a wide range of growth conditions (Zunger and Mahajan 1995). The ordering strongly influences the optical properties of this material and can be used to tune the band gap without changes in composition. An inherent drawback of ordering is the occurence of antiphase boundaries (APB) which have been found to significantly lower the performance of ordered lasers (Geng et al 1997). It is therefore important to gain information about the physical mechanisms by which the APBs form and evolve during growth. The formation of APBs on the substrate is generally discussed within the model of Suzuki and Gomyo (1991) which is based on step-flow growth and assumes phase-locking across steps to be mediated by surface reconstruction. Concerning the evolution of APBs during further growth, it has been suggested that the APBs may be located at supersteps where the coherence of the surface reconstruction is likely to be lost (Stringfellow and Su 1996).

This paper studies the APBs directly on the vicinal GaAs buffer layer by plan-view transmission electron microscopy (TEM). The APB-patterns are not consistent with the model of Suzuki and Gomyo (1991), but indicate that the growth starts by the formation of ordered islands. Furthermore, the connection between the APBs and the surface undulations is investigated. Atomic force microscopy (AFM) is used to determine the surface structure of the layers whereas TEM is used to simultaneously image the surface structure and the APBs. We do not find any correlation between the APBs and the surface undulations.

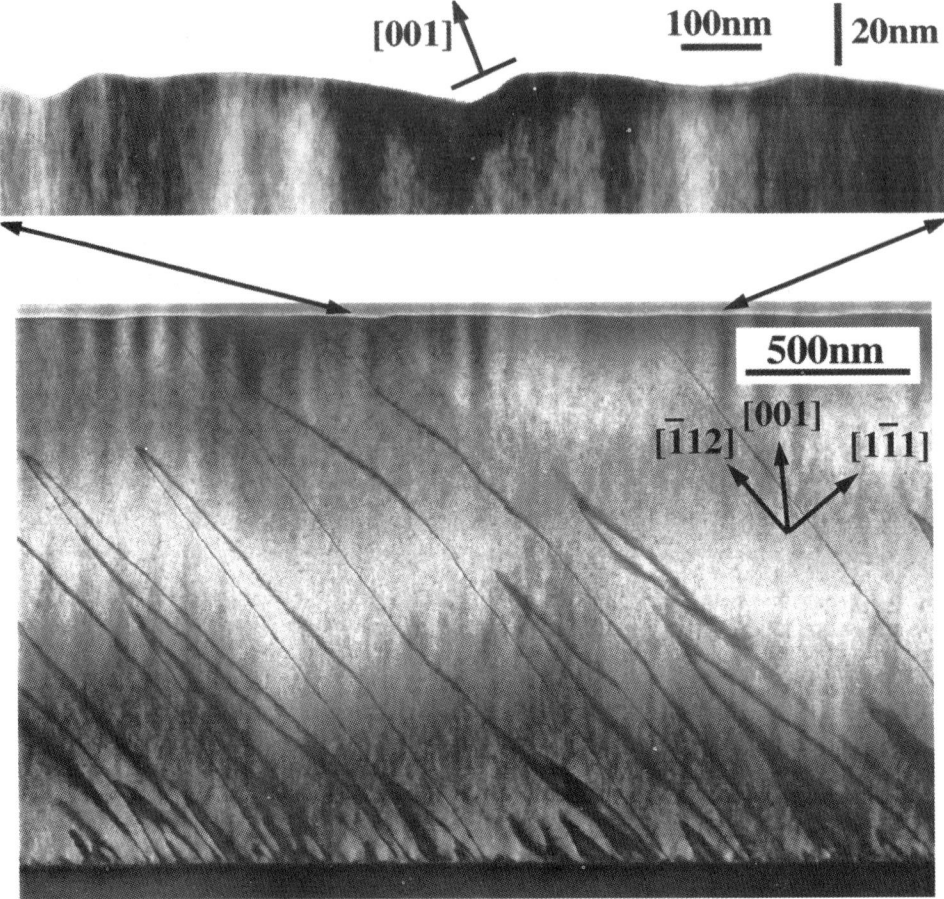

Figure 1: [110]cross-section $\frac{1}{2}(1\bar{1}1)$DF-image of a 1.7μm thick (GaIn)P-layer grown at 750°C on the 6° misoriented substrate; top: surface section of the corresponding BF-image, the vertical direction is stretched by a factor 4 with respect to the horizontal direction in order to improve the visibility of the surface undulations (see scale bars). The indicated [001]-direction has been determined by high-resolution lattice imaging.

2. EXPERIMENTAL

The Ga$_{0.5}$In$_{0.5}$P epitaxial layers were grown by low pressure metalorganic vapor phase epitaxy (MOVPE) on GaAs(001) substrates misoriented by 2° and 6° towards [1$\bar{1}$0]. Trimethylgallium, trimethylindium, and phosphine were used as sources. Two different growth temperatures were used, 750°C and 690°C, the growth rate and V/III ratio were 2.1μm/h and 400 respectively. (As pointed out by Geng et al (1997) the actual substrate temperatures are about 20°C lower than the nominal temperatures stated in the text.)

Thin foils for cross-section and plan-view TEM were prepared by mechanical polishing followed by Ar$^+$ ion milling. TEM examination was performed using a Philips EM400 working at 120kV and a CM200 working at 200kV. The surface was characterized using a Nanoscope III AFM in the tapping mode. Scan rates of 2 lines per second were used and data were taken at 512 points/line and 512 lines per scan area.

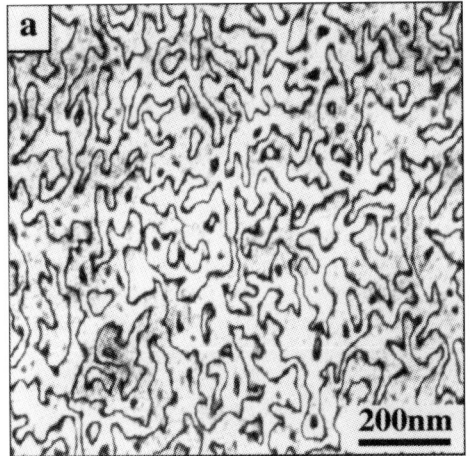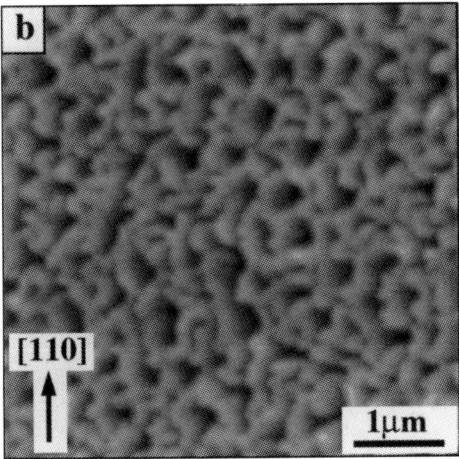

Figure 2: a) Plan-view TEM $\frac{1}{2}(\bar{1}5\bar{1})$DF-image of the APBs in the 30nm thick (GaIn)P-layer grown at 750°C on the 6° misoriented substrate, b) AFM-image of the same layer (black: 0nm, white: 15nm). Both images are equally oriented, but note the two different magnifications.

3. RESULTS AND DISCUSSION

3.1 Antiphase boundaries on the substrate

Fig.1 shows a [110] cross-section $\frac{1}{2}(1\bar{1}1)$TEM dark-field image of a 1.7μm thick (GaIn)P-layer grown at 750°C on the 6° misoriented substrate. While the overall evolution of the APBs can be determined from the image, it does not reveal the morphology of the APBs at or near the substrate, because the density of APBs is too high in this region and the APBs are mostly inclined to the electron beam resulting in blurred contours. To study this region a thin uncapped layer (thickness 30nm) of (GaIn)P was grown at identical growth conditions and investigated by plan-view TEM. The chosen layer thickness produces enough intensity in superstructure reflections to enable imaging of the APBs and at the same time ensures that the interface to the buffer layer is contained in the foil. It is seen from the APB-contours in Fig.1a that tilting towards the [$\bar{1}$12] zone axis is favourable in order to image the APBs edge-on. Fig.2a shows a $\frac{1}{2}(\bar{1}5\bar{1})$DF image taken near the [$\bar{1}$16] zone axis. The morphology of the APBs at the substrate is clearly revealed: they possess a curved irregular shape, which is somewhat elongated in the [110]-direction (corresponding to the preferred direction of step-edges on the vicinal substrate).

To study the dependence of the shape of the APBs on growth temperature and step density, similar TEM images have been obtained from layers grown at 690°C on the 6° miscut substrate and 750°C on the 2° miscut substrate. The overall appearances of the APB-patterns are similar to that shown in Fig.2a. A detailed analysis of the patterns has been performed after digitizing the APB-contours in 1μm × 1μm areas (referred to the specimens). Some extracted data are collected in Tab.1: the APB-densities ρ_{apb} (length per area), the APB-anisotropies $l_{[110]}/l_{[1\bar{1}0]}$ (relative elongation in the direction of the step edges on the substrate), and the mean distances $\overline{d}_{[1\bar{1}0]}$ between neighbouring APBs in the direction perpendicular to the step edges. For convenience the mean terrace widths $\overline{d}_{ter}$ on the vicinal substrate assuming monolayer steps are also indicated. The results can be summerized as follows: The reduction of the growth temperature from 750°C to 690°C increases the density and the anisotropy of the APBs by about 30% and 10%, respectively. Reduction of the misorientation angle from 6° to 2° at 750°C has the opposite effect, whereby the APB-density is only slightly reduced. Especially the mean distance between

layer	ρ_{apb} $[10^5\mathrm{cm}^{-1}]$	$l_{[110]}/l_{[1\bar{1}0]}$	$\overline{d}_{[1\bar{1}0]}$ [nm]	$\overline{d}_{\mathrm{ter}}$ [nm]
750°C, 6°B	3.31(3)	1.55(4)	40	2.7
690°C, 6°B	4.20(6)	1.68(5)	30	2.7
750°C, 2°B	3.16(5)	1.45(3)	43	8.1

Table 1: Results of the analysis of the APB contours in three different (GaIn)P-layers at the GaAs buffer layer. The growth conditions (temperature, misorientation) are indicated in the first column. Further meanings are: ρ_{apb}: APB-density (length/area) referred to (001), $l_{[110]}/l_{[1\bar{1}0]}$: ratio of the integrated components of the APB-contours in the respective directions, $\overline{d}_{[1\bar{1}0]}$: mean distance between neighbouring APBs in the $[1\bar{1}0]$-direction (perpendicular to the step edges), $\overline{d}_{\mathrm{ter}}$: mean terrace width on the substrate assuming monolayer steps.

neighbouring APBs in the direction perpendicular to the step edges is almost not affected by the reduced step density due to lower miscut. For all layers this distance is much larger then the mean terrace width on the substrate in accordance with the results of Suzuki and Gomyo (1991). Like these authors we did not find any indication for step-bunching on the GaAs buffer layer by high-resolution lattice imaging of the interface.

The APB-patterns on the substrate certainly store some information about the state of the GaAs-surface and the processes occuring on it at the beginning of the layer growth, however any extraction of information is based on model assumptions. The step-terrace-reconstruction model (Suzuki and Gomyo 1991) for ordering assumes a step-flow growth. The ordering starts by preferential sticking of one group III atom at the step edges. The preferred arrangement of alternating In-rich and Ga-rich [110] rows than completes the ordering of the first layer. To explain the long distances between neighbouring APBs compared to the one order of magnitude smaller terrace widths, a 'phase-locking' mechanism has been proposed (Suzuki and Gomyo 1991): the reconstruction of the As-terminated surface favours the formation of terraces with an even number of atoms, which garantees that growth at neighbouring step edges starts with the same phase. Within this model, APB-patterns on the substrate mainly reveal the coherence length of the reconstruction. However, the coherence length should strongly depend on the step density, because the steps are the main locations where the coherence can be destoyed. The weak dependence of the APB-density on the step density found in our experiments therefore clearly contradicts the step-terrace-reconstruction model.

It is more natural to conclude from the weak dependence of the APB-pattern on the step density that the migrating group III atoms can easily cross the steps. Indeed there are clear indications for the weak bonding of group III atoms at steps with edges in the [110] direction (Asai 1987, Hiramoto et al 1994). In this case the growth of (GaIn)P may start by the nucleation of two-dimensional ordered islands on the vicinal substrate. Within this scenario the APB-contours represent the boundaries between merging islands of different phase. The physical quantities that could be probably extracted (cf. Kasu and Kobayashi 1997) are surface diffusion coefficients and/or sticking coefficients of group III atoms. The experimentally observed increase of the APB-density with decreasing temperature is expected because of the reduced surface diffusion. Furthermore, the increased anisotropy of the APBs at lower temperatures and higher step densities indicates that the steps act as barriers for the diffusion.

While it has been unambiguously shown that the reconstruction of the P-terminated (GaIn)P surface plays an important role in the evolution of ordering (Murata et al 1997), the significance of reconstruction at the beginning of the layer growth is unclear. In this respect, it would be very interesting to study the dependence of the APB-patterns on the V/III-ratio and the growth rate.

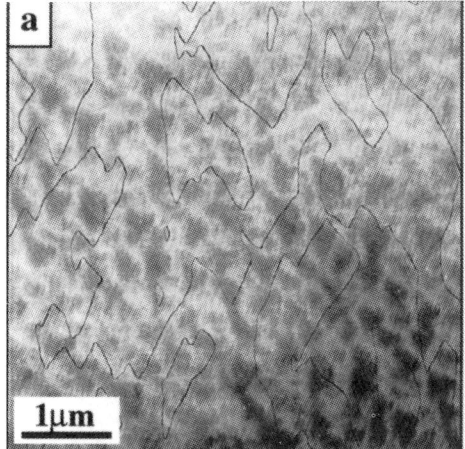
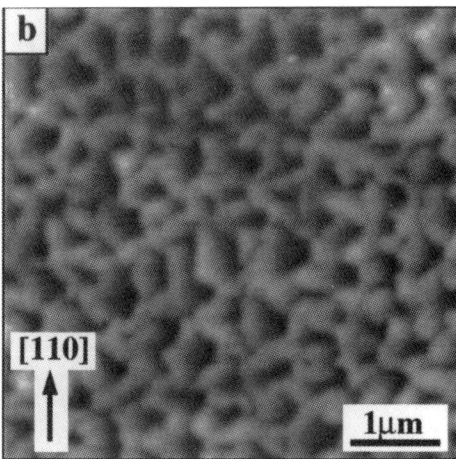

Figure 3: a) Plan-view TEM $\frac{1}{2}(\overline{1}5\overline{1})$DF-image of the APBs at the top of the (GaIn)P-layer shown in Fig.1 with (220) strongly excited, b) AFM-image of the same layer (black: 0nm, white: 15nm). The orientation of the images are the same as in Fig. 2. Note that the TEM-image reveals both the APBs and the surface undulations simultanously.

3.2 Antiphase boundaries and surface undulations

During the growth of (GaIn)P on misoriented GaAs(001) the surface does not remain flat but becomes wavy. The occurence of short (001)-facets has been reported which are compensated by large vicinal regions with a misorientation angle slightly higher than that of the substrate (Gomyo et al 1994) or by shorter vicinal regions with an even higher misorientation angle (Stringfellow and Su 1996), sometimes referred to as 'supersteps'. Stringfellow and Su (1996) pointed to the common dependence of the density of supersteps and the density of APBs on the growth temperature and growth rate, and discuss the supersteps as possible locations of APBs. Saß et al (1997) reported on the correlation of APBs and supersteps in thin (GaIn)P-layers, while Nasi et al. (1997) found correlations between the domain shape imaged by plan-view TEM and the surface step structure imaged by AFM.

We have studied the relation between the APBs and the surface undulations at the layers grown at 750°C on the 6° substrate. From the following three observations we conclude that the APBs are not directly correlated to recognizable surface features (i.e. no spatial coincidence):

1. By cross-section TEM (Fig.1) the surface undulations can be clearly revealed. In order to increase their visibility a surface section taken from the BF-image and stretched in the vertical direction by a factor of 4 is shown at the top of the figure. Analysis of the surface (the absolute orientation was obtained by high resolution lattice imaging) indeed reveals the features described above, whereby the local misorientation angle varies between 10° and −2°. However by inspection of long surface regions, we did not find a systematic spatial coincidence of surface features (facets, hillocks, valleys) and antiphase boundaries.

2. The peak-to-valley distance of the surface undulations is found from Fig.1 to be about 15nm. In plan-view TEM such thickness changes can lead to clearly visible intensity variations when a low index reflexion is excited. Fig.3a) shows a $\frac{1}{2}(\overline{1}5\overline{1})$DF image taken at the [$\overline{1}16$] zone axis with (220) strongly excited. Both the APBs and the surface undulations are visible at the same time. The AFM image (Fig.3b)) clearly proves that the background intensity variations in the TEM image stem from surface undulations. The APBs show moderate facetting on ($1\overline{3}2$) and ($3\overline{1}2$) planes (cf. Fig.1), which corresponds to [310] and [$\overline{1}30$] directions on the growing surface. These types of facets have been also found by McFadden (see Mahajan 1997). Also the

surface undulations reveal features (valleys and facets) aligned in directions inclined to [110], as can be seen from both the TEM and AFM image. Thus there might be a common cause for the preferred occurence of these directions on the growing surface. However a closer inspection of the TEM image confirms the lack of a systematic spatial coincidence of recognizable surface features and APBs.

3. Fig.2b) shows an AFM image of the 30nm thick (GaIn)P layer grown at 750°C on the 6° miscut substrate. Even at this small thickness apparent surface undulations quite similar to that at the top of the thick layer (Fig.3b) are present. However when going from the thin to the thick layer the density of the hillocks is reduced by a factor of about 2. In contrast to this, the APB-density decreases by a factor of 15 (Fig.2a and 3a, note the different magnifications). In the thin layer the mean distance between neighbouring APBs is much smaller than the distance between the hillocks on the surface whereby the relation is reversed at the top of the thick layer. Because of this totally different density dependence it is very unlikely that a direct spatial coincidence of the surface features seen with the AFM and the APBs exists at any stage of the layer growth, at least for the growth conditions investigated.

4. SUMMARY

We have studied the pattern of antiphase boundaries (APB) in partially ordered (GaIn)P directly on the vicinal GaAs buffer layer by plan-view transmission electron microscopy (TEM). The APB-patterns reveal only a weak dependence on the step density which contradicts the step-terrace-reconstruction model of Suzuki and Gomyo (1991). A mechanism for the formation of the APBs is proposed that assumes (GaIn)P-growth to start by nucleation of two-dimensional ordered islands on the vicinal substrate rather than by step-flow. The observed dependence of the APB-pattern on growth temperature and step density can be well interpreted. Furthermore, we have investigated the relation between APBs and surface undulations using TEM and atomic force microscopy. No spatial coincidence between recognizible surface features (facets, hillocks, valleys) and the APBs has been found.

ACKNOWLEDGEMENTS

We thank K. Sauthoff and K.J. Engel, two members of our STM group, for technical support with the AFM.

REFERENCES

Asai H 1987 J. Cryst. Growth **80**, 425

Geng C, Moritz A, Heppel S, Mühle A, Kuhn J, Ernst P, Schweizer H, Phillipp F, Hangleiter A and Scholz F 1997 J. Cryst. Growth **170**, 418

Gomyo A, Hotta H, Miyasaka F, Tada K, Fujii H, Fukagai K, Kobayashi K and Hino I 1994 J. Cryst. Growth **145**, 126

Hiramoto K, Tsuchiya T, Sagawa M and Uomi K 1994 J. Cryst. Growth **145**, 133

Kasu M and Kobayashi N 1997 J. Cryst. Growth **170**, 246

Mahajan S 1997 Inst. Phys. Conf. Ser. **160**, 367

Murata H, Ho I H and Stringfellow G B 1997 J. Cryst. Growth **170**, 219

Nasi L, Fermi F, Ferrari C, Francesio L, Lazzarini L, Zanotti-Fregonara C, Pellegrino S and Salviati G 1997 Inst. Phys. Conf. Ser. **157** 269

Saß T, Pietzonka I, Franzheld R, Gottschalch V and Wagner G 1997 Inst. Phys. Conf. Ser. **160** 377

Stringfellow G B and Su L C 1996 J. Cryst. Growth **163**, 128

Suzuki T and Gomyo A 1991 J. Cryst. Growth **111**, 353

Zunger A and Mahajan S 1995 Handbook of Semiconductors, edited by Mahajan S (Elsevier, Amsterdam, 1995) Vol.3, 1399

Inst. Phys. Conf. Ser. No 164
Paper presented at Microsc. Semicond. Mater. Conf., Oxford, 22–25 March 1999
© 1999 IOP Publishing Ltd

Contrast analysis in TEM images of InGaAs/GaAs strained layers grown on non-planar substrates

A M Condó*, K Leifer, A Rudra, J Michler[#], E Blank[#] and E Kapon

Department of Physics, Swiss Federal Institute of Technology (EPFL), 1015 Lausanne, Switzerland.
* Present Address: Centro Atómico Bariloche, 8400 S.C. de Bariloche, Argentina.
[#] Laboratoire de metallurgie physique, Dept. Materials EPFL, 1015 Lausanne, Switzerland.

ABSTRACT: Strained $In_xGa_{1-x}As$ layers (x ~ 0.15, 15-30nm thick) grown by organometallic chemical vapor deposition on V-grooved and planar (100) GaAs substrates were investigated by dark field (DF) imaging in transmission electron microscopy (TEM). The effect of strain and composition on the contrast in (200) DF images was studied. The contrast features not varying with slightly different orientations and not observed in (400) DF reveal differences in composition with the spatial resolution of DF imaging. Higher In content was found at the bottom of the V-grooves.

1. INTRODUCTION

In lattice matched $Al_xGa_{1-x}As$ layers grown by organometallic chemical vapor deposition (OMCVD) on V-grooved GaAs substrates, different incorporation rates of the elements in the different facets lead to local variations in composition of the $Al_xGa_{1-x}As$ alloy giving rise, for example, to vertical quantum well formation (Biasiol et al 1996). In the $In_xGa_{1-x}As$ system, the lattice misfit with GaAs (1% for x= 0.15) gives rise to a strain field, and the resulting segregation of the In and Ga elements has not yet been well studied. Chemical analysis of non-strained layers by (200) dark field imaging (DF) is based on the intensity dependence of the structure factor on the composition. Bithell and Stobbs (1989) have developed a procedure for quantification of the composition of AlGaAs from the (200) DF technique. In strained layers two more effects influence the intensity of the (200) DF image: the new lattice parameters (bulk property) and the surface relaxation of lattice planes. For planar InGaAs, variations in composition have been investigated by McCaffrey et al (1997) with the assumption that the surface relaxation does not affect the contrast. Surface relaxation has been investigated in the cladding GaAs layers by Jacob et al (1998) in (400) BF and DF images.

In the present work, we study InGaAs/GaAs non-planar epitaxial layers by DF imaging. Since in these V-grooved structures a high spatial resolution is required, we are unable to tilt the sample to an orientation with only one systematic row excited (Bithell and Stobbs 1989). Therefore, we have to find an orientation which is as close as possible to the zone axis parallel to the grooves but which, on the other, hand limits the influence of multiple beam excitation on the contrast. The effects of strain are analysed by comparing (200) with (400) DF images since they are sensitive to the same type of deformation and the latter are much less sensitive to composition. The influence of the variation of the lattice parameters caused by strain was studied by calculating its effect on the intensity. Preliminary finite elements calculations of the strain on complex V-grooved structures were performed.

2. EXPERIMENTAL

The nanostructures were grown by low-pressure OMCVD on (100) GaAs substrates patterned with a grating of V-grooves oriented in the [01$\bar{1}$] direction and with a 500nm pitch (see for example

Biasiol et al 1996, Gustafsson et al 1997). $In_xGa_{1-x}As$ layers of 15nm and 25nm were grown simultaneously on V-grooved and planar substrates at temperature of 550°C and with nominal In concentration of x=0.15. Cross-sectional specimens of the V-grooved samples were mechanically thinned and ion milled with Ar^+ at an angle of 7 degrees and voltage of 4kV. Cleaved wedge specimens were prepared to study the planar layers (Ganière et al 1989). They were observed in a PHILIPS EM430 microscope operating at 300 kV.

3. RESULTS AND DISCUSSION

Figures 1a and 1b show a (200) DF and (400) DF image of the InGaAs layer of 25nm nominal thickness. Three regions are evident: the ridges, the sidewalls of the groove and the bottom of the groove (QWR). In the (200) DF image two dark vertical branches are observed in the QWR region and a horizontal central clear region is observed at the ridges. In the (400) DF image the contrast changes from the centre of the InGaAs layer to the interface with GaAs. This is clearly observed at the bottom of the groove.

Finite elements calculations were performed for the InGaAs layer grown on V-grooved substrate in Fig. 1 considering the elastic constants of GaAs and $In_xGa_{1-x}As$: C_{11}=119-35.7x, C_{12}=53.8-8.5x, C_{44}=59.5-19.9x (10^{10} dyn cm^{-2}) (Jain et al 1996); a lattice misfit of 0.01, a constant In composition and a sample thickness of 100nm. Figure 2 shows the distortion of the lattice planes with a cut at the centre of the QWR. The deformation has been amplified 100 times to be visible. The (100) planes inside the InGaAs layer are appreciably bent in the last 15nm near the surfaces and at the InGaAs/GaAs interface. At the ridges they are bent by 4 mrad and at the QWR by about 6 mrad.

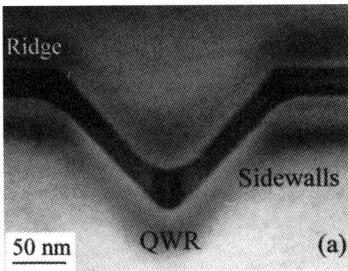 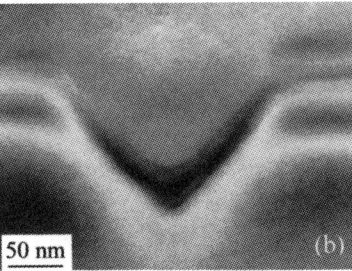

Fig. 1. DF images observed from near [01$\bar{1}$] for a sample thickness at QWR of 115nm. a) (200) DF image. b) (400) DF image.

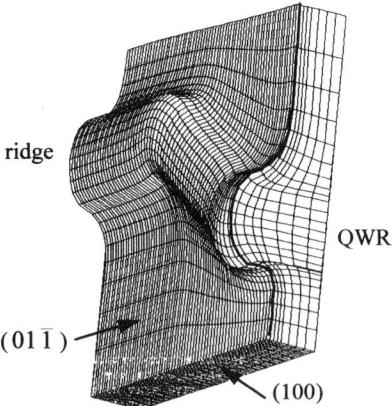

Fig. 2. Finite elements simulated deformation of the structure corresponding to half the image in Fig. 1. The deformation has been amplified 100 times to be visible.

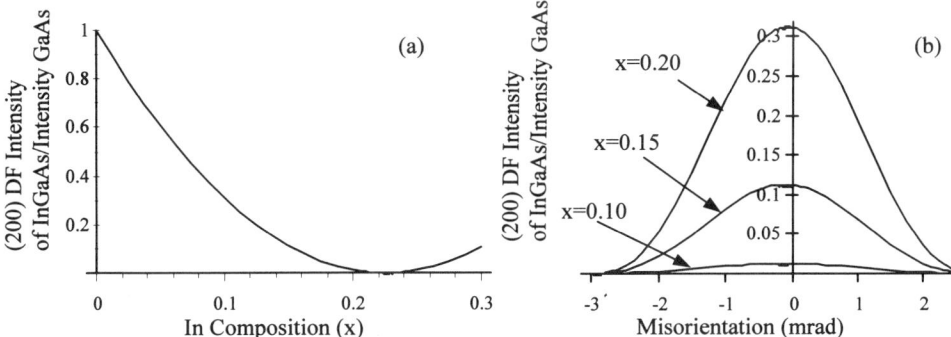

Fig 3. Calculated dynamical intensity of the strained In$_x$Ga$_{1-x}$As planar (100) layer: a) as a function of composition, b) as a function of the misorientation (see text).

Figures 3a and b show the calculated intensity in (200) DF images of a strained In$_x$Ga$_{1-x}$As planar (100) layer as a function of x (Fig. 3a) and as a function of misorientation (deviation angle of the (100) planes from the exact Bragg condition, Fig. 3b). Fig. 3b was calculated to study the effect of surface relaxation. The intensity, plotted relative to the intensity of GaAs, was calculated using the dynamical theory of electron diffraction with no absorption for an acceleration voltage of 300kV, a sample thickness of 100 nm and isotropic elastic constants. For a given composition or misorientation we found that the intensity is almost unaffected by the change of the lattice parameters caused by strain. In Fig. 3b, the intensity was found to be reduced to about 50% for a misorientation of 1.5 mrad.

Figure 4 shows the DF images of the cleaved planar samples observed from a [010] zone axis. In the (200) DF image (Fig. 4a, tilt ~4degrees) the intensity across the InGaAs layer was uniform. For different layer thickness (8-30nm) the contrast showed the same qualitative behaviour. This means that the effect of strain on planar layers is not visible on the (200) DF images. Since the intensity is very weak contrast may be lost by background intensity due to inelastic scattering which was found to be around 10% of the intensity of GaAs (Bithell and Stobbs 1989). In the (400) DF image (Fig. 4b), the intensity variation at the centre of the InGaAs layer as a function of thickness is similar to the thickness fringes in non-strained GaAs. This means that the contrast in the centre of the InGaAs layer is not affected by surface relaxation. For a thickness value around 75nm (dashed line) the contrast behaviour across the layer is similar to the one observed at the right ridge in Fig. 1b for which the thickness is around 75nm. This means that the ridges can be considered as planar layers.

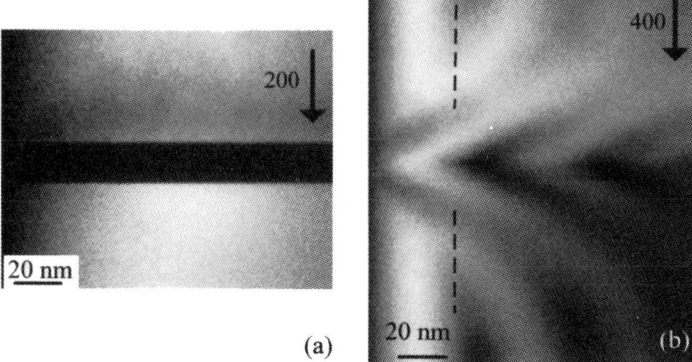

Fig. 4. DF images of the planar layer of 15nm near [010] zone axis. The thickness in each image increases linearly from 0 at the left border to 300nm at the right border. a) (200) DF image, b) (400) DF image. The position of the InGaAs layer is indicated by the (200) DF image.

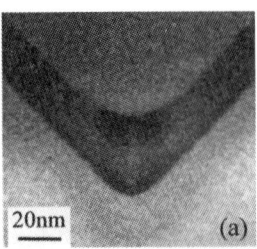 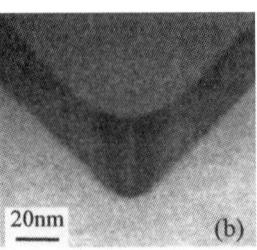  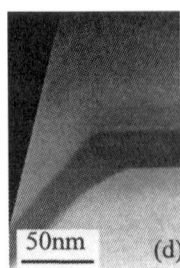

Fig. 5. (200) DF images from near [01$\bar{1}$]. a) and b) QWR at sample thickness 50nm. a) Tilt of 4 degrees. b) Tilt of 6 degrees. c) and d) Ridge and sidewall. c) Tilt of 4 degrees. d) Tilt of 5 degrees.

The (200) DF contrast in the InGaAs non planar layers was found to be non uniform (Fig. 1a) and to change with small variations of the tilt angle from [01$\bar{1}$] zone axis (Fig. 5). At small angles (around 4 deg) a central clear contrast parallel to the GaAs/InGaAs interface was observed (Fig. 5a and 5c) which disappears at higher angles (Fig. 5b and 5d, respectively). The contrast feature in the ridge in Fig. 5c was not observed near [010] and disappears with increasing tilt angle near [01$\bar{1}$] so, it was attributed to multi-beam interactions. The contrast features in Fig. 5a changing with tilt also disappear for higher angles. Thus, the effect of multi-beam interaction is considered to be suppressed in Figs. 5b and 5d. The contrast is rather uniform in the planar facets in Fig. 5d as in the planar layer in Fig. 4a. Thus, we consider there is no effect of strain on the ridges and the sidewall being only a chemical contrast. A contrast difference between the sidewalls and the QWR that was as strong as the one in the (200) DF image (Fig. 5b) was neither observed on the (400) DF image in Fig. 1b nor on other (400) DF images from a tilt series. Because of this, we interpret the contrast in Fig. 5b also as a chemical one.

Considering the effect of composition (Fig. 3a) darker intensity means higher In content. This is valid if the composition is lower than 0.22. Since the composition in the planar sample was measured by photoluminescence to be 0.15 we assume variations around this value. From Fig. 5b and 5d the intensity is lower at the ridges and at the QWR and thus higher In content is expected at the ridges and at the QWR. Moreover, two dark vertical branches are observed at the QWR and thus two In rich branches are present.

4. CONCLUSION

Near the [01$\bar{1}$] zone axis a compositional image of In$_x$Ga$_{1-x}$As layers can be obtained if care is taken to avoid multi-beam interactions. For x<0.22, the relative In concentration can be deduced. In the case of 25nm thick layers (nominal x=0.15) grown on V-grooved substrates we found a higher In content at the ridges and two In rich branches at the QWR.

We would like to acknowledge Dr. F. Lelarge for PL measurements and discussions and the Centre Interdépartemental de Microscopie Electronique (EPFL) where samples were analysed.

REFERENCES

Biasiol G, Reinhardt F, Gustafsson A, Martinet E and Kapon E 1996 Appl. Phys. Lett. **69**, 2710
Bithell E G and Stobbs W M 1989 Phil. Mag. **60**, 39
Ganière J D, Reinhart F K, Spycher R, Bourqui B, Catana A, Ruterana P, Stadelmann P A and Buffat P A 1989 J. Microsc. Spectrosc. Electron. **14**, 407
Gustafsson A, Biasiol G, Dwir B, Reinhardt F and Kapon E 1997 Proc. Microscopy of Semiconducting Materials, Inst. Phys. Conf. Ser. **157**, 373
Jacob D, Androussi Y, Benabbas T, François P and Lefevre A 1998 Phil. Mag. A **78**, 879
Jain S C, Willander M and Maes H 1996 Semicond. Sci. Technol. **11**, 641
McCaffrey J P, Wasilewski Z R, Robertson M D and Corbett J M 1997 Phil. Mag. A **75**, 803

Inst. Phys. Conf. Ser. No 164
Paper presented at Microsc. Semicond. Mater. Conf., Oxford, 22–25 March 1999
© 1999 IOP Publishing Ltd

Microcharacterisation of MOVPE-grown AlAsSb/GaAsSb superlattices by STEM

C Mendorf, F Schulze-Kraasch, S Li, X G Xu*, C Giesen*, K Heime* and E Kubalek

Werkstoffe der Elektrotechnik, Gerhard-Mercator-Universität Duisburg, Bismarckstr. 81, 47048 Duisburg, Germany
*Institut für Halbleitertechnik, RWTH Aachen, Templergraben 55, 52056 Aachen, Germany

ABSTRACT: Structural properties of $AlAs_{0.14}Sb_{0.86}/GaAs_{0.1}Sb_{0.9}$ superlattices grown on InAs substrate (SL_{InAs}) and on GaAs substrate (SL_{GaAs}) by low-pressure MOVPE were studied using CL combined with SE imaging, AFM and STEM. The surfaces exhibit structural defects like islands and dislocation lines. Cross-sectional specimens examined by STEM show good crystalline quality for the SL_{InAs} whereas defects on the $\{111\}$ planes due to misfit occur in the SL_{GaAs}. For the quantitative examination of local strain within the SLs CBED has been applied. The investigation of the interface properties of the SLs yields an asymmetric behaviour of the SL_{GaAs}. Moreover, an interlayer in the $AlAs_{0.14}Sb_{0.86}$ layers of SL_{InAs} has been proved.

1. INTRODUCTION

There has been an increasing interest on III-V compounds based on antimonides as alternative materials for devices working in the near and mid-infrared region, e.g. for applications in communication, medicine or environmental analytical techniques. Most of the work on Sb-based compounds that has been published was done on MBE-grown structures, but only little is known of material quality with respect to defects and chemical homogeneity of MOVPE-grown heterostructures. Therefore, we investigated the structural and compositional properties of $AlAs_{0.14}Sb_{0.86}/GaAs_{0.1}Sb_{0.9}$ superlattices (SL) applying methods like cathodoluminescence (CL) combined with secondary electron (SE) imaging as well as atomic force microscopy (AFM) and scanning transmission electron microscopy (STEM).

2. GROWTH AND APPLIED CHARACTERISATION TECHNIQUES

The $AlAs_{0.14}Sb_{0.86}/GaAs_{0.1}Sb_{0.9}$ SLs studied here were grown on InAs substrate (SL_{InAs}) or on GaAs substrate (SL_{GaAs}) in a horizontal low-pressure (20 mbar) metalorganic vapour-phase epitaxy (LP-MOVPE) apparatus at a temperature of $T_g = 560°C$. Triethylgallium (TEGa), Tritertiarybutylaluminium (TTBAl), Triethylantimony (TESb) and Tertiarybutyl-arsine (TBAs) were used as precursors. TTBAl was used to reduce strong aluminium-carbon bonds which typically leads to a substantial carbon incorporation into the epilayers.

CL was excited by a modified Camscan scanning electron microscope (SEM) operating with primary electron energies in the range of 5-20 keV. The luminescence emitted from the samples was collected by an ellipsoidal mirror. The panchromatic CL images were detected by a PbSe infrared detector operating at a temperature of -30°C. For AFM investigations a conventional TopoMetrix Explorer AFM microscope was used. The images were taken in contact mode using forces in the range of 10^{-8}N. Cross-sectional specimens have been investigated in a cold field-emission STEM working at 100 keV (VG Microscopes: HB 501) which is described in detail elsewhere (Lakner et al 1996). Bright-field imaging was used to investigate the structural properties of the samples. The interface abruptness was qualitatively determined by atomic number (Z-) contrast imaging. Convergent beam electron diffraction (CBED) and corresponding simulations have been carried out for the quantitative examination of local strain within the superlattices. For the simulations we used the Bloch-wave dynamical program developed by Spence and Zuo (1992).

3. RESULTS AND DISCUSSION

3.1 Topography and surface defects

On the surfaces of the investigated SLs structural defects in the forms of islands and dislocation lines were observed. Figure 1 represents a typical panchromatic CL surface image which has been recorded from the SL_{GaAs}. One can see that on the defect positions the luminescence is generally higher than elsewhere which has been reported by Liu et al (1998). The density of the 3D islands is around 7.5×10^5 cm^{-2} for the SL_{GaAs}, respectively 3.5×10^5 cm^{-2} for the SL_{InAs}.

In order to determine the size of the structural defects and the surface roughness, AFM images were also recorded. Size and shape of most of the defects are quite similar for both samples. Figure 2 shows an AFM image of the surface of the SL_{GaAs}. The lateral dimension of the islands is around 3-4 µm and the height is around 1 µm. Beneath single islands one can see two smaller islands in the middle of the AFM image (see markings) which are so close together that the 3D growth process has started to replace the two islands by a single one (Tersoff et al 1996).

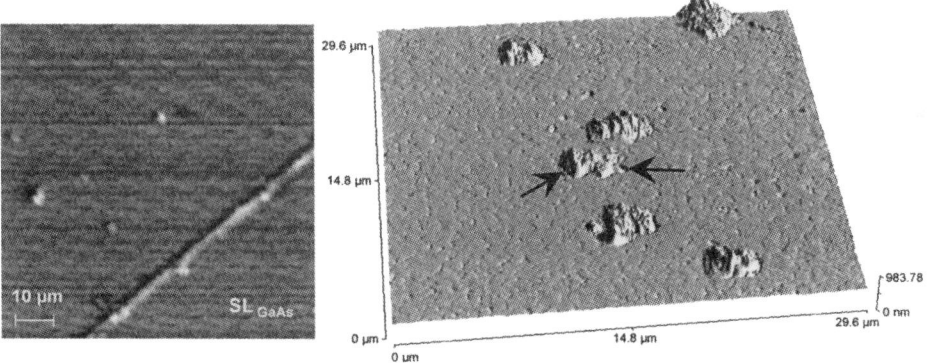

Fig. 1: Panchromatic CL image of SL_{GaAs} displaying islands and a dislocation line

Fig. 2: AFM image of SL_{GaAs} showing typical 3D islands

3.2 Cross-sectional STEM characterisation

In order to achieve additional information about the described structural properties cross-sections of the superlattices have been characterised by STEM. By the use of techniques like bright-field imaging, Z-contrast imaging and CBED the local structural properties and local chemical composition fluctuation could be examined on a nanometer scale.

Figure 3 a) and b) shows bright-field images of the superlattices. Whereas the SL_{InAs} (see fig. 3a) exhibits good crystalline quality, defects on the {111} planes are clearly visible due to misfit in the SL_{GaAs} (fig. 3b). Additionally, it can be observed how threading dislocations propagate through the superlattice, originating from a thick $AlAs_{0.15}Sb_{0.85}$ buffer.

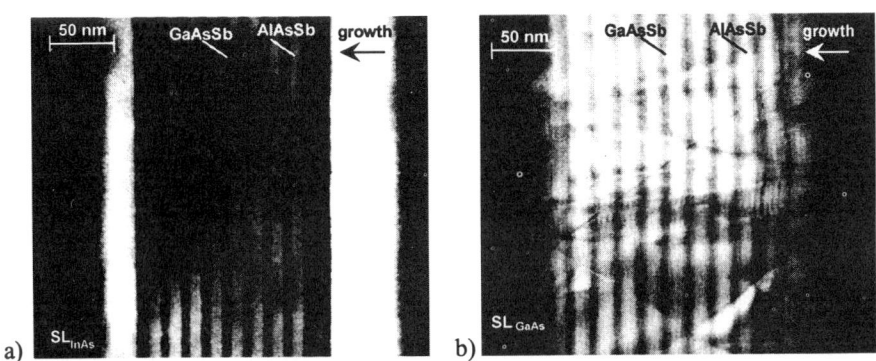

Fig. 3: Bright-field images of SL_{InAs} (a) and SL_{GaAs} (b)

The SLs have been grown with a nominal periodicity of $\Lambda=20$ nm. The evaluation of the STEM results yields a periodicity of $\Lambda_{InAs}=16$ nm for the SL_{InAs} with thicknesses of 6.5 nm for wells ($GaAs_{0.1}Sb_{0.9}$) and 9.5 nm for barriers ($AlAs_{0.14}Sb_{0.86}$) and $\Lambda_{GaAs}=19$ nm for the SL_{GaAs} with thicknesses of 8 nm for wells and 11 nm for barriers, respectively. Figure 4 shows Z-contrast images and linescans extracted from the Z-contrast micrographs of the SLs. Due to Z-contrast the $GaAs_{0.1}Sb_{0.9}$ layers appear bright. The linescan of the SL_{InAs} displays a nearly symmetric behaviour at the interfaces. However, the linescan clearly reveals the existence of interlayers in the barriers of the SL_{InAs}. In comparison, the linescan of the SL_{GaAs} indicates an asymmetric grading of the chemical composition at the interfaces which may be explained by an Al carry-over into the wells (Giesen et al 1998).

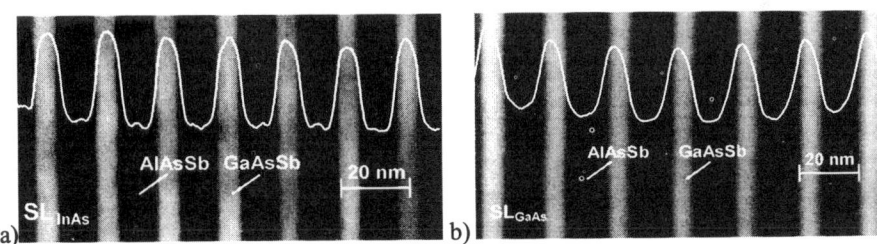

Fig. 4: Z-contrast images and embedded linescans of SL_{InAs} (a) and SL_{GaAs} (b)

In addition, the layers of the SLs have been analysed regarding to strain by application of CBED technique. Figure 5 shows a detail of CBED diffraction patterns recorded from the $AlAs_{0.14}Sb_{0.86}$ and the $GaAs_{0.1}Sb_{0.9}$ layers of the SL_{GaAs}. The patterns are easily distinguishable because the 006-diffraction-line appears in the $AlAs_{0.14}Sb_{0.86}$ diffraction pattern but not in the $GaAs_{0.1}Sb_{0.9}$ pattern. From the distances measured between the 008-diffraction-lines, we obtained a change in the lattice constant c of 7.4% for the $GaAs_{0.1}Sb_{0.9}$-layer and 8.0% for the $AlAs_{0.14}Sb_{0.86}$ layer related to the GaAs-substrate. Thus, the relative change in the lattice constant c between the $GaAs_{0.1}Sb_{0.9}$ and the $AlAs_{0.14}Sb_{0.86}$layers is -0.6%. The latter value

192

corresponds to a tetragonal distortion in a bulk material (Perovic et al, 1991), as it is twice the theoretical lattice constant difference $\Delta a/a_{GaAs}$ between a single crystal $GaAs_{0.1}Sb_{0.9}$ and a single crystal $AlAs_{0.14}Sb_{0.86}$ ($\Delta a/a_{GaAs}= -0.3\%$ according to Vegard's Law). Thus, we can conclude that strain relaxation has taken place between the superlattice and the substrate.

Figure 6 shows the experimental CBED patterns obtained from the SL_{InAs} (left column). The evaluation of the 008-diffraction lines yields a much smaller change in the lattice constant c related to the InAs-substrate: -0.3% for $AlAs_{0.14}Sb_{0.86}$ and 0.3% for $GaAs_{0.1}Sb_{0.9}$. The theoretical values $\Delta a/a_{InAs}$ for single crystals are -0.11% and -0.17%. The corresponding simulations where a tetragonal distortion of +/-0.3% was assumed are shown in the right column. They are in a good agreement with the experiment (left column), especially visible in a shift of the [117]-intersection (see marking). The overlap of experimental CBED discs is not considered in the simulation, which leads, especially in the (004) systematics orientation, to some differences in the fine structure of the diffraction patterns (Lakner et al 1995).

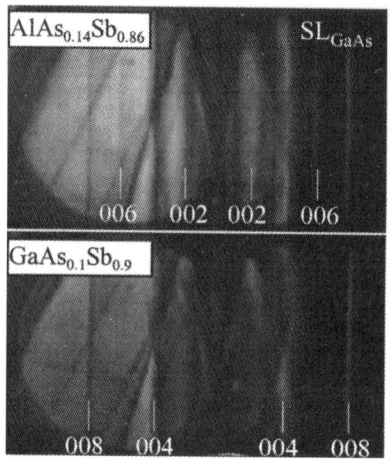

Fig. 5: Experimental CBED-pattern of $AlAs_{0.14}Sb_{0.86}$ and $GaAs_{0.1}Sb_{0.9}$ of the SL_{GaAs} in (006) systematics orientation

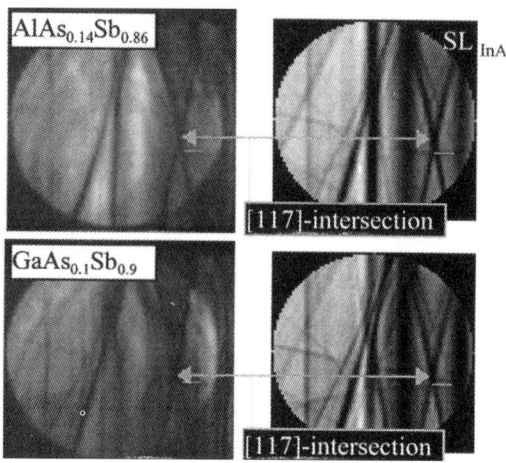

experiment   simulation

Fig. 6: Experimental (left) and simulated (right) CBED-pattern of $AlAs_{0.14}Sb_{0.86}$ and $GaAs_{0.1}Sb_{0.9}$ of the SL_{InAs} in (004) systematics orientation

ACKNOWLEDGEMENT

Dr. Li is a Scholarship Holder of the Alexander von Humboldt Foundation.

REFERENCES

Giesen C, Beerbom M M, Xu X G and K Heime 1998 J. Crystal Growth **195**, 85
Lakner H, Bollig B, Ungerechts, S and Kubalek E 1996 J. Phys. D: Appl. Phys. **29**, 1767
Lakner H, Ungerechts, S and Kubalek E 1997 Inst. Phys. Conf. Ser. **146**, 267
Liu Q, Brockt G, Meinert A, Kalisch H, Heuken M and Lakner H 1998 J. Crystal Growth **184/185**, 139
Perovic D D, Weatherly G C and Houghton D C 1991 Phil. Mag. A **64**, 1
Spence J C H and Zuo J M 1992 Electron Microdiffraction (New York: Plenum)
Tersoff J, Teichert C and Lagally M G 1996 Phys. Rev. Lett. **76**, 1675

Inst. Phys. Conf. Ser. No 164
Paper presented at Microsc. Semicond. Mater. Conf., Oxford, 22–25 March 1999
© *1999 IOP Publishing Ltd*

Growth characterisation of InAlAs/InGaAs structures on InP non_(001) index substrates

J Arbiol, F Peiró, A Cornet and A Christou[*]

EME Electronic Materials and Engineering, Dept. Electronics, University of Barcelona, Martí i Franquès 1, 08028 Barcelona, Spain. E-mail: paqui@el.ub.es

[*]**Dept. of Materials and Nuclear Engineering,** University of Maryland, Building 090, Room 2135. College Park, MD 20742-2115, U.S.A.

ABSTRACT: In the present work we analysed InAlAs/InGaAs samples grown by molecular beam epitaxy (MBE) on InP substrates with different orientations: (100), (110)A, (110)B, (111)B and (112). Our aim is to assess the influence of substrate orientation on the appearance of alloy decomposition and ordering phenomena. The specimen characterisation was accomplished by Transmission Electron Microscopy (TEM) and Atomic Force Microscopy (AFM). Whereas specimens grown on (100) and (110)B InP are rather homogeneous, for the rest of the growth directions significant features related to lateral layer decomposition or order appear. The formation of oval and hexagonal defects in samples grown on (111) and (112) is also described.

1. INTRODUCTION

The pseudomorphic growth of III-V heterostructures on InP substrates allows a wide spectrum of possibilities for the design of semiconductor devices in the field of optoelectronics and high frequency applications. Moreover, the growth of these compounds on novel substrate orientations generates great attraction because of the interesting physical properties.

Epitaxial layers grown on (110) substrates present a polarisation anisotropy of the optical transition as well as a maximum in the optical gain, all these favouring the implementation of optical polarizers and modulators (Mitsuhara 1994, Okuno 1997 and Bhat 1992). On the other hand, (111) substrates show an important piezoelectric effect (PE) along the growth direction, and this characteristic allows us to obtain important electric fields when a strained epitaxial growth is achieved. The appearence of the PE field may be exploited for improved optical and electro-optical device performance (Chen 1995). Moreover, layers grown on (112) substrates also present an important piezoelectric effect, although not as high as in those grown on (111) substrates. The reason of this effect's reduction lies in the misorientation of the polarisation vector in relation to the growth direction (Rees 1997 and Smith 1997). However, these electrical and optical properties of the grown layers are strongly influenced by defects, compositional variations and ordering (Zakharov 1993).

In the present work we studied the epitaxial growth by molecular beam epitaxy (MBE) of InGaAs and InAlAs layers on InP substrates with non conventional orientations (100), (110)A, (110)B, (111)B and (112). Our aim is to assess the influence of substrate orientation on the appearance of alloy decomposition and ordering phenomena.

2. EXPERIMENTAL DETAILS

We grew a double layer system consisting of 50 nm of $In_{0.518}Al_{0.482}As$ and 50 nm of $In_{0.53}Ga_{0.47}As$ over an InP substrate. All samples were grown by molecular beam epitaxy (MBE) at a growth temperature of $T_g = 544$ °C. The InP substrate indexes used were (100), (110)A, (110)B, (111)B and (112). The beam equivalent pressure of As used during the growth was $P_{As} = 1.24 \cdot 10^{-5}$ Torr which corresponded to a beam equivalent pressure ratio (V/III) of 17.2. The In molar fraction for both InAlAs and InGaAs was intended for lattice matching to InP.

Specimen analysis was made using transmission electron microscopy (TEM), in planar view (PVTEM) and cross section (XTEM) orientations. Atomic force microscopy (AFM) complemented the surface morphology characterisation, and finally, the study of transmission electron diffraction (TED) patterns and high resolution images (HRTEM) confirmed the existence of ordering in some of our samples.

3. RESULTS AND DISCUSSION

From TEM and AFM observations of samples with (100) and (110)B index substrates we can conclude that the epitaxial growth was homogeneously done. These samples had high crystal quality, no defects were generated during the growth. Neither layer decomposition nor ordering phenomena were observed. Both samples had smooth surfaces and no roughness appeared.

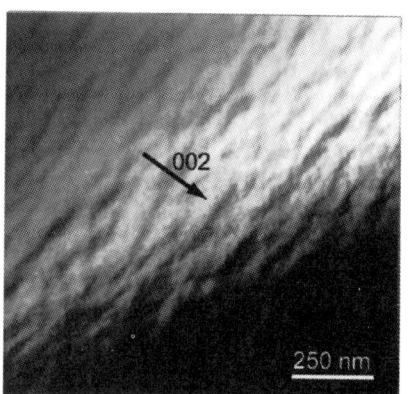

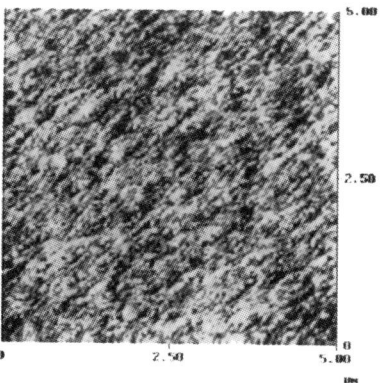

Figure 1a. (002) bright-field two-beam micrograph of sample with (110)A InP substrate. We can appreciate a roughness in the [$\overline{2}20$] direction .

Figure 1b. AFM image where roughness on (110)A sample surface is clearly shown.

Conversely, planar observation of the sample with the (110)A InP substrate shows a pronounced roughness oriented along the [$\overline{2}20$] direction (Figures 1a and 1b). From AFM analysis (Figure 1b) we found that the distance between hillocks along the [002] direction was about 95 nm. In figure 2 we can appreciate the presence of contrast stripes in the InAlAs layer inclined about 8-12 degrees with respect to the horitzontal. The distance between hillocks on the surface sample coincided with the distance between stripes observed in XTEM micrographs (d=95 nm). This result confirms that surface (110)A sample roughness was due to those contrast inhomogeneities showed in figure 2. These contrast stripes also appeared in InGaAs layer, but now with weaker intensity and they were perpendicular to growth direction. In our opinion, these contrast stripes must be due to compositional modulation (Guyer 1998). This compositional modulation is expected to be present with higher strength in the InAlAs layer because of the presence of Al. As suggested by Zakharov et al (1993) this compound is more

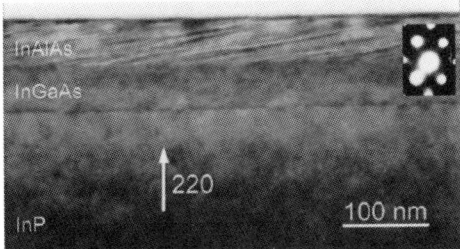

Figure 2. XTEM micrograph of sample grown on (110)A. We can see compositional nonuniformities in InAlAs and InGaAs layers.

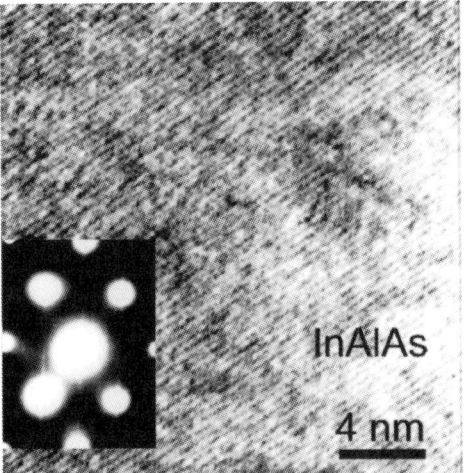

Figure 3. HRTEM micrograph with the corresponding diffraction pattern, exhibiting satellite spots due to the existence of ordering.

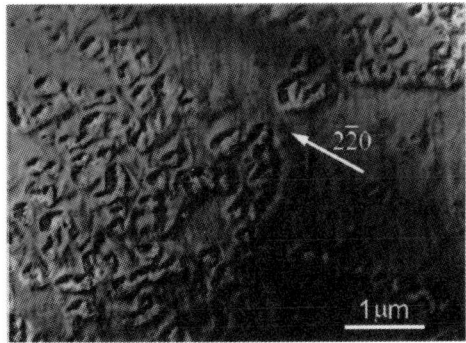

Figure 4. ($2\bar{2}0$) bright-field micrograph of (111)B InP substrate sample showing faceted craters on the surface.

susceptible to present compositional nonuniformities (Zakharov 1993). Our observations also confirmed their results. Figure 3 shows a HRTEM micrograph of the InAlAs layer with its corresponding TED pattern. We found two diffuse intensity maximums between 220 and 440 diffraction spots which revealed the existence of ordering of CuAu I-type on (110) planes (Zakharov 1993, Nakata 1991 and Norman 1993). This division of the 220 diffraction vector in three parts suggests that on average each third atomic plane was occupied by monoatomic platelets. Our results are in agreement with Zakharov et al (1993). The distances between the matrix and the diffuse spots were slightly larger than $1/3g_{220}$. In direct space, it corresponds to a smaller interplanar [110] distance between monoatomic platelets. They conclude that each third plane, then, must be occupied by Al atoms, if it is taken into account the atomic radii of the elements, $r_{Al} < r_{Ga} < r_{In}$.

Sample with (111)B InP substrate presented a high density of oval defects on surface as well as a faint roughness on [$20\bar{2}$] direction (Figure 4). These oval structures present faceted faces, forming inverted pyramids whose edges' projections on planar surface lie along A=[$2\bar{1}\bar{1}$], B=[$\bar{1}2\bar{1}$] and C=[$\bar{1}\bar{1}2$] directions. Figure 5 represents the inverted pyramid structure, where letters A, B and C label the three edges' projections on sample surface. Sample with (112) InP substrate also had a high density of crater defects (Figure 6), having an inverted triangular pyramid faceted shape (Figure 5). The edges' projections were now on [$1\bar{1}\bar{1}$], [$\bar{3}11$] and [$13\bar{1}$] directions corresponding to A,B,C directions in figure 5, respectively. XTEM analysis allowed us to observe the faceted crater formation (Figure 7). These defects were formed in the InGaAs/InP interface, and grew up to the sample surface crossing both (InGaAs and InAlAs) layers. The origin of these faceted defects was due to the growth inhibition of InGaAs layer over the InP substrate. Since the growth velocity is different on inclined surfaces than on plane, the consequent growth of InAlAs upper layer over these inhibited InGaAs sharp holes created these faceted structures. Both samples presented an important lateral contrast modulation as we deduced from XTEM observation (Figure 7). This contrast modulation started in the InGaAs/InP interface and was

196

extended to the rest of the layer, continuing also along the InAlAs upper layer. The lateral contrast modulation observed commonly appears in III-V epitaxial growth and it can be due to lateral decomposition or to strain contrast (Guyer 1998).

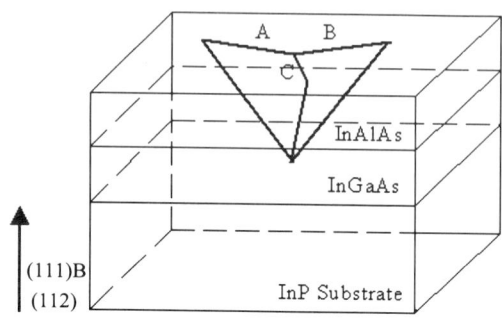

Figure 5. Scheme of the inverted pyramid structures in samples (111)B and (112). Letters A, B and C label the three edges' projections on sample surface.

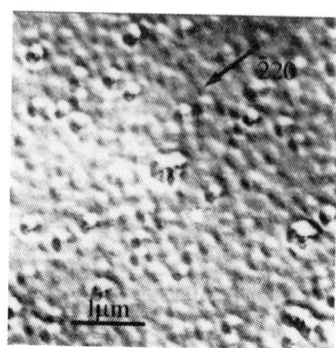

Figure 6. ($\overline{2}20$) bright-field two-beam micrograph of sample with (112) InP substrate showing faceted craters.

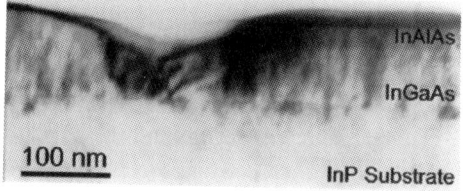

Figure 7. XTEM micrograph showing lateral contrast modulation in (112) InP substrate sample as well as the formation of one crater faceted structure growing from the InP/InGaAs interface.

4. CONCLUSIONS

In this work, we studied the most common effects in InAlAs/InGaAs layers grown on non_(001) index InP substrates. While in (100) and (110)B InP substrate samples no noticeable effects occurred, for the rest of the substrate orientations remarkable phenomena appeared. In (110)A substrate samples, we found the presence of CuAu I-type ordering and compositional modulation in the InAlAs layer due to the presence of Al. Moreover, in layers with (111)B or (112) index substrates, we observed the presence of lateral contrast modulation as well as faceted craters formed all over the surface. These faceted defects originated due to the InGaAs layer growth inhibition over the InP substrate.

REFERENCES

Bhat R, Koza M A, Hwang D M, Brasil M J S P and Nahory R E 1992 J. Crystal Growth **124**, 311
Chen X, Molloy C H, Woolf D A, Someford D J, Blood P, Shore K A and Sarma J 1995 Appl. Phys. Lett. **67**, 1393
Guyer J E and Voorhees P W 1998 J. Crystal Growth **187**, 150
Mitsuhara M, Okamoto M, Iga R, Yamada T and Sugiura H 1994 J. Crystal Growth **136**, 195
Nakata Y, Ueda O and Fujii T 1991 Jap. J. Appl. Phys. **30**, 249
Norman A G, Seong T-Y, Philips B A, Booker G R and Mahajan S 1993 Inst. Phys. Conf. Ser. **134**, 279
Okuno Y, Tsuchiya T and Okai M 1997 Appl. Phys. Lett. **71**, 1918
Rees G J 1997 Microelec. J. **28**, 957
Smith D L 1997 Microelec. J. **28**, 707
Zakharov N D, Liliental-Weber Z, Swider W, Washburn J, Brown A S and Metzger R 1993 J. Elec. Mat. **22**, 1495

Inst. Phys. Conf. Ser. No 164
Paper presented at Microsc. Semicond. Mater. Conf., Oxford, 22–25 March 1999
197

X-ray diffraction tools and methods for semiconductor analysis

Paul F Fewster

Philips Research Laboratories, Cross Oak Lane, Redhill, UK

ABSTRACT: X-ray diffraction methods have been used to extract information on the microstructure of GaN and Quantum Dot structures. Various techniques and geometrical arrangements are combined with a new theoretical model and direct interpretation to obtain domain sizes in GaN and the asymmetry of quantum dots. The principal method employed is reciprocal space mapping and it is the simulation of these maps using an extended dynamical theory that yields the domain sizes and their distribution in a GaN sample. High-resolution reciprocal space maps with in-plane scattering geometry have been used to measure the distribution of the diffuse scattering from quantum dots and from this distribution the asymmetry in the shape of the quantum dots has been determined.

1. INTRODUCTION

The analysis of semiconductor thin films generally requires several approaches to build a full picture of a sample. X-ray diffraction is a very versatile tool and by choosing the appropriate method a large amount of detail can be extracted. X-ray diffraction is very strain sensitive with comparatively large averaging probe volumes, however this does not prevent the analysis of samples with fine scale features. Topography will obtain images of features on the micron scale by using high spatial resolution photographic film or imaging cameras. However inhomogeneity on a far smaller scale can be observed because of the reciprocal nature of the diffraction process with that of spatial dimensions. This allows us to study quantum dots and very small correlation lengths in imperfect material. In general these methods will detect the deviation from perfection that makes X-ray diffraction a very powerful analysis method for semiconductor materials.

To illustrate some of these techniques, GaN and quantum dot structures have been analysed. New opportunities in light emitting diodes and lasers have arisen from recent developments in GaN, and this material has now become an important material to the semiconductor industry. GaN is in general of moderate quality and hence knowledge of the composition and the structural quality are important to aid improved growth control. Quantum dot structures also offer considerable potential for opto-electronic devices and understanding the growth of these materials is becoming very important. These offer considerable challenges to all analysis techniques and X-ray diffraction can have a very important role to play.

2. EXPERIMENTAL METHODS

Most of the analyses described in this paper are performed in high-resolution mode, that is high angular or high reciprocal space resolution. The experiments described were performed on the Philips X'Pert Materials Research Diffractometer that has been described in detail (Fewster, 1989, 1991a, 1991b, 1997). The instrument is composed of a monochromator made from two Ge "U" shaped blocks reflecting off the 220 crystallographic planes parallel to the surface cuts, this gives a probing beam divergence of ~ 4"arc and a wavelength divergence $\Delta\lambda/\lambda \sim 10^{-5}$. The analyser is just in front of the detector and is a grooved crystal of Ge set to reflect three times from the 220 crystallographic planes that are parallel to the groove surfaces. This allows us to obtain very high reciprocal space resolution in the plane of scattering. With this arrangement and an X-ray mirror we can achieve a

198

dynamic range of ~10^8 in intensity with 10 second counting times. The intensity ranges from ~30,000,000 counts per second at the maximum to a background intensity of ~0.2 counts per second.

The high precision optically encoded axes allow the lattice parameter to be determined routinely to within 1ppm. The diffraction profiles and reciprocal space maps obtained with this arrangement but without the X-ray mirror can therefore be placed on an absolute scale using the method described by Fewster and Andrew (1995). Obtaining strain or orientation images with one micron resolution is possible by placing a photographic emulsion in the scattered beam from the sample. These images, as with most topography, is partially swamped by the Bragg scattering, however placing the emulsion after the analyser will create the equivalent of a dark field image, Fewster and Andrew (1993a). This latter topography is very powerful for imaging dislocations by placing the probe within the diffuse scattering region.

When the quality of the material is poor the scattering can be significantly broadened and it is not always necessary to perform experiments in high reciprocal space resolution. There are many advantages in using low-resolution reciprocal space mapping, Fewster and Andrew (1993b) and this method has been applied to aspects of the study on GaN. It is clear that we have to suit the instrumental configuration to the problems we wish to solve.

In this paper it will become clear that instrument versatility greatly assists the analysis. For example to obtain precision lattice parameters for the measurement of relaxation (and for topography which is not discussed further) we require high resolution (monochromator + analyser) but no X-ray mirror. However to obtain high-resolution reciprocal space maps of weak features we require an X-mirror + monochromator + analyser. If the features are very weak and the crystal quality is poor then we should change to slit optics without an X-ray mirror for low-resolution work. Clearly the instrument should be reconfigured to suit the problem and this is the advantage of the modular nature of the optics with this instrument.

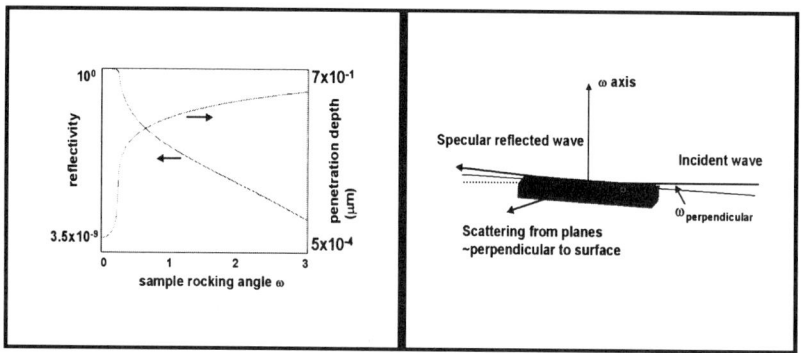

Fig.1. (left) The variation of specular reflectivity of X-rays from a perfect silicon surface and the variation in penetration depth with incident beam angle.

Fig.2. (right) The geometry of in-plane scattering. The small penetration of the X-ray beam into the sample close to total external reflection is used to scatter from atomic planes normal and close to the surface.

Generally work on semiconductor thin films is limited to reflection geometry, however this can be extended through the use of in-plane scattering. Access to reciprocal space is a function of X-ray wavelength (this defines the outer bounds of the scattering sphere) and sample absorption (this limits the transmission possibilities). If the sample is highly absorbing as in most thin layer substrates then the simplest analyses are best performed in reflection mode. This can be restricting for some studies since several reflections are unobservable, however we can make use of the fact that X-rays can undergo specular reflection from the surface. The refractive index of X-rays is less than unity and therefore will undergo total external reflection. Because the sample will be absorbing the X-rays incident close to the critical angle for total external reflection will penetrate into the sample. The strength of the specular reflectivity and the X-ray energy penetration depth are given in Fig.1.

Because the X-rays penetrate the surface they can be used to reflect off planes perpendicular to the surface plane and also the penetration depth can be controlled, Marra et al (1979). The experimental configuration is given in Fig.2. Both high resolution and low-resolution optics are used with this geometry in this study. The beam is incident on the sample at angles close to the critical angle ~0.3⁰ and the exit beam emerges at a similar angle, but is scattered at the appropriate scattering angle for the planes studied.

3. SIMULATION OF IMPERFECT STRUCTURES

Dynamical diffraction theory of X-rays is well-established (Takagi 1962, 1969: Taupin 1964), however the incorporation of defects is less straightforward. Kato (1980) has included the breakdown of wave-field coherence in a statistically based model. Isolated threading dislocations do not significantly contribute to the scattering except in terms of the diffuse scattering, which has a characteristic spherical shape. Misfit dislocations on the other hand are strongly located at the interfaces and can dramatically change the scattering process, Holy et al (1995), Fewster (1992). The former model is based on kinematical theory. The latter model treats the problem dynamically with a column approach that has now been extended to all interfaces and as many columns as the computer can cope with. The strain fields associated with the interface distortion are all included and the breakdown of the coherence in the X-ray wave-field. The wave-field in each column is then determined and combined in various ways. The ideal situation is to include all the contributions as though they are coherent, however the number of photon paths in many inhomogeneous samples is such that this is impractical. We can approximate the large range of beam paths to phase averaging which is roughly equivalent to an incoherent sum of the scattering. We therefore include an optional partial coherence parameter that relates to the problem to be solved.

Reciprocal space mapping can extend the limited information available to single diffraction profiles. Much of the data can be interpreted directly, however a more complete analysis can be obtained by modelling the scattering in reciprocal space. In conventional dynamical theory the strength of the scattering is simply related to the incident wave direction and no account is made of the scattered wave direction. However the scattered wave direction can be obtained from solving the boundary condition of the internal and external wave-fields. We can therefore include shape factors associated with any lateral inhomogeneity to derive lateral sizes of features less than the coherence length. The simplest way is to consider the phase coherence for all possible paths for a single photon and in this way simulate a reciprocal space map of an imperfect structure. A further complication in all this analyses is that most thin layers will distort in a complex way and therefore we have developed the model to accommodate any symmetry and distortion.

Of course what has been described so far is the interaction of photons with the sample. Aberrations from monochromators, analysers and detectors, etc., must be and are all included. These functions are very complicated and will not be discussed here except to state that the whole process from X-ray generation to X-ray capture must be considered. There is no simple instrument function. The full three-dimensional effect, sample size, type of scan, etc., must all be included for a precise analysis. A full account of the theory of scattering from imperfect structures and the instrumental aberrations are given in Fewster (1999).

4. GaN ANALYSIS

The example structure of interest here is a basic structure. It is composed of a 2 μm GaN layer on a sapphire substrate followed by a 5nm $In_{0.3}Ga_{0.7}N$ active region and a 30nm $Al_{0.3}Ga_{0.7}N$ layer and thin, 30nm, GaN cap. The thicknesses and compositions were only known approximately. A standard rocking curve close to the 0002 reflection of GaN revealed a broad profile with little helpful information content. A reciprocal space map close to this region produced the complex shape given in Fig.3a. A central scan through the map along a radial reciprocal space direction produced a profile with strong fringing, Fig.4. The broadening was similar for all directions normal to this radial scan and therefore the intensities extracted were considered realistic.

If the GaN was matched to the sapphire at the interface (they are both 0001 orientated) the ratio of the atomic spacings depend upon the relative orientations. Akasaki et al (1989) have found that the two cells are orientated and matched along the 10-10 and the 1-210 directions. In this study we found the orientation of GaN and sapphire also to register along these directions, although this is by no means conclusive as will be described later. Measurement of the in-plane spacing from the 0006 and 22-49 reflections of sapphire and the 0002 and 10-15 reflections of GaN indicate values close to the unstrained parameters and therefore for this study we have worked on the basis that the relaxation is close to 100%. Defining a more precise value for the quality of this epitaxy is unnecessary since it will not significantly influence the composition measurement. With this knowledge we can model the scattering close to the GaN 0002 profile and determine the composition and thicknesses in the structure. The structural information that gave the closest fit to the profile suggested is that the In had segregated into the AlGaN layer giving a complex quaternary phase of $In_{0.08}Al_{0.4}Ga_{0.52}N$. This phase accounts for the broad peak underneath the strong GaN peak. The alloy composition is almost certainly a much more complex graded layer.

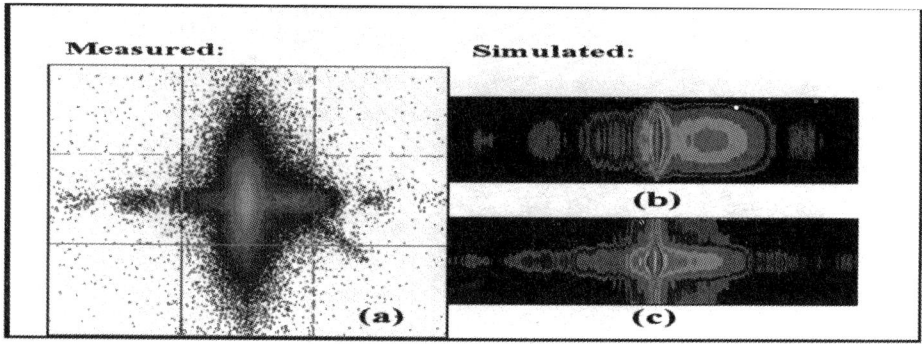

Fig.3. The measured and simulated 0002 reciprocal space maps from a relaxed GaN sample with lateral domains defined by defects. (a) is the measured map and (b) is the simulated map assuming a single 92 nm lateral correlation length and (c) shows the shape change when a distribution of correlation lengths is included.

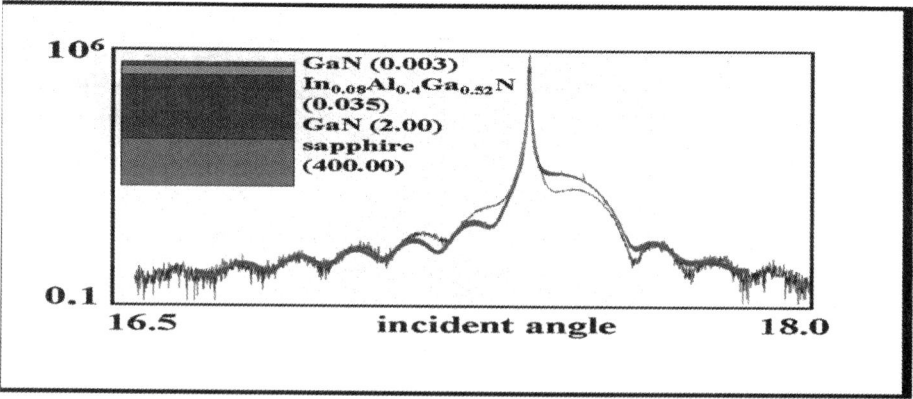

Fig.4. The fit (darker line) of the structure (thickness in microns) to the measured profile of the 0002 reflection.

Other important parameters for these materials relate to the quality of the microstructure. The reciprocal space map close to the 10-15 reflection for the GaN layer indicated that the broadening was almost entirely due to lateral correlation lengths parallel to the surface of the sample. This probably represents the distance between defects. Undertaking a simple analysis yielded a correlation length of 92nm, (Fewster, 1997). This value was included into the simulation of the reciprocal space map, however the shape did not match that of the measured data, Fig.3b. Only by creating a distribution of correlation lengths did this show agreement, Fig.3c. Because of the averaging effects of the distribution of length scales the average distance between defects had to be increased to 150nm, with a range of 50nm assuming an even distribution. This will require refining for a more precise fit. Also these length scales are duplicated through the whole structure above the sapphire interface. The sapphire on the other hand indicated a curvature of ~6m, this is evident from the 0006 and 22-49 reciprocal space maps where the intensity distribution is normal to the radial direction only.

Isolating single reflections does not always indicate epitaxial quality and often a more extensive search is worthwhile. With this sample the 10-15 reflection appeared rather indistinct and the layer structure information was difficult to see. The epitaxy parallel to the growth direction is clear but orientation conformity in the surface plane is far from perfect. This was clearly shown, by measuring the scattering from planes perpendicular to the surface plane with the geometry of Fig.2. The collimation and detector acceptance is perfectly adequate with slits and the scan indicated a multitude of peaks indicative of very poor matching of the atomic planes at the surface, leading to almost random orientations. This makes precise relaxation values inappropriate.

We can summarise our analysis in being able to obtain the composition and thicknesses, including segregation of In in this sample, the degree of relaxation at the GaN / sapphire interface, a distribution of distances between defects and the epitaxial quality.

5. SELF ASSEMBLED QUANTUM DOT STRUCTURES

InGaAs / GaAs quantum dot structures offer considerable advantages for the manufacture of lasers. The small size of the dots limit the number of recombination (mainly Auger process) possibilities for the electrons, remove the necessity of cleaving perfect facets for stimulating emission (vertical cavity emission). Also there are considerable advantages in temperature insensitivity compared with conventional and multiple quantum well lasers. Consequently the efficiency can be increased significantly. The ability to tune the emission wavelengths depends on the dot size and composition and these are the challenges in analysis. Depositing 1.7 monolayers of InAs on (001) GaAs creates the structures considered here. The growth is governed by the Stranski-Krastanow process, i.e. a two-dimensional wetting layer and island growth. The dots (islands) are then overlaid with GaAs and the process repeated to create a periodic array of dots in three dimensions. The samples also included AlGaAs layers either side of this 10 layer repeat sequence. Although these islands appear well formed, almost pyramidal soon after deposition, the overgrowth of GaAs causes the InAs to alloy with the GaAs and the shape changes dramatically, Xie, Chen, Ramachandran, Nayfonov, Konkar and Madhukar (1995). The analysis of Siverns et al (1998) by STEM suggests that the composition in the buried dots may be about 30% [In]. Since all device structures will be capped it is important to determine the shape and composition after growth. Other useful parameters are the degree of vertical correlation, the dot spacing, etc.; in fact anything that will help interpret the structure and link the optical properties to what has been grown.

A reciprocal space map close to the 004 reflections of two samples with different GaAs spacer layers covering the dots are given in Fig.5. The sample with the thicker spacer layers (22.8nm) is given in Fig.5a and differs from the sample with thinner spacer layers (11.8nm) given in Fig.5b. The periodicity of the satellite reflections gives the average repeat, whereas the spreading normal to the modulation direction is indicative of vertical correlation in composition or strain or both. It is clear that the degree of vertical correlation is strongest when the separation between quantum dots is reduced to levels below about 20nm or less. To obtain an indication of the vertical correlation in composition alone, maps of the intensity close to the strong specular scattering (reflectometry) were measured, Zabel (1994). These clearly indicated that the chemical modulation follows the evidence from the 004 reciprocal space maps. However both samples create significant diffuse scattering that can be related to the presents of quantum dots. The 444 and 044 reflections were used to obtain the average spacing and dispersion between dots in the plane of the interfaces along the <110> and <010> directions respectively, Darhuber et al (1997). A typical measurement is shown in Fig.6. For the sample with the stronger vertical correlation these separations (dispersions in brackets) were determined to be 216(97-400)nm along <110>, 99(67-200)nm along <1-10>, 100(62-200)nm along <100> and 104(81-191)nm along <010>. Clearly the separations along the principle axes are similar whereas there is a clear difference along the two [110] type directions.

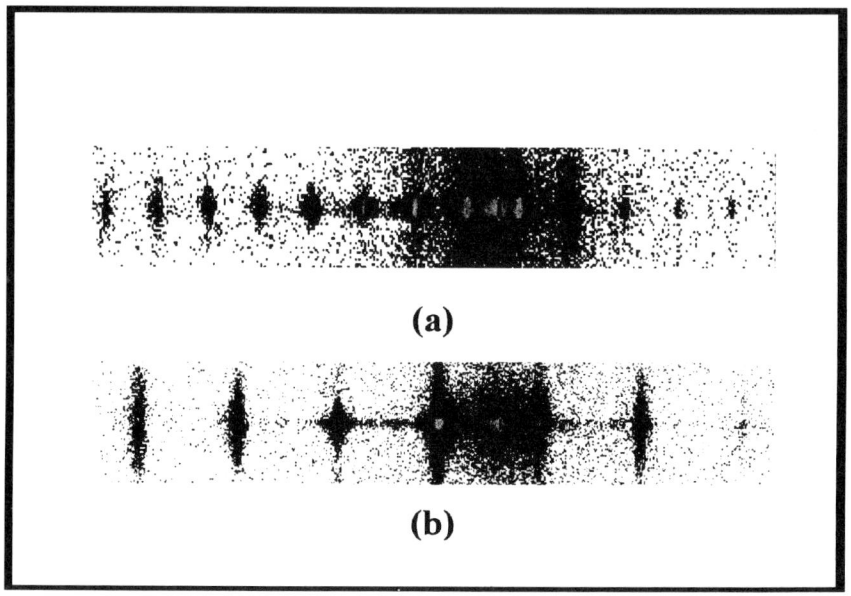

Fig.5. The reciprocal space maps (004 reflections) from periodically deposited Quantum Dot structures; (a) small vertical correlation and (b) significant vertical correlation of the dots from layer to layer.

To obtain the composition in the dots we firstly have to know the dimensions and relate this to the measurement over which we are averaging. An approximate fit to the 004 and 002 reflections (to separate the strain and composition correlation) has been obtained based on the assumption of two coherently scattering columns of GaAs + wetting layer and GaAs + wetting layer + dot. What is needed though is the area distribution, which will relate to a strain balance provided no relaxation takes place. The reciprocal space map around the 044 reflection shows that the substrate and layer peaks all lie on the normal to the surface plane and therefore their in-plane lattice parameters are identical. This was also confirmed with high resolution reciprocal space maps of the diffuse scattering of the 220 and 400 type reflections, since their scattering angles were within a few arc seconds of the value expected for bulk GaAs, Fewster & Andrew (1995).

To obtain an indication of the shape of the dots high-resolution reciprocal space maps using the in-plane scattering geometry (point focus + monochromator + analyser: Fig.2). Reciprocal space maps were recorded for the 400, 220, 040 and the –220 reflections, Fig.7. The differentiation of the <110> and <1-10> type directions was carried out as described in Fewster (1991c). The shape of the diffuse scattering is very different in each azimuth and clearly indicates asymmetry. The direction horizontal to the figures relates to strain whereas vertical directions indicate positive rotations in omega (the angle of incidence). The average strain as discussed before corresponds to that of GaAs, and intensity to the left of each shape in Fig.7 corresponds to larger lattice parameters. Consider the 400 and 040 reflections first. Clearly for 400 the higher lattice parameters appear at lower incident angles and this will correspond to the dot region, whilst the surrounding GaAs will be compressed, the latter appears at higher incident angles. The 040 reflection on the other hand creates higher strains at higher incident angles, etc. If we now work on the assumption that the dot shape is asymmetric then we can consider the directions to match those given in Fig.8. The strain variation must be at a maximum along the longer directions of the dot since there are more unit cells to contain the In and therefore this direction would match that of the <-110>. The smaller strain variation with more gradual bending of the crystallographic planes will match the diffuse scattering associated with the <110> direction. These results on the shape asymmetry agree with those found by Steans et al (private communication) for similar samples with thin GaAs overgrown layers on InGaAs dots analysed by STM.

This work is at a preliminary stage and the data obtained so far will be subject to more interpretation since much is only described at a qualitative level. However the data contains significant material for quantitative analysis to be able to extract the density of dots, asymmetry parameters and approximate limits on the range composition.

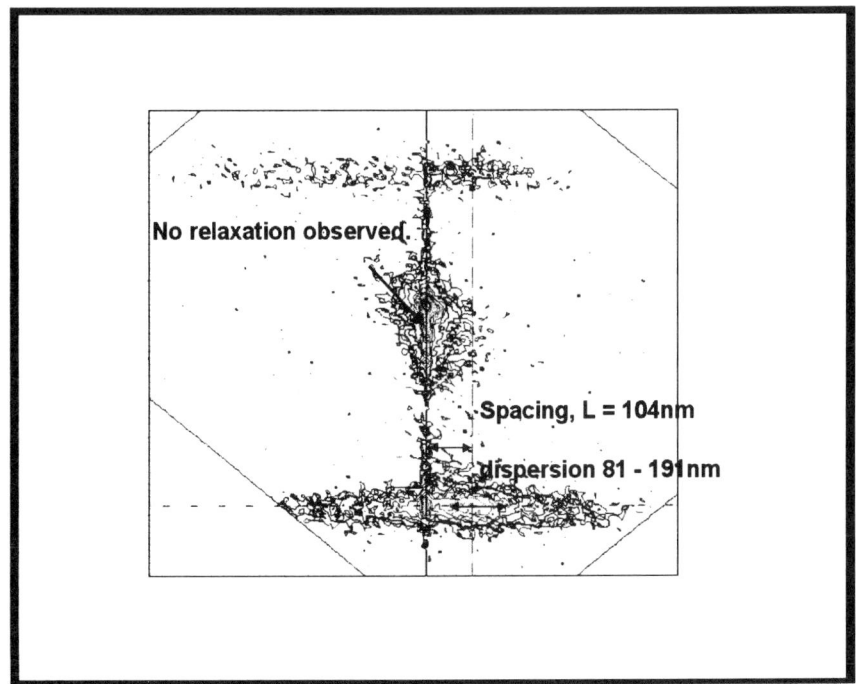

Fig.6. Reciprocal space map close to the 044 reflection, showing the lateral satellite peaks associated with the −1 satellite of the superlattice. The spacing of this lateral satellite and breadth of the peak gives an estimate of the average dot spacing and the dispersion.

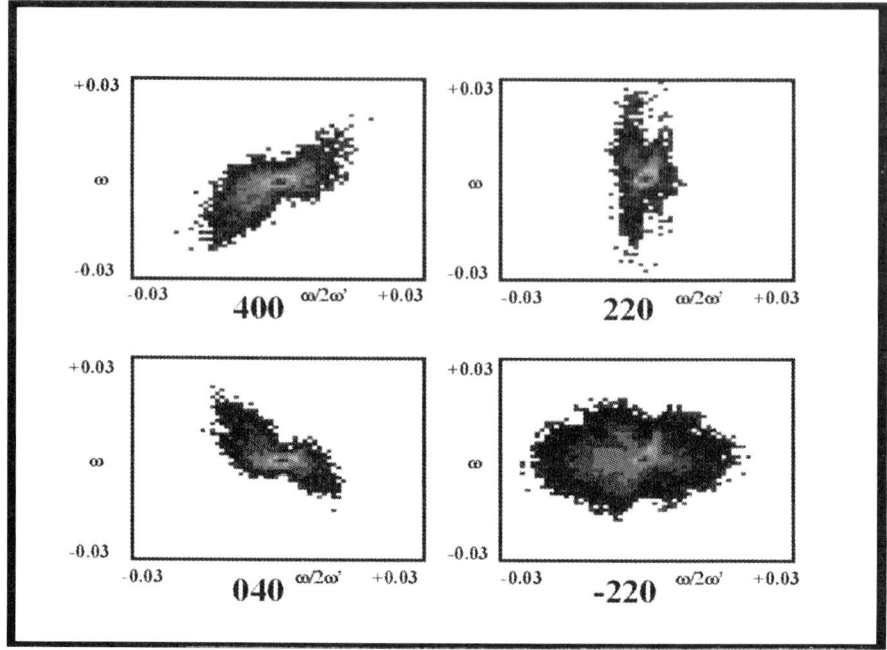

Fig.7. The diffuse scattering associated with the strain around the dots indicate a clear asymmetry. This data was collected with the geometry of Fig.2 with multiple crystal optics. The sharp central peak arises from the lattice plane spacings of the average lattice and corresponds precisely to that expected for GaAs, i.e. there is no lattice relaxation.

6. CONCLUSIONS

Recent advances in X-ray diffraction methods allow detailed analysis of semiconductor materials. These methods include; reciprocal space mapping, precision determination of interatomic spacings, diffuse scattering topography and the simulation of the scattering process. This paper has shown how a mixture of reciprocal space mapping combined with the simulation of imperfect relaxed structures can yield useful microstructural information. Also the use of in-plane scattering has been shown to give a good indication of epitaxial quality in GaN and the anisotropy of the shapes in Quantum Dot structures. Since these types of analyses require numerous geometries from low to high-resolution optics, conventional and in-plane scattering, the advantages of modular optics that are all pre-aligned is obvious. X-ray diffraction clearly offers a range of experimental configurations, using a large variation of optical combinations, to suit the analysis and it make possible to create quite a detailed picture of the samples under investigation. The development of an extended dynamical scattering model has also made it possible to simulated reciprocal space maps of relaxed and laterally inhomogeneous samples. Hence the scattering can now be modeled of a GaN structure for example, that consists of a range of domain sizes and a heavily dislocated interface to the underlying sapphire substrate. Simulating the scattering in a reciprocal space map indicates a high sensitivity to the dispersion in the size of domains.

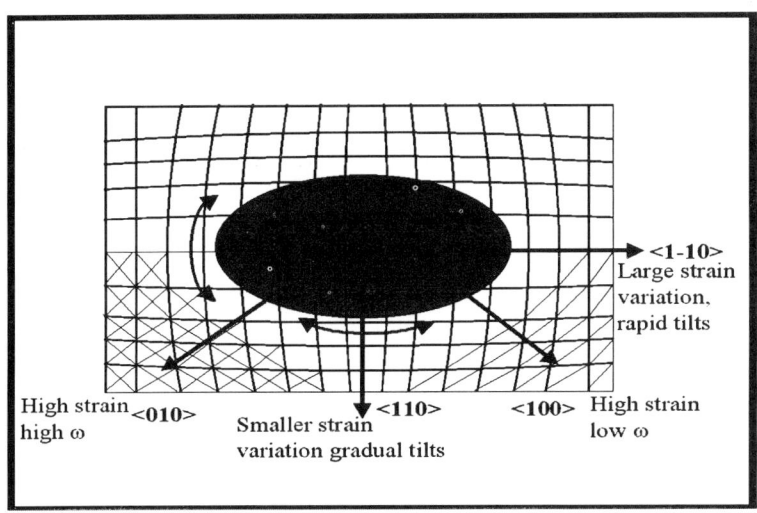

Fig.8. The qualitative interpretation of the diffuse scattering of Fig.7, indicating the anisotropy of the average dot shape.

ACKNOWLEDGEMENTS

I would like to thank N Andrew (Philips Research Laboratories) for carrying out the experimental work. Also I thank K Saito (Philips Analytical, Japan) for obtaining the GaN sample and colleagues at the Centre for Electronic Materials and Devices (Imperial College, London) for many fruitful discussions; B A Joyce, J H Neave, T Jones, P H Steans, D W Pashley and D Zhi.

REFERENCES

Akasaki I, Amano H, Koide Y, Hiramatsu K and Sawaki N 1989 J. Cryst. Growth **98**, 209
Fewster P F 1989 J. Appl. Cryst. **22**, 64
Fewster P F 1991a J. Appl. Cryst. **24**, 178
Fewster P F 1991b Appl. Surf. Sci. **50**, 9
Fewster P F 1991c Analysis of Microelectronic Materials and Devices, eds Grasserbauer and Werner (New York: Wiley) Chapter 3.1.1 *Note: the directions indicated in the figure of this chapter are reversed, whereas the text is correct.*
Fewster P F 1997 Critical Reviews in Solid State and Materials Science **22**, 69
Fewster P F 1992 J. Appl. Cryst. **25**, 714
Fewster P F 1999 *Scattering from Semiconductors* (Singapore: World Scientific Publishing)
Fewster P F and Andrew N L 1995 J. Appl. Cryst. **28**, 451
Fewster P F and Andrew N L 1993a J. Appl. Cryst. **26**, 812
Fewster P F and Andrew N L 1993b Materials Science Forum, eds E J Mittemeijer and R Delhez (Switzerland: Trans Tech) **133-136**, pp 221-230
Darhuber A A, Schittenhelm P, Holy, V, Stangl J, Bauer G and Abstreiter G 1997 Phys. Rev. B **55**, 15652

Holy V, Li J H, Bauer G, Schaffler F and Herzog H–J 1995 J. Appl. Phys. **78**, 5013

Kato N 1980 Acta Cryst. **A36**, 763

Marra W C, Eisenberger P and Cho, A Y 1979 J. Appl. Phys. **50**, 6927

Takagi S 1962 Acta Cryst. **15**, 1311

Takagi S 1969 J. Phys. Soc. Japan **26**, 1239

Taupin D 1964 Bull. Soc. Fran. Miner. Cryst. **87**, 469

Siverns P D, Malik S, McPherson G, Childs D, Roberts, C, Murray R, Joyce B A and Davock H 1998 Phys. Rev. **B58**, R10217

Xie Q, Chen P, Ramachandran T R, Nayfonov A, Konkar A and Madhukar A 1995 J. Cryst. Growth **150**, 357

Zabel H 1994 Appl. Phys. **A58**, 159

Inst. Phys. Conf. Ser. No 164
Paper presented at *Microsc. Semicond. Mater. Conf., Oxford, 22–25 March 1999*
© 1999 IOP Publishing Ltd

The influence of NaCl on the microstructure of CdS films and CdTe solar cells

K Durose[1], D Albin, R Ribelin, R Dhere, D King, H Moutinho, R Matson, P Dippo, J Hiie[2] and D H Levi

National Renewable Energy Laboratory, 1617 Cole Boulevard, Golden, Colorado 80401-33393, USA
[1]On leave from: Department of Physics, University of Durham, South Rd, Durham, DH1 3LE, UK
[2]Inst. of Materials Technology, Tallin Technical Univ, 5 Ehitajate Rd, 19086 Tallin, Estonia

ABSTRACT: NaCl treatments of CdTe/CdS solar cells are of interest because of the possibility that group I ions may enhance photovolatic performance. In this work the influence of NaCl on the microstructure of both CBD CdS films and CSS CdTe/CdS cells was studied by AFM, SEM and GIXRD. Morphological and microstructural changes to the films occurring on heating are described together with optical transmission and device results. Although addition of high levels of NaCl degrades the device performance, evidence is presented which suggests that impurity diffusion from glass substrates is beneficial.

1. INTRODUCTION

Heterojunction solar cells comprising CdTe/CdS/SnO₂/glass have now been studied widely and the behaviour of the materials is known to be complex (Durose et al 1999b, Durose et al 1999c). Issues known to be important include the polycrystallinity of the material, interdiffusion at the CdTe/CdS interface, and the nature of the active electrical centres. All of these are influenced by CdCl₂ 'activation'; a processing step in which the material is exposed to CdCl₂ and annealed - often in air - at around 400°C. This encourages recrystallisation and grain growth (the pseudobinary phase diagram has a eutectic at ~508°C), influences the doping and also interdiffusion. Although such processing is becoming better understood, little attention has been paid to the possible importance of other impurities in these solar cells. In this work the influence of added Na⁺ ions on the microstructure of CdS layers and CdTe/CdS solar cells has been examined: It has been suggested (Romeo 1998) that Na diffusing from glass substrates may enhance device performance. Na is already known to improve the crystal quality of epitaxial CdTe/GaAs (Hails et al 1998), and it also acts as an acceptor in CdTe (Hoffman et al 1992). In the somewhat parallel field of CuInSe₂ solar cell research, out-diffusion of Na from glass substrates has been shown to have a significant influence on microstructure and device performance (Contreras et al 1997). This is a preliminary study in which the Na⁺ was added as NaCl to the films at various stages in the processing, the chloride being chosen on account of the role of this anion in existing processing routes.

2. EXPERIMENTAL

In this work all films and devices were deposited on low alkali glass (Corning 1737F) which had first been coated with 0.5μm of SiO₂, followed by 0.5μm of SnO₂, the silica being included to eliminate uncontrolled out-diffusion from the glass.

CdS films deposited on such substrates by chemical bath deposition (Kaur et al 1980) were used to make 3 sample types:- a) as-deposited, b) heated in the presence of 1torr of O₂ using the same heating profile as is usually used prior to the deposition of CdTe at 600°C, and c) heated as for the previous sample but after coating with NaCl. Throughout this work the NaCl was added by dipping the samples in a saturated solution of NaCl in methanol (~1 wt%) and blowing dry. The CdS films were examined by AFM, grazing incidence XRD (CuKα), XPS and optical transmission/reflectance.

Four CdTe/CdS/SnO$_2$/SiO$_2$/glass samples were also prepared on the same substrates with the addition of NaCl being as follows:- a) no added NaCl - 'control' sample, b) NaCl added after CdTe deposition, prior to CdCl$_2$ treatment, c) NaCl added after CdS deposition, d) NaCl added after both the CdS and the CdTe as in b) and c) above. In each case ~8μm of CdTe was deposited close space sublimation (CSS) and CdCl$_2$ treatment was done in air at 400°C for 15 mins. Portions of samples a) and d) above were made into devices with graphite contacts. These devices were examined by SEM, AFM, PL and their AM1.5 I-V characteristics were measured. The I-V characteristics of a 'standard' cell grown on soda lime glass (no diffusion barrier) were also measured.

3. RESULTS AND DISCUSSION

Fig. 1 shows AFM micrographs of the 'heated' (NaCl-free) and 'Na-treated' CdS films. The morphology of the 'heated' sample is comparable to that of the 'as-deposited' one (not shown) - both show texture on the scale of ~0.2μm, this being typical of the underlying SnO$_2$, and both have average grain sizes in the range 230-250Å. Heating was also accompanied by a decrease in the optical band gap by 0.05eV, as has been reported elsewhere (Albin et al 1997). GIXRD taken with incident angles of between 0.3 and 2.0° showed strong CdS peaks, with SnO$_2$ peaks being present for scans with higher penetration. Of the low index CdS peaks, the 002$_{hex}$ was strongest with shoulders of intensity being present in the 100 $_{hex}$ and 101 $_{hex}$ positions. Rietvelt analysis has shown this to indicate that the material is neither purely cubic nor hexagonal, but of a mixed polytypic type (Gibson et al 1998). Heating in the presence of NaCl causes the same band-gap reduction as heating alone, and does not change the structure apparent from GIXRD. However it does cause grain growth by almost a factor of 3, as shown in Fig. 1 and Table 1. GIXRD also revealed the presence of halite NaCl, this being confirmed by XPS and by the appearance of cubic, hopper-faceted crystals ~1μm in size in AFM micrographs. It is possible that NaCl could influence CdS if oxidation to CdSO$_4$ occurs: in that case NaCl would react with the sulphate to give Na$_2$SO$_4$ + CdCl$_2$, the latter being a known sintering aid. No XRD lines due to Na$_2$SO$_4$ were observed in these samples. Similar reactions with CdTeO$_3$ might also be possible.

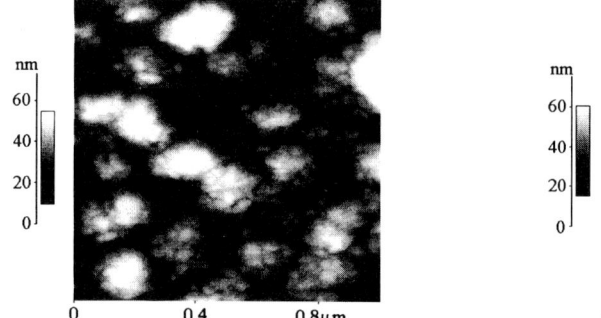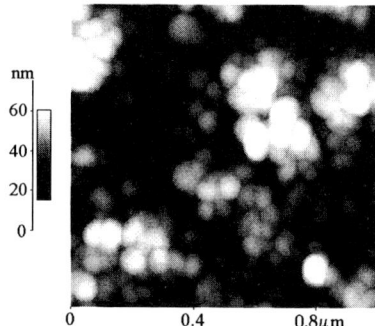

Fig. 1   AFM of CdS/SnO$_2$/SiO$_2$/glass surfaces. Heated (left) and heated in the presence of NaCl (right). Grain structure in the CdS is superimposed on the ~0.2μm texture of the SnO$_2$.

CdS/SnO$_2$/SiO$_2$ glass sample	Thickness / Å	Grain size / Å	RMS roughness / Å	Bandgap / eV
a) as-deposited	1002	233	133	2.42
b) heated	864	246	108	2.37
c) NaCl-treated	943	724	142	2.37

Table 1. CdS/SnO$_2$/SiO$_2$/glass films and their properties.

Fig. 2 shows SEM micrographs of the four device structures fabricated on SnO$_2$/SiO$_2$/glass substrates. The sequence a-d in the figure indicates the effect of increasing incorporation of NaCl in the devices. NaCl has no noticeable effect on the grain size in the CdTe, and this is expected since

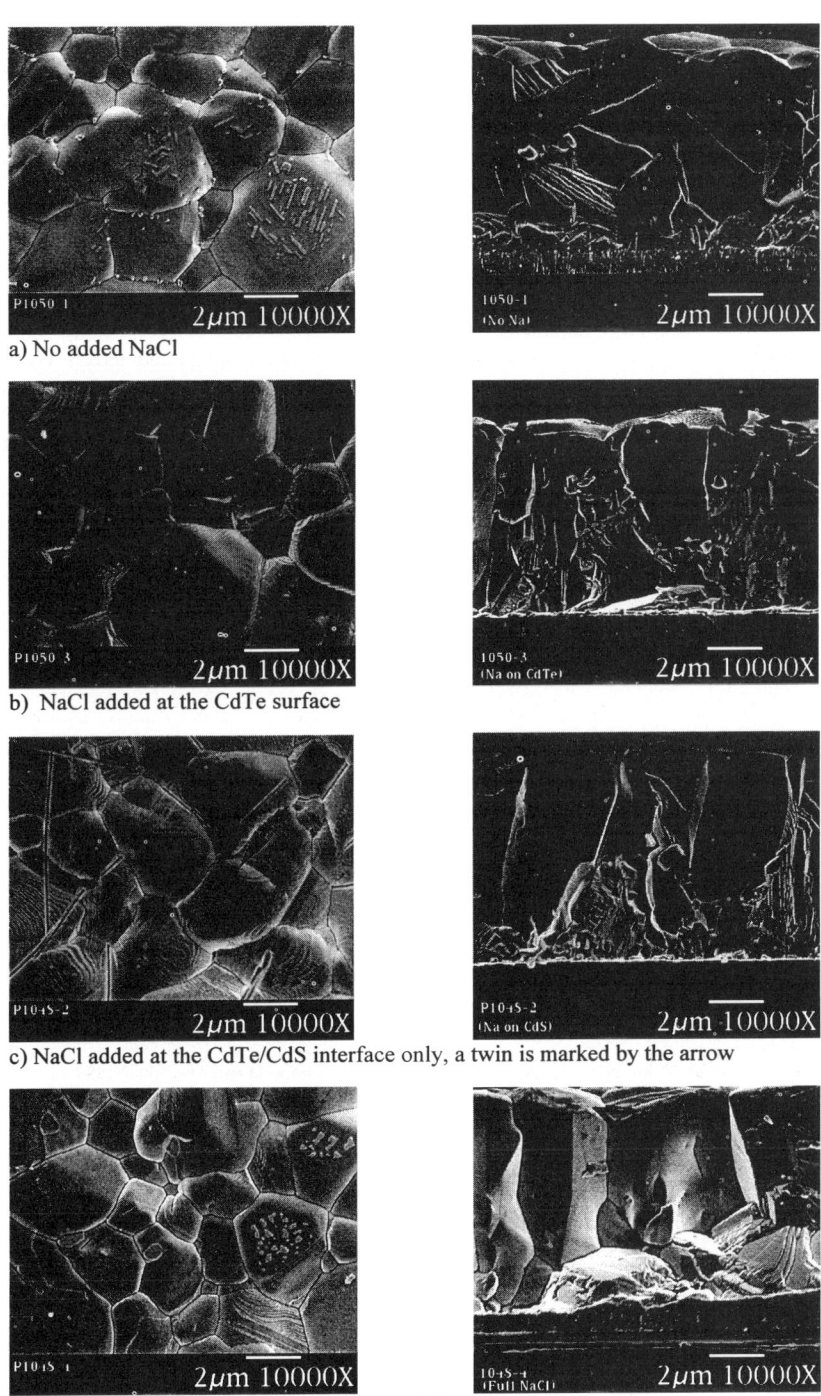

a) No added NaCl

b) NaCl added at the CdTe surface

c) NaCl added at the CdTe/CdS interface only, a twin is marked by the arrow

d) NaCl added at both the CdTe surface and the CdTe/CdS interface

Fig. 2 Plan view and cross-section SEM of NaCl-treated CdTe/CdS/SnO$_2$/SiO$_2$/glass structures

CSS CdTe does not usually undergo grain growth, even in the presence of $CdCl_2$ (Durose et al 1999b). The plan view micrographs show residues discussed by Waters et al (1998) and terracing. For sample c, AFM showed the terraces to be $\sim$270 Å in height and their interruption by a twin (see arrow in fig. 2) indicates their crystallographic nature. Samples were prepared for cross section SEM by clipping fragments off them with pincers to ensure that the layer failed at its weakest points. It can be seen that the increasing incorporation of NaCl has the effect of weakening the grain boundaries. Hence while failure occurs through the grains in the NaCl-free sample (fig. 2a), the grain boundary interfaces themselves fail in the fully treated sample (fig. 2d). Segregation phenomena are likely to be responsible for this.

With regard to the opto-elecronic influences of Na, photoluminescence of the CdTe revealed complex near-edge and DAP spectra which might be attributed to Na acceptors. This is reported in (Durose et al 1999a). Table 2 shows the photovoltaic parameters of the devices including those of a 'standard' cell grown on soda lime glass without a diffusion barrier. Addition of NaCl caused a very marked decline in all parameters. However *exclusion* of diffused species in the control sample acts to reduce V_{oc} *below* that routinely achievable in standard processing in which substrate out-diffusion may occur. It may be inferred from this that the substrate type has an influence on cell performance, and moreover that diffusion is likely to be a key factor. However without further study it is not possible to state that any particular diffusing species is responsible for enhancing device response.

Sample	V_{oc}/V	J_{sc}/mAcm^{-2}	FF/%
Standard	0.813	22.2	72.5
Control (Na-free)	0.775	21.4	58.1
NaCl-treated	0.202	6.4	29.9

Table 2. Photovoltaic parameters of the solar cells. Inclusion of an SiO_2 diffusion barrier in the 'control' cell reduces V_{oc} from the value obtained in 'standard' material.

4. CONCLUSIONS

The influence of adding NaCl to CdS and CdTe/CdS films prior to annealing has been studied. For CdS films the NaCl promotes grain growth, while for CdTe it weakens grain boundary interfaces. Inclusion of high levels of NaCl in device structures causes a serious decline in photovolatic performance. However, (nominal) exclusion of species diffusing from glass substrates by means of an SiO_2 barrier reduces V_{oc} from what is achievable in standard cells which have no such barrier. Impurites may therefore be beneficial. Further studies of the effect of intentionally added species and their quantification by SIMS will be of value. With regard to explaining the effects of NaCl on microstructure it would be helpful to determine the phase diagrams of NaCl with CdTe and CdS.

ACKNOWLEDGEMENTS

The authors thank N Romeo, Parma Univ, for drawing our attention to the importance of Na.

REFERENCES

Albin D, Rose D, Dhere R, Niles D, Swarzlander A, Mason A, Levi D, Moutinho H and Sheldon P 1997 NREL/SNL Photovoltaics Programme Review (AIP New York) p 665

Contreras M A, Egaas B, Dippo P, Webb J, Grantna J, Ramanathan K, Asher S, Swartzlander A and Noufi R 1997 26th IEEE PVSEC (Anaheim, Calafornia, USA: IEEE)

Durose K, Albin D, Ribelin R, Dhere R, King D, Moutinho H, Matson R, Dippo P and Levi D H 1999a Submitted to J Phys D: Applied Physics

Durose K, Edwards P R and Halliday D P 1999b J Cryst Growth **197**, 733

Durose K, Potter M D G, Cousins M A and Halliday D P 1999c Applied Physics (in press).

Gibson P N, Ozsan M E, Sherborne J M, Lincot D and Cowache P 1998 2nd World Conference on Photovoltaic Solar Energy Conversion (Vienna: IEEE) p 1089

Hails J E, Cole-Hamiliton D and Geiss J 1998 Journal of Electronic Materials **27**, 624

Hoffman D M, Omling P, Grimmeiss H G, Meyer B K, Benz K W and Sinerius D 1992 Phys Rev B **45**, 6247

Kaur I, Pandya D K and Chopra K L 1980 J Electrochemical Society **127**, 943

Romeo N 1998 Personal Communication

Waters D, Niles D, Gessert T A, Albin D, Rose D H and Sheldon P 1998 2nd World Conference on Photovoltaic Solar Energy Conversion (VIenna: WIP) p 1031

Inst. Phys. Conf. Ser. No 164
Paper presented at Microsc. Semicond. Mater. Conf., Oxford, 22–25 March 1999
© *1999 IOP Publishing Ltd*

Limitations to homoepitaxial silicon growth in plasma-enhanced chemical vapour deposition at low temperatures

L Houben, M Luysberg*, R Carius, P Hapke, F Finger and H Wagner

Institut für Schicht- und Ionentechnik, Forschungszentrum Jülich GmbH, D-52425 Jülich, Germany
*Institut für Festkörperforschung, Forschungszentrum Jülich GmbH, D-52425 Jülich, Germany

ABSTRACT: Low-temperature epitaxial growth of Si thin films on Si(100) was studied by transmission electron microscopy. The films were prepared by very-high frequency plasma-enhanced chemical vapour deposition at 200 °C. Large-scale epitaxial growth, which is restricted to pyramidal regions, is followed by a columnar growth with column diameters ≈50 nm, similar to the growth on non-crystalline substrates. The appearance of columnar growth is accompanied by the formation of {111}-facets, microtwins and stacking fault tetrahedra. These structural features bear a close similarity to those observed in low temperature MBE, which indicates similar mechanisms in the formation of growth defects.

1. INTRODUCTION

Plasma-enhanced chemical vapour deposition (PECVD) using silane diluted in hydrogen is the preferred technique for the deposition of thin film microcrystalline silicon (μc-Si:H) on various substrates at low temperature (≈200-400 °C). μc-Si:H provides enhanced light absorption in the near IR compared to amorphous silicon (a-Si:H) as well as increased stability against light induced degradation when compared to amorphous silicon-germanium alloys. It has been successfully applied as absorber layers in thin film devices like solar cells (Meier et al 1996) or colour sensors (Stiebig et al 1998). A number of studies have been devoted to the structural properties of PECVD grown μc-Si:H on non-crystalline substrates (e.g. glass, SiO_2, a-Si:H) as a function of the plasma conditions to elucidate the growth mechanism and to correlate transport behaviour and structure. Details of the structure like the crystal volume fraction and the typical column sizes are affected strongly by the hydrogen dilution (Tsai 1988, Houben et al 1998) and also the plasma excitation frequency (Luysberg et al 1997), but a common feature of all films is the columnar structure with faulted single crystal columns smaller than ≈50 nm in diameter and coherent twin domains with an extent of a few nm. In this paper we focus on homoepitaxial growth on clean Si(100) wafers in order to clarify whether the columnar fine-grain structure is related to the details of film nucleation or is an inherent property of the PECVD growth of crystalline silicon at low temperature.

2. EXPERIMENTAL DETAILS

The films were prepared in a conventional diode-type PECVD reactor using silane strongly diluted in hydrogen at very-high plasma excitation frequency of 95 MHz and a total pressure of 300 mTorr. The flow rates of silane and hydrogen were 2(3) sccm and 98(97) sccm, respectively. The samples were prepared at a substrate temperature of 200 °C and an externally applied HF-power of 5 W. Standard and patterned Si(100)-wafers were used as substrates. The patterned Si substrates were produced by

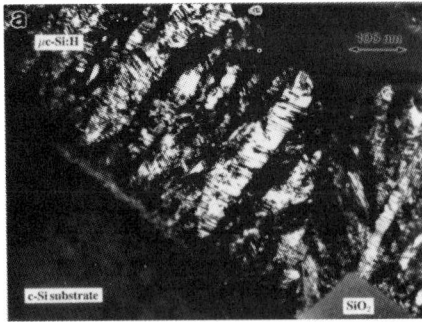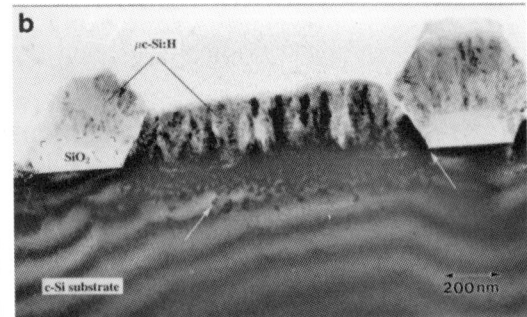

Figure 1: TEM diffraction contrast images of a cross sectional sample with patterned substrate. (a) Dark-field image recorded in a two beam case with a (400) substrate reflection. (b) Bright-field image in [011]-projection. See text for details.

photolithography and have a regular grid of stripes of thermally grown SiO_2. The substrate cleaning procedure involved a 60 s dip in AF91, a short rinse in water and immediate transfer to the deposition chamber. Prior to deposition, the substrates were cleaned additionally for 60 s in a pure hydrogen plasma in order to remove residual C and O contamination.

The microstructure was characterized by TEM using a JEOL 2000EX microscope operated at 200 kV and a JEOL 4000EX microscope operated at 400 kV. Additionally, X-ray pole-figures were recorded using a conventional Cu K_α radiation source.

3. RESULTS

Figure 1 shows diffraction contrast images of a cross-sectional specimen of an intrinsic film on a patterned substrate. The SiO_2 stripes are the bright regions with trapezium shape. Figure 1(a) shows that the long term evolution of film growth results in columnar growth. The single crystal columns are less than 70 nm in diameter. They are mostly lamellar twinned with nm sized coherent domains. The growth rates estimated from the film thickness on the Si substrate and on the SiO_2 are equal, indicating that there is no long term effect of the substrate on the film growth. Figure 1(b) shows that the large-scale epitaxial growth in the early stages is restricted to regions with pyramidal shape, bounded by {111}-planes (the right arrow in figure 1(b) points in ⟨111⟩-direction). Due to the hydrogen radical impingement during deposition, hydrogen platelets (HIP) are formed on {111}- and {100}-planes within the p-type silicon substrate (left arrow in figure 1(b)). The length of the HIP is mostly less than 20 nm and the most probable depth within the top surface layer of the substrate is about 40 nm. The HIP were identified by their closed Burgers circuits on lattice images as well as by the contrast reversal upon changing from underfocus to overfocus, which reveals their void-like nature.

Figure 2(a) shows a high-resolution image of the film structure in the very early stages of film nucleation. Microtwins and stacking faults are formed right at the beginning of film growth. Stacking fault tetrahedra are often observed (see arrow in figure 2(a)). Moreover, the epitaxially grown pyramidal regions also show planar defects like HIP or stacking faults. However, there is no evidence for a connection between the HIP in the substrate and the nucleation sites for stacking faults or microtwins.

The formation of higher order twins (for illustration some first- and second-order twins are labelled in figure 2(a)) leads to incoherent grain boundaries and columnar growth and gradually turns the layer polycrystalline. The top surface of the film is shown in the lattice image in figure 2(a). Besides a rough (100) interface the protruding part of a crystalline column shows {111}-facets (arrows), which were also found on the pyramidal regions adjacent to the SiO_2 stripes (left arrow in figure 1(b)).

X-ray pole figures (not shown here) reveal a significant contribution of first order twins. This confirms the results from the high-resolution TEM images that the randomization of crystal orientation is not due to the formation of nuclei with random orientation but due to the successive formation of

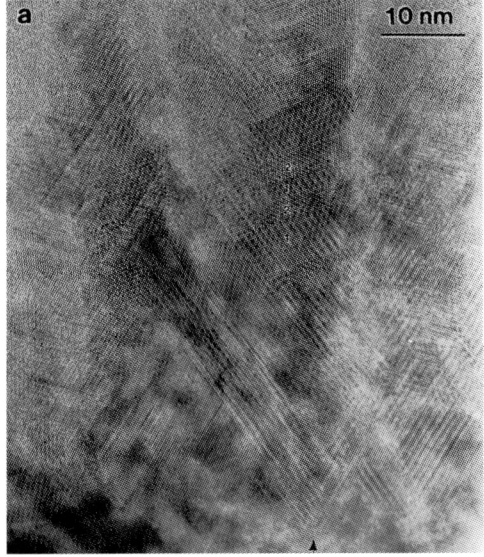
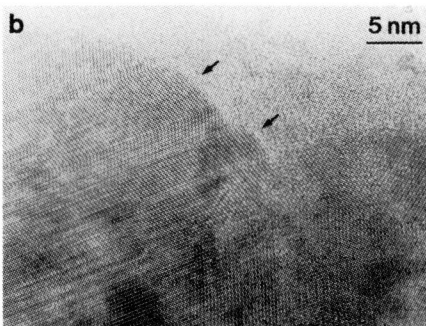

Figure 2: High-resolution TEM micrographs showing (a) the initial film nucleation and (b) the film surface. The arrow in (a) points to the nucleation site of a stacking fault tetrahedron. For illustration some first- and second-order twins are labelled within the region of the onset of columnar growth. In (b) {111}-facets adjacent to a rough (100)-plane are marked by arrows.

twin boundaries during grain-epitaxy. Moreover, the probability for twin formation, estimated from the intensity ratio of twin reflections and substrate reflections, was found to be equal for all four {111}-planes.

4. DISCUSSION

Homoepitaxial growth of silicon in radio-frequency PECVD has been reported by Tsai et al (1991). These authors also found a limited epitaxial thickness at low temperature and interpreted their results in terms of excessive hydrogen etching leading to 'microcrystalline-like' structures. In this study we found a close similarity of this microcrystalline structure with the microstructure of films grown under similar plasma conditions on non-crystalline substrates (see Luysberg et al 1997, Houben et al 1998). Whereas this means that there is no long term effect of the substrate on the film structure in PECVD, the mechanisms by which large scale epitaxial growth is hampered under these growth conditions are not known in detail. Lattice misfit due to a strained substrate in the presence of the HIP is not likely to be the cause of disordered growth. Firstly, the strain field is highly localized around the HIP and their separation is too large to result in a considerable lattice compression. Secondly, there is no evidence for a coincidence between the HIP and the nucleation sites for stacking-faults or microtwins.

Additionally, remarkable similarities can be found when comparing PECVD grown Si films with films that were grown by low-temperature MBE of semiconductors (see Eaglesham 1995, and references therein), despite the considerably lower (background) hydrogen concentrations in MBE. For instance Varhue et al (1996) investigated the temperature dependence of Si homoepitaxy in ECR-PECVD and found a thermally activated epitaxial thickness and pointed out the similarity to the limited thickness epitaxy in low-temperature MBE. Similarities can also be found in film roughening and facet formation. The standard model of Karder et al (1986) describes the film roughness w emerging with film thickness h as $w = h^\beta$, where β takes its maximum value $\beta_{max} = 0.5$ in the zero-diffusion limit. Both the MBE grown samples (see Eaglesham 1995) as well as PECVD-samples on Si(100) and Si(111) (see Kondo et al 1998) grown under plasma conditions similar to those used in this study show roughening exponents $\beta > 1$, considerably larger than β_{max}. The kinetic roughening is believed to be enhanced by the presence of hydrogen. Hydrogen covered surfaces lower the mobility of Si on Si(100) in MBE (Vasek et al 1995) and were also found to lower the mobility of SiH_2-precursors on Si(100) and promote a multilayer-growth during gas-source CVD with disilane (Owen et al 1997). The importance of roughening arises from the observation, that local extrema in the surface morphology provide nucleation sites for growth defects (Eaglesham 1995, Luysberg et al 1999).

The growth on {111}-facets at local cusps in the surface, as shown for the PECVD growth in this paper and for the MBE growth of Si by Eaglesham et al (1990) and Ramana Murty and Atwater (1994), could provide a simple picture for the defect formation since the {111}-planes are susceptible to the formation of stacking faults or microtwins, similar to the solid phase crystallisation of amorphous silicon (see Batstone 1993, and references therein). As described before, PECVD growth under strong hydrogen dilution then leads to columnar structures due to the formation of incoherent grain boundaries between higher order twins. Successive twinning gradually turns the film polycrystalline. There is no indication for a link between the typical column diameter and the typical length scale of local epitaxy since the bulk of the columns also show a high density of planar defects. Therefore, shadowing effects as in amorphous silicon deposition may play a significant role for the typical length scale in the stationary columnar growth.

5. CONCLUSIONS

Homoepitaxial growth of Si on Si(100) by PECVD at 200 °C has a local nature and turns to a columnar growth with structural features similar to growth on non-crystalline substrates. The columnar growth results from the formation of grain boundaries between higher order twins. Successive twinning gradually turns the films polycrystalline. No evidence could be found for any influence of hydrogen induced platelets within the substrate on the formation of growth defects. A comparison with the low-temperature MBE of Si reveals important similarities: Kinetic roughening as well as the formation of {111}-facets are observed for both processes and may be responsible for the local nature of epitaxial growth in PECVD due to promotion of the formation of growth defects.

REFERENCES

Batstone J L, 1993, Phil. Mag. A **67**, 51.
Eaglesham D, Gossman H-J and Cerullo M, 1990, Phys. Rev. Lett. **65**, 1227.
Eaglesham D J, 1995, J. Appl. Phys. **77**, 3597.
Houben L, Luysberg M, Hapke P, Carius R, Finger F and Wagner H, 1998, Phil. Mag. A **77**, 1447.
Karder M, Parisi G and Zhang, Y-C, 1986, Phys. Rev. Lett. **56**, 889.
Kondo M, Ohe T, Saito K, Nishimiya T and Matsuda A, 1998, J. Non-Cryst. Solids **227–230**, 890.
Luysberg M, Hapke P, Carius R and Finger F, 1997, Phil. Mag. A **75**, 31.
Luysberg M, Specht P and Weber E R, 1999, *this conference*.
Meier J, Torres P, Platz R, Dubail S, Kroll U, Selvan J A, Pellaton Vaucher N, Hof Chr, Fischer D, Keppner H, Shah A, Ufer K-D, Giannoulès P and Koehler J, 1996, Mat. Res. Soc. Symp. Proc. **420**, 3.
Owen J, Miki K, Bowler D, Goringe C, Goldfarb I and Briggs G, 1997, Surf. Sci. **394**, p. 79 and p. 91.
Ramana Murty M and Atwater H, 1994, Phys. Rev. B **49**, 8483.
Stiebig H, Knipp D, Hapke P and Finger F, 1998, J. Non-Cryst. Solids **227–230**, 1330.
Tsai C C, 1988, In Amorphous Silicon and Related Materials, ed. by Fritzsche H (Singapore: World Scientific Publishing Co.), p. 123.
Tsai C C, Anderson G and Thompson R, 1991, J. Non-Cryst. Solids **137&138**, 673.
Varhue W, Andry P, Rogers J, Adams E, Kontra R and Lavoie M, 1996, Solid State Techn. **39**, 163.
Vasek J, Zhang Z, Salling C and Lagally M, 1995, Phys. Rev. B **51**, 17207.

Inst. Phys. Conf. Ser. No 164
Paper presented at Microsc. Semicond. Mater. Conf., Oxford, 22–25 March 1999

An investigation of segregation-induced interface broadening in *p*-channel SiGe/Si MOSFET device structures by electron energy-loss imaging

D J Norris, A G Cullis, T J Grasby[*] and E H C Parker[*]

Department of Electronic and Electrical Engineering, Sheffield University, Mappin Street, Sheffield, S1 3JD, UK.
[*] Department of Physics, University of Warwick, Gibbet Hill Rd, Coventry, CV4 7AL, UK.

ABSTRACT: SiGe *p*-channel heteroepitaxial MOS transistor test structures, fabricated by molecular beam epitaxy, have been investigated using high resolution transmission electron microscopy and electron energy-loss imaging. Using these techniques, we have been able, for the first time, to quantitatively determine the nanoscale Ge distribution across a typical SiGe alloy channel. The Ge profile across the alloy channel was found to be asymmetrical due to the occurrence of segregation, with an exponential-like distribution directed towards the surface. The results match closely with the Ge profile predicted by segregation theory.

1. INTRODUCTION

Developments towards a more advanced Si-based device technology have been driven by the need for increasingly high-speed analogue and digital electronics. One particular device structure that we are currently investigating is the strained *p*-channel field-effect transistor (FET), a component of which consists of a pseudomorphic layer of SiGe grown by molecular beam epitaxy (MBE) on Si. A crucial requirement for good electrical performance is sharpness of the interfaces bounding the SiGe layer, since small scale interface roughness leads to deleterious carrier scattering (Whall and Parker 1999). Interface sharpness can be degraded by wave-like undulations which may form on the final growth surface due to thermodynamically driven growth instability (Cullis et al 1992). It is possible to produce more uniform alloy layers by performing growth at a relatively low-temperature, capping with Si, and then annealing at high-temperature to restore the required electrical properties (Grasby et al 1999). However, even after implementing the above growth recipe, interfaces may still not be optimally sharp (Norris et al 1999). It is crucial, therefore, that the interfaces can be characterised by chemically sensitive techniques, which can analyse the structure and composition of these layers with sub-nanometre resolution.

Interfaces of this type have been analysed (Fukatsu et al 1991, Grasby et al 1999) using secondary ion mass spectroscopy (SIMS); however, this method tends to sample large areas, to give averaged results, and consequently it is difficult to obtain nm-scale resolution in the high-concentration regime (Norris et al 1999). Small electron-probe methods using transmission electron microscopy (TEM) (Walther et al 1997a, Benedetti et al 1999) have involved electron energy-loss spectroscopy (EELS) and energy dispersive X-ray spectroscopy (EDS) and have been used to profile such Ge distributions. Moreover, image-based techniques include energy-filtered TEM (EFTEM) and high-angle annular dark-field (HAADF) studies. Previous EFTEM work on SiGe layer structures includes Si-L deficit profiling of relatively thin multiple layers (Mayer et al 1998) and Ge-L profiling of morphologically distorted SiGe layers (Walther et al 1997b). In addition, HAADF quantitative analysis has, for example, been performed on multiple quantum wells (Walther et al 1999).

In this work, we present a study using combined HREM and energy-loss filtered imaging for the analysis of the Ge distribution within a ~9nm SiGe channel heterostructure grown by MBE. For EFTEM, images were formed using the Ge-L energy-loss electrons from which the Ge profile across the layer could be determined with nm-scale resolution.

2. EXPERIMENTAL DETAILS

Layers were grown on RCA-cleaned, n-Si(001) substrate in a VG V90S MBE system. After an *in-situ* clean, a 250 nm of Si was deposited whilst ramping the heater down to 450°C.
A 9 nm layer of SiGe (Ge~30%) was next deposited, followed by a 25nm Si capping layer, both at 450°C, this temperature being low enough to suppress SiGe undulation formation (Grasby et al 1999). The temperature was then ramped up to 700°C for 30min for a post-growth *in-situ* anneal, before being reduced to 600°C for the growth of a final layer of boron doped Si. All layers were deposited at a growth rate of 0.1nm/s.

Cross-sectional TEM specimens, with [110] surface normals, were prepared by mechanical polishing and argon ion milling to electron transparency. These were examined using a JEOL 2010F transmission electron microscope with a field emission gun; this operated at 200kV and was equipped with a Gatan Imaging Filter (GIF), which provided electron energy-loss spectroscopy (EELS) and imaging.

3. RESULTS

Conventional TEM performed on cross-sectional samples, and a typical region of the channel layers is shown in figure 1. The 200 bright-field (Fig. 1a) and HREM (Fig. 1b) images show no large scale roughening or defects within the SiGe channel region. However, there is a clear difference in the relative sharpness of the buffer/channel and channel/cap interfaces. An intensity profile, taken from Fig. 1a along the growth direction (A to B), is plotted in Fig. 1c and shows that the buffer/channel interface is significantly sharper than the channel/cap interface. However, the presence of strain contrast in such images makes a quantitative comparison of the Ge gradation difficult; therefore, we have used energy filtered TEM.

In figure 2a, an EELS spectrum is shown, obtained using a 3nm electron probe focused into the central region of the SiGe channel. Both the Ge-L and Si-K energy loss edges are shown in the range 1-2keV of the spectrum. For EFTEM imaging, the delayed Ge-L signal was used with the post-edge window placed at 1300eV and pre-edge windows at 1125eV and 1175eV. A series of filtered images were acquired at each of these windows using a 30eV slit width and exposure times of 40s: the specimen drift was no more than 0.3nm during this exposure period. The three-window background subtraction method was used to generate Ge maps of the channel area. In figure 2b, an example of a Ge map is shown derived using the Ge-L edge, from which the profile distribution of Ge across the channel could be determined. The latter, along the growth direction A to B, is shown in Fig. 2(c) and exhibits a central plateau region with asymmetrically sloping sides. The Ge elemental distribution across the channel was independently calibrated by determination of the Ge peak concentration using the spectrum shown in Fig. 2a. The 3nm electron probe was incident at the relatively flat central region of the distribution and, from the relative Ge and Si edge amplitudes, a Ge concentration of 29±4% was obtained.

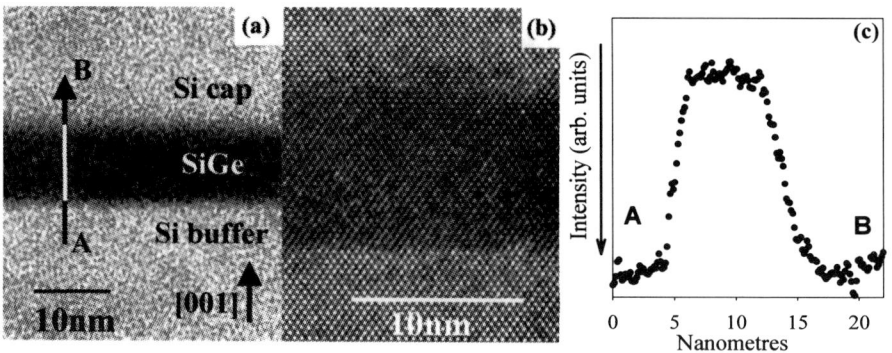

Fig. 1 (a) 200 BF image and (b) HREM image of the SiGe alloy channel layer; (c) inverted profile of region A-B as indicated in (a).

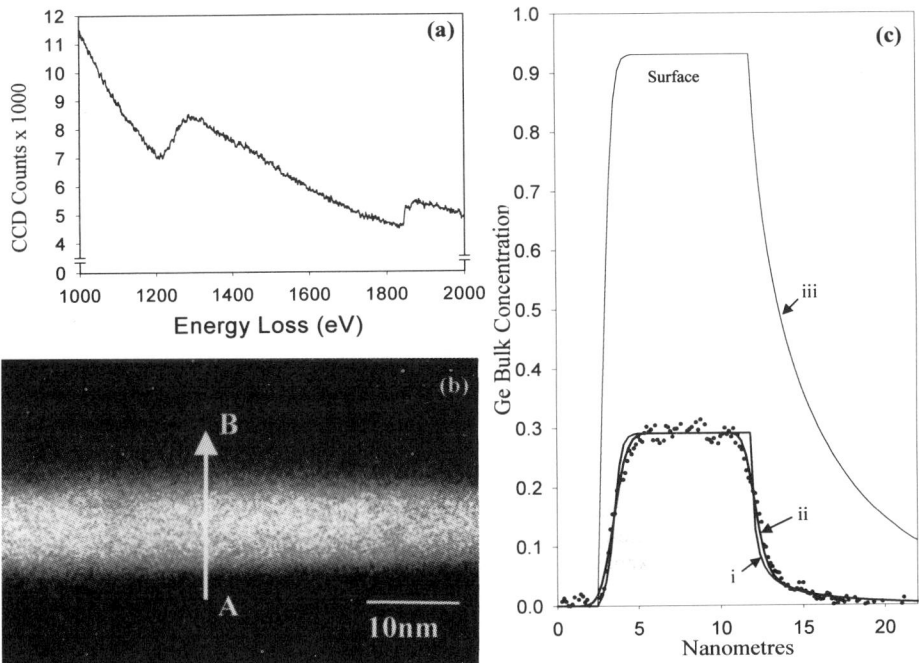

Fig. 2 (a) electron energy-loss spectrum from within the SiGe channel; (b) Ge elemental map acquired using the Ge-L edge (1217eV) shown in (a); (c) Ge profile from region A-B outlined in (b) where (i) is the theoretically derived bulk Ge profile, (ii) is the same profile but subjected to a 1nm resolution limitation and (iii) is the profile of the segregated Ge concentration in the surface monolayer.

4. DISCUSSION

Although the theoretically expected spatial resolution of the Ge distribution is still unclear (Dehaese et al 1995, Fukatsu et al 1991), for the present work, we estimate the spatial resolution in two ways. Firstly, by comparing the measured Ge profile with that (Fig. 1c) across an elementally sensitive 200-type image of the channel, a match to significantly better than 1nm is found in both full-width-half-maximum and overall shape. Secondly, segregation theory was found to correlate closely with the experimentally derived profile (see below). Both of these methods give significant confidence that actual resolution is achieved on the 1nm scale.

From the EFTEM Ge-L profile in Fig 2(c), it is clear that the Ge distribution exhibits a 6-7nm relatively flat central region, across which the measured Ge concentration is close to the nominal 30% value. However, the channel interfaces are clearly asymmetrical. The buffer/channel interface is relatively sharp, with a width of ~1.5nm (from 10% to 90% of maximum Ge concentration). In comparison, the channel/cap interface is significantly broader, having an exponential-like tail, in the Ge concentration range 10%-1%, which extends several nanometres from the body of the channel. These main features of the channel shape were a general characteristic of other regions of the SiGe alloy channel investigated. Ge profiles obtained using SIMS, on similar layers, have been found to be insensitive to the extended Ge-distribution in the high-concentration regime (Norris et al, to be published).

The predicted Ge profile derived from segregation theory has been fitted directly to the asymmetrical Ge profile. For this, we have used the kinetic, two-state exchange segregation model, with self-limitation (Dehaese et al 1995, Fukatsu et al 1991), to predict the expected Ge distribution for our layer growth temperature. For the present calculations, the rate equation (1) and mass conservation relation (2) for the bulk (B) and surface (S) concentration were:

$$dS(t)/dt = \Phi_{Ge} + P_1 B(t)(1 - S(t)) - P_2 S(t)(1 - B(t)), \qquad (1)$$

$$S(t) + B(t) = S(0) + B(0) + \Phi_{Ge} t \qquad (2)$$

where $P_1 = \nu_1 \exp(-E_a/kT)$ and $P_2 = \nu_2 \exp(-(E_a + E_s)/kT)$.

Parameters for fitting included a lattice vibration frequency ($\nu_1 = \nu_2$) of 5×10^{12}/s; an incident Ge flux concentration (Φ_{Ge}) of 0.29 (as determine from EELS); an activation energy (E_a) of 1.8eV; and a segregation energy (E_s) of 0.2eV (based on previous work with adjustment to fit the long range, low concentration Ge tail measured in our samples by SIMS): k is the Boltzmann constant and T the temperature in Kelvin. Solving these equations numerically leads to the profiles (i) and (iii), for the bulk and surface concentration respectively, in Fig. 2(c). These theoretical profiles (i and iii) make clear the way in which a large segregated surface Ge concentration controls incorporation of Ge into the bulk, thus giving the characteristic shape expected for the segregation-induced interface broadening. However, to account for the estimated resolution limitation, a Gaussian convolution function of 1nm FWHM was imposed upon the resultant bulk profile (i) to give the final curve (ii) in Fig. 2(c). It is evident that there is an excellent correlation with the experimental data at the 1nm resolution level.

5. CONCLUSIONS

The present work has demonstrated that the Ge distributions across very narrow SiGe MBE layers can be quantitatively determined with approximately nanometre resolution by electron energy-loss imaging employing the Ge-L edge. The Ge distribution within the present layer structure exhibited a surface-directed, exponential-like tail which extended several nanometres from the body of the channel. This is shown to be in agreement with theoretical modelling. Such segregation-induced interface broadening has significance for the performance of FET devices fabricated using these layers.

ACKNOWLEDGEMENTS

The authors would like to thank Prof. M.G. Dowsett and T.E. Whall for helpful discussions, Drs C.P. Parry and P.J. Phillips for the growth of SiGe/Si layer structures and the EPSRC for financial provision.

REFERENCES

Benedetti A, Norris D J, Hetherington C J D, Cullis A G, Armigliato A, Balboni R, Robbins D J and Wallis D J 1999 Microscopy of Semiconducting Materials 1999 eds A G Cullis and R Beanland (Bristol: IOPP) pp 219-222
Cullis A G, Robbins D J, Pidduck A J and Smith P W 1992 J. Crystal Growth **123** 333
Dehaese O, Wallart X and Mollot F 1995 Appl. Phys. Lett. **66** 52
Fukatsu S, Fujita K, Yaguchi H, Shiraki Y and Ito R 1991 Appl. Phys. Lett. **59** 2103
Grasby T J, Hammond R, Parry C P, Phillips P J, McGreggor B M, Morris R J H, Braithwaite G, Whall T E, Parker E H C, Knights A and Colman P G 1999 Appl. Phys. Lett. **74** 1848
Jager W and Mayer J 1995 Ultramicroscopy **59** 33
Norris D J, Cullis A G, Dowsett M G and Parker E H C (to be published)
Norris D J, Cullis A G, Grasby T J and Parker E H C 1999 J. Appl. Phys. (in press)
Walther T and Humphreys C J 1999 J. Crystal Growth **197** 113
Walther T, Humphreys C J, Cullis A G and Robbins D J 1997a Microscopy of Semiconducting Materials 1997 eds A G Cullis and R Beanland (Bristol: IOPP) pp 47-54
Walther T, Humphreys C J, Cullis A G and Robbins D J 1997b Materials Science Forum **196-201** 505
Whall T E and Parker E H C 1998 J. Phys. D: Appl. Phys. **31** 1397

Inst. Phys. Conf. Ser. No 164
Paper presented at Microsc. Semicond. Mater. Conf., Oxford, 22–25 March 1999

219

Strain and Ge concentration determinations in SiGe/Si multiple quantum wells by TEM methods

A Benedetti, D J Norris, C J D Hetherington, A G Cullis, A Armigliato[+], R Balboni[+], D J Robbins* **and D J Wallis***

Department of Electronic and Electrical Engineering, Sheffield University, Mappin Street, Sheffield S1 3JD, UK
[+] CNR-Istituto LAMEL, Via P. Gobetti 101, 40129 Bologna, Italy
*Defence and Evaluation Research Agency, St Andrews Road, Malvern, Worcs WR14 3PS, UK

ABSTRACT: A range of transmission electron microscopy (TEM) techniques has been employed to determine the Ge concentrations in, and profiles across, low pressure CVD-grown SiGe/Si MQWs. In addition to conventional imaging techniques, we have used EDS and EFTEM for direct chemical analysis. From these measurements, we have found that the wells exhibit a segregation-induced asymmetry in interface sharpness. Furthermore, we have used other indirect TEM methods to determine the average Ge concentration from analysis of the strain distribution across the wells. These methods include analysis of the relative strain-induced lattice displacements from HREM images, and also the strain-induced shift in the <111> Bragg contours observed in LACBED patterns. We will show that the measurements of Ge fraction determined using these four different techniques are in good agreement.

1. INTRODUCTION

SiGe/Si multiple quantum well (MQWs) structures, with SiGe layer thicknesses approaching a few nanometres, are currently being developed for use in infra-red (IR) radiation detection. However, their electronic properties depend critically on the composition, thickness, strain and interface morphology of the layers (Karunasiri et al 1994). Hence, there is a strong need for characterisation techniques that can analyse such thin layers with both high spatial resolution and high chemical sensitivity. High-angle annular dark-field (HAADF) imaging and fine-probe analysis can be used for elemental mapping (Walther et al 1997, 1999); however, we have used energy-filtered TEM (EFTEM) imaging which can potentially provide spatial resolution of chemical species on the nanoscale.

In this paper, we present analysis by four different TEM methods that have been used to determine the compositional profiles across ~4nm SiGe quantum wells (QW). These include the use of X-ray energy-dispersive spectroscopy (EDS) to measure the Ge fraction. In addition, both Si and Ge elemental maps have been obtained using EFTEM imaging, where high spatial resolution to better than 1nm can be achieved. The second pair of techniques involves analysis of local strain fields using the large angle convergent beam electron diffraction (LACBED) technique and high resolution electron microscopy (HREM). In the former case, the strain-induced average lattice displacement inside a QW has been measured through the shift of a <111> Bragg contour, whereas in the latter the phase shift between two pure Si regions has been calculated from HREM images and related to the well composition.

2. EXPERIMENTAL DETAILS

SiGe MQWs were grown in Si by low-pressure chemical vapour deposition (LPCVD) on (001) Si substrates using mixtures of silane and germane at 650°C and at a pressure of ~140 mTorr. TEM cross-sectional specimens, with <110> surface normals, were prepared by face-to-face gluing, mechanical grinding and low energy argon ion-beam thinning to electron transparency. The specimens were examined using two different electron microscopes. For HREM, EFTEM and EDS work, a JEOL 2010F TEM with a field emission gun operating at 200 kV was employed. Important accessories attached to the microscope include a Gatan imaging filter (GIF) and an Oxford Instruments ISIS energy dispersive X-ray detection system. LACBED analysis was performed using a Philips CM30 TEM, operating at 300 kV with specimen cooling at liquid nitrogen temperatures.

3. RESULTS AND DISCUSSION

3.1 EDS analysis

MQW analysis was carried out using a 0.5nm focussed electron probe. A stepped-scan was performed across, and perpendicular to, a SiGe QW and EDS spectra were taken at each location. The Ge composition was extracted from each of these to yield a Ge concentration profile, such as that shown in Fig.1. Whilst the measured peak Ge concentration of 19% is close to that intended during growth, the Ge profile across the well shows a clear asymmetry in the relative sharpness of the two QW interfaces: the SiGe-on-Si interface appears much sharper than the Si-on-SiGe interface.

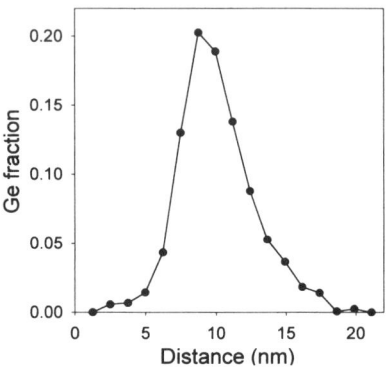

Fig. 1 *EDS Ge profile across a single QW*

Despite the analysed area being thin (~20 to 50 nm), the composition profile is expected to be partially smeared, most likely due to the effects of beam broadening. However, using a LINK software routine, the effect of beam broadening determined for our probe size and specimen thickness was found to be between 0.5 and 1nm suggesting a spatial resolution limitation (Reed 1982) of ~1 nm. As a consequence, the true peak Ge concentration is expected to be a few percent higher, and the full-width-at-half maximum (FWHM) should be wider than corresponding profiles derived from BF/HREM images and/or EELS (see Table 1). Nevertheless, the FWHM of 5.3± 0.5 nm was taken to be a reasonable measure of the QW width, whilst the peak Ge concentration of 19±3% was taken to be that at the centre of the well.

3.2. EFTEM analysis

Ge concentration profiles were also determined from energy-loss filtered images, in particular from those images derived using the Ge-L edge (1217eV). Both the three-window (3W) and jump-ratio (JR) background subtraction methods were employed, with pre-edge windows at 1175eV (for 3W) and 1210eV (for 3W and JR) and a post-edge window at 1300eV. Images such as those in Figs. 2 and 3 were obtained. A slit-width of 30eV and an exposure time of 40s were used throughout and no significant drift could be observed during image acquisition.

A comparison of a 3W Ge image with a conventional (200) BF image (Fig. 2) shows good agreement both in the well-width and spacing, suggesting that resolution better than 1nm is being achieved, as should be expected for this technique (Egerton 1996). In a similar manner to Fig. 1, a typical Ge profile (Fig. 3b) taken across the region outlined in the JR Ge image (Fig. 3a) shows further the asymmetry in

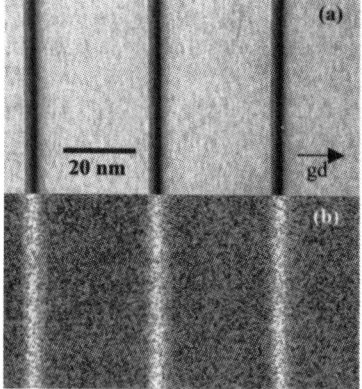

Fig. 2 *BF image (a) compared with a three windows Ge map (b) (growth direction gd)*

the Ge concentration profile. Again, the SiGe-on-Si interface appears sharper than the Si-on-SiGe interface.

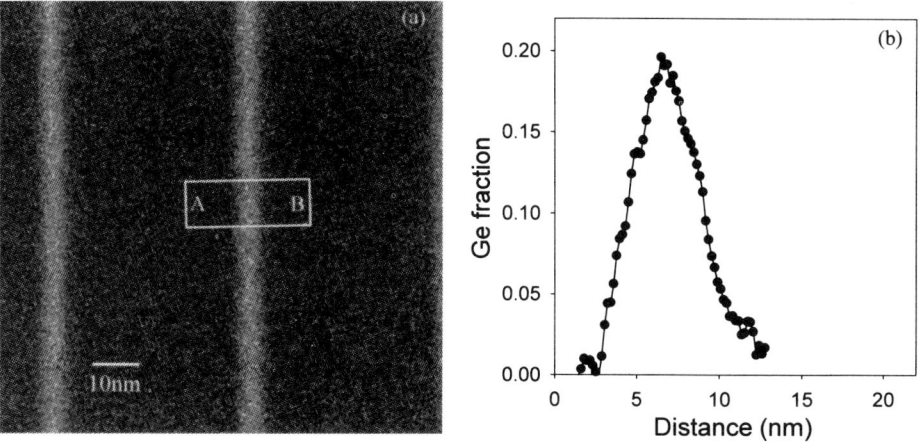

Fig. 3 *(a)Ge jump ratio image of two wells ; (b) intensity profile of region A-B (growth direction) outlined in (a)*

3.3 HREM imaging analysis

HREM images of the QWs were obtained and a typical example is shown in Fig. 4(a). The QW width was measured from such images to be (4.2 ± 0.5) nm. A (200) phase shift image is shown in Fig. 4(b), which was derived from the Fourier transform of the HREM image in (a).

From the intensity profile (Fig. 4c) the phase shift and then the lattice displacement between the two Si regions surrounding the QW were calculated as $\psi = (1.5 \pm 0.2)$ rad and $\phi_{(100)} = \phi_{(200)} / 2 = (0.064 \pm 0.006)$ nm, respectively. The strain field inside the QW can then be determined. If we define

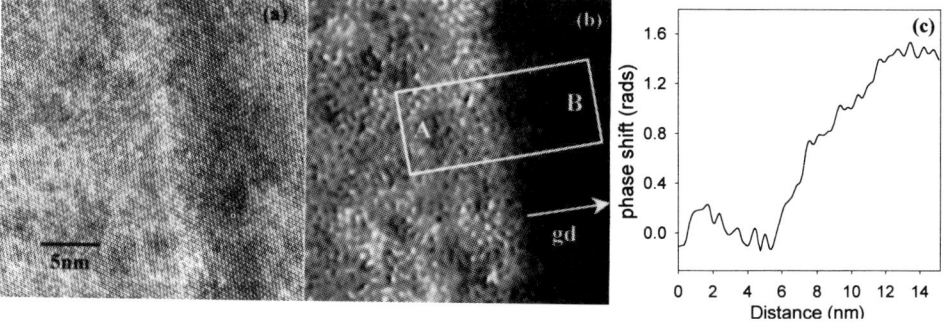

Fig. 4 *(a) HREM image of a single well; (b) 200 phase image of (a); (c) phase-shift intensity profile of region A-B outlined in (b)*

$\delta = a_{SiGe} - a_{Si}$ and N as the number of $\{100\}$ planes within the well, the width W of a QW will be given by $W = (a_{Si} + \delta)N$. At the same time, W can be written as $W = (a_{Si} \times N) + \phi$, where ϕ is the total displacement of the $\{100\}$ planes as measured above. Hence, δ can be expressed as the following:

$$\delta = a_{Si}\phi / (W - \phi).$$

The strain along the growth direction is $\varepsilon^{th} = \delta /a_{Si} = (8.4\pm0.8)\times10^{-3}$. Since the sample thickness is estimated to be an order of magnitude greater than the QW width, negligible surface relaxation is expected (Treacy and Gibson 1986). The Ge concentration, x, is equal to (Armigliato et al 1992) $\varepsilon^{th}/ (1+\alpha) \times N \times \beta = (13 \pm 2)$ %, with $\alpha =2\nu / (1 - \nu)$. N is the Si atomic density, β is the expansion coefficient, x is the Ge fraction and ν is the Poisson ratio.

3.4 LACBED analysis

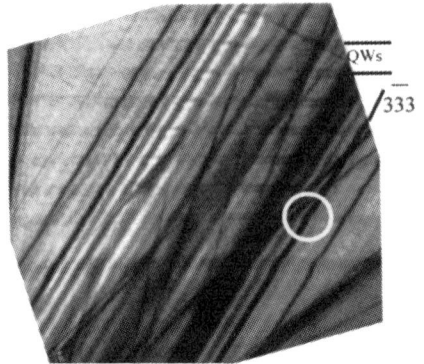

Fig. 5 *LACBED pattern*

An indirect method of determining the Ge composition is to measure the strain-induced distortions of reciprocal Bragg contours in LACBED patterns. A typical LACBED pattern of the QW region, taken in projection a few degrees off the [110] zone axis, is shown in Fig. 5. A shift in the position of the <111> contours is clearly visible (see white circle), caused by a strain-induced rotation of the corresponding planes. The rotation angle is $\Delta\theta = d / r$, where d is the measured shift (see Fig. 5) and r is the angular distance from the pole of [110] zone axis. The unrelaxed strain is given by $\varepsilon^{th} = 2 \times \Delta\theta / \sin2\theta$ (Duan et al 1994), where θ is the angle between the (333) and (001) planes, $\theta = 54.94°$. Again, a negligible relaxation is expected, since the analysed specimen area is relatively thick. Hence, the Ge fraction, x, given by $x = \varepsilon^{th} / (1+\alpha) \times N \times \beta$ (α, β and N are defined previously) was found to be 15±4%.

4. CONCLUSIONS

Our comparison of the four different TEM techniques for the analysis of the concentration and profile of Ge within ~4nm CVD-grown SiGe QWs has yielded results which are in satisfying agreement. Both EDS and EFTEM Ge profiles have been found to exhibit a clear asymmetry in the relative sharpness of the two SiGe QW interfaces: the Si-on-SiGe interface is clearly broader than the SiGe-on-Si interface. This behaviour may, in part, be dependent upon the effects of Ge segregation (Norris et al 1999): further details with quantitative correlation will be published in Benedetti et al (1999). This is the first time that this behaviour has been studied by combining BF, EDS and EFTEM analysis. We are currently investigating the relationship between such segregation-induced broadening and the electrical performances of devices.

The measurements of the Ge concentration made by indirect methods, namely LACBED and HREM phase-shift imaging, as summarised in Table 1, are in good agreement. It is important to note, however, that these indirect methods give averaged QW concentrations, rather than an absolute Ge profile as for the EDS and EELS methods. Hence, this average composition would be expected to be lower than the peak concentration, as indeed we observe.

	EDS	EFTEM	HREM	LACBED
W	(5.3 ± 0.5) nm	(4.7 ± 0.5) nm	(4.2 ± 0.5) nm	
ε^{th}			$(8.4 \pm 0.8) \times 10^{-3}$	$(1.0 \pm 0.2) \times 10^{-2}$
x_{ave} (%)			(13 ± 2) %	(15 ± 4) %
x_{pk} (%)	(19 ± 2) %			

Table 1 : W=well width; ε^{th} =lattice strain; x_{ave}/x_{pk}=average/peak Ge concentration

AB, DJN and CJDH would like to acknowledge EPSRC for financial support.

REFERENCES

Armigliato A, Servidori M, Cembali F, Fabbri R, Rosa R, Corticelli F, Govoni D, Drigo A, Mazzer M, Romanato F, Frabboni S, Balboni R, Iyer S S and Guerrieri A 1992 Micr. Microan. Microstr. **3**, 363
Benedetti A, Hetherington C J D, Norris D J, Cullis A G, Robbins D J and Wallis D J 1999, to be published
Duan X F, Cherns D and Steeds J W 1994 Phil Mag A **70**, 1091
Egerton R F 1996 Electron Energy-Loss Spectroscopy in the Electron Microscope, New York: Plenum Press, Ch 1
Karunasiri R P G, Park J S, Kang L W and Sang K C 1994 Optic. Engin. **33**, 1468
Norris D J, Cullis A G, Grasby T J and Parker E H C 1999 J. Appl. Phys., in press
Reed S J B 1982 Ultramicroscopy **7**, 405
Treacy M M J and Gibson J M 1986 J. Vac. Sci. Technol. B **4**, 1458
Walther T and Humphreys C J 1997 Inst. Phys. Conf. Ser. No. **157**, 47
Walther T and Humphreys C J 1999 J. Crystal Growth **197**, 113

Inst. Phys. Conf. Ser. No 164
Paper presented at Microsc. Semicond. Mater. Conf., Oxford, 22–25 March 1999
© 1999 IOP Publishing Ltd

Study of the relaxation in InGaAs SQW grown on (111)B substrates

M Gutiérrez, D González, G Aragón, J J Sánchez*, I Izpura* and R García

Departamento de Ciencia de los Materiales e Ingeniería Metaúrgica y Química Inorgánica. Universidad de Cádiz. Apdo 40, Puerto Real, 11510-Cádiz. Spain.
*Departamento de Ingeniería Electrónica. Universidad Politécnica de Madrid. Ciudad Universitaria, 28040-Madrid. Spain.

ABSTRACT: The critical layer thickness was experimentally determined by transmission electron microscopy through direct observation of misfit dislocations in an InGaAs single quantum well grown on (111)B GaAs substrates. For high In-content, a new configuration of misfit dislocation network was found. This consists of <112> dislocation lines and $1/2$<1$\bar{1}$0> Burgers vectors which are contained in the growth plane. The strain relief intensity of this new network is larger than that of the classical network for low In-content. This implies two different relaxation mechanisms depending on In-content.

1. INTRODUCTION

Strained InGaAs/GaAs Quantum Wells grown on (111) substrates present strong piezoelectric fields (Smith 1986) due to biaxial stress resulting from the lattice mismatch. Such built-in piezoelectric fields make this system suitable for potential applications in new electro-optical devices operating within 1.3-1.5 μm wavelength.

Concerning strain relaxation, a larger Critical Layer Thickness (CLT) for the (111) orientation with respect to (001) one is expected (Anan et al 1992, Calle et al 1995). This larger CLT permits the introduction of a higher In-content and to reach wavelengths beyond 1 μm. Up to date, few strain relaxation studies have been done. Dislocations in InGaAs heterostructures using different misorientated (111)B substrate types have been studied by Mitchell and Unal (1991), Sacedón et al (1994) and Edirisinghe et al (1997) but the In-content was always below 20%. Nevertheless, the strain relaxation mechanism is still not well studied for high In-content in strained InGaAs on GaAs (111)B.

2. EXPERIMENTAL

A Single Quantum Well (SQW) series has been grown on GaAs (111)B misoriented 1° off towards [211] direction by Molecular Beam Epitaxy (MBE). The structure consisted of a 0.3 μm of GaAs Si-doped followed by a layer 10 nm-thick of $In_xGa_{1-x}As$ and finally other 0.3 μm of GaAs Be-doped. The In-content was 10, 15, 20, 25, 30 and 35%.

Samples were prepared for Transmission Electron Microscopy (TEM) by mechanical thinning followed by chemical etching ($H_2SO_4+H_2O_2+H_2O$) for planar view observation (PVTEM). The TEM studies were performed using a JEOL 1200EX transmission electron microscopy operating at 120kV.

3. RESULTS

A misfit dislocation (MD) network was observed for SQW with In-content beyond 30% with two outstanding characteristics as is shown in Fig. 1. First, the short MD segments followed the three $<11\bar{2}>$ directions. Second, the **g.b** analysis with 220 and 224 reflections revealed Burgers vectors of $1/2<1\bar{1}0>$ type which are contained in the (111) growth plane. Thus, the glide plane of MDs is the growth plane and the MD character is 30°. Hereafter, this new MD network will be called 30°-in.

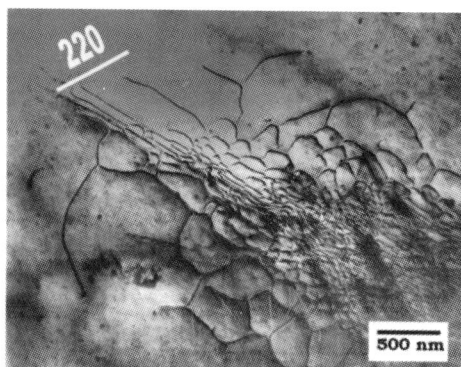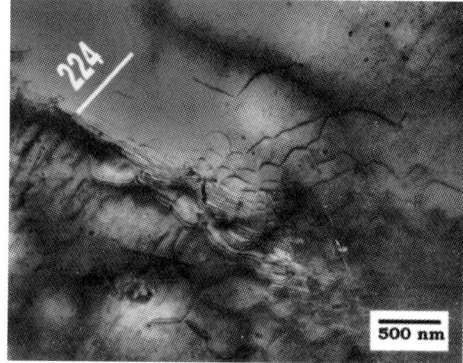

Figure 1. Bright field PVTEM image of $In_{0.35}Ga_{0.65}As$ SQW with 220 reflection (left micrograph) and 224 reflection (right micrograph) showing the new MD network. Note that the MD segments are along $<11\bar{2}>$ directions and under one 224 reflection, one of the three $<11\bar{2}>$ MD lines is out of contrast.

On the contrary, the MD network which has been observed for In-content below 20% (Edirisinghe et al 1997, Sacedón et al 1994, Mitchell and Unal 1991) was parallel to all three $<1\bar{1}0>$ directions. The Burgers vectors were $1/2<110>$ type which were out the (111) growth plane at 60° with MD lines. Hereafter, this classical MD network will be called 60°-out.

4. DISCUSSION

Two different approximations to work out the CLT in (111) epitaxy have been used by Anan et al (1992) and Colson and Dunstan (1997). For a capped strained layer with lattice mismatch f, Anan´s equation neglecting its surface energy term can be written as:

$$h_c = \frac{G_{111}b^2(1-\upsilon_{111}\cos^2\theta)}{2\pi(1-\upsilon_{111})M_{111}b_r f}\ln\left(\frac{\alpha h_c}{b}\right) \qquad (1)$$

and Colson's equation as:

$$h_c = \frac{G_{001}b^2(1-\upsilon_{001}\cos^2\theta)}{2\pi(1-\upsilon_{001})M_{111}b_r f}\ln\left(\frac{\alpha h_c}{b}\right) \tag{2}$$

where G, υ, and M are the shear modulus, Poisson's ratio and biaxial Young's modulus respectively, b, θ and α, the Burgers vector modulus, angle between the Burgers vector and the MD line and MD core radius respectively and b_r, the misfit-relieving component of Burgers vector defined as the projection of **b** onto a line in the interface at right angles to the MD.

The main difference between both equations is the used G and υ values. Anan et al (1992) have taken into account the orientation of the growth plane in the dislocation energy through G_{111} and υ_{111}. On the contrary, Colson and Dunstan (1997) have considered the dislocation energy independent of the growth plane orientation, this is done through G_{001} and υ_{001}. However, both Anan and Colson have worked out the CLT for (111) epitaxy considering a 60°-out MD network. This network has a b_r equal to $1/\sqrt{12}\,b$.

Our experimental results put forward a 30°-in MD network for high In-content (x>25%). This new network (30°-in) has a b_r equal to 1/2b which is larger than that of the classical network (60°-out). The different strain-relieving components of both 30°-in and 60°-out networks will have implications in the (111) epitaxy CLT.

The CLT for 30°-in and 60°-out networks in (111) substrates according to equation 1 and 2 is plotted in fig. 2. Beside, the CLT in (001) substrates for comparative purpose is plotted as well. Moreover, in the Fig. 2, the studied SQW are also drawn to compare the two CLT approximations. It can be seen that the SQW 10 nm-thick with 30% and 35% In-content overcome the CLT according to Anan approximation much better than Colson one.

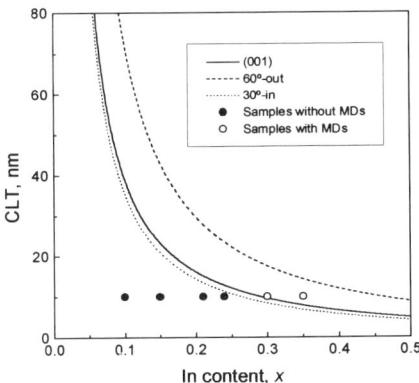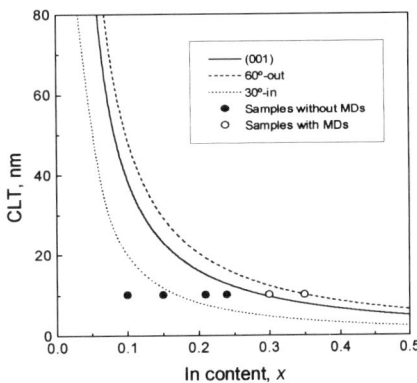

Figure 2. Anan's CLT (left plot) and Colson's CLT (right plot) for 30°-in (dotted line) and 60°-out (dashed line) networks in (111) orientation together the CLT in (001) epitaxy (full line). The studied SQWs with MDs are indicated as open circles while the ones without MDs as full circles. Note that better agreement is found for Anan's CLT of 30°-in MD network.

From the Fig. 2, the CLT for the 30°-in network is smaller than that for 60°-out network due to its misfit-relieving component being larger. This fact explains the lower CLT found experimentally for high In-content. Therefore, one expects a lower CLT for high In-content ($x > 25\%$) than one for low In-content ($x < 25\%$) due to a change in the MD network from a 60°-out to a 30°-in type.

5. CONCLUSIONS

InGaAs SQWs 10 nm-thick grown on (111)B GaAs substrates have been studied by TEM. A new MDs network was found for In-content beyond 30%. In this network, MDs can glide on the (111) growth plane. This implies a larger misfit-relieving component than that of the classical 60° MD network typical for SQW below 25% In-content. A comparison between our experimental CLT data and those worked out according to the Anan and Colson approximations has given a better agreement with the Anan approximation which takes into account the substrate orientation in the shear modulus and Poisson's ratio. A CLT transition due to a change in the MD network type depending on In-content is expected.

ACKNOWLEDGEMENTS

The present work was supported by the CICYT project TIC98-0826 and the Andalusian government (PAI TEP-0120). The TEM study was carried out at The Electron Microscopy Facilities of the Universidad de Cádiz.

REFERENCES

Anan T, Hishi K and Sugou S 1992 Appl. Phys. Lett. **60**, 3159
Calle F, Álvarez A L, Sacedón A and Calleja E 1995 Phys. Stat. Sol (a) **152**, 201
Colson H G and Dunstan D J 1997 J. Appl. Phys. **81**, 2898
Edirisinghe S, Staton-Bevan A E and Grey R 1997 J. Appl. Phys. **82**, 4870
Mitchell T E and Unal O 1991 J. Electron. Mat. **20**, 723
Sacedón A, Calle F, Álvarez A L, Calleja E, Muñoz E, Beanland R and Goodhew P 1994 Appl. Phys. Lett. **65**, 1
Smith D L 1986 Sol. St. Comm. **57**, 919

Inst. Phys. Conf. Ser. No 164
Paper presented at Microsc. Semicond. Mater. Conf., Oxford, 22–25 March 1999

TEM study of an anti-correlation relation in corrugated layers of Si$_{1-x}$Ge$_x$(C)/Si superlattices

E Müller, R Hartmann and D Grützmacher

Labor für Mikro- und Nanostrukturen, PSI Würenlingen u. Villigen, CH-5232 Villigen/PSI, Switzerland

ABSTRACT: In Si(C)/Si and Si$_{1-x}$Ge$_x$(C)/Si superlattices a pronounced undulation of the individual layers is observed when a critical growth temperature and/or C-concentration is exceeded. This undulation is locally correlated from one layer to the next one by an anti-correlation relationship. The onset of a similar anti-correlation is found in stacks of C-induced Ge-dots provided the thickness of the Si-spacer between the dot layers is large enough. If its thickness is below 10 nm, the dots are vertically aligned.

1. INTRODUCTION

Over many years, attempts have been made to modify the bulk properties of silicon, e.g. by compositional variation or by a reduction of the dimensionality. Aiming at an integration of optoelectronics into Si-technology, the incorporation of Ge has been studied for many years. More recently, the addition of C-atoms to either pure Si or to SiGe has been proposed (Soref 1993). Besides the growth of quantum wells, the synthesis of quantum dots and the occurrence of self-organization phenomena are intensively studied. For SiGe layered structures it is known that corrugations are transmitted from one layer to the next one. This has been applied to the growth of Ge quantum dots on Si: If dot layers are vertically stacked at a relatively short distance, the dots are found to vertically align (Schittenhelm et al 1998). According to Schmidt et al (1998) this is still the case if prior to the first layer of Ge dots some tenths of a monolayer (ML) carbon are deposited in order to reduce the dot size and to increase the dot density (C-induced Ge-dots). When depositing C-atoms prior to every layer, however, no ordering could be observed anymore.

The present study aims at a description of two new ordering phenomena in the SiGe(C) system: an anti-correlation relationship in corrugated Si(C)/Si and SiGe(C)/Si superlattices (SLs), and the transition from vertical alignment to the onset of the respective anti-correlation relationship in a sequence of layers containing C-induced Ge-dots on Si.

2. EXPERIMENTAL

Two types of samples were grown by molecular beam epitaxy: a) "quantum well samples": Si(C)/Si and Si$_{1-x}$Ge$_x$(C)/Si SLs and b) "dot samples": stacks of layers containing Ge dots, respectively C-induced Ge dots on Si. Details of the growth procedure were given by Gruetzmacher et al (1998) and Leifeld et al (1999). A maximum C-concentration of 2.5 % could be incorporated into Si by the MBE-system used for the growth of the samples being investigated in this study. The optimum growth parameters for flat Si(C) and Si$_{1-x}$Ge$_x$(C) layers have been found to be T$_{Growth}$ ≈ 450 - 470°C at a maximum C-concentration ≈ 1.5 %.

For the XTEM investigation cross-sectional samples were mechanically pre-thinned and finally ion-etched to electron transparency using Ar-gas (accelerating voltage: 4.3 kV; etching angle: 4°). They were investigated with a Philips CM30ST transmission electron microscope at an acceleration voltage of 300 kV and a point to point resolution of 1.9 Å. All XTEM images were taken along a [110] direction with an objective aperture of 20 nm-1.

3. RESULTS

3.1 Quantum well samples

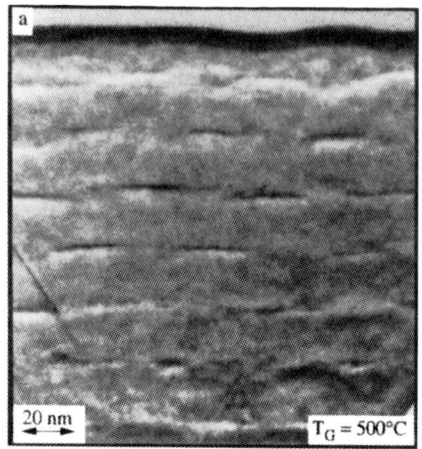

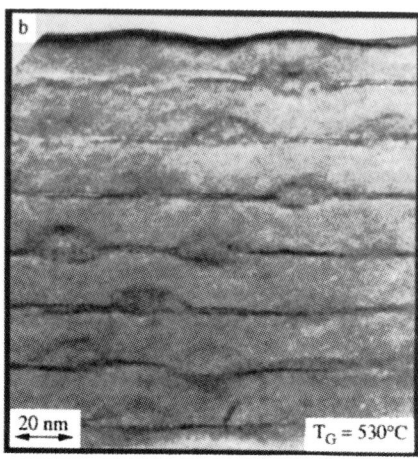

Fig. 1

Si(C)/Si SLs grown at a) 500°C and b) 530°C, respectively. The carbon concentration in the Si(C) layers is about 1.5 %.

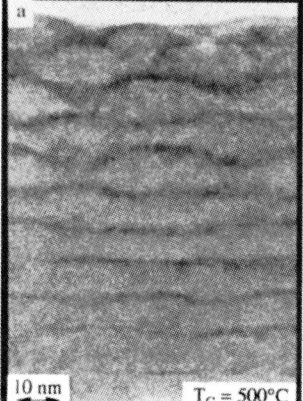

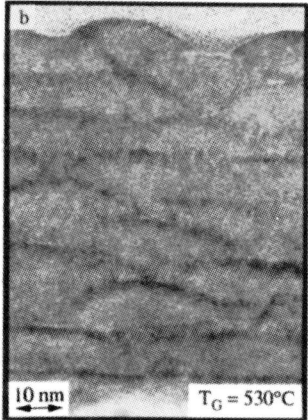

Fig. 2

Si(C)/Si SLs grown at a) 500°C and b) 530°C, respectively. The carbon concentration is as high as 2.5 %.

At a C-concentration of 1.5 % and growth temperatures of 500°C and 530°C, respectively, the Si(C) layers (being dark in contrast possibly due to lattice distortions) appear no longer to be flat. While they seem to be periodically interrupted at $T_{Growth} = 500°C$, faint ellipsoidal contrast features being periodically aligned can be recognized at $T_{Growth} = 530°C$ (Fig. 1). If, additional to the elevated growth temperature, the C-content is increased to a concentration of 2.5 %, a pronounced undulation of the Si(C) layers is visible in the upper half of the SL (500°C), respectively in the whole SL (530°C) (Fig. 2). It is concluded that the ellipsoidal features of Fig.1 can be described as a small buckling of the layers, resembling bubbles. They are smaller than the thickness of the TEM samples and are therefore imaged as a whole giving ellipsoidal contrast, while in Fig. 2 only sections of them are seen due to their larger size. They are also large enough that they can interfere with the upper layer, compared to being well separated in Fig. 1. Both this undulation as well as the ellipsoidal contrast features, are not vertically aligned in the SL. They appear to be coupled via an anti-correlation relationship: The features observed in the 2nd, 4th, 6th etc. layer are shifted by one half of their lateral period compared to those of the 1st, 3rd, 5th etc. layer.

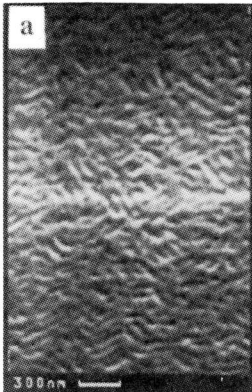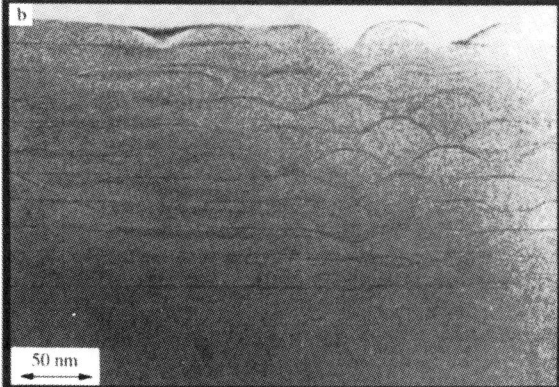

Fig. 3   a) SEM image of the surface roughening which occurs due to the presence of carbon. b) XTEM image showing on its left part a section seen across the "hills" and on its right part a projection along the "hills" visible in the SEM image.

SEM images show that the surface roughening is not a real long range order phenomenon. Probably due to the presence of two equivalent orientations, a complicated system of "hills", each being elongated along one of the <110> directions, is observed (Fig. 3a). Usually only 3-5 hills are lying parallel to each other. On TEM images this results in regions showing the above discussed anti-correlation relationship alternating with areas being characterized by flat Si(C)-layers. The latter represent sections imaged across the hills seen by SEM (Fig. 3b: left part), while those showing surface roughening are seen in a projection along the hills (Figs. 1, 2; Fig. 3b: right part).

3.2 Dot samples

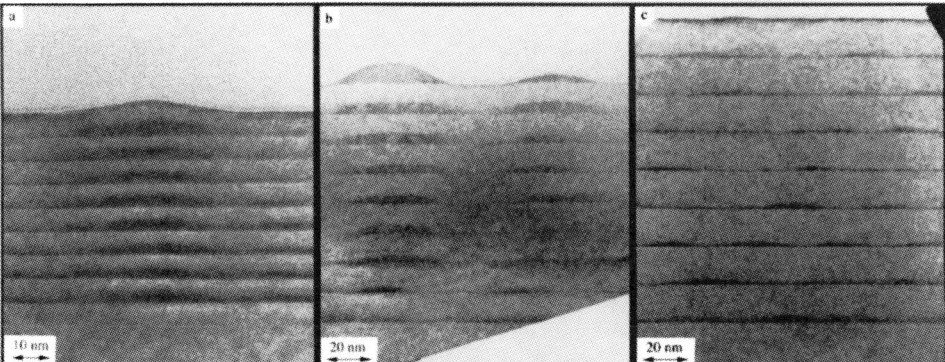

Fig. 4   A series of layers containing pure Ge dots is found to grow vertically aligned on top of a starting layer which consists of C-induced Ge dots (a). If a small amount of carbon (0.05 -0.1 ML) has been deposited prior to all layers, the Ge dots grow b) in the form of stacks, provided the Si-spacer layer is thin enough. c) For a Si-spacer layer of more than about 15 nm thickness, the onset of an anti-correlation relationship comparable to that in quantum well samples is observed.

For comparison a set of samples containing Ge-dots and/or C-induced Ge-dots was investigated. As described by Schmidt et al (1998) the dots were found to vertically order if only the first layer consisted of C-induced Ge dots (Fig. 4a). If carbon had been deposited prior to every dot layer, an additional parameter revealed to be crucial to the kind of ordering: At a thickness of the Si-spacer of less than 15 nm the dots exhibited perfect vertical ordering at temperatures of 500 – 530°C (Fig. 4b). When the spacer thickness was raised above 15 nm, however, the onset of an ordering comparable to the anti-correlation relationship in quantum well samples was observed (Fig. 4c).

4. DISCUSSION

In quantum well samples the presence of an anti-correlation relationship appears to be a rather general phenomenon being caused by the incorporation of C-atoms: It is not restricted to $Si(C)/Si$ SLs but it is observed in $Si_{1-x}Ge_x(C)/Si$ SLs (Fig. 5) and, as can be concluded from Fig. 5 in the publication of Yang et al (1997), also in $Si_{1-x}Ge_x(C)/Ge$ SLs.

As in the dot samples, the thickness of the Si-spacer layer is an important parameter in the quantum well samples. If it is thinner than about 10 - 15 nm the ordering is replaced by a chaotic intermixing of the layers. Increasing the undulation by choosing either a still higher growth temperature or a higher C-concentration would therefore ask for a thicker Si-spacer. For both types of samples, further work is necessary to optimize the parameters (distance between layers, T_{Growth}, C-concentration, respectively amount of C-predeposition). It would be especially interesting to know what causes the change in ordering state of the dot samples. One critical point might be the distribution of the C-atoms compared to that of the Ge-dots: Do the Ge islands grow where there is no carbon or is the surface regularly covered by carbon (compare Leifeld et al. 1999).

Fig. 5 $Si_{1-x}Ge_x(C)/Si$ SL grown at 530°C. The carbon concentration in the $Si_{1-x}Ge_x(C)$ layers is about 1.1 %.

5. CONCLUSIONS

It has been shown that under certain conditions (elevated growth temperature or C-concentration) the undulations of the layers proceed in a $Si(C)/Si$ or a $Si_{1-x}Ge_x(C)/Si$ SL proceed from one layer to the next one not in the form of a direct vertical alignment. It can be described by an anti-correlation relationship instead. This is contrary to the vertical alignment observed in the pure SiGe-system. In a sample consisting of several layers with C-induced Ge-dots a transition from vertical alignment to an anti-correlation relationship comparable to that in layered systems has been reported. It depends on the thickness of the Si-spacer.

ACKNOWLEDGEMENTS

We would like to thank P. Wägli very much for the SEM work. Financial support by the Swiss National Science Foundation is greatfully acknowledged.

REFERENCES

Gruetzmacher D, Hartmann R, Schnappauf P, Gennser U, Mueller E, Baechle D, Dommann A 1998 Thin solid films **321**, 26
Leifeld O, Mueller E, Gruetzmacher D, Mueller B, Kern K 1999 Appl. Phys. Lett. **74**, 994
Schittenhelm P, Engel C, Findeis F, Abstreiter G, Darhuber A A, Bauer G, Kosogov A O, Werner P 1998 J. Vac. Sci. and Techn. **B16**, 1575
Schmidt O G, Schieker S, Eberl K, Kienzle O, Ernst F 1998 Appl. Phys. Lett. **73**, 659
Soref R A 1993 Proc. IEEE **81**, 1687
Yang B-K, Krishnamurthy M and Weber W H 1997 J. Appl. Phys. **82**, 3287

Inst. Phys. Conf. Ser. No 164
Paper presented at Microsc. Semicond. Mater. Conf., Oxford, 22–25 March 1999
© 1999 IOP Publishing Ltd

Hall photovoltage imaging of the carrier flux in a 2DEG using an optical fibre

A Böhm[1], B Özyilmaz[1,3], J Heil[1], W C van der Vleuten[2], L W Molenkamp[3], U Beyer[1] and P Wyder[1]

[1] Grenoble High Magnetic Field Laboratory, Max-Planck-Institut für Festkörperforschung and Centre National de la Recherche Scientifique, 25 Avenue des Martyrs, BP 166, F-38042 Grenoble Cedex 9, France
[2] Einhoven University of Technology, Department of Physics, PO Box 513, NL 5600 MB Eindhoven, The Netherlands
[3] Rheinisch-Westfälische Technische Hochschule, 2. Physikalisches Institut, D-52056 Aachen, Germany

ABSTRACT: We report experiments using a Hall photovoltage imaging technique to investigate the carrier transport in a two-dimensional electron gas (2DEG). A laser-beam is coupled into an optical fibre, which can be scanned by a mechanical cryogenic micropositioning device. This technique allows a resolution better than 5μm to visualise the potential profile of the 2DEG in a standard Hall bar device at low temperatures and in high magnetic fields. We see flux channels (10μm in size) due to the edge confining potential and in agreement with the theory of skipping orbits. So-called light-induced Shubnikov-de Haas oscillations have been investigated near the edges.

1. INTRODUCTION

Experiments aimed at measuring the electrostatic potential distribution were initially carried out by attaching electric contacts to the interior of the 2DEG and measuring the voltage difference between adjacent contacts. These electric contacts however disturb the system under investigation. Real space-resolved contactless measurements are one alternative to interior contacts. Standard real-space resolved techniques (STM, SFM, SEM) have resolutions in the order of 0.1μm but are difficult to apply since the 2DEG is buried tens of nm under a semiconducting layer and high magnetic fields represent a complicated environment.

Fontain et al (1988) used, for the first time, the photoelectric effect in GaAs to inject carriers into the 2DEG. The sample is locally illuminated by a light spot with a diameter of ~100μm and this region of illumination acts as a current injection contact (photo-induced electrons) and can be scanned across the sample. These measurements were restricted to room temperature. In 1995 van Haren et al extended the technique to measurements at liquid helium temperatures and high magnetic fields with a spot size of 25 μm. In 1997 Shaskin et al improved the resolution to 5 μm. The method presented here represents an extension of this work. Here, the light spot is not controlled by means of scanning mirrors from the exterior of a glass cryostat, but the positioning takes place within the cryostat by means of an optical fibre, as used for scanning near field optic microscopes (SNOMs).

2. EXPERIMENTAL SET-UP

Fig. 1 shows a diagram of the setup. A He-Ne laser beam is chopped with frequency of ~100Hz and coupled into an optical fiber, the end of which is brought to within ~20μm of the sample surface. An area of ~5μm^2 is illuminated with a power of ~0.1mW. All experiments are performed with the sample immersed in liquid He[4] at temperatures of 1.1K. At the point of illumination electron-hole pairs are generated in the GaAs layer. Electrons enter the 2DEG by diffusion, where they can propagate with the mobility of $\mu = 3 \cdot 10^6$ cm^2/Vs (mean free path $l* \sim 15\mu$m). To record an image of the carrier flux, the fiber is scanned across the Hall bar. The so-called lateral photovoltage (LPV) V_{12} between the contacts 1 and 2 is recorded as a function of the fibre position. A purely mechanic scanning device realises the positioning of the fibre. In 1995 Heil showed that the resolution and reproducibility of such a versatile cryogenic scanning unit is 1μm over a scanning range of several mm.

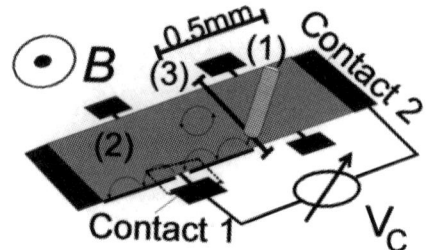

Fig. 1. Experimental Set-up.
(1) optical fibre, (2) Standard Hall bar, (3) position of line-scans

3. LINESCANS

Linescans have been performed across the sample as indicated in Fig. 1. The results for different magnetic fields from 0 to 6.3 T are shown in Fig. 2. The magnetic field corresponds to integer and non-integer filling factors. In both cases the lateral photoelectric voltage has opposite polarities at opposite edges. Together with the fact that the polarities reverse upon reversing the magnetic field, this points to the Hall-effect origin of the signal. At magnetic fields above 3T a strong response is observed only at the edge. In this field range at the edge a peak of approximately 10μm HWFM is observed. These experimental results were modeled by means of an integral equation for the self-consistent Hall voltage profile in an ideal impurity-free sample with i completely filled Landau levels. Beenakker (1991) has shown that in a Hall bar with width W the potential distribution as a function of the position x perpendicular to the current flux is given by

$$V_{12} = \frac{IR_H}{2} \ln\left|\frac{x-W/2}{x+W/2}\right| \left(1+\ln\frac{W}{\varsigma}\right)^{-1} \text{ with } \varsigma = \frac{il_B}{\pi a*} \text{ and } a* = \frac{\varepsilon\hbar}{m*e^2},$$

where l_B is the magnetic length, a^* is the effective Bohr radius and m^* is the effective mass. The

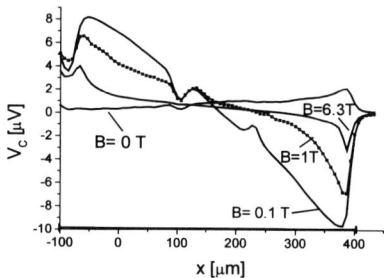

Fig. 2. Line-scans across the sample (line (3) in Fig.1) for different magnetic fields B. A strong response at high magnetic fields is only observed at the edges of the sample. With increasing magnetic field the lateral photo-voltage changes from linear space dependence to a logarithmic one. The breakdowns in the middle of the sample are caused by an optical gate structure, which partly covers the sample surface.

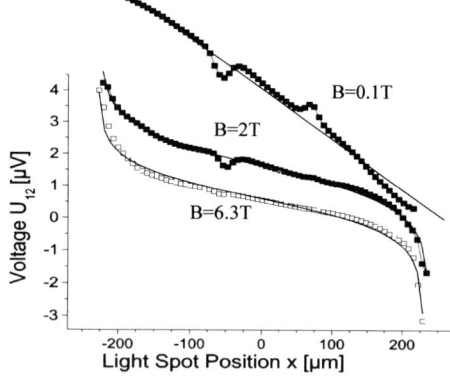

Fig. 3. The region between the two peaks at opposite edges of the line-scans shown in Fig.2 is fitted with the theoretical self-consistent Hall voltage profile in an ideal impurity-free sample (see text). The line-scan at $B = 0.1$ T is fitted with a linear function.

fitting parameter is IR_H, where I is the current and R_H is the Hall resistance. IR_H is in the range of 12 μV (6.3 T) to 45 μV (1 T). Assuming that $R_H = ih/e^2$, this corresponds to a current in the order of 1 nA.

From Fig.3 it is can be seen that the model is in a good agreement with the experimental observations at magnetic fields $B \geq 2T$. This proves that the measured signal corresponds to the confining potential in the 2DEG. However, the observed potential profile does not depend on the filling factor v. In high magnetic fields the measurements always agree with the model calculation whether the filling factor is an integer value or not. As a consequence this would mean that at high magnetic fields the photo-induced electrons accumulate at the edges even if $\sigma_{xx} \neq 0$.

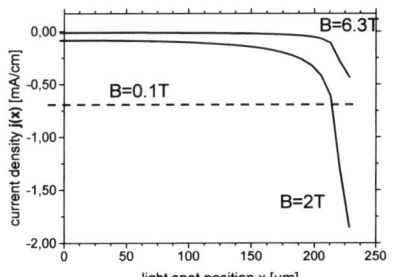

Fig. 4: The current density distribution at the right half of the sample obtained by deriving the fit function of Fig.3 and dividing by the corresponding magnetic fields. At $B = 0.1$ T the current density distribution is constant. With increasing field the edge contributes more and more to the total current.

In 1991 Beenaker estimated the nonequilibrium current density $j(x)$ from the potential distribution V_{12} by

$$j(x) = \frac{en}{B} \frac{dV_{12}(x)}{dx}$$

In Fig. 4 the $j(x)$ for the right half of the sample are plotted. From this it can be estimated that, beyond filling factor 2, 75% of the total current flows within a 50μm region next to the edge.

4. IMAGING THE CONFINING POTENTIAL NEAR THE CONTACTS

The potential distributions at the edge at filling factor $v \leq 1$ have been imaged. Ballistic carriers move inside the 2DEG on circular orbits (cyclotron radius $r_c \sim$ nm) and thus cannot contribute to the signal. Only scattering processes permit them to propagate towards the contacts, where they can be detected. The carriers in vicinity of the 2DEG-edge (B) can skip along the border towards the contact 1. This produces the bright structures in Fig. 5(b) (next page). This phenomenon is in agreement with the theory of so-called "skipping orbits". Carriers from edge (A) skip away from contact 1 and cannot contribute to the signal. In Fig. 5(c) the magnetic field (6.3 T) is inverted and thus the carriers can skip only in the A-B direction. The bright signal corresponds to the carriers moving from edge (A) into the contact 1. The dark structures (negative signal) are created from carriers moving out of the contact I and skipping into contact II. Thus they create a negative signal.

5. LIGHT INDUCED QUANTUM EFFECTS

If V_{12} is recorded as a function of the magnetic field at position x indicated in Fig. 5, oscillations periodic in $1/B$ are clearly visible. The frequency of these so-called Light-Induced Shubnikov-de Haas Oscillations (LISHO) is direct proportional to the electron density of the 2DEG. The quantum limit is reached at $B = 4T$. Spin splitting is visible above 1 T. The origin of the oscillations can be explained by Landau quantization of the electron states and the occurrence of back-scattering at non-integer filling factors.

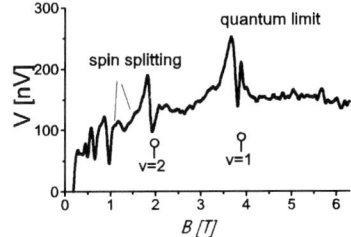

Fig. 6: V_{12} is recorded at position x in Fig. 5 as a function of the magnetic field B. The oscillations are proportional to $1/B$.

6. CONCLUSION

A simple technique to image the confining potential in a standard Hall bar has been presented. A purely mechanical device scans an optical fibre. Carriers are injected locally due to the photoelectric effect without disturbing the system. The potential distribution near the edge and in vicinity of the contacts was nicely imaged. Line scans can be well described by the theoretical Hall voltage profile proving that the measured signal presents directly the potential distribution in the sample. The current density can be obtained by the derivative. At higher fields the potential profile is peaking near the edge with an FWHM of 10µm. This means that 75% of the total current flows within a 50µm region next to the edge. So-called light-induced Shubnikov-de Haas oscillations have been observed near the edges.

REFERENCES

Beenakker C W J and van Houten H 1991 'Quantum transport in semiconductor nanostructures' in
    Solid State Physics **44**

Fontain P F, Hendriks P, Peat R, Williams D E and Andre J P 1988 J. Appl. Phys. **64**, 3085

Fontain P F, Kleinen J A, Hendriks P, Blom F A P, Wolter J H, Lochs H G M, Driessen F A J M,
    Giling L J, and and Beenakker C W J 1991 Phys. Rev. B **43**, 12090

Heil J, Böhm A, Primke M and Wyder P 1995 Rev. Sc., Instrum.**74**, 146

Shashkin A A, Kent A J, Owers-Bradley J R, Cross A J, Hawker P and Henini M 1997 Phys. Rev.
    Lett. **79**, 5114

Van Haren R J F, Blom F A P and Wolter J H 1995 Phys. Rev. Lett. **74**, 1198

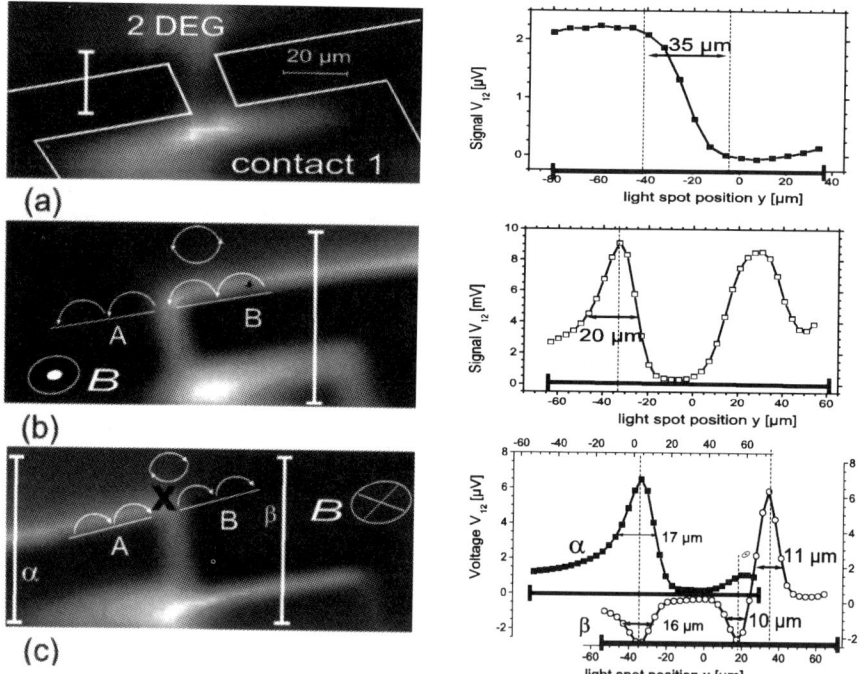

Fig. 5: Image of the lateral photoelectric voltage at (a) B = 0 T, (b) B = 6.3 T and (c) B = -6.3 T. The images cover an area of 270 x 120µm with 40 x 18 data points. Line-scans are taken vertically through the images (a) – (c). The electrons injected in the bulk 2DEG are localised. Only at the vicinity of the edge the electrons can move in skipping orbits along the edge and reach the contact1, where they generate a positive signal (white). For B = -6.3 T electrons can move out of contact 1 and skip towards contact 2, where they generate a negative signal (black).

Inst. Phys. Conf. Ser. No 164
Paper presented at Microsc. Semicond. Mater. Conf., Oxford, 22–25 March 1999

A cross-sectional HRTEM study of particle defects in an epitaxial diamond film

H Sawada, H Ichinose, D Takeuchi[1] and H Okushi[1]

Department of Material Science, The University of Tokyo: 7-3-1 Hongo, Bunkyo-ku, Tokyo 113-8656, Japan
[1]Electrotechnical Lab.: 1-1-4 Umezono, Tukuba, Ibaraki 305-8568, Japan

ABSTRACT: Homoepitaxial single crystal diamond films grown by the CVD method suffer from spontaneous growth of small polycrystalline particles that degrade the quality of the film. In these polycrystalline particles, which are not epitaxially aligned, the most dominant defects are (111) $\Sigma3$ CSL boundaries. The $\Sigma3$ boundary, which is not parallel to the (111) plane, is curved in macroscopic appearance. An asymmetrical (111)/(115) $\Sigma3$ boundary is straight and long.

1. INTRODUCTION

Diamond has received considerable interest for various applications. Chemical vapour deposition (CVD) of diamond on diamond substrates is the most promising way to obtain a high quality large-scale single crystal diamond film (Hayashi et al 1996). However, in practice growth of small polycrystalline particles in the film degrades the film quality. Therefore, the desired quality of diamond epitaxial films is still far beyond the actual quality level achieved. The quality degradation of the artificial diamond is mainly attributed to the nature of grain boundaries, characterized by local breakdown in crystal periodicity which disturbs the electronic structure.

The boundary structure of diamond is different from that of other covalently bonded materials (Ichinose and Nakanose 1998). This characteristic feature is related to some unique features of dangling bond reconstruction in diamond. In the present study, the origin and grain boundary structure of the polycrystalline particles is investigated using high-resolution electron microscopy.

2. EXPERIMENTAL

Diamond films (1.5 - 2.0µm) were deposited on synthetic Ib diamond (001) substrates (4.0×4.0×0.3mm³) by the CVD method. Many diamond crystallites ~2.0µm in diameter were present in the deposited epitaxial films. Transmission electron microscope (TEM) specimens of these particles were prepared by cleaving the substrate into small pieces, 300µm×300µm×1.5mm in size, followed by FIB sectioning through the centre of a particle. The plane of section was perpendicular to the (110) plane of the film. Atomic structure observation was carried out employing the JEM-ARM1250 high voltage atomic resolution TEM at the University of Tokyo.

3. RESULTS AND DISCUSSION

3.1 Structure and origin of the diamond polycrystalline particle

A cross section through the centre of a particle showed it to have an inverted pyramidal morphology (Figure 1c). The particle/substrate interface at the bottom of the particle was not parallel to any low index crystal plane of the matrix in most cases. Two different origins of the particle appear likely. One is the nucleation at the end of a screw dislocation originating from the substrate. A dislocation terminating at the point of the pyramid was actually observed. The other is a nm-sized diamond particle. A polygonal crystal, a few nm in size was sometimes observed at the point of the pyramid (Figure 1d). Lattice dislocations in this case were rarely observed and many grain boundaries were present in the particle.

3.2 Diamond grain boundary in a polycrystalline diamond particle

A pyramidal particle consisted of fine grains several nm in size. The main change in orientation between the fine grains, measured from electron diffraction patterns, were found to be 70.5° and 39° around the [110] axis, corresponding to $\Sigma 3$ CSL and $\Sigma 9$ CSL relation respectively.

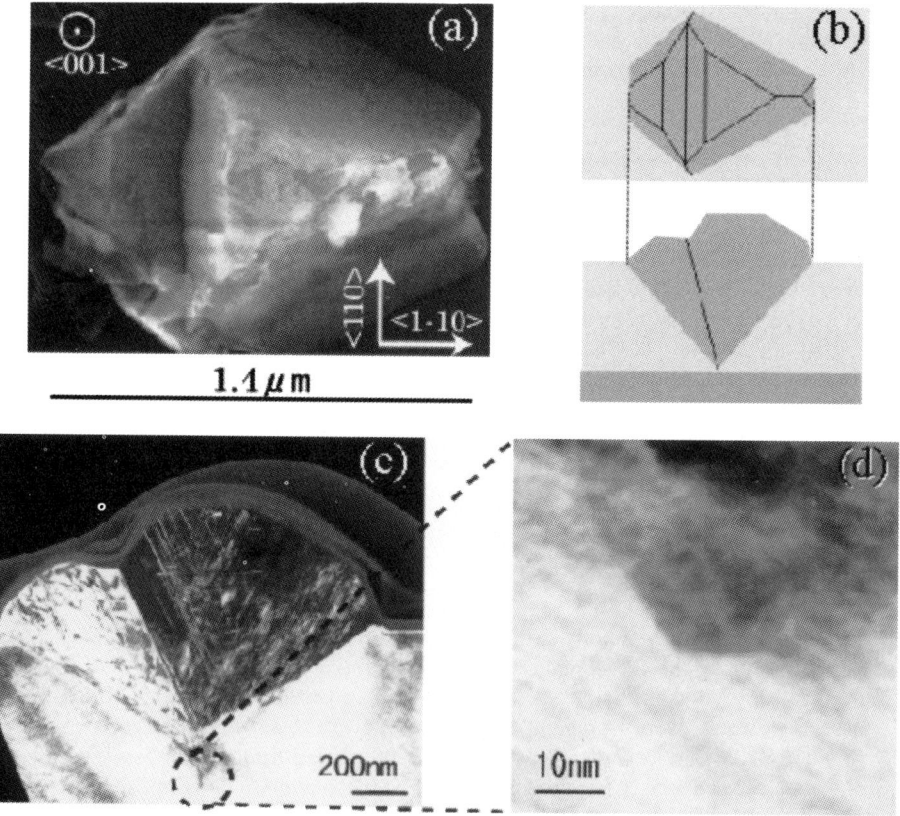

Figure 1. (a) A diamond particle on the homoepitaxial diamond film; (b) Schematic view of the particle and the (110) plane; (c) Cross section, dark field ($g = 111$) image of the diamond particle in (a); (d) Bright field image of the nucleation site.

3.3 (111) Σ3 grain boundary

The (111) Σ3 CSL boundary was most common. Often, Σ9 CSL boundaries were observed to connect to the (111) Σ3 CSL boundary, although the (111) Σ3 boundary was much longer than the others (Figure 2). The (111) Σ3 boundary in diamond was much longer than that in silicon.

3.4 (112) Σ3 grain boundary

Incoherent Σ3 boundaries were curved even on an atomic scale (Figure 3). In silicon an incoherent Σ3 boundary shows a rigid translation in the <111> direction in order to relax the repulsive stress. This would result in an asymmetrical atomic structure. The curved Σ3 boundary consisted of many short (111) Σ3 and (112) Σ3 boundaries connected with each other. The high-resolution TEM image of the (112) Σ3 boundary structure was consistent with asymmetrical 5-7 atom rings.

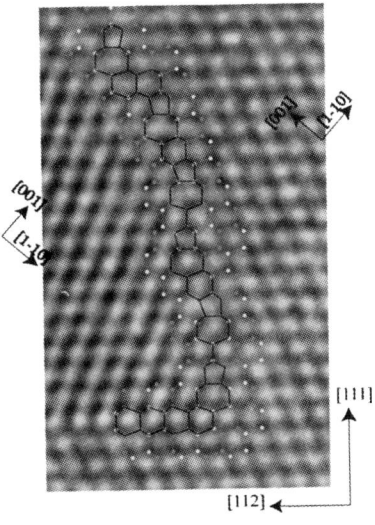

Figure 2. Symmetrical structure of the {111} Σ3 grain boundary. The dark contrast in the image corresponds to the atomic column position.

Figure 3. Superimposing a geometrical model on the HRTEM image of the series of the {111} Σ3 and {112} Σ3 grain boundaries

3.5 Σ9 grain boundaries

A Σ9 boundary was produced by the meeting of two Σ3 boundaries. In silicon, Σ9 boundaries are mostly parallel to {221} planes, in which atoms can exist without any dangling bonds. The most stable boundaries in silicon correspond to those with the smallest number of dangling bonds. In diamond, Σ9 boundaries were found to be parallel to {221} and {114} planes. If the Σ9 boundary is parallel to the {114} plane, four dangling bonds for one CSL period of the boundary are inevitably generated. Our observation of these planes shows that dangling bonds do not affect the grain boundary internal energy in diamond, as is shown in the diamond polycrystalline film (Ichinose and Nakanose 1998).

3.6 Asymmetrical boundary parallel to (111)

Asymmetrical high angle boundaries such as (111)/(115) Σ3 boundaries were also observed (Figure 4). Their structure models were constructed by a combination of 5-7 membered atom rings and boat-shaped 6-membered atom rings. These rings have the same shape as those which constitute the structural units of (111) and (112) Σ3 grain boundaries.

Figure 4. HRTEM image of the asymmetrical (111)/(115)Σ3 grain boundary

4. SUMMARY

We have performed a TEM study of the small crystallites which form in epitaxial diamond films grown on single crystal diamond substrates by CVD. The crystallites were observed to nucleate at a point on the surface corresponding with the emergence of a substrate dislocation or a small (nm-sized) diamond particle. The defects present in the crystallites were investigated using high resolution TEM. The most common defects were Σ3 and Σ9 CSL grain boundaries. The Σ3 boundaries were the most common, and lay on {111} and {112} planes. Σ9 boundaries were formed at the junction of two Σ3 boundaries, and lay on {221} and {114} planes. Their atomic structure was described in terms of 5- 6- and 7-member atom rings.

REFERENCES

Hayashi K, Yamanaka S, Okushi H and Kajimura K 1996 Appl. Phys. Lett. **68**, 1220
Ichinose H and Nakanose M 1998 Thin Solid Films **319**, 87

Inst. Phys. Conf. Ser. No 164
Paper presented at Microsc. Semicond. Mater. Conf., Oxford, 22–25 March 1999
© *1999 IOP Publishing Ltd*

CBED to determine the lattice parameters of strained SiC films on 6H-SiC substrates

U Kaiser, K Saitoh[1], K Tsuda[1], M Tanaka[1] and W Richter

Institut für Festkörperphysik, Friedrich-Schiller University Jena, 00743 Jena, Germany.
[1]Research Institute for Scientific Measurements, Tohoku University, Sendai 980-8577, Japan.

ABSTRACT: Many-beam Bloch wave CBED pattern simulation is used to determine the lattice parameter of thin SiC films on 6H-SiC substrates from experimental on-axis CBED patterns at [320] and [331] zone axes incidence, which show lattice parameter change sensitive HOLZ line shifts. The agreement of three ratios between defined HOLZ lines in each zone axis pattern ensures a lattice parameter determination with high accuracy. The transition from the cubic to the rhombohedral symmetry was found in one specimen and explained as a result of layer stress.

1. INTRODUCTION

One way to measure the strain in a thin film is to determine its lattice parameter. For small layer thickness (<500nm) determination by X-ray methods (e.g. Kräußlich et al 1998), usually used for a lattice parameter determination of high accuracy, cannot be applied because it averages over the whole crystal volume and includes the substrate as well. Here, convergent-beam electron diffraction (CBED) is the method of choice as the information is obtained from a very small crystal volume and the position of higher order Laue zone (HOLZ) lines appearing in the 000 disk of CBED patterns at a definite incident electron energy are very sensitive to lattice parameter changes.

In this paper, we describe the CBED analysis of thin epitaxial SiC films on 6H-SiC substrates grown by molecular beam epitaxy using different growth conditions. Calculations in order to fit the experimental patterns were performed using the many-beam Bloch method.

2. EXPERIMENTAL DETAILS

2.1 Thin film growth

The SiC films were grown on 6H-SiC substrates by solid-source molecular beam epitaxy as is described more fully by Fissel et al (1997) and Kaiser et al (1999). Some important growth parameters and the main characteristics of the thin film defect structure are summarized in the Table:

	Specimen 1	**Specimen 2**
substrate	hexagonal (6H)	hexagonal (6H)
growth temperature	1300°C	1050°C
growth mode	alternate supply of C, constant Si flow	simultaneous supply of C and Si high Si excess in the flux
thin film defect structure	very few defects	defective areas containing big Si and SiC hillocks
thin film thickness	100nm	160nm

Table I. Some important growth parameters and main characteristics of Specimens 1 and 2.

2.2 Transmission electron microscopy

TEM foils were prepared in cross-section using standard techniques. The TEM studies were carried out using a JEOL 2010 electron microscope operating at 200kV. As seen in Fig. 1 they revealed the presence of strain at the substrate-layer interface (marked S) in both specimens but large hillocks, marked H, consisting of Si single crystals were only apparent in specimen 2.

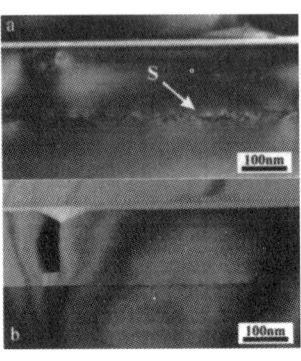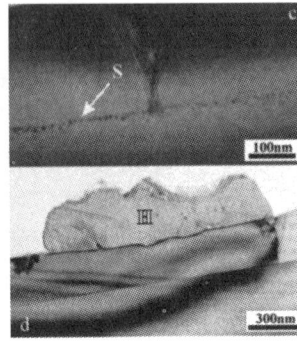

Fig. 1 Cross-sectional bright-field images of Specimens 1 (a, b) and 2 (c, d) showing strain marked S at the layer-substrate interface. In d) hillocks (marked H) are seen which were only found in specimen 2.

For the CBED studies the microscope was operated at about 100kV. To determine the exact accelerating voltage, large angle CBED (LACBED) patterns of Si [111] were obtained always before the experiments on SiC were started (for more details see Kaiser et al (1999b)).

CBED experiments on SiC were carried out at room temperature at two zones ([331] and [320]) at defect-free positions in the middle of the layers.

2.3 CBED pattern simulation and fitting procedure

The [331] and [320] zones of cubic SiC at about 100kV accelerating voltage were found to be suitable because ratios of distances between HOLZ lines of higher order reflections could be found which moved in different directions when the lattice parameter varied.

For the dynamical calculations at Si [111], SiC [331] and SiC [320] zones, the program of Tsuda and Tanaka (1995) has been used. The input values were: atom co-ordinates of Si and 3C SiC respectively, Debye-Waller factors: $B_{Si}= 0.20A^2$, $B_C= 0.25A^2$ [9]; Laue zones included: 1st to 5th order; number of beams: at Si [111] zone: 86, at SiC [331] zone: 134, at SiC [320] zone: 104; number of pixels at Si [111] zone: 201, at SiC [331] zone: 401, at SiC [320] zone: 201.

Three ratios were calculated from distances measured between the sensitive HOLZ lines in the CBED patterns. The distances between HOLZ lines were measured in the same way from the experimental CBED patterns and from the dynamically calculated CBED patterns using an image processing program. Each length was measured carefully at a high magnification 10 times to determine the standard deviation of the values. The ratios determined from the calculated patterns were plotted as a function of the lattice parameter. The cross point of this function with the ratio determined from the corresponding experimental pattern is the measured lattice parameter.

3. RESULTS

Figure 2 shows calculated CBED patterns at the cubic [320] and [331] SiC zone axes, together with schematic HOLZ lines. The distances used for the determination of the three ratios are marked by dotted lines. Reflections from the 1st to the 4th order Laue zone are included.

When the experimental CBED patterns of specimens 1 and 2 are compared, the striking fact is the difference in the symmetry. Figures 3a) and 3b) show the experimental CBED patterns obtained at the [320] zone for specimens 1 and 2, respectively. It is possible to see that HOLZ lines arrowed in b) (specimen 2) are asymmetric about the transparent vertical line. In contrast, the pattern in a) (specimen 1) is fully symmetric. At the [331] zone, the CBED pattern of specimen 2 remains symmetric about the transparent line discussed. This behaviour at [320] and at [311] zone of specimen 2 could be fitted by the assumption of a rhombohedrally strained cell.

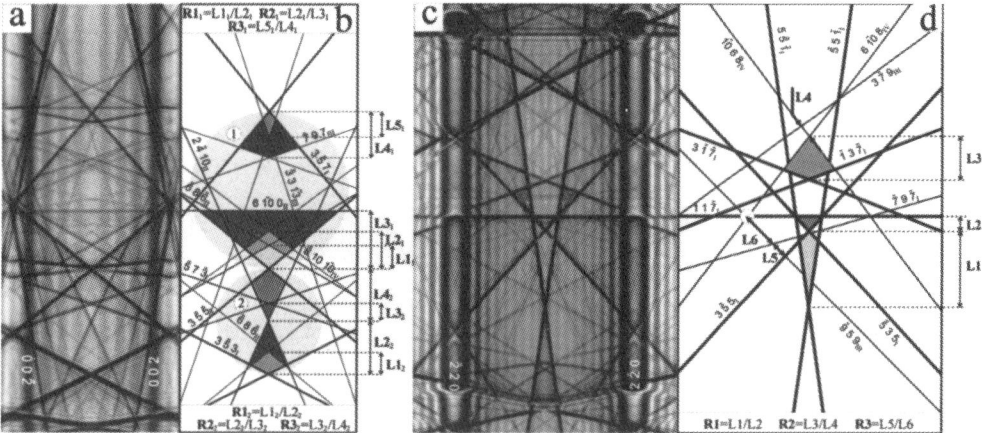

Fig. 2   Simulated CBED pattern of cubic SiC at the [320] (a) and at the [331] zone (c) zone axis together with the scheme of HOLZ lines in b, d, respectively. The HOLZ lines are indexed and the suffix shows the order of Laue zone reflections. In b) parts 1 and 2 are covered by two CBED patterns at two orientations. 3 ratios in each pattern were determined of the distances marked L_x as inserted in the figures.

The transformation from the cubic space group F-43m to the rhombohedral space group R3m is accompanied by the loss of the mirror symmetry of (002) planes. This results in an asymmetry of the pattern at the [320] zone axis incidence ([002] is the ZOLZ base vector) in the case of a rhombo-hedrally strained cell as is illustrated in the schematic in Fig. 3 d. At the [331] zone axis the ZOLZ base vector is [2-20] and no change of the symmetry occurs parallel to the (2-20) planes when the layer is strained in [111] direction.

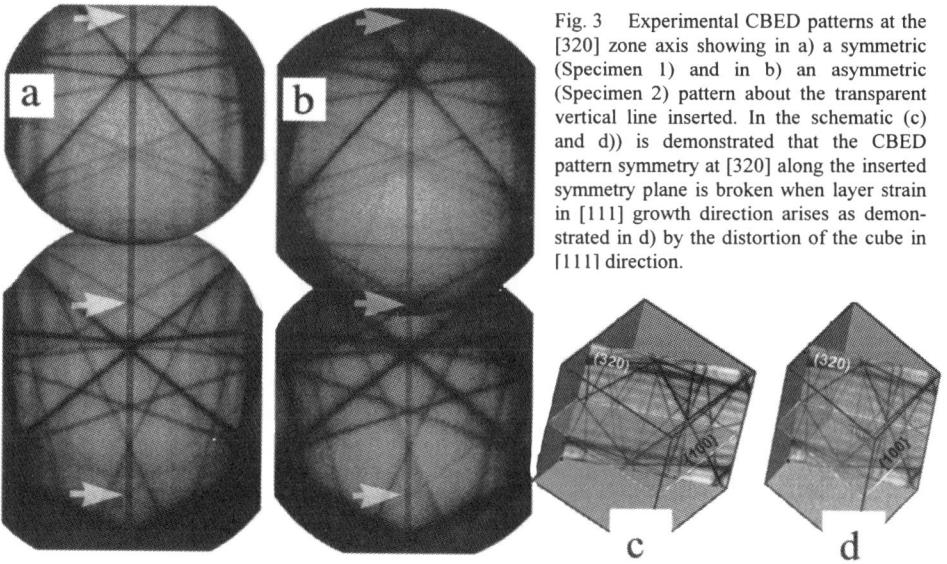

Fig. 3   Experimental CBED patterns at the [320] zone axis showing in a) a symmetric (Specimen 1) and in b) an asymmetric (Specimen 2) pattern about the transparent vertical line inserted. In the schematic (c) and d)) is demonstrated that the CBED pattern symmetry at [320] along the inserted symmetry plane is broken when layer strain in [111] growth direction arises as demon-strated in d) by the distortion of the cube in [111] direction.

Simulations have been carried for the lattice parameter a from 0.4359nm to 0.4378nm in 0.0001nm steps and for rhombohedral angles α from 89.95° to 89.75° in 0.05° steps. The values for the lattice parameters obtained from the fit (see 2.3) between the experimental and simulated patterns for Specimen 1 and 2 are summarised in Tab. II.

Zone axis	lattice parameter Specimen 1	lattice parameters Specimen 2
[320] RT	a=0.4374nm	a=0.4369 nm, α=89.9°
[331] RT	a=0.4375nm	a=0.4368 nm, α=89.9°

Table II Lattice parameters of Specimen 1 and 2. The accuracy of the values was found to be for a: ± 0.0001 nm and for α: ±0.05°. The literature value for cubic SiC bulk material determined by Kräußlich at al (1998) at RT is 0.4359nm.

The values for the lattice parameters a between the two specimens differ significantly. We suggest that the reason is connected with compressive strain arising in the layer due to the large hillock on top of the layer (see Fig. 1d).

Although the accuracy determined is high, the values differ greatly from the bulk 3C-SiC values, which were recently determined by X-ray analysis by Kräußlich at al (1998). The main reason should be associated with the strong layers strain arising at the interface (see Figure 1a, c). As the theoretical lattice parameter of the a value of 6H-SiC differs only slightly by 0.04% from the corresponding a-value of the cubic SiC, a cubic layer grown pseudomorphically on the 6H substrate is expected to grow less strained. Investigation to clarify this problem are in progress. We will prove in more detail, whether the local temperature increase possible under the electron beam gives rise to this difference. As a first estimation under the condition used showed, the temperature rise ΔT for SiC could be about 100K whereas for the reference Si it could be increased by the factor of about 4 to 400K (due to different thermal conductivity), which then would result in a lattice parameter enlarged by 0.1%.

4. SUMMARY AND CONCLUSIONS

The lattice parameter was shown to change significantly with growth parameters. The appearance of Si hillocks on top of the layer may be the reason for compressive stress explaining the measured transformation from cubic to rhombohedral symmetry in specimen 2. The lattice parameters a and α of the SiC films have been determined with high accuracy of ± 0.0001nm and ± 0.05° (for the rhombohedral cell) respectively in defect-free positions of thin SiC films after determining exactly the high voltage of the microscope. We used the ratios of HOLZ line distances sensitive to small lattice parameter change determined in dynamically simulated and experimental CBED patterns at [320] and [331] zone axes of cubic and rhombohedral cells. The values of the lattice parameter were found to be up to 0.4% larger than the literature value for bulk SiC determined by X-ray analysis. The question of whether this increase can be explained only by the strain in the thin film will be answered in due course.

ACKNOWLEDGEMENTS

The authors are grateful to F. Sato for the skilful maintenance of the microscope, to J. Jinschek for help in TEM specimen preparation and to Dr. A. Fissel for providing the SiC material. The work was carried out during U. Kaiser's stay as associate Professor at the Tohoku University in Sendai, Japan.

REFERENCES

Fissel A, Pfenninghaus K, Kaiser U, Schröter B and Richter W 1997 Mater. Sci. Eng. **B46**, 324
Kaiser U, Khodos I, Brown P D, Humphreys C J, Chuvilin A, Albrecht M, Fissel A and Richter W 1999a J. Mat. Res. accepted
Kaiser U, Saitoh K, Tsuda K and Tanaka M 1999b J. Electron Microscopy, accepted
Kräußlich J, Bauer A and Goetz C 1998 Phys. Stat. Sol. **82**, 284
Tsuda K and Tanaka M 1995 Acta Crystallogr. **A 51**

Inst. Phys. Conf. Ser. No 164
Paper presented at Microsc. Semicond. Mater. Conf., Oxford, 22–25 March 1999

Structure of SiC layers grown by LPE in microgravity and on-ground conditions

B Pécz, R Yakimova*, M Syväjärvi*, C Lockowandt**, G Radnóczi and E Janzén*

Research Inst. for Technical Phys. and Matl. Sci. H-1525 Budapest, P.O.Box 49, Hungary
* Department of Physics and Measurement Technology, Linköping University, S-581 83 Linköping, Sweden
** Swedish Space Corporation, P.O.Box 4207, S-171 04 Solna, Sweden

ABSTRACT: High quality, hexagonal SiC layers have been grown in microgravity conditions and on-ground as well. The surface of the layers is always stepped. The dislocation density of the layers is increased closer to the surface. Scandium carbide precipitates, nanopipes and cavities were found in the SiC layers grown on-ground, but none of them were traced in the layers grown under microgravity conditions.

1. INTRODUCTION

Liquid Phase Epitaxy (LPE) is a suitable method to grow SiC layers and especially advantageous when thick epitaxial layers are needed as for example for fabrication of high power SiC devices. Pure SiC layers were grown by LPE from silicon solvent (Koga and Yamaguchi 1991), at a growth rate of a few micrometers per hour due to low solubility of carbon in a silicon melt. By introducing scandium into the system the solubility of carbon is increased and growth rates as high as 350 μm/h have been achieved (Syväjärvi et al 1997). However, growth from solution with high growth rate may result in morphological instabilities and defect formation due to gravitation-induced convection. LPE growth under reduced gravity gives the possibility of eliminating a growth parameter that it is not possible to control under normal conditions.

2. EXPERIMENTAL PROCEDURE

Liquid phase epitaxy of 4H- and 6H-SiC was performed by using a modified travelling solvent method (Yakimova et al 1995). The growth utilised a container-free sandwich configuration with a Si (40% Sc and C)-melt between a SiC source material and the (0001) oriented, Si face of 6H-SiC (misoriented by 3.5°) and 4H-SiC (misoriented by 3.5° or 8°) substrates. The SiC layers were grown at 1750°C for 360 seconds (the samples were preheated to 1200°C to be able to reach the processing temperature as fast as possible). Control samples were grown on-ground in the same LPE unit shortly after the space experiments.

Cross sectional samples for TEM analysis were prepared by ion beam milling. The thinned samples were examined using a Philips CM20 transmission electron microscope operating at 200 kV. Energy Dispersive X-ray microanalysis (EDS) was performed by a NORAN Voyager system attached to the TEM.

3. RESULTS AND DISCUSSION

3.1 Common features in layers grown in microgravity and on-ground

The thickness of the layers obtained was about 20 μm during the 6 minutes growth time. The grown layers followed the polytype of the substrate in all cases, which was established by both TEM and X-ray diffraction. No other polytypes have been found in the 4H layers grown on 4H substrates or in the 6H layers grown on 6H substrates. As a consequence the original substrate/layer interface can hardly be detected by TEM. The surfaces of the grown layers are always faceted showing larger and smaller steps (Fig. 1). The dislocation density of the layers is increased closer to the surface as can be seen in Fig. 2, which is a dark field image of a 4H-SiC layer showing the near surface region.

Fig.1.a, b: Cross sections of SiC layers grown on ground showing steps on the surface.

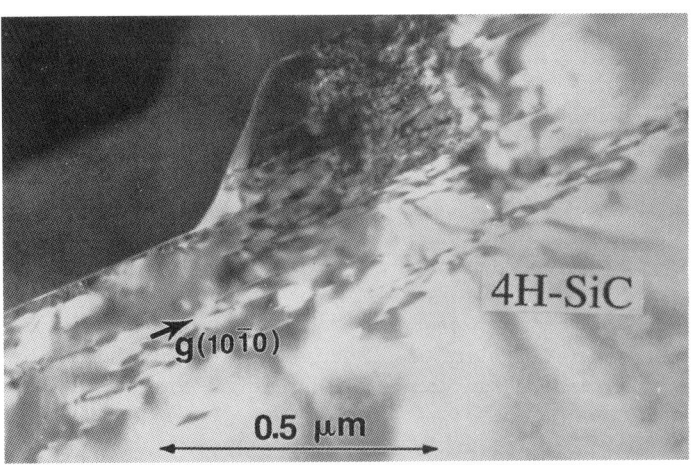

Fig. 2 Dark field image of a 4H-SiC layer showing the near surface region with high defect density.

3.2 Macrodefects in layers grown on-ground

While the layers grown in microgravity are free of macrodefects, the layers grown on-ground under identical conditions (except the gravity) differ from this point of view.

In 4H-SiC layers grown onto 4H-SiC substrates (misoriented by 8°) some holes were found at a depth of about 7 μm measured from the top surface. One of those holes is shown in Fig. 3 at low magnification. In all cases an elongated contrast resembling impressions/dents in the sample was observed which was directed towards the sample surface (Fig. 3). This suggests that the holes are cross sections of macro/nanopipes, which lay in the TEM foil, directed in the [0001] direction and pierce the foil at the depth of some μm. High resolution TEM images of the nanopipe sections did not give any indication of the presence of any other polytype at the walls of the hole. This agrees with the results of Heindl et al (1997) who gave a general description of the micropipe formation.

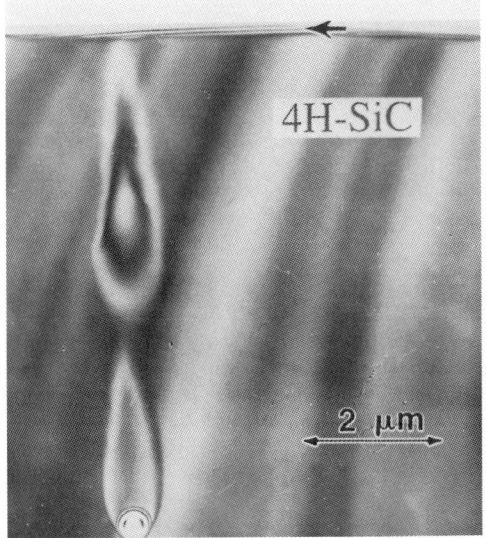
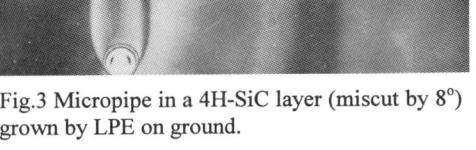
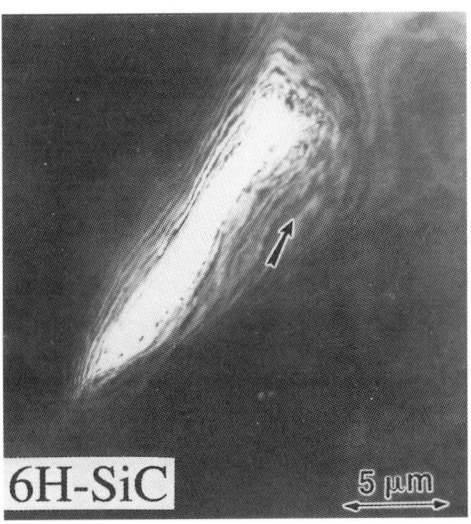

Fig.3 Micropipe in a 4H-SiC layer (miscut by 8°) grown by LPE on ground.

Fig. 4. Large cavity in a 6H-SiC layer grown by LPE on ground.

In a 6H-SiC layer grown onto a 6H-SiC substrate (miscut by 3.5°) a cavity has been found which is shown in Fig. 4. The arrow in Fig. 4. indicates the growth direction. This cavity in the grown layer is probably connected to a micropipe having a configuration similar to that already observed by Heindl and Strunk (1996). Taking selected area electron diffraction patterns at the edge of this cavity we have observed that the internal wall of the cavity is covered by nanocrystalline 3C-SiC. The 3C crystallites are not responsible for the formation of the cavity. We suggest that the cavity was formed during the growth of the 6H-SiC layer and the 3C polytype nucleated at lower temperature during the cooling down process and caused by the residues enclosed in it. This can be supported by two arguments: First, the nucleation of 3C-SiC is not expected under the applied growth conditions because they were selected for preferred growth of the substrate structure, and also Sc is thought to stabilise the growth of the hexagonal phase. Secondly, once 3C-SiC is nucleated at high temperature it will grow further forming larger grains.

A large cavity containing a grain, clearly differing from the SiC matrix, has been found in a 4H layer grown onto 4H-SiC (Fig. 5.a). The reasons for formation of this cavity are not clear. We focused on the identification of the grain situated at the top edge of the cavity, at a depth of 6.5 μm measured from the surface. EDS analysis of the grain was carried out in the TEM and the spectrum of the above grain clearly showed that it is composed of scandium and carbon. Selected area electron diffraction (SAED) patterns of the grain have been taken at different tilts and the obtained diffraction patterns (e.g. Fig. 5.b) could be indexed as tetragonal $Sc_{15}C_{19}$ (JCPD card 38-0797). Considering the equilibrium phase binary diagram of scandium and carbon (Sidorko et al 1995) this is one of the two scandium carbide phases forming at the temperature while Sc_4C_3 can be excluded in this case according to the SAED patterns. The investigated grain is large being about 600 nm long and 200 nm wide (Fig. 5.a). We have to note that there is a second small scandium carbide grain just beside the large one (left side of the large grain in Fig. 5.a) in this cavity.

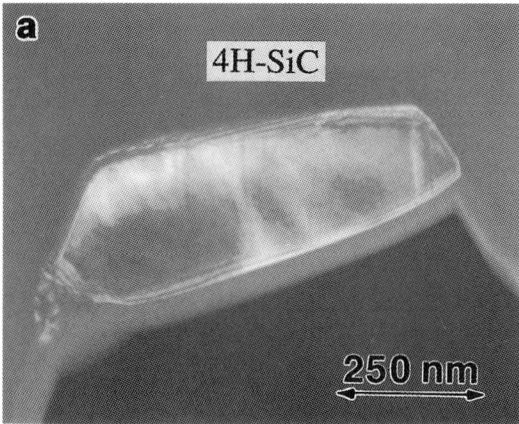
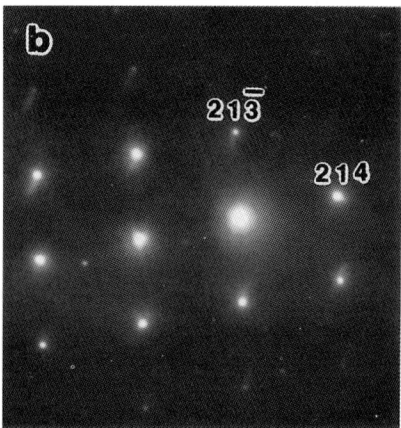

Fig. 5.a Dark field image of the $Sc_{15}C_{19}$ precipitate taken using the 214 reflection.

Fig. 5.b Selected area diffraction pattern of the precipitate.

The macrodefects described above were not discovered by TEM in the layers grown in microgravity.

ACKNOWLEDGMENT

Support from the National Swedish Space Board and from the European Space Agency is acknowledged. OTKA project No. T 030447 is also acknowledged for financial support.

REFERENCES

Heindl J and Strunk H P 1996 phys. stat. sol. (b) **193**, K1
Heindl J, Strunk H P, Heydemann V D and Pensl G 1997 phys. stat. sol. (a) **162**, 251
Koga K and Yamaguchi T 1991 Prog. Crystal Growth and Charact. **23**, 127
Sidorko V R, Goncharuk L V, Gordiychuk O V and Antonchenko R V 1995 J. of Alloys and Compounds **228**, 159
Syväjärvi M, Yakimova R and Janzén E 1997 Diamond Rel. Mat. **6**, 1266
Yakimova R, Tuominen M, Bakin A S, Fornell J-O, Vehanen A and Janzén E 1995 Inst. Phys. Conf. Ser. **142**, 101

Inst. Phys. Conf. Ser. No 164
Paper presented at Microsc. Semicond. Mater. Conf., Oxford, 22–25 March 1999
© 1999 IOP Publishing Ltd

TEM and XRD analysis of CuInS$_2$ layers for solar cell applications

J Marcos-Ruzafa, A Romano-Rodríguez, J Álvarez-García, A Pérez-Rodríguez, J R Morante, J Klaer[*] and R Scheer[*]

Dept. of Electronics, University of Barcelona, Martí i Franquès 1, E-08028 Barcelona, Spain
[*]Hahn-Meitner-Institut, Dept. of Physical Chemistry, Glienicker Str. 100, D-14109 Berlin, Germany

ABSTRACT: CuInS$_2$ layers deposited by two different evaporation methods have been analysed by TEM, XRD and EDS. For processing temperatures above 500°C the grain size is about 1µm and a surface CuS layer has been observed for Cu-rich layers. For In-rich layers, In-rich phases have been detected, but a mixture with the CuInS$_2$ seems to exist. With increasing temperature, voids appear at the interface between the substrate and the CuInS$_2$ layer. The CuInS$_2$ shows chalcopyrite cation ordering and, in some areas, also CuAu type ordering.

1. INTRODUCTION

A^{I}B^{III}X$^{VI}_2$ (A=Cu; B=In, Ga, Al; X=S, Se) are compounds that crystallise in the chalcopyrite structure and which show very interesting semiconducting properties well suited for photovoltaic applications. Solar cells based on this type of absorber layer have been proven to have efficiencies up to about 17% (Wada 1997), which has been achieved using Cu(InGa)Se$_2$ layers. Theoretically efficiencies up to 20% have been projected for optimum materials and process parameters.

Among the different absorber layers, those free of Se (CuInS$_2$, hereafter called CIS, and related materials) are gaining interest in the last years because of environmental reasons and because of the larger bandgap, about 1.5 eV, which is better suited for the solar spectrum at the sea level. Furthermore, solar cells having absorber layers of this material have shown efficiencies in excess of 11% (Klaer et al 1998) and their efficiencies are continuously increasing. This improvement is strongly related to the better knowledge and control of the structural, optical and electrical properties of the absorber layers.

In this work the structural characterisation of CIS absorber layers deposited on glass substrates is presented.

2. EXPERIMENTAL

CIS absorber layers have been deposited on soda-lime glass substrates either covered or uncovered with a sputtered Mo layer, which acts as back contact in solar cell production. The CIS layers have been deposited by two different methods. The first one uses sequential sputtering of Cu and In layers, followed by sulphurisation of the bilayer, which is performed at 550°C, and which promotes the formation of CuInS$_2$. In the experiment the layers are very

Cu-rich, with a Cu-to-In ratio of about 1.8, which leads to the formation of Cu-related secondary phases. The second method uses simultaneous evaporation of the reacting elements from Cu, In and S sources. For this experiment the substrate is kept at temperatures in the range 400-600°C and due to the geometry of the evaporation sources, Cu- and In-rich layers are formed on each sample, which allows to study a broad range of compositions on one particular sample. In this case the maximum Cu-to-In ratio is estimated to be about 1.2.

Cross-section specimens have been prepared by ultrasonically cutting a rectangular piece from the sample, gluing it to a similar piece of Si, cutting a disc out of the sandwich and embedding the whole in a metallic cylinder. Next discs of 500 µm are cut from the filled cylinder and the samples are mechanically ground to 50µm and dimpled to a final thickness of about 15µm. Finally Ar ion milling is performed at angles between 6° and 13°, either at room temperature or cooling the sample at liquid nitrogen. Cross-section TEM analysis of these samples has been performed in Philips CM30 microscope, equipped with a EDS detector.

2θ /θ XRD spectra have been taken using a Philips MRD diffractometer.

3. RESULTS AND DISCUSSION

3.1 Sequentially evaporated samples

Fig. 1 shows two XRD spectra of the sequentially evaporated sample, as-obtained and after etching using KCN. The as-obtained sample is characterised by the presence of reflections corresponding to tetragonal $CuInS_2$ (roquesite), the Mo contact (peak at about 40°) and a secondary phase, hexagonal CuS (covellite), which is formed because of the excess copper. From this spectrum a (112) texture of the CIS layer can be observed. After etching the sample,

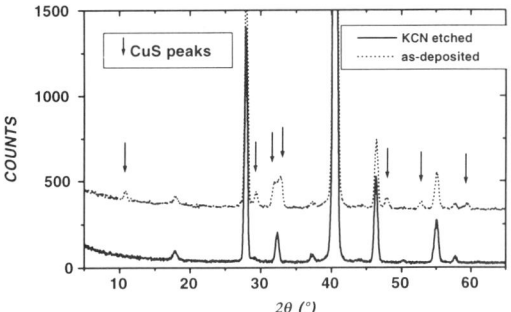

Fig.1: XRD spectra of the sample sequentially evaporated and sulphurised: as-obtained and KCN etched.

the CuS peaks completely disappear and, thus, only the CIS layer is remaining. This procedure of etching the remaining CuS is a step routinely used in the production of solar cells based on this material.

TEM analysis performed on this sample shows (fig. 2a) that the sample is polycrystalline and that the grains have dimensions in the order of 0.7-1.5µm. The surface roughness is very large, of about 1µm (compared to the thickness, about 3.5µm), and the Mo-CIS interface shows the presence of a large number of voids, which probably did develop during the sulphurisation step due to Cu migration. In this image the grains that are at the top of the image are slightly darker and correspond to the CuS, detected by electron diffraction. Two EDS spectra taken from the top and from the centre of the deposited layer from this sample are presented in fig. 2b. The main difference between both is the almost absolute absence of In peaks at the top area, which confirms that the CuS is formed at the surface. After etching this sample using KCN, the whole CuS layer is removed from the sample and only the CIS layer remains, which gives rise to a reduction in thickness of about 40%, which fits well with the excess Cu in the sample. No additional CuS has been detected after etching, although it has been speculated that during the sulphurisation and due to Cu diffusion segregation could occur at the grain boundaries.

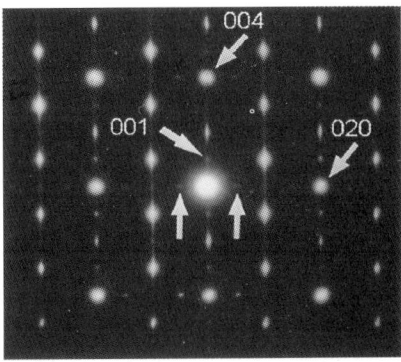

Fig. 3: [100] zone axis a tetragonal CuInS$_2$ grain, showing CuAu-related additional spots ((001)) and additional reflections not yet identified (arrows).

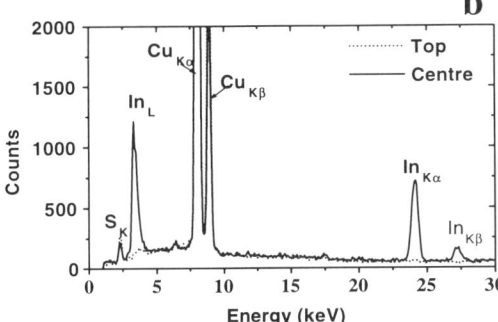

Fig. 2: a) Cross-section TEM image of the Cu-rich region of a coevaporated sample and b) EDS spectra from the top and central regions.

Furthermore, CuInSe$_2$ absorber layers obtained using a similar method on Mo contacts have been investigated (Wada 1997) and the formation of MoSe$_2$ has been detected due to selenisation of the Mo contact. In a similar way, the possible formation of MoS$_2$ in these samples due to the sulphurisation step has been investigated, but TEM results did not show the presence of this phase.

Several grains in this sample show a very particular diffraction pattern, as shown in fig. 3, which is characterised by the presence of additional spots forbidden by the CuInS$_2$ space group ($I\overline{4}2d$). CuAu ordering of the cations in this compound have been observed and simulated (Su et al 1999) and some of the spots could be accounted for. However other additional spots (arrowed) cannot be accounted for by this ordering and simulations are now ongoing in order to identify the crystalline structure. Furthermore a high density of planar defects can be deduced from the streaking in the pattern, which occurs along [001] directions.

Solar cells made using this layer as absorber (after removal of the CuS) have proven to have efficiencies in excess of 11%.

3.2 Simultaneously evaporated samples (coevaporated)

XRD spectra have been taken at different regions of the sample, covering from Cu- to In-rich regions. Similarly to the sequentially evaporated samples, the Cu-rich region shows the presence of peaks corresponding to CuInS$_2$ and CuS. While proceeding towards In-rich regions, the CuS peaks decrease strongly in intensity and for the In-rich region CuInS$_2$ remains and new In-rich phases appear, which depend on the substrate temperature. For a

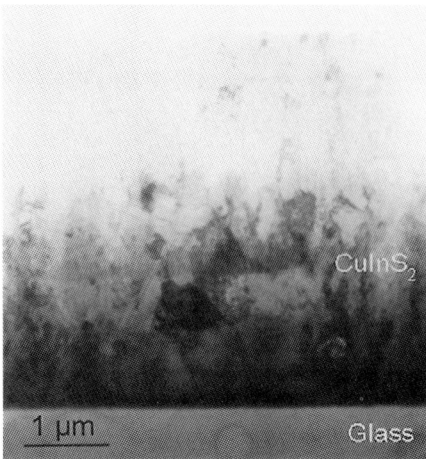

Fig. 4: TEM image of the Cu-rich region of the sample coevaporated at about 450°C.

deposition temperature of about 450°C, the main phase is In_2S_3, while for a temperature of about 520°C, $CuIn_5S_8$ or $CuIn_7S_{11}$ are the primarily observed secondary phases. This result fits with the measurements performed using micro-Raman spectroscopy of the surface region. XRD also shows that again the Cu-rich region presents a (112) texture, while no preferential orientation is observed in the In-rich region.

Fig. 4 shows the TEM image of the Cu-rich region of the sample deposited at about 450°C. The In-rich region is similar to the Cu-rich and both present a much smaller grain size than before, of the order of 0.3-0.6μm. The growth mode tends to be rather columnar, but at the centre of the layer this growth seems to be stopped and larger grains are formed. Furthermore, no voids are observed at the glass-CIS interface, the surface presents a low roughness and no CuS-areas have been observed. These facts are most probably related to the low deposition temperature, which strongly reduces the Cu mobility, and could also be partly due to the reduced Cu-to-In ratio, as compared to the sequentially evaporated sample. However, as mentioned, XRD confirms the CuS existence.

For a growth temperature of 520°C the grain size is strongly increased to about 0.7-1μm, which approaches the value for the sequentially evaporated sample. In the XRD spectra from this sample no texture in the CIS layer has been observed. Again, as for the sequentially evaporated sample, voids are formed at the interface Mo-CIS.

4. CONCLUSIONS

The structural characterisation of the formation of $CuInS_2$ layers by deposition on glass substrates has been investigated as a function of the Cu and In concentrations. An evolution of the grain size and surface roughness has been presented and the formation of voids at the interface between the CIS layer and the substrate has been observed. Ordering of the cations has been detected in the CIS layer, probably being of CuAu type.

ACKNOWLEDGEMENTS

This work has been financially supported by the European Union through the JOULE contract JOR3-CT98-0297, SULFURCELL.

REFERENCES

Klaer J, Bruns J, Henninger R, Siemer K, Klenk R , Ellmer K and Bräunig D 1998 Semicond. Sci. Technol. **13**, 1456
Su DS, Neumann W, Hunger R, Schubert-Bischoff P, Giersig M, Lewerenz HJ, Scheer R and Zeitler E 1998 Appl. Phys. Lett. **73**, 785
Wada T 1997 Proc. 11th Int. Conf. On Ternary and Multinary Compounds, eds. R.D. Tomlinson, A.E. Hill, R.D. Pilkington (Bristol: IOP Publishing Ltd) p 903

Inst. Phys. Conf. Ser. No 164
Paper presented at Microsc. Semicond. Mater. Conf., Oxford, 22–25 March 1999
© 1999 IOP Publishing Ltd

Electron microscopy study of indium oxide thin films.

J Ratajczak, A Maląg, W Sobkowicz[1] and J Kątcki

Institute of Electron Technology, al. Lotników 32/46, 02-668 Warsaw, Poland
[1] Institute of Electronic Materials Technology, ul. Wólczyńska 133, 01-919 Warsaw, Poland

ABSTRACT: Thin films of undoped In_2O_3 have been produced by a conventional vacuum evaporation of In in oxygen ambient. These films have been designed as the cladding layers in a new developed folded cavity surface emitting laser with a prism-like resonator made by selective area MOCVD of GaAs. By means of electron microscopy (SEM and XTEM) we have investigated the structure of the layers deposited with different deposition rates and of different thickness. We have found strong dependence of the granularity of the layer deposited on the prism on the evaporation velocity. Differences in the structure of the layers deposited on various substrates, such as GaAs, masking SiO_2 and GaAs treated with S_2Cl_2 have been also observed.

1. INTRODUCTION

Indium oxide films have a wide range of applications. Their high transparency in the visible range and good electrical conductivity allows them to be applied as transparent electrodes in many optoelectronic devices, such as: solar cells, flat display panels or vertical cavity lasers.

Recently an oportunity has arisen to use In_2O_3 layers as the n-type cladding layers in a newly developed vertical folded cavity surface emitting laser. An active region of this laser is produced in a prism-like shape by GaAs selective area MOCVD. Operation of such a device is based on total internal reflection at the mirrors, folded at the angle of 90°, of a corner reflector, and radiation is emitted from the bottom side of an etched-out substrate. The theoretical operation analysis of this device was given by Maląg (1998) and the selective MOCVD process of obtaining GaAs prism-like resonators with lateral planes bent with an angle of $\pi/4$ to the surface was described by Strupiński et al (1998). To satisfy the total reflection conditions at the $\pi/4$ angle of incidence in the GaAs active region with refractive index $n_a \approx 3.6$, the value for the n-cladding layer has to be $n_c \leq n_a/\sin(\pi/4) \approx 2.55$. The material which is potentially to be used to form n-type cladding is indium oxide which has a value of refractive index below 2.5 over a wide range of wavelengths (Woollam et al, 1994). Due to the high sensitivity of laser parameters to losses at the GaAs-In_2O_3 heterointerface the quality of this interface is of great importance and the deposition process of indium oxide should be carefully optimised. In this study we have used SEM and cross-sectional TEM to characterise the structure of indium oxide made with different deposition rates. This structure is correlated with optical and electrical features of In_2O_3 layers.

2. EXPERIMENTAL PROCEDURES

Undoped indium oxide layers have been produced by means of vacuum evaporation of metallic indium in an oxygen atmosphere in a conventional, resistively heated system. It has been reported that such a method gives the possibility of obtaining nearly stoichiometric, polycrystalline In_2O_3 films under various oxygen pressures (8×10^{-5}Torr - 9×10^{-4}Torr) and substrate temperatures (25°C - 330°C) (Korobov et al., 1994). In order to minimise the number of variable parameters all evaporation processes have been performed at: oxygen pressure - 5×10^{-4}Torr, substrate temperature - 200°C and time - 1h. The only variable was deposition rate controlled by the current to the evaporator. Each deposition process has been performed on two types of surfaces: *i)* as to be used in the laser structure i.e. GaAs covered with 100nm thick SiO$_2$ mask in which 4μm × 16μm windows have been etched and GaAs resonator prisms have been grown by the MOCVD process, *ii)* clean flat GaAs substrates, which were assigned to be used as the TEM cross-sectional samples, and to perform characterisation with other methods such as ellipsometry or conductivity measurements. The thickness of the layers has been measured in a scanning electron microscope on cleaved samples and with Alpha-Step on oxide step etched out by HCl.

3. RESULTS

The indium oxide layers deposited with different rates on laser structures are presented in Fig.1. The deposition rates for samples shown in Fig.1 (a), (b) and (c) were 1,2μm/h, 0.8μm/h and 0.15μm/h respectively. The sample presented in Fig.1d has been obtained in a three step process in which evaporation was performed with rates: 0.2μm/h(15min), 0.3μm/h (25min) and 0.4μm/h(20min). Total evaporation time was also 1h in this case and the thickness of such a layer was 0.25μm

As shown in Fig.1a the evaporation with the highest rate produces indium oxide on a GaAs resonator surface with large grains separated one from another resulting in a non continuous layer. It is worthy of note that this part of the In$_2$O$_3$ layer which is deposited on the silicon oxide mask is continuous and has quite a different surface structure. The layer obtained with lower deposition rate (0.8μm/h – Fig.1b) has still very large grains but is continuous on the resonator surface. The lowest rate layer (Fig.1c) has a fine structure, but sizes of some of its grains are bigger. The finest surface

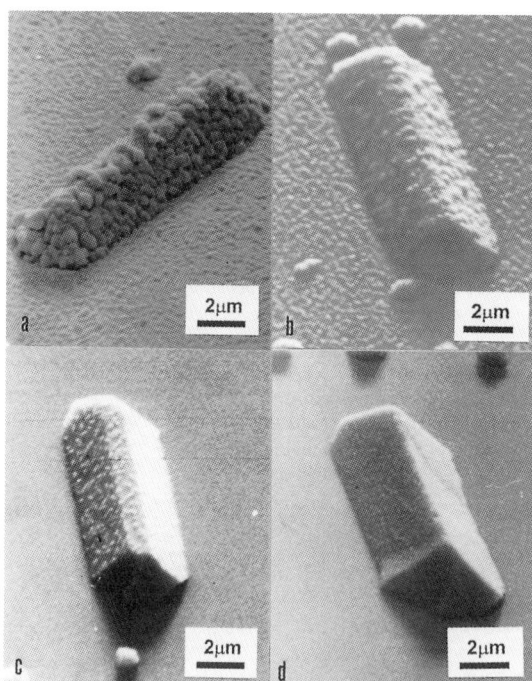

structure has been obtained in a three step process in which the deposition rate was increased during evaporation (Fig.1d). The layers deposited with low rates are continuous and well reflect the shapes of the selectively grown prism. One can conclude that they perfectly stick to the resonator surface.

Electrical measurements have shown that resistivity of all obtained indium oxide layers was in the range of 5×10^{-4}-$10^{-3}\Omega$cm independent of the grain size. The refracive index (n) as measured by ellipsometry was 2.07 (λ=546.1nm) for layers with the lowest granularity, in good agreement with data given by Woollam et al. (1994), and n=1.62 for layers with large grain sizes. For λ=632nm, which is closer to GaAs emission, the refractive index of flat films was 1.97 and these films were lossless, while the layers with bigger grains had bigger losses.

On each micrograph in Fig.1 differences between the surface structures of indium oxide deposited on a SiO$_2$ mask and the part which is positioned on the GaAs surface are visible. These differences were caused by various growth conditions on different types of substrate.

Fig.1. Surface topography of In$_2$O$_3$ films deposited with rates: a) - 1.2μm/h, b) - 0.8μm/h, c) - 0.15μm/h, d) – three step process: 0.2μm/h(15min), 0.3μm/h(25min), 0.4μm/h (20min)

The structure of the layer deposited on a silicon oxide mask is presented in Fig.2. It can be seen that the grains are highly misoriented. This is the result of the amorphous nature of the surface. The grains have small sizes and rounded shapes. This is the reason for characteristic surface topography observed in Fig.1a in a flat region.

In turn in Fig.3 a cross-section of the In_2O_3 layer obtained on a GaAs surface with low deposition rate (0.15μm/h) is shown. This layer mainly consists of well oriented flat grains. Its thickness is equal to the grain height. Since the heights of the majority of the grains are similar, the layer is flat, but there are singular grains with shapes extended above the top surface. These grains have also been observed on the surface of the GaAs resonator in Fig.1c. The most important fact is that the interface between GaAs and In_2O_3 is flat, without any voids which could have given rise to additional optical loses on this mirror.

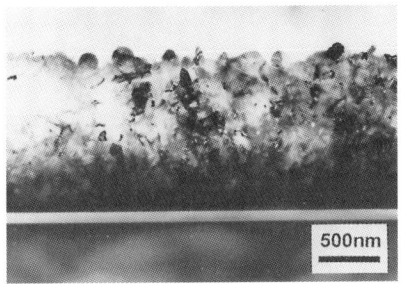

Fig.2. The structure of In_2O_3 deposited on silicon oxide mask.

Fig.3. The structure of thin In_2O_3 layer deposited on a GaAs surface

The importance of the semiconductor-oxide interface in the laser with the prism-like, folded cavity resonator concerns not only its optical properties but also electrical ones. Both of them can be improved by a special chemical treatment, passivation. This process reduces the density of surface states, resulting in the reduction of surface recombination velocity and improvement of device performance.

We have used a relatively new sulphur passivation method: a S_2Cl_2 treatment (Li et al., 1994). GaAs surfaces treated with S_2Cl_2 has exhibited a remarkable increase of photoluminescence intensity. In Fig.4 electrical characteristics of devices made from wafers with and without sulphur passivation are compared. These devices were made by deposition of ohmic contacts (Ag-Te from substrate and Cr-Au from the oxide side) and cleaving into small chips. While characteristics of the chips made from the unpassivated structure exhibit "leaky" Schottky junctions those made after sulphur passivation are similar to nonlinear "ohmic" contacts.

The influence of the passivation of a GaAs surface with S_2Cl_2 on the structure of the indium oxide films has also been observed. These layers looked mat under visual examination. Electron microscopical investigations have shown that it is caused by a very rough oxide surface.

Fig.4. I-V characteristics of GaAs-In_2O_3 heterojunctions.

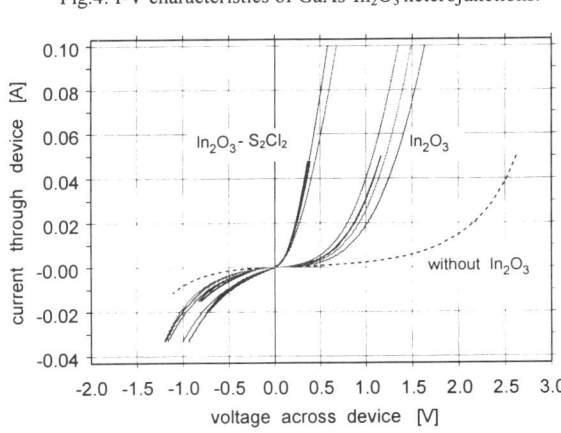

In Fig.5 cross-sections of the In_2O_3 layers deposited on the S_2Cl_2 treated surface are presented. Grain complexes grown above the surface have various orientations as observed in a lower magnification in Fig.5a. These grains are the reason for the surface roughness. In Fig.5b one of the grain complexes is shown in higher magnification. It consists of small grains deposited one on another, finally giving a dendrite-like shape elongated nearly normal to the surface. But directions of other grains are different. Such a random growth can be explained as an effect of the etching of GaAs by the S_2Cl_2 which occurs during the passivation process. Sulphur treatment produces a clean, native oxide free, but a little rough surface of GaAs. Some of its uncovered crystal directions allow for the accelerated deposition of indium oxide grains.

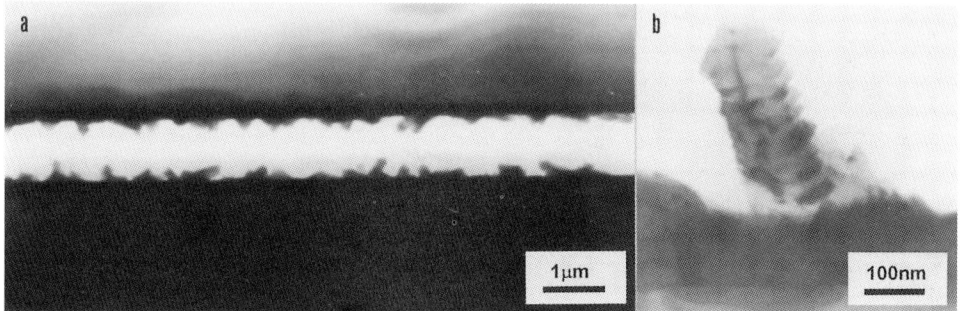

Fig.5. Cross-sections of In_2O_3 layers deposited on GaAs passivated with S_2Cl_2.

4. CONCLUSIONS

It has been shown that careful choice of evaporation rate of indium oxide films results in the production of layers of good structural and optical quality, which allows them to be applied as cladding layers in the vertical folded cavity surface emitting laser with a prism-like corner reflector. The difficulties with obtaining good quality In_2O_3 layers on GaAs passivated with S_2Cl_2 exclude application of this passivation process for improvement of the GaAs-In_2O_3 heterointerface in such a device.

ACKNOWLEGMENT

The authors are much indebted to Mrs A.Kamińska for the ellipsometry measurements.

REFERENCES

Korobov V, Shapira Y, Ber B, Faleev K, Zushinskiy D, (1994) J.Appl.Phys. **75** 2264
Li Z S, Cai W Z, Su R Z, Dong G S, Huang D M, Ding X M, Hou X Y, Wang X, (1994)
 Appl.Phys.Lett **64** 3425
Maląg A, (1998) IEE Proc.-Optoelectron. **145** 151
Strupiński W, Maląg A, Ratajczak J, (1998) J.Cryst.Growth **195** 474
Woollam J A, McGahan W A, Johs B, (1994) Thin Solid Films **241** 44

Inst. Phys. Conf. Ser. No 164
Paper presented at Microsc. Semicond. Mater. Conf., Oxford, 22–25 March 1999
© 1999 IOP Publishing Ltd

Microstructures of Y_2O_3 films grown on Si (111) by ionized cluster beam

D-H Lee,[1] T-Y Seong,[1] M-H Cho[2] and C-N Whang[2]

[1]Department of Materials Science and Engineering, Kwangju Institute of Science and Technology (K-JIST), Kwangju 500-712, Korea
[2]Department of Physics, Yonsei University, Seoul 120-749, Korea

ABSTRACT: Y_2O_3 films were grown on SiO_2-covered Si (111), and H-terminated Si (111) substrates at a temperature of 500 °C by ultrahigh vacuum ionized cluster beam deposition (UHV-ICB). The microstructures and growth behaviour of these films have been investigated by transmission electron diffraction (TED) and high resolution electron microscopy (HREM). The TED results show that the Y_2O_3 film grown on the SiO_2-Si has the epitaxial relationship of $(11\text{-}1)_{Y2O3}//(111)_{Si}$ and $[\text{-}110]_{Y2O3}//[\text{-}110]_{Si}$. The film on the H-Si contains $YSi_{2\text{-}x}$ and amorphous YSi_xO_y layers at the interface. For the $YSi_{2\text{-}x}$ and the Si substrate, orientation relationship is $(0001)YSi_{2\text{-}x}//(111)Si$ and $[1\text{-}210]YSi_{2\text{-}x}//[\text{-}110]Si$. For the Y_2O_3 and the $YSi_{2\text{-}x}$, the relationship is as follows: $(11\text{-}1)Y_2O_3//(0001)YSi_{2\text{-}x}$ and $[\text{-}110]Y_2O_3//[1\text{-}210]YSi_{2\text{-}x}$; $(111)Y_2O_3//(0001)YSi_{2\text{-}x}$ and $[\text{-}110]Y_2O_3//[1\text{-}210]YSi_{2\text{-}x}$. The formation mechanisms of the interfacial phases of SiO_x, YSi_xO_y and $YSi_{2\text{-}x}$ are presented. It is shown that the crystallinity of the Y_2O_3 film on the SiO_2-Si (111) is better than that of Y_2O_3 on H-Si (111).

1. INTRODUCTION

Extensive research has been carried out to obtain materials with a high dielectric constant, which could be used in the storage capacitors of Giga bit dynamic random access memory. SiO_2 films, having a dielectric constant of 3.9, have been commonly used in the fabrication of metal-insulator-semiconductor devices. The scaling-down of the lateral dimension of the insulator-semiconductor devices requires reduction in the thickness of thermally grown SiO_2 films (Sharma and Rastogi 1993). However, the SiO_2 films thinner than 3 nm suffer from reliability problems, such as a high pin-hole density, an enhanced tunneling current conduction and a low dielectric breakdown strength (Hu 1984). Therefore, different oxide films such as Ta_2O_5, ZrO_2, HfO_2 and Y_2O_3 have been investigated to overcome such problems. Manchanda and Gurvitch (1988) showed that an $Al/Y_2O_3/Si$ capacitor is a potential candidate for a high storage capacity insulator for ultra-large-scale-integration (ULSI) devices.

Yttrium oxide has a cubic bixbyite Mn_2O_3 structure, in which the unit cell consists of eight unit cells of an incomplete fluorite structure. There is a lattice mismatch of ~2.45 % between Y_2O_3 and silicon, where the lattice constant of Y_2O_3 is 1.06 nm and (that of Si)×2 is 1.086 nm (Fukumoto et al 1989). Fukumoto et al (1989), investigating structural properties of Y_2O_3 films deposited on Si (001) and (111) (4° off), showed that the Y_2O_3 films were epitaxially grown on Si (111) at 800 °C.

Recently, Choi et al (1997) investigated Y_2O_3 films grown on Si (100) at 800 °C by ultra-high vacuum ionized cluster beam system. They illustrated the successful growth of epitaxial films having orientation relationship of $(110)Y_2O_3//(100)Si$.

In this work, effects of differently treated surface conditions (e.g., SiO_2-covered and H-terminated Si (111) substrates) on the growth behaviour and microstructure of Y_2O_3 films have been investigated using TED and HREM. The films were grown at a relatively low temperature, 500 °C, as compared to the previously reported works (Manchanda and Gurvitch 1988, Fukumoto et al 1989,

Choi et al 1997). The formation mechanisms of interfacial phases are discussed.

2. EXPERIMENTS

The deposition of the Y_2O_3 films was carried out by an UHV-ICB system. The ICB system and growth conditions have been described elsewhere (Choi et al 1997). The base pressure of the growth chamber was 5×10^{-10} Torr and the substrate temperature was 500 °C. The substrates were SiO_2-covered and hydrogen-terminated Si (111) wafers, cut 4° off toward the [-1-12] direction. The SiO_2-covered substrate (SiO_2-Si) was prepared by chemical cleaning using the RCA method (Kern and Puotinen 1970). For the preparation of the H-terminated surface (H-Si), a Si (111) substrate was dipped into a dilute HF solution after chemical cleaning using the RCA method. The Y_2O_3 films were deposited using an Y metal cluster beam generated by an ICB source and oxygen gas that was supplied into the chamber through a gas inlet line installed below the sample holder.

For electron microscope examination, [110] cross-section specimens were prepared using standard procedures and finishing by Ar+ ion thinning with the specimen cooled to ~77K and examined in a JEM 2010 instrument operated at 200 kV.

3. RESULTS AND DISCUSSION

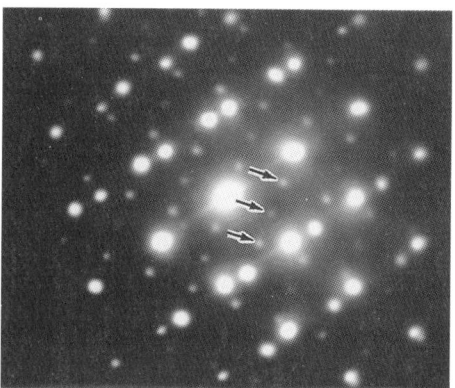

Fig. 1 shows a [-110]$_{Si}$ TED pattern from a region including the Y_2O_3 film and the SiO_2-Si (111) substrate.

Figure 1 shows a [-110]$_{Si}$ TED pattern obtained from a region including the Y_2O_3 film and the SiO_2-Si (111) substrate. To characterise the structural information of the [-110] TED pattern of the Y_2O_3 film, we have performed computer simulations using NCEMSS (Kilaas 1987). The details about the electron microscope used and crystallographic information (a bixbyite Mn_2O_3 structure) have been taken into account in the simulations. The calculated [-110] TED pattern of the Y_2O_3 film reproduced diffraction spots observed in Fig. 1. The TED results indicate that the Y_2O_3 film is epitaxially grown on the Si (111) substrate. The orientation relationship between the Y_2O_3 and the substrate is found to be $(11-1)_{Y2O3}//(111)_{Si}$ and $[-110]_{Y2O3}//[-110]_{Si}$. This shows that the film has twin relation to the substrate; (11-1) of Y_2O_3 is parallel to (111) of Si. This result is in good agreement with the XRD results of Fukumoto et al (1989) who investigated structural characteristics of Y_2O_3 films which were electron-beam-deposited on Si (111) at 800 °C.

To investigate the microstructure of the film/substrate interface, [-110] cross-section HREM examination was carried out. Fig. 2 shows a [-110]$_{Si}$ HREM image taken from a region including the Y_2O_3/SiO_2-Si interface. The image reveals the presence of an interfacial amorphous SiO_x layer at the interface. The SiO_x/Si interface appears to be reasonably planar with an undulation of ~±3 monolayers. However, the Y_2O_3/SiO_x interface is found to be rather wavy. It should be noted that despite the presence of the amorphous SiO_x layer ~2 nm thick, the Y_2O_3 film has epitaxial relation to the substrate (Fig. 1). The presence of the amorphous layer can be attributed to the indiffusion of oxygen from the surface during growth, as discussed below (Inoue et al 1991, Fenner et al 1991, Choi et al 1997). As expected from the TED results, it can be seen that (11-1) of the Y_2O_3 film is parallel to (111) of the Si substrate. There are bright and dark bands lying parallel to (110) of the Y_2O_3 film (marked 'B'). The bands have a periodicity (peak to peak: 0.75 nm) four times larger than the spacing of the (440) lattice planes, which is responsible for the presence of the additional weak spots (as indicated by the arrows) in the [-110]$_{Si}$ pattern (Fig.1).

Figure 3 shows a [-110]$_{Si}$ TED pattern obtained from the interface region of the Y_2O_3 film

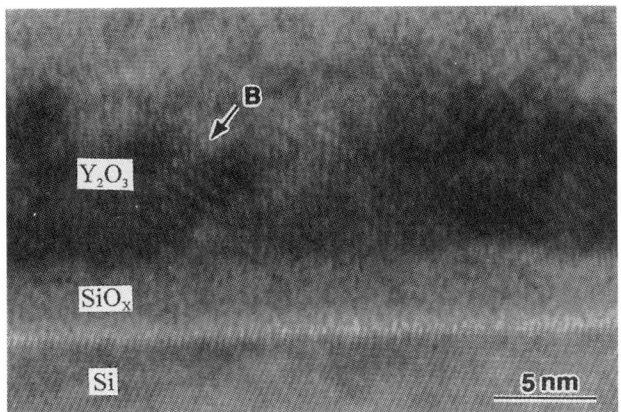

Fig. 2 shows a $[-110]_{Si}$ HREM image taken from a region including the Y_2O_3/SiO_2-Si interface.

grown on the H-Si (111) substrate. The pattern indicates that the film contains different materials having different crystallographic orientations. The simulated results and HREM results show that there are two sets of the <110> patterns of the Y_2O_3 film. In addition, there are weak elongated diffraction spots (or diffuse intensity) (indicated by the small arrows) that can be attributed to the presence of Y-silicide, as discussed below. The Y-silicide (YSi_{2-x}) is known to have a hexagonal structure (defected AlB_2 type, but with 15~20 % vacancies on the Si sublattice). Assuming that the Si substrate spots correspond to undistorted materials with the bulk Si lattice parameter of 0.543 nm, the lattice parameters of the YSi_{2-x} were calculated from the measured spacings of the spots. The result shows that the silicide has the bulk YSi_{2-x} lattice parameters of 0.385 ± 0.002 nm (a) and 0.415 ± 0.002 nm (c), which is in agreement with the reported values of Knapp and Picraux (1986), and Gurvitch et al (1987).

Figure 4 shows a $[-110]_{Si}$ cross-section HREM image obtained from a region including the Y_2O_3/H-Si interface. The image shows the presence of the crystalline YSi_{2-x} layer and an additional intermediate amorphous YSi_xO_y layer. The TEM result showed that the YSi_{2-x} layer varied in thickness from ~1 to ~5 nm, whilst that of the amorphous layer was in the range of $0.5 - 2$ nm. There are also bright and dark bands lying parallel to (110) of the Y_2O_3 film (marked 'B'), having a periodicity four times lager than that of the (440) lattice planes.

The pattern (Fig. 3) shows that there is orientation relationship between the Y_2O_3 film, the YSi_{2-x} layer, and the H-Si(111) substrate. The relationship between the YSi_{2-x} layer and the Si substrate is $(0001)YSi_{2-x}//(111)Si$ and $[1-210]YSi_{2-x}//[-110]Si$. For the Y_2O_3 and YSi_{2-x} layers, the relationship is $(11-1)Y_2O_3$-A$//(0001)YSi_{2-x}$ and $[-110]Y_2O_3$-A$//[1-210]YSi_{2-x}$; $(111)Y_2O_3$-B$//(0001)YSi_{2-x}$ and $[-110]Y_2O_3$-B$//[1-210]YSi_{2-x}$. The higher-order spots are elongated slightly and/or split into several spots

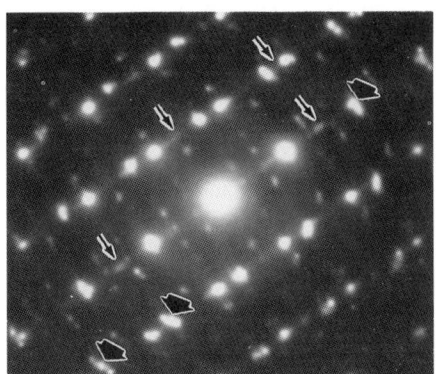

Figure 3 shows a $[-110]_{Si}$ TED pattern obtained from the interface region of the Y_2O_3 film grown on the H-Si (111) substrate.

(as indicated by the larger arrows), implying that although the Y_2O_3 film is a single-crystal, there are patches of grains having different orientations. Such Y_2O_3 grains are rotated ±1~4° about $[-110]_{Si}$ with reference to the (0001) plane of the YSi_{2-x} (and also the (111) plane of the H-Si substrate).

The TED and HREM results showed that the formation of the interfacial layers depends on the surface conditions. For the SiO_2-Si substrate, the formation mechanism of the amorphous SiO_x layer can be explained as follows. It is worth noting that in the deposition process, a Y metal cluster beam was first supplied for 3 min before the introduction of oxygen gas. The ionized Y clusters could favourably react with the SiO_2 on the top surface of the Si (111) substrate to form Y_2O_3, since the heat of the formation of Y_2O_3 (1758 kJ/mol) is larger than that of SiO_2 (911 kJ/mol) (Cowley 1995).

258

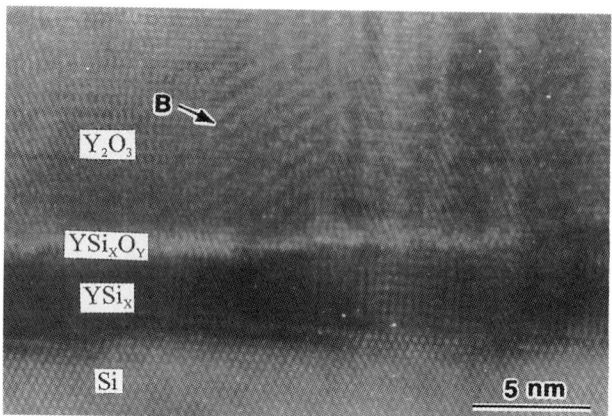

Fig. 4 shows a $[-110]_{Si}$ HREM image taken from a region including the Y_2O_3/H-Si interface.

Thus, this reaction process leads to the growth of the Y_2O_3 film without an interfacial oxide layer, since the native oxide layer is completely consumed by the ionized Y clusters at the early stage of the film growth (Inoue et al 1991, Choi et al 1997). During growth, however, oxygen may diffuse into the Si substrate through the Y_2O_3 film and react with the Si substrate to form the intermediate amorphous SiO_x layer. This explanation is in good agreement with the TED and HREM results (Figs. 1 and 2), showing that despite the presence of the interfacial amorphous layer, there exists the epitaxial relationship between the film and the SiO_2-Si substrate. Similar reactions have been also reported to occur in different material systems such as Y_2O_3/Si (100) (Choi et al 1997), yttria-stabilised zirconia (YSZ)/Si (Fenner et al 1991), and CeO_2/Si (100) (Inoue et al 1991).

For the H-Si (111) substrate, there are two types of the intermediate layers, YSi_{2-x} and YSi_xO_y. The formation mechanism for the intermediate layers can be speculated as follows. The ionized Y clusters could also react with the Si substrate to form YSi_{2-x} until oxygen gas is introduced into the chamber. The Y clusters then react with the supplied oxygen to form Y_2O_3 on the YSi_{2-x} layer. In a similar manner, during growth, oxygen may diffuse into the YSi_{2-x} layer through the Y_2O_3 film and may react with the YSi_{2-x} to form the YSi_xO_y layer. The absence of an amorphous oxide layer at the YSi_{2-x}/Si substrate interface suggests that the YSi_{2-x} layer act as a diffusion barrier for oxygen diffusing into the Si substrate.

ACKNOWLEDGMENT

This work was supported by the Korea Science and Engineering Foundation (KOSEF) through the Atomic-scale Surface Science Research Centre (ASSRC) at Yonsei University.

REFERENCES

Choi S C, Cho M H, Whangbo S W, Whang C N, Kang S B, Lee S I and Lee M Y 1997 Appl. Phys. Lett. **71**, 903
Cowley J M 1995 Handbook of Oxide (Oxford University Press, New York) pp 270
Fukumoto H, Imura T and Osaka Y 1989 Appl. Phys. Lett. **55**, 360
Fenner D B, Viano A M, Fork D K, Connel G A N, Boyce J B, Ponce F A and Tramontana Geballe T H 1991 J. Appl. Phys. **69**, 2176
Gurvitch M, A F J Levi, R T Tung and S Nakahara 1987 Appl. Phys. Lett. **51**, 311
Hu C 1984 IEDM Technical/Digest (IEEE, New York) pp 6
Inoue T, Ohsuna T, Luo L, Wu X D, Maggiore C J, Yamamoto Y, Sakurai Y and Chang J H 1991 Appl. Phys. Lett. **59**, 3604
Kern W and Puotinen D A 1970 RCA Rev. 187
Knapp J A and Picraux S T 1986 Appl. Phys. Lett. **48**, 466
Kilaas R 1987 Proc. 45th EMSA (San Francisco Press, San Francisco) pp 66
Manchanda L and Gurvitch M 1988 IEEE Electron Dev. Lett. **9**, 180
Sharma R N and Rastogi A C 1993 J. Appl. Phys. **74**, 6691

Inst. Phys. Conf. Ser. No 164
Paper presented at Microsc. Semicond. Mater. Conf., Oxford, 22–25 March 1999

Dislocation dynamics in graded relaxed SiGe/Si, InGaAs/GaAs, and InGaP/GaP

E A Fitzgerald, A Y Kim, M T Bulsara and M T Currie

MIT Materials Science and Engineering, 77 Massachusetts Ave., Cambridge, MA 02139

ABSTRACT: The concept of relaxed graded layers is applied to the SiGe/Si, InGaAs/GaAs, and InGaP/GaP systems in order to achieve new flexibility in semiconductor integration. In graded composition layers, the threading dislocation density should remain constant once an equilibrium density is achieved. Deviations from this ideal case, resulting in higher threading dislocation densities, are due to microstructural features that inhibit dislocation flow. By eliminating the obstruction in each material system, a steady-state residual threading dislocation density can be achieved resulting in high quality lattice mismatched materials on commercially available binary compound substrates.

1. INTRODUCTION

As lattice-matched materials systems become fully exploited commercially, the continued drive towards system level improvements in electronic systems has forced a re-evaluation of the utility of lattice-mismatched materials. Lattice-mismatched materials in this context refer to the vast majority of semiconductor compounds that do not possess the lattice constant of a conventional substrate, such as Si, GaAs, and InP. The mismatched materials often can offer new design freedom at the system level, or enhanced performance of device technologies. For example, the SiGe/Si system can enhance the mobility of carriers in Si (Mii et al 1991, Schaffler et al 1992, Ismail et al 1995, Sugii et al 1998), the InGaP/GaP system can create an all epitaxial transparent substrate LED technology (Kim and Fitzgerald 1998, Kim et al 1999a), and the InGaAs/GaAs system can increase the performance of high electron mobility transistors (Grider et al 1990) as well as create 1.3μm devices on GaAs substrates (Bulsara et al 1998). Although most device engineers have been traditionally cautious about exploring purposely-dislocated material, it is useful to note that high volume visible red light emitting diodes (LEDs) are manufactured using thick graded GaAsP layers on GaAs substrates (Casey and Panish 1978).

Despite the success in the GaAsP system, relaxed mismatched epitaxy was not implemented in other materials systems. There are three main reasons for this lack of further development. First, the lattice-matched systems had not yet been fully exploited. With the development of semiconductor lasers in the AlGaAs/GaAs system, it was clear that lattice matching was the most expedient path to high performance devices. Thus, the demand for lattice mismatched materials would not increase until lattice matched materials systems were fully exploited. Second, the knowledge base of understanding dislocations in lattice mismatched materials was not extensive. Thus, even though GaAsP/GaAs success was achieved through empirical understanding, the basics of lattice mismatched epitaxy were just beginning to appear (Matthews 1975). Third, and most relevant to this publication, is the fact that other materials systems did not result in high quality material like the GaAsP/GaAs system (Abrahams et al 1975).

Threading dislocation densities in significantly relaxed material in the SiGe/Si, InGaAs/GaAs, and InGaP/GaP materials systems are shown in Figure 1. It is clear that as the degree of plastic deformation increases with greater graded buffer thickness, the threading dislocation density in the top relaxed layer increases sharply. If we were to simply apply an empirical relationship to graded layer growth as is done for bulk deformation, it would appear that an intrinsic dislocation multiplication is occurring in all systems, and that composition grading is a poor strategy.

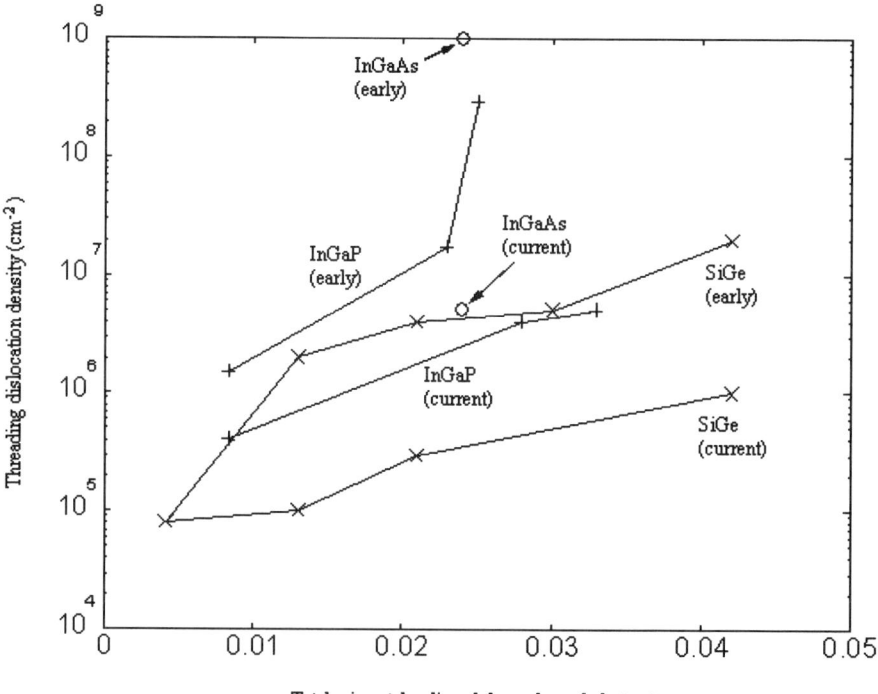

Figure 1. Plot of threading dislocation density in relaxed graded layers for SiGe/Si, InGaAs/GaAs, and InGaP/GaP materials systems. The terms 'early' and 'current' refer to data before and after current graded layer improvements were applied.

2. DISLOCATION DYNAMICS MODEL

The fact that some very relaxed graded layers have low threading dislocation densities suggests that the empirical 'dislocation multiplication' is not active under all conditions. For example, GaAsP alloys graded to 1% mismatch on GaP, and SiGe alloys graded to 1% mismatch on Si substrates possess threading dislocation densities in the 10^5 cm^{-2} range. Such layers are extremely dislocated, and therefore one would think the multiplication source should be active.

An alternative point of view is that a certain threading dislocation density is required to prevent excessive strain from accumulating in the graded layer (Fitzgerald et al 1992). For example, if we ignore nucleation, then we can estimate the threading dislocation density required to relax the graded layer sufficiently to avoid a large increase in strain. This calculation is possible because the threading dislocation velocity is proportional to the rate at which the misfit dislocation is lengthened, which in turn is the rate of strain relief. We can then obtain the following relationship:

$$\dot{\delta} = \frac{\rho_t b}{2} B Y^m \varepsilon_{eff}^m e^{\frac{-U}{kT}} \tag{1}$$

where $\dot{\delta}$ is the rate of plastic strain (mismatch strain relief), ρ_t is the threading dislocation density at the surface, b is the Burgers vector, B is a constant, Y is the elastic constant for a biaxial system, ε_{eff} is the effective strain on the threading dislocation, U is the activation energy for dislocation glide, and T is the temperature. Note that the majority of the right hand side of Eqn. 1 is the empirical expression for dislocation velocity in materials. A more useful expression transforms Eqn. 1 into an expression in terms of the growth parameters. At graded layer thickness much greater than the critical thickness for dislocation introduction, the plastic strain rate is approximately linear with thickness (Fitzgerald et al 1992), leading to:

$$\rho_t = \frac{2 R_g R_{gr} e^{\frac{U}{kT}}}{b B Y^m \varepsilon_{eff}^m} \tag{2}$$

where R_g is the growth rate, and R_{gr} is the grading rate. Eqn. 2 shows us that we can not expect to lower threading dislocation densities easily if the system is near equilibrium. This limit is due to the fact that the variables in Eqn. 2 can not be changed easily by multiple orders of magnitude. An interesting result of Eqn. 1 and 2 is that the expected threading dislocation density in different semiconductor alloy systems is on the order of 10^5-10^6 cm^{-2} for the most common growth conditions. Thus, GaAsP/GaAs graded layers and SiGe/Si layers graded to about 1% mismatch possess the expected threading dislocation density of about 10^5-10^6 cm^{-2}.

3. SiGe/Si

Viewing SiGe graded layers within the context of the model, we must question why higher threading dislocation densities are observed in layers graded to higher Ge concentration. Samavedam and Fitzgerald have shown that threading dislocation pile-ups begin to form along <110> directions beginning at final Ge concentrations near 25-30% in graded SiGe layers (Samavedam and Fitzgerald 1997, Fitzgerald and Samavedam 1997). They attribute this phenomenon to the blocking of threading dislocations via misfit dislocation strain fields and surface morphology. The surface morphology is also a consequence of the buried misfit dislocation strain fields since the strain affects local growth rate on the surface. Such pile-ups increase the overall threading dislocation density since once the dislocation is trapped, it can no longer contribute to strain relief, and the system will then nucleate a replacement.

Currie et al. proved the importance of surface morphology by planarising the graded layer surface in the middle of the epitaxial growth (Currie et al 1998). Removing the surface morphology nearly eliminated the threading dislocation pile-ups, reduced the overall threading dislocation density, and produced high quality Ge (Currie et al 1998) and GaAs on Si (Sieg et al 1998).

An additional relationship revealed in the work of Currie et al. is that the pile-up density and threading dislocation density away from the pile-ups (referred to as field density) both improve upon surface planarisation. We can now explain this behaviour by noting in general that Eqn. 2 suggests that the effective stress on the threading dislocations be lowered by the increase in surface roughness. Therefore, the field threading dislocation density must increase. The pile-up density will also increase, since the pile-ups occur at infrequent regions in which the surface cross hatch is deep enough to start arresting dislocation motion. The surface roughness increase can then be expected to increase both the field dislocation density as well as the pile-up dislocation density.

Because the parameters for dislocation velocity in the SiGe system are well known (Houghton 1991), we can create a predictive expression for dislocation density as a function of the two most important parameters in Eqn. 2, the effective strain and the temperature of growth. Such a threading dislocation density map is shown in Figure 2. As expected from the Eqn., high temperature growth and high effective strain acting on the threading dislocation at the surface promote low threading dislocation densities.

We note here that the effective strain can be a function of the growth temperature. Three regimes can be envisioned. At the lowest temperatures, the overall surface roughness is expected to be low due to surface kinetics, i.e. the equilibrium cross hatch pattern from the underlying misfit dislocations can not be obtained. In the SiGe system, such low temperatures also result in very low dislocation velocities, and create numerous pile-ups since a few nucleation sources are very active but dislocation travel is limited. Even though the effective stress is high in this case, the low temperatures create a kinetically limited situation for the dislocations as well. At intermediate temperatures, the cross hatch pattern can be quite rough, since the surface can respond to the underlying dislocation strain fields, but the misfits are not evenly distributed since high dislocation nucleation rates occur at a few heterogeneous sources and cross slip is not active enough. In this regime, the graded structures may be nearing complete relaxation since the dislocation velocities are high enough, but the surface roughness ensures that the effective strain is less than the equilibrium strain value. Finally, at the highest temperatures, many heterogeneous sources are active and cross slip is very active, so even though the surface can respond easily to the underlying misfit dislocation strain fields, the misfit dislocations are more evenly distributed. This distribution results in a lower surface roughness and a higher effective strain at the surface, which should be very near the equilibrium strain. It is in this regime that the lowest threading dislocation densities are expected in the SiGe/Si system.

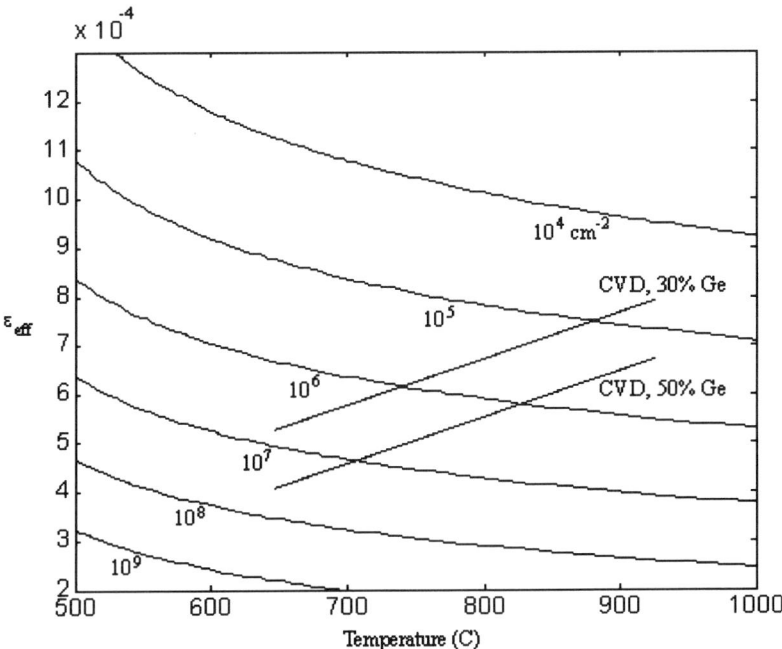

Figure 2. Threading dislocation density map for graded structures as a function of growth temperature and effective strain on the dislocation. The dashed lines represent data for graded SiGe layers graded to 30 and 50% Ge. The fact the lines are sloped suggests that the effective strain is a function of the surface roughness, which decreases at higher growth temperatures.

In Figure 2, we have superimposed interpolated threading dislocation data for graded layers with final compositions of 30% and 50% Ge. All layers are graded at approximately 10% Ge per micron thickness. First note that the lines do not have zero slope, supporting the idea that the effective strain at the surface of the layers is a function of the growth temperature. Also note that the effective strain increases as the temperature is increased. This fact, along with the experimental observation that the surface roughness decreases with higher temperature, suggests that we are in regimes 2 and 3 above. The higher temperatures are promoting a more even misfit dislocation distribution and lower surface roughness, allowing the effective strain to approach the true equilibrium residual strain value.

The 50% Ge alloys show a lower effective strain value. This lower value is expected since the surface roughness increases with further grading (no planarisation is used in the data shown in Figure 2). The trend in higher effective strain with increasing temperature is also seen in this series.

It is logical that the results from the SiGe/Si system can be applied to the III-V materials systems as well. As we see below, a different mechanism is responsible for the impediment to threading dislocation motion in these materials.

4. InGaAs/GaAs

Based on the results from the SiGe/Si system, and the data in Figure 1, we might assume that the same mechanism is responsible for the increase in the threading dislocation density as InGaAs is graded to higher In concentration. However, investigations into low mismatched layers with only 6% In show that there is a surface roughness mechanism taking place which is different than the SiGe/Si mechanism (Bulsara et al 1998). This study shows that the InGaAs surface roughness appears before the threading dislocation pile-ups form, and the roughness appears to be originating from planar boundaries, which we refer to as branch defects after their appearance in the InGaP/GaP system (Kim and Fitzgerald 1998). The number of these branch defects is a strong function of temperature.

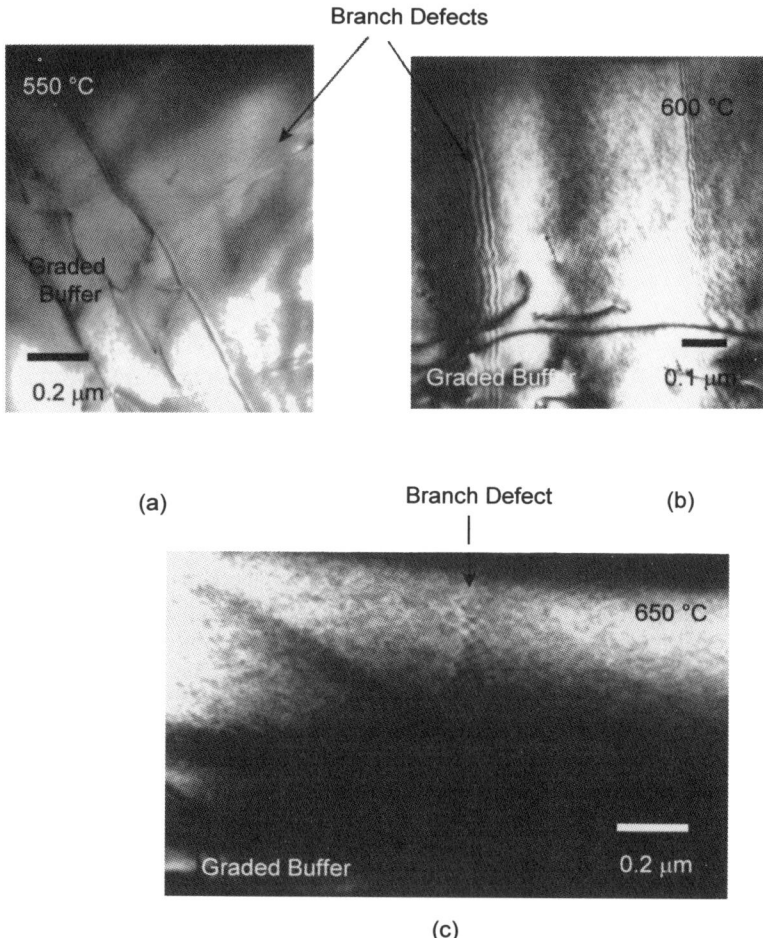

(a)   (b)

(c)

Figure 3. TEM cross-sections showing branch defects in graded InGaAs films on GaAs. The films are graded to 6% In, at temperatures of 550, 600, and 650C, as indicated.

Figure 3 is a collage of TEM cross sections of InGaAs alloys graded to 6% In and grown at different growth temperatures. The InGaAs alloys grown at 550C show the most numerous branch defects, whereas at 650C, the spacing of the defects is great enough that observing them in cross section TEM is difficult. The decrease in branch defect density with increasing temperature correlates directly to the decrease in surface roughness with increasing growth temperature in Bulsara et al 1998). It is reasonable to conclude that the two effects are correlated, and one can observe that the branch defects tend to have surface features associated with them in cross section TEM.

With further grading at 550C, threading dislocations become blocked, and there is a rapid rise in threading dislocation density with increasing In concentration. These results clearly show that graded InGaAs layers must be grown at temperatures less than 500C or greater than 700C for the In compositions investigated (<33% In). Most of the graded InGaAs relaxed buffer work has been performed with molecular beam epitaxy, in which the higher temperature regime can not be accessed. However, with the higher growth temperatures afforded by MOCVD, high optical quality, low threading dislocation density relaxed buffers can be achieved (Bulsara et al 1998).

5. InGaP/GaP

The InGaP/GaP system appears to be similar to the InGaAs/GaAs system in that branch defects are observed in this system as well. However, the branch defects in the InGaP/GaP system are much sharper in plan view TEM.

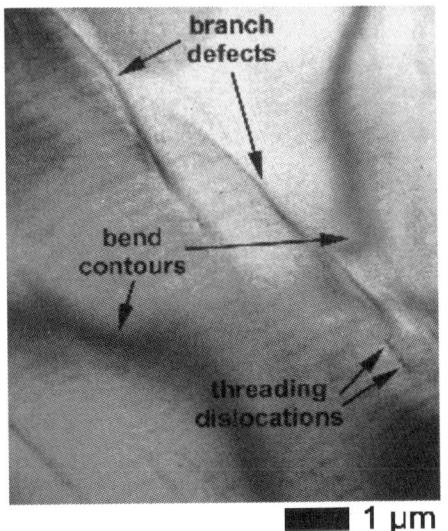
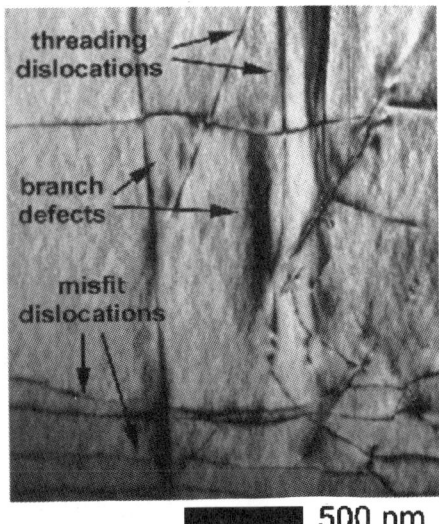

Figure 4 TEM plan view (left) and cross section (right) of the uniform cap region in an InGaP/GaP graded structure, graded to 26% In. The growth temperature was 760C, leading to branch defect formation as the In concentration was increased.

Figure 4 is a plan view TEM micrograph and a cross section of a TEM micrograph revealing the branch defects and their ability to block threading dislocations. Note that the bend contours in the TEM foil have a high curvature as they cross the branch defects, indicating a high strain at the branch defects. The plan view shows two threading dislocations that have been arrested near the branch defect, and the cross section shows numerous threading dislocations trapped in branch defects creating a pile-up.

Because the TEM evidence suggests that the branch defects are a major impediment to dislocation flow, we have roughly mapped the presence of branch defects in graded relaxed films as a function of growth temperature in Figure 5. At any given growth temperature, the branch defects are present at a high enough In composition. Since the InGaP alloy becomes direct gap somewhere in the vicinity of 25% In, and the maximum growth température in our MOCVD is approximately 800C, we can see that it is impossible to get to the direct band gap composition without creating branch defects.

However, it has been observed that the apparent strength of blocking threading dislocations at the branch defect decreases with growth temperature. Thus, at low enough temperature, the density of branch defects is high but their strain field appears to be much weaker (Kim et al 1999a). Therefore, a strategy for achieving relaxed InGaP layers with direct band gaps can be to deposit the graded layer with a substrate temperature of 800C, but as the 20% In composition is reached, the temperature can be dropped to 650C. At first, it may appear that this strategy is inconsistent with Eqn. 2, since the lower temperature will slow the dislocations and increase the threading dislocation density. But as the In concentration increases, the velocities of the dislocations in InGaP increase for a given temperature. Thus, by 20-30% In, the dislocation velocities are high enough that relatively low threading dislocations can still be expected. Recently, visible yellow and red LEDs with transparent GaP substrates were fabricated using this methodology (Kim and Fitzgerald 1999b).

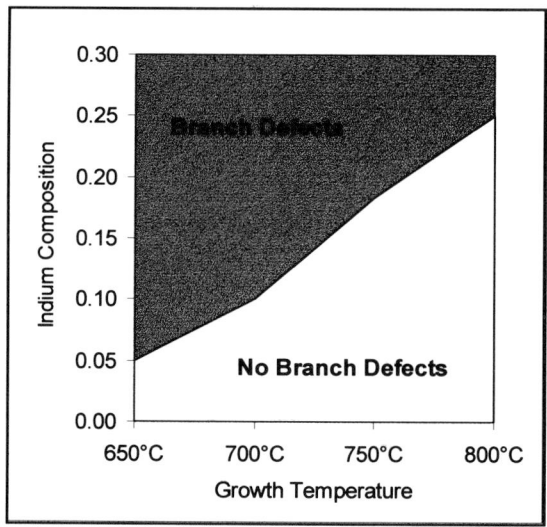

Figure 5 A rough estimate of when branch defects form based on TEM micrographs of graded InGaP/GaP samples. As the In concentration increases in the graded buffer, branch defect formation is eventually unavoidable.

Our understanding of the phase diagram in Figure 5 suggests that at 800C, InGaP layers graded to 10% In should not contain any branch defects, and at such temperatures we might expect perfect threading dislocation flow if there are no serious surface obstructions. Atomic force microscopy of a series of $In_{0.10}Ga_{0.90}P$ graded layers on GaP grown at different temperatures between 650 and 800C show that the surfaces display the expected cross hatch and that the overall surface roughness is quite low. The expected perfect dislocation flow is confirmed by analysis of the threading dislocation density vs. growth temperature. Using Eqn. 2, we can extract the best fit when using an activation energy of about 2eV when m=1. 2eV is the expected activation energy for dislocation glide in this system if we interpolate between the GaP and InP activation energies (Fitzgerald et al 1999).

6. SUMMARY AND CONCLUSIONS

We have shown that if there are no impediments to dislocation flow during graded layer epitaxy, the threading dislocation density will be determined by growth temperature and the effective strain on the threading dislocation. The residual effective strain can be approximated by the residual equilibrium strain in the relaxed structure at the growth temperature. In each materials system, different mechanisms can exist that can contribute to inhibiting dislocation flow. In SiGe/Si, the surface roughening created by the buried misfit dislocation strain fields can create the barrier. In the ternary III-V compounds, branch defects most likely arising from a connection to the ordering that can occur in these compounds are impediments, and lead to surface roughening as well, which can create further impediments. Control of these factors has led to relaxed SiGe alloys of any composition for application in electronics and integrated optoelectronics, as well as to relaxed InGaAs and InGaP alloys for novel optoelectronic devices. The current dislocation densities achieved in the materials systems with the improvements are shown in Figure 2. The general result of such research is that traditional bulk substrates serve as a starting template for the growth of a rich variety of new virtual semiconductor substrates.

ACKNOWLEDGEMENTS

We gratefully acknowledge research support for this work under a grant from the Army Research Office, ARO DAAG55-97-1-0111, and partial support under NSF DMR-940034. We would also like to acknowledge the DOD fellowship program and Steve Stockman and Glen Carey of Hewlett-Packard Optoelectronics for donation of GaP and GaAsP wafers.

REFERENCES

Abrahams M S, Buiocchi C J, Olsen G H 1975 J. Appl. Phys. **46**, 4259
Bulsara M T, Leitz C W and Fitzgerald E A 1998 Appl. Phys. Lett. **72**, 1608.
Casey Jr H C and Panish M B 1978 Heterostructure Lasers: Part B (Boston: Academic Press) p. 52
Currie M T, Samavedam S B, Langdo T A, Leitz C W, Fitzgerald E A 1998 Appl. Phys. Lett. **72**, 1718
Fitzgerald E A, Xie Y-H, Monroe D, Silvennan P J, Kuo J-M, Kortan A R, Thiel F A, Weir B E and Feldman L C 1992 J. Vac. Sci. Tech. B **10**, 1807

Fitzgerald E A and Samavedam S B 1997 Thin Solid Films **294**, 3108

Fitzgerald E A, Kim A Y, Currie M T, Langdo T A and Taraschi G 1999 Mater. Sci. and Eng. B, Proceedings from the Lawrence Epitaxy Conference, Mesa, AZ, to be published

Grider D E, Swirhun S E, Narum D H, Akiwande A I, Nohava T E, Stuart W R, Joslyn P and Hsieh K C 1990 J. Vac. Sci. Tech. B **8**, 301

Houghton D C 1991 J. Appl. Phys. **70**, 2136

Ismail K, Arafa M, Saenger K L, Chu J O and Meyerson B S 1995 Appl. Phys. Lett. **66**, 1077

Kim A Y and Fitzgerald E A 1998 MRS Proceedings **510**, 131

Kim A Y, McCullough W S and Fitzgerald E A 1999a J. Vac. Sci. Tech. B, July

Kim A Y and Fitzgerald E A 1999b SPIE Proceedings 3621, **179**

Matthews J W 1975 Epitaxial Growth Part B (Boston: Academic Press)

Mii Y J, Xie Y H, Fitzgerald E A, Monroe D, Thiel F A, Weir B E and Feldman L C 1991 Appl. Phys. Lett. **59**, 1611

Samavedam S B and E A Fitzgerald 1997 J. Appl. Phys. **81**, 3108

Schaffler F, Tobben D, Herzog H J, Abstreiter G and Hollander B 1992 Semi. Sci. Tech. **7**, 260

Sieg R M, Carlin J A, Boeckl J J, Ringel S A, Currie M T, Ting S M, Langdo T A, Tarasci G, Fitzgerald E A and Keyes B M 1998 Appl. Phys. Lett. **23**, 3111

Sugii N, Nakagawa K, Kimura Y, Yamaguchi S and Miyao M 1998 Semicond. Sci. Technol. **13** A140

Inst. Phys. Conf. Ser. No 164
Paper presented at Microsc. Semicond. Mater. Conf., Oxford, 22–25 March 1999
© 1999 IOP Publishing Ltd

Climb/glide dislocation sources at low misfit $Ge_x Si_{1-x}$/Si(001) interfaces

S Jiao[+], P B Hirsch[*] and D D Perovic[+]

[+] Department of Metallurgy and Materials Science, University of Toronto, Toronto M5S 3E4, Canada
[*] Department of Materials, University of Oxford, Parks Road, Oxford OX1 3PH, U.K.

ABSTRACT: Detailed analysis of Burgers vectors and line directions has been carried out on the "double half loops", formed during growth, and other small loops, formed during growth and on annealing, observed by Perovic and Houghton (1992) in low misfit $Ge_x Si_{1-x}$ / (001) Si heterostructures. Both defect types originate from small vacancy loops, near the interface, the former visibly associated with precipitates > 10 nm in size; both have to transform into glissile configurations by stress driven vacancy absorption and glide before generating misfit dislocations. This corresponds to the nucleation stage for misfit dislocation generation (Houghton 1991).

1. INTRODUCTION

Houghton (1991) measured the rate of nucleation of misfit dislocations in low misfit $Ge_x Si_{1-x}$ (0.03 < x < 0.25) / (001) Si heterostructures directly by Nomarski interference microscopy. The activation energy was found to be 2.5 ± 0.2 ev, independent of the method of growth (MBE, RTCVD, UHVCVD), and was attributed in 1995 to a vacancy controlled mechanism of nucleation of vacancy type prismatic loops at Ge rich platelets at the strained layer interface (Perovic and Houghton 1995). Reduction of nucleation rates during rapid thermal anneals after Si irradiation at room temperature of the as-grown samples have provided support for this suggestion (Stirpe et al 1997).

TEM studies identifed two possible types of sources: "Double half loops" and simple shear loops (Perovic and Houghton 1992, 1995), which were thought to be formed from the prismatic loops nucleated around the interfacial Ge rich precipitates. Hirsch (1997) suggested that a likely configuration of the prismatic loops is normal to the interface and platelet plane, and proposed a dissociation mechanism for the formation of the "double half loops". An important conclusion was that for a low misfit epitaxial layer, say $Ge_{0.1} Si_{0.9}$, the size of the precipitate parallel to the interface would have to exceed ~ 10 nm in order for the loops to be able to expand by glide to the surface. A more detailed characterisation of the "double half loops" and of the single loops has now been carried out with a view to elucidating the nucleation mechanism. The results are described in this paper.

2. EXPERIMENTAL

$Ge_x Si_{1-x}$ / (001) Si heterostructures of low misfit strain (< 2 %) were grown by MBE as described by Houghton (1991). Specimens included single layers of $Ge_x Si_{1-x}$ about 100 nm thick, and multilayers of 5 nm GeSi and 20 nm Si, total thickness 400 nm, with a 700 nm Si capping layer (SLS specimens). The work reported in this paper was carried out on the SLS specimens, grown at 350 – 500 °C. Some samples were annealed for 30 s at 750 °C or 800 °C. Specimens were examined in plan view, i.e. parallel to (001); others in cross-section parallel to ($\bar{1}$10). Samples were prepared by mechanical grinding followed by ion-beam milling, for the (001) specimens on the Si substrate side. Electron microscopy was carried out on a Hitachi H800 TEM, operating at 200 Kv.

3. RESULTS

3.1 Double half loop (X and Y) defects

Figs 1a, b are weakbeam images of two "double half loop" defects, consisting of 4 (X defect) and 3 (Y defect) dislocations respectively, in an unannealed (001) SLS specimen. (Variants rotated 90° about [001] are also observed.) Burgers vector analysis was carried out by imaging in $\underline{g}$ = 220, 2$\bar{2}$0, 040, 400, 20$\bar{2}$, 022. The Burgers vectors of the various dislocations are indicated on the figures. Contrary to the preliminary results published previously (Perovic and Houghton 1992), all the 4 dislocations in the X defect have different Burgers vectors. By tilting about [110] the planes containing AB, CD and BE, CF for the X defect in Fig. 1a were found to be approximately (1$\bar{1}$1) and ($\bar{1}$11) respectively. The line directions of these dislocations are approximately [$\bar{2}$13], [$\bar{1}$23], [1$\bar{2}$3] and [2$\bar{1}$3] respectively, in each case about 19° away from the screw direction, tilted towards the (110) plane. Similar analysis for the Y defect in Fig. 1b shows the Burgers vectors to be as shown in the figure. PG is an edge dislocation with $\underline{b}$ = $\frac{1}{2}$[110], lying on (1$\bar{1}$1); its line direction is [$\bar{1}$12].

Under some diffraction conditions, particles 20 – 50 nm in diameter can be observed at the apex of the X and Y defects; an example is P in Fig. 1a. Under other imaging conditions the contrast due to the local dislocation configuration predominates. In Fig. 2a the X defect of Fig. 1a is imaged in $\underline{g}$ = 040; this shows that AB and DC in Fig. 1a actually originate from point S in Fig. 2a, and EB, FC from R, and AS, DS and ER, FR cross in the (001) projection. RS is an edge dislocation with $\underline{b}$ = $\frac{1}{2}$[110], confirmed by imaging with $\underline{g}$ = 400. The detailed geometry of the dislocations near the apex varies from defect to defect, but sometimes RS and the crossing-over can be clearly seen as in the $\underline{g}$ = $\bar{2}$ $\bar{2}$ 0 image in Fig. 2b, from an annealed SLS sample. This crossing-over is also apparent for the $\bar{Y}$ defect in Fig. 1b. The X and Y defects observed in annealed SLS specimens are very similar to those in unannealed specimens.

In order to elucidate the three-dimensional structure of the X and Y defects, annealed ($\bar{1}$10) cross-section SLS specimens were studied. Fig. 3a shows an X defect and two single loop defects imaged with $\underline{g}$ = 220 in such a specimen. This and similar images from other X defects show that one pair of dislocations originates at the SLS/substrate interface, and the other some distance above it. The connecting edge dislocation RS has $\underline{b}$ = $\frac{1}{2}$[110] similar to RS in Figs 2a, b, but its length is parallel to [001] in this projection. The points marked B', C' in Fig. 3a correspond very probably to the extremities of the line through B, C along [110] in Fig. 1a, i.e. that line is not along the interface, but is in the form of a V in the ($\bar{1}$10) plane.

The cross-section samples also show images of X and Y defect variants rotated through 90° about [001] with respect to those shown in Figs 1a, b. Some of these show only two arms, e.g. corresponding to AB, BE in Fig. 1a, the other two arms having been lost through thinning. These defects confirm the conclusions reached from the other sections; in particular that AB, BE lie on the (1$\bar{1}$1) and ($\bar{1}$11) planes respectively, and that B'RC' in Fig. 3a lies approximately on ($\bar{1}$10).

Figs 4b, c are models of the X and Y defects which encapsulate the results of the contrast experiments on the plan view and cross-section specimens. The nucleus of these defects appears to be a $\underline{b}$ = $\frac{1}{2}$[110] prismatic vacancy loop formed during growth, to relieve the misfit around an interstitial type precipitate at the substrate/SLS interface (Fig. 4a). In the stress field of the precipitate, the prismatic loop dissociates at R and S according to

$$\tfrac{1}{2}[110] = \tfrac{1}{2}[10\bar{1}] + \tfrac{1}{2}[011] \text{ and } \tfrac{1}{2}[110] = \tfrac{1}{2}[101] + \tfrac{1}{2}[01\bar{1}] \tag{1}$$

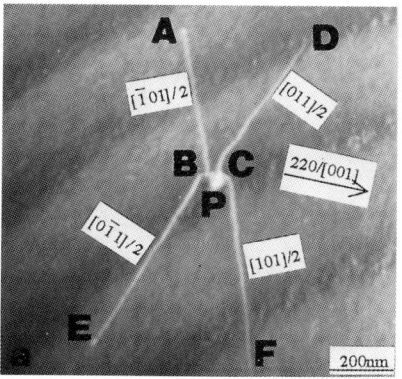
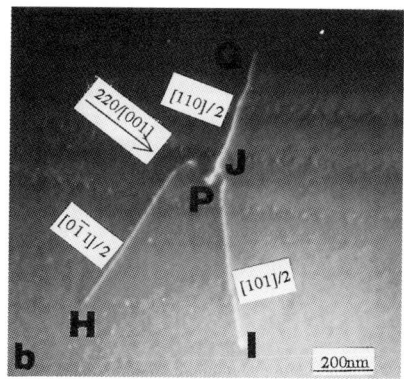

Fig.1 Plan view weak beam images of a) X and b) Y defects in an unannealed capped SLS Ge$_x$ Si$_{1-x}$ / (001) Si specimen, taken with $\underline{g}$ = 220.

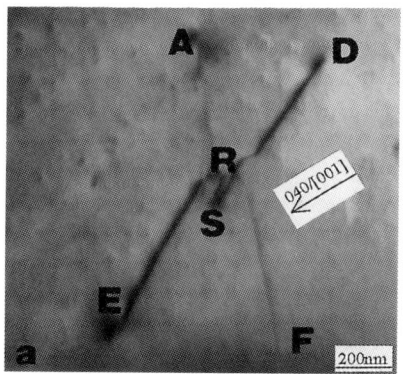
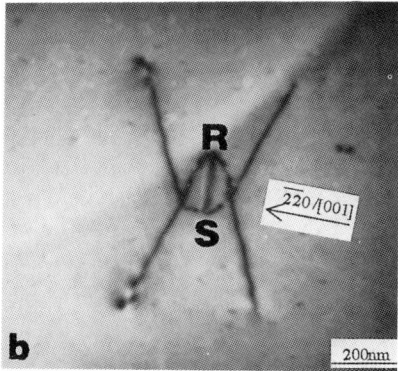

Fig.2 a) Same defect as in Fig. 1a, but with $\underline{g}$ = 040;
b) X defect in an annealed SLS sample ($\underline{g}$ = $\overline{2}$ $\overline{2}$ 0).

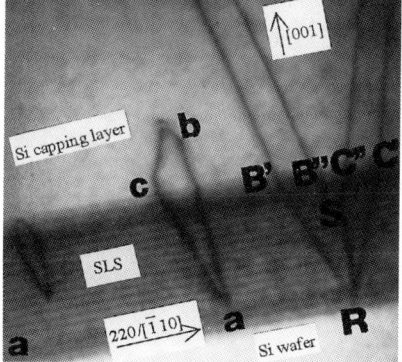
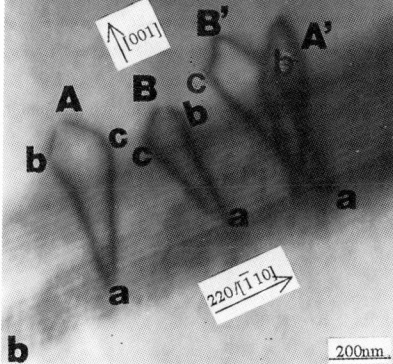

Fig.3 a) X defect and 2 simple loop defects and b) 4 simple loop defects in a ($\overline{1}$ 10) section of an annealed, capped SLS specimen, with $\underline{g}$ = 220.

respectively. In the SLS region the dissociation occurs on ($\bar{1}$ 10) by climb and glide, but in the capping layer the dislocations fan out on the (1 $\bar{1}$ 1) and ($\bar{1}$ 11) glide planes respectively. The position of S in Fig. 4b varies somewhat from defect to defect, as is clear from Figs 2a, b. B'RC' and B"SC" are parallel to ($\bar{1}$ 10), while EB', FC' and AB", DC" lie in the ($\bar{1}$ 11) and (1 $\bar{1}$ 1) planes respectively. B'C', B"C" are at the top of the SLS layer; R is at the SLS/ substrate interface. The Y defect shown in Fig. 4c forms when the dissociation at S does not occur.

The directions of the dislocations in the capping layer, e.g. EB', FC', are ~ 19° away from the screw directions in the ($\bar{1}$ 11) glide plane. This corresponds to the minimum energy direction; deviation towards an edge orientation between the two dislocations shortens their length, but increases the energy per unit length of dislocation line. It is easy to show that for a Poisson's ratio of 0.215 for Si, the deviation should be 20°, in excellent agreement with the observations.

3.2 Simple loop defects

Fig. 5 shows two small loops in an annealed (001) SLS specimen, imaged with $\underline{g}$ = 220. Loop A vanishes with $\underline{g}$ = 400, and loop B with $\underline{g}$ = 040. This implies that A could have $\underline{b}$ = $\frac{1}{2}$ [011] or $\frac{1}{2}$ [0$\bar{1}$ 1], and $\bar{B}$, $\underline{b}$ = $\frac{1}{2}$ [101] or $\frac{1}{2}$ [$\bar{1}$ 01]. Because of the thickness of the specimens it was not possible to obtain images in other reflections to distinguish between these possibilities. Instead similar loops were identified by their corresponding shapes in ($\bar{1}$ 10) cross-section specimens. To do this it was noted that the directions ab and ac on Fig. 5 are approximately parallel to the[$\bar{1}$ 10] and [110] directions respectively. Also intensity oscillations suggest that ab and ac are inclined to (001).

Fig. 3b shows 4 loops in an annealed ($\bar{1}$ 10) cross-section SLS specimen, imaged with $\underline{g}$ = 220; two other similar loops are shown in Fig. 3a. Most, but not all, of the loops appear to be nucleated at the Si wafer SLS interface. They all have similar shapes with ab lying in the (110) plane, and ac at about 20° away from [001] towards [110] in the ($\bar{1}$ 10) projection. These directions for defect A in Fig. 3b are consistent with ab, ac in the (001) projection for defect A in Fig. 5, assuming that ab for defect A in Fig. 3b is tilted towards [$\bar{1}$ 10]. This was confirmed by tilting the specimen with defect A in Fig. 3b about the [110] axis; thus the images of defect A in Figs 3b and 5 come from the same type of defect. By imaging defect A in Fig. 3b with $\underline{g}$ = 220, 004, $\bar{2}$ $\bar{2}$ 2 and 222, the Burgers vector $\underline{b}$ was found to be either $\frac{1}{2}$ [101] or $\frac{1}{2}$ [011], but for consistency with the Burgers vector analysis for A in Fig. 5a, $\underline{b}$ = $\frac{1}{2}$ [011].

Thus the defects imaged in plan view in Fig. 5 have steeply inclined sides ab, ac originating from an apex a, usually at the wafer SLS interface, and at about 20° to [001] in the (110) and ($\bar{1}$ 10) planes; this is illustrated in Fig. 6a. There are 4 variants of this shown in Fig. 6b in (001) projection; A and B in Fig. 5 correspond to two of these; three different types appear in Fig. 3b; ab for B and B' are inclined in opposite directions to the ($\bar{1}$ 10) plane. The actual nucleus of the loop at a is not known, but the resultant shape is consistent with it being a small prismatic vacancy loop, with $\underline{b}$ one of $\frac{1}{2}$ [011], $\frac{1}{2}$ [101], $\frac{1}{2}$ [0$\bar{1}$ 1], $\frac{1}{2}$ [$\bar{1}$ 01], around a small interstitial precipitate. The misfit stress then causes the part of the loop away from the interface to glide along its glide cylinder and expand by cross-slip and climb to form the observed defect (Fig. 6a). Unlike the X and Y growth defects described in section 3.1, the simple loop defects appear to be formed on annealing as well as during growth. They do not penetrate the capping layer in any specimens.

4. DISCUSSION

The experiments show that X and Y are growth defects, nucleated from a prismatic vacancy loop normal to the interface, and lying around a large precipitate (> 10 nm). Although this step and the conditions for nucleation are as expected (Hirsch 1997), the dissociation turns out to be different. So far it has not been possible to identify the precipitates. The prismatic loops generated during growth

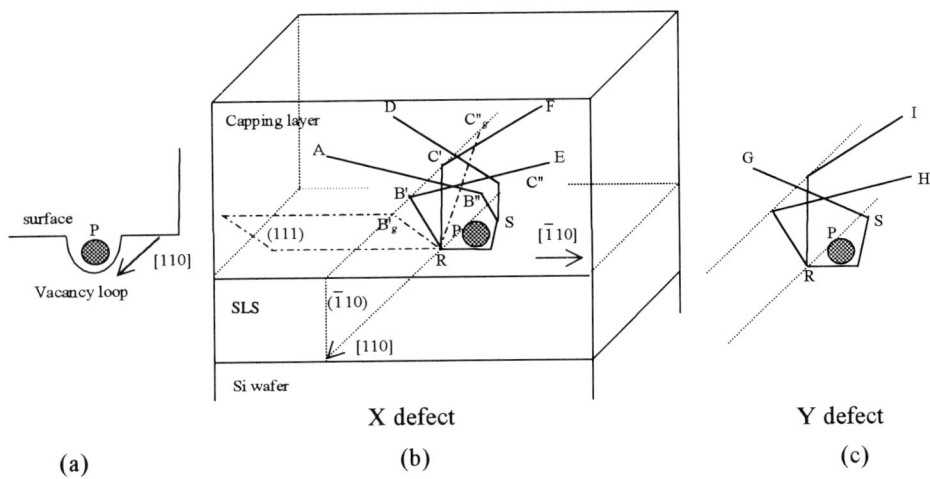

Fig. 4 a) Initial vacancy loop nucleated around precipitate P during growth, and models of resulting a) X and b) Y defects.

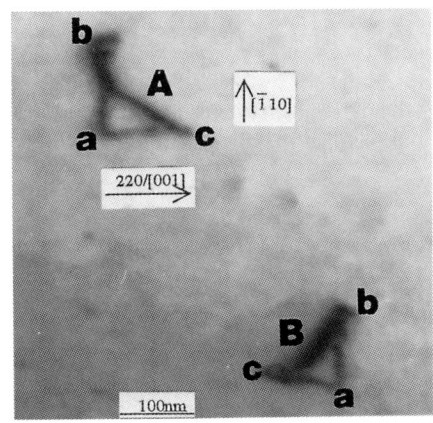

Fig. 5 Two small loops in an annealed, capped SLS specimen viewed along [001], with $\underline{g}$ = 220.

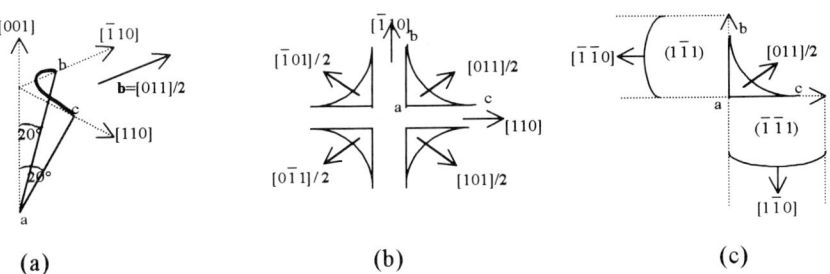

Fig. 6 a) Model of small loop defect; b) 4 variants viewed along [001]; c) rotation of ab, ac into $(1\bar{1}1)$ and $(\bar{1}\,\bar{1}1)$ planes when glide can occur.

and on annealing may also have been nucleated from small vacancy loops, but no evidence for such precursor loops, or for possible precipitates associated with them has been found so far.

Although all the detailed TEM has been carried out on SLS specimens, similar defects have been observed previously in single layers of $Ge_x Si_{1-x}$, both capped and uncapped. Fig. 3b in Perovic and Houghton (1995) shows X-type and 3 variants of the prismatic loops in an annealed specimen of a capped single layer of $Ge_x Si_{1-x}$, and Fig. 4a of Perovic and Houghton (1992) shows examples of X and Y defects in an uncapped single layer of $Ge_x Si_{1-x}$.

It is likely therefore that these defects contribute to or control the nucleation rates in all these types of specimens (Houghton 1991). The effective stress acting on any glide dislocation threading through the misfit layers is positive, so why are the observed defects apparently stable? The answer presumably lies in their detailed structure. For the X and Y defects the dislocations in the capped SLS lie approximately in the $(\bar{1} 10)$ plane for a defect originating from a $\underline{b} = \frac{1}{2}[110]$ prismatic loop (see Fig. 4b). Under the influence of the stress these dislocations can under suitable conditions rotate by vacancy absorption and glide until they lie in the $\{111\}$ glide planes, indicated as B'_g, C'_g in Fig. 4b, whereupon they will glide and generate misfit dislocations at the interfaces.

Ideally, for a SLS with a thick capping layer, if both $B'_g R$ and $C'_g R$ glide to the left in Fig. 4b, two different 60° dislocations will be generated along the dashed lines at the top SLS interface, and a pure edge dislocation or two closely spaced 60° dislocations at the bottom interface. The effective stress in this case would be rather larger than predicted by the Houghton (1991) expressions, which apply if only one of the two dislocations were to move. Related mechanisms will operate for the single $Ge_x Si_{1-x}$ layer specimens. Under certain conditions these defects can act as regenerative sources. As to the prismatic loops, the dislocations ab, ac also have to rotate by vacancy absorption and glide until they lie in the glide planes, whereupon they can form 60° interface dislocations by glide along two orthogonal <110> directions at the interface, indicated in the (001) projection as dashed lines in Fig. 6c.

Thus climb by vacancy absorption and glide of mixed dislocations are controlling the rate at which the X and Y defects and the prismatic loop defects turn into configurations which can glide and generate misfit dislocations. But for the prismatic loop defects climb and glide processes also control the initial nucleation and growth for those formed on annealing. This supports therefore the suggestion made by Perovic and Houghton (1995) that the activation energy of 2.5 ± 0.2 ev for the rate of nucleation of misfit dislocations is determined by a mechanism involving vacancies, and explains the effects observed by Stirpe et al (1997).

Finally it should be noted that these defects are more complex than conventionally assumed in calculations of effective stress or critical thickness. There has to be a positive driving force to transform the initial defects into glissile configurations, and the critical thickness becomes defect specific.

REFERENCES

Hirsch P B 1997 Inst. Phys. Conf. Ser. **157**, 121
Houghton D C 1991 J. Appl. Phys. **70**, 2136
Perovic D D and Houghton D C 1992 Mat. Res. Soc. Symp. Proc. **263**, 391
Perovic D D and Houghton D C 1995 Inst. Phys. Conf. Ser. **146**, 117
Stirpe M B, Perovic D D, Lafontaine H L and Goldberg R D 1997 Inst. Phys. Conf. Ser. **157**, 127

Inst. Phys. Conf. Ser. No 164
Paper presented at Microsc. Semicond. Mater. Conf., Oxford, 22–25 March 1999
© *1999 IOP Publishing Ltd*

Dislocation blocking in strained layers grown on vicinal substrates

P J Goodhew

Materials Science & Engineering, University of Liverpool, Liverpool L69 3GH

ABSTRACT: The blocking of gliding threading dislocations by perpendicular misfit dislocations laid down earlier has been reported to restrict the relaxation of strained layers. There is some evidence that this blocking is reduced in layers grown on vicinal substrates. The results of a simple computer model show that this effect (if true) cannot be accounted for simply by the different dislocation configurations generated following growth on a vicinal substrate.

1. INTRODUCTION

The development of arrays of misfit dislocations frequently controls the relaxation of heteroepitaxial semiconductor layers. Recently interest has focused on the dynamics of the relaxation process, since it is evident from many experimental results that relaxation by movement and/or multiplication of dislocations usually starts long after the critical layer thickness has been exceeded (e.g. Beanland et al 1996).

For layers grown on III-V substrates, with threading dislocation densities of at least 10^4 lines cm^{-2}, the first relaxation mechanism to operate is the movement by glide of threading dislocations (TDs) across the epilayer, laying down lengths of 60° misfit dislocation (MD) in the interface. The extent of "primary" relaxation is proportional to the total length of MD and if the strain is substantial this will be limited by exhaustion of TDs. Exhaustion will occur when all the TDs initially present have either slipped out of the epilayer at the edge of the wafer or have had their glide blocked by some means. For further relaxation to occur, new dislocation sources must be created but these secondary mechanisms are not considered here.

For layers grown on (001) substrates TDs can glide on one of two glide planes in each of two orthogonal <110> directions. Each gliding TD therefore encounters previously-laid-down MDs orthogonal to its path, which may have Burgers vectors parallel or non-parallel to it. Freund (1990) has analysed many of the possible interactions and has demonstrated that in general the gliding dislocation will be resisted by orthogonal MDs and that blocking of its passage may occur. In addition, closely spaced MDs of the same Burgers vector should act as stronger blocks. There seems to be plenty of evidence, from transmission electron microscope (TEM) images and surface striations (cross-hatch), that blocking does occur (e.g. Freund 1990; MacPherson and Goodhew 1997) and must reduce misfit relief by limiting the length of the MDs which can be laid down by gliding TDs. There is some evidence that the rate of strain relaxation is different when epilayers are grown on vicinal substrates (see Goodhew and Giannakopoulos (1999a) for a summary) and this is not clearly understood. Here we show some results of a computer model in which the development of an array of MDs is simulated, taking account of the effect of blocking. The details of the model are given elsewhere (Goodhew 1999).

2. MISFIT DISLOCATION ARRAYS

Primary MD arrays between III-V epilayers and exact (001) substrates consist of two orthogonal sets of essentially straight 60° dislocations along <110> (see e.g. Fig. 1). The arrays are not regularly spaced and the only detailed measurements of their spacing (in the InGaAs/GaAs system) have shown that they follow a Poisson distribution consistent with random location of the original TDs (MacPherson et al, 1995). When the epilayer is grown on a vicinal substrate with offcut angle θ, each <110> set of MDs becomes two almost-parallel sets with line directions at an angle of θ√2 (Kightley et al, 1991). This might have two effects on the blocking behaviour. Firstly the inevitable frequent crossing of MD lines should lead to accelerated interaction between nearly-parallel MDs to form segments of edge dislocation (Kightley et al, 1991) which might offer a different blocking strength from 60° MDs. Secondly the close proximity of MDs as they approach their intersections might provide a greater length of closely-spaced MD acting as stronger blocks than in the parallel arrays on exact substrates.

A statistical analysis shows that, if the MDs are Poisson distributed, then the double array on the vicinal substrate should offer almost the same blocking resistance to gliding TDs as does the single array on an exact (001) substrate (Goodhew 1999). It is more difficult to perform the statistical analysis when the MDs are not distributed at random and therefore a computer model has been constructed to verify the exact/vicinal comparison and to extend it to TDs which are not randomly distributed.

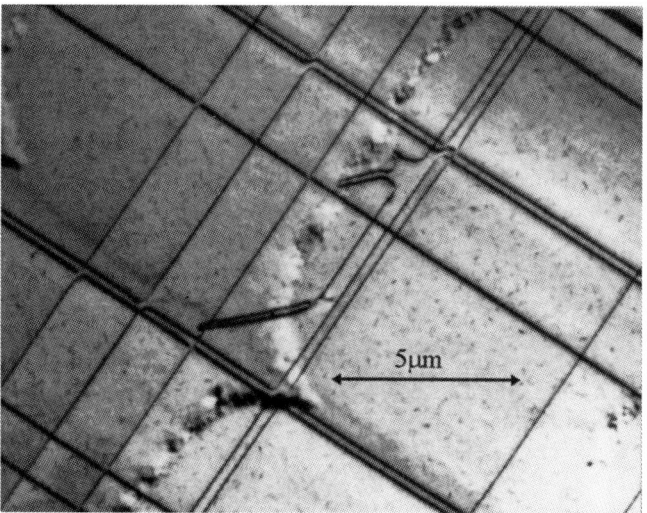

Figure 1. TEM image of orthogonal arrays of 60° misfit dislocations in the interface between an In$_{0.1}$Ga$_{0.9}$As layer and an exact (001) GaAs substrate.

3. A BLOCKING SIMULATION

The basis for the computer simulation is as follows:

1. Threading dislocations are assumed to exist at a fixed density per unit area and to begin to glide in random sequence. This models the situation where a substrate has a significant density

of TDs which begin to move as the layer thickness increases beyond the critical thickness.

2. Each TD moves from its initial position in one of eight directions corresponding to the two glide planes available in each of the <110>. On an exact (001) substrate these eight directions reduce to four. The direction for each dislocation is selected at random to simulate the choices available to the TDs, which will actually have four different Burgers vectors.

3. As the moving TD reaches each previously laid-down orthogonal MD (or nearly-orthogonal MD for the vicinal case) it has a probability B_1 of being blocked by the single dislocation and a different probability B_2 of being blocked by the next two MDs if they are spaced closer than a selected distance x. If it is blocked the TD moves no further, but if it not blocked its next intersection with an orthogonal MD is examined and so on until the edge of the defined area is reached. In the work reported here the blocking probabilities B_1 and B_2 have been set to 0.2, recognising that blocking is not certain to occur even when (in about 1 in 4 cases) the Burgers vector combination is appropriate.

4. When all TDs have been allowed to glide, the number of blocks of each type and the average MD segment length are calculated.

Some arrays produced by the model are shown in Fig. 2. The main conclusion, drawn from many runs of the simulation, is that there is no difference between the blocking behaviour of exact and vicinal MD arrays. The total number of MDs is the same in the two cases, while the increase in closely-spaced dislocations resulting from their crossings is almost exactly balanced by the increased average spacing in each array, as predicted by the statistical analysis. Additionally the simulation reveals that there is no statistically significant difference between the blocking behaviour in random arrays (e.g. Fig. 2c) and extremely non-random arrays (e.g. Fig 2d, in which only two groups of TD sources operate). This is an important conclusion because the distribution of TDs in real substrates is usually far from random.

A key experimental result from etch pit studies of TD densities in partially relaxed InGaAs/GaAs layers is that the TD density in vicinal layers is almost an order of magnitude smaller than that in exact layers (Goodhew and Giannakopoulos 1999b). It is tempting to explain this in terms of easier (or more extensive) glide of TDs on the vicinal layers, allowing more TDs to be eliminated at the edge of the wafer. However the results of the present model run counter to this analysis and the reason for the apparently different relaxation behaviour of vicinal layers remains to be explained.

ACKNOWLEDGEMENTS

The author would like to acknowledge useful discussions with Drs Glyn MacPherson and Kostas Giannakopoulos and advice on the statistical analysis from Professor Raj Bhansali and Dr Paula Williamson.

REFERENCES

Beanland R, Dunstan D J and Goodhew P J 1996 Adv. Physics **45**, 87
Freund L B 1990 J. Appl. Phys. **68**, 2073
Goodhew P J 1999 to be published
Goodhew P J and Giannakopoulos K 1999a Micron **30**, 59
Goodhew P J and Giannakopoulos K 1999b TMS Proc., in press

278

Kightley P, Goodhew P J, Bradley R R and Augustus P D 1991 J. Crystal Growth **112**, 359
Macpherson G, Goodhew P J and Beanland R 1995 Phil. Mag. **72**, 1531
Macpherson G and Goodhew P J 1997 Appl. Phys. Lett. **70**, 2873
MacPherson G, Goodhew P J and Beanland R 1995 Phil Mag **A72**, 1531

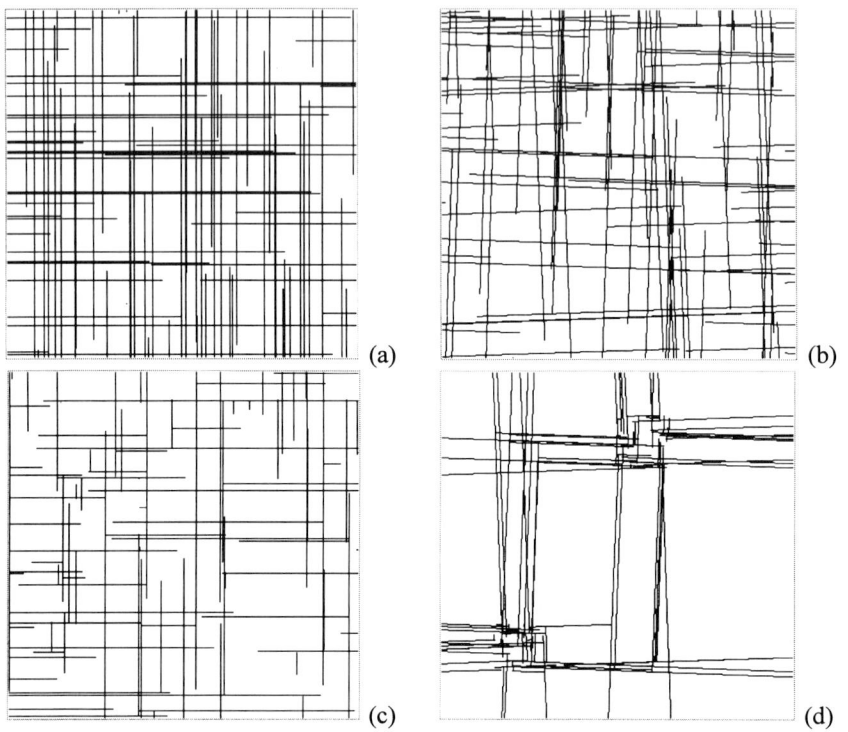

Figure 2. Simulations of MD arrays. (a) Random origins on exact substrate with no blocking; (b) random origins on a 2° offcut substrate with no blocking; (c) random origins on an exact substrate with blocking; (d) grouped origins on a 2° offcut substrate with blocking.

Inst. Phys. Conf. Ser. No 164
Paper presented at Microsc. Semicond. Mater. Conf., Oxford, 22–25 March 1999

Mechanisms of the formation of structural defects during low temperature growth of GaAs

M Luysberg, P Specht* and E R Weber*

Institut für Festkörperforschung, Forschungszentrum Jülich, 52425 Jülich, Germany
* Department of Materials Science, University of California, Berkeley, CA 94720

ABSTRACT: GaAs grown by MBE at low temperatures (LT-GaAs) is highly non-stoichiometric with an excess in As of up to 1.5%. At a certain thickness depending on the growth parameters a change in growth mode from perfect epitaxial growth to columnar, polycrystalline growth is observed. The breakdown of single-crystallinity is governed by the formation of "pyramidal" defects consisting of microtwins, stacking faults and dislocations. The nucleation of the defects is found to be clearly correlated with hillocks on the growing surface. Furthermore, an anisotropy of the shape of the hillocks and of the type of defects is observed. The mechanisms leading to the surface roughening and the subsequent breakdown of the epitaxial growth are discussed.

1. INTRODUCTION

The unique properties of GaAs grown by MBE at low temperatures (LT-GaAs) make this material suitable for device applications like terahertz photodetectors or radiation-hardened satellite communication devices. The ultrashort carrier lifetimes in as-grown mayerial and the high resistivity in annealed samples can be attributed to the incorporation of point defects during growth. The layers contain excess As mostly in the form of As antisite defects, As_{Ga} (Liu et al 1994). The larger bond length of the As-As_{Ga} compared to As-Ga bond induces a dilation, which leads to a mismatch of up to 0.16% between the LT-GaAs layer and the GaAs substrate (Lilental-Weber at al 1994). Despite the low growth temperatures, the as-grown LT-GaAs layers are free of structural defects such as dislocations or precipitates. However, the perfect epitaxial growth extends only up to a certain thickness, where a change in growth mode is observed. As pointed out earlier (Lilental-Weber 1992) this epitaxial thickness sensitively depends on the growth parameters. In this paper we investigate the mechanisms leading to the breakdown of epitaxial growth.

2. EXPERIMENT

Undoped and Be-doped (10^{20} cm^{-3}) LT-GaAs layers were grown by molecular beam epitaxy on (100) GaAs wafers (for details see Luysberg et al (1998)). The growth temperature measured by diffuse reflectance spectroscopy was varied from 155°C to 200°C. Different beam equivalent pressure (BEP) ratios from 11 to 40 were used. In one case, LT-GaAs layers (thickness: 50 nm) and LT-AlAs layers (thickness: 1.7 nm) were grown alternately on top of a 600 nm LT-GaAs layer. These AlAs marker layers enable the investigation of the surface morphology by cross sectional transmission electron microscopy (TEM). The structural properties were investigated by JEOL

4000EX and JEOL 4000FX transmission electron microscopes operated at 400 kV. Samples were prepared in cross section by conventional techniques including Ar ion milling.

3. RESULTS AND DISCUSSION

The breakdown of epitaxial growth is governed by the formation of "pyramidal" defects consisting of a high density of microtwins, stacking faults and dislocations, which will be discussed in detail below. Figure 1 summarizes the epitaxial thickness D_{epi} to which the layers can be grown single-crystalline for different growth conditions. We define the epitaxial thickness as the distance from the substrate/epilayer interface where the single-crystalline volume fraction is 50%. D_{epi} increases with increasing growth temperature and decreasing As/Ga flux ratio as measured by the BEP ratio. In particular at low growth temperatures (155°C) the epitaxial thickness varies within a factor of 2, which may be indicative

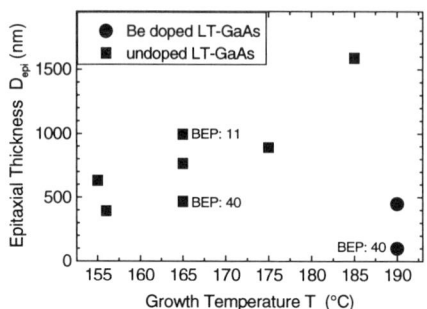

Figure 1: Epitaxial thickness plotted for the different growth conditions. All samples were grown at a BEP ratio of 20 unless indicated differently.

of local inhomogeneities and/or different impurity levels. In case of Be doping, a considerably smaller epitaxial thickness is observed. Again, a larger thickness is found for smaller BEP ratios. The lowering of the growth temperature and/or the increase of the BEP ratio results in a higher As_{Ga} concentration incorporated during growth (Luysberg et al 1998). The increase of the non-stoichiometry may contribute to the instability of the perfect single crystalline growth (see below).

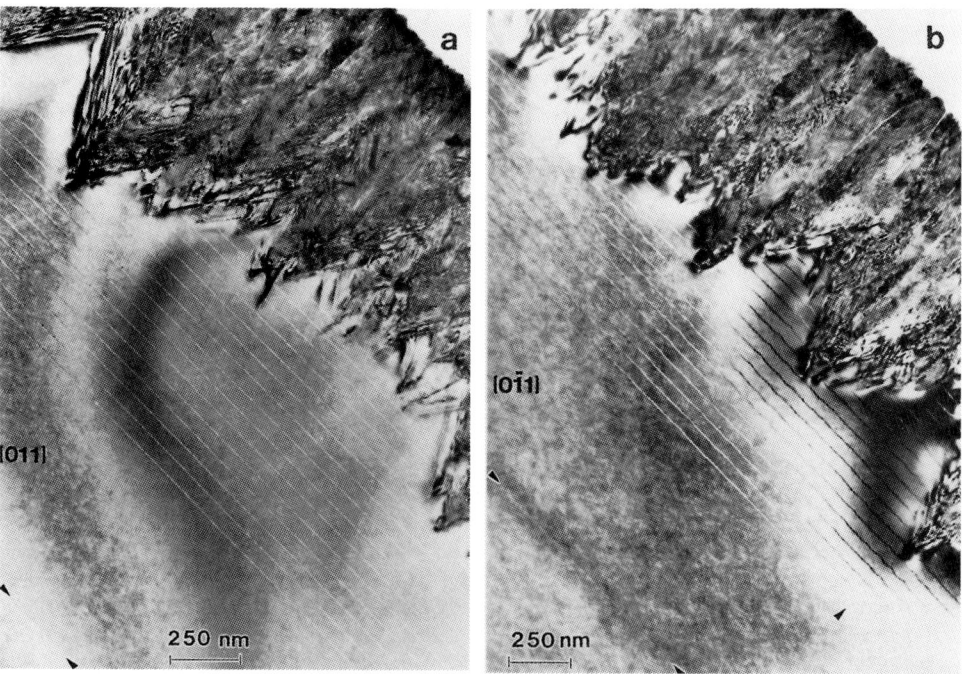

Figure 2: Bright field electron micrographs obtained with (004) two beam conditions (s>0). The cross sections in the (011) orientation (a) and (0-11) orientation (b) reveal the transition from epitaxial to polycrystalline growth. AlAs marker layers appear as bright lines. The substrate/epilayer interface is marked by arrows.

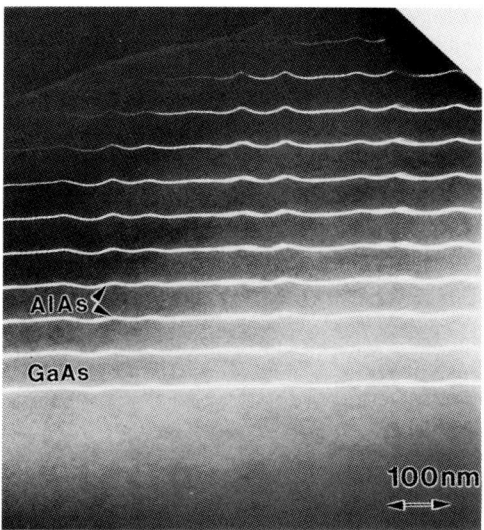

Figure 3: Dark field electron micrographs obtained with (002) reflection. AlAs layers appear bright.

The formation of defects leading to the breakdown of single-crystalline growth is illustrated in the bright field TEM images of a LT-GaAs sample grown at 200°C and a BEP ratio of 20 (Fig. 2). The sample is shown in two different {110} orientations. A defect-free, epitaxial layer extends from the GaAs substrate (marked by arrows) up to a thickness of about 1.3 μm. The AlAs marker layers reveal an anisotropic morphology of the growing surface, i.e. hillocks elongated along the [0-11] direction. In the (011) orientation the AlAs layers are perfectly flat and perpendicular to the [100] growth direction (Fig. 2a). In contrast, the (0-11) projection reveals increasing surface roughness with increasing thickness (Fig. 2b), which is seen more clearly in the (002) dark field image showing the AlAs marker layers as bright lines (Fig. 3). The high resolution micrograph (Fig. 4b) demonstrates that the defects nucleate at the facets of the hillocks, as soon as the hillocks have reached a height of about 10 nm.

Besides the surface morphology, the defect structure is also observed to be anisotropic. In both {110} projections planar defects like twins and stacking faults as well as dislocations are observed. However, the density of primary twins is much higher in the (011) orientation, where the planar defects lie on As terminated {111} planes. By contrast the (0-11) orientation reveals predominantly dense dislocation networks. This anisotropy is demonstrated in Fig 4. The contrast in Fig 4b can be identified as Moiré fringes, which are caused by the superposition of twinned areas (Neumann et al. 1994). The core of a pyramidal defect seen in Fig. 4a is surrounded by twin lamellae (T) and stacking faults (SF). As two primary twins meet, the formation of higher order twins occurs (marked by arrow). The role of twinning in low temperature growth of GaAs was pointed out by Claverie et al (1993), who did not report, however, on any anisotropy of the defect formation.

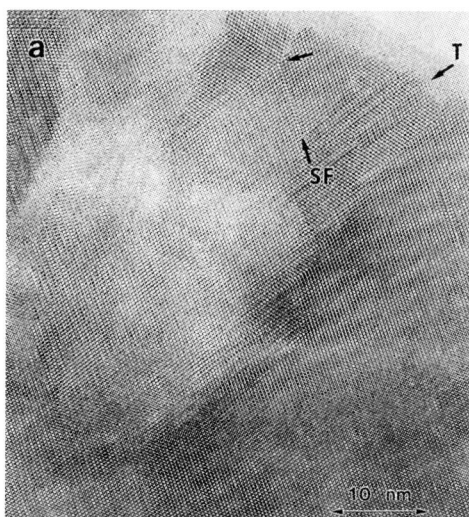
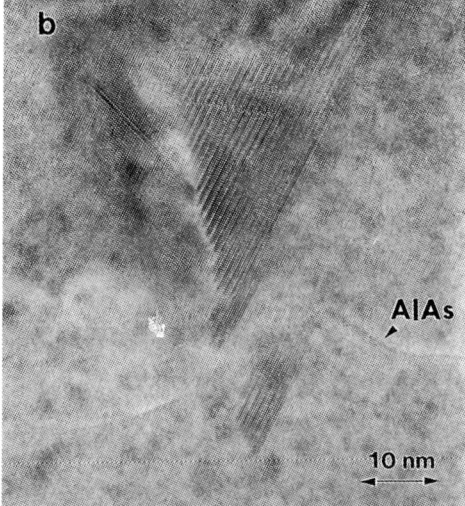

Figure 4: High resolution electron micrographs in the (011) orientation (a) and (0-11) orientation (b) reveal the anisotropy of the "pyramidal" defects.

The pyramidal defects act as nuclei for the subsequent columnar, polycrystalline growth. The formation of higher order twins, particular in the later stages of the growth turns the LT-GaAs layer more and more polycrystalline, i.e. with grains of random orientation. The individual columns of a typical width of about 50 nm consist of heavily twinned crystalline grains. It has to be stressed, that the polycrystalline layer as well as the cores of the pyramidal defects are free of any As precipitation, which we concluded from the lack of any evidence in the high resolution micrographs and in EDX analyses. This is in contrast to previous studies (Liliental-Weber et al 1992).

The results of the structural investigation clearly demonstrate the influence of surface roughening on the breakdown of epitaxy. Similar results were reported on the transition from epitaxial to amorphous growth of Si and GaAs (Eaglesham 1995). In the case of low-temperature epitaxy thermally activated processes with high activation energies are suppressed. Therefore, relaxation mechanisms, like the formation of misfit dislocations, are not the dominant processes. The same argument applies to stress induced surface diffusion, which has been evoked to explain surface roughening. It has to be pointed out that the formation of surface undulations and of the defects is not induced by a global strain. This can be concluded from the fact, that the Be doped samples, which have a nearly vanishing lattice mismatch with respect to the substrate (Luysberg et al 1999), show even smaller epitaxial thickness than its undoped counterparts with a mismatch of about 0.16% (see Fig. 1). Therefore, kinetic roughening, i.e. statistical fluctuations during growth, has to be considered as dominant mechanism of the buildup of surface undulations. However, according to Eaglesham (1995) this effect may not be intrinsic, but induced by impurities like oxygen or, in our doped samples, by Be. Recently, simulations of the kinetic roughening has been successfully applied to describe atomic force microscopy images of LT-GaAs samples (Apostolopoulos 1999). The anisotropy was explained by anisotropic surface diffusion. On the other hand the role of excess As incorporation may have to be considered as well. The formation of As_{Ga} defects accounting for the non-stoichiometry imposes locally a dilation of the bonds. Therefore, the formation energy of As_{Ga} defects may be considerably lower on top of the hillocks, where the bonds are stretched. Although there is no evidence of inhomogeneous As_{Ga} incorporation on a macroscopic scale (Liliental-Weber 1992, Luysberg et al 1998), microscopically the incorporation can be favored on top of the hillocks, which can gradually increase the buildup of surface undulations.

ACKNOWLEDGMENTS

This work was supported by the AFOSR grant No. F49620-95-1-0091. MBE growth was performed in the NSF supported Integrated Materials Laboratory of UC Berkeley.

REFERENCES

Apostolopoulos G, Herfort J, Däweritz L, and Ploog, K H 1999 Proccedings og the XII Panhellis Solid State Conf, Ioamina, 1998, accepted

Claverie, A, Liliental-Weber Z 1993, Mat. Sci Eng, **B22**, 45

Eaglesham D J 1995, J. Appl. Phys. **77**, 3597

Liliental-Weber Z 1992 Mat. Res. Soc. Symp. Proc. **241**, 101

Liliental-Weber Z, Ager J, Look D, Lin X W, Liu X, Nishio J, Nichols K, Schaff W, Swider W, Wang W, Washburn J, Weber E R , and Whitaker J 1994, in: *Proc. of the 8th Conf. on Semiinsulating III-V Materials*, ed. M Godlewski (World Scientific) p. 305

Liu X, Prasad A, Chen W M , Kurpiewski A, Liliental-Weber Z, and Weber E R 1994, Appl. Phys. Lett. **65**, 3002

Luysberg M, Sohn H, Prasad A, Specht P, Liliental-Weber Z, Weber E R , Gebauer J, and Krause-Rehberg R 1998, J. Appl. Phys. **83** 561

Luysberg M, Specht P, Thul K, Liliental-Weber Z, Weber E R 1999 Proc. of the SIMC X, Berkeley, CA (1999), will appear

Neumann W, Hofmeister H, and Heydenreich J 1994, phys. stat sol. (a) **146**, 437

Inst. Phys. Conf. Ser. No 164
Paper presented at Microsc. Semicond. Mater. Conf., Oxford, 22–25 March 1999
© 1999 IOP Publishing Ltd

TEM Study of ordered domain structures in Te-doped GaInP layers grown by organometallic vapour phase epitaxy

T-Y Seong,[1] C-J Choi,[1] R Spirydon,[1] S H Lee [2] and G B Stringfellow[2]

[1]Department of Materials Science and Engineering, Kwangju Institute of Science and Technology, Kwangju 500-712, Korea
[2]Department of Materials Science and Engineering, University of Utah, Salt Lake City, Utah 84112, USA

ABSTRACT: Organometallic vapour phase epitaxial (OMVPE) GaInP layers were grown on (001) GaAs singular and vicinal substrates at 670 °C to investigate the effects of Te doping on ordering and antiphase boundaries (APBs). The density of APBs in the vicinal samples is increased by a factor of 2, while that of the singular samples is slightly increased, as the Te concentration increases. APBs are inclined 9–57° from the (001) growth surface. For the singular samples, the angle remains virtually unchanged with increasing Te concentrations. However, for the vicinal samples, the angle decreases significantly with increasing Te concentration.

1. INTRODUCTION

TEM and TED have been widely used to assess CuPt ordering in (001) layers of III-V ternary alloys (Stringfellow and Chen 1991, Seong et al 1994a, and Seong et al 1994b) grown by OMVPE and molecular beam epitaxy (MBE). CuPt ordering is known to be a surface related phenomenon occurring during epitaxial growth (Stringfellow and Chen 1991, Zunger and Mahajan 1994, Philips et al 1994). TEM dark field (DF) examination has revealed ordered domains in MBE and OMVPE III-V ternary layers (Seong et al 1994b). The ordered domains usually contain a density of APBs. Such APBs are reported to adversely influence the optical and electrical properties of the layers. Therefore, it is important to be able to control such defects to enhance device performance. In this work, TEM and TED studies of APBs in Te-doped GaInP layers grown on GaAs (001) singular and vicinal substrates are presented.

2. EXPERIMENTAL PROCEDURE

Te-doped GaInP layers were grown in a horizontal OMVPE reactor using trimethylindium, trimethylgallium, and tertiarybutylphosphine with diethyltelluride as the dopant precursor. The substrates were Cr-doped (001) GaAs and (001) GaAs misoriented by 3° in $[111]_B$. A GaAs buffer layer ~30 nm thick was grown first. This was followed by the GaInP layer grown at 670 °C with a rate of 0.5 μm/h. The Te concentrations were varied from 4.80×10^{16} to 8.40×10^{18} cm^{-3}.

<110> cross-section thin foil films were prepared by mechanical polishing followed by Ar$^+$ ion milling using a LN$_2$ cold stage and examined in a JEM 2010 instrument operated at 200 kV. Convergent beam electron diffraction technique was employed to determine the thin foil thickness (Kelly et al 1975). The thicknesses of the thin foil films examined by TEM were mostly in the range 115 – 400 nm.

The surface morphology was examined using a Nanoscope III atomic force microscope (AFM) in the tapping mode. Etched single-crystalline Si tips were used with an end radius of ~5 nm with a sidewall angle of ~35°. The samples were measured in air, so covered by a thin, conformal oxide layer.

3. RESULTS AND DISCUSSION

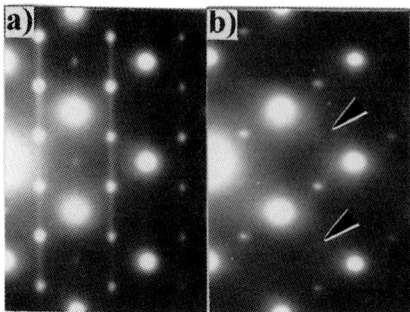

Fig. 1 shows [110] TED patterns from the GaInP layer grown on (a) singular and (b) vicinal (001) GaAs at 670 °C.

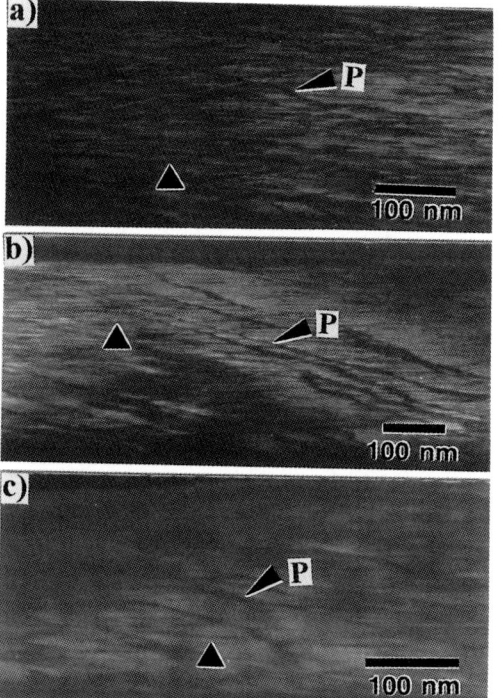

Fig. 2 shows ½(-33-1) DF images from the singular samples with concentrations of (a) 4.8×10¹⁶, (b) 2.1×10¹⁷ and (c) 4.7×10¹⁷ cm⁻³.

Fig. 1(a) shows a [110] TED pattern from the GaInP layer grown on singular (001) GaAs at 670 °C, with a carrier concentration of 2.15×10^{17} cm⁻³. The pattern shows the main zinc-blende spots and the superlattice spots at ½(1-11) and ½(-111), indicating the occurrence of CuPt ordering on the (1-11) and (-111) planes. Fig. 1(b) shows a [110] pattern from the layer grown on vicinal (001) GaAs at 670 °C, with 1.23×10^{17} cm⁻³. The pattern reveals the main spots and $\frac{1}{2}\{111\}_B$ superlattice spots, indicating that ordering occurs on the (-111) plane with much weaker ordering on the (1-11) plane (indicated by the arrows). The observation of both variants in the vicinal sample is attributed to the presence of different types of steps on the buffer layers, i.e., thermal steps occurring during growth (Pashley et al 1988). TED results showed that for both the singular and vicinal samples, the degree of order gradually decreases with increasing Te concentrations and appears to be completely disordered at concentrations exceeding $\sim1\times10^{18}$ cm⁻³.

Figs. 2(a), (b) and (c) show ½(-33-1) DF images from the singular samples with concentrations of 4.8×10^{16}, 2.1×10^{17} and 4.7×10^{17} cm⁻³, respectively. The images reveal the ordered domains as the bright regions. The ordered regions vary from ~8 to ~35 nm in width and contain fine modulations 0.6~1.2 nm across (indicated by the arrows) lying parallel to the (001) surface. The ordered regions of each variant are separated by APBs (marked 'P'). The APBs are inclined counterclockwise 9~18° from the (001) growth surface. The ½(-331) DF images (other variant) from the singular samples showed that the APBs were inclined clockwise by about the same angles. A comparison of the ½(-33-1) and ½(-331) images showed that the two variants are complementary with alternating laminae, 0.8~2 nm across, of each variant. The lamina structures are better defined in the samples with lower doping concentrations.

Figs. 3(a), (b) and (c) show ½(-33-1) DF images from the vicinal samples with concentrations of 4.2×10^{16}, 1.2×10^{17} and 3.1×10^{17} cm⁻³, respectively. The ordered regions also contain APBs 5-16 nm across. The APBs are inclined counterclockwise 10~57° to the (001) growth surface. It is noted that the ordered regions contain no fine modulations that were observed in the singular samples.

There are characteristic features in the DF images obtained from the singular and vicinal

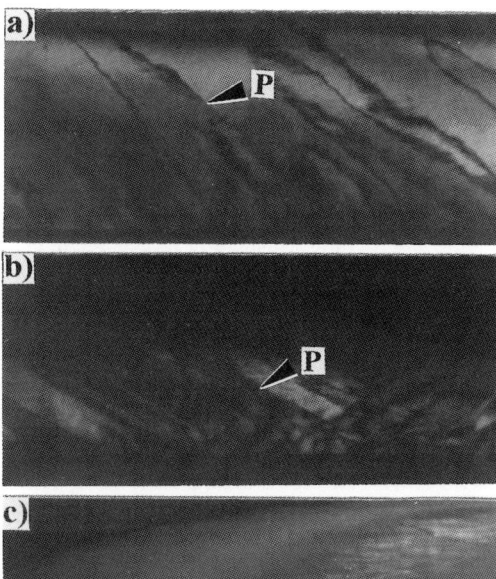

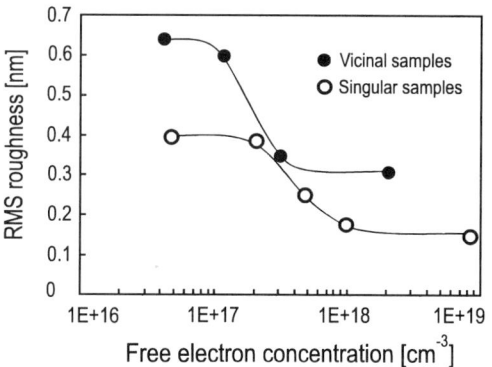

Fig. 3 shows ½(-33-1) DF images from the vicinal samples with concentrations of (a) 4.2×10^{16}, (b) 1.2×10^{17} and (c) 3.1×10^{17} cm^{-3}.

Fig. 4 shows the plot of root mean square roughness versus carrier concentration.

samples. Firstly, the density of APBs in both the samples increases with increasing Te concentration. For the singular samples, it is ~8.7×10^9, ~8.5×10^9 and ~9.5×10^9 cm^{-2} for concentrations of 4.8×10^{16}, 2.1×10^{17} and 4.7×10^{17} cm^{-3}, respectively. For the vicinal samples, the density is measured to be ~7.2×10^9, ~8.1×10^9 and ~1.3×10^{10} cm^{-2} for concentrations of 4.2×10^{16}, 1.23×10^{17}, and 3.1×10^{17} cm^{-3}, respectively. Secondly, APBs are inclined some degrees from the (001) growth surface. For the singular samples, the tilting angle seems to remain virtually unchanged with increasing Te concentrations. The angle is in the range of 10–17° for a concentration of 4.8×10^{16} cm^{-3}, 10–18° for 2.15×10^{17} cm^{-3}, and 9–16° for 4.7×10^{17} cm^{-3}. However, for the vicinal samples, the angle decreases significantly with increasing Te concentration. It is in the range of 36–57° for a concentration of 4.2×10^{16} cm^{-3}, 22–30° for 1.23×10^{17} cm^{-3}, and 10–15° for 3.1×10^{17} cm^{-3}. Thirdly, the fine modulations are present in the singular samples, but not observed in the vicinal samples. For the singular samples, the modulations became ill-defined with increasing Te doping level. It is worth noting that the surface of the singular samples became smoother, as the Te concentration increased. Therefore, the TEM and AFM results indicate that the fine modulations are associated with the presence of hillocks introduced by surface undulations.

AFM was used to investigate the surface structures of the singular and vicinal samples having carrier concentrations in the range of 4.2×10^{16} – 8.4×10^{18} cm^{-3}. Fig. 4 shows the plot of root mean square roughness versus carrier concentration. It is clear that the addition of Te results in smoother surfaces. For the singular samples, the step structure changes from a mixture of monolayer and bilayer steps to atomically flat surface for carrier concentrations exceeding ~1×10^{18} cm^{-3}. As for the vicinal samples, the step structure changes mainly from bilayers to monolayers. Similar doping behaviour was observed in OMVPE GaInP (Lee et al 1998).

The mechanism for the carrier concentration dependence of the behaviour of APBs in the present GaInP layers may be related to a change in the surface structures. Several models (Stringfellow and Chen 1991, Zunger and Mahajan 1994, Philips et al 1994) have been proposed to interpret the formation mechanism of CuPt ordering in III-V alloy layers, which includes the important role of surface atomic steps running along the [110] direction ([110] steps). [110] steps were observed to assist the ordering process, whereas [-110] steps retard ordering (Murata et al 1996).

Even when starting with substrate wafers with surfaces cut close to the (001) orientation, the surface of the buffer layer will probably have surface undulations and hence surface atomic steps. The AFM results of the singular samples (Lee et al 1998, Seong et al 1999) showed that due to the anisotropic step velocity, the hillocks change from a circular shape to an elliptical shape with the longer sides running along [110]. This change yields a slight increase in the C regions (in which step edges are not parallel to [110]) of the doped sample as compared to the undoped layers. Seong et al (1994b), investigating domain structures in ordered GaInAs OMVPE layers grown at a wide range of growth conditions, showed that undimerised atoms could form at the step edge when the steps were not along the [110] direction and such atoms could lead to the formation of APBs. Therefore, a slight increase in the density of APBs observed in the singular samples may be due to the increase in the C regions.

The AFM and TED results of the vicinal samples (Lee et al 1998, Seong et al 1999) showed that the surface of the undoped sample consists of [110] steps, while the doped sample contains numerous kinks at step edges (Seong et al 1999). In other words, the addition of Te causes more kinks to form at step edges. Similar kink behaviour was also observed in GaAs (Pashley et al 1991). Pashley et al (1991) showed that during MBE growth of GaAs, Si doping led to the loss of coherency of (2×4) cells, resulting in a high density of kinks. Such kinked terraces may contain numerous undimerised atoms that could lead to the formation of APBs (Seong et al 1994b). Therefore, the behaviour of APBs in the vicinal samples may be attributed to the presence of kinks at step edges associated with thermal steps.

The Te doping dependence of tilting angle of APBs may be explained as follows. If APBs form at the step edges, they are expected to follow the step edges during growth. Since the presence of Te changes step velocity, the higher doping levels leads to higher velocity and hence lower angle. Details of the Te doping dependence of the behaviour of APBs will be published elsewhere (Seong et al 1999).

ACKNOWLEDGMENT

The authors wish to thank MOST (Korea) for financial support. One author (GBS) wishes to thank the Department of Energy for partial support of his work.

REFERENCES

Kelly P M, Jostons A, Blake R G and Napier J G 1975 Phys. Stat. Sol. **31**, 771

Lee S H, Hsu T C and Stringfellow G B 1999 J. Appl. Phys. (submitted)

Murata H, Ho I H, Hosokawa Y and Stringfellow G B 1996 Appl. Phys. Lett. **68**, 2237

Pashley M D, Haberern K W, Friday W, Woodall J M and Kirchner P D 1988 Phys. Rev. Lett. **60**, 2176

Pashley M D and Haberern K W 1991 Phys. Rev. Lett. **67**, 2697

Philips B A, Norman A G, Seong T-Y, Mahajan S, Booker G R, Skowronski M, Harbison J P and Keramidas V G 1994 J. Cryst. Growth **140**, 249

Seong T-Y, Booker G R, Norman A G and Ferguson I T 1994a Appl. Phys. Lett. **64**, 3593

Seong T-Y, Norman A G, Booker G R and Cullis A G 1994b J. Appl. Phys. **75**, 7852

Seong T-Y, Choi C-J, Spirydon R, Lee S H and Stringfellow G B 1999 Semicond. Sci. Technol. (submitted)

Stringfellow G B and Chen G S 1991 J. Vac. Sci. Technol. **B 9**, 2182

Zunger A and Mahajan S 1994 Handbook on Semiconductors, edited by Mahajan S (Elsevier Science, Amsterdam) Vol.3, Chap. 19

Inst. Phys. Conf. Ser. No 164
Paper presented at Microsc. Semicond. Mater. Conf., Oxford, 22–25 March 1999
© 1999 IOP Publishing Ltd

Formation model for microstructures in a (Al, Ga, In)P natural superlattice

Y Kuno, S Takeda, M Hirata, Y Ohno, N Hosoi* and K Shimoyama*

Department of Physics, Graduate School of Science, Osaka University,
1-16, Machikane-yama, Toyonaka, Osaka 560-0043, Japan
*Opto-Electronics Research & Technology Development Center Mitsubishi Chemical Co.,
Tsukuba Plant, 1000 Higashi-mamiana, Ushiku City, Ibaraki, 300-12, Japan

ABSTRACT: Antiphase boundaries (APBs) in a natural superlattice in (Ga, In)P and (Al, Ga, In)P have been observed by means of transmission electron microscopy. The extending behavior of APBs depends on the substrate misorientation angle. The existing models of crystal growth such as step flow growth fail to interpret the behaviour. We discuss the extending mechanism of APBs based on the experimental results.

1. INTRODUCTION

A natural superlattice is spontaneously formed in (Ga, In)P and (Al, Ga, In)P during metal organic vapor phase epitaxial (MOVPE) growth (Gomyo *et al* 1988). The ordered structure is described as a monolayer superlattice with the atoms alternately stacked on (Al)GaP and InP layers along [$\bar{1}$11] and [1$\bar{1}$1]. Surface reconstruction and thermodynamics relating to the cause of atomic ordering have been developed (Suzuki and Gomyo 1991, Froyen and Zunger 1996, Zhang *et al* 1995) , while little insight into growth kinetics exists as yet. In fact, the epitaxial layers contain many antiphase boundaries (APBs) and the behavior of APBs depends on crystal growth conditions. In addition, APBs give rise to undesirable influence on the optical and the electrical properties of epitaxial layers. Therefore, the study of APBs leads to information on the growth kinetics, with which a better understanding of the epitaxial process is obtained. APBs in ordered GaInP layers and the related ordered structures have been experimentally investigated (Su *et al* 1994a, 1994b, Ernst *et al* 1996, Seong *et al* 1997) . However, no specific model is given for the behaviour of APBs. In this paper, we investigate APBs in GaInP and AlGaInP using the transmission electron microscope (TEM), and discuss the origin of microstructures associated with APBs.

2. EXPERIMENTS

GaInP and AlGaInP layers were grown by MOVPE on GaAs substrates misoriented from the (001) plane in the [1$\bar{1}$0] direction by a range of angles. A GaAs buffer layer about 0.1 μm-thick and a $Al_{0.7}Ga_{0.3}As$ buffer layer about 0.5 μm-thick were grown first. The V/III ratio was constant at 370. The growth temperature was 660 °C. The sample had 50 nm thick layers of $Ga_{0.5}In_{0.5}P$ and $(Al_{0.7}Ga_{0.3})_{0.5}In_{0.5}P$ alternately. The thickness of $(Al_{0.7}Ga_{0.3})_{0.5}In_{0.5}P$ layers were, from bottom to top, 5 nm, 10 nm, 20 nm, and 50 nm. Substrates were misoriented by angles θ_m, of 2°, 6° and 10°. TEM studies have been performed using [110] cross-sectional samples. APBs were observed using a 200 kV TEM (JEM2010). The dark-field images have been taken using the ($\bar{1}$/2 1/2 3/2) diffraction spots.

3. RESULTS

Figure 1 shows the epitaxial layers which were grown on the substrates misoriented by different angles. We first define the extending angle of an APB, θ_{APB} as the angle between the APB and the (001) plane. The extending angles for AlGaInP layers differ from those for GaInP layers. The interface between GaInP and AlGaInP layers does not influence the extending behaviors of APBs. We summarized the measured θ_{APB} as a function of θ_m in Fig. 2. θ_{APB} in GaInP layers is much greater than θ_{APB} in AlGaInP layers. θ_{APB} increases markedly as θ_m increases. It is believed that Al atoms are less moveable on a terrace than the other species, so we relate the smaller extending angle in AlGaInP layers to the effect of Al.

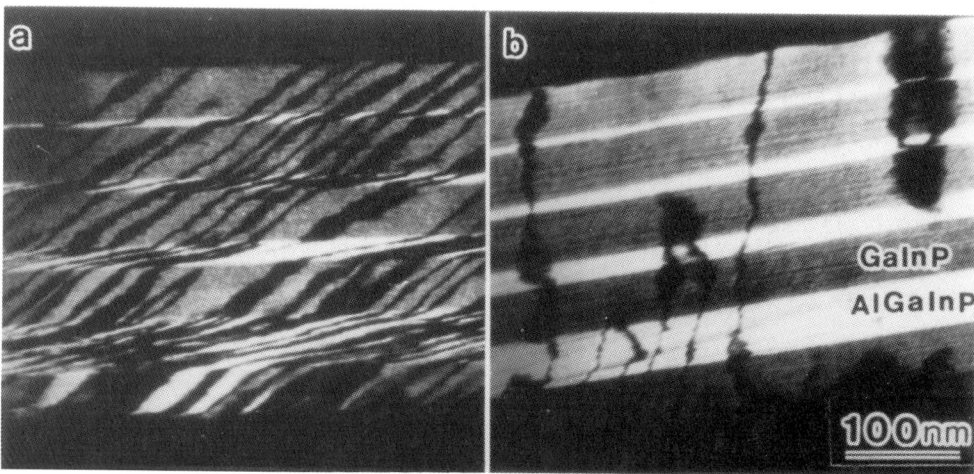

Fig. 1   Dark field TEM images of microstructures in the epitaxial layers grown on the substrates misoriented by angles of (a) 2° and (b) 10°. AlGaInP layers appear brighter than GaInP layers in the imaging condition. The growth direction is along the upward direction in the images. APBs extend against the step flow direction.

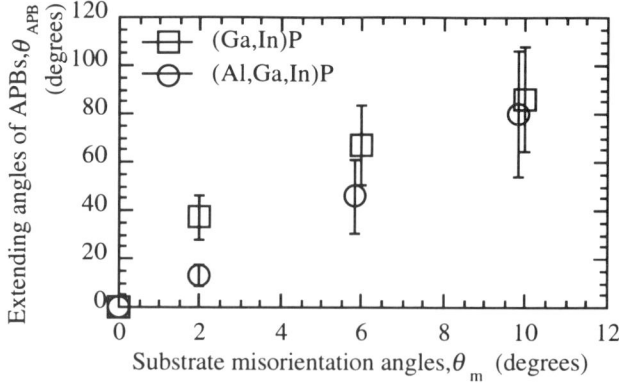

Fig. 2   Extending angle of APBs, θ_{APB} as a function of θ_m in GaInP and AlGaInP layers.

Figure 3 shows a high resolution TEM image of APBs in AlGaInP layers and its Fourier filtered image. The minor variant of the ordered structure is localized at the laterally extending part of the APBs.

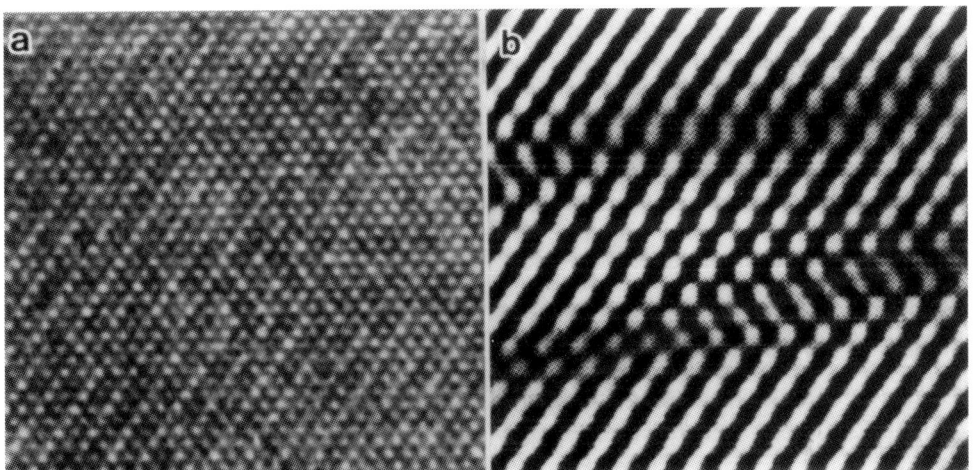

Fig. 3 (a) High resolution TEM images of APBs in AlGaInP layers, $\theta_m = 2°$. (b) Fourier filtered image with the ordered spots.

4. DISCUSSION

We conclude from the experiments that θ_{APB} is small if the terrace width is large and the mobility of adatoms is small. This result implies that two-dimensional (2D) nucleation of islands occurs during epitaxial growth. Only step-flow growth mode has been used to account for the formation of the ordered structures but apparently it can not explain the behavior of APBs.

We assume the 2D nucleation on a terrace (Fig. 4). In Fig. 4, the ordered structure is projected onto the (110) plane. For simplification, P atoms and surface reconstructions are not drawn, and one-molecular-layer-high steps are assumed. We also assume the interplanar ordering during the epitaxial growth: Ga-atom lines (or Al-atom lines) and In-atom lines are deposited alternately along [$\bar{1}$10] or [1$\bar{1}$0] (Suzuki and Gomyo 1991). If θ_m is small, the terrace-widths are wide, then 2D nuclei become large, and θ_{APB} will be small (Fig 4(a) to (e)). The small θ_{APB} in AlGaInP implies the larger sticking rate of Al at the APBs on a surface. These explanations interpret the experimental results in part. A detailed view will be given elsewhere about the extension of APBs with the island nucleation at the intersection of APBs and the terrace (Takeda et al 1999) and shown in Fig. 4 (b') to (e'). The other variant (1$\bar{1}$1) appears when θ_{APB} is small. Since most of these APBs are parallel to the (001) planes, thin lamellae of the minor (1$\bar{1}$1) variant are formed there.

The atomic force microscopy study suggests the spacing of the supersteps is nearly equal to the distance between APBs measured using electron microscopy (Stringfellow and Su 1995). Applying our model, we understand that the phenomenon is as follows. If the APBs at a terrace shift laterally to other APBs, then pair annihilation takes place, while the unmovable APBs at supersteps can escape from the annihilation. In consequence, the density of APBs decreases to that of supersteps.

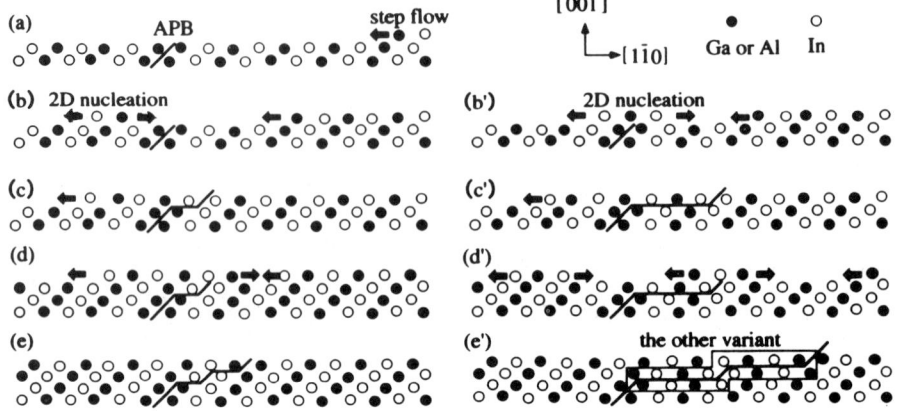

Fig. 4 Development of microstructures during crystal growth associated with the extension of APBs on the substrates. The 2D nucleation occurs on a terrace in (a) to (e), and at the APB exposed on a terrace in (b') to (e'). A terraced (001) substrate has an APB. A surface step moves to the left. The APB extends to the position where the 2D nucleus and the step-flow edge encounter one another. The other variant of the ordered structure appears in the laterally shifted APB in (e').

5. CONCLUSIONS

APBs in GaInP and AlGaInP were investigated by means of the TEM. Taking account of the experimental results, we suggested that the microstructures in the epitaxial layers are formed by both the step-flow mode and the 2D nucleation mode.

REFERENCES

Ernst P, Geng C, Hahn G, Scholz F, Schweizer H, Phillipp F and Mascarenhas A 1996 J. Appl. Phys. **79**, 2633.
Froyen S and Zunger A 1996 Phys. Rev. **B 53**, 4570.
Gomyo A, Suzuki, T and Iijima S 1988 Phys. Rev. Letter. **60**, 2645.
Seong T-Y, KimJ H, Chun Y S and Stringfellow G B 1997 Appl. Phys. Lett. **70**, 3137.
Stringfellow G B and Su L C 1995 Appl. Phys. Lett. **66**, 3155.
Su L C, Ho I H and Stringfellow G B 1994a J. Appl. Phys. **75**, 5135.
Su L C, Ho I H and Stringfellow G B 1994b J. Appl. Phys. **76**, 3520.
Suzuki T and Gomyo A 1991 J. Cryst. Growth **111**, 353.
Takeda S, Kuno Y, Hosoi N and Shimoyama K, to be published in J. Crystal Growth.
Zhang S B, Froyen S and Zunger A 1995 Appl. Phys. Lett. **67**, 3141.

Inst. Phys. Conf. Ser. No 164
Paper presented at Microsc. Semicond. Mater. Conf., Oxford, 22–25 March 1999

291

Ordering and domain structures in Mg-doped GaInP layers

S-M Lee,[1] S W Jun,[2] T-Y Seong[1] and G B Stringfellow[2]

[1]Department of Materials Science and Engineering, Kwangju Institute of Science and Technology, Kwangju 500-712, Korea
[2]Department of Materials Science and Engineering, University of Utah, Salt Lake City, Utah 84112, USA

ABSTRACT: The effects of Mg-doping on ordering and domain structures in organometallic vapour phase epitaxial (OMVPE) GaInP layers grown on (001) GaAs singular and vicinal substrates at 670°C have been investigated by TEM and TED. It is shown that the degree of order decreases insignificantly with increasing Mg concentrations up to ~3.0×10^{19} cm^{-3}. The TEM results show that for both the singular and vicinal samples the addition of Mg results in an insignificant change in the density of APBs. The TED results show that for vicinal samples, both of the two ordered variants are observed, although one variant is much stronger than the other. The intensity of the other variant increases slightly with the Mg levels. It is shown that the Mg-doped layers contain threading dislocations, stacking faults and microtwins, which originate from the GaInP/GaAs interface. A simple comparison is made to interpret the behaviour of APBs.

1. INTRODUCTION

GaInP layers are of increasing interest for their application in visible light-emitting diodes and lasers. Atomically ordered {111} superlattice, i.e., CuPt-type ordering, generally occurs in these layers (Bellon et al 1989, Stringfellow and Chen 1991, Baxter and Stobbs 1994). TEM and TED techniques have been widely utilised to characterise such ordering in III-V ternary layers. CuPt ordering is believed to be a surface related phenomenon occurring during growth (Stringfellow and Chen 1991, Zunger and Mahajan 1994, Philips et al 1994). Models based on the (2×n) surface reconstruction and the presence of [110] atomic steps have been suggested to interpret the mechanisms by which the ordering occurs in III-V (001) alloy layers. CuPt ordering is significantly affected by growth conditions such as V/III ratio, growth temperature, growth rate, and impurity diffusion. It has been shown that impurities such as Zn, Si, Te, and Mg cause the ordered structure to become a random alloy. In this work, we describe effects of Mg-doping on ordering, ordered domains and APBs in GaInP layers grown on GaAs (001) singular and vicinal substrates.

2. EXPERIMENTS

Mg-doped GaInP layers were grown in a horizontal OMVPE reactor using trimethylindium, trimethylgallium, and tertiarybutylphosphine with CP$_2$Mg as the dopant precursor. The substrates were Cr-doped (001) GaAs and (001) GaAs misoriented by 3° in [111]$_B$. A GaAs buffer layer ~30 nm thick was grown first. This was followed by the GaInP layer grown at 670 °C with a rate of 0.5 μm/h. The Mg concentrations were in the range of $3.0 \times 10^{16} - 3.0 \times 10^{19}$ cm^{-3}.

<110> cross-section thin foil films were prepared by mechanical polishing followed by Ar$^+$ ion milling using a LN$_2$ cold stage and examined in a JEM 2010 instrument operated at 200 kV. Convergent beam electron diffraction technique was employed to determine the thin foil thickness (Kelly et al 1975). The thicknesses of the thin foil films examined by TEM were mostly in the range 120 – 410 nm.

3. RESULTS AND DISCUSSION

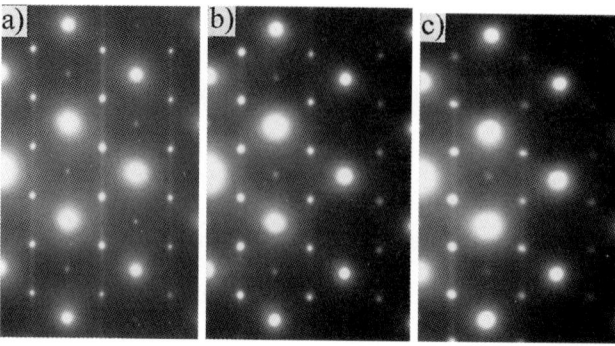

Fig. 1 shows [110] TED patterns from the GaInP layer grown on singular (001) GaAs, with concentrations of ~3.0×10¹⁶, ~7.0×10¹⁷ and ~3.0×10¹⁹ cm⁻³, respectively.

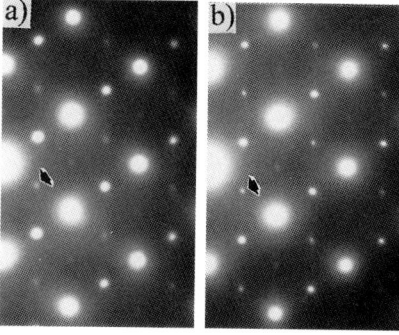

Figs. 2(a) and 2(b) show [110] patterns from the layer grown on vicinal (001) GaAs, with concentrations of ~7.0×10¹⁷ and ~3.0×10¹⁹ cm⁻³, respectively.

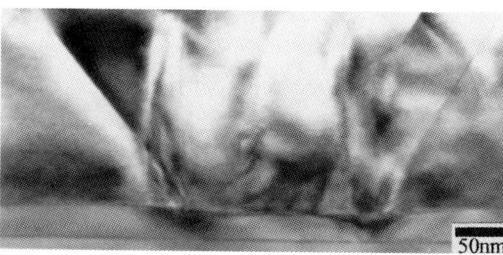

Fig.3 shows a (002) DF image from the vicinal layer with a concentration of 7.0×10¹⁷ cm⁻³.

Figures 1(a), 1(b) and 1(c) show [110] TED patterns obtained from the GaInP layer grown on singular (001) GaAs at 670 °C, with carrier concentrations of ~3.0×10¹⁶, ~7.0×10¹⁷ and ~3.0×10¹⁹ cm⁻³, respectively. The patterns show the main zinc-blende spots and the superlattice spots at ½(1-11) and ½(-111), each with its associated family of spots, indicating that CuPt ordering occurs on the (1-11) and (-111) planes. Figs. 2(a) and 2(b) show [110] patterns taken from the layer grown on vicinal (001) GaAs at 670 °C, with concentrations of ~7.0×10¹⁷ and ~3.0×10¹⁹ cm⁻³, respectively. The patterns show the main spots and ½{111}ᴮ superlattice spots, indicating that ordering occurs on the (-111) plane with weaker ordering on the (1-11) plane (indicated by the arrows). The observation of both variants in the vicinal sample is attributed to the presence of different types of steps on the buffer layers, i.e., thermal steps occurring during growth (Pashley et al 1988). TED results show that for both the singular and vicinal samples, the degree of order seems to be insignificantly decreased with increasing Mg concentrations. This is different from the addition of Te, in which the ordered layer was found to be completely disordered at concentrations exceeding ~1×10¹⁸ cm⁻³ (Seong et al 1999).

TED DF examination showed that the Mg-doped samples contain threading defects, although no defects are observed in the undoped layers. For example, a (002) DF image obtained from the vicinal layer with a concentration of ~7.0×10¹⁷ cm⁻³ is shown in Fig.3. There are threading dislocations and microtwins originating from the GaInP/GaAs interface. Such defects are likely to be associated with the presence of magnesium oxides (Mg_xO_{1-x}) that formed during growth.

Figs. 4(a), 4(b) and 4(c) show ½(-33-1) DF images obtained from the singular samples with concentrations of ~3.0×10¹⁶, ~7.0×10¹⁷ and ~3.0×10¹⁹ cm⁻³, respectively. The images reveal the ordered domains as the bright regions. The ordered regions vary from ~7 to ~33 nm in width and contain fine modulations 0.6~1.2 nm across (indicated by the arrows) lying parallel to the (001) surface. The ordered regions of each variant are separated by APBs (marked 'P'). The APBs are inclined counterclockwise 9~55° from the (001) growth surface. The ½(-331) DF images (other variant) from

Fig. 4 shows ½(-33-1) DF images obtained from the singular samples with concentrations of ~3.0×10¹⁶, ~7.0×10¹⁷ and ~3.0×10¹⁹ cm⁻³, respectively.

the same samples showed that the APBs were inclined clockwise by about the same angles. A comparison of the ½(-33-1) and ½(-331) images showed that the two variants are complementary with alternating laminae, 0.6~2.1 nm across, of each variant.

Figs. 5(a), 5(b) and 5(c) show ½(-33-1) DF images from the vicinal samples with concentrations of ~3.0×10¹⁶, ~7.0×10¹⁷ and ~3.0×10¹⁹ cm⁻³, respectively. The ordered regions contain APBs 6-15 nm across. The APBs are inclined counterclockwise 42~82° to the (001) growth surface. It is interesting to note that the ordered regions contain no fine modulations that were observed in the singular samples. Fig. 5(d) shows a ½(-331) DF image obtained from the layer with a concentration of ~3.0×10¹⁹ cm⁻³. The image reveals that the APBs were inclined clockwise 12 – 28° to the (001) growth surface.

The DF images obtained from the singular and vicinal samples illustrate a few characteristic features. Firstly, for both the singular and vicinal samples, the addition of Mg results in an insignificant change in the density of APBs. For the singular sample, it is ~2.5×10⁹, ~2.7×10⁹ and ~3.0×10⁹ cm⁻² for concentrations of ~3.0×10¹⁶, ~7.0×10¹⁷ and ~3.0×10¹⁹ cm⁻³, respectively. As for the vicinal samples, the density is ~3.2×10⁹, ~3.4×10⁹ and ~3.4×10⁹ cm⁻² for concentrations of ~3.0×10¹⁶, ~7.0×10¹⁷ and ~3.0×10¹⁹ cm⁻³, respectively. Secondly, the fine modulations are present in the singular samples, but not observed in the vicinal samples, which were attributed to the presence of hillocks introduced by surface undulations (Seong et al 1999). Thirdly, for the vicinal samples, the intensity of the ½(1-11) superlattice spots (other variant) increases with increasing Mg concentrations. This indicates that a high density of kinks involving [110] steps facing [1-10] was introduced in the doped samples. This result is consistent with the occurrence of patches of the (1-11) ordered domain, as shown in Fig. 5(d). (It is worthwhile to note that the (1-11) variant is not expected to occur on the surface cut a few degrees toward [111]ᵦ.) In addition, Seong et al (1999) argued that such kinked terraces contain numerous undimerised atoms.

AFM examination showed that the root mean square (rms) roughness varies with the Mg concentrations. As for the singular sample, it ranged from 0.44 nm to 2.32 nm, while for the vicinal samples, the roughness varied from 0.95 nm to 3.24 nm. It is clear that the addition of Mg leads to rougher surfaces having bunched steps. This is in contrary to the doping behaviour of Te (Seong et al 1999). The AFM results showed that there are small crystallites (white dots) on the surface of the Mg-doped samples, but not observed in the undoped samples. The density of the dots depends on the Mg concentrations. The size of the dots varied from 12 to 40 nm in diameter and from 20 to 30 nm in height.

Different mechanisms have been proposed to describe the behaviour of APBs in ordered III-V alloy layers. Su and Stringfellow (1995), investigating ordering and step structures in GaInP OMVPE layers, suggested that APBs arise from bunched steps. Seong et al (1994), investigating ordered domain structures in GaInAs OMVPE layers, argued that undimerised atoms could form at the step edge when the steps were not along the [110] direction and such atoms could lead to the formation of APBs. Thus, a comparison of these mechanisms and our results indicates that the addition of Mg

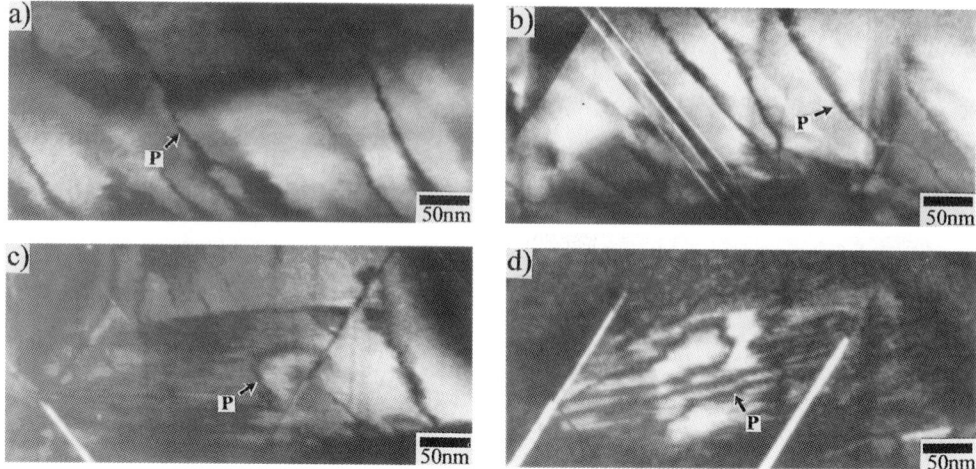

Fig. 5 shows ½(-33-1) DF images obtained from the vicinal samples with concentrations of ~3.0×10^16, ~7.0×10^17 and ~3.0×10^19 cm^{-3}, respectively. (d) A ½(-331) DF image obtained from the layer with a concentration of ~3.0×10^19 cm^{-3}.

could result in an increase in the density of APBs. However, the DF results show that this is not the case for the present system; there are little changes in the density of APBs throughout the doping range. This discrepancy may be attributed to the ill-defined surface structures of the Mg-doped samples. As shown by TEM and AFM, for the doped samples, there are a density of defects propagating through the full layer thickness of ~280 nm and white dots on the surface. These defects affect the surface step structures during growth.

ACKNOWLEDGMENT

The authors wish to thank MOST (Korea) for financial support. One author (GBS) wishes to thank the Department of Energy for partial support of his work.

REFERENCES

Bellon P, Chevalier J P, Augarde E, André J P and Martin G P 1989 J. Appl. Phys. **66**, 2388
Baxter C S and Stobbs W M, 1994, Phil. Mag. **69**, 615
Kelly P M, Jostons A, Blake R G and Napier J G 1975 Phys. Stat. Sol. **31**, 771
Pashley M D, Haberern K W, Friday W, Woodall J M and Kirchner P D 1988 Phys. Rev. Lett. **60**, 2176
Philips B A, Norman A G, Seong T-Y, Mahajan S, Booker G R, Skowronski M, Harbison J P and Keramidas V G 1994 J. Cryst. Growth **140**, 249
Seong T-Y, Norman A G, Booker G R and Cullis A G 1994 J. Appl. Phys. **75**, 7852
Seong T-Y, Choi C-J, Spirydon R, Lee S H and Stringfellow G B 1999 Semicond. Sci. Technol. (submitted)
Stringfellow G B and Chen G S 1991 J. Vac. Sci. Technol. **B 9**, 2182
Zunger A and Mahajan S 1994 Handbook on Semiconductors, edited by Mahajan S (Elsevier Science, Amsterdam) Vol.3, Chap. 19
Su L C and Stringfellow G B, 1995 J. Appl. Phys. **78**, 6775

Inst. Phys. Conf. Ser. No 164
Paper presented at Microsc. Semicond. Mater. Conf., Oxford, 22–25 March 1999
© 1999 IOP Publishing Ltd

Formation of edge-type misfit dislocations in $In_{0.15}Ga_{0.85}As$/GaAs heterostructures during thermal processing

X W Liu and A A Hopgood

Faculty of Technology, The Open University, Milton Keynes MK7 6AA, UK

ABSTRACT: The formation of edge-type misfit dislocations in GaAs/$In_{0.15}Ga_{0.85}As$/GaAs heterostructures which have already been relaxed by 60° misfit dislocations is presented. Dislocations with Burgers vectors of $a/2<101>$ perpendicular to their $<010>$ directions are formed at the interfaces during thermal processing at 1040K. A model is proposed in which the formation of these edge dislocations is ascribed to vacancy-producing jogs on pre-existing 60° dislocations, whose motion by climb leaves a trailing dislocation dipole.

1. INTRODUCTION

Strained-layer structures are commercially important, particularly for semiconductor lasers. The reliability of such devices is dependent on the stability of the strained layer. In contrast, where a misfitting layer is to be used as a buffer, complete relaxation of the layer is desired. In either case it is vital to be able to predict reliably the level of dislocation introduction and consequent strain relaxation.

The accommodation of misfit between an epitaxial film and its substrate by misfit dislocations has been intensively investigated. In heteroepitaxial semiconductor systems with the zinc-blende structure, the dominant type of misfit dislocation is the so-called 60° dislocation. Orthogonal arrays of 60° misfit dislocations at interfaces have been assumed to be the general configuration, and the investigation of nucleation, propagation and multiplication of misfit dislocations has been based on them, e.g. Matthews et al (1976) and Hagen and Strunk (1978).

Beanland et al (1997) observed that the rotation of 60° dislocations into 90° dislocations produced additional misfit relief by the conversion of screw components into edge components. We report here a new mechanism for the relaxation of strained-layer structures by the introduction of a new kind of misfit dislocation, viz. pure edge dislocations formed by jogs on pre-existing 60° dislocations.

2. EXPERIMENTAL DETAILS

All specimens were grown by molecular beam epitaxy (MBE). A 50 nm AlAs layer, which was used to release the heterostructures from the substrate by the epitaxial lift-off (ELO) technique, was first grown on a GaAs (001) substrate. This was followed by epitaxial growth of a 25 nm layer of $In_{0.15}Ga_{0.85}As$ between two 200 nm GaAs layers. The growth temperature was 800K for the $In_{0.15}Ga_{0.85}As$ layer and 870K for other layers. The growth rate was 1μm/hour. The composition and layer thickness were calibrated by means of reflection high-energy electron diffraction (RHEED).

The bulk specimens (approx. 5 mm × 5 mm) were annealed in a Carbolite furnace. Annealing was performed in a nitrogen atmosphere at temperatures ranging from 773K to 1273K and times of up to 300 s. The film specimens for transmission electron microscope

(TEM) observation were prepared using the ELO technique (Liu and Hopgood, 1999). TEM observation was on a JEOL 2000FX operated at 200 kV.

3. RESULTS

At 25 nm, the thickness of the $In_{0.15}Ga_{0.85}As$ layer was above its critical value. Fig. 1 is the TEM plan-view image of the original 60° dislocations in $GaAs/In_{0.15}Ga_{0.85}As/GaAs$. These misfit dislocations are formed at the strained-layer interfaces along the intersections of {111} slip planes and (001) interfaces, i.e. in <110> directions. The Burgers vectors of these dislocations are $a/2<101>$, inclined at 45° to (001) and at 60° to their line directions (Matthews et al, 1976).

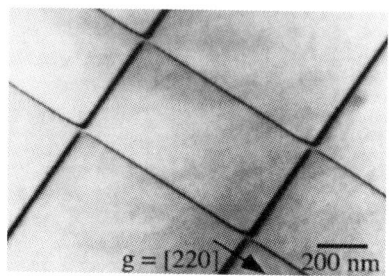

Fig. 1. 60° misfit dislocations in $GaAs/In_{0.15}Ga_{0.85}As/GaAs$.

Fig. 2 shows TEM images of dislocations in the same type of heterostructure after thermal processing at 1040K. The dislocations in <110> directions, e.g. *M1* and *M2*, are 60° dislocations that pre-existed in the as-grown structure. The dislocation lines which are at 45° to the original 60° dislocations, such as *P1* and *P3*, are newly-formed. These newly-formed dislocations are all in <100> directions. All dislocations lying in [010] directions, e.g. *P3*, are invisible in Fig. 2(b) when viewed using **g** = [040]. From the criterion of **g.b** = 0, the invisibility of these dislocations shows that their Burgers vectors must lie in (010) planes and that they are therefore pure edge. Similarly, dislocations lying in [100] directions are also edge type because they are invisible in Fig. 2(c) when viewed using **g** = [400], showing their Burgers vectors to lie in (100) planes.

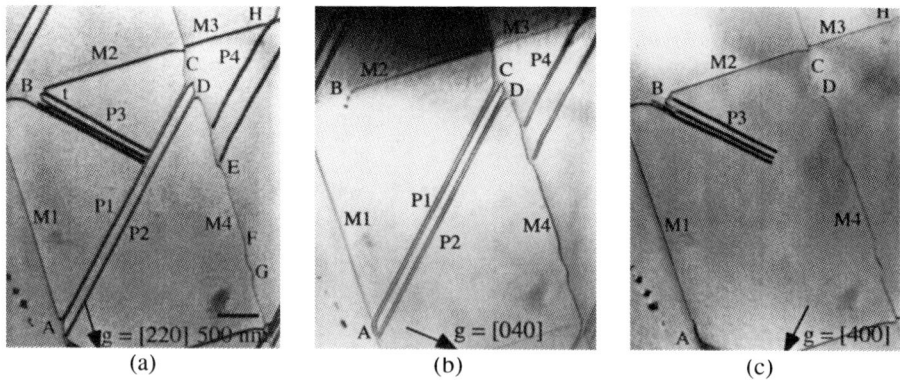

(a)   (b)   (c)

Fig. 2. Misfit dislocations in thermally processed $GaAs/In_{0.15}Ga_{0.85}As/GaAs$.

All of the newly-formed dislocations are in pairs protruding from the original 60° dislocations. They are extensions of the original dislocations; there are no breaks. In Fig. 2(a), two pairs of dislocations, *P1* and *P2*, come from *C* and *D* respectively on the dislocation *M4*.

Fig 2(c) shows that dislocation segment *M4* is broken at *C* and *D* while dislocation segment *M1* is continuous at *A*. Thus *P1* and *P2* have been formed by extension of *M4*, not of *M1*. It can be seen from Fig. 2(b) that the pair *P1* interacts with dislocation *M1* at *A*. Similarly, three pairs come from the line *M2* and have been stopped by the newly-formed dislocation pair *P1* as can be seen in Fig. 2(a) and Fig. 2(c).

So it is clear that all of the newly-formed dislocations come from the pre-existing 60° dislocations. Some of them end at the pre-existing 60° dislocations, while others end at other newly-formed dislocations. The newly-formed dislocations must therefore be at the interfaces, i.e. they are interfacial or misfit dislocations.

Another feature of the newly-formed dislocations is that all dislocation pairs generated from the same 60° dislocation line are oriented in the same direction. *M2* and *M3* can be regarded as one dislocation line; the pair *P3* can be identified as coming from *M2* but the pair *P4* do not come from *M3*. If the pair *P4* had come from the line *M3* there would be a break at *H* as occurs at *B* in Fig. 2(b). The distortion of dislocation *M3* at *H* is further evidence that the pair *P4* did not come from the line *M3* but have interacted with it.

4. DISCUSSION

Hopgood (1994) has pointed out that vacancies in a compressively strained $In_xGa_{1-x}As$ layer can relieve the strain energy. The strain energy per unit volume relieved by a vacancy concentration C_v can be expressed as

$$\Delta Q_v = -4G\left(\frac{1+v}{1-v}\right)\left(\frac{xC_v\Delta d_v\Delta d_{In}}{d_{GaAs}^2}\right),$$

where d_{GaAs} is the lattice constant of GaAs, d_v is the lattice constant of a GaAs unit cell containing one vacancy, $\Delta d_{In} = d_{InAs}-d_{GaAs}$, and $\Delta d_v = d_v-d_{GaAs}$. Thus ΔQ_v is the difference in strain energy per unit volume between vacancy-free $In_xGa_{1-x}As$ on a GaAs substrate and $In_xGa_{1-x}As$ containing vacancies on a GaAs substrate.

At high temperature, thermal activation assists the formation and movement of vacancies. The movement of vacancies enables pre-existing dislocations to climb from their slip planes along the interfaces. The distortion of the dislocation line *M4* in Fig. 2(a) is evidence of this. The vacancy movement results in a jog on the original dislocation line. If a small segment *l* of dislocation undergoes a small displacement *s*, the local change in volume, ΔV, is

$$\Delta V = \mathbf{b}.\mathbf{l}\times\mathbf{s} = \mathbf{b}\times\mathbf{l}.\mathbf{s},$$

since the two sides of the area element $\mathbf{l}\times\mathbf{s}$ are displaced by $\mathbf{b}$ relative to each other during the movement. The slip plane of the element is by definition perpendicular to $\mathbf{b}\times\mathbf{l}$. When either $\mathbf{s}$ is perpendicular to $\mathbf{b}\times\mathbf{l}$ or $\mathbf{b}\times\mathbf{l} = 0$, so the element is pure screw, ΔV is zero. Otherwise, volume is not conserved ($\Delta V \neq 0$) and the motion is climb. For a 60° dislocation $\mathbf{b}\times\mathbf{l} \neq 0$, and any movement away from its slip plane will produce an $\mathbf{s}$ which cannot be perpendicular to $\mathbf{b}\times\mathbf{l}$. Therefore any movement of 60° dislocations will be non-conservative. This is the mechanism for the formation of vacancy-producing jogs on a 60° dislocation. One of the most important processes of their formation is the absorption and emission of vacancies, i.e. jogs are sources and sinks for vacancies. The jogs will move away from the 60° dislocations under the effects of misfit stresses. A moving jog will trail a dislocation dipole and a string of vacancies.

As the jog moves through the lattice, the dislocation dipole trailed by it becomes longer until stopped by another dislocation. Because the jog is merely a segment of the dislocation itself and has the same Burgers vector as the remainder of the dislocation, any jogs on the pre-existing 60° dislocations must have a Burgers vector of $a/2<101>$. As they lie in (001) planes and they are formed by climb, they must be pure edge type lying in $<010>$ directions.

Subsequent climb of those new edge dislocation segments must continue along (001) in the direction normal to their Burgers vectors. Thus all these new edge dislocations are observed to grow in the same direction, at 45° to the pre-existing 60° dislocations. This may explain why bowing out of the point G and the section $E-F$ on dislocation line $M4$ in Fig. 2(a) could not develop further to form dislocation dipoles, or even jogs, in other directions.

5. CONCLUSIONS

Pure edge-type misfit dislocations have been found to be introduced into GaAs/In$_{0.15}$Ga$_{0.85}$As/GaAs strained-layer heterostructures during thermal processing at 1040K for up to 300 seconds. These edge dislocations are produced by vacancy-producing jogs protruding from pre-existing 60° dislocations. The jogs move out from the original dislocations by climb, trailing dislocation dipoles behind. The resultant dislocations have Burgers vectors of $a/2<101>$ perpendicular to their $<010>$ directions. They form new orthogonal arrays of dislocations in additional to the pre-existing 60° dislocation network. In this way, the relaxed strained-layer structure is relaxed further.

ACKNOWLEDGEMENTS

Financial support from the Open University Research Committee is acknowledged. The authors are grateful to H.Wang and B.F.Usher for the provision of specimens, to N.Williams for technical support, and to N.Braithwaite for his comments and suggestions.

REFERENCES

Beanland R, Lourenco M A, and Homewood K P 1997, Microscopy of Semiconducting Materials, Oxford

Hagen W and Strunk H 1978, Appl. Phys., **17**, 85

Hopgood A A 1994, J. Appl. Phys., **76**, 4068

Liu X W and Hopgood A A 1999, "Formation of misfit dislocations in In$_x$Ga$_{1-x}$As/GaAs strained-layer heterostructures" to be published

Matthews J W , Blakeslee A E, and Mader S 1976, Thin Solid Film, **33** , 253

Inst. Phys. Conf. Ser. No 164
Paper presented at Microsc. Semicond. Mater. Conf., Oxford, 22–25 March 1999
© 1999 IOP Publishing Ltd

Study of plastic relaxation of layer stress in ZnSe/GaAs(001) heterostructures

J Schreiber, U Hilpert, L Höring, L Worschech*, B König*, W Ossau*, A Waag* and G Landwehr*

Fachbereich Physik, Martin-Luther-Universität Halle-Wittenberg, Friedemann-Bach-Platz 6, 06108 Halle (Saale), Germany
*Physikalisches Institut, Universität Würzburg, Am Hubland, 97074 Würzburg, Germany

ABSTRACT: A series of ZnSe/GaAs(001) samples grown with various epilayer thickness from 50nm to 1800nm is investigated to study the role of polar misfit dislocations, in respect of their nucleation and propagation behavior in the strain relaxation process. Photoluminescence and SEM-cathodoluminescence, both applied at low temperatures, are exploited to investigate misfit dislocation configurations and related stress relief in ZnSe epilayers. Expected effects concerning the asymmetry in the dislocation misfit arrangements and optical anisotropy could be proved and analyzed in the framework of a modified Dodson-Tsao-Model.

1. INTRODUCTION

ZnSe/GaAs (001) MBE heterostructures play a key role for the realization of II/VI-based short-wavelength emitting optoelectronic devices. Despite of low lattice mismatch ($f=-2.7 \cdot 10^{-3}$) and employment of almost dislocation-free substrates for the heterostructural MBE layer growth, the generation of dislocations responsible for misfit accommodation is a subject of continued attention (Behr 1997). The misfit-caused biaxial compressive layer stress is found to be completely relieved for epilayer thicknesses of about 1µm resulting in misfit dislocation densities up to 10^6-10^8cm^{-2}. The epilayers may exhibit misfit dislocation configurations that are clearly asymmetric in respect of the [110] and [1$\bar{1}$0] directions (Petruzzello et al 1988, Rosenauer et al 1996). This asymmetry of the dislocation distribution is thought to originate from differences in the nucleation and propagation of polar A(g) and B(g) glide dislocations in the zinc-blende lattice realizing the major part of plastic stress relaxation. The asymmetric stress relaxation is found to be correlated with optical anisotropy unexpected in the matrix of (001) layers. Recent photoluminescence studies (Worschech et al 1996) could prove linear polarization $\bar{E} \| [110]$ and $\bar{E} \| [1\bar{1}0]$. The polarization degree varies with layer thickness suggesting distinct dependence on the amount of anisotropic strain relief (Worschech et al 1998).

In this paper we are studying by means of SEM cathodoluminescence and PL spectroscopy the relationship of the evolution of asymmetric misfit dislocation configurations to the strain relaxation as monitored by the optical anisotropy. Using CL microscopy enables one to reveal the distinct contribution of A(g)- and B(g)- related dislocation arrangements to strain reduction in the epilayers. For this purpose, the dislocation-induced Y luminescence is utilized to identify Se(g) dislocation segments occurring in the misfit dislocation arrangements. The observed evolution of separated [110] and [1$\bar{1}$0] misfit dislocation subsystems will be explained in the framework of a geometrical model for the activation of particular {111}⟨1$\bar{1}$0⟩ slip systems producing at surface nucleation sites Zn(g)- and Se(g)-related dislocation half loops, respectively.

2. MODEL OF STRAIN RELAXATION

Although the ZnSe/GaAs(001) heterostructures are being grown under optimized conditions with regard to the lattice mismatch and density of grown-in defects in the substrates, the strain relaxation proceeds involving a relatively high density of misfit dislocations.

Current understanding of the stress relief in the ZnSe epilayers is based on the Dodson-Tsao-Model (Dodson and Tsao 1987). It implies particular nucleation and multiplication processes of polar glide dislocations being necessary for the formation of the final misfit dislocation configurations.

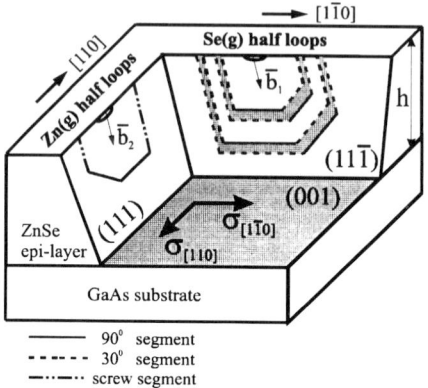

Fig. 1: Scheme of evolution of misfit defect configuration in ZnSe epilayer by nucleation and propagation of Zn(g)- and Se(g)-related glide dislocations.

Dislocation nucleation at surface sites is considered by various authors (Fitzgerald et al 1989, Guha et al 1993, Hull and Bean 1992, Petruzzello et al 1988) a predominant defect generation mode because grown-in threading dislocations are not available in adequate quantities from the high-quality GaAs substrates. In Fig. 1 a schematic model proposed for nucleation and propagation of Zn(g)- and Se(g)-related dislocation half loop structures is illustrated. Independent creation of the Zn(g)- and Se(g)-related dislocations for layer thickness, h, just above the critical thickness, h_c, is assumed. The dislocations generated have Burgers vectors $\overline{b} = a/2 \cdot \langle 101 \rangle$ and spread by slipping in $(\overline{1}1\overline{1})[011]$ and $(11\overline{1})[011]$. The resulting orthogonal [110] and $[1\overline{1}0]$ dislocation subsystems form an in-plane network consisting of the polar misfit segments, and yield a certain density of threading polar and screw line segments ending up in the layer surface, also.

Taking into account differences in the nucleation and migration behaviours for the dislocations with opposite polarity, this model predicts asymmetric strain relaxation due to an unequal density and distribution of the defects in the independent [110] and $[1\overline{1}0]$ dislocation subsystems. The asymmetric strain relaxation may be regarded as the origin of the optical anisotropy occurring in the (001)ZnSe layers as proved by PL polarization (Worschech 1996). The outlined model is supported by the experimental findings reported here, which seem to indicate dislocation nucleation at the epilayer surface.

3. EXPERIMENTAL

Samples used were grown by MBE at the Department of Physics of the University of Würzburg. The nominally undoped layer material obtained on almost dislocation-free (001) GaAs showed high n-type resistivity. A detailed description of this material is given by Fischer (1995).

The PL studies were performed applying Argon-Ion-Laser excitation at 364nm. During the measurements the samples were cooled down to 2°K in an optical He-bath cryostat. The luminescence was analyzed using an YOBIN-YVON spectrometer HR1000 with a resolution of 0.008nm. For the purpose of spectra recording, an optical multichannel analyzer LN/CCD-1100-pf/UV (PRINCETON INSTRUMENTS Inc.) was utilized.

The samples were investigated by panchromatic and spectrally resolved CL-microscopy in a SEM (Schreiber and Hergert 1989) equipped with the low-temperature cryostat sample stage CF302 and the MONO-CL systems (OXFORD Instr.). The sample temperature applied in all CL experiments was 72K. The CL image recording was performed by means of the lock-in technique. Quasi-monochromatic CL maps could be obtained using spectral filters with cut-off wavelengths for separating matrix emission from defect contrasts caused by the dislocation-bound luminescence.

4. EXAMINATION OF STRAIN RELAXATION

Utilization of polarization-resolved PL spectroscopy yielded very instructive information on the stress state of the epilayers. The registered PL spectra proved alteration of stress with layer thickness. The observed spectral shift, splitting and polarization of the layer luminescence peaks have been analyzed according to the response of the electronic valence band structure on strain. The expected shift and corresponding splitting of the excitonic peak are given by

$$\Delta E_{lh} = \left(-2a\frac{C_{11}-C_{12}}{C_{11}} - b\frac{C_{11}+C_{12}}{C_{11}} \right)\varepsilon \quad \text{and} \quad \Delta E_{hh} = \left(-2a\frac{C_{11}-C_{12}}{C_{11}} + b\frac{C_{11}+C_{12}}{C_{11}} \right)\varepsilon, \quad (1)$$

where a is the hydrostatic potential, b is the sheer deformation potential, C_{nm} are elastic constants and ε is the layer strain.

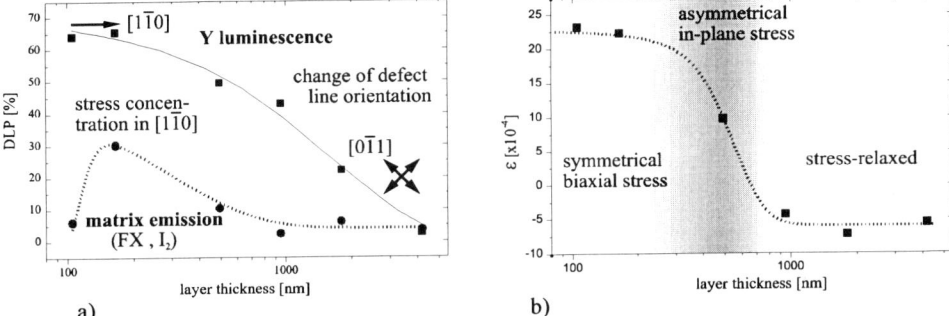

Fig. 2: Degree of linear polarization (DLP) for Y luminescence and matrix emission in dependence on layer thickness (a). Relaxation of strain ε with increasing layer thickness (b) as derived from the PL spectra.

To the stress-induced polarization applies $\overline{E} \perp \overline{\sigma}$ as a good approximation, with $\overline{E}, \overline{\sigma}$ as polarization and stress vectors, respectively. The registered PL spectra included a pronounced Y luminescence peak at about 2.6eV proving Se(g) dislocations in the epilayer. The Y peak specific for this type of polar glide dislocation originates from excitons bound in the defect-induced electronic states (Rebane and Shreter 1991). Contrary, Zn(g) dislocations are believed to cause non-radiative recombination.

In particular, the Y luminescence have very recently been found to show relatively strong polarization parallel to the direction of the defect lines (Mitsui and Yamamoto 1997). This feature is exploited to characterize the orientation of Se(g) dislocation line segments in the epilayers.

In Fig. 2.a the graphs of the degree of linear polarization (DLP) as determined for the Y luminescence and matrix emission, respectively, are given as functions of layer thickness. The DLP of the matrix emission exhibits a maximum pointing at anisotropical stress concentration in [1$\bar{1}$0], whereas in both very thin layers and for increased layer thickness the stress is likely to be more isotropical.

The DLP of Y luminescence is seen to decrease steadily with increasing layer thickness. That should illustrate the change of the orientation of Se(g) dislocation segments under consideration from [1$\bar{1}$0] to [0$\bar{1}$1].

The graph in Fig. 2.b shows the relaxation of strain ε upon the epilayer thickness as obtained by evaluating the exciton peak splitting.

The thickness-dependent strain behavior demonstrates good correspondence with the Dodson-Tsao-Model and appears to relate clearly to the misfit dislocation configurations recognized in the ZnSe/GaAs(001) samples.

5. ANALYSIS OF MISFIT DISLOCATION CONFIGURATION

The sample series studied covers the range of layer thicknesses (h) between the pseudo-morphic state and full stress relaxation of the ZnSe/GaAs(001) heterostructure.

CL micrographs presented in Fig. 3 show typical dislocation configurations as observed for the various epilayers from 50nm up to 1800nm. No defect contrasts are revealed for the pseudomorphic layer (Fig. 3.a), indicating missing interface- as well as misfit-related defects. Thus, no stress relaxation is expected. Fig. 3.b shows the CL micrograph taken from a 165nm thick epilayer; that is close to the critical layer thickness (h_c). The picture contains bright CL defect contrasts arranged along $[1\bar{1}0]$ only and represents the initial formation of misfit dislocations. The bright defect contrasts correspond to the appearance of the Y luminescence strongly polarized $\parallel$ $[1\bar{1}0]$. This gives clear evidence for Se(g) dislocation segments occurring within the $[1\bar{1}0]$ subsystem in misfit position.

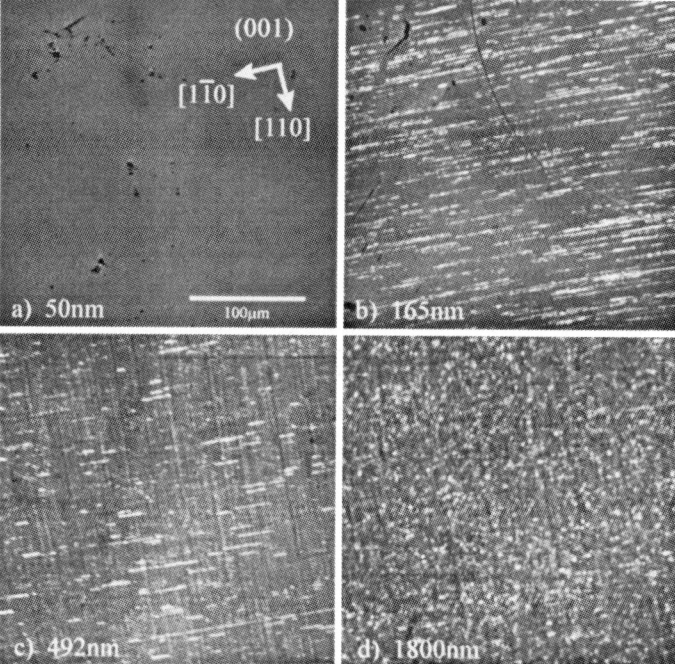

Fig. 3: Dislocation configuration in dependence on ZnSe epilayer thickness comprising Zn(g) and Se(g) dislocations as revealed by CL microscopy discriminating the polar defects by dark and bright contrasts. No misfit dislocation for sub-critical layer thickness (a), Se(g)-related $[1\bar{1}0]$ subsystem formed just above critical thickness (b), appearance of Zn(g)-related $[110]$ subsystem (c), arrangement of threading dislocation segments at the surface of thicker layer (d) experienced almost full stress relief.

The preferential emergence of the $[1\bar{1}0]$ misfit dislocation configuration results in asymmetrical partial stress relief causing optical anisotropy as seen from the PL measurements.

The increase of the layer thickness ($h>h_c$) is found to be accompanied by the development of the second dislocation subsystem. In Fig. 3.c dark CL defect contrasts aligned along $[110]$ can be seen, which have been formed in addition to the $[1\bar{1}0]$ oriented bright contrasts. The dark contrast lines may be attributed to Zn(g) dislocations arranged in the relevant $[110]$ dislocation subsystem. The entire dislocation configuration under consideration as imaged by the bright and dark contrasts illustrates a transient stage for the formation of a misfit dislocation network appearing in the final relaxation state.

Fig. 3.d is the CL micrograph of a thicker epilayer. The sample surface has a sufficient distance from the interface to contain the dislocation configuration as characteristic of the upper region of a stress-relaxed layer. The picture reproduces mainly the arrangement of threading dislocation segments. There is no preferred contrast distribution with respect to $[110]$ or $[1\bar{1}0]$. The prevailing densities of dark contrasts appearing in layers about 4μm thick (not shown here in the CL micrographs) point at an increasing content of Zn(g) and/or screw dislocation segments. In this case, the total defect density can be estimated at $10^6..10^7 cm^{-2}$.

6. COMMENT ON DISLOCATION NUCLEATION SITE

The crystallographically determined misfit dislocation configurations deduced from the CL microscopy confirms the important role of the polar glide dislocations within the process of plastic strain relaxation. The preferential nucleation of Zn(g)- and Se(g)-related dislocation half-loops at surface site, as supposed in the model above, seems to be confirmed by the conclusions to be drawn from the two CL images given in Fig. 4. The micrographs were obtained with spectrally discriminated CL signals of the ZnSe epilayer and GaAs substrate, thus the defects situated in these regions should be imaged separately. The layer thickness of the chosen sample corresponds to the initial stage of plastic relaxation.

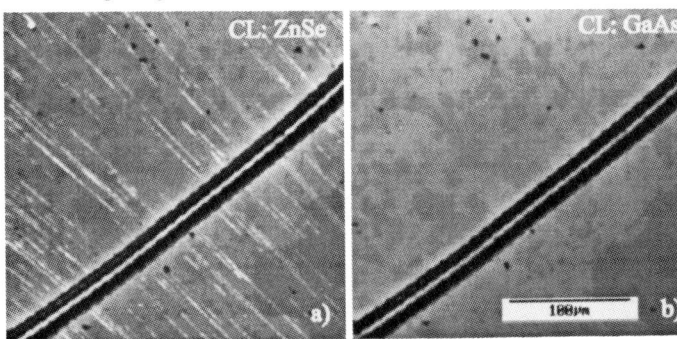

Fig. 4: Defect recognition in ZnSe epilayer (a) and GaAs substrate (b), strong dark contrasts belong to dust particles at layer surface and a scratch.

There is no indication of any interface defects such as grown-in substrate dislocations or stacking fault pyramids (Rosenauer et al 1996) in the CL pictures. From the CL image of the ZnSe layer the Se(g)-related dislocation configuration is deduced to be situated within the near-surface region of the epilayer. Additional results of beam voltage dependent CL investigations support the idea of active nucleation centers at the layer surface.

7. SUMMARY AND CONCLUSIONS

The plastic relaxation of layer stress in the ZnSe/GaAs(001) heterostructure has been studied with regard to the Dodson-Tsao-Model assuming dominant contribution of polar glide dislocations nucleating during the relaxation process without interfacial defects.

PL spectroscopy and CL microscopy were successfully utilized to reveal the stress states and misfit dislocation configurations in the examined ZnSe epilayers. Furthermore, the luminescence technique reflected the optoelectronic activity of the observed misfit dislocations that is not accessible by TEM yielding structural information only. The dislocation-related spectroscopical data based on the Y luminescence allowed the reliable recognition of Se(g) dislocation segments within the occurring dislocation configurations.

A series of ZnSe/GaAs(001) samples with various layer thicknesses corresponding to the range from the compressively stressed pseudomorphic to the completely stress-relaxed layer state has been investigated. The results obtained establish distinct contributions of the Zn(g)- and Se(g)-related dislocations arranged in [110] and [1$\bar{1}$0] subsystem, respectively. The corresponding asymmetrical strain relaxation results in an anisotropical stress relief as clearly demonstrated by the degree of linear polarization of the layer luminescence.

The evolution of the misfit dislocation configuration realizing stress relief in dependence on layer thickness could be analyzed. The strain relaxation was found to start close to the critical layer thickness by the preferential generation of Se(g)-related dislocation structures only, which form the $[1\bar{1}0]$ subsystem. This asymmetric defect configuration was shown to relieve stress in $[110]$.

With increasing layer thickness accompanied by stress concentration in $[1\bar{1}0]$ direction initiating the appearance of the $[110]$ subsystem related to the formation of Zn(g) dislocation, a network configuration of the misfit dilocations was brought out, which tends to produce more isotropic stress relief. The thicker epilayers considered exhibited almost complete strain relaxation as diagnosed in the near-surface region by the PL polarization measurements. In this case, CL microscopy yielded high-density threading dislocation configurations including mainly Zn(g) and/or screw dislocations, and to lesser extent Se(g) dislocation segments, as well. The observed misfit dislocation configuration may be explained by the geometrical model proposed for dislocation glide in the $\{111\}\langle1\bar{1}0\rangle$ slip system operating in the stress field originating from the interface lattice mismatch. The experimental findings concerning the evolution of the $[110]$ and $[1\bar{1}0]$ dislocation subsytems support the idea of independent nucleation of Zn(g) and Se(g) dislocations at surface sites.

Finally, it is worth mentioning that as for the thick stress-relaxed epilayer the CL micrographs emphasize Zn(g) and screw dislocation segments causing non-radiative carrier recombination as dominating electronically active defects which may extend in subsequent epitaxial layers.

ACKNOWLEDGEMENT

The author would like to thank Prof W Ossau, University of Würzburg, for using lab and equipment to perform the PL measurements. A special thank goes to Dipl Phys B König for friendly help. The valuable assistance of Dipl Ing H Mähl during the scanning electron microscope experiments and the useful discussions with Dr S Hildebrandt are acknowledged. This work was supported by the 'Graduiertenkolleg 415' of the 'Deutsche Forschungsgemeinschaft' and by the 'Kultusministerium des Landes Sachsen-Anhalt' under project number: 2437A/0086B.

REFERENCES

Behr T W 1997 dissertation, University of Würzburg

Dodson B W and Tsao J Y 1987 Appl. Phys. Lett. **51**, 1325

Fischer C 1995 diploma work, University of Würzburg

Fitzgerald E A, Watson G P, Proano R E, Ast D G, Kirchner P D, Pettit G D and Woodall J M 1989 J. Appl. Phys. **65** (6) 2220

George A and Rabier J 1987 Revue Phys. Appl. **22**, 941

Guha S, Munekata H and Chang L L 1993 J. Appl. Phys. **73** (5), 2294

Hull R and Bean J C 1992 Sol. St. and Mat. Sci. **17** (6), 507

Mitsui T and Yamamoto N 1997 J. Appl. Phys. **81** (11) 7492

Petruzzello J, Greenberg B L, Cammack D A and Dalby R 1988 J. Appl. Phys. **63** (7) 2299

Rebane Yu T and Shreter Yu G 1991 Springer Proc. Phys. **54**, Polycrystalline Semiconductors II, 28

Rosenauer A, Reisinger T, Franzen F, Schütz G, Hahn B, Wolf K, Zweck J and Gebhardt W 1996 J. Appl. Phys. **79** (8) 4124

Schreiber J and Hergert W 1989 Inst. Phys. Conf. Ser. **102**, 97

Worschech L, Fischer C, Schenk H, Ossau W, Kurtz E, Schäfer H, Faschinger W, Waag A and Landwehr G 1996 Int. Symp. On Blue Laser and LED's 421

Worschech L, Ossau W, Lugauer H-J, Behr T, Waag A and Landwehr G 1998 J. of Crystal Growth **184/185** 500-504

Inst. Phys. Conf. Ser. No 164
Paper presented at Microsc. Semicond. Mater. Conf., Oxford, 22–25 March 1999

Anisotropic lattice mismatch accommodation in the epitaxial MnAs/GaAs-system

A Trampert, F Schippan, L Däweritz and K H Ploog

Paul-Drude-Institut für Festkörperelektronik, Hausvogteiplatz 5–7, D-10117 Berlin, Germany

ABSTRACT: We investigate the epitaxial growth of hexagonal MnAs on cubic GaAs. The layers were deposited by solid-source molecular beam epitaxy. Despite the different symmetry of the adjacent planes at the hetero-interface and the large lattice mismatch, which is additionally orientation dependent, the hexagonal MnAs grows epitaxially on (001) GaAs with its prism plane parallel to the cubic substrate. A detailed TEM analysis of the interface structure as well as of the lattice mismatch accommodation process explains these unexpected results of real heteroepitaxy.

1. INTRODUCTION

The heteroepitaxial growth of magnetic layers on semiconducting substrates has drawn considerable attention because of its potential to develop novel device structures (Prinz 1990). The crystalline quality of these heterostructures is most important because of its critical role for the system's physical properties. The different structure and the different atomic bonding of the materials involved in these newly developed heterosystems form interfaces which are of particular importance because they do not only determine the epitaxial alignment, but also the way of lattice mismatch accommodation and, in addition, the character and density of extended defects. MnAs-GaAs is considered as a promising candidate in this field (Tanaka 1998), although the growth of MnAs is impeded by the appearance of structural phase transitions (Okamoto 1989).

2. EXPERIMENTAL DETAILS

Thin films of hexagonal MnAs are deposited on cubic GaAs(001) substrates by solid-source molecular beam epitaxy (MBE). A 100 nm thick GaAs buffer layer is grown in order to realize well-defined substrate surfaces for the following heteroepitaxy. The growth temperature and growth rate of MnAs is about 250°C and in the range of 9 to 21 nmh^{-1}, respectively (Schippan et al 1999). In spite of the marked difference in both the crystal symmetry and the lattice constant, MnAs shows the following epitaxial orientation relationship which was observed by in-situ reflection high-energy electron diffraction (RHEED): $(1\bar{1}.0)$ MnAs $\parallel$ (001) GaAs and [00.1] MnAs $\parallel$ [1$\bar{1}$0] GaAs, i.e., the hexagonal prism plane is parallel to the cubic plane. In order to get a better understanding of this unexpected result, the interface structure and the lattice mismatch accommodation is investigated by conventional as well as high-resolution transmission electron microscopy (TEM). The TEM specimen were prepared using standard techniques involving mechanical grinding, dimpling and ion beam milling in a cold stage (liquid nitrogen) in order to minimize sample damage. The TEM observations are carried out in Jeol JEM 4000FX microscope operating at 400 kV.

3. RESULTS AND DISCUSSION

Figure1 shows selected area diffraction (SAD) patterns obtained from three different directions with respect to the GaAs substrate: (a) plan-view along the [001] GaAs-direction, (b) cross-sectional view parallel to the [110] and (c) parallel to the $[\bar{1}10]$ GaAs-direction. These diffraction patterns reveal the same orientation relationship between both crystal lattices as that obtained by the in-situ RHEED experiments. Besides, the SAD patterns clearly demonstrate the asymmetric character of the interface resulting in misfit parameters which strongly depend on the inspected in-plane direction. The lattice misfit f corresponding to the $[\bar{1}10]$ GaAs-direction, i.e. between the {00.2} MnAs and the {110} GaAs planes, amounts to 30%, based on the following definition for f:

$$f = \frac{d_{epi}\{hkl\} - d_{sub}\{h'\,k'\,l'\}}{d_{sub}\{h'\,k'\,l'\}}$$

Here, $d_{epi}\{hkl\}$ and $d_{sub}\{h'k'l'\}$ denotes the corresponding lattice plane distance on both sides of the interface in epilayer and substrate, respectively. The value f for the GaAs [110]-direction, i.e. between the {11.0}MnAs and the {110}GaAs planes, is only about 7.5%. This large direction dependence of lattice misfit and the differences in crystal symmetry result in anisotropic lattice mismatch accommodation.

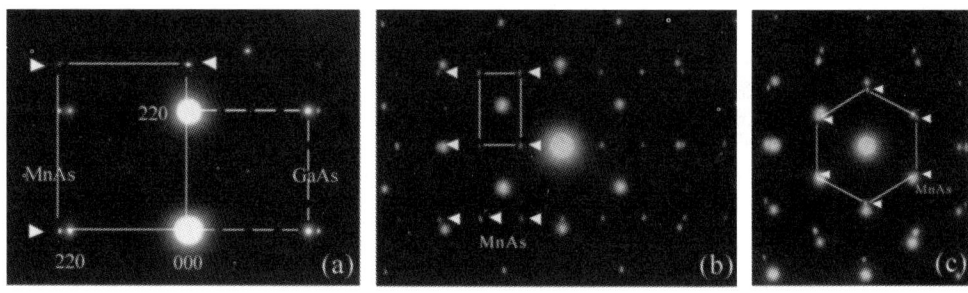

Fig. 1: (a) SAD pattern in plan-view taken along [001] GaAs ∥ [1 $\bar{1}$.0] MnAs; (b) SAD pattern in cross-sectional direction along [110] GaAs ∥ [11.0] MnAs and (c) along [1 $\bar{1}$ 0] GaAs ∥ [00.1] MnAs.

Figure 2(a) shows a cross-sectional dark field micrograph of the MnAs/GaAs-heterostructure imaged along the [00.1] MnAs ∥ [1 $\bar{1}$ 0] GaAs-direction (lattice misfit 7.5%). The abrupt change in contrast reflects a smooth and chemically sharp boundary with no indication of an extended interfacial reaction phase. A periodic array of strain contrast features along the interface (marked by arrows) is clearly observed which is expected for an array of misfit dislocations. In fact, the high-resolution image [Fig. 2(b)] confirms the semi-coherent description of the interface where regions of perfect lattice matching are separated by localized misfit dislocations, which are characterized by a strong lattice plane bending perpendicular and parallel to the interface. A Burgers circuit around such a dislocation core is used to determine the Burgers vector to be **b** = 1/3 [11.0], a typical lattice dislocation in hexagonal materials. This Burgers vector is located parallel to the boundary plane and, therefore, most efficient in strain relief. However, by measuring the mean distance D between the dislocations, a residual strain ε in the epilayers is calculated if applying the equation $\varepsilon = f - b / D \approx - 0.01$, where f defines the above introduced lattice misfit. Assuming that the mismatch between the MnAs layer and the GaAs substrate is completely relaxed at the growth temperature, a residual compressive strain arises during sample cooling to room temperature, mainly because of a discontinuous increase of the lattice constant a of about 1% at the structural transition temperature $T_c \approx 45°C$ (Willis and Rooksby 1954). This increase in the MnAs lattice constant results in a decrease of the lattice mismatch value and, thus, in larger misfit dislocation distances. However, because of the relative low temperature of the structural transition, the dislocations should be immobile to follow this change in lattice constant resulting in residual epilayer strain. In fact, this compressive stress state caused by the too closely spaced misfit dislocations is also verified by local lattice distortions in perfectly matched interfacial regions. In the high-resolution image [Fig. 2(b)] the angle Θ between the misaligned {111} GaAs and the {1$\bar{1}$.0} MnAs planes is increased from the unstressed value of 5.26° to 6.5 ± 0.5°, corresponding to a compressively stressed state.

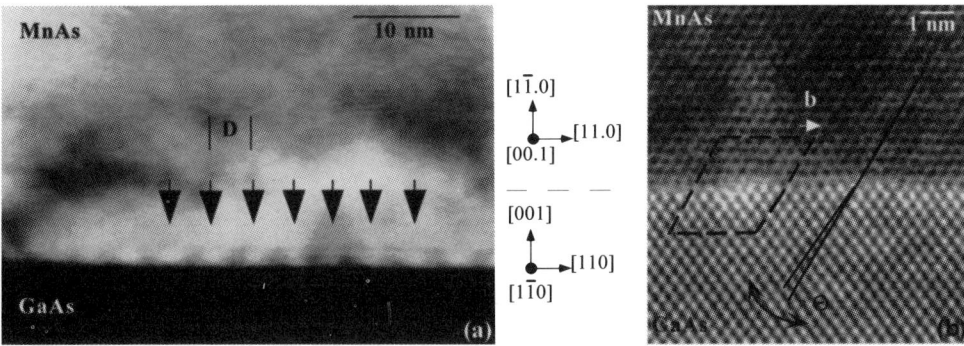

Fig. 2: Cross-sectional dark field micrograph (a) and the HRTEM image (b) of the MnAs/GaAs interface with the incident beam parallel to [00.1] MnAs ∥ [1$\bar{1}$0] GaAs. Note the array of periodic strain contrast along the interface (arrows) in (a) and the Burgers circuit around a dislocation core in (b).

Along the GaAs [110]-projection, where the lattice mismatch is about 30% between the {110} GaAs and the {00.2} MnAs planes, the HRTEM micrograph [Fig.3(a)] reveals a completely different interface character. The HRTEM contrast of the MnAs lattice imaged in [11.0]-direction is wavy-like with a period corresponding to the hexagonal lattice constant c in agreement with image simulations. No localized misfit dislocation or strong coherence strain features, respectively, are visible in the image. At a first glance, the interface appears completely incoherent, as one would expect for heterosystems with large lattice mismatch and weak interfacial bond strength. However, a more careful inspection reveals an interface structure where the mismatch accommodation becomes plausible by employing the near coincidence site lattice (CSL) model (see, e.g., Trampert et al 1997): every fourth {00.2} MnAs plane fits every sixth {220} GaAs plane forming a commensurate interface region [see Fig.

2(b)]. This 4/6-ratio reduces the natural lattice mismatch to about 5%, a reasonable value to guarantee epitaxial growth. The deviation from the exact coincidence F is accommodated by secondary or coincidence lattice misfit dislocations, an example is shown in Fig. 3(c). The result is then an interface structure with commensurate domains separated by localized CSL-dislocations. The measured spacing of these secondary dislocations agrees with the calculated value given by $D_F = b_F/F$, where b_f denotes the Burgers vector of the coincidence lattice misfit dislocation.

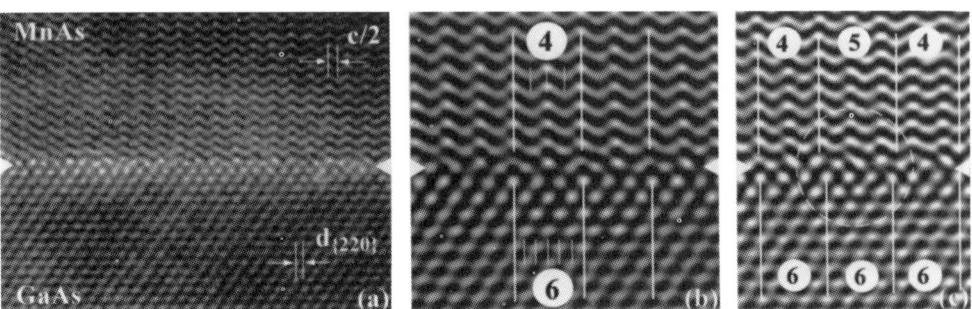

Fig. 3: Cross-sectional HRTEM image of the MnAs/GaAs interface in [11.0] MnAs ‖ [110] GaAs projection, (a) overview and (b) magnified part after Fourier filtering; (c) coincidence lattice misfit dislocation indicated by the white circle .

4. CONCLUSION

We have demonstrated that the differences in crystal symmetry and lattice constant lead to an anisotropic lattice mismatch accommodation. Along the $[1\bar{1}0]$-direction, the misfit strain is relieved by regularly arranged misfit dislocations, whereas along the perpendicular interface direction, the four times larger lattice misfit produces no localized misfit strain. The interfacial atomic arrangement is explained by a near coincidence site lattice model.

ACKNOWLEDGEMENTS

The authors would like to thank the Forschungszentrum Jülich, Institut für Festkörper-forschung, for providing their microscope facilities.

REFERENCES

Okamoto H 1989 Bulletin of Alloy Phase Diagrams **10**, 549
Prinz GA 1990 Science **250,** 1092
Schippan F, Trampert A, Däweritz D and Ploog KH 1999 J. Cryst. Growth (in press)
Tanaka M 1998 Physica E **372**, 380
Trampert A, Brandt O, Yang H, Ploog K H 1997 Appl. Phys. Lett. **70**, 583
Willis BTM and Rooksby HP 1954 Prov. Phys. Soc. (London) B **67**, 290

Inst. Phys. Conf. Ser. No 164
Paper presented at Microsc. Semicond. Mater. Conf., Oxford, 22–25 March 1999

Structural properties of compressive and tensile strained InGaAs/InP heterostructures

L Lazzarini, G Salviati, M Natali°, M Berti°, D Cerolini°, D De Salvador°, AV Drigo°, G Rossetto^ and G Torzo^

CNR-MASPEC Institute, Parco Area delle Scienze 37/A, I-43010 Fontanini- Parma, Italy
° INFM, Department of Physics, University of Padova, Via Marzolo 8, I-35131 Padova, Italy
^ CNR-ICTIMA Institute, Corso Stati Uniti 4, I-35127 Padova, Italy

ABSTRACT: Different extended defects affect the InGaAs alloy when tensile or compressively stressed. In tensile stressed samples, grooves, planar defects and cracks are present in addition to the interfacial network of misfit dislocations. Here we report a systematic study of the structural properties of MOVPE grown InGaAs/InP tensile and compressively strained epilayers carried out by means of by TEM, CL, RBS-Channeling, X-ray diffraction and SFM techniques. The correlation between the observed defects and the mechanisms of strain relaxation in both cases is discussed.

1. INTRODUCTION

The strain relaxation mechanism for lattice mismatched epitaxial layers under compressive strain has been extensively studied, in particular for the InGaAs/GaAs system (Drigo et al 1989, Lavoie et al. 1995). For layers under tensile strain such as InAlSb/InSb it has recently been suggested (Maignè et al 1995) that the relaxation mechanism might be different since the residual strain is significantly larger than in InGaAs/GaAs layers with comparable misfit. However a quantitative comparison between the two systems is not necessarily appropriate because the materials exhibit different physico-chemical, mechanical and thermal properties.

The $In_xGa_{1-x}As/InP$ material system is ideal for the direct comparison of the relaxation processes involved in compressive and tensile strains. In fact, varying the In composition below or above x=0.53, the epitaxial layer shows negative or positive lattice mismatch. Furthermore the role of the Indium composition itself can be studied by comparing compressively strained InGaAs/GaAs and InGaAs/InP layers having the same misfit.

2. EXPERIMENTAL PROCEDURE

Thirtyseven epitaxial films of InGaAs were grown on (001) semi-insulating InP substrates via metalorganic vapour phase epitaxy (MOVPE) in order to cover large intervals of Indium concentration (0.2<x<0.73) and film thickness (8 nm<t<2400 nm).

TEM bright field and dark field images of <001> plan view and <110> cross-sections were recorded with a JEOL 2000 FX microscope at 200 kV on mechano-chemically thinned samples finished by argon ion-milling (Gatan 600 Duo Mill).

Cathodoluminescence (CL) investigations were carried out in a 360 Cambridge Stereoscan SEM in the range 5-300 °K using a nitrogen cooled Germanium detector.

The indium composition and strain of the layers were determined by high-resolution X-ray diffraction (Philips MRD) measuring rocking curves of symmetric (004) and asymmetric (444) reflections along four <110> in-plane directions. RBS measurements were performed with 2MeV 4He+

beams at the AN-2000 Van der Graaf accelerator (LNL Legnaro) to determine the layer thickness and to cross-check the indium compositions.

Scanning Force Microscopy measurements were performed with a Park CP model operated in contact-mode, using ultra-lever tips with nominally 10 nm tip radius.

3. RESULTS AND DISCUSSION

3.1 Compressively strained layers

The strain relaxation along the two 110 directions is symmetric in the compressive case (In concentration varying in the interval 0.61<x<0.73). The difference in the relaxation coefficients, defined as $\Delta R = R[110] - R[1\bar{1}0]$ (where $R = (m-\varepsilon)/m$, ε=strain, m=misfit), is relatively small and of the same sign and order of magnitude as found for the InGaAs/GaAs system by Lavoie et al (1995). In Fig. 1 it is shown that he residual strain as a function of the layer thickness follows the empirical curve previously found for MBE grown InGaAs/GaAs layers by Drigo et al (1989) relaxing at a thickness much larger than predicted by the Matthews-Blakeslee curve (1974). It is worth to note that those samples had Indium concentrations x<0.2.

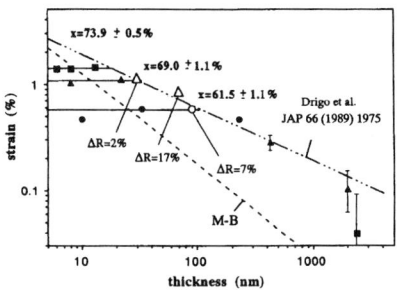

Fig. 1: Residual strain for compressive InGaAs/InP layers vs layer thickness

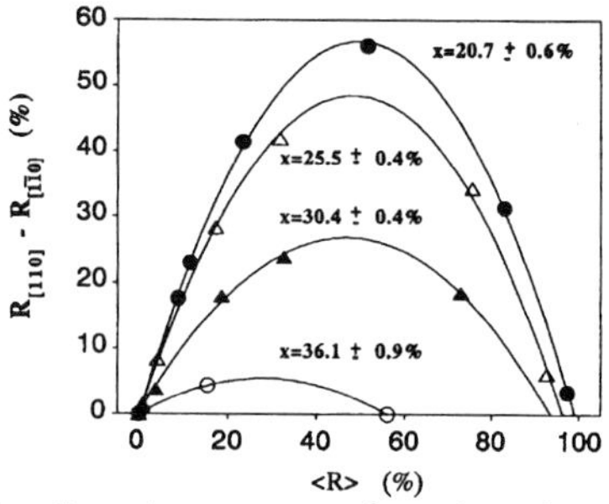

Fig. 2: Asymmetry of strain relaxation in tensile InGaAs/InP layers vs average relaxation

The surface morphology of compressive InGaAs/InP layers having thicknesses between the above mentioned relaxation curves is characterized by striations parallel to [1$\bar{1}$0] associated with MDs. Striations parallel to [110] develop for samples along the relaxation curve, leading to the typical cross-hatch pattern. The only extended defects present in these samples are misfit dislocations, 60° mixed in character, arranged in the usual orthogonal network at the interface between the epilayer and the substrate. The density of the misfit dislocations accounts for the measured strain relaxation. This behaviour is similar to that reported for InGaAs/GaAs by Lavoie et al (1995). The main result is that strain relaxation in compressive InGaAs layers is independent of both growth conditions and Indium composition. The present samples have larger Indium composition and higher growth temperature than those previously studied and MBE grown on GaAs (Drigo et al 1989) so that a significant increase in dislocation glide velocity is expected (Sumino and Yonegana 1992). From their equal relaxation behaviour it can be inferred that strain relaxation in compressive InGaAs layers is not kinetically limited by dislocation glide but by activation of MD nucleation or multiplication mechanisms.

3.2 Tensile strained layers

For tensile InGaAs/InP layers (In concentration varying in the interval 0.2<x<0.36) the strain relaxation is highly asymmetric. Fig. 2 shows the behaviour of the asymmetry parameter ΔR as a function of the average relaxation coefficient for four series of samples of different composition. The [110] direction relaxes first and the strain relaxation asymmetry increases with increasing tensile misfit. For layers with Indium composition higher than 38% we expect that the asymmetry would disappear. The critical thickness for strain relaxation along [1$\bar{1}$0] is about one order of magnitude larger than for compressive layers. Once initiated, however, the strain relaxation proceeds much faster than for compressive layers. Although the strain relaxes along [110] at smaller thicknesses than along [1$\bar{1}$0] the relaxation occurs at larger thicknesses than for compressive layers. Samples that appear pseudomorphic from XRD analysis contain extended defects along both <110> directions as shown by CL images (Fig. 3). The TEM analysis reveals that, at this stage of growth, all the defects are long (some mm) stacking faults (SF) or twins (TW), suggesting that all the dislocations are dissociated into partials.

One of the main morphology differences with respect to compressive samples is the development of grooves along [1$\bar{1}$0] and [110], which start in correspondence with the onset of measurable strain relaxation along the respective perpendicular directions. While grooves deepen and widen as the layer thickness increases, their density doesn't change significantly. This fact indicates that the density of grooves cannot be correlated directly with the amount of relaxed strain.

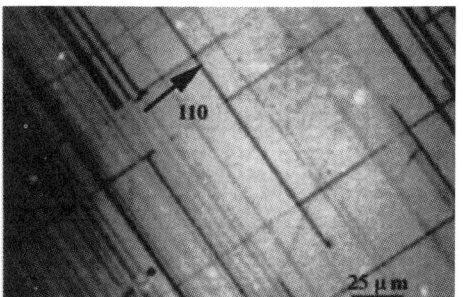

Fig. 3: Panchromatic CL image of an unrelaxed sample, showing a network of extended defects.

Fig. 4: (001) plan view TEM image of a 450 nm thick layer where the interface region has been removed.

Figure 4 shows a plan-view TEM image of a network of SF and/or TW which reproduces the groove in-plane arrangement as it is observed by SFM. We thus believe SF/TW are correlated to the formation of grooves. The noncrystallographic surface step associated to the SFs could prevent step-flow growth in the neighbourhood of the SF as suggested by Van Nostrand et al (1998) for tensile SiGe/Ge layers. SF and TW coexist with a network of misfit dislocations whose density roughly corresponds to that required to relax the measured (XRD) amount of relaxed misfit, but only in low-misfit samples. The groove side-walls often coincide with the {111} planes but faceting occurs also on higher index planes such as {112}, {113}, and so on, as shown in Fig 5. A careful inspection of Fig. 4 shows faceting (lozenge features) also along the groove length.

Fig. 5: (110) TEM cross section showing a V-groove defect with faceted side-walls

Fig. 6: (110) TEM cross-section showing a crack crossing a SF or TW.

Other defects peculiar to tensile layes are cracks which in the present samples form mainly along the [110] in-plane direction. The cracks we have observed always pass through the layer and go into the substrate at a depth ranging between once and twice the layer thickness. As it is shown in Fig. 6, the cracks have a very narrow V-shaped profile, lie on vertical {110} planes but, at some depth in the substrate, they deviate onto inclined {111} planes. The density of cracks starts to increase at a thickness which is about twice the theoretical critical thickness predicted by the model of Murray et al (1995), reaches a maximum at a thickness which corresponds to the onset of strain relaxation along [1$\bar{1}$0] and decreases towards zero with further increase of the layer thickness. This behaviour is clearly incompatible with the formation of cracks during growth. On the contrary we have several evidences that cracks form after growth: cross-sectional TEM images usually show cracks cutting extended defects such a SF and TW propagating from the interface to the top of the layer as for instance shown in figure 6. Furthermore SFM images show that atomic growth steps are correlated on the two sides of the crack, implying that no preferential growth has taken place in the relaxed region around the crack. The decrease of the crack density at high thickness on the other hand can be related to the decrease of the residual strain in the layer below the threshold for crack formation. Similarly the absence of cracks along [1$\bar{1}$0] for the higher misfit samples can be understood in terms of formation of cracks after growth and the strong asymmetry of the residual strain of the layers. The model of crack formation of Murray et al (1995) has thus been refined to take into account the asymmetry of the residual strain (Natali 1999). The predicted thickness interval for crack formation reasonably agrees with our experimental data.

Simple calculations show that in our samples the maximum amount of misfit relaxed by cracks turns out to be of the order of 0.1%, which is a negligible quantity, close to the amount of tensile thermal strain added to the layer during the post-growth cooling-down. This fact indicates that strain relaxation by misfit dislocations and by crack formation are competing mechanisms, the latter prevailing at low temperature, where MD nucleation or glide is frozen, eventually being triggered by the thermal strain.

4. CONCLUSIONS

Layers under compression present the same defects and morphology and relax following the same thickness dependence as that found for MBE grown InGaAs/GaAs showing no composition or growth technique effect.

Contrary to the compressive case, in tensile InGaAs/InP layers the strain relaxation is highly asymmetric, being much faster along [110] than along [1$\bar{1}$0]. The thickness for strain relaxation is larger than for the compressive case even for the fast relaxing [110] direction. In the relaxed layers, in addition to the interfacial misfit dislocation network, V-shaped grooves are present, most likely correlated to stacking faults and twins. A significant density of cracks is found only parallel to [110]: this is related to the pronounced asymmetry of the strain relaxation. It has been assessed that cracks form after the growth has been completed.

REFERENCES

Drigo AV, Aydinli A, Carnera A, Genova F, Rigo C, Ferrari C, Franzosi P and Salviati G 1989 J. Appl. Phys. **66**, 3334
Lavoie C 1995 Appl.Phys.Lett. **67**, 3744
Maignè P, Lockwood DJ and Dharma-Vardana C, Webb JB 1995 J.Appl.Phys. **77**, 1466
Matthews JW and Blakeslee AE 1974 J.Cryst.Growth **27**, 118
Murray RT, Kiely CJ and Hopkinson M 1995 Inst.Phys.Conf.Ser.**146**, 207
Natali M 1999 Strain relaxation, defect formation and surface morphology in tensile and compressive InGaAs/InP layers, PhD Thesis, University of Padova
K. Sumino and I. Yonegana (Paper presentedat the 7th Semi-insulating III-V materials Conference) ed C.J. Miner,W.Ford, and E.R. Weber(Bristol, Institute of Physics 1992), 29
Van Nostrand JE, Cahill DG, Petrov I and Greene JE 1998 J.Appl.Phys. **83**, 1096

Inst. Phys. Conf. Ser. No 164
Paper presented at Microsc. Semicond. Mater. Conf., Oxford, 22–25 March 1999

313

TEM study of defects formed in InGaAs/GaAs and InGaAs/InP heterostructure devices

J Kątcki, J Ratajczak, J Muszalski, F Phillipp[+] and N Y Jin-Phillipp[+]

Institute of Electron Technology, Al. Lotników 32/46, 02-668 Warsaw, Poland
[+] Max-Planck Institut für Metallforschung, Stuttgart, Germany

ABSTRACT: Cross-sectional transmission electron microscopy (TEM) was used to find the best way of growing photodetector heterostructures with a low dislocation density InAs active layer. The growth of a thick InAs layers on a GaAs or InP substrate was compared to the growth of a graded $In_xGa_{1-x}As$ buffer layer on a GaAs substrate.

1. INTRODUCTION

For some applications there is a need of good quality InAs layers. The electron mobility in good layers of InAs is equal to 30000 cm^2/Vs and is more than three times higher than that for GaAs. The size of the InAs bandgap equals 0.354 eV and is four times smaller than that for GaAs. This makes InAs and InGaAs compound based devices exhibit good electronic performance. Layers of InAs and InGaAs are grown both on GaAs and InP substrates.

Since the lattice constants of InAs, GaAs and InP are 0.60583, 0.56533 and 0.58688 nm respectively a lattice misfit occurs during the epitaxial growth of InAs layers both on GaAs and InP substrates. Complete miscibility of InAs and GaAs means that by adjusting the mole fraction of InAs the desired properties of $In_xGa_{1-x}As$ alloy can be obtained. The lattice constant, a, of $In_xGa_{1-x}As$ can be determined from the equation: $a = 0.60583-0.04050(1-x)$ [nm](Stringfellow 1993). The higher the indium content, the higher the electron mobility but also the larger the lattice constant. The extreme case of this alloy is InAs.

Heterostructures with an InAs active layer are used for photodetectors. Its guarantees the best performance of the devices but only when dislocation density in an area of the device is small enough. The purpose of this work is to find the best way of growing by means of molecular beam epitaxy (MBE) the low dislocation density InAs layers on GaAs and InP substrates. Cross-sectional transmission electron microscopy was used to observe defects in heterostructures. TEM results were compared with results obtained by other characterisation techniques such as PL spectroscopy and Hall measurements.

2. EXPERIMENTAL DETAILS

Heterostructures discussed in our paper were grown by means of MBE on the (100)-oriented GaAs and InP semi-insulating substrate in a Riber 32P system.

Three sets of heterostructures were grown: *i)* step-graded and index-graded $In_xGa_{1-x}As$ layers on a GaAs substrate; *ii)* thick InAs layers on a GaAs substrate and *iii)* thick InAs layers on an InP substrate. In the first set of samples *a)* in step-graded $In_xGa_{1-x}As$ layers x was increased every 300 nm by 0.11±0.01 and *b)* in index-graded $In_xGa_{1-x}As$ layers x was increased linearly from 0.01 to 0.53. In order to determine the best growth conditions thick InAs layers were grown at temperatures 450°C, 475°C and 500°C on a GaAs substrate and at 480°C on an InP substrate.

Heterostructures were observed in transmission electron microscopes JEM-200CX operating at 200 kV (Institute of Electron Technology) and JEM-4000FX operating at 400 kV (Max-Planck

Institut für Metallforschung). Cross-sectional specimens for the TEM observations were prepared using a technique developed in our laboratory described by Ratajczak (1990) and Kątcki et al (1995).

3. RESULTS AND DISCUSSION

The large difference between lattice constants of InAs and GaAs and InAs and InP causes the lattice misfit between the layers and is the reason for the formation of numerous dislocations in a layer. The performance of InAs photodetectors strongly depends on the dislocation density in active areas of the devices. The lower the dislocation density the better the device performance. Hence, the aim of our work was to find the best way of growing InAs layers of low-dislocation density in crucial areas of the devices i.e. in a 1 μm thick subsurface area of the heterostructure. Three solutions were tested.

3.1 Graded $In_xGa_{1-x}As$ Layers on a GaAs Substrate

In order to slow down the dislocation formation process a buffer graded $In_xGa_{1-x}As$ layer was grown on a GaAs substrate. Fig. 1 shows two examples of such a layer. In Fig. 1a a heterostructure consisting of six $In_xGa_{1-x}As$ sublayers of different indium content and almost equal thickness grown on a GaAs semi-insulating substrate in shown. The exact thicknesses of the sublayers (counting from the substrate) were 250, 300, 300, 300, 300, 300 nm, respectively and exceeded the critical thickness causing the formation of dislocation networks at the interfaces. For those sublayers where the indium content is lower than 0.3, dislocations which formed at interfaces of sublayers moved down to the bulk. In those layers where the indium content exceeded the value of 0.3, threading dislocations were formed. The number of dislocations in the high indium content sublayers is a few orders of magnitude higher than in the layers where x is below 0.3. In top buffer sublayers the number of dislocations was so high that it was difficult to identify the interface between the sublayers.

Another example of an $In_xGa_{1-x}As$ buffer layer is shown in Fig. 1b. In the index-graded $In_xGa_{1-x}As$ layer grown on a semi-insulating GaAs substrate the value of x changed almost linearly from 0.01 to 0.53. On the top of this layer an $In_{0.53}Ga_{0.47}As$ layer was grown. No threading dislocations were observed in the index-graded layer. Due to relaxation of the crystalline lattice during epitaxial growth dislocation networks parallel to the interface were formed. This process took place several times during the growth of this layer. The mechanism of this process is discussed elsewhere (Kątcki 1999). At the top interface of an index-graded $In_xGa_{1-x}As$ layer with $In_{0.53}Ga_{0.47}As$ numerous threading dislocations were formed.

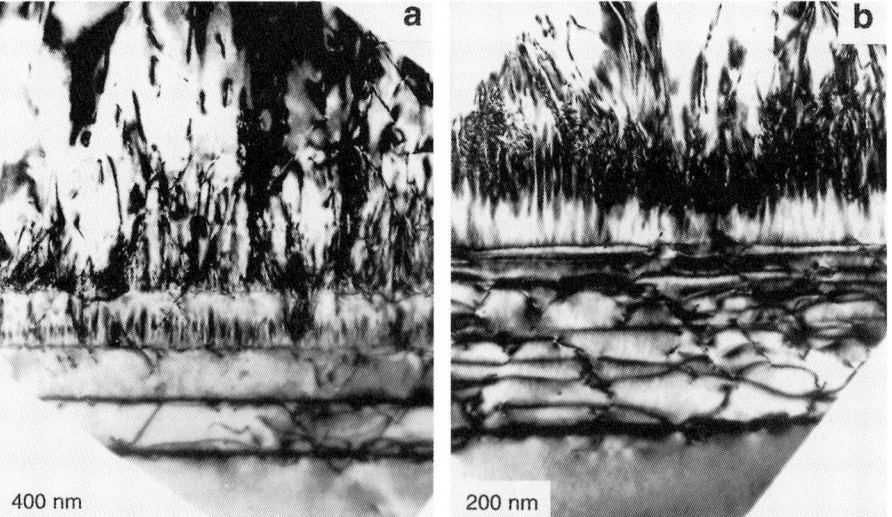

Fig. 1. Cross-sectional TEM view of a) step-graded and b) index-graded $In_xGa_{1-x}As$ layer grown on a GaAs substrate.

3.2 InAs Layers on a GaAs Substrate

Thick InAs layers were grown on a GaAs substrate at different temperature (Fig. 2). A large difference between the lattice constants of InAs and GaAs (misfit coefficient is equal 7.16%) caused the formation of numerous dislocations at the InAs/GaAs interface. The number of dislocations decreases with the distance from the interface. In the subsurface area the dislocation density is a few orders of magnitude smaller than in the bottom part of an InAs layer. Observations of a wide range of samples proved that this result is not caused by changes of the sample thickness in its central part. The layers grown at 450°C, 475°C and 500°C are shown in Fig. 2a, 2b and 2c, respectively. Surfaces of InAs layers grown at 450°C and 475°C are perfectly flat. In Fig. 2c a rough surface of an InAs layer grown at 500°C can be observed. This is the feature of MBE growth. Comparing cross-sectional TEM views of InAs layers grown at various temperatures one can conclude that the least density of dislocations was observed in the InAs layer grown at 475°C and 500°C, but the second one should be eliminated because its rough surface. This conclusion was confirmed both by PL spectroscopy and Hall measurements. InAs layers grown at 475°C exhibit the narrowest FWHM. The carrier concentration for these layers equalled 1.11×10^{16} cm^{-3} (at room temperature) and 6.76×10^{15} cm^{-3} (at liquid N$_2$ temperature). Their electron mobility was equal 14500 cm^2/Vs (at room temperature) and 32800 cm^2/Vs (at liquid N$_2$ temperature). The area of high dislocation density in heterostructures grown at 475°C and 500°C was about 2 µm thick.

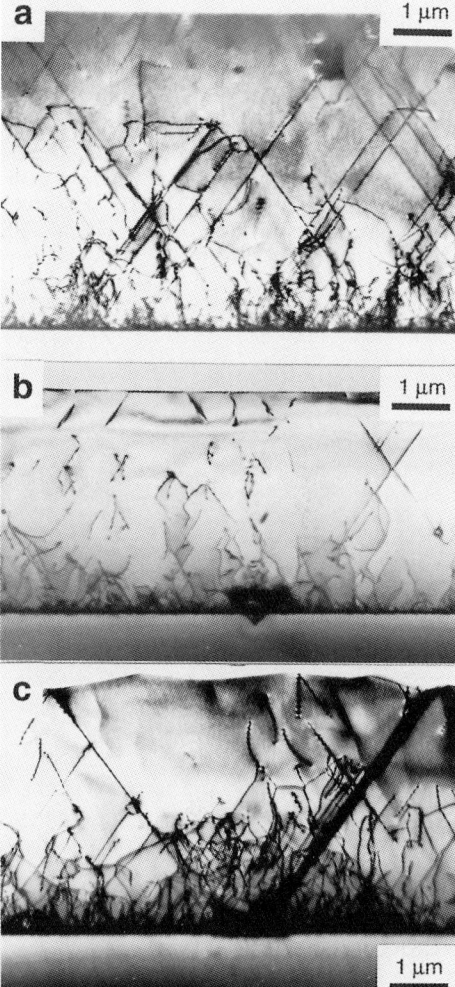

Fig. 2. Cross-sectional TEM view of an InAs layer grown on a GaAs substrate at **a)** 450°C **b)** 475°C and **c)** 500°C.

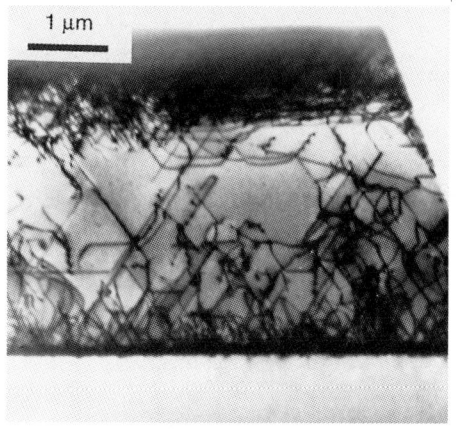

Fig. 3.Dislocation clusters in an InAs layer grown on a GaAs substrate.

The thickness of the same area in heterostructures grown at 450°C was above 3 µm. As a consequence, for low dislocation density area in an InAs layer to be wide enough, the layer must be thicker.

The presence of dislocation clusters in a subsurface area of an InAs layer can be a reason for worsening electrical properties. The layer shown in Fig. 3 exhibits worse transport properties (higher carrier concentration and lower electron mobility).

3.3 InAs Layers on an InP Substrate

In Fig. 4 a cross-sectional TEM view of an InAs layer grown on InP substrate is shown. The misfit coefficient for this system is lower than for the InAs/GaAs system and is equal to 3.81%. This explains why the number of dislocations formed at the InAs/InP interface is smaller than in the case of heterostructures discussed before. Hence, an area of the low-dislocation density is wider. An InAs layer was grown at 480°C. The carrier concentration for these layers was equal 6.70×10^{16} cm^{-3} (at room temperature) and 5.84×10^{16} cm^{-3} (at liquid N$_2$ temperature). The electron mobility equalled 8500 cm^2/Vs (at room temperature) and 11600 cm^2/Vs (at liquid N$_2$ temperature). Values of electron mobility were smaller and of carrier concentration were higher than in the case of an InAs layer grown on a GaAs substrate at 475°C. This can be explained by the presence of numerous long vertical dislocations penetrating the InAs layer.

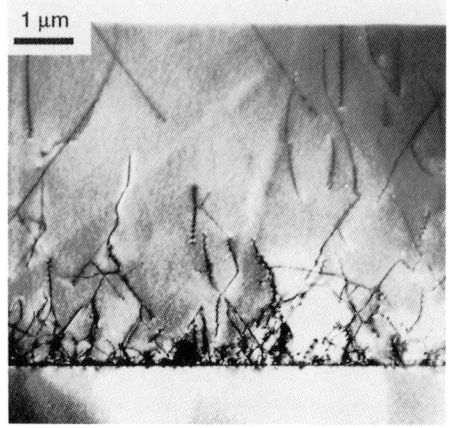

Fig. 4. Cross-sectional TEM view of an InAs layer grown on an InP substrate at 480°C.

3. CONCLUSIONS

Techniques of growing In$_x$Ga$_{1-x}$As based heterostructures from the point of view of obtaining InAs photodetectors were discussed. High performance photodetectors can be obtained from the heterostructures with a low dislocation density InAs layer. This was confirmed by the results of Hall measurements and cross-sectional TEM observations. Application of step-graded and index-graded buffer layers does not give satisfactory results. Despite using graded buffer layers on the top of the buffer numerous threading dislocations were formed. Total thickness of the buffer layer of step graded layer is about 1.8 μm. The thickness of the high dislocation density area in heterostructures grown at 475°C is about 2.5 μm. Comparing heterostructures with and without In$_x$Ga$_{1-x}$As graded buffer layer one can conclude that growing thick InAs layers on a GaAs or InP substrate is a much easier way of obtaining good heterostructures for device application. Despite a bigger lattice misfit InAs/GaAs heterostructures can be used for photodetectors as successfully as heterostructures of InAs/InP.

ACKNOWLEDGEMENTS

This publication is based on the work sponsored by the Polish Government under the project #8T11B.064.14. The authors are very much indebted to Prof. M Bugajski and Drs K Regiński, M Wesołowski and K Klima for collaboration in the investigations, Ms D Szczepańska for assistance with specimen preparation and Ms J Wiącek for careful preparation of micrographs.

REFERENCES

Kątcki J, Ratajczak J, Maląg A and Piskorski M 1995 *Microscopy of Semiconducting Materials 1995*, eds A G Cullis and A E Statton-Bevan, IOP, Bristol, pp. 273-6.
Kątcki J, Ratajczak J, Adamczewska A, Phillipp F, Jin-Phillipp N Y, Regiński K and Bugajski M 1999 *phys. stat. sol.(a)* **171**, 275
Ratajczak J 1990 *IET Rep.* **9**, 51
Stringfellow G B 1993 Indium Gallium Arsenide , eds O Bhattacharya, INSPEC, p.3

Inst. Phys. Conf. Ser. No 164
Paper presented at Microsc. Semicond. Mater. Conf., Oxford, 22–25 March 1999
© 1999 IOP Publishing Ltd

Structural study of GaAs/GaAs twist-bonded compliant substrates

G Patriarche, C Mériadec, G Le Roux, I Sagnes, J-C Harmand and F Glas

France Télécom, Centre National d'Etudes des Télécommunications, Laboratoire Concepts et Dispositifs pour la Photonique, 196 avenue Henri Ravéra, BP 107, 92225 BAGNEUX Cedex, France

ABSTRACT: We investigate by TEM the structure of the interface fabricated by twist-bonding two GaAs wafers in order to obtain a compliant substrate. The interface contains a dense network of pure screw dislocations even for twist angles larger than 15°. Mismatched $In_xGa_{1-x}As$ layers have subsequently been grown on such substrates. X-ray diffraction shows that they contain significantly fewer structural defects than similar layers grown on standard GaAs substrates.

1. INTRODUCTION

The fabrication of electronic and optoelectronic devices requires semiconductor materials of high structural quality. Their active region is often a complex structure involving heteroepitaxial layers grown on a substrate. If the layers are lattice-matched to the substrate or if, when lattice-mismatched, they are kept below their critical thickness, no structural defects generally appear. Above the critical thickness, which decreases rapidly with increasing misfit, the epitaxial layers relax plastically through the introduction of misfit dislocations at the interface. Unfortunately, this relaxation process also generates a large density of threading dislocations extending between the interface and the surface. Not only are these dislocations unnecessary to relaxation, they also deleteriously affect the optical and electrical properties of the structure, and the various procedures tried in order to eliminate them in standard epitaxy have so far remained insufficient.

The use of a 'compliant' substrate should overcome this problem (Lo 1991). Such a substrate must accommodate the substrate/layer misfit by deforming elastically or plastically in a zone localised near its top surface, below the heteroepitaxial layer. This superficial region is the actual compliant layer. Several methods of fabrication of compliant substrates have been described. Here, we investigate the scheme first proposed by Ejeckam et al (1997), which makes use of a thin compliant layer bonded to a bulk substrate with a large twist disorientation (several degrees to several tens of degrees). Ejeckam et al (1997) tested their method by growing on their substrates thick misfitting layers whose cross-sectional Transmission Electron Microscopy (TEM) images showed no evidence of threading dislocation. Zhu et al (1998) grew misfitting multi-quantum wells on similar substrates and measured a photoluminescence efficiency significantly improved compared with growth on a standard GaAs substrate. The mechanisms involved are however not well understood.

Here, drawing upon our previous detailed investigation of the GaAs/InP interface bonded with a slight non-intentional twist (Patriarche et al 1997), we investigate by TEM the structure of the interface between the twist-bonded bulk substrate and compliant layer. We then assess the structural quality of the heteroepitaxial layers grown on such substrates by x-ray diffraction (XRD).

2. FABRICATION OF THE COMPLIANT SUBSTRATES

The compliant substrates were obtained by first bonding two GaAs bulk wafers twisted through a large angle θ (between 6 and 25°) with respect to each other around their common normal. The bonded surfaces, slightly above 1 cm^2, were nominally exact (001) planes. Bonding temperatures between 500°C and 730°C and bonding times of about one hour (unless indicated otherwise) have been used. After bonding, one side of the assembly was thinned, first down to a few tens of micrometers by mechanical polishing, and then down to a few nanometers by chemical etching, making use of etch-stop layers previously deposited on the wafer.

3. STRUCTURE OF THE TWIST-BONDED INTERFACE

At least up to $\theta=16°$, the TEM plan-view images of the twist-bonded interface taken in a weak beam (WB) condition (usually $(\mathbf{g},4\mathbf{g})$) show a square dislocation network (observed only for low twist by Ejeckam et al (1997)). For $\mathbf{g}=220$ we observe half the dislocations, with lines all nearly parallel to $\mathbf{g}$, and for $\mathbf{g}=2\bar{2}0$ the other half (Fig. 1). These are thus pure screw dislocations with Burgers vectors $\mathbf{b}$ of the a/2<110> type. The dislocations lines are in fact parallel to the bisector of the pair of <110> directions closest to each other in the two crystals. In addition to the expected double diffraction spots, the corresponding diffraction patterns display specific spots (Fig. 2) produced by the very regular dislocation network, as already discussed in the case of the GaAs/InP bonded interface (Patriarche et al 1997). Quantitative analysis confirms that the observed network does indeed accommodate the twist. For instance, the spacing measured in Fig. 1 is 2.02 nm, the spacing deduced from diffraction for the same specimen (Fig. 2) is 2.0 nm. The calculated spacing between screw dislocations accommodating twist θ is $D=|\mathbf{b}|/2\sin(\theta/2)$. The spacing so calculated for the twist $\theta=11°$ measured directly in Fig. 2 is 2.04 nm, very close to the latter measurements. For $\theta\sim25°$ however, WB images do not show individual dislocations anymore, as predicted by Jesser et al (1999).

A second network is also observed. It is a single set of roughly parallel dislocations which appear in the strong beam images as broken lines with a spacing varying between samples (about 160 nm in Fig. 3). Although they produce no contrast, these dislocations also appear in the WB images because they interact with the screw dislocations of the first network: the latter are all shifted by half their spacing upon crossing a line of the second set. This shows that the dislocations of the second network have Burgers vectors of the a/2<101> type not lying in the interface. Similar observations have already been made in the cases of Si/Si twist-bonding (Benamara et al 1992) and of GaAs/InP heteroepitaxial bonding (Patriarche et al 1997). Here again, this second set is very likely accommodating the slight disorientation of the bonded surfaces from nominal (001) planes.

The plan-view images show microscopic non-bonded areas. In a given sample, the density and size of these areas vary with the zone observed (Fig. 3). No systematic difference is observed between samples bonded at different temperatures. From a practical point of view, the quality of the interface probably depends more on a low density of macroscopic non-bonded areas. We observe more of these at low bonding temperatures, whereas no significant difference is noted in the 600-730°C range.

(110) cross-sectional views have also been studied (Fig. 4). In $\mathbf{g}=004$ dark field images (note that [001] is the only direction common to the two twisted crystals), small segments tilted with respect to the interface appear. In Fig. 4 (from the same specimen as Figs. 1 and 2), these segments (about 2.2 nm long) are spaced by 2.04nm, precisely the spacing of the dislocations of the first network. These screw dislocations should however then be out of contrast, since $\mathbf{g}.\mathbf{b}=0$. The contrast observed might be due to a stacking fault associated with the dissociation of the screw dislocations.

4. STRUCTURAL QUALITY OF THE LAYERS GROWN ON COMPLIANT SUBSTRATES

In order to test the efficiency of the method, $In_xGa_{1-x}As$ layers were grown by molecular beam epitaxy on the compliant substrates. A composition compatible with layer-by-layer growth in standard epitaxy ($x \leq 0.2$, corresponding to about 1.5% misfit) was selected. The layers were not deposited in the same epitaxy run, which explains the slight unintentional composition differences between samples (Table). A layer thickness of 500 nm, well above the critical thickness (~20 nm), was chosen.

We tested the structural quality of the deposited layers by measuring the full width at half maximum (FWHM) β of the peaks of the double-crystal XRD rocking curves, which increases in presence of structural defects, such as threading dislocations. We believe that this method (to be shortly complemented by plan-view TEM studies) is more reliable than cross-sectional TEM, which is sensitive only to very high densities of extended defects. The table gives the FWHM of the 004

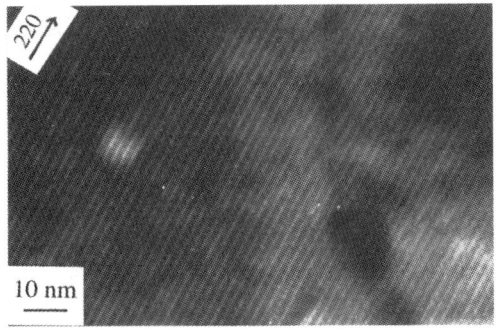

Fig. 1 : Plan –view image in g=220 weak beam condition.

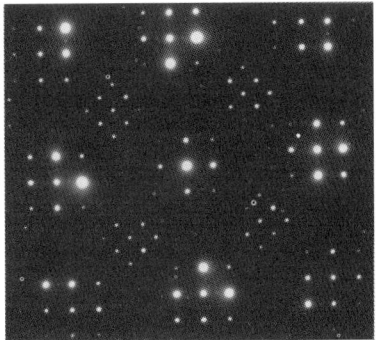

Fig. 2 : (001) diffraction patern of a plan –view specimen.

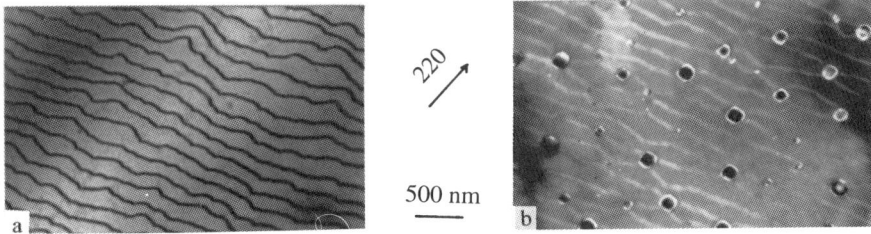

Fig. 3 : 220 dark field plan –view images of two zones of the same specimen showing the lines of the second network. Roughly square non-bonded areas are present in (b) but not in (a).

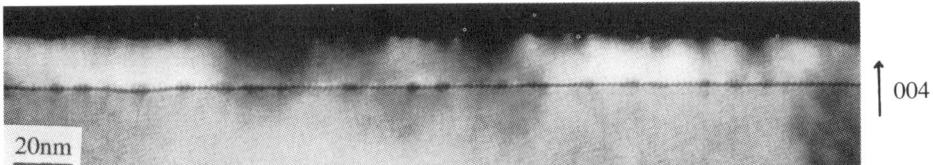

Fig. 4 : (110) cross-sectional view in g=004 dark field condition.

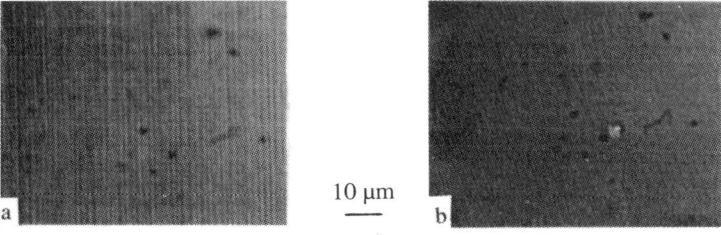

Fig.5 : Optical micrographs of (a) sample No 2 and (b) sample No 6.

peaks associated with the $In_xGa_{1-x}As$ layers grown on various compliant layers. These FWHMs are compared with values for similar thick mismatched reference layers grown directly on a standard GaAs substrate, with or without insertion of a GaAs buffer layer. The alloy layers grown on the 20 nm thick GaAs compliant layers display a worse structural quality than the reference layers. On the other hand, the alloy layers deposited on a 5 nm thick compliant layer exhibit a significantly better quality than the reference layers, manifested by a decrease by more than 200" of β (more than could be expected from the non-intentionally slightly reduced In content). To estimate roughly the density d of threading dislocations per unit area in the alloy layer, we use the simple formula $d=\beta^2/9b^2$, with b the Burgers vector modulus for the dislocations involved and β in radians (Hirsch 1956). We find $d=6.3 \times 10^8$ cm^{-2} for the best layer grown on a compliant substrate (sample No 6), to be compared with $d=1.1 \times 10^9$ cm^{-2} for the best layer grown on a standard substrate (sample No 2). No significant difference in FWHM is observed upon changing the twist angle.

Moreover, the optical micrographs of the surfaces of layers grown on compliant and standard substrates differ deeply: the cross-hatch patterns usually observed on plastically relaxed misfitting layers is present in the latter but is absent for all layers grown on twist-bonded substrates (Fig. 5).

Sample No	Substrate			$In_xGa_{1-x}As$ layer	
	Twist angle (°)	Bonding temperature (°C)	Compliant layer thickness (nm)	x (from XRD)	FWHM β of the 004 XRD peak (")
1 (reference)	none: epitaxy on bulk GaAs	—	—	0.182	860
2 (reference)	none: same as 1 with GaAs buffer	—	—	0.178	825
3	10.5	730	20	0.177	897
4	13.5	568 (for 3 hours)	20	0.18 (estimated)	1134
5	16	636	5	0.147	653
6	6.5	636	5	0.142	624

Table: Characteristics of the substrates and of the deposited misfitting heteroepitaxial layers.

5. DISCUSSION AND CONCLUSIONS

We have shown that the description of the twist-bonded interface as a dense network of pure screw dislocations remains valid at least up to $\theta=16°$, but not for $\theta=25°$. Regarding the structural quality of the layers grown on the compliant substrates, the thickness of the actual compliant layer seems to be of paramount importance. We intend to reduce it even more to improve further the deposited layers, where the defect density is currently still high. On the contrary, changes of twist angle (and thus of the screw dislocation density) appear to have no effect on the structural quality of the grown layers. We have undertaken a TEM study of the complete structure (after growth) in order to investigate the behaviour and evolution of the compliant and deposited layers and of the dislocation network, and to compare them with theoretical predictions (Kästner et al 1998, Jesser et al 1999).

REFERENCES

Benamara M, Rocher A, Laporte A, Sarrabayrousse G, Lescouzères L, PeyreLavigne A, Fnaiech M and Claverie A 1995 Mater. Res. Soc. Symp. Proc. **378**, 863
Ejeckam F E, Lo Y H, Subramanian S, Hou H Q and Hammons B E 1997 Appl. Phys. Lett. **70**, 1685
Hirsch P B 1956 Progress in Metal Physics, eds B Chalmers and R King **6**, 236
Jesser W A, van der Merwe J H and Stoop P M 1999 J. Appl. Phys. **85**, 2129
Kästner G, Gösele U and Tan T Y 1998 Appl. Phys. A **66**, 13
Lo Y H 1991 Appl. Phys. Lett. **59** 2311
Patriarche G, Jeannès F, Oudar J-L and Glas F 1997 J. Appl. Phys. **82**, 4892
Zhu Z H, Zhou R, Ejeckam F E, Zhang Z, Zhang J, Greenberg J, Lo Y H, Hou H Q and Hammons B E 1998 Appl. Phys. Lett. **72**, 2598

Inst. Phys. Conf. Ser. No 164
Paper presented at Microsc. Semicond. Mater. Conf., Oxford, 22–25 March 1999
© *1999 IOP Publishing Ltd*

Structural analysis of epitaxial Si layers grown on porous silicon

S Jin, H Bender, L Stalmans, J Poortmans and M Caymax

IMEC, Kapeldreef 75, B-3001 Leuven, Belgium, sing@imec.be

ABSTRACT: Transmission electron microscopy (TEM) is used to study the microstructural properties of porous silicon layers (PS) and of the epitaxial Si layers grown on top of the PS. A more dense silicon layer exists in the upper part of the porous silicon. The defect density of the epitaxial layers is found to depend strongly on the morphology of the porous Si layers, and on the deposition temperature. A defect-free epi-Si layer with 800nm thickness can be obtained when depositing on a low porosity layer at 725°C.

1. INTRODUCTION

Porous silicon is formed during electrochemical anodization of silicon single crystals in hydrofluoric acid (Canham 1990, Lehmann and Gösele 1991 and Cullis et al 1997) and is regarded as a very promising material for crystalline Si photovoltaics applications (Jain et al 1981 and Rohatgi et al 1993). Epitaxial growth of thin crystalline layers on porous silicon can provide opportunities for applications in silicon-on-insulator (SOI) and Si-based solar cells if the epitaxial layer has a low defect density and if the porosity of the initial porous layer is preserved (Oules et al 1992, Sato et al 1995 and Stalmans et al 1997).

In order to optimize the processing conditions for the fabrication of the Si film grown epitaxially on a porous silicon layer, transmission electron microscopy (TEM) is performed, as discussed in this paper.

2. EXPERIMENTAL

The porous layers are prepared by conventional electrochemical anodization in a HF/ethanol electrolyte. These studies are concentrated on highly-doped (< 0.02 Ω·cm) p-type silicon substrates because the porous etching of highly-doped materials results in mesoporous Si which is the preferred type of porous material for further epitaxial growth.

The epitaxial Si layers are grown on porous Si in an Epsilon-One chemical vapour deposition (CVD) system. Various deposition temperatures are used in the range of 700°C to 850°C. The reactor is operated at a pressure of 40 torr. SiH_2Cl_2 (DCS) is used as Si source gas, which is diluted in a flow of H_2 as carrier gas. The time between porous silicon formation and the loading into the CVD-system is kept as short as possible to avoid ageing.

The samples are characterised with JEOL 200CX electron microscope. Cross-sectional specimens for TEM studies are prepared by mechanical thinning and subsequent ion milling. The ion energy is set at 3.8 keV in order to minimise the ion damage in the layers. During the final thinning the incidence angle is lowered from the initial setting of 15° to 7°. The normal to the specimen is parallel to the [011] direction of silicon.

3. RESULTS AND DISCUSSION

3.1 Morphology of porous silicon formed on highly-doped p-type silicon

Figure 1 shows a cross-sectional transmission electron microscopy (XTEM) image of a porous silicon (PS) layer formed on a p-type silicon substrate. The speckled contrast of dark and bright regions represents Si crystallites and pores, respectively. The porous surface is flat, while the interface between the porous layer and the silicon substrate is quite rough. The total thickness of the porous silicon layer is about 220 nm. The top of the porous silicon consists of a more dense layer with a thickness of about 20 nm.

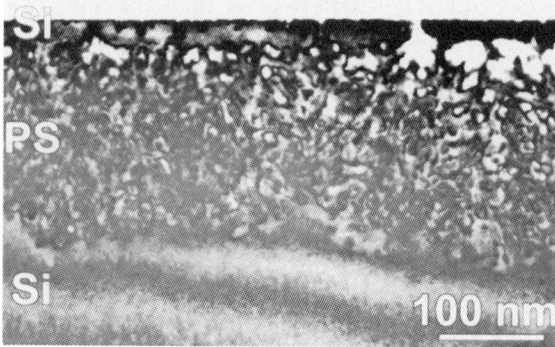

Fig.1   XTEM image of the porous silicon layer formed on a p-type silicon substrate.

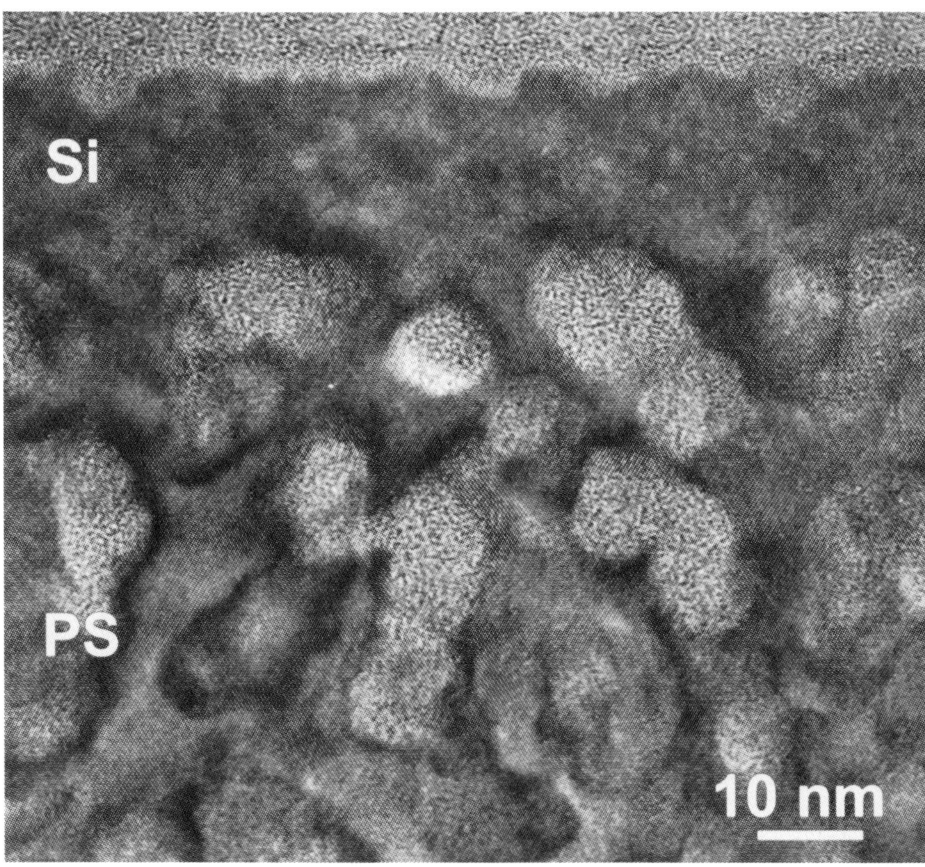

Fig.2   XHREM images near the top of the PS layer.

Figure 2 shows a cross-sectional high resolution electron microscopy (XHREM) image taken near the top of the PS layer showing the dense top layer. The crystalline skeleton of the PS has the same orientation as the silicon substrate over the full thickness of the layer. Neither the pore propagation nor the branch growth follows exactly the <100> crystallographic directions. The porous structure has a crystalline feature size, defined as the mean distance between neighbouring pores, of about 10 to 20 nm. The pores have a radius of 20 nm. In the thinnest regions of the TEM specimen where the thickness is on the order of the pore and crystallite dimensions, the pores are visualised as an amorphous material. This is due to the amorphisation of the very thin silicon walls around the pores by the ion milling used for the specimen preparation. In thicker specimen areas the projection of the silicon skeleton and pores over the thickness of the TEM specimen gives rise to an image as for a crystalline material with variable density.

3.2 Epitaxial growth on porous silicon

As discussed above, a dense surface layer is formed on the top of the porous silicon, which is an advantage for the further epitaxial growth. Figure 3 is a XTEM image of an epitaxial silicon film over a porous silicon layer formed on a highly doped p-type silicon substrate. The porosity of the initial porous silicon is measured to be 20% by spectroscopic ellipsometry (SE) measurements. During the deposition, the temperature is kept at 725°C. A bake is done at 800°C for 10min after deposition. The epitaxial silicon film and porous silicon layer have thicknesses of 800nm and 160nm respectively, and a dense thin layer is clearly observed on the top of the porous silicon. The epi-Si layer is uniform, the surface is flat and no defects are observed in the epitaxial layer within the statistics of the TEM observation.

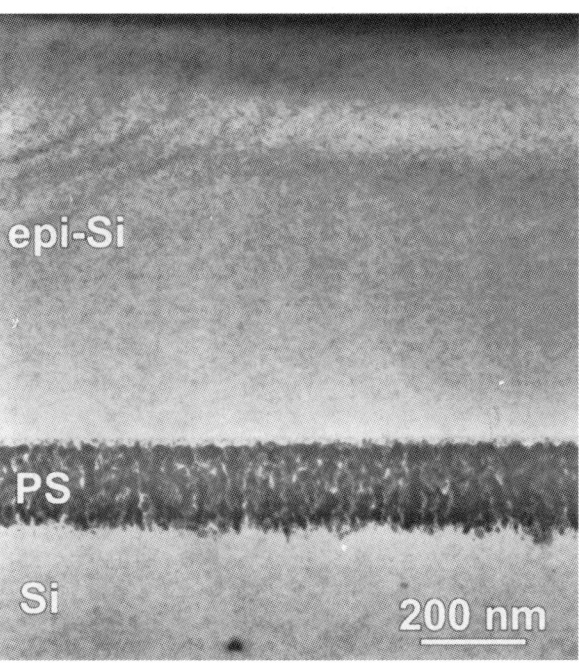

Fig.3 XTEM image of the epitaxial Si layer on the porous silicon. The CVD process is performed at 725°C for 20min, and a post-bake after the deposition is done at 800°C.

Studies of the optical properties of the Si/PS/Si structures have been made to assess the effects of the heat treatment on the reflectance behaviour (Stalmans et al 1998). The porous Si reflectance behaviour is dependent on the temperature of the epi-Si growth. Figure 4 shows the XTEM images corresponding to different growth temperatures and porosities. One can see that the epitaxial quality strongly depends on both the porosity of the initial porous silicon layer and the deposition temperature. When depositing at 725°C, an increase of the porosity from 20% to 60% increases the number of defects, as shown in Fig. 3 and Fig. 4(a). For the highly porous layer, the defect density further increases when the deposition temperature is increased to 800°C (Fig. 4(b)), resulting in a highly defective Si layer and destruction of the porous Si underlayer.

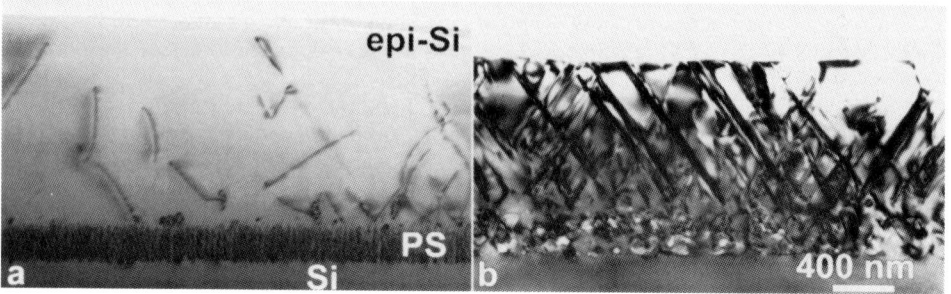

Fig.4   XTEM images corresponding to the different deposition temperatures and PS porosities: (a) 725°C, 60%, (b) 800°C, 60%.

4. CONCLUSION

The defect density of the epitaxial layers depends strongly on the morphology of the porous Si layers and on the deposition temperature. The deposition temperature also influences the porosity of the porous Si underlayer. The porous structure is completely destroyed after thermal CVD deposition of Si at too high temperature, resulting in a highly defective epitaxial layer. A defect-free epi-Si layer with 800nm thickness is obtained when depositing on a low porosity layer at 725°C.

ACKNOWLEDGEMENT

The transmission electron microscopy investigations were performed with the microscopes at EMAT, University Antwerpen (RUCA), Belgium. L. Stalmans is supported by a fellowship from IWT (Flemish Institute for Promotion of Scientific-Technological Research in the Industry).

REFERENCES

Canham L T 1990 Appl. Phys. Lett. **57**, 1046
Cullis A G, Canham L T and Calcott P D J 1997 J. Appl. Phys. **82**, 909
Jain G C, Singh S N, Kotnala R K and Arora N K 1981 J. Appl. Phys. **52**, 482
Lehmann V and Gösele U, 1991 Appl. Phys. Lett. **58**, 856
Oules C, Halimaoui A, Regiolini J L, Perio A and Bomchil G 1992 J. Electrochem. Soc. **139**, 3595
Rohatgi A, Weber E R and Kimerling L C 1993 Journal of Electronic Materials **22**, 65
Sato N, Sakaguchi K, Yamagata K, Fujiyama Y and Yonehara T 1995 J. Electrochem. Soc. **142**, 3116
Stalmans L, Poortmans J, Caymax M, Bender H, Jin S, Nijs J and Mertens R 1998, 2ⁿᵈ World Conference on Photovoltaics, Vienna, Austria, Conference Proceedings pp 124-7

Inst. Phys. Conf. Ser. No 164
Paper presented at Microsc. Semicond. Mater. Conf., Oxford, 22–25 March 1999

X-ray diffraction measurements of strained and relaxed SiGe epitaxial layers on Si

D J Wallis, A M Keir, D J Robbins, J C Jones, G M Williams, A J Pidduck, R Carline and J Russell

DERA, St Andrews Road, Malvern, Worcestershire, WR14 3PS

ABSTRACT: SiGe epitaxy is being introduced into Si integrated circuit technology for high-frequency low-power applications. This new development brings to the industry a requirement for new techniques to characterise the deposited layers. High resolution X-ray diffraction techniques have high measurement precision and are non-destructive, and thus are well suited to the task. We describe examples showing the high sensitivity of X-ray diffractometry to SiGe layer strain state, thickness, composition and profile. Additionally we use X-ray diffraction space mapping and single crystal topography to gain information about the relaxation mechanisms occurring in layers deposited to form SiGe virtual substrates.

1. INTRODUCTION

Incorporation of SiGe epitaxial layers within Si-based devices can be used to enhance the performance of, and add new functionality to, Si integrated circuits (ICs). SiGe heterojunction bipolar transistors (HBTs) employing a thin strained SiGe base layer are currently going into production for high-frequency low-power applications such as mobile communications. Multiple quantum well (MQW) strained SiGe structures for use in IR photodetectors, and fully relaxed SiGe layers for use as "virtual substrates" for high mobility Si field effect transistors (HEMTs) are at the development stage. These advances in Si technology bring new measurement requirements to achieve controlled reproducible SiGe layer growth and device processing. In this paper we summarise the use of high resolution X-ray diffraction (HRXRD) rocking curves, reciprocal space mapping and topography techniques to measure the strain state, composition, and profiles of strained SiGe epilayers, and to study relaxation mechanisms of SiGe layers on Si.

2. EXPERIMENTAL

The SiGe layers studied were grown by low pressure chemical vapour deposition (Robbins and Young 1987) (LP-CVD) onto chemically cleaned 100mm diameter (001) Si wafers after in-situ oxide desorption at 850-900°C in 130Pa H_2. For the strained SiGe structures the SiGe growth rate was approx. $0.4\mu mh^{-1}$ at a growth temperature of 605-610°C. This growth temperature was maintained during the growth of Si spacers for quantum well structures. In the case of the virtual substrate samples, a range of growth conditions have been explored from slow grade rate (approx. 5% Ge μm^{-1}) and low growth temperature

(560-610°C) to high grade rate (approx. 50% Ge μm^{-1}) and high growth temperature (800°C) to examined how relaxation of the layers is affected. The results presented are for layers with a multilayer structure grown on them consisting of a constant composition buffer layer, a modulation-doped Si quantum well for electron mobility measurements and a SiGe cap.

X-ray measurements were made on a Bruker high resolution diffractometer equipped with a Goebel mirror and a Bede D3 diffractometer with a Cu rotating anode source. All simulations were carried out using a commercial software package, RADS by Bede Scientific.

3. RESULTS AND DISCUSSION

3.1 Strained Single SiGe Epitaxial Layers

Figure 1 shows a (004) double crystal X-ray rocking curve taken from a single, constant composition SiGe layer deposited on to a Si Substrate (Sample 8B9). If required, the strain state of the layer may be determined by comparing the lattice parameters determined from symmetric and asymmetric reflections. However, the presence of strong thickness fringes gives a good indication that the layer is fully strained. In principle, the composition of a fully strained layer may be simply estimated by measuring the lattice parameter of the layer from the separation between the layer and substrate peaks. In practice, the rocking curve must be simulated using a full dynamical simulation package in order to get an accurate assessment of the composition due to effects such as diffraction pulling

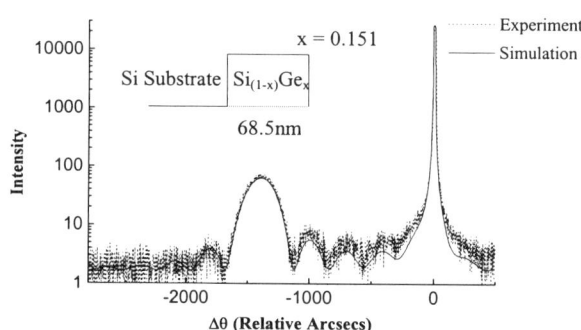

Figure 1. (004) Double crystal rocking curve for a single SiGe layer on Si (Sample 8B9).

where the peak position is changed due to overlap with adjacent features in the rocking curve. Additionally, a deviation from the linear relationship, known as Vegard's law, between lattice parameter and composition has been reported for the SiGe system (Dismukes et al 1964) and must be considered. Whilst this is a systematic error in the Ge fraction and may be simply corrected, not all commercial packages include this correction and it is important that it is allowed for, if the values estimated are to be compared with those derived from other techniques. Accurate simulation of the rocking curve also allows the average thickness of the layer to be measured to an accuracy of ±0.2nm by fitting the spacing of the subsidiary fringes known as thickness fringes. For the layer shown in Fig. 1 the measured Ge fraction is 0.151±0.002 (corrected) and the thickness is estimated as 68.5±0.2nm.

3.2 Multiple Quantum Wells

Figure 2 shows an (004) double crystal X-ray rocking curve collected from a 30 period SiGe/Si superlattice (Sample 5C37) and two models fitted to this data. For a superlattice structure, the position of the zeroth order peak, relative to that of the substrate, is determined by the mean composition of the entire structure. Therefore, the peak position is sensitive to not only the Ge fraction of the wells, but also their thickness and the thickness of the Si spacers. These parameters may be separated however, since the spacing of the superlattice peaks is determined by the total period of the superlattice and their intensity contains information about the Ge fraction of the wells. The width of the superlattice peaks also contains information about any variations in the superlattice period. Another important parameter to determine when trying to predict device characteristics is the composition profile of the quantum wells. Information about the well profile is contained in the damping envelope which controls the relative intensity of the superlattice peaks. For a "top hat" shaped well profile, the damping envelope is a $Sinc^2$ function and results in enhanced intensity of some of the higher order satellites as shown in Fig. 2b where the experimental data is modelled with no well period variation and a "top hat" well profile. However, a non "top hat" well profile results in a modification of the $Sinc^2$ damping function and a redistribution in intensities of the higher order superlattice peaks. In order to model accurately the rocking curve for the multiple quantum well shown in Fig. 2 a graded profile must be added to the well leading and trailing edges Fig. 2a. Note that although the best fit with the two models is achieved with the same total Ge content (mean composition x well thickness), the correct peak Ge concentration (13.4%) is only obtained when the graded well profile is taken into account (Fig. 2a).

a)
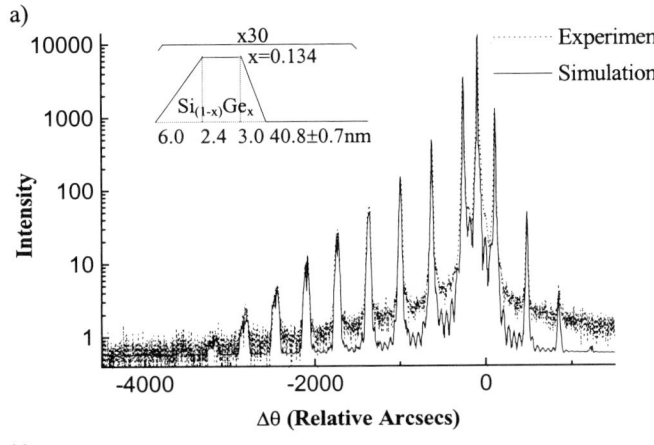

b)
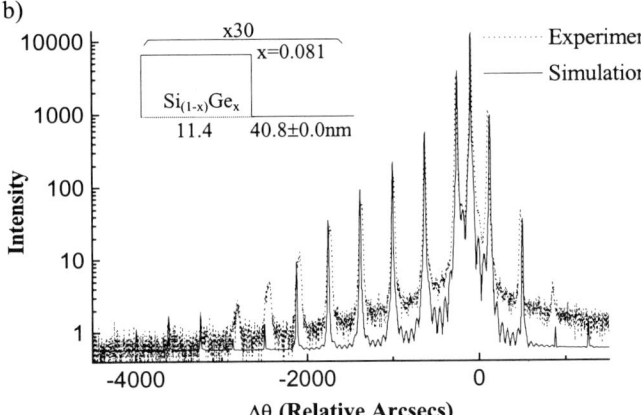

Figure 2. (004) double crystal rocking curve for a 30 period superlattice (Sample 5C37), a) modelled using graded well profile and period variations, and b) modelled using a "top hat" well profile and no period variations.

The rocking curves shown in Figs. 1 and 2 may be repeated on different regions across a wafer with a lateral resolution of about 0.5mm to give information about the uniformity across a wafer during growth.

3.3 SiGe Virtual Substrates

A "virtual substrate" consists of relaxed layers deposited onto a Si wafer which provide a variable lattice parameter for subsequent epitaxial growth. This allows the use of strain effects to enhance device performance such as those seen for stained Si 2-Dimensional electron gasses (Churchill et al 1997) and also offer the possibility of epitaxial growth of dissimilar materials onto cheap Si substrates. In order to allow such applications, the relaxation mechanism of the SiGe layers and the effect that these mechanisms have on surface morphology must be understood (LeGoues et al 1993). Figure. 3a shows a single crystal topograph taken using a grazing incidence (422) reflection from a constant composition buffer layer above a virtual substrate. The intensity of this image contains information about the size of tilts present in the crystal due to dislocations. A scanning optical microscope image, which contains information about the slope of the surface of this layer, is shown in Fig. 3b. From Fig. 3 the close relationship between the strain field surrounding misfit dislocations and the "cross-hatch" surface morphology can be clearly seen.

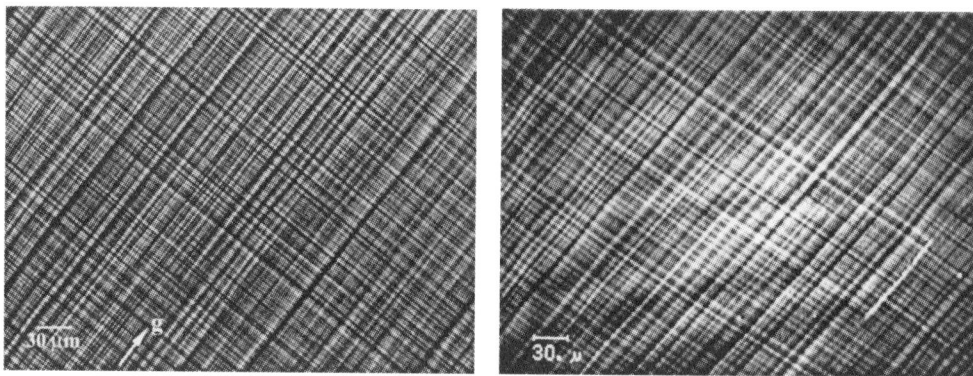

Figure 3a) (422) topograph and b) scanning optical microscope image of a SiGe Virtual substrate (Sample 6B35).

X-ray diffraction space maps provide more detailed information about the size and nature of the pile-up of these dislocations. Two (004) diffraction space maps taken with the X-ray beam parallel to one of the [110] directions for two virtual substrates (6A58 and 6B35) are shown in Fig. 4. Along the ordinate, such maps show changes in Bragg angle, and thus the lattice parameter of these fully (>95%) relaxed epilayers. By combining this information with a knowledge of the composition profile obtained via SIMS, an approximate depth scale may be applied to the ordinate axis of the diffraction space maps. This may be used to draw conclusions about the depth in the structure where relaxation occurs. The abscissas of the space maps are sensitive to the tilt of the epilayer relative to the substrate and can thus be interpreted in terms of a preference for dislocations of a particular Burgers vector since a tilt of the layer necessarily implies the presence of more dislocations with a particular tilt component. Contributions to the width of the peaks across the ω-axis from crystal size effects are considered to be negligible.

For the two virtual substrates shown, which are typical for high and low temperature growth irrespective of grade rate, the constant composition buffer layer is seen to have a net average tilt relative to the substrate. From Fig. 4, tilting of the layers occurs during growth of the initial third of the graded Ge region. This is confirmed using cross-sectional TEM (not shown) which show that the majority of the large dislocation pileups occur in this region and have a characteristic separation of 0.5-2μm (Wallis et al 1998).

a)   b)

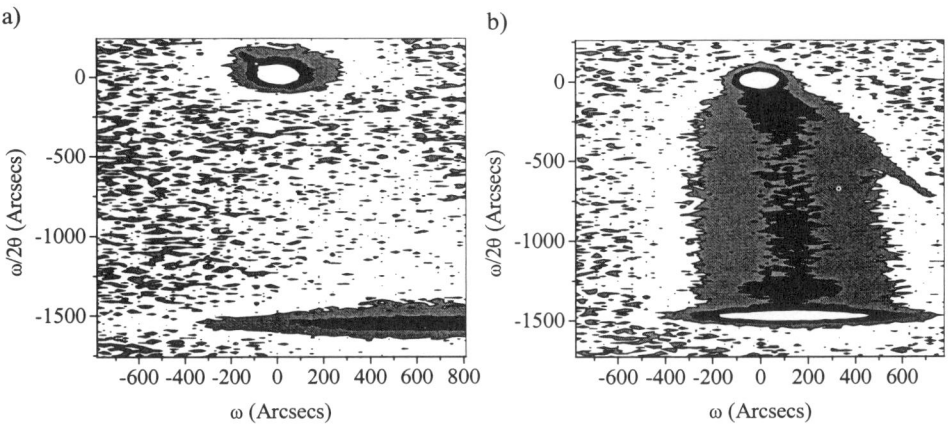

Figure 4. (004) X-ray diffraction space maps typical of virtual substrates grown at a) low temperatures (Sample 6A58) and b) high temperatures (Sample 6B35).

The width of the tilt distribution is related to the mosaic spread of tilts in the crystal and contains information about the number of dislocations with a particular Burgers vector in each pileup. This can be simply understood since each dislocation can only give rise to a specific maximum value of tilt and so the large tilts seen must be due to the accumulation of many dislocations with identical Burgers vectors. If a pileup were to occur with a random Burgers vector distribution, this would lead to a very narrow range of mosaic tilts as their Burgers vectors would cancel. Presumably, the pileup of dislocations with the same Burgers vectors results from the operation of heterogeneous dislocation sources in these samples. The width of the tilt distributions for the low temperature layers is much larger (1300 arcsecs FWHM) than for the high temperature layers (400 arcsecs FWHM). This indicates that at low temperatures a relatively small number of sources are active which leads to large pileups of dislocations with identical Burgers vectors. At higher temperatures the pileup size is reduced as the number of active sources and the average length of dislocation misfit segments increases.

For the two growth regimes illustrated, the effects on the resulting surface morphologies are shown in Fig. 5 using Atomic Force Microscopy (AFM). The forms of the surface roughness seen in Fig. 5 may be correlated with the growth conditions and the size of the dislocation pileups which result. Lutz et al (1995) have shown that each 60° dislocation is associated with a 2.5A height surface step, thus pileup of many dislocations with the same Burgers vectors can result in a tiled surface morphology as seen for the low temperature layers (Fig. 5a). For layers grown at higher temperatures the size of the pileups is significantly reduced and the tiling at the surface is expected to be much weaker. However, for high temperature growth, the mobility of surface atoms is higher which encourages the onset of strain driven roughening in the presence of residual strain fields. Such strain fields are associated with the pile up of misfit dislocations whose Burgers vectors do not cancel (Fitzgerald et al 1992) and this results in the undulating surface morphology seen for the high temperature layers (Fig. 5b).

330

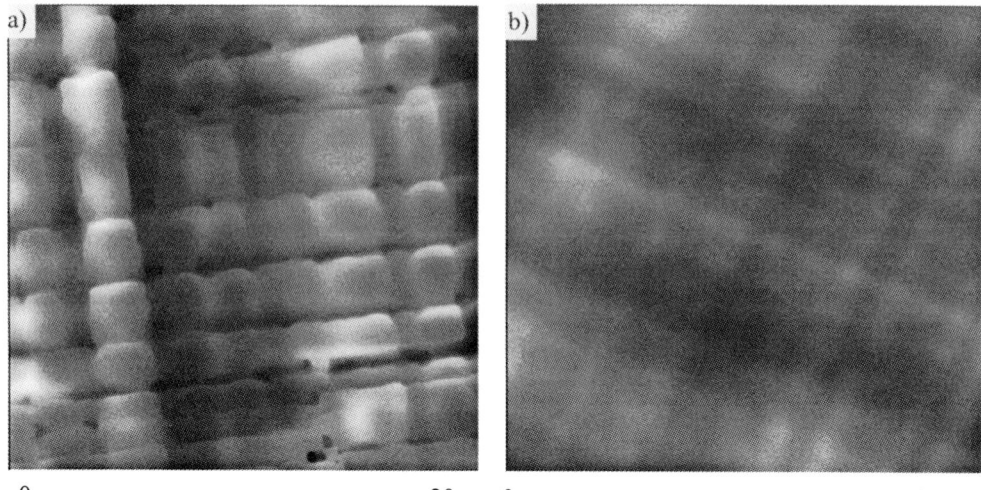

0   20μm 0   20μm

Figure 5. 20μm square AFM images typical of virtual substrate surfaces grown at a) low temperatures (Sample 6A58) and b) high temperatures (Sample 6B34). Grey scale 0-20nm.

4. CONCLUSIONS

High resolution X-ray diffraction techniques have played a major role in bringing SiGe epitaxy to the point of technological application. In this paper we have demonstrated the use of HRXRD techniques to make high precision measurements of strain state, Ge composition (within ±0.2%, given an assumption about deviations from Vegard's Law behaviour), layer thickness (within ±1nm) and profile in strained SiGe layers, and also, in conjunction with other techniques, to learn details of the strain relaxation process. The non-destructive nature of HRXRD means, furthermore, that the method is well suited to the support of process development and control in a production environment.

REFERENCES

Churchill A C Robbins D J Wallis D J Griffin N Paul D J and Pidduck A J 1997 Semicond. Sci. Technol. **12**, 943
Dismukes JP, Ekstrom L and Paff RJ 1964 J. Phys. Chem. **68**, 3021
Fitzgerald E XieYH Monroe D Silverman PJ Kuo JM Kortan AR Thiel FA and Weir BE 1992 J. Vac. Sci. Technol. B **10**, 1807
LeGoues FK, Mooney PM and Chu JO 1993 Appl. Phys. Lett. **62**, 140
Lutz MA Feenstra RM LeGoues FK Mooney PM and Chu JO 1995 Appl. Phys. Lett. **66**, 724
Robbins DJ and Young IM 1987 Appl. Phys. Lett. **50**, 1575
Wallis DJ Robbins DJ Pidduck AJ Williams GM Churchill A and Newey J 1998 MRS Symposium Proceedings **533**,77

Inst. Phys. Conf. Ser. No 164
Paper presented at Microsc. Semicond. Mater. Conf., Oxford, 22–25 March 1999
© *1999 IOP Publishing Ltd*

Characterization of surface steps on heteroepitaxial 3C-SiC thin films by TEM

W Bahng[1], H Matsuhata[1], T Takahashi[1], Y Ishida[1], H Okumura[1], S Yoshida[1,2], H Sawada[3], and H Ichinose[3]

[1]Electrotechnical Laboratory, Tsukuba, Ibaraki, 305-8568 Japan
[2]Faculty of Engineering, Saitama University, 255, Shimo-Okubo, Urawa, Saitama 338-8570 Japan
[3]University of Tokyo, 7-3-1 Hongo, Bunkyo-ku, Tokyo, 113-8654 Japan

ABSTRACT: The microstructures beneath the large surface steps aligned to two orthogonal <110> directions on 3C-SiC thin films were investigated by transmission electron microscopy. The films were grown on exactly oriented (001) Si substrates. Most of the steps were 20μm in length, although some of them were shorter. The surface steps were identified to be twin bands, which originated at some points approximately 5μm above the heterointerface, forming triangular plates on the (111) crystallographic plane. The three-dimensional structure of the surface steps linked with each other is also discussed.

1. INTRODUCTION

SiC is a potential semiconducting material for high power and high temperature applications. However, defects such as twins and anti-phase boundaries (APBs) are often found in SiC on Si substrates. It is generally accepted that these defects are produced at the interface between the Si substrate and SiC epilayer due to large lattice mismatch (~20%) and the large difference of thermal expansion coefficients (~6%). Usually, the density of defects introduced at interface decreases as the thickness increases. Shibahara et al (1986) proposed that APBs could be annihilated by encountering two APBs. A Si (001) substrate slightly tilted towards <110> was used to reduce APBs and a successful result was reported (Kong et al 1988). However, there were no reports about epilayers free from APBs grown on an exact (001) Si substrate. Ishida et al (1997) succeeded in growing SiC epilayers with very smooth surface on exactly oriented (001) Si substrates by low pressure CVD. In their case, the APBs were produced at the interface and their density was reduced with an increase of thickness. On the surface of these epilayers, large steps aligned to two orthogonal <110> directions were found (Takahashi et al 1998). In this paper, we report on the investigation of surface steps by cross-sectional transmission electron microscopy (XTEM).

2. EXPERIMENTAL

SiC thin films were grown at low pressure using a horizontal cold-wall type CVD system. Before the growth, a Si substrate was heated to remove the surface oxide under atmospheric pressure of hydrogen. Then the substrate was heated quickly and was carbonized for 2min in the flow of H_2 and C_3H_8 at 10 torr. Thereafter, SiH_4 was introduced and 3C-SiC epilayers were grown at 1150°C to 1350°C (Ishida et al 1997). The films grown for 10 hrs were approximately 16μm thick and few APBs were detected at the surface by chemical etching (Ishida 1999).

The focused ion beam (FIB) method was used for TEM specimen preparation. A W film was deposited on the surface to protect against damage and then non-deposited surface regions were

332

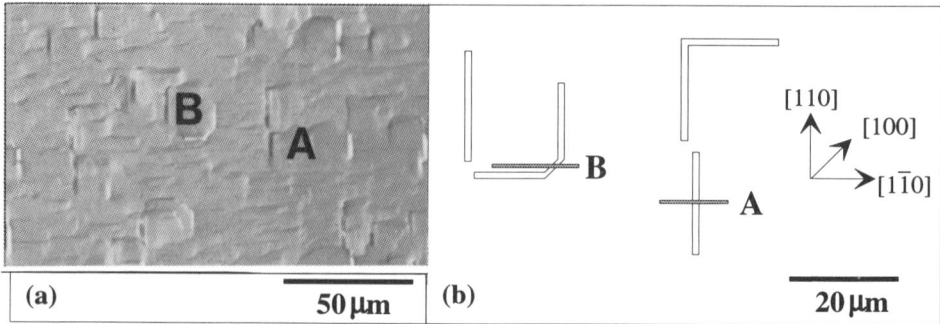

(a) 50 μm (b) 20 μm

Fig. 1 Typical optical micrograph of surface steps(a) and magnified schematic diagram of it(b). The section A and B were observed by cross-sectional TEM.

etched out by Ga$^+$ ions. A JEM-4000FX was used and the accelerating voltage was 200keV for observation.

3. RESULTS AND DISCUSSION

Fig. 1a is an optical micrograph of typical large steps on the surface of epitaxially grown SiC film on a Si (001) substrate. These steps are a few hundred nm in height and 10~20μm in length, although most of the steps are 20μm in length. All of these steps are aligned to two orthogonal <110> directions on (001) surface. They have a crystallographic relationship with matrix, which we will investigate in this report. There are two typical features marked A and B as shown in Fig. 1b. In the section B, there are two perpendicular <110> aligned steps linked with a small <100> aligned step, while the section A has a single surface step.

A cross-sectional TEM micrograph of section A in Fig. 1b is shown in Fig. 2a. One can see that the step is a large twin band whose width is approximately 300nm. This is in accord with the result of former etching experiment that these surface steps have a fixed angle of 54.7° with respect to surface plane (Takahashi et al 1998). In addition, the large twin band consists of multiple twin boundaries (marked MTB in Fig. 2b). The geometrical relationships of crystallographic directions between

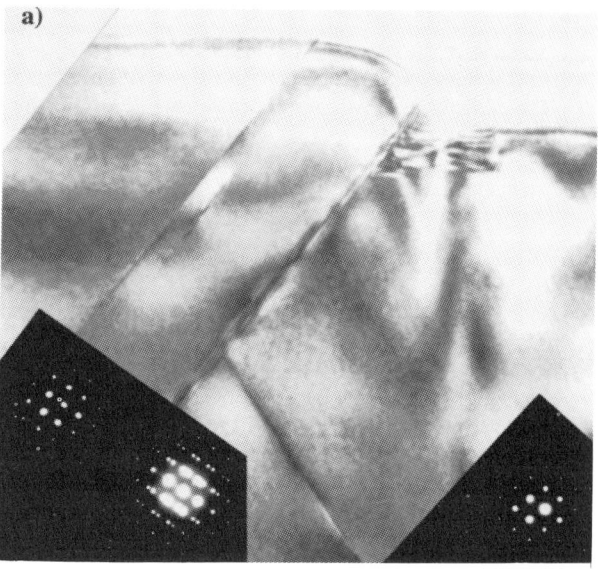

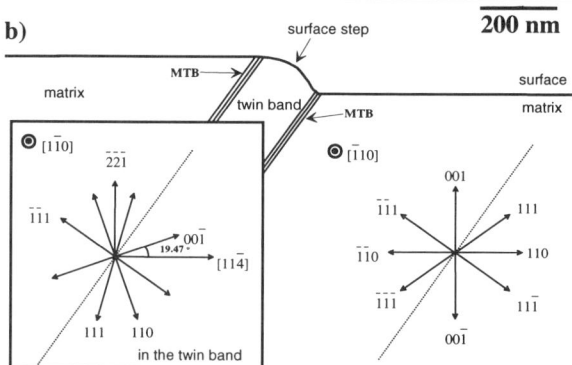

Fig. 2 Cross-sectional TEM micrograph of surface step(a) and schematic diagram of geometrical relationship. The inserted SAD patterns, from the left, were taken from twin band, boundary between twin and matrix, and matrix, respectively.

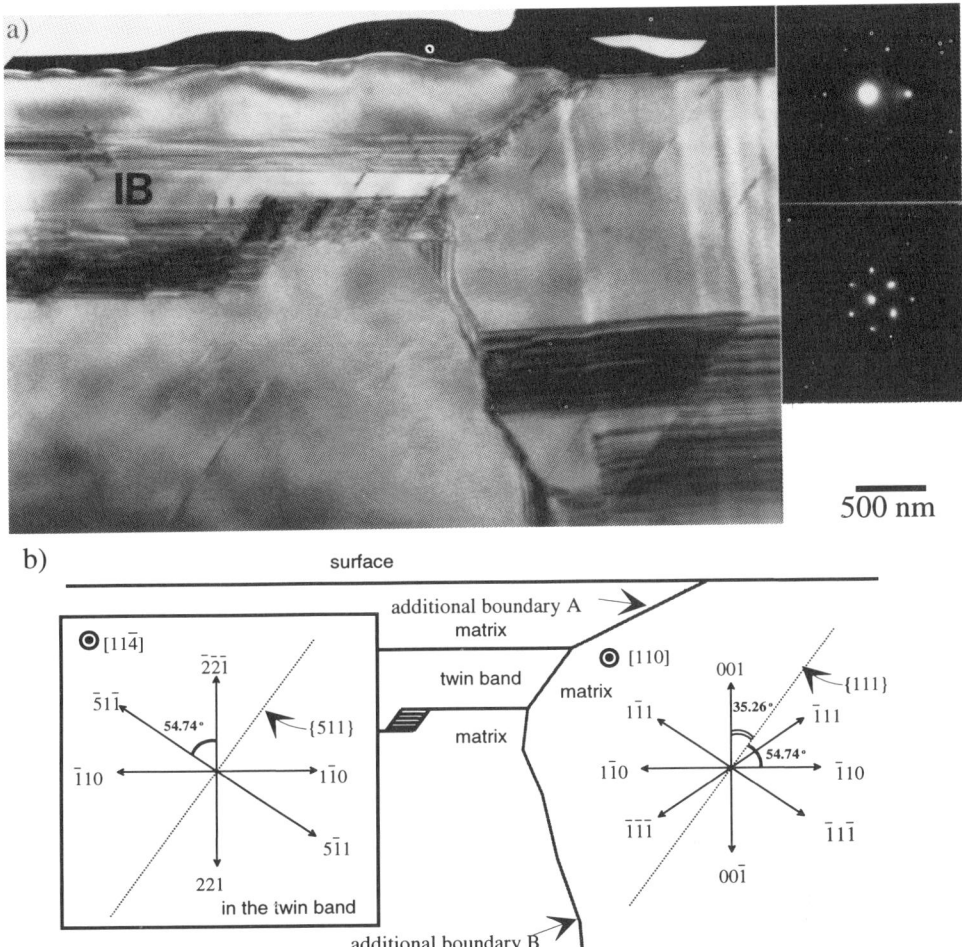

Fig. 3 TEM micrograph(a) and schematic diagram(b) observed parallel to the surface step. The inserted SAD patterns were taken from the region marked IB (upper one) and matrix (lower one). The black deposit on surface is tungsten film as an ion damage protecting layer.

the twin band and matrix are illustrated. These show the twin band has changed its growth plane to {221} with respect to (001) of the matrix. There is no (001) plane on and/or near the surface of twin band, which may result in the difference in height. The twin band and matrix will have [11$\bar{4}$] and [110] directions, respectively, in the same geometrical direction.

Fig. 3a is an electron micrograph of the section beneath the surface step, which was observed parallel to the surface step similar to the section B in Fig. 1b. This complex image shows an inserted band (IB), additional boundaries A, B, and stacking faults. The selected area diffraction (SAD) patterns show that the electron beam direction for the inserted band is the [11$\bar{4}$] direction (upper one) where the matrix is [110] (lower one). These SAD patterns confirm that the inserted band is a twin, as illustrated in Fig. 2b. This micrograph shows that the side edge of twin band is (5$\bar{1}$1) which corresponds to (1$\bar{1}$1) of the matrix, suggesting the twin band forms a triangular plate on {111} plane of the matrix. This formation of a triangular plate may be caused by fact that {111} is the most stable plane. We have calculated the expected depths of the origins of twins considering the triangular plate shape. The calculation shows that the most of the twins are introduced at 5μm above the

334

heterointerface. This indicates that the origins of twins are neither experimental instabilities during film growth nor the stress induced from the heterointerface.

We could not identify the additional boundary A, B because we could not change diffraction conditions arbitrarily for the present samples made by FIB. Ernst and Pirouz (1989) had observed termination of a microtwin by an APB in a GaP epilayer, which has the same zinc blend structure as for SiC. Considering their report, the APB may be responsible for the termination of the twin band and the additional boundaries A, B seem to be APBs due to their irregular shapes. A calculation based on fringe width and thickness of the TEM sample confirms that the additional boundary A is a

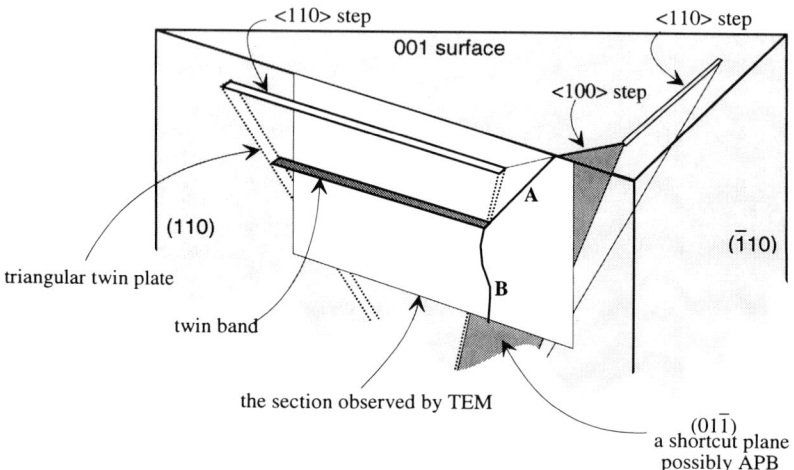

Fig. 4 Schematic diagram of the whole defect structure of the two combined <110> aligned surface steps linked with a <100> aligned step.

$(01\bar{1})$ plane, this is a shortcut plane which connects the two plates of twin bands. Thus, the whole structure of the defects combined with two twin bands will be a part of a {011} truncated {111} pyramid, as shown in Fig. 4.

4. CONCLUSION

We have investigated the surface steps on heteroepitaxial 3C-SiC films on Si substrates with TEM. The surface step had a large twin band whose side edge was a {511} plane of the twin band. The twin band was shown to be triangular plate-like in shape. These twins were introduced 5μm above the interface but not exactly at the interface. Some twin bands were linked with a (110)-type plane which is a shortcut plane of the two twin bands: on film surface the two perpendicular <110> aligned steps were linked with a <100> aligned small step.

REFERENCES

Ernst F and Pirouz P 1989 J. Mater. Res. **4**, 834
Ishida Y 1999 private communication
Ishida Y, Takahashi T, Okumura H, Yoshida S and Sekigawa T 1997 Jpn. J Appl. Phys. **36**, 6633
Kong H S, Wang Y C, Glass J T and Davis R F 1988 J. Mater. Res. **3**, 521
Shibahara K, Nishino S and Matsunami H 1986 J. Cryst. Growth **78**, 538
Takahashi T, Ishida Y, Okumura H, Yoshida S and Sekigawa T 1998 Materials Science Forum **264-268**, (Switzerland: Trans Tech Publications) pp 207-210

Inst. Phys. Conf. Ser. No 164
Paper presented at Microsc. Semicond. Mater. Conf., Oxford, 22–25 March 1999
© *1999 IOP Publishing Ltd*

Lateral- and pendeo-epitaxial growth and characterization of low defect density GaN thin films

Robert F Davis, O-H Nam, T S Zheleva, M D Bremser, K J Linthicum, T Gehrke, P Rajagopal and D B Thomson

Department of Materials Science and Engineering, North Carolina State University, Raleigh, NC 27965,

ABSTRACT: Lateral- and pendeo-epitaxy were used to grow GaN and $Al_xGa_{1-x}N$ films on SiC/AlN and Si/SiC/AlN substrates either from GaN stripes contained within windows of an SiO_2 mask or from side walls of GaN seed structures containing Si_3N_4 top masks. Scanning and transmission electron microscopies and atomic force microscopy were used to evaluate the microstructure, the dislocation distribution and the surface roughness of the films. These regions contained a low density of dislocations. The RMS roughness of the $(11\overline{2}0)$ sidewall plane of the pendeo-epitaxial structures was approximately 0.1 nm.

1. INTRODUCTION

The III-Nitride materials have been recognized for several decades for their potential, and, recently, for their commercial viability in wide band gap optoelectronic device applications including green and blue light emitting diodes and laser diodes (Nakamura et al 1994, Nakamura et al 1996 and Doverspike et al 1998). Ultraviolet light detectors and high-frequency, -power and –temperature microelectronic devices with promising characteristics have also been achieved. Heteroepitaxial process routes have been employed to grow films of these materials because of the dearth of bulk substrates on which homoepitaxial films could be grown. The resulting films contain dislocation densities of 10^8-10^{10} cm^{-2} because of the mismatches in the lattice parameters and the coefficients of thermal expansion between the buffer layer and the film and/or the buffer layer and the substrate. These high concentrations of dislocations compromise the optical and electrical properties of the aforementioned devices.

The present authors (Nam et al 1997abc and Zheleva et al 1997) as well as several other groups of investigators (Underwood et al 1995, Kato et al 1994, Sakai et al 1998, Marchand et al 1998, Zhong et al 1998) have conducted research regarding the use and efficacy of selective area growth (SAG) and lateral epitaxial overgrowth (LEO) techniques to achieve GaN films with significantly reduced densities of dislocations. Both techniques employ the deposition of a double layer sequence of a GaN seed layer and an overlying mask layer, with the latter containing etched stripes or other patterns. Transmission electron microscopy (TEM) has shown that these approaches yield GaN films containing dislocation densities of $\approx 10^5$ cm^{-2} in the areas of overgrowth. Increased emphasis in this research topic was fueled in part by the announcement by Nakamura, et al. (1997) of the dramatic increase in projected lifetime of their GaN-based blue light-emitting laser diodes fabricated on one layer of LEO material. However, to benefit from this reduction in defects in the first LEO layer, the placement of devices incorporating LEO technology requires careful alignment with respect to the underlying mask stripes to take advantage of the superior quality material. This limits the placement of the devices to the regions on the final GaN device layer that are located on the overgrown regions and thus limits the device size.

Recently we have pioneered a new approach for the selective epitaxy of materials, namely, pendeo- (from the Latin: to *hang* or be *suspended*) epitaxy (PE) (Zheleva et al 1999ab, Linthicum et al 1999ab, Gehrke et al 1999, and Thomson et al 1999), as a promising new process route leading to a single, continuous, large area layer, multilayer heterostructures or discrete platforms of these materials. In our research it has been specifically applied to GaN and $Al_xGa_{1-x}N$ layers. It incorporates mechanisms of growth exploited by the conventional LEO process by using an amorphous mask to prevent vertical propagation of threading dislocations; however, it extends beyond the conventional LEO approach to employ the substrate itself as a *pseudo-mask*. This unconventional approach differs from LEO in that growth does not initiate through open windows on the (0001) surface of the GaN seed layer; rather, it is forced to selectively begin on the sidewalls of a tailored microstructure comprised of forms previously etched into this seed layer. Continuation of the pendeo-epitaxial growth of GaN or the growth of the $Al_xGa_{1-x}N$ layer until coalescence over and between these forms results in a complete layer of low defect-density material. This is accomplished in one (GaN), two ($Al_xGa_{1-x}N$)) re-growth steps. And the need to align devices or masks for the growth of the subsequent layers over particular areas of overgrowth is eliminated, unless deposition only in certain areas is desired.

The following sections describe the experimental parameters necessary to achieve GaN films via LEO and PE and $Al_xGa_{1-x}N$ layers via PE on SiC-based substrates. The microstructural evidence obtained for the resulting films is also described, discussed and summarized.

2. EXPERIMENTAL PROCEDURES

The LEO of GaN stripes patterned in SiO_2 masks deposited on a GaN film/AlN buffer layer/6H-SiC(0001) substrates has been achieved in the manner as shown schematically in Figure 1. The GaN was deposited on the underlying GaN layer through the windows in the SiO_2 mask. The deposited material grew vertically to the top of the mask and then both laterally over the mask and vertically until the lateral growth fronts from many different windows coalesced and formed a continuous layer. The underlying 1.5-2.0 μm thick GaN films were grown at 1000°C on high temperature (1100°C) AlN buffer layers previously deposited on 6H-SiC(0001) substrates in a cold-wall, vertical and RF inductively heated OMVPE system. Additional details of the growth experiments have been presented by Weeks et al (1995). Each 100 nm thick SiO_2 mask layer was subsequently deposited on each GaN/AlN/6H-SiC(0001) sample via low pressure chemical vapor deposition at 410°C. Patterning of the mask layer was achieved using standard photolithography techniques and etching with a buffered HF solution. The pattern contained 3μm and 5μm wide stripe openings, both spaced parallel at various distances (3-40μm), and oriented along <11$\overline{2}$0> and <1$\overline{1}$00> in the GaN film. Prior to lateral overgrowth, the patterned samples were dipped in a 50% buffered HCl solution to remove the surface oxide of the underlying GaN layer. The lateral overgrowth was achieved at 1000 -1100°C and 45 Torr. Triethylgallium (13-39 μmol/min) and NH_3 (1500 sccm) precursors were used in combination with a 3000 sccm H_2 diluent.

Each pendeo-epitaxial GaN and $Al_xGa_{1-x}N$ film, underlying GaN seed layer and AlN buffer layer were grown in the MOVPE system used for the LEO research. Each seed layer consisted of a 1 μm thick GaN film grown on a 100 nm thick AlN buffer layer previously deposited on a 6H-SiC(0001) substrate. A 100 nm silicon nitride growth mask was deposited on each seed layer via plasma enhanced chemical vapor deposition. A 150 nm nickel etch mask was subsequently deposited using e-beam evaporation. Patterning of the nickel mask layer was achieved using standard photolithography techniques followed by dipping in

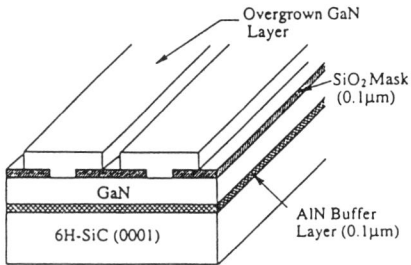

Figure 1. Schematic diagram showing lateral epitaxial overgrowth of GaN layer on SiO₂ mask from GaN deposited within striped window openings on GaN/AlN/6H-SiC substrates.

HNO₃ for approximately five minutes. The samples were subsequently cleaned by consecutive dips in trichloroethylene, acetone, methanol, and HCl for five minutes each and blown dry with nitrogen. The final, tailored, microstructure consisting of seed forms was fabricated via removal of portions of the nickel etch mask via sputtering and by inductively coupled plasma (ICP) etching of portions of the silicon nitride growth mask, the GaN seed layer and the AlN buffer layer. The etching of the seed-forms was continued into either the 6H-SiC substrate or the 3C-SiC layer, thereby removing all III- nitride material from the areas between the sidewalls of the forms. This step was critical to the success of the pendeo-epitaxial growth. The seed-forms used in this study were raised rectangular stripes oriented along the <1 1̄ 00> direction, thereby providing a sequence of GaN sidewalls (nominally (11 2̄ 0) faces). Seed form widths of 2 and 3 µms coupled with separation distances of 3 and 7 µms, respectively, were employed. The remaining nickel mask protecting the seed structures during the ICP etching process was removed using HNO₃. Immediately prior to pendeo-epitaxial growth, the patterned samples were dipped in a 50% HCl solution to remove the surface oxide from the walls of the underlying GaN seed structures.

Schematics of the pendeo-epitaxial growth of GaN(0001), $Al_xGa_{1-x}N(0001)$ and layered structures of these materials are illustrated in Figure 2. There are three primary stages associated with the pendeo-epitaxial formation of the first layer of each of these materials: (i) initiation of lateral homoepitaxy (GaN) or heteroepitaxy ($Al_xGa_{1-x}N$) from the sidewalls of a GaN seed layer, (ii) vertical growth and (iii) lateral growth over the silicon nitride mask covering the seed structure. Pendeo-epitaxial growth of GaN and $Al_xGa_{1-x}N$ was achieved within the temperature range of 1050-1100°C and a total pressure of 45 Torr. The precursors (flow rates) of triethylgallium (23 -27 µmol/min) and NH₃ (1500 sccm) were used in combination with a H₂ diluent (3000 sccm). The introduction of triethylaluminum at flow rates of 2.5 µmol/min and 5.8 µmol/min into the growth chamber produced $Al_xGa_{1-x}N$ layers containing Al concentrations of approximately 5% and 10%, respectively. Additional experimental details regarding the pendeo-epitaxial growth of GaN and $Al_xGa_{1-x}N$ layers are given in Refs. [11] - [16].

The morphologies and defect and surface microstructures for both types of samples were investigated using scanning electron microscopy (SEM) (JEOL 6400 FE), transmission electron microscopy (TEM) (TOPCON 0002B, 200 KV) and atomic force microscopy (AFM) (Digital Instruments, Inc. Dimension 3000). The AFM was operated in the tapping mode with an Olympus tapping mode etched silicon probe.

3. RESULTS AND DISCUSSION

Figure 3 shows the surface and cross-sectional morphologies of the homoepitaxial GaN stripes grown for different growth times on stripe openings oriented along <11 2̄ 0> and <1 1̄ 00>. The morphologies of the stripes were very similar after the 3 min of growth

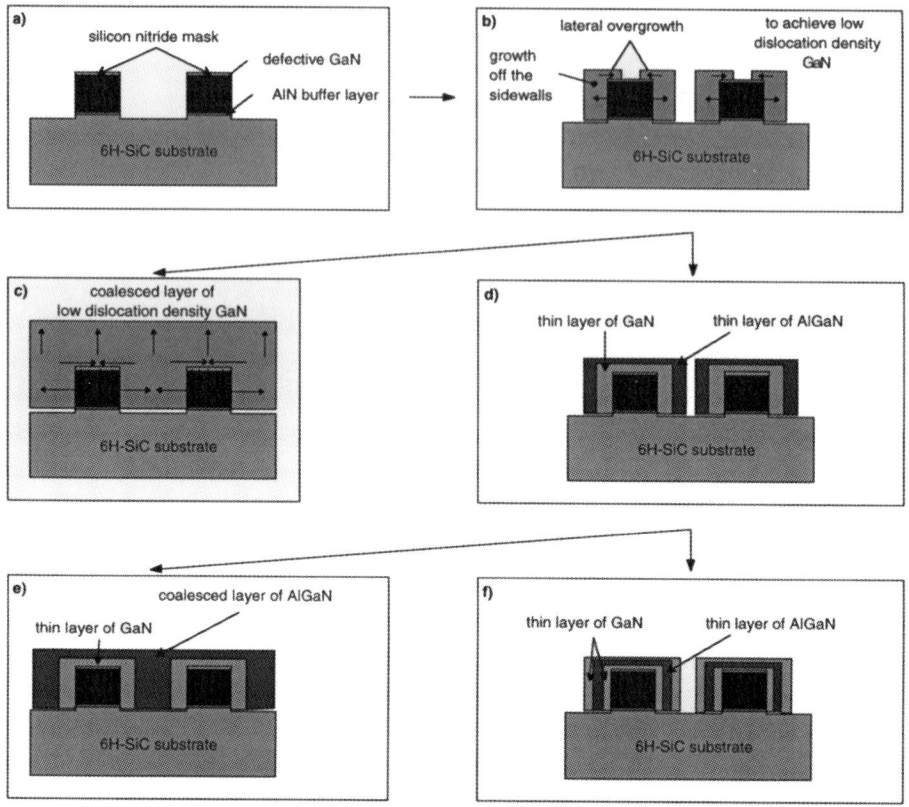

Figure 2. Schematic diagrams of the steps necessary to form pendeo-epitaxial layers and structures. The process routes move from (a) the etched columnar forms in the GaN seed layers to (b) lateral growth from the side walls of the seed layer and lateral overgrowth over the silicon nitride mask, to the growth of either (c) a continuous coalesced GaN film, or (d) a discrete bi-layer of GaN and $Al_xGa_{1-x}N$ from which further growth results in either (e) a continuous coalesced layer of $Al_xGa_{1-x}N$ or (f) a layered structure of GaN and $Al_xGa_{1-x}N$.

regardless of stripe orientation, as shown in Figures 3 (a)-(d). The stripes subsequently developed into different shapes as the growth proceeded. Triangular stripes having $\{1\bar{1}01\}$ side facets were observed for window openings along $<11\bar{2}0>$, while rectangular stripes having a (0001) top facet and $\{11\bar{2}0\}$ side facets developed for those along $<1\bar{1}00>$. The amount of lateral growth also exhibited a strong dependence on stripe orientation. Results obtained under various growth conditions showed that the lateral growth rates of the $<1\bar{1}00>$ oriented stripes were much faster than those along $<11\bar{2}0>$. Additional studies regarding the dependence of the morphological evolution and the lateral and vertical growth rates of the GaN stripes on growth conditions as well as stripe orientation are underway.

Parallel 3 μm wide stripe openings spaced 3μm apart were used to achieve continuous GaN layers by lateral overgrowth. Continuous 5 μm thick GaN layers were obtained, as shown in Figure 4. Atomic force microscopy of the surfaces revealed an average RMS roughness of the pit-free overgrown layers to be 0.25 nm which is similar to the RMS roughness values obtained for the underlying GaN films. Each black spot in the overgrown GaN layer shown in Figure 4 (a) is a void that forms when two growth fronts coalesce.

Stripe Orientation

$<11\bar{2}0>$ $<1\bar{1}00>$

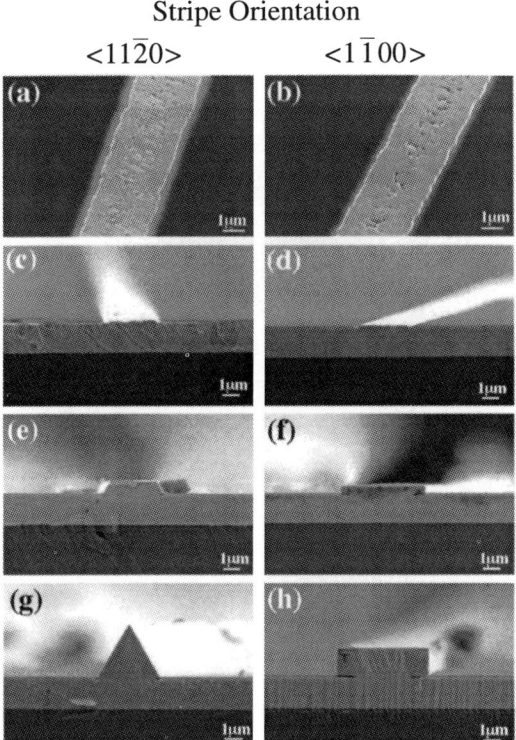

Figure 3. Scanning electron micrographs of GaN layers grown on 3 μm wide stripe openings oriented along $<11\bar{2}0>$ and $<1\bar{1}00>$ for the growth times of (a)-(d) 3 min, (e)(f) 9 min, and (g)(h) 20 min.

These voids were most often observed under the lateral growth conditions wherein rectangular stripes having vertical { $11\bar{2}0$ } side facets developed.

A cross-sectional TEM micrograph showing a typical laterally overgrown GaN layer is presented in Figure 5. Threading dislocations, originating from the GaN/AlN buffer layer interface, propagate to the top surface of the regrown GaN layer within the window regions of the mask. By contrast, there were no observable threading dislocations in the overgrown layer. Microstructural studies of the areas of lateral growth obtained using various growth conditions indicated that the overgrown GaN layers contained only a few dislocations. These dislocations formed parallel to the (0001) plane via the extension of the vertical threading dislocations after a 90° bend in the regrown region. These dislocations did not subsequently propagate to the surface of the overgrown GaN layers.

Cathodoluminescence of selected regions using a scanning electron microscope indicates that yellow emission originates from the regrown regions having high dislocation densities; however, only strong band edge emission was observed from the overgrown layers.

The pendeo-epitaxial phenomenon is made possible by taking advantage of growth mechanisms identified by Zheleva et.al. (1997) in the conventional LEO technique, and by

340

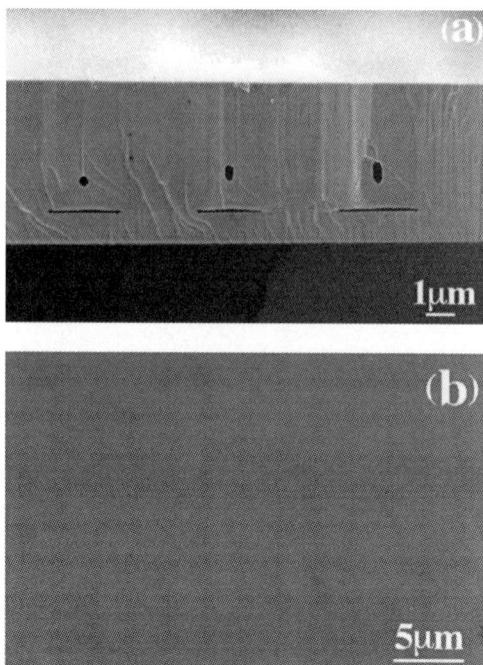

Figure 4. (a) Cross-section and (b) surface SEM micrographs of the coalesced GaN layer grown on 3 μm wide and 3 μm spaced stripe openings oriented along <1 1̄ 00>.

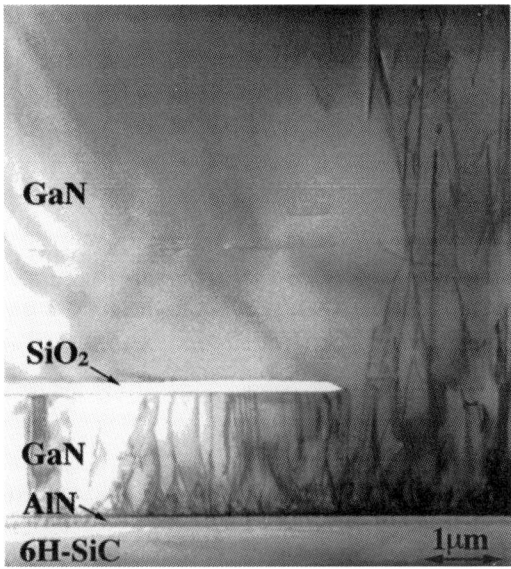

Figure 5. Cross-section TEM micrograph of a laterally overgrown GaN layer on a SiO_2 mask.

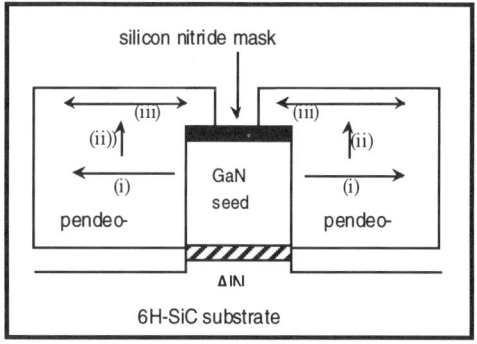

Figure 6. Schematic of the growth of GaN both laterally and vertically over a Si₃N₄ mask over the seed.

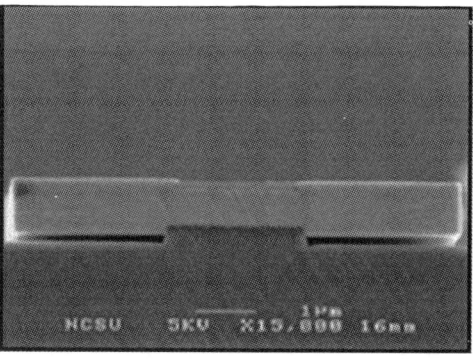

Figure 7. Cross-sectional SEM of a GaN pendeo-epitaxial growth structure with limited vertical growth from the seed sidewalls and no growth on the seed mask.

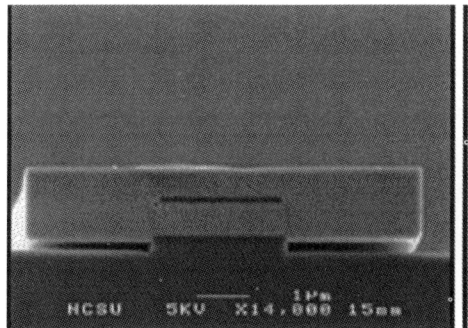

Figure 8. Cross-sectional SEM of a GaN/Al$_x$Ga$_{1-x}$N pendeo-epitaxial growth structure showing coalescence over the seed mask.

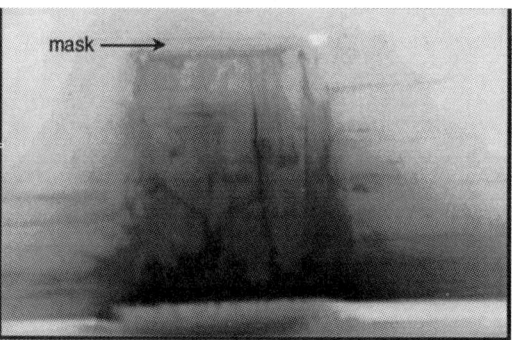

Figure 9. Cross-sectional TEM of a GaN pendeo-epitaxial structure showing confinement of threading dislocation under the seed mask, and a reduction of defects in the regrown areas.

using two additional key steps, namely, the initiation of growth from a GaN face other than the (0001) and the use of the substrate (in this case SiC) as a mask. By capping the seed-forms with a growth mask, the GaN was forced to grow initially and selectively only on the GaN sidewalls, as shown in Figure 6. Common to conventional LEO, no growth occurred on the silicon nitride mask covering the seed forms. Deposition also did not occur on the exposed SiC surface areas at the higher growth temperatures employed to enhance lateral growth. The Ga- and N-containing species more likely either diffused along the surface or evaporated (rather than having sufficient time to form GaN nuclei) from both the silicon nitride mask and the silicon carbide substrate. The pronounced effect of this is shown in Figure 7 wherein the newly deposited GaN has grown truly suspended (*pendeo-*) from the sidewalls of the GaN seed structure. During the second PE event (ii), vertical growth of GaN occurred from the advancing (0001) face of the laterally growing GaN. Once the vertical growth became extended to a height greater than the silicon nitride mask, the third PE event

(iii) occurred, namely, conventional LEO-type growth and eventual coalescence over the seed structure, as shown in Fig. 8. A cross-sectional TEM micrograph showing a typical pendeo-epitaxial growth structure is shown in Fig. 9. Threading dislocations extending into the GaN seed structure, originating from the GaN/AlN and AlN/SiC interfaces, are clearly visible.

The silicon nitride mask acted as a barrier to the further vertical propagation of these defects into the laterally overgrown pendeo-epitaxial film. Since the newly deposited GaN is suspended above the SiC substrate, there are no defects associated with the mismatches in lattice parameters between GaN and AlN and between AlN and SiC. Preliminary analysis of the GaN seed/GaN pendeo-epitaxy interface revealed evidence of threading dislocations or stacking faults within the (0001) planes. This indicates evidence of the lateral propagation of the defects; however, there is yet no evidence that the defects reach the (0001) surface where device layers will be grown. As in the case of LEO, there is a significant reduction in the defect density in the regrown areas.

The continuation of the pendeo-epitaxial growth results in coalescence with adjacent growth fronts and the formation of a continuous layer of GaN, as observed in Figure 10. This

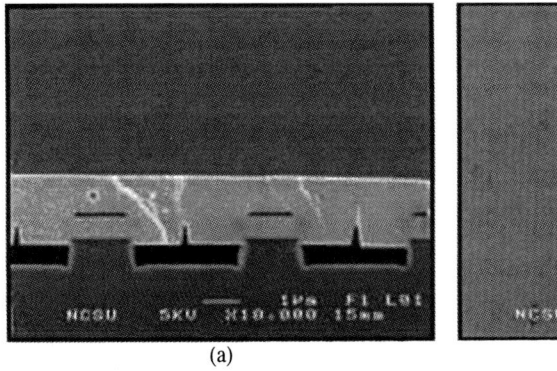
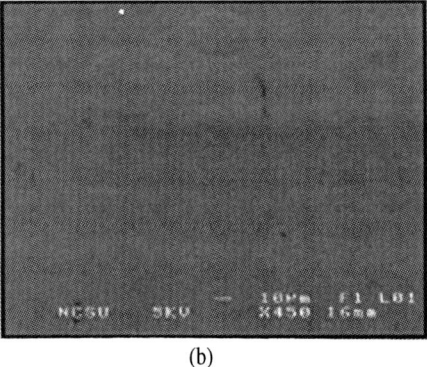

(a)   (b)

Figure 10. Micrographs taken via (a) cross-sectional SEM and (b) plan-view SEM of examples of pendeo-epitaxial growth with coalescence over and between the seed forms resulting in a single GaN layer.

also results in the practical elimination of all dislocations stemming from the heteroepitaxial growth of GaN/AlN on SiC. Clearly visible in Fig. 10(a) are the voids that form when adjacent growth fronts coalesce. Optimization of the pendeo-epitaxial growth technique should eliminate these undesirable defects.

Each of the microstructures depicted in Figures 2(d), (e) and (f) has been realized in this research. It was necessary to refine the experimental procedure described above to achieve the different microstructures. To produce the heterostructure shown in Figure 2(e) a thin GaN layer was grown for 2 min at 1070°C and for 2 min at 1090°C. The $Al_{10}Ga_{90}N$ was deposited using three growth steps at the susceptor temperatures and times of 1090°C and 2 min., 1110°C and 50 min and 1090°C and 30 min. The second step was employed to force the $Al_{10}Ga_{90}N$ to grow laterally and coalesce; the third step was used to grow the film vertically. The growth rate of the $Al_{10}Ga_{90}N$ films was markedly lower at both 1090 and 1100°C than that of the GaN at any temperature employed in this study. The multi-layered structure shown in Figure 2(f) was also realized using several different growth parameters. The first layer of GaN was grown at susceptor temperatures of 1075°C for 3 min and 1090°C for 2 min. The subsequent layer of $Al_{10}Ga_{90}N$ was grown using susceptor temperatures of 1090°C for 10 min and 1075°C for 5 min. The temperature during the $Al_{10}Ga_{90}N$ growth was not increased to 1110°C, to limit the lateral growth. The additional layer of GaN was grown at a susceptor temperature of 1075°C for 10 min.

4. SUMMARY

Continuous GaN films having very low dislocation densities have been achieved via lateral growth and coalescence of GaN homoepitaxial stripes over SiO_2 masks. The results suggest that lateral overgrowth of GaN via OMVPE is a promising technique for obtaining low defect density layers and could be useful for optoelectronic and microelectronic device applications. The pendeo-epitaxial technique has been developed as an alternative and more simple approach of growing uniformly low-defect density thin films over the entire surface of a substrate. In particular, we have demonstrated the growth of both discrete structures and coalesced GaN and $Al_xGa_{1-x}N$ films and multilayer heterostructures using pendeo-epitaxy on etched GaN seed layers previously grown on AlN/6H-SiC substrates.

ACKNOWLEDGEMENTS

The authors acknowledge Cree Research, Inc. and Motorola for the SiC and the silicon wafers respectively. This work was supported by the Office of Naval Research under contracts N00014-96-1-0765 (Colin Wood, monitor) and N00014-98-1-0654 (John Zolper, monitor).

REFERENCES

Doverspike K, Bulman G, Sheppard S, Kong H, Leonard M, Dieringer H, Weeks Jr. W, Edmond J, Brown J, Swindle J, Schetzina J, Song Y, Kuball M and Nurmikko A 1998 Materials Res. Soc. Symp. Proc **482**, (Pittsburgh, PA: Materials Research Society) pp 1169-1174

Gehrke T, Linthicum K, Thomson D, Rajagopal P, Batchelor D and Davis R 1999 to be published in Materials Res. Soc. Symp. Proc **537**, eds S J Pearton, C Kuo and T Uenoyama (Pittsburgh, PA: Materials Research Society)

Kato Y, Kitamura S, Hiramatsu K and Sawaki N 1994 J. Cryst. Growth **144**, 133

Linthicum K, Gehrke T, Thomson D, Tracy K, Carlson E, Smith T, Zheleva T, Zorman C, Mehregany M and Davis R 1999b to be published in Materials Res. Soc. Symp. Proc **537,** eds S J Pearton, C Kuo and T Uenoyama (Pittsburgh, PA: Materials Research Society)

Linthicum K, Gehrke T., Thomson D, Carlson E, Rajagopal P, Smith T Batchelor D and Davis R 1999a (submitted to Applied Physics Letters).

Marchand H, Wu X, Ibbetson J, Fini P, Kozodoy P, Keller S, Speck J, Denbaars S and Mishra U 1998 Appl. Phys. Lett. **73**, 747

Marchand H, Zhang N, Zhao L, Golan Y, Rosner S, Girolami G, Fini P, Ibbetson J, Keller S, DenBaars S, Speck J and Mishra U, 1999 MRS Internet J. Nitride Semicond. Res. **4**, 2.

Nakamura S, Mukai T, and Senoh M, Appl. Phys. Lett 1994 **64**, 1687

Nakamura S, Senoh M, Nagahama S, Iwasa N, Yamada T, Matsushita T, Kiyoku H and Sugimoto Y 1996 J. Appl. Phys. **35**, L74

Nakamura S, Senoh M, Nagahama S, Iwasa N, Yamanda T, Matsushita T, Kiyoku H, Y, Sugimoto, Kozaki T, Umemoto H, Sano M and Chocho 1997 K, Proc. of the 2nd Int. Conf. On Nitride Semicond., Tokushima, Japan (unpublished)

Nam O, Bremser M, B Ward, R Nemanich, and R Davis 1997a Mat. Res. Soc. Symp. Proc., **449**, (Pittsburgh, PA: Materials Research Society) pp 107-112

Nam O, Bremser M, Ward B, Nemanich R and Davis R 1997b Jpn.J. Appl. Phys. Part 1 **36**, L532

Nam O, Zheleva T, Bremser M and Davis R 1997c Appl. Phys. Lett. **71**, 2638

Sakai A, Sunakawa H and Usui A 1998 Appl. Phys. Lett., **73**, 481

Thomson D, Gehrke T, Linthicum K, Rajagopal P, Hartlieb P, T Zheleva and Davis R 1999 to be published in Materials Res. Soc. Symp. Proc **537**, eds S J Pearton, C Kuo and T Uenoyama (Pittsburgh, PA: Materials Research Society)

Underwood R, Kapolnek D. Keller B, Keller B, Denbaars S and Mishra U 1995 Topical Workshop on Nitrides, Nagoya, Japan, September (unpublished)

Weeks T, Bremser, M, Ailey S, Carlson E, Perry W and Davis R 1995 Appl. Phys. Lett. **67**, 401

Zheleva T, Nam O-H, Bremser M and Davis R 1997 Appl. Phys. Lett. **71**, 2472

Zheleva T, Smith S, Thomson D, Linthicum K, Gerhke T, Rajagopal and R Davis 1999a (to be published in the Journal of Electronic Materials)

Zheleva T, Thomson D, Smith S, Rajagopal P, Linthicum K, Gehrke T and Davis R 1999b to be published in Materials Res. Soc. Symp. Proc **537,** eds S J Pearton, C Kuo and T Uenoyama (Pittsburgh, PA: Materials Research Society)

Zhong H, Johnson M, McNulty T, Brown J, Cook Jr J and Schezina J 1998 Materials Internet Journal, Nitride Semiconductor Research, **3**, 6

Inst. Phys. Conf. Ser. No 164
Paper presented at Microsc. Semicond. Mater. Conf., Oxford, 22–25 March 1999
© 1999 IOP Publishing Ltd

Microscale luminescence from ELOG specimens on (0001) sapphire

A Amokrane, S Dassonneville, B Sieber*, K Jacobs, Z Bougrioua and I Moerman**

*Laboratoire de Structure et Propriétés de l'Etat Solide, ESA CNRS 8008, Bât. C6,
Université des Sciences et Technologies de Lille, 59655 Villeneuve d'Ascq Cédex, FRANCE
** IMEC-INTEC, University of Gent, Sint-Pietersnieuwstraat 41, B-9000 Gent, Belgium.

ABSTRACT: The spatial variations of the luminescence properties of an epitaxially overgrown gallium nitride specimen are studied by means of spectrally resolved plan-view cathodoluminescence. GaN material grown over the stripes exhibits intrinsic and extrinsic luminescence intensities which both can be twice higher than those recorded in the windows. The columnar structure of the epilayer is made with cells of larger size above the stripes than in the window areas.

1. INTRODUCTION

Wide band-gap semiconductors of the group III nitride family such as Gallium Nitride (GaN) are still receiving much attention due to their manifold commercial interestssuch as violet and blue/green light-emitting diodes (Nakamura et al 1995), and blue laser diodes (Nakaruma et al 1996). Nitride compounds have also demonstrated their ability in the fabrication of electronic devices based on MESFET or HFET (Fan et al 1997). But its further development requires now the growth of epilayers where the intrinsic violet luminescence dominates the extrinsic yellow one. In that respect, the epitaxially laterally overgrown GaN (ELOG) is promising since the first results obtained on such layers have shown a reduced density of dislocations (Marchand et al 1998) which are thought to be partly responsible for the yellow band (Ponce et al 1996), although this yellow luminescence might also be related to the presence of Ga vacancies (Neugebauer et al 1996). In this paper, we present the results obtained by means of spectrally resolved plan-view cathodoluminescence (CL) on such a specimen. The recombination properties of the ELOG specimen are compared to those of an epitaxial GaN layer, which was grown without SiO2 stripes.

2. EXPERIMENTS

The ELOG specimen has been grown by organometallic vapor phase epitaxy, as follows: after the deposition of a 50 nm GaN buffer layer on top of a c-plane sapphire substrate at 630°C, a 2.4 μm thick undoped GaN film was grown at 1080°C. Then 150 nm SiO2 was deposited and stripes were defined along the [1$\bar{1}$00] direction, using conventional photolithography and dry etching. Afterwards, regrowth of a 5 μm thick undoped GaN epilayer took place at 1060°C. The width of SiO2 stripes is 3 μm and the distance between two stripes (window width) 5 μm. The layer was n-type with a carrier density close to 1×10^{17} cm^{-3}. A reference sample consisting of a 50 nm GaN buffer and an epitaxially undoped 2.4 μm thick GaN film grown on (0001) sapphire was also investigated and compared to the ELOG specimen. Plan-view CL spectra and monochromatic images were recorded at 87K on a Cambridge 250 MK3 scanning electron microscope (SEM) fitted with an Oxford CL system, a H10 UV Jobin Yvon monochromator and a GaAs R 636 photomultiplier. A H10 IR Jobin Yvon monochromator was used to record the yellow luminescence. The CL spectra presented in this paper are not corrected for the system response. The accelerating voltage was varied from 10 to 20 kV. Thus, the electron beam does create electron-hole pairs only in the top GaN layer since its penetration depth is of 0.4 to 1.4 µm respectively.

3. LUMINESCENCE PROPERTIES OF ELO AND EPITAXIAL GaN

3.1. Epitaxial GaN

Figures 1 and 2 show typical CL spectra and monochromatic CL images recorded on the GaN reference sample. The intrinsic band, which is more intense than the extrinsic blue and yellow bands, peaks at 358 nm (3.464 eV). The experiments being performed at 87 K, it could correspond to the radiative recombination of the A free exciton (Tchounkeu et al 1996). The low-energy shoulder of the intrinsic peak, located at 365 nm (3.397 eV) could be a phonon replica of the UV peak (Beaumont 1999).

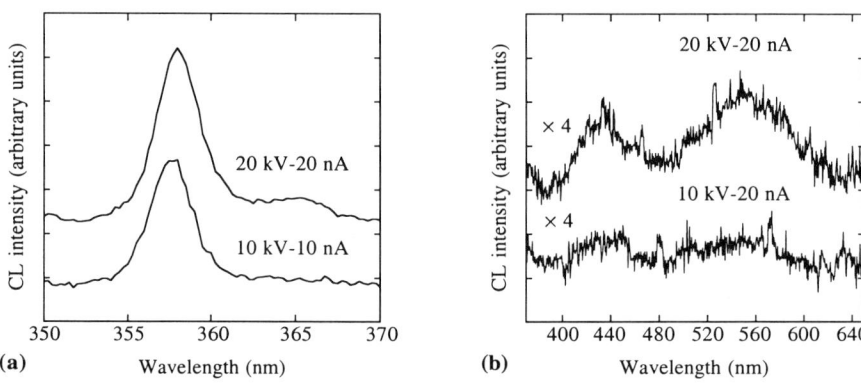

Fig.1: Typical CL spectra recorded at 87 K on GaN reference sample for two values of the accelerating voltage. a) UV peak . b) extrinsic bands.

A broad blue band is easily detected in the epilayer. This shows that residual magnesium atoms present inside the reactor, and used in previous experiments, have been incorporated in the layer during its growth. From the energy position of the band (2.884eV;

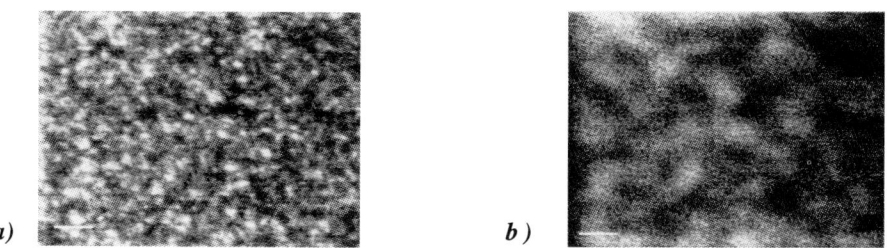

Fig.2: 87 K typical CL monochromatic images recorded at 20 kV on GaN reference sample; a) 346 nm, b) 556 nm . The bar scale is 4 µm.

430 nm), it can be concluded that the Mg acceptor is in a deep configuration and that the concentration in the volume of the epilayer probed by the electron beam can be estimated in the

range 0.03 % - 0.3 %. (Eckey et al 1998). The yellow band (YB at 556 nm) is more visible on the spectra recorded at 20kV than at 10 kV (Figure 1b). As the YB is not being absorbed by the material, most of the defects responsible for it are likely to be situated close to the interface than to the free surface of the epilayer. CL images recorded with the UV peak and the blue band (not shown) are quite similar, but the spatial resolution of the bright spots decreases from about 1 μm to about 4 μm when the wavelength increases from 346nm to 556 nm (Figure 2). A low signal to noise ratio could explain such a trend.

3.2. ELO specimen

The CL spectra recorded on the stripes <u>and</u> in the windows of the ELOG specimen exhibit a strong increase of the UV to YB intensities ratio, compared to that observed in the reference sample (figure 3a). Furthermore, the UV luminescence efficiency of the ELO specimen in the UV range is about 10 times larger than that of the reference sample, since a 10 times lower beam current was necessary to get a similar CL signal. It has to be noticed that the increase of the UV luminescence intensity occurs for the material located above as well as in-between the stripes. When measured at 20 kV and 10 nA, the YB intensity has been found to be about 4 times higher in the windows of the ELOG specimen than in the reference sample. A DAP band located at 377 nm (3.289 eV) with its phonon replica (390nm: 3.1795 eV) shows that the Mg concentration is close to 0.03 %. In few areas of the ELO, the DAP transition and a blue broad band located close to 430 nm are both visible on the spectra.

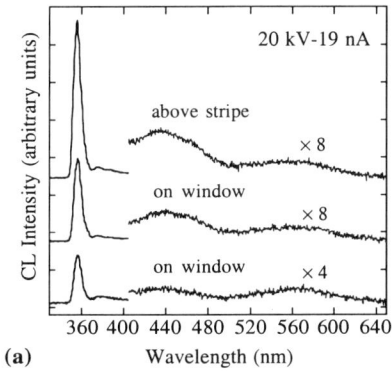
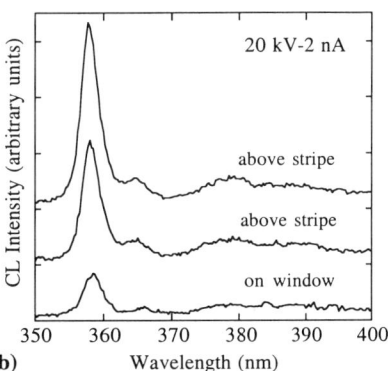

(a)   **(b)**

Fig. 3: CL spectra recorded at 87 K on the stripes and windows of the ELOG specimen at 20 kV a) whole spectra, recorded with the IR monochromator; the beam current (I_b) is 19 nA b) UV and blue parts of the spectra recorded with the UV monochromator; I_b = 2 nA.

Monochromatic CL images show that the UV luminescence as well as the extrinsic blue (not shown) and yellow luminescence bands are higher above the stripes than in-between them (figure 4). The enhancement of the UV peak on the stripes has been already quoted in the literature (Li et al 1998, Yu et al 1998) and has been associated with a reduction of the dislocation density as observed in ELOG samples (Marchand et al 1998). A preliminary TEM analysis of an equivalent sample confirms that the dislocation density of threading dislocations is much lower above the stripes than in the windows (Jacobs et al 1999). But, usually, a decrease of the YB above the stripes with respect to the windows has been quoted in the literature (Li et al 1998, Yu et al 1998). Thus, it could be suggested that dislocations and the point defects segregated around them are not associated with the YB luminescence. This is also supported by the larger YB measured on the ELOG sample. Recording of the UV luminescence shows that, in the studied sample, its intensity is about five to six times larger on the stripes than in the openings. The CL intensity of the UV peak can vary by a factor two along the stripes as well as in the windows. The size of the luminescent cells is larger above the stripes than in-between them; their width is about 1 μm in the openings, and can reach 3-4 μm within the stripes. A dark dashed line in the middle of the stripes can be observed, although no contrast of the GaN surface could be detected in secondary electrons. This cannot be explained by uncompleted coalescence of 5 μm GaN above the stripes. These dark lines may be related to regions where impurities have segregated and possibly also to the presence of dislocations (Marchand et al 1998). The UV and YB CL images of the stripes and of the openings are identical and display the same spatial resolution (figure 4).

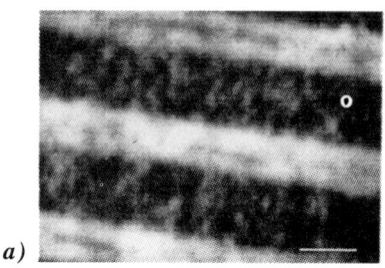 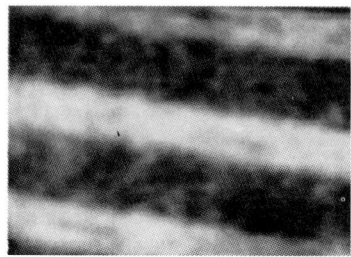

a)   *b)*

Fig. 4: 87 K monochromatic CL images of the ELOG specimen recorded at 20 kV. a) 358nm b) 556nm. The noise has been canceled by Fourier transformation. The bar scale is 4 μm.

4. CONCLUSION

Intrinsic and YB intensities given by an ELOG specimen have been compared with those of a GaN epilayer. The ELOG specimen, despite a higher YB luminescence, displays a higher intrinsic UV luminescence intensity. In this sample, both the UV and YB luminescences are higher above the stripes than in-between them. The ELOG process has also enlarged the size of the cells of the columnar structure above the SiO_2 stripes.

ACKNOWLEDGEMENT

We should like to thank C. Vanmansart for his help in the CL experiments.

REFERENCES

Beaumont B 1999 private communication

Eckey L, von Gfug U, Holst J, Hoffmann A, Kaschner A, Siegle H, Thomsen C, Schineller B, Heime K, Heuken M, Schön O and Beccard R 1998 J. Appl. Phys. **84**, 5828.

Fan Z, Lu C, Botchkarev AE, Mohammad SN, Roth M, Jenkins T, Kehias L and Morkoç H 1997 Electron. Lett. **33**, 814

Jacobs K, Van der Stricht W, Moerman I, Demeester P, Verstuyft S, De Nayer J, Van Daele P, Amokrane A, Dassonneville S, Sieber B and Thrush EJ, Submitted to Electronic Materials Conference (EMC '99)

Li X, Bishop SG and Coleman JJ 1998 Appl. Phys. Lett. **73**, 1179

Marchand H, Ibbetson JP, Fini PT, Kozodoy P,Keller S, DenBaars S, Speck JS and Mishra UK1998 MRS Internet J. Nitride Semicond. Res., **3**, 3

Marchand H, Wu XH, Ibbetson JP, Fini PT, Kozodoy P, Keller S, Speck JS, DenBaars SP and Mishra UK 1998 Appl. Phys. Lett. **73**, 747.

Nakamura S, Senoh M, Iwasa N and Nagahama S 1995 Jpn. J. Appl. Phys. 2, Lett. **34**, L797

Nakaruma S, Senoh M, Nagahama SI, Iwasa N, Yamada T, Matsushita T, Kiyoku H and Sugimoto Y 1996 Appl. Phys. Lett. **68**, 3269

Neugebauer J, Van de Walle CG 1996 Appl. Phys. Lett. **69**, 503

Ponce FA, Bour DP, Götz W and P.J. Wright PJ 1996 Appl. Phys. Lett. **68**, 57.

Tchounkeu M, Briot O, Gil B, Alexis JP and Aulombard RL 1996 J. Appl. Phys. **80**, 5352.

Yu Z, Johnson MAL, Mcnulty T, Brown JD, Cook JW and Schetzina JF 1998 MRS Internet J. Nitride Semicond. Res, **3**, 6.

Inst. Phys. Conf. Ser. No 164
Paper presented at Microsc. Semicond. Mater. Conf., Oxford, 22–25 March 1999
© *1999 IOP Publishing Ltd*

STEM investigations of GaN

U Bangert, A J Harvey, C Dieker[a], A Rizzi[b], D Freundt[b] and H Davock[c]

Department of Physics, UMIST, Manchester M60 1QD, UK
[a]Mikrostrukturanalytik, Technische Fakultät der CAU, Kaiserstr 2, 24143 Kiel, Germany
[b]Institut fuer Schicht und Ionentechnik, Forschungszentrum Juelich, 52425 Juelich, Germany
[c]Department of Materials Science and Engineering, University of Liverpool, Liverpool L69 3BX, UK

ABSTRACT   The low energy loss region of electron energy loss (EEL) spectra of hcp and cubic GaN is investigated. The joint density of states times the matrix element curves show differences between the hcp and cubic phase. Possible effects of individual dislocations on the energy loss intensity in the interband scattering regime in EEL spectra will be highly localized. Relying on the high spatial resolution of this technique and in spite of the low intensities in the interband scattering regime we investigate this possibility. For the case of screw dislocations in hcp and of perfect dislocations in cubic material distinct differences in the intensity of pre-bandedge states could be detected.

1. INTRODUCTION

The III-V nitrides are among the most promising semiconductors for blue and ultra-violett device applications. Device quality wurzite-GaN can be grown by metal organic vapor deposition on (0001) sapphire substrates. The large misfit results in dislocation tangles near the interface, but also in isolated threading dislocations parallel to the c-axis with densities of ~ 10^9 cm^{-2}, which cross the active regions of devices. There is currently some controversy concerning their effects on the device performance: an unexpected finding is that these dislocations do not lead to a reduction in the lifetime of light emitting devices (e.g. Lester et al 1996); on the other hand, atomic force microscopy in combination with cathodoluminescence (CL) has shown that threading dislocations act as non-radiative recombination centres and degrade the luminescence efficiency in the blue light spectrum of the epilayers (Rosner et al 1997). More specifically there are arguments that it is the screw dislocations, which act as non-radiative recombination centres. CL studies have also shown that the undesired yellow luminescence, which is ascribed to a V_{Ga}-O complex, is spatially non-uniform and can be correlated with extended defects, in particular low angle grain boundaries, which contain dislocations (Ponce et al 1996).

Plan view specimens can conveniently be prepared in this material with the threading dislocation perpendicular to the thin foil, so that the electron beam can be channelled down the dislocation line. The volume ratio of the dislocation core to the matrix can be 1:10 in such a geometry and hence there is a fair chance to separate the effects of the dislocation from the matrix contribution. Cross sectional specimen can be used for studies of dislocations running parallel to interfaces and edge-on stacking faults. The bandgap of ~3.4 eV in hcp GaN makes it an ideal material for electronic bandstructure studies in a highly localized fashion via EEL spectroscopy (EELS). The purpose of this study was to investigate whether dislocations introduce midgap or pre-bandedge states in GaN.

2. EXPERIMENTAL

All investigations were carried out in a VG cold emission FEG STEM at 100 keV. The electron probe size was typically 0.8 -1 nm in diameter. The energy spread of the electron beam was 0.45-0.55 eV with a tungsten filament and 0.3 - 0.4 eV when a zirconiated tungsten filament with a lower work function was used. The GATAN 666 PEEL spectrometer was set at an energy dispersion of 0.1 eV per channel. Samples were made from hcp GaN films grown by MBE on sapphire or on hcp SiC for inspection in plan view. In this orientation the film normal was parallel to the c-axis and also parallel to the majority of the dislocations in the film. Hence the electron beam could be aligned to encompass individual dislocations during EELS. Strong yellow luminescence was exhibited by the material grown on sapphire but not by the material grown on SiC. (110) cross-sections were made from cubic GaN films grown on cubic SiC (on Si) such that EELS could be carried out with the electron probe along [110] dislocations, and using box or line scan geometries on stacking faults which could be viewed edge-on in this orientation. The spectrum acquisition conditions are described elsewhere (Bangert et al 1998). Spectra of the low loss region up to 50 eV were recorded. A modelled zero loss function, fitted to each respective spectrum zero loss peak, was subtracted from the spectra.These were then subjected to Kramer-Kronig analysis in order to enable the extraction of the electronic joint density of states. The use of a modelled zero loss function was preferred to a zero loss peak measured without a specimen, since the shape of an unscattered beam was found to differ from an elastic scattering distribution, which is part of an energy loss spectrum.

In a second approach an energy broadening function was deconvolved from the raw spectra before employing Kramer-Kroening analysis. This procedure sharpened the spectrum peaks to a certain extent, but did not reveal any new information.

EEL spectra were taken of a number of dislocations and stacking faults and compared to spectra taken from crystallographically perfect parts of the matrix next to the respective defects.

3. RESULTS AND DISCUSSION

Plan view micrographs reveal different dislocation types: screw dislocations with the typical inverse contrast double-lobes when viewed edge on, and also dislocations with bow-tie or V-contrast, indicative of an edge component (Fig. 1a). No nano-pipes were observed. EEL analysis was undertaken on both types of dislocations. Though each EEL spectrum is the result of the summation of many hundred spectra, it becomes clear from the low loss spectra that the low scattering cross

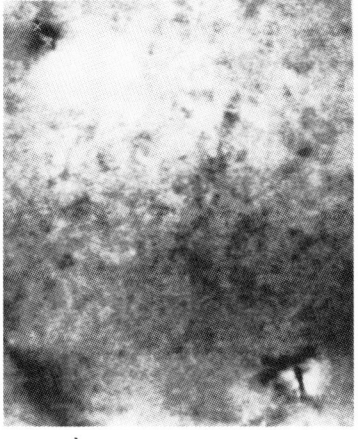 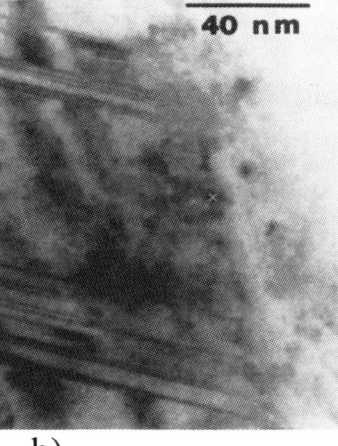

a)   b)

Fig.1. a) plan view of hcp GaN film, showing edge-on dislocation contrast of a pure screw (x) and of dislocations with edge component. b) cross-sectional view of cubic GaN film, showing edge-on stacking faults and dislocations.

sections for interband scattering events are the limiting factor in this kind of experiment.

The value of the Y-axis in all spectra is the product of the joint density of states and the transition matrix element. Under the assumption that the latter is constant we call the Y-value freely the 'joint density of states'.The joint density of states spectra shown in Fig.2 are representative of crystallographically perfect parts (solid lines) and of dislocations (dashed lines), and are taken of a hcp GaN film which exhibited strong yellow luminescence. The image contrast suggested that the dislocation had edge components. The distinct features in all hcp spectra are a broad peak or double peak at around 4 eV, a dip at around 6 eV and a second major rise with a peak at around 7.5 eV. The peaks can be attributed to transitions around the Γ point (including $\Gamma \rightarrow A$) and at the H-point in the Brillouin zone (since inter-zone transitions and such with non zero initial k-vectors are possible). The ratio of these peaks undergoes variations from spectrum to spectrum, predominantly because the spectrum statistics are rather poor. It is therefore extremely difficult to draw conclusions about differences between perfect and dislocated parts: often spectra taken at different perfect locations vary as much as between perfect and dislocated locations. If we concentrate on the bandedge and pre-bandedge region, it emerges that there is a pronounced rise in the joint density of states at around 3.4 eV in perfect parts though there is also some scattering intensity at energies below 3.4 eV, whereas the dislocation spectra show an earlier and slower rise. It could be argued that this effect is marginal, but it seems to be consistant throughout all our studies. It therefore appears that near bandgap states are present throughout the material with an increase in spectra taken of small volumes including dislocation cores.

The spectra shown in Fig.3 a), b) and c) are of a hcp GaN film, which exhibited no significant yellow luminescence, and

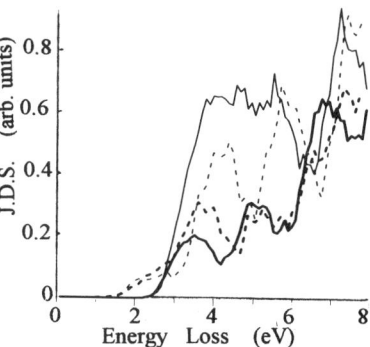

Fig.2 Joint density of states in GaN/Al$_2$O$_3$, on dislocations (dashed) and in perfect regions besides the dislocations (solid)

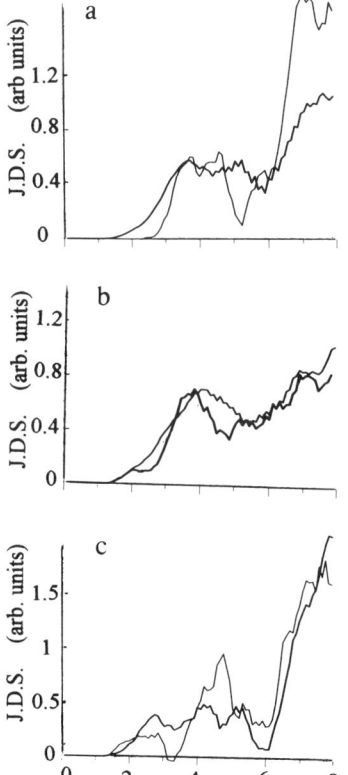

Fig. 3) Joint density of states curves for GaN/hcp-SiC a) in perfect regions, b) on dislocations with edge components, c) on screw dislocations

are taken in perfect crystal parts and on close-by dislocations with edge components and on screw dislocations, respectively. Again there is a pronounced rise at 3.4 eV in the perfect material and some pre-bandedge intensity. The dislocation curves in Fig.3b of mixed dislocations are similar to those of the perfect material in Fig.3a and arguments of stronger pre-edge intensity in the former are disputable. The screw dislocations in Fig.3c, however, show a more pronounced peak between 2 and 3 eV in relation to the bandedge rise than any of the other spectra.

Fig. 4 shows joint density of states curves representative of perfect crystal parts (curve a), dislocations (curve b) and stacking faults (curve c) in a cubic GaN film. A corresponding image is shown in Fig. 1b). Other than that they are of the perfect type, the nature of the dislocations, strongly visible as circular dark contrast spots, has not been established in this study.

The spectra are generally flatter with shallower peaks than those of the hcp structure, reflecting the 'simpler' bandstructure of the cubic phase, with Brillouin zone transitions at Γ-point and along the shallow valley from Γ to L. It should be noted that the steepest rise in the perfect material is just after 3 eV, which is in accordance with the bandgap at 3.2 eV. There is some pre-edge intensity in all spectra. The dislocations investigated in this study introduce a distinctive shift in the rise of the spectrum down to about 2 eV. The contrast suggests that the dislocations are decorated by impurities, either native to the GaN or introduced by contamination during TEM sample preparation. (No contamination was observed during STEM investigation!) The rise in the stacking fault spectra, too, appears to start earlier and to proceed more slowly than in the perfect material. The stacking fault spectra were taken via line scans and it is estimated that in this geometry planar defects can contribute up to 50% to the scan volume in samples of about 50 nm thickness. Nevertheless this is an unexpected observation; at

352

least in hcp GaN stacking faults are not expected to introduce localized states in the bandgap (Stampfl et al, 1998). This suggests that there might be decoration of the stacking faults also.

To draw conclusions from this investigation, though carried out with great care on many dislocations and stacking faults, is tricky. The low intensities in the interband scattering regime make statements about the effect of dislocations unreliable in most cases. Pre-bandedge states between 2 and 3 eV can be detected everywhere in the materials with varying intensities. The finger prints of the bandstructure in perfect parts, characteristic of the hcp and the cubic phase are, however, clearly noticeable. Also there is a distinct difference in comparison with all the other spectra, in the intensity of the pre-bandedge states of the screw dislocation in the hcp and of the perfect dislocation in the cubic material. This intensity is much greater than expected from a 10% contribution towards the low loss signal (in fact if it was only 10% we would not see the effects over the statistical error as in the other cases), and it is therefore reasonable to conclude that this must be due to a change in the transition matrix element at the dislocation. The limited energy resolution of the instrument does not, however, allow us to study the pre-bandedge intensity in greater detail.

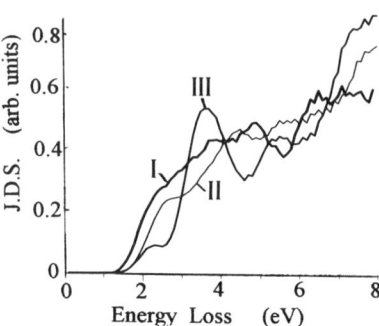

Fig.4 Joint density of states curves for c-GaN/c-SiC on a dislocation (curve I), on a stacking fault (curve II) and of a perfect portion (III).

To reliably detect the effects of dislocations at all is extremely useful, in order to enable the identification of microstructural features, which might contribute to non-radiative recombination or to undesired optical emission bands. In this light the improvement of the energy resolution would not appear to be a prime requirement: the spectra in Fig.2 were acquired with a zirconiated tip of significantly improved energy resolution, the zero loss function, however, exhibited similar tails as that of a 'normal' tungsten tip. Hence we conclude that it is not the emission function, but the detector resolution, which contributes to the high background in the 0.5 to 5eV loss regime. To improve the strength of the low loss signal over the background in this region is the key issue, if the low loss technique is to become a viable spectroscopic tool. To this end sophisticated deconvolution procedures are not the answer either: the tails of the zero loss function have to be reduced and this is only possible with the use of an improved detection system. New detectors based on CCD camera technology, which are now available for PEELS, might contribute to the solution of the problem.

ACKNOWLEDGEMENTS

We thank the Anglo German foundation for financial support through ARC project #857

REFERENCES

Bangert U, Harvey A J, Davidson J, Dieker C and Keyse R 1998 J. Appl. Phys. **83**,7726
Lester D, Ponce F A, Cranford M G and Steigerwald D A 1996 Appl. Phys. Lett. **66**, 1249
Ponce F, Bour D B, Gotz W and Wright P J 1996 Appl. Phys. Lett. **86**, 57
Rosner S J, Carr E C, Ludowise M G, Girlami G and Erikson H I 1997Appl. Phys. Lett. **70**, 420
Stampf C and Van der Walle C G 1998 Phys. Rev. B **57**, R15052

Inst. Phys. Conf. Ser. No 164
Paper presented at Microsc. Semicond. Mater. Conf., Oxford, 22–25 March 1999
© 1999 IOP Publishing Ltd

Spatially resolved low-loss EELS analysis of optical properties of GaN and related alloys

G Brockt, H Lakner* and E Kubalek

Werkstoffe der Elektrotechnik, Gerhard-Mercator-Universität Duisburg, 47048 Duisburg, Germany
*Fraunhofer Institut für Mikroelektronische Schaltungen und Systeme, 01109 Dresden, Germany

ABSTRACT: Interest in the optical properties of GaN compounds remains strong. Information deduced from optical measurements lacks spatial resolution, thus representing macroscopic properties. Here, the local dielectric properties are measured by spatially resolved EELS performed in a cold field-emission STEM at an energy-resolution of 0.35 eV. The low-loss EEL spectra reveal information on plasmon excitations, bandgap transitions on a nanometre scale and the local dielectric properties are deduced via Kramers-Kronig transformation. These data show excellent agreement with optical ellipsometric results. The superior spatially resolution of EELS is applied to the characterization of individual defects.

1. INTRODUCTION

Although the recent progress in growth of GaN-based optoelectronic devices has already resulted in commercial success, the interest in understanding and controlling the dielectric properties remains high. To characterize these properties, usually optical measurements are applied which offer high energy resolution of 1-500 meV. For example, ellipsometry has been used successfully, where synchrotron radiation must be applied in order to cover the necessary spectral range between 2 and 25 eV (Wethkamp et al 1997). But the spatial resolution of optical methods is limited in the range of micrometers, or even worse, which restricts its application. In contradistinction to this, low-loss EELS in a transmission microscope offers spatially resolved information, but mostly is handicapped by a lower energy-resolution. Nevertheless, improved experimental methods for EELS already show promising results on the characterization of properties of GaN (Natusch et al 1997 and 1998, Bangert et al 1997, Brockt et al 1997). But their accuracy and expressiveness is limited and varies with the energy-resolution achieved and the data processing methods applied.

In this work it is demonstrated how dedicated methods of electron energy-loss spectroscopy (EELS) in a scanning transmission electron microscope (STEM) using subnanometre electron probes yield improved data on the characterization of local dielectric properties near the bandgap region of group III-nitrides at a spatial resolution on a nanometre scale.

2. EXPERIMENTAL METHODS

Cross-sectional specimens from heterostructures of wurtzite group III-nitrides grown by MOVPE on sapphire substrates have been investigated in a STEM. The microscope used is a cold field-emission STEM (VG Microscopes: HB501). The minimum diameter of the electron probe is 0.3 nm at an energy of 100 keV.

For the EELS investigations we made use of a dedicated parallel detection system (Mc Mullan et al 1992) providing an experimental energy-resolution of 0.35 eV (Brockt et al 1999). To separate the onset of the bandgap signal from the zero-loss background the zero-loss peak has been removed using a Fourier-ratio deconvolution (Egerton 1996) with an experimentally recorded vacuum zero-loss peak followed by a convolution with a suitable gaussian peak to avoid the domination of high-frequency artifacts.

In the low-loss spectrum the measured energy-loss function depends on the joint density of states where maxima occur at the critical points in the bandstructure. As the energy-loss function is proportional to Im(-1/ε), the local properties of the real and imaginary part of the dielectric function ε can be deduced via Kramers-Kronig transformation (Egerton, 1996). Further optical data like the reflectivity R can be calculated from this and compared to results from optical measurements.

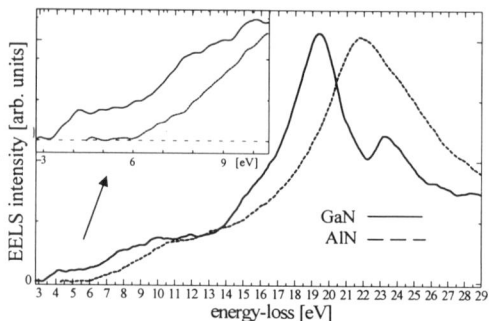

Fig.1: Low-loss spectra of GaN and AlN

3. RESULTS AND DISCUSSION

In the case of III-nitride alloys, which typically show high defect density, the spatial resolution of the STEM technique allows the characterization of defect-free areas. Fig.1 shows the measured energy-loss functions of defect free wurtzite GaN and AlN after the removal of the zero-loss peak. As a result of the described methods the EEL-spectra directly reveal the onset of the bandgap itself and the characteristic shape of the joint density of states. Thus without the use of any extrapolating fitting-routines the bandgap energies can be observed at 3.3 eV for GaN and 6.1 eV for AlN which confirms theoretically expected values.

In the energy-loss function recorded from GaN the region between 15 eV and 25 eV is strikingly dominated by two characteristic peaks at 19.4 eV and 23.3 eV. This double peak, which can not be observed for the case of AlN, originates from the influence of

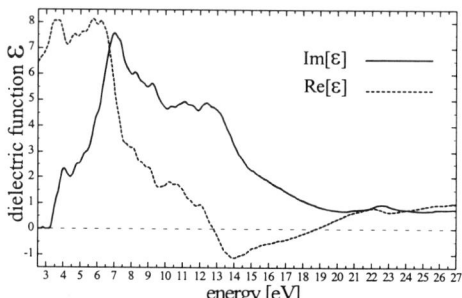

Fig.2: Real and imaginary part of dielectric function of wurtzite GaN

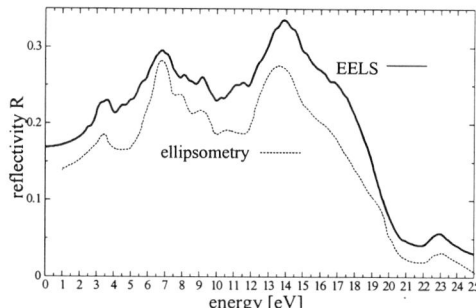

Fig.3: Reflectivity of GaN by EELS and synchrotron ellipsometry (Logothetidis, 1994)

transitions from the Ga d-band and has been observed by synchrotron ellipsometry and electron spectroscopy as well (Olson et al 1981 and Hedman et al 1980).

In the lower energy region the used energy-resolution and dispersion enable a detailed investigation of the electronic structure. Several smaller rises in the lower eV range of the loss-function, indicating interband-transitions, become more enhanced as peaks in the plot of the calculated imaginary part of the dielectric function, which is shown in Fig.2 for the case of GaN. In contradistinction to recent work (Natusch et al 1997 and 1998), here emphasis was put on features in the lower energy scale between the fundamental bandgap and the first collective exitations (2 - 25 eV), which is essential for optoelectronic applications. Peaks in the imaginary part of the dielectric function give rise to the presence of critical points in the band structure. In the energy range investigated characteristic, peaked features at 7, 8, 9, 11 and 13 eV could be resolved which have been observed in optical measurements as well (Logothetidis et al 1994, Olson et al 1981, Wethkamp et al 1998).

From the dielectric function the reflectivity has been calculated and compared to the results of synchrotron ellipsometric results of Logothetidis et al. (1994) as pointed out in Fig.3. The data do not only agree in their overall shape but in the finer resolved

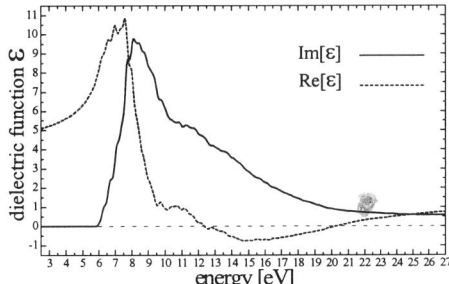

Fig.4: dielectric function of AlN

features as well. Thus the excellent agreement between the achieved EELS results and the optical measurements can be demonstrated.

In comparison to the properties of GaN Fig.4 shows the measured real and imaginary parts of the dielectric functions of AlN. Again the data show several peaked features in the dielectric function with a bandgap onset at 6.1 eV. In contradiction to GaN, due to the absence of d-bands in Al, here the dielectric function does not show a hump shaped feature near 23 eV (Phillip et al 1962).

On characterizing the properties of highly localized features one has to consider the spatial resolution of EELS measurements in a STEM. As a cold field-emission STEM enables the use of subnanometre electron probes the spatial resolution of the measurements no longer is limited by the electron probe diameter, but by the physical localization of the scattering process itself. Here the size of the energy depending impact parameter determines the localization which is in the range of nanometres for low energy-losses (< 10 eV) and subnanometric for ionization energies (Pennycook 1982). These facts were verified on the mentioned experimental setup and

materials. EELS linescans at low-loss and core-loss energies across an AlN/GaN interface were recorded. Fig.5 shows the recorded loss functions at distances of 0.8 and 2.3 nm to both sides of the interface. In the low-loss spectra the plasmon's shape and energy does not change abruptly when crossing the interface. But near the interface it shows a mixed behavior between GaN and AlN where the major peak of the plasmon region starts to shift and a certain influence of the GaN d-band gives a double peak simultaneously. A corresponding behavior can be observed in the bandgap region of the spectra. In addition to that N-K ionization

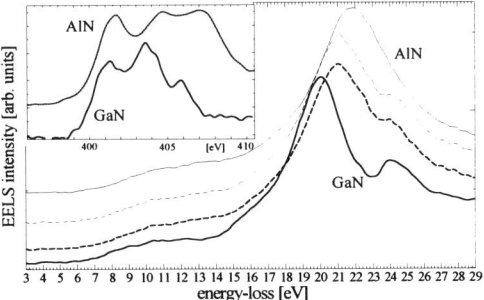

Fig.5: low-loss spectra and N-K edge mapping across a AlN/GaN interface in 1.5 nm steps

edges were recorded on the same sites. They reveal material induced near-edge fine structure. In contrast to the low-loss results the ionization edges change abruptly across the interface. Even at the positions close to the interface, plotted in the inset of Fig.5, they show either the fine-structure of GaN or that of AlN but no mixed characteristic. Thus the experimental energy resolution of our measurements is demonstrated to be in the range of nanometres for low-loss EELS and on a subnanometre scale for core-loss excitations.

Besides the investigation of heterostructures, (Brockt et al 1999) the superior spatial resolution can be used to characterize the properties of highly localized features like defects. The dielectric properties within threading dislocations, typical for wurtzite GaN layers, have been investigated. The measured imaginary part of the dielectric functions is plotted in Fig.6 in which a third plot from a defect free area is added for comparison. The spectrum from

356

dislocation A differs significantly from the defect free one between 5 and 8 eV, but the characteristic peaks are still visible. In comparison the spectrum from dislocation B shows a rise in the imaginary part of its dielectric function already at 2.6 eV which gives a hint of defect states within the bandgap. The extent in the differences of the shown dielectric functions can be visualized even more enhanced in the Cole-Cole plot of Fig. 7. Though deviations in the dielectric properties of individual dislocations within wurtzite GaN could be detected, currently, no common systematic features of defect properties can already be deduced.

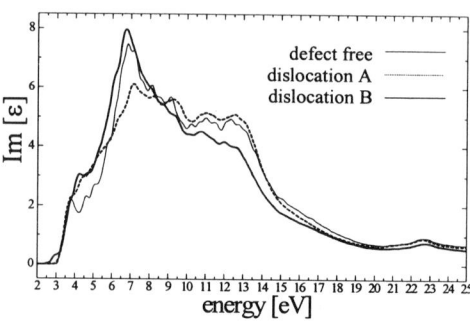

Fig.6: imaginary part of dielectric function of wurtzite GaN on dislocations and defect free sites

4. CONCLUSIONS

In summary it can be concluded that dedicated low-loss EELS methods give access to bandgap properties and the dielectric function of group III-nitrides. The achieved data show detailed structure in the lower energy scale and are in excellent agreement with synchrotron ellipsometric results. But in comparison with these techniques, the described advanced EELS method is essentially less influenced by surface features and yields a superior spatial resolution of better than 7 nm. This enables detailed study to be undertaken of the effect of local defects and boundaries on the optical properties.

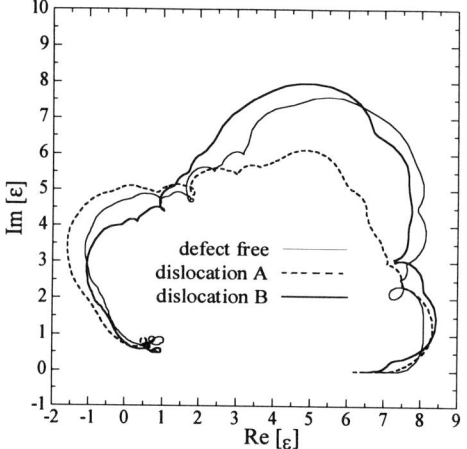

Fig.7: Cole-Cole diagram of wurtzite GaN on dislocations and defect free sites

REFERENCES

Brockt G and Lakner H 1999 in press at Material Science and Engineering B
Brockt G, Mendorf, C, Scholz, F and Lakner, H 1997 Inst. Phys. Conf. Ser. **157** 221
Bangert U, Harvey A, Keyse R and Freundt D 1997 Inst. Phys. Conf. Ser. **157** 209
Egerton R F 1996 Electron Energy Loss Spectroscopy. 2nd ed., Plenum Press New York
Ehrenreich, H. 1966 The optical properties of solids. Academic Press New York
Hedman J and Martensson N. 1980 Phys. Scr. **22** 1276
Logothetidis S Petalas J and Cardona M 1994 Phys. Rev. B, **50** 18017S
McMullan D, Fallon P, Ito Y and McGhibbon A J 1992 Electron Microscopy I 103
Natusch M K H, Botton E A and Humphreys C T 1997 Inst. Phys. Conf. Ser. **197** 213
Natusch M K H, Botton E A, Broom R F, Brown P D, Tricker D M and Humphreys C T 1998 Mat. Res. Soc. Symp. Proc. Vol. **482** 763
Olson C G, Lynch D W and Zehe A 1981 Phys. Rev. B **24** 4629
Wethkamp T, Wilmers W, Cobet C, Esser N, Richter W, Ambacher O, Brandt O, Müllhäser J R, Ploog K H, and Cardona, M 1998 Thin Solid Films **745** 313
Pennycook S T, 1982 Contemp. Phys. **23** 371
Phillip H and Ehrenreich H 1963 Phys. Rev. **129** 1550

Inst. Phys. Conf. Ser. No 164
Paper presented at Microsc. Semicond. Mater. Conf., Oxford, 22–25 March 1999
© *1999 IOP Publishing Ltd*

Electron holography studies of piezoelectric fields in InGaN/GaN layers

D Cherns and J Barnard

H H Wills Physics Laboratory, University of Bristol, Tyndall Avenue, Bristol BS8 1TL, UK.

ABSTRACT: Electron holography has been used to measure the piezoelectric field across a 1.5 nm $In_{0.52}Ga_{0.48}N$ quantum well in α-GaN. The results show an increase in the inner potential $\Delta V_o = 0.6 \pm 0.2V$ in the [0001] direction implying a piezoelectric field of 4MV cm^{-1} along [0001]. The accuracy and limitations of the method are discussed.

1. INTRODUCTION

The last few years have seen the development of light-emitting diodes and lasers based on hexagonal GaN/InGaN structures in (0001) orientation. One factor of importance is the existence of large piezoelectric fields across strained layers. These fields can have marked effects, such as the red shifted emission from InGaN quantum wells in GaN caused by field-induced band bending (Takeuchi et al 1997). In this paper we show that electron holography can be used to directly measure the local piezoelectric field across a strained InGaN quantum well.

2. RESULTS AND INTERPRETATION

Studies were carried out on a 7 nm GaN/1.5 nm $In_{0.52}Ga_{0.48}N$/2 μm GaN/(0001) sapphire sample grown by metal-organic chemical vapour deposition (MOCVD) at temperatures of 1050°C (GaN) and 800°C (InGaN). Electron microscope specimens were prepared in cross-sectional orientation using mechanical polishing, followed by ion thinning using 4kV Ar^+ ions at 10-12° incidence in a liquid nitrogen cooled stage. The growth direction, or polarity, was confirmed as [0001], i.e. the Ga-N bond direction, by convergent beam electron diffraction as described by Ponce at al (1996). Electron holography studies were then carried out in a Hitachi HF2000 cold field emission gun TEM equipped with a quartz fibre biprism, and a Gatan imaging PEELS system.

Figure 1 shows a low magnification micrograph in which the InGaN layer is close to edge-on. The image shows a threading dislocation passing undeviated through the InGaN layer. The InGaN layer, which is mostly pseudomorphic and, therefore, has an in-plane compressional strain of –0.055, shows some contrast due to surface relaxation of misfit strains.

Electron holograms were taken using the off-axis method in which a beam passing through the specimen is combined with a reference beam passing through the neighbouring vacuum. Figure 2 shows an experimental hologram with the trace of the InGaN layer indicated. The 0.14 nm fringes were produced with a voltage of 100V on the biprism.

To reduce diffraction effects, holograms were formed with the (0001) planes ~1-2° off the edge-on orientation, such that the InGaN/GaN interface was slightly inclined; holograms were recorded with no objective aperture present and were energy-filtered. The zero loss images were recorded on the slow scan 1024 x 1024 CCD of the Gatan Imaging Filter (GIF). Amplitude and phase maps were derived from the holograms using standard computing methods (Midgley et al 1997). Figure 3 illustrates phase and "τ/λ" profiles along the growth direction. To reduce noise, these profiles were averaged over a section of interface ~20 nm long. Two pairs of profiles are

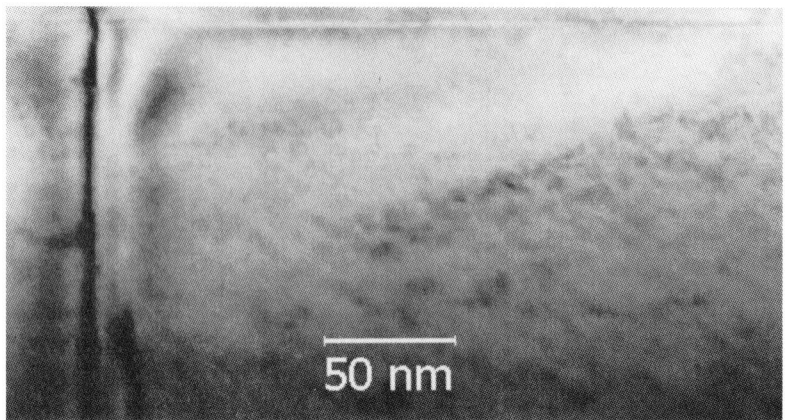

Fig. 1. The InGaN/GaN cross-section showing some contrast from the InGaN quantum
well, and a threading dislocation.

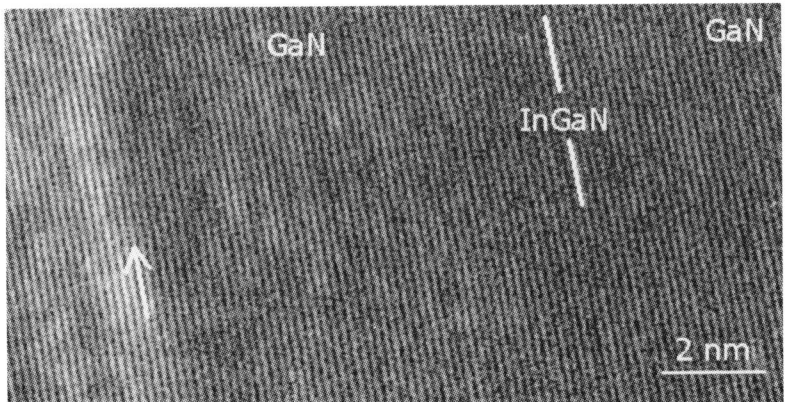

Fig. 2. Hologram showing the positions of the InGaN layer and the sample edge
(arrowed).

shown representing opposite tilts of the (0001) planes from the edge-on orientation. The phase profiles show a general increase in phase ϕ, left to right, due to an increase in the foil thickness. In both cases, however, there is also a small phase offset, $\delta\phi$, across the InGaN layer. The corresponding τ/λ profiles were taken from the amplitude maps using the relationship $I = I_0 e^{-\tau/\lambda}$ where I/I_0 is the ratio of fringe intensities with and without the specimen, τ is the foil thickness and λ is the effective inelastic mean free path. The τ/λ curves are much noisier than the phase showing an increase at the InGaN layer which can be ascribed to increased inelastic scattering in the InGaN compared to the GaN (giving a reduced λ). For one of the profiles there is a strong Fresnel fringe oscillation at the sample edge (not present in the phase profile). There are also offsets across the InGaN layer.

In a perfect crystal under diffracting conditions where channelling is unimportant, the phase change $\phi = CV_0\tau$ where C is a constant and V_0 is the inner potential. Thus, assuming V = O outside the specimen and τ is known, V_0 can be obtained from the phase profile. Unfortunately, the results are complicated by surface relaxation of misfit strains at the InGaN layer, which leads to some local diffraction contrast (c.f. Fig. 1). The change in diffracting conditions affects the τ/λ profiles in particular. Experiments on InP single crystals have shown these effects more clearly; for example,

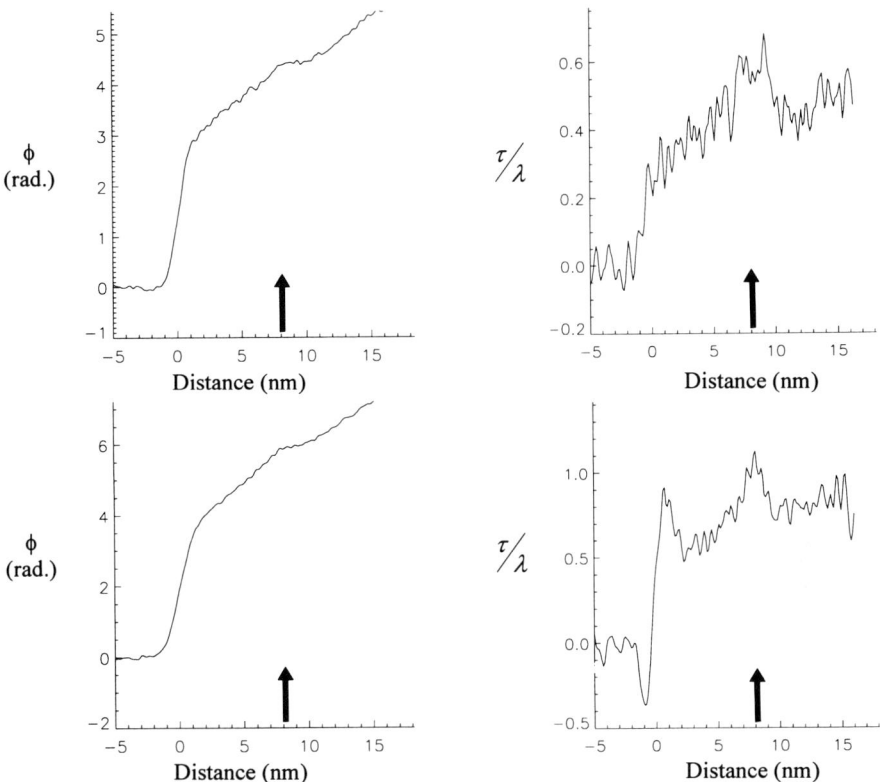

Fig. 3. Pairs of phase ϕ and thickness τ/λ profiles derived from experimental holograms. Distance is measured from the edge of the sample with the position of the quantum well arrowed. The top and bottom profiles are from different holograms with the (0001) planes tilted in opposite senses about the edge-on orientation.

comparison of a weakly diffracting condition (orientation 1) with 2-beam diffracting conditions (g = 040, orientation 2) using similar experimental conditions to those in Fig. 3 gave $(\tau/\lambda)_2:(\tau/\lambda)_1 = 2.21:1$ and $\phi_2:\phi_1 = 1.13:1$ (Barnard, to be published). Although such effects should be less important in GaN/InGaN since (a) weakly diffracting conditions were selected and (b) relaxation affects mainly the surface layers (to a depth comparable to the layer thickness d = 1.5 nm), it is clear that τ/λ profiles do not give foil thicknesses reliably. In contrast, since the phase profiles are much less affected, we have chosen, for a preliminary analysis, to compare phase profiles for different areas under a range of diffracting conditions and to assume a smoothly varying thickness profile in each case. Figure 4 shows measurements of $\delta\phi$ against ϕ obtained in this manner.

The results in Fig. 4 show that, despite some scatter, there is a consistent drop in phase on passing from the top to the bottom GaN layer. We propose that this is due to a change ΔV_0 in V_0 due to the piezoelectric field where ΔV_0 = Ed where E is the field in the InGaN assumed constant. Taking $\delta\phi \propto \phi$, i.e. ignoring effects due to fringing fields (Pozzi 1996) and surface relaxation (Treacy and Gibson 1986) both of which should be small for d<<τ ($\tau \leq 30$ nm in Fig. 4) gives $\Delta V_0 = (0.56 \pm 0.10)V_0$. The value of V_0 is not known experimentally; preliminary holography experiments give $V_0 = 11 \pm 3V$ (H Mokhtari, private communication), much less than the calculated neutral atom value of 19.6 V (Rez et al 1994). Using $V_0 = 11 \pm 3V$ gives $\Delta V_0 = 0.6 \pm 0.2V$ and a field E = 4MV cm^{-1} along the [0001] direction in good agreement with bulk estimates (e.g. Takeuchi et al 1997).

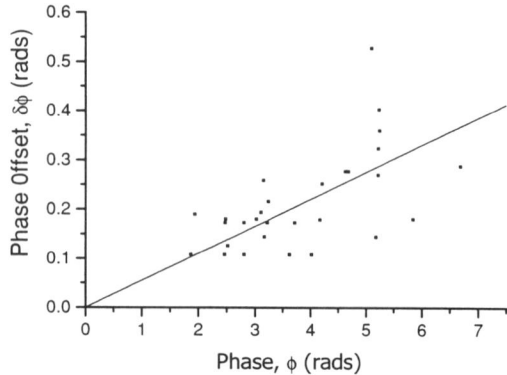

Fig. 4. The phase offset across the InGaN layer measured in the growth direction plotted against the total phase ϕ measured at the well. The points represent measurements from different areas for a range of weakly diffracting conditions.

ACKNOWLEDGEMENTS

The authors are grateful to Dr F A Ponce (ASU) for provision of the sample and for financial support from NATO (grant CRG 960690). JB is grateful to the EPSRC and Motorola, AZ, USA for financial support.

REFERENCES

Midgley P A, Barnard J and Cherns D 1997 Inst. Phys. Conf. Ser. **157**, 75
Ponce F A, Bour D A, Young W T, Saunders M and Steeds J W 1996 Appl. Phys. Lett. **69**, 337
Pozzi, G 1996 J. Appl. Phys. **29**, 1807
Rez D, Rez, P and Grant I 1994 Acta Cryst. **A50**, 481
Takeuchi T, Sota S, Katsuragawa M, Komori M, Takeuchi H, Amano H and Akasaki I 1997 Jpn. J. Appl. Phys. **36**, L382
Treacy M M J and Gibson J M 1986 J. Vac. Sci. and Technol. **B4**, 1458

Inst. Phys. Conf. Ser. No 164
Paper presented at Microsc. Semicond. Mater. Conf., Oxford, 22–25 March 1999
© 1999 IOP Publishing Ltd

Spontaneous polarisation of AlN measured by electron holography

M Albrecht, S Christiansen, T Remmele, H P Strunk, H Banzhof[1], R Goldberg[1] and H Lichte[1]

Universität Erlangen-Nürnberg, Institut für Werkstoffwissenschaften – Mikrocharakterisierung, Cauerstrasse 6, D-91058 Erlangen, Germany.
[1]Institut für Angewandte Physik, Technische Universität Dresden, Zellescher Weg 16, D-01062 Dresden, Germany.

ABSTRACT: We show that the spontaneous polarisation of AlN can be measured by electron holography at the interface between cubic and wurtzite material (i.e. a stacking fault). The stacking fault represents a capacitor of 1.5 nm thickness, the surface charge density of which is given by the spontaneous polarisation. The measured value agrees well with theoretical values.

1. INTRODUCTION

According to theoretical work by Bernardini et al. (1997), group III-nitrides in the wurtzite phase exhibit a strong spontaneous polarisation in addition to the piezoelectric polarisation. This spontaneous polarisation exists due to the non-cubic crystal symmetry and the high ionicity of the bonds. While the existence of piezoelectric fields can be concluded indirectly from the red shift of excitonic transitions of strained InGaN- and AlGaN-quantum wells (Martin et al 1996, Takeuchi et al 1997, Im et al 1998), no experimental proof exists up to now for the spontaneous polarisation. In principle the spontaneous polarisation manifests itself as interface charges at interfaces where the bulk symmetry is broken. Such an interface is of course the surface/vacuum interface that may compensate the charges due to polarisation. However, a high number of dangling bonds occur at free surfaces that usually tend to reconstruct and may induce states in the band gap. Even more drastic, for GaN(0001) it has been shown by both theoretical and experimental work that the (0001) and (000-1) surfaces are Ga terminated, independent on the polarity (Smith et al 1997). Thus the surface is metallic in character and screens the polarisation field.

A more favourable interface is the solid interface between the zincblende and the wurtzite phase, that occur intrinsically at stacking faults on (0001) planes (eg Albrecht et al 1997). First of all the cubic lattice for reasons of higher symmetry shows no spontaneous polarisation as does the vacuum (Bernardini et al 1997), secondly contrary to the free surface all bonds are saturated and no states in the forbidden gap occur (Stampfl and van de Walle 1998). Thus a stacking fault in wurtzite material is an ideal candidate for measuring the spontaneous polarisation as it represents a plate capacitor, the field of which is induced by the polarisation charge induced at the interfaces.

In this work we utilise transmission electron holography (for an overview see Lichte 1997) to directly image and quantify the electric field and the charge density distribution across the interfaces in wurtzite heterostructures. Electron holography is based on the interference of a reference wave with a scattered wave and is directly sensitive to the inner potentials in solids. Using a Möllenstedt-type biprism in a field emission transmission electron microscope allowed us to interfere a reference wave that has passed the vacuum with the scattered object wave. The resulting interference fringe pattern contains information on the phase and the amplitude of the scattered wave.

2. EXPERIMENTAL DETAILS

AlN samples are grown 1 µm thick by plasma induced molecular beam epitaxy (Angerer et al 1997) at 1050°C onto Al_2O_3(0001) substrates. Cross-sectional samples of these layers were prepared with standard techniques including mechanical grinding and polishing to a thickness of 2µm followed by 4 keV Ar$^+$ ion etching until a hole appears. Off-axis holograms were taken at 200 keV in a Philips CM200 FEG electron microscope equipped with a Schottky emitter electron source and an electrostatic biprism. The potential applied to the biprism determines the spacing of the interference fringes and thus the spatial resolution, which is twice the fringe spacing for weak objects. To measure the phase shift induced by the piezoelectric field it is important to choose a spacing broader than that of atomic distances to avoid the superposition of atomic and piezoelectric potentials. All holograms of the layers were taken under [11-20] multibeam conditions with the incident electron beam perpendicular to the surface normal of the interface. A tilted interface/stacking fault might affect the measurement of the inner potential. Electron holograms and reference holograms of the vacuum reference wave were taken with a 1kx1k CCD camera. The digital hologram is reconstructed numerically.

3. EXPERIMENTAL RESULTS

Fig. 1 shows the reconstructed phase of a hologram of the stacking fault of an AlN layer, shown in the corresponding high resolution transmission electron micrograph in Fig 1b. The stacking fault is seen edge on in the [11-20] projection as a thin black and white line. This is the only projection where stacking faults lying parallel to the incident electron beam can be revealed in HRTEM micrographs. The decreasing magnitude of phase from the upper left to the lower right edge of the image corresponds to a decreasing thickness of the wedge shaped sample. A line scan integrating several nm along the stacking fault is shown in Fig. 1c. The wedge shape of the sample is seen as a linearly varying background. Subtracting this background, we find a phase difference across the stacking fault of $\Delta\Phi$ = 0.4 rad. To exclude possible effects due to Fresnel fringes at the interfaces between the cubic and the wurtzite phase we performed defocus series. These defocus series showed no change in the phase difference.

4. QUANTITATIVE EVALUATION OF THE SPONTANEOUS POLARISATION

In the following we will quantitatively analyse our experimental data to determine the spontaneous polarisation from the phase shift of the electron wave between the top and bottom interface of the stacking fault. The phase of the scattered wave can be changed due to (i) variations of the mean inner potential (related to fluctuations in composition or atomic density) (ii) a change of the specimen thickness (iii) electrostatic and magnetic fields. In the following we concentrate on our case of the interfaces between cubic and wurtzite phase material. In this case we have to consider a potential difference in the inner potential between the wurtzite phase and the cubic phase and the electrostatic field induced by the charge at the interfaces.

4.1. The methodological approach

The phase difference of the electron wave passing the vacuum to that of the wave that passed the object is given by

$$\delta\Phi(x,y) = \frac{\pi e}{\lambda E} \int V(x,y,z) \, dz \qquad \textbf{(1)}$$

where x,y are the object coordinates, z is the distance the wave travels along in the sample and E is the kinetic energy of the electron and V is the inner potential. The usual relativistic correction adds a factor 1.164 to the right term of eq.1. The inner potential V(x,y,z) includes two terms: V_{el} is the potential

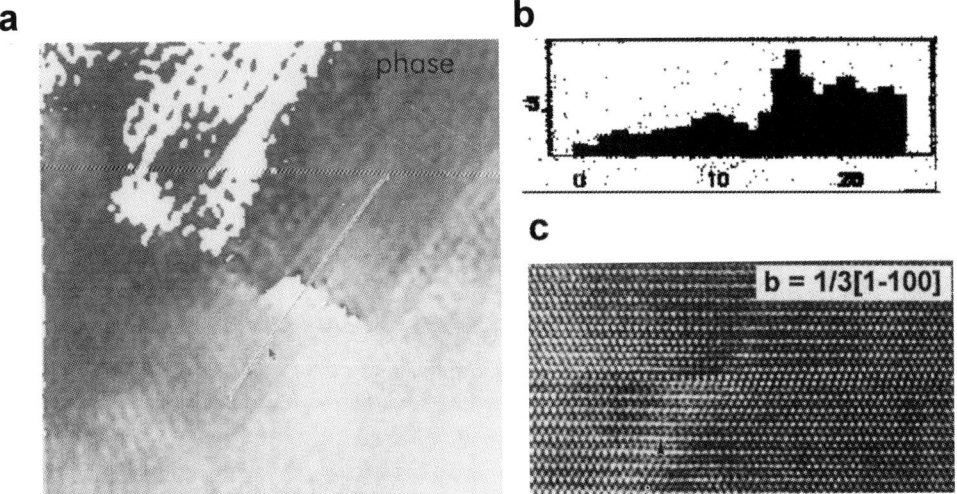

Fig. 1. (a) Phase reconstructed from a hologram of a cross-sectional sample of AlN: The hologram was taken along the <11-20> projection. The stacking fault is visible in the upper right quadrant as a thin black line with seams. It crosses the box introduced there. The line indicated by an arrow crosses an inversion domain boundary, which is not discussed here. (b) Phase integrated along several nm across the stacking fault. Integration parallel to the stacking fault within the box. (c) High-resolution transmission electron micrograph of a stacking fault in AlN. The stacking fault is a 3 monolayer thick cubic inclusion in the wurtzite lattice. The inserted [1-100] lattice plane is indicated by an arrow.

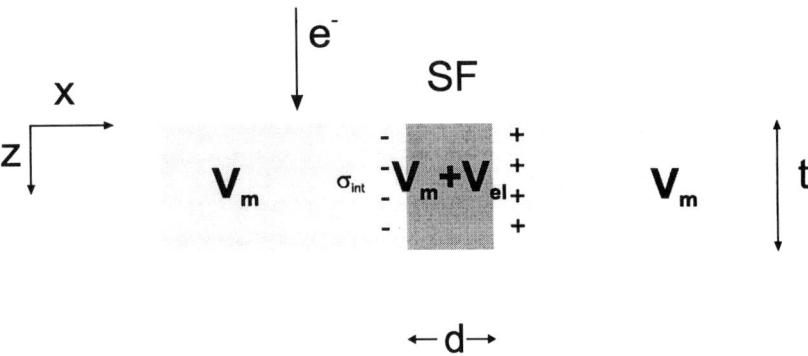

Fig. 2. Schematic showing the geometry of the measurement. The stacking fault (marked SF, dark grey) is embedded in wurtzite material with a mean inner potential V_m. X-axis corresponds to the (0001) direction, the z-axis corresponds to the [11-20] direction that is parallel to the incident electron beam. At the interfaces between the stacking fault and the host wurtzite lattice positive and negative charge is accumulated according to the spontaneous polarisation. This results in an electrostatic potential V_{el} within the stacking fault.

related to the electrostatic field and V_m is the mean inner potential of the cubic or the wurtzite phase respectively. In the following we consider the interfaces between both materials to lie perpendicular to the electron beam incident along the <11-20> direction (a schematic is shown in Fig.2). The slight difference in the atomic densities between zincblende and wurtzite material (due to the deviation from an ideal c/a value of wurtzite material) is negligible over the whole range of thicknesses considered here. This can be seen in the phase shifts of the zero beam (multislice technique, EMS program package) for AlN and GaN, see Fig. 3. An important difference between GaN and AlN as regards the phase shift/thickness relationship has to be pointed out: While in AlN the phase in the zero beam shifts almost linearly with thickness within one oscillation period, superimposed oscillations can be seen for GaN. These oscillations may cause severe problems in determining piezoelectric fields in GaN, InN and InGaN alloys, since small changes in thickness may cause strong phase differences that may erroneously be interpreted as potential fluctuations or alloy fluctuations. An experimental way to determine the phase thickness relationship is the measurement of samples with a defined geometry. For this purpose the sample has to be cleaved along defined crystallographic planes. This is impossible for heteroepitaxial material grown on sapphire due to the special epitaxial relationship (30° rotation of the layer normal and substrate). An analysis of cleaved AlN and GaN single crystals will be presented elsewhere.

4.2 The evaluation of the spontaneous polarisation

The spontaneous polarisation $\mathbf{P}$ leads to an areal interface charge density at the two interfaces of the stacking fault that form two plates of a capacitor given by

$$\sigma_{int} = \pm n \cdot P \qquad (2)$$

where $\mathbf{n}$ is the interface orientation ([0001]), and the sign of σ_{int} depends on the polarity of the respective surface (P is also parallel [0001]). According to Bernardini and Fiorentini (1998) and Majewski and Vogl (1998) this polarisation-induced charge is localised in an interface region less than 0.5 nm and results in macroscopic electrostatic field. In our case we have to consider two interfaces with different polarity. This results in a negative (positive) charge at the top interface and a positive (negative) charge at the bottom interface (Fig 2). In consequence the field between the two interfaces of our stacking fault can be described as a 1.5 nm thin capacitor. The potential drop between these interfaces is given by

$$\Delta V = \frac{\sigma_{int}}{\epsilon\epsilon_0}d \qquad (3)$$

where d is the thickness of the Capacitor (=cubic inclusion), ϵ is the dielectric constant and ϵ_0 is the dielectric constant of vacuum. The decay of this field into the crystal can be described by that of the classical space charge layers and is expected to be over-exponentially. Assuming a constant polarisation induced potential along z throughout the film and neglecting surface effects in a first approach the phase shift is given by

$$\Phi = \frac{\pi}{\lambda E}\left(\frac{\sigma_{int}}{\epsilon\epsilon_0}d + V_m\right)t \qquad (4)$$

If the sample thickness is identical across the stacking fault, which is a reasonable assumption, the phase difference between both interfaces of the stacking fault is given by

$$\Delta\Phi_{max} = \frac{\pi}{\lambda E}\frac{\sigma_{int}}{\epsilon\epsilon_0}d\,t \qquad (5)$$

a   **b**

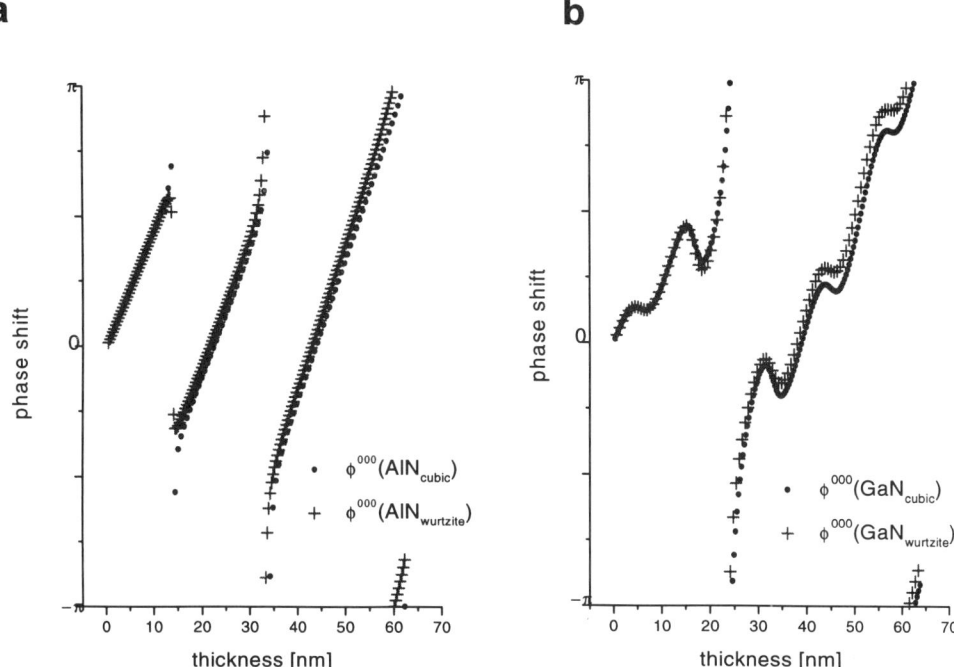

Fig. 3. Phase shift of the zero-beam as dependent on the transmission sample thickness in cubic and wurtzite (a) AlN and (b) GaN, respectively. Cubic material is represented by dots, wurtzite material by crosses. The difference in phase between cubic and wurtzite materials is almost vanishing for thicknesses up to 50nm. Oscillations in phase of the zero-beam are seen for GaN. Multislice calculation with the EMS program package used experimental values for the lattice constants listed in Table 1.

From our theoretical calculation of the phase shift by the multislice technique, we can deduce a sample thickness of around 50 nm. The thickness of the stacking fault according to structural images (see Fig. 1c) is 1.5nm (three atomic layers). If we now substitute the wavelength and the kinetic energy of 200keV electrons (λ=0.00251nm), the dielectric constant of AlN (ε=8.5 Edgar 1994) and the observed phase shift of 0.4rad, we obtain an interface charge induced by the spontaneous polarisation of 0.055Cm^{-2} (with relativistic correction). This value is in reasonable agreement with the theoretically predicted value of 0.081Cm^{-2} (Bernardini et al 1997). The two sources of inaccuracy have to be considered here, in addition.

	wurtzite		cubic
	a	c	a
GaN	0.319 nm	0.5185 nm	0.4503 nm
AlN	0.311 nm	0.4978 nm	0.438 nm

Table 1: Lattice constants of the cubic and the wurtzite polytypes of AlN and GaN (Edgar 1994). Data used for the multislice simulation (see Fig. 3).

The first source is the noise in the phase of the electron wave (which is ±0.2) which results in an inaccuracy of ±0.032 Cm^{-2} in determination of the polarisation. The second source is the screening of the potential induced by the free carriers generated by the electron beam. Based on the Bethe- Bloch formula, considering an electron- hole formation energy of three times the energy gap of AlN (18.6 eV), a beam current 1 nA and a sample thickness of 50 nm, we obtain a generation rate of $0.6 \cdot 10^{10}$ s^{-1}. From experimental measurements of the lifetime of minority carriers we know that the recombination time is in the range of 0.1 ns. The steady state carrier density is thus $\approx 0.2 \cdot 10^{15}$ cm^{-3}. This is about 1% of the doping level of 10^{17} cm^{-3}, which means that we have low injection conditions, i.e. no noticeable screening. A detailed analysis of the screening as dependent on the injection conditions would require the diffusion equation of carriers to be solved, which is beyond the scope of this work.

5. CONCLUSIONS

Based on the evaluation of transmission electron holograms of stacking faults we have measured the spontaneous polarisation of AlN. The notion of this measurement is to consider the stacking fault, which is a planar cubic inclusion of 1.5 nm thickness within the host wurtzite lattice, a plate capacitor. The two electrodes are charged by a charge density representing the spontaneous polarisation of the wurtzite host. We obtained 0.055 Cm^{-2} for the polarisation which is in satisfactory agreement with the theoretical value of 0.081 Cm^{-2}. These first experiments which are currently being extended show that electron holography is a powerful tool to analyse polarisation related phenomena in many materials, not only in nitrides. Possible applications are the analysis of piezoelectric fields of dislocations and their influence on scattering of charge carriers. Other fields of interest are the analysis of piezoelectric fields in strained quantum wells in devices, and all phenomena related to polarity.

REFERENCES

Albrecht M, Christiansen S, Salviati G, Zanotti-Fregonara C, Rebane Y, Shreter Y G, Mayer M, Pelzmann A, Kamp M, Ebeling K J, Bremser M D, Davis R F, and Strunk H P 1998 Mat.Res.Soc.Symp.Proc. **468**, 293

Angerer H, Ambacher O, Stutzmann M, Metzger T, Höpler R, Born E, Bergmaier A, Dollinger G 1997 Mat.Res.Soc.Symp.Proc. **468** 305

Bernardini F, Fiorentini V and Vanderbilt D 1997 Phys.Rev.**B 56**, R10024

Bernardini F and Fiorentini V 1998 Phys.Rev.**B 57**, R9427

Edgar J H (Ed.) 1994 Properties of Group III-Nitrides, emis datareview series (London: INSPEC)

Im J S, Kollmer H, Off J, Sohmer A, Scholz F and Hangleiter A 1998 Mat.Res.Soc.Symp.Proc. **482** 513

Lichte H 1997 in Handbook of Microscopy Vol I, Eds:. S.Amelinckx, D van Dyck, J van Landuyt, G van Tendeloo, (Weinheim:VCH), pp 515-36

Martin G, Botchkarev A, Rockett A, Morkoç 1996 Appl.Phys.Lett. **68**, 2541

Majewski J A, and Vogel P 1998 MRS Internet J. Nitride Semiconductor Res. 3 Art. 21

Smith A R, Feenstra R M, Greve D W, Neugebauer J, and Northrup J E 1997 Phys.Rev.Lett. **79**, 3934

Stampfl C and van de Walle C G 1998 Phys.Rev.**B 57** R15052

Takeuchi T, Sota S, Katsuragawa M, Komori M, Takeuchi H, Amano H, and Akasaki I 1997 Jpn.J.Appl.Phys. **36** L382

Inst. Phys. Conf. Ser. No 164
Paper presented at Microsc. Semicond. Mater. Conf., Oxford, 22–25 March 1999
© 1999 IOP Publishing Ltd

The structure of GaN films grown on a-plane sapphire by MOCVD

M E Twigg, R L Henry, A E Wickenden, D D Koleske, M Fatemi, and J C Culbertson

Electronics Science and Technology Division, Naval Research Laboratory, Washington, D.C. 20375-5320

ABSTRACT: Using cross-sectional transmission electron microscopy (XTEM), we have determined the structure of GaN films grown on a-plane sapphire by metal-organic chemical vapor deposition (MOCVD). We have assessed the effects of substrate nitridation and vicinality, reactor pressure, and multiple nucleation layers on grain size and orientation as well as on dislocation density.

1. INTRODUCTION

Recent attention to growing GaN on a-plane sapphire by MOCVD (metal-organic chemical vapor deposition) has made this material competitive with GaN grown on c-plane sapphire (Wickenden et al 1999). Typically these GaN films have dislocation densities between 10^8 and $10^9/cm^2$. Advances in the understanding of the effects of substrate nitridation and vicinality, reactor pressure, and dislocation filtering have led to strategies for reducing dislocation density and increasing grain size. These strategies, in turn, have contributed to the growth of uniform GaN films with carrier mobilities in excess of 500 $cm^2/V/s$ and carrier densities of less than $2x10^{17}/cm^2$.

2. EXPERIMENTAL

MOCVD growth of GaN films has been conducted in two types of vertical reactors: a water-cooled, inductively-heated, quartz tube (QT) reactor with the gas inlet located 10 cm above the sample; and a resistively-heated showerhead (SH) reactor with the gas inlet 1 cm above the sample. In both reactors, trimethylgallium (or trimethylaluminum for AlN) was used as the group III precursor, ammonia as the group V source, and hydrogen as the carrier gas. Silane or disilane serve as the dopant source for Si-doped films. Prior to growth of the high temperature (1000-1060°C) GaN layer, a ~20 nm nucleation layer (NL) is deposited. For the QT reactor, only AlN NLs were grown, whereas in the SH reactor both AlN and GaN NLs have been used.

3. RESULTS AND DISCUSSION

Using XTEM to determine the preferred crystallographic configuration for nitride growth on a-plane sapphire, we have found that films with carrier mobilities above 100 $cm^2/V/s$ are characterized by the orientation relationship $GaN[2\bar{1}\bar{1}0]/sap[1\bar{1}00]$; $GaN(0001)/sap(\bar{1}\bar{1}20)$. The other observed configuration, $GaN[1\bar{1}00]/sap[1\bar{1}00]$; $GaN(0001)/sap(\bar{1}\bar{1}20)$, suffers from poor grain alignment, with a resulting large dislocation density ($>10^{10}/cm^2$) at the grain boundaries, as shown in Fig.1. The latter configuration occurs for GaN films grown in the SH reactor following nitridation with ammonia at 675°C for 10 minutes and the growth of a GaN NL at 550°C (Wickenden et al 1999). The former orientation is obtained by nitriding at the higher temperature of 1085°C, resulting in a corresponding decrease in dislocation density to less than $10^9/cm^2$.

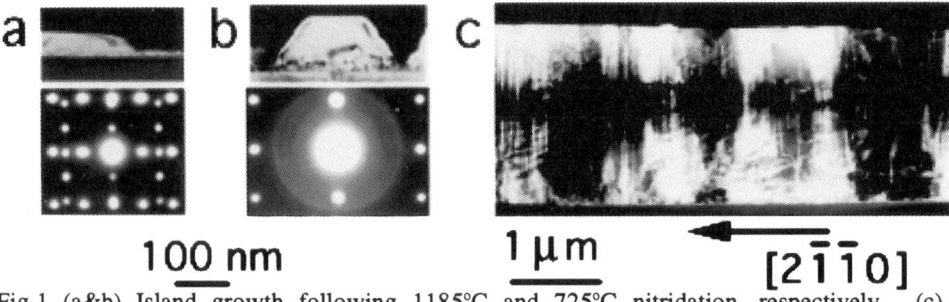

100 nm   **1 μm**   ⟵ **[2$\bar{1}\bar{1}$0]**

Fig.1 (a&b) Island growth following 1185°C and 725°C nitridation, respectively. (c) Coalesced GaN film following 675°C nitridation.

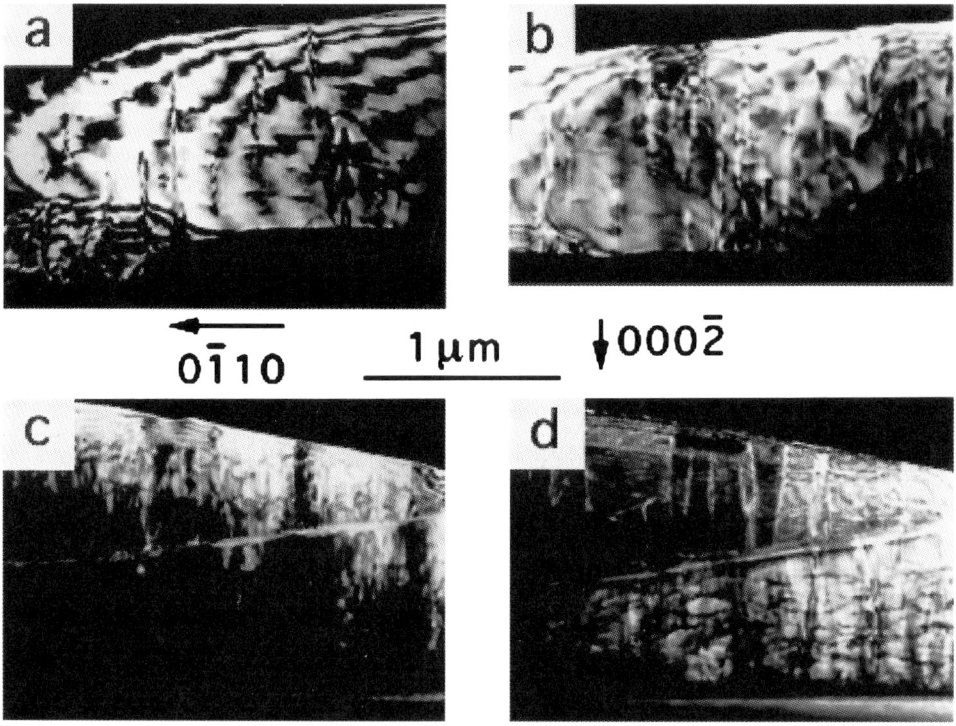

⟵ **0$\bar{1}$10**   **1 μm**   ↓**000$\bar{2}$**

Fig.2 Dark-field XTEM images showing the effect of vicinal growth on a-plane sapphire. (a) On-axis growth; g=0$\bar{1}$10. (b) On-axis growth; g=000$\bar{2}$. (c) Viscinal growth; g=0$\bar{1}$10. (d) Viscinal growth; g=000$\bar{2}$.

Further improvements in grain alignment may be brought about by growing on vicinal a-plane sapphire (Fatemi et al 1998). For samples grown in the QT reactor at 50 torr (as shown in Fig.2), the density of edge dislocations in vicinally-grown samples is less than $10^8/cm^2$, as compared with an edge dislocation density of $5x10^8/cm^2$ for films deposited upon on-axis substrates. Because the density of screw dislocations is $5x10^8/cm^2$ for both vicinal and on-axis films, the dislocation density in the former ($5 \times 10^8/cm^2$) is half that of the latter ($10^9/cm^2$). It is our conjecture that the presence of steps on the vicinal sapphire surface provides a better template for grain alignment, which in turn leads to a lower density of the edge dislocations at the low-angle grain boundaries between adjacent grains.

We have also observed that growing at reactor pressures over 100 torr improves film quality by increasing grain size in both QT and SH reactors. Typically, reactor pressures of less than 50 torr result in an average grain size of 0.5 μm. For reactor pressures greater than 100 torr, the average grain size is greater than 1 μm (as shown in Fig.3). According to Koleske et al (1998), higher hydrogen pressure promotes GaN decomposition. Thus, enhanced desorption at higher pressures may retard grain nucleation, thereby resulting in larger grain size (Wickenden et al 1999). For the lower-pressure (49 torr) film in Fig.2a, the mobility μ and carrier density n were measured as 150 $cm^2/V/s$ and $1.5 \times 10^{17}/cm^2$, respectively. For the higher-pressure (250 torr) film in Fig.1b, however, μ=518 $cm^2/V/s$ and n=$9.4 \times 10^{16}/cm^2$. These results agree with previous observations that larger grain films have higher carrier mobilities (Wu et al 1998; Twigg et al 1999).

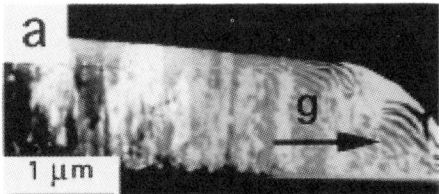

Fig.3 Dark-field XTEM images recorded using the GaN $01\bar{1}0$ reflection. (a) GaN film grown at 49 torr. (b) GaN film grown at 250 torr.

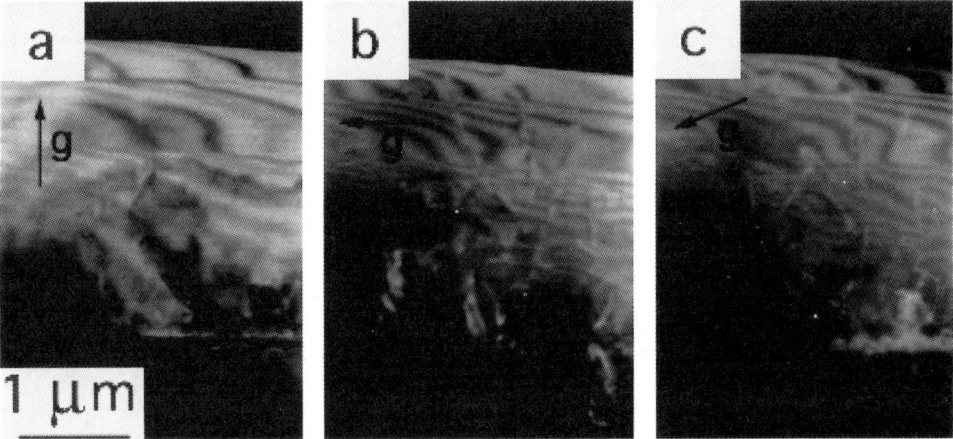

Fig.4 Weak-beam g(3g) XTEM images recorded using GaN (a) 0002, (b) $0\bar{1}10$, and (c) $0\bar{1}1\bar{2}$ reflections for a film grown with multiple nucleation layers.

A more novel approach to improving film quality is that of interrupting high-temperature growth with a series of low-temperature NLs (Iwaya et al 1998). Weak-beam XTEM images of a GaN film, grown in the SH reactor at 130 torr with AlN NLs, are shown in Fig.4. Diffraction contrast (**g·b**) analysis of the XTEM images indicates that the NLsprimarily consist of dislocations with the Burgers vector perpendicular to the growth plane. This array of NL dislocations then acts to annihilate threading screw dislocations, thereby reducing screw dislocation density to less than $10^8/cm^2$. Despite the large (2-6 μm) grain size in this film, however, the edge dislocation density is $\sim 10^9/cm^2$. Nevertheless, mobilities in excess of 700 cm^2/V/s and carrier concentrations of ~2x10^{17}/cm^2 were obtained in n-type Si-doped GaN films grown using this multiple-NL approach.

4. CONCLUSION

High-temperature nitridation, and growth on vicinal a-plane sapphire substrates results in better grain alignment, with a corresponding decrease in the density of edge dislocations. Growing at higher pressures results in larger GaN grains, while multiple-NLs filter screw dislocations. All four strategies result in fewer extended defects and higher carrier mobilities.

ACKNOWLEDGEMENTS

This work was supported by the Office of Naval Research. We thank Larry Ardis and Bob Gorman for expert technical assistance.

REFERENCES

Fatemi M, Wickenden A E, Koleske D D, Twigg M E, Freitas J A, Jr., Henry R L, and Gorman R J, 1998 Appl. Phys. Lett., **73,** 608.

Iwaya M, Takeuchi T, Kanaguchi, Wetzel C, Amano H, Akasaki I, 1998 Japn. J. Appl. Phys. **37,** L316.

Koleske D D, Wickenden A E, Henry R L, Twigg M E, Culbertson J C, and Gorman, R J, 1998 R J, Appl. Phys. Lett. **73,** 2018.

Twigg M E, Henry R L, Wickenden A E, Koleske D D, and Culbertson J C, 1999 Appl. Phys. Lett, in press.

Wickenden A E, Koleske D D, Henry R L, Gorman R J, Culbertson J C, and Twigg M E, 1999 J. Electron. Mater., **28,** 229.

Wu X L, Fini P, Tarsa E J, Heyring B, Keller S, Mishra U K, DenBaars S P, and Speck J S, 1998 J. Cryst. Growth. **189/190,** 231.

Inst. Phys. Conf. Ser. No 164
Paper presented at Microsc. Semicond. Mater. Conf., Oxford, 22–25 March 1999
© 1999 IOP Publishing Ltd

The Stranski-Krastanov growth mode of GaN on AlN

J-L Rouviere, M Arlery, B Daudin and G Feuillet

CEA-Grenoble/ Département de Recherche Fondamentale sur la Matière Condensée /SP2M
17 rue des Martyrs - 38054 Grenoble Cedex 9, France

ABSTRACT : We report on a Transmission Electron Microscopy study of GaN dots embedded in an AlN matrix. Using "High Resolution Transmission Electron Microscopy (0002) interplanar profiles", also called lattice fringe profiles, the thickness and the chemical composition of the wetting layer (2 GaN monolayers spread on 4 planes) as well as the height of the GaN dots were measured. The approximate shape of the GaN dots was determined to be a truncated hexagonal pyramid. We found that even the bigger dots (diameter of 15.3nm, height of 3.3nm) are coherently grown on AlN : the growth of GaN dots on AlN does not produce additional defects, but big GaN dots nucleate at the edge of threading edge dislocations, which propagate in AlN.

1. INTRODUCTION

GaN and its alloys with AlN and InN have emerged as an important family of semiconductors for commercial optoelectronic devices such as diodes and laser diodes. The commercial diodes with active GaInN layers do work in spite of a high density of dislocations. It is now supposed that this is possible because In composition fluctuations localise the electron-hole pairs avoiding their recombination on defects (for a review of GaN research, see Nakamura (1995) or Gil (1998)). The use of controlled dots that would also localise the pairs could enhance the optical emission of the devices. This is why we investigated GaN dots in AlN and studied their optical and structural properties. In this paper, we will report the structural study performed by transmission electron microscopy. More information on the results of the other techniques can be found in Widmann et al (1998).

GaN/AlN is becoming a new semiconductor quantum dot system in addition to the more studied ones, which are Ge/Si and InAs/GaAs. Tanaka et al. (1998) have also produced GaN dots. GaN dots have many specific properties such as their wurtzite structure, their strong piezoelectric coefficients and, as we will see, can have a rather small size.

2. EXPERIMENTAL PROCEDURES

Depending on the growth temperature T_g, two dimensional GaN layers (T_g=600°C) or three dimensional GaN islands (T_g=710°C) have been grown by Molecular Beam Epitaxy (MBE) on thick (0001) AlN layers deposited on (0001) sapphire, and have been studied in situ by Reflection High Energy Diffraction (RHEED) and ex situ by Atomic Force Microscopy (AFM), photoluminescence and Transmission Electron Microscopy (TEM). Conventional and High Resolution Transmission Electron Microscopy (HRTEM) studies were realised on a JEOL4000EX electron microscope, specially equipped for HRTEM (Scherzer resolution about 0.17nm, tilt ±20°). Specimens for TEM were prepared using the standard techniques: mechanical polishing and Argon ion milling. HRTEM images were simulated using the EMS software (Stadelmann, 1987) with the multislice method. Interplanar profiles, also called lattice fringe profiles, have been measured using the SEMPER software.

In this paper we present a TEM study of a stack of 20 GaN dot layers each separated by 5.5nm of AlN (Fig. 1). Each GaN dot layer has been grown at 710°C and underwent a growth interruption of 10 seconds under vacuum before being recovered by AlN.

3. THE WETTING LAYER STUDY

Figure 1 is a general view of five GaN dot layers of the studied sample. The dots, are clearly seen and well separated from each other. The wetting layer (WL) in between these dots has a uniform appearance. This WL has been analysed by measuring on HRTEM the apparent distances between the (0002) planes perpendicular to the growth axis: we obtain what we call here an "HRTEM interplanar distance profile" (Fig. 2a), indicating that this profile is not the real profile of the structure. Indeed this profile has distances (extrema m, m' and M) that are higher or lower than the (0002) interplanar distance of the pure materials (AlN and GaN). These extrema come from intensity oscillations that occur at potential discontinuities. These intensity fluctuations are generally called Fresnel fringes and are an example of the non linearity of HRTEM image formation. One has to do HRTEM image simulations, which reproduce these oscillations, to extract detailed information from these profiles.

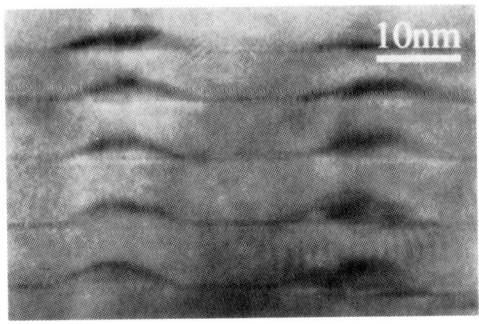

Fig. 1 HRTEM images of five GaN dot layers observed along a $<01\overline{1}0>$ direction

The HRTEM interplanar distance profiles were obtained using the following steps (Rouviere, 1994) :

(i) The digitised image is first projected along the $[2\overline{1}\,\overline{1}0]$ direction (direction perpendicular to the (0002) direction and the $[01\overline{1}0]$ observation axis).

(ii) The position of the peaks are then determined by fitting the extrema neighbourhood (generally 2 points on each side of the extrema) with a polynomial function of degree 2.

(iii) The distances between the successive (0002) planes are then obtained (Fig. 2). The average lattice parameter measured in the AlN matrix far from the interface is taken as reference for the distance calibration.

The quantitative analysis of these profiles is carried out in two main steps. Firstly, the experimental parameters (defocus Δz of the objective lens and thickness t of the specimen) have to be determined. The defocus (here 85nm) was determined from diffractograms of an amorphous region at the edge of the sample. The thickness (t=8.6nm) was adjusted by comparing the experimental images with defocus-thickness maps of perfect AlN and more precisely with defocus-thickness map of an abrupt GaN WL (corresponding to the WL of fig. 2b having 2 GaN MLs) : the thickness was determined by reproducing both the contrast in the AlN material and the extrema m, m' and M. Secondly, the composition profile is determined by "trial and error comparisons" between experiment and simulation.

One loop of this "trial and error procedure" involves the four following steps.

(i) A trial composition profile x_i is given. "i" represents the integer number of the ith (0002) plane of composition $Al_{xi}Ga_{1-xi}N$ (i runs generally from 0 to about 20).

(ii) By assuming a linear variation with x of the lattice parameters a(x) and c(x) and elastic constants of $Al_xGa_{1-x}N$ alloys and assuming a biaxial deformation of GaN on the AlN substrate, one obtains from the composition profile x_i an interplanar profile d_i that allows one to construct an atomistic box containing all the atoms with the correct concentrations and positions :
$$d_i = 0.5\ c(x_i)\ (1\text{-}2\ C_{13}(x_i)\ /C_{33}(x_i)\ (a_{substrate} - a(x_i))/a(x_i))$$
(iii) These atomic positions are used to simulate HTREM images.

(iv) An HRTEM interplanar profile is obtained and compared to the experimental profiles.

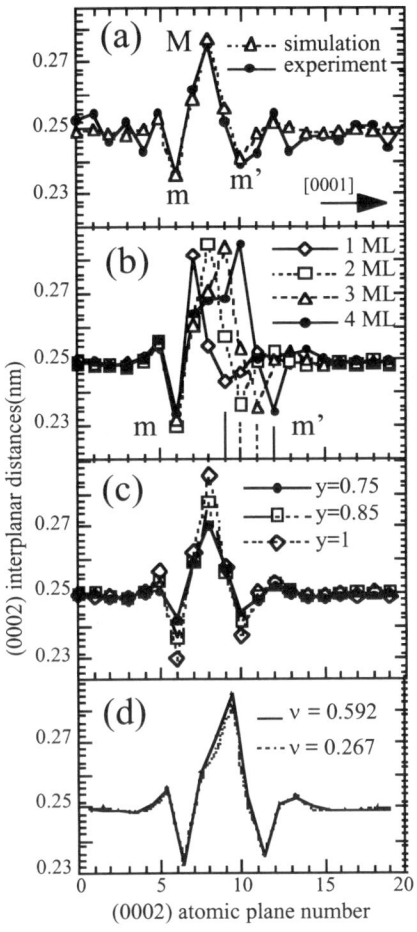

(0002) interplanar distances(nm)

(0002) atomic plane number

This "trial and error procedure" was facilitated by the observed results that the relative position of m and m' (i.e. the number of atomic planes between m and m') depends exclusively on the total number n of pure GaN in the WL (Fig. 2b) and that the amplitudes of the extrema m, M and m' depend on the chemical dilution y between Ga and Al (Fig. 2c). The best fit was obtained with a WL consisting of 2 monolayers of pure GaN spread out on 4 monolayers, which have Ga compositions of respectively 15%, 85%, 85% and 15%.

It must be pointed out that the published values of the ratio $v=2C_{13}(x=1)/C_{33}(x=1)$ for GaN are relatively dispersed (Gil 1998), but that the simulated HRTEM interplanar profiles obtained with the lower and higher values of the literature are very similar (Fig. 2d). This comes from the fact that Fresnel fringes are more sensitive to potential discontinuities than to distance variations.

Fig. 2 : a) Experimental and simulated HRTEM interplanar distances ("c/2") between the successive (0002) planes. The projection was obtained with a box of 7 nm lateral size.
b) Simulated HRTEM interplanar distance profiles corresponding to GaN WLs of different thicknesses: 1,2,3 or 4 GaN monolayers (ML).
c) Three simulated profiles of three WLs containing a total amount of 2 GaN MLs that are spread differently on 4 planes. The Al compositions of these 4 planes are respectively y, 1-y, 1-y and y (with y=1, y=0.85 and y=0.75).
d) Two simulated HRTEM interplanar profiles calculated using two $v=2C_{13}/C_{33}$ GaN values.

4. THE GaN DOT STUDY

The GaN dots were observed in three different directions in order to determine their shapes, their coherency with AlN and their deformations.

Figure 3a and figure 3b are two HRTEM images of the same dot observed in two different directions that make an angle of 30°: [2$\overline{1}$$\overline{1}$0] and [10$\overline{1}$0]. The Fourier filtering (Figs. 3c and 3d) of these images enables to outline the approximate shape of the dot (Figs. 3e and 3f). This shape is compatible with a truncated pyramid with a hexagonal base that is bounded by {01$\overline{1}$0} facets (Figs. 3g, 3h, 3i and 3j). In particular, the measured ratios between the apparent diameters of the dots in the [2$\overline{1}$$\overline{1}$0] and [10$\overline{1}$0] directions agree rather well with the geometrical value of $\sqrt{3}/2=0.866$. On figure 3, these ratios were measured equal to $D_b'/D_b=0.9$ and $D_t'/D_t=0.78$ for respectively the base and the top diameter ratios. The Fourier filtering also determines that the GaN dots are coherently grown on AlN : the growth of GaN does not introduce any additional planes i.e. edge dislocations.

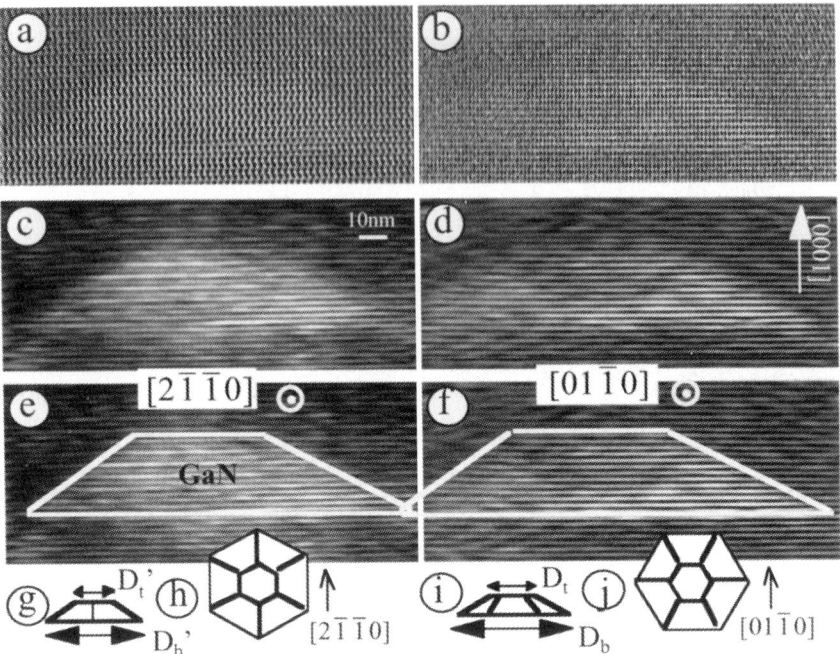

Fig. 3 : The same GaN dot has been observed in two directions $[2\bar{1}\bar{1}0]$ and $[10\bar{1}0]$, which make an angle of 30° from each other. a) HRTEM image taken along the $[2\bar{1}\bar{1}0]$ direction. b) HRTEM image taken along $[10\bar{1}0]$. c)Fourier filtering of image 3a. All the frequencies along the (0001) directions lower than the AlN frequencies have been taken : GaN appear brighter than AlN. d) Fourier filtering of image 3c obtained in the same condition as image 3c. e) Same image as figure 3d, but the dot has been outlined by visual aspect. f) Same image as figure 3d, but the dot has been outlined. g) Scheme of the GaN dot viewed along the $[2\bar{1}\bar{1}0]$ direction. h) The GaN dot ofFig. 3g observed from the top that is to say along [0001]. i)Scheme of the GaN dot viewed along the $[10\bar{1}0]$ direction. j) The GaN dot of figure 3i observed from the top.

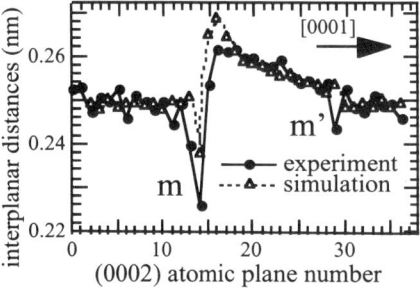

Fig. 4 Experimental and simulated HRTEM interplanar distance profiles of a GaN. They have been obtained at the center of a dot.

The exact height of the pyramid was determined by making an HRTEM interplanar profile at the centre of the pyramid (Fig. 4). The minima m and m' allowed the determination of the exact height of the dot. Indeed, the WL study has shown that the distance between m and m' is equal to the number of planes of GaN plus two. In the case of Fig. 4, we found that the GaN dot height contains thirteen (0002) planes. More precisely, we realised HRTEM simulation of the GaN dot image and found a dot height of 3.3nm. We had to take into account the position of the dot within the cross-section sample to reproduce the decreasing part of the profile (Arlery et al., 1999). In a first approximation the centre of the pyramid seems to undergo a biaxial deformation.

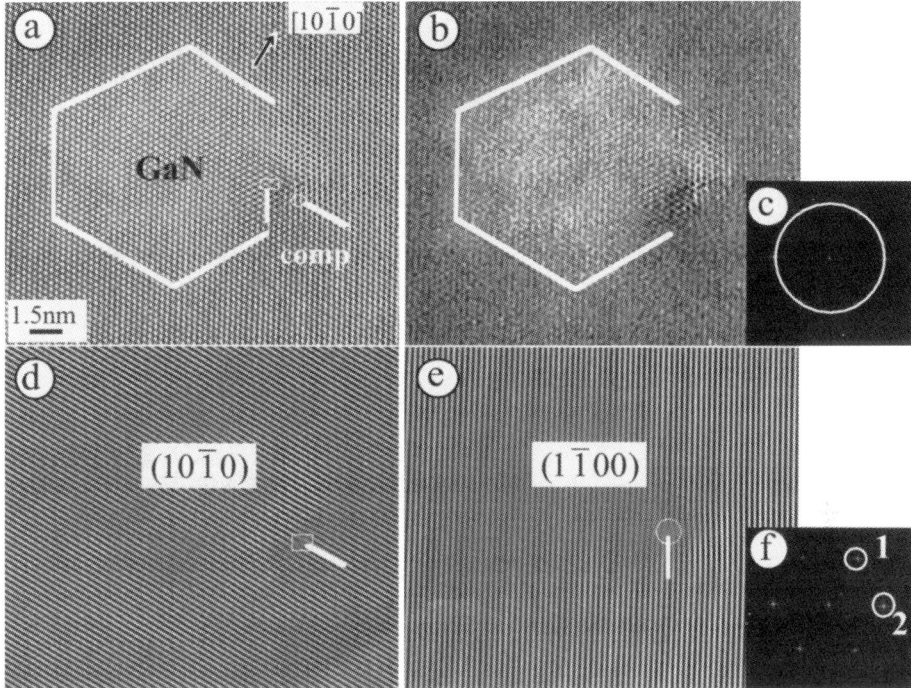

Fig. 5 a) HRTEM image of a GaN dot viewed along the [0001] direction. The hexagonal shape of the dot has been determined from the filtered image 5b. The two additional {01$\overline{1}$0} planes, obtained from images 5d and 5e, are marked with a short white line ended either by a circle or a rectangle that indicate the end of the additional planes.

b) The Fourier filtered image of image 5a obtained by using the mask shown in 5c. This mask includes all the frequencies whose modulus is strictly inferior to the AlN (01$\overline{1}$0) frequency. The GaN dot is thus outlined.

c) The modulus of the Fourier transform of image 5a. The mask that has been used for obtaining image 5b is outlined.

d) The Fourier filtered image of image 5a obtained by using the mask marked 1 in image 5f. This mask cuts all the crystal frequencies but the (01$\overline{1}$0) one. The additional (01$\overline{1}$0) plane and its end are outlined respectively by a white line and a white rectangle. By cutting and pasting, this information was added to image 5a.

e) Fourier filtered image of 5a using a mask that is marked 2 in image 5f and that cut all the frequencies but the (01$\overline{1}$0) one.

f) Modulus of the Fourier transform of image 5a. The masks used for obtaining images 5d and 5e are respectively marked 1 and 2.

The contrasts of the plan view images are weak (Fig. 5a), but again by Fourier filtering the images we could outline the approximate hexagonal shape of the dots (Fig. 5b). By making Burgers circuits, we determined that every dot grows at the edge of a threading edge dislocation whose dislocation line in AlN is [0001] and Burgers vector is 1/3<2$\overline{1}$$\overline{1}$0>. An easier way to determine this Burgers vector and the exact core location of the dislocation was to Fourier filter the HRTEM image three times, by selecting alternatively the three {01$\overline{1}$0} frequencies. Two of these three filtered images are shown in figures 5d and 5e. In each of these filtered image, an additional {01$\overline{1}$0} plane can be seen. The third filtered image does not contain any additional plane. The two {01$\overline{1}$0} additional planes are equivalent to one additional $(2\bar{1}\bar{1}0)$ plane, i.e. to an edge dislocation of Burgers vector $1/3[2\bar{1}\bar{1}0]$. It is interesting to note that the neighbourhood of the dislocation that contains the additional planes is in compression (marked comp in figure 5a) and has locally a lattice parameter smaller than AlN. GaN, whose lattice parameter is larger than the AlN one by about 2.5%, tends to grow at the opposite side of the dislocation that is in tension and where the AlN lattice parameter is locally wider than the AlN bulk value. The GaN dots with their associated edge dislocations could also be seen in two-beam images (Rouviere et al., 1999).

We think that this is the first observation of the nucleation of an island (dot) at the edge of an edge dislocation. This was possible because the density of edge dislocations was of the same order of magnitude as the density of GaN dots (about 10^{+11} cm^{-2}).

5. CONCLUSION

HRTEM study has been a powerful tool to analyse the wetting layer and the GaN dots. By producing HRTEM interplanar profiles, the thickness and the composition of the wetting layer have been determined (GaN is spread on 4 planes of respective GaN composition equal to 0.15, 0.85, 0.5 and 0.15). It has been determined that a stacking of GaN dot layers, associated with a growth interruption after each GaN dot layer deposition, produces stacks of correlated truncated hexagonal pyramids of diameter and height respectively equal to 15.3nm and 3.3nm. The GaN dots grow coherently on the AlN buffer layer and are dislocation free, but they grow at the edge of threading AlN edge dislocations. Photoluminescence studies have shown that these GaN dots do emit light (Rouviere et al., 1999).

ACKNOWLEDGEMENT

We would like to thank Dr H. Mariette and N T Pelekanos for discussions.

REFERENCES

Arlery M, Rouviere J L, Widmann F, Daudin B, Feuillet G, Mariette H 1999 accepted in Appl. Phys. Lett. (June issue)
Gil B. (editor) 1998 *Group III Nitride Semiconductor Compounds : Physics and Applications*, (Clarendon Press, Oxford)
Nakamura S and Fasol G 1997 *The blue laser diode* (Springer, Heidelberg)
Rouviere J L 1994 in *Electron Microscopy 1994* Proceeding of the 13th international congress on electron microscopy (les éditions de Physique) vol. **2A** p123
Rouviere J L, Daudin B, Feuillet G, Pelekanos N T and Simon J 1999 submitted to Appl. Phys. Lett.
Stadelmann P A 1987 Ultramicroscopy **38** 265
Tanaka S, Hiramaya H, Aoyagi Y, Narukawa Y, Kawakami Y, Fujita S 1997 Appl. Phys. Lett. **71** 1299
Widmann F, Daudin B, Feuillet G, Samson Y, Rouviere JL and Pelekanos N T 1998 J. Appl. Phys. **83** 7618

Investigation of the atomic structure of [0001] tilt grain boundaries in GaN/sapphire epitaxial layers

V Potin, G Nouet, P Ruterana and RC Pond[*]

Laboratoire d'Etudes et de Recherches sur les Matériaux , UPRESA 6004 CNRS, Institut des Sciences de la Matière et du Rayonnement, 6 boulevard Maréchal Juin, 14050 Caen Cedex, France.
E-mail: potin@lermat8.ismra.fr
[*]Department of Materials Science and Engineering, the University of Liverpool, Liverpool L69 3GH, England.

ABSTRACT: A grain boundary rotated around [0001], with a misorientation corresponding to $\theta = 21.8°$ has been studied in GaN. In order to identify interfacial defects, the micrographs were characterised within the formalism of topological theory using circuit mapping. It is found that the period of the boundary consists of three edge $1/3 <11\overline{2}0>$ dislocations. This boundary plane is seen to shift from one plane to another without need of any additional defect. Although the interface is made of only one type of dislocation, its reconstruction is shown to necessitate the use of several atomic configurations for the cores: 4, 5, 7 and 8 atom rings.

1. INTRODUCTION

The III-V nitride semiconductors are characterised by their large direct bandgap in the range 1.89 – 6.2 eV, which covers most of the visible spectrum and extends out into the ultraviolet range. InN, GaN and AlN form a continuous alloy system which makes available any band gap within this range. Among them, GaN has been extensively studied and several technological breakthroughs have led to the commercial availability of high brightness light emitting diodes and laser diodes (Nakamura et al 1996, 1998).

These results are surprising if the large defect densities observed in the active layers are considered. For example, densities of 10^{10} cm^{-2} have been reported which are at least 10^5 times larger than those acceptable for GaAs devices (Lester et al 1995). The defects are mainly threading dislocations, basal and prismatic stacking faults, inversion domain boundaries and nanopipes (Vermaut et al 1997, Potin et al 1998a, 1999a) . Among them, we have focussed our attention on the threading dislocations which form high-angle grain boundaries (Potin et al 1998b). The grains present special misorientations corresponding in particular to $\Sigma = 7$, $\Sigma = 19$, $\Sigma = 31$ in the coincidence site lattice (CSL) notation (Potin et al 1999b). The atomic structure of the grain boundaries has been determined by high resolution electron microscopy (HREM).

2. EXPERIMENTAL DETAILS

The GaN layers were grown on the (0001) sapphire surface by NH$_3$ gas source molecular beam epitaxy (MBE). The active layer was deposited at 800°C on top of a low-temperature (550°C) GaN buffer layer of 40 nm thickness. The plan–view samples were prepared as usual by mechanical grinding and ion milling. HREM experiments were carried out along the [0001] zone axis on a Topcon 002B electron microscope operating at 200 kV.

3. CRYSTALLOGRAPHY

In the topological theory of line defects in interfaces developed by Pond (1989), it was shown that all the admissible defects between crystallographically equivalent surfaces can be known a-priori. In this formalism, one crystal is designated as white (λ) and the other as black (μ). The configuration formed by the interpenetrating lattices and crystals are called dichromátic pattern and complex, respectively (Pond et al 1983). The Burgers vectors of some admissible interfacial dislocations are given by :

$$\mathbf{b}_{ij} = \mathbf{t}\,(\lambda)_j - \mathrm{Pt}\,(\mu)_i$$

where $\mathbf{t}\,(\lambda)_j$ and $\mathbf{t}\,(\mu)_i$ represent the ith and jth translation vectors in the λ and μ crystals, respectively and P is a matrix which re-expresses $\mathbf{t}\,(\mu)_i$ in the λ coordinate frame. The cores of such defects are associated with steps whose heights are given by $h(\lambda) = \mathbf{n}_\lambda \cdot \mathbf{t}\,(\lambda)_j$ and $h(\mu) = \mathbf{n}_\mu \cdot \mathbf{t}\,(\mu)_i$ where $\mathbf{n}$ are unit vectors normal to the interface, oriented towards the λ crystal.

These interfacial defects can be characterised in HREM images by circuit mapping as proposed by Pond (1995). A closed circuit is constructed on the micrograph with segments in the λ and μ crystals called $\mathbf{c}(\lambda)$ and $\mathbf{c}(\mu)$, respectively. When the circuit is mapped into a reference space, any closure failure is equal to the total defect content and using the RH/FS convention, the defect is given by $\mathbf{c}\,(\lambda,\,\mu)^{-1}$. If the circuit is mapped into a single crystal, the primary dislocations are exhibited whereas in the dichromatic pattern, secondary dislocations are determined. In the simplest case, we obtain :

$$\mathbf{b}_{ij} = -\,\mathbf{c}(\lambda,\mu) = \mathbf{c}(\lambda) + \mathrm{P}\,\mathbf{c}(\mu)$$

4. RESULTS

A low magnification image of a high-angle grain boundary is shown in Fig. 1a and the corresponding diffraction pattern (Fig. 1b) indicates a misorientation equal to $\theta = 22° \pm 0.2°$.

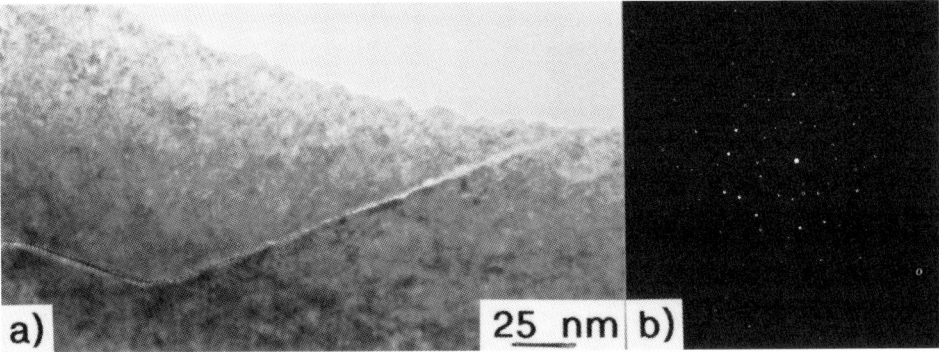

Fig 1a and b: low magnification plan view of a high angle grain boundary with its corresponding diffraction pattern.

This value is very close to that of $\Sigma = 7$ misorientation in the CSL notation where the rotation angle is 21.79° about [0001]. The grain boundary is characterised by facets and the studied boundary is asymmetric and corresponds to $(\bar{3}\,8\,\bar{5}\,0)_\lambda\,/\,(0\,1\,\bar{1}\,0)_\mu$. On Fig. 2, the large black dots (such as X_2, S_2/F_2, X/F_3, S/F) define periods which are broken by the step $S_2/F_2 - X/F_3$. By using circuit mapping

around them, we determined their dislocation content. We found for SXF : $c(\lambda) = SX =$
$\bar{a}_2 - 4\bar{a}_1 + 4\bar{a}_3 = \dfrac{1}{3}\left[\bar{1}3\ 2\ 11\ 0\right]_\lambda$ and $c(\mu) = XF = -\bar{a}_2 + 6\bar{a}_1 - \bar{a}_3 = \dfrac{1}{3}\left[14\ \bar{7}\ \bar{7}\ 0\right]_\mu$.

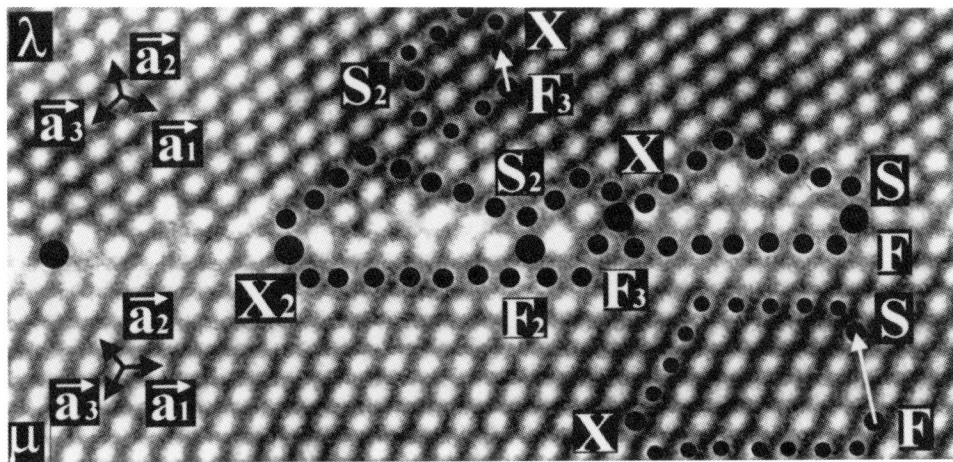

Fig. 2: HREM image of the boundary separating λ and μ crystals, several closed circuits have been constructed and mapped in the single crystals. The large black dots define periods of $\Sigma = 7$ (21°79 / [0001]).

By reporting the circuit SXF in the μ crystal, the closure failure is $\mathbf{FS} = 2\bar{a}_2 - \bar{a}_3$, corresponding to the primary dislocations. When the same circuit is mapped into the dichromatic pattern rather than into the lattice of a single crystal, secondary dislocations may be determined. In this case, this leads to

$$c(\lambda) + P_7\,c(\mu) = 0.$$

Therefore, in this period, the interface appears to be completely flat, without any additional defect. The result is exactly the same if the circuit $S_2X_2F_2$ is mapped.

A step is clearly present between S_2 and X. We have tried to determine if a defect was associated with it. Thus, the circuit XS_2F_3 was mapped and we found $\mathbf{c'}(\lambda) = \mathbf{XS_2} = 1/3\ \left[\bar{4}\ \bar{1}\ 5\ 0\right]_\lambda$ and $\mathbf{c'}(\mu) = \mathbf{S_2F_3} = 1/3\ \left[5\ \bar{1}\ \bar{4}\ 0\right]_\mu$. In fact, the circuit XS_2F_3 corresponds to one period of the $\left(\bar{2}\ 3\ \bar{1}\ 0\right)_\lambda / \left(\bar{1}\ 3\ \bar{2}\ 0\right)_\mu$ interface (one side of the $\Sigma = 7$ unit cell) which is symmetric (Fig. 3a). In the λ crystal, it exhibits a closure failure equal to $\mathbf{F_3X} = \bar{a}_2$ and in the dichromatic pattern, no secondary dislocation was found. Therefore, we may conclude that the analysed part of the grain boundary corresponds exactly to the particular coincident orientation $\Sigma = 7$ (21°79 around [0001]).

We have studied the atomic structure of the boundary using some models of dislocation cores previously observed for isolated dislocations (Ruterana et al 1998). This part of the grain boundary is made of three periods corresponding to the $\left(\bar{3}\ 8\ \bar{5}\ 0\right)_\lambda / \left(01\bar{1}0\right)_\mu$ interface and a step corresponding to one period of the $\left(\bar{2}\ 3\ \bar{1}\ 0\right)_\lambda / \left(\bar{1}\ 3\ \bar{2}\ 0\right)_\mu$ interface (Fig. 3b).

Along this interface, we have identified similar features like the 5/7 atom rings which are located at ends of each period. Inside the periods, the contrast is rather complex and the cores of the two remaining dislocations appear to be different. From left to right of the micrograph, the contrast of one of the dislocations is similar in period 1 and 4 and it exhibits 5/4/7 atom rings. The other has an 8 atom ring core in period 1 and 5/7 atoms in period 4. In period 2, the two inner dislocations

380

have 8 and 5/7 atom rings respectively. For period 3, which corresponds to the step, the unique dislocation core is made of one 5/7 atom ring unit.

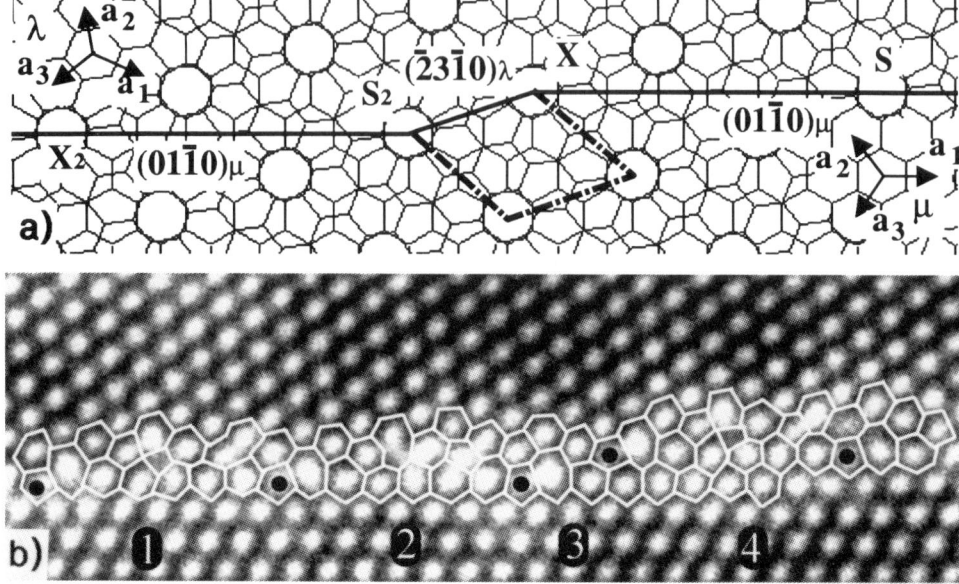

Fig. 3: similar image as Fig.2 a) the corresponding dichromatic pattern is shown with the $\Sigma = 7$ unit cell underlined in bold and b) a reconstruction is shown for the dislocation cores of the four different periods.

5. CONCLUSION

A $\Sigma = 7$ GaN grain boundary has been analysed using circuit mapping. In this asymmetric boundary, we have identified one period of the $\Sigma = 7$ symmetric interface which acted as a step and allowed it to change from one plane to another without any additional defects. The atomic structure of the boundary is based on several $1/3 <11\overline{2}0>$ dislocation cores made of 4, 5, 7 and 8 atom cycles.

REFERENCES

Lester S D, Ponce F A, Craford M G and Steigerwald D.A 1995 Appl. Phys. Lett. **66**, 1249.
Nakamura S, Senoh M, Nagahama S, Iwasa N, Yamada T, Matsushita T, Kiyoku K and Sugimoto Y 1996 Jpn. J. Appl. Phys. **35**, L74.
Nakamura S, Senoh M, Nagahama S, Iwasa N, Yamada T, Matsushita T, Kiyoku H, Sugimoto Y, Kozaki T, Umemoto H, Sano M and Chocho K 1998 Appl. Phys. Lett. **6**, 832.
Pond RC et Vlachavas DS 1983 Proc. Roy. Soc. London A **386**, 95.
Pond RC 1989 Dislocations in Solids, Ed. F.R.N. Nabarro, North Holland, Amsterdam **8**, 1.
Pond RC 1995 Interface Science **2**,1.
Potin V, Vermaut P, Ruterana P and Nouet G 1998a J. Elec. Mat. **4**, 266.
Potin V, Ruterana P, Hairie A and Nouet G 1998b Mat. Res. Soc. **482**, 435
Potin V, Ruterana P and Nouet G 1999a Phil. Mag. A (in press).
Potin V, Ruterana P, Nouet G and Pond RC 1999b Phil. Mag. A (to be published).
Ruterana P, Potin V and Nouet G 1998 Mat. Res. Soc. **482**, 459
Vermaut P, Ruterana P, Nouet G and Morkoc H 1997 Phil. Mag. A **75**, 239.

Inst. Phys. Conf. Ser. No 164
Paper presented at Microsc. Semicond. Mater. Conf., Oxford, 22–25 March 1999
© *1999 IOP Publishing Ltd*

Inversion domain nucleation in homoepitaxial GaN

P.D. Brown[1]*, J.L. Weyher[2,3], C.B. Boothroyd[1], D.T. Foord[1], A.R.A. Zauner[2], P.R. Hageman[2], P.K. Larsen[2], M. Bockowski[3] and C.J. Humphreys[1].

[1]Department of Materials Science and Metallurgy, University of Cambridge, Pembroke Street, Cambridge, CB2 3QZ, UK.
[2]University of Nijmegen - RIM, Exp. Solid State Physics III, Toernooiveld 1, 6525 ED Nijmegen, The Netherlands.
[3]High Pressure Research Centre, Polish Academy of Sciences, ul. Sokolowska 29, 01-141 Warsaw, Poland.

ABSTRACT: Homoepitaxial GaN grown by metal organic chemical vapour deposition on the N-polar surface of bulk GaN commonly exhibits gross hexagonal pyramidal features, typically 5 to 50μm in size, depending on layer thickness. The evolution of these defects is dominated by the growth rate of an emergent core of inversion domain (typically 100nm in size). The associated nucleation events are considered to be thin bands of amorphous gallium hydroxide (2 to 5nm in thickness) remnant from the standard KOH chemo-mechanical substrate polishing procedure used.

1. INTRODUCTION

Interest in homoepitaxial GaN continues, driven by its potential for high power optoelectronic device applications. In this context, the absence of microstructural defects afforded by homoepitaxial growth is considered beneficial, particularly in view of recent definitive evidence confirming that dislocations do indeed exhibit non-radiative recombinative properties (e.g. Sugahara et al, 1997). However, metal organic chemical vapour deposition (MOCVD) grown homoepitaxial GaN on chemo-mechanically polished (000$\bar{1}$), N-polar, substrates commonly exhibit gross hexagonal shaped hillocks that could be somewhat problematic for subsequent device processing. Here we use a range of microscopical techniques to characterise the nature of the cores of these hillock structures and the precise form of their associated nucleation events.

2. EXPERIMENTAL

The homoepitaxial GaN samples examined in this study were grown by MOCVD at 1050°C as detailed elsewhere (De Theije et al, 1999, Weyher et al, 1999a). The bulk GaN substrate material was grown under a high hydrostatic pressure of nitrogen (15 – 20kbar) from liquid Ga at 1600°C (Porowski, 1996). Prior to growth, the (000$\bar{1}$) surfaces were mechanically polished using 0.1μm diamond paste and then chemo-mechanically polished in an aqueous KOH solution (Weyher et al, 1997). Epitaxial growth was performed using trimethylgallium and NH$_3$ precursors with H$_2$ as the carrier gas, under a total pressure of 50mbar. Electron transparent samples were prepared in plan-view using conventional sequential mechanical polishing and ion beam thinning procedures, whilst cross-sectional samples were prepared using a Ga-source focused ion beam (FIB) workstation. The site

---

* Now at the Division of Materials Engineering and Materials Design, School of Mechanical, Materials, Manufacturing Engineering and Management, University of Nottingham, University Park, Nottingham, NG7 2RD.

382

Fig. 1 Secondary electron image of homoepitaxial GaN/GaN(000$\bar{1}$) with an FIB prepared membrane through the core of one of the growth hillocks.

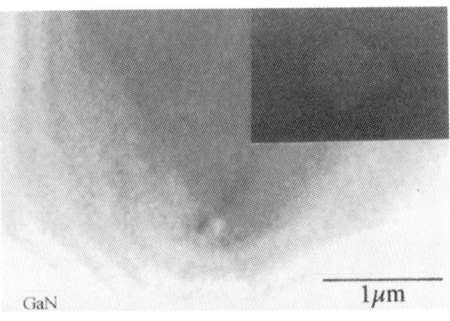

Fig. 2 Slightly tilted plan-view TEM image through a growth hillock revealing the central defect core. Enlarged <0001> image of hillock core inset.

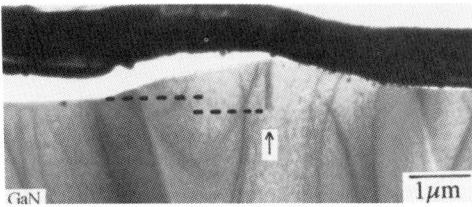

Fig. 3 Low magnification cross-sectional TEM image showing a growth hillock in profile (defect core arrowed).

selectivity of the FIB technique enabled cross-sections through the emergent cores of the hillocks to be obtained, thereby allowing the nucleation events to be isolated and characterised.

3. RESULTS AND DISCUSSION

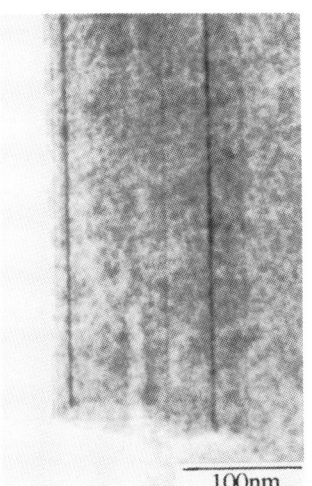

Fig. 4 TEM image showing the faceted structure of a hillock core in cross-section.

Fig. 1 shows a secondary electron image of the homoepitaxial GaN/GaN(000$\bar{1}$) growth hillocks recorded during FIB cross-sectional sample preparation. The hillocks are typically 5 to 50µm in size depending on layer thickness (time of growth). When prepared in plan-view geometry (Fig. 2), such hillocks exhibit a small faceted core structure at the centre, but otherwise the layers are generally found to be defect free. Fig. 3 shows a low magnification image of one such growth hillock from a very thin layer imaged in cross-section. The protective Pt-stripe deposited on a thin Au gold coating layer prior to FIB milling has become detached from the top surface of the sample and slightly obscures the hillock. Nevertheless, the hillock core structure can clearly be seen (arrowed). It is presumed that the defect originates at the line of the original epilayer / substrate interface since no other contrast delineating the region of this interface could be discerned. The hillock structure is enlarged and slightly tilted in Fig. 4 to emphasise its faceted structure. The core emanates from a thin platelet nucleus that is only ~100nm in size. Convergent beam electron diffraction (CBED) analysis confirmed the N-polar growth surface of the surrounding matrix material, but no clear CBED patterns could be obtained from this particular core structure since it was embedded within matrix material.

Fig. 5 shows a cross-section through a core structure from a thicker layer, although the nucleating event has been sectioned away during sample foil preparation in this instance. In

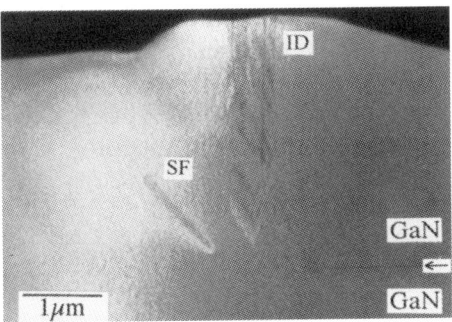 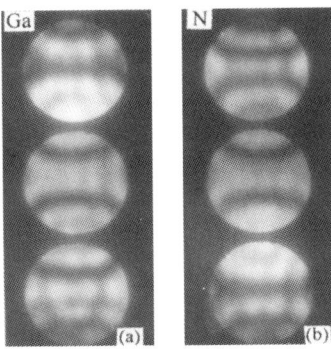

Fig. 5 Thinner section through a hillock core with two associated CBED patterns (a,b) recorded either side of the core boundary, corresponding to core and matrix material respectively. The reversal in contrast within the {0002} diffraction discs confirms the defect to be an inversion domain.

addition to the central core there is also an inclined stacking fault and a horizontal screw dislocation (arrowed) which, again, presumably delineates the line of the original epilayer / substrate interface. CBED patterns recorded either side of the boundary wall of the core immediately reveal a reversal of the contrast within the {0002} diffraction discs and as such confirm the defect core to be an inversion domain (ID) (e.g. Cherns et al, 1998). Correction for image rotation introduced by the electron microscope (using a polarity calibrated sample) demonstrates that the central core has a Ga-polar growth surface, while the surrounding matrix has a N-polar growth surface, commensurate with the polarity of the substrate used for growth. When compared with Fig. 3, it is clear that the cores of the inversion domains have a much higher growth rate (approx. x3) than the surrounding N-polar matrix, leading to the production of the 'circus-tent' hillock structures around them. Competition between the growth and desorption rates of Ga and N-polar surfaces allows the gross hexagonal pyramids to evolve. Interestingly, the sense of Ga-polar surface having a faster growth rate than the N-polar surface is contrary to that found for thin film growth of wurtzite CdS, where the S-polar surface has a higher growth rate than the Cd-polar surface (Halsall et al, 1988). The implication, therefore, is that the N-polar surface has a much higher desorption rate than the Ga-polar surface for the MOCVD growth conditions used here.

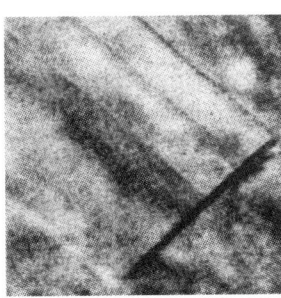 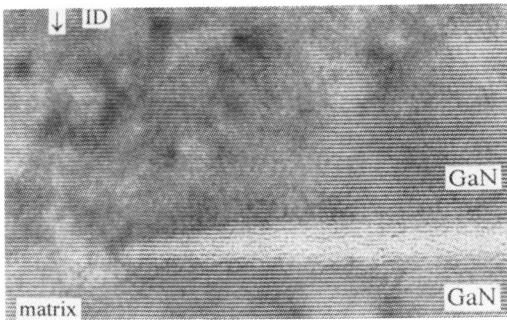

Fig. 6 HAADF image, confirming the presence of low atomic number material at the ID source.

Fig. 7 HREM image confirming the presence of a narrow band of amorphous material at the ID source.

High angle annular dark field imaging (HAADF) imaging (Fig. 6) of the sample shown in Figs. 3,4, tilted slightly off a systematic row orientation to minimise the effects of diffraction contrast, shows dark contrast at the position of the inversion domain, and therefore confirms the presence of a low atomic number material associated with the nucleation event. Preparing samples thin enough for high resolution electron microscopy (HREM) confirmed the nature of such nucleation events to be narrow bands of amorphous material, 2 to 5nm in thickness (Fig. 7). Electron energy loss spectroscopy (EELS) confirms the presence of

384

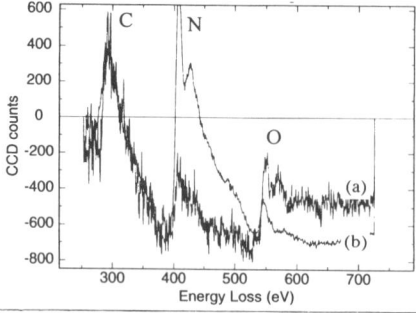

Fig. 8 EEL spectra confirming the presence of oxygen at the nucleating event. (a) Nucleation event. (b) Matrix.

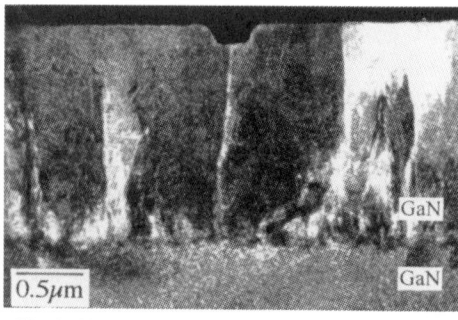

Fig. 9 Homoepitaxial GaN grown on a Ga-polar GaN substrate, showing the association of a grain boundary with a surface flat-bottomed pit.

oxygen (Fig. 8) within these narrow bands of amorphous material, and as such these features are attributed to remnant contamination from the chemo-mechanical polishing technique used to prepare the substrates prior to growth. The oxygen-containing residue is presumed to be gallium hydroxide, a likely reaction product of the KOH etchant with GaN. An improved surface preparation method incorporating a short deoxidising polishing procedure in an aqueous solution of NaCl (Ohkubo, 1998) led to a dramatic reduction of these nucleation sources and thus allowed N-polar homoepitaxial GaN films to be grown largely free of these gross hillock structures (Weyher et al, 1999a).

Preparation of (0001)Ga-polar substrates suitable for homoepitaxy proved to be a difficult problem since chemo-mechanical polishing procedures were found not to be effective. Thus, substrates were mechanically polished with 0.1μm diamond paste and received a reactive ion etching (RIE) treatment prior to growth. Such samples were found to contain a very high density of threading dislocations, with FIB prepared cross-sectional samples showing the association of grain boundaries, rather than inversion domains, with the location of flat bottomed surface pits (e.g. Fig. 9). Such highly faulted layers were again attributed to remnant surface roughness and contamination after the substrate preparation process. However, improved RIE procedures have recently enabled the preparation of GaN homoepitaxial layers on Ga-polar surfaces free of structural defects, and this will be reported on elsewhere (Weyher et al, 1999b).

ACKNOWLEDGEMENTS

PDB wishes to acknowledge EPSRC for funding under contract number GR/L21686. JLW wishes to thank the NATO Scientific Affairs Division for the grant HTECH: LG972924. This work was financially supported by the Dutch Technology Foundation (STW).

REFERENCES

Cherns D, Young WT, Saunders M, Steeds JW, Ponce FA and Nakamura S, 1998, Phil. Mag. A77, 273
De Theije FK, Zauner ARA, Hageman PR, van Enckevort WJP, Larsen PK, 1999, J. Crystal Growth, 197, 37
Halsall MP, Davies JJ, Nicholls JE, Cockayne B, Wright PJ and Russell GJ, 1988, J. Crystal Growth 91,135
Ohkubo M, 1998, J. Crystal Growth 189/190, 734
Porowski S, 1996, J. Crystal Growth, 166, 583
Sugahara T, Sato H, Hao M, Naoi Y, Kurai S, Tattori S, Yamashita K, Nishino K, Romano LT and Sakai S, 1997, Jpn. J. Appl. Phys. 37, L398
Weyher JL, Müller S, Grzegory I and Porowski S, 1997, J Crystal Growth 182, 17
Weyher JL, Brown PD, Zauner A, Müller S, Boothroyd CB, Foord DT, Hageman PR, Humphreys CJ, Larsen PK, Grzegory I and Porowski S, 1999a, submitted to J. Crystal Growth.
Weyher JL, Zauner A, Brown PD, Karouta F, Wysmolek A and Porowski S, 1999b, to be presented at, ICNS3, Montpelier.

Inst. Phys. Conf. Ser. No 164
Paper presented at Microsc. Semicond. Mater. Conf., Oxford, 22–25 March 1999
© 1999 IOP Publishing Ltd

HREM analysis of planar defects in GaN layers grown on (0001) Al₂O₃ vicinal surfaces

B Barbaray*, P Ruterana*, G Nouet*, M A di Forte Poisson**, F Huet** and M Tordjman**

*Laboratoire d'Etudes et de Recherches sur les Matériaux, UPRESA CNRS 6004, Institut des Sciences de la Matière et du Rayonnement, 6 Boulevard du Maréchal Juin, 14050 CAEN Cedex – France.
nouet@labolermat.ismra.fr
**Thomson-CSF/Laboratoire Central de Recherches, Domaine de Corbeville, 91404 ORSAY Cedex – France

ABSTRACT: HREM analysis was carried out on GaN layers grown on 3° misoriented (0001) surface of Al₂O₃. In the layers, numerous basal stacking faults were observed and identified as I1 type, with R_{I1} fault vector equal to $=1/6<20\bar{2}3>$. They are connected to substrate steps by prismatic stacking faults which fold into the basal plane. A comparison is made with defects found in GaN layers grown on non-vicinal (0001) sapphire substrates.

1. INTRODUCTION

Gallium nitride thin films contain high densities of defects such as threading dislocations, stacking faults, inversion domains, pinholes and nanopipes. A non-exhaustive review of these defects has been recently published by Potin *et al.* (1998). Although these defects have been identified and characterized, the role of the interface structure on their formation is still under discussion. It is now clear that the very high density of threading dislocations ($10^{10} cm^{-2}$) is due to the island growth mode studied by Ning (1996). It has been shown by Vermaut *et al.* (1997) that prismatic stacking faults in AlN or GaN grown over silicon carbide (SiC) are generated on interface steps. Such defects are present in GaN or AlN layers independently of the substrate: SiC or sapphire (Al₂O₃).

A geometric analysis of the anionic stacking showed that steps on substrate could bring about stacking faults and inversion domain boundaries in the epitaxial layer (Barbaray, 1999). In this work, we investigated the role of the numerous steps present on the surface of a 3° misoriented sapphire substrate on the formation of extended defects.

2. CRYSTALLOGRAPHY

The substrate, Al₂O₃, belongs to the rhombohedral R$\bar{3}$c space group. The lattice parameters are a=0.476 nm and c=1.2991 nm. The anion framework forms a hexagonal closed packed (hcp) structure that is distorted by the Al³⁺ which occupy 2/3 of the octahedral sites. By using the approximation (x, y, z)≈(1/3, 0, 1/4) instead of (x, y, z)=(0.306, 0, 1/4) for the O²⁻ location, one can consider the anion framework as a perfect hcp lattice. In the following analysis, due to the largest size of the anions, the (0001) sapphire surface is supposed to be oxygen terminated. Thus, the steps that can occur on the substrate have a height which is a multiple of $c_{substrate}/6$, which corresponds to the distance between two oxygen basal planes.

GaN layers crystallize in the wurtzite structure, P 6_3mc, with lattice parameters a=0.319 nm and c=0.518 nm. The Ga^{3+} cations occupy one of the two families of tetrahedral sites β_1 or β_2 of the anions hcp lattice. Conventionally, the polarity of a layer is given by the direction of the Ga^{3+}–N^{3-} bond parallel to the c axis. Therefore, by occupying β_1 or β_2 tetrahedral sites, the polarity can be reversed, giving rise to inversion domains. Geometrically, an inversion domain is constructed by applying an operator that can be a symmetry centre or a mirror parallel to the basal plane (which is equivalent due to the 6_3 axis parallel to [0001]). In non-centrosymmetric materials, several inversion domain boundaries have been described and observed (Austerman, 1966, Holt, 1969, Northrup, 1996). The stacking faults in the wurtzite structure are the same as in the hcp structure. Their fault vectors are $R_{I1}=1/6<20\bar{2}3>$, $R_{I2}=1/3<10\bar{1}0>$ and $R_E=1/2<0001>$, and their stackings are hch, hcch and hccch, respectively (h=hexagonal sequence, c=cubic sequence).

GaN layers grow on the Al_2O_3 substrate according to the well-known epitaxial relationship:

$$(0001)_{GaN}//(0001)_{Al_2O_3},$$

$$[0\bar{1}10]_{GaN}//[11\bar{2}0]_{Al_2O_3}$$

This relationship leads to the continuation of the anion compact stacking: O^{2-} in the substrate and N^{3-} in the layer. The difference between the parameters leads to a mismatch along the $<11\bar{2}0>$ and $<0001>$ directions: 13.9 % and 16.9 % respectively.

3. EXPERIMENTAL

Two kinds of epitaxial GaN layers have been studied: (i) growth on a non-vicinal (0001) surface substrate and (ii) on a 3° misorientated (0001) surface substrate. For (i), the sapphire substrate is chemically cleaned and treated in-situ by a hydrogen plasma. The GaN is grown at 800°C by ECR MBE, without a buffer layer at 40 nm/h, up to a thickness of 3μm. (ii): the (0001) substrate surface is 3° +/-0.5° misorientated. The epitaxial layer is grown by MOCVD. After nitridation of the substrate, a 15 nm nucleation layer is deposited at 500°C. The active GaN layer is grown at 1000°C at a rate of 0.4 μm/h.

The specimens for electron microscopy were prepared in the conventional way by mechanical polishing followed by ion milling down to electron transparency. The experimental micrographs were recorded in the 002B Topcon microscope operated at 200 kV, with a point to point resolution of 0.18 nm.

4. RESULTS ON NON-VICINAL SURFACES

By studying the crystallography and the epitaxial relationship, we showed that the operator that relates two GaN crystallites grown on two adjacent terraces may exhibit a R_{SF} translation and a [0001] residual translation T_{res}. The R_{SF} translation, which can be R_{I1}, R_{I2} or R_E, is due to the several ways to stack the nitrogen framework of GaN. During coalescence, this translation creates a prismatic stacking fault with R_{SF} as displacement vector. The residual translation is due to the mismatch along c between the deposit and the layer and has to be included in the operator when there is a step. With the approximation $2c_{substrate}\sim5c_{deposit}$, T_{res} can take four values $T_{Res}=0$, $T_{Res}=1/12$, $T_{Res}=1/6$ and $T_{Res}=1/4$, in c GaN units. This translation can be elastically relaxed or minimised by the formation of an inversion domain boundary. The HREM observations confirm that on non-vicinal substrate, small terraces limited by $c_{Al2O3}/3$ steps give rise to inversion domain boundaries in $\{10\bar{1}0\}$ planes (Barbaray et al. 1999).

5. RESULTS ON VICINAL SURFACES

The misorientation can be measured directly on the micrographs and is evaluated to 3.5°-4°, as seen on figure 1. The interface is made by a succession of steps difficult to locate due to the poor resolution. In these conditions, the minimum height of step which can be detected is $1/3c_{Al2O3}$. Actually, the heights of these steps are evaluated to be c_{Al2O3}, $2/3 c_{Al2O3}$ and $1/3 c_{Al2O3}$.

These GaN layers do not contain any inversion domain boundaries but numerous basal stacking faults are present close to the interface. These faults can be identified by counting the number of cubic sequences (c) into the hexagonal stacking (h), and all the analysed ones correspond to I1. Figure 2 shows an example of I1 basal stacking faults.

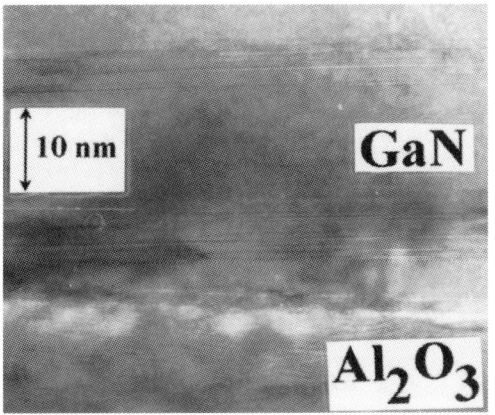

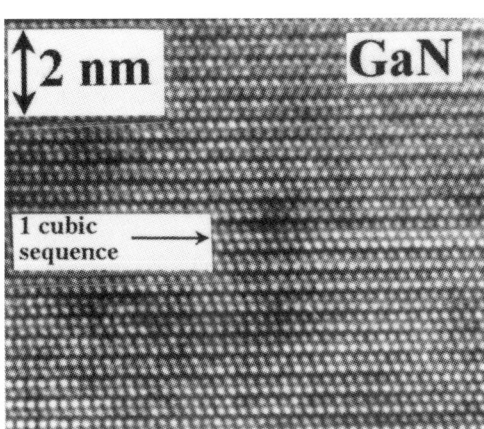

Fig. 1: HREM micrograph of the epitaxial layer near the interface. The basal stacking faults are parallel to (0001) planes, thus the misorientation of the interface is measured : $\alpha = 3.5° - 4°$.

Fig. 2: Detail of a basal stacking fault. The cubic sequence allows to identify it as a I1 stacking fault.

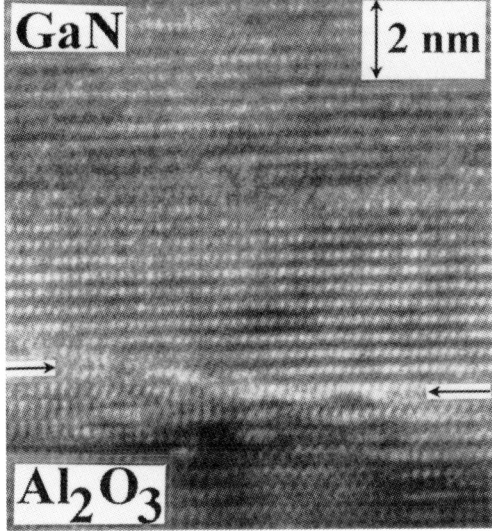

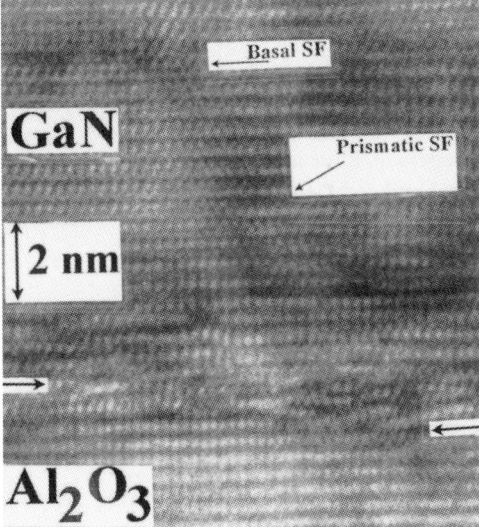

Fig. 3: HREM micrograph of the interface that shows a step which does not create a stacking fault in the epitaxial layer. Its height is evaluated to be 2/3 c_{Al2O3} (arrows indicate the interface position on each terrace).

Fig. 4: HREM micrograph of the interface that shows a c_{Al2O3} step which creates a stacking fault in the epitaxial layer. Arrows indicate the interface position on each terrace. A prismatic stacking fault, that could be due to the different stacking sequences on each terrace, folds into the basal plane (basal SF).

Approximately, we can count one step every 10 nm at the interface. If each step created a basal stacking fault, the spacing of faults along [0001] in the GaN layer would be: $d=10$. tan $3.5° = 0.6$ nm. In fact, in the region where there is a high density of stacking faults, their average distance is larger than 5 nm. This means that there is no direct connection between the stacking faults and the steps.

However, concerning the relationship between surface steps and layer defects, two distinct cases were observed. In the first case, illustrated by figure 3, the GaN stacking around the step is not disturbed and no basal stacking fault is formed. The residual translation induced by the step is probably elastically compensated. In the other case, the step is connected to a basal stacking fault by an area of blurred contrast (figure 4). A model for the formation of such a defect can be proposed in comparison to work already published in GaN/SiC (Vermaut *et al.* 1997). The stacking at the interface is not the same on either adjacent terrace: ABA/babab and ABA/cbcbcb leading to a $R_{I2}=1/3<10\overline{1}0>$ translation in the basal plane. This vector, added to the translation from the step, which is in this case $\frac{1}{2}[0001]$, leads to the formation of a prismatic stacking fault characterised by $R_{I2}+1/2[0001] = R_{I1}$. This fault folds into the basal plane.

6. CONCLUSION

We have investigated the extended defects in the epitaxial layer due to steps at the substrate surface. On non-misoriented substrates, the small hexagonal terraces may give rise to inversion domain boundaries in order to minimise the mismatch along the c axis. On vicinal surfaces, the steps are large and they create elastic deformations or stacking faults in the epitaxial layers. These faults are identified as I1 basal stacking faults and are explained by the several stacking possibilities that can occur at the interface.

REFERENCES

Austerman S B and Gehman W G 1966 J. Mater. Sci., **1**, 249
Barbaray B, Potin V, Ruterana P and Nouet G 1999 Phil. Mag. A, submitted
Holt D B 1969 J. Phys. Chem. Solids, **30**,1297
Ning X J, Chien F R, Pirouz P, Wang J W and Khan M A 1996 J. Mater. Res., **3**, 580
Northrup J E, Neugebauer J and Romano L T 1996 Phys. Rev. Lett., **77**, 103
Potin V, Ruterana P and Nouet G 1998 J. Elect. Mat., **27**, 266
Vermaut P, Ruterana P, Nouet G and Morkoç H 1997 Phil. Mag. A, **75**, 239

Inst. Phys. Conf. Ser. No 164
Paper presented at Microsc. Semicond. Mater. Conf., Oxford, 22–25 March 1999
© 1999 IOP Publishing Ltd

Growth of GaN layers onto misoriented (0001) sapphire by MOCVD

B Pécz, M A di Forte Poisson*, G Radnóczi and L Tóth

Research Inst. for Technical Phys. and Mat. Sci. H-1525 Budapest, P.O.Box 49, Hungary
* Thomson CSF, Laboratoire Central de Recherches, Domaine de Corbeville, 91404 Orsay,
Cedex, France

ABSTRACT: The growth processes of GaN layers grown by MOCVD on sapphire have been characterised by TEM. GaN buffer layers grown at 510°C consist of hexagonal single crystals with a mosaic structure already in their as-grown state. Annealing of the buffer layers leads to substantial smoothing of their surfaces due to the coalescence of the grains. The GaN layers themselves are single crystalline, hexagonal and epitaxial to the substrate. Layers grown on exactly oriented (0001) sapphire substrates and on miscut substrates are compared.

1. INTRODUCTION

GaN layers are grown by different techniques, the most promising ones being gas source MBE (Molecular Beam Epitaxy) and MOCVD (Metal Organic Chemical Vapour Deposition). Although several materials are used as substrates to grow hexagonal GaN (6H-SiC, ZnO and Hf), sapphire is still a reasonable choice for that purpose as it is available widely for relatively low cost. Growth of GaN on sapphire is usually carried out by a two step procedure (Amano et al 1986). In this procedure a thin AlN or GaN buffer layer is deposited at low temperature before the growth of the thick GaN at high temperature. Very often the surface of sapphire is nitrided before two step growth in the NH_3 stream, while in other cases the substrate is exposed to H_2 gas at high temperature. The formation of an AlN layer is detected on nitrided sapphire surfaces by diffraction techniques showing the crystalline layers formed. Our earlier Transmission Electron Microscopy (TEM) results (Pécz et al 1997) showed the formation of a 4 nm thick AlN layer on the substrate due to the nitridation process.

Little attention has been paid in the literature to the structure of the GaN buffer layer grown at relatively low temperature (400-600°C). In some works the growth parameters are optimized based on the final quality of the thick GaN layers, but very confusing data as well as some speculations are published on the structure of the thin GaN buffer layers. There are results showing that the buffer GaN layers are cubic in their as-grown state (Wu et al 1996) while other authors are of the opinion that the buffer layers are amorphous when deposited (Kuznia et al 1993 and Sugiura et al 1997). The structure of the as-grown and annealed GaN buffer layers are revealed in this paper to clarify this important step of the GaN growth.

Usually GaN/sapphire heteroepitaxial layers are grown on exact cut (0001) sapphire surfaces. Very few studies can be found in the literature on miscut substrates (Hiramatsu et al 1991) despite the fact that this is a general technique in the MOCVD growth of GaAs for example. Very little improvement of the layer structure is reported: only one paper is known to us (Grudowski et al 1996) which shows better optical properties for GaN layers grown on misoriented c-plane sapphire, although no structural information on the grown layers is given. In this paper GaN layers grown on miscut sapphire with specular surfaces are characterised and their defect density is shown to decrease by the optimisation of the growth parameters.

2. EXPERIMENTAL PROCEDURE

(0001) sapphire substrates were treated in ammonia at 1000°C before the deposition of the layers. This nitridation process leads to the formation of a thin AlN layer on the sapphire substrate (Pécz et al. 1997). Then a thin GaN nucleation layer was deposited at 510°C on both exact and miscut (3.5° off toward the [-2,1,1,0] axis) substrates by MOCVD. This was followed by about 1.5 μm thick GaN layers were grown at 600 mbar pressure. The growth temperature of the GaN layer varied between 960 and 1000°C. Triethylgallium (TEG) and ammonia (NH_3) were used as Ga and N sources.

Cross sectional and plan view specimens were prepared for TEM by ion milling.

3. RESULTS AND DISCUSSION

A cross section of a typical GaN layer is shown in Fig. 1. The layer was grown on a miscut sapphire substrate at 990°C. The nominal thickness of the buffer layer deposited at 510°C was 15 nm. The thick GaN layer is hexagonal and epitaxial to the substrate (see Fig. 1.c). The layer fulfils the most important technological requirement, as the surface is flat, which is observed even at lower deposition temperature (Pécz et al 1997) when miscut substrates are used. However, the defect density depends strongly on the growth temperature as well as on the buffer layer thickness. Therefore buffer layers with three different nominal thickness (11, 15 and 22.5 nm) were grown.

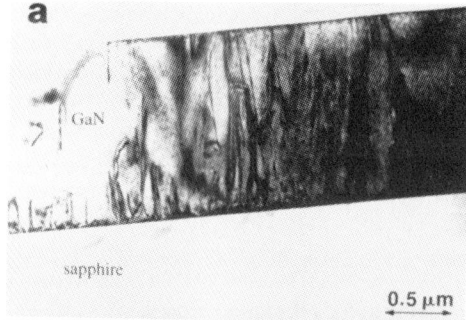
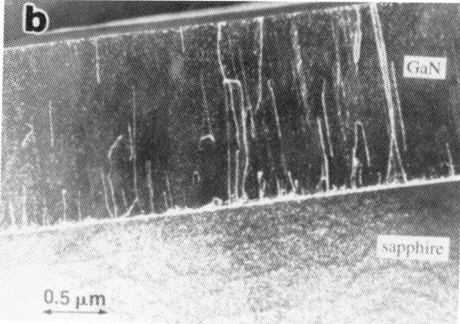
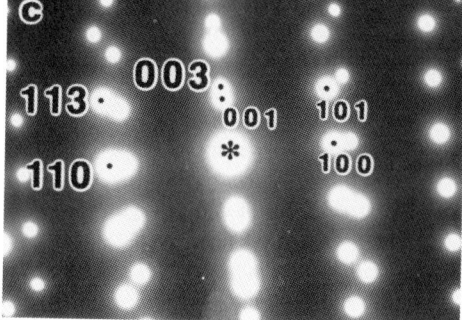

Fig. 1. GaN layer grown by MOCVD on miscut sapphire (0001) a: bright field, b: dark field (g=0004) micrographs, c: selected area diffraction pattern (SAED) showing the orientation relationship. (The larger indices belong to the substrate.)

The as grown buffer layer (nominal thickness is 22.5 nm) is continuous, but the surface is very rough (Fig. 2.a) and the mean thickness is less than the nominal one. The layer is single crystalline, hexagonal GaN (see Fig. 2b) with a mosaic structure. Despite the streaks in the diffraction pattern (Fig. 2b) the as grown buffer layer obeys the orientation relationship observed in thick GaN layers (see Fig. 1.c). The layer is composed of slightly misoriented grains. For the determination of the in-plane misorientation angle the rotation in Moiré images on plan view samples has been measured. The calculated misorientation is 3.3°, while the SAED patterns taken on the same sample gave one-two degrees higher value. This can be explained by the fact that SAED patterns average a larger area.

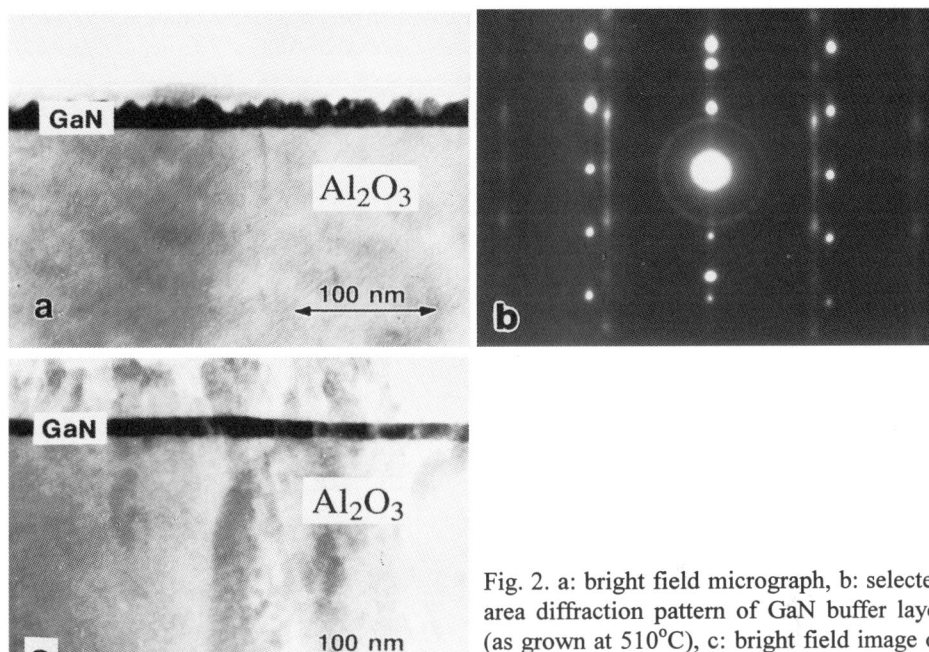

Fig. 2. a: bright field micrograph, b: selected area diffraction pattern of GaN buffer layer (as grown at 510°C), c: bright field image of the layer after annealing at 980°C.

After annealing the buffer layer up to 980°C (to the growth temperature of the thick GaN film) the surface became substantially smoother (Fig. 2.c). This is explained by the coalescence of the grains. A smoother layer provides an appropriate buffer to grow thick GaN layers with specular surfaces. Our experiments have shown that the thinnest buffer layer (with nominal thickness of 11 nm) is the smoothest after annealing. This means that the optimum thickness of the buffer layer is 11 nm.

Defect density of the thick GaN layers is usually measured by high resolution X-ray diffraction. Plan view TEM images are also informative from this point of view. Fig. 3.a shows the top region of a thick GaN layer grown at relatively low temperature (960°C) and onto a thick (22.5 nm, far from the optimum) buffer layer. A high defect density is revealed (Fig. 3.a). When the growth temperature was increased to 998°C and the thick GaN layer was deposited onto a thin (11 nm thick) buffer layer, the plan view TEM image (Fig. 3.b) shows substantial improvement in the defect (dislocations) density in the near surface region. When the whole layer was probed by high resolution X-ray technique, decreasing of the defect density to one third was observed.

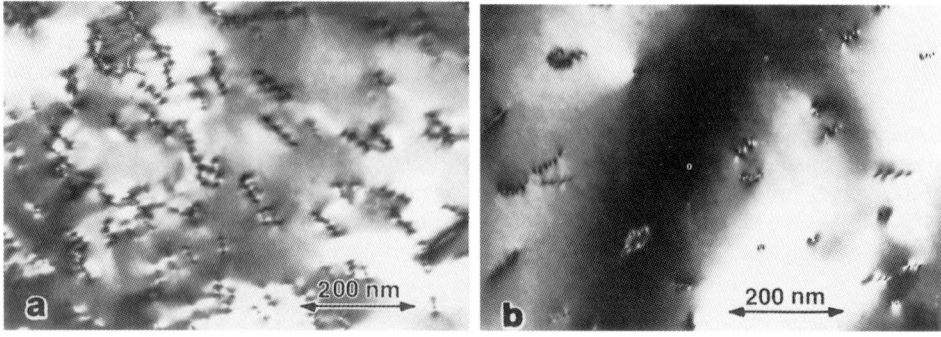

Fig. 3. Comparison of the defect structure of the top surface region of GaN layers grown on miscut sapphire a: at 960°C with 22.5 nm thick buffer and b: at 998°C with an 11 nm thick buffer layer.

4. CONCLUSIONS

GaN buffer layers grown at 510°C are hexagonal single crystals with a mosaic structure already in their as-grown state. Annealing of the buffer layers at 980°C leads to substantial smoothing of their surfaces due to the coalescence of the grains.

The defect density of the thick GaN layer substantially decreases when thinner (11 nm) buffer layers are used.

ACKNOWLEDGMENT

EU-98-B4-144, OTKA T 030447 projects and French-Hungarian bilateral co-operation (F-36/97) are acknowledged for financial support.

REFERENCES

Amano H, Sawaki N, Akasaki I and Toyoda Y 1986 Appl. Phys. Lett. **48**, 353
Grudowski P A, Holmes A L, Eiting C J and Dupuis R D 1996 Appl. Phys. Lett. **69**, 3626
Hiramatsu K, Amano H, Akasaki I, Kato H, Koide N and Manabe K 1991 J. Crystal Growth, **107**, 509
Kuznia J N, Khan M A, Olson D T, Kaplan R and Freitas J 1993 J. Appl. Phys., **73**, 4700
Pécz B, di Forte-Poisson M A, Tóth L and Radnóczi G 1997 Inst. Phys. Conf. Ser. **157**, 227
Sugiura L, Itaya K, Nishino J, Fujimoto H and Kokubun Y 1997 J. Appl. Phys. **82**, 4877
Wu X H, Kapolnek D, Tarsa E J, Heying B, Keller S, Mishra U K, DenBaars S P and Speck J S 1996 Appl. Phys. Lett. **68**, 1371

Inst. Phys. Conf. Ser. No 164
Paper presented at Microsc. Semicond. Mater. Conf., Oxford, 22–25 March 1999
© 1999 IOP Publishing Ltd

Comparison of the interfacial relaxation between AlN/Al$_2$O$_3$(0001) and AlN/6H-SiC(0001) studied with TEM

S Kaiser, M Jakob, J Zweck and W Gebhardt

Institute of Experimental and Applied Physics, University of Regensburg, D-93040 Regensburg, Germany

ABSTRACT: We investigate the interfacial relaxation of AlN on Al$_2$O$_3$ and on 6H-SiC by TEM. We find a typical dislocation density of 10^9cm^{-2} in a subsequently grown GaN layer for both substrates despite a clearly different lattice mismatch. HRTEM images of the AlN/Al$_2$O$_3$ interface reveal confined misfit dislocations which relax the mismatch efficiently to a residual strain $\varepsilon_r \approx -0.7\%$. For AlN/6H-SiC we only find non-confined dislocations as a result of the smaller misfit f(AlN/SiC)=-1.0%. Since ε_r and f(AlN/SiC) are almost equal, we expect a very similar defect density in both systems.

1. INTRODUCTION

GaN and the related III-V-nitrides are useful semiconducting materials for the fabrication of optoelectronic and high temperature electronic devices. Recently, GaN-based light emitting devices for the violet to green spectral range could be successfully realized using sapphire as substrate material (Nakamura 1995, 1997). On the other hand 6H-SiC(0001) seems to be an alternative substrate for the III-V-nitride epitaxial growth, because the in-plane lattice parameter a_{SiC}=0.308nm has a much smaller lattice mismatch f to GaN (f=-3.4%) and AlN (f=-1.0%) compared with f=-13.9% (GaN/Al$_2$O$_3$) and f=-11.7% (AlN/Al$_2$O$_3$) for the sapphire substrate. Meanwhile, there are many reports on TEM investigations of III-V-nitride samples grown on Al$_2$O$_3$ and 6H-SiC. They show dislocations and defects which appear either in both systems or in only one of them (Akasaki 1997, Chien 1996, Tanaka 1996). All these authors observe a very similar dislocation density of 10^8-10^9cm^{-2} in both systems, although there is a large difference in the corresponding lattice mismatch. We have already discussed the interfacial relaxation of GaN grown without any buffer layer on Al$_2$O$_3$(0001) and studied with TEM (Kaiser 1998a). We found in this case an efficient strain reduction by dislocations which are confined at the interface and which do not contribute to the defect density in the epilayer itself. In the present study we discuss the interfacial relaxation of AlN on Al$_2$O$_3$ as well as on 6H-SiC which seems to be a well-suited buffer material for the growth of III-V-nitrides on both substrate systems. High resolution TEM (HRTEM) investigations reveal confined dislocations at the interface of AlN/Al$_2$O$_3$ which do not extend into the layer. This relaxation reduces the mismatch efficiently down to a residual strain ε_r which is estimated from the observations. For the system AlN/6H-SiC, however, we did not find an equivalent relaxation process by confined dislocations. In this paper a model for the initial stage of growth is discussed which gives a plausible explanation for the almost equal defect density of 5-8·10^8cm^{-2} observed in Plan View TEM of these samples.

2. EXPERIMENTAL PROCEDURE

Epitaxially GaN films were grown on Al$_2$O$_3$(0001) as well as on 6H-SiC(0001) by molecular beam epitaxy (MBE) after the deposition of a 15-20nm thick AlN buffer layer in both cases. The growth direction is <0001> without any misorientation of the substrate surface. The AlN buffer layers were deposited at a substrate temperature of T$_s$=550-600°C and a growth rate of typically 60-100nm/h. For the subsequent GaN layer growth the temperature and the growth rate were increased to 780°C and 0.6μm/h. All investigated samples have a GaN layer thickness of about 500nm. For TEM investigations cross sectional samples were prepared in a conventional sandwich technique. The specimens were thinned mechanically to a thickness of about 20μm before ion milling is applied to reach electron transparency. Furthermore, Plan View TEM samples were prepared by mechanical polishing and ion milling only from the back side. All TEM observations were carried out with a Philips CM30 electron microscope (EM) operated at an accelerating voltage of 300kV.

3. RESULTS AND DISCUSSION

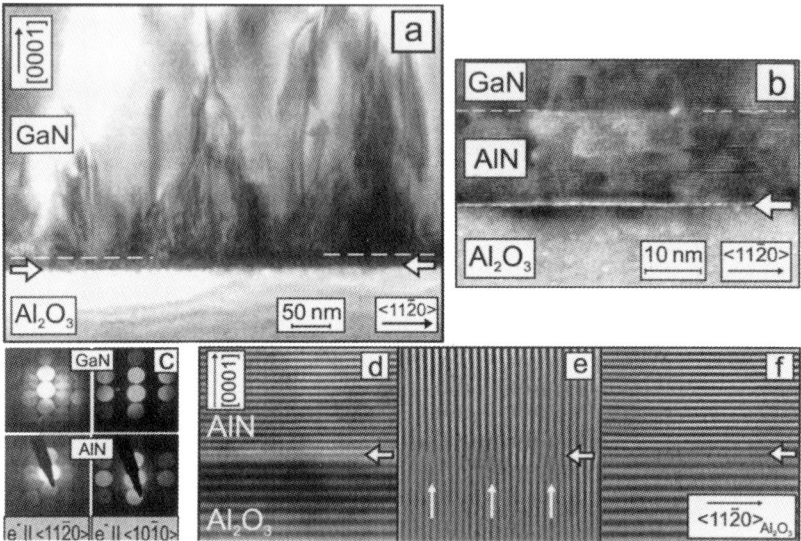

Fig.1: Cross sectional TEM images of GaN/AlN/Al₂O₃ showing almost the whole epilayer (a), the AlN buffer layer (b), corresponding diffraction patterns (c), a HRTEM image section of the interface (d) and corresponding Fourier filtered images visualizing the lattice planes (e, f).

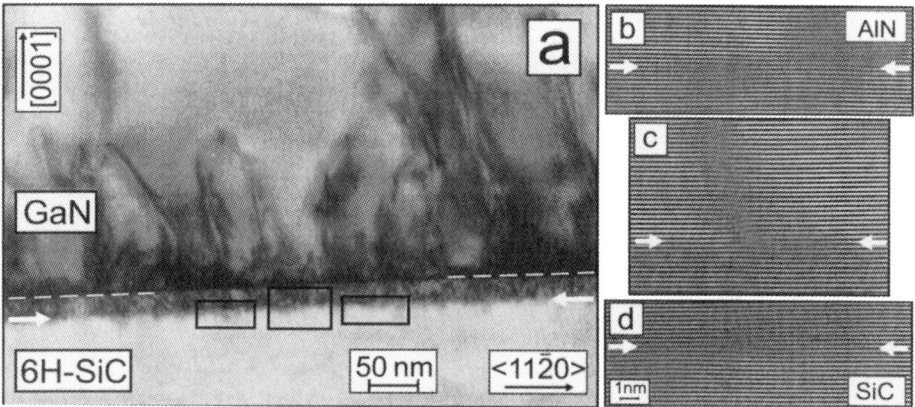

Fig.2: Cross sectional TEM images of the system GaN/AlN/6H-SiC showing almost the whole layer (a) and the (0001) lattice fringes of substrate and buffer at different positions (b-d).

Fig.1 shows TEM images of the cross sectional sample of GaN/AlN/Al₂O₃, whereas Fig.2 contains photographs taken from GaN/AlN/6H-SiC. The comparison of Fig.1a with Fig.2a gives the impression of a similar defect density in the upper region of both samples which could be confirmed by Plan View TEM investigations (without figures). Notice that these Plan View TEM results are comparable since the whole layer thickness is almost the same in both cases. In this way we determine a dislocation density of $8 \cdot 10^8 cm^{-2}$ for the system on Al_2O_3 and $5 \cdot 10^8 cm^{-2}$ for 6H-SiC. The observed dislocations are mainly threadings along the <0001> growth direction. Fig.1b shows a HRTEM image of the AlN buffer and of GaN. The corresponding convergent beam electron diffraction pattern in Fig.1c were taken separately in buffer and GaN layer using the nanoprobe mode of the EM. In order to visualize the lattice planes in buffer layer and substrate a Fourier filtering procedure was applied (Fig.1e,f) to a HRTEM image section of the AlN/Al$_2$O$_3$ interface (Fig.1d). The image reconstruction using the in-plane Fourier coefficients reveals terminating substrate fringes in regular distances of about 9.0±0.7 substrate spacings (Fig.1e). The perpendicular Fourier filtering (Fig.1f) visualizes the (0001) lattice planes in AlN and Al$_2$O$_3$ running parallel to the interface without terminating or bending of the fringes. A contrasting result is obtained from images taken from the interface between AlN and 6H-SiC (Fig.2b-d) which show terminating (0001)-AlN fringes with distances of 5-8nm.

3.1 GaN/AlN/Al$_2$O$_3$(0001)

For a correct interpretation of the Fourier filtered HRTEM images (Fig.1e,f) it is necessary to determine the structural quality of the buffer layer and its orientation relative to the Al$_2$O$_3$ substrate. This is quite important, because there are some reports about MOVPE grown low temperature AlN buffers on sapphire which show strongly faulted areas as well as misorientated and tilted grains (Akasaki 1997). In our case of MBE grown AlN the HRTEM images (Fig.1b) and the nanoprobe diffraction pattern (Fig.1c) reveal a monocrystalline structure of the buffer layer. Additionally, the correspondence of the diffraction pattern taken from AlN and GaN separately for both in-plane electron beam incidences shows the same in-plane orientation of AlN and GaN relative to Al$_2$O$_3$ (Fig.1c). Thus, we may interpret the Fourier filtered images Fig.1e,f in the same way as in the case of GaN/Al$_2$O$_3$ without buffer layer (Kaiser 1998a): The terminating $11\overline{2}0$ substrate fringes in Fig.1e show the edge components of misfit dislocations perpendicular to the electron beam. However, these misfit dislocations cannot have a Burgers vector component inclined to the interface since no bending or terminating of (0001) lattice fringes is observed (Fig.1f). Thus, the extension of these dislocations into the AlN layer seems to be suppressed and the dislocations are confined at the interface. In order to determine the efficiency of the relaxation, we count an averaged value of substrate lattice spacings between two terminating fringes n=9.0±0.7. It is now reasonable to divide the lattice mismatch f in two components $f=\delta+\varepsilon_r$ with δ the degree of relaxation and ε_r the residual strain. Since ε_r is a function of n (Kaiser 1998a) we estimate $\varepsilon_r=(-0.7\pm1.0)\%$. Therefore, we suggest the following model for strain reduction during the initial growth stage (Fig.3): The relaxation of AlN on Al$_2$O$_3$ starts with the growth of the first monolayer (ML) since the critical thickness due to the large mismatch of f=-11.7% is $h_c\leq1ML$. As long as the substrate surface is not completely covered with AlN, dislocations nucleate at island borders. They glide beneath the islands along the (0001) interface which acts as primary slip plane in the hexagonal crystal (Fig.3a,b). These dislocations relax the mismatch of the buffer layer, but do not contribute to the defect density. This results in a residual strain of -0.7%. A possible reason for the remaining strain could be a slight twist between adjacent islands, which depends on the growth temperature and growth rate (Kaiser 1998b). Another possibility maybe vertical steps at the interface which inhibit a complete relaxation by confined dislocations. After the island coalescence (Fig.3c) the strain energy of the layer increases with proceeding layer growth due to ε_r, until at a critical thickness

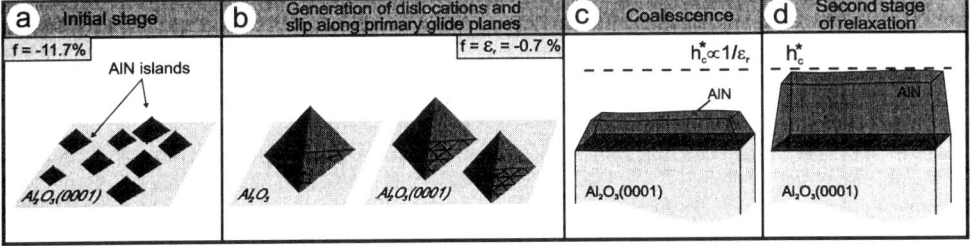

Fig.3: Schematic drawing of the relaxation of AlN on Al$_2$O$_3$(0001) during the initial stage of growth.

h_c^* dislocations are again generated (Fig.3d). In this second stage the relaxation by dislocations with glide planes parallel to the interface seems to be less effective, because they do not reduce the strain below the growth surface. Therefore, secondary glide planes inclined to the interface are activated to relax ε_r. The corresponding dislocation lines possess threading segments which are also inclined to the interface and contribute to the observed defect density of $8 \cdot 10^8 cm^{-2}$. We also note that the generation of basal stacking faults, which are frequently observed near the interface, is a further relaxation process along the (0001) primary glide planes. However, their contribution to the strain reduction is only small since their associated partial dislocations have only a small in-plane displacement.

3.2 GaN/AlN/6H-SiC(0001)

For AlN grown on 6H-SiC we do not observe any misorientation since they have the same crystal symmetry and a smaller lattice mismatch of f=-1.0%. The terminating (0001) AlN fringes in Fig.2b-d represent edge components of misfit dislocations with Burgers vectors inclined to the interface. These terminating (0001) fringes have distances of typically 5-8nm at the interface AlN/6H-SiC. Obviously, the dislocation confinement discussed above does not take place in this case. A possible reason may be the much larger critical thickness h_c due to a smaller mismatch: If the whole mismatch would be relaxed by confined dislocations only, we would expect terminating SiC substrate fringes perpendicular to the interface similar to Fig.1e. The distance between two terminating vertical fringes, given by n=-1/f=100 substrate lattice spacings, would in this case correspond to about 27nm. This is much larger than the observed distances of 5-8nm between two terminating (0001) fringes (Fig.2b-d). Using the relation of Matthews and Blakeslee (1974) we estimate a critical thickness of h_c=4.6nm for AlN/6H-SiC. Tanaka et al (1996) report on TEM investigations of pseudomorphic AlN buffer layers with a thickness of approximately 1.5nm which is smaller than h_c. These AlN layers possess a closed two-dimensional layer surface. With regard to our growth model discussed above the first relaxation of AlN on 6H-SiC starts when the substrate is completely covered with layer material (see Fig.3d). Again, the strain reduction at this stage of growth is only effective when secondary glide planes are activated. In contrast to AlN/Al₂O₃ the whole misfit f=-1.0% for AlN/6H-SiC is relaxed by non-confined dislocations which cause the observed terminating (0001) AlN fringes at the interface and contribute to the defect density of $5 \cdot 10^8 cm^{-2}$. Furthermore, the role of basal stacking faults is the same as in AlN/Al₂O₃.

4. CONCLUSION

TEM is used to investigate the interfacial relaxation of MBE grown AlN on Al₂O₃ as well as on 6H-SiC. We find an efficient relaxation by confined dislocations in the first case, whereas for AlN/6H-SiC the whole mismatch is relaxed by non-confined defects. The difference is discussed in a growth model which mainly depends on the critical thickness of each system. The comparison of the determined residual strain ε_r=(-0.7±1.0)% at the interface AlN/Al₂O₃ after formation of confined dislocations with the misfit f =-1.0% for AlN/6H-SiC reveals a very similar strain condition in both systems. Consequently, the subsequent relaxation causes an almost equal density of threading segments of $5-8 \cdot 10^8 cm^{-2}$, although there is a large difference in the corresponding mismatch.

REFERENCES

Akasaki I and Amano H 1997 Jpn. J. Appl. Phys. Vol.**36**, pp 5393
Chien F R, Ning X J, Stemmer S, Pirouz P, Bremser M D and Davis R F 1996 Appl. Phys. Lett. **68**, 2678
Kaiser S, Preis H, Gebhardt W, Ambacher O, Angerer H, Stutzmann M, Rosenauer A and Gerthsen D 1998a Jpn. J. Appl. Phys. Vol.**37**, pp 84
Kaiser S, Reinwald M, Riechert H, Averbeck R, Zweck J and Gebhardt W 1998b Proc. 14th Int. Conf. on Electron Microscopy (ICEM14), Cancun, Mexico, Vol.**III**, pp 447
Matthews J W and Blakeslee A E 1974 J. Cryst. Growth **27**, 118
Nakamura S, Senoh M, Iwasa N and Nagahama S 1995 Jpn. J. Appl. Phys. **34**, 794
Nakamura S, Senoh M, Nagahama S, Iwasa N, Yamada T, Matsushita T, Sugimoto Y and Kiyoku H 1997 Appl. Phys. Lett. **70**, 868
Tanaka S, Kern R S, Bentley J and Davis R F 1996 Jpn. J. Appl. Phys. Vol.**35**, pp 1641

Inst. Phys. Conf. Ser. No 164
Paper presented at Microsc. Semicond. Mater. Conf., Oxford, 22–25 March 1999
© 1999 IOP Publishing Ltd

Structural characterisation of AlGaN/AlN/Si(111) heterostructures by transmission electron microscopy

A M Sánchez, S I Molina, F J Pacheco, R García, M A Sánchez-García[+] and E Calleja[+]

Departamento de Ciencia de los Materiales e I.M. y Q.I., Universidad de Cádiz, Apdo. 40, 11510 Puerto Real (Cádiz), Spain
[+]Departamento de Ingeniería Electrónica, ETSI Telecomunicación, U.P.M., Ciudad Universitaria s/n, 28040 Madrid, Spain

ABSTRACT: The crystalline structure of AlGaN epilayers grown, with varying Al content, by Molecular Beam Epitaxy on AlN buffered (111)Si substrates was studied using Transmission Electron Microscopy techniques. The wurtzite AlGaN layers exhibited a mosaic structure where the presence of Al in the films produced a decrease of the subgrain size. A strong increase in threading dislocation density and a large improvement of surface quality has been observed in AlGaN epilayers, in relation to GaN epilayers grown under similar conditions.

1. INTRODUCTION

Ternary $Al_xGa_{1-x}N$/GaN heterostructures are the basis of HEMTs, high power and high temperature FETs, photodetectors and laser diodes emitting in the ultraviolet spectrum from 3.4 to 6.2 eV by varying the Al content (Nakamura et al 1993). UV LEDs are of interest due to their use in white light conversion, medical applications, solar UV monitoring systems and solar blind photodetectors (Morkoç et al 1994). AlGaN is also a promising material for device applications in the visible range when it is associated to InGaN.

On the other hand, the epitaxial growth of GaN on Si is attractive because the Si technology is well established. It has been found that by growing GaN directly on a Si substrate, the formation of an amorphous layer at the interface can take place, which is not desirable for good epitaxy (Othani et al 1994, Bourret et al 1998). However, by growing on AlN-buffered (111)Si substrates the crystal quality clearly improves (Meng et al 1994). Similarly, $Al_xGa_{1-x}N$ can be used as buffer layer for this kind of growth.

In this paper AlGaN/AlN/Si(111) heterostructures have been characterised by Transmission Electron Microscopy (TEM) and High Resolution Transmission Electron Microscopy (HREM). The influence of the Al content on the structural quality of the epilayer has been studied by comparing it with the GaN/AlN/Si(111) heterostructure.

2. EXPERIMENTAL

A series of samples consisting of AlGaN films have been grown by plasma-assisted Molecular Beam Epitaxy (MBE) on Si(111) substrates. An optimised AlN buffer layer (Sánchez-García et al 1998a, Calleja et al 1997 and Sánchez-García et al 1998b) has been grown on the Si substrate before the AlGaN growth. For comparison purposes, we also carried out measurements on a GaN film grown with the same conditions. The growth conditions of the studied samples are shown in table I.

TABLE I

Sample	N1	N2	N3
Growth Temperature (°C)	740	760	780
Thickness of AlGaN layer (μm)	1	1	0.6
Al content (%)	0	10	17

The described heterostructures have been characterised by Transmission Electron Microscopy (TEM) in a Jeol 1200-EX transmission electron microscope operating at 120kV and the High Resolution Transmission Electron Microscopy (HREM) work was performed on a Jeol 2000-EX working at 200kV. The specimens were mechanically thinned and then ion milled with Ar^+ at 4.5 kV. The Al content was deduced from photoluminescence (PL) and X-Ray Diffraction (XRD) measurements. Low temperature PL was measured with the 334 nm line of an Ar^+ laser using a Jobe-Yvon THR 1000 monochromator, a GaAs photomultiplier and a lock-in amplifier. XRD data were obtained with a Huber D5000HR.

3. RESULTS AND DISCUSSION

TEM, HREM and Selected Area Electron Diffraction (SAED) techniques have been used to investigate the crystal structure of the AlGaN films. These studies reveal that AlGaN has a hexagonal structure with the following relationship between the AlGaN layer and the Si substrate: $[1120]_{AlGaN}$ // $[110]_{Si}$ and $[0001]_{AlGaN}$ // $[111]_{Si}$. SAED patterns registered from areas of about 65 μm^2 of PVTEM specimens (figure 1) show the existence of a mosaic structure.

It is clear that the spots of the pattern of figure 1 are curved and elongated due to the in-plane misorientation between the subgrains of this mosaic structure. Such structure is typical of epitaxial films having a large lattice mismatch with their substrates (Srikant et al 1997).

The grain size was measured by the software "AMT^{TM}-VIDs V Semiautomatic Image Analysis System of Synoptics". The obtained grain sizes are presented in table II. The AlGaN grain size is clearly smaller than that of the GaN grains. The average size

Fig. 1. <0001> SAED pattern of the AlGaN layer.

grain of AlGaN layers was measured to be below 150 nm while it is larger than 200 nm in the GaN layer. Hence the grain size decreases as the Al content increases.

TABLE II

Sample	N1	N2	N3
Al content(%)	0	10	17
Average grain size (nm)	>200	<150	<100
Dislocation density (10^{10} cm^{-2})	0.01	≥1.0	≥1.0

The analysis of PVTEM images permits to obtain the density of dislocations that reach the free surface of the grown layer. Three types of perfect Burgers vectors can occur in wurtzite layers: $1/3<2\bar{1}\bar{1}0>$, $<0001>$ and $1/3<2\bar{1}\bar{1}3>$. The dislocation density measurements were carried out by accounting dislocations in PVTEM images registered in two beam conditions with $g=(11\bar{2}0)$ (figure 2). Hence the dislocations considered have Burgers vectors $b=1/3<11\bar{2}3>$ or $b=1/3<11\bar{2}0>$. In III-N layers, screw dislocations tend to form half loops, and therefore they are greatly reduced (Rouviére et al 1997). Table II shows the obtained dislocation density for the series of studied samples. It is clear that the dislocation density at the GaN layer free surface is smaller than that measured in the AlGaN layer. These results are in agreement with XRD results, as the AlGaN samples (N2 and N3) present XRD-FWHM values around 20 arcmin while in GaN sample (N1) the FWHM value from XRD measurements is 8.5 arcmin.

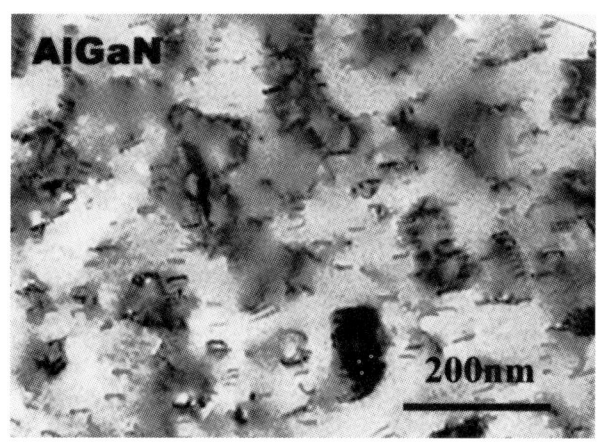

Fig. 2. Weak-beam PVTEM image registered under two beams condition with $g=(11\bar{2}0)$.

A high density of planar defects parallel to the (0001) growth plane has been obtained in sample N1. Some of these planar defects have been observed by XTEM to stop threading dislocations. This factor contributes to decrease the number of dislocations reaching the free surface of the film.

The HREM analysis shows that the interface between AlGaN and AlN is better defined than the GaN/AlN interface, probably due to the larger lattice mismatch of the latter interface. The $<11\bar{2}0>$ HREM images show some grains at the AlGaN/AlN interface that exhibit the sphalerite crystalline structure. These grains do not reach the free surface and often they are surrounded, in the AlGaN layer, by an area containing stacking faults.

The surface quality of AlGaN samples strongly improves relative to that of GaN layer (figure 3). The improvement in the surface smoothness makes $Al_xGa_{1-x}N$ a promising buffer layer to enhance the growth of GaN on Si.

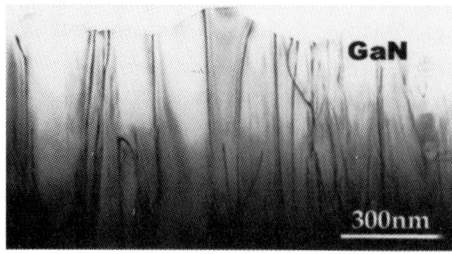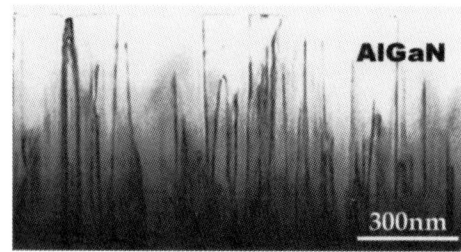

Fig. 3. XTEM images showing the smoothness of the GaN and AlGaN surfaces.

4. CONCLUSIONS

The Al content of the AlGaN epilayer grown by MBE on (111) Si substrates with an AlN buffer layer determines the structural quality of the grown film. SAED studies have shown the existence of mosaic structure in the AlGaN/AlN/Si(111) heterostructure. The average subgrain size of such mosaic structure decreases with the Al content. In relation to this AlGaN heterostructure, a decrease of two orders of magnitude in the dislocation density and a higher density of planar defects have been observed in a GaN/AlN/Si(111) heterostructure obtained using similar growth conditions.

ACKNOWLEDGEMENTS

This work has been supported by CICYT Project MAT98-0823-C03-02 and Junta de Andalucía (Group PAI TEP0120)

REFERENCES

Bourret A, Barski A, Rouvière J-L, Renaud G and Barbier A 1998 J. Appl. Phys. **83**, 2003

Calleja E, Sánchez-García M A, Monroy E, Sánchez F J, Muñoz E, Sanz-Hervás A, Villar C and Aguilar M 1997 J. Appl. Phys. **82**, 4681

Meng W J and Perry T A 1994 J. Appl. Phys. **76**, 7824

Morkoç H, Strite S, Gao G B, Lin M E, Sverdlov B and Burne M 1994 J. Appl. Phys. **76**, 1363

Nakamura S, Senoh M and Mukai T 1993 Appl. Phys. Lett. **62**, 2390

Rouvière J-L, Arlery M, Daudin B, Feuillet G and Briot O 1997 Mat. Sci. Eng. **B50**, 61

Sánchez-García M A, Calleja E, Sánchez F J, Calle F, Monroy E, Basak D, Muñoz E, Villar C, SanzHervás A, Aguilar M, Serrano J J and Blanco J M 1998a J. Electron. Mater. **27**, 276

Sánchez-García M A, Calleja E, Monroy E, Sánchez F J, Calle F, Muñoz E and Beresford R 1998b J. Cryst.Growth **183**, 23

Srikant V, Speck J S and Clarke D R 1997 J. Appl. Phys. **82**, 4286

Inst. Phys. Conf. Ser. No 164
Paper presented at Microsc. Semicond. Mater. Conf., Oxford, 22–25 March 1999
© 1999 IOP Publishing Ltd

Crystalline quality of $In_xGa_{1-x}N$ samples assessed by SEM, Raman and PL

R Correia, R Seitz, C Gaspar, T Monteiro, E Pereira, M Heuken[1], O Schoen[1] and H Protzmann[1]

Department of Physics, University of Aveiro, Campus Universitário de Santiago, 3810 Aveiro, Portugal
[1]AIXTRON, Kackertstr. 15-17 - D-52072 Aachen, Germany

ABSTRACT: In this work a series of InGaN samples 0.1 μm thick grown by metalorganic vapour phase epitaxy are analysed by SEM, micro-Raman and photoluminescence. The mole fraction of indium is determined by microanalysis. Sample morphology varies widely in different samples, ranging from good uniformity to cellular structure, where hexagonal pits are observed inside the cells. The composition dependences of experimental PL peak position and band gap energy (E_g), are studied and compared with the results found in the literature. Raman spectra show that all samples are strained affecting the band gap edge emission.

1. INTRODUCTION

The group III nitrides (AlN, GaN and InN) form a complete series of ternary alloys, with a band gap varying from 1.95-6.2 eV. The fact that they also generate efficient luminescence has been the main driving force for recent technological development. In particularly, $In_xGa_{1-x}N$ can be tuned over a wide range from visible red to ultraviolet by changing the alloy composition and by forming heterostructures such as quantum wells (Nakamura and Fasol 1997). However several factors influence the optical properties of the layers. Namely it is known that high indium mole fraction causes phase separation and poor crystal quality (Keller et al 1998).

GaN-based devices as blue light-emitting diodes (LED´s) and laser diodes (LDS) require an $In_xGa_{1-x}N$ active layer. In order to improve the performance of devices high quality layers must be obtained.

2. EXPERIMENTAL PROCEDURE

$In_xGa_{1-x}N$ layers of 0.1 μm thickness (S255#) were grown on 1.8 μm thick GaN:Si on c-plane sapphire by metal-organic vapour epitaxy (MOVPE). The growth temperature of the GaN:Si base layer was 1200 °C, the $In_xGa_{1-x}N$ layer was grown at 700-850 °C, with an indium mole fraction ranging from 3.5% to 20.3 % as indicated by electron probe microanalysis (EPMA) measurements. The indium fraction was determined using analyses performed on a homogeneous area of the film. Another set of samples of $In_xGa_{1-x}N$ (S20#), with 0.2 μm, was used for comparison purposes.

Steady state photoluminescence (PL) spectra were obtained after excitation above the band gap by using a HeCd laser as the excitation source (325 nm). Absorption measurements at 11K were carried out using an experimental set up with a Tungsten arc lamp as light source.

Raman spectra were taken at room temperature (RT) using a 514.5 nm line of an argon ion laser in backscattering Z (XX) Z geometry. The laser beam was focused on the sample with a spot of 2 μm. Peak positions are determined with an accuracy of 1 cm^{-1}.

3. RESULTS AND DISCUSSION

3.1 Observations by SEM

Typical scanning electron microscopy (SEM) images are shown in Fig. 1. These images show that the thicker samples have a much smoother morphology (Fig. 1a), while in the thinner samples (Figs. b to f) besides smooth continuous films hexagonal pits can be observed. Some "islands" are also observed on the surface, with diameters varying from 1 μm to 10 μm.

The 0.1μm series shows a progressive increase of pit number density with increasing indium concentration.

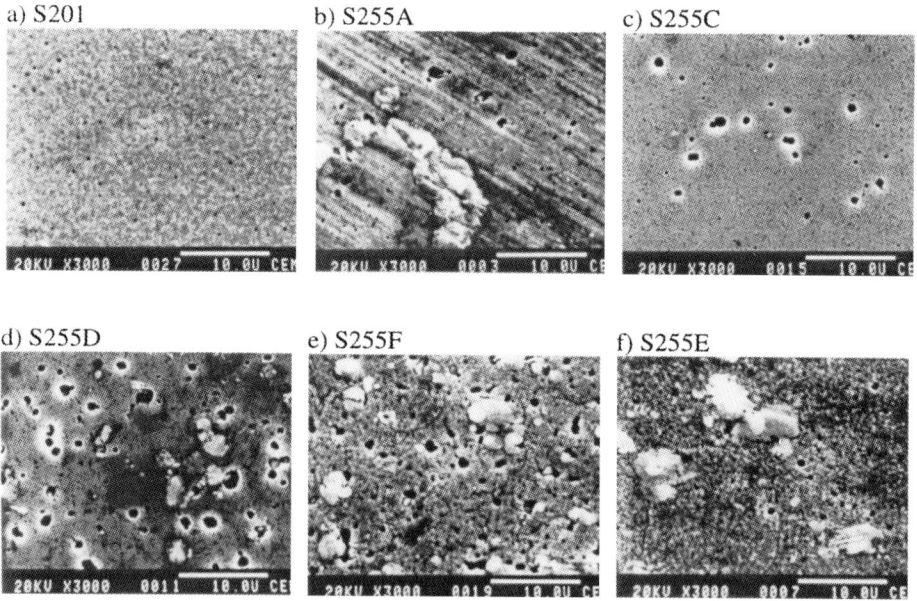

Fig. 1. Surface SEM images of In$_x$Ga$_{1-x}$N grown on a GaN: Si epitaxial layer.

By EPMA it is shown that the islands are indium droplets, attributed to indium precipitation in the final stage of growth. The precipitates are clearly indium rich as compared to the surrounding matrix (Fig. 2).

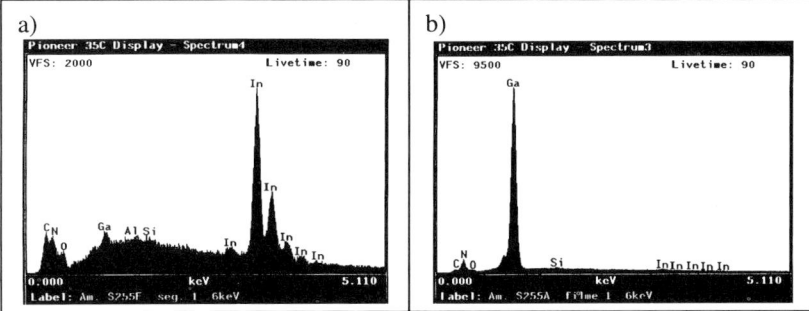

Fig. 2. a) X-ray spectra in the "islands regions" (Sample S255F); b) X-ray spectra in " film region"(Sample S255F).

Cross-sectional TEM observations carried out by Hiramatsu et al (1997) describe the hexagonal pits (Fig. 3 a), suggesting they are due to threading dislocations which penetrate into the $In_xGa_{1-x}N$ layer. With increase of indium content a cellular structure appears, as is apparent from Fig. 3 b. Inside the cells, small pits (~100 nm) are observed.

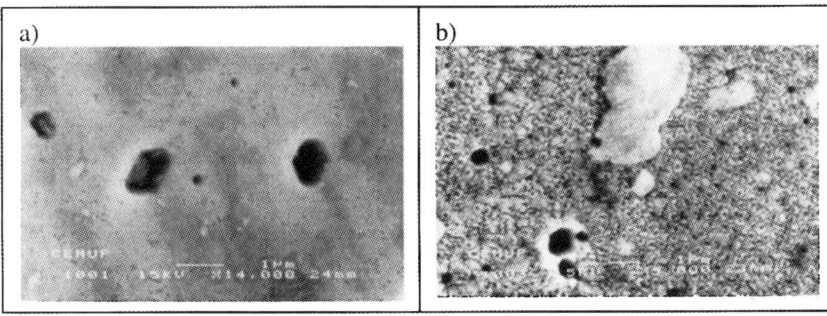

Fig. 3. SEM images of $In_xGa_{1-x}N$ grown on a GaN:Si epitaxial layer (a) Sample-S255A with 3.6% In; (b) Sample-S255E with 13.4% In.

EPMA has been used to evaluate the indium concentration in the different samples. A correct measure of the actual indium content is an open question. Rutherford backscattering spectroscopy is considered to be the most reliable method to measure the indium fraction. It is known that XRD overestimates the indium fraction, since this estimation is obtained assuming that Vegard's law is valid (Nakamura and Fasol 1997). Nevertheless the contraction of the unit cell with changing composition may not be the same for all three axes in the presence of biaxial strain that results from the lattice mismatch between $In_xGa_{1-x}N$ and GaN. The biaxial compression in the $In_xGa_{1-x}N$ layer leads to an increase in the lattice spacing along the c-axis (McCluskey et al 1998).

EPMA may underestimate the indium fraction if a pit is included in the interaction volume, or overestimate it if the area includes regions where pulling effects took place (Hiramatsu et al 1997). Actually, when we compare XRD and EPMA data on the same sample (0.2 μm thick) we found a higher value for EPMA (28.9% for EPMA versus 25.7% for XRD with a sample of the S20# series).

3.2 Observations by Raman and PL

Due to the strain in the films the use of optical measurements, such as absorption, Raman shift of modes and excitonic luminescence to determine the indium content of layer may lead to incorrect results.

The $In_xGa_{1-x}N$ films have been confirmed to have the Wurtzite crystal structure (Osamura et al 1975). Raman active phonons belong to $A_1(Z)$, $E_1(X,Y)$ and E_2 modes. Several studies refer to the way in which the linewidths and the frequencies of the phonon lines and selection rules for scattering may contain information on the quality of the films and strain in the films (Kirillov et al 1996). The literature concerning Raman spectra from $In_xGa_{1-x}N$ films and InN is poor.

In order to evaluate the effect of strain in the films, Raman spectra have been taken for both sets of samples. Taking account of the geometry used we may observe only the allowed E_2 and A_1 (LO) modes, but a small misalignment enables forbidden modes also to be observed, namely the E_1 (LO) mode. In spite of the E_2 (high) mode being more intense, no shift was observed with increase of molar fraction of indium. The dispersion observed in frequency was within the measurement precision.

For comparison we have obtained the linear interpolation between GaN and InN for the E_1 (LO) mode, using the values 742 cm^{-1} (Siegel et al 1997) and 694 cm^{-1} (Edgar 1994), respectively. The observed trend and the lack of InN localised modes suggested this alloy exhibits one-mode behaviour, in agreement with the study carried out by Hayashi et al (1991) for $Al_xGa_{1-x}N$ films. The 0.1 µm samples are systematically shifted to higher energies, indicating that in these films strain is more important, as expected. However it is expected that the shift in band gap is not larger than 10 meV (Hiramatsu et al 1997).

Typical PL spectra are shown in Fig. 4 for samples S255C and S255F. It is apparent that when the morphology is good the peaks are narrow and all the emission is fast. However when the film morphology shows more pits and grain boundaries, the peak becomes broader and a second slow transition can be identified. In Fig. 5 we present a plot of the PL peak position of the higher energy emission as a function of concentration.

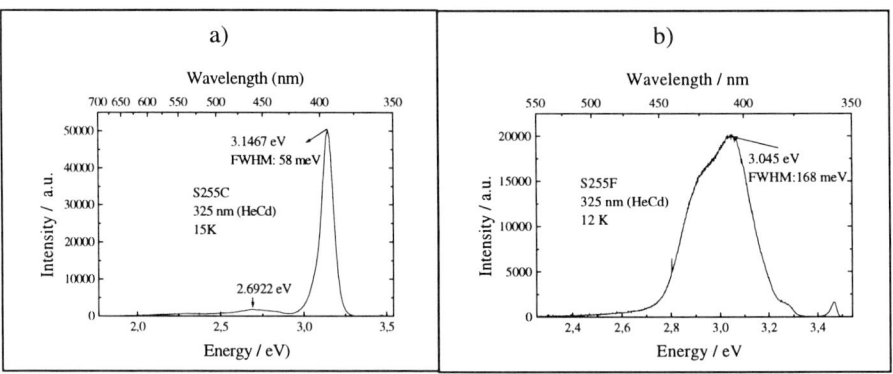

Fig. 4. PL spectra at 11K after above band gap excitation: a) S255C; b) S255F.

Absorption spectra were also measured at 11K. However, due to band tailing attributed to inhomogeneity in sample composition, it is difficult to use the data to derive the band gap. Band gap values derived from absorption spectra are shown in Fig. 5, assuming an error of

± 20 meV. It is also noticed that the best morphology gives a steeper increase of the band gap. Photoreflectivity could not be measured in these samples, as no free exciton is observed.

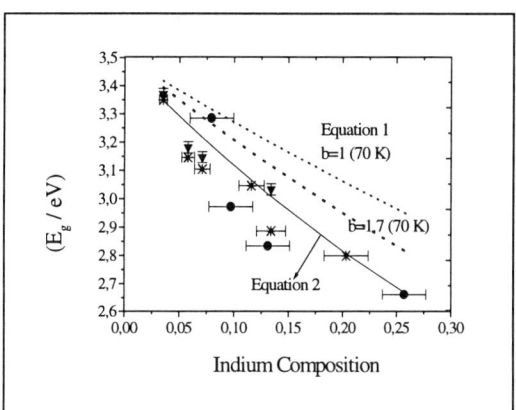

Fig. 5. Composition dependence of PL peak position and band gap obtained by absorption measurements. ● PL (S20#); ▼- absorption (S255#) ; ✱-PL (S255#).

Band gap values for $In_xGa_{1-x}N$ have been calculated by several authors, assuming an empirical bowing parameter (Nakamura and Fasol 1997; Shan et al 1998). A shift of the band gap with temperature has also been reported (Osamura et al 1975; Orton et al 1997). Higher bowing parameters are associated with disorder (Shan et al 1998).

(1) (Nakamura and Fasol 1997) $E_g(x) = xE_{gInN} + (1-x)E_{gGaN} - bx(1-x)$

(2) (Shan et al 1998) $E_g(x) = 3,42 - 3,91 E_{gGaN} - 2,39x^2 + 0,06,$

where 60 meV was taken as the difference in band gap energies between RT and 70 K.

By analogy with GaN we have assumed that the temperature dependence of the band gap energy of InN, between 11 K and 78 K, is affected to a small extent. Therefore, we used an E_{gInN} of 2.11eV obtained by Osamura et al (1975), to estimate the band gap at 11K by equation (1). The 0.1 μm series shows a near systematic shift of PL peak position and band gap (Fig. 5). This is particularly observed when the curve given by equation (2) is taken into account. The difference between the $E_g(11K)$ calculated, and the $E_g(RT)$ given by equation (2) is of the order of 200 meV. Although the 0.2 μm samples have in general narrower lines, they also have a larger dispersion between band gap and peak position. This problem is now being analysed in more detail.

4. CONCLUSIONS

Films of $In_xGa_{1-x}N$ with a thickness of 100 nm have a morphology that shows pits, hexagonal in shape, while thicker films present a more continuous structure. Cellular structure becomes important for higher indium concentrations. EPMA gives values for the indium concentration, which agree within 11% with the values measured by XRD. Raman spectra confirm that strain decreases with film thickness. Samples with the best morphology give narrower luminescence peaks closer to the band gap.

ACKNOWLEDGMENTS

This work was partially supported by the European communities under contract BRPR-CT96-034. Also financial support by JNICT, contract PRAXIX XXI PBIC/C/CTM/1925/95 is acknowledged. The authors are grateful to Professor Carlos Sá from CEMUP for helpful discussions about SEM measurements.

REFERENCES

Edgar J H 1994 Properties of Group III Nitrides (London: INSPEC) pp 254
Hayashi K, Itoh K, Sawaki N and Akasaki I 1991 Solid State Commun. **77**, 115
Hiramatsu K, Kawaguchi Y, Shimizu M, Sawaki N, Zheleva T, Robert F D, Tsuda H, Taki W, Kuwano N and Oki K 1997 MIJ-NSR **2**, article 6
Keller S, Keller B P, Minsky M S, Bowers J E, Mishra U K, DenBaars S P and Seifert W 1998 J. Cryst. Growth **189/190**, 29
Kirillov D, Lee H and Harris J S, Jr 1996 J. Appl. Phys. **80**, 4058
McCluskey M D, Romano L T, Krusor B S, Bour D P, Johnson N M and Brennan S 1998 Appl. Phys. Lett. **72**, 1730
Nakamura S and Fasol G 1997 The Blue Laser Diode – GaN Based Light Emitters and Lasers (Berlin: Springer) pp137
Orton J W and Foxon C T 1998 Rep. Prog. Phys. **61**, 1
Osamura K, Naka S and Murakami Y 1975 J. Appl. Phys. **46**, 3432
Shan W, Walukiewicz W, Haller E E, Little B D, Song J J, McCluskey M D, N M Johnson, Feng Z C, Schurman M and Stall R A 1998 J. Appl. Phys. **84**, 4452
Siegle H, Kaczmarczyk G, Filippidis L, Litvinchuk A P, Hoffmann A and Thomsen C 1997 Phys. Rev. B **55**, 7000

Inst. Phys. Conf. Ser. No 164
Paper presented at Microsc. Semicond. Mater. Conf., Oxford, 22–25 March 1999

Investigation of composition fluctuations in In$_x$Ga$_{1-x}$N

B Neubauer, E Hahn, A Rosenauer, D Gerthsen and M Heuken*

Laboratorium für Elektronenmikroskopie, Universität Karlsruhe, Kaiserstraße 12, 76128 Karlsruhe, FRG
*AIXTRON AG, Kackertstr. 15-17, 52072 Aachen, FRG

ABSTRACT: The chemical compositions of In$_x$Ga$_{1-x}$N/GaN quantum wells are studied with high-resolution transmission electron microscopy (HRTEM) on an atomic scale. The samples were grown by metal organic chemical vapour deposition (MOCVD) on Al$_2$O$_3$(0001) substrates. The In contents were varied by lowering the substrate temperature but with otherwise unaltered growth conditions. Composition fluctuations and average In contents are assessed by the analysis of the local lattice parameters with the evaluation software DALI (digital analysis of lattice images) on the basis of HRTEM images.

1. INTRODUCTION

InGaN/GaN heterostructures are the basis of optoelectronic devices like light emitting diodes and lasers (Nakamura et al 1997). For InGaN as the active layer in a laser structure the potential emitted wavelength is tunable nearly across the entire visible range because the bandgap energy can be lowered from 3.39 eV to 1.89 eV with increasing In content. The difficulty of growing homogeneous InGaN layers with high In concentrations is induced by a miscibility gap predicted for $x_{In} >$ 10% by thermodynamic calculations (Ho et al 1996, Matsuoka 1997). Therefore, composition inhomogeneities occur which affect the device properties. In order to optimise the efficiency, suitable growth conditions to suppress the decomposition have to be found and the relation between the composition profile of the active layer and the emitted wavelength has to be understood. HRTEM images of cross-section samples evaluated with the DALI procedure (Rosenauer et al 1996) allow the analysis of composition fluctuations in ternary layers on an atomic scale. Three samples with different In contents were investigated.

2. EXPERIMENTAL

All samples were grown by metal organic chemical vapour deposition (MOCVD) on Al$_2$O$_3$(0001) substrates. First, a 2 µm GaN:Si (n = 1.5 x 10^{18} cm^{-3}) buffer layer was deposited at a substrate temperature of 1170°C to obtain a low defect density. The subsequent quantum well structure consists of about 3 nm In$_x$Ga$_{1-x}$N embedded in two 7 nm thick In$_{0.02}$Ga$_{0.98}$N layers, which is capped with 10 nm GaN. The growth temperature of the In$_x$Ga$_{1-x}$N layer was 840°C, 820°C and 800°C whereas other growth conditions as the growth temperature of the cladding layers (T= 1110°C) and the gas fluxes were unaltered. The emitted wavelengths measured with room temperature photoluminescence (PL) with a HeCd laser (325 nm) were 416 nm, 449 nm and 484 nm. Cross-sectional samples for the HRTEM investigations were prepared in a conventional manner by grinding, dimpling and Ar$^+$-ion milling at 6 kV, 1 mA and an incident angle of 7° in a specimen holder cooled with liquid nitrogen.

The samples were viewed close to a $<1\bar{1}00>$-zone axis in a Philips CM 200 FEG/ST electron microscope at an acceleration voltage of 200 kV by imaging conditions as follows: the samples were tilted out of the zone axis orientation about $5°$ in $<11\bar{2}0>$ direction in such a way that only beams of type $\{0002n\}$ are excited. In order to reduce the influence of the sample thickness and defocus on the image pattern the objective aperture contained the (0000) and the $\pm(0002)$ beams only. The photographic negatives were digitised with an off-line CCD camera at a resolution of 1024 x 1024 pixels with a grey scale depth of 12 bits per pixel. Only dislocation free areas were selected for the further evaluation.

3. EVALUATION PROCEDURE

The HRTEM images were evaluated with the DALI program package including the following steps: First, the images are Fourier-transformed and the noise is reduced by a Wiener filtering procedure. Only the regions surrounding the $\pm(0002)$ reflections and the zero-frequency pixel were used for the inverse Fourier-transform whereas other regions were deleted. Assuming a completely strained layer these $\{0002\}$-plane distances are sufficient for the calculation of the In content. The next step is the generation of a two-dimensional grid with one set of grid lines running along the $\{0002\}$ fringes and the second one perpendicular to them. For that purpose, the DALI procedure detects intensity maxima positions along linescans perpendicular to the $\{0002\}$ fringes. The distance of the linescans is arbitrary and in our case approximately equal to the $\{0002\}$-fringe distances. Note that the actual positions of the $\{0002\}$-lattice planes need not be known with respect to the intensity maxima positions along the fringes, if the sample thickness does not significantly change in the investigated area. Reference $\{0002\}$-lattice plane distances d_r are computed in a region of pure GaN. Local $\{0002\}$-plane distances d_j are measured and normalized with respect to d_r yielding $D = d_j/d_r$. The lattice mismatch of the $\{0002\}$ planes of the binary materials is defined by $f_0:= (d_{InN} - d_{GaN})/d_{GaN}$. According to Vegard's law $d_{InGaN} = d_{GaN} + x \cdot (d_{InN} - d_{GaN})$, the relation between the normalized distance D and the In content x is given by

$$x = \frac{1}{f_0} \cdot \frac{1}{1+K} \cdot (D-1), \tag{1}$$

where K is the factor for the tetragonal distortion (Eq. 2) that depends on the elastic constants c_{ij}. Along the electron beam direction, a complete relaxation of the tetragonal distortion occurs for very thin TEM samples which modifies K according to Eq. (3).

$$K = 2 \cdot \frac{c_{13}}{c_{33}} \tag{2}$$

$$K = \frac{c_{13}}{c_{33}} \cdot \left(1 - \frac{c_{12}c_{33} - c_{13}^2}{c_{11}c_{33} - c_{13}^2}\right) \tag{3}$$

We find $K = 0.19 \pm 0.035$ in the limit of a very thin specimen ("thin foil approximation") and $K = 0.55 \pm 0.067$ in the limit of a bulk sample ("thick foil approximation"). The calculated values for K are almost independent of the In content (e.g. van Schilfgaarde et al 1997).

For the calculation of local displacements in growth direction, a reference lattice is calculated inside the reference area and superimposed on the whole image. The local displacement then is given by the distance between the local intensity maximum position and the corresponding reference lattice position. The maximum displacement u_{max} induced by the InGaN layer is linked with the total amount C of deposited InN by:

$$C = u_{max} \cdot \frac{1}{f_0} \cdot \frac{1}{1+K} \quad \text{in monolayers (ML) of InN.}$$

4. RESULTS

Figure 1 shows normalized distances D averaged along the {0002} fringes as well as a grey scale coded map of local In concentrations calculated according to Eq. (1) for the InGaN quantum well with an emitted wavelength of $\lambda = 416nm$.

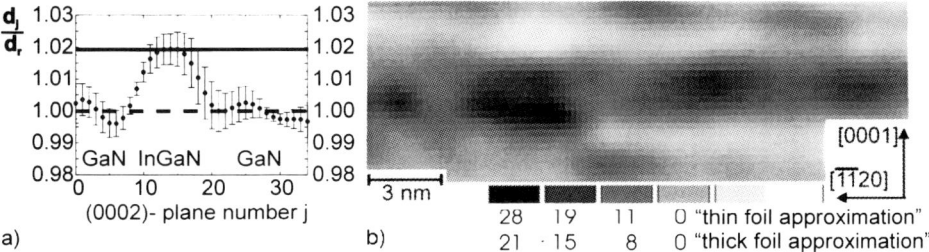

a)   b)

28	19	11	0	"thin foil approximation"
21	15	8	0	"thick foil approximation"

Fig. 1. *InGaN quantum well with* $\lambda = 416nm$; *a) averaged {0002}-plane distances as a function of the ML number in growth direction; b) local In- concentrations in [%].*

The maximum of the averaged normalized lattice distances shown in Fig. 1a corresponds to an In content of $x_{max} = (14 \pm 6)\%$. The average content calculated inside the region that is marked as InGaN in Fig. 1a) yields a mean concentration of $x_{mean} = (12 \pm 4)\%$. The maximum local value obtained from Fig. 1b is $x_{max}^{local} = (25 \pm 14)\%$. The errors are calculated by considering the unknown sample thickness and deviations of the elastic constants in the literature. The total amount of deposited In is (1.2 ± 0.3) ML InN. The mean In content varied between $(17 \pm 6)\%$ and $(3 \pm 4)\%$ with $x_{mean} = 9\%$ averaged over all images.

The InGaN layer with $\lambda = 449nm$ (Fig. 2) showed larger In concentrations.

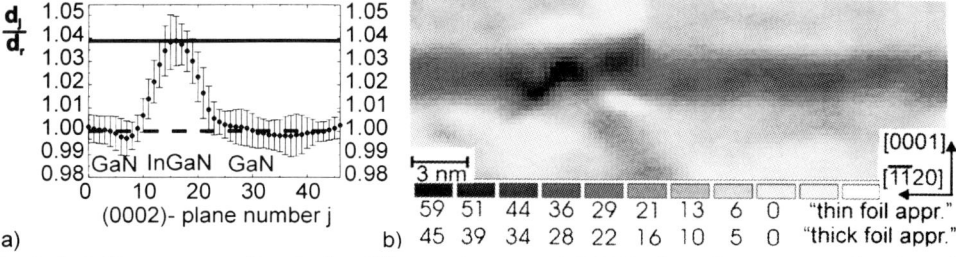

a)   b)

59	51	44	36	29	21	13	6	0 "thin foil appr."
45	39	34	28	22	16	10	5	0 "thick foil appr."

Fig. 2. *InGaN quantum well with* $\lambda = 449nm$; *a) averaged {0002}-plane distances as a function of the ML number in growth direction; b) local In-concentrations in [%].*

In Fig. 2, the maximum In content is $x_{max} = (29 \pm 7)\%$. The average In content is $x_{mean} = (24 \pm 5)\%$, and the maximum local In content reaches $x_{max}^{local} = (52 \pm 17)\%$ which is exceptionally high for this sample. The evaluation of the displacements yields a total amount of deposited In of (2.5 ± 0.3) ML InN. From the evaluation of all images we obtain an averaged In content of $x_{mean} = 18\%$ with a minimum and maximum of $(6 \pm 2)\%$ and $(29 \pm 7)\%$, respectively.

Figure 3 shows an example for the quantum well emitting the longest wavelength of $\lambda = 484nm$ which would indicate a higher In content than in the other two samples. However, the measured In contents are smaller than those found in the sample with $\lambda = 449nm$. The selected area shows a maximum averaged In content of only $x_{max} = (15 \pm 3)\%$ and a mean value of $x_{mean} = (10 \pm 1)\%$. The maximum local In contents in this region did not exceed $x_{max}^{local} = 23\% \pm 10\%$.

410

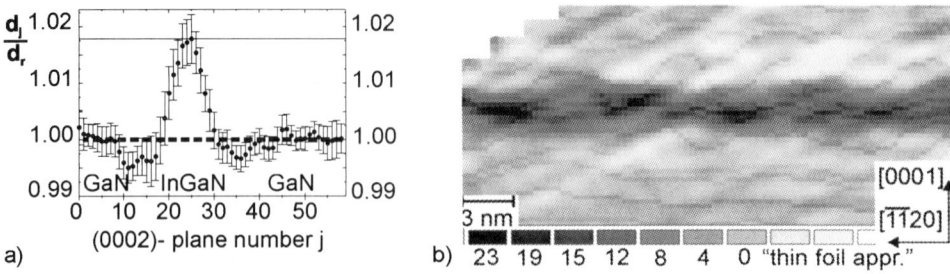

Fig. 3. *InGaN quantum well with* $\lambda = 484nm$ *; a) averaged {0002}-plane distances as a function of the ML number in growth direction; b) local In-concentrations in [%].*

The total amount of deposited In is equivalent to 1.1 ML InN as calculated from the displacements. An average In concentration of $x_{mean} = 17\%$ was determined from all HRTEM images with minimum and maximum values of $(10 \pm 3)\%$ and $(36 \pm 8)\%$. A decrease of the GaN lattice parameter compared to the bulk lattice parameter is observed in the white areas of Figs. 1b)-3b) which can be explained by the distortion close to regions with particularly high In contents.

5. DISCUSSION

The experimental results show that the In incorporation can be influenced by only changing the substrate temperature under otherwise unaltered growth conditions - in particular unaltered gas fluxes. All samples show significant composition fluctuations as predicted in the literature. A crude estimation of the In concentration can also be obtained from PL data assuming homogeneous and unstrained InGaN which yields $x_{In} = 19\%$ for T= 840°C, $x_{In} = 30\%$ for T= 820°C and $x_{In} = 40\%$ for T= 800°C. However, the average In concentrations determined by DALI are at least 10% lower. A possible explanation for this discrepancy is a red shift of the emitted wavelength due to the piezoelectric field caused by the strain of the quantum well (quantum confined Stark effect, Takeuchi 1998). This field tilts the valence and conduction bands which yields a bandgap transition with decreased emission energy.

6. CONCLUSION

In conclusion, we observed significant fluctuations of the In content in all investigated samples. The strain state analysis was facilitated by the application of an off-axis imaging condition that decreases the influence of variations of sample thickness and defocus.

ACKNOWLEDEGMENT

This work is supported by the Deutsche Forschungsgesellschaft under contract number Ge 841/5.

REFERENCES

Nakamura S and Fasol G 1997 "the blue laser diode", Springer
Ho I-hsiu and Stringfellow G B 1996 Appl. Phys. Lett. **69**, 2701
Matsuoka T 1997 Appl. Phys. Lett. **71**, 105
Rosenauer A, Kaiser S, Reisinger T, Zweck J, Gebhardt W 1996 Optik (Stuttgart) **102**, 63
van Schilfgaarde M, Sher A and Chen A-B 1997 J. Cryst. Growth **178**, 8
Takeuchi T 1998 J. Cryst. Growth **189/190**, 616

Inst. Phys. Conf. Ser. No 164
Paper presented at Microsc. Semicond. Mater. Conf., Oxford, 22–25 March 1999
© 1999 IOP Publishing Ltd

Energy selected elemental analysis of InGaN multiple quantum wells in GaN

M W Fay, D J Norris, C J D Hetherington, A G Cullis, P J Parbrook, C R Whitehouse and M Schurman[1]

University of Sheffield, Department of Electronic and Electrical Engineering, Mappin Street, Sheffield S1 3JD UK
[1] Emcore Inc. Elizabeth Avenue, Somerset, New Jersey, NJ 08873, USA

ABSTRACT: Multiple quantum wells of InGaN have been grown within GaN epitaxial layers by metal-organic vapour phase epitaxy. Cross-sectional EFTEM imaging and EDX analysis have been used to determine the spatial distribution of In and Ga within these well structures. We shall show how these techniques may be used to evaluate elemental distributions on the nm-scale. Fluctuations in EFTEM contrast are observed in In elemental maps, but it is necessary to remove effects of observed point defect clustering in order to quantify In concentration variations.

1. INTRODUCTION

GaN is becoming acknowledged as possibly one of the most important semiconductor materials other than silicon. This is substantially due to the wide band-gap and efficient optical emission properties of this material that enable optoelectronic devices to operate in the blue and UV regions of the electromagnetic spectrum (Nakamura et al 1997).

Alloying GaN with In makes it possible to tune devices to shorter wavelengths. Laser devices have been produced from these materials employing multiple InGaN quantum wells in GaN (Nakamura et al 1997). The high quantum efficiency seen in these layers is believed to be related to large localisation of electrons (Chichibu et al 1997) which may originate from In rich regions of the wells, which act as quantum dots (Narukawa et al 1997). The self-formation of these structures may be due to interface fluctuations, but might also be a result of the nature of the InGaN alloy (Ho et al 1996). It is therefore of vital interest to determine the extent of such roughness and compositional fluctuations.

In this work, we have used high-resolution energy-filtered transmission electron microscope imaging and X-ray analysis to examine the In and Ga elemental distributions in a series of InGaN/GaN multiple quantum wells.

2. EXPERIMENTAL METHODS

Samples consisting of 10 InGaN wells (3nm wide), separated by 10nm thick GaN layers were grown by metal-organic vapour phase epitaxy (MOVPE) on sapphire. Cross-sectional specimens were prepared by mechanical polishing and argon ion milling to electron transparency. These were examined using a JEOL 200CX transmission electron microscope operating at 200KV for lower magnification imaging. A JEOL 2010F field emission gun TEM operating at 200KV, equipped with a Gatan Imaging Filter (GIF) and an Oxford Instruments Energy Dispersive X-ray (EDX) detection system was used for high resolution TEM, electron energy loss spectroscopy (EELS), energy filtered TEM (EFTEM) and EDX analyses.

3. RESULTS AND DISCUSSION

Low magnification bright-field [1010] images showing the entirety of the layer structures, from the GaN layer/Al$_2$O$_3$ substrate interface (a) to the InGaN/GaN multiple quantum wells (b) have been taken, and a typical region is shown in Fig. 1. Significant defect clustering occurs at the Al$_2$O$_3$/GaN interface, with some threading dislocations (~10^8 cm^{-2}) extending to the free surface. A higher magnification bright field image, showing more clearly the MQW stack, is shown in Fig. 2. A series of 10 InGaN QWs (~3nm) are separated by GaN (~10nm) spacer layers. During initial imaging the layers appeared uniform in contrast; however, after some exposure (~20s), fluctuations in the contrast became increasingly visible within the InGaN

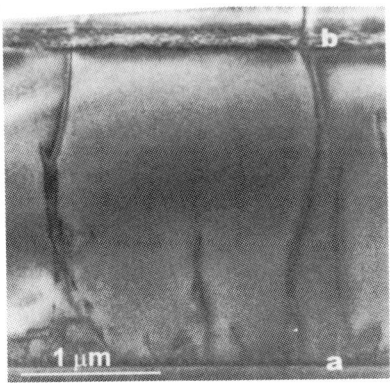

Fig.1 Bright field image showing the InGaN quantum wells in the GaN.

Fig.2 High magnification bright field image of the InGaN quantum wells.

layers. Such fluctuations are believed to be associated with the formation of point defect clusters in the InGaN wells. This also explains the strain field variations observed within the surrounding GaN spacer layers.

For a semi-quantitative analysis of the In distribution within the InGaN QWs, EELS and EFTEM imaging have been performed. One major difficulty associated with these methods, however, is the close proximity of the N-K (401eV) and In-M (443eV) edges within the EEL spectrum and this complicates EFTEM imaging. For the present work, we have performed EELS upon a 3nm InGaN QW, and an example background-subtracted spectrum (in the 200-1200eV energy loss range) is shown in Fig. 3, which clearly exhibits appreciable In-M$_{4,5}$ signals from which EFTEM images could be derived.

The In distribution has been obtained using EFTEM jump-ratio imaging and a typical image is shown in Fig. 4. The In distribution within the QWs appears fairly uniform; however, we have observed small-scale fluctuations in the In concentration on the nanometre-scale.

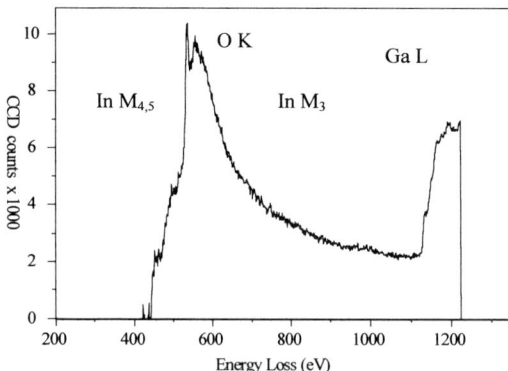

Fig.3 Background subtracted EELS spectrum showing the onset of the In M edges.

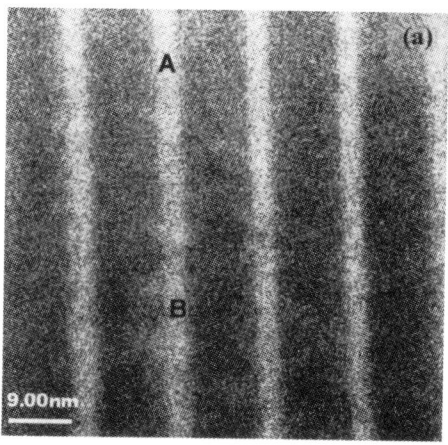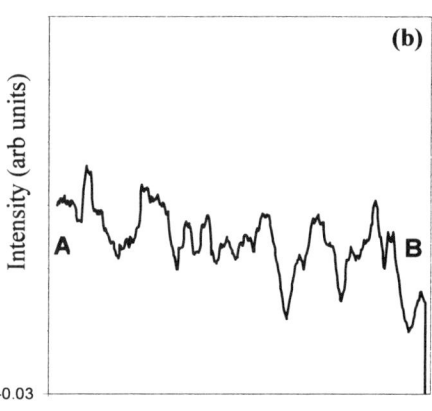

Fig.4 (a) - EFTEM image of InGaN quantum wells: Higher In concentration appears bright. (b) - Profile of the In distribution along the well as labelled in (a).

In Fig. 4(b), an intensity profile of the In signal - taken along the well (A-B) - shows more clearly these small-scale fluctuations: their spacing appears to be of the order of the QW width. These are consistent with the presence of In clusters; however, it is necessary to remove possible contrast features associated with beam induced point defect clustering, as observed in Fig. 2.

As an alternative to EFTEM, small-probe (~0.5nm) EDX elemental mapping has been performed to determine the In distribution. Typical elemental maps using the N-K, In-L and Ga-K X-ray peaks are shown in Fig. 5. While the N elemental map (Fig 5a) appears fairly featureless, the Ga map (Fig. 5c) shows a Ga-deficit at the region of the QWs. The In map (Fig 5b) shows clearly a series of bright lines that correspond to the In concentration within the InGaN QWs. However, the distribution appears broader than that observed in EFTEM maps and therefore does not provide sufficient resolution to enable precise composition fluctuations to be determined.

To assess the spatial resolution of these chemically-sensitive techniques, In elemental profiles across the wells derived from EFTEM and EDX elemental maps have been compared. The EFTEM In profile, shown in Fig 6a, exhibits a series of well-defined In-peaks having QW widths of ~4nm which are comparable to the widths obtained from BF images. In addition, there are plateau regions of ~10nm between the In well profiles, which clearly indicate that resolution at the nm-scale has been achieved. The corresponding EDX profile across the same wells, shown in Fig 6b, is less well resolved with no discernible central plateaux region between the In wells.

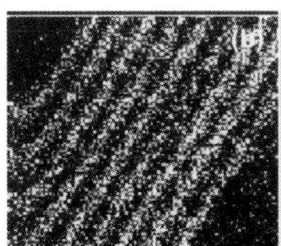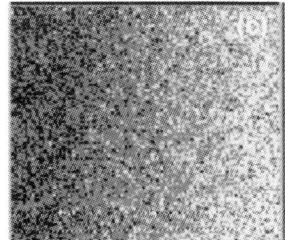

Fig.5 EDX maps of InGaN quantum wells, showing the distribution of (a) N, (b) In and (c) Ga

414

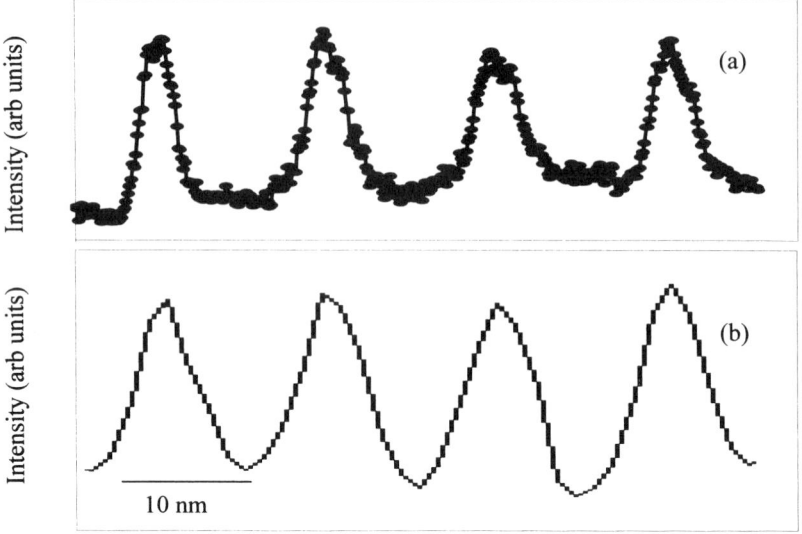

Fig.6 (a) - Profile of the In distribution across the wells, derived from the EFTEM image seen in Fig. 4a (b) - EDX line profile of InGaN quantum wells using the In-L peak for X-ray linescan.

5. CONCLUSIONS

EFTEM and EDX imaging techniques have been used to determine the spatial distribution of In within InGaN/GaN MQWs. In EFTEM images, small-scale fluctuations have been observed in the In distribution laterally along the QW regions. However, it is necessary to remove from these images possible artefacts associated with electron beam damage. EDX imaging is not able to resolve the In distribution within the well to the same extent as the EFTEM method due to beam broadening effects.

ACKNOWLEDGEMENTS

MWF, DJN and CJDH would like to acknowledge the EPSRC for financial support.

REFERENCES

Chichibu S, Wada K and Nakamura S 1996 Apply Phys. Lett. **71,** 2346
Ho I-H and Stringfellow G B 1996 Appl. Phys. Lett. **69**, 2701
Nakamura S and Fasol G 1997 "The Blue Laser Diode" (Springer: Berlin)
Narukawa Y, Kawakami Y, Funato M, Fujita S, Fujita S and Nakamura S 1997 Appl. Phys. Lett. **70**, 983

Inst. Phys. Conf. Ser. No 164
Paper presented at Microsc. Semicond. Mater. Conf., Oxford, 22–25 March 1999
© 1999 IOP Publishing Ltd

A TEM assessment of GaN/SiC layers grown by MBE

A N Bright, P D Brown, D M Tricker, N Jeffs[a], C T Foxon[a] and C J Humphreys

Department of Materials Science and Metallurgy, University of Cambridge, Pembroke Street, Cambridge, CB2 3QZ, UK
[a] Department of Physics, Nottingham University, University Park, Nottingham, NG7 2RD, UK

ABSTRACT: We report TEM and AFM observations of GaN films deposited by plasma assisted molecular beam epitaxy directly on to 6H-SiC substrates without buffer layers. When substrate cleaning procedures are optimised, single crystal wurtzite GaN films can be grown, with dislocation densities of between $4*10^{10}$ and $3*10^{11} cm^{-2}$. The dislocations are primarily edge in character. The films also contain prismatic stacking faults. The GaN/SiC interface is free of amorphous phase, as evidenced by the presence of misfit dislocations. The effects of nucleation temperature and substrate vicinality on film quality and morphology are studied.

1. INTRODUCTION

The properties of GaN, which include a wide band gap and a large electron velocity, make it promising as a material for high power, microwave frequency amplifiers. To fully exploit these properties, a substrate with high thermal conductivity is needed to transmit heat away from the device, as thermal effects are expected to be limiting. A candidate substrate is SiC, and high quality GaN films grown by Molecular Beam Epitaxy (MBE) directly on SiC substrates have already been reported as reaching state of the art quality even without a buffer template (Ploog et al 1998). MBE is thus a potential source of GaN for device applications, though its development is still some way behind that using Metal Organic Chemical Vapour Deposition. The initial nucleation stage is of crucial importance in the successful growth of GaN films on SiC substrates, and is still not well understood. We are growing and characterising films of varying thicknesses down to the nanometre-scale with a view to elucidating the mechanisms of nucleation and growth as a function of temperature, substrate preparation and vicinality. In this paper we describe the results for continuous films of intermediate thicknesses (approx 100nm) deposited on 6H-SiC directly by plasma-assisted MBE.

2. EXPERIMENTAL

Initial growth experiments were performed on chemically prepared on-axis (0001) SiC substrates (Si face), aiming to optimise the first few monolayers of growth. A standard three step chemical etch was used ($NH_4OH:H_2O_2$, then $HCl:H_2O_2$, then HF). To prevent nitridation of the SiC surface by the active nitrogen, and the consequent formation of amorphous Si_3N_4 at the interface, Ga alone was deposited initially, and the nitrogen plasma was started only when several monolayers of Ga were present on the SiC surface. The nitrogen shutter was then opened to react with the Ga to form GaN. The temperature of nucleation was varied (table 1) to find a compromise between high temperature nucleation, where the Ga immediately desorbs from the SiC surface, and low temperature nucleation, where the initial film quality is poor. RHEED was used in the MBE system to help determine when Ga coverage had been successful, and when initial GaN had been formed. In every case, growth subsequent to the nucleation stage was performed at 810°C (the nominal temperatures quoted in this paper are thought to be accurate to within +50°C).

Once the nucleation conditions had been optimised, the effect of substrate vicinality was investigated by growing on 3.5° and 8° off-axis as well as on-axis SiC substrates (table 1). For all these, the standard chemical etch was used. As a further experiment, a different substrate cleaning procedure was attempted, where a general degrease was performed using standard solvents and an ultrasonic bath, and removal of the SiO$_2$ layer on the SiC was attempted by deposition of Ga without the nitrogen plasma, then heating to allow the Ga to desorb, taking the oxide layer along with it and leaving a clean SiC surface behind.

The films were characterised using Bright Field and Dark Field Transmission Electron Microscopy, Diffuse Dark Field Imaging, High Resolution Imaging (HREM) and Atomic Force Microscopy (AFM).

Specimen	Substrate vicinality	Substrate preparation	Nucleation temperature	Cell size/nm PVTEM /AFM	Dislocation density/cm^{-2}
MS40	On-axis 6H-SiC	chemical etch	no nucleation step	45 / 118	8*10^{10}
MS41	On-axis 6H-SiC	chemical etch	low N power at 810°C	60 / undefined	3*10^{11}
MS42	On-axis 6H-SiC	chemical etch	570°C	71 / undefined	7*10^{10}
MS43	On-axis 6H-SiC	chemical etch	620°C	80 / 133	5*10^{10}
MS44	On-axis 6H-SiC	chemical etch	700°C	64 / 130	6*10^{10}
MS45	3.5° off axis 6H-SiC	chemical etch	700°C	130 / 200	4*10^{10}
MS46	8° off axis 4H SiC	chemical etch	700°C	75 / 133	not measured
MS47	3.5° off axis 6H-SiC	Ga desorption	700°C	130 / undefined	not measured

Table 1. List of observed samples

3. EFFECT OF SUBSTRATE PREPARATION

The films grown on substrates prepared by the standard chemical etch were found by TEM to be uniformly single crystalline wurtzite GaN, adopting the following orientation relationship with the SiC: $(0001)_{GaN} // (0001)_{SiC}$ and $[11\bar{2}0]_{GaN} // [11\bar{2}0]_{SiC}$. The success of the chemical cleaning procedure is demonstrated by the absence of amorphous phase at the interface (fig. 1) and by the presence of misfit dislocations. Not only can these be surmised from HREM images such as fig. 1, but their strain fields can, with care, be imaged in dark field plan view, when the specimen contains the interface itself (fig. 2). Fringes running parallel to $<1\bar{2}10>$ can be seen, of spacing around 80Å. We are unable to match this to any possible moiré fringe spacing in this system. The expected misfit dislocation spacing for a relaxed interface is 78Å, which corresponds very well with the observed spacing of 80Å above. This is also the approximate spacing of dislocations measured from HREM images such as fig. 1.

By contrast, the film grown on the substrate that had been cleaned by Ga desorption (MS47) contained significant amounts of cubic material, as seen in HRTEM (fig. 3) and as evidenced by the inclined {111} microtwins and stacking faults seen in cross section (fig. 4).

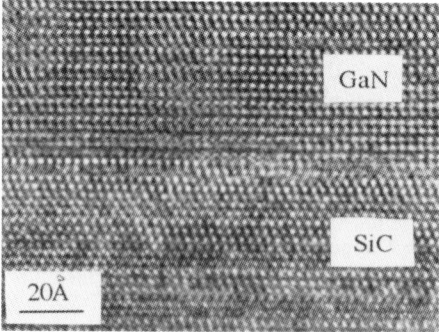

Fig. 1. HREM of the interface region of MS42 showing a clean interface with misfit dislocations.

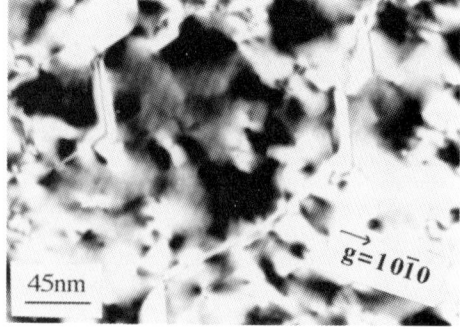

Fig. 2. Dark Field TEM image showing misfit dislocations in plan view from MS44.

These inclined stacking faults were not seen in the other samples (fig. 5), and along with high densities of basal plane stacking faults throughout the film (not just at the interface), suggest that the cleaning process had adversely affected the film microstructure. It is worth noting, however, that no amorphous layer could be unambiguously identified at the interface of this sample either by HREM or by diffuse dark field imaging.

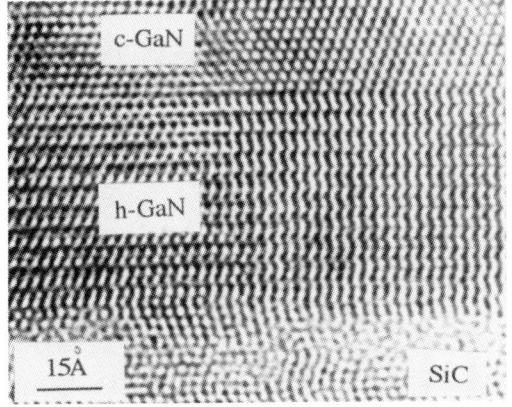

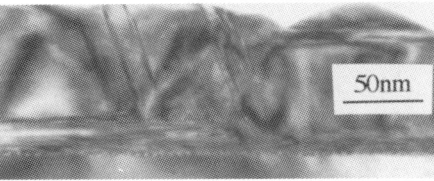

Fig. 4. Bright field image of MS47 with basal plane & inclined stacking faults.

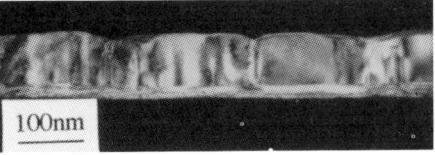

Fig. 3. HREM of MS47 showing a region of cubic GaN growing on wurtzite in the near-interface region.

Fig. 5. Dark field, MS44 cross section showing a cusped surface morphology.

4. DEFECT MICROSTRUCTURE AND FILM MORPHOLOGY

Threading dislocations in all the films were characterised by standard **g.b** analysis, and were shown to be predominantly of edge character. Measured dislocation densities were high, from $4*10^{10}cm^{-2}$ to $3*10^{11}cm^{-2}$, with the majority of dislocations located in cell boundaries. Prismatic stacking faults on the $\{11\bar{2}0\}$ planes were present in all the films, with varying densities (figs. 6 and 7), and basal plane stacking faults were commonly seen in HREM images of the interface region. These microstructures were similar to those previously observed in MBE GaN grown on GaAs, GaP and sapphire substrates (Tricker et al 1997).

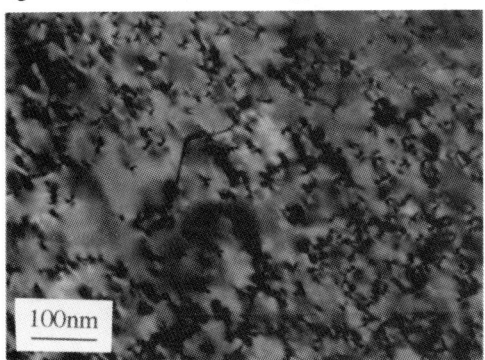

Fig. 6. Bright field image of MS41 with an ill-defined cell structure and high dislocation density.

Fig. 7. Plan view bright field TEM of MS45 showing dislocations and prismatic stacking faults.

The surface morphology, as observed by AFM, shows variation from sample to sample, between very bumpy structures (fig. 8) and a surface characterised by flat regions, surrounded by trenches or cusps (fig. 9). The change in scale of this surface morphology from specimen to specimen as nucleation and growth conditions are changed is found approximately to correlate with the change in cell size as viewed in plan view TEM (table 1), although the 'cell size' of surface morphology is consistently larger than the TEM cell size. The reason for this consistent

418

difference has not yet been established. Cross-sectional TEM observations suggest that the surface cusps correspond to the emergence of dislocations and stacking faults (fig. 5) which are thought to be linked to the initial coalescence of islands in the nucleation layers. Preferential re-evaporation at dislocations or stacking faults could account for much of the observed surface morphology. The cells in fig. 9 contain small craters whose density (around $5*10^9 cm^{-2}$) suggests that these may be the emergence points of dislocations not located in the cell walls.

Fig. 8. AFM image of the surface of MS41.   Fig. 9. AFM image of the surface of MS45.

5. EFFECT OF NUCLEATION TEMPERATURE AND SUBSTRATE VICINALITY

5.1 Nucleation Temperature: Nucleation was optimised by observing RHEED patterns in the MBE system to determine what layers (amorphous layers or Ga layers/crystalline GaN etc.) were present on the SiC surface. Although for samples MS40-44 an amorphous layer was observed by RHEED, no amorphous phase was detectable at the GaN/SiC interface either by HREM or by diffuse dark field imaging. The cell size observed in the final films is expected to correspond to island size during the nucleation process. Although it was difficult to accurately define a cell size for many of these samples, samples MS44-46 were generally of larger cell size and lower dislocation density than MS40-42. The conclusion from the RHEED observations was that 700°C was the optimum nucleation temperature, although subsequent observations by TEM showed that 620°C nucleation led to a similar cell size and dislocation density.

5.2 Substrate Vicinality: Three GaN films were grown under similar conditions except for a change in the substrate vicinality (MS44-45-46, see table 1). The 3.5° off-axis sample (MS45, figs 7 and 9) has the largest cell size and the lowest dislocation density, but was found to have a high density of stacking faults in the HREM and plan-view images. However, the variation in growth morphology between these samples is not large.

6. CONCLUSIONS

Eight 100nm GaN films have been grown by plasma-assisted MBE on vicinal and on-axis (0001)6H-SiC and examined using TEM and AFM with a view to understanding the nucleation process for GaN on SiC. Effective substrate cleaning is essential to ensure single crystal GaN growth. No amorphous phase was detectable at the GaN/SiC interface using TEM. All the films contain stacking faults and high densities of threading dislocations, mostly of edge-type, which define cell boundaries. The lateral dimension of surface morphology as measured by AFM scales with cell size as viewed by TEM, but is consistently larger. No clear relationship between microstructure and substrate vicinality has been established.

REFERENCES

Ploog K H, Brandt O, Yang H, Yang B and Trampert A 1998 Journal of Vac. Sci. and Tech. B, **16(4)** 2229
Tricker D M, Natusch M K H, Boothroyd C B, Xin Y, Brown P D, Cheng T S, Foxon C T and
    Humphreys C J 1997 Inst. Phys. Conf. Ser. **157** 217
Vennegues P, Beaumont B, Vaille M and Gibart P 1997 J.Crystal Growth **173** 249

AB acknowledges Marconi Electronic Systems and the EPSRC for funding.

Inst. Phys. Conf. Ser. No 164
Paper presented at Microsc. Semicond. Mater. Conf., Oxford, 22–25 March 1999
© 1999 IOP Publishing Ltd

Excitonic transitions in cubic AlGaN

G Salviati, C Zanotti-Fregonara†, M Albrecht°, N Armani, S Christiansen°, H P Strunk°, H Angerer^, O Ambacher^ and M Stutzmann^

Istituto MASPEC-CNR, Parco Area delle Scienze 37/A, I-43100 Parma, Italy
† also with Centro Interdipartimentale Materiali e Tecnologie dell'Informazione, Parco Area delle Scienze, I-43100 Parma, Italy
°Università Erlangen, Institut für Werkstoffwissenschaften, Mikrocharakterisierung Cauerstr.6 D-91058 Erlangen, Germany
^Walter Schottky Institut, Technische Universitaet Muenchen Am Coulombwall, D-8574 Garching, Germany

ABSTRACT: This paper reports on the study by means of cathodoluminescence and high resolution transmission electron microscopy of both structure and excitonic transitions in zincblende $Al_xGa_{1-x}N$ (0.5<x<1) layers. The layers are grown by plasma induced molecular beam epitaxy by purpose at 810°C onto sapphire (0001) layers. It is shown that excitonic transitions can be observed in the whole range of compositions. In the cubic phase they are characterised by at least two main optical transitions separated by up to ~ 200 meV, of energy ranging from 4.33 and 4.45 eV (x=0.52) to 4.62 and 4.85 eV (x=0.81). By plotting the near band edge emission vs the alloy composition, a band gap bowing of about 1 eV is found. From this plot, a band gap for zincblende AlN at ~ 5.41eV is estimated instead of the theoretically calculated 4.9 eV. The full width at half maximum of the optical transitions vs Al concentration for hexagonal and cubic materials are also compared. A sharp increase of the full width at half maximum between x=0.52 and x=0.58 is found; it is ascribed to the increase of the intervalley scattering probability resulting from the occurrence of an indirect band gap.

1. INTRODUCTION

Optical and electronic   properties of group-III nitrides and their alloys, which are interested for the production of optoelectronic devices (Nakamura 1997), essentially depend on composition and crystal structure. The two main polytypes that have been reported for single crystalline heteroeptaxial layers are (i) the wurtzite and (ii) the zincblende structure. Despite the large interest for optoelectronic devices working in the ultraviolet region of the spectrum (Al concentration > 30%), at present few experimental data exist on excitonic transitions in AlGaN with high Al concentration (see for instance for cubic AlGaN   Okumura et al 1998 and for hexagonal AlGaN Perry et al 1998) and, to the best of our knowledge, nothing is known about excitonic transitions in cubic AlGaN with Al concentrations larger than 50%. In this paper we will show how the complementary use of structural and optical investigation techniques having high spatial and analytical resolution, like transmission electron microscopy and cathodoluminescence in the scanning electron microscope, can give information on the correlation between the onset of an indirect band-gap and excitonic transitions in cubic AlGaN epitaxial layers.

We will show by high resolution transmission electron microscopy (HRTEM) that $Al_xGa_{1-x}N$ (0.5<x<0.8) layers grown by plasma induced molecular beam epitaxy result in a predominant cubic phase. We will also show that excitonic transitions can be observed in the whole range of compositions. From the strong increase of the FWHM compared to that of the wurtzite phase we will conclude the bandgap is indirect for x>0.52.

2. EXPERIMENTAL

Not intentionally doped cubic $Al_xGa_{1-x}N$ (x=0.58<x<0.8) with a thickness of 1μm has been grown by purpose at growth temperatures as low as 810 °C by plasma induced molecular beam epitaxy on (0001) oriented sapphire substrates (Angerer et al 1997). The Al concentrations of the layers were determined by high resolution X-ray diffraction by reciprocal space maps of asymmetrical reflections. The data were confirmed by elastic recoil detection analyses.

The optical properties of the layers have been assessed by CL investigations performed between 5K and 300 K, using different energies of the incident electron beam and different beam currents. We used a commercial MonoCL system from Oxford Instruments equipped with a multialkali photomultiplier fitted in a 360 Stereoscan SEM from Cambridge Instruments (Albrecht et al 1997). Conventional and HRTEM investigations were performed on cross sectional samples (cut in the <11-20> projection) in a Philips CM300 and in a Philips CM300 UT (point to point resolution d = 0.165 nm) operating at 300 KV respectively.

3. RESULTS AND DISCUSSION

Conventional TEM cross sectional bright field micrographs of the layer with x=0. showed the usual high density of threading dislocations (~ 1.5 10^{10} cm^{-2}). HRTEM investigations of sample with x=0.58 revealed the layer had a predominant cubic phase (see the 2 sets of {111} lattice planes in (Fig. 1a). A high density of stacking faults (SFs), consisting of a single layer of

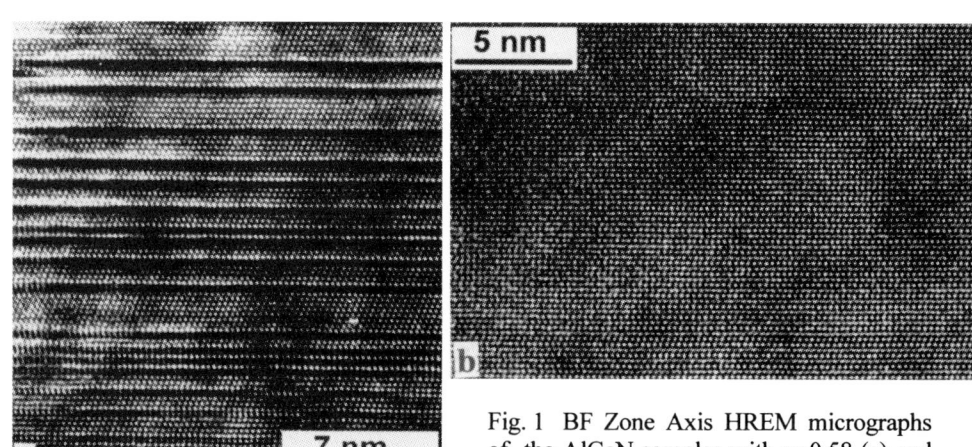

Fig. 1 BF Zone Axis HREM micrographs of the AlGaN samples with x=0.58 (a) and x=0.50 (b)

wurtzite AlGaN embedded in the host cubic lattice parallel to the (0001) substrate surface has been found to be present in the whole layer. The SFs end at threading dislocations that have a Burgers vector with a c-component (a=1/3<11-23>, a=<0001>). The most striking result from TEM investigations is that in all the samples with x >0.58 the cubic phase is predominant (85%).

Wurtzite phase is exclusively present in form of stacking faults. The total amount of wurtzite phase according to the HRTEM images evaluation was of about 15%.

Fig. 2 shows CL spectra of these predominant cubic AlGaN samples with Al contents x= 0.58, 0.75, 0.82.

The spectra have been taken at 5 K under the same experimental conditions reported in the inset. It can be clearly seen that the dominant transition of the cubic layers shifts to higher energy and gets broader with increasing the Al content. The dominant transitions are comparatively broad bands and show several shoulders in the high and low energy regions. As a comparison, a spectrum from a wurtzite AlGaN sample with x=0.5 is also shown. As expected, the CL peak emission of the wurtzite sample with x =0.5 is higher in energy than the cubic sample with x=0.58.

Deconvolution of the spectra revealed at least five different emissions underneath the bands for all the samples in the concentration range investigated here. In the following, we will discuss more in detail the sample with x=0.58 (data for all samples will be reported elsewhere). As an example, the deconvolution of a spectrum taken at 5 K on the above mentioned sample gives five transitions at energy values of 4.01 eV, 4.06 eV, 4.17 eV, 4.23eV, 4.35 eV. Changing the excitation power by five orders of magnitude (Fig. 3), by varying the electron beam current, does not affect the positions of the four peaks at higher energies. In contrast, a blue shift

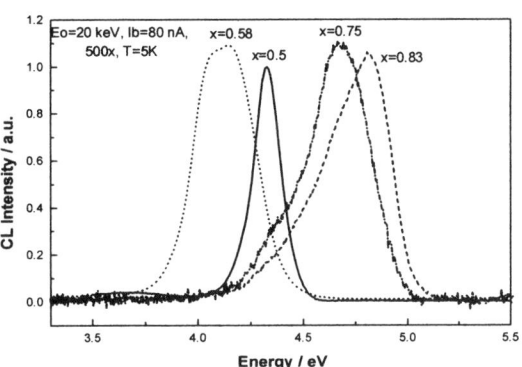

Fig. 2   5 K CL spectra of AlGaN samples with different Al content. The CL emission from a sample with x=0.5 is also reported for comparison.

of the lowest energy peak emission (3.89 eV) with increasing   excitation conditions and a saturation at a beam current of about $5 \cdot 10^{-9}$ A were found. From this observation we conclude that this last emission can be interpreted as due to a Donor Acceptor Pair (DAP) recombination.

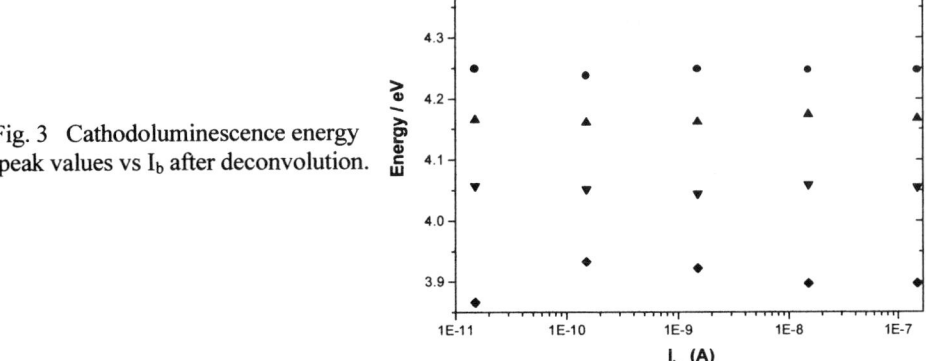

Fig. 3 Cathodoluminescence energy peak values vs I_b after deconvolution.

This is also confirmed by depth dependent CL investigations performed by changing the electron beam energy that evidenced a strong increase of the DAP band and its phonon replicas by approaching the interface. XTEM investigations actually evidenced a high density of structural defects at the interface.

Temperature dependent measurements (Fig. 4) show a decrease in energy of the four higher energy transitions with increasing temperature. This is consistent with an excitonic nature of these emissions. The peak at about 4.35 eV disappears at T=150 K. We assume this faint emission represents the near band edge emission of the wurtzite phase in our layers. The two main transitions are separated by up to ~ 200 meV; a similar splitting of the dominat line has also been observed for wurtzite AlGaN material with an Al content x<0.25 (Kan et al 1986). In these samples these emissions have been ascribed to Near Band Edge (NBE) and Bound Exciton (BE) transitions (the last ones being the so-called I_1 and I_2-lines (Kan et al 1986)

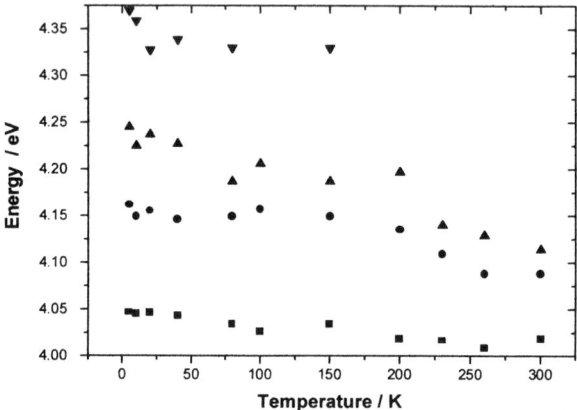

Fig.4 Cathodoluminescence energy peak positions vs T after deconvolution.

A comparison of absorption data with the energy position of the two main excitonic transitions observed in our layers has also been done. For the wurtzite layers with x=0.5 and x=0.15 a good agreement between absorption and CL was obtained, i.e. the excitonic emission occurs at a slightly lower energy value than the absorption data.

For the dominant cubic material, the main absorption lines are significantly lower in energy than the respective absorption data. This can be explained taking into account that a significant volume fraction of the material is wurtzite-type and thus may contribute to the absorption. On the other hand, emission from the wurtzite phase can easily be absorbed by the cubic phase that has the lower band gap and therefore doesnot contribute significantly to the emssion although it represents 0.15 of the total volume fraction.

Fig. 5 shows the shift for the dominant transition of our cubic samples with increasing the Al content. For comparison, recent experimental data for low Al content cubic layers have also been included from (Wu et al 1998).

From these data we estimate a bandgap bowing according to the following equation:

$$E_g = E_g^{GaN} x + E_g^{AlN} x(1-x) + bx(1-x)$$

with E_g^{GaN} =3.29 eV, E_g^{AlN} =5.41 eV and the bowing parameter b=1.02 eV. Thus we obtain an experimental value of cubic AlN of 5.41 eV. From the theoretical work by Rubio et al (1993), a value of 4.9 eV has been considered till now. It is worth noticing that Rubio et al estimated an error of about ±0.3 eV in the predicted band gap of cubic AlN due to the use of the theoretical lattice constants instead of the experimental ones.

Fig. 5 Experimental band-gap values for the samples studied obtained by CL spectra (black squares). The bowing parameter has been obtained by using as a comparison the data from J. Wu et al 1998 (red circles).

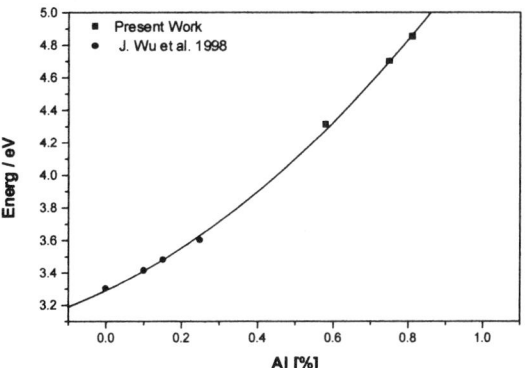

The full width at half maximum (FWHM) of the excitonic transitions increases with increasing   Al content showing a maximum at x=0.75. Comparing the FWHM of pure wurtzite and zincblende materials with comparable Al content (x=0.50 (cubic); x=0.58 (wurtzite)), we observe the zincblende AlGaN to have a FWHM that is two times larger that that of the corresponding wurtzite layer (Figure 6). This is consistent with an increase of the intervalley scattering probability resulting from the occurrence of an indirect band gap (Bignazzi et al 1998).   Rubio et al (1993) predicted an indirect gap for zincblende AlN, while zincblende GaN has a direct gap. A corresponding change from a direct to an indirect bandgap has been observed for zincblende AlGaAs at x=0.38 at low temperature. So we may conclude that our material with an Al content x>0.58 exhibits an indirect energy gap. The values of the FWHM from PL measurements (Wu et al 1998) reported for zincblende AlGaN with an Al content x<0.25 are lower than 35 meV, which is comparable to that of corresponding wurtzite material.

Fig.6 Experimental FWHM values of excitonic transitions for 0.15<x<0.82; the max value occurs at x=0.75.

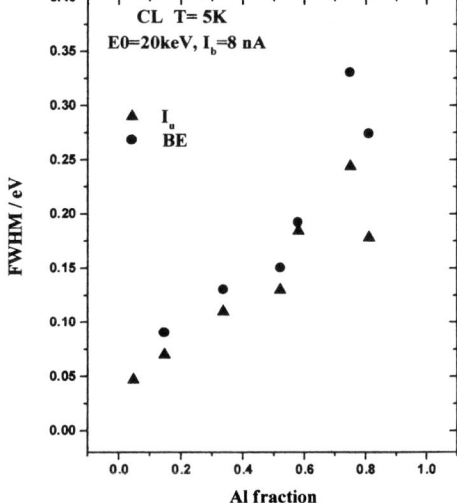

4. CONCLUSIONS

This paper reports on the study by means of cathodoluminescence and high resolution transmission electron microscopy of both structure and excitonic transitions in zincblende $Al_xGa_{1-x}N$ (0.5<x<1) layers. The layers are grown by plasma induced molecular beam epitaxy by purpose at 810°C onto sapphire (0001) layers.

It is shown that excitonic transitions can be observed in the whole range of compositions. In the cubic phase they are characterised by at least two main optical transitions separated by up to ~ 200 meV, of energy ranging from 4.33 and 4.45 eV (x=0.52) to 4.62 and 4.85 eV (x=0.81).

By plotting the near band edge emission vs the alloy composition, a band gap bowing of about 1 eV is found. From this plot, a band gap for zincblende AlN at ~ 5.41eV is estimated instead of the theoretically calculated 4.9 eV. The full width at half maximum of the optical transitions vs Al concentration for hexagonal and cubic materials are also compared.

A sharp increase of the full width at half maximum between x=0.52 and x=0.58 is found; it is ascribed to the increase of the intervalley scattering probability resulting from the occurrence of an indirect band gap.

REFERENCES

Albrecht M, Christiansen S, Salviati G, Zanotti-Fregonara C, Rebane Y T, Shreter Y G, Mayer M, Pelzmann A, Kamp M, Ebeling K J, Bremser M D, Davis R F, Strunk H P 1997 MRS Symp. Proc **468**, pp 293-296

Angerer H, Brunner D, Freudenberg F, Ambacher O, Stutzmann M, Hopler R, Metzger T, Born E, Dollinger G, Bergmaier A, Karsch S, Korner H J 1997 Appl. Phys. Lett. **71**, 1504

Bernardini F and Fiorentini V 1997 Phys. Rev. B **56**, R10024

Bignazzi A, Grilli E, Guzzi M, Bocchi C, Bosacchi A, Franchi S and Magnanini R 1998 Phys. Rev. B **57**, 2295

Nakamura S 1997 MRS Symp. Proc. **482**, pp 1145-1150

Kan M R H, Kiode Y, Itoh H, Sawaki N and Akasaki I 1986 Solid State Commun. **60**, 509

Perry W G, Bremser M B and Davis R F 1998 J. Appl. Phys. **83**, 469

Rubio A, Corkhill J L, Cohen M L, Shirley E and Louie S G 1993 Phys. Rev. B **48**, 11810

Wu J, Yaguchi H, Onabe K and Shiraki Y 1998 Appl. Phys. Lett. **73**, 193

Inst. Phys. Conf. Ser. No 164
Paper presented at Microsc. Semicond. Mater. Conf., Oxford, 22–25 March 1999
© 1999 IOP Publishing Ltd

Electrical activity of dislocations in GaN epilayers

A Castaldini[a,b], A Cavallini[a,b], L Polenta[a,b], M Avella[c], E de la Puente[c] and J Jimenez[c]

a) INFM, b) Dipartimento di Fisica, Universita' di Bologna, I-40127 Bologna, Italy, c) Fisica de la Materia Condensada, ETSII, E-47011 Valladolid, Spain

ABSTRACT: Analyses by electron beam induced current (EBIC) mode of scanning electron microscopy, photo-luminescence (PL) and cathodoluminescence (CL) on hydride vapour phase epitaxial films of gallium nitride on sapphire have been performed to obtain information on the role extended defects play in carrier transport properties. EBIC, CL and PL findings have been cross-related to get insight into the question about the actual dislocation electrical activity and whether they do act as non-radiative recombination centres. The minority carrier diffusion length results equal to 0.5- 0.8 μm. The influence of the dislocations on EBIC and PL results is discussed.

1. INTRODUCTION

Group III nitrides are considered predestined materials to improve the performance of electronic devices. In this context gallium nitride (GaN) is a highly promising material because of its wide direct band gap and high thermal conductivity that make it appealing for, respectively, very efficient blue/UV lasers and diodes and high-power and high-temperature devices. Despite the very large body of work being done on GaN, we still have insufficient experimental insight into the actual electrical activity of dislocations, their spatial distribution and influence on the minority carrier lifetime. In this work we investigated radiative and non-radiative recombination in GaN epilayers and determined inhomogeneous distributions of defects and minority carrier diffusion lengths.

2. EXPERIMENTAL

The GaN sample (289D) examined in this study was grown by hydride vapour phase epitaxy (HVPE) on a (0001) Al_2O_3 substrate to a thickness of ≈ 35 μm. The substrate was first coated with a 200-300 nm thick RF-sputter-deposited ZnO pre-treatment layer, which was thermochemically desorbed at the beginning of the GaN HVPE overgrowth. The main function of this pre-treatment was to promote the dense, homogeneous nucleation of GaN on the sapphire substrate (Molnar et al 1995). The film was grown at 1050°C at a growth rate of 15-20 μm/hour. The unintentionally doped sample was n-type with a room temperature carrier concentration of n= 5×10^{16} cm^{-3}, an electron mobility of μ_n = 780 cm^2 /Vs, and shallow donor and acceptor concentrations of N_D = 2.1×10^{17} cm^{-3} and N_A = 5×10^{16} cm^{-3}, respectively (Look et al 1997, Molnar et al 1995).

The sample was cleaned in acetone and then methanol to remove organic contaminants. Subsequently, the sample was put into boiling aqua regia for 10 minutes and finally rinsed in distilled water. A Schottky diode 200 Å thick was formed on the GaN film by gold evaporation in a vacuum chamber. Its barrier height Φ_B was equal to 0.64 eV. The ohmic contacts were prepared using an In-Ga eutectic solution or by soldered indium at a distance equal to 2mm from the Schottky diode.

426

Scanning photoluminescence (PL) and cathodoluminescence (CL) measurements were carried out both in plan view and cross-section. Electron beam induced current (EBIC) investigations were performed in planar collector configuration since this geometry can be suitably used in the study of defects and simultaneously to measure the diffusion length in the same sample region. The EBIC measurements reported here were carried out at room temperature using a primary beam accelerating potential ranging from 14 to 20 kV. Correspondingly, the maximum electron range R_e, calculated by the Gruen formula (Gruen 1957), varied from 0.65 to 0.86 µm. By conveniently selecting the beam energy and, in turn, the primary electron penetration depth, the spatial resolution for imaging the defect recombination activity was optimised. The minority carrier diffusion length L and surface recombination velocity s were evaluated by the EBIC data using the method of Kuiken and van Opdorp (1985).

3. RESULTS AND DISCUSSION

Cross-section photoluminescence observations showed optical emission corresponding to the parasitic yellow luminescence (YL) of the GaN epilayer. Figure 1a shows a PL micrograph of the GaN film on the sapphire substrate taken at a wavelength equal to 537 nm, thus in the yellow band centred at 560 nm. The presence of this YL has been observed only in n-type material and its origin has been associated with the presence of extended crystalline defects, such as dislocations (Ponce et al 1996). Further studies by Neugebauer and Van de Walle (1996) indicated that the extended defects themselves are not the source of the YL but native defects gettered at extended defects are responsible for such luminescence. In more detail, the Ga vacancy (V_{Ga}) has been indicated as the key defect responsible for yellow emission.

Figure 1 b) shows the CL spectrum of the same sample as in Fig. 1a). Two main spectral bands were observed, a shorter wavelength emission consisting of the transitions due to the free exciton at 357 nm, the donor bound exciton at 364 nm and the DAP (Donor Acceptor Pair) at 375 nm, and the yellow band (YL) centred at 560 nm. Earlier reports (Lester et al 1995, Rosner et al 1997) were contradictory on both the band-to-band and YL emissions from GaN epilayers (Cremades et al 1998, Rosner et al 1997, Ponce et al 1996). The yellow luminescence band was reported to be more intense around extended defects (Ponce et al 1996). In this context we observed an increase of the recombination yielding YL intensity near the substrate/film interface (see Fig.1 a) where higher density of dislocations is expected to relieve the strain. Yellow luminescence PL images reproduce the extended defect distribution in the film.

In gaining a better understanding of the electrical recombination activity, its spatial distribution and correlation with dislocations, electron beam induced current observations are a very powerful tool. In EBIC images the dark areas correspond to regions depleted of minority charge

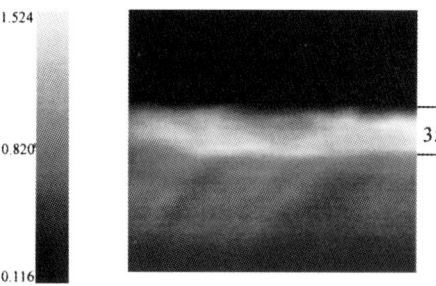

Fig.1a) Room temperature photoluminescence image of the GaN film on sapphire substrate taken in cross-section geometry at wavelength λ equal to 537 nm.

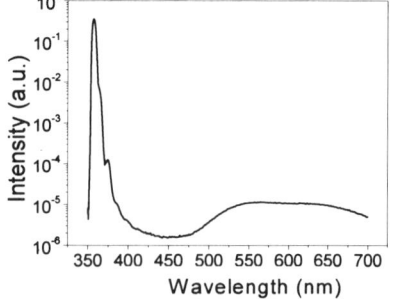

Fig.1b) Cathodoluminescence spectrum of the same sample as in a) at 80 K.

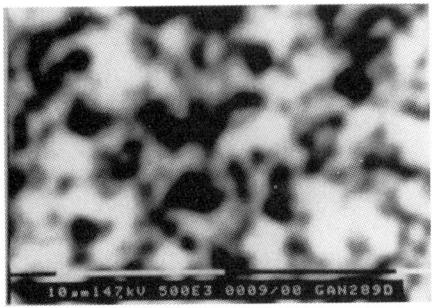

Fig. 2 EBIC micrograph of the spatial distribution of recombining defects. Accelerating energy of the electron beam $E_b = 15$ keV.

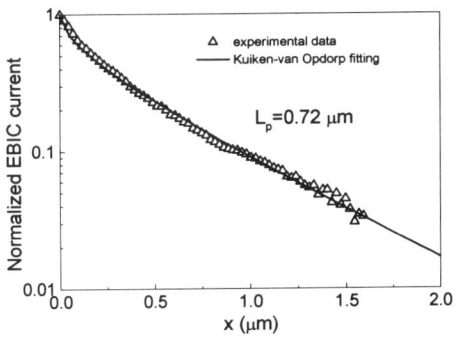

Fig. 3 Normalised semilogarithmic EBIC vs. beam-to-diode distance d.

carriers due to the presence of nonradiative recombining centres, and the bright areas to regions where these centres are absent.

The plan view micrograph of the GaN epilayer imaged in Fig. 2 was obtained by raster scanning the diode area. This micrograph exhibits a highly inhomogeneous distribution of electrically active defects, the smallest feature of which is about 0.3 μm. Furthermore, it may reveal features consisting of defect-free grains, at the boundaries of which a strong recombination activity occurs. Similar features, but relevant to surface morphology, have been observed by atomic force microscopy (AFM) by Bandić et al (1998a, 1998b). In comparing our results to the AFM results, it is worth remembering that EBIC shows the defects in terms not of their structure size but of the extent of their actual recombination activity. Because of this, a certain discrepancy in defect dimensions can be justified.

Thus, EBIC observations are consistent with the already reported (Bandić et al 1998a) presence of extended defect-free grains, which were demonstrated to have a columnar structure with dislocations at low angle grain boundaries associated with yellow luminescence (Ponce et al 1996). Indeed, our results provide additional information about the electrical activity of the dislocations at the boundaries of the grains, since the EBIC maps clearly demonstrate that they show a strong recombination.

Unfortunately, cross-section EBIC images can not be obtained to check the correlation between columnar structure and recombination activity. Indeed, the thin, degenerate n-type region at the GaN/sapphire interface (Look and Molnar 1997) produces an effect similar to a "short-circuit" of the charged threading dislocations, that are generated at this interface to relieve the strain (Look and Sizelove 1999).

The EBIC contrast of the dislocations is 20%. On average, the full width at half maximum of the dislocation contrast profiles as well as the diameter of most of the defect-free grains are on the order of 1-1.5μm. However, also a few grains 3-4 μm wide are present (see Fig.2).

The minority carrier diffusion length L and surface recombination velocity s were evaluated by fitting the curves of the normalised semilogarithmic induced current using the method of Kuiken and van Opdorp (1985) (Fig. 3). Values of L ranging from 0.52 to 0.86 μm were obtained, with an average value $L_a = 0.72$ μm, while the value of the recombination velocity s approaches 0. Table I summarises majority and minority carrier transport properties.

It is of considerable interest to observe that the diffusion length is significantly shorter than the grain dimensions. This excludes the possibility that the limiting factor is the subgrain size length (Bandić et al 1998b). An alternative explanation could be the presence of point defects (impurities/complexes) inside the grains, which behave as recombination centres. Due to the low

Table 1.

Thickness	Carrier density	Mobility	Minority carrier diffusion length	Surface recombination velocity
37 μm	$n = 5 \times 10^{16}$ cm^{-3}	$\mu_n \simeq 780$ cm^2/Vs	$L_p = 0.72$ μm	s = 0 cm/s

value of the free carrier concentration n, dislocation scattering dominates over scattering by native or impurity defects (Look and Sizelove 1999). However, these latter centres would control the carrier diffusion length.

Surface recombination velocity s ~ 0 cm/s results from the fitting procedure by Kuiken and van Opdorp (1985) without any "a priori" assumption. This finding is in agreement with literature results (Chernyak et al 1996, Cremades et al 1998, Bandić et al 1998b). However, it should be regarded with due reservation because of the high density of recombining dislocations terminated at the surface. This discrepancy can be accounted for by assuming that the large photoluminescence efficiency actually prevails over the defect recombination and, thus, determines s.

4. CONCLUSIONS

Luminescence and electron beam induced current analyses of an HVPE GaN epilayer have been performed to determine the electrical activity of extended defects, its spatial distribution and minority carrier diffusion length.

An increase of the YL intensity near the substrate/film interface has been found and the yellow luminescence PL images reproduce the extended defect distribution in the film cross-section, while plan view EBIC maps show grains bounded by strong recombination activity.

By comparing grain size and minority carrier diffusion length, it can be deduced that charge carriers mainly recombine at point defects (impurities/complexes) inside the grains since scattering by dislocations dominates over other scattering mechanisms.

ACKNOWLEDGEMENTS

R. Molnar of MIT, Lincoln Laboratory, Lexington, MA, USA is kindly acknowledged for providing the sample. This work was supported by NATO (grant CGR 961123). The Spanish group was funded by the Spanish government (project: MAT 97-0686)

REFERENCES

Bandić Z Z, Bridger P M, Piquette E C and McGill T C 1998 a Appl. Phys. Lett. **72**, 3166
Bandić Z Z, Bridger P M, Piquette E C and McGill T C 1998 b Appl. Phys. Lett. **73**, 3277
Chernyak L,Osinsky A,Temkin H,Yang J W,Chen Q and Asif Khan M, 1996 Appl. Phys. Lett 69, 2531
Cremades A,Albrecht M,Voigt A,Krinke J,Ambacher O, Stutzmann M 1998 Sol. St. Phen.**63-64**, 139
Gruen A E, 1957 Z. Naturforsch, **12A**, 89
Kuiken H K and van Opdorp C, 1985 J. Appl. Phys. **57**, 2077
Lester S D, Ponce F A, Craford M G and Steigerwald D A 1995 Appl. Phys. Lett. **66**, 1249
Look D C, Reynolds D C, Hemsky J W, Sizelove J R, Jones R L and Molnar R J, 1997 Phys. Rev. Lett. **79**, 2273
Look DC and Molnar R J 1997b Appl. Phys. Lett. 70, 3377
Look D C and Sizelove J R 1999 Phys. Rev. Lett. 82, 1237
Molnar R J,Nichols KB,Makai P,Brown E R and MenIngailis I,1995 Mater. Res. Soc. Symp.**378**, 479
Nuegebauer J and Van de Walle CG 1996 Appl. Phys. Lett. **69**, 503
Ponce F A, Bour D P, Götz W and Wright P J 1996 Appl. Phys. Lett. **68**, 57
Rosner S J, Carr E C, Ludowise M J, Girolami G and Erikson H I 1997 Appl. Phys. Lett. **70**, 420

Inst. Phys. Conf. Ser. No 164
Paper presented at Microsc. Semicond. Mater. Conf., Oxford, 22–25 March 1999

TEM for process development of silicon devices

H Cerva, A Rucki, V Klüppel and T Ohnemus

Siemens AG, Corporate Technology, Otto Hahn Ring 6, D-81730 München, Germany

ABSTRACT: The high-resolution imaging and analytical capabilities of transmission electron microscopy (TEM) are applied to problems in process development and production of silicon integrated circuits. Conventional and high-resolution imaging are used to study the initial stages of dislocation generation at vertical bird beaks. Another topic is the application of electron spectroscopic imaging (ESI) to materials reactions at the bottom of contacts to silicon. The investigation of interfacial layers requires the small electron probe of a TEM with a field emission gun and electron energy loss spectroscopy (EELS) to identify light elements. Problems occuring during the integration of $Ba_{1-x}Sr_xTiO_3$ dielectric layers into memory devices are addressed.

1. INTRODUCTION

Transmission electron microscopy has been in use for silicon material and device studies at least for three decades. The steadily increasing need and interest over the last twenty years is documented best in the proceedings of the Oxford Conference on "Microscopy of Semiconducting Materials" (Cullis 1979-1999). Two developments had a major impact on the application of TEM for silicon studies:

The preparation of thin TEM cross sections (Petit and Booker 1971) was quickly employed in many labs and resulted finally in the development of a special mechanical polishing tool which allows specific site preparation of sub-micron device features (Benedict et al 1990). Cross-sectional preparation with a focused gallium ion beam was only in its beginnings (Kirk et al 1989) ten years ago and today is a standard technique indispensable in an industrial TEM lab. These preparation tools are a prerequisite for studying materials effects in small device structures.

The continuous improvement of the microscopes led to various commercially available dedicated TEMs for e.g. high-resolution imaging, small probe analytical analysis (with a field emission gun and various detectors for x-ray (EDX) and electron energy loss (EELS) analysis) in the early eighties. Today analytical TEMs equipped with a field emission gun, an energy dispersive x-ray detector, an energy filtering system and high-resolution capabilities are a standard for the investigation of sub-micron device structures.

Here we present some case studies which are typical for process development and optimization. All TEM investigations were carried out in small device structures. The analytical TEM used is a Philips CM200FEG with a Noran HPGe-EDX detector and a Gatan imaging filter system.

2. DISLOCATION FORMATION AT A VERTICAL BIRD´S BEAK

Capacitor trenches in dynamic random access memories (DRAM) are surrounded either by the dielectric (e.g. oxide or oxide-nitride-oxide) or oxide isolation collars close to the substrate surface (Adler et al 1995). Thus, such configurations are prone to vertical bird´s beak formation because these areas are laterally oxidized during high temperature anneals. Bird´s beak formation is known to produce high mechanical stress and defects. Dislocation formation at vertical bird´s beaks in 4Mbit DRAMs has already been reported by Dellith et al (1996). The formation of the defects is thought to proceed as follows: the oxidation creates silicon interstitials which agglomerate under the mechanical stress field, then the agglomerates form dislocation nuclei at their interfaces with the silicon matrix. These dislocations may finally glide, under the stress fields, deep into the substrate.

Here high-resolution imaging (HREM) is applied to investigate the initial stages of such a process. In a non-optimized process scheme the oxide collar may be exposed to a high temperature treatment in oxygen through an adjacent trench isolation. This leads to the formation of a vertical bird´s beak formation at the oxide collar which lies below a local poly-Si layer. Figures 1a,b show a badly processed sample where dislocations have been generated. The oxide collar is wider at the top than at its lower parts. Dislocations and stacking faults emerge from the tip of the oxide collar lying on {111} planes. They are marked by "1" in Fig. 1a. Moreover, each cell exhibited a dark dot contrast (arrow in Fig. 1a). Dislocations on {111} planes lying perpendicular to the oxide collar are shown in Fig. 1b. The majority of these dislocations have Burgers vectors of the type $a/2<110>$ parallel to the surface and perpendicular to the oxide collar. HREM imaging reveals a further small defect: at the tip of the oxide collar an extrinsic stacking fault, 5 nm long, terminated by a Frank partial dislocation has formed in the substrate (between A, A´, B´, B in Fig. 2). The poly-Si on top of the oxide collar has to some extent recrystallized epitaxially on the Si-substrate. The set of {111} planes marked by the long "black on white" lines is seen to be bent when it crosses the original poly-Si/Si interface just above the oxide collar. The way they bend may be understood by the upward expansion of the collar oxide during the oxidation. Then the small area (about 100 nm^2) in the substrate - just above the collar oxide tip - must be tensile strained. Si-interstitials which are generated during oxidation are injected into this area. They may precipitate in the form of the observed extrinsic stacking fault. A Lomer dislocation is observed at L, L´ in Fig. 2. with a {110} plane missing between the dislocation line and the substrate surface. The position of the Lomer dislocation corresponds to the dark dot contrast mentioned above. Due to bird`s beak formation the poly-Si layer over the oxide collar compresses the substrate (to the right in Fig. 2). Most probably the Lomer dislocation was nucleated at an irregularity at the poly-Si/substrate interface. Under the high compressive stress this dislocation may have glided on a {100} plane and stopped approx. 50 nm away from the interface. Weak-beam images of planview and strongly tilted cross sections show that the Lomer dislocations "bend back" to the poly-Si/Si interface (Figs. 3a,b). These dislocation segments have pure screw character. Under particular stress conditions these may glide deep into the substrate as observed in Fig. 1b.

A measure for the extent of stress which builds up with the vertical bird´s beak formation at the collar oxide are: the bending of the {111} lattice planes and the length of the tiny extrinsic stacking fault (Fig. 2). Indeed, with a reduced bird´s beak formation (an optimized oxidation) lattice bending, extrinsic stacking fault growth and Lomer dislocation formation cease. This is demonstrated in HREM images of samples which have been subjected to three different oxidations (Figs. 4a-c).

3. ALUMINUM SPIKING IN CONTACT HOLES USING HOT AL-PROCESSES

Aluminium contact hole and via fill processes have received new attention particularly for submicron technology with advanced multilevel metallizations. Advantages of Al over the established standard W plug-fill are its lower resistivity and a reduced process complexity. Moreover, a complete hole-fill may be achieved. A critical issue, however, is the reliability of the Al filled contact holes with respect to Al spiking in the Si substrate, since all the new Al fill techniques (Al-reflow, cold/hot Al sputtering, high-pressure Al (Forcefill)) require temperatures up to 500°C. To prevent spiking the standard Ti/TiN barrier layer processes (collimated, long throw sputtering) have to be optimized with respect to these Al fill processes (Barth 1996). Some critical material aspects which can be observed best by electron spectroscopic imaging (ESI) are described below - for details see Cerva et al (1998).

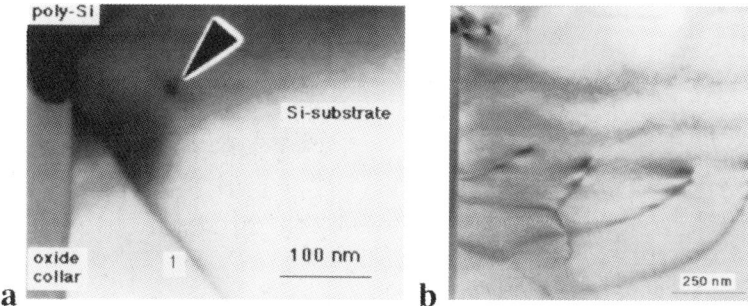

Fig. 1. Dislocation generation at the tip of the oxide collar due to vertical bird´s beak formation. (a) Dislocation on {111} glide plane. Arrow points to dislocation seen end on. (b) Several dislocations generated by a multiplication process.

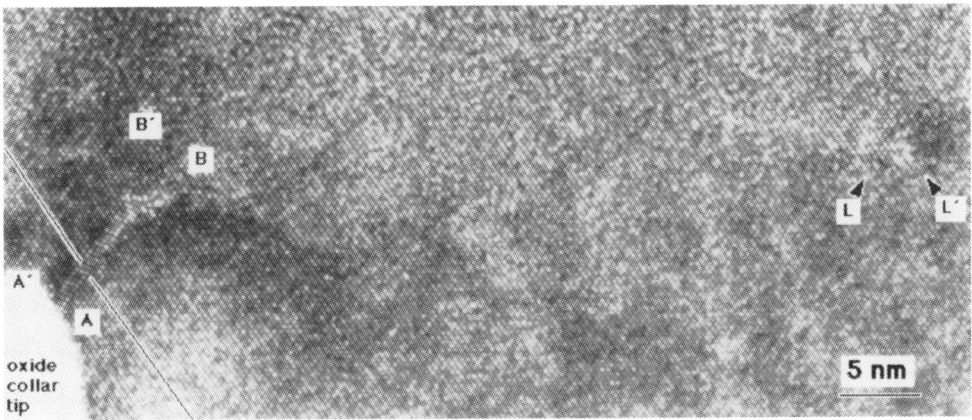

Fig. 2. High-resolution TEM cross section. Extrinsic stacking fault in area AA´BB´ bounded by a Frank partial dislocation. L and L´denote extra {111} planes of a Lomer dislocation.

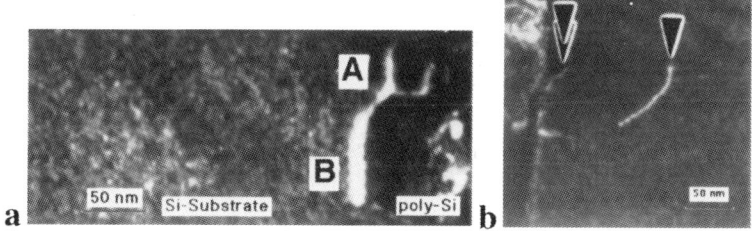

Fig. 3. Lomer dislocations generated at the oxide collar tip. (a) Planview TEM. (b) Tilted cross section.

432

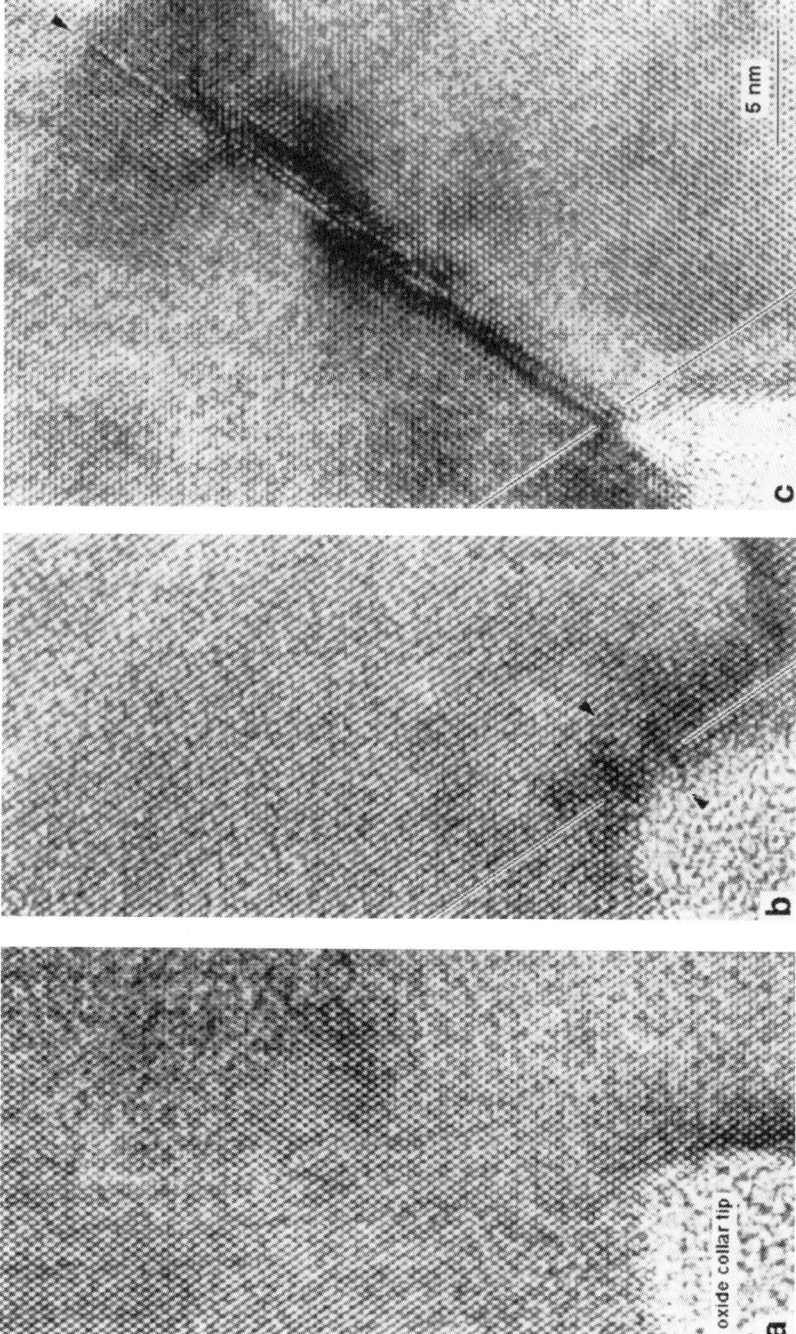

Fig. 4. HREM cross sections. Samples with bird´s beaks after increasingly vigorous oxidations. (a) No {111} plane bending and no extrinsic stacking fault. (b) {111} lattice plane bending and stacking fault in its initial stage. (c) Fully developed stacking fault and strong bending of {111} planes.

The set of samples which showed Al spiking comprises arrays of contact holes, approx. 1μm in diameter, which were filled with a Ti/TiN/Ti (20/100/60 nm) barrier by collimated sputtering. After the TiN deposition an airbreak "stuffs" the grain boundaries with oxygen which improves the barrier's resistance to spiking. The final, thick Ti layer serves as a wetting layer for the successive depositions of a thin cold-sputtered AlSi(1%)Cu(0.5%) layer (seed layer) and the hot-sputtered AlSi(1%)Cu(0.5%)-fill (520°C).

Figure 5a shows a typical cold/hot Al filled contact hole. The TiN barrier at the sidewall is very thin and tapers from the top to the bottom of the hole. The pronounced grain structure everywhere on top of the TiN represents Al-Ti containing grains (mainly Al_3Ti) resulting from the reaction of the thick Ti layer with the thin cold sputtered Al seed layer. The barrier at the bottom is well below the original surface of the Si substrate and the arrow at the edge marks the onset of an Al-spike. Figure 5b is a higher magnification of a contact bottom edge. The columnar orientations of the TiN grains at the bottom and the sidewall are different. In particular the sidewall barrier is less dense than on the bottom. Below the contact a massive Al-spike has formed already. The Al-K ratio map in Fig. 5c shows clearly that an Al-diffusion path exists at the edge of the contact hole through a weak spot in the sidewall barrier. The Ti-L ratio map (Fig. 5d) on the other hand shows the barrier leak at the edge and also the rough and inhomogeneous TiN layer at the sidewall.

The bright-field image in Fig.6a shows the bottom of a contact hole which was analyzed by ESI. Jump ratio images from a selected area are shown in Figs. 6b-e. The Al_3Ti grains above the TiN layer can be seen clearly. The Si-images suggest that there is even a small Si contribution. The spike in the substrate consists of two kind of grains: Al-grains and Ti-silicide grains. In the Ti-images two fine dark lines (2-4 nm thick) parallel to the TiN layer may be observed. There is no continuous $TiSi_2$ layer as found in regular contact holes, but large $TiSi_2$ grains have formed. Inspection of the Ti-, Al-, N-, and Si-images makes clear that the thin upper layer consists mainly of Al and N whereas the thin lower layer contains mainly Al. The former result is in agreement with a recent EDX analysis (Sobue et al 1995, Okihara 1995).

Two features lead to spiking in these samples: firstly the open grain structure of the TiN sidewall layer and the strong Al_3Ti formation, and secondly the fact that the contact hole was etched so deep into the substrate that the sidewall barrier touches the substrate.

4. INTERFACIAL LAYERS

Silicon device technology is strongly concerned with interfacial layers in the nanometer range because they may cause contact resistance problems. The nature of such layers is diverse, comprising electrically poor conducting oxides, polymer layers, and damage zones. The occurence of oxygen and carbon in such layers calls for TEM analysis with electron energy loss spectroscopy (EELS and ESI): analysis of EEL spectra obtained with a 1 nm electron probe yield both compositional and structural information whereas ESI displays the elemental distribution down to the nanometer scale. This will be demonstrated by three examples.

A typical tungsten filled contact hole with a Ti/TiN barrier layer is shown in Fig. 7a. The thin Ti layer serves as a "glue" layer to the silicon by dissolving the native silicon oxide. In this case an air-break before the subsequent CVD TiN barrier deposition and a final Ti-liner anneal at 400°C were applied. Three layers - denoted by letters a-c in Fig7b - can be distinguished below the tungsten plug. The EELS analyis at point "a" reveals a thin titanium silicide (Si-, Ti-$L_{2,3}$ edges) which was formed by the reaction of silicon with the Ti-layer. Moreover a small O-K signal is observed. At point "b" a thin oxygen rich Ti layer is identified which is responsible for the poor contact resistance.

Polymeric interfacial layers can originate either from non-volatile carbon-hydrogen reactants during CVD deposition or dry etching. Figure 8a shows an interfacial layer, a few nanometers thick, with bright contrast. The C-K jump ratio image clearly shows a continuous carbon-rich layer at the contact hole bottom (Fig. 8b).

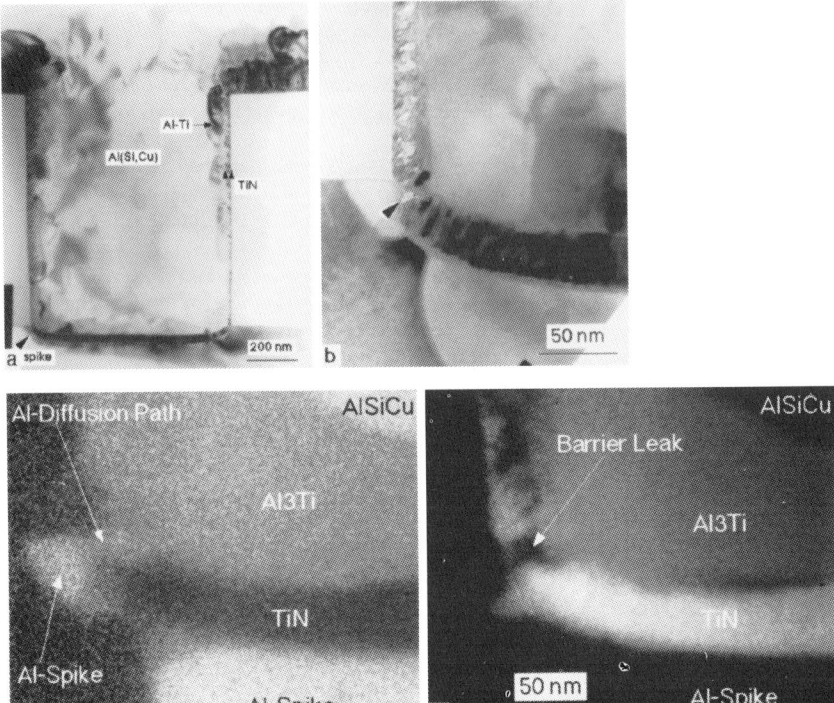

Fig. 5. (a) Al contact hole with an Al-spike. (b) Bottom corner of the contact hole. (c,d) Al-K and Ti-L jump ratio image (electron spectroscopic imaging) of (b)

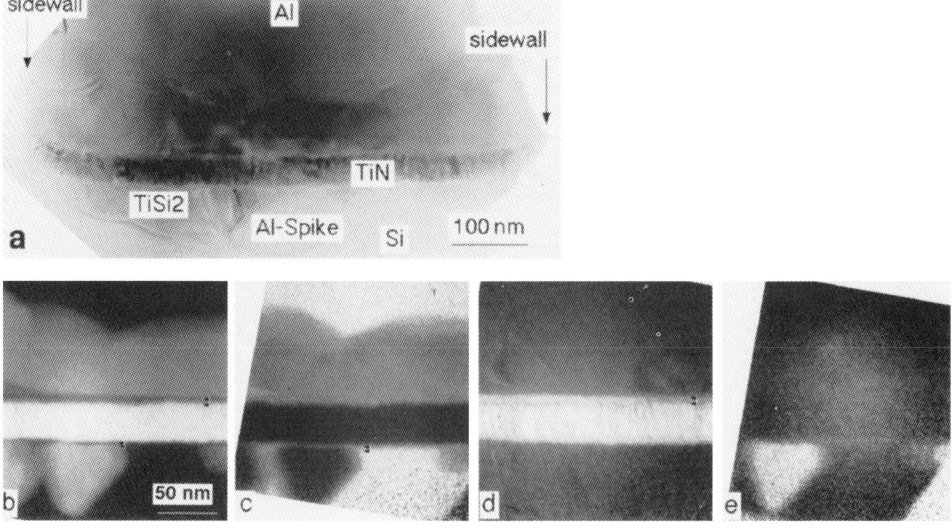

Fig. 6. (a) Bright-field image of contact hole with Al-spike. Jump ratio images of the middle of the contact hole of (a):(b)Ti, (c) Al, (d) N, (e) Si.

Sometimes interfacial layers may be observed between poly-Si layers (Fig. 9a). They are often due to insufficient or ineffective cleaning steps. The EEL spectrum recorded from the interfacial layer shows, apart from a strong silicon signal, small amounts of carbon and oxygen (spectrum b in Fig. 9b). The slightly different shape of the Si-$L_{2,3}$ edge in spectrum "b" reflects the structural change of the silicon atom environment in the interfacial layer as compared to that in poly-Si. The analysis clearly shows that a carbonaceous silicon oxide layer is present.

5. Ba$_{1-x}$Sr$_x$TiO$_3$ AS A HIGH-ε DIELECTRIC IN DRAM STORAGE CAPACITORS

Ba$_{1-x}$Sr$_x$TiO$_3$ (BSTO) thin films will replace nitride/oxide multilayer systems as dielectric in capacitors of Gbit DRAMs (Hwang 1998, Kotecki 1997). Sticking to DRAM cell concepts with even thinner nitride/oxide dielectric layers would cause high dielectric leakage currents and would require even more complicated three-dimensional (stacked or trench) capacitors in order to increase capacitor area. Since a minimum capacitor charge of about 25fC has been found to be indispensable for safe information storage, a new dielectric with a high-ε value must be introduced. Though BSTO thin films with ε up to 600 have been produced by metal-organic chemical vapour deposition (in order to ensure good step coverage), ε becomes smaller with decreasing film thickness and deposition temperature (Hwang 1998).

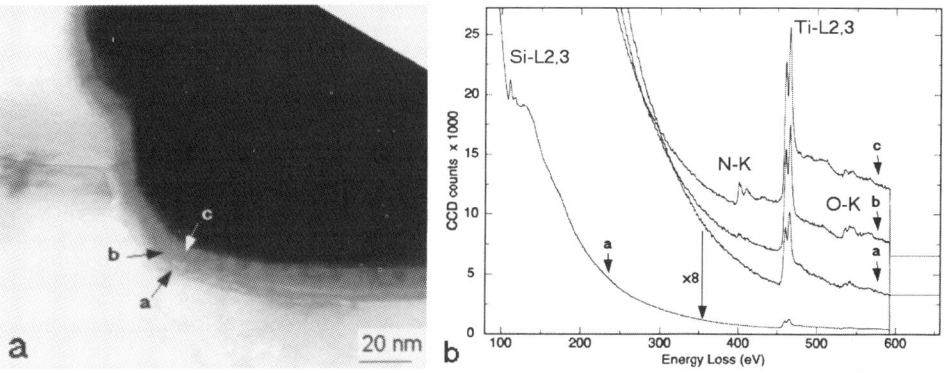

Fig. 7. (a) Cross section through bottom of a tungsten filled contact hole with a Ti/TiN barrier. (a) EEL spectra from the positions marked in (a).

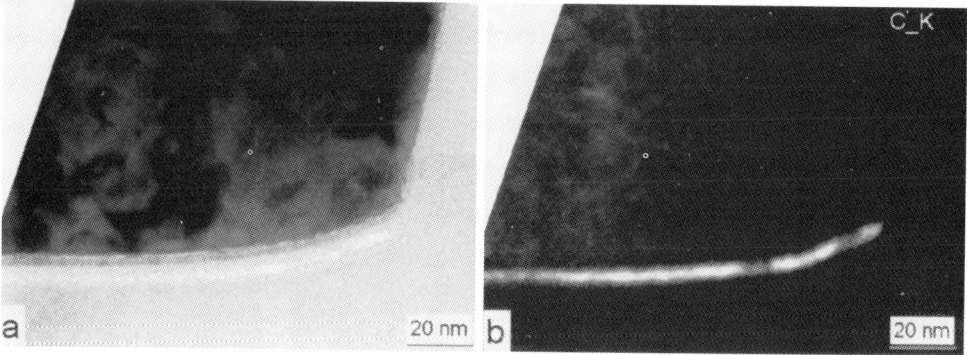

Fig. 8. (a) Cross section through bottom of a tungsten filled contact hole with a Ti/TiN barrier revealing a thick interfacial layer. (b) C-K jump ratio image of (a).

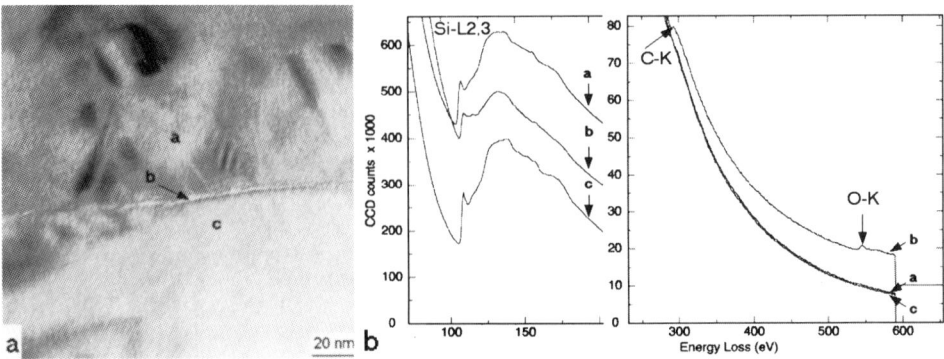

Fig. 9. (a) Poly-Si/poly-Si interface with a 1 nm thick intermediate layer. (b) EEL spectra from the positions marked in (a).

The integration of BSTO into a silicon device process requires new electrodes such as Pt or conductive RuO_2 and diffusion barrier materials like TiN, TaN, TiSiN, TaSiN, or TiAlN. The barrier has to prevent the reaction of a) oxygen from the BSTO deposition or post-capacitor anneal processes and b) the electrode material with the silicon plug and substrate.

To meet these challenges a cell concept as schematically drawn in Fig. 10a is needed. Figure 10b displays a TEM cross section of one such functional cell realized in a 4Mbit DRAM test device. The capacitor is located above the bitline and the transistor and the storage node are connected to the transistor via a poly-Si plug. In order to obtain functional bits the barrier must not become oxidized and hence process optimization becomes necessary. Figure 11a shows the sidewall of the capacitor stack at higher magnification. While the BSTO layer is about 20 nm thick between the two Pt electrodes on top of the capacitor stack, it becomes thicker at the sidewall where the BSTO is in direct contact with the TiN barrier. The TiN reveals the typical columnar grains below the electrode and the BSTO at the sidewall shows homogeneous contrast typical for amorphous material. At the sidewall there is a zone approximately 20 nm thick with a different contrast. This contrast extends like a wedge down the Pt electrode/TiN interface. The letters "a-f" and the marker "ref" denote the spots where EELS analysis with a 2 nm electron probe was carried out. Comparing the BSTO reference spectrum "ref" with the spectra at "a", "b", and "c" shows that the $Ba-M_{4,5}$ edges have been strongly reduced and at "c" only nitrogen and titanium are detected (Fig. 11b). The same is true for the spectra at "d", "e", and "f" (Fig. 11c). In spectrum "f" only titanium and oxygen are measured which indicates that isolating Ti-oxide has been formed. The oxidation of the barrier is found to increase with BSTO deposition temperature and with post-capacitor annealing. At the sidewall the oxygen diffuses directly into the TiN whereas at the Pt electrode/TiN interface the Pt may catalyze the oxidation of the barrier by producing atomic oxygen (Hwang 1998). A possible solution is to cover the barrier sidewall with the Pt bottom electrode as shown in Fig. 12a. The EEL spectra in Fig. 12b indicate that the TiN was not oxidized after the complete capacitor process.

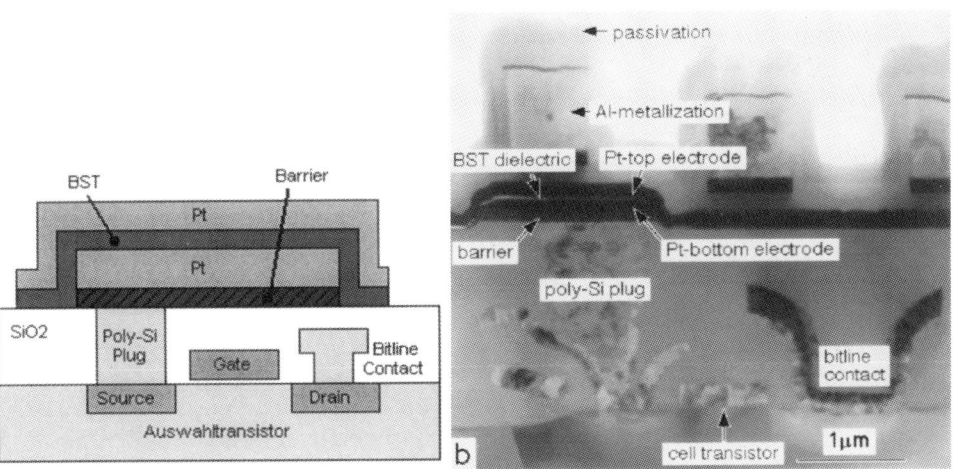

Fig. 10. DRAM cell with a stacked capacitor. (a) Schematic drawing, (b) TEM cross section.

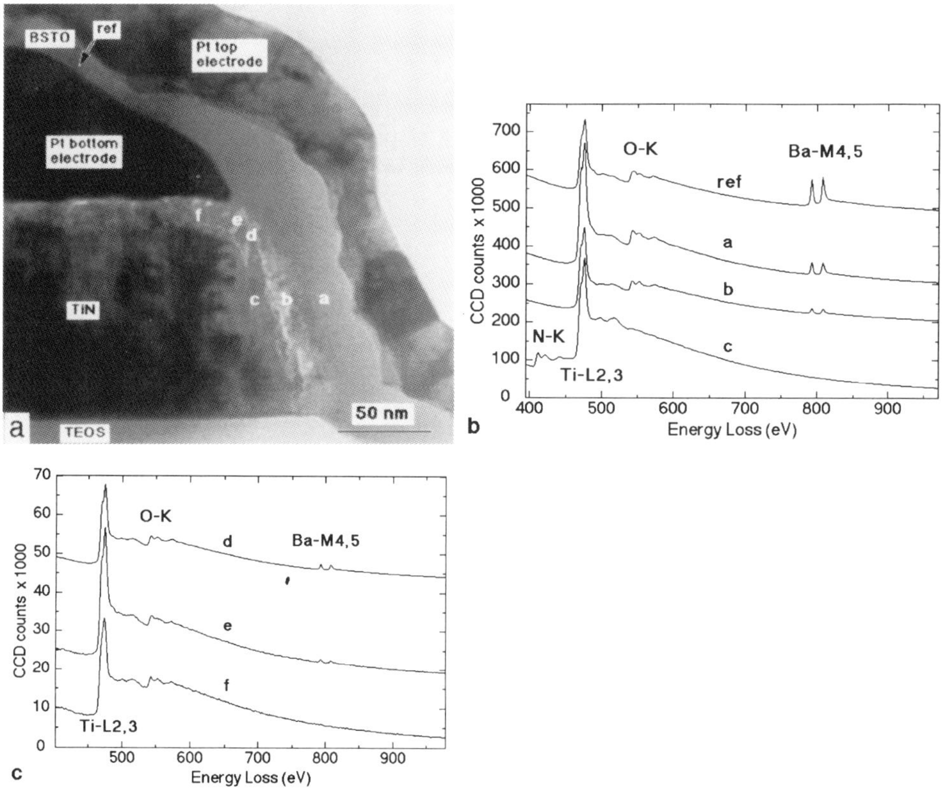

Fig. 11. (a) Sidewall of the stacked capacitor: BSTO dielectric in direct contact with TiN barrier, (b,c) EEL spectra recorded at the positions of the letters in (a).

438

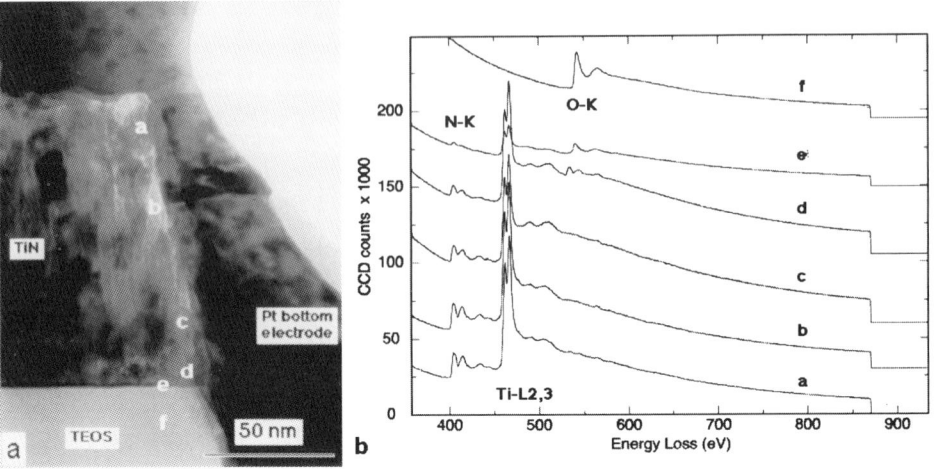

Fig. 12. (a) Sidewall of the stacked capacitor: Pt bottom electrode in direct contact with TiN barrier, (b,c) EEL spectra recorded at the positions of the letters in (a).

ACKNOWLEDGMENT

The authors wish to thank E. Hammerl, H.-J. Barth, H. Helneder, A. Ruf, and C. Dehm from *Infineon Technologies* (former Siemens Semiconductor Group) for the fruitful collaboration.

REFERENCES

Adler E, DeBrosse J K, Geissler S F, Holmes S J, Jaffe M D, Johnson J B, Koburger C W III, Lasky J B, Lloyd B, Miles G L, Nalos J S, Noble W P Jr., Voldman S H, Armacost M and Ferguson R 1995 IBM J. Res. Develop. **39**, 167.

Barth H J 1996 Mat. Res. Soc. Symp. Proc. **427**, 253

Benedict J P, Anderson A, Klepeis S J and Chaker M 1990 Mat. Res. Soc. Proc. **199**, 189

Cerva H, Klüppel V, Barth H J and Helneder H 1998 Mat. Res. Soc. Symp. Proc. **523**, 115

Cullis A G (editor) 1979 J.Microscopy **118**, 1981-1997 Inst. Phys. Conf. Ser. **60, 67, 76 , 87 ,100 117, 134, 146, 157**

Dellith M, Booker G R, Kolbesen B O, Bergholz W and Gelsdorf F 1996 J. Electrochem. Soc. **143**, 210

Hwang C S 1998 Mater. Sci. Engin. **B 56**, 178

Kirk E C G, Williams D A and Ahmed H 1989 Inst. Phys. Conf. Ser. **100**. 501

Kotecki D E 1997 Integr. Ferroelectr. **16**, 1

Okihara M, Hirashita N, Hashimoto K and Onoda H 1995 Appl. Phys. Lett. **66**, 1328

Petit H R and Booker G R 1971 Proc. 25th Anniversary Meeting of EMAG (editor W C Nixon), IOPPubl, London, p. 290

Ruf A, Rucki A, Cerva H, Gehring O and Pahlitzsch J 1999 J. Electrochem. Soc. submitted

Sobue S, Mukainakano S, Ueno Y and Hattori T 1995 Jpn. J. Appl. Phys. **34**, 987

Inst. Phys. Conf. Ser. No 164
Paper presented at Microsc. Semicond. Mater. Conf., Oxford, 22–25 March 1999
© 1999 IOP Publishing Ltd

Void formation at the interface of bonded hydrogen-terminated (100) silicon wafers

R Scholz, L F Giles, S Hopfe, A Plößl and U Gösele

Max-Planck-Institut für Mikrostrukturphysik, Weinberg 2, D - 06120 Halle, Germany

ABSTRACT: Hydrogen-terminated (100) wafers of FZ-grown silicon were joined through wafer direct bonding at room temperature. After various temperature treatments of bonded wafer pieces between 520°C and 1300°C their interfaces were investigated by TEM. Various imaging variants were applied to cross section and quasi plan-view specimens for detection and characterization of minute voids at the interfaces in the presence of emerging dislocation networks.

1. INTRODUCTION

Two solids with sufficiently smooth and flat surfaces adhere to each other when brought into contact in ambient air at room temperature. This phenomenon is the basis of wafer direct bonding, a joining technology aptly reviewed by Tong and Gösele (1999). In the case of hydrogen-terminated silicon, the assumption of van der Waals interaction suffices to account for the adhesion. The energy of adhesion increases through heating, above about 500°C eventually being comparable with the energy of cohesion of silicon. The increase in adhesion commonly is attributed to the internal desorption of hydrogen and diffusion along the interface or into the Si lattice, and the accompanying formation of Si-Si bonds across the interface.

Because of the inevitable roughness of real surfaces, initial contact will be made only at the highest asperities, which in turn will be elastically deformed by the adhesion, so that further asperities can make contact. Hitherto the completeness of the bonding has been investigated through infrared transmission, scanning acoustic microscopy, X-ray topography or defect etching; voids larger than 1 μm, usually associated with contamination, can be detected with these methods. TEM investigations aimed at dislocation networks at the interface of bonded wafer pairs annealed at high temperatures (Benamara et al. 1995). Oxide precipitation in the wake of high-temperature annealing of the commonly used Czochralski (Cz) Si complicated the TEM detection of any minute voids. In agreement with the observed leak-tightness, the general notion has been that bonding of state-of-the-art Si wafers, meticulously cleaned and contacted in dust-free environment, implies void-free intimacy of contact. However, the question remains: are the elastic deformations sufficient to preclude unbonded "islands" in the "sea" of intimate contact? Here we set out to investigate this question with TEM techniques.

2. EXPERIMENTAL

Because FZ-grown Si contains appreciably less oxygen than Cz-Si, FZ-Si was used. The 100 mm diameter wafers were n-type, nominally of (100) +1° orientation. After standard wet chemical cleaning, the native oxide layer was dissolved in a 2 vol.% aqueous solution of HF, rendering the surface hydrogen-terminated. Immediately after retracting the hydrophobic surfaces, they were brought into contact. The bonded wafer pairs, or pieces thereof, were annealed in inert atmosphere by RTA or in a tube furnace at temperatures between 520°C and 1300°C for various times. The results reported here were obtained on parts of two bonded wafers which were heated in a first step at 700°C for 10 min (wafer pair 1) and at 520°C for 2 h (wafer pair 2). Cross section specimens were prepared in the usual way by final Ar ion milling. For plan-view investigations 25°-cuts were used, finally thinned either chemically or preferentially by ion milling. The TEM investigations were carried out at CM20T and JEM-4000EX microscopes.

3. RESULTS

On wafer pair 1, after the initial treatment at 700°C, the results, as depicted in Fig. 1 (a-c), were obtained. The plan-view survey image in Fig. 1a shows a dislocation network consisting of square meshes of screw dislocations representing a twist component of misorientation, inevitable in bonding experiments. An additional set of dislocations (usually for short termed 60°-dislocations), accommodating a tilt component caused by the miscut, crosses the square network almost diagonally and causes well-known half-distance shifts. Twist and tilt component can be determined to be about 0.5° and 0.1°, resp. Subsequent investigations of a cross section specimen additionally revealed a high density of minute voids (or gas filled cavities, Fig. 1b and 1c). They are partly superimposed in an edge-on view (Fig. 1b). When tilting a cross section specimen in the microscope goniometer by larger angles, they can be observed separated (Fig. 1c, about 40° tilted from a usual <011> transmission). Kinematical conditions and Fresnel contrast underfocus settings were used for Figs. 1b and 1c in order to suppress strong diffraction contrasts and making the voids visible. The section in Fig. 1c shows irregular forms of voids of different sizes. Using ion-milled 25°-cut specimens, it was also possible to image the small interface voids in plan-view together with contrasts of dislocations. At overfocus setting, in Fig. 1d, the voids are visible as dark patches among dislocations. The smallest voids on such images could be estimated to be 1.5 to 2 nm in diameter. Maximum extensions of voids up to 5 nm perpendicular to the interface were measured in HREM images of cross sections.

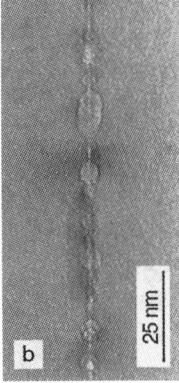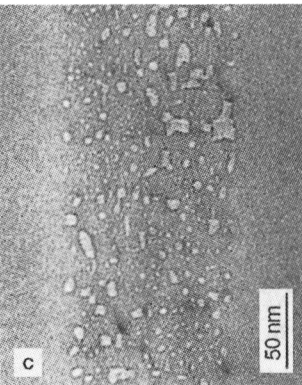

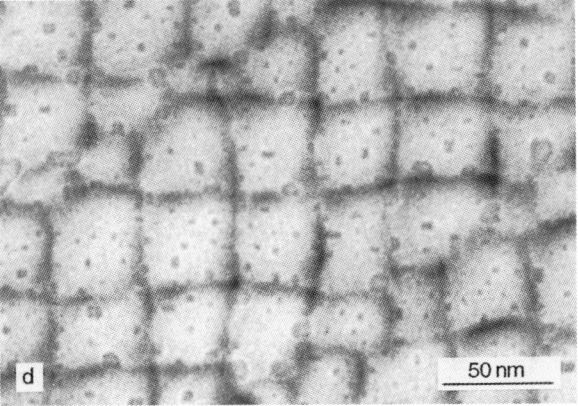

Fig. 1. TEM images of a wafer bonding interface (wafer pair 1) after RTA at 700°C for 10 min.
(a) Plan-view survey image of the dislocation network revealing twist and tilt component;
(b) Cross section edge-on image at underfocus showing minute voids at the interface (Fresnel-contrast);
(c) Distribution of voids in a thicker cross section region after tilting by about 40°;
(d) Enlarged section of the interface (plan-view) at overfocus setting with the voids dark among dislocations.

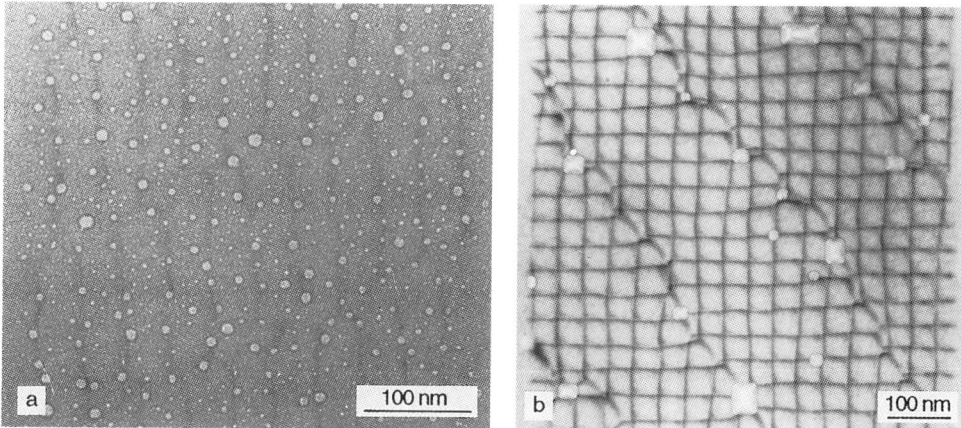

Fig. 2. Plan-view images obtained under different imaging conditions of the same bonding interface as in Fig. 1 after additional RTA for 10 min (a) at 900°C and (b) at 1100°C.

Figures 2a and 2b give an impression of the interface structures obtained after further heating of pieces of wafer pair 1 at higher temperatures. After RTA at 900°C for 10 min a similar high density of voids as before is observed which are now mainly spherical (Fig. 2a). The largest voids have diameters up to 13 nm and the smallest ones detectable were in the range of 2 nm in diameter. Large voids were found preferably located at or near the positions of 60°-dislocations. This was also the case after the initial treatment at 700°C, not visible in small sections like Figs. 1c and 1d. It points to more free volume initially present near and along former steps at the wafer surfaces. After heating at 1100°C (Fig. 2b) a small number of large voids is being left in the interface, mainly in contact with 60°-dislocations. The largest ones (up to 80 nm in their long extension) are facetted polyhedra preferably bound by Si {111} and {100} planes as confirmed by cross section imaging.

In a first series, pieces of wafer 2 were heated at 520°C for 2 h, 10 h, 50 h and 245h. No remarkable differences in interface microstructure were observed, neither in plan-view nor cross section. Distinct Fresnel contrasts of three-dimensional voids as shown before could not be obtained and only first indications of dislocation network formation has been locally discernible as shown in Fig. 3a. Several square meshes of dislocation contrasts among weak Moiré fringes are visible, the periodicities of which indicate a twist misorientation of about 1.65°. In cross sections at high resolution it became obvious that at the interface a very thin (1 nm and less) discontinuous layer of higher electron transparency exists (Fig. 3b). At large

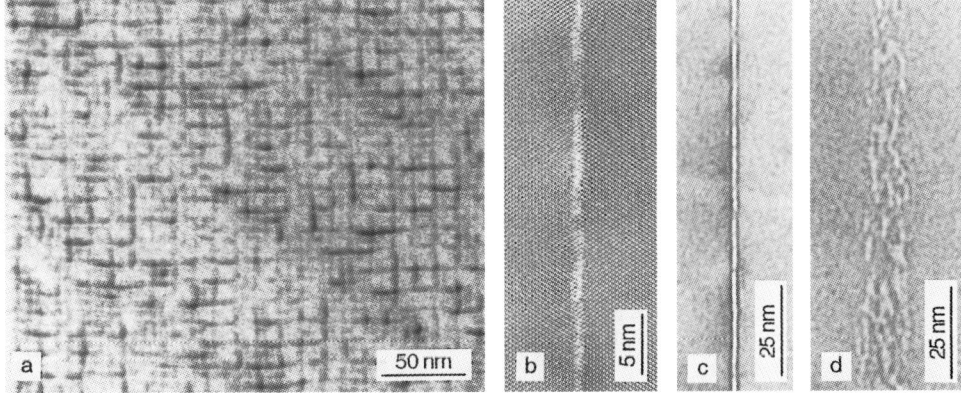

Fig. 3. TEM images of bonded wafer pair 2 after heating at 520°C for 50 h. Plan-view (a); Cross section edge-on at high resolution (b), at underfocus (c) and 20° tilted, underfocus (d).

442

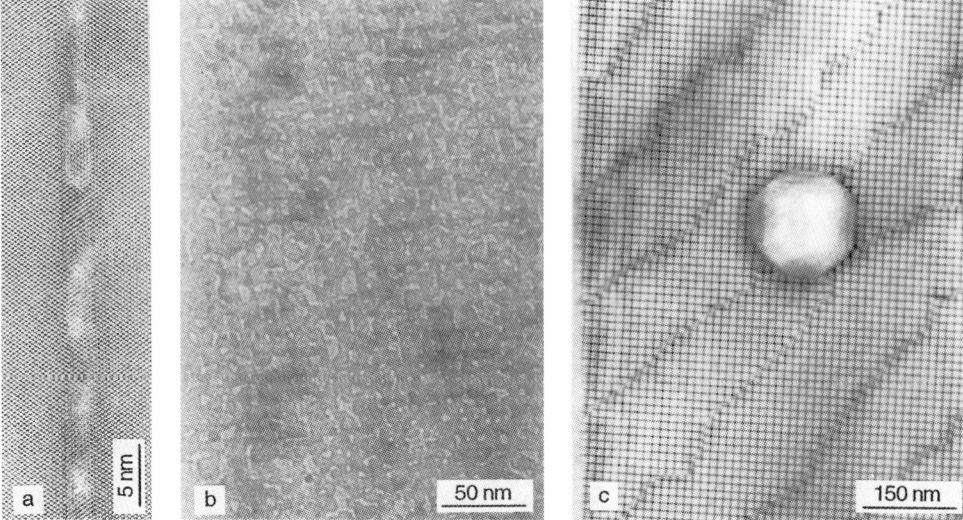

Fig. 4. TEM images of wafer pair 2 after heating at 650°C for 2h, (a) cross section, (b) plan-view at underfocus; and after 1300°C for 2h (c).

defocus settings Fresnel contrasts were observed in edge-on position along the interface (Fig. 3c). By tilting a cross section up to 20°, Fresnel contrast features as in Fig. 3d became visible, that were not attainable after tilting to higher angles or under plan-view conditions. It is assumed the contrasts in Fig. 3d reflect a distribution of very shallow voids at the interface. They can only be imaged by Fresnel contrasts if the void thickness in beam direction, which changes by specimen tilting, exceeds a certain limit.

After further treatment at 650°C for 2h the voids at the interface were observed to be up to 2 nm thick (Fig. 4a) and their island shape became now visible in plane-view if diffraction contrast from dislocations is suppressed (Fig. 4b). Subsequent heating of pieces of wafer pair 2 at still higher temperatures caused transformation of the shallow void islands to more three-dimensional shape, easily detectable within the well-developed dislocation network as in case of wafer pair 1. Fig. 4c shows a section of the interface after a treatment at the highest temperature applied. In 25°-cut specimens of samples annealed at 1300°C, only a few large polyhedra (diameters up to 190 nm) were found.

4. SUMMARY AND CONCLUSIONS

Contrary to the general notion, bonded hydrogen-terminated Si wafers clearly contain excess volume at the interface, as shown here through TEM. The number of voids drastically decreased for higher annealing temperatures. There was a large scatter in void volume per unit interface area, between specimens of the same annealing experiment and even within the same TEM specimen, rendering a quantitative evaluation problematic. However, the assumption that the ripening process conserved the void volume per area is consistent with the observed data.

ACKNOWLEDGEMENTS

The authors thank the Bundesministerium für Bildung, Wissenschaft, Forschung und Technologie for financial support under grant number 13 N 6758

REFERENCES

Tong Q-Y and Gösele U 1999 Semiconductor wafer bonding: Science and Technology (New York: Wiley)
Benamara M, Rocher A, Laporte A, Sarrabayrouse G, Lescouzères L, PeyreLavigne A, Fnaiech M and Claverie A 1995 Mat. Res. Soc. Symp. Proc. **378**, 863

Inst. Phys. Conf. Ser. No 164
Paper presented at Microsc. Semicond. Mater. Conf., Oxford, 22–25 March 1999
© *1999 IOP Publishing Ltd*

Morphology and defects in shallow trench isolation structures

C Stuer, J Van Landuyt, H Bender[1], R Rooyackers[1] and G Badenes[1]

EMAT, University of Antwerp, Groenenborgerlaan 171, B-2020 Antwerp, Belgium
[1] IMEC, Kapeldreef 75, B-3001 Leuven, Belgium

ABSTRACT: Transmission electron microscopy (TEM) is used in this study to examine the evolution of the morphology and the defect formation in shallow trench isolation (STI) structures with various line-widths and spacings. This research is performed to optimise the STI processing, for further scaling to smaller dimensions. Bright field and high resolution imaging along the $[011]_{Si}$ cross-section allow geometry studies to determine the optimal processing conditions and to investigate defect formation. Plan-view imaging is used for further study of the defects.

1. INTRODUCTION

Due to the impressive progress in microelectronics, devices have reached submicron dimensions. An accurate identification of the morphology and defects calls for TEM observations. Crystallographic defects such as dislocations may form during particular steps in device processing and it is well known that these have an important influence on the electrical behaviour and the reliability of the final devices.

Shallow trench isolation (STI) is a promising technology for isolation structures in the new generations of ULSI devices with dimensions below 0.18 μm (Nandakumar et al 1998). Nowadays, a lot of attention is paid to optimisation of STI processing, because of its large scalability and high planarity.

2. STI PROCESSING

The experimental procedure is pictorially demonstrated in Fig. 1. The starting substrates are 150 mm (100)-orientation, p-type silicon substrates.

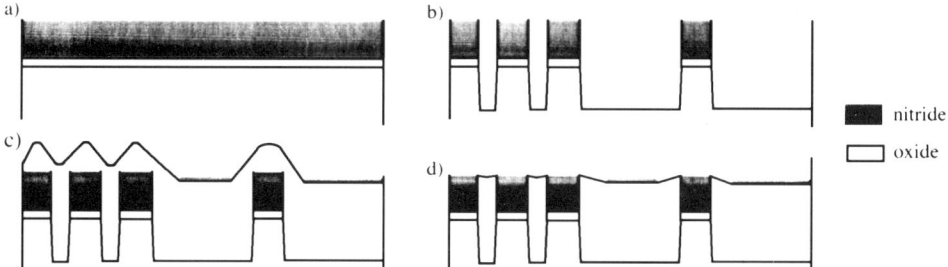

Fig. 1. STI process flow. a) Pad oxide growth and nitride deposition. b) Definition and etching of the trenches. c) Filling of the trenches with a HDP oxide and nitride deposition on large active areas. d) Chemical-mechanical polishing (CMP).

The isolation mask is prepared by depositing a 200 nm low-pressure chemical vapour deposited (LPCVD) nitride film on a 20 nm thermal pad oxide (Fig. 1.a). After definition of the active area using a standard photolithographic step, the nitride/oxide stack is etched. Subsequently, 0.4 μm deep trenches in the silicon substrate are etched (Fig. 1.b). After cleaning, a sidewall oxidation of 20 nm at high temperature is performed to round the top corners of the trenches. Next, the trenches are refilled using a 500 nm undoped high-density plasma (HDP) oxide and a 70 nm thick LPCVD nitride. This second nitride layer is defined as a field protecting nitride by using a complementary mask of the active area with a size of 0.4 µm. The nitride is etched selectively towards the HDP-oxide (fig. 1.c).   Finally, the wafers are planarised by removing the excess oxide using a chemical-mechanical polishing (CMP) process with slurry selective towards nitride.

3. TRENCH RE-OXIDATION

The trench formation and the re-oxidation parameters are studied using bright field $[011]_{Si}$ cross-section images to obtain the optimal conditions before refilling the trenches with an oxide. No defects are found in the examined structures after re-oxidation.

The influence of creating a cavity into the pad oxide before re-oxidation and the influence of the re-oxidation temperature on the morphology of the trench are analysed with bright field $[011]_{Si}$ cross-sections. On the one hand, rounding the silicon corner on the top of the trench without resulting in a negative slope is necessary to avoid gate oxide thinning.   On the other hand, the re-oxidation conditions can play an important role in the stress behaviour and hence in the defect formation because the later filling of the trench with an oxide will cause high stresses. A rounded corner spreads the stress while a sharp corner keeps it localised. This may cause defects in the silicon substrate.

3.1 Influence of a cavity etching before re-oxidation

The size of the trenches and the active area is a dominating parameter in the STI processing. The smaller the dimensions, the more difficult it is to obtain the desired morphology by etching. Therefore, in any experiment to optimise the processing, it is very important to compare structures with the same dimensions.

A HF solution is used to undercut the nitride in order to create a cavity of 12 nm into the pad oxide.   This allows oxygen diffusion in the cavity, but due to its small size, only a small part of the silicon substrate underneath is exposed to the oxygen, which implies less oxidation. This results in rounding the silicon top corner as is observed by bright field $[011]_{Si}$ cross-sections (Fig. 2). In this figure, the size of both the active area and the trench is equal to 0.9 µm.

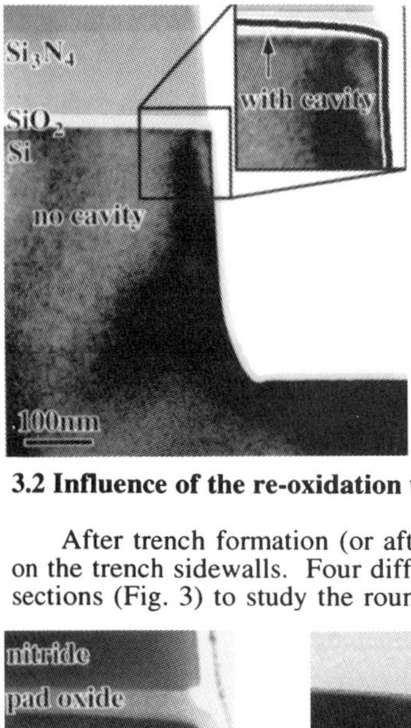

Fig. 2. Bright field [011] cross-section of an 1150°C RTP re-oxidised structure without cavity. The outline of the structure with a cavity is overlaid on it.

3.2 Influence of the re-oxidation temperature

After trench formation (or after a cavity etching in the pad oxide), a 20 nm oxide is grown on the trench sidewalls. Four different temperatures are analysed with bright field $[011]_{Si}$ cross-sections (Fig. 3) to study the rounding of the trench top corners.

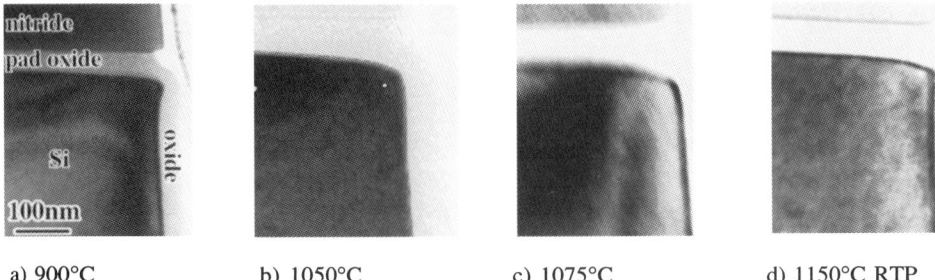

a) 900°C   b) 1050°C   c) 1075°C   d) 1150°C RTP

Fig. 3. Bright field $[011]_{Si}$ cross-sections, which show the influence of the re-oxidation temperature on the morphology of the silicon corner under the nitride. All structures got a cavity etching before re-oxidation.

On the one hand, low temperature CMOS processing is preferred to reduce radial temperature gradients, which induce thermal stresses in the substrate. On the other hand, re-oxidation at low temperatures leads to a silicon corner with a negative slope (Fig. 3.a), which can not be tolerated. The oxidation rate differs with the type of crystallographic plane. The use of higher temperatures allows faster oxidation, even at the so-called slow oxidation planes. This results in a more or less equal oxidation of the [100] and the [011] planes. An optimal corner is obtained with an oxidation temperature of 1075°C (Fig. 3.c).

Rapid thermal processing (RTP), where an oxidation takes place at a high temperature in a short time, keeps the morphology of the etched trench (Fig. 3.d). A sharper corner is obtained in comparison with a 1075°C re-oxidation.

4. TRENCH FILLING

Starting with the optimal re-oxidation conditions, i.e. a 1075°C re-oxidation after a 12 nm HF-dip in the pad-oxide, the trenches are filled with a 500 nm or 550 nm HDP oxide.

The defect formation depends, as the morphology of the trenches, in a large measure on the structure density. Defects are mainly observed at the bottom of the isolated trenches, i.e. trenches with dimensions of 0.35 μm surrounded by 50 μm wide active areas, whereas the defect formation on the bottom of 0.35 μm trenches surrounded by active areas of 2 μm or smaller is extremely reduced. In the latter trenches, dislocations are not observed but some cracks can appear in the corner at the bottom of the trench. STI structures with 0.4 μm trenches or larger show no defect formation.

The presence of defects near the bottom of the isolated structures is related to the shape of the isolated trenches and the stresses induced by the filling with the HDP oxide. A supplementary stress from the nitride on the 50 μm active areas may cause the defects. A different structure and expansion coefficient of the nitride and the silicon substrate causes stress in the silicon. The larger the active areas and hence the larger the nitride layer on top of it, the more stress will be caused by this reason. The stress from the nitride on the 2 μm and smaller active areas is less. Added to the stress due to the filling of the trench, it does not lead to a large amount of defects. Further experiments on structures with a grading density will be performed to substantiate this hypothesis. The formation of defects is similar for the samples with both HDP oxide thicknesses.

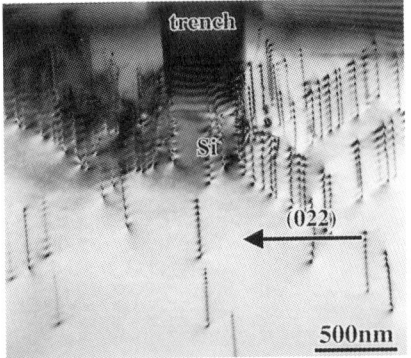 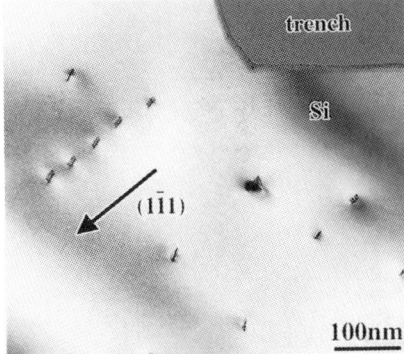

Fig. 4. Dislocations observed in a bright field cross-section of isolated trenches tilted to two beam conditions, as is indicated on the picture. The trenches are filled with a 550 nm (a) and a 500 nm (b) HDP oxide.

From cross-section measurements (Fig. 4), dislocations pointing in the [011] direction, parallel to the nitride mask are observed. The dislocation arrays lie mainly under the nitride masks (Fig. 4.a). This releases the stress in the substrate. In the opposite case, when the array of dislocations lies in the field oxide region, as can occasionally occur, the tensile stresses are lowered. To reduce the energy, the dislocations in the silicon dissociate in two Shockley partials with the formation of an extrinsic stacking fault as can be observed in Fig. 4.b. From extinctions, it follows that the two partials have a different Burgers vector of the type 1/6[121].

The HREM image (Fig. 5) shows that the stacking fault lies in a (111) plane and is terminated by two partials. Due to the high stresses around the defect, it is not possible in this case to identify the exact Burgers vectors of the partials.

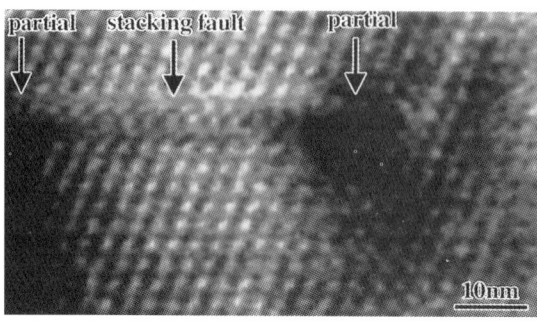

Fig. 5. HREM image in $[011]_{Si}$ of an extrinsic stacking fault terminated by two partials.

In the plan-view image shown in fig. 6, the main type of dislocations are straight 60° dislocations in [011] direction parallel to the film surface with Burgers vector b = 1/2 [110]. These are the most common undissociated dislocations in silicon and are already extensively studied (Amelinckx 1979). Dislocation networks are also observed.

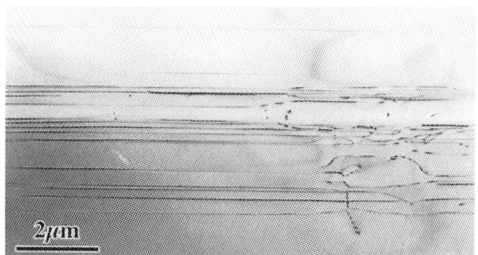

Fig. 6. Plan-view image of the isolated trenches, showing straight 60° dislocations and dislocation networks.

4. CONCLUSION

TEM measurements on unfilled STI structures show no defect formation. From bright field $[011]_{Si}$ cross-sections, it can be concluded that a 12 nm cavity-etching before the re-oxidation and the use of a 1075°C re-oxidation temperature result in an optimal rounding of the silicon corner under the nitride. Later filling of these trenches with an oxide induces stresses in the substrate. A supplementary stress due to the nitride mask on active areas may cause defects at the bottom of the trenches. Most defects are dissociated dislocations with the formation of an extrinsic stacking fault, as is observed by bright field and HREM imaging along $[011]_{Si}$.

ACKNOWLEDGEMENT

This work is supported by the Flemish institute for the encouragement of scientific and technological research in industry (IWT).

REFERENCES

Amelinckx S 1979 Dislocations in particular solids, ed. Nabarro, pp 288-300
Nandakumar M, Chatterjee, A, Sridhar S, Joyner K, Rodder M and Chen I C 1998, International Electron Device Meeting, IEDM Technical Digest, pp 133-136

Inst. Phys. Conf. Ser. No 164
Paper presented at Microsc. Semicond. Mater. Conf., Oxford, 22–25 March 1999
© 1999 IOP Publishing Ltd

Strain determination in submicron isolation structures by TEM/CBED

A Armigliato[1], R Balboni[1], S Balboni[1], S Frabboni[2], A Tixier[3], G P Carnevale[4], P Colpani[4], G Pavia[4] and A Marmiroli[4]

[1] CNR-Istituto LAMEL, Via P Gobetti 101, 40129 Bologna, Italy

[2] Istituto Nazionale di Fisica della Materia (INFM) and Dipartimento di Fisica Università di Modena e Reggio Emilia, Via G Campi 213/A, 41100 Modena, Italy

[3] ISEN-IEMN, Avenue Poincaré, B.P. 69, 59655 Villeneuve d'Ascq, France

[4] ST Microelectronics s.r.l., Via C Olivetti 2, 20041 Agrate Brianza, Italy

ABSTRACT:. The convergent beam electron diffraction (CBED) technique in a transmission electron microscope (TEM) has been employed to determine the strain distribution along a cutline parallel to the padoxide/Si interface in a 0.80 μm wide recessed-LOCOS (LOCal Oxidation of Silicon) structure. The values of the components of the strain tensor so obtained have been compared with those computed by two process simulation codes. It has been found that both the LOCOS morphology and the strain distribution, experimentally determined, were in agreement with the simulation results, if some oxidation-related parameters were modified.

1. INTRODUCTION

It has been recently recognized that the presence of localized stress fields at the perimeter of the components of a submicrometric integrated circuit, yet below the yield stress value, adversely affects the device characteristics (Hu 1991, Smeys et al 1996). Therefore, the problem of mechanical stresses must be controlled in order to improve the device reliability and the production yield. An experimental technique like CBED is then needed, since it is capable of measuring the strain distribution in a region of the silicon substrate close to the isolation structure, with a sufficient spatial resolution and sensitivity. This technique has been previously applied with success to determine the lattice strain in a silicon region underlying micrometer-sized isolation structures (Armigliato et al 1993). From the stress simulation viewpoint the LOCOS structures can be studied by the currently available two-dimensional process codes like ATHENA (Silvaco Int.'l 1996) and IMPACT (Baccus et al 1991), as they can calculate the stress/strain distribution in devices. The accuracy of the simulation is however critically dependent on the choice of the different physical parameters implemented into the codes. In order to make a suitable choice of the simulator parameters, in this work the computed profiles of the components of the strain tensor are compared with the corresponding values obtained by CBED patterns taken in different points of TEM cross sections of a 0.80 μm LOCOS structure.

2. EXPERIMENTAL

A recessed-LOCOS sample has been prepared for CBED analysis. The process flow for the sample preparation includes the deposition on a CZ silicon substrate of (i) a stress relief oxide (10 nm) called *padoxide*; and (ii) a Si_3N_4 film (117 nm) by means of a low pressure chemical vapour deposition (LPCVD) system. Then a lithographic step plus an etching step are used to define Si_3N_4

stripes (0.8 μm wide) and to reduce by 50 nm the level of silicon in the region not covered by nitride (recession). Finally, a 570 nm thick oxide, called *field oxide*, is grown.

TEM <110> cross-sections of the LOCOS structures have been analysed. They have been prepared according to a procedure which involves glueing, sawing, mechanical lapping down to 20 μm, and ion-beam milling to perforation.

The specimens have been investigated by using a Philips CM 30 TEM. The accelerating voltage was 300 kV for imaging, and 100 kV for the CBED experiments, which have been performed in a <130> orientation. A Gatan liquid-nitrogen cooled double tilt holder was employed, in order to improve the HOLZ line visibility. The spot size at the specimen level was 10 nm, obtained in the nanoprobe mode.

3. TEM/CBED RESULTS

In Fig. 1b is reported a TEM micrograph of the LOCOS structure investigated in this work. From a morphological viewpoint, the following features should be noted:

a) the lateral oxidation of silicon near the nitride edge called *bird's beak*; b) the growth of the pad oxide thickness during the field oxidation called *padoxide punch through*; c) the abrupt oxide/silicon interface discontinuity near the nitride edge; d) the reduction of the final field oxide thickness in narrow space called *field oxide thinning*.

The values of the parameters are generally in good agreement with the expected ones, except for the pad oxide thickness which is 52 nm instead of the expected 11 nm and for the field oxide thickness (474 nm instead of 570 nm). This is due to the well known *bird's beak punch through* and *field oxide thinning* phenomena. This discrepancy will strongly affect the simulation of the technological process, as it will be discussed in the next section.

On the structure reported in Fig. 1b, CBED patterns were recorded at nine points located along a cut line, at a depth of 200 nm from the pad/substrate interface. It must be noted that the strain values, as measured by CBED, are given in the cubic unit cell system (indices in caps), where the X,Y and Z axes are directed along the [100], the [010] and the [001] crystallographic directions, respectively. Instead, as will be seen in the next section, the process simulation codes assume a 'sample-based' frame of reference (x,y,z), where z≡Z=[001], x and y are orthogonal <110> directions in the wafer plane (y being parallel to the length of the nitride stripe).

From the comparison between the experimental CBED patterns and the simulated ones, the different components of the strain tensor in the nine selected points have been determined. In Figs. 1d and 1e the plots of the ε_{XX}, ε_{ZZ} and ε_{XZ} components are reported. The experimental values are drawn as diamonds, whose diagonals represent the error bars.

4. PROCESS SIMULATION OF THE LOCOS STRUCTURE

The morphological simulations of the structure have been performed using both ATHENA and IMPACT codes, with the same set of coefficients calibrated during previous work (Tixier 1998). The most relevant difference between the models implemented in the two simulators regards the stress calculation modules: in ATHENA the stress field inside the silicon substrate is calculated only at the end of the oxidation step, while for IMPACT this computation is performed during the growth of the field oxide. The set of equations implemented consists of an equilibrium relationship calculated for the bulk substrate (Silvaco Int.'l 1996) and a boundary condition applied to each node of the silicon oxide interface proportional to the velocity of the interface inside the substrate.

We observed first a better agreement in the morphology and between the measured and the simulated strain obtained with ATHENA, but there was an unsatisfactory asymmetric behaviour of the three strain components along the cutline. On the other hand, the simulations obtained with IMPACT showed a qualitatively correct behaviour of the strain, but at the same time an overestimation of the same quantities (see Fig. 1d). Moreover, significant morphological differences with respect to the real sample (Fig. 1b), particularly for the *bird's beak punch through* phenomenon, are observed (Fig. 1a).

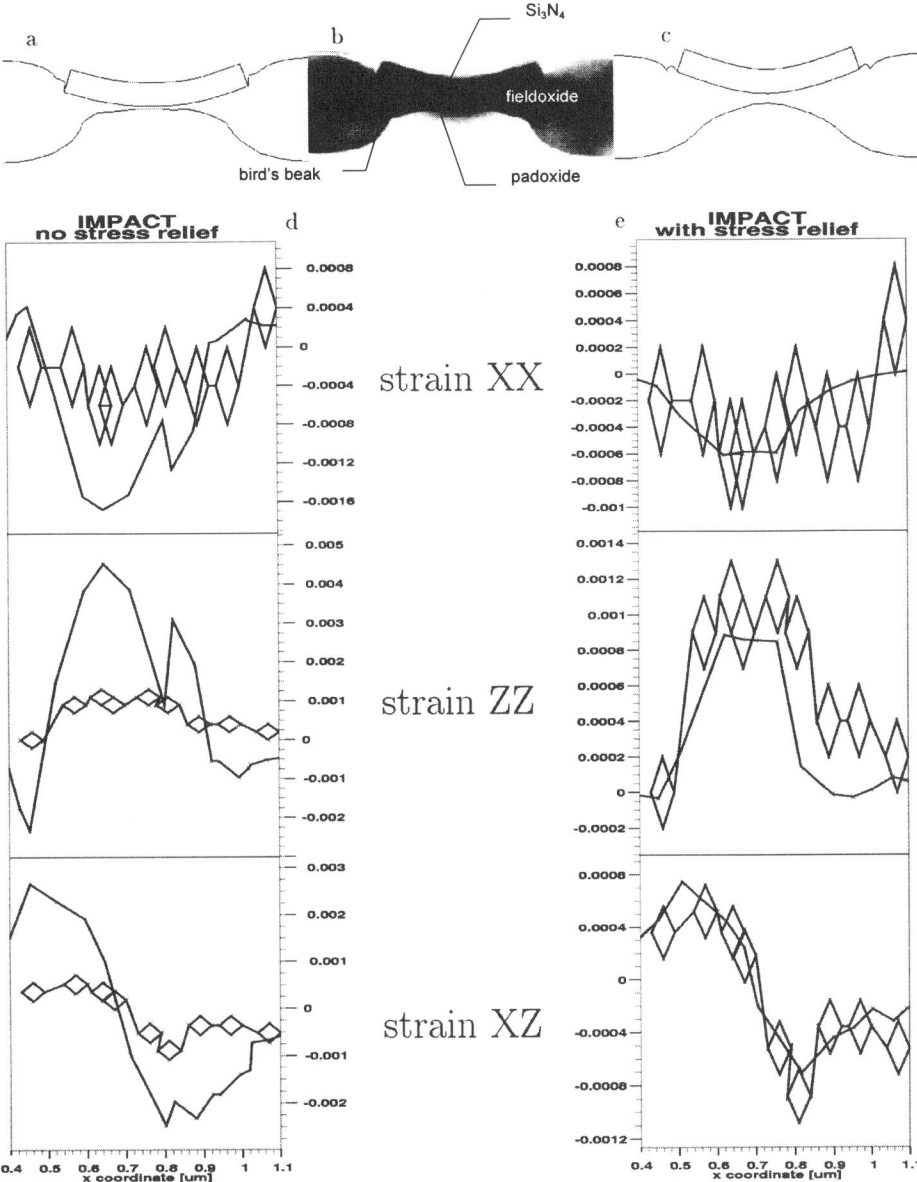

Fig. 1. Top: a comparison of the IMPACT simulated profiles (a) without stress relief and (c) with stress relief and TEM image (b). The plots (bottom) refer to the XX, ZZ and XZ components of the strain tensor (d) without stress relief and (e) with stress relief. The experimental CBED points are marked with diamonds, whose diagonals represent the associated errors.

450

As a consequence the internal code of IMPACT was modified in order to neglect the accumulation of the stress during the simulation of the oxide growth. In fact the implementation of the model of stress relief produces an effect also on the morphological profile of the simulated structure. This effect may be ascribed directly to the different stress relief models and in particular to the calculation of the reaction rate at the silicon-oxide interface which is reduced by the stress normal to the interface (Kao et al 1988). If the model of stress relief is not activated (Fig. 1a) the value accumulated during the oxidation time will be able to inhibit the reaction rate, then a thinner oxide is expected in this case. On the other hand, when the stress in the silicon is allowed to relax, the value normal to the interface will never reach values able to effectively reduce the reaction rate, then the oxide is grown freely at the interface (Fig. 1c). The latter case, in terms of strain tensor components, is reported in Fig. 1e, where a good agreement between the experimental points and the simulated ones can be observed. Also the *bird's beak punch through* phenomenon is accurately replicated.

5. CONCLUSIONS

In this work it has been demonstrated that the TEM/CBED technique can be successfully applied to the determination of the strain distribution, as well as of the morphology of a sub-micron LOCOS isolation structure. These results have been compared with the ones deduced from two dimensional process simulators (ATHENA, IMPACT). It has been found that the assumption of full stress relaxation leads us to obtain a better agreement between the simulated and experimental profiles: this hypothesis must be confirmed by other measurements performed on wafers processed at different oxidation temperatures.

ACKNOWLEDGEMENTS

This work was partially supported by CNR Applied Research Project *Microelectronics*, Theme 6: *Modellization and simulation of processes, structures and devices*.

REFERENCES

Armigliato A, Balboni R, De Wolf I, Frabboni S, Janssens K G F and Vanhellemont J 1993 Inst. Phys. Conf. Ser. **134**, 229

Baccus B, Collard D, Ferreira P, Senez V and Vandenbossche E 1998 IMPACT Isen Modelling Package For Integrated Circuit Technology, Version 4.8.1

Hu S M 1991 J. Appl. Phys. **70**, R53-R80

Kao D B, McVittie J P, Nix W D and Saraswat K C 1988 IEEE Trans. Electron Devices **ED-35**, 25

Silvaco International 1996 ATHENA User's Manual, 2D Process Simulation Software

Smeys P, Griffin P B, Rek Z U, De Wolf I and Saraswat K C 1996 Proc. IEDM 96 (San Francisco, 1996) p. 709

Tixier A 1998 PhD Thesis (University of Lille)

Inst. Phys. Conf. Ser. No 164
Paper presented at Microsc. Semicond. Mater. Conf., Oxford, 22–25 March 1999
© *1999 IOP Publishing Ltd*

Measurement of thin silicon oxide/nitride/oxide layers using transmission electron microscopy

R Beanland

Marconi Materials Technology Ltd., Caswell, Towcester, Northants NN12 8EQ

ABSTRACT: We present a method of measuring layer thicknesses in silicon oxide/nitride/oxide structures using the Fresnel fringes surrounding each interface when the specimen is out of focus. A simple electron-optical model shows that the interface lies mid-way between bright and dark fringes. If the specimen is thin enough for mass-thickness contrast differences between the materials to be negligible, it is possible to measure average layer thicknesses with an accuracy of about 0.3 nm. Experimental data showing the viability of the approach is presented.

1. INTRODUCTION

Ultrathin silicon oxide/nitride/oxide layers are now in widespread use as high dielectric constant materials in ULSI devices (e.g. Cheng et al 1995). The layer thickness in such structures is typically a few nanometres. Since these layers are amorphous, and have only slight differences in density, neither high resolution TEM or diffraction contrast can be used to distinguish between the layers. An obvious solution is to use a high resolution STEM or FE-TEM with imaging electron energy loss spectroscopy or energy-dispersive x-ray spectroscopy, which allows the nitrogen content to be observed on a nanometre scale (e.g. Sekiguchi et al 1998). This is particularly suitable when the nitrogen profile is unknown. Here, we present a means of characterising such structures using the Fresnel fringes which appear about the interfaces when the specimen is out of focus. Although this cannot quantify any gradation in nitrogen content, it does provide a rapid and easy way to measure layer thickness, with a spatial resolution of about 0.3 nm.

The basis of the technique is the observation that Fresnel fringes are present around the interface between materials of different electron refractive index. Such fringes have been used to measure, for example, layer thickness in thin oxide layers (Ross and Stobbs 1991). In cases where strong diffraction may occur, a series of experimental through-focus images must be compared with simulated images based on various models of the physical structure. In the present case, a far simpler approach is sufficient to enable layer thicknesses to be measured.

2. THEORY

The appearance of fringes in optical images at the edges of opaque objects illuminated with coherent light - i.e. near field diffraction - can be described reasonably well by the Huygens-Fresnel diffraction theory (e.g. Hecht and Zajac 1974). The intensity and position of the fringes can be calculated simply by adding up all the contributions to a given point in the image P from each scattering centre in the object, taking account of the amplitude and phase of each contribution. The amplitude is proportional to $1/r$, where r is the distance between the scattering centre and P, and also decreases with increasing scattering angle, described by the obliquity factor $K(\theta) = \frac{1}{2}(1 + \cos\theta)$. A similar approach can be applied to electron optics, the main difference being the more rapid decrease in the intensity of an electron wave with scattering angle. This is described by the scattering factor for electrons, $f(\theta)$. Here, we use

$$f^2(\theta) = \frac{4\gamma^2 Z^2}{a_0^2 k_0^4} \frac{1}{\left(\theta^2 + \theta_0^2\right)^2},$$

(1)

where $\gamma = (1 - v^2/c^2)^{-\frac{1}{2}}$, Z is the atomic number, $\theta_0 = 1/k_0 r_0$, $r_0 = a_0 Z^{1/3}$, a_0 is the Bohr radius, and $k_0 = 2\pi/\lambda$, where λ is the electron wavelength (Egerton 1996).

We now use the most basic model that can describe the behaviour of Fresnel fringes at an interface. The simplest approach is to consider the phase shift of the electron wave as it propagates through the specimen. The two materials have different electron refractive indices, and therefore there will be a difference in electron wavelength inside the two materials. This will give a difference in phase between electrons that have passed through the two materials. The problem thus reduces to the near-field diffraction image when there is a difference in phase ϕ between electrons that emerge from each side of the interface between the two materials. Since the difference in phase is purely relative, we may say that the wave emerging from material 1 has a phase shift of $-\phi/2$, and that from material 2 has a phase shift of $+\phi/2$. We ignore the slight difference in $f(\theta)$ and Z between the different materials, the effect of finite specimen thickness, microscope aberrations, and far-field diffraction effects from an objective aperture. We also consider the materials to be infinite in lateral extent, making the problem one-dimensional. The amplitude of the electron wave at a point P as a function of position, x, in the image is thus

$$A(x) = \int_{-\infty}^{0} \frac{K(\theta) f(\theta)}{r} \exp\left(ikr - \frac{\phi}{2}\right) d\theta + \int_{0}^{\infty} \frac{K(\theta) f(\theta)}{r} \exp\left(ikr + \frac{\phi}{2}\right) d\theta.$$

(2)

Inspection of Eqn. (2) reveals that $A(x)$ is an odd function, i.e. $A(-x) = -A(x)$. Thus a bright fringe on one side of the interface is mirrored by a dark fringe on the other. Performing the integration in Eqn. (2) numerically gives the intensity distributions shown in Fig. 1. As electron refractive indices are always close to 1 (typically $1+10^{-4}$), the phase shift ϕ will be small; we have chosen $\pi/50$ as a representative value; the intensity has been normalised to be equal to 1 away from the interface. It can be seen that the bright and dark fringes are equidistant from the interface, and that the distance between them increases with defocus F ($F^2 = r^2 - x^2$).

The above discussion shows that it is possible to accurately measure the position of an interface from a defocused image simply by taking the midpoint between the bright and dark Fresnel fringes, provided the assumptions mentioned above are reasonable. The assumptions concerning the specimen are best met when the specimen is very thin. Those concerning the microscope will change the intensity distribution, but will not change the antisymmetric nature of the fringes. Thus if Fresnel fringes are observed in very thin specimens, they will be antisymmetric about the interface. Also, since the electron intensity decreases very rapidly with scattering angle, the method should be applicable even to very thin layers. Other points worth noting are that the fringes converge to lie on the interface at perfect focus, and that the relative phase shift reverses when the defocus changes sign.

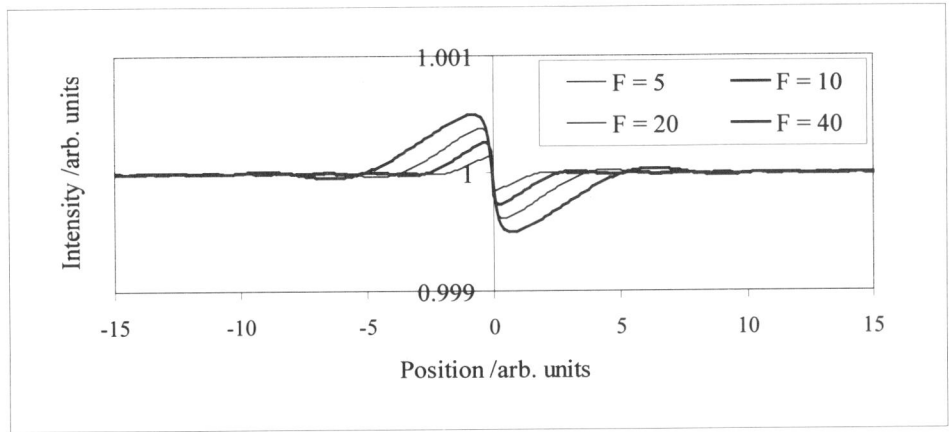

Figure 1. Fresnel fringe intensities calculated numerically from Eqn. 2.

3. EXPERIMENT

Figure 2 shows a TEM image of a test silicon oxide/nitride/oxide structure deposited on (001) silicon taken on a JEOL 120CX analytical TEM. Strong diffraction in the silicon was avoided by tilting approx. 0.5° from a 004 diffraction condition. Note that mass-thickness contrast is present in this image, since this region of the specimen is relatively thick. Fresnel fringes can be seen at each interface in the structure. The intensity, measured from digitised negatives, of a through focus series is shown in Fig. 3. Contrast due to thickness variations was removed by subtracting the intensity of an image taken at exact focus. The fringes are not as sharp as in the simple model (spherical aberration and far field diffraction at the objective aperture give a nominal point-to-point resolution of 0.8 nm). However, the basic characteristics of the model are observed, i.e. the fringes grow more intense with greater defocus and the sign reverses from overfocus to underfocus. The apparent position of the upper nitride/oxide interface can be seen to shift with focus. This may be due to a gradient in nitrogen content, which would give a gradual change in the phase shift with position. The fringes overlap in the lower oxide, which makes it difficult to obtain an accurate measure of its thickness. However, since the fringes move closer to the interface with decreasing defocus, the layer thickness can be obtained by simply by measuring the distance between the outer fringes and extrapolating to zero defocus.

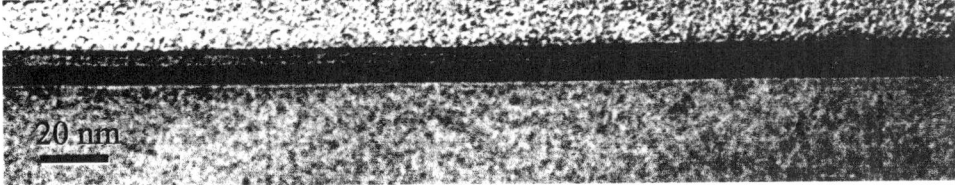

Figure 2. TEM image of an oxide/nitride/oxide structure on Si.

454

4. SUMMARY AND CONCLUSIONS

We have described a simple and rapid method for measuring the dimensions of amorphous layers that are only a few nanometres in thickness. Using simple electron-optical wave optics, we have shown that the interface lies midway between the bright and dark Fresnel fringes in the image of a very thin specimen. Experimentally we find that this midpoint varies by less than 0.3 nm in silicon oxide/nitride/oxide dielectric layers with thicknesses of a few nm. This allows a rapid and reproducible means of measuring these structures without recourse to elemental analysis. However, gradients in nitride content produce a shift in the position of the midpoint, and it is therefore important to take a through-focus series in order to be confident that the measurement is reproducible.

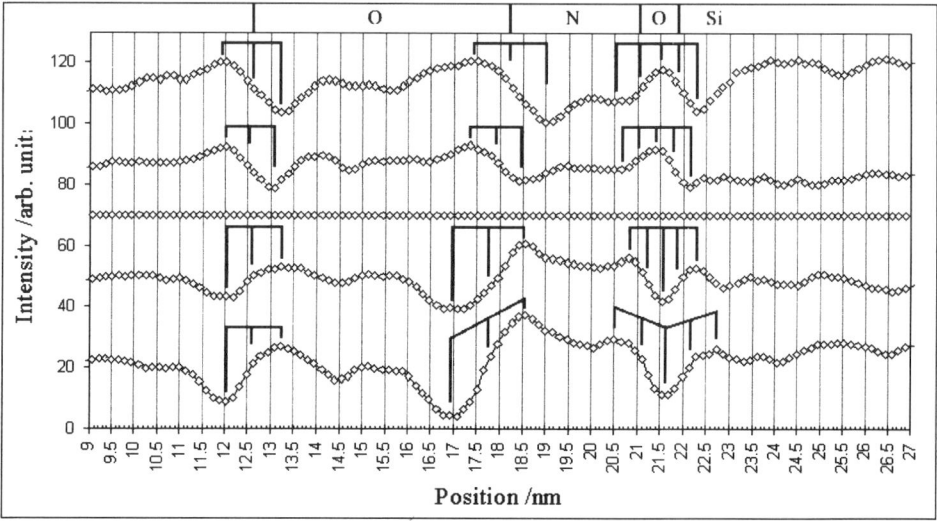

Figure 3. Measured Fresnel fringes in a silicon oxide/nitride/oxide structure. The outer lines mark the position of the bright and dark fringes, and the central line marks the position of the interface.

ACKNOWLEDGEMENTS

The samples used to develop this technique were supplied by Richard Rodrigues and Peter Kay of Seagate Microelectronics Ltd. I would like to thank Dr. Chris Boothroyd and Dr. Paul Hurley for interesting and valuable discussions.

REFERENCES

Cheng H C, Liu H W, Su H P and Hong G 1995 IEEE Electron Device Letters **16**, 509
Sekiguchi T, Kimoto K, Aoyama T and Mitsui Y 1998 Jpn. J. Appl. Phys. **37**, L694
Ross F M and Stobbs W M 1991 Phil. Mag. **A63** 37
Hecht E and Zajac A 1974 Optics (Addison-Wesley, Reading, Mass) pp 364-88
Egerton R F 1996 Electron energy-loss spectroscopy in the electron microscope (Plenum Press, New York) p 134

Inst. Phys. Conf. Ser. No 164
Paper presented at Microsc. Semicond. Mater. Conf., Oxford, 22–25 March 1999
© 1999 IOP Publishing Ltd

Transmission electron microscopy studies of silicon-based electron waveguides fabricated by oxidation of etched ridges on buried oxide layers

A Gustafsson, J Ahopelto[1], M Kamp[2], M Emmerling[2] and A Forchel[2]

Solid State Physics/nm Structure Consortium, Lund University, Box 118, S-221 00 Lund, Sweden
[1]VTT Microelectronics Centre, Box 1101, FIN-02044 VTT, Finland
[2]Technische Physik, Universität Würzburg, Am Hubland, D-97074 Würzburg, Germany

ABSTRACT: We have used transmission electron microscopy to study the structural properties of etched and oxidised Si electron waveguides in SiO_2. A variety of ridge widths was studied. The Si/SiO_2 interfaces are defined on a 1-2 monolayer scale. The smalest structures we have observed have an effective diameter <50 nm.

1. INTRODUCTION

Electron waveguides can be used for the realisation of novel semiconductor devices for low power and fast switching applications. The conventional way to fabricate the waveguides, is to use a 2-dimensional electron gas (2DEG) in a III-V semiconductor system (Weissbuch and Winter 1991). The waveguides and other devices are subsequently formed by etching and re-growth, or simply by gating and depletion of the 2DEG via metal contacts on the surface of the structure (Weissbuch and Winter 1991). An important test structure for the fabrication of this type of structure is the quantum point contact (QPC) (van Wees et al 1988). This structure is a very narrow constriction in the 2DEG, where the conductance through becomes quantized. The conductance through the constriction will be an integer number times $2e^2/h$ ($=13k\Omega^{-1}$). This typically requires a constriction with a diameter of <50 nm at helium temperature. In the ideal waveguide, the conductance is independent of the length, provided that the diameter is constant. An important device, based on electron waveguiding, is the Y-branch switch (Palm and Thylén 1992). This is an electrical device, based on the idea of an optical switching device. In the Y-branch switch, an electron waveguide is split into two branches. By applying different bias voltages on gate electrodes, the electron wave can be lead into either one of the two branches, thereby switching between two states.

One alternative way to fabricate these devices is to make them in Si instead of using III-V. A simple way would be to use reverse-biased pn-junctions to create the barriers. A different approach is to use Si-oxide as barriers. The fabrication of this type of structure is compatible with conventional CMOS technology. The only additional step needed is the definition of the waveguides, where the dimensions require electron-beam (e-beam) lithography. The compatibility also means that it is possible to incorporate conventional CMOS devices and waveguides on the same chip.

2. FABRICATION AND EXPERIMENTAL DETAILS

The fabrication of the Si waveguides incorporates a number of processing steps. The starting point was commercially available (001) separation by implantation of oxygen (SIMOX) and bonded silicon on insulator (SOI) substrates (Colinge and Bower 1998). The bonded SOI wafers were fabricated using a Smart-Cut process (Colinge and Bower 1998). The SIMOX wafer had a 380 nm thick buried SiO_2 layer (BOX) below a 200 nm single crystal Si layer. The bonded wafer had a 200 nm thick SOI layer on top of a 400 nm thick BOX film. The samples were patterned by UV-lithography and dry-etched down into the buried oxide layer. In the case of sub-250 nm features, e-beam lithography was used. An acceleration voltage of 100 kV was used to obtain significantly high resolution in the e-beam lithography. The width and thickness of the waveguides were further reduced by oxidation. In this report, we have studied two types of structures. i) samples testing the process. The samples in this category generally contain waveguides along the [110] direction with different initial widths, typically 400 nm - 2000 nm. The period (ridge:trench) of the test structures is 1:1 or 1:10. The final reduction of the ridge size was done by wet oxidation to reach a high oxidation rate. In sample A, the final oxidation stage was one hour and sample B two hours. ii) samples with device structures, i.e., QPCs and Y-branch switches (templates only). On this wafer, the size reduction after e-beam lithography and etching was done by dry oxidation, which gives a lower oxidation rate, but a much higher control on the oxidation rate. For structural analysis, dedicated cross-sectional transmission electron microscopy (XTEM) patterns were incorporated into these wafers. These patterns have four different periods (pitch:ridge): 1) 400:240 nm, 2) 400:144 nm, 3) 200:152 nm and 4) 200:120 nm, as well as broad-area ridges to determine the vertical oxidation rate. In this sample, the waveguides were oriented along both the [110] and the [1-10] directions. This will be referred to as sample C.

We have used XTEM to study the structure of the waveguides in both categories of samples. The samples in this study were prepared by conventional methods of mechanical thinning and dimpling, followed by ion milling. The TEM work was performed at either 200 kV or 400 kV under either high resolution (HR) conditions or in multi-beam bright-field (BF) conditions with the sample oriented on the [110] zone axis. All the images were obtained with the electron beam oriented along the waveguides.

3. RESULTS

3.1 The test wafer

The starting point was the SIMOX wafer. The oxide was initially 380 nm thick, with a 200 nm thick single crystal Si layer. During the oxidation process, the rectangular shape of the waveguide is modified, because the corners of the ridges oxidise more slowly. This is probably caused by a strain-induced reduction of the oxygen diffusion around these corners (Namatsu H, Horiguchi et al 1997). Concentrating on 500 nm wide ridges, the original width of the ridge is reduced to 350 nm after 1 hour (see figure 1(a)) and 325 nm after 2 hours of oxidation (see figure 1(b)). At the same time the thickness at the centre is reduced from 100 nm to 40 nm during the additional hour of oxidation. As a comparison, 1 hour of oxidation of a larger flat area results in 145 nm of material remaining below a 135 nm thick oxide layer. This is

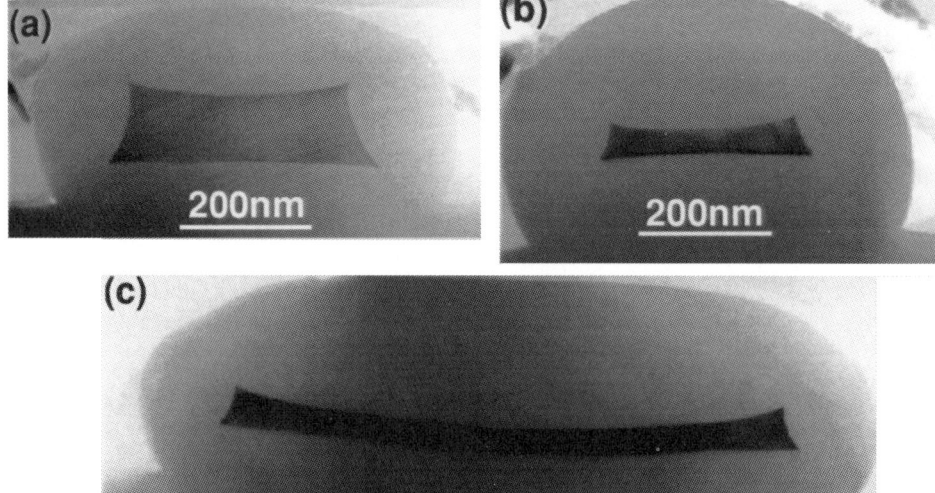

Figure 1. Images of the test structures, sample (a) A - 500 nm, (b) B - 500 nm, and C - 1250 nm.

consistent with a stoichiometric oxide, where the volume of SiO_2 is slightly more than twice (2.25 times) that of Si (Kao et al 1987). Since the oxide can only expand in one dimension, its thickness is roughly twice the thickness of the Si. The major part of the oxidation takes place on the top and side of the ridge, but some oxidation takes place at the lower interface. This can be observed in the wider ridges as a significant bending of the structure at the edges (see figure 1(c)). We also found that the oxide thickness on the top of the ridges equals that of the centre of the sides after 1 hour of oxidation. During the second hour of oxidation, the thickness reduction is larger than the width reduction. indicating that there can be some type of self limitation on the oxidation rate (Liu et al 1993). In the test samples, we do not find any specific planes defining the structure of the remaining crystalline Si. The sides of the ridges are concave rather than straight or trapezoidal (Nakajima et al 1994). After the oxidation, the interface between the oxide and the remaining crystalline material is defined within a few monolayers of Si, on the scale of TEM sample thickness. This was determined by HR imaging (not shown here).

3.2 The device wafer

This sample was made from a wafer-bonded SOI substrate. In this case, the buried oxide was 400 nm thick, below a 200 nm thick Si layer. The 200 nm thick layer was first reduced to 100 nm by thermal oxidation at 1080°C for 35 minutes, and the oxide was removed before the wafer was patterned. After patterning the remaining SOI layer, the final size reduction was done by dry oxidation for 20 minutes at 1000°C. Dry oxidation was used because of its well defined oxidation rate. In this case the broad areas reveal an oxide thickness of 95 nm on top of a 60 nm thick Si layer. Again consistent with the original Si thickness of ~100 nm.

In the patterned areas, the wider ridges (400:240, figure 2(a)) show a similar profile to the ridges of the test wafer, with a slightly thinner centre and sloping, concave sides. The lower interface is mainly straight. The narrower ridges (400:144, figure 2(b)) with the same spacing, shows a different profile, with no flat upper part but a rounded tip. The shape of the sides is still concave, but there is a tendency to form crystallographic planes at the lower part, near {111}, giving the lower part of the structure a triangular shape. The narrowest ridges (200:120, figure 2(c)) have a similar shape, but with a wider base, and not as pronounced concave sides. There are also more variations in the shape between the different ridges in this part of the sample. For most of these structures, the major part of the variations between the adjacent ridges was caused by irregularities in the e-beam lithography. The thickness of the PMMA mask was 300 nm and the resulting high aspect ratio after patterning could cause wiggling of the narrow PMMA stripe. In the close-pitch areas, there is an additional effect in the swelling of the ridges as they oxidise. The individual ridges touch, or almost touch each other after oxidation. As a comparison, the 400:144 nm ridges expand to just above 200 nm, which is equal to the pitch of the narrowest ridge. The touching of the sides of the ridges can cause local tilt of the Si cores. We observed various tilts of the Si core of the ridges. This is illustrated in the low-magnification image of figure 2 (d). The variations in the shape can also be observed in this image. The local tilt can be

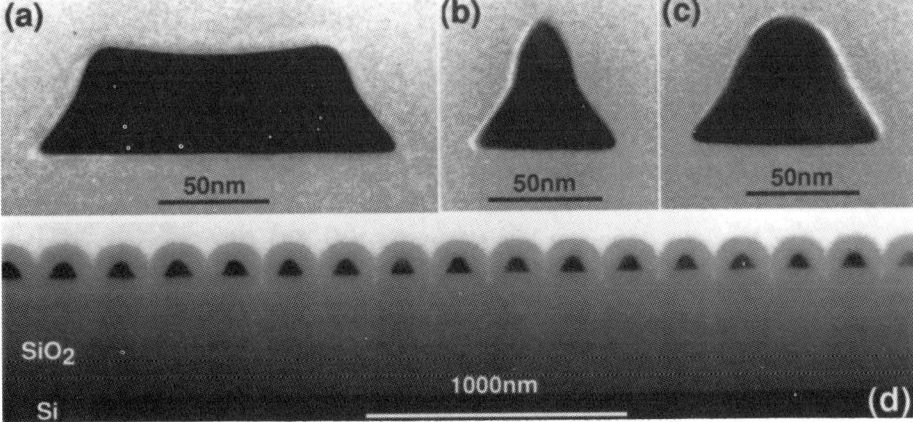

Figure 2. Images of the device structure. a) 400:240 nm, b) 400:144 nm, c) 200:120 nm and a lower magnification image of d) 200:120 nm.

seen as a contrast variation between the different cores depending on how close they are to the zone axis. One clear consequence of the smaller pitch is the larger size of these cores despite of the smaller starting size. This is due to hindered flow of the oxidizing ambient around the Si ridges.

In the device structure, the interfaces are defined on a scale of a 1-2 monolayers of Si, as determined from HRTEM images. The only exceptions are the short-pitch ridges, that are less well-defined, as discussed above.

4. DISCUSSION

The aim of these types of structures is to produce new types of devices, based on quantum effects. For silicon this will require a diameter of the waveguide below 50 nm even at helium temperature. We have demonstrated a triangular cross-section of the Si core of $60{\times}60$ nm^2. Not taking the shape of the waveguide into account, the area is equivalent to a diameter of below the required 50 nm. This is reflected in the fact that the QPCs of this device wafer show quantized conductance.

The difference between the narrow-pitch and wide pits areas of figure 2 (b) and (c) illustrates the importance of free space surrounding the ridges before the oxidation (Namatsu et al 1997). The narrowest structure was produced by a slightly wider ridge than the narrowest ridge on the sample, where the difference can be found in the problem in feeding the oxygen to the sides of the ridges. In addition there are differences in the oxidation rate in different crystallographic directions (Kao et al 1987). Furthermore, there is a tendency for the smaller cores to self-limit on {111} planes (Nakajima et al 1994). The design of the Y-branch type devices will take some significant design work, where the differences in oxidation rate of the ridges in the centre of the Y will have to be taken into account, but it could be done.

In summary, we have studied Si-based waveguide structures where we have found a well-defined Si/SiO$_2$ interface and structures with cores as small as triangular cross-sections of $60{\times}60$ nm^2. This size of the core is sufficient for a QPC to show quantized conductance at liquid helium temperatures.

ACKNOWLEDGEMENTS

The TEM studies were performed on the TEMs of Inorganic Chemistry 2, where J-O Malm is thanked for his assistance. S Eränen, H Kattelus, N Pirilä and J Pekkala are acknowledged for technical assistance. This work was done as part of the Esprit project Q-SWITCH, project number 30960, and partially funded also by Technology Development Centre of Finland (TEKES) through a grant no. 4422/96.

REFERENCES

Colinge J P and Bower R W 1998 MRS Bulletin **23**, 13 (and references therein)

Kao D-B, McVittie J P, Nix W D and Saraaswat K C 1987 IEEE Trans. Electron Devices **ED-34**, 1008

Liu H I, Biegelsen D K, Johnson N M, Ponce F A and Pease R F W 1994 J. Vac. Sci. Technol. **B11**, 2532

Palm T and Thylén L 1992 Appl. Phys. Lett. **60**, 236

Nakajima Y, Takahashi Y, Horiguchi S, Iwadate K, Namatsu H, Kurihara K and Tabe M 1994 Appl. Phys. Lett. **65**, 2833

Namatsu H, Horiguchi S, Nagase M and Kurihara K 1997 J. Vac. Sci. Technol. **B15**, 1688

van Wees B J, van Houten H, Beenakker C W J, Williamson J G, Kouwenhoven L P, van der Marcel D and Foxon C T 1988 Phys. Rev. Lett. **60**, 848

Weisbuch C and Winter B 1991 Quantum Semiconductor Structures (Academic, Boston)

Inst. Phys. Conf. Ser. No 164
Paper presented at Microsc. Semicond. Mater. Conf., Oxford, 22–25 March 1999

Studies of low-energy ion implantation in silicon

T-S Wang, A G Cullis, E J H Collart*, A J Murrell*, M A Foad* and J A Van Den Berg[+]

Dept of Electronic & Electrical Engineering, University of Sheffield, Sheffield, S1 3JD, UK
*Applied Materials, Thermal Processing and Implant Division, Foundry Lane, Horsham, West Sussex, RH13 5PY, UK
[+]Joule Laboratory, Dept of Physics, University of Salford, Salford M5 4WT, UK

ABSTRACT: We have used high resolution transmission electron microscopy, secondary ion mass spectrometry and energy filtered transmission electron microscopy to study low energy B^+ implanted Si wafers, at ion doses from 5E14 to 5E15 cm^{-2}. Some implants were carried out in Ge$^+$ preamorphised Si wafers. Selected implanted layers were annealed over a range of temperatures from 800°C to 1050°C for 1sec or 10sec. We have found the Ge$^+$ pre-amorphisation did not suppress transient enhanced diffusion during annealing. The evolution of the implantation damage structures has been assessed and the highest implantation dose resulted in the formation of disorder clusters which could trap most of the dopant.

1. INTRODUCTION

Future CMOS (complementary metal oxide semiconductor) devices, with enhanced performance, will require ultra shallow source and drain extensions to avoid the short-channel effects (Crabbé et al 1996) and, in turn, these will necessitate the use of low energy ion implantation to provide the required dopant distributions. It has also been demonstrated that Ge$^+$ preamorphisation is an effective process to suppress ion channeling and can give excellent regrowth (Ozturk et al 1988) by solid phase epitaxy (SPE). However, many features of these implants are not yet fully understood, including dopant clustering, segregation behaviour and the effect upon transient enhanced diffusion (TED) (Eaglesham et al 1994, Zhang et al 1995, Liu et al 1998).

For this paper, 8 Si wafers have been implanted with B^+ ions, at low energies of 500eV and 1keV, with various ion doses. Some implantations were carried out in Si layers which had been preamorphised with Ge$^+$. Selected implanted layers were then subjected to rapid thermal annealing (RTA) treatment at different temperatures. The structure of the samples has been analysed by using high resolution transmission electron microscopy (HRTEM) while secondary ion mass spectrometry (SIMS) studies have given information on dopant distributions. Indeed, for 500eV 5E14cm^{-2} B^+ implantation, the effect of the Ge$^+$ preamorphisation has been studied. Also, energy filtered transmission electron microscopy (EFTEM) has been used to detect boron in the highest dose samples. Boride clustering has been examined for 1keV 5E15cm^{-2} B^+ implanted samples subjected to annealing at various temperatures.

2. EXPERIMENTAL

Low energy B^+ ion implantation was carried out in an Applied Materials xRLEAP system with implant energies of 500eV and 1keV. In the case of 500eV implantation, two sets of wafers have been prepared using different implantation conditions as shown in Table 1. Two wafers, as-implanted and 1050°C RTA treated, of each set have been analysed. Both sets of samples had the same B^+ implantation ion dose of $5E14cm^{-2}$ but set-A had Ge^+ preamorphisation at 10keV with an ion dose of $5E14cm^{-2}$. Another four wafers of set-C have been implanted with B^+ ions at a dose of $5E15cm^{-2}$ at 1keV implant energy and three of them have been annealed at various temperatures between 800°C and 1000°C, as indicated in Table 1. The RTA treatment has been carried out by using a Centura system. The SIMS measurements were performed by Evans East on a Physical Electronics 6600 instrument using an O_2^+ primary beam. The beam energy was 1.5keV, with an impact angle of 65° from the surface normal. The HRTEM analyses have been carried out primarily in a JEOL 2010F field-emission TEM at 200kV, together with a Gatan imaging filter. The cross-sectional TEM specimens were thinned mechanically and then ion milled with low energy Ar^+ ions at 10° incidence in a Technoorg Linda miller with samples at LN_2 temperature: the samples were finished using 100V ion bombardment to minimise residual surface damage.

Table 1

Implant energy (eV)	500				1000		
Total Dose (ion/cm²)	5E14		5E14		5E15		
Ge⁺ preamorphisation	10keV/5E14		non		Non		
Annealing (°C)-Time(sec)	no	1050-1	no	1050-1	no	800-10	900-10	1000-10
Name of Set	Set-A		Set-B		Set-C		

3. RESULTS

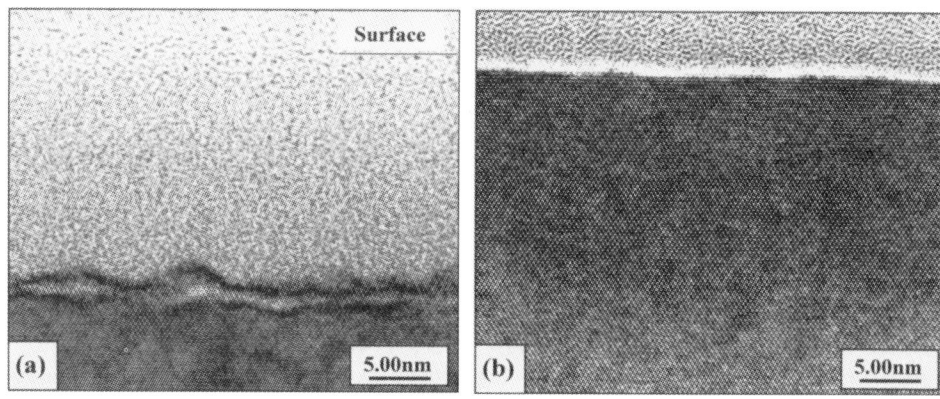

Fig. 1. *500eV B^+ implantation at a dose of 5E14 cm^{-2} with Ge preamorphisation (a) as-implanted (b)1050°C annealed.*

In Fig. 1(a), the thickness of the amorphous layer due to Ge^+ preamorphisation is shown to be ~20nm and the amorphous/crystalline interface is highly strained and exhibits nonuniform dark or bright contours. As inferred from TRIM simulation, the end of range (EOR) of 500eV B^+ ions is at a depth of ~9nm. Therefore, essentially all the implanted B^+ ions should be stopped in the amorphous region but the SIMS profile, Fig. 3, shows that some implanted boron has escaped into deeper regions. The amorphous layer was regrown with high quality after 1050°C RTA for 1sec as illustrated in Fig. 1(b). In the case without Ge^+ preamorphisation, there was no detectable structural difference between the as-implanted sample and the annealed sample under XHRTEM as shown in Fig. 2(a) and Fig. 2(b), but boron diffusion was found after annealing as illustrated in Fig. 3.

A comparison of the SIMS profiles of set-A and set-B in Fig. 3 shows that Ge^+ pre-amorphisation did suppress B^+ ion channeling beyond 9nm (EOR) during implantation, but the difference between the final diffusion profiles after annealing of the two sets of samples was minor. It seems the implanted B in the Ge+ preamorphised sample can diffuse faster in the tail region. This might be because the Ge+ preamorphsation produced more interstitials which resulted in faster TED. Also the boron concentration at the EOR was 1E20cm^{-3}, which was the crossover point of the as-implanted and the annealed profiles (Fig. 3). This suggests that the concentration of implanted B can almost reach the solid solubility at the EOR.

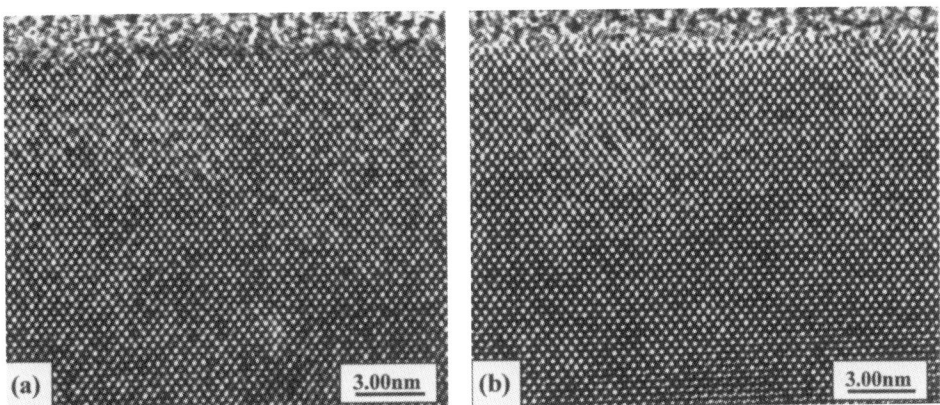

Fig. 2. *500eV B+ implantation at a dose of 5E14cm^{-2} without preamorphisation (a) as-implanted, (b) 1050 ℃ annealed.*

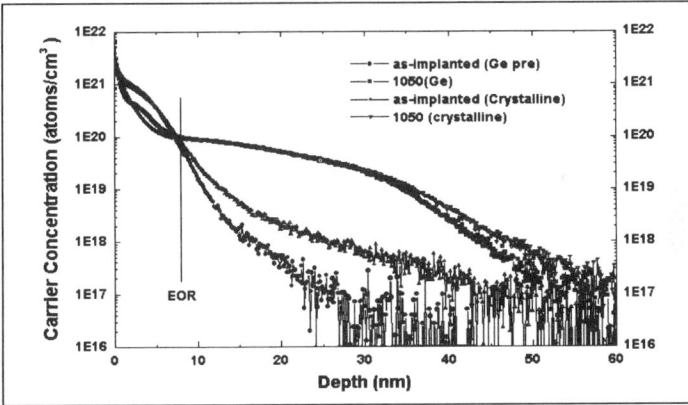

Fig. 3. *SIMS depth profiles of 5E14cm^{-2} B+ implantation at 500eV in the both cases of with and without Ge+ preamorphisation.*

For 1keV B+ ion implantation at a dose of 5E15cm^{-2}, Fig. 4(a) shows a XHRTEM image of the as-implanted sample. The surface was amorphised to a depth of not more than 5nm, which is approximately the ion projection range (Rp) of 5.8nm, as inferred from TRIM simulation work. Also, beyond the amorphous layer, the implanted zone was heavily damaged to a depth of 14nm, which extends almost to the EOR (~15nm). In the 800°C annealed sample, a distribution of disordered zones is formed extending to a depth of ~11nm, as shown in Fig. 4(b). These zones were present in the form of nonuniform defect clusters giving rise to strain fields, which produced strong contrast features in the images.

462

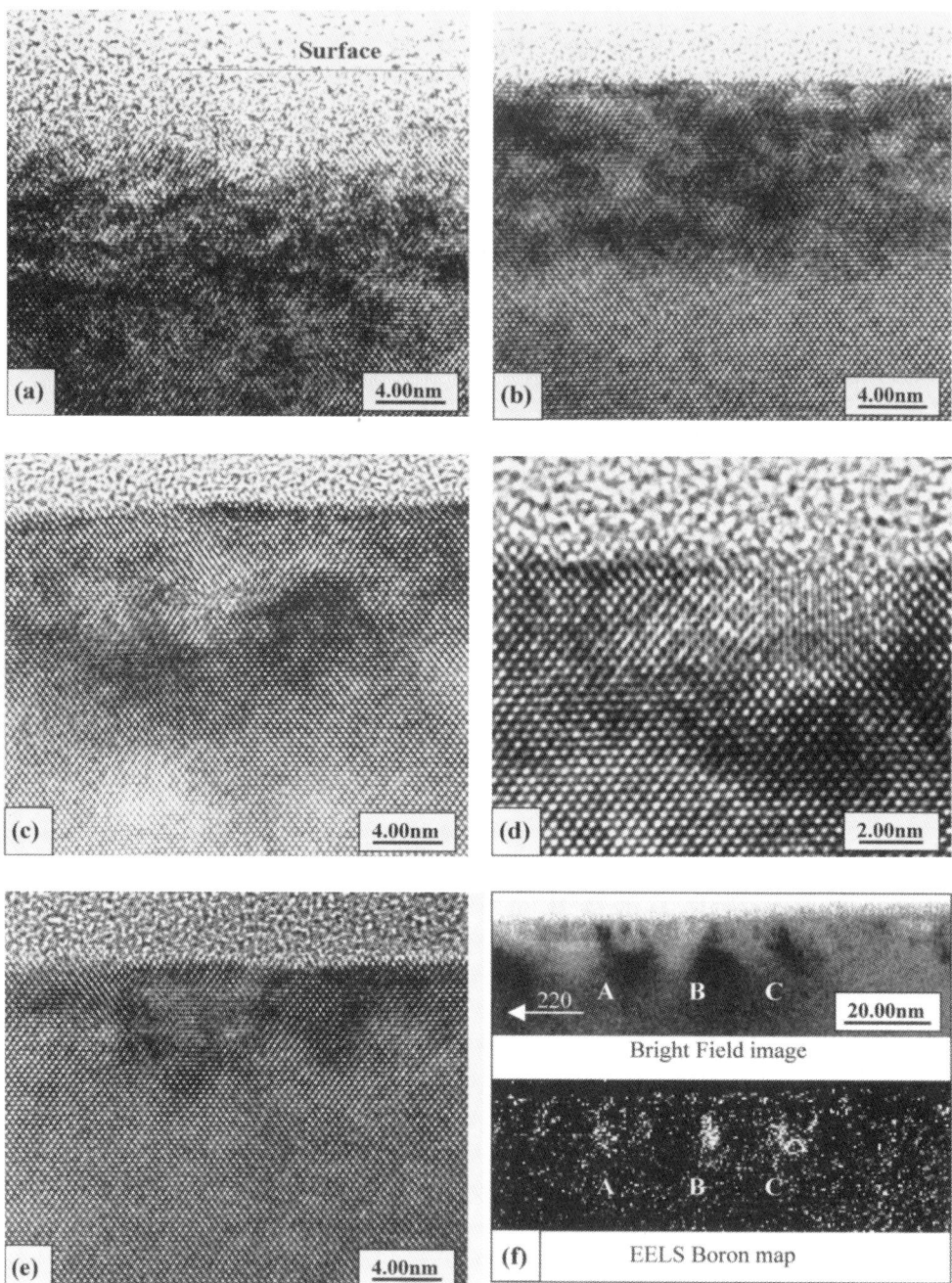

Fig. 4. *XHRTEM images of Si implanted with 1keV B^+ ions to a dose of $5E15cm^{-2}$ (a) as-implanted (b) 800°C annealed (c) 900°C annealed (d) a cluster reached the surface after 900°C anneal (e) 1000°C annealed (f) an EFTEM boron map with a bright field image.*

The clusters increased in size as the annealing temperature increased, as illustrated in Fig. 4(c), which shows a XHRTEM image of the 900°C annealed sample. Most of the clusters were of elongated shape and located at the depth of ~6nm from the surface, slightly below the Rp. In the region above the Rp, some small round clusters were found and few of them even reached the surface as shown in Fig. 4(d). This is likely to be because the implantation damage can disperse faster in the near surface region, as the surface is an excellent sink for point defects. Also boron diffusion was observed on the SIMS profile at below solid solubility level of ~1E20cm^{-3}, as shown in Fig. 5.

After 1000°C RTA treatment, the number density of clusters was less than for the 900°C annealed sample. The mean diameter of the clusters was ~4nm and they were located at a depth of not more than 7nm, as shown in Fig. 4(e). The sample was also chemically analysed using EFTEM. In order to avoid electron channeling and any unwanted contribution due to strain contrast effects, the specimen was tilted a few degrees off the [110] zone axis, along the 400 Kikuchi band. Consequently, the clusters showed little strain contrast and were almost invisible in a bright field image, Fig. 4(f) upper. EFTEM boron maps were obtained using the three-windows background subtraction method with a post-edge window placed at 210eV and pre-edge windows at 168eV and 190eV. Exposure times of 10s were used for each image with a slit width of 20eV. The boron map, Fig. 4(f) lower, shows that there is a direct correlation between structurally disordered clusters and elemental boron concentrations. This suggests that the disordered clusters comprised Si boride, since the boron concentration in clusters was extremely high. Furthermore, strain contrast could be observed around the clusters since the high boron concentration in the clusters resulted in significant lattice mismatch. Only few discrete crystallographic defects could be observed but, at a higher annealing temperature or for longer annealing time, more discrete defects might be found as a result of boron cluster decomposition.

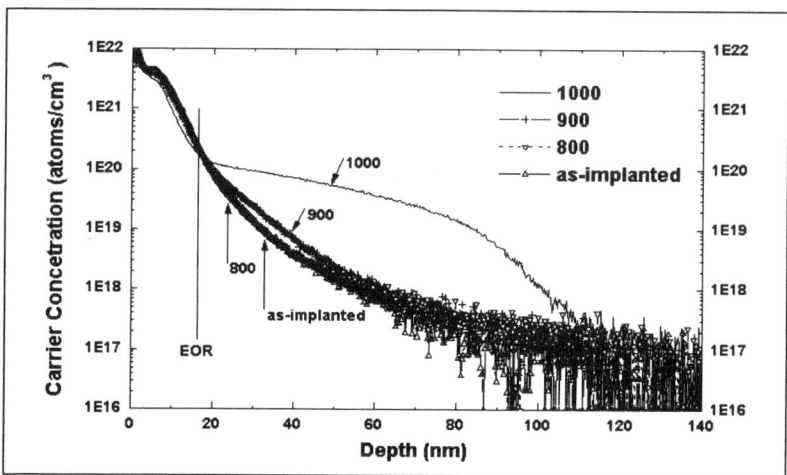

Fig. 5. *SIMS profiles of 1keV B$^+$ implantation with a dose of 5E15cm^{-2}.*

In Fig. 5, there was a 'shoulder' in each of the SIMS profiles at between depths of 5nm and 10nm, where there was also the region with a high cluster density. On the 1000°C profile, the shoulder was less well defined than the for 900°C annealed profile due to pronounced boron TED. From a comparison of Fig. 3 and Fig. 5, a less identifiable shoulder on a SIMS profile indicates more boron diffusion, especially above the solid solubility level after annealing. This suggests that for the high dose implant, the damage/clusters around the Rp region can trap most of the dopant which results in temperature-sensitive boride clustering during high temperature annealing.

4. CONCLUSION

For B^+ implantation doses of 5E14cm^{-2} at 500eV and 5E15cm^{-2} at 1keV, the concentration of the implanted boron at the EOR was ~1E20cm^{-2} which almost reached the solid solubility level. Also beyond the EOR, the implanted boron diffused significantly during the RTA treatment at 1050°C due to boron TED.

For 500eV 5E14cm^{-2} B^+ implantation, TEM studies showed that excellent regrowth has been achieved in samples both with and without Ge$^+$ preamorphisation. It also has been demonstrated that preamorphisation can reduce incident B^+ ion channeling during 500eV B^+ implantation at a dose of 5E14cm^{-2}, but it did not suppress boron TED during RTA treatment at 1050°C.

We have also found that for 1keV B^+ implantation at a dose of 5E15cm^{-2}, disorder clusters formed during annealing and, as shown by EFTEM boron mapping, boron dopant was trapped in these clusters. The clusters/damage around the Rp region reduced dopant diffusion in the region above the EOR. Also the 'shoulder' on the SIMS profiles at around the Rp region was likely to be a signature of boride clustering and was well defined as the implantation dose increased.

ACKNOWLEDGEMENTS

The authors would like to thank Dr. C J D Hetherington and Dr. D J Norris for their helpful suggestions. The authors also wish to acknowledge A Walker for technical support.

REFERENCES

Crabbé E, Logan R, Snare J, Agnello P, and Sun J 1996 Proc. IEEE International Electron Devices Meeting pp 571-574

Eaglesham D J, Stolk P A, Gossmann H-J and Poate J M 1994 Appl. Phys. Lett. **65**, 2305

Liu J, Krishnamoorthy V, Johes K S, Law M E, Shi J and Bennett J 1998 Proc. IEEE Ion Implantation Technology pp 626-629

Ozturk M C, Wortman J J, Osburn C M, Ajmera A, Rozgonyi G A, Frey E, Chu W-K and Lee C 1988 IEEE Trans. Electron Devices **35**, 659

Zhang L H, Jones K S, Chi P H and Simons D S 1995 Appl. Phys. Lett. **67**, 2025

Inst. Phys. Conf. Ser. No 164
Paper presented at Microsc. Semicond. Mater. Conf., Oxford, 22–25 March 1999
© 1999 IOP Publishing Ltd

Chemical bevelling and SIMS linescan analysis of low energy boron implanted silicon

K P Johansen, D S McPhail and S Fearn

Department of Materials, Imperial College, Prince Consort Road, London, SW7 2BP, UK

ABSTRACT: High magnification bevels were produced in samples of low energy boron implanted silicon by chemical etching. Reproducible etching results were obtained for a range of doses for 1keV implanted annealed and non-annealed samples. Implant profiles of boron were obtained by linescanning along the bevel and depth calibrating the measurements using light interference 3D surface profiling. The results were found to be in very good agreement with low energy Secondary Ion Mass Spectrometry depth profiles.

1. INTRODUCTION

Recent developments in VLSI technology have led to sub-micron characteristic device lengths and the use of implantation energies less than 1 keV. Predictions from theoretical models become inaccurate at these low energies, and new characterisation techniques are needed in order to be able to quantify the implant profiles with a depth resolution of a few nm. The analysis time for Secondary Ion Mass Spectrometry (SIMS) depth profiling increases rapidly for beam energies below 1keV, magnifying the effect of instrumental factors such as drift. Physical damage introduced by the use of SIMS depth profiling, such as beam induced mixing and the development of surface topography in the crater base, also underlines the need for the development of alternative techniques.

The idea of chemical bevelling aims to enable in-depth analysis of multilayer structures by a bevel and linescan approach without the physical damage introduced by sputtering the material away. The method also offers the advantage of rapid data acquisition compared to the other high resolution SIMS techniques, which in turn reduces the cost of the analysis and minimises the apparatus related problems. The technique was initially developed for GaAs structures by Hsu and McPhail (1995). The etching of silicon by the $HF:HNO_3:CH_3COOH$ system has been extensively investigated by several authors (Sze (1985), Turner (1960) and Robbins and Schwartz (1961)). The effect of dopant concentration on the rate of etching has also been studied by Ukraintsev et al (1998), Gong et al (1989 and 1995) and others. The theoretical bevel magnification M_t for a constant dipping rate R has been expressed by Hsu and McPhail (1995) as a function of the rate of etching r as

$$M_t(r) = R/r \qquad [1]$$

Based on experimental observations suggesting that the depth resolution of SIMS linescan implant profiles is controlled by two different mechanisms at low and high bevel magnifications, a model was developed by Hsu et al (1995). For very large bevel magnifications the three major depth resolution-limiting factors were considered to be upward beam induced mixing, the information depth and the microtopography. A term accounting for the finite width of each linescan point on the bevelled surface must be added for low values of bevel magnification or large beam diameters.

2. EXPERIMENTAL PROCEDURE

Non-annealed and annealed samples of ^{11}B implanted silicon were bevelled using a mixture of HF (40%), HNO_3 (69%) and CH_3COOH (glacial) with a $HF:HNO_3:CH_3COOH$ ratio of 5:70:17. SIMS linescans were then conducted along the bevel and the data were subsequently depth calibrated using a light interference surface profiler. A bevelling apparatus (Fig. 1) with a dip rate of 15 mm/min was used, and the samples were then lowered into the solution (25^0C) at a constant rate.

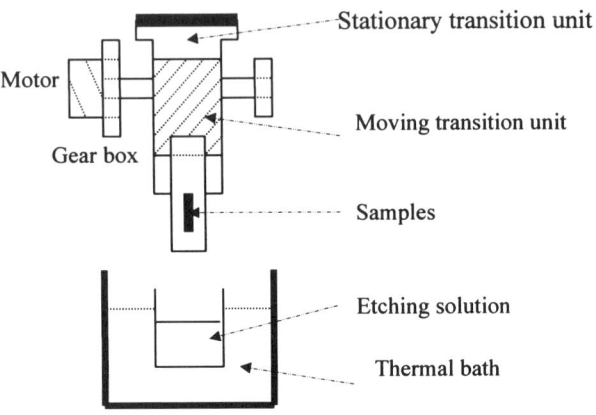

Figure 1 *Bevelling apparatus.*

The linescans (1000 μm long) and depth profiles were done on a FLIG-based Atomika 4500 SIMS instrument using a 500eV primary O_2^+ beam at normal incidence. The beam focus was estimated from the surface profiles of the linescan trenches to be 50μm and the information depth to be 4-5nm. A Zygo® NewView 3D light interference surface profiler was used to assign a depth value to each of the data points from the SIMS linescan.

3. RESULTS AND DISCUSSION

The bevelled surfaces were found by atomic force microscopy to have a roughness similar to that of the original surface. The etch rate was found to be higher in the implant region in accordance with previous studies by Spinella et al (1995). Typical bevel magnifications were in the region 5000-10000.

The depth profiles from the bevel and linescan approach for non-annealed silicon samples implanted with ^{11}B doses of $5x10^{14}$ and $1x10^{15}$ atoms.cm^{-2} and 1keV implantation energy are compared with conventional SIMS depth profiles in Figs. 2.1 and 2.2. The results for the other materials that were investigated are not illustrated here because of the limited space available, and because they display the same characteristic differences between the bevel-and-linescan profile and the conventional depth profile as discussed below.

The general shape of the linescan and depth profile curves is the same, albeit with a number of differences. The linescan data appear to oscillate slightly around the depth profile data, especially at depths above 15nm, and the conditions used for the initial linescan data acquisition reported on here appear not to be ideal for picking up the peaks of the shallow boron implants.

The differences between the depth profile and linescan results for each material can in general be explained by a number of factors, and these factors can be divided into three categories: bevel magnification, linescan conditions and surface profiling conditions.

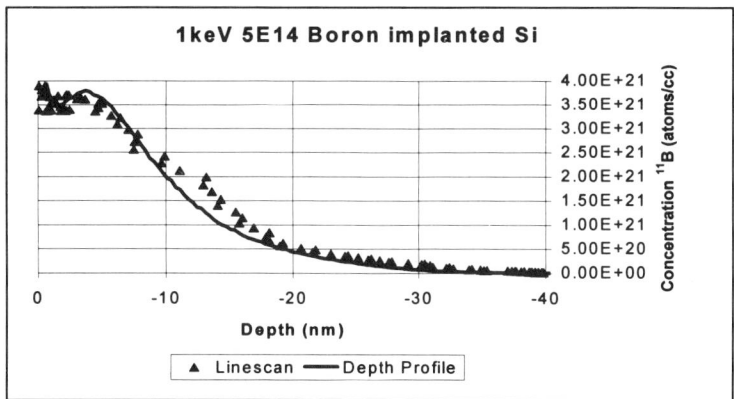

Figure 2.1 *Depth profile and linescan results for non-annealed boron implanted silicon (dose 5E14 atoms.cm^{-2}).*

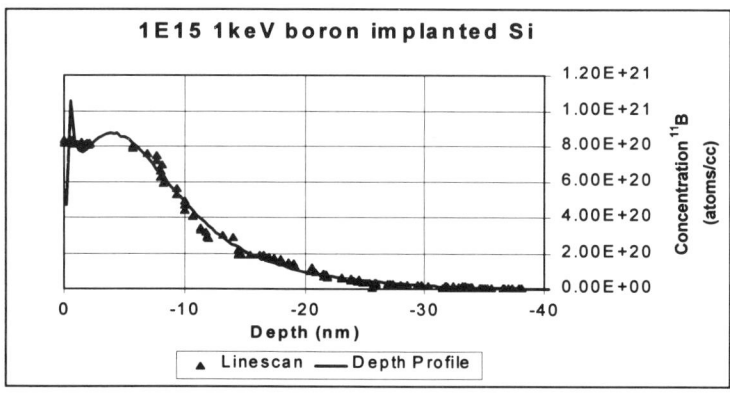

Figure 2.2 *Depth profile and linescan results for non-annealed boron implanted silicon (dose 1E15 atoms.cm^{-2}).*

All the samples analysed were found to have an apparent surface waviness of 1.5-2nm, and this clearly influences the results by making the determination of the zero depth difficult and to a certain extent arbitrary. The cluster of data points close to the assigned zero depth level for the bevel-and-linescan profile shows how this surface waviness influences the results obtained by this method. The apparent surface waviness could arise from either a genuine unevenness of the surface or from vibrations during the light interference profiler measurements. Further experiments are needed in order to confirm the origin of this waviness. The light interference surface profiling technique would also benefit from an investigation into the effect of a variation in the refractive index across the surface, caused by e.g. oxide layers, to quantify the error introduced in the measurements as a result of this factor.

The importance of a high bevel magnification in the region of interest can be seen in Fig. 2.2. The etch rate was found to be strongly dependent on the boron concentration in agreement with previous studies by Muraoka et al (1973). The rapid etch rate in the region of the implant peak, i.e. the low bevel magnification, meant that the spacing between adjacent linescan data points was too large to pick up the peak of the implant. Clearly this is undesirable. The problem can be overcome by decreasing the length of the linescan, hence reducing the separation between adjacent linescan data points. Shorter linescans can also be profiled at higher magnifications on the light interference profiler, which means that the camera resolution improves accordingly. With the current bevel magnifications for ^{11}B implanted silicon the linescan lengths can be decreased to and still cover the total exposed length of the implant.

The profiling of linescans after SIMS analysis requires that the trench created by the sputtering of material can be observed on the screen of the light interference profiler. This depends on the apparent pixel size, and hence on the objective magnification used. A linescan 300μm long can be profiled on the Zygo® surface profiler in one image using a magnification of 15. The apparent pixel size using at this magnification is 0.73μm and the lateral resolution is 1.29μm. A trench width of 1.5-2μm can hence be detected. A good beam focus is also required to obtain independent data from closely spaced linescan points, and the smaller the beam diameter the smaller the depth range of the information contained in one linescan data from the bevelled surface. For a typical bevel magnification in the implant peak region of 5000 neglecting the effects of analytical depth removed and upward mixing, a 50μm diameter linescan data point contains information from a 10nm depth range whilst a 2μm beam limits the data point depth range to 0.4nm. An improved beam focus is therefore expected to contribute to a significantly improved depth resolution of the bevel and linescan results.

However, the depth resolution ultimately depends on the analytical depth, i.e. the amount of material removed by the linescan. To obtain a depth resolution of ultimately about 1nm the analytical depth must be reduced to a similar value. The dynamic range of the linescan results would be expected to decrease somewhat as a result of the reduction in the analytical volume per data point.

4. CONCLUSIONS

High magnification bevels have been successfully produced by chemical etching. SIMS linescans have been performed along the bevel to produce results in good agreement with low energy depth profiles. Further improvement in the depth resolution to values in the range 1-2nm is expected once the SIMS analysis conditions have been optimised. Further work is needed to characterise the influence of variations in refractive index of the surface of the bevelled samples on the light interference profiling measurements.

ACKNOWLEDGEMENTS

We are grateful to Professor Mark Dowsett and Dr Graham Cooke for giving us access to their Atomika 4500, to facilitate the low energy linescan analyses of the bevels.

REFERENCES

Gong L, Barthel L A, Lorenz J and Ryssel H 1989 Proc. ESSDERC'89, eds A Heuberger, H Ryssel and P Lange (Springer: Berlin) p 198
Gong L, Petersen S, Frey L and Ryssel H 1995 Nucl. Instr. Meth., B96, 133
Hsu C M and McPhail D S 1995 Nucl. Instr. Meth. Phys. Res., B 101, 427
Hsu C M, Sharma V K M, Ashwin M J and McPhail D S 1995 Surf. Interf. Anal. 23, 665
Muraoka H, Ohhashi T and Sumitomo Y 1973 2nd Intl. Symp. on Silicon Materials, (Electro. Chem. Soc.: Pennington), pp 327-338
Robbins H and Schwartz B 1961 J. Electrochem. Soc. 108, 365
Spinella C, Raineri V, Saggio M, Privitera V and Campisano S U 1995 Nucl. Instr. Meth. B96, 139
Sze SM 1985 Semiconductor Devices, Physics and Technology (Wiley: New York) pp 451-456
Turner D R 1960 J. Electrochem. Soc. 107, 810
Ukraintsev V A, McGlothin R, Gribelyuk M A and Edwards H 1998 J. Vac. Sci. Technol. B16, 476

Inst. Phys. Conf. Ser. No 164
Paper presented at Microsc. Semicond. Mater. Conf., Oxford, 22–25 March 1999
© *1999 IOP Publishing Ltd*

Indium implantation in silicon: TEM, RBS and SIMS characterization

G Pavia[1], A Losavio[1], G Queirolo[1], F Zanderigo[1] and G Ottaviani[2]

[1]STMicroelectronics, Central R&D, via C. Olivetti 2, I-20041 Agrate Brianza, Italy
[2]Università di Modena, Via Campi 213A, I-41100 Modena, Italy

ABSTRACT: Silicon samples have been characterized after indium implantation at different doses, energies and incidence angles by Transmission Electron Microscopy (TEM), Rutherford Backscattering Spectroscopy (RBS) and Secondary Ion Mass spectroscopy (SIMS). TEM analysis has been performed both in cross-section and plan view, RBS characterization has been performed in channeling geometry. Indium produces strongly damaged regions, which are not completely amorphized up to 150 keV, 5×10^{13} at/cm^2. The precipitation phenomena expected due to the low solid solubility have been studied on samples thermally treated for different times at different temperatures.

1. INTRODUCTION

Device size reduction in ultra large-scale integration (ULSI) technology requires junctions shallower than about one hundred nanometres. Boron exhibits relatively large diffusion and eventually will produce short channel effects. Indium is an acceptor dopant in silicon with low diffusion coefficient, about a factor 5 less than boron at the same temperature, and for this reason it is a valued candidate for the ultra-shallow junctions in the next generation devices. In addition, boron implantation leaves in the implanted region a number of lattice defects while indium is expected to induce a more recoverable damage. Limits to the application of indium are imposed by its low solid solubility. Only a moderate doping can be obtained, and precipitation phenomena could appear due to the various thermal treatments included into a typical CMOS process flow. Nowadays, indium has already been used to obtain a retrograde channel profile in deep sub-micron nMOSFET devices (Taur et al 1993). Indium is expected to improve the carrier mobility close to the surface, to reduce the appearance of short channel effects, and to increase the junction capacitance and the body effect.

2. EXPERIMENTAL

Indium has been implanted in (100) Cz silicon wafers at doses ranging from 10^{12} to 5×10^{13} at/cm^2, energies between 50 and 200 keV and at 0 and 7° implant angles. All the samples have been implanted at room temperature through a 10 nm silicon oxide layer. The split scheme is reported in Table 1 together with the TEM results. The work has been carried out around the best-suited condition to produce an ultra shallow junction and avoid precipitate formation. A LEO 922 Omega TEM, at 200kV accelerating voltage, has been used to obtain the micrographs. TEM analyses have been performed on both Si <110> cross sections and <100> planar view. RBS measurements have been carried out with a 2 MeV ^{4}He$^+$ ion beam within a Van De Graaf accelerator in the channeling geometry configuration and using simultaneously two detectors placed at 20° and 73°. Before analysis the SiO$_2$ layer was removed with buffered HF. SIMS has been performed with a CAMECA IMS-4f. An oxygen primary beam with impact energy of 5.5 keV has been used to determine indium profiles both before and after thermal treatments.

3. EXPERIMENTAL RESULTS

The TEM results are schematically reported in the following table, for each sample analyzed. In the "amorphized" column, data are reported only if a clearly near-continuous layer is found in TEM images.

Table 1: split scheme and TEM measurements

Energy	Dose	Tilt	Damage	Amorphized
50 keV	$7x10^{12}$	7°	30 nm	-
100 keV	$7x10^{12}$	7°	55 nm	-
150 keV	$7x10^{12}$	7°	80 nm	-
200 keV	$7x10^{12}$	7°	110 nm	50 -100 nm
150 keV	$7x10^{12}$	0°	80 nm	-
150 keV	$1x10^{12}$	7°	u.d.l.	-
150 keV	$3x10^{12}$	7°	60 nm	-
150 keV	$1x10^{13}$	7°	80 nm	30 - 50 nm
150 keV	$5x10^{13}$	7°	80 nm	10 - 65 nm

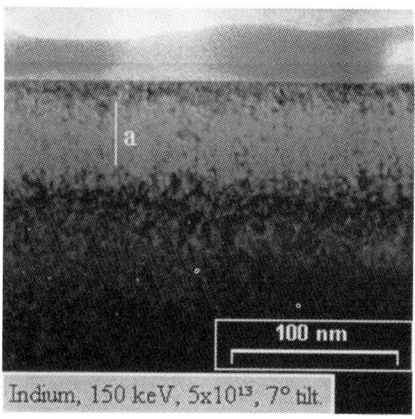

Figure 1. X-TEM micrograph showing amorphized region near the sample surface.

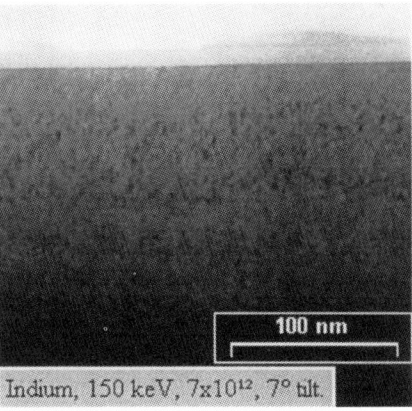

Figure 2. X-TEM image showing damage but not amorphization at lower implant dose.

For all samples a damaged but still crystalline region is found at the surface. Deeper into the samples, RBS data show that a damaged region is present. The amount of damage increases as expected with the implanted dose, but complete amorphization is not obtained even for the higher dose as also shown by the TEM image (Fig. 1).

The dose $7x10^{12}$ at/cm^2 is not generally enough to produce even localized amorphization (Fig. 2); only at 200 keV, the number of recoils is large enough to induce a local amorphization.

The two lower doses ($1x10^{12}$ and $3x10^{12}$), with primary beam energy of 150 keV, show respectively a non-detectable damage, and a damage extending to about 60 nm.

The plan views confirm that there is not noticeable damage at the $7x10^{12}$ at/cm^2 fluence, while at $5x10^{13}$ at/cm^2 the amorphized regions are clearly distinguishable from the still crystalline ones (Fig. 3).

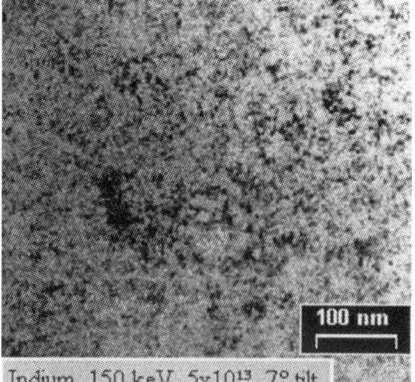

Figure 3. Plan view of amorphized and crystalline regions in the highest dose

After thermal annealing (900°C, 30 min.) of the sample implanted at the highest dose (Fig. 4), indium precipitates heterogeneously at the End Of Range (EOR) defects. Other precipitates are found closer to the surface possibly due to precipitation at the external EOR defects, at the location of the implant peak, or where the two crystallization fronts meet. After further treatments, to simulate the full process thermal budget, only a few precipitates at EOR location are still present. At the lowest dose $(7x10^{12} at/cm^2)$ no indium precipitation occurs, either after one or many thermal treatments.

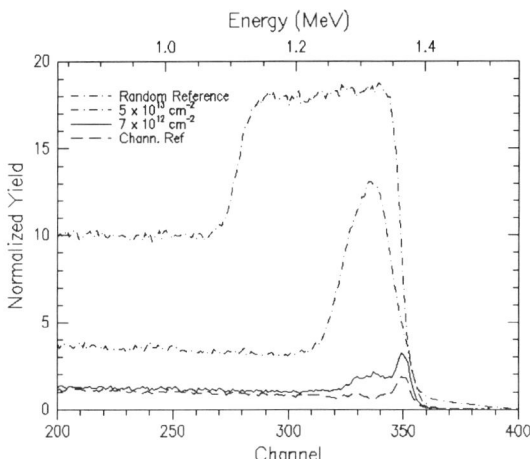

Figure 4. RBS spectra in channeling condition obtained in two samples implanted at 150 keV 7° tilt with 2 different doses. Spectra from reference samples are also reported.

Figure 4 shows the RBS channeling spectra obtained from the samples implanted at 150 keV 7° tilt with doses of $7x10^{12}$ and $5x10^{13}$ at/cm^2 with the detector placed at 73°. In the same picture are also reported as reference the spectra from an Ar amorphized single crystal silicon sample and from <100> unimplanted silicon. The

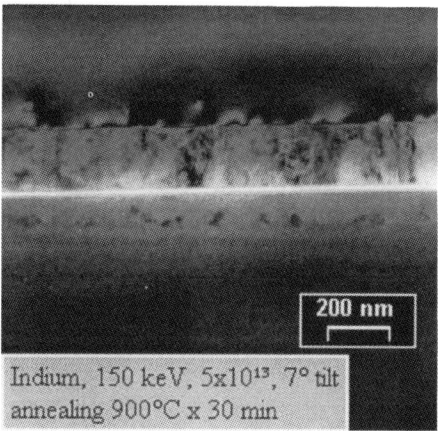

Figure 5. indium precipitates in silicon after thermal treatment at 900°C 30 min.

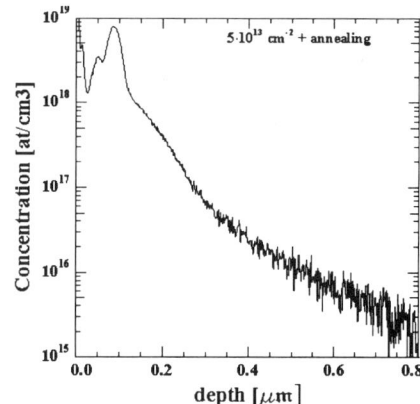

Figure 6. SIMS profile of the sample implanted at $5x10^{13}$ at/cm² and annealed at 900°C 30 min.

damage in both the implanted samples is localized between 10 and 70 nm, in agreement with TEM data. The shape of the curve of the sample implanted at $5x10^{13}$ at/cm^2 is consistent with presence of an amorphous layer and a narrow crystalline region still present at the surface. The other sample is substantially undamaged with an amorphous region which is about only 10% of the total one.

SIMS measurements too suggest the presence, in the most doped sample, of precipitates at the same locations found in the TEM images. This has been found both after the initial thermal treatment

472

at 900°C 30 min (Fig. 6) and at the end of the process. In the last case, not shown, there are noticeably fewer precipitates, in agreement with TEM results. Annealing at 900° 30 min of the sample implanted at $7x10^{12}$ at/cm^2 does not give rise to precipitate formation as the In concentration is below the solid solubility. This has been verified both with TEM (not reported) and SIMS profile (Fig. 7).

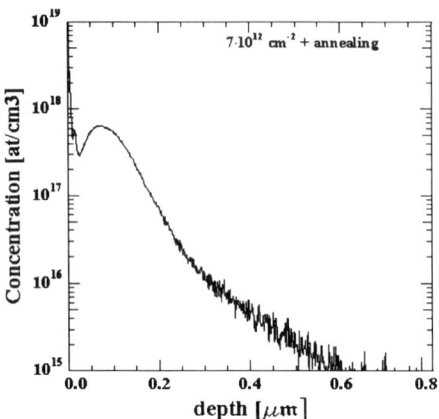

Figure 7. SIMS profile of the sample implanted at $7x10^{12}$ at/cm^2 and annealed at 900°C 30 min.

4. CONCLUSIONS

Indium implantation and diffusion has been studied at low and high fluence. At low doses, as typically used for channel doping, the indium concentration is below the solid solubility limit and good quality of silicon is found after thermal diffusion, without formation of any precipitate. In contrast, for higher doses, precipitation phenomena are present, as shown by SIMS and TEM data. From SIMS profiles, a value of $1.5x10^{18}$ at/cm^3 for the solid solubility at 900°C is found.

AKNOWLEDGEMENTS

The authors want to thank Dr. Maurizio Sbetti and Dr. Massimo Bersani, ITC-IRST, Trento for acquisition of the SIMS profiles.

REFERENCE

Taur Y, Wind S, Mii Y J, Lii Y, Moy D, Jenkins K A, Chen C L, Coane P J, Klaus D, Bucchignano J, Rosenfield M, Thomson M G R and Polcari M 1993 IEDM Tech. Dig. 127

Inst. Phys. Conf. Ser. No 164
Paper presented at Microsc. Semicond. Mater. Conf., Oxford, 22–25 March 1999
© 1999 IOP Publishing Ltd

A novel method for studying the regrowth of implanted silicon

K D Vernon-Parry[+], I Brough[#] and J H Evans-Freeman[+]

[+]Centre for Electronic Materials and Dept. of Electrical Engineering & Electronics, UMIST, PO Box 88, Manchester M60 1QD
[#]Centre for Electronic Materials and Manchester Materials Science Centre, UMIST, PO Box 88, Manchester M60 1QD

ABSTRACT: A new application of electron back-scattered diffraction (EBSD) pattern analysis has been developed to study the regrowth of implanted silicon. EBSD patterns are produced by elastically scattered electrons and are intimately related to the local crystal orientation and quality. In this work, a Philips XL-30 FEG SEM equipped for EBSD has been used to study self-implanted silicon in cross-section. An abrupt boundary was observed between the crystalline substrate and the amorphous layer. The results agree well with cross-sectional transmission electron microscopy data on the same sample.

1. INTRODUCTION

Implantation amorphization with subsequent regrowth is a standard VLSI process, widely used to prevent channelling when doping for shallow junctions and for more specific applications such as Er implanted into Si for optoelectronic devices (Seidel 1983, Evans-Freeman et al 1998). It is important to ensure that the amorphous layer is fully regrown without using overly long annealing times, as these will cause excessive diffusion.

Studies of recrystallization of amorphous silicon have traditionally used techniques such as RBS (Csepregi et al 1975), cross-sectional transmission electron microscopy (XTEM) and time-resolved reflectivity (Olson and Roth 1988). None of these methods is particularly quick and easy, and they are not attractive if the only information that is required is whether or not the sample is fully recrystallized.

Electron back-scattered diffraction patterns arise from the coherent scattering of elastically back-scattered electrons by the ordered arrangement of atoms in the surface layer of the sample. Dingley and Randle (1992) have reviewed the primary use of EBSD, which is the determination of crystal orientation in polycrystalline material on a grain-by-grain scale.

Silicon is routinely used as a calibration standard for EBSD systems: however, EBSD has rarely been applied to the systematic study of semiconductors (Wilkinson 1996). To date, all EBSD work on silicon has been done on plan-view samples.

2. SPECIMEN PREPARATION

2.1 Details of samples

A wafer of n-type Si with (100) orientation and a resistivity of 0.6 Ωcm was self-implanted at liquid nitrogen temperature using doses of 1×10^{16} cm^{-2} ions at an energy of 1.2 MeV followed by 6×10^{15} cm^{-2} ions at an energy of 0.5 MeV. This was planned to give an amorphous layer approximately 1.5 μm thick. This is deeper than usual for device fabrication (eg Franzò et al 1997) and was chosen to ensure that the end of range damage from the implantation was beyond the depletion width at 10V reverse bias to enable electrical characterisation (Scholes 1998).

474

2.2 EBSD specimen preparation

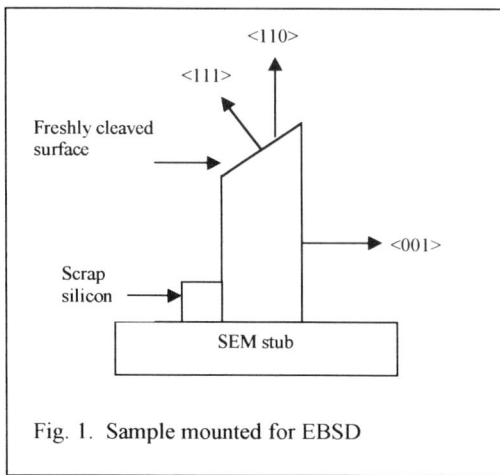

Fig. 1. Sample mounted for EBSD

Pieces a few mm square were cleaved from the wafer, degreased and the native oxide removed. They were then given conventional furnace anneals of between 0 and 6 hours at 550°C under dry nitrogen, a typical regrowth temperature (Csepregi et al 1975, Huda et al 1997).

After annealing, the samples were scribed lightly on the back and cleaved. The specimen was mounted vertically on a SEM stub using RS silver conductive paint with the freshly cleaved surface uppermost. A small piece of scrap silicon was found useful to give the sample a little extra stability. After drying at room temperature overnight the samples did not drift in the SEM, even at high magnifications. (To collect an area scan of the sample, the specimen must be stable for upwards of 10 minutes at magnifications of 40 – 50 thousand times.)

3. EBSD ANALYSIS

Humphreys and Brough (1999) describe the system used in this work. The samples were put into the Philips XL-30 FEG SEM and tilted to 70°. This angle optimises the quality of the EBSD patterns, which are collected by a camera inserted into the chamber (specimen to camera distance is 25 mm). The Channel+ program was used to gather the data and index the EBSD patterns. Further analysis of the data is possible using the VMAP software. A flat region of the edge was selected at a magnification of 40,000x, and the beam (20kV, beam current 5×10^{-8}A) scanned across it in 80 nm steps. At least 10 such scans were performed on each sample. There is an abrupt change in the EBSD pattern at the amorphous-crystalline boundary, which is highly reproducible. Fig. 2 illustrates EBSD patterns from (a) an amorphous region and (b) from a crystalline region (indexed using Channel+; VMAP was used to discover that <110> was normal to the SEM stub). Note that no lines are apparent in the EBSD pattern from the amorphous region, while they are readily visible in the pattern from the crystalline silicon.

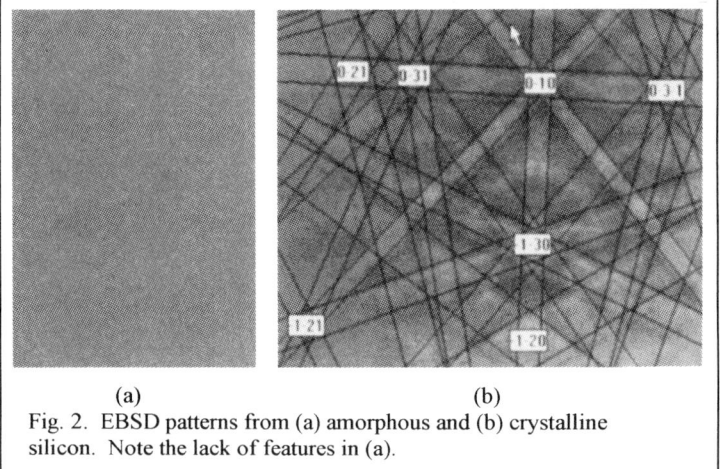

(a)   (b)

Fig. 2. EBSD patterns from (a) amorphous and (b) crystalline silicon. Note the lack of features in (a).

4. RESULTS

The regrowth results are summarised in Fig. 3. The average rate of regrowth is 0.67Å/s. This

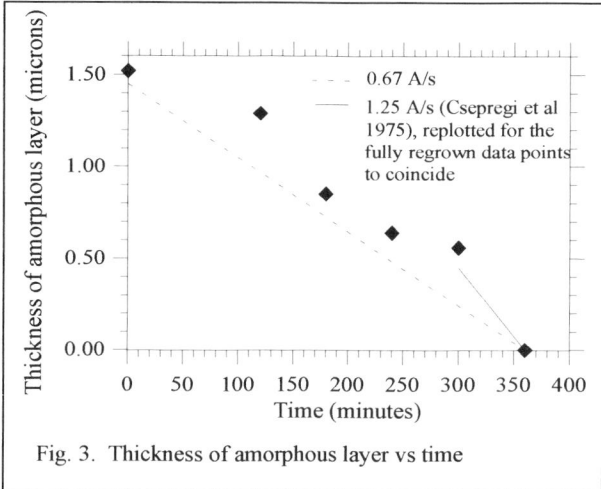

Fig. 3. Thickness of amorphous layer vs time

agrees reasonably well with the data of Custer et al (1994), who measured a regrowth velocity of 0.29Å/s at 525°C (equivalent to 0.76Å/s at 550°C). Regrowth studies (e.g. Olson and Roth 1988, Csepregi et al 1975) suggest an interface velocity in (100) Si at 550°C that is approximately twice that measured in this work. However, such studies have traditionally been carried out on thinner layers of amorphous silicon, and it may be that the sheer thickness of the amorphous silicon is sufficient to cause the difference in growth rate. Fig. 3 shows that when the amorphous layer becomes thinner, our data do indeed

approach the rate of 1.25Å/s reported by Csepregi et al (1975).

5. COMPARISON WITH TEM

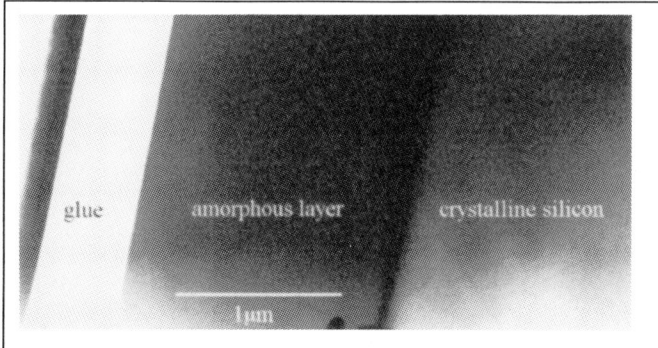

Fig. 4. Bright field TEM micrograph of the unannealed sample.

The unannealed material was also examined by cross sectional TEM, and a bright field electron micrograph is shown in Fig. 4. The depth of the amorphous layer is revealed to be 1.57 ± 0.01μm, which is in good agreement with the results from EBSD analysis of 1.52 ± 0.04μm, thus confirming the usefulness of this technique.

6. SUMMARY

EBSD is a quick and simple method of determining significant changes in crystal structure. With a state-of-the-art system, point-to-point resolutions of 80 nm or less are possible. While this method does not reveal the same amount of extra information as XTEM, the ease of sample preparation and the comparatively non-destructive nature of the technique make it a useful tool to confirm that implanted silicon has been regrown, and predict required annealing schedules.

ACKNOWLEDGEMENTS

This work was supported by the EPSRC and the implantations reported in this work were done by SCRIBA (Surrey Centre for Ion Beam Applications), University of Surrey.

Channel+ is available commercially from HKL Technology
VMAP is developed and distributed by F J Humphreys, Manchester Materials Science Centre, UMIST.

REFERENCES

Csepregi L, Mayer J W and Sigmon T W 1975 Phys.Letters A **54**,157

Custer J S, Polman A and van Pinxteren H M 1994 J. Appl. Phys. **75**, 2809

Dingley D J and Randle V 1992 J. Mater. Sci. **27**, 4545

Evans-Freeman J H, Naveed A T, Huda M Q, Peaker AR, Houghton D C and Wright A C 1998 Mat. Res.Symp.Proc.**533**, 133

Franzò G, Coffa S, Priolo F and Spinella C 1997 J. Appl. Phys. **81**, 2784

Huda M Q, Peaker A R, Evans-Freeman J H, Houghton D C and Gillin W P 1997 Electronics Letters **33**, 1182

Humphreys F J and Brough I 1999 Proc. Microscopy and Analysis 99, Portland USA, in press

Olson G L and Roth J A 1988 Materials Science Reports **3**, 1

Scholes A 1998 PhD thesis, UMIST

Seidel T E 1983 in: VLSI Technology ed. Sze S M (McGraw-Hill) p 260

Wilkinson A J 1996 Ultramicroscopy **62**, 237

Inst. Phys. Conf. Ser. No 164
Paper presented at Microsc. Semicond. Mater. Conf., Oxford, 22–25 March 1999
© 1999 IOP Publishing Ltd

Enhanced interfacial oxide break up and polysilicon regrowth using a methanol-last wafer preclean and a fluorine implant

C D Marsh, N E Moiseiwitsch*, G R Nash*, G R Booker and P Ashburn*

Dept. of Materials, University of Oxford, Parks Rd, Oxford OX1 3PH, UK.
*Dept. of Electronics & Computer Science, University of Southampton, Southampton SO17 1BJ, UK.

ABSTRACT: Si wafers were given a methanol-last or a HF preclean, LPCVD poly-Si layers were deposited and implanted with As^+ plus F^+ or As^+ only, the specimens were annealed and the structures determined by TEM. For break up of the interfacial oxide plus initial epitaxial regrowth of the poly-Si to occur, the 850°C anneal times were - for an HF preclean without F, >480min; for a methanol-last preclean without F, 120mins; and for a methanol-last preclean with F, <30min. The corresponding 950°C anneal times are 1800, 60 and 30s respectively. These thermal budgets for a methanol-last preclean and F^+ implant are the lowest yet reported for interfacial oxide break up and initial poly-Si regrowth and are suitable for the fabrication of advanced poly-Si bipolar transistors.

1. INTRODUCTION

In the fabrication of poly-Si emitter bipolar transistors a thin oxide layer (6-14Å) is invariably present at the poly-Si/Si interface (Patton et al 1986). To reduce the emitter resistance (Marsh et al 1998) and variations in the base current (Crabe et al 1987) it is an advantage to completely break up the interfacial oxide layer during the anneal used to drive in the As^+ dopant. A temperature of 1055°C is typically used for this anneal (Pontcharra et al 1997) but reducing the anneal thermal budget while retaining the oxide break up would give better device performance. We have shown that during As^+ drive-in anneals at 850°C (Marsh et al 1995 & Moiseiwitsch et al 1995) or at 950°C (Marsh et al 1998) oxide break up and initial regrowth occurred more rapidly when $1x10^{16}/cm^2$ F^+ was implanted into the poly-Si. Recently we have shown (Moiseiwitsch et al 1998) that using a Si wafer methanol-last preclean (Lai et al 1994) gives a thin 1-2Å interfacial oxide compared to an HF preclean that gives a 6-8Å oxide. In this paper we investigate for the first time the combined role of a methanol-last preclean and a $1x10^{16}/cm^2$ F^+ implant on the break up of the interfacial oxide layer and the regrowth of the poly-Si during an As^+ drive-in anneal. The results are compared to a methanol-last preclean without a F^+ implant and a HF clean with and without a F^+ implant.

2. EXPERIMENTAL

Cz Si (100) wafers, p-type resistivity 5-35Ωcm, were given either a HF or a methanol-last wafer preclean. The HF preclean consisted of a 10s etch in 7:1 buffered HF, followed by a 90s rinse in deionised water and a spin dry. The methanol-last preclean consisted of a 10s etch in 7:1 buffered HF, followed by a standard RCA clean (which produces a surface oxide layer 14Å thick), followed by a 15s etch in 40:1 H_2O:HF and a 30s etch in 10:1 methanol:HF. The wafers were spun dry with no additional water rinse. The precleans were immediately followed by low-pressure chemical vapour deposition (LPCVD) at 610°C of 350nm of undoped poly-Si. The samples were then implanted with As^+ ($1x10^{16}/cm^2$, 70keV) and some also with F^+ ($1x10^{16}/cm^2$, 50keV). The samples were then capped with 600nm of LPCVD oxide to prevent subsequent As loss, and furnace annealed at 850°C, or rapid thermally annealed at 950°C, for times between 4s and 480mins.

3. RESULTS AND DISCUSSION

Fig. 1 shows the epitaxial regrowth of the poly-Si layer to single crystal Si (average fractional volume determined by TEM) for samples given a HF or methanol-last preclean, with or without a F^+ implant and annealed at 850 or 950°C as a function of the anneal time. In samples given a methanol-last preclean, no F^+ implant and annealed at 850°C the interfacial oxide layer had roughened and 1% regrowth had occurred after 120mins (figs. 2a&b). After annealing for 180 and 480mins the oxide layer had completely broken up forming oxide particles, size 40-80Å (fig. 2), and the regrowths were 55 and 98% respectively (fig. 1a). In the equivalent samples annealed at 950°C for 30, 90 and 120s no oxide break up, break up with 1% (figs 3a&b) and 100% regrowth respectively had occurred (fig. 1b).

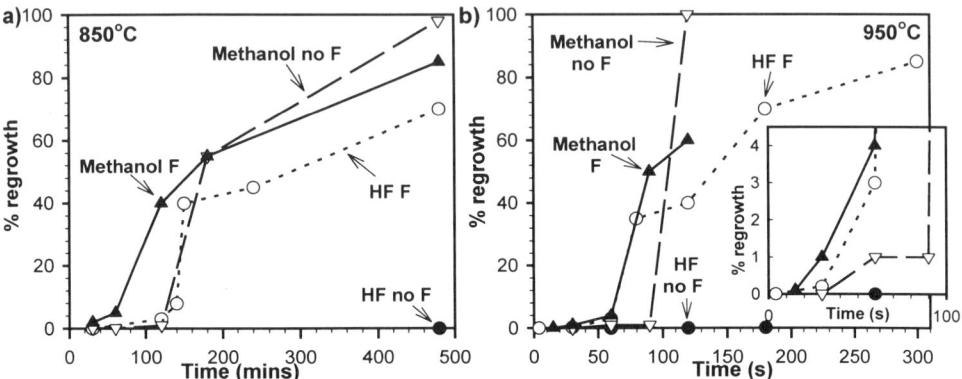

Fig. 1: Regrowth (%) of the poly-Si layer to single crystal Si for anneals at a) 850°C & b) 950°C.

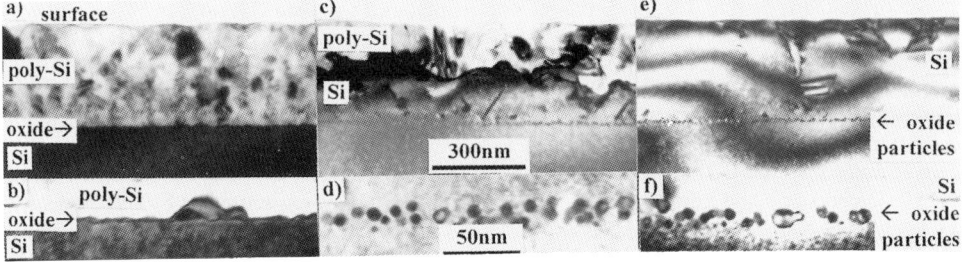

Fig 2: Samples with methanol-last preclean, no F, anneal 850°C for a)&b) 120, c)&d) 180 and e)&f) 480mins.

In the samples given a HF preclean, no F^+ implant and annealed at 850°C for 480mins (figs. 3c&d) or 950°C for 120s (figs. 3e&f) the interfacial oxide layer had roughened slightly but ≤0.1% regrowth had occurred. In the samples given a methanol preclean, no F^+ implant and the same anneals the regrowths were 98 (fig. 2e) and 100% respectively. Hence oxide break up and regrowth occurs more quickly for the methanol (figs. 2&3a) than for the HF preclean (fig. 3c&e). This may be because if the oxides are stoichiometric SiO_2 the thickness after a methanol preclean (1-2Å) is less than after a

Fig. 3: Samples with a)&b) methanol-last preclean, no F, annealed at 950°C for 90s, c)&d) HF preclean, no F, annealed at 850°C for 480mins and e)&f) HF preclean, no F, annealed at 950°C for 120s.

Fig. 4: Samples with methanol-last preclean, F, anneal 850°C for a)&b) 30, c)&d) 60 and e)&f) 120mins.

HF preclean (6-8 Å) and hence breaks up more easily. Alternatively, the precleans result in oxides of different stoichiometries with the stoichiometry from a methanol preclean being less thermally stable.

In the samples given a methanol-last preclean, a F^+ implant and annealed at 850°C, the interfacial oxide layer had broken up and 2 and 5% regrowth had occurred after 30 and 60mins (figs. 4a-d). After annealing for 120 (figs. 4e&f), 180 and 480mins, the oxide layer was completely broken up forming oxide particles, size 40-80Å, and the regrowths were 40, 55 and 85% respectively (fig. 1a). In the equivalent samples annealed at 950°C for 30, 60 and 90s the regrowths were 1, 4 and 50% respectively (figs. 5 & 1b). For samples given a methanol preclean, after annealing at 850°C for 120mins and 950°C for 90s, regrowths of 40% (fig. 4e) and 50% (fig. 5e) were observed when F was present, compared to only 1% (figs. 2 & 3) when F was not present. Hence with a methanol-last preclean oxide break up and initial regrowth occurs more quickly when F is present. F may be enhancing oxide break up because F is very electro-negative and hence readily breaks and forms bonds.

In the methanol precleaned samples containing oxide particles after anneals at 850°C (120, 180 and 240mins with F and 180 and 240mins without F) the sizes of the oxide particles (40-80Å) were similar in samples with and without F but the particle densities were higher without F ($\sim4\times10^{10}$/cm^2 (figs. 2d&f) compared to $\sim6\times10^9$/cm^2 (fig. 4f)). Hence after annealing the total volume of the oxide particles is less in the samples with F than in those without F. In the equivalent samples annealed at 950°C with F (90 and 120s) the oxide particles (size 40-120Å) were present in a band 550Å wide (fig. 5f). This is much wider than the 80Å wide band of oxide particles (size 40-60Å) in the sample without F (anneal 120s). Both these results suggest that as well as breaking oxide bonds F increases the mobility of the oxide during the anneals and so further enhances oxide break up.

Fig. 5: Samples with methanol-last preclean, F, and annealed at 950°C for a)&b) 30, c)&d) 60 and e)&f) 90s.

In the samples given a HF preclean and a F^+ implant and annealed at 850°C for 30mins the interfacial oxide layer had roughened and <0.1% regrowth had occurred. After annealing for 120, 140, 150 and 480mins oxide break up and regrowths of 3, 8, 40 and 70% respectively had occurred (fig. 1a). In the equivalent samples annealed at 950°C, no oxide break up or regrowth had occurred after 4s. After annealing for 30, 60, 120, 180 and 300s the regrowths were 0.2, 3, 40, 70 and 85% respectively (fig. 1b). After annealing samples with F at 850°C for 120mins (figs 4e&f) and at 950°C for 30s (figs. 5a&b), for a methanol preclean regrowths of 40 and 1% had occurred compared to only

480

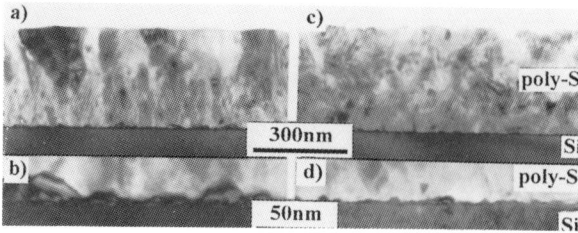

Fig. 6: Samples with HF preclean, F, annealed at a)&b) 850°C for 120mins and c)&d) 950°C for 30s.

3% (figs 6a&b) and 0.2% (figs. 6c&d), respectively, for a HF preclean. Hence F more quickly breaks up the oxide after a methanol preclean than after a HF preclean. This may also be because after a methanol preclean the oxide is thinner or the stoichiometry of the oxide differs, making it less thermally stable, than after a HF preclean.

The increase in the oxide break up and initial regrowth due to F is greater at 850°C (fig. 1a) than at 950°C (fig. 1b inset). The greater effect of F at 850°C may be because oxide becomes more viscous on going from 850°C to 950°C. Hence at 850°C fewer of the oxide bonds are broken due to their thermal energy and the potential for F to speed up the oxide break up by chemically breaking the oxide bonds is greater.

Oxide break up and regrowths of ~1% (figs. 5a&b) and ~40% (figs. 4e&f) are important for the fabrication of poly-Si bipolar transistors. The results show that for oxide break up and ~1% regrowth the times (t) for annealing at 850°C are, for a HF preclean t >480mins, for a HF preclean and a F+ implant 30< t <120mins, for a methanol-last preclean 60< t <120mins and for a methanol-last preclean and a F+ implant t <30mins. The corresponding 950°C anneal times are 1800, 30< t <60, 60 and 30s respectively. The results show that for oxide break up and ~40% regrowth the times for annealing at 850°C are, for a HF preclean t >480mins, for a HF preclean and a F+ implant 150mins, for a methanol-last preclean 120< t <180mins and for a methanol-last preclean and a F+ implant 60< t <120mins. The corresponding 950°C anneal times are t >1800, 80< t <120, 90< t <120 and 60< t <90s respectively. Hence at both temperatures the shortest anneals to achieve oxide break up and regrowths of 1 and 40% are for a methanol-last preclean and a F+ implant. Annealing such samples at 950°C rather than 850°C enables the anneal times for regrowths of 1 and 40% to be reduced to 30 and 90s, from 30 and 120mins, respectively. The shorter anneals are more suitable for device fabrication.

4. CONCLUSIONS

Enhanced interfacial oxide break up and poly-Si regrowth have been demonstrated when a methanol-last Si wafer preclean is combined with a F implant into poly-Si layers. The thermal budget required to break up the interfacial oxide and initiate epitaxial regrowth of the poly-Si was just 30s at 950°C for a methanol preclean and a F implant compared to 1800s at 950°C for a HF preclean and no F implant. This thermal budget for the methanol-last preclean is the lowest yet reported for interfacial oxide break up and initial poly-Si regrowth and is suitable for the fabrication of advanced poly-Si emitter bipolar transistors. The presence of the F at the poly-Si/Si interface has been shown to increase the mobility of the interfacial oxide during anneals at 850 and 950°C.

ACKNOWLEDGEMENTS

The authors would like to thank the EPSRC for funding and the staff of the Microscopy facility, University of Oxford and the Clean Room, University of Southampton for their assistance.

REFERENCES

Crabe E, Hoyt J L, Moslehi M, Pease R & Gibbons J F, 1987 IEEE Tech. Dig.: VLSI Tech. Symp.
Pontcharra J, Behouche E, Ailloud L, Thomas D & Chantre A, 1997 IEEE Trans. Electron Dev. **ED-44** 2091
Lai K, Hao M, Chen W, Lee J C, 1994 IEEE Electron Dev. Lett. **EDL-15** 446
Marsh C D, Moiseiwitsch N E, Booker G R & Ashburn P, 1995 Inst. Phys. Conf. Ser. **146** 457
Marsh C D, Moiseiwitsch N E, Booker G R & Ashburn P, 1998 Proc. MRS **Vol 523** 195
Moiseiwitsch N E, Marsh C D, Ashburn P & Booker G R, 1995 Appl. Phys. Lett. **66** 1918
Moiseiwitsch N E, Ashburn P, Marsh C D, & Booker G R, 1998 Electrochem. & Sol. State Lett. **1** 91
Patton G L, Bravman J C & Plummer J D, 1986 IEEE Trans. Electron Dev. **ED-33** 1754

Inst. Phys. Conf. Ser. No 164
Paper presented at Microsc. Semicond. Mater. Conf., Oxford, 22–25 March 1999
© *1999 IOP Publishing Ltd*

Deactivation and diffusion of boron in ion-implanted silicon: dopant mapping through secondary electron imaging

M R Castell[1], T W Simpson[2], I V Mitchell[2], D D Perovic[3] and J-M Baribeau[4]

[1] Department of Materials, University of Oxford, Parks Road, Oxford OX1 3PH, U.K.
[2] Department of Physics and Astronomy, University of Western Ontario, London N6A 3K6, Canada.
[3] Department of Metallurgy and Materials Science, University of Toronto, Toronto M5S 3E4, Canada.
[4] Institute for Microstructural Sciences, National Research Council, Ottawa K1A 0R6, Canada.

ABSTRACT: Secondary electron (SE) imaging in a scanning electron microscope is used to map electrically active dopant distributions of B-doped superlattices in Si. By comparing SE contrast profiles with secondary ion mass spectroscopy data, it is shown that B is electrically deactivated when the damage caused during Si implantation falls onto a doped region. Following a 450 °C anneal, the effect of the implantation damage is severely reduced in the SE profiles and the B is partially reactivated. An 815 °C anneal results in transient enhanced diffusion of some of the B with the remainder trapped in an inactive immobile peak.

1. INTRODUCTION

The incorporation of dopant atoms into semiconductors is widely achieved through implantation techniques. After B implantation into crystalline Si, an annealing period is necessary to heal ion-implant damage and place the dopants on electrically active substitutional sites. The anneal gives rise to transient enhanced diffusion (TED) where some of the B displays anomalously high diffusion rates for a limited time. The remaining B is trapped in immobile and electrically inactive B clusters. These two processes place a major restriction on the successful implementation of shallow ion-implantation technology in advanced Si-based devices and have stimulated many investigations in this area (Stolk et al 1997).

TED not only affects implanted dopants but also substitutional dopants incorporated during growth. It is therefore possible to study the effects of ion-implantation and TED by implanting Si ions into a structure containing narrow dopant marker layers (Stolk et al 1997, Huizing et al 1996, Cowern et al 1996, Simpson et al 1997). Annealing of ion implantation damage produces a transient flux of Si interstitials (I) which form BI pairs (one substitutional B atom and one silicon interstitial). The BI complex can be activated into the highly mobile interstitial B configuration leading to enhanced B diffusion rates. These processes have been well described by Fahey et al (1989). However, there is little understanding of the mechanisms which lead to the formation of the *immobile* B component. As in TED, B deactivation is also driven by the Si interstitial flux, through a mechanism believed to be associated with the formation of small B clusters (Pelaz 1997 et al, Caturla et al 1998).

Secondary ion mass spectroscopy (SIMS) has been used extensively to measure changes in the B concentration profiles for various implantation and annealing conditions but progress in this area has been limited by the inability to distinguish between the electrically active and inactive B in such profiles. This is especially problematic in the regime following implantation and before or early in the anneal where little or no diffusion has occurred. Spreading resistance profiling has also been used to characterise B deactivation after implantation and following an anneal (Cowern et al 1996, Larsen et al 1996). Simulations suggest that B deactivation due to Si implantation alone is accomplished through the formation of BI or BI2 complexes which dissolve rapidly in the very early stages of annealing (Pelaz et al 1997, Caturla et al 1998, Larsen et al 1998).

This paper presents results of a new study where secondary electron (SE) imaging in a scanning electron microscope (SEM) has been used in combination with SIMS to investigate B deactivation and TED of B in a dopant superlattice. Modern field emission SEMs can now routinely resolve sub-nanometre features and detect small changes in work function (Castell et al 1997). This high sensitivity to the local band structure has led to the development of dopant contrast imaging in the SEM (Perovic et al 1995, Castell et al 1995, Howie 1995, Turan et al 1996, Sealy et al 1997, Venables et al 1998). In practice, an SE generated in a p-type region experiences a lower local work function than an SE generated in an adjacent n-type region. The bright regions in an SE micrograph of dopant distributions therefore correspond to p-type doping and the dark regions to n-type. Because dopant contrast is purely due to electronic effects, SE imaging reveals only the electrically active component of the B (Turan et al 1996), so deactivated B can be identified by comparing SE data with B concentration profiles from SIMS measurements.

2. EXPERIMENTAL PROCEDURES

The B doped heterostructure used in the experiments was grown on a Si (001) substrate at 650 °C by molecular beam epitaxy (MBE) and consisted of four 25 nm layers of Si doped with B (5×10^{19} cm^{-3}) separated by 240 nm of undoped Si and capped with 350 nm undoped Si. Implantation of Si ions to a dose of 5×10^{13} cm^{-2} was carried out at 90 °C at energies of 80 keV ($R_p \sim 120$ nm) and 200 keV ($R_p \sim 300$ nm). Annealing was performed under flowing N_2 at 450 °C for 900 s, or at 815 °C for 100 s. SIMS was performed in a Cameca 3f facility with 5 keV O_2^+ primary ions.

Active dopant distributions were imaged from air-cleaved sample cross-sections in a Hitachi S4500II SEM operated at 1 keV (~3 nm beam width). This SEM is equipped with a field emission electron source and was operated using the minimum accessible working distance of 3 mm. The energy of the primary electrons where the observations described in this paper were most clearly seen were at 1 keV for SE imaging. All SEM images were digitally recorded as 1024 × 768 pixels with 256 greyscales per pixel. Image analysis was performed using the NIH Image software package (version 1.61).

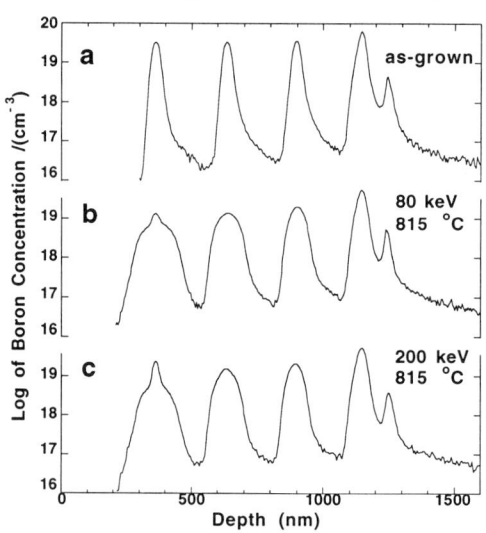

Fig. 1 SIMS profiles showing the B concentration of the as-grown dopant superlattice (a); after Si implantation at 80 keV (b) or 200 keV (c) and a 100 s anneal at 815 °C.

3. RESULTS AND DISCUSSION

We begin the discussion of our results with the SIMS data shown in Fig. 1. The spectrum of the as-grown structure (Fig. 1a) shows the four B layers together with a deeper smaller spike that was created anomalously at the early stages of MBE growth. There was no change of the SIMS profile following implantation or after a 450 °C anneal, as expected, because no B diffusion is observed under these conditions. However, for the implanted samples annealed at 815 °C, significant B diffusion is found as shown in Fig. 1(b,c). The characteristic broadening of the B peaks is due to TED, and the smaller spike on top of the shallowest layer is associated with electrically inactive and clustered B. An estimate of the fraction of the inactive B can be made through fitting of gaussian curves to the small inactive peak and the large TED profile. This analysis yields 0.12 and 0.37 for the shallowest layer of the 80 keV and 200 keV implanted samples, respectively.

We now turn to an analysis of the active dopant distributions of the same samples using SE imaging. Fig. 2 shows a progression of SE images from cleavage cross-sections of the samples as-grown (a), 80 keV implanted (b), implanted and annealed at 450 °C (c), and implanted and annealed at 815 °C (d). Corresponding SE profiles are shown in Fig. 3.

The SE data from the as-grown sample (Fig. 3a) should be compared with the SIMS profile of Fig. 1a. The good agreement between the two curves confirms the close relationship between SE contrast and the logarithm of the active dopant concentration. As the micrographs are recorded electronically it is possible to use image analysis software to integrate the SE signal intensities along lines parallel to the B-doped layers. The result of this integration gives an average SE intensity profile perpendicular to the layers. Profiles of this type show the variation of the SE signal due to the active B content but there is also a varying background

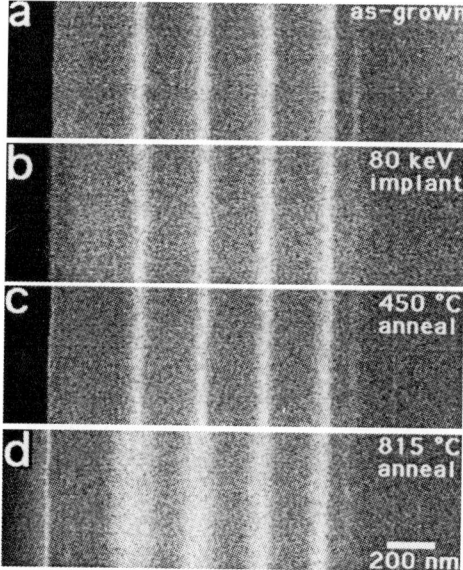

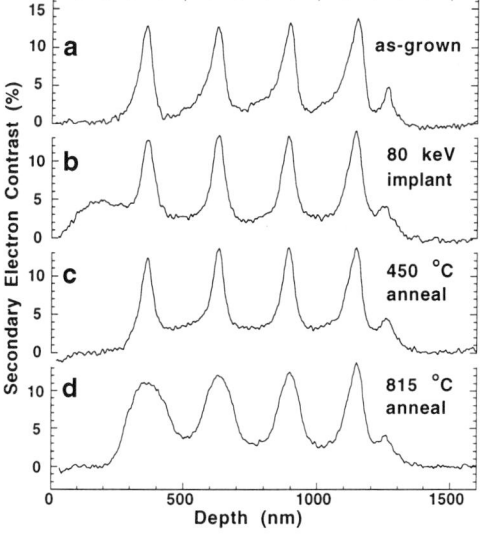

Fig. 2 SE images of cleavage cross-sections through samples that were as-grown (a), 80 keV implanted (b), implanted and 450 °C annealed (c), and implanted and 815 °C annealed (d). Integrated SE profiles of these images are shown in Fig. 3.

Fig. 3 The SE profile of the active B distribution in the as-grown sample is shown in (a). Following implantation with 80 keV Si ions the SE signal increases in the damaged region (b), and is removed again through a 450 °C anneal (c). An 815 °C anneal results in broadening of the dopant peaks through TED (d). The electrically inactive peak on the shallowest layer is not seen in (d).

level that requires subtraction. In practice the variation of the background is usually around 4 % contrast across a 1 μm scan region. In order to quantitatively compare different SE profiles all data is presented in absolute contrast units. The SE level equivalent to 0 % contrast is the intensity of the undoped Si. Further details of how SE profiles can be obtained through image integration are outlined in Castell et al (1997).

Figs 2b and 3b show a SE image and profile after ion implantation at 80 keV. The effect of implantation damage is more easily seen in the SE profile (Fig. 3b) than in the image (Fig. 2b). The implantation damage has caused an increase in the SE emission from the near-surface region, at a depth that correlates with the damage layer depth determined through a TRIM91 simulation (Biersack and Haggmark 1980, 120 nm range, 44 nm longitudinal straggle). No significant portion of the implantation damage falls on the shallowest B layer. Presumably the increase in SE emission from the damaged region can be attributed to implantation related defect states (Fahey et al 1989, Watkins 1997) in the band gap which cause a reduction of the local work function and result in an increased SE signal. Consistent with this hypothesis, the SE signal returns to its normal level (Fig. 3c) after a 900 s, 450 °C sample anneal which is known to heal point defects. When the implanted sample is annealed for 100 s at 815 °C the SE profiles broaden in the usual manner associated with TED. This is shown in the SE image in Fig. 2d and the corresponding SE profile of Fig. 3d, and should be compared with the SIMS profile of Fig. 1b of the same sample. Again, there is good correspondence between the SE and SIMS data, apart from the expected absence of the electrically inactive peak on the shallowest B layer in the SE profile.

Fig. 4 shows a set of SE images from the sample that was implanted at 200 keV and then annealed. The corresponding SE profiles are shown in Fig. 5. The broad SE signal around the

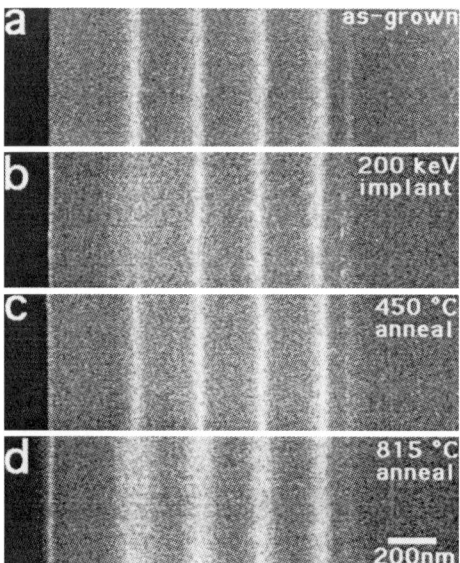

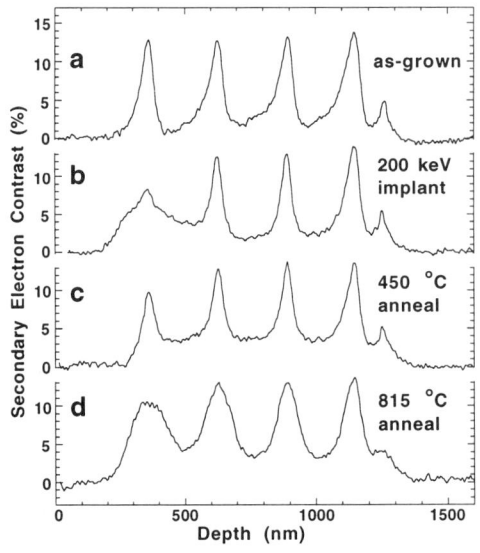

Fig. 4 SE images of cleavage cross-sections through samples that were as-grown (a), 200 keV implanted (b), implanted and 450 °C annealed (c), and implanted and 815 °C annealed (d).

Fig. 5 Sequence of SE profiles, showing the active B distribution in the as-grown sample (a), and after 200 keV Si ion implantation (b) where the broad peak around the shallowest layer is due to implant damage which results in electrical deactivation of the B layer. A 450 °C anneal (c) heals the damage and partially reactivates the B dopants. The broadening of the dopant peaks in (d) is due to TED following an 815 °C anneal.

shallowest B layer following Si implantation (Fig. 5b) can be attributed to point defects. The difference between this profile and the 80 keV implanted profile (Fig. 3b) is that the implant damage falls directly onto the shallowest B layer. This observation is confirmed through a TRIM91 simulation which predicts that the primary damage layer should be located at a depth of 300 nm with 91 nm straggle. Another significant difference in Fig. 5b is that the shallowest B layer has been partially deactivated by the implantation process alone. The effect of deactivation is particularly marked in the SE image (Fig. 4b). This electrical deactivation is not due to B clustering, as no broadening of the B SIMS peaks was observed at this stage. There cannot have been sufficient movement of B to diffuse and form B pairs as this would result in a measurable increase in the B SIMS peak widths (Pelaz et al 1997). Also, there is not nearly enough collisional displacement of the B atoms to explain the deactivation. Presumably the B is still located on substitutional sites, but has been deactivated by pairing with mobile Si interstitials to form BI or BI2 complexes (Pelaz et al 1997, Caturla et al 1998). This deactivation of individual B atoms reduces the active dopant concentration and hence the SE emission from the B marker layer. Further reduction in the SE signal occurs because the effect of the active dopants on the electronic structure is reduced by the high defect density.

After point defect annealing at 450 °C for 900 s the B is partially reactivated (Fig. 5c) and most of the SE signal due to implantation damage has been extinguished. Full dopant reactivation has not occurred, which indicates that some electrically inactive B configurations remain. After an 815 °C anneal for 100 s TED broadening of the B profiles is seen (Fig. 5d). When comparing these SE data with the SIMS profile of Fig. 1c the anticipated absence, through electrical inactivity, of the B clustering (B3I) peak should be noted (Pelaz et al 1997, Caturla et al 1998).

4. CONCLUSION

We have demonstrated the successful application of SE imaging to the study of diffusion and electrical activity of B dopants in self-implanted Si. We show that electrical deactivation of B marker layers occurs during Si implantation at 90 °C if the primary damage region is located on the B layer. This deactivation is thought to be due to BI and BI2 defects and can be partially reversed following a 450 °C anneal. Independent of the location of the primary damage region, B deactivation is also observed following TED after an 815 °C anneal, and can be explained through B3I clustering mechanisms.

ACKNOWLEDGEMENTS

We are grateful to R. Turan for preliminary investigations, and C. Marsh and B. Henry for useful discussions. This work was funded through the Ontario Centre for Materials Research and Natural Sciences and Engineering Research Council of Canada.

REFERENCES

Biersack J and Haggmark L 1980 Nucl. Instr. and Meth. **174**, 257
Castell M R, Perovic D D and Lafontaine H 1997 Ultramicroscopy **69**, 279
Castell M R, Perovic D D, Howie A, Ritchie D A, Lavoie C and Tiedje T 1995 Inst. Phys. Conf. Ser. **146**, 281
Caturla M J, Johnson M D and Diaz de la Rubia T 1998 Appl. Phys. Lett. **72**, 2736
Cowern N E B, Huizing H G A, Stolk P A, Visser C C G, de Kruif R C M, Larsen K K, Privitera V, Nanver L K and Crans W 1996 Nucl. Instr. and Meth. B **120**, 14
Fahey P M, Griffin P B and Plummer J D 1989 Rev. Mod. Phys. **61**, 289
Howie A 1995 J. Microsc. **180**, 192
Huizing H G A, Visser C C G, Cowern N E B, Stolk P A and de Kruif R C M 1996 Appl. Phys. Lett. **69**, 1211

Larsen K K, Privitera V, Coffa S, Priolo F, Campisano S U and Carnera A 1996 Phys. Rev. Lett. **76**, 1493

Pelaz L, Jaraiz M, Gilmer G H, Gossmann H-J, Rafferty C S, Eaglesham D J and Poate J M 1997 Appl. Phys. Lett. **70**, 2285

Perovic D D, Castell M R, Howie A, Lavoie C, Tiedje T and Cole J S˙ W 1995 Ultramicroscopy **58**, 104

Sealy C P, Castell M R, Reynolds C L and Wilshaw P R 1997 Inst. Phys. Conf. Ser. **157**, 561

Simpson T W, Goldberg R D, Mitchell I V and Baribeau J-M 1997 Mat. Res. Soc. Symp. Proc. **438**, 15

Stolk P A, Gossmann H-J, Eaglesham D J, Jacobson D C, Rafferty C S, Gilmer G H, Jaraiz M, Poate J M, Luftman H S and Haynes T E 1997 J. Appl. Phys. **81**, 6031

Turan R, Perovic D D and Houghton D C 1996 Appl. Phys. Lett. **69**, 1593

Venables D, Jain H and Collins D C 1998 J. Vac. Sci. Technol. B **16**, 362

Watkins G D 1997 Mat. Res. Soc. Symp. Proc. **469**, 139

Inst. Phys. Conf. Ser. No 164
Paper presented at Microsc. Semicond. Mater. Conf., Oxford, 22–25 March 1999
© 1999 IOP Publishing Ltd

TEM and AFM studies of selectively etched Si specimens to determine 2-D dopant profiles associated with n+-p junctions

K D Yoo, C D Marsh and G R Booker

Department of Materials, Oxford University, Parks Road, Oxford, OX1 3PH

ABSTRACT: The etching/TEM and etching/AFM methods for determining 2-D dopant concentration depth profiles are applied to n+-p junctions in silicon. The work includes the use of an independent etch rate vs. dopant concentration calibration curve, comparisons by using the two methods on the same specimens and the application of the methods to junction depths down to 60nm. The TEM method gave good results while the AFM method requires further development.

1. INTRODUCTION

Methods have previously been developed to determine experimentally 2-D dopant concentration depth profiles associated with p-n junctions in silicon wafers, e.g. as occur in the sources and drains of MOS transistors. Wet etchants sensitive to dopant concentration were applied to cross section specimens and 2-D etching profiles were determined using TEM (Roberts et al 1985, Gold et al 1989, Maher and Zhang 1994, Alvis et al 1996), STM (Takigami and Yanimoto 1991) and AFM (Barrett et al 1995). These results were converted to 2-D dopant profiles using calibrations provided by either a 1-D SIMS profile obtained from a large area region similarly processed or an etch rate vs. dopant concentration curve obtained using measurements from separate specimens with different dopant concentrations. The junction depths were typically in the range 2.0 to 0.15μm and the results revealed, for example, the important 2-D dopant distributions at the mask edges. As devices for micro-electronic circuits become smaller, there is a need for p-n junctions in the typical depth range 150 to 50 nm. In this paper, we have investigated the etching/TEM and etching/AFM methods to obtain 2-D dopant concentration profiles for As+ implanted and annealed silicon specimens with junction depths of 60 or 100nm.

2. ETCH RATE CALIBRATION

For the calibration specimens, five p-type 8-10 Ω cm (001) Cz silicon wafers were implanted with As+ and annealed. Conditions were chosen so that for each specimen a constant dopant concentration extended from the surface to a depth of $\sim 0.5\mu m$. 1-D As SIMS depth profiles showed five different surface concentrations covering the range 2×10^{16} to $8 \times 10^{19} cm^{-3}$. The specimen surfaces were etched using 0.5%HF / 99.5%HNO3 at room temperature with the etchant stirred, followed by a rinse in deionised water. Etch rates were determined by removing typically 120 to 500nm. The etch rate vs. dopant concentration curves obtained for stirring rates of 1.7, 5.0 and 8.3Hz are shown in Fig. 1. For each dopant concentration and stirring rate, two specimens were separately etched and all of the results are shown by points on the curves. The small spreads indicate good etching reproducibility. Etch rate is not significantly dependent on stirring rate for the lower dopant concentrations but increases markedly with increasing stirring rate for the higher concentrations. The selectivity, e.g. the ratio of the etch rate at the highest and the lowest dopant concentrations, increases from 2.4 at 1.7Hz to 4.2 at 8.3Hz. The corresponding selectivities obtained previously by Roberts et al (1985) and Takigami and Yanimoto (1991) were 2.0 and 3.0 respectively.

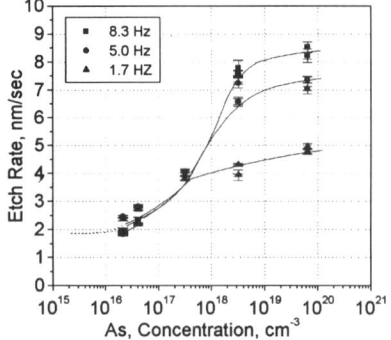

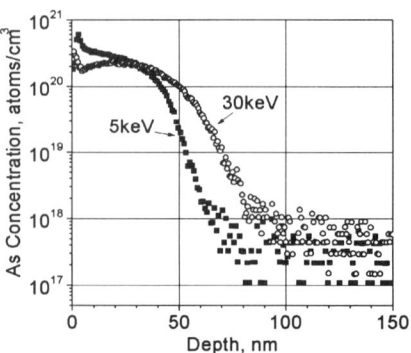

Fig. 1 Etching calibration curves for As doped silicon.

Fig. 2 1-D SIMS As concentration profiles for 2-D profile specimens.

3. ETCHING/TEM

For the 2-D profile specimens, p-type 8-10 Ω cm (001) Cz silicon wafers were used. A 5nm surface thermal oxide layer was formed and a 250nm amorphous silicon layer was deposited. Stripe windows typically 0.4μm wide were etched, implanted with 1×10^{15}cm^{-2} As^{+} at either 5 or 30keV and given a rapid thermal anneal at 1050°C for 20s. A 0.7μm plasma oxide was deposited to cover the whole specimen. 1-D As SIMS depth profiles obtained from large area regions on separate wafers that had been similarly processed are shown in Fig. 2. If these specimens had been fabricated as MOS devices instead of test structures, similar n^{+}-type As dopant distributions would have been obtained but the underlying p-type B dopant concentration would have been 10^{18}cm^{-3} rather than 10^{15}cm^{-3}. Consequently, SIMS indicates that junctions would have occurred at depths of ~65nm and ~95nm for the 5 and 30keV As^{+} implants respectively (subsequently referred to as 60nm and 100nm junction specimens).

Pairs of (110) cross-section thin slices were cleaved from these wafers and stuck together with their original (001) surfaces face-to-face. The resulting specimen was polished flat on both sides, finishing with colloidal silica. For the TEM specimen assessment method, the specimen was ion beam thinned to give areas of electron transparent silicon at the edge of a hole. Reflection optical microscope examination using sodium light revealed bright and dark interference fringes arising from the wedge-shaped specimen and enabled the foil thickness adjacent to each As doped region to be determined. The specimen was then etched for 25s on both sides simultaneously using the procedure described above with a 8.3Hz stirring rate. Optical microscope examination revealed new fringes and enabled the new thicknesses of the foil at regions adjacent to each As doped region to be determined.

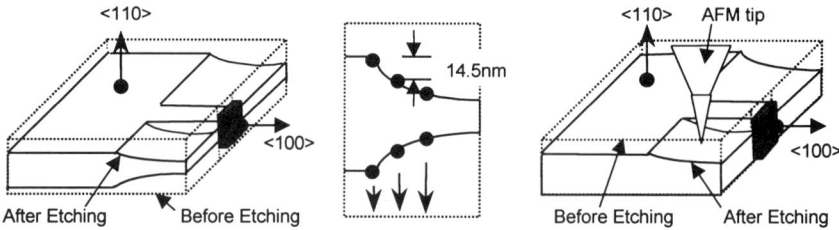

Fig. 3a Etching/TEM method.

Fig. 3b Position of TEM thickness fringes.

Fig. 3c Etching/AFM method.

Diagrams illustrating the etching/TEM method are shown in Figs. 3a and b. The etched foil is examined in the TEM at 200keV with the specimen tilted to the [110] on-axis many-beam diffraction condition. Closely spaced thickness fringes occur in the As doped regions due to the rapidly varying foil thickness after etching. The fringe distributions and contrasts that occur in each of these regions depend on the foil thickness before etching. An optimum region showing typically eight fringes on going from the wafer surface to deeper into the wafer is selected for analysis (Fig. 4a). These fringes correspond to 2-D thickness contours with a foil thickness change of 29nm (electron extinction distance for 200keV electrons and these diffraction conditions) on going from dark-to-dark fringes. These results enable the fringes to be converted to 2-D etch rate contours and then with the aid of the calibration curve to 2-D As dopant concentration contours (Fig. 4b). These etching/TEM results obtained for the 60nm junction specimen show the 1-D dopant distribution perpendicular to the wafer surface in the middle of the implanted region (line xx'). This profile is plotted in Fig.5 and is in good agreement with the 1-D As SIMS profile obtained for this specimen. It indicates a junction depth of 65nm. The etching/TEM results also show the 2-D dopant profiles in the region under the edge of the polysilicon gate where the dopant has diffused laterally. Parallel to the wafer surface, the dopant contours are closest at a depth of 18nm, corresponding to the highest lateral electric field (line yy').

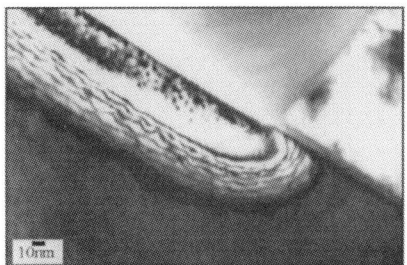

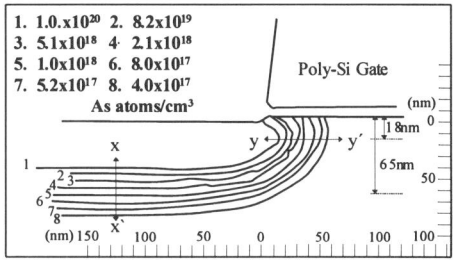

Fig. 4a XTEM micrograph of 60nm junction specimen etched for 25s.

Fig. 4b 2-D As doping profiles deduced from etching/TEM results of Fig. 4a.

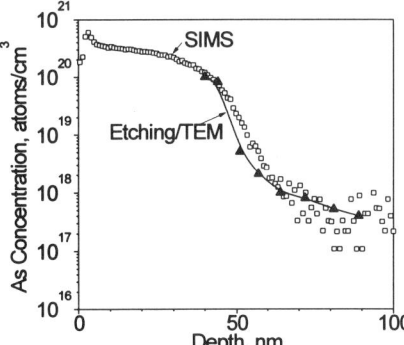

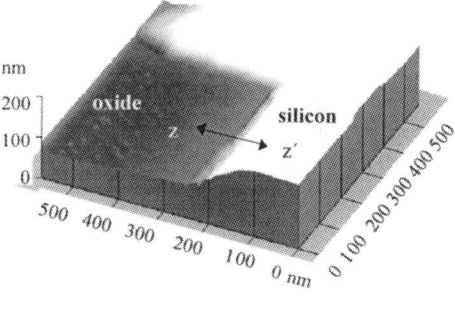

Fig. 5 1-D As doping profiles for 60nm junction specimen by etching/TEM and SIMS.

Fig. 6 3-D etching/AFM image for 60nm junction specimen etched for10s.

4. ETCHING/AFM

The AFM assessments were performed on specimens that had been given precisely the same processing as those used for the TEM assessments. For the etching/AFM method and the 60 and 100nm bulk specimens, the silicon was etched for 10s on one side using the procedure described above (Fig. 3c). The surface plasma oxide layer (not shown in Fig. 3) was then etched with buffered HF so that on going from the etched silicon region to the etched oxide region, there was a step down of 10nm. The latter etch does not affect the silicon. The AFM 3-D image in Fig. 6 shows an etched region for the 60nm specimen. An AFM 1-D profile corresponding to the line zz' of Fig. 6 is shown in Fig. 7a as the continuous line. The zero positions on the ordinate and abscissa scales are arbitrary.

On going from right to left, for the silicon region, the depth is constant. For the As doped silicon region, the depth progressively increases and is then constant for a short distance. The approximate depth profile expected for this specimen is shown as the dashed line of Fig. 7a. The reason why, for the doped silicon region, the AFM depth is less than the actual depth is because of the relatively large size of the AFM tip (nominally 18 degrees semi-angle and 20nm radius). As the tip moves across this region, contact is progressively made with the right side of the tip, rather than the bottom of the tip, and so shallower depths are recorded. The bottom of the tip does not make contact with the silicon surface again until it is a short distance from the silicon/oxide interface. The depth in this latter region is constant because the dopant concentration is high and so the etchant is insensitive to dopant changes. An analogous AFM 1-D depth profile and its expected profile for the 100nm specimen are shown in Fig. 7b. These AFM 1-D profiles for 60 and 100nm specimens underestimate the junction depths because of the relatively large tip size. Sharper tips would give more accurate junction depths.

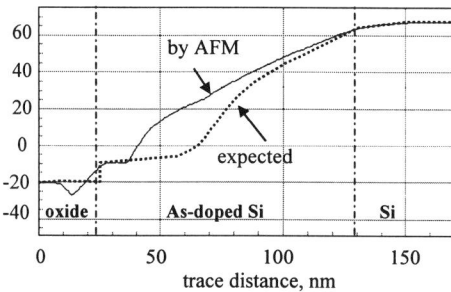

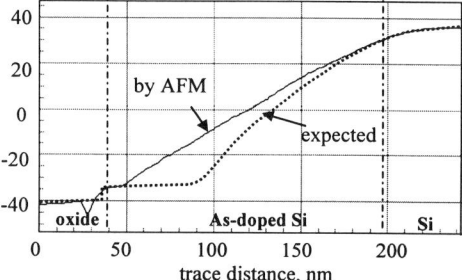

Fig. 7a 1-D AFM profile and expected profile for 60nm junction specimen etched for 10s.

Fig. 7b 1-D AFM profile and expected profile for 100nm junction specimen etched for 10s.

5. CONCLUSIONS

The selectivity of the HF/HNO$_3$ etchant to dopant concentration changes is significantly increased by stirring the etchant. The etching/TEM method in conjunction with an independent etch rate/dopant concentration calibration curve can be used to obtain 1-D and 2-D dopant concentration profiles over the complete arsenic doped region for n$^+$-p junction depths down to 60nm. The etching/AFM method is capable of giving 1-D dopant profiles for shallow junctions but underestimates the junction depths. More accurate results would be obtained if sharper tips were used.

REFERENCES

Alvis R, Luning S, Thompson L, Sinclair R and Griffin P 1996 J. Vac. Technol. B **14**, 231
Barrett M, Dennis M, Tiffin D, Li Y and Shih C K 1995 IEEE Electron Device Lett. **16**, 118
Gold D P, Wills J H, Booker G R, Wilson M C and Godfrey D J 1989 Inst. Phys. Conf. Ser. **100**, 537
Maher D M and Zhang B 1994 J. Vac. Sci Technol. B **12**, 347
Roberts M C, Yallup K J and Booker G R 1985 Inst. Phys. Conf. Ser. **76**, 483
Takigami T and Yanimoto M 1991 Appl. Phys. Lett. **58**, 2288

Inst. Phys. Conf. Ser. No 164
Paper presented at Microsc. Semicond. Mater. Conf., Oxford, 22–25 March 1999
© *1999 IOP Publishing Ltd*

Two-dimensional dopant profiling in shallow junctions using TEM and scanning capacitance microscopy

C-J Choi and T-Y Seong

Department of Materials Science and Engineering, Kwangju Institute of Science and Technology, Kwangju 500-712, Korea

ABSTRACT: TEM and scanning capacitance microscopy (SCM) were used to assess two-dimensional (2D) dopant profiles in MOSFETs with source/drain where arsenic implantation was performed for various conditions, i.e., energy ranging from 35 to 100 keV As^+ with doses of 2×10^{13} and 1×10^{14} cm^{-2}. As for TEM, major technique is based on the selective chemical etching of doped regions in silicon using a solution of HF and HNO_3. This etching results in the local variations in crystal thickness giving rise to the appearance of thickness fringes. Such fringes could be interpreted as 2D isoconcentration contours that map the dopant distribution. Thickness fringes corresponding to a concentration of $\sim 10^{16}$ cm^{-3} could be observed with accuracy better than 5 nm. The results are compared with SCM results and simulation data provided by SUPEREM-IV.

1. INTRODUCTION

Very light implant conditions have been employed to reduce semiconductor device dimensions down to the submicron level (Hill 1992). Advance in this field consequently requires accurate information concerning dopant distribution and junction depth, especially the lateral diffusion of dopants in the gate area of metal-oxide-semiconductor (MOS) devices. A variety of techniques such as secondary ion mass spectroscopy (SIMS) (Osburn and Reisman 1987), TEM (Spinella et al 1995a and 1995b, Cerva 1992), SEM (Gong et al 1995), atomic force microscopy (AFM) (Choi et al 1998), and scanning capacitance microscopy (SCM) (Williams et al 1990) have been successfully used to obtain experimental 2D dopant profiles in semiconductor devices. In this work, TEM was used to assess 2D dopant profiles in MOSFETs with gate lengths of 2, 1.5, and 1 μm. The TEM results have been compared with SCM results and simulated data provided by SUPREM-IV.

The principle of chemical etching used for dopant profiling is expressed by the following chemical reaction:

$$3Si + 4HNO_3 + 18HF \rightarrow 3H_2SiF_6 + 4NO + 8H_2O$$

The HNO_3 oxidizes the silicon surface, while HF removes the oxide layer. The addition of dopants leads to a decrease in the activation energy of the oxidation process, so that the layers of different dopant concentrations can be etched with different rates; the etching rate is higher at regions with higher dopant level. By calibrating the thickness variation of the etched sample using SIMS measurements, the dopant distribution can be spatially mapped out. Thus, in the TEM images, the thickness fringes could be interpreted as 2D isoconcentration contours.

2. EXPERIMENT

The patterned samples were implanted with energies ranging from 35 to 100 keV As^+ to doses of 2×10^{13} and 1×10^{14} cm^{-2} and then annealed at 950 °C for 30 min in a nitrogen ambient. In order to realize a spatial resolution better than 5 nm, the preparation of a highly flat cross-section surface is required before the etching process. Cross-section specimens were prepared using a conventional 'sandwich' technique. A sandwich structure was prepared by gluing one sample on a piece of scrap silicon face to face. After mechanical polishing, the glued specimens were thinned by Ar$^+$ ion milling using a liquid N$_2$ cold stage. Finally, the ion-milled samples were chemically etched using a mixture of 0.5 % HF and 99.5 % HNO$_3$. The etching was performed at a constant temperature of 5 °C, since the etching rate is known to be highly sensitive to temperature variations.

3. RESULTS AND DISCUSSION

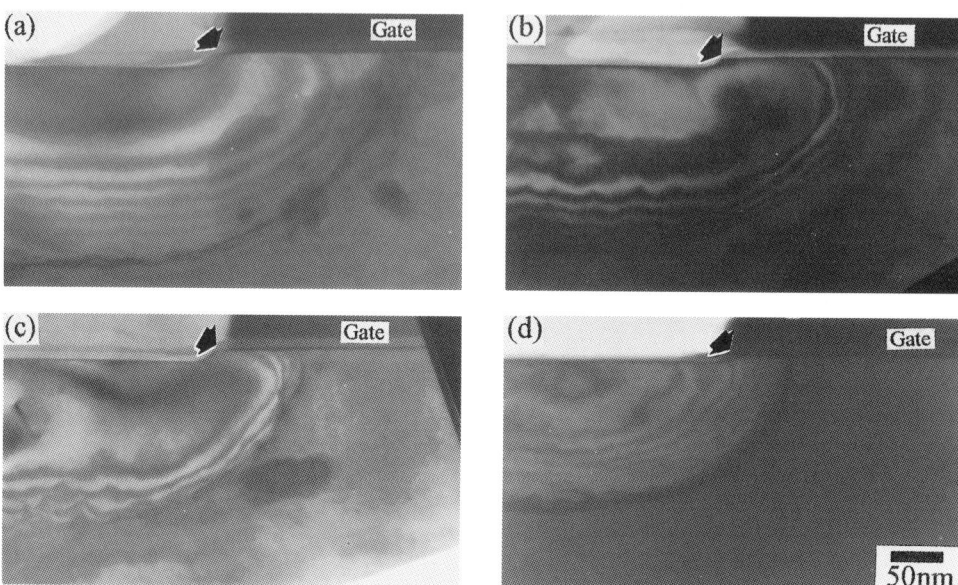

Fig. 1 TEM images from the samples implanted with (a) 35 keV As$^+$ to a dose of 2×10^{13} cm^{-2}, (b) 35 keV As$^+$ to 1×10^{14} cm^{-2}, (c) 60 keV As$^+$ to 1×10^{14} cm^{-2} and (d) 100 keV As$^+$ to 1×10^{14} cm^{-2}.

Figure 1 shows TEM images obtained from the etched samples that were ion-implanted at various conditions. The images clearly illustrate the presence of vertical and lateral isoconcentration lines. (SIMS dopant profile measurements were compared with the TEM results to calibrate the vertical doping concentrations.) For the sample implanted with 35 keV As$^+$ to a dose of 2×10^{13} cm^{-2} (Fig.1(a)), the deepest fringe in the image is located at a depth of 189 nm from the oxide film/silicon interface. The calibration shows that this fringe corresponds to a concentration of 1.3×10^{16} cm^{-3}. The corresponding lateral isoconcentration line is located at 150 nm from the edge of the mask (indicated by the arrow). For the sample implanted with 35 keV As$^+$ to a dose of 1×10^{14} cm^{-2} (Fig. 1(b)), the deepest fringe corresponding to the concentration of 6.23×10^{16} cm^{-3} is located at 147 nm from the interface. The corresponding lateral isoconcentration line is extended up to 89 nm from the edge of the mask. Examination of the samples implanted with 60 and 100 keV As$^+$ to a dose of 1×10^{14} cm^{-2} (Fig. 1(c) and 1(d), respectively) shows that the deepest fringes lie at 151 and 138 nm from the interface and correspond to 8.74×10^{16} and 1.84×10^{18} cm^{-3}, respectively. The corresponding lateral fringes are located at 64 and 49 nm from the mask edge, respectively. It is shown that the location of the deepest fringes increase as the energy and dose increase.

Fig. 2(a) shows a (400) weak beam TEM image obtained from the sample implanted with 100 keV As$^+$ to 1×10^{14} cm^{-2}. There are dislocation loops and stacking faults. Fig. 2(b) exhibits a bright field image of the same region of Fig. 2(a) and shows three thickness fringes (f_1, f_2, f_3). The fringes f_1 and f_2 obtained from the region, where defects are present, represents non-uniformity, while the fringe f_3 obtained from the region away from the defects shows a reasonably uniform contour. It was shown that the selective etching is dependent upon hole concentrations (Spinella et al 1995a, Maher

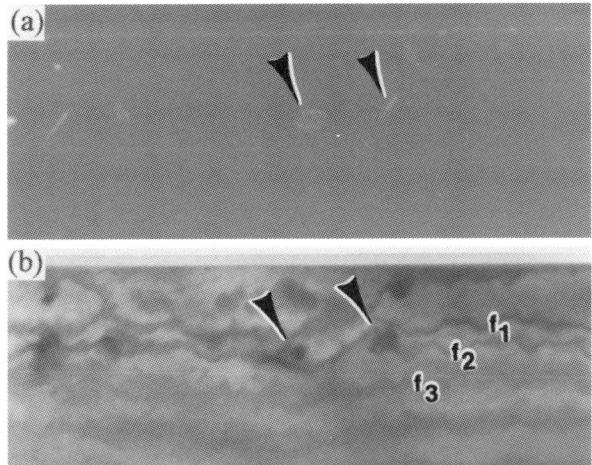

Fig. 2 (a) A (400) weak beam TEM image. (b) A bright field image of the same region of Fig. 2(a).

and Zhang 1994). Since crystallographic defects act as recombination centers for free carriers, the presence of defects would reduce current density and hence result in a decrease in the etching rate.

Fig. 3 shows 2D SCM images of the cross-section samples implanted with energies of 35 and 100 keV As⁺ to doses of 1×10^{13} and 1×10^{14} cm⁻². The images are exhibited with an alternating black and white scale. Edward et al (1998) proposed a simple calibration method of scanning capacitance spectroscopy (SCS) data, which allows the SCM contours to be converted into doping concentrations. It is shown that SCM can be used to map the dopant concentration down to ~10^{16} cm⁻³. Some contour lines extend into the regions beneath the gate oxides. These contours may be related to either a charge trap effect or the limited lateral resolution. Another possible explanation is that the masking during implantation may not be sufficient and so the masked area was implanted to some extent in the near surface region. Despite the effects, however, the SCM technique can resolve 2D carrier concentrations on cross-section and is relatively easy to perform. This indicates that the technique is potentially useful for the design and characterisation of semiconductor devices.

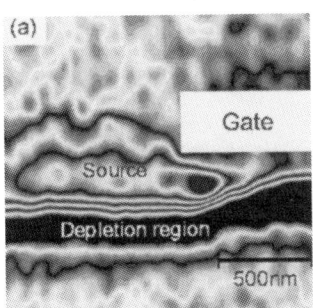 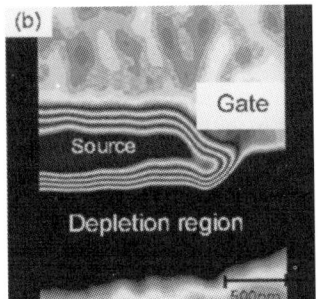 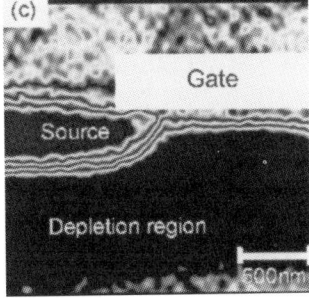

Fig. 3. SCM measured 2D dopant profiles of the samples implanted with (a) 35 keV As⁺ to 1×10^{13} cm⁻², (b) 35 keV As⁺ to 1×10^{14} cm⁻² and (c) 100 keV As⁺ to 1×10^{14} cm⁻².

Fig. 4 presents a comparison of the lateral doping profiles obtained using SUPREM-IV simulations and the TEM measured data. In the SUPREM-IV calculations, the behaviour of implantation and diffusion was simulated using the Dual Pearson and Fermi models, respectively. The lines and solid squares indicate the simulation results and the experimental data, respectively. It is noted that there is a discrepancy between the two results. This discrepancy in the experimental and simulation data may be caused by either measurement uncertainty, which might be introduced by TEM measurements, or more likely the nature of the simulator. The 2D simulation programme such as SUPREM-IV has a number of process-modeling parameters. In fact, the simulator is an

494

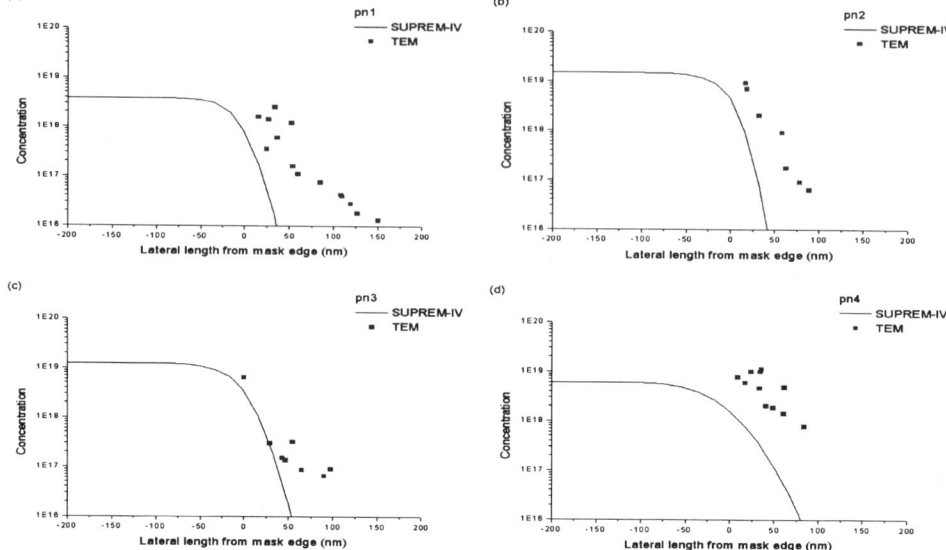

Fig. 4. A comparison of the lateral results obtained using SUPREM-IV simulations of the experimental conditions and the TEM measured data. The samples implanted with (a) 35 keV to 1×10^{13} cm^{-2}, (b) 35 keV to 1×10^{14} cm^{-2}, (c) 60 keV to 1×10^{14} cm^{-2}, and (d) 100 keV to 1×10^{14} cm^{-2}.

engineering model platform, in which the user applies his own calibration algorithm to get the computed results to match up with available process measurements (Fair and Shen 1997). Therefore, a lack of quantitative 2D lateral information in the junctions leads to such discrepancy presented in this work.

ACKNOWLEDGMENT

The authors wish to thank MOST (Korea) for financial support.

REFERENCES

Choi K K, Seong T-Y, Lee S H, Hwang H and Sohn Y S 1998 MRS Symp. Proc. 490, 53
Edwards H, McGlothlin R, Martin R S, Elias U, Gribelyuk M, Mahaffy R, Shih C K, List R S and Ukraintsev V A 1998 Appl. Phys. Lett. 72, 698
Gong L, Peterson S, Frey L and Ryssel H 1995 Nucl. Instrum. Methods B96, 133
Cerva H 1992 J. Vac. Sci. Technol. B 10, 491
Fair R B and Shen M, paper presented at 4th International Workshop on the Measurement, Characterisation and Modelling of Ultra-Shallow Doping Profiles in Semiconductors, Research Triangle Park, USA, April 6-9, 1997
Hill C 1992 J. Vac. Sci. Technology. B 10, 289
Maher D H and Zhang B 1994 J. Vac. Sci. Technol. B 12, 347
Osburn C M and Reisman A 1987 J. Electron. Mater. Vol. 16, p223
Spinella C, Raineri V and Campisano S U 1995a J. Electrochem. Soc. 142,1601
Spinella C, Raineri V, Saggio M, Privitera V and Campisano S U 1995b Nucl. Instrum. Methods in Phys. Res. B96 139
Williams C C, Slinkman J, Hough W P and Wickramasinghe H K 1990 J. Vac. Sci. Technol. A 8, 895

Inst. Phys. Conf. Ser. No 164
Paper presented at Microsc. Semicond. Mater. Conf., Oxford, 22–25 March 1999
© 1999 IOP Publishing Ltd

Clustering of vacancies on {113} planes in Si layers close to Si-Si₃N₄ interfaces and further aggregation of self-interstitials inside vacancy clusters during electron irradiation

L Fedina [1], **A Gutakovskii** [1], **A Aseev** [1], **J Van Landuyt** [2] **and J Vanhellemont** [3,4]

1 Institute of Semiconductor Physics, pr. Lavrentyeva 13, 630090 Novosibirsk, Russia
2 University of Antwerp, Groenenborgerlaan 171, B2020 Antwerpen, Belgium
3 Interuniversity Microelectronics Center (IMEC), B3001, Leuven, Belgium
4 Present address: Wacker Siltronic AG, PO Box 1140, D-84479, Burghausen, Germany

ABSTRACT: In situ HREM irradiation of (110) FZ-Si crystals covered with thin Si₃N₄ films was carried out in a JEOL-4000EX microscope, operated at 400 keV at room temperature. It is found that clustering of vacancies on {113} planes is realised in a Si layer close to the Si-Si₃N₄ interface at the initial stage of irradiation. Further aggregation of self-interstitials inside vacancy clusters is considered as an alternative way of point defect recombination in extended shape, to be accomplished with the formation of the extended defects of interstitial type upon interstitial supersaturation.

1. INTRODUCTION

It is well-known that in Si crystals, irradiated with a high dose of ions or with intense electron beams, large interstitial supersaturation exists which results in the formation of interstitial type extended defects. Detailed investigations of the kinetic growth of interstitial type defects during in situ electron irradiation in a HVEM in a wide temperature range (20-1150°C) led to an important conclusion concerning the difference between the coefficients of point defect interaction with the real surface (Aseev et al 1994). The physical reason for this is not yet clear but a larger vacancy capture coefficient with the surface results in an interstitial supersaturation close to the surface of Si specimens during irradiation. The supersaturated self-interstitials cluster into metastable defect types with preferential {113} habit planes at temperatures not exceeding 500°C. The atomic structure of {113} defects produced by electron irradiation at 450°C has first been revealed by Takeda et al (1994). However, a much more complex amorphous-like structure of the {113} defect, originating from clustering of vacancies close to the surface capped with SiO₂ or Si₃N₄ films and further interaction with interstitials, has been found during electron irradiation at room temperature (Fedina et al 1997a). These results suggest that capping layers and the slightly slowed down mobility of vacancies at room temperature, may both play an essential role in the accumulation of vacancies close to the surface. In situ HREM irradiation of strained GeSi /(001) Si heterostructures also shows that small compressive strains within 0.2-0.5 % promote an accumulation of vacancies on {113} planes inside the GeSi layers (Fedina et al 1997b). The discovery of vacancy clustering during electron irradiation allows us to study this effect straightforwardly in a high resolution electron microscope and our paper presents this kind of investigation concerning the aggregation of point defects on {113} habit planes in a Si layer close to the Si-Si₃N₄ interface.

2. RESULTS AND DISCUSSION

Figure 1 shows a typical HREM image of the {113} defects as created in thin irradiated areas (about 10 nm) of Si crystal covered with 3-5 nm Si₃N₄ films. Because the HREM image was taken near the optimum defocus condition, every dark dot in the image corresponds to a chain of atoms in [110] direction which is parallel to the electron beam. One can see that the image of the defect edge marked by a white rectangle differs from the central part of the defect. This fact shows the structure transformation of the defect during

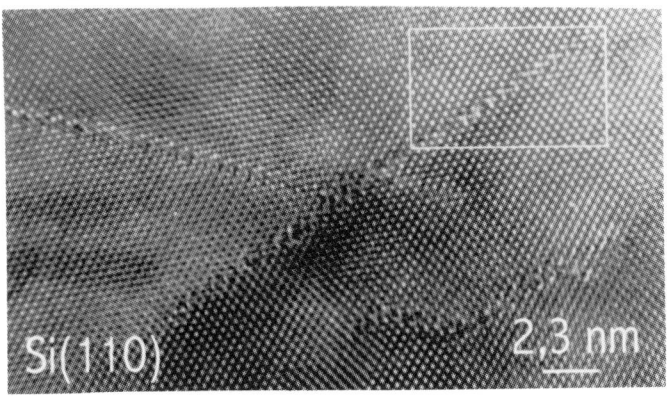

Fig.1. [110] HREM image of {113} defects in Si crystal produced after 15 minutes of irradiation in JEOL-4000EX at room temperature.

irradiation. An enlarged HREM image of this part of the defect is shown in Fig.2a,b with an atomic model superimposed on the image. There is no visible atomic column displacement around the defect plane, but very small hardly visible distortion (within 0.02±0.01 nm) of {113} planes parallel to the defect plane, which is of vacancy type, can be observed. We assume that white dots arising in the place of dark ones on {113} planes are due to vacancies in the atomic chain in [110] direction, parallel to the electron beam. An arrangement of vacancies along this direction can not be deduced straightforwardly from

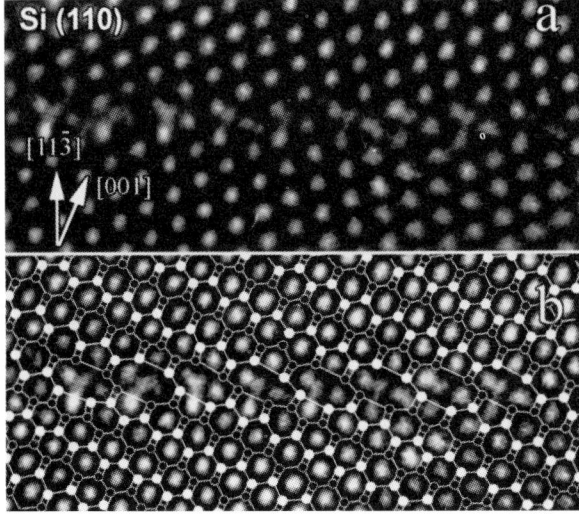

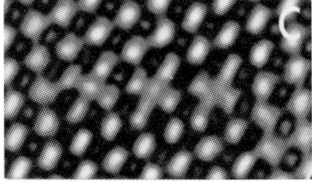

Fig.2. An experimental (a,b) and simulated (c) HREM images of the {113} defect obtained by using the model as superimposed on the image in b). Parameters for the calculation: defocus value dF=-35 nm and specimen thickness 8 nm.

the HREM image which is sensitive to the missing atoms only. In our tentative model for the {113} defect shown in Fig.2b each   eight-membered ring located on the defect plane corresponds to a continuous chain of vacancies in which no dangling bonds are involved along the chain (excluding two dangling bonds at the edges of the chain). This is a simple way of minimizing the defect energy, which is to minimize the number of broken bonds (Tan 1981). Based upon this model, the computer simulated HREM image of the {113} defect is in good agreement with the experimental one (see Fig.2c). The extended {113} defect of vacancy type has never been predicted with theory and it was only recently detected by irradiation of strained GeSi alloy as well as by irradiation of Si crystal covered with a Si_3N_4 film (Fedina et al 1997b and 1999). This is a new type of extended defect in Si crystal with the smallest displacement vector and it originates from the aggregation of vacancies in the shape of a chain-like configuration located in the {113}planes.

Figures 3a to d show the experimental HREM images with corresponding structural models of the same aggregate of point defects taken at successive stages of growth in a thicker Si specimen where a larger interstitial supersaturation is created (Fedina et al 1999). The model of the defects in their initial stage in Fig.3c reveals the formation of three-dimensional clusters consisting of two parallel {113}defects connected with each other by a sequence of five-membered atomic rings. Large white spots in the defect plane corresponding to vacancy chains are observed in the simulated image with 60 nm of defocus (not shown here). The shape and the size of white spots in the HREM image of a

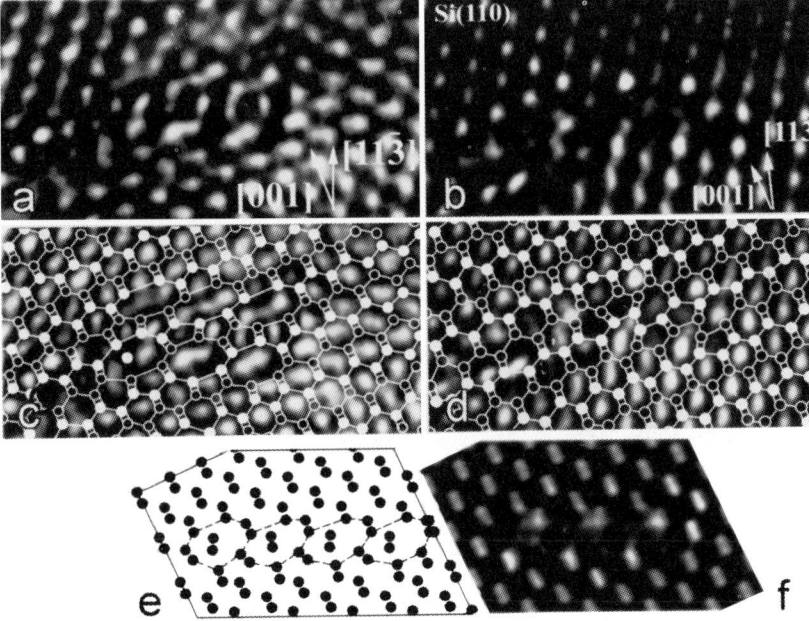

Fig.3. HREM images of one and the same point defect aggregate after (a) 20 and (b) 35 minutes of electron irradiation. c), d) Corresponding structural models superimposed on the experimental images. c) Geometric atomic model of {113} defect in a further formation stage obtained by insertion of double self-interstitials inside initially formed eight-membered rings shown with the broken lines. f) Computer simulated image of a {113} defect based on the model presented in (e). Parameters for the calculation are: a defocus value dF=-60 nm and specimen thickness 15 nm.

vacancy chain depends strongly on the number of vacancies in the chain as well as on the amount of inward-introduced relaxation. More calculation and experimental work is required in order to elucidate in detail the structure of these large aggregates of vacancies. The structure of aggregates is dramatically changed during further irradiation (Fig.3b) and a short sequence of double five- and double seven-membered atomic rings can be found on the atomic model of the defect superimposed on the image (Fig.3d). The same model can be easily designed by geometrical modelling of vacancy chains (eight-membered rings) on a {113} plane followed by further insertion of double interstitials (interstitial chain) inside each of these rings and rearrangement of the Si-Si bonds in the defect plane (Fig. 3e). Based upon this model the simulated HREM image presented in Fig.3f is in satisfactory agreement with the experimental one in Fig.3b. This is because the real chain structure in the [110] direction is changed during irradiation, which was not taken into account in the simulations. Clearly, this kind of {113} defect is very metastable. The model of the {113} defect at this stage is close to the one of interstitial type predicted by Tan (1981); however, this defect is not of interstitial nor of vacancy type since a small rotation of the interstitial chain inside the eight-membered ring and an additional rearrangement of the Si-Si bonds leads to the restoration of a perfect crystal structure, i.e. to the recombination of the vacancy chain. This takes about 20-30 minutes. The formation of a metastable configuration of interstitials inside a vacancy aggregate prevents fast recombination of the point defects in extended shape at room temperature. But it will provide an additional effective way for recombination of point defects at higher temperature close to the interface where an accumulation of vacancies forms. However the dominant interstitial supersaturation in Si crystal upon long irradiation time for the recombination of vacancy type {113} defects at a low temperature, enlarges the possibility for insertion of other interstitials into vacancy clusters so that the transformation of vacancy type {113} defects into the interstitial ones becomes possible (Fedina et al 1999). The results obtained show that the combined aggregation of point defects on {113} planes close to the Si-Si_3N_4 interface can be considered as an alternative method for their recombination through an extended shape, which is accomplished with the formation of the defects of interstitial type upon interstitial supersaturation.

ACKNOWLEDGEMENTS

This work was supported by grant No 98-02-17798 from the Russian Foundation of Basic Research. L.F. is grateful to the Belgian Science Policy Office (DWTC) for her fellowship at RUCA EMAT (Antwerp) in 1995-1996.

REFERENCES

Aseev AL, Fedina LI, Hoehl D and Barsch H 1994 Clusters of Interstitial Atoms in Silicon and Germanium, Academy Verlag, Berlin
Fedina L, Gutakovskii A, Aseev A, Van Landuyt J and Vanhellemont J 1997a In: « In situ Electron Microscopy in Material Research», Kluwer International Academic Publishers, p. 63
Fedina L, Lebedev O, Van Tendeloo G, Van Landuyt J 1997b Inst. Phys. Conf. Ser. No 157, p. 55
Fedina L, Gutakovskii A, Aseev A, Van Landuyt J and Vanhellemont J 1999 Phys. Stat. Sol. (a), **171**, p. 147
Takeda S, Kohyama M and Ibe K 1994 Phil. Mag. A, **70**, p. 287
Tan TY 1981 Phil.Mag. A, **44**, p. 101

Inst. Phys. Conf. Ser. No 164
Paper presented at Microsc. Semicond. Mater. Conf., Oxford, 22–25 March 1999
© 1999 IOP Publishing Ltd

Polycrystalline silicon – silicon nitride multilayers

D A Williams, S B Newcomb* and K Nakazato

Hitachi Cambridge Laboratory, Cavendish Laboratory, Madingley Road, Cambridge CB3 0HE U.K.
* Department of Materials Science and Metallurgy, University of Cambridge, Pembroke Street, Cambridge CB2 3QZ U.K.

ABSTRACT: Cross-sectional and planar transmission electron microscopy, as well as high-resolution scanning electron microscopy, were used to characterise stacked polycrystalline silicon – silicon nitride multilayers, designed for single-electron and few-electron device applications which are insensitive to background charge effects and co-tunnelling.

1. INTRODUCTION

In today's memory circuits, one bit of information is represented by $10^4 - 10^5$ electrons. The minimum feature size of semiconductor devices has been consistently reducing for several decades, and the trend is set to continue, which means that the number of electrons per bit is expected to reduce to the order of 100 in 15 years time. It is therefore useful to devise structures which use small numbers of electrons for future electronic applications. One such approach is to use the Coulomb blockade of electrons in quantum-dot devices, where the charging of a dot by one electron prevents subsequent electron flow(Geerligs et al 1993). This has been demonstrated in many materials and geometries, and has led to the development of single-electron and few-electron memory and logic circuits(Nakazato et al 1994, Tsukagoshi et al 1998).

However, whilst demonstrating the principle that single-electron types of circuit are possible, these experiments have shown that there are many difficulties to be overcome in their successful implementation as commercial circuits. Some of the main problems are temperature of operation, background charges and co-tunnelling. The purpose of the experiments described in this paper is to investigate a structure which is robust to the latter effects, and is also constructed using silicon-based technology.

2. MULTIPLE TUNNEL JUNCTIONS

Multiple tunnel junctions were used for the first successful single-electron memory cell using a random-potential approach. Fig. 1 shows a single – island Coulomb blockade structure, which is the basic element of single-electronics, and a multiple tunnel junction array. The charging energy of a single electron on the quantum dot has to be overcome by a voltage applied to the lead for charge to flow, and this is called the blockade voltage. For successful device operation, the blockade voltage must be many times greater than ambient thermal fluctuations, and for circuit operation the spread of blockade voltages over the devices in the circuit must be small. The single island is very sensitive to local charge distributions, which may capacitatively induce charge on the island in the range $-e/2$ - $+e/2$, altering the blockade voltage and in principle reducing it to zero. If the background charge is randomly distributed, as is usually the case in solid-state structures (both spatially and temporally), this will lead to a random spread of blockade voltages which means that such a structure is useless for circuit applications. For the case of the multiple junction, however, for a range of background charge distributions, the overall behaviour of the tunnel array is not so sensitive, as it is always the junction with largest resistance that dominates. This is currently the subject of much debate and has recently been simulated for the first time(Mizuta et al 1998).

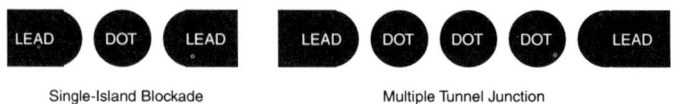

Single-Island Blockade   Multiple Tunnel Junction

Figure 1: Schematic picture of the single and multiple tunnel junction arrays.

The single island also has the difficulty that there is always a small but finite chance that two electrons will tunnel simultaneously on or off the island, called co-tunnelling, which leads to increased noise and difficult circuit implementation, but this probability becomes negligibly small over several junctions in series.

Such multiple junctions have been extensively studied using a number of different material systems, and most implementations use a random one or two-dimensional array formed in metal or semiconductor devices. For potential implementation, it is preferable to use silicon-based technology, and for repeatability and understanding of the device operation it is preferable to use an ordered structure.

3. FABRICATION

Multiple tunnel junctions have been formed in many material systems, and in this study the aim was to produce a structure with silicon-based technology. For this to be effective, it is necessary to build a layered structure of silicon and silicon nitride with controllable and uniform layer thicknesses, barrier heights and dopant levels.

The polycrystalline silicon layers, with thicknesses in the range 20-60nm, were formed by chemical vapour deposition. They were grown without intentional doping, on a heavily doped polycrystalline silicon substrate. The silicon layers were interspersed with thin (1-2nm) silicon nitride layers which were grown by a self-limiting process at a temperature of 950 °C. The samples were then capped with another layer of heavily-doped polycrystalline silicon, patterned, etched and oxidised.

The advantage of using nitridation is that it is inherently a self-limiting process and so in principle will give a tunnel barrier of known width and height for a given growth temperature. However, most existing experimental investigations of this process have been on single-crystal silicon for gate insulator applications. It was therefore necessary to ensure that the nitridation process was insensitive to surface orientation so that growth on a polycrystalline surface would be of uniform thickness. The application in question also requires that the 'lead' layers of silicon have high doping, whereas the layers that will form the dots have low to intrinsic dopant levels.

4. ANALYSIS

Specimens for transmission electron microscopy were prepared by focused ion-beam milling, and examined in a JEOL 2000F. The silicon layers in contact with the large grain-size heavily-doped layers were found to have a large degree of epitaxial linkage with those layers, despite the presence of a residual native oxide layer. SIMS analysis also showed that the nominally-undoped layers had also become heavily doped. However, the layers which were separated by thin silicon nitride had completely different morphology, and had a much smaller grain size. There was also evidence of voiding in several of these internal layers, but the thin nitride layers were found to have grown conformally on the silicon surfaces, with an even thickness. SIMS analysis shows that the nitride barriers are efficient diffusion barrier, retaining intrinsic dopant concentrations in intermediate layers, with heavy doping in adjacent layers

Bright- and dark-field micrographs of the whole layer structure are shown in Fig. 2. An immediate distinction can be made between the heavily doped regions with large grain size, as at A, and the isolated, nominally undoped regions between the nitride layers (B) as well as the rather thicker amorphous oxide above the n+ poly-Si upper layer.

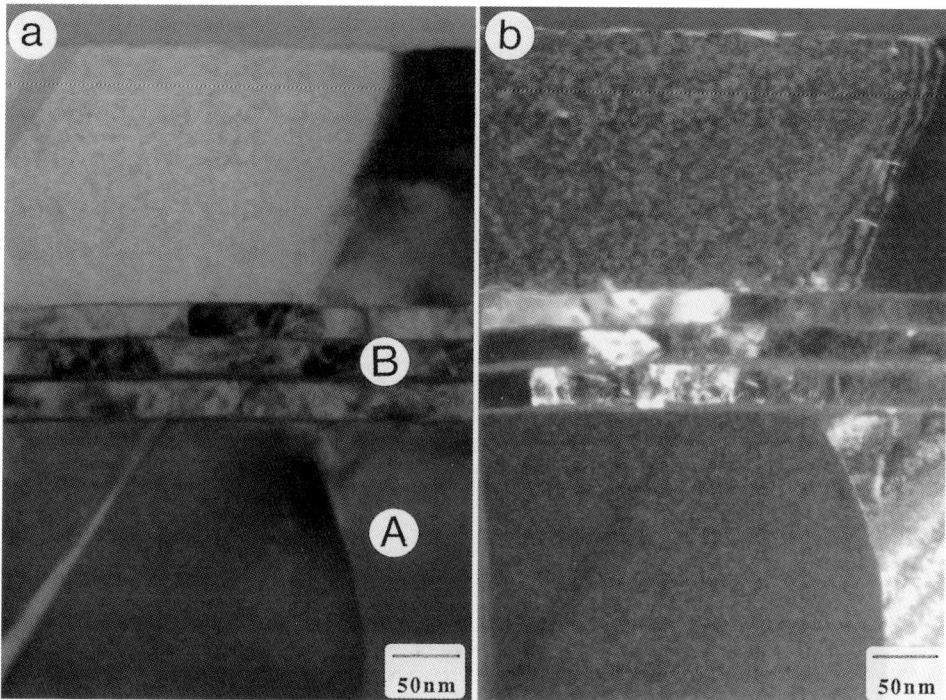

Figure 2: a)Bright and b)dark field micrographs of the overall array, showing the upper and lower highly doped polysilicon sandwiching the inner polysilicon – silicon nitride stack.

The different layers were examined in further detail and the region of interest is shown at higher magnification in Fig. 3, which is a through focal series of bright field images where the Fresnel effects associated with the changes of scattering potential at the different layer boundaries have been used to enhance their visibility. The sense of the changes in contrast as a function of the defocus conditions provides clear evidence for the fact that the poly-Si layers marked at 1-4 are interleaved with fine layers which are of relatively low scattering potential. These layers, which are marked X,Y and Z, are fully amorphous and the thickness of Y is rather higher than that of either X or Z. The nitride layers were found to undulate because of the associated roughness of the poly-Si underlying them and for this reason only limited regions were found to be lying edge-on in the through focal series of images shown. However, measurements of the thicknesses of the three nitride layers were obtained and those of X,Y and Z were found to be 1.2, 2.9 and 0.97nm respectively.

502

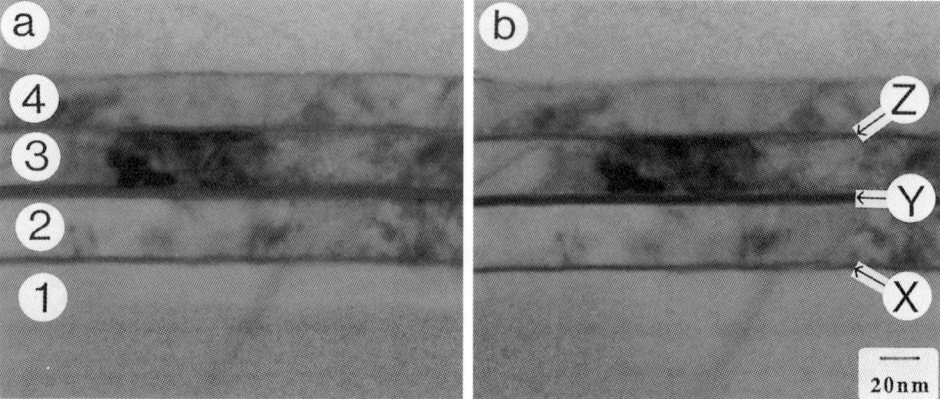

Figure 3: Through-focus pair of the polysilicon – nitride stack.

The microstructures of the different poly-Si zones were examined in further detail. In general both the highly-doped regions are of low defect content and relatively high grain size although there is a tendency for the upper parts of the n+ first gate to contain rather more defects than elsewhere. The positions of the intervening i-Si layers are marked (1-4) in Fig. 3a, and the boundary between the n+ and i-Si for the first polysilicon layer can be seen below layer X. The nominally i-Si is of higher defect content than the n+ Si underlying it, but is generally epitaxial. SIMS observation confirmed that the i-Si region has the same dopant level as the underlying region. This was also true of the upper nominally undoped layer in contact with the upper doped polysilicon, although in this case there was an abrupt change in crystallinity between the two layers.

In conclusion, the use of stacked layers of polycrystalline silicon and silicon nitride is a promising approach for the manufacture of vertical multiple-tunnel junctions, with applications for single-electron and few-electron devices.

REFERENCES

Geerligs L J, Harmans C J P M and Kouwenhoven L P 1993 *The Physics of Few-Electron Nanostructures*, Physica B, **189**

Mizuta H, Williams D, Katayama K, Müller H-O, Nakazato K and Ahmed H 1998 *Proc. 2nd Int. Workshop on Physics and Modelling of Devices Based on Low-Dimensional Structures*, pp 67-72 (IEEE Computer Society)

Nakazato K, Blaikie R J and Ahmed H 1994 J. Appl. Phys. **75**, 5123

Tsukagoshi K, Alphenaar B W and Nakazato K 1998 Appl. Phys. Lett. **73**, 2515

Inst. Phys. Conf. Ser. No 164
Paper presented at Microsc. Semicond. Mater. Conf., Oxford, 22–25 March 1999
© 1999 IOP Publishing Ltd

Defect characterization of multicrystalline silicon for solar cell applications

M Werner, A Lawerenz*, M Ghosh* and H J Möller*

Max-Planck-Institut für Mikrostrukturphysik, Weinberg 2, 06120 Halle, Germany; *Institut für Experimentelle Physik, Technische Universität Bergakademie Freiberg, Silbermannstr. 1 09599 Freiberg, Germany

ABSTRACT: The main material for photovoltaic applications is multicrystalline silicon (mc-Si). The efficiency of corresponding solar cells depends on lattice defects such as grain boundaries, dislocations, small precipitates and point defects. Oxygen and carbon are the main impurities in mc-Si. The concentrations of oxygen are comparable to Cz-silicon, whereas the carbon concentrations are usually higher in mc-Si. Their agglomeration depends on the different growth techniques and subsequent technological processing steps. The present investigations include the precipitation behaviour of mc-Si grown by directional solidification in an ingot. The material is characterized by FTIR spectroscopy and TEM. The precipitation behaviour of mc-Si is compared to Cz-material.

1. INTRODUCTION

Multicrystalline silicon is a low-cost material for photovoltaic applications. The solar conversion efficiencies of commercial cells are typically in the range of 12 to 15%. The efficiency of multicrystalline solar cells is mainly limited by minority carrier recombination at dislocations and intragrain defects such as certain impurities, small clusters or precipitates. Recombination at pure dislocations is known to be relatively weak but the decoration of the latter with precipitates or metallic impurities is responsible for the enhanced recombination (e.g. Mc Hugo et al 1997, Cabanel et al 1990). The main impurities in mc-Si are oxygen and carbon. The concentration of oxygen is comparable with Cz-silicon (up to $2x10^{18}$ cm^{-3}), whereas the carbon concentration is usually higher in mc-Si (up to 10^{18} cm^{-3}).

Oxygen is known to affect the conversion efficiency of solar cells. A degradation of the solar cell performance has been reported both for Cz- and mc-Si (e.g. Glunz et al 1998, Möller et al 1999). Oxygen may form various defects that differently affect the electrical behaviour. Clusters of a few oxygen atoms and larger SiO_2 precipitates of different size and crystal structure have been observed (Borghesi et al 1995). Thermal donors are well-known defects, consisting of clusters of a few oxygen atoms; new donors are considered to be SiO_2 precipitates (Michel et al 1994). In addition, oxygen which also precipitates at grain boundaries and dislocations can also change the electrical behaviour of the latter two. The evolution of the oxygen defects strongly depends on the thermal history of the material. Besides, each thermal step from the crystal growth to the subsequent solar cell processing has to be considered. Carbon forms defects that are electrically less active, but it is important because of its influence on the oxygen precipitation. Present investigations include the precipitation behaviour of multicrystalline silicon grown by directional solidification in an ingot. The as-grown material has been investigated by Fourier-transform infrared (FTIR) spectroscopy and transmission electron microscopy (TEM).

2. EXPERIMENTAL PROCEDURE

As-grown wafers were selected from different places in the ingot, which was boron-doped using a concentration of about $2x10^{16}$ cm^{-3}. Typical grain sizes were in the centimetre range. The

504

dislocation densities determined by selective etching of mechanically polished surfaces varied between 10^5 - 10^6 cm^{-2}. The concentrations of dissolved oxygen and carbon were determined by FTIR spectroscopy, described in more detail by Möller et al (1998). For conventional and high-resolution electron microscope investigations a microscope JEM 4000 EX was used, operating at an acceleration voltage of 400 kV. To determine the elemental distribution with a high lateral resolution (< 5nm) electron energy loss spectroscopy (EELS) was performed in an CM 20 FEG equipped with a Gatan imaging filter system (GIF).

3. RESULTS

3.1 FTIR measurements

The typical crystallization process of a multicrystalline ingot takes about 40 - 60 hours. During solidification the planar melt interface moves from the bottom to the top of the ingot. Due to this process the oxygen concentration decreases from bottom to top from 10^{18} to 10^{16} cm^{-3}. In contrast to oxygen, the concentration of carbon and many other impurities (metals) increases from bottom to top. Both oxygen and carbon may partly precipitate during cooling, depending on their concentrations and the cooling rates. Oxygen is expected to precipitate to a larger extent in the bottom where a higher oxygen concentration is measured. An example of FTIR spectra is given in Fig. 1. In addition to the peak of interstitial oxygen there is a small peak at wave number 1224 cm^{-1}, which is ascribed to plate-like SiO_2 precipitates (Hu 1980). Both peaks decrease with increasing height in the ingot. It is obvious that more precipitates are formed in such as-grown samples with a higher interstitial oxygen concentration.

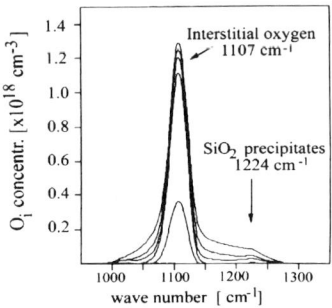

Fig.1. FTIR spectrum of oxygen in various heights within a mc-Si ingot.

3.2 TEM investigations

Different types of defects within the bulk of the grains could be identified by TEM. Dislocations decorated with precipitates are the most frequently observed defects. They occurred in all height positions of the ingot which have been investigated up to now. The top region is still under investigation. The dislocation density varies between 10^5 and 10^6 cm^{-2}, locally up to 10^7 cm^{-2}. A typical example of decorated dislocations is shown in Fig. 2a. The particles visible by their strain fields have a non-homogeneous amorphous structure. Energy filtered images were taken as to identify

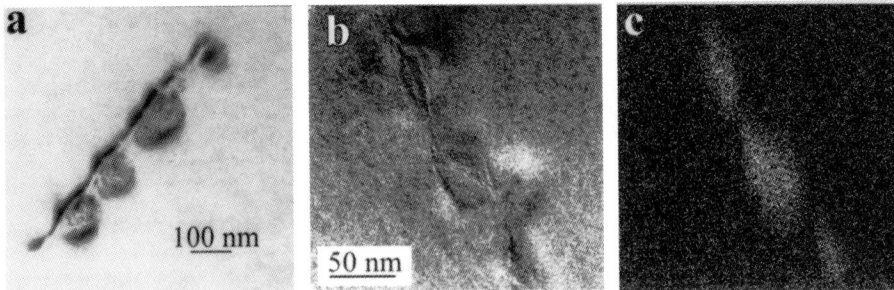

Fig.2. Bright-field image of a dislocation decorated with precipitates (a,b). The energy-filtered image is generated by the oxygen content of the precipitates (c).

these precipitates (see Fig. 2c). The image is generated by the oxygen content of the precipitates. A quantitative determination of the oxygen concentration is not possible, as the sample surfaces themselves are additionally covered with SiO_2 layers.

In the following the defects at the bottom are analysed. Besides decorated dislocations, polyhedral SiO_2 particles were identified, cf. Fig. 3. This HREM image clearly shows the amorphous structure besides the lattice planes of the matrix. The particles reach sizes up to 50 nm. Owing to their sporadic occurrence it is not possible to determine accurately their low density.

Fig.3. HREM image of an amorphous, polyhedral SiO_2 particle

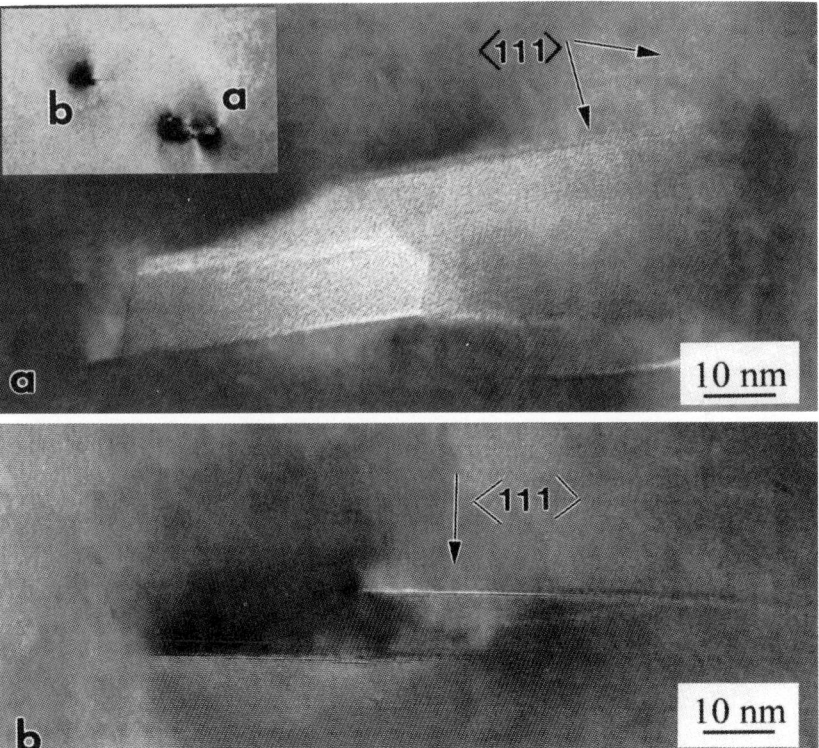

Fig.4. Conventional diffraction contrast TEM and HREM images of defects in mc-Si: a) plate-like SiO_2 precipitate, b) agglomeration of Si self-interstitials on {111} planes. The inset in a) is a low magnification of the defects in a) and b).

As indicated by the above FTIR measurements, there should be plate-like SiO_2 precipitates, which are indeed detected, especially at the bottom of the ingot (Fig.4a). However, their density determined by TEM is rather low ($<10^5$ cm^{-2}). It is known from Cz-silicon that the SiO_2 precipitates are lying on {100} planes. It is not unambiguously clear whether the plate-like precipitate in Fig. 4a lies on a {100} or {111} plane. Precipitates lying on {100} planes have been identified very rarely in cast Si. In addition, we have also identified plate-like defects on {111} planes which seem to have a thickness of 1-2 monolayers (Fig. 4b). These defects are up to 100 nm long and show a strong local strain field. Their density is higher than that of plate-like SiO_2 precipitates. Concerning the analysis of HREM images we assume that these defects arise from agglomerations of silicon self-interstitials.

In addition to the above described defects complex defect structures have been observed, too. Respective examples are shown in Fig. 5, where the defect consists of parallel-arranged agglomerates on {111} planes surrounded by very strong strain fields.

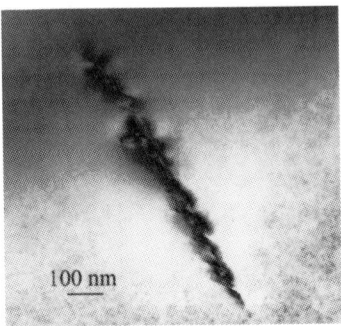 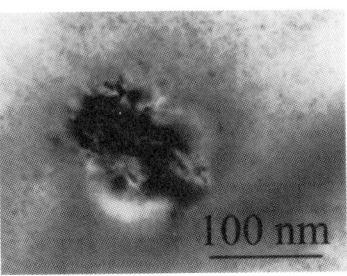

Fig.5. Bright-field images of a complex structure surrounded by strong local strain field

4. SUMMARY

The precipitation behaviour of mc-Si grown by directional solidification has been investigated by FTIR and TEM. The precipitation of oxygen in mc-Si is stronger than in Cz-Si due to the higher density of nucleation centres as, e.g., of dislocations and carbon. The precipitates have more complex structures and their densities are higher than in Cz-Si. The majority of the defects occur at the bottom of the ingot, where the oxygen content is highest. The following kinds of precipitates have been identified: i) SiO_2 precipitates at dislocations, ii) polyhedral amorphous SiO_2 precipitates, iii) plate-like SiO_2 precipitates, and iv) agglomerations of Si self-interstitials on {111} planes. The densities of the precipitates decrease with increasing height in the ingot (with a decreasing oxygen content). Dislocations decorated with precipitates have been identified in all regions of the ingot investigated up to now.

ACKNOWLEDGEMENTS

This work has been supported by the BMBF under contract number 0329743D.
We would like to thank Dr. E. Pippel for providing the energy filtered TEM images.

REFERENCES

Borghesi A, Pivac B, Sasella A and Stella A 1995 J. Appl. Phys. **77**, 4169
Cabanel C and Laval J 1990 J. Appl. Phys. **67**, 1425
Glunz S W, Rein S, Wartha W, Knobloch J and Wettling W 1998 Proc. 2nd World Conf. on Photovoltaic Solar Energy Conversion, Vienna, 1343
McHugo S A, Hieslmair H and Weber E R 1997 Appl. Phys. A **64**, 127
Michel J and Kimmerling L C 1994 Semiconductors and Semimetals **42**, 251
Möller H J, Long L, Werner M and Yang D 1999 phys. stat. sol. (a) **171**, 175

Inst. Phys. Conf. Ser. No 164
Paper presented at Microsc. Semicond. Mater. Conf., Oxford, 22–25 March 1999
© *1999 IOP Publishing Ltd*

New insights into the formation processes of macropores in n-Si(001) and p-Si(001)

C Jäger, C Dieker, W Jäger, M Christopherson*, J Carstensen* and H Föll*

Center for Microanalysis, Faculty of Engineering, Kaiserstr. 2, D-24143 Kiel, Germany
*Materials Science, Faculty of Engineering, Kaiserstr. 2, D-24143 Kiel, Germany

ABSTRACT: Morphology and interfaces of macropores in (001)-oriented n- and p-type silicon were studied by analytical and high-resolution transmission electron microscopy (TEM) for different stages of their evolution during electrochemical etching. In n-Si etched by oxidizing HF-electrolytes dendritic pores along <100>-directions (dimensions ~ few nm) were found to be the precursors of macropores (dimensions ~ 1 μm). The pore morphology during the early stages of pore formation consists of a periodic arrangement of truncated octahedra with {111} facets, and clear indications exist that oxide inclusions are formed at the interfaces. The interfaces of macropores possess {111} facets of similar size. In p-Si etched by non-oxidizing organic electrolytes, nanopipes along <100> (dimensions ~ few nm) were found to be precursors of macropores. During the later formation stages macropores possess {111} interface facets and are filled with nanoporous crystalline Si. These results indicate an oscillatory mode of pore formation as a result of oxidation and subsequent dissolution of SiO_2 for n-Si and a predominance of a direct dissolution mechanism for p-Si.

1. INTRODUCTION

It was first demonstrated by Canham (1990) that microporous Si can emit photo-luminescence with high quantum efficiency, and it has been proposed that the luminescence mechanism is the recombination of quantum-confined carriers in crystalline Si nanostructures (Canham 1990, Lehmann and Gösele 1991). Strong confirmation for this mechanism was provided by experimental work adressing the structural and luminescence properties of porous silicon (Cullis et al 1997 for review).

In contrast to microporous Si, the processes of macropore formation in Si during electrochemical etching are not yet understood, and a comprehensive explanation of the formation of porous structures in silicon is still missing (Smith and Collins 1992). Depending on the experimental conditions used for etching of Si, various types of pore structures can be created in n-type Si (Lehmann 1993, Osaka et al 1997). An orientation-dependent macropore formation involving {113} crystallographic directions was reported recently for n-type silicon (Rönnebeck et al 1999). The influences of electrolyte type and crystal orientation on the pore formation were investigated also for p-type Si and led to a proposal that H termination may influence the chemical reactivity during pore etching (Rieger and Kohl 1995).

This paper summarizes first results of TEM investigations of macropore formation in n- and in p-type Si for etching with HF electrolytes in aqueous solution and with organic acetonitrile solution. The results will show that the fraction of oxidizing electrolytes in the etch solution is essential in controlling the microscopic formation processes of macropores.

2. EXPERIMENTAL DETAILS

For the comparison of electrochemical pore etching in polished n-type and p-type Si(001) wafers two types of anodic etching conditions have been applied. For n-type Si (specific resistivity 4-5 Ωcm) etching was performed under backside illumination using an HF electrolyte (4 wt.-% HF in aqueous solution) and a current density of 4 mA/cm^2. For p-type Si (specific resistivity 10-16 Ωcm) etching was performed using an organic HF electrolyte (4 wt.-% HF in in acetonitril and a current density of 2 mA/cm^2. Microscopic investigations have been performed on cross-section and plan-view specimens using a Philips CM 30 transmission electron microscope at 300 kV. The chemical compositions of inclusions at macropore interfaces were analysed by spatially-resolved X-ray microanalyses (XEDS).

3. EXPERIMENTAL RESULTS

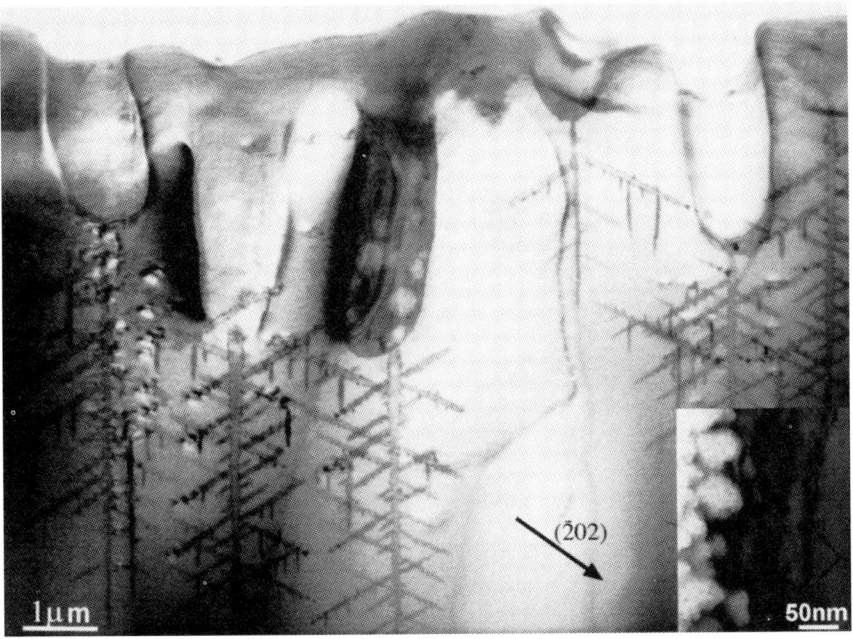

Fig.1 Macropores with {111} wall facets (insert) and dendritic pores in n-type Si. TEM bright-field micrographs.

The observations of macropores in n-type Si and in p-type Si show that the temporal evolution of pore formation can be subdivided into three stages: (a) homogeneous etching (b) roughening of the substrate surface and (c) formation of pores. These observations explain also the different regimes in Si etch rates measured by profilometry. For n-Si surface roughening is followed by the formation of dendritic pores and of macropores connected with dendritic pores (Fig. 1). Macropores and dendritic pores have preferred <100> orientations of the pore axes. Pore diameters are approximately 1 μm for macropores and 0 25 μm for dendritic pores. The pores extend to depths of about 7 μm for 15 min etch time and about 100 μm for 30 min etch time. For p-Si a layer of nanoporous crystalline Si with preferential <100> fiber orientation develops at the substrate surface before the onset of macropore

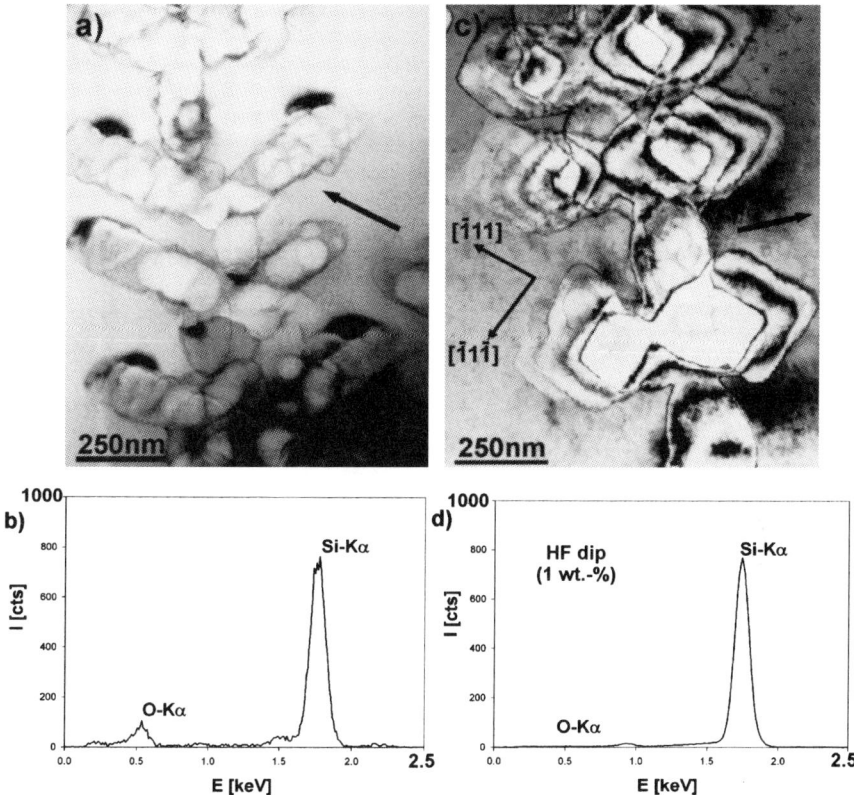

Fig.2 Dynamical bright field images of dendritic pores before (a) and after (c) an HF dip treatment of the TEM specimen. Imaging vectors $\vec{g} = (2\bar{2}0)$. Note thickness contours due to {111} facets of pore walls.(b,d) Energy-dispersive X-ray spectra, I(E) with O-K_α and Si-K_α lines taken from a wall region with strain contrast before applying an HF dip (b) and a pore wall region after applying an HF dip (d).

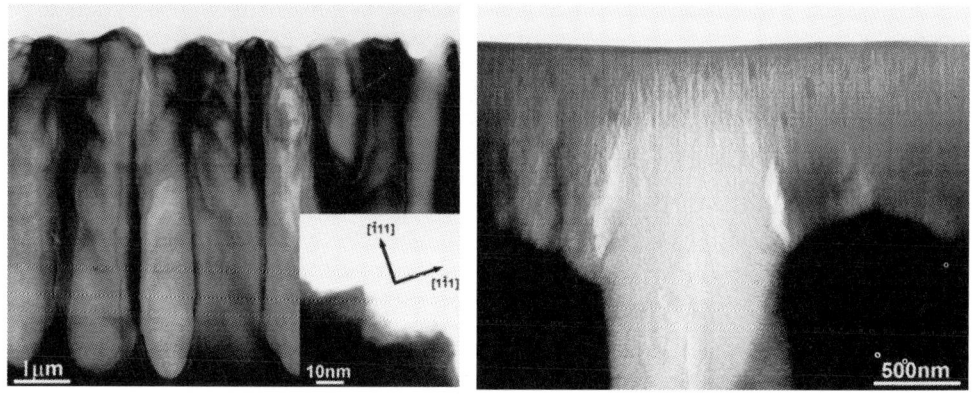

Fig.3 Macropores with {111} wall facets (insert) in p-type silicon (left). Surface layer and macropore with nanocrystalline Si fibres (right). TEM bright-field micrographs.

510

formation (Fig. 3). Plan-view observations indicate that also the macropores are completely or partially filled with Si fibres (Fig. 6). Investigations of macropore wall sections in cross-sectional specimens show that the walls at some distance to the pore tips are facetted on (111) planes for n-type and p-type Si (Figs.1, 3, inserts).

The morphology of dendritic pores has been determined from imaging in <001> and <110> zone axes. From such experiments it can be concluded that dendritic pores consist of a quasiperiodic arrangement of truncated tetrahedral voids with large fractions of {111} faces which are aligned along <001> directions (Fig. 2c). The perfection of this morphology decreases as etching proceeds (Fig. 2a).

Under dynamic imaging conditions pronounced strain contrasts are observed at the tips of the individual branches indicating the presence of inclusions causing distortions of the surrounding Si lattice (Fig. 2a). Spatially-resolved x-ray microanalyses of such regions show clearly that Si and

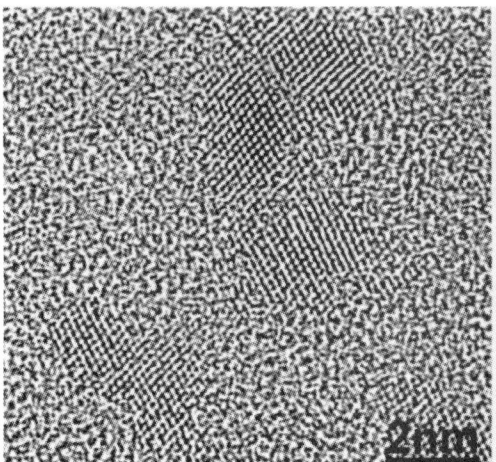 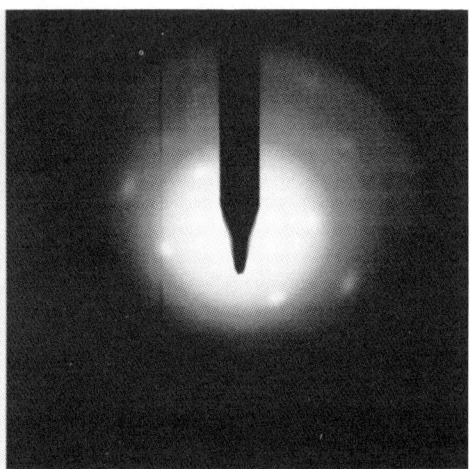

Fig.4 <110> lattice image of nanoporous crystalline Si fibres in p-type Si.

Fig.5 Selected area diffraction pattern taken from nanoporous surface layer showing individual diffraction spots.

oxygen are present (Fig. 2b) suggesting that the inclusions consist of an SiO_x phase. This result was confirmed by measurements taken from wall sections of dendritic pores after applying an HF dip (using a 1wt.-% HF etch for 30 min) which is known to remove oxides. Oxygen is absent in the corresponding XEDS spectra (Fig. 2d), and the strain contrast phenomena attributed to oxide inclusions disappear (Fig. 2c).

Macropores in p-type Si (diameters ≈ 1 μm) are completely or partially filled with dense arrays of nanocrystalline Si fibres. The Si fibres are preferentially oriented along [001] near the pore axes and emerge from the pore walls at angles of about 90°. Lattice images taken along <110> zone axes show that the Si fibres are crystalline with diameters on the scale of up to a few nanometers (Fig. 4). The crystalline nature of the fibers could also be confirmed by the presence of distinct diffraction spots in selected-area diffraction patterns (Fig. 5). The interfaces of macropore tips are roughened (imaged as contrast fluctuations in Figs. 6, 7). Furthermore, straight tubes or pipes along <100> (lengths ≤ 50 nm, diameters $\leq$ few nanometers) are found to be connected to the tip regions of macropores. Examples of the defocus contrasts of such nanopipes formed at the tip of macropores are shown in Fig. 7.

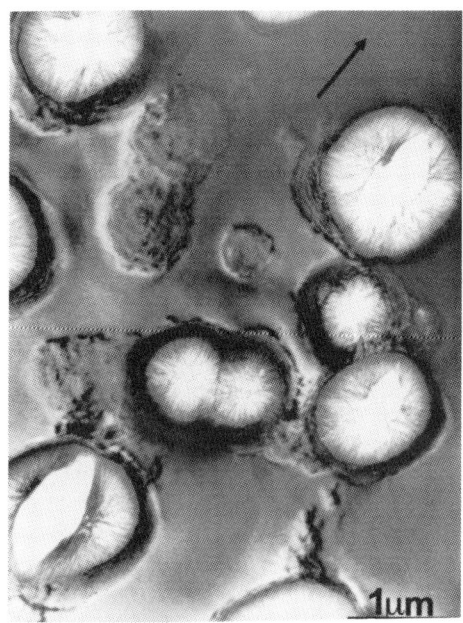

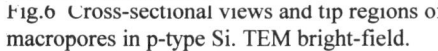

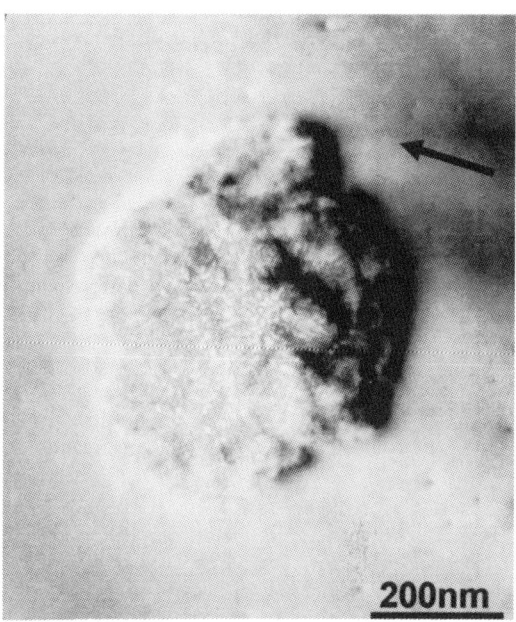

Fig.6 Cross-sectional views and tip regions of macropores in p-type Si. TEM bright-field.

Fig.7 Defocus contrast of nanopores (bright) at tip of macropore in p-type Si. TEM bright-field, g = (220).

3. DISCUSSION

Our TEM observations in n-type and p-type Si provide new insights into the microscopic processes of macropore formation during electrochemical etching. A detailed model will have to consider the following aspects which we can address here only in a summarizing discussion.

Macropore formation in n-Si and p-Si appears to occur via precursor stages under the etching conditions chosen: Dendritic pores are precursors or competitors of macropores in n-Si whereas nanopipes appear to act as precursors of macropores in p-Si. Characteristic of the later stages are empty pores for n-Si and macropores filled with nanoporous crystalline Si for p-Si. An important point is that macropore walls both in n-type and p-type Si possess {111} facets. In n-type Si also the dendritic pores possess {111} facets during the early stages.

These microstructures are intimately connected with the differences in the mechanisms of pore formation for the two cases considered here. For etching of n-Si using an aqueous solution of HF electrolyte the pore formation obviously proceeds via formation of localized oxide inclusions at the Si-electrolyte interfaces and their subsequent dissolution. The quasiperiodic arrangements of truncated octahedra with {111} facets are indicative of an oscillatory formation mode in the electrochemical dissolution of Si whereby {111} pore walls become increasingly passivated against dissolution and {100} pore tips favour dissolution (Fig. 8). It is expected that the dynamics of oxide growth and dissolution cause current oscillations (Carstensen et al 1998) The anisotropy of pore growth indicates that passivation of interfaces by hydrogen termination hinders the electrochemical reaction at the pore walls in certain crystallographic directions. The efficiency for passivation of {111} facets is higher than that for {100} facets at the current densities used. Hydrogen termination has been proposed to be important also in the anisotropic etching of pipe arrays in silicon (Santos and Teschke 1998). Preferred growth of dendritic pores along <100> directions has been reported by Osaka (1997), however, no detailed analyses of pore structures have been published. The spatial distribution and the

512

branching of dendritic pores suggest that local differences in the transport of electric charge carriers and / or ions are affecting the dissolution processes during the etching under backside illumination. These results are consistent with earlier proposals (Lehmann 1993).

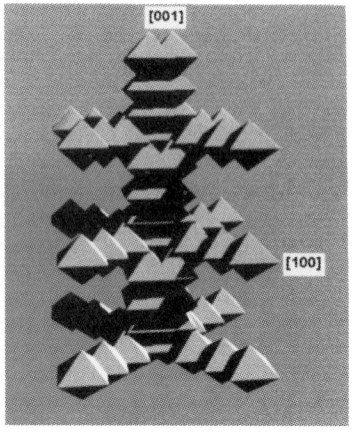

Fig. 8 Schematic illustration of the oscillatory mode of pore formation in n-Si.

For etching of p-Si with a non-oxidizing organic electrolyte the pore formation appears to proceed locally via direct dissolution of silicon resulting in the formation of nanopipes at the interface between Si substrate and electrolyte. Si is dissolved in a highly correlated fashion resulting in a thin surface layer with oriented nanocrystalline Si fibres (Fig. 3). At a later stage instabilities of the substrate-electrolyte-interface occur which favour macropore formation by dissolution of Si along nanopipes. Again anisotropy appears to favour pore growth along <100> crystallographic directions. However, nanocrystalline Si fibres emerge also from side walls of macropores at right angles indicating that nanopipe formation can also occur along other directions. The preferred growth of nanopipes in the <100> directions of macropores is suggestive of an efficiency for interface passivation by hydrogen termination which is smaller for {001}interfaces as compared to other interface planes.

4. CONCLUSIONS

The transmission electron microscopy investigations of morphology and interfaces of macropores in n-Si(001) and p-Si(001) at different stages of their evolution have led to new insights into the formation processes. The dominating microscopic mechanisms of macropore formation appear to be governed by the fraction of oxidizing electrolyte present in the etch solution and may lead to different dissolution reactions. Etching of n-Si with oxidizing HF-electrolytes leads to an oscillatory mode of pore formation as a result of oxidation and subsequent dissolution of SiO_x. Etching of p-Si for etching with non-oxidizing organic electrolytes leads to a direct dissolution of silicon.

REFERENCES

Canham L T 1990 Appl. Phys. Lett. **57**, 1046
Carstensen J, Prange R, Popkirov G S and Föll H. 1998 Appl. Phys. **A 67**, 459
Cullis A G, Canham L T and Calcott P D J 1997 J. Appl. Phys. **82**, 909
Lehmann V and Gösele U 1991 Appl. Phys. Lett. **58**, 856
Lehmann V 1993 J. Electrochem. Soc. **140**, 2836
Osaka T, Ogasawara K and Nakahara S 1997 J. Electrochem. Soc. **144**, 3226
Rieger M M and Kohl P A 1995 J. Electrochem. Soc. **142**, 1490
Rönnebeck S, Carstensen J, Ottow S and Föll H 1999 Electrochem. Solid-State Lett. **2**, 3
Santos M C and Teschke O 1998 J. Vac. Sci. Technol. **B16**, 2105
Smith R L and Collins S D 1992 J. Appl. Phys. **71**, R1

Inst. Phys. Conf. Ser. No 164
Paper presented at Microsc. Semicond. Mater. Conf., Oxford, 22–25 March 1999
© 1999 IOP Publishing Ltd

Transmission electron microscopy studies of photochemically etched porous silicon

A Wellner, R E Palmer, L Koker[1], K W Kolasinski[1] and M Aindow[2]

Nanoscale Physics Research Laboratory, School of Physics and Astronomy,
The University of Birmingham, Edgbaston, Birmingham B15 2TT, UK
[1] School of Chemistry, The University of Birmingham, Edgbaston, Birmingham B15 2TT, UK
[2] School of Metallurgy and Materials, The University of Birmingham, Edgbaston,
Birmingham B15 2TT, UK

ABSTRACT: The morphology of photochemically etched n-type [100] porous silicon has been investigated using transmission electron microscopy. The porous silicon was found to form square macropores with edges aligned along the <011> directions and edge lengths of several microns. These macropores contain nanoporous silicon which still exhibits the [100] orientation. The nanoporous silicon exhibits significant lattice strain, estimated to be in the range of 1- 2%. Since the samples have been exposed to air the expected oxide layer may account for this strain.

1. INTRODUCTION

The discovery of strong visible photoluminescence from porous silicon (por-Si) by Canham (1990) has stimulated intensive studies of this material (for a review see Cullis et al 1997) and various models of por-Si formation have been proposed (for a review see Smith and Collins 1992). Four basic methods of production can be distinguished. (i) Electrochemical etching is the most well-established of these techniques and the morphology of electrochemically grown por-Si has been investigated using a wide variety of techniques, one of the most powerful being Transmission Electron Microscopy (TEM) (Beale et al 1984). (ii) For lightly doped n-type silicon, illumination is necessary during the anodization. This technique is generally referred to as photoelectrochemical etching and the morphology of this material has been studied by Lévy-Clément et al (1996) using SEM and by Albu-Yaron et al (1994) applying TEM. (iii) Porous silicon can also be produced by chemical (stain) etching without applying light or electrical current, and structural investigations using TEM have been performed by Shih et al (1993). (iv) The formation of porous silicon by irradiating n-type silicon immersed in aqueous HF with a laser, referred to as photochemical etching, has been reported by Noguchi and Suemune (1992) and Dubbelday et al (1993). Ngan et al (1998) used SEM for investigating the surface structure.

Photochemical processing of n-type Si produces por-Si films that luminesce efficiently in the visible. The wavelength of the photoluminescence (PL) peak maximum is found to depend upon the wavelength of the fabrication laser and exposure of the por-Si to air (Kolasinski et al, in preparation). Here we report the first structural investigations of photochemically etched porous silicon using TEM. We have developed a different preparation technique for TEM specimens which ensures minimal damage and fewer artefacts.

2. EXPERIMENTAL

Preparation of TEM samples of por-Si often suffers from the introduction of artefacts which affect the fine internal structure observed. In this study we take advantage of the photochemical etching production process to minimise this problem.

514

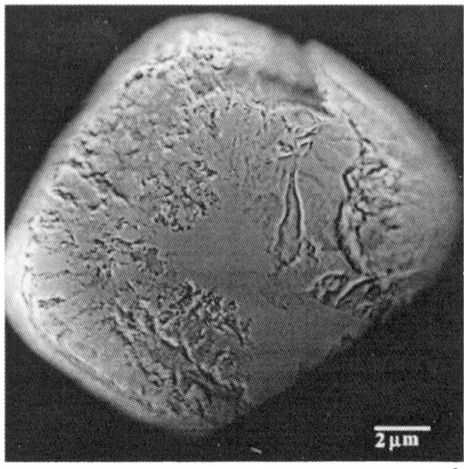

a   b

Fig. 1 Axial bright field images of macropores in photochemically etched porous silicon,
a) macropore containing silicon which is still intact (not porous) and exhibits many bend contours;
b) macropore exhibiting a hole at its centre and containing nanoporous silicon.

Porous silicon was prepared by photochemical etching of n-type Si [100] using a 15 mW HeNe laser operating at 633 nm (for more details see Koker and Kolasinski 199X). Irreproducibility in fabrication is eliminated by insuring the cleanliness of the initial Si surface by ultrasonically cleaning the Si substrate in acetone and ethanol prior to etching. The teflon container in which etching is performed must also be cleaned ultrasonically in ethanol. The samples were irradiated in 48% HF (aq) solution at near normal incidence. The gaussian laser intensity profile leads to a gaussian shape of the bulk silicon-porous silicon interface while at the same time the porous silicon at the HF interface is slowly dissolved away. Taken together these two processes result in the etching of a hole into the wafer with porous silicon at the edges. Applying this technique, a plan view TEM specimen was produced within roughly 3 h exposure time using a thin (10-20 micron) [100] wafer. After etching to perforation, the sample was immediately removed from the HF, rinsed in de-ionised water and transferred onto a copper support grid. The samples were examined in a top-entry JEOL 200 CX TEM ($C_s = 1.1$ mm) operating at 200 kV.

3. RESULTS

Figures 1a) and 1b) show plan view axial bright field electron micrographs (taken at lower magnification) illustrating the general morphology of the photochemically etched porous silicon. The etched n-type silicon was found to form square macropores with edges aligned along the <011> directions and edge lengths of several microns. The macropore in Fig. 1a) is a typical example and has an edge length of 14–15 microns. It can be seen that the silicon within the macropore is still intact and not porous but suffers from stress which leads to many bend contours. Bend contours arise from diffraction contrast and indicate that the lattice planes are oriented under the bragg condition with respect to the electron beam. Furthermore it should be noted that the edges are not smooth but serrated. Other macropores, like the macropore shown in Fig. 1b), have a hole in the middle, smooth edges and clearly contain nanoporous material. This suggests that Fig. 1a) shows the bottom of a macropore which has not been etched through.

Images of the nanoporous silicon taken at higher magnification show crystalline material containing irregular nanopores, but no evidence of 'nanowires'. The axial bright field image of Fig. 2 shows a typical area. The selected area diffraction pattern (inset to Fig. 2) demonstrates clearly that the porous silicon still has the [100] orientation of the wafer and no additional reflections occur. It can be seen from Fig. 2 that the thickness of the crystal is not uniform and bend contours, indicating

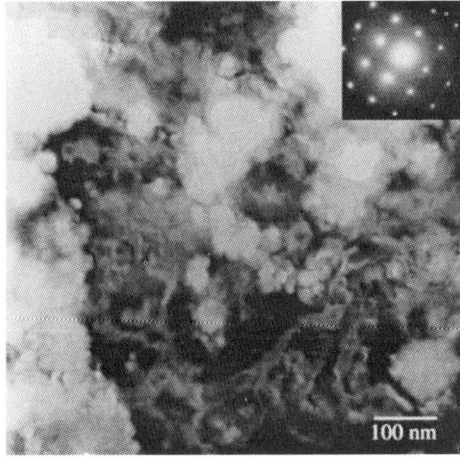

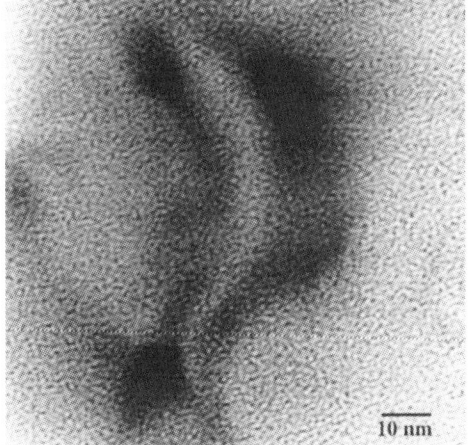

Fig. 2 Axial bright field image of nanoporous silicon contained in a macropore. The silicon still exhibits the [100] orientation as the selected area diffraction pattern (inset) shows.

Fig. 3 Axial bright field image of nanoporous silicon showing bend contours at high magnification. The speckled background indicates additional amorphous material.

significant lattice strain, are also evident. This strain was observed in all samples and on very different length scales. Figure 3 is an axial bright field image showing bend contours (dark lines) at high magnification, a frequently observed feature in these specimens. These bend contours display strain fields and reveal that the crystal experiences lattice strain. Assuming that the bragg-reflecting lattice planes are the {220} planes, the radius of curvature can be estimated to be 1530 nm. For a crystal thickness of 50 nm this value corresponds to a surface strain of 1.6%. Since Young's modulus for silicon is of the order of $1.2 \ 10^{11}$ Pa, this strain corresponds to a surface stress of $1.9 \ 10^9$ Pa. It should be pointed out further that the 'speckled' background visible in Fig. 3 indicates the existence of additional amorphous material.

4. DISCUSSION

The formation of square macropores in photoelectrochemically etched n-type porous silicon has been reported previously by several authors (for a summary see Cullis et al 1997). The pores propagate along the [100] direction and for [100] silicon the edges are crystallographically aligned with the <011> directions. Lévy-Clément et al (1994) have observed the formation of square macropores with edge lengths of up to several microns and containing nanoporous silicon in photoelectrochemically etched silicon. The structure of these pores is strikingly similar to our observations. The complex composition of por-Si, and the way it is being formed, are still under discussion and various models have been proposed (for a review see Smith and Collins 1992). Our results indicate that photochemically etched porous silicon appears to resemble photoelectrochemically etched porous silicon. This result will have to be taken into account in explaining the mechanism of production.

Lattice strain in porous silicon, as commonly observed in our samples, has been reported previously by other authors. Sugiyama and Nittono (1990) reported x-ray diffraction data which shows a lattice expansion of up to 0.3% of the porous silicon compared with the silicon lattice; their explanation was based on hydrogen atom chemisorption on the porous silicon. However, this lattice expansion is too small to explain the strain found in our samples. Buttard et al (1996) have pointed out that the oxide layer on porous silicon must also be taken into account. The Si-Si bond length in SiO_2 is up to 30% larger than in Si. This lattice mismatch should lead to considerable strain at the Si-SiO_2 interface. Since our samples have been exposed to air they are expected to be covered with a thin oxide layer. This is confirmed by the 'speckled' background observed in images taken at high magnification, which is consistent with an amorphous oxide layer. As porous silicon possesses a large surface area, the formation of an oxide layer should result in considerable oxide stress which may account for the strain seen in our samples.

4. CONCLUSIONS

We have reported the first structural investigations of photochemically etched porous silicon using TEM. The porous silicon consists of large crystallographically aligned macropores, with edge lengths of several microns. The macropores contain nanoporous silicon which still exhibits the [100] orientation of the wafer. The structure of these macropores is similar to that of photoelectrochemically etched porous n-type silicon observed previously by Lévy-Clément et al (1994).

Our samples also show considerable lattice strain which is consistent with the formation of a native oxide layer covering the porous silicon.

ACKNOWLEDGEMENTS

Support for this research was provided by the Engineering and Physical Sciences Research Council and the Royal Society. A. W. would like to thank Dr. Peter Nellist for valuable discussions.

REFERENCES

Albu-Yaron A, Bastide S, Bouchet D, Brun N, Colliex C and Lévy-Clément C 1994 J. Phys. I France **4**, 1181
Buttard D, Bellet D and Dolino G 1996 J. Appl. Phys. **79**, 8060
Canham L T 1990 Appl. Phys. Lett. **57**, 1046
Cullis A G, Canham L T and Calcott P D J 1997 J. Appl. Phys. **82**, 909
Dubbelday W B, Szaflarski D M, Shimabukuro R L, Russell S D and Sailor M J 1993 Appl. Phys. Lett. **62**, 1694
Koker L and Kolasinski K W submitted
Kolasinski K W, Koker L, Ganguly S, Barnard J C and Palmer R E in preparation.
Lévy-Clément C, Lagoubi A and Tomkiewicz M 1994 J. Electrochem. Soc. **141**, 958
Ngan M L, Lee K C and Cheah K W 1998 J. Appl. Phys. **83**, 1637
Noguchi N and Suemune I 1993 Appl. Phys. Lett. **62**, 1429
Shih S, Jung K H, Qian R -Z and Kwong D L 1993 Appl. Phys. Lett **62**, 467
Sugiyama H and Nittono O 1990 J. Cryst. Growth **103**, 156

Inst. Phys. Conf. Ser. No 164
Paper presented at Microsc. Semicond. Mater. Conf., Oxford, 22–25 March 1999
© 1999 IOP Publishing Ltd

Oxide layers on 6H silicon carbide substrates

R T Murray[1] S Taylor[2] G.P.Kennedy[2], L S Riley[2] and S Hall[2]

[1] Dept. of Material Science & Engineering
[2] Dept. of Electrical Engineering and Electronics, University of Liverpool, L69 3BX

ABSTRACT: Electronic grade wafers of 6H SiC have been given a surface oxide layer by either thermal or plasma processing. TEM, EDX and PEELS have been used to characterise the layer and its interface with the parent carbide.

1. INTRODUCTION

The properties of insulating layers which can be grown on Si by high temperature thermal processing or by low temperature plasma methods are crucial to the performance of silicon based microelectronics. Many studies have therefore been undertaken to understand the interface between the almost amorphous oxide and the underlying lattice on which it has been formed (Gibson and Lanzarotti 1989, Ourmadz and Taylor 1987). It has been found that interface planarity and sharpness are important features for high performance circuits (Krivanek and Mazur 1980). Silicon carbide offers many of the advantages of silicon but because of its wider band gap (~3eV), enhanced thermal stability and mechanical robustness it can be operated at higher temperatures such as may be encountered in automobile or industrial applications. In this paper we will present an investigation into the oxide/substrate interface for 6H-SiC.

2. TECHNIQUES AND METHODOLOGY

Various forms of electron beam analysis have been used to build up a picture of the oxide layer and its interface including.

2.1 High resolution TEM

High resolution TEM on both conventional (JEOL 2000EX) and field emission (JEOL 2010F) instruments to give information on the sharpness of the transition from crystalline to amorphous texture and to analyse surface ripple.

2.2 Nanoprobe EDX

Nanoprobe EDX performed on both the JEOL 2010F and the VG HB601 scanning transmission microscope to determine elemental profiles from the substrate through the oxide layer.

2.3 Point analysis by electron spectroscopy

Point analysis by electron spectroscopy on the VG microscope equipped with a Gatan parallel electron energy loss spectrometer has enabled the bonding state of carbon to be determined in the semiconductor and its oxide.

3. OBSERVATIONS ON OXIDIZED SiC

Oxides on [0001] wafers of 6H-SiC supplied by Cree Inc. have been grown to approximately the same thickness by both thermal and plasma means. Thermal oxidation was performed at 1000°C in a dry oxygen atmosphere for 40 minutes. The use of low temperature plasma anodization involved the inverted suspension of the sample positively biased in the presence of an inductively coupled RF oxygen plasma (Chappel et al 1997). Low temperature (350°C) growth of the oxide is then achieved by the drift of O⁻ ions through the growing film to react at the SiC-SiO₂ interface.

3.1 TEM Observations

Specimens for cross-section TEM were prepared by dicing the wafer into 5mm x 3mm fingers with the short direction parallel to the [10-10] lattice orientation. Four fingers were epoxied together in the usual way and cut into slices using a diamond impregnated wire saw. Before ion milling a small depression across the inner two interfaces of the slices was machined on a Southbay dimpler using 6μ and 1μ diamond suspensions. The specimens were then orientated in the TEM, using converging beam electron diffraction, so the oxide to semiconductor interface was parallel to the electron beam; thus a sharp image of the interface could be realised. However, due to relaxation of strain during the preparation there was considerable distortion of the thinned regions and orientation was not perfect. The results obtained by both field emission and conventional microscopes were normalised by reference to images of the polished but unoxidized face of the wafer. Two artefacts have been identified by the process:

1. Where the protective overlayer of epoxy has been almost etched away, the surface of SiC shows corrugation with a wavelength of ~3nm and a depth of 0.5nm. This we attribute to damage caused by the ion milling.
2. Particularly in the highly coherent field emission microscope, focussing can be difficult and this can lead to information on the SiC lattice being recorded within the oxide layer. It has been found best to focus on the epoxy-vacuum interface as close as possible to the region of interest.

The results averaged over a number of micrographs are presented in Table 1.

TABLE I Dimensions of the thermal and plasma oxide layers on SiC

Sample	Oxide Thickness nm	Interface Roughness nm	Ripple wavelength nm	Ripple Depth Peak to Trough nm
Plasma oxide	25 ± 5	0.5 – 2	30 ± 5	2.5 ± 1
Thermal oxide	20	0.7 –3	31 ± 5	2 ± 2
Polished SiC/Epoxy	--	0.5	40 ± 10	1 ± 0.5

1. The sharpest interface is that between SiC and epoxy (Fig.1.)
2. This interface is far from flat displaying an undulation with average wavelength of 40nm and a peak to trough depth of ~1nm
3. The ripple retains its wavelength upon oxidation but appears to double in depth. To some extent the oxide layer is conformal to the initial surface (Fig. 2)

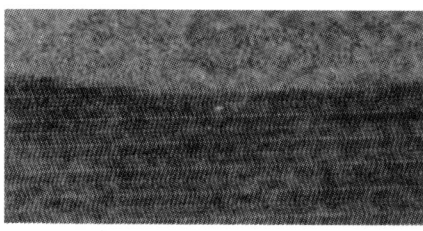

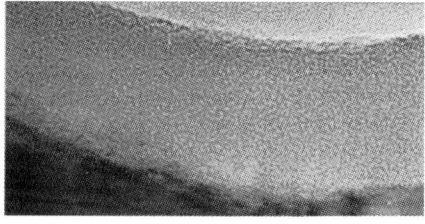

━━━ 5nm

━━━ 10nm

Fig. 1. The sharp interface between polished SiC and epoxy. Note the surface ripple which can transfer to the oxide interface after processing.

Fig. 2. A thermally oxidised layer showing the conformal action of the oxide on this defect.

4. The transition zone from the SiC lattice to the amorphous oxide varies from 0.5nm to 3nm for both oxidation processes (Fig. 3).

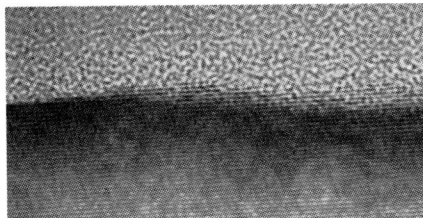

—— 5nm

Fig. 3. A plasma oxide layer showing the variation of "sharpness" over a distance of tens of nanometers.

3.2 Interface analysis by EDX

With the interface orientated parallel to the electron beam, EDX line scans for C, O and Si were obtained on either the 2010F or the VG HB601. Excitation was provided either by 1nm or 3nm diameter electron probes at 200 and 100 keV respectively. Despite excellent vacuum, better than 10^{-7} Torr at the specimen, some contamination was noted. Of greater interest perhaps was the severe damage caused to the SiC substrate by the 3nm probe whereby the boundary has become more diffuse as seen in Table II.

Table II: Chemical Analysis of oxidised SiC, * Plasma oxide, T_{ox} = oxide thickness.

Microscope Specimen	Element	Phase		Edge definition (nm)	T_{ox} (nm)	$[Si]_{SiCO}/$ $[Si]_{SiC}$	$[C/Si]_{SiCO}/$ $[C/Si]_{SiC}$
		SiC (counts)	'SiCO' (counts)				
PO* SiC in VG HB601 with 1nm probe	C	540	190	2.5	---------	-------------	0.57±0.2
	O	55	1810	2.2	---------	-------------	
	Si	2620	1620	2.4	---------	0.62	
PO SiC in JEOL 2010 1nm probe	C	21	6	3.5	---------	-------------	0.75±0.2
	O	3	18	8.0	36	-------------	
	Si	85	32	6.5	36	0.38	
Thermal oxide JEOL 2010 3nm probe	C	270	42	8.5	18	-------------	0.51±0.2
	O	70	207	5.0	21	-------------	
	Si	2057	635	16.0	---------	0.31	

The oxidation process could have several outcomes,
1. SiO_2 is produced and all C is lost as CO or CO_2 , giving rise to no detectable C in the O rich band and $[Si]_{oxide}/[Si]_{SiC} = 0.51$ using $\rho_{oxide} = 2.45$ and $\rho_{SiC} = 3.2$ gcm^{-3}.
2. Only some of the C is burnt off and the layer is a mixed oxide represented forthwith as 'Si_2CO_6' $[C/Si]_{SiCO} / [C/Si]_{SiC} < 1$
3. No C is lost and O, Si and C form a continuous network highlighted by the above ratio being ≈ 1
Again as shown in Table II for all oxide measurements, Carbon is detected in the oxide rich layer and the count ratio of C/Si in the layer is always about half that in the SiC. We thus conclude that the composition of the oxide in the vicinity of the interface is close to Si_2CO_6.

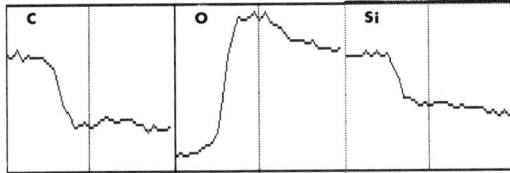

Fig. 4. EDX spectra acquired on the VG STEM microscope from SiC and plasma oxide interface. Note the significant C signal in the oxide layer

520

The edge definition as measured by the EDX represents the concentration gradient of the elements rather than the crystalline: amorphous transition. In addition the actual profile is convoluted with the probe diameter including any side lobes so we should not expect EDX to give the same values for roughness as found by imaging. The edge roughness quoted in Table II is obtained by drawing a tangent to the profile at the centre point of the edge and measuring the spatial separation of the two points where the tangent intercepts the extrapolated lines representing the concentration of that element on either side of the interface. Comparing Tables I and II we conclude that EDX overestimates the boundary zone by up to a factor of 2 and the STEM gives closer correspondence with high resolution TEM than does the 2010F.

3.3 Nanoprobe analysis using PEELS

The plasma oxidised wafer has been analysed by PEELS in the region of the Si L_{23} edge and the carbon K edge.; so as to give information on the bonding within the layer. Fig. 5 shows that in SiC the L_{23} edge occurs at 101eV whereas within the oxide layer it is at 107eV as commonly found in SiO_2 (Papworth and Fox 1997). This indicates the dominant Si bond is to O. The carbon K edge within the SiC is very similar to that obtained from tetrahedral amorphous carbon showing that the majority of carbon bonding is sp^3 hybridized. Within the oxide layer the π bonding signature at 285eV is strengthened indicating that in this region bonding is more diverse (fig. 6).

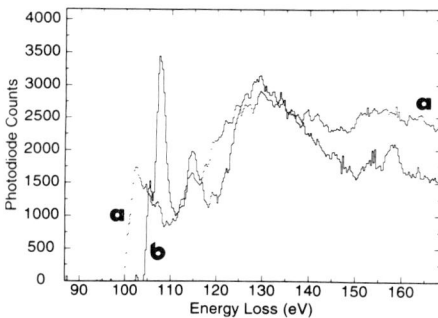

Fig. 5. The Si Electron Energy Loss peak from a) the SiC and b) the oxide. In the oxide the edge shifts towards 107eV as commonly found in SiO_2.

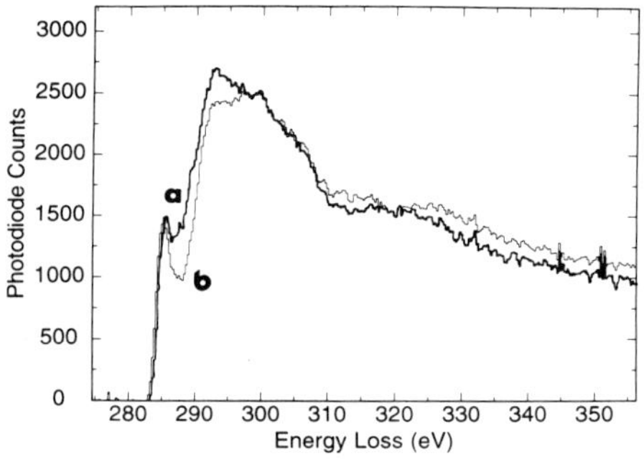

Fig. 6. The carbon loss spectra from a). SiC and b) plasma grown oxide. Note the higher π bonding at 285eV in the oxide layer.

4. CONCLUSIONS

The sharpness and planarity of the SiC-to-oxide interface has been observed. The edge roughness varies from 0.5 to 3nm as seen by TEM. EDX, which measures the elemental profile, indicates levels within a factor of 2 of this. The oxide is close to Si_2CO_6 in composition, but contains a higher proportion of π bonding carbon than does SiC.

REFERENCES

Chappel D C, Taylor S, Smith J P, Eccleston W, Das M, Cooper J and Melloch M 1997 Electron. Lett. **33**, 97
Gibson J M and Lanzarotti M Y 1989 Ultramicroscopy **31**, 29
Krivanek O L and Mazur J 1980 Appl. Phys. Lett. **37**, 392
Ourmazd A, Taylor D W, Rentschler J A and Bevk J 1987 Phys. Rev. Lett. **59**, 213
Papworth A and Fox P 1997 Mater. Sci. Tech. **13**, 912

Inst. Phys. Conf. Ser. No 164
Paper presented at Microsc. Semicond. Mater. Conf., Oxford, 22–25 March 1999
© *1999 IOP Publishing Ltd*

Effect of the temperature ramp rate during carbonization of Si (111) on the crystalline quality of SiC produced

F J Pacheco, A M Sánchez, S I Molina, D Araújo, R García and A J Steckl[*]

Departamento de Ciencia de los Materiales e I.M. y Q.I., Universidad de Cádiz. Apdo. 40, 11510 Puerto Real (Cádiz), Spain
[*]Nanoelectronics Laboratory, University of Cincinnati, 899 Rhodes Hall, P.O.Box 210030, Cincinnati, OH 45221-0030, USA

ABSTRACT: 3C-SiC thin films obtained by carbonization of Si (111) are characterized by Scanning Electron Microscopy and Transmission Electron Microscopy techniques. The high lattice mismatch between SiC and Si leads to a mosaic structure in the SiC layers. Voids in the Si substrate as well as misfit dislocations at the SiC/Si interface were present in the thin films. Differences in temperature ramp rate during carbonization of (111) Si are shown to induce changes in the sub-grain size as well as in the void size distribution and morphology in the SiC layers produced.

1. INTRODUCTION

The need of electronic devices able to operate at high temperatures, high frequencies and high power levels, makes wide band gap semiconductor materials such as II-N or SiC very attractive. Among them, SiC exhibits interesting physical and electronic properties for microwave, high temperature and radiation resistant device applications. On the other hand, SiC is a suitable substrate for III-N growth. Indeed, carbonization before growing III-N layers on Si substrates has been shown to raise the crystalline quality of the nitride layers. In this study we investigated the impact of the thermal sequence on the resulting SiC crystalline quality.

3C-SiC thin films, obtained by carbonization of Si (111), are characterized by Scanning Electron Microscopy (SEM) and Transmission Electron Microscopy (TEM). The influence of temperature ramp rate during the carbonization process on the structural quality of the SiC layers is analyzed.

2. EXPERIMENTAL TECHNIQUE

Two samples with a 40 nm thick 3C-SiC layer have been obtained by propane Rapid Thermal Chemical Vapor Deposition (RTCVD) on Si(111) substrates at 1250°C with a flow rate of 15 sccm of propane (diluted to 5% in H_2) and 1 slpm of H_2 at atmospheric pressure. The carbonization process was carried out using temperature ramp rates of 25°C/s and 5°C/s for samples S25 and S5, respectively. More details about the growth technique and the RTCVD system have already been published by Steckl and Li (1992).

TEM characterization was performed on JEOL 1200-EX and JEOL 2000-EX transmission electron microscopes operating at 120 and 200 kV, respectively. SEM observations were carried out on a JEOL 830-JSM scanning electron microscope. The TEM specimen preparation was achieved by Ar^+ ion milling after mechanical thinning.

To investigate the crystal structure of the SiC films and the voids in the Si substrate at the Si/SiC interface, SEM, plan-view TEM (PVTEM), cross-sectional TEM (XTEM), transmission electron diffraction (TED) and high resolution TEM (HREM) techniques were used.

3. EXPERIMENTAL RESULTS AND DISCUSSIONS

Selected area TED (SAED) patterns registered from areas of about 65 μm^2 of plan-view specimens of both samples reveal the presence of a mosaic structure in the SiC layers (Pacheco et al. 1999). The existence of cubic SiC crystals with a very small angle of disorientation is shown by curved shape elongated spots in the above mentioned SAED patterns. This result can be expected because epitaxial films with a large lattice mismatch with respect to the substrate may present a mosaic structure of slightly disoriented sub-grains (Srikant et al. 1997). The lattice mismatch between the 3C-SiC (a_{SiC}=0.4359 nm) and Si (a_{Si}=0.5430 nm) is extremely high, 19.7% in tension. The grain diameters were estimated by PVTEM and SEM as about 170 and 100 nm for samples S25 and S5, respectively.

Fig. 1. <110> cross-sectional HREM micrograph of sample S5 showing the high misfit dislocation density at the SiC/Si interface. The {111}- planes associated to a misfit dislocation are drawn in the center of the HREM image.

From cross-sectional TEM studies of both samples is clear that the SiC/Si interfaces present high misfit dislocation densities necessary to accommodate the high lattice mismatch between the two materials. In the <110> HREM image shown in Fig. 1, an extra SiC plane appears every four Si {110} interplanar spacings as a consequence of those misfit dislocations. This fact explains how the SiC layer lattice parameter corresponds to the bulk SiC lattice constant even for monolayers close to the SiC/Si interface, as it has previously been observed by Li et al. (1993). This is in good accordance with the SiC lattice parameter estimated from the diffracted spots corresponding to SiC reflections in SAED patterns of cross-sectional and plan-view specimens of both samples (Pacheco et al. 1999). To increase the precision of the SiC lattice constant determination, an analysis of intensity profiles in HREM images, acquired with a Gatan slow scan CCD camera, has been used. The value obtained was 0.441 ± 0.011 nm.

The presence of voids in the Si substrate just below the SiC/Si interface has been shown in XTEM, PVTEM and SEM images (Fig. 2). Through observations by the former techniques there are two different possible shapes: (i) truncated tetrahedral geometry voids faceted by {111} planes with their sides along <110> directions and (ii) quasi-spherical voids. The void side size distribution has been studied from SEM images shown in Fig. 2. The void size distribution histograms are also shown in Fig. 2. Sample S25 presents a bimodal void distribution whereas of sample S5 has a single peak distribution. The first peak in sample S25, resulting from a multiple gaussian fit, approximately corresponds to the single peak of sample S5 is at a void size (lateral dimension) of ~0.35 μm while the second peak, with about half the area of the first one, is centered at void size which is roughly twice as large (~0.8 μm).

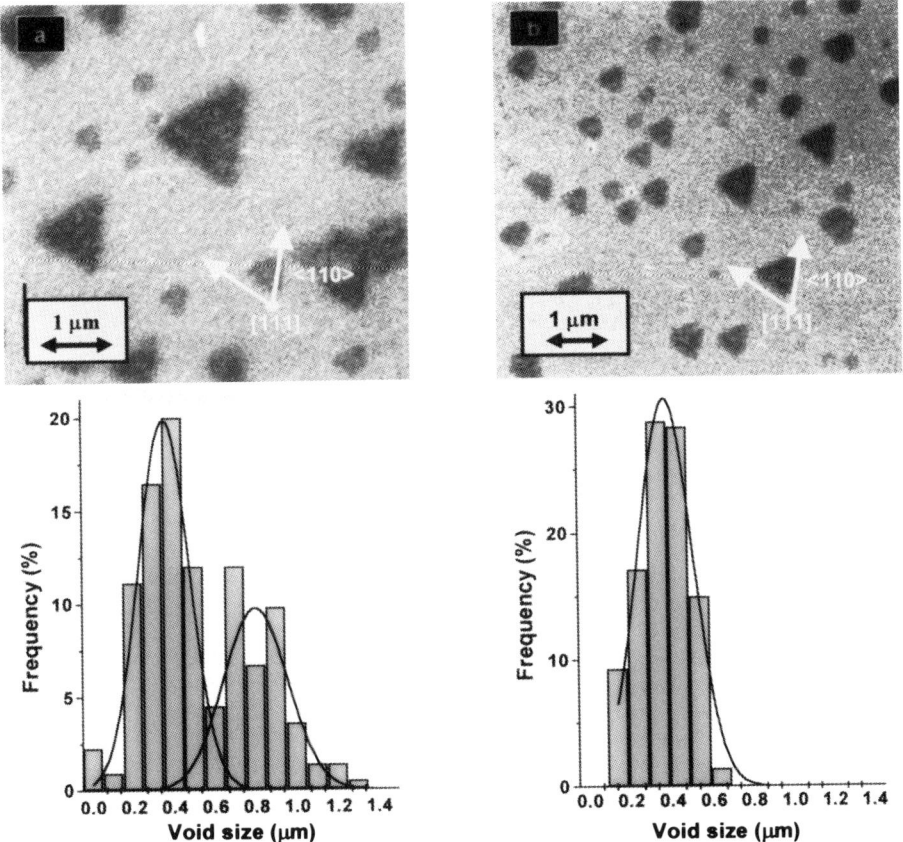

Fig. 2. Scanning electron microscopy images revealing the presence of equilateral triangular and irregular shape voids. Void size (lateral side length) histograms are shown with their corresponding gaussian fit curves for samples S25 (a) and S5 (b).

The void depths were estimated, from a vast extension of the SiC/Si interfaces in low magnification XTEM images (Fig. 3). The average depths are approximately 0.16 for the small voids and 0.25 μm for the large voids of sample S25 and 0.23 μm for sample S5. Table I summarizes the measures above mentioned: the void side sizes estimated by gaussian fits in the distribution histograms of Fig. 2, and void depth and grain diameter for both samples.

Table I

Sample	Void size (μm)	Void depth (μm)	Sub-grain diameter (μm)
S25	0.35, 0.80	0.16, 0.25	0.17
S5	0.37	0.23	0.10

A smaller grain size means a larger number of nuclei during the first stages of the Si carbonization. Therefore, assuming the SiC layer growth model proposed by Li and Steckl (1995), the number of voids should be higher and their lateral extension smaller as the grain size decreases.

524

This is in good agreement with our experimental observations. Sample S5 presents smaller grains and voids than sample S25.

The presence of a second peak in sample S25 void size distribution could indicate the existence of voids coalescence, a phenomenon that was not considered in our void growth scheme proposed in Pacheco et al. (1999).

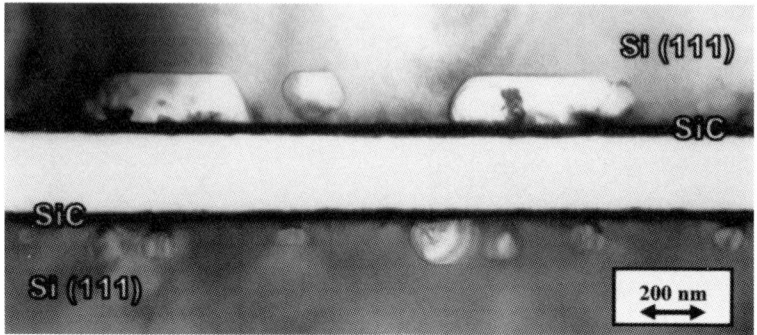

Fig. 3. Low magnification XTEM image on a <110> projection showing a general view of the SiC layer and facetted and non-facetted voids in the Si (111) substrate of sample S25.

Since the grown SiC layer thicknesses are approximately identical in both samples and all the Si atoms, necessary to grow the SiC, come from the Si substrate, the total volume occupied by the voids should be the same. Indeed, in sample S5 the void size is smaller but the surface void density is considerably higher. The existing differences between the two samples must be related to kinetic and thermodynamic aspects of the Si out-diffusion and SiC island nucleation mechanism during the carbonization process.

4. CONCLUSIONS

The high lattice mismatch between SiC and Si leads to a mosaic structure in the SiC layer. Voids in the Si (111) substrate as well as misfit dislocations at the SiC/Si interface are present in the thin films.

In summary, the temperature ramp rate affects the SiC grain size as well as the morphology and void size distribution. The average grain and void sizes increase with the temperature ramp rate while the SiC grown layer thickness obtained is roughly the same.

ACKNOWLEDGEMENTS

The present work has received financial support from the *Comisión Interministerial de Ciencia y Tecnología*, CICYT, Project MAT 98-0823 and form the *Junta de Andalucía* under the group TEP-0120.

REFERENCES

Li J P and Steckl A J 1995 J. Electrochem. Soc. **142**, 634
Li J P, Steckl A J, Golecki I, Reidinger F, Wang L, Ning X J and Pirouz P 1993 Appl. Phys. Lett. **62**, 3153
Pacheco F J, Sánchez A M, Molina S I, Araújo D, Devrajan J, Steckl A J and García R 1999 Thin Solid Films, in press
Srikant V, Speck J S and Clarke D R 1997 J. Appl. Phys. **82**, 4286
Steckl A J and Li J P 1992 IEEE Trans. Electron Devices ED-**39**, 64

Inst. Phys. Conf. Ser. No 164
Paper presented at Microsc. Semicond. Mater. Conf., Oxford, 22–25 March 1999
© 1999 IOP Publishing Ltd

HREM investigation of structural defects in Al- and B- implanted 4H and 6H SiC

P O A Persson[1], E Olsson[2] and L Hultman[1]

[1] Thin Film Physics, Linköping University, S-581 83 Linköping, Sweden
[2] Ångström Laboratory, Uppsala University, P.O. Box 534 S-75121 Uppsala Sweden

ABSTRACT: Ion implantation is currently the tool for selective area doping of SiC. However, ion implantation results in crystal damage. High-temperature implantation and annealing ($\geq 1600°C$) are used to restore the crystal order. High resolution electron microscopy and electron-energy loss elemental mapping were used to study the nature of defects found in both Al and B implanted SiC. Microscopy revealed small loops on the basal (0001) planes when the dopant concentration is higher than 10^{17} cm^{-3}. Defect size increases with annealing temperature and time. Elemental mapping revealed an elevated level of dopants in the regions of the defects.

1. INTRODUCTION

Silicon carbide is a suitable material for electron devices operating at high temperatures, high powers and high frequencies. In order to achieve the full potential of SiC, device processing technology such as oxidation ion implantation and metallization plays an important role. Ion implantation in particular is the key technique for selected area doping in SiC due to the extremely low diffusion rates in this material. Donor implantation such as N and P doping into Al or B doped epilayers has been shown to give reasonably low sheet resistances and quite high electrical activation ratios. When it comes to Al and B implanted N doped epitaxial layers, few results were published until recently. These reports show a wide range of measured electrical properties, see e.g. Kimoto et al (1996) and Troffer et al (1997). Unfortunately, only a few reports have been published on transmission electron microscopy (TEM) and especially of high resolution electron microscopy (HREM) studies of implanted and annealed samples which could give valuable information about larger clusters of point defects, see e.g. Pensl et al (1996), Persson et al (1998) and Grisolia et al (1999).

In this paper we use TEM and HREM to investigate Al and B implanted high quality 4H and 6H SiC epilayers, the implantation induced structural damage after annealing and the effects of implant concentration.

2. EXPERIMENT

The samples were prepared from epitaxially grown 4H and 6H-SiC layers on (0001) substrates oriented 3.5° off towards (11-20). Al and B ions were implanted to produce single peaks and box profiles with implant concentration ranging from 10^{17}-10^{20} atoms/cm^3 using energies ranging from 20 keV to 200 keV at elevated temperatures (600-800°C). After that, the samples were annealed in a high temperature furnace for 30 min. at temperatures 1600°C, 1700°C and 2000°C.

Cross sectional XTEM samples were prepared by clamping two pieces of the material together in a Ti grid, face to face, followed by mechanical thinning, polishing and ion milling. A BalTec RES 010 precision ion mill operated at 8.0 kV and 4.2 mA, with Ar as the sputter gas, was used to make the samples electron transparent. Before the TEM studies, all samples were cleaned using the BalTec operated at 2 kV and 1.4 mA for 15 min on both sides. Microscopy was performed using a Philips CM20 UT electron microscope operated at 200 kV (resolution 1.9Å) equipped with a LaB$_6$ filament. The distribution of B, Si and C in the B-doped XTEM samples was also investigated using a Gatan imaging filter (GIF) attached to a Philips CM200 supertwin field emission gun (FEG) microscope.

3. RESULTS

Microscopy of the B and Al as-implanted samples shows dark contrast in the region of the implantation. High-resolution imaging of the lattice is still possible although the resulting image is confused due to the point defects created by the high energy implanted species.

After annealing at high temperatures the crystal order is restored. The only visible evidence of the implantation are "black dots" in the region where the implanted species reside. The black dot is interpreted as a strong strain field around a defect, and has been observed before, for instance by Pensl (1996). As can be seen in fig. 1 the distribution of defects, seen from <11-20> in a B implanted sample, is evenly distributed in the region with the highest dopant concentration as measured by SIMS.

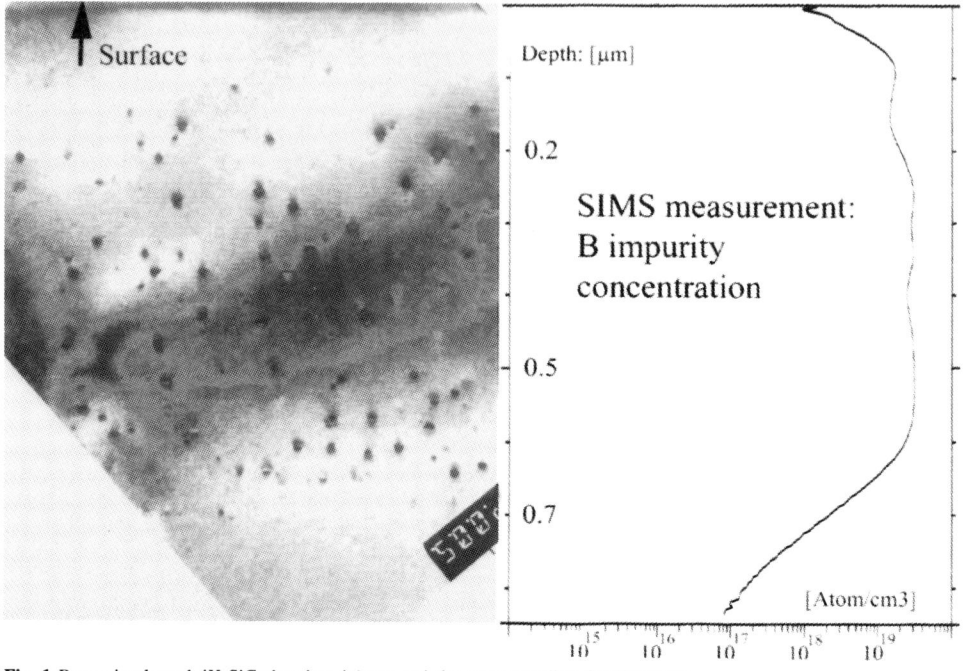

Fig. 1 Boron implanted 4H SiC showing defects and the corresponding SIMS profile.

When studied under two-beam conditions these defects exhibit a contrast which is typical for a precipitated layer of interstitials surrounded by a Frank-loop. Samples with shallow doping were studied in plan-view, but no defects could be found, probably due to low contrast when studied from this direction. To see if the defects were rodlike or indeed 2-D planar precipitates, a sample was also studied from <1100>. This sample showed the presecence of planar precipitates of the same density as for the cross-sectional view. Typical B-related defect size after 30 min. anneal at 1700°C is about 5-10 nm.

The density of defects in the samples is dependent on the dopant concentration in the implanted region. At concentrations of 10^{17} cm^{-3}, there are very few of these defects and they would probably not form at lower concentrations.

Both Al and B implanted samples show black dots in both 4H and 6H. The contrast from the defects is easily seen in implanted 4H material, but is very weak in 6H. Al-related spots are typically much smaller than the B-related spots, but the volume density of defects is higher. This could be seen using two samples which were implanted and annealed under the same conditions, and cut from the same wafer.

Elemental mapping using electron energy filtering was also used to study a B-related defect. The sample was tilted so that the <11-20> direction would not be parallel to the optical axis in order to prevent diffraction effects. The studied defect is shown in fig. 2a. Elemental mapping of boron showed an elevated level of boron in the region of the defect, see fig. 2b. Si and C maps were also recorded, but any fluctuations in the concentration of these elements were below detection level.

High-resolution studies provided additional information abouth these defects. The samples shown in figs. 3 and 4 have been annealed at 1700°C for 30 min. As seen from the figures the defects reside on the basal planes in the crystal structure. This type of defect has been shown to exist also in neutron and Xe irradiated hexagonal SiC by Yano (1990) and Lehrmitte-Sebire (1994).

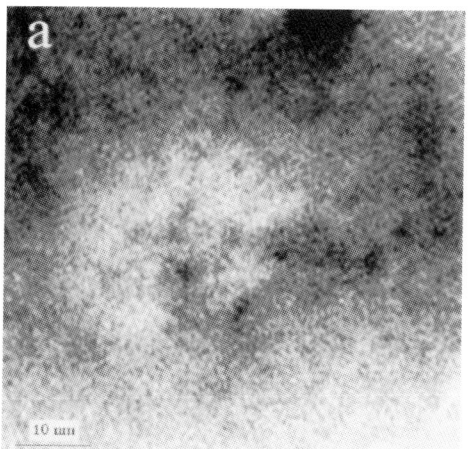 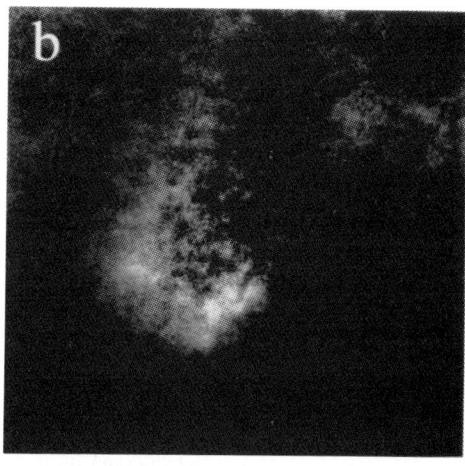

Fig. 2 Images of a B-related defect in 6H-SiC. a) Bright field image of the area used for elemental mapping containing one defect (bright strain contrast). b) B elemental map.

The Al-related defect seen in fig. 3 clearly shows the interstitial plane even though the loop diameter is not very wide, typically less than 5nm. This inserted plane results in an extrinsic stacking fault. A typical B-related defect as seen in fig. 4, however, does not show any clear indication of an inserted plane, lattice planes around the defect are bent. From the figure we can also see a difference in stacking sequence inside the defect. An Al implanted sample, annealed in a high temperature furnace at 2000°C for 30 min. is shown in fig. 5. The defects are now about 20-50nm wide. From high resolution imaging it can be seen that the defect consists of a plane of interstitials with the same stacking sequence as the sample in fig. 3. What is more striking is their tendency to cluster. The clusters typically contain 10-20 defects of varying size and are separated by an average distance of 0,5μm. Inbetween these clusters few defects are found. SIMS measurements of this sample reveal that the depth dependence of the Al-concentration was the same as before annealing.

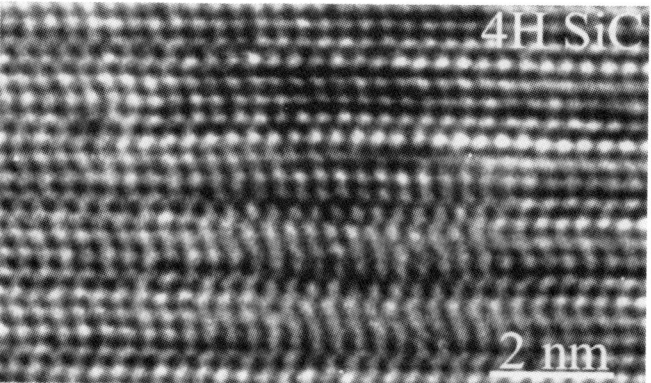

Fig. 3 Image of Al-doped 4H SiC sample with clear indications of an inserted plane. The highlighted stacking sequence through the fault is typical for the Al-related defect.

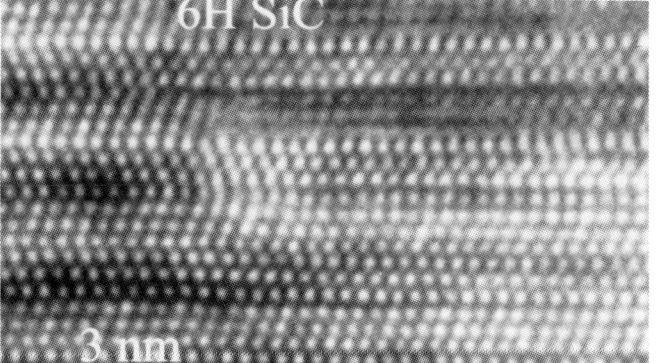

Fig. 4 High-resolution image of the edge of the B-related planar defect in 6H SiC. Compare the difference in stacking sequence is observed inside and outside the defect.

4. DISCUSSION

All defects found in this study of implanted SiC were of the same nature, an extrinsic plane of interstitials surrounded by a dislocation loop. No prismatic faults or 3D-impurity precipitates could be seen as reported by Lehrmitte (1994) and Suvorov (1997).

The gathering, during annealing, of interstitials on basal planes in hexagonal SiC is probably due to that basal planes are the higher diffusivity paths in hexagonal materials, and because of this the interstitials can move around more freely. These interstitials are then suggested to interact with other interstitals and form extended planar defects. After further annealing more interstitials will add to the planar defect. When the region around each defect is low on interstitials, larger defects, like those seen in the high temperature annealed sample in fig. 5, will form when smaller defects merge. Large defects then continue to grow at the expense of smaller defects which is the principle of Ostwald ripening. This is evident from looking at fig. 5 where large areas are free of interstitial planar defects.

As is obvious from the difference in Al and B related planar defects, the composition and possibly the structure of the defects differ. For the case of B, we can connect elevated levels of this impurity to the region of the defect. An extra plane is clearly visible in the Al doped samples but not convincingly visible in the B samples.

5. CONCLUSION

Planar interstitial defects surrounded by a Frank loop are formed on the basal planes in 4H and 6H SiC after B and Al single impurity implantation. These defects appear at annealing temperatures higher than 1600°C and when the impurity concentration is higher than 10^{17} cm^{-3}. These defects then evolve according to the principles governed by the Ostwald ripening process to form clusters.

ACKNOWLEDGEMENTS

Asea Brown Boveri (ABB), the Swedish Research Council for Natural Sciences (NFR) and the Swedish Foundation for Strategic Research (SSF) are acknowledged for financial support. The Elemental Mapping was performed, using the Philips CM 200 FEG instrument, at Chalmers University of technology.

REFERENCES

J Grisolia, B de Maudit, J Gimbert, Th Billon, G Assayag, C Bourgerette and A Claverie (1999) Nucl. Instr. and Meth. B **147** 62-67.

T Kimoto, A Itoh, H Matsunami, T Nakata and M Watanabe (1996) J. Electron. Mater, **25**.

I Lehrmitte-Sebire, J Vicens, JL Chermant, M Levalois and E Paumier (1994) Phil. Mag. A. **69**, 2, 237.

G Pensl, VV Afanasev, M Bassler, M Schadt, T Troffer, J Heindl, HP Strunk, M Maier and WJ Choyke (1996) Inst. Phys. Conf. Ser. No 142: Chapter 2, 275.

PO Persson, Q Wahab, L Hultman, N Nordell, A Schöner K Rottner E Olsson and MK Linnarson (1998) Mater. Sci. Forum **264-268** 413-416

A V Suvorov, OI Lebedev, AA Suvorova, J Van Landuyt and IO Usov (1997) Nucl. Instr. and Meth. B **127/128**, 347.

T Troffer, M Schadt, T Freank, H Itoh, G Pensl, J Heindl, HP Strunk and M Maier (1997) Phys. Stat. Sol (a) **162**, 277.

T Yano and T Iseki (1990) Phil. Mag. A. 62, 4, 421.

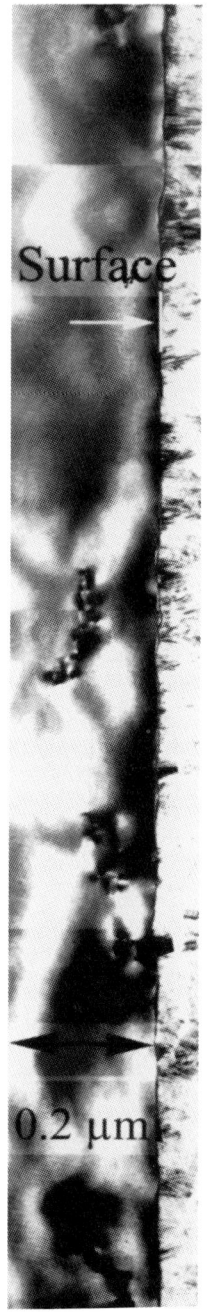

Fig. 5 Al-doped, 2000°C annealed sample showing clusters of planar defects. Inbetween these clusters no defects can be seen.

Inst. Phys. Conf. Ser. No 164
Paper presented at Microsc. Semicond. Mater. Conf., Oxford, 22–25 March 1999
© 1999 IOP Publishing Ltd

Comparison of defect formation in O+- and Ne+-implanted 6H-SiC

B Pécz, T Dalibor*, G Pensl* and J Stoemenos**

Research Inst. for Technical Phys. and Matl. Sci. H-1525 Budapest, P.O.Box 49, Hungary
* Inst. of Applied Phys., University of Erlangen-Nürnberg, Staudtstr. 7, D-91058 Erlangen, Germany
** Aristotle University of Thessaloniki, Physics Department, 54006, Thessaloniki, Greece

ABSTRACT: Oxygen introduced by implantation into 6H-SiC forms oxygen-related shallow centres, which act as shallow donors or deep acceptors in a certain analogy with oxygen-related defects in silicon. In order to separate damage-induced defects from those related to the chemical bonding of oxygen, implantation of the noble gas neon was also performed with similar profiles as for oxygen. The structural characteristics of these defects were studied.

1. INTRODUCTION

It has been shown recently that oxygen introduced into 6H-SiC by implantation forms two types of oxygen-related centres: shallow donors below the conduction band ($O_{I/II}$) and deep acceptor-like defects (O_{III}, O_{IV}, O_V) near the midgap; see Dalibor et al (1998a) and Dalibor et al (1998b). The shallow oxygen-related centres are sensitive to heat treatments, like the thermal donors in silicon, which form larger agglomerations with increasing annealing time and show smaller ionisation energies (Pensl 1987). Annealing of the specimens at 1650 °C to 1800 °C leads to a decrease of the ionisation energy. In order to separate damage-induced defects from those related to the chemical bonding of oxygen, implantation of the noble gas neon was also performed with similar profiles as for oxygen. Electrical characterisation of these specimens does not show any donor-like behaviour. The difference is attributed to the types of defects, which are formed in the two cases. The defects were studied by cross sectional and plan view Transmission Electron Microscopy (TEM).

2. EXPERIMENTAL PROCEDURE

N-doped 6H-SiC CREE wafers 3.5° off-axis along [11$\bar{2}$0] were used. The net doping concentration $N_D - N_A$ was $1.2 \times 10^{17} \text{cm}^{-3}$. Part of the 6H-SiC wafer was implanted with O^+ at 100 keV with a dose of 5×10^{15} $O^+\text{cm}^{-2}$ at 300°C in order to avoid amorphization. Neon was implanted as a reference, under identical conditions as oxygen resulting in a similar profile to the O^+ ions. In this way, damage-induced defects can be separated from those related to the chemical bonding. Due to the slightly higher mass of Ne the generated damage was also slightly higher.

The implanted samples were annealed at 1000°C for two hours. Part of the annealed specimens was further annealed at 1650°C for two hours.

3. RESULTS AND DISCUSSION

3.1 Samples implanted with oxygen

A defect zone is observed in the specimen implanted and subsequently annealed at 1000 °C for 2 hours. This defect zone extends down to a depth of 225 nm from the surface. No well-defined defects can be distinguished in this zone. Obscure contrast variations reveal the formation of the defects, as shown in the cross-section TEM (XTEM) micrograph of Fig. 1a.

Further annealing at 1650°C results in the formation of a defect zone, which extends from 110 to 250 nm in good accordance with computer simulation, denoted by the letter D in Fig. 1b. Above this zone and up to the surface the crystal is defect free. The defect zone is shown at higher magnification in Fig. 1c. The defects are mainly small clusters with a density of $2 \times 10^{11} \text{cm}^{-2}$ and with a mean diameter of 4 nm. No stacking faults are observed in this case. The defect zone was also studied by plan view TEM (PVTEM) after thinning the specimen from both sides so that only the implanted zone was left, as shown in Fig. 1d. The presence of small clusters is evident; however, no other defects were formed during the annealing at 1650°C. The related diffraction pattern with the electron beam parallel to the c-axis is shown in Fig. 1e. Very faint forbidden $10\bar{1}0$ reflections are observed. These originate from higher order Laue zones due to elongation of the diffraction spots in reciprocal space as has already been discussed for 6H-SiC by Suvorov et al (1997).

Electrical admittance spectra reveal that the oxygen-related donors exhibit an ionisation peak in the energy range of 144-160 meV. This peak is shifted to lower ionisation energies after a subsequent annealing at 1700 °C and 1800 °C as shown in Table I. The reduction of the ionisation energy is correlated with an increase of the electrical capture cross-section σ_n. Two cases are considered: $\sigma_n \propto T^0$ (multiphonon capture) and $\sigma_n \propto T^{-2}$ (cascade capture), as shown in Table I.

Table I

Ionisation energies ΔE of oxygen-related donors subsequent to different heat treatments as obtained from an Arrhenius analysis of admittance spectra. The implanted oxygen concentration is $4 \times 10^{18} \text{cm}^{-3}$.

Heat Treatment	ΔE (meV)	Capture cross-section $\sigma_n(\text{cm}^2)$ $(\sigma_n \propto T^0 / \sigma_n \propto T^{-2})$
1000°C/30min, 1400°C/15min, 1650°C/15min	144/160	3×10^{-13} / 2×10^{-12}
1700°C/15min	132 /147	6×10^{-13} / 2×10^{-12}
1800°C/30min	132/145	8×10^{-13} / 6×10^{-12}

3.2 Samples implanted with neon

The low temperature annealing at 1000 °C for 2 hours does not result in the formation of a well-developed defect zone. The effect of implantation can be seen only as a general blackening of the implanted area (denoted by the letter B in Fig. 2a). The same specimen subsequently annealed at 1650 °C for 2 hours exhibits a high density of defects, mainly Stacking Faults (SFs). These SFs have a mean diameter of 80 nm, as shown in Fig. 2b. The micrograph was taken with the electron beam exactly parallel to the $[11\bar{2}0]$ direction. Lattice fringes reveal the 6H periodicity. The SFs are situated on the basal planes in a zone denoted by the letter S in Fig. 2b. Close to the SFs bright dots are observed in Fig. 2b.

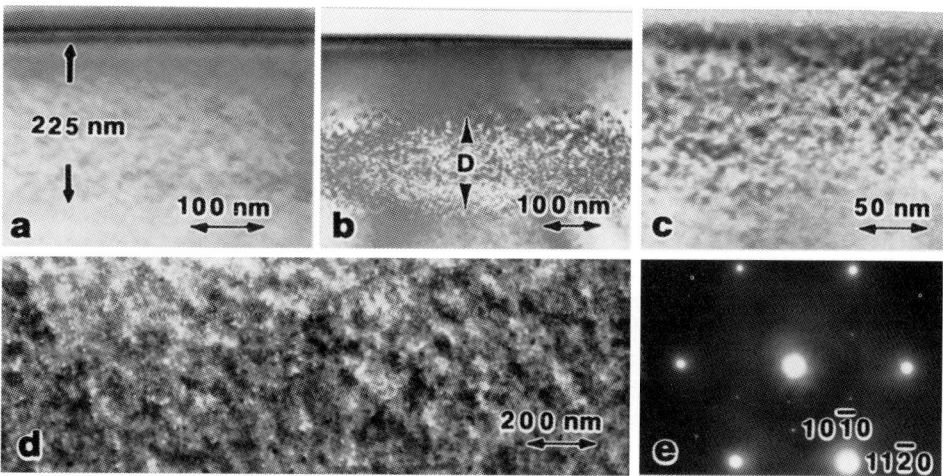

Fig.1 Specimens implanted by oxygen and annealed a) at 1000°C for 2h, b) at 1000°C for 2h plus 1650°C for additional 2h, c) same as (b) at higher magnification, d) PVTEM micrograph from the specimen annealed at 1000°C for 2h plus 1650°C for 2h.

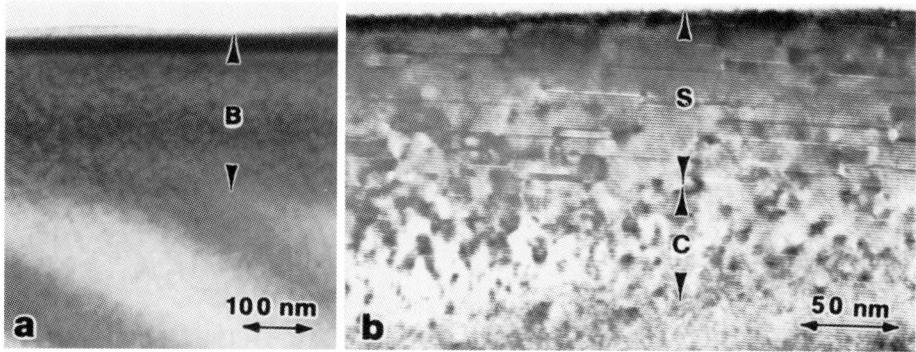

Fig.2 Specimens implanted by neon and annealed a) at 1000°C for 2h, b) at 1000°C for 2h plus 1650°C for 2h.

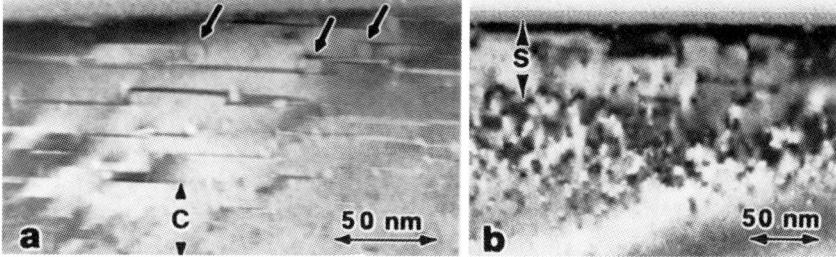

Fig.3 Specimen implanted by neon and annealed at 1000°C for 2h plus 1650°C for 2h a) under two beam case with $g_{11\bar{2}0}$ b) under two beam case with g_{0006}.

532

The dots are preferentially located at the edges of the SFs. Contrast analysis suggests that these are neon gas bubbles trapped in the SiC lattice which agglomerated during the second annealing process. Deeper in the implanted zone a high density of clusters is evident (denoted by the letter C in Fig. 2b.). The SFs exhibit strong contrast under two-beam conditions with $g_{11\bar{2}0}$. Under these conditions, all the clusters in the zone C are not invisible as shown in Fig. 3a. The two-beam case with the g_{0006} reflection makes the SFs invisible, as shown in Fig. 3b, revealing that these are of the Shockley type. In this case the clusters exhibit strong contrast revealing that their displacement vector is parallel to the c-axis.

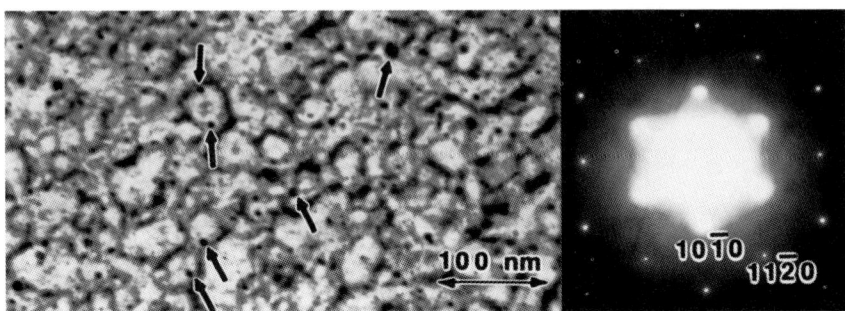

Fig.4 PVTEM micrograph from the specimen implanted with neon and annealed at 1000°C for 2h and at 1650°C for another 2h.

PVTEM observations of the defect zone S reveal a high density of dislocation loops and bubbles denoted by arrows in Fig. 4. The related diffraction pattern with the electron beam parallel to the c-axis is shown in the inset of Fig. 4. The forbidden $10\bar{1}0$ reflections are now stronger than the principal diffraction spots of the [0001] zone, revealing the high density of the overlapping SFs on the basal planes (compare the diffraction spots in Fig. 1e and Fig. 4). It is evident that the defects produced by the oxygen and the neon implantation are significantly different in spite of the very close atomic masses of the two elements. It is evident that, apart from radiation damage, the oxygen ions also react with the 6H-SiC resulting in the formation of precipitates. The exact nature of these precipitates is not known. Annealing of the implanted specimens at 1900°C for 2 hours reveals the coarsening of the precipitates, which are apparently amorphous suggesting the formation of SiO_2. Further studies are in progress for a better understanding of the oxygen precipitates in SiC.

ACKNOWLEDGMENTS

Two of the authors B. P. and J. S. would like to thank the NATO Science Fellowships Programme and the bilateral Greek-Hungarian scientific action for partial support of this work. B.P. also acknowledges the financial support of OTKA project No. T 030447.

REFERENCES

Dalibor T, Pensl G, Yamamoto T, Kimoto T, Matsunami H, Sridhara S G, Nizhner D G, Devaty R P, Choyke W J 1998a Mater. Sci. For. **264-268**, 257

Dalibor T, Trageser H, Pensl G, Kimoto T, Matsunami H, Nizhner D G, Schigiltchoff O and Choyke W J 1998b European Conference on Silicon Carbide and Related Materials, Montpellier, abs. J1

Pensl G 1987 Proc. 5th Int. School on Physical Problems in Microelectronics, ed J. Kassabov (World Scientific, Singapore) p 155

Suvorov A V, Lebedev O L, Suvorova A A, Van Landuyt J, Usov L O 1997 Nuclear Instr. Meth. Phys. Res. **B 127/128,** 347

Inst. Phys. Conf. Ser. No 164
Paper presented at Microsc. Semicond. Mater. Conf., Oxford, 22–25 March 1999

533

A crystalline (amorphous) silicon bubble 3-D lattice in a synthetic opal matrix

V N Bogomolov, N A Feoktistov, V G Gobulev, J L Hutchison*, D A Kurdyukov, A B Petsov, J Sloan* and L M Sorokin

Ioffe Physico-Technical Institute, RAS. 194021 St. Petersburg, Russia
*Department of Materials, Oxford University, Parks Road, Oxford OX1 3PH, UK

ABSTRACT: Silicon has been until now the most important material in modern solid state electronics. Regular systems of silicon nanoclusters, effectively "bubbles" containing up to 10^{14}cm^{-3} elements have been fabricated in a sublattice of opal voids. Structural studies of samples by TEM, HREM and also Raman measurements have been carried out. Following this, regular lattices of Pt-Si junctions have also been obtained, and the current-voltage characteristics (CVC) investigated.

1. INTRODUCTION

Contemporary solid state electronics is based on essentially 2-D planar technology. Transition to 3-D systems of semiconductor devices is necessary for further increase of the volumetric density of elements. By using 3-D dielectric matrices such as in opal, it may be possible to obtain 3-D ensembles of semiconductor nanodevices with density of elements as high as 10^{14}cm^{-3}.

2. EXPERIMENTAL

To fabricate semiconductor nanocomposites we used "monocrystals" of synthetic opals having optically perfect structure (Bogomolov et al. 1996). The opals consist of 250 nm diameter, close packed amorphous silica spheres and have a regular sublattice of voids, tetrahedral, of size 45 nm, and octahedral, of size 90 nm. Bogomolov and Pavlova (1995) have shown that up to 26% of the void volume is accessible to filling with guest materials. In order to incorporate silicon into opal samples the thermal CVD technique was used, as developed by Bogomolov et al. (1998). The CVD reactor consisted of a quartz tube with an external heater, through which a gas mixture of SiH$_4$ (5%) and Ar was flowing. A plate-shaped opal sample was placed normal to the gas flow. The reactor design precluded gas flow around the sample. As a result of the thermal decomposition of the silane, a silicon film was deposited on the inner surfaces of the opal cavities, i.e. on the surface of the silica spheres. The decomposition conditions were isothermal. To increase the volume fraction of nanocrystalline silicon phase the samples were subsequently annealed at 800°C and gas pressure about 1 Torr. To investigate the final products of silane thermal decomposition, TEM, HREM investigations and also Raman spectrum measurements were undertaken. The electron microscopes were a JEM 4000EXII and JEM 2010 equipped with EDX for analysis of object areas down to 3 - 5 nm in diameter. Electron microscope images in diffraction contrast, phase contrast and high resolution modes were obtained. Specimens were prepared for electron microscopy by mechanical grinding down to 70 - 80 micrometers followed by Ar^+ ion milling to electron transparency. The Raman spectroscopy was carried out in the backscattering geometry at a spectral resolution of 5 cm^{-1}, with a scattering accuracy of about 1 cm^{-1}. The spectra were excited by the 4888 AA-line of an argon laser. The distribution of material across the thickness of the sample was investigated by X-ray absorption, as shown by Ratnikov et al. (1999). In order to create metal-semiconductor-metal (MSM) junctions, the samples were then filled with platinum. This was achieved by impregnating with a solution of $PtCl_4$ in ethanol, followed by reduction of the $PtCl_4$ with hydrogen.

3. RESULTS AND DISCUSSION

Figure 1 shows the Raman spectrum of an annealed sample. The narrow peak, associated with the Raman-active TO phonon mode of crystalline silicon is seen to be shifted to a lower frequency range, compared to that of bulk silicon. Such a transformation in the spectrum is evidence of formation of a nanocrystalline Si phase, see Igbal and Veprek (1992). Analysis of the Raman spectra within the framework of the model of a strong spatial confinement of optical phonons has allowed us to estimate the average size ($L \sim 4$ nm) and volume fraction ($X = 52\%$) of crystallites in the amorphous-crystalline silicon system, Campbell and Fauchet (1986); Pevtsov et al (1995); Golubev et al (1997).

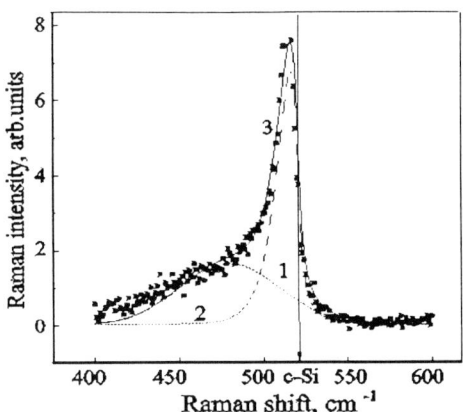

Fig. 1. Raman spectra of a opal-silicon nanocomposite.
1: amorphous component; 2: nanocrystalline component; 3: "total" spectrum. The phonon frequency corresponding to crystalline Si is marked.

The TEM investigations of annealed samples showed that as-deposited Si films were amorphous, with only occasional nanocrystallites in the 3 - 5 nm range. In the annealed samples, the silica spheres were covered uniformly with a 20 - 25 nm thick layer of mixed amorphous and crystalline material, as shown in Figure 2a, recorded in diffraction contrast mode. The composition of the layers was identified by EDX analysis, in which it was established that only silicon was present, compared to silicon and oxygen within the spheres themselves. The crystalline state of the silicon was determined by electron diffraction, which showed mainly <110> oriented Si grains. In Figure 2a the dark contrast within the Si layers represents separate grains of microcrystalline Si in exact Bragg reflecting conditions.

The HREM image of a section of Si layer (Figure 2b) shows that separate grains have single crystal structure of most of the thickness of the layer. The interface between the crystalline Si layer and the amorphous silica sphere is well defined (arrowed). In most instances we also found an amorphous layer on the surface of the Si layers; we interpreted this as silicon oxide (also confirmed by EDX) arising as a result of oxidation of the Si film surface, due to local oxidation during specimen preparation. In some cases the thickness of this oxide extended to about one third of the total Si film thickness. We note that as the thickness of the Si layers (including the oxide skins) is in the 20 - 25 nm range, this becomes close to the size of the voids themselves. Thus the voids are very nearly, but not completely, filled by silicon. This is seen clearly in Figure 2b.

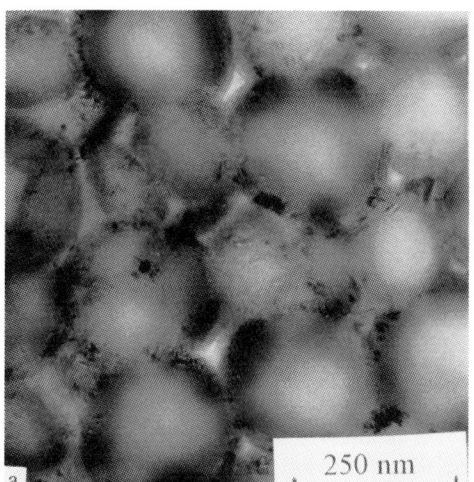
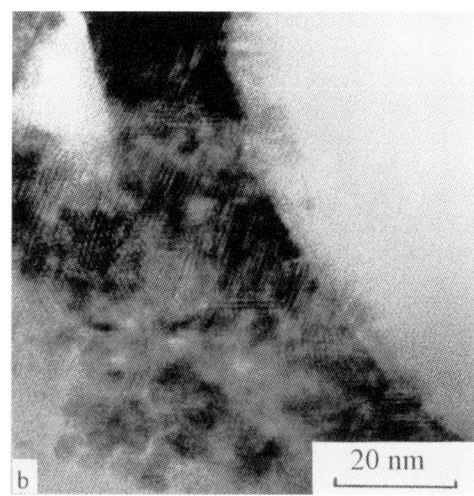

Fig.2a. TEM image of an annealed sample of opal-silicon, showing the formation of crystalline Si layers on silica spheres. 2b. HREM image of Si film on silica sphere, showing a high degree of ordering.

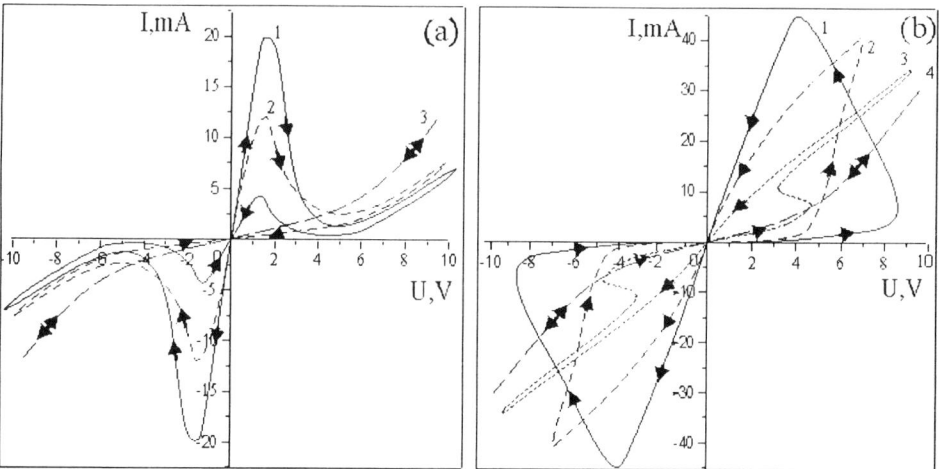

Fig. 3. CVC of opal-Si-Pt nanocomposites with different Pt fill factors.
   a: N-shaped (1: 0.1 Hz, 2: 100 Hz, 3: 10 kHz)
   b: N-and S-shaped (1: 0.01 Hz, 2: 1Hz, 3: 100 Hz and 4: 10kHz)

The X-ray absorption described by Ratnikov et al (1999) was used to gain an estimate of the depth of opal which was impregnated successfully with Si; this was found to be 200 μm. The remaining empty volume is in principle accessible to filling with other materials. In our experiments, we succeeded in impregnating the opal with platinum clusters. Current-voltage characteristics (CVC) were measured for opal-Si-Pt nanocomposites, and examples are shown in Figure 3a,b. The type of CVC depends on the fill factor for the Pt, the behaviour of CVC being defined by the redistribution of carriers on the Pt-Si interface.

4. CONCLUSION

It is shown that the thermal CVD technique can be used to deposit 20 - 25 n thick, uniform Si films on the inner surfaces of the void sublattice of synthetic opals. As a result of this, a 3-D bubble lattice of silicon can be fabricated. The thickness of the Si film as well as the degree of filling can be simply varied by both the duration of the thermal CVD process, as well as by varying the thickness of the opal sample in the reactor. 3-D multilayer, quasiplanar structures can then be envisaged, and hence Pt-Si junctions were designed and their properties measured. The resulting structures were found to have S-like or N-like CVC.

ACKNOWLEDGEMENTS

This work was supported by a Royal Society Travel Grant to one of us (LMS) and the Russian R&D "Nanostructures" Program (Grant No. 97-2016) and RFBR Grant No. 98-02-17350.

REFERENCES

Bogomolov V N, Kurdyukov D A, Prokofiev A V and Samoilovich S M 1996 JETP Lett. **63,** 496
Bogomolv V N and Pavlova T M 1995 Semiconductors **29,**826
Bogomolov V N, Golubev V G, Kartenko N F, Kurdukov D A, Pevstov A B, Prokofiev A V, Ratnikov V V, Feoktistov N A and Sharenkova N V 1998 Tech.Phys. Lett. **24**, 326
Ratnikov V V, Kurdyukov, D A and Sorokin L M 1999 Phys.Solid State **40** 1249
Igbal Z and Veprek N 1992 Solid State Physics **15**, 377
Campbell L H and Fauchet P M 1986 Solid State Commun. **56** 377
Pevstov A B, Davydov V Yu, Feoktistov N A and Karpov V G 1995 Phys. Rev. **B52,** 955
Golubev V G, Davydov V Yu, Pevstov A B and Feoktistov N A 1997 Phys. Solid State **38,** 1197

Inst. Phys. Conf. Ser. No 164
Paper presented at Microsc. Semicond. Mater. Conf., Oxford, 22–25 March 1999
© *1999 IOP Publishing Ltd*

EFTEM study of Ti/TiN and Co-silicide thin films at cross-sections of device structures

W Blum, H J Engelmann, J Rinderknecht and E Zschech

AMD Saxony Manufacturing GmbH, Dresden, Germany

ABSTRACT: EFTEM was used for cross-sectional on-product characterization in CMOS structures. The investigations of Ti/TiN adhesion layers and diffusion barriers reveal the possibility of distinguishing clearly between these functional layers with high lateral resolution. Defect analysis concerning the formation of Ti-Al intermetallic phases at grain boundaries and at TiN/Ti/Al interfaces were performed by EFTEM. The investigation of the influence of a cap layer on Co silicide formation reveals the formation of SiO_x at the $CoSi_2$ surface for samples without a cap layer on top of Co, which verifies the benefit of a cap layer for $CoSi_2$ formation.

1. INTRODUCTION

EFTEM (Energy-Filtering Transmission Electron Microscopy) has been introduced only very recently to characterize semiconductor manufacturing processes (Egerton 1996, Hofer et al 1995). This technique allows one to determine the qualitative chemical composition of thin films with nearly the same lateral resolution as TEM bright field imaging. Particularly the ongoing scaling-down of device structures and growing aspect-ratios result in the deposition of extremely thin layer stacks. Specifications are usually tested on blanket test wafers using X-ray methods and TEM bright field imaging. However, the more realistic cross-sectional on-product characterization of thin film layers has been a serious problem so far, thus jeopardizing successful attempts to measure layer thickness with high accuracy by means of TEM bright field imaging and to determine the elemental distribution in thin films by EDS, as well.

In this EFTEM study, Ti/TiN layers and $CoSi_2$ thin films in the 100 Å range were investigated. Ti/TiN layers are well established for application to local interconnects (LI) and metal lines. The layer is acting as an adhesion and low contact resistance layer, as an anti reflective coating layer and as a via etch stop on top of aluminum metal lines. $CoSi_2$ films have proven to be appropriate as low resistance contact layers in active areas for 0.25μm CMOS technology. Their physical integrity is crucial to product yield and their function can only be ensured if defined film thickness ratios, compositions and morphologies are met.

2. EXPERIMENTAL

The EFTEM study was performed on a Philips CM 200FEG microscope with Gatan Imaging Filter system (GIF 200). The background correction of the elemental maps was done with the conventional three-window method using a smooth power law approximation (Egerton 1996). TEM samples were prepared by conventional cross-section preparation as well as by FIB-cut preparation. The Ti/TiN layers were formed by Ti Physical Vapor Deposition (PVD) sputtering in different atmospheres. Cobalt silicide is formed as a self-aligned silicide: After Co PVD deposition, a low temperature Rapid Thermal Annealing (RTA) reaction creates a Co-Si compound on the exposed silicon of the gate and source/drain regions as well. The unreacted Co over the oxide regions is removed subsequently. A high temperature RTA transforms the Co-Si compound into $CoSi_2$. The influence of an additional cap layer on top of the Co layer, which is removed after the low RTA step, is investigated and compared to the case of samples without a cap layer.

3. RESULTS

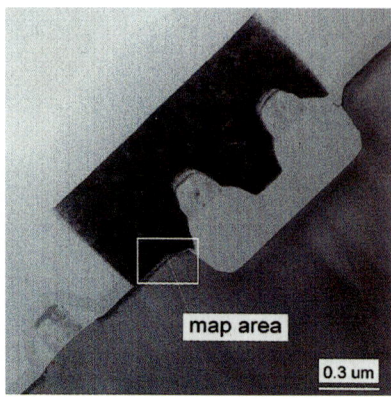

Fig.1: TEM bright field of a LI Structure

Fig. 1 shows an overview TEM bright field image of a LI structure. The area where the Ti/TiN barrier deposition was characterized is marked by 'map area'. This map area is shown in Fig. 2 at higher magnification. The TEM bright field image does not allow us to distinguish clearly between the Ti and TiN layers. A sufficient Ti layer thickness is important for the performance, because Ti significantly reduces contact resistance. Particularly, the thickness of the Ti layer can not be determined. The application of EDS line scans also fails due to the overlapping of the N K_α- peak with the Ti $L_{\alpha,\beta}$- peaks. EFTEM allows us to measure the film thickness: Fig. 3 represents a colored overlay (RGB image) of the Ti and N maps.

The Ti layer (red) can be well separated from the TiN layer (yellow: red from Ti and green from N yields yellow for TiN). A film thickness ratio of 5:2 was determined using the scale of the corresponding TEM bright field image.

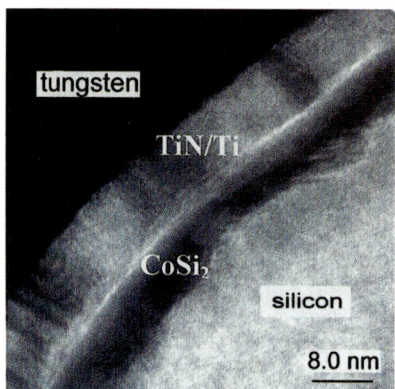

Fig. 2: Bright field of the map area

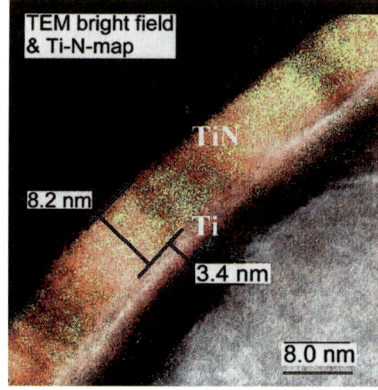

Fig.3: Overlay of TEM bright field image and Ti-N map

EFTEM was also used to analyze the chemical composition of defects in reliability test metal lines. Fig. 4 is a bright field image of an Al metal line. At a grain boundary within the Al line, one can observe dark grains of probably unknown phases growing from the bottom and the top into the metal line. Fig. 5 shows the colored overlay (RGB image) of the Ti map (red), the N map (green), the Al map (blue) and the bright field image. EFTEM as well as EELS (Energy Loss Spectroscopy) allow us to distinguish clearly between the TiN layer on top (yellow) and a small underlying Ti rich layer (red). Figure 7 shows the corresponding EELS-spectra, which are taken from the circled areas in figure 6. At the bottom of the metal line the former Ti layer is reacted to form a uniform Ti-Al alloy layer (pink color). This uniform alloying is not visible on top. The formation of Ti-Al intermetallic phases occurs at a grain boundary, which gives a strong indication of grain boundary diffusion of Ti into the Al. The Ti rich top layer reveals a Ti depleted area (green) at the point where the Ti-Al alloy is formed (Fig. 6). Al metal lines are alloyed with Cu (about 1 wt %) to improve electromigration resistance.

In comparison to the surrounding Al, the dark and triangle shaped grain on the bottom reveals a higher Cu content (Fig. 6), which indicates a significant Cu-enrichment at the Al grain boundary and probably a $CuAl_2$ phase formation at the grain boundary.

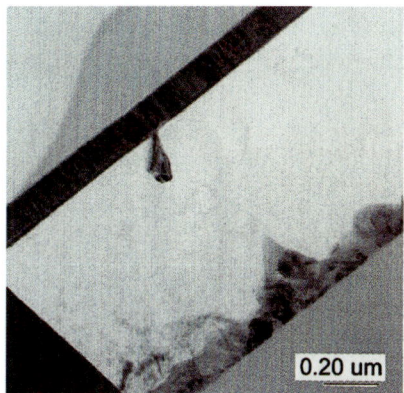

Fig. 4: TEM bright field of a Al metal line

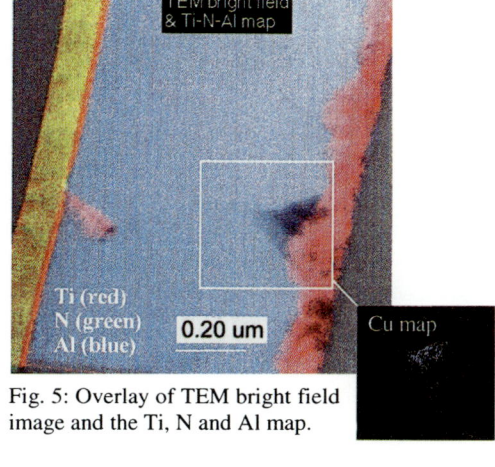

Fig. 5: Overlay of TEM bright field image and the Ti, N and Al map.

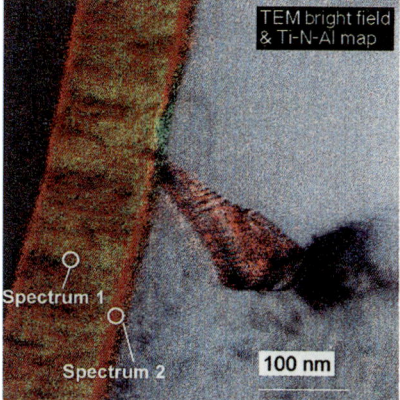

Fig. 6: Overlay of TEM bright field image and the Ti, N and Al map

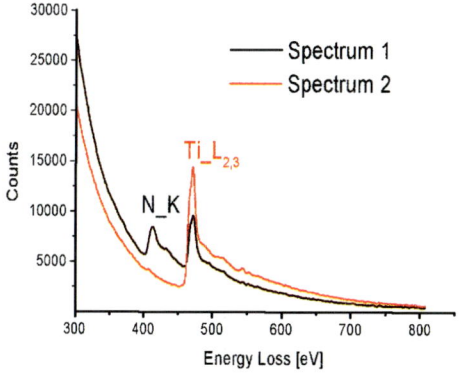

Fig. 7: EELS-spectra referring to circles in Fig. 6

Cobalt silicide samples were investigated by EFTEM, as well. Samples with a Ti-cap layer reveal perfect $CoSi_2$ formation. Elemental maps, taken from samples without a cap layer on top of Co, reveal the formation of very small, disk shaped phases with a diameter of about 50 Å, and of a thin surrounding layer with a thickness of about 20 Å at the surface. This is shown in the bright field image (Fig. 8). The elemental mapping of Co, Si, and O verifies clearly the formation of SiO_2 at the surface. Fig. 9 and Fig. 10 show the colored overlay (RGB image) of the Co map (red), the Si map (green), the O map (blue). Fig. 11 and Fig. 12 show the cobalt and the oxygen maps. These complementary maps prove that little, if any, cobalt oxide is formed. SiO_2 is formed at the surface but also underneath the disk shaped Co-Si phases, which might be due to grain boundary diffusion of oxygen into the CoSi layer during low temperature RTA. The RGB image (Fig. 10) shows a higher Co content in the disk shaped phases. This can be seen by the orange color of the disks in comparison to the yellow color of the underlying $CoSi_2$. This experimental result gives a strong indication of incomplete $CoSi_2$ formation during high temperature RTA. This could be explained by SiO_2 formation at the CoSi / Co interface during low temperature RTA, as reported in the literature (Besser et al 1998).

540

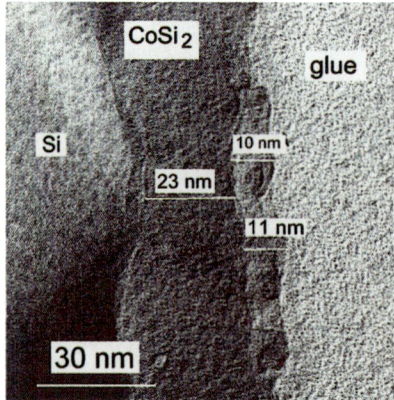

Fig. 8: TEM bright field, CoSi₂ formation without a cap layer.

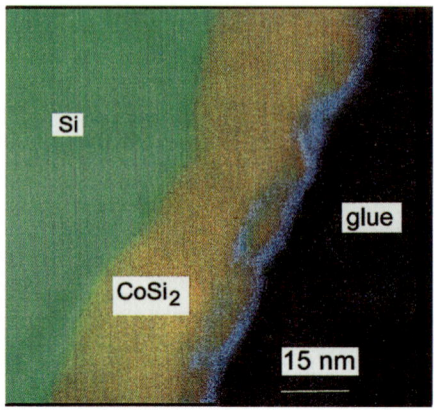

Fig. 9: Overlay (RGB image) of the Si map (green), Co map (red), and the O map (blue).

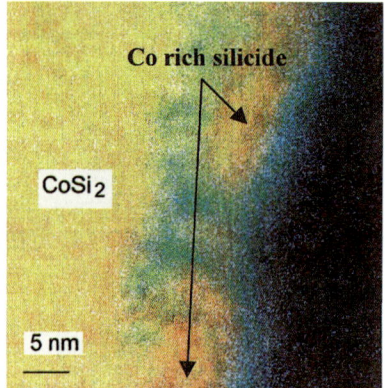

Fig. 10: Detail (RGB image) of Fig. 9

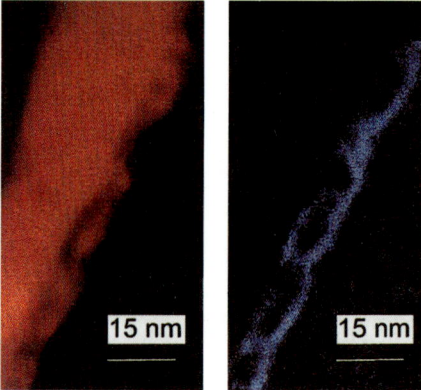

Fig. 11: Co map   Fig. 12: O map

3. CONCLUSIONS

The superior capabilities of EFTEM for the analytical characterization of CMOS devices have been confirmed. It has been shown that EFTEM can be used for cross-section on-product characterization of Ti/TiN layer stacks. A detailed defect analysis in Al metal lines to study Ti-Al phase formation and very small Cu enrichments at grain boundaries is possible. The investigation of the Co-silicide formation without a cap layer reveals the possibility to detect the formation of SiO₂ and of a Co-rich silicide phase at the surface over a range of a few Angstroms.

REFERENCES

Besser P, Lauwers A, Roelandts N, Maex K, Blum W, Alvis R, Stucchi M and de Potter M 1998 Mat. Res. Soc. Symp. Proc. **514**, 375

Egerton R F 1996 Electron Energy Loss Spectroscopy in the Electron Microscope (Plenum Press: New York)

Hofer F, Warbichler P and Grogger W 1995 Ultramicroscopy **59**, 15

Inst. Phys. Conf. Ser. No 164
Paper presented at Microsc. Semicond. Mater. Conf., Oxford, 22–25 March 1999
© *1999 IOP Publishing Ltd*

Characterisation of tungsten nitride barrier layer for copper metallisation

S Jin, H Li, H Bender, I Heyvaert and K Maex

IMEC, Kapeldreef 75, B-3001 Leuven, Belgium, sing@imec.be

ABSTRACT: Transmission electron microscopy (TEM), focused ion beam (FIB) and Auger electron spectroscopy (AES) are used to study tungsten nitride films. In the present work, tungsten nitride (W_xN) films with different W/N ratios are deposited, by PECVD using $WF_6/N_2/H_2$ chemistry, on Si (100) wafers covered with blanket or structured PECVD SiO_2 layers. The as-deposited W_xN is amorphous and the crystallisation temperature is dependent on the stoichiometry of the tungsten nitride films. W_2N is found to be the main phase after annealing. The grain size is about a few tens of nanometres. TEM studies on blanket and trench samples with $Cu/W_xN/SiO_2$ structure show excellent W_xN step coverage and no significant interfacial reactions even after RTP annealing at 700°C, indicating that W_xN layer is an effective barrier layer between Cu and SiO_2.

1. INTRODUCTION

The demands of manufacturing integrated circuit devices require the shrinking of semiconductor interconnections. Recently, much effort has been devoted to Cu metallisation due to its lower resistivity and better resistance to electromigration than aluminium alloys (Murarka 1997). Due to the fast diffusion of copper into dielectrics, a barrier layer is introduced against the Cu penetrations. Tungsten nitride (W_xN) has attracted considerable attention as a potential barrier material (Wang 1994 and Galewski and Seidel 1999).

A clear picture of the structure of thin films can only emerge from a combination of several analytical techniques. Therefore, transmission electron microscopy (TEM), focused ion beam (FIB) and Auger electron spectroscopy (AES) are used, in the present work, to study the Cu metallisation films and the W_xN diffusion barrier layers.

2. EXPERIMENTAL

The deposition of W_xN films is carried out in a plasma enhanced chemical vapour deposition (PECVD) chamber of a commercial W CVD system using a $WF_6/N_2/H_2$ chemistry. All W_xN films reported in this paper are deposited at a susceptor temperature of 400°C on 150 mm Si (100) wafers covered with SiO_2.

Blanket and trench samples with $Cu/W_xN/SiO_2$ structure are annealed at different temperatures up to 700°C. The Cu is deposited by sputtering (blanket samples) or by electroplating on a sputtered seed layer (trench samples), respectively.

542

The samples are characterised with JEOL 200CX electron microscope, FIB 200 workstation and PHI 600 Auger analysis system. Cross-sectional specimens for TEM studies are prepared by focused ion beam. For the preparation of Cu samples, special care is exercised to avoid the corrosion of copper with I_2 (Bender 1999). The normal of the TEM thin foil is parallel to the [011] direction of silicon. Plan-view TEM specimens are produced by mechanical thinning from the substrate side and milling from the backside with a single ion beam only.

3. RESULTS AND DISCUSSION

The initial stage of the formation and evolution of the W_xN film is investigated by AES, as shown in Fig. 1. They are obtained with the electron beam scanning over an area of $500\mu m^2$, so that the electron beam induced oxide reduction is minimal. At the deposition time of one-second, strong Si but no W or N peak is observed, indicating that within the AES sensitivity no film is deposited on the SiO_2 yet. For the two-second sample, the W and N peaks are clearly visible, while the intensity of the Si peak decreases, indicating the formation of a W_xN layer. The AES spectrum of the three-second sample reveals that a continuous coverage probably starts to form at this time, as the Si signal from the substrate has almost vanished. Further increase of the deposition time to 10 seconds shows no further increase of the intensity of the W and N signals.

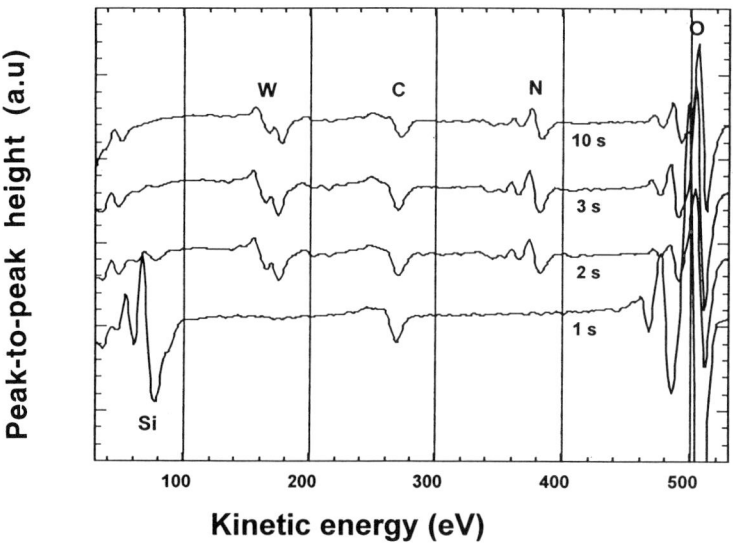

Fig. 1 AES spectra from the as-deposited samples. The deposition time is 1, 2, 3 and 10 seconds

The results show that probably some Si is incorporated into the W_xN films during the initial stage of the W_xN growth. The mechanism is discussed in detail in another report (Li et al 1999).

The TEM micrographs of the W/N=2 and W/N=3 films are shown in Fig. 2. In the as-deposited samples, both films have a very fine granular-like structure with size 15-25 nm (Fig. 2(a) and (d)). The electron diffraction patterns exhibit only a few broad diffraction rings, indicating that the W_xN films did not crystallise. Annealing at 500°C does not change the

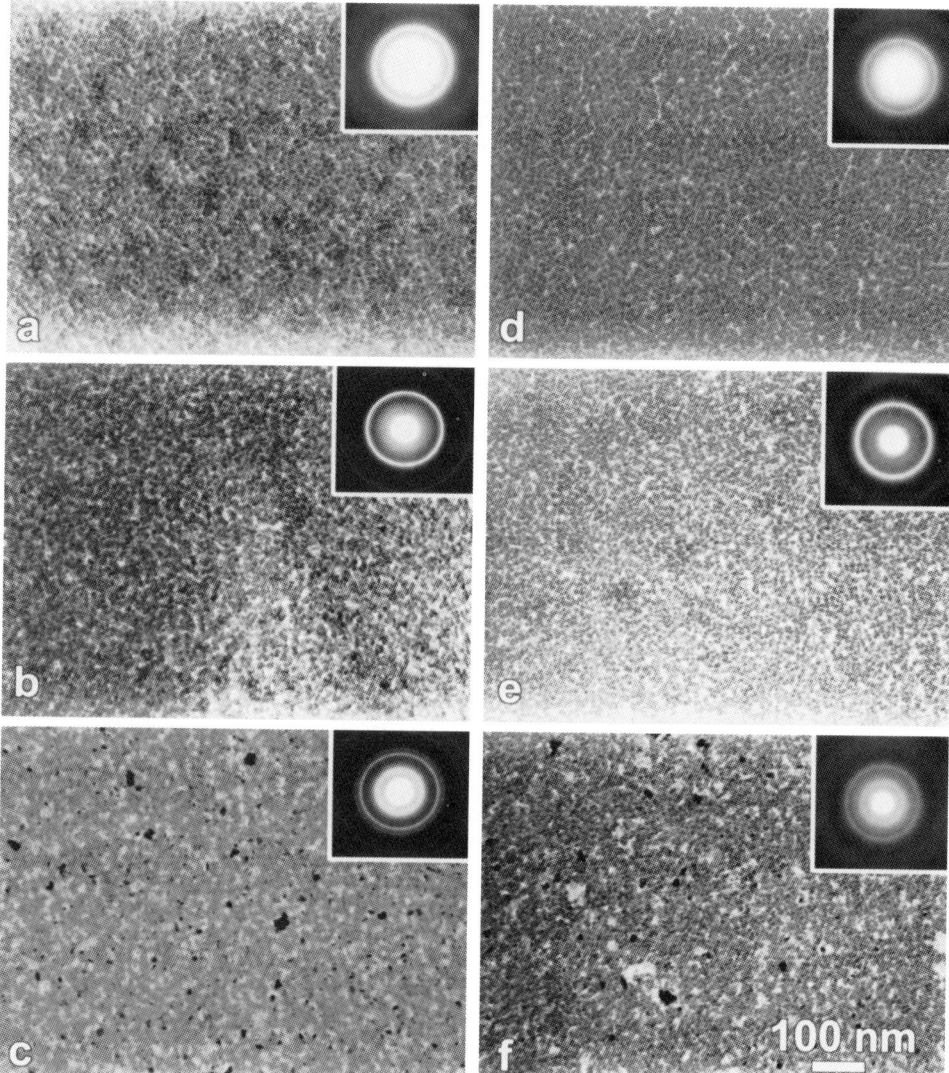

Fig. 2 Plan-view TEM images from the samples: (a) as-deposited (W/N=2), (b) 500°C (W/N=2), (c) 600°C (W/N=2), (d) as-deposited (W/N=3), (e) 500°C (W/N=3), (f) 600°C (W/N=3).

granular-like structure in both films (see Figs. 2(b) and (e)). The electron diffraction pattern of the 500°C W/N=2 sample shows several sharp diffraction rings, indicating the formation of a crystalline phase. These rings can be indexed as belonging to the W_2N phase (cubic, a= 0.4126nm). For the W/N=3 film, the electron diffraction rings remain diffuse after annealing at 500°C. Further increase of the temperature to 600°C results for both films in the appearance of a second phase with a grain size of 5-30 nm and which shows a dark contrast in the TEM micrographs in both films (Fig. 2 (c) and(f)). The sharp electron diffraction rings found in both films can be assigned to the W_2N phase. The irregularly shaped grains with the dark contrast are α-W. These results on the phase evolution are consistent with the XRD studies (Li et al. 1999).

Grain boundaries are the dominant diffusion path. For this reason, the preferred structure is an amorphous one thus the grain boundary diffusion is completely blocked. To evaluate the quality of W_xN as Cu diffusion barrier layer, samples with $Cu/W_xN(W/N=3)/SiO_2$ structure are annealed up to 700°C. Figure 3 is a cross-sectional TEM (XTEM) image showing sharp interfaces between the W_xN and the Cu and the SiO_2. The W_xN layer has also a good conformal step coverage when it is deposited on SiO_2 trench, as shown in Fig. 4. These results reveal that the W_xN is still an effective barrier material even when the W_xN layer is fully crystallised.

Fig.3 XTEM image of the blanket sample after annealing at 700°C for 15min.

Fig.4 XTEM image of the structured sample after annealing at 700°C for 15min.

4. CONCLUSION

The as-deposited W_xN is amorphous and crystallisation temperature is dependent on the stoichiometry of the tungsten nitride films. W_2N is found to be the main phase after annealing. The grain size is about a few tens of nanometres. TEM studies on blanket and trench samples with $Cu/W_xN/SiO_2$ structure show excellent W_xN conformal step coverage and no significant interfacial reactions even after RTP annealing at 700°C, indicating that W_xN layer is an effective barrier layer between Cu and SiO_2.

ACKNOWLEDGEMENT

The transmission electron microscopy investigations are performed with the microscopes at EMAT, University Antwerpen (RUCA), Belgium.

REFERENCES

Bender H 1999 this conference
Galewski C and Seidel T 1999 European Semiconductor, January, 31
Li H, Jin S, Bender H, Lanckmans F, Heyvaert I, Maex K and Froyen L 1999 J. Appl. Phys. (submitted)
Murarka S P 1997 Mater. Sci. Eng. Rep. **19**, 87
Wang S Q 1994 MRS Bulletin **17**, 30

Inst. Phys. Conf. Ser. No 164
Paper presented at Microsc. Semicond. Mater. Conf., Oxford, 22–25 March 1999
© 1999 IOP Publishing Ltd

HRTEM and EFTEM studies of the evolution of Cu/Ta/SiO₂/Si interfaces in ULSI devices

Kai-Min Yin, Li Chang,* Fu-Rong Chen, Ji-Jung Kai, Cheng-Cheng Chiang Graham Chuang, Chun-Feng Shen, Shen-Chuan Lo, Peijun Ding,[+] Barry Chin, [+] Hong Zhang[+] and Fusen Chen[+]

Department of Engineering and System Science, National Tsing Hua University, Hsinchu, Taiwan 300, Republic of China
*Department of Material Science and Engineering, National Chiao Tung University, Hsinchu, Taiwan 300, Republic of China
[+]Metal Deposition Products Group, Applied Materials, Santa Clara, California 95054, USA

ABSTRACT: The Cu/Ta/SiO₂/Si films have been annealed at temperatures from 500°C to 600°C in various vacuum conditions. Transmission electron microscopy has been used to characterize the microstructure of the films after annealing. It shows that different thicknesses of amorphous interlayer were formed between Cu and Ta under various vacuum conditions. Energy dispersive spectroscopy confirmed this interlayer to be tantalum oxide. It has also found that tantalum oxidation is caused by oxygen from the outside atmosphere diffusing along grain boundaries in copper films.

1. INTRODUCTION

Copper has a lower bulk electrical resistivity and better resistance to electromigration than aluminum and aluminum alloys. Though copper metallization has been recently realized in commercial ULSI device fabrication, there are a number of issues in processing with copper for future devices of smaller sizes that have not been clarified. Among these, a diffusion barrier against copper plays an extremely important role in the copper metallization. Tantalum has been shown to be a promising material as diffusion barrier layer for copper metallization because Ta and Cu are mutually insoluble (Jang et al 1996; Wang et al 1998; Holloway et al 1990; Catania et al 1992; Ono et al 1994). The thermal stability of the barrier is one of the most important properties in microelectronics, particularly in the damascence process (Price et al 1997). Although the thermal stability of the Cu/Ta/SiO₂/Si structure has been studied in the past (Olowolafe et al 1993; Vogt et al 1995), few investigations show the details of interfacial structures between the layers on the nanometre scale, which will affect the barrier properties. Furthermore, the effects of thermal environments so far have not been reported in published works.

In this report, microstructures of barriers and Cu after thermal annealing in various environments are studied to understand the effect of oxygen on the barriers. The high affinity of Ta for oxygen will promote oxide formation, resulting in structural changes of Ta. The microstructural and compositional evolution at the Cu/Ta/SiO₂ interfaces are investigated by cross-sectional transmission electron microscopy (TEM). Cu/Ta/SiO₂/Si films have been annealed at temperatures from 400°C to 600°C in different atmospheres and/or pressures.

2. EXPERIMENT

Cu/Ta/SiO₂/Si structures as blanket films were used for the present study. The Cu and Ta films were deposited on 8-inch silicon wafers in an Applied Materials Electra™ system which utilized ion metal plasma (IMP) processing technology (Dixit et al 1996). The base pressure of the IMP chamber

546

was typically ~ 10^{-8} mbar. The stacking sequences and thicknesses of the deposited films were Cu (150nm)/Ta (25-30nm)/SiO$_2$(1000nm) on Si wafers.

To investigate the thermal stability with temperature, a sample was annealed at 500°C in a vacuum furnace of a pressure of 10^{-2} mbar. A couple of samples were treated at 500°C and 600°C in a pure argon (4N) atmosphere with a pressure of 3 mbar. Before feeding with Ar gas, the furnace was purged several times with Ar, and evacuated to the 10^{-2} mbar range. The annealing time of all treatments was 30 min.

A JEOL 2010F field-emission-gun TEM, equipped with an energy dispersive spectrometer (EDS) and a Gatan Imaging Filter (GIF) was used for studying the structures and chemical compositions. Electron energy loss spectroscopy (EELS) was also used for compositional analysis because of its higher spatial resolution and better energy resolution than EDS.

3. RESULTS

Figure 1 shows a typical cross-sectional high-resolution TEM (HRTEM) image from an as-deposited Cu/Ta/SiO2/Si sample, taken immediately after TEM specimen preparation. There is no evidence of reaction or formation of any interlayer between stacked layers. Thus, it is certain that the as-deposited film is clean and has no contamination or oxidation from the system environment. From the image, it can be seen that the Ta is in the form of nanocrystals. A selected area diffraction pattern (SADP) from the Ta, shown as the inset in Fig. 1, confirms that the Ta is a beta-phase (bct, body-center-tetragonal structure). This TEM sample was re-examined after it was kept in air at room

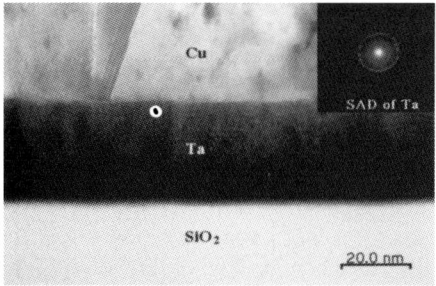

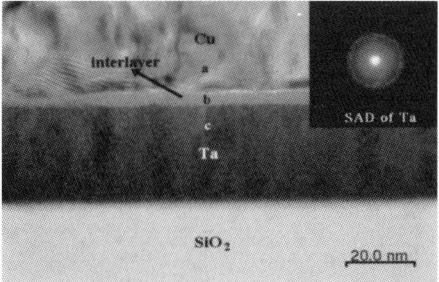

Fig. 1 Cross-sectional TEM image from an as-deposited Cu/Ta/SiO$_2$/Si sample with inset of a SAD pattern from the Ta layer.

Fig. 2 Cross-sectional TEM image of the as-deposited TEM sample with inset of a SAD pattern from the Ta layer was re-examined after it was kept in air at room temperature for one week.

temperature for one week. Figure 2 shows an interlayer of thickness from 2 nm to 4 nm between Cu and Ta. No diffraction spots can be observed from this interlayer. Thus, it is an amorphous phase. The Ta film is still beta-phase as confirmed from the diffraction pattern in Fig. 2. The thickness of the Ta reduced from 30 nm in the as-deposited condition to 28 nm, which is in excellent agreement with formation of a 2 nm thick interlayer. EELS maps in Fig. 3, obtained from another area, show the distribution of Cu, Ta, and O. An intensity profile across the interface clearly reveals strong oxygen signals at the interlayer. The above results clearly show that an amorphous tantalum oxide interlayer has been formed between the Cu and Ta layers held at room temperature for one week. The oxidation occurs at the interface between the Cu and Ta layers rather than between the Ta and SiO$_2$. After heat treatment at 500°C for 30 min all metallic Ta transformed into amorphous tantalum oxide (Fig. 4). The thickness of tantalum oxide has increased to about 70 nm.

Samples were treated in an Ar atmosphere with a pressure of 3 mbar to determine whether they could be protected from oxidation. Annealing at 500°C greatly reduced the oxide thickness. As shown in

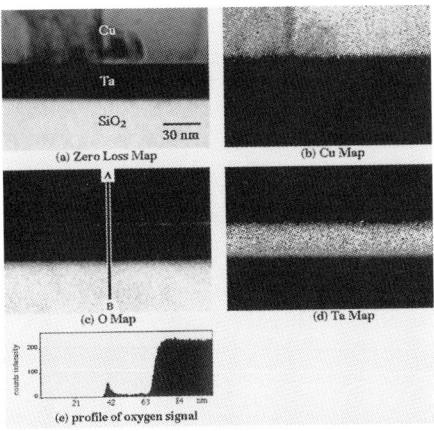

(a) Zero Loss Map

(b) Cu Map

(c) O Map

(d) Ta Map

(e) profile of oxygen signal

Fig. 3 EELS maps of Cu/Ta/SiO₂/Si sample after one week aging at room temperature. (a) the zero loss map, (b) copper map, (c) oxygen map (d) tantalum map and (e) an intensity profile of oxygen map.

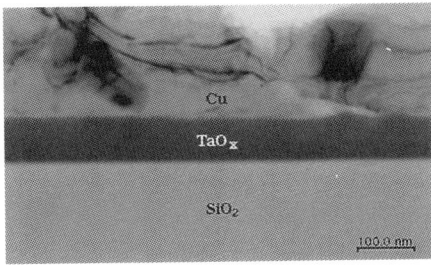

Fig. 4 Cross-sectional TEM image of as-deposited Cu/Ta/SiO₂/Si sample after thermal annealing at $500^{\circ}C$ for 30 min in a vacuum of 10^{-2} mbar.

Fig. 5, an oxide layer of 5-7 nm thick exists between the Cu and the Ta. For those treated in Ar atmosphere at 600°C, TEM in Fig. 6 shows that TaO_x is in contact with underlying SiO_2, and along a Cu grain boundary. A thin oxide layer parallel to the interface is also found between the Cu and the Ta. Most of the metallic Ta layer remains between Cu and SiO_2. The interface between the Ta and the SiO_2 is still very sharp. The results strongly suggest that oxygen from an outside source diffuses along grain boundaries of Cu, and gradually oxidizes the Ta. Under this situation, copper oxidation is less favorable. The diffraction patterns indicate that Ta oxide is amorphous, and Ta remains in the bct phase.

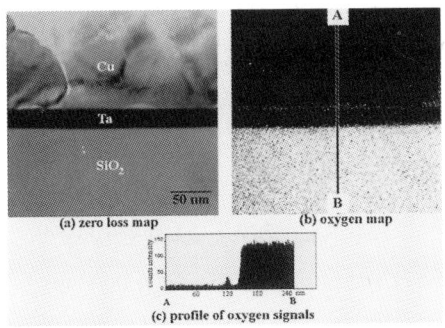

(a) zero loss map

(b) oxygen map

(c) profile of oxygen signals

Fig. 5 EELS maps of Cu/Ta/SiO₂/Si sample after thermal annealing at $500^{\circ}C$ for 30 min in argon atmosphere. (a) the zero loss map, (b) oxygen map and (c) an intensity profile of oxygen map.

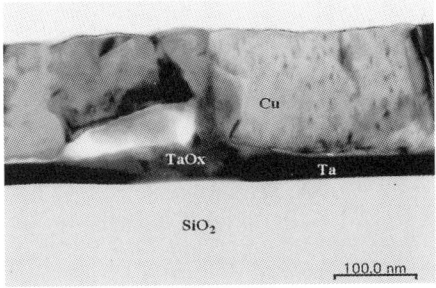

Fig. 6 Cross-sectional TEM image of as-deposited Cu/Ta/SiO₂/Si sample after thermal annealing at $600^{\circ}C$ for 30 min in argon atmosphere.

4. DISCUSSION

Ta has a higher oxygen affinity than Cu. Therefore, Ta oxidation retards the copper oxide formation at low temperatures. However, at high temperatures both Cu and Ta are oxidized. The reason why tantalum oxide formed at the interface between the Cu and the Ta rather than between the Ta and the SiO_2 is simply because Si has a greater oxygen affinity than Ta (Beyers et al 1984). From thermodynamic calculations, it is found that all the reactions of Ta with SiO_2 in which products can be $TaSi_2$, Ta_2Si, Ta_5Si_3 or Ta_2O_5 have positive free energies. Hence, it can be understood why there is no reaction between Ta and SiO_2 even after annealing at temperatures up to 600°C. Oxygen from ambient environments probably diffuses through the Cu and reacts with the Ta. It is believed that oxygen as residual gas in the vacuum furnace can diffuse along grain boundaries in Cu films to form Ta oxide on the Ta film. With the use of an environment of Ar gas, the oxygen concentration is reduced, resulting in retardation of oxidation of the Ta film.

The above results should have significant implications for device fabrication by thermal processing. It can not be ignored that the annealing environment has a strong effect on the barrier properties of Ta films as thin oxide can form even in low vacuum with an Ar atmosphere at 500°C. Also, it has been shown that *in situ* oxygen dosing during deposition of Ta and Cu increases the failure temperature by 30-250°C during annealing in comparison with those of without dosing (Clevenger et al 1993). Though the effectiveness of amorphous tantalum oxide in retarding Cu diffusion has not been investigated in the present work, it is likely that diffusion of copper might be slowed down by a thin oxide layer due to the amorphous structure of the latter. The oxide formation certainly results in an increase of apparent resistance of Cu. On the other hand, it is questionable whether the failure temperature of the device would be raised. Thick oxide as insulator, formed after annealing in air and low vacuum conditions, certainly will not be helpful for the performance of the interconnects.

5. CONCLUSIONS

The results are summarized as the following.
1. Oxidation starts from an ambient environment at room temperature. An interlayer of amorphous tantalum oxide was found between the Cu and the Ta.
2. Under low vacuum condition, heat treatments at higher temperatures enhance growth of the oxide. At 500°C, metallic Ta films transform completely into amorphous tantalum oxide. An argon atmosphere in a base vacuum of 10^{-2} mbar can significantly reduce the extent of oxidation.
3. Oxidation occurs mainly at the interface between the Cu and the Ta, suggesting that oxygen comes from outside sources rather than from SiO_2.
4. Therefore, for copper metallization, it is important to understand the effect of oxygen from the surrounding atmosphere on barrier structures and their properties during thermal treatments.

REFERENCES

Beyers R, Sinclair R, and Thomas M E 1984 J. Vac. Sci. Technol. B2, 781
Catania P, Doyle J P, and Cuomo J J 1992 J. Vac. Sci. Technol. A10, 3318
Clevenger L A, Bojarczuk N A, Holloway K, Harper J M E, Cabral C, Schad R G, Cardone F, and Stolt L 1993 J. Appl. Phys. 73, 300
Dixit G A, Hsu W Y, Konecni A J, Kirshnan S, Luttrner J D, Havemann R H, Forster J, Holloway K and Fryer P M 1990 Appl. Phys. Lett. 57, 1736
Jang S Y, Lee S M and Baik H K 1996 J. Mater. Sci. Electron. 7, 271
Ono H., Nakano T, and Ohta T 1994 Appl. Phys. Lett. 64, 151
Olowolafe J O, Mogab C J, and Gregory R B 1993 Thin Solid Films 227, 37
Price D T, Gutmann R J and Murarka S P 1997 Thin Solid Films 308, 523
Vogt M and Drescher K 1995 Applied Surface Science 91, 303
Wang M T, Lin Y C, and Chen M C 1998 J. Electrochem. Soc. 145, 2538

Inst. Phys. Conf. Ser. No 164
Paper presented at Microsc. Semicond. Mater. Conf., Oxford, 22–25 March 1999

Titanium disilicide nanoislands on Si(001) surfaces

G Medeiros-Ribeiro, D P Basile, T I Kamins, D A A Ohlberg, G A D Briggs* and R Stanley Williams

Hewlett-Packard Laboratories, 3500 Deer Creek Road, MS 26U-12, Palo Alto, CA 94304-1392, USA
*on sabbatical leave from Oxford University, Department of Materials, Parks Road, Oxford OX1 3PH , England.

ABSTRACT: Both common phases of titanium disilicide form islands on Si(001) surfaces. At temperatures up to about 850°C the C49 phase forms, with two related (4×4) reconstructions on (010) faces. At higher temperatures the C54 phase forms, with unreconstructed (110) faces. Current imaging tunneling spectroscopy reveals straight parallel fringes crossing flat C49 islands. By comparing with plan and cross sectional view TEM, and with Fourier analysis of STM images, it is possible to relate these fringes to the incommensurate $TiSi_2(010)/Si(001)$ interface.

1. WHY TITANIUM DISILICIDE NANOISLANDS?

A key parameter in the size distribution of self assembled structures for quantum technology is the lattice mismatch between the islands and the substrate. A small enough mismatch can give layer-by-layer growth, with the mismatch eventually being taken up by dislocations. If the mismatch is larger, then the associated elastic strain energy can promote island growth, with the equilibrium island volume depending on the inverse sixth power of the mismatch. Therefore a higher mismatch may be expected to give a smaller island size distribution. But this is true only if the interface remains commensurate; if it is incommensurate then the mismatch strain may be relaxed from the beginning of island formation, and then the elastic energy constraint will not apply. Titanium disilicide is used for interconnects. It has two phases (of which the higher temperature phase is preferred for interconnects because of its higher conductivity and stability), each of which offers a large lattice mismatch to silicon (Jeon et al 1992, 1997). The growth of silicide islands on silicon can be studied within a UHV growth system by STM (Stephenson and Welland 1995a, b, Goldfarb and Briggs 1999), and we have investigated the phases and surface structures of titanium disilicide islands (Medeiros-Ribeiro et al 1999). In the process, we have discovered that it is also possible to observe periodicity in the interface structure.

2. TWO PHASES AND THEIR SURFACES

2.1 Sample preparation

Samples for study in UHV STM were grown on n-doped (2×10^{18} Sb cm^{-3}) Si(001) wafers that had been prepared to give acceptable (2×1) dimer rows in STM. They were exposed at 650°C to submonolayer equivalent coverage of titanium from an electron beam evaporator. They were then annealed for one minute at temperatures in the range 650 – 950°C. For TEM studies islands were grown by CVD on lightly p-doped substrates at 670°C, and subsequently annealed for 30 minutes at 920°C. The calibration of the temperature in the growth chambers is not necessarily the same, but both methods gave, as assessed using AFM, islands with similar sizes, orientation, and morphology, suggesting that they may be not far from equilibrium.

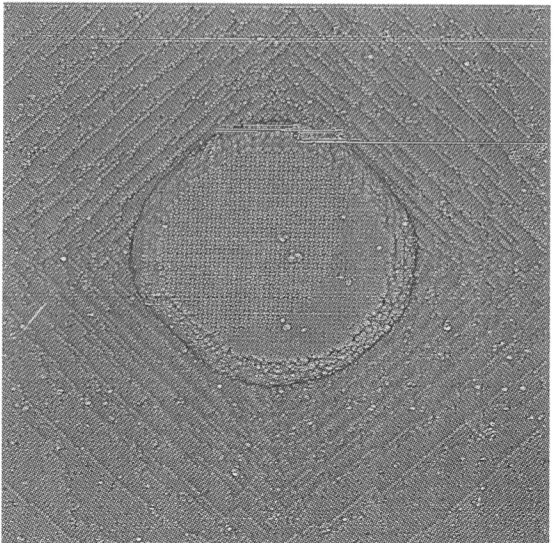

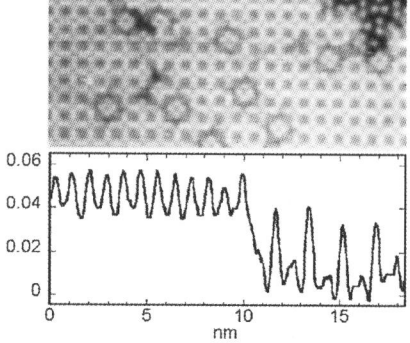

Fig. 1(a) 100 nm × 100 nm STM scan of a flat TiSi₂ island grown at 850°C. The grayscale corresponds to the laplacian of the topographic data, aiding in the visualization of the reacted Si surface and island.

Fig. 1(b) 18.3 nm × 18.3 nm STM scan of an island with two different surface reconstructions, and corresponding line scan, $V = -2$ V and $I = 0.1$ nA.

2.2 C49 islands

Figure 1 shows an island grown at 850°C in UHV. Larger area scans of the surface revealed islands spaced 20 nm apart or more, with a variety of orientations. Some of the islands had tops that were flat and parallel to the Si(001) surface, as in this case. The island shown in Fig. 1 had dug a hole in the surrounding substrate surface, presumably to provide silicon for the reaction with the titanium to form the silicide, and so in order to provide the maximum fine scale detail Fig. 1(a) has been processed with a laplacian filter (so that the contrast is determined by local surface curvature). Two different kinds of reconstruction are present on the TiSi₂ surface, and these are seen at higher magnification in Fig. 3(b), together with a line scan across the surface. The reconstructions are related to each other, and we find that they can be accounted for in terms of an (010) surface of the C49 phase of TiSi₂, which has a mean lattice parameter of 0.361 nm. The origin of the reconstruction remains to be explained, as does the remarkable fourfold symmetry of the defects seen to the left in Fig. 1(b). For islands like this one that are approximately equiaxed, the C49⟨100⟩ direction in the surface is parallel to Si⟨110⟩, but for more elongated islands the orientation is at 45° to this, with C49⟨101⟩ ∥ Si⟨110⟩.

2.3 C54 islands

Figure 2 shows an island that formed after an anneal at 850°C. The top is flat and parallel to the Si(001) substrate surface, and although the near side is not straight, the far edge of the island is parallel to the Si⟨110⟩ dimer directions. An enlargement of the region indicated by the circle in Fig. 2(a) is shown in Fig. 2(b), where the spacing of the rows in 0.85 nm. This is the length of the c-axis of the C54 phase of TiSi₂, which in this direction consists of four stacked hexagonal planes of Ti and Si atoms. While unreconstructed C54(100) and (010) surfaces would give a period of half the observed value, a C54(110) surface would be expected to give a period equal to the length of the c-axis, with either Ti or Si atoms prominent depending on the nature of the termination. We believe that this accounts for the observed structures, and that the dark and light areas in Fig. 2(b) correspond to these two cases. Non-epitaxial C54 islands also had other facets, some of them reconstructed.

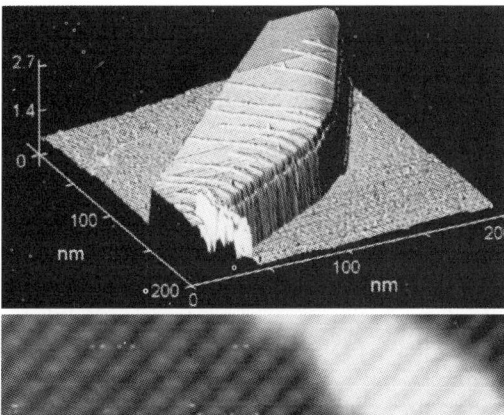

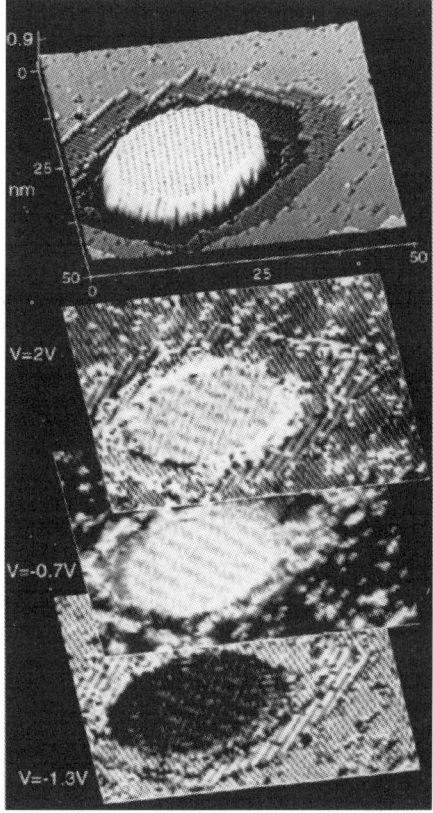

Fig. 2 (a) TiSi₂ island in the C54 phase, grown at 850°C and then annealed at 950°C; (b) 20 nm × 20 nm scan of part of the surface of the island in (a), sample bias $V = -2$ V, $I = 0.1$ nA.

Fig. 3. Topography image (top, $V = -2$ V) of a C49 island, and current images at three indicated sample bias voltages.

3. THE INCOMMENSURATE INTERFACE

3.1 Scanning tunneling microscopy

Fourier analysis of the type of island surface seen in Fig. 1, calibrated by the spacing of the surrounding silicon dimers, enables the lattice mismatch to be determined. We find a mean period for the C49(010) surface structure of 0.676 nm, compared with the silicon dimer spacing of 0.768 nm, giving a 12% lattice mismatch. Using the current imaging tunneling spectroscopy (CITS, Hamers et al 1986) technique we can image the current at a series of different voltages for a given height of the tip above the surface. Fig. 3 shows a series of such current images, crossed by parallel dark fringes, whose contrast reverses at different tip bias voltages.

3.2 Transmission electron microscopy

Fig. 4(a) shows plan view TEM of several islands. The contrast seen in the islands is due to translational Moiré fringes from the lattice mismatch with the substrate. The contrast variation across individual islands and the barrel and pincushion distortion are due to the imaging conditions. Analysis of the diffraction patterns from the islands confirms that they are indeed in the C49 phase. Both the spacing and the orientation of the fringes are the same as in the CITS images, and they correspond to the difference between the TiSi₂(220) and the Si(220) interplanar spacings.

A cross sectional TEM (XTEM) image through an island is shown in Fig. 4(b). The fringes parallel to the island surface may be the (010) stacking structure of C49. The islands are buried like icebergs, with more below the surface than above it. The bottoms of the islands are not flat, so that through a

552

given section the electron beam may sample both silicide and silicon. The periodic nature of the interface is readily seen, but it has not proved possible to determine the structure of the interface or of the dislocations, even after using Fourier image processing to remove the localised contrast by subtracting out the Moiré contribution.

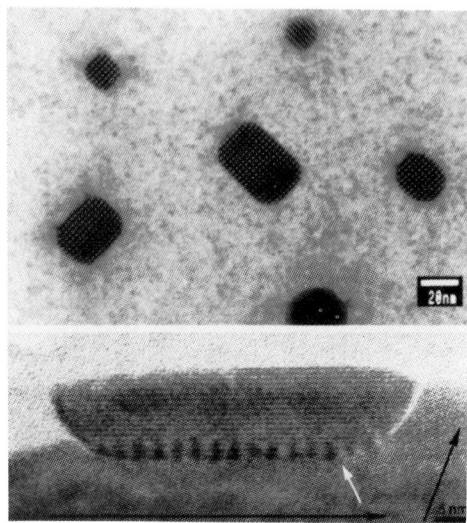

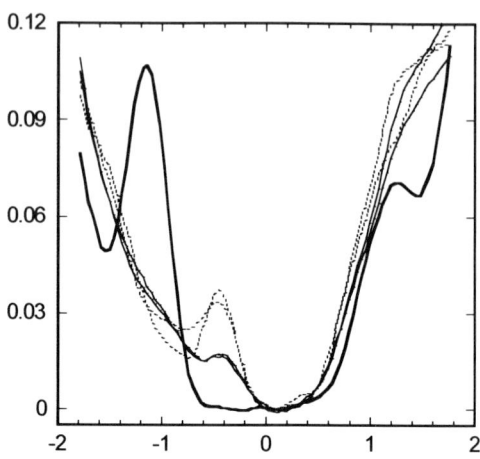

Fig. 4. Upper (a) plan view TEM of C49 islands. Lower (b) cross section in high resolution; dark arrows indicate decreasing sample thickness. Moiré contrast is reduced in very thin regions of the sample (white arrow).

Fig. 5. *dI/dV* curves calculated from the same data set as the current images in Fig. 3. The curve with the large peak at sample bias –1.1 V is obtained from the silicon; the other curves are averaged from either light or dark bands in Fig. 3.

3.3 Scanning tunneling spectroscopy

How are we to account for the contrast observed in the CITS images? That it corresponds to the periodic mismatch at the interface is not in doubt, but what is the mechanism? We have considered several candidate explanations, some of which we find more plausible than others, and the one that we prefer attributes the contrast to electrons travelling ballistically through the silicide to or from the interface (Sirringhaus et al 1994, Meyer and von Känel 1997). From the CITS data we can extract tunneling spectra averaged from the different fringes, and from the surrounding silicon. Such spectra are presented in Fig. 5, as *dI/dV* curves to facilitate discussion in terms of local densities of states. The peak in the Si curve at –1.2 V is readily identified with surface states characteristic of Si(001) dimers. We believe that the peak in *dI/dV* measured periodically in the island may similarly be associated with a local density of states, but in this case localised at the interface. We have seen that the interface must have a periodic nature, both from geometrical necessity and from direct observation, and it is entirely reasonable that there should be associated periodic electronic properties.

REFERENCES

Goldfarb I and Briggs G A D 1999 *Microscopy of Semiconducting Materials XI* (this volume)
Hamers R J, Tromp R M and Demuth J E 1986 Phys. Rev. Lett. **56**, 1972
Jeon H, Sukow C A, Honeycut J W, Rozgonyi, G A and Nemanich R J 1992 J. Appl. Phys **71**, 4269
Jeon H, Yoon G and Nemanich R J 1997 Thin Solid Films **299**, 178
Medeiros-Ribeiro G, Ohlberg D A A, Bowler D R, Tanner R E, Briggs G A D and Williams R Stanley 1999 Surf. Sci. **431**, 116
Meyer T and von Känel H 1997 Phys. Rev. Lett. **78**, 3133
Sirringhaus H, Lee E Y and von Känel H 1994 Phys. Rev. Lett. **73**, 577
Stephenson A W and Welland M E 1995a J. Appl. Phys **77**, 563
Stephenson A W and Welland M E 1995b J. Appl. Phys **78**, 5143

Inst. Phys. Conf. Ser. No 164
Paper presented at Microsc. Semicond. Mater. Conf., Oxford, 22–25 March 1999
© *1999 IOP Publishing Ltd*

Nucleation and growth of CoSi$_2$ dots on Si(001)

I Goldfarb and G A D Briggs

University of Oxford, Department of Materials, Parks Road, Oxford OX1 3PH, UK

ABSTRACT: Nucleation and growth of reactively-deposited CoSi$_2$ were analysed in-situ by scanning tunneling microscopy and surface electron diffraction. Nucleation on flat Si(001) terraces leads to a Volmer-Weber growth, resulting in a mixture of faceted (221)- and flat-topped (001)-oriented three-dimensional nanocrystals. It is imperative to understand the growth modes of this technologically-important silicide, as, when grown with a smooth morphology, it is a primary candidate for epitaxial metallization in ULSI. However the apparent crystalline perfection of the nanocrystals implies another possibility, i.e. of application in quantum-dot devices. Therefore the growth kinetics of CoSi$_2$/Si(001) nanocrystals (dots) was analysed from the evolution of dot size distribution with annealing time. It was found that the dot growth obeys a power law with 1/5 exponent, and is dominated by coalescence. Some parallels with Ge/Si(001) dots can be drawn.

1. INTRODUCTION

It is well known that smooth epitaxial growth of CoSi$_2$ on Si(001) substrates is hampered by misoriented grains (Bulle-Lieuwma *et al* 1992, Adams, Yalisove and Eaglesham 1994). This has been the major obstacle for epitaxial CoSi$_2$ metallization in ultra-large scale integration (ULSI) technology. On the other hand scanning tunneling microscopy (STM) observations of nanometer-size, cristallographically-perfect crystallites (Scheuch, Voigtlander and Bonzel 1997, Goldfarb and Briggs 1999) suggest potential applications in quantum-dot devices. Internal photoemission sensor for infrared radiation, based on CoSi$_2$ nanoparticles embedded in Si, with the six times higher quantum efficiency than in planar CoSi$_2$ Schottky diodes, has been demonstrated (Fathauer *et al* 1990).

Low electrical resistivity and high thermal stability of CoSi$_2$ (Murarka 1983, K. Maex 1993) make it especially attractive as contact and interconnect material, and thus the vast majority of efforts have been devoted to the development of methods for obtaining smooth epitaxial morphology. Far less attention has been paid to the exploration of the nanocrystallites, and the reasons and mechanisms underlying their nucleation and growth. Goldfarb and Briggs (1999) have emphasised the importance of the Si(001)-substrate morphology, and the resulting types of nucleation site, on the CoSi$_2$ morphology. In particular, reactive deposition epitaxy (RDE) of Co onto a flat Si(001) surface results in three-dimensional (3D) nanocrystallites, while RDE onto a vicinal Si(001) surface leads to flatter, 2D morphology. Due to its simplicity, the latter can be an attractive alternative to the popular "template" technique (Stalder *et al* 1992, Adams, Yalisove and Eaglesham 1994, Buschmann *et al* 1998), which involves more processing steps.

In this work we discuss the evolution of the reactively-deposited CoSi$_2$/Si(001) nanocrystals (dots) with annealing time at the growth temperature, and draw parallels with the somewhat analogous Ge/Si(001) dots.

2. EXPERIMENTAL

The nominally flat Si wafers used for this study were n-doped, cut into 1×7 mm^2 pieces and chemically degreased *ex-vacuo*. The samples were handled with ceramic tweezers and clamped to the Ta support on the holder by Ta clamps. In UHV, the samples were degassed for several hours, repeatedly flashed at 1150°C (keeping the pressure below 10^{-7} Pa), quenched, and slowly cooled to the desired temperature of 500°C, to produce well-ordered (2×1) Si surfaces.

A JEOL elevated-temperature STM (base pressure 1×10^{-8} Pa), equipped with Low-Energy-Electron-Diffraction (LEED/Auger) spectrometer and Reflection-High-Energy-Electron-Diffraction (RHEED), and capable of operation up to 1200°C was used to monitor the surface evolution. The constant-current images were obtained using electrochemically etched W tips both in real-time, i.e. during exposure to a Co flux at 500°C, and, occasionally, during anneal, to follow the morphological development of the growing surface. Co was supplied from a water-cooled four-element e-beam source at 45° to the sample mounted in the STM stage. The sample was heated by a direct current. Temperatures were measured by an infrared pyrometer with an accuracy of 30°C.

3. RESULTS AND DISCUSSION

Terrace-nucleation, and even step-edge-nucleation, of cobalt silicide on flat Si(001) substrates leads to a nucleation of 3D nanocrystals (Scheuch, Voigtlander and Bonzel 1997). This is contrary to the step-edge-nucleation of the silicide at the double-height steps on a vicinal Si(001) substrate, where 2D silicide platelets reflect a close-to-equilibrium step separation, as determined by the interactions between force-multipoles at the step-edges (Goldfarb and Briggs 1999). In other words, elastic relaxation of the silicide, which is under tension due to -1.2 % lattice mismatch with silicon, is achieved at the double-height step-edges. On the other hand, when the silicide nucleates on terraces of a flat Si(001) surface, i.e. at some distance from the nearest single-height step edges, the strain can be elastically relieved at the crests of 3D pyramidal islands (see Fig. 1). This relaxation prevails over additional, facial surface energy, and some compression (tension) at the troughs of compressively (tensely) strained islands. The thickness gradient of a pyramid strains the underlying substrate, reducing the original difference in the layer-substrate lattice constants (Tersoff and LeGoues 1994).

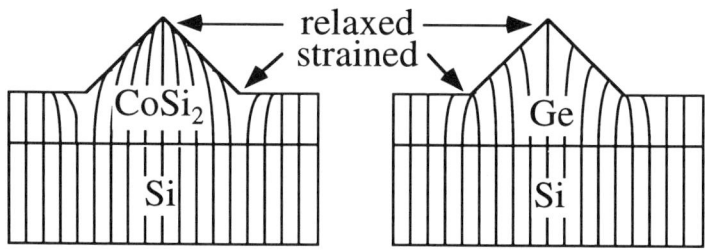

Fig. 1: Sketch showing the strain relaxation (crests) and enhancement (troughs) due to three-dimensional islands. $CoSi_2$/Si is under 1.2 % tension, and Ge/Si is 4.2 % compressed.

An STM image of $CoSi_2$ dots on a flat Si(001) surface is shown in Fig. 2(a), and consists mostly of prismatic {221}-oriented nanocrystals with {111} faces (see inset), but also of small number of flat-topped epitaxial (001)-nanocrystals, which cannot relax strain in the way illustrated in Fig.1. To be useful in quantum-dot devices the nanocrystals must fulfil the requirements of (i) small size, as well as (ii) uniformity of size and shape. Fig. 2(b) shows that, having 7-8 nm mean size, $CoSi_2$ dots fulfil the first requirement. However the control over uniformity is more difficult to achieve in self-assembled processes. Anisotropic elongation of the dots along [110] and [$\bar{1}$10] Si-substrate crystallographic directions, in Fig. 2(a), explains the lack of size uniformity as reflected in the distribution width in Fig. 2(b). Tersoff and LeGoues (1994) have shown that a square-base pyramid represents the equilibrium shape of a prismatic stress-induced nanocrystal at any size, if it is allowed to grow both laterally and vertically. Even assuming the height-invariance, minimum energy conditions require laterally-isotropic growth (Tersoff and Tromp 1993) of $CoSi_2$ nanocrystals up to almost 200 nm size (Brongersma et al 1998). It thus can be concluded that the dot array in the present experiment does not represent an equilibrium configuration.

To learn more about the kinetic factors governing the anisotropic elongation of the dots, we annealed them at the growth temperature, measuring their size distribution and lateral aspect ratio, as a function of annealing time. The anneal caused a significant increase in the both dot mean size and lateral anisotropy, up until 70 hours, after which the mean size and anisotropy sharply decreased (see Fig. 2(c)). As during anneal the mass of the deposited cobalt is conserved, the growth can be described by power law of the form:

where $r(t)$ is the time-dependent dot size, r_0 is the critical nucleus size, k is the kinetic rate constant, t is the annealing time, t_0 is the time to form critical nucleus, and n depends on the dimensionality of the system and the limiting growth mechanism (Goldfarb *et al* 1997). The best fit to the first three data points in Fig. 2(c) yielded $n = 5$.

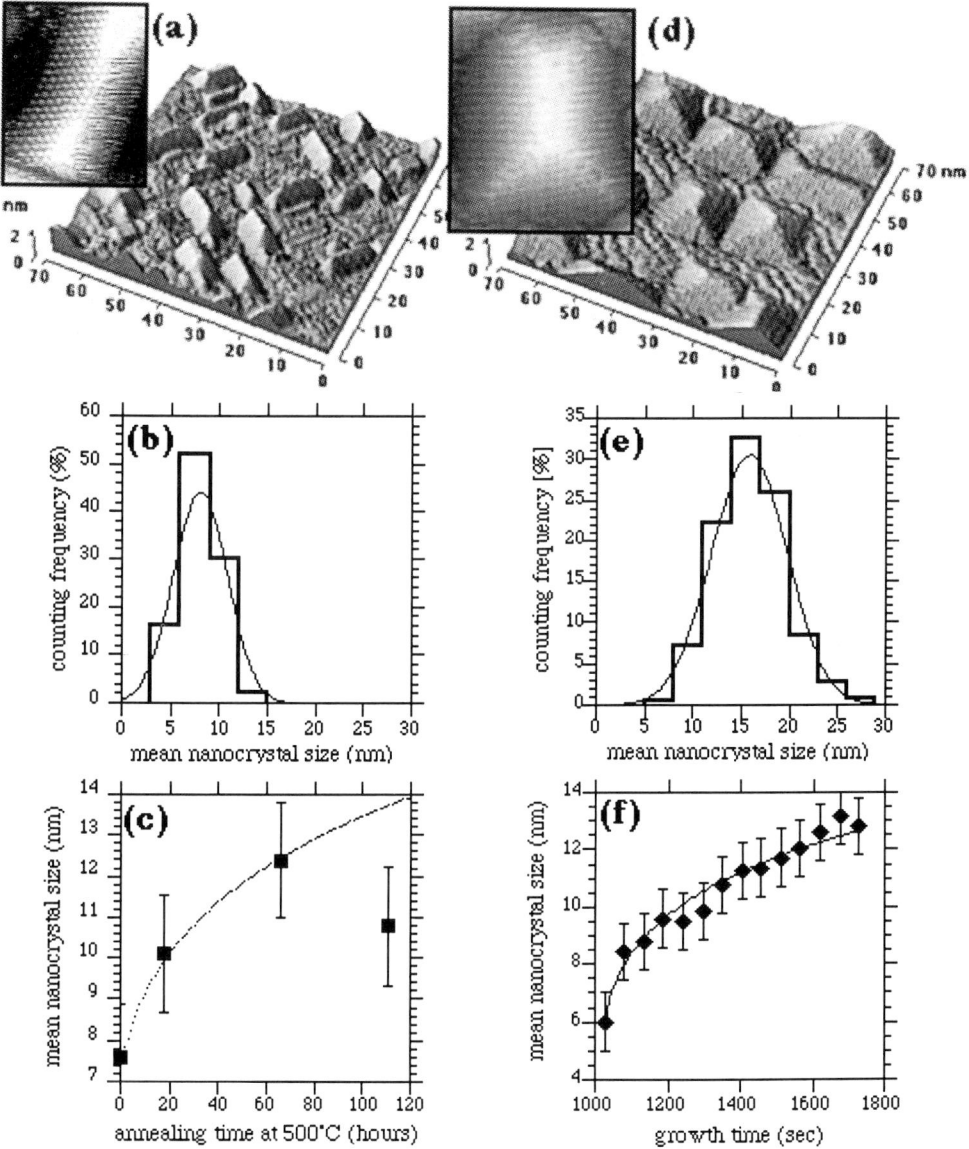

Fig. 2: Typical appearance of $CoSi_2/Si(001)$ (a) dots, (b) dot size distribution, and (c) growth curve. $Ge/Si(001)$ (d) dots, (e) dot size distribution, and (f) growth curve are shown for comparison.

4. CONCLUSIONS

In summary, the resemblance in geometric shapes, size distributions, and growth curves, between $CoSi_2$ (Fig. 2(a)-(c)) and Ge dots (Fig. 2(d)-(f)) on Si(001) invites almost inevitable comparison. The most important observation is that, inherently, $CoSi_2$ dots are half the size of the Ge ones, which makes them even better candidates for quantum-dot devices. Similarity of the growth curves is another striking observation. The similar time exponents, around 1/5, imply similar growth mechanisms, from which it follows that the growth of $CoSi_2$ dots is not controlled by the Co-Si reaction. The fact that the elongated $CoSi_2$ dots are initially metastable, but reach a closer to equilibrium state during anneal (Fig. 2(c)), is also consistent with the elongated Ge dots (so-called "huts"), which transform into stable square-based pyramids upon anneal (Medeiros-Ribeiro et al 1998).

However, there are also some differences. While the Ge dots grow in epitaxial (001) orientation, and their {501} facets are actually 11.2° vicinal (001) surfaces, most of the $CoSi_2$ dots are misoriented {221}-nanocrystals bound by {111} faces. All four {501} facets are well defined in Ge dots, unlike {111} faces in the $CoSi_2$ case. Ge dots are aligned in parallel to the elastically-"soft" [100] and [010] crystallographic directions of the substrate. Although these directions are also the softest in the fluorite lattice of $CoSi_2$, these dots align themselves along [110] and [T10] substrate directions. In spite of the apparent similarity in growth mechanisms, there is an order of magnitude difference between their respective growth rate constants (k_{CoSi_2} = 0.006 sec^{-1}, and k_{Ge} = 0.06 sec^{-1}). In other words, Ge dots elongate much faster than the $CoSi_2$ ones, which may indicate higher activation barriers for $CoSi_2$ dot growth. Alternatively, this difference may be explained by the fact that the $CoSi_2$ curve was obtained during isothermal anneal, i.e. with a negligible Co supersaturation, and that of Ge during deposition (Goldfarb et al 1997).

REFERENCES

Adams D P, Yalisove S M and Eaglesham D J 1994 J. Appl. Phys. **76**, 5190
Brongersma S H, Castell M R, Perovic D D, and Zinke-Allmang M 1998 Phys. Rev. Lett. **80**, 3795
Bulle-Lieuwma C W T, van Ommen A H, Hornstra J, and Aussems J 1992 J. Appl. Phys. **71**, 2211
Buschmann V, Rodewald M, Fuess H, van Tendeloo G and Schäffer C 1998 J. Cryst. Growth **191**, 430
Goldfarb I, Hayden P T, Owen J H G, and Briggs G A D 1998 Phys. Rev. B **56**, 10459
Goldfarb I and Briggs G A D 1999 submitted
Fathauer R W, Iannelly J M, Nieh C W and Hashimoto S 1990 Appl. Phys. Lett. **57**, 1419
Medeiros-Ribeiro G, Kamins T I, Ohlberg D A A and Williams R S 1998 Phys. Rev. B **58**, 3533
Murarka S P 1983 Silicides for VLSI applications (Academic Press, New York)
Maex K 1993 Mater. Sci. Eng. R **11**, 53
Scheuch V, Vöigtlander B and Bonzel P 1997 Surf. Sci. **372**, 71
Stalder R, Schwarz C, Sirringhaus H and von Känel H 1992 Surf. Sci. **271**, 355
Tersoff J and LeGoues F K 1994 Phys. Rev. Lett. **72**, 3570
Tersoff J and Tromp R M 1993 Phys. Rev. Lett. **70**, 2782

Inst. Phys. Conf. Ser. No 164
Paper presented at Microsc. Semicond. Mater. Conf., Oxford, 22–25 March 1999
© 1999 IOP Publishing Ltd

Superconducting contacts to a two-dimensional electron gas

D A Williams, T D Moore* and S B Newcomb**

Hitachi Cambridge Laboratory, Cavendish Laboratory, Madingley Road, Cambridge CB3 0HE U.K.
*Microelectronics Research Centre, University of Cambridge, Cavendish Laboratory, Madingley Road, Cambridge CB3 0HE U.K.
** Department of Materials Science and Metallurgy, University of Cambridge, Pembroke Street, Cambridge CB2 3QZ U.K.

ABSTRACT: Metallic tin-based contacts to a GaAs:AlGaAs heterostructure were made using rapid thermal annealing. The contacts were superconducting at low temperature, and had a relatively high critical magnetic field, allowing the observation of Andreev reflection at high field. The microstructures of the contacts have been examined using cross-sectional transmission electron microscopy.

1. INTRODUCTION

GaAs:AlGaAs heterostructures are attractive substrates for superconductor - semiconductor junctions because of the high electron mobilities obtainable, potentially making the study of ballistic and geometric effects in such junctions more easily fabricable. However, the high-mobility electron channel is inevitably buried, usually 70-100 nm below the semiconductor surface, which makes contact formation difficult. The problem of contact formation to buried GaAs:AlGaAs channels is not limited to the formation of superconducting contacts, but is also of intrinsic importance in the fabrication of devices for the investigation of the quantum Hall effects and other transport physics experiments, and in the fabrication of commercial electronic and optoelectronic devices. Although in the latter case contact resistances remain problematic, in the former, variations in contact behaviour are typically removed by performing four-terminal measurements, and the microscopic nature of charge transport at the interfaces has not been extensively studied. In the case of superconducting contacts, however, and the study of Andreev reflection processes in particular, the microscopic nature of the contact and charge transport across it are of crucial importance(For a comprehensive review of superconductor – semiconductor transport physics, see Lambert and Raimondi 1998).

Charge transport through superconductor - semiconductor junctions depends upon the superconductor energy gap, the elastic and inelastic scattering lengths of electrons in the semiconductor, and the interface structure. Andreev reflection, whereby electrons from the semiconductor pair and tunnel into the superconductor ground state, is of particular current interest, as it can be used to couple the pair state to high-mobility mesoscopic semiconductor structures. The probability of Andreev reflection is strongly dependent on the interface properties, and variations lead to a large variety of electrical characteristics.

2. CONTACT FORMATION

The samples are formed by the deposition of tin-gold contacts, separated by a gap defined by electron-beam lithography, and constrained by a silicon nitride and a chromium cap. The sample is then sintered in an electron beam rapid annealer. This allows the precise optimisation of the annealing process to form electrically transparent contacts for the study of interface charge transport. The channel length between contacts is typically 2µm and width is 200 µm. The substrate is a high-mobility GaAs:AlGaAs heterostructure, with an electron density of $3.32 \times 10^{11} \text{cm}^{-2}$ and a low-temperature mobility of $3.34 \times 10^{6} \text{cm}^{2}\text{V}^{-1}\text{s}^{-1}$. A layer of tin is evaporated and then capped in-situ with gold to prevent oxidation during specimen transfer. A layer of silicon nitride is then sputtered, a refractory layer of chromium and gold evaporated, and the structure is sintered using rapid electron-beam annealing with a time at peak temperature of ~10ms. A thin layer of aluminium oxide, not wetted by tin, is used to prevent contact flow across the gap during sintering.

This structure was designed to improve the repeatability of the electrical characteristics of the sintered contacts. The combination of refractory chromium and silicon nitride above the tin, and aluminium oxide between the contacts, keeps the tin in place during annealing and ensures uniformity of alloying. The layer of silicon nitride is used as a diffusion barrier to prevent chromium, which can be responsible for pair-breaking scattering, mixing and alloying with the tin and the semiconductor during annealing. The critical temperature, T_C and the critical magnetic field, H_{CII} of the superconducting alloy contact material vary between specimens, lying in the ranges 4-7.5 K and 3-6.5 T respectively.

3. ELECTRICAL CHARACTERISTICS

Electrical contacts were made to the tin layer to allow four-terminal transport measurements. These were made at temperatures down to 300 mK in a Helium III refrigerator, using lock-in techniques, with a perpendicular magnetic field of up to 6T. The samples were found to show Andreev reflection, with a probability increasing with magnetic field at moderate fields. Figure 1 shows the current-voltage characteristics of such a device, taken at 300mK with an applied field of 2T. The differential resistance is also shown, and the dip around zero bias is characteristic of the excess conductance due to Andreev reflection, and corresponds to the super-ohmic kink in the I-V curve. A plot of the magnitude of this dip, which is related to the Andreev reflection probability, shows that the probability first *increases* then decreases with magnetic field. This was the first observation of Andreev reflection at these field strengths, and the result contradicts previous predictions of a monotonic *decrease* in Andreev reflection probability with increasing field. A full set of electrical characteristics and a detailed discussion of the mechanisms involved is given elsewhere(Moore and Williams 1999).

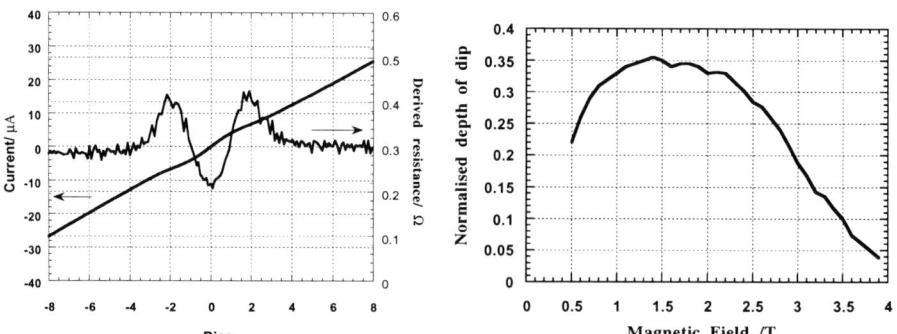

Figure 1. Left – Current-voltage characteristics and differential resistance of a device at 2T and 300mK. **Right** – magnitude of the dip as a function of applied magnetic field.

4. MICROSTRUCTURE

The contacts were thinned to electron transparency using focused ion beam milling, and examined in cross-section by transmission electron microscopy. Several distinct features were noted, and it was found that the microstructure differed somewhat from similar contacts formed with a less complex layer structure(Marsh and Williams 1996). Figure 2 shows a typical micrograph of the contact cross-section. The AlGaAs layer was found to be intact, except in two regions at the very edges of the metallic cover. The capping GaAs layer was not observed after annealing, but pyramidal GaAs structures were seen spread over the AlGaAs layer. The first metal layer was found to be Sn_4Au, as expected, but was highly ordered, rather than being the expected polycrystal. The capping layer of silicon nitride was intact, and the upper layers of chromium and gold were polycrystalline with a similar morphology to the unannealed state.

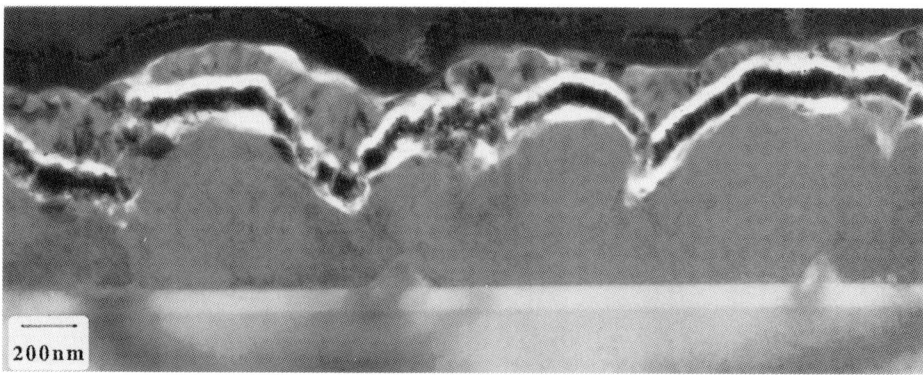

Figure 2. Cross-sectional transmission electron micrograph of the metallic part of the contact structure after annealing. The uneven morphology can easily be seen, along with the capping layer of silicon nitride, and the pyramidal GaAs structures.

The formation of the microstructures described above can be explained as follows: On deposition, the tin forms an uneven polycrystalline layer, which is covered conformally by the upper layers. During annealing, the refractory silicon nitride and chromium layers retain this shape and mould the underlying metal. The tin melts, and alloys with the gold to form Sn_4Au. The two layers were deposited to give this stoichiometric ratio, as it had been noted in previous experiments that Sn_4Au is a high H_{CII} superconductor with a higher T_C than pure tin. These features have all been observed in previous experiments with simpler layer structures(Marsh and Williams 1996)

A typical region of the boundary formed between the Sn and the AlGaAs is shown in a bright field micrograph in Figure 3. Here Fresnel effects associated with local changes in scattering potential have been used to enhance the visibility of the pores formed at the interface between the two layers. A mix of large and small pores can be seen, as indicated by the regions marked A and B. It was also observed that the upper surface of the metal layer where it is considerably thinner than elsewhere exhibits the formation of a relatively well developed "finger" of porosity that extends for a significant depth into the metal. Although there is some evidence for small localized misorientations in the metal at its upper surface, in general it is clear that it is not poly-crystalline. Both spherical and other pores, which are rather more loosely defined in shape, can be seen in the bulk regions of the Sn_4Au.

Figure 4 shows a micrograph of the $GaAs:Sn_4Au$ interface, and the pyramidal structures at the upper AlGaAs interface. The former interface forms at a break in the AlGaAs layer, and is very abrupt. Its formation is a result of the heat flow and heat loss in vacuo during and immediately after the ultra-rapid annealing cycle, and is believed to be the origin of epitaxial seeding of the Sn$_4$Au layer, leading to a near-single crystal metal layer on the surface. A detailed analysis of this process will be published elsewhere.

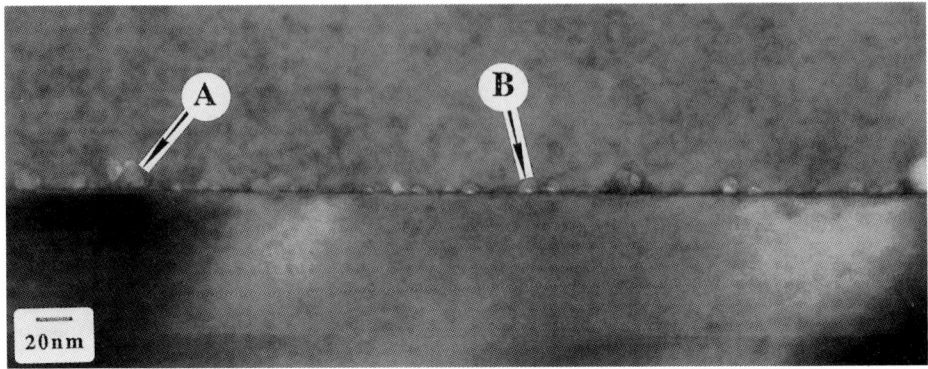

Figure 3. Porosity at the interface between the AlGaAs insulator and the Sn4Au metal layer.

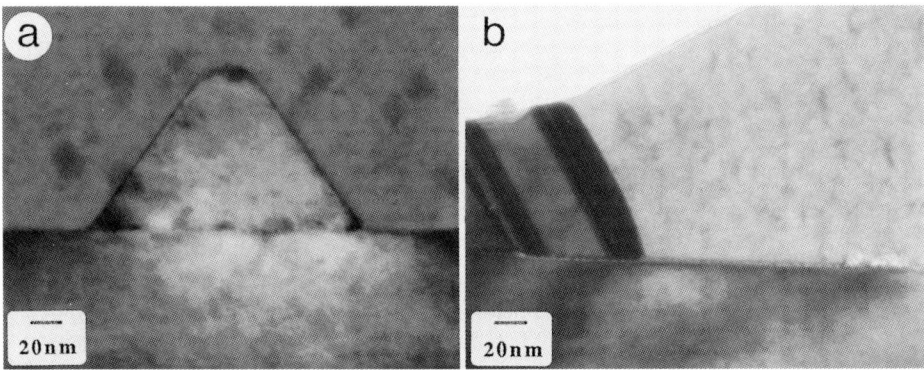

Figure 4. a) A pyramidal GaAs region above the AlGaAs layer and below the metallic region. b) The interface between the metallic and semiconductor regions of the contact.

Cross-sectional and planar scanning electron microscopy was also used to verify that the microstructures identified in TEM were typical of all such devices. Although detailed crystallographic information is not obtainable, a reasonably detailed match of morphology is usually possible between the two methods, and the simpler specimen preparation and faster observation in SEM allows a statistically representative sampling of many specimens

REFERENCES

Lambert C J and Raimondi R 1998 J. Phys: Cond. Mat. **10**, 901
Marsh A M and Williams D A 1996 J. Vac. Sci. Technol. A **14**, 2577
Moore T D and Williams D A 1999 Phys. Rev. B **59**, 7308

Inst. Phys. Conf. Ser. No 164
Paper presented at Microsc. Semicond. Mater. Conf., Oxford, 22–25 March 1999
© 1999 IOP Publishing Ltd

561

Microstructural investigation of oxidized Ni/Au ohmic contact to p-type GaN

Li-Chien Chen[1], Fu-Rong Chen[1], Ji-Jung Kai[1], Li Chang[2], Jin-Kuo Ho[3], Charng-Shyang Jong[3], Chien C Chiu[3], Chao-Nien Huang[3], Chin-Yuen Chen[3] and Kwang-Kuo Shih[3]

[1]Department of Engineering and System Science, National Tsing Hua University, Hsinchu, Taiwan 300, Republic of China.
[2]Department of Material Science and Engineering, National Chiao Tung University, Hsinchu, Taiwan 300, Republic of China.
[3]Opto-Electronics and Systems Laboratories, Industrial Technology Research Institute, Chutung, Hsinchu, Taiwan 310, Republic of China

ABSTRACT: The phase evolution of oxidized (10nm) Ni/(5nm) Au films on p-GaN was examined with a field emission gun transmission electron microscope in conjunction with composition analyses to explore the mechanism of formation of low resistance ohmic contact to p-GaN. The p-GaN/Ni/Au sample heat treated above 500°C in air was mainly composed of a mixture of crystalline NiO, Au and amorphous Ni-Ga-O phases. Small voids adjacent to the p-GaN film were also found. Moreover, NiO partially contacts to p-GaN as well as Au islands and the amorphous Ni-Ga-O phases. The results suggest that the crystalline NiO and/or amorphous Ni-Ga-O phases may play a significant role in the formation of a low resistance ohmic contact to p-GaN.

1. INTRODUCTION

Gallium nitride based semiconductors are among the most promising material for optoelectronic applications such as light emitting diodes (LEDs) and laser diodes (LDs) in the blue and ultraviolet ranges. These optoelectronic devices need low resistance ohmic contacts to both n- and p-type GaN to enhance their performance. In contrast to extensive attention paid to the n-type GaN contact, a low resistance ohmic contact to p-type GaN has achieved less success. Multilayer metallic films with high work function such as Ni/Au (Mohs et al 1995; Nakamura et al 1996; Ishikawa et al 1997; Sheu et al 1998; Ho et al 1999a, 1999b), Pt/Au (King et al 1997) and Pt/Ni/Au (Jang et al 1998) are comprehensively used as contacts to p-type GaN nowadays. However, the specific contact resistance to p-GaN has been restricted to 10^{-2} to 10^{-4} Ω–cm^2. Such a high value is insufficient for high performance device applications.

Recently Ho et al (1999a, 1999b) have successfully developed a novel contact to p-type GaN with specific contact resistance lower than 1×10^{-4} Ω–cm^2 by oxidizing a (10nm)Ni/(5nm)Au metallization scheme at 500°C (Ho et al 1999a). A higher or lower annealing temperature will degenerate the electric properties. Later, they use the same technique to achieve an ultra low specific contact resistance as 5×10^{-6} Ω–cm^2 (Ho et al 1999b). This ultra low contact resistance is approximately two to three orders of magnitude lower than ever previously reported (Mohs et al 1995; Ishikawa et al 1997; King et al 1997; Jang et al 1998; Sheu et al 1998). The excellent electrical properties have led us to investigate in detail the microstructure of as-deposited and oxidized Ni/Au films on p-GaN, to analyse the phase evolution of alloys correlating with low contact resistance. The reaction mechanism and diffusion behaviour are also discussed.

2. EXPERIMENTAL

The sapphire/undoped GaN(2 μm)/Mg-doped GaN(2 μm) samples were prepared by metal-organic chemical vapour deposition (MOCVD). Hall measurements showed that the hole concentration of p-GaN was $2x10^{17}cm^{-3}$. Prior to metal deposition, the samples were etched with $HCl:H_2O = 1:1$ for one minute to remove native oxide, and then 10 nm Ni and 5 nm Au were deposited on the p-type GaN as a metal contact by electron beam evaporation under a base pressure lower than 3×10^{-6} torr. Finally, the samples were oxidized in air at 300°C, 500°C, and 600°C for 10 minutes to induce the interdiffusion of bi-layer metals (Ni and Au) and the reaction with oxygen. Microstructural and compositional analyses were performed by using a JEOL 2010F field emission gun TEM (FETEM) equipped with an Oxford energy dispersive X-ray spectrometer (EDS). Crystal structure identification and EDS analysis were carried out using a 0.5 nm nanobeam probe. Cross-sectional TEM samples were prepared by the tripod polishing method and ion milling.

3. RESULTS AND DISCUSSION

The as-deposited Ni and Au films were polycrystalline with small grain size of a few nanometers, as shown in Fig. 1. The Ni layer was grown in columnar-like structure on p-GaN with the preferred orientation $(001)_{Ni} //(0001)_{GaN}$. The subsequently deposited Au layer epitaxially aligned with the Ni layer with an ambiguous interface. No interlayer between Ni and p-GaN was observed. This finding indicates that the sample pre-cleaning and subsequent deposition brought no contamination or reaction between the interface of Ni and p-GaN. However, in current-voltage measurements, the as-deposited sample showed a rectifying characteristic (Ho et al 1999a). Heat treatment of the as-deposited samples in N_2 or forming gas did not modify the rectifying characteristic.

Fig. 2 (a), (b) and (c) demonstrate the typical microstructure of the oxidized sample annealed at 300°C, 500°C, and 600°C in cross-section view, respectively. As Fig. 2 (a) demonstrates, the Ni/Au bi-layer oxidized at 300°C retained the same structure as the as-deposited sample in most of the areas. However, some reaction products were observed on the surface of the metal layer. This result indicates that the 300°C sample showed the initial stage of the oxidation reaction and interdiffusion of Ni and Au.

With oxidizing treatment at 500°C in air, the bilayer metals transformed to a new microstructure as shown in Fig. 2(b), which made the contact ohmic with a low interface resistance. Discrete crystalline islands with darker contrast were observed, and an amorphous phase and small voids existed among the crystalline grains. A non-uniform crystalline thin film with uneven surface covered these features. The 600°C heat treatment results in a similar structure to that at 500°C However, a different feature of the 600°C sample is the distribution of increasingly large voids adjacent to the p-GaN, as shown in Fig. 2(c).

In this study, high-resolution TEM, EDS analysis and nanobeam diffraction were performed to examine the composition and microstructure of the 500°C sample in detail, regarding its superior electrical properties. The results are shown in Fig. 3 and Fig. 4. Fig. 3(a) is a magnified micrograph of the typical oxidized sample. The EDS spectra corresponding to the points marked "a", "b" and "c" in Fig. 3 are shown in Fig. 4. Fig. 4(a) is the EDS spectrum from point "a" of the discontinuous island of Fig. 3 (a), which indicates that the island primarily consisted of Au with a small amount of Ni. The diffraction pattern from the same point is shown in Fig. 3 (b) and is consistent with the Au structure. The results identify the discrete islands as Au-rich grains. On the other hand, the crystalline film above the Au islands was mainly composed of Ni and O as shown from the EDS spectrum in Fig.4 (b), which was obtained from the point 'b' in Fig. 3 (a). The film was further confirmed as face-centred-cubic NiO phase with the nanobeam diffraction pattern shown in Fig. 3 (c). The EDS analysis, as illustrated in Fig.4 (c) from point 'c' of Fig. 3 (a), shows that the amorphous regions consisted of a relatively large amount of Ga as well as Ni and O. The orientation relationships among the NiO film, the gold islands and the p-GaN are determined as $(111)_{NiO} //(11\bar{1})_{Au} //(0001)_{GaN}$ and $[1\bar{1}0]_{NiO} //[1\bar{1}0]_{Au} //[1\bar{1}20]_{GaN}$.

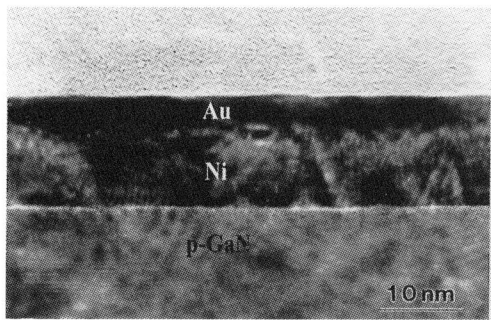

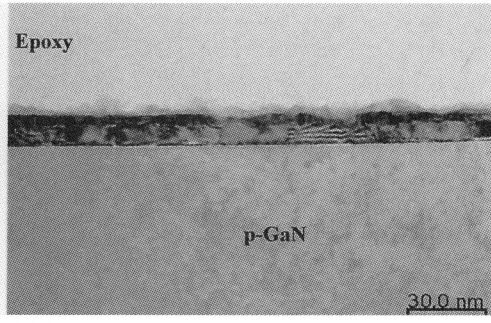

(a)

Fig. 1 A typical cross-sectional TEM image of the as-deposited p-GaN/Ni/Au sample.

The above results have the following implications. During annealing at 500°C in air, interdiffusion of Ni and Au atoms occurred. It is conceivable that grain boundaries (Richards and McCann 1969) of the as-deposited Au film may have served as quick diffusion channels for out diffusion of Ni atoms onto the surface at this temperature. The driving force for the Ni atom diffusion out through the Au layer results from the stronger affinity of Ni to O than of Au to O. During the out diffusion of Ni atoms to the surface, interdiffusion between Ni and Au atoms at the interface also occurred and could lead to the formation of Ni-Au alloy. According to the Ni-Au phase diagram (Okamoto and Massalski 1990), the equilibrium concentration of Au-Ni alloy at 500°C is 5%NI-Au or 93%Ni-Au as a result of a miscibility gap. The EDS quantification shows that the concentration of Ni in Au-rich islands was much less than 5% with essentially no Ni being detected. It indicates that the Ni in the Ni-Au alloy was steadily consumed due to oxidation of Ni at the surface. Thereby little Ni was left in the Au-rich islands and a NiO film was constructed on top of them.

After oxidation, the Au-rich islands were sandwiched by the NiO and GaN, which resulted in formation of two new interfaces of Au/NiO and GaN/Au. The average thickness of Au-rich islands also changed to approximately 10 nm, that is double to the as-deposited thickness. Small voids appeared as a result of conservation of volume of Au. The formation of voids reduced the contact area of metal to p-GaN after oxidation and could have degraded contact performance.

(b)

(c)

Fig. 2 Cross-section TEM micographs of the oxidized p-GaN/Ni/Au samples annealed at (a) 300°C, (b) 500°C and (c) 600°C for 10 min in air.

Consequently the authors believe that, during oxidation, some Ni atoms first diffused into the Au layer enriching the grain boundaries and then propagated to the surface to react with oxygen to form NiO. At the same time, the Ni left behind reacted with GaN to form Ni-Ga alloy, Ga_4Ni_3 and Ga_3Ni_2 (Sheu et al 1998; Bermudez et al 1993). When the NiO grew down to contact the Ni/GaN reaction products such as Ni-Ga alloy, Ga_4Ni_3 and Ga_3Ni_2, a new oxidation reaction took place and amorphous Ni-Ga-O phase formed with the liberation of nitrogen to the air or the voids.

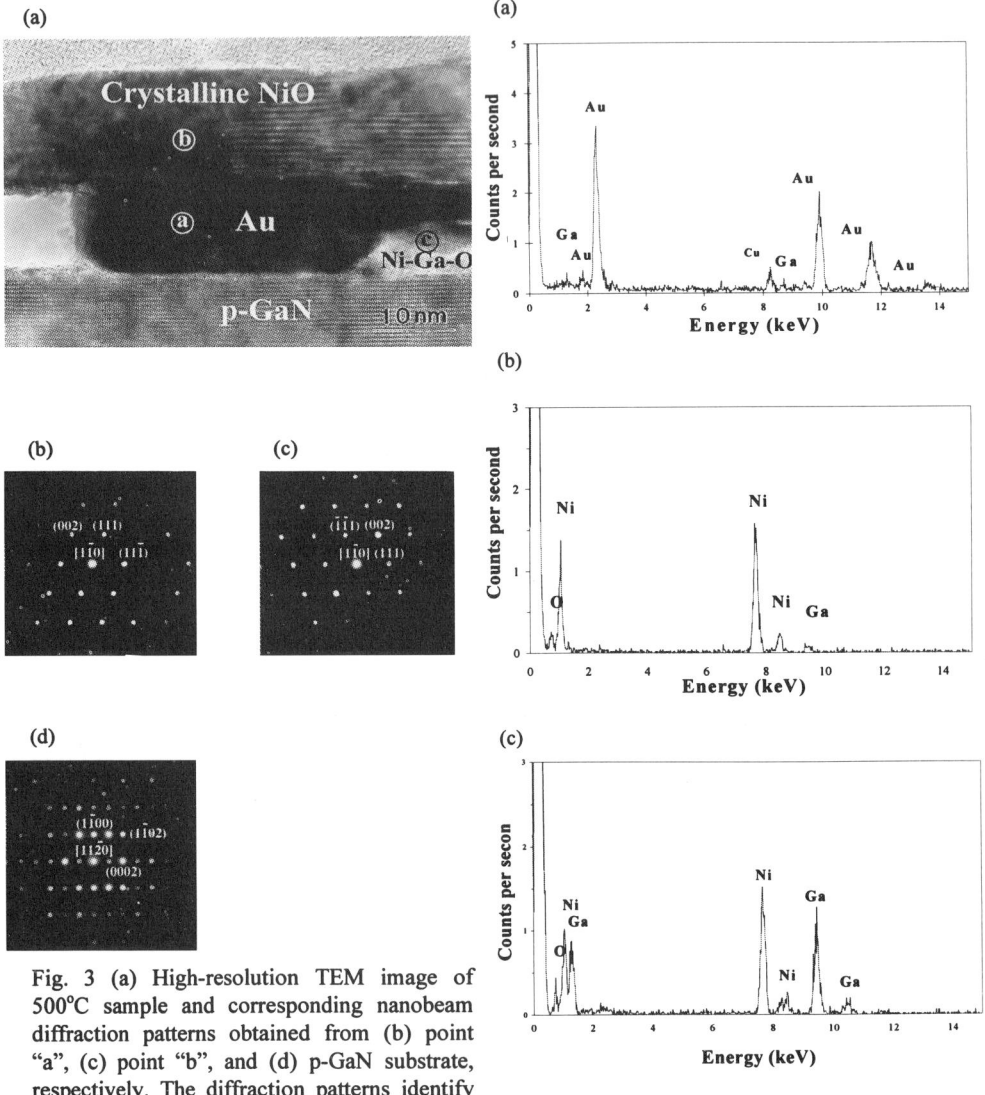

Fig. 3 (a) High-resolution TEM image of 500°C sample and corresponding nanobeam diffraction patterns obtained from (b) point "a", (c) point "b", and (d) p-GaN substrate, respectively. The diffraction patterns identify the phases as (b) Au, (c) NiO in $[1\bar{1}0]$ zone axis, and (c) GaN in $[11\bar{2}0]$ zone axis.

Fig. 4 EDS spectra obtained from (a) point "a", (b) point "b" and (c) point "c" of Fig. 3 (a), respectively.

Although the stoichiometric crystalline NiO is an insulator, the NiO was found to be a p-type semiconductor while increasing Ni^{3+} ions by introducing nickel vacancies and/or interstitial oxygen (Adler 1968; Sato et al 1993). The low specific contact resistance of the oxidized Ni/Au metallization scheme was attributed to the high conductivity of Au islands and low contact barrier of p-NiO to p-GaN. However, in this investigation, we found two products, the amorphous Ni-Ga-O phase and crystalline NiO, contacting with p-GaN. Whether either one or both of these products were responsible for reducing the interface resistance is still unclear at this moment. Further studies will continue to reveal the mystery of formation of low resistance ohmic contacts to p-GaN.

4. CONCLUSIONS

Very low specific contact resistance to p-type GaN (lower than 1×10^{-4} Ω cm^2) was attained by oxidizing (10nm)Ni/(5nm)Au bilayer metallization. The microstructural evolution of this ohmic contact to p-type GaN has been investigated. During oxidation, nickel diffused out through the gold layer to react with oxygen and thus formed crystalline NiO. Nickel was steadily consumed in the oxidizing reaction. The epitaxial orientation relationship of crystalline NiO, Au-rich island and p-type GaN were determined as $(111)_{NiO} // (11\bar{1})_{Au} // (0001)_{GaN}$ and $[1\bar{1}0]_{NiO} // [1\bar{1}0]_{Au} // [11\bar{2}0]_{GaN}$. Besides crystalline NiO, an amorphous Ni-Ga-O phase was also found. The results suggest the crystalline NiO and/or amorphous Ni-Ga-O phase may play a significant role in lowering ohmic contact resistance to p-type GaN.

REFERENCES

Adler D 1968 Solid State Physics (Academic, New York,) p. 21
Bermudez V M, Kaplan R, Khan M A and Kuznia J N 1993 Phys. Rev. B **48**, 2436
Ho J K, Jong C S, Huang C N, Chen C Y, Chiu C C and Shih K K 1999a Appl. Phys. Lett. **74**, 1275
Ho J K, Jong C S, Huang C N, Chen C Y, Chiu C C and Shih K K 1999b Submitted to Appl. Phys. Lett.
Ishikawa H, Kobayashi S, Koide Y, Yamasaki S, Nagai S, Umezaki J, Koike M and Murakami M 1997 J. Appl. Phys. **81**, 1315
Jang J S, Kim H G, Park K H, Um C S, Han I K, Kim S H, Jang H K and Park S J 1998 Mat. Res. Soc. Symp. Proc. **482**, 1053
King D J, Zhang L, Ramer J C, Hersee S D and Lester L F 1997 Mat. Res. Soc. Symp. Proc. **468**, 421
Mohs G, Fluegel B, Giessen H, Tajalli H and Peyghambarian N 1995 Appl. Phys. Lett. **67**, 1515
Nakamura S, Senoh M, Nagaham S, Iwasa N, Matsushuta T, Kiyiku H and Sugimoto Y 1996 Appl. Phys. Lett. **68**, 3269
Okamoto H and Massalski T B 1990 Alloy Phase Diagrams (ASM, Ohio,) p. 289
Richards J L and McCann W H 1969 J. Vac. Sci. Technol. **6,** 644
Sato H, Minami T, Takata S and Yamada T 1993 Thin Solid Films **236,** 27
Sheu J K, Su Y K, Chi G C, Chen W C, Chen C Y, Huang C N, Hong J M, Yu Y C, Wang C W and Lin E K 1998 J. Appl. Phys. **83**, 3172

Inst. Phys. Conf. Ser. No 164
Paper presented at Microsc. Semicond. Mater. Conf., Oxford, 22–25 March 1999
© 1999 IOP Publishing Ltd

Microstructure of TiN layers used as ohmic contacts on GaN

P Ruterana, G Nouet, Th Kehagias[1], Ph Komninou[1], Th Karakostas[1], M A di Forte Poisson[2], F Huet[2], M Tordjman[2] and H Morkoç[3]

Laboratoire d'Etudes et de Recherches sur les Matériaux , UPRESA 6004 CNRS, Institut des Sciences de la Matière et du Rayonnement, 6 boulevard Maréchal Juin, 14050 Caen Cedex, France.
1. Aristotle University, Physics Department, 54006 Thessaloniki, Greece
2. Thomson-CSF/Laboratoire Central de Recherches, Domaine de Corbeville, 91404 Orsay Cedex , France
3. Virginia Common Wealth University, Department of Electrical Engineering, Richmond, VA 23284-3072, USA
email: ruterana@lermat8.ismra.fr

ABSTRACT: TEM investigations have been carried out on annealed Al/Ti/GaN and directly deposited TiN films on GaN. In the Al/Ti/GaN system, the annealing step gave rise to a ~ 20 nm rough TiN film. The ohmic contact is made by TiN crystallites which are epitaxially related to GaN, the orientation relationship being (0001)GaN//(111)TiN. In between, amorphous patches of 1 to 2 nm extension can be found. When stoichiometric TiN was deposited directly on GaN, we obtained columnar TiN grains of 10-20 nm section, which cross the whole film thickness and are rotated mostly around the [111] axis. The epitaxial relationship is similar and no amorphous patches are observed at the interface.

1. INTRODUCTION

Successful fabrication of GaN based devices needs reliable ohmic contacts to be attained. The performance of the optoelectronic devices and transistors depends critically on the contact resistance. Hence, the development of reliable ohmic contacts on GaN is of great practical importance. Until now, the best results include Ti as the first layer deposited on the GaN surface (Lin et al 1993, Fan et al 1996). In the case of Ti/Al, a thin layer of titanium (20 nm) is first deposited and covered by aluminium to a total thickness of 100-200nm. The ohmic contact forms during a subsequent anneal at temperatures which can be as high as 900°C (Ruvimov et al 1996). The ohmic contact has been shown to be due to a thin TiN film from the reaction of Ti with GaN. In this process, resistivity in the range of 10^{-5} ohm.cm is obtained upon 10^{18} cm^{-3} silicon doped GaN layers. During the annealing steps many reactive phases form between Al, Ga, Ti and N which may be a problem for the quality of the contact. Moreover, the formation of TiN may lead to a loss of N from the GaN surface layer, as well as to an excess of Ga in the reactive area. One of the ways to avoid these drawbacks would be to deposit TiN directly on the GaN surface. In this work, a comparative TEM investigation has been carried out on annealed Al/Ti/GaN and directly deposited TiN films on GaN.

2. EXPERIMENTAL

Two types of specimens were investigated. In one case, the Ti/Al films were deposited on clean GaN surfaces first as 20 nm Ti covered by a 80nm Al film in order to avoid mostly the interaction of

568

titanium with air. They were then submitted to a rapid thermal annealing for 10 seconds at 500°C. In the other case, TiN_x contacts were deposited at room temperature on the GaN films by dc reactive magnetron sputtering using an Alcatel SCM 600 deposition system equipped with an *in situ* ultrafast spectroscopic ellipsometer. Details of the deposition system are presented elsewhere (Dimitriadis et al. 1995) First, the samples were cleaned in air with a $HCl:H_2O$ (1:1) solution and then, after loading them into the deposition chamber, a very low energy (~ 5 eV) Ar^+ ion dry etching was used to remove the native surface oxide just before the deposition. The working gas was Ar (99.999%) and N_2 (99.999%) was used as the reactive gas. The deposition condition of the substrate bias voltage V_b was -40 or -120 V and the N_2 flow rate was about 2 sccm. The Ar flow was kept constant at 15 sccm and at a partial pressure of 7.5×10^{-3} mbar. In-situ spectroscopic ellipsometry investigations have shown that TiN films deposited at $V_b = -120$ V are stoichiometric, while for $V_b = -40$ V the TiN_x films are non-stoichiometric (Dimitriadis et al. 1995). TEM samples were prepared in the conventional way by mechanical grinding followed by ion milling. HREM observations were carried out on an Topcon 002b electron microscope operated at 200 kV with a point to point resolution of 0.18 nm

3. RESULTS

In the annealed Ti/Al films, the ohmic resistivity was measured to be 10^{-5} ohm.cm which is consistent with the silicon doping level of $10^{18} cm^{-3}$. Conventional TEM shows a polycrystalline metallic layer which has a total thickness of nearly 150nm.

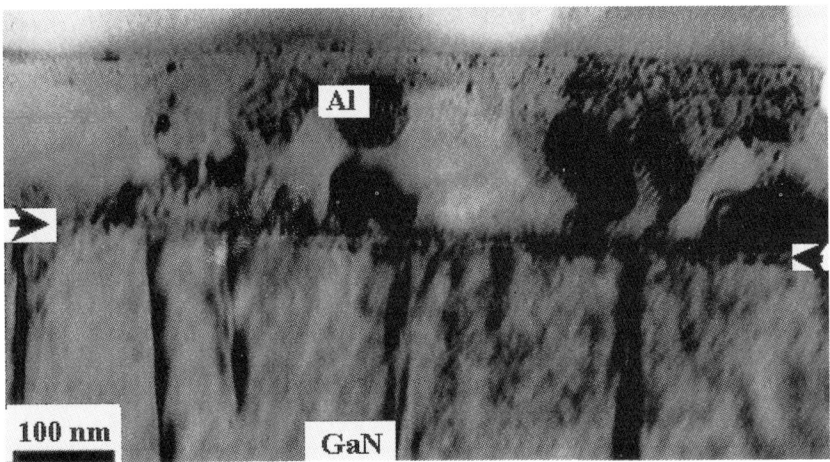

Fig. 1 Morphology of low temperature annealed Ti/Al contact.

At this stage, the interfacial layer is not visible and the Ti containing layer can be estimated as about 10 nm (fig. 1, as shown by arrows). The thick metal layer exhibits only Al composition, so there has been no reaction between the two metals, probably due to the low temperature (500°C) anneal. At higher magnification, one starts to notice the reative area. The GaN surface is flat, whereas the TiN layer is 2 nm thick and is rough. In areas with large crystallites (5-10 nm), only the {111} TiN lattice fringes are parallel to (0001) GaN, meaning that the grains have only the <111> direction parallel to the [0001] GaN (fig. 2).

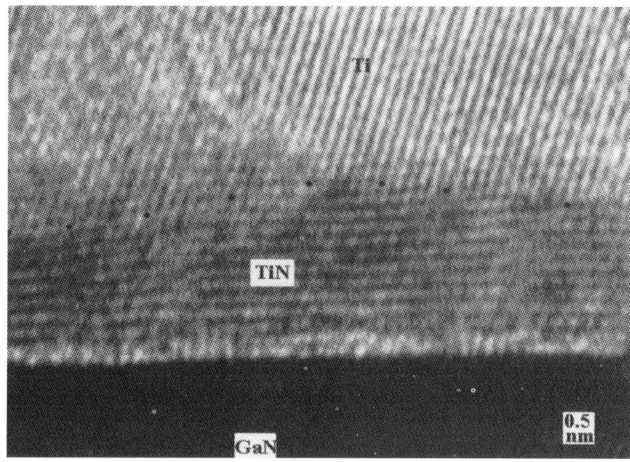

Fig. 2 Interfacial area with a continuous and rough TiN layer.

It was then possible to find nanometric areas where the $<11\bar{2}0>$ GaN and $<1\bar{1}0>$ of TiN directions are parallel, leading to the now well known epitaxial relationship.

Unfortunately, these areas are always small and can be separated by patches which exhibit amorphous contrast in the HREM images. If they are made of amorphous material due to surface contamination or non completed reaction between GaN and Ti, one may suspect them to constitute points of non optimum ohmic contact, which may lead to discontinuities in current flow. For the directly deposited TiN layers, we obtained a perfect columnar structure. As seen on the diffraction pattern (fig.4) recorded along the $<11\bar{2}0>$ zone axis of GaN, the 111 streaks are quite narrow. The TiN crystallite section is 3-5 nm and they are strongly twinned as can be seen by most of the vertical fringes. They mainly have their {111} planes parallel to the (0001) GaN as in the conventional epitaxial relationship (fig. 5).

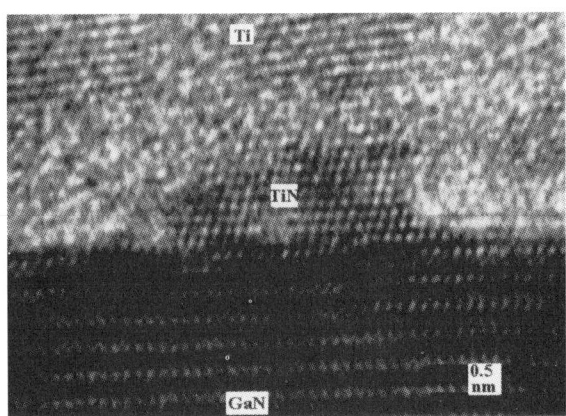

Fig. 3 Nanometric TiN crystal, with (111) lattice fringes parallel to GaN (0001).

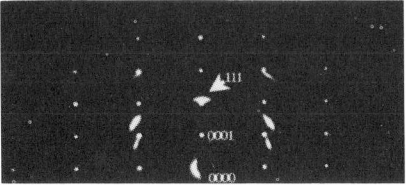

Fig. 4 Diffraction pattern along $[11\bar{2}0]$ showing the small misorientation of the {111} TiN atomic planes (arrow).

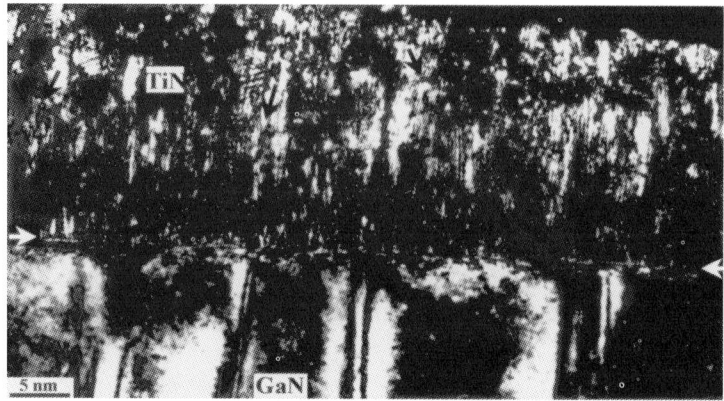

Fig. 5 TiN layer, directly obtained by deposition. Interface is shown (white arrows) . Twin fringes are visible in the TiN layer (black arrows), as well as the columnar growth.

HREM observations of the interface indicate that the transition between GaN and TiN crystallites is flat, with the (0001) GaN // (111) TiN atomic planes as in the case of RTA samples. However no amorphous patches could be noticed at this interface.

Electrical I/V measurements show that this deposition gives rise to the formation of an ohmic contact even at the deposition temperature (Dimitriadis et al. 1999). The next step is to investigate the effect of low temperature RTA and to make resistivity measurements upon this system.

4. CONCLUSION

Although it has been known that the best ohmic contacts to n type GaN is made by TiN formation, the latter is usually made by deposition af Ti and Al, followed by a 900°C anneal. This results in the formation of more compounds that the necessary TiN, which may constitute a drawback for device performance. In this work we have shown that, although the low temperature annealing may be an alternative, one is left with amorphous patches at the interface. Therefore the deposition of stoichiometric TiN may be one solution and work is going on in order to optimize this material.

ACKNOWLEDGEMENT

This work was done within a collaboration of CNRS and NHRF (Greece) under project number 6919.

REFERENCES

Dimitriadis C A, Logothetidis S and Alexandrou I 1995 Appl. Phys. Lett. **66**, 502
Dimitriadis C A private communication
Fan Z, Mohammad S N, Kim W, Actas O, Botchkarev AE and Morkoç H 1996 Appl. Phys. Lett. **68**, 1672
Lin M E, Ma F, Huang F Y, Fan Z F, Allen L H and Morkoç H 1994 Appl. Phys. Lett. **64**, 1003
Ruvimov S, Liliental-Weber Z, Washburn J, Dustad K J, Haller EE, Fan Z F, Mohammad S N, Kim W, Botchkarev A E and Morkoç H 1996 Appl. Phys. Lett. **69**, 1556

Inst. Phys. Conf. Ser. No 164
Paper presented at Microsc. Semicond. Mater. Conf., Oxford, 22–25 March 1999
© 1999 IOP Publishing Ltd

Comparison of the growth mode of intimate In contacts deposited at room and cryogenic temperatures on (100) InGaAs and InAlAs self-decomposed layers

F Peiró[1], A Vilà[1], A Cornet[1], D S Cammack[2] and S A Clark[2]

[1]EME Electronic Materials and Engineering, Dept. Electronics, University of Barcelona, Martí i Franquès 1, 08028 Barcelona, Spain.
[2]Materials Research Institute, Sheffield Hallam University, Sheffield S1 1WB, UK.

ABSTRACT: In this work we present the characterisation of the growth mode of a metallic layer from high purity In metal evaporated on top of (3x1) reconstructed $In_{0.52}Al_{0.48}As$ and $In_{0.53}Ga_{0.47}As$ layers grown by Molecular Beam Epitaxy on InP(100) substrates. These ternary compounds exhibit both vertical and lateral decomposition, the latter inducing a significant surface roughness prior to metallisation. The formation of a homogenous In layer, the nucleation of In islands and the appearance of mixed InAs-In pure islands depending on the state of the ternary buffer is discussed based upon TEM, HREM, AFM, and XPS measurements.

1. INTRODUCTION

The investigation of the intimate configuration of metal–semiconductor contacts is of special relevance for ultimate device performance. The morphological features and chemical reactions at the interface level play a dominant role in achieving the low resistance and rectifying properties required for Ohmic and Schottky contacts, respectively (Baca 1997 and Swenson 1998). Among others, the temperatures involved in the whole technological process are a key parameter for the final electrical properties of the contact to be optimized. Moreover, thermal stability and interfacial reactions during further processing will be highly dependent on the range of temperatures used (Pilkington 1998). Concerning the effects of the temperature during metallisation itself (T_m), in previous work (Clark 1996) the electrical characterisation of I-V curves of In contacts on InGaAs revealed an improvement of Schottky contact behaviour when metallisation is done at low T_m.

In the present work we use TEM to examine the configuration of intimate contacts between In metal and atomically clean InGaAs (InAlAs)/InP semiconductor layers depending on the temperature, cryogenic or room temperature, at which the sample is held during the metallisation process. TEM has provided significant information, in agreement with previous X-ray photoelectron spectroscopy (XPS) and I-V measurements, to understand the mechanisms responsible for the Φ_b values encountered.

2. EXPERIMENTAL DETAILS

Four types of samples have been studied, corresponding to In metallisation deposited onto $In_{0.52}Al_{0.48}As$ and $In_{0.53}Ga_{0.47}As$ layers 1 μm thick (samples In#Al-.. and In#Ga-.. respectively), held at T_m=294 K (room temperature -RT) and 124K (cryogenic temperature –LT). The growth of the ternary compounds was performed onto InP (100) n+S-doped substrates using a V80H MBE system at a growth temperature of 500°C and growth rate of 1μm/h under As_4 stabilised conditions. The samples were capped in the growth chamber with a thick amorphous As layer for protection during transport to the chamber for XPS experiments. This cap was removed by annealing the samples *in situ* at 380°C. The reconstruction of the decapped InAlAs and InGaAs (100) surfaces was monitored by LEED replicating the (3x1) symmetry observed by RHEED after the MBE growth. High purity In metal was evaporated from well-outgassed tungsten filaments onto decapped samples. The total thickness of the In layer was in the range 70-220Å. More details of sample growth and metallisation, and XPS and electrical measurements have been already published elsewhere (Cammack 1997 and 1998).

The Transmission Electron Microscopy (TEM) observations have been performed in a Hitachi 800 NA microscope operated at 200 kV, in bright-field (BF) two-beam diffraction conditions for in plan (PVTEM) and cross-sectional (XTEM) views and in a Philips CM-30 at 300kV for High resolution mode (HREM). Surface morphology studies have been complemented by Atomic Force Microscopy (AFM) using a Nanoscope III in tapping mode.

3. RESULTS AND DISCUSSION

Concerning the growth of the semiconductor layers, both InGaAs and InAlAs films present a vertical superlattice in strong contrast for $g=200$ (Fig. 1). Selected area diffraction patterns (SADP) revealed satellite spots along the direction [100], corresponding to a period of the superlattice of 5 nm. At the moment we do not know whether the origin of such structure is a spontaneous decomposition of the alloy or is related to the experimental set-up of the MBE system. The configuration of the InAlAs layer is still more complicated. In figure 1c we distinguish the superlattice but also a lateral corrugation linked to columnar domains alternating along the [022] direction and in strong contrast for $g=022$.

Recent models (Guyer 1998) have described the development of lateral composition modulations and/or morphological corrugation in III-V compounds even for alloy compositions latticematched to the substrate and growth conditions well above the spinodal temperature for alloy decomposition. In our case this phenomenon has driven an anisotropic surface undulation of the InAlAs

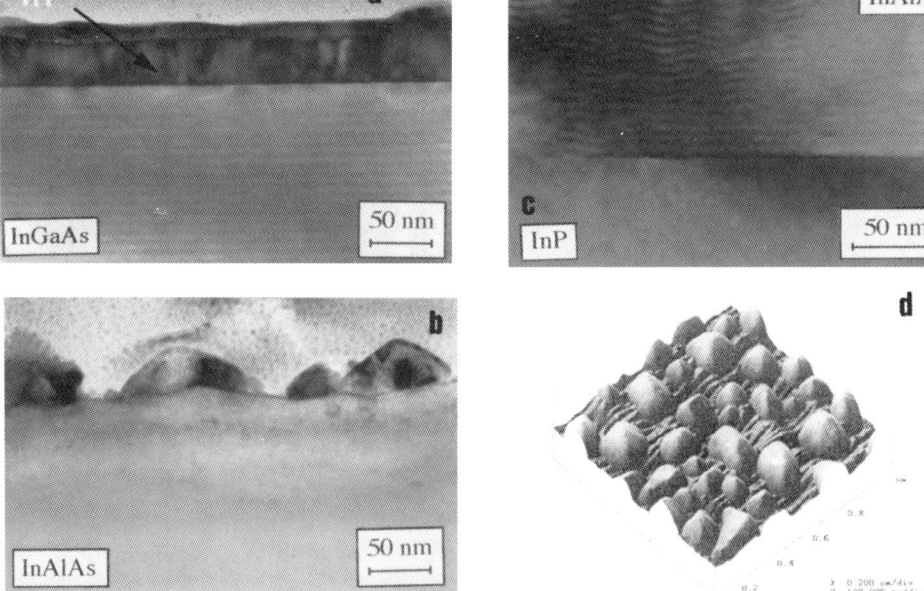

FIG. 1. Comparison of the growth mode of the In metallic layer depending on the state of the ternary layer: **(a)** Sample In#Ga-RT: continuous layer of In grown on InGaAs with vertical superlattices; **(b)** Sample In#Al-RT: nucleation of In islands; **(c)** Detail of the state of the InAlAs compound. Notice the vertical superlattice but also the lateral corrugation driving InAlAs surface roughness; **(d)** AFM image of sample In#Al-LT: notice the anisotropic roughness of the InAlAs surface underneath the In islands.

layer as revealed by AFM (Fig, 1d). The distance between hillocks is ≈40 nm and the height between the crests and the valleys is 4.7 nm near the InAlAs/InP interface and reaches up to 18 nm at the top.

The presence of the vertical superlattice does not seem to have affected the metallisation on InGaAs. Samples In#Ga-RT and In#Ga-LT, both exhibit a continuous In layer over the InGaAs. For sample In#Ga-RT (Fig. 1a) the thickness of the In layer is very irregular ranging between 20 and 50 nm. The distance between maximum undulations is not constant, oscillating between 130 and 430 nm. However, the metal layer in sample In#Ga-LT is much more uniform, with thickness oscillations in the range 38-43 nm. The In film has become polycrystalline with a noticeable texture following the relationship $[100]_{In}//[011]_{InGaAs}$. Conversely, the metallisation onto InAlAs has been strongly influenced by the corrugation of the surface: In islands have been nucleated instead of a continuous metallic film in both In#Al-RT (fig. 1b) and In#Al-LT samples with a nominal amount of In of 220 Å and 150 Å respectively. The morphology of the islands is very irregular although most of them are facetted with {111} lateral planes and slightly elongated towards the $[02\bar{2}]$ direction (Fig.1b and 1d). The island size is bigger for the sample metallised at low T (Fig. 2), with a mean diameter of the islands of $\varnothing_{LT} = 90$ nm and $\varnothing_{RT} = 58$ nm, and mean island height of $h_{LT}= 33$ nm and $h_{RT}= 30$nm, for LT and RT samples respectively. The island density (ρ) is higher in the sample held at high temperature, being $\rho_{LT} = 3.2 \cdot 10^9$ cm^{-2} and $\rho_{RT}= 9.1 \cdot 10^9$ cm^{-2}. These results are explained by the enhanced diffusion of ad-atoms during metallisation at RT. For sample In#Al-LT we distinguish a uniform contrast of In islands in bright field images. Conversely, for sample In#Al-RT, quite a lot of islands present a mixed configuration, with a central region of weak contrast and a dark region at the border (Fig. 2b).

In previous work concerning XPS measurements, the evolution of As3d, Ga3d and In4d core levels was analysed as a function of the In coverage (Cammack 1998). The results from the As3d core evolution for both In#Ga-LT and RT samples revealed out-diffusion of As towards the surface. Accounting for the signal of In, a transition from a covalent bonded In to a In metal emission was found for the In#Ga-LT sample, being the component relative to the covalent bonded In more significant in In#-Ga-RT sample. HREM images of the interface have confirmed the presence of an As-poor layer underneath the In metal, with higher strength in In#Ga-RT sample (Fig. 3a). This region of As deficiency extends up to 6 monolayers under the interface. The electrical measurements of Schottky barriers gave $\Phi_b=0.3$ eV and 0.45 eV for samples metallised at 294K and 125K respectively (Cammack 1998). The chemical reaction at the interface level influences the density of gap states and therefore plays an important role in determining the pinning position of the Fermi level. The inhibited interface reaction at LT may explain the enhancement of the Schottky barrier.

For metallisation on InAlAs no indication of chemical reactions were detected by XPS experiments. However, the shift of Fermi level as increasing In deposition resulted $\Phi_b=0.3$ eV, much lower than that previously measured for intimate In/n-InAlAs contacts ($\Phi_b=0.91$ eV) (Cammack 1998). In this case, HREM images combined with EDX measurements have demonstrated that As out-diffusion during metallisation has also occurred in InAlAs samples grown at RT. At the first stages of metallisation, In nuclei, with a lattice

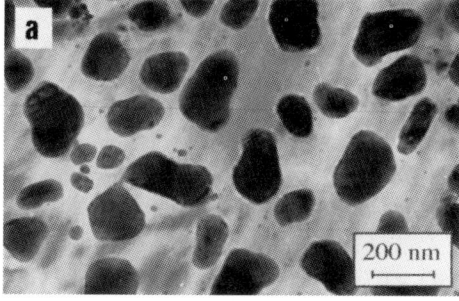

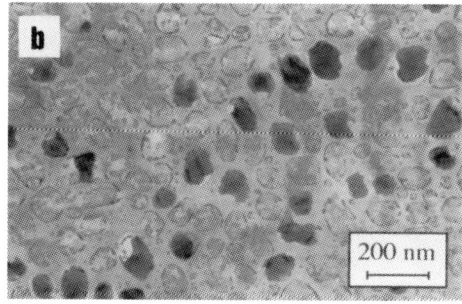

Fig. 2. Bright field (100) images of the In Islands in sample In#Al-LT held at cryogenic temperature **(a)**, and sample In#Al-RT held at room temperature **(b)**.

parameter smaller than that of InAlAs (3.252 Å and 5.868Å respectively), are preferentially deposited at the valleys of the InAlAs surface rippling, where it is expected to find highly compressive sites after alloy lateral decomposition (Cullis et al 1995). The As out-diffusion develops InAs islands as further In is added. Higher In coverage results in the formation of a pure In metallic layer covering the InAs islands or, in some cases, to a metallic deposit at the lateral faces of the islands (Fig. 3b), corresponding to the dark regions noticed in plane view images (Fig. 2b). Moreover, the islands are highly defective, with stacking faults and twins in the core and border of the islands, as usually observed in the Stransky-Krastanov growth mode of strained compounds. The presence of the lateral decomposition inside the $In_{0.52}Al_{0.48}As$ layer, the defective interfacial configuration due to the induced semiconductor surface roughness and the structural defects inside the islands, are all factors that contribute to the formation of surface and gap states that pin the Fermi level at an energy far below the Schottky limit.

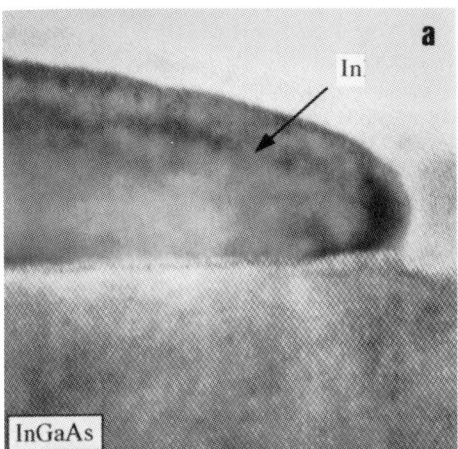 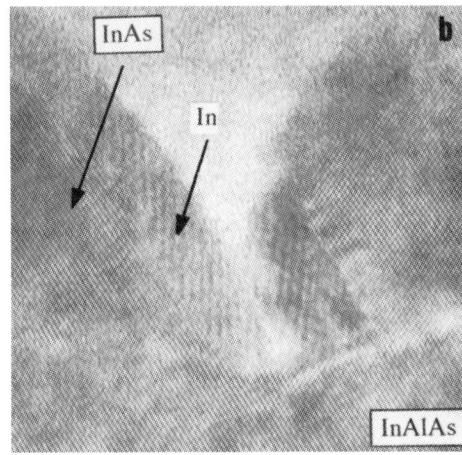

FIG. 3. HREM images of the In#Ga-RT (**a**) and In#Al-RT (**b**) samples. Mixed InAs-In islands are observed in (**b**).

4. CONCLUSIONS

Structural properties of the interfaces formed at room and cryogenic temperatures between In and atomically clean $In_{0.52}Al_{0.48}As$ and $In_{0.53}Ga_{0.47}As$ layers have been studied. The growth mode of the In layer has been described depending on the state of the ternary compound. We have shown that the presence of a vertical superlattice does not affect the In deposition. Conversely, the metallisation onto laterally self-decomposed InAlAs layers driving surface undulations leads to island growth. High temperatures favour As out-diffusion from the semiconductor. For metallisation on InGaAs this chemical reaction at the interface level is responsible for a reduced value of the Schottky barrier as measured by I-V. For InAlAs, the As loss from the ternary results in the nucleation of mixed In and InAs-In islands. The irregularities of the InAlAs and the high density of defects in the islands determine the pinning position of the Fermi level below the usual Schottky barrier.

REFERENCES

Baca A G, Ren, F, Zolper J C, Briggs R D and Pearton S J 1997 Thin Sol. Films **308-309**, 599
Cammack D S, McGregor S M, McChesney J J, Dharmadasa I M, Clark S A et al. 1997 J. Appl. Phys. **81**, 7876
Cammack D S, Clark S A, Dunstan P R, Pan M, Wilks S and Elliot M 1998 J. Appl. Phys. **84**, 4443
Clark S A, Wilks S P, Kestle A, Westwood D I and Elliot M 1996 M Surf. Sci. **352-354**, 850
Cullis A G, Pidduck A J and Emeney M T 1995 Phys. Rev. Lett. **75**, 2368
Guyer J E and Voorhees P W 1998 J. Cryst. Growth **187**, 150
Pilkington S J and Missous M 1998 J. Appl. Phys. **83**, 5282
Swenson D, Jan C H and Chang Y A 1998 J. Appl. Phys. **84**, 4332

Inst. Phys. Conf. Ser. No 164
Paper presented at Microsc. Semicond. Mater. Conf., Oxford, 22–25 March 1999
© 1999 IOP Publishing Ltd

Applications and problems for TEM of semiconductor products

A John Mardinly

Materials Technology Dept., Intel Corporation, 2200 Mission College Blvd., SC2-24, Santa Clara, CA 95052-8119

ABSTRACT: The demand for higher performance semiconductors has stimulated industry to respond by producing devices with smaller transistors, more layers of metal, dielectric and interconnects, and more interfaces. As features are being shrunk in the quest for higher performance, the role of Transmission Electron Microscopy as a characterization tool takes on a continually increasing importance. As the feature size shrinks below one quarter micron, the thickness of the specimens must shrink as the square root of the feature size reduction. Moreover, the center-targeting of these specimens must improve so that the center-targeting error shrinks linearly with the feature size reduction.

1. INTRODUCTION

Transmission electron microscopy (TEM) provides for the characterization of materials at high spatial resolution and high contrast through the use of diffraction contrast, micro-chemical analysis, and other techniques. These capabilities have led to an increasingly essential role for TEM-based analysis in process development, defect identification, yield improvement and root-cause analysis within the electronics industry. With continuing reductions in semiconductor device dimensions, the high spatial resolution of TEM-based techniques will be required to an even greater extent.

2. IMAGING EXPERIMENTS

The high-resolution mode of imaging is useful also for the analysis of amorphous materials even though they exhibit no long range order. Fig. 1 shows the familiar example of a gate dielectric between the single crystal silicon substrate and the polycrystalline gate electrode. The amorphous gate dielectric shows a mottled contrast characteristic of amorphous materials. However, the thickness of the gate dielectric and the roughness of the top and bottom interfaces can still be assessed.

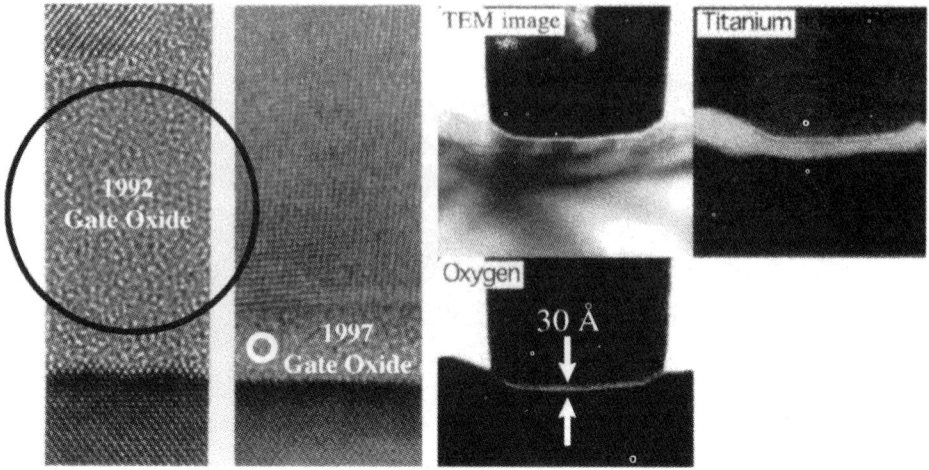

Fig. 1. Montage of TEM imaging and analysis situations. Left two images are gate oxides from two generations of process technologies with superimposed circles showing analytical probe sizes for LaB₆ and FEG TEMs. Right images are Energy Filtered TEM images of a contact to salicide that had been oxidized before tungsten deposition.

Again, the lattice image in the silicon substrate can serve as a direct and accurate calibration of the magnification of the image. On the right side of Fig. 1 are Energy Filtered TEM images of a contact to salicide that had been accidentally oxidized before tungsten deposition. These examples also illustrate quite vividly how device scaling is driving the microstructure of the devices further into a realm where the highest spatial resolution analytical tools, such as TEM, are required.

Traditional use of TEM as a materials science tool is based on the philosophy that the specimen is homogeneous, and any materials science observations in one part of the specimen should be repeatable anywhere else in the specimen. In semiconductor products, the opposite is frequently the case. The structure of wires and insulators above the silicon resembles a tiny city, and it may be necessary to perform the TEM analysis on a selected part of the structure, due to a known defect. Moreover, we find that much of the necessary data must be collected from this structure of wires and insulators, and thus we find much of the TEM of semiconductor based products is performed on materials that are not semiconductors. Fig. 2 illustrates this with a cross-section of a microprocessor with four levels of metallization, dielectric layers, interconnects, and transistors on the silicon substrate.

Fig. 2. Cross-section of a microprocessor showing four levels of metallization, dielectric layers, interconnects on the silicon substrate.

3. SPECIMEN PREPARATION

As shrinking process geometries push the size of features close to that of traditional electron transparent TEM specimens, special requirements are placed on the preparation of cross-sectional samples. Since typical interconnects are cylindrical, there are simple geometrical effects that become significant as the radius of curvature scales with the shrinking of devices. Geometrical blurring reduces resolution and becomes significant when the specimen is too thick, or off-center, even though it may well be outstanding in its electron transparency and freedom from surface amorphization, as shown in Fig. 3. Quantitative assessment of the effects of shrinking radius of curvature on a centered specimen can be made with a relatively simple model, as illustrated in Fig. 4. Define the "interfacial layer" as the blurring due to geometrical effects. Geometrical blurring becomes significant in a cross-section when the outermost surface of an interfacial layer can be directly above or below the innermost surface of that layer, thus convoluting the image information from the inside and outside of that interfacial layer. Equation (1) describes a simple right triangle containing R, the interconnect radius, and equation (2) describes the solution for T, the specimen thickness. The conclusion is that if cross-sections can be perfectly centered, routine specimen thickness must continue to scale downwards as the square root of the scaling of semiconductor process geometries, which is reduced by a factor of 70% with each new generation.

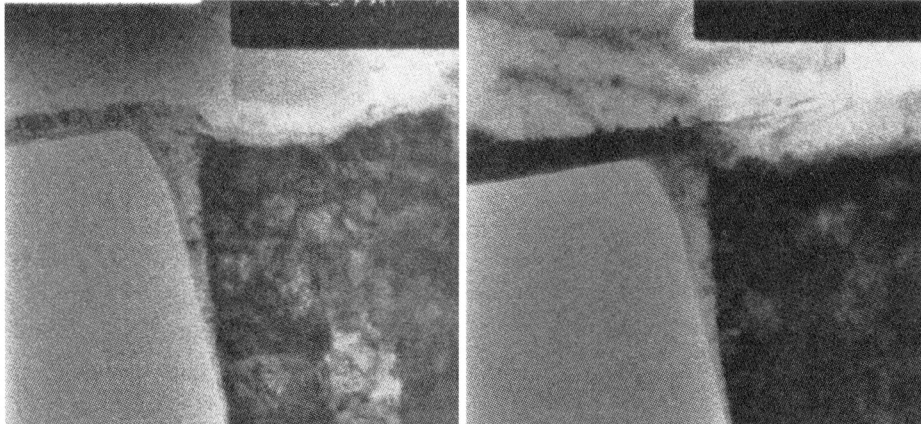

Fig. 3. Two images of tungsten interconnects from the same specimen. On the left is a thin, well centered specimen, while on the right is the same structure not centered, with the view of the structure not properly represented.

Fig. 5 illustrates a more typical case where the cross-section may miss the exact center of the feature. Equations (3) and (4) describe two right triangles, and equation (5) describes the solution. In this case, the requirement is more severe than for centered specimens, and this suggests a requirement that specimen thickness must scale linearly with shrinking process geometries in order to avoid geometrical blurring.

Fig. 6 shows a plot of equation 5 solved for T, the specimen thickness, with R from three different process geometries, D, the centering error equal to 0, 100Å, 200Å and 400Å, and Δ, the blurring, equal to 30 Ångstroms. Note that these specimen thickness requirements are all less than the extinction distance for silicon. It can be seen that TEM of anticipated future devices will require stringent control of centering, since specimens thickness requirements for poorly centered cross-sections (200-400Å error) are so severe, these specimens will be very difficult to produce, provide poor contrast in the TEM, and are frequently dominated by surface amorphization.

One favored method of providing well centered specimens is the gallium FIB. The milling and imaging capability, particularly for dual beam FIBs makes achieving excellent center targeting almost routine. However, high energy gallium ions cause surface amorphization and implantation several hundred angstroms in thickness that seriously degrade the imaging and microanalysis qualities of the specimen[2].

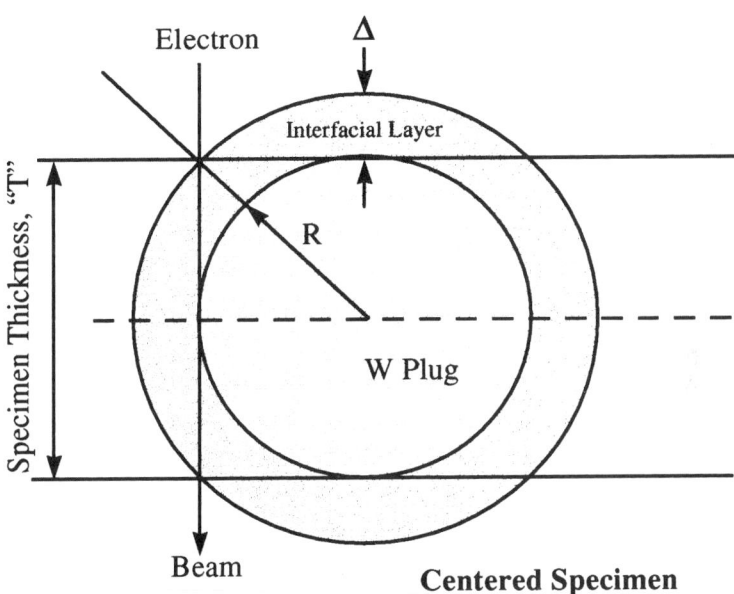

Fig. 4. Construction of the model for calculating specimen thickness required to avoid geometrical blurring for thick centered specimens.

$$R^2+(T/2)^2=(R+\Delta)^2 \quad (1)$$

$$T=2(2R\Delta+\Delta^2)^{0.5} \quad (2)$$

Fig. 7. shows a Monte Carlo simulation of 30 kV Ga+ ion milling. The Monte Carlo calculations show implantation and vacancy formation are present at depths approaching 300Å. Fig. 8 shows experimental verification of this simulation in which a FIB cross-section was itself cross-sectioned, and the amorphised layer was measured to be 280Å. This damaged material accounts for a large fraction of a TEM sample that is thin enough to avoid geometrical blurring, as described in Fig. 6. The quality of TEM lattice images is impaired by the presence of so much amorphous and damaged material in the sample. In addition, ion-beam induced compositional mixing of adjoining layers of material severely degrades analytical data obtained by EDS or EELS. Clearly, there is an emerging industry-wide need

580

to develop methods for precise, center-targeted sample preparation that produce significantly less surface damage.

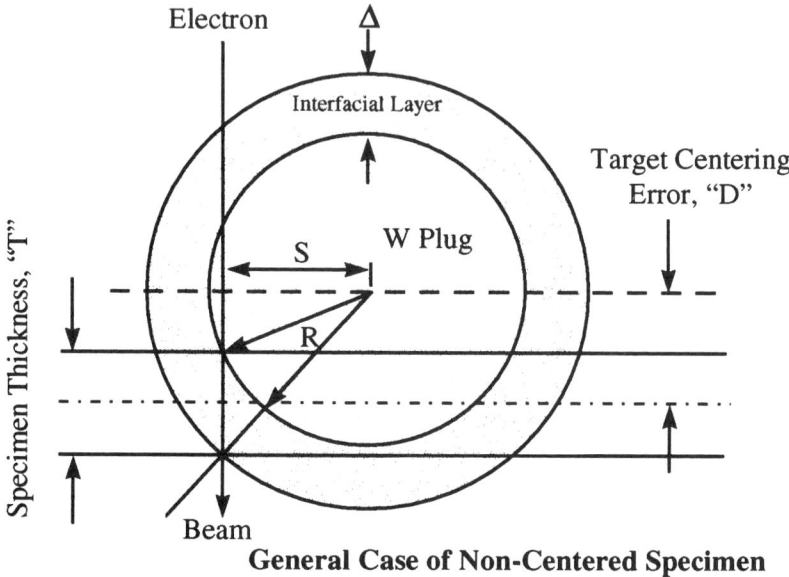

General Case of Non-Centered Specimen

Fig. 5. Construction of the model for calculating specimen thickness required to avoid geometrical blurring for non-centered specimens.

$$R^2=S^2+(D-T/2)^2 \qquad (3)$$

$$(R+\Delta)^2=S^2+(D+T/2)^2 \qquad (4)$$

Describe two right triangles, and the solution for "T" is:

$$T=(2R\Delta+\Delta^2)/2D \qquad (5)$$

Traditional argon ion milling is the only technique now known to provide thin specimens with significantly less surface amorphization than that provided by gallium FIB systems. However, the endpoint control that can provide thin, centered cross-sections of small structures is limited. Traditional argon ion mills are equipped with a low-power optical microscope that is inadequate for this task because the features cannot be resolved by the microscope. The only alternative to the slow and tedious procedure of iterative milling and examination with electron optics is to have the ion mill and SEM integrated into a single tool. We have constructed a new tool utilizing an argon ion mill mounted inside the specimen chamber of a field emission scanning electron microscope to monitor the surface of a specimen in order to achieve a level of control of the ion milling process never before possible. The resolution of the field emission SEM allows monitoring the milling process with sufficient resolution and sensitivity to find the correct endpoint within 100 Ångstroms. Furthermore, the use of the 3KeV electron beam causes no damage to the specimen.

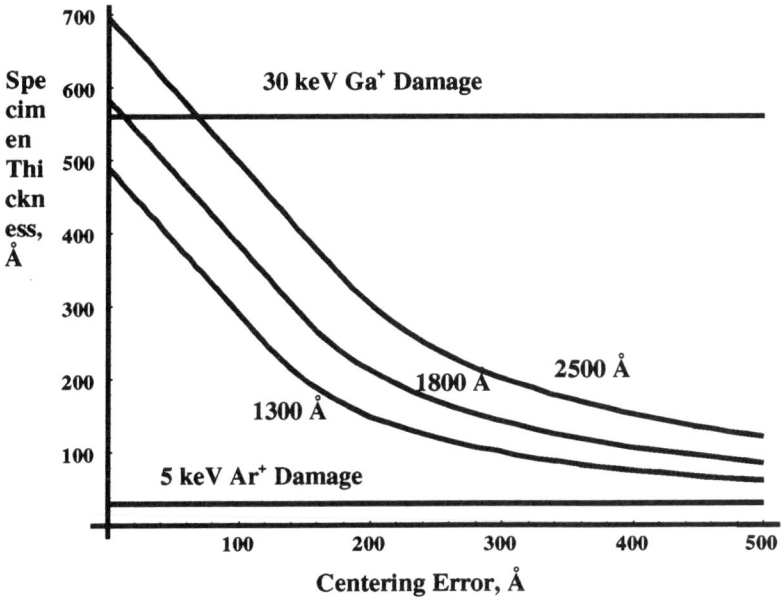

Fig. 6. Plot of TEM specimen thickness requirements for current, and two future process technologies as a function of centering error required to keep geometrical blurring below 30Å. Note that the thickness is always less than one silicon extinction distance, and errors greater than 100Å lead to a requirement of extremely thin specimens much of which would be consumed by amorphization.

This tool was constructed using a Hitachi S-800 cold field emission SEM converted to accept a side entry goniometer specimen holder from a JEOL 1200EX TEM. Fig. 9 shows the VCR saddle-field ion gun with a viewport for observing the specimen and the ion beam, and a custom designed two-axis aiming mechanism for aiming the ion beam to the center of the specimen inside the SEM. Feed-throughs for high voltage, argon, and monitor current were embedded in a custom made flange designed to fit one of the many ports available on the Hitachi SEM. Fig. 10 shows the result of argon ion milling tungsten interconnects to their precise center. The interconnects and all interfaces are imaged sharply.

Monte Carlo 30 kV Ga⁺ Simulation

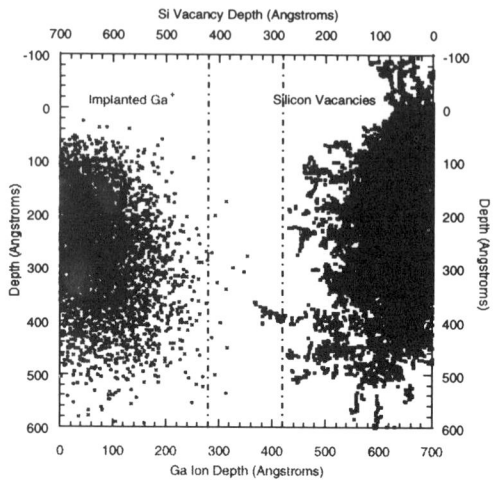

• Incident milling angle of 6°

•Projected onto X-Y plane

•Impact location is a point source

•Ion data from 10,000 Ga⁺

•X-axis full range represents realistic FIBXTEM sample thickness

The simulations are from SRIM package of software by James Ziegler at IBM Research, Yorktown, NY (see: www.research.ibm.com/ionbeams/)

Fig. 7. TRIM calculation of gallium implantation and vacancy creation in silicon.

4. CONCLUSIONS

Transmission electron microscopy encompasses a range of high spatial resolution imaging and analytical techniques for the characterization of semiconductor products. Currently, TEM plays an essential role in process development, defect identification, yield improvement and root-cause analysis within the semiconductor industry. As semiconductor device dimensions continue to decrease, TEM will become a critical analytical tool for ULSI technology. Improvements in the design of field emission electron sources, lenses etc. and in the development of new, sophisticated analysis modes have led to the commercial availability of TEMs with an unprecedented combination of resolving power and analytical capability. However, a practical limitation to an expanded role for TEM-based techniques is the need for producing an extraordinarily thin, artifact-free specimen, precisely centered on a particular sub-quarter micron feature, in a timely and cost- effective manner. To meet this need, further improvement in existing techniques, and the development of novel specimen preparation techniques is urgently required.

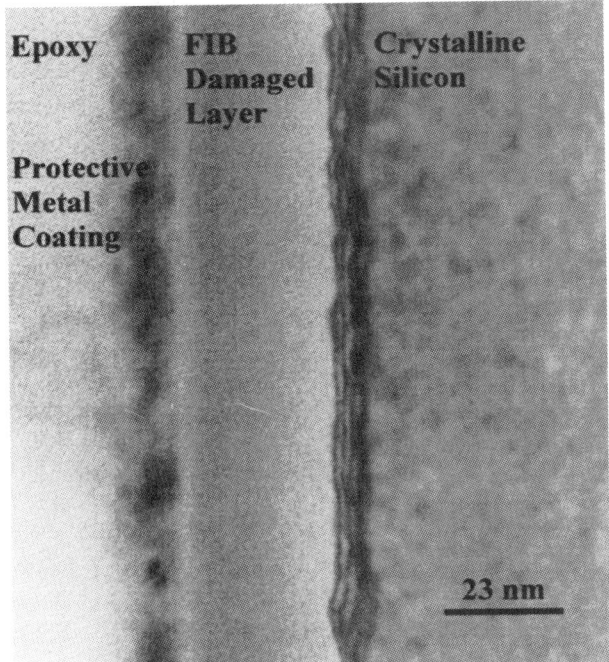

Fig. 8. Cross-section of FIB cross-section showing ~280Å of amorphization caused by the gallium beam.

Fig. 9-View of the ion gun installed inside the SEM specimen chamber with gimbal control mechanism.

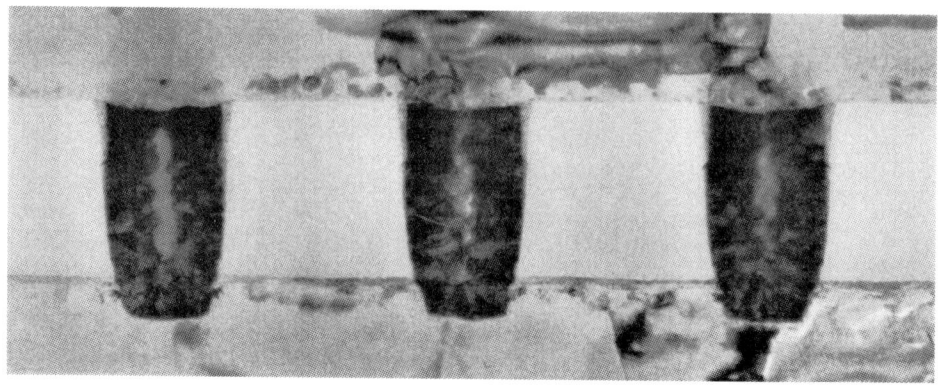

Fig. 10-TEM Image of tungsten interconnects after ion milling with ultra high precision endpoint detection.

ACKNOWLEDGEMENTS

The author gratefully acknowledges the contributions of Kevin Johnson for the TRIM calculations, and Robert Jamison for constructing an Argon Ion Mill inside a SEM.

REFERENCES

Barna A, Pecz B and Menyard M 1998 Ultramicroscopy **70**, 161
Basile D, Boylan R, Baker B, Hayes K and Soza D 1992 MRS Symp. Proc. **254**, pp 23-41
Bravman J and Sinclair R 1984 J. Electron Microsc. Techn. **1**, 53
Mardinly J and Jamison R 1998 Proc. ICEMXIII **3**, pp 403-404
Schurke T, Mandl M, Zweek J and Hoffman H 1992 Ultramicroscopy **41**, 429
Susnitzky D and Johnson K 1998 Proc. MSA pp 656-657

Inst. Phys. Conf. Ser. No 164
Paper presented at Microsc. Semicond. Mater. Conf., Oxford, 22–25 March 1999

585

Focused ion beam microscopy investigation of InGaP/GaAs heterojunction bipolar transistors

A J Pidduck, C Reeves, G M Williams, M A Crouch, D G Hayes, K Hilton, P Parmiter, J Birbeck and A Schertel*

DERA, St Andrews Road, Malvern, Worcestershire, WR14 3PS, UK
*Micrion GmbH, 85622 Feldkirchen, Munich, Germany

ABSTRACT: We describe the application of focused ion beam (FIB) microscopy to the direct structural characterisation of specific completed InGaP/GaAs heterojunction bipolar transistor devices. FIB is used firstly for direct sectioning and imaging of devices, and secondly for the preparation of transmission electron microscopy specimens. In the latter case, a "lift-off" process, requiring no pre-preparation of the device wafer, is demonstrated.

1. INTRODUCTION

Heterojunction bipolar transistors (HBTs) based on epitaxial n-InGaP/p$^+$-GaAs/n-GaAs are very promising for high-power microwave frequency (>10GHz) applications (Ueda et al 1997, Pan et al 1998). Some devices can however exhibit a long-term degradation in gain after prolonged electrical stressing at elevated temperatures, limiting their useful lifetime as power transistors. Investigations of the origin of such phenomena require techniques able to probe specific devices to allow correlation of structural detail with measured electrical characteristics. We show in this paper that high-resolution focused ion beam (FIB) microscopy, alone and in combination with transmission electron microscopy (TEM), is very well suited to this task.

2. EXPERIMENTAL

The devices studied are n-p-n transistors employing a lattice-matched $Ga_{1-x}In_xP$ (x=0.48) wide band-gap emitter layer and a highly C-doped base layer. The epitaxial layer structure is shown in Fig. 1(a), and a schematic of the device construction in Fig. 1(b). Epitaxial layer growth was by MOVPE. Features of the device construction include polyimide dielectric, device isolation by proton implantation, alloyed metal contacts to the emitter (E) and collector (C) and a non-alloyed metal contact to the base layer (B), formed by self-alignment to the E-contact. An emitter finger geometry is used as shown in Fig. 2(a).

FIB microscopy was carried out using a Micrion 2500 instrument with a 50keV Ga$^+$ ion beam. Ion currents used were in the nA-μA range for milling, down to 5pA for the highest resolution imaging (base resolution ≤10nm).

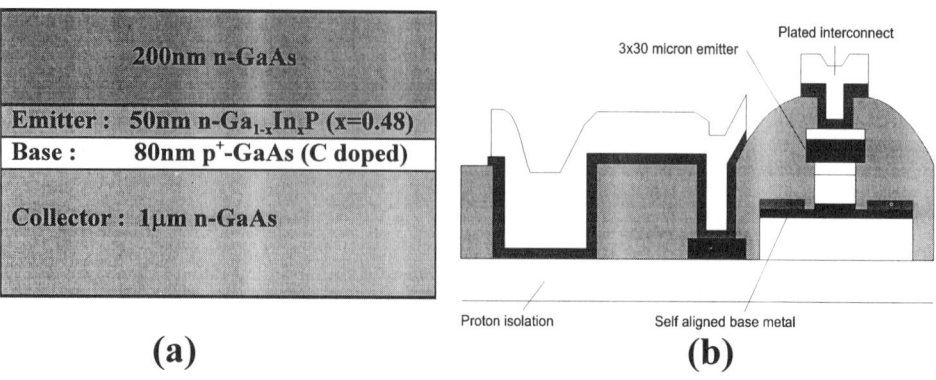

(a)   **(b)**

Figure 1. (a) HBT epitaxial layer structure (b) schematic of device structure

3. RESULTS AND DISCUSSION

3.1 Direct FIB imaging of devices

Figure 2(b) shows a typical FIB trench milled across a device emitter finger. Secondary electron FIB images obtained from the FIB-polished vertical face are shown in Fig. 3. These reveal the device construction in considerable detail. Metal grain structure shows up with high contrast in the Au emitter metallisation, and alloying is evident at the semiconductor / metal interface. The InGaP layer appears as a dark stripe in the semiconductor. Unexpected voiding is apparent in the polyimide dielectric in the emitter recess, and cracking is evident at the interface with the Au interconnect.

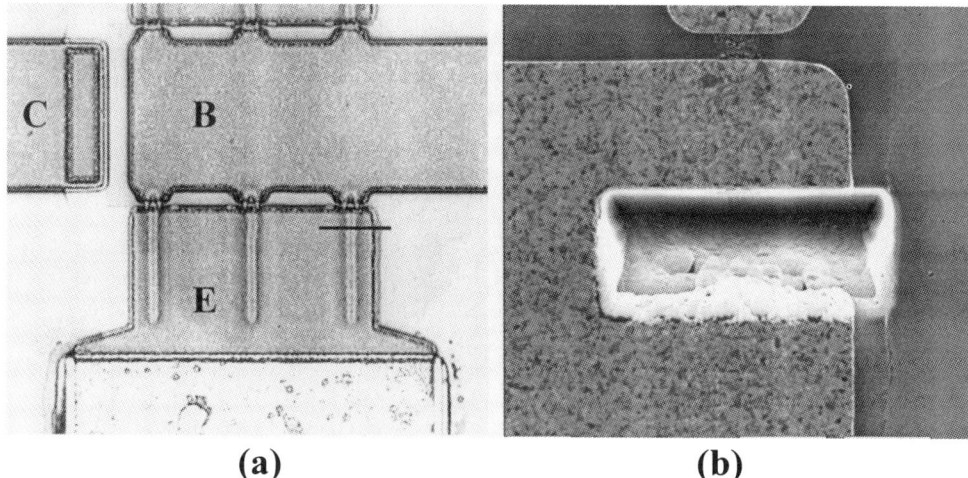

(a)   **(b)**

Figure 2. (a) optical micrograph of a multi-finger HBT. (b) secondary electron contrast plan-view FIB image of a tapered FIB trench milled through an individual emitter finger. The widths of the bar marker in (a) and the trench in (b) are about 30μm.

(a)   (b)

Figure 3. Secondary electron contrast FIB images from the polished face of the trench in Fig.2(b). The vertical scale is 1.4x greater than the horizontal due to tilting through 45°.

3.2 Preparation of FIB-TEM specimens ("Lift-off" technique)

Tapered trenches were ion-milled to leave a free-standing membrane of 100nm nominal thickness after careful ion-polishing of each face, as shown in Fig. 4. The sides and base of this membrane were then cut with the ion beam to leave an isolated foil. This was then picked up with an electrostatic silica probe, using a micromanipulator, and transferred to a holey-carbon grid for TEM analysis.

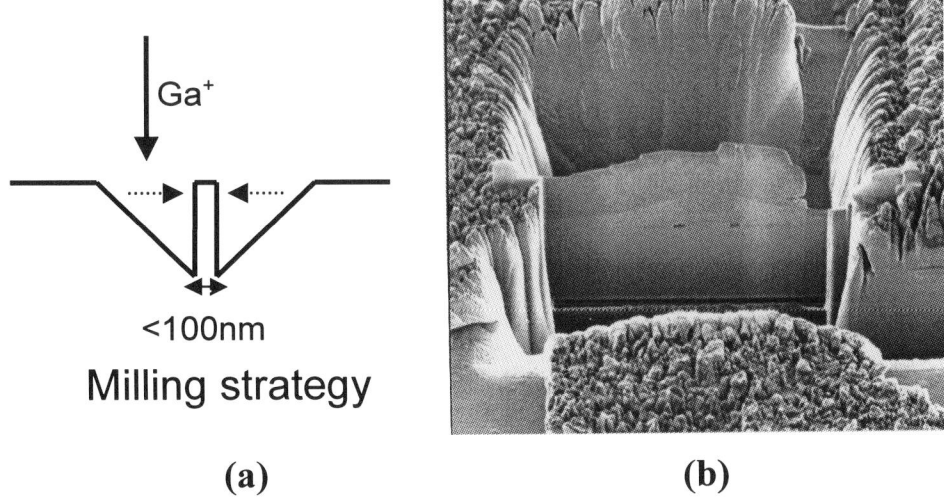

(a)   (b)

Figure 4. (a) Schematic showing the FIB milling strategy resulting in an electron-transparent lamella, shown in (b), which was then isolated and transferred to a holey carbon grid for TEM analysis. The lamella is 18μm wide.

3.3 FIB-TEM imaging of devices

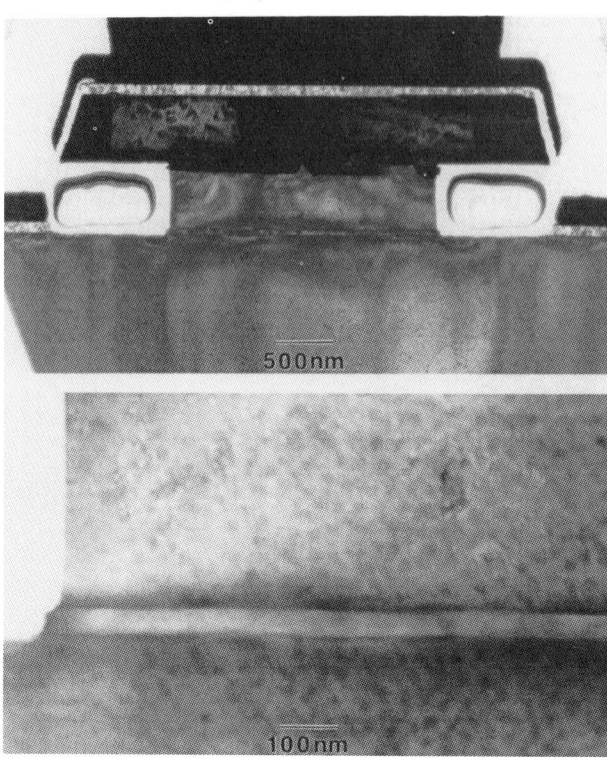

500nm

100nm

TEM images of the device (Fig. 5) show up very clearly the self-aligned base contact layer and the InGaP emitter, in addition to those features observed by direct FIB imaging. There are signs of weak strain fluctuations and thickness variations in the InGaP layer, and over-etching into the base layer by 10-20nm. Background mottling, believed to result from Ga droplet formation during FIB preparation of the GaAs, masks information on the possible presence of precipitates in the base layer, which are one proposed transistor gain degradation mechanism (Ueda et al). Optimised FIB milling conditions to avoid this artefact clearly need to be developed.

Figure 5 (top and bottom). Bright field diffraction contrast TEM images from the foil in Fig. 4.

4.   CONCLUSIONS

High-resolution FIB microscopy, applied with and without TEM for the direct structural characterisation of specific completed HBT devices, allows detailed correlation of device structure with process history on the one hand, and with measured electrical performance on the other. A "lift-off" technique for FIB preparation of TEM specimens, requiring no pre-preparation of the device wafer, enables TEM investigation of processed wafers in an essentially non-destructive manner.

REFERENCES

Pan N, Elliott J, Knowles M, Vu D, Kishimoto K, Twyman J, Sato H, Fresina M and Stillman G 1998 IEEE Electron Device Lett. **19**, 115

Ueda O, Kawano A, Takahashi T, Tomioka T, Fujii T and Sasa S 1997 Solid State Electronics **41**, 1605

Inst. Phys. Conf. Ser. No 164
Paper presented at Microsc. Semicond. Mater. Conf., Oxford, 22–25 March 1999

TEM investigation of wet oxidised AlAs in VCSEL structures

M A Al-Khafaji, H Meidia, A G Cullis, K Y Chang, J Woodhead and J S Roberts

Department of Electronic and Electrical Engineering, University of Sheffield, Sheffield S1 3JD, UK

ABSTRACT: The microstructure and composition of laterally oxidised AlAs layers in a vertical cavity surface emitting laser (VCSEL) were studied using analytical TEM and nano-scale EDS analysis. This work revealed that there is no trace of As in the oxidised AlAs layer, while high concentrations of Al and Oxygen were detected indicating the formation of Al oxide. The phases of the reaction products were identified using high resolution electron microscopy and image processing. In addition, the chemical segregation and sharpness of the oxide-semiconductor interfaces with surrounding layers of different Al contents were also studied. This revealed that the sharpness may be degraded by the oxidation process. Narrow areas of high porosity were observed at these interfaces, which were explained in term of the AlAs oxidation mechanism.

1. INTRODUCTION

The technique of lateral sidewell wet oxidation of selected layers (Dallesasse et al 1990, Sugg et al 1993) has gained considerable interest in recent years due to improvements which it offers in the optical and electrical properties of various electronic devices. Vertical cavity surface emitting laser (VCSEL) (Choquette et al 1994, 1996) devices are particularly important due to their wide applications, for example in optical communication and optical recording (Ebeling 1996). The properties of these devices are expected to be sensitive to chemical composition and interface microstructure between the oxidised and unoxidised/partially-oxidised layers. Hence understanding of the oxidation mechanism, chemical homogeneity and interface structure is important. In this work these properties were investigated using various TEM techniques, such as high resolution electron microscopy (HREM), nanoprobe energy dispersive spectroscopy (EDS) and electron energy loss spectroscopy (EELS).

2. EXPERIMENTAL

The VCSEL structure was grown using the MOVPE method with metalorganic sources and arsine: the trenches were obtained by reactive ion etching (RIE). The depth of these trenches was controlled via in-situ optical monitoring and terminated in the first few pairs of the lower distributed Bragg reflector (DBR). The oxidation of the AlAs layer was done by exposing the structure to H_2O vapour in a wet nitrogen ambient at an elevated temperature of 400°C. A schematic illustration of three copies of the top emitting VCSEL (each containing a single oxide aperture) is shown in Fig. 1a. A detailed magnified section of the oxide interfaces, with surrounding materials, is shown in Fig. 1b. TEM specimens were prepared by cleaving along directions perpendicular to the etched trenches and then glued face to face. In

590

order to prevent deformation of the top DBRs, care was taken not to excessively press the stripes together or slide them sideways. The samples were then mechanically thinned from both sides to a thickness of about 30μm and finally ion beam thinned using Ar⁺ until perforation. These were examined using a field emission JEOL JEM 2010F TEM equipped with an EDS detector and Gatan Imaging Filter (GIF) system.

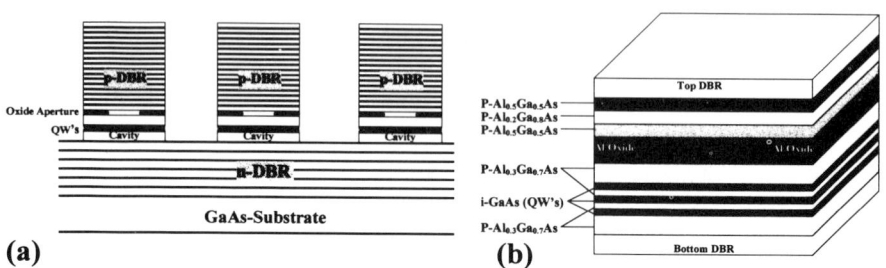

(a)   **(b)**

Figure 1. (a) Schematic of the oxide confined top emitting VCSEL, (b) details of the oxidised/unoxidised AlAs interface and surrounding layers.

3. RESULTS AND DISCUSSION

In Fig. 2a, a low magnification image, acquired along the [110] zone axis, shows the termination of the oxidised AlAs layer (white contrast). In many samples examined in this work there were a number of cracks propagating through the multilayers. These cracks must be generated during the oxidisation process, possibly as a mechanism for release of strain, which is generated due to the change in the volume fraction of the oxide layer during the oxidation process (Guha et al 1995). This effect was utilised in this study to examine the interface structure.

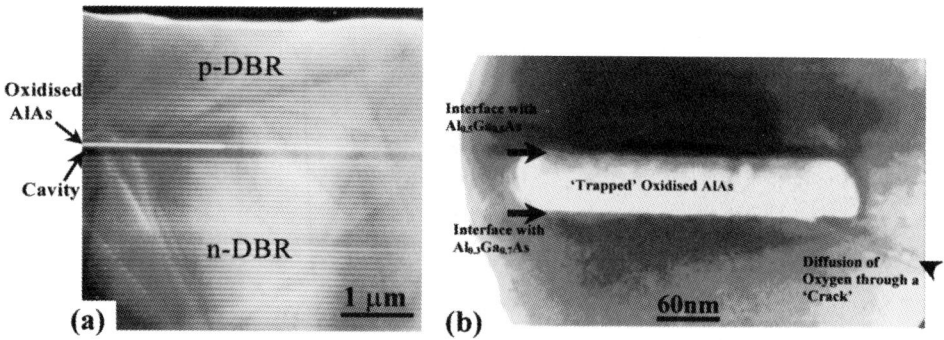

Figure 2. [110] zone axis bright field images showing (a) the termination of the oxide front and (b) entrapped oxidised AlAs region formed by leakage of oxygen through a crack.

In Fig. 2b, a crack can be seen propagating through the lower DBR and through the AlAs layer. This crack has behaved as a channel for oxygen migration to oxidise the AlAs layer and hence a 'trapped oxidised AlAs' area was generated as seen in Fig. 2b where oxidation fronts can be seen on either side. This area was used to study the formation of the oxidised AlAs and map the composition of the various elements. X-ray maps of this region, acquired from both ends using an electron probe with nominal diameter of 0.5nm, are shown in Fig. 3a. These compositional images were obtained using the K lines from Al, Ga, O and As. Note the presence of oxygen in the crack confirming that this crack is a source of oxygen migration. The most obvious feature is that the oxidation process has led to the complete removal of the As (see Fig. 3b). The maps were acquired with the sample oriented along the [110] zone axis, which ensured that the scanning electron beam is mapping parallel to the epitaxial layers. Therefore the sharpness of the interface is determined by the scanning probe size (including beam spreading) and the interface physical dimensions.

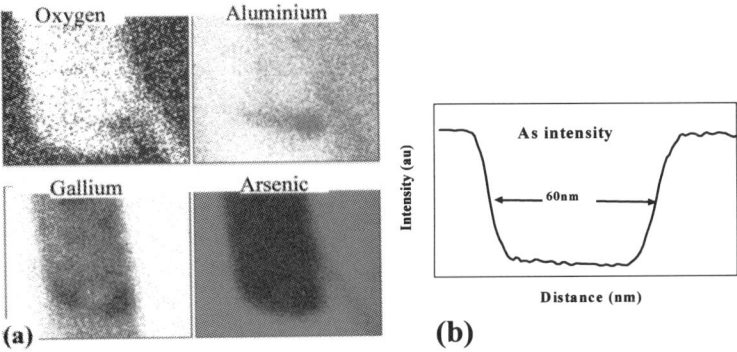

Figure 3. X-ray elemental mapping of oxidised AlAs Layer acquired from one end of the entrapped area shown in Fig. 2. In (b) the arsenic line profile is shown.

Oxygen maps were also obtained using EELS imaging from the same region and the result is shown in Fig. 4. Line profiles along AA and BB are shown in Figs 4b and c, respectively. The oxygen intensity profile is almost symmetric about the centre of the oxidised region and peaks at the interface between the semiconductor (AlGaAs in this case) and the

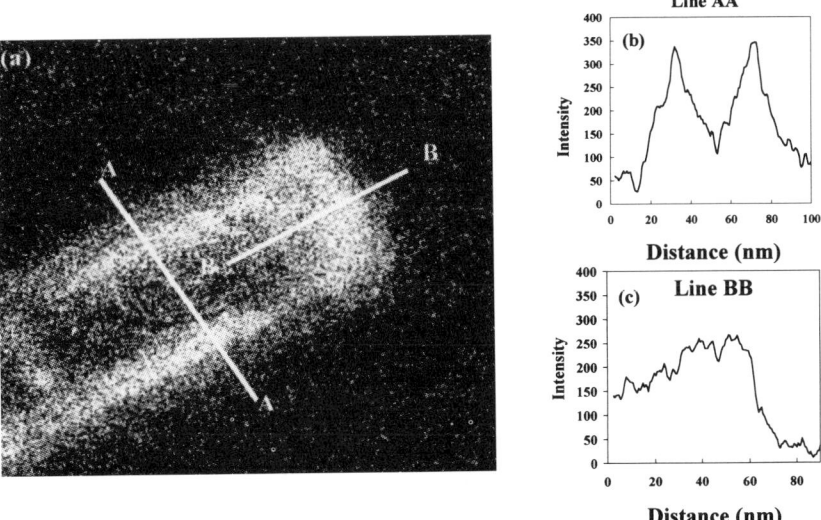

Figure 4. Oxygen EELS map and intensity line profiles from the same area shown in Fig. 3.

oxidised AlAs region. This symmetry indicates that the Al fraction in the AlGaAs layers (0.5 in layer on top and 0.3 in layer below) does not affect the profile, presumably because the oxidation rate for these low Al fraction layers is negligible. However, high resolution TEM does show a roughening of the interface. The symmetry in the EELS oxygen profile can be explained in terms of an effect of thickness on the observed EELS spectra. It is known in EELS that as the sample thickness is increased, plural scattering tends to dominate (Egerton 1996). This effect is basically a subsequent multiple scattering of an electron which will result in either deflecting the electron outside the collection aperture (elastic) or assigning a new energy loss and deflection to the electron (inelastic). In either case the detected edges will have large background and hence the basic background subtraction routines will fail due to poor signal to noise ratio and hence thicker regions will appear darker than thinner regions of the same materials. In the present case the chemical reaction transformation of AlAs to Al_2O_3 is (Guha et al 1995):

$$2AlAs + 3H_2O \rightarrow Al_2O_3 + 2AsH_3.$$

Since the volume of the AlAs unit cell in the zinc blend structure is 181.5 Å^3 and that of the Al_2O_3 in the hexagonal structure is 254.8 Å^3, then the above formula implies that there is a 30% shrinkage during formation of the reaction solid product. This yields a porous region near the Al_2O_3/semiconductor interface and hence less material in these regions results in a higher intensity in the EELS elemental maps.

4. CONCLUSIONS

For a zone axis orientation and fixed probe size, composition mapping can be used to determine the interface sharpness and the extent of the AlAs oxidation process into the surrounding layers. AlGaAs layers with Al percentage less than 50% show very little oxidation. EDS elemental maps show a complete transformation of As into a gas form and the remaining aluminium is oxidised to form the Al_2O_3 phase. This chemical reaction results in the shrinkage of the total volume by 30% and hence produces a porous thin layer at the interface with the surrounding semiconductor. This result explains the symmetry observed in the oxygen EELS elemental map.

ACKNOWLEDGEMENTS

MAA acknowledges financial support from the EPSRC.

REFERENCES

Choquette K D, Geib K M, Chui H C, Hammons B E, Hou H Q Drummond T J and Hull R 1996 Appl. Phys. Lett. **69**, 1385
Choquette K D, Schneider R P, Lear K L and Geib K M 1994 Electron. Lett. **30**, 2043
Dallesasse J M, Holonyak J N Jr., Sugg A R and Richard T A 1990 Phys. Lett. **57**, 2844
Ebeling K J, Fiedler U, Michalzik R, Reiner G and Weigl 1996 Proc. Of 22nd European Conference on Optical Communications ECOC 96 81.
Egerton R 1996 Electron Energy Loss Spectroscopy in the Electron Microscope (New York: Plenum Press)
Guha S, Agahi F, Pezeshki B, Kash J A, Kisher D W and Bojarczuk N A 1995 Appl. Phys. Lett. **68**, 906
Sugg A R, Chen E I, Richard T A, Holonyak J N Jr and Hsieh J 1993 J. Appl. Phys. **74**, 797

Inst. Phys. Conf. Ser. No 164
Paper presented at Microsc. Semicond. Mater. Conf., Oxford, 22–25 March 1999
© *1999 IOP Publishing Ltd*

Focused ion beam specimen preparation for transmission electron microscopy studies of ULSI devices

H Bender

IMEC, Kapeldreef 75, B-3001 Leuven, Belgium, hugo.bender@imec.be

ABSTRACT: The different strategies used for TEM specimen preparation with focused ion beam are reviewed. Experimental data for the damage layer on the FIB prepared sidewalls and in the top layers of the structures are discussed for different types of materials. The amorphized layer on the silicon can be etched in-situ by a short exposure to XeF_2. The possibilities for high resolution structure imaging of FIB prepared specimens are discussed. In-situ Cu corrosion leading to the formation of CuI crystals on the FIB prepared specimens can occur due to a low I_2 background in the FIB after gas-assisted etching. The outlook for the future needs for FIB preparation of device materials is discussed.

1. INTRODUCTION

High quality specimen preparation is a key issue for any transmission electron microscopy work. For semiconductor materials the major routes for specimen thinning can be classified as (in order of historical emergence) : chemical thinning, ion beam milling, wedge polishing with the tripod tool and focused ion beam (FIB) milling. Whereas the chemical thinning technique is less suitable for device materials, the latter 3 methods are widely applied nowadays. The FIB technique for TEM specimen preparation was first reported just ten years ago (Kirk et al 1989) and has since then become an established method which is extensively used for microelectronics development and failure analysis. This is clearly due to its major strengths : site-specific preparation, plan-parallel specimens, high yield and reproducibility, speed and relative ease to use.

Several authors, e.g. Anderson and Klepeis (1997) and Hunt (1997), have recently discussed the strengths and disadvantages of the major specimen preparation techniques. De Veirman and Weaver (1999) have discussed a comparison of the ion milling, tripod polishing and FIB techniques as applied to the same type of sample and they showed that FIB produces the best results with a nearly 100% yield and high reproducibility.

This paper reviews in outline the major routes as used for FIB preparation and then focuses on the damage layer created by the Ga ion beam, the implications for high-resolution structure imaging and the corrosion effects observed during the FIB preparation of Cu metallization specimens.

2. FIB PREPARATION TECHNIQUES

2.1 Conventional milling procedure

The standard milling procedure as initially outlined by Kirk et al (1989) and subsequently further discussed by e.g. Park (1990), Young et al (1990) and Basile et al (1992) is still the most used FIB preparation method. A detailed procedure was recently discussed by Su et al (1997). A thin strip of silicon is cut from a device directly by sawing to a thickness of ~<50 µm, by thicker sawing combined with shallow sawing along the region of interest, or by a combination of sawing and grinding or tripod polishing. Subsequently the strip is mounted on a large slot Cu grid and with the Ga ion beam orthogonally incident on the top surface of the device, two trenches are milled step-wise with decreasing beam currents so that a thin wall of 50-200 nm remains (Fig. 1a). Before the milling the surface on the region of interest is protected by an in-situ deposited metal layer (Pt or W). The total transparent area is generally on the order of 10×5 to 20×10 μm². In order to reduce the slope of the thinned wall some tilting is often done during the final steps of the preparation (see further). To speed up the milling of the large trenches and to reduce the redeposition, gas-assisted etching with a halogen gas (I_2 or Cl_2) can be used.

2.2 Lift-out technique

The lift-out technique for FIB preparation was developed by Overwijk et al (1993) and has become popular in the last few years (Herlinger et al 1996, Shaapur et al 1997, Giannuzzi et al 1997, Stevie et al 1998). It has the major advantage that no sawing and grinding pre-thinning steps are involved so that the preparation can directly be done on device wafers or on packaged devices. The preparation involves: the milling of two craters near the region of interest, tilting the sample once the thinned slice is 0.5-1 μm thick and then cutting the bottom and one side fully and the other side partially, further thinning the slice from the top to the required thickness and finally cutting the partially cut side from the top (Fig. 1b). The free slice will then fall down in the milled crater and must be picked up with a fine needle and deposited on an amorphous C film which is deposited on a TEM-grid. For the pick-up, metal (W) needles (Overwijk et al 1993, Herlinger et al 1996, Shaapur et al 1997) or glass rods (Shaapur et al 1997, Giannuzzi et al 1997, Stevie et al 1998) are used. Whereas a high humidity is favourable for the metal needles, dry conditions are required with the glass rods as adhesion in that case is based on electrostatic attraction. An alternative extraction method is also proposed based on a replica method (Sheng et al 1997).

The thinned area is similar in size as with the conventional technique. The absence of any sidewalls due to the trenches makes the lift-out specimens more suitable for analytical TEM. The lift-out technique is also more useful than the conventional method for samples with layers which easily delaminate (e.g. Cu, PZT) during the sawing and grinding. As the milling is done on a large sample the orthogonal incidence of the ion beam is well guaranteed, whereas the tilt alignment in the FIB of the strips glued on the Cu grids for the conventional milling procedure is less accurate. For thick structures (e.g. 5-layer metallization) this can pose the problem that the thinned slice is not perfectly perpendicular to the device.

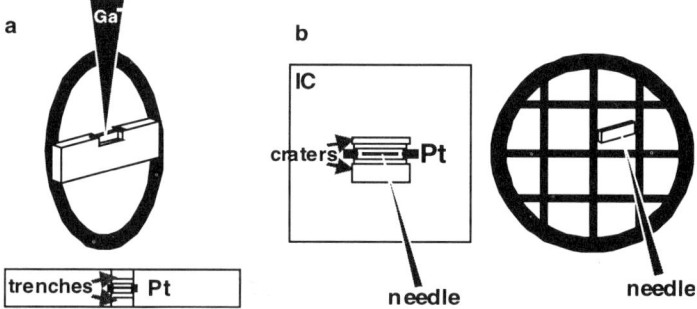

Fig. 1 : Conventional FIB milling (a) of large trenches in a strip (2000×50μm²×wafer thickness) cut out of the device, and lift-out technique (b) by milling craters, pick up of the 20×5×0.1 μm³ thinned slice and deposition on a TEM grid.

2.3 Wedge technique

In the earliest reports on FIB preparation (Kirk et al 1989, Young et al 1990) the sample was rotated a few degrees around the surface normal for the final linescans so that a wedge shaped specimen was obtained. Also Yamaguchi et al (1993) applied this method to get GaAs specimens ultimately thin. This rotation technique is not very suitable for the investigation of small features on device structures where a uniform specimen thickness is preferable. However, it is an excellent method to get specimens ultimately thin for HREM or EELS analysis in cases that hard layers are involved which unevenly mill during conventional ion milling or which are only present in relatively small regions on the devices.

2.4 Plan-view preparation

The majority of the TEM investigations of device structures involves cross-sectional preparation. With the increase in number of metallization layers, there is a growing demand for the study of the device structures at a specific level in plan-view. Young et al (1990) first reported the use of FIB milling for plan-view preparation of GaAs. They used Cl_2 gas-assisted milling from the backside through the 150 μm thick substrate. Obviously this method poses a problem of localization of the area of interest and of end-point detection. Leslie et al (1995), Anderson and Klepeis (1997) and Stevie et al (1998) discussed a far better plan-view preparation method which involves (tripod) polishing from the backside, and FIB milling from the sidewall. In this way the thinning can be done at any required depth in the structure. It can be combined with the pick-up method (Stevie et al 1998) or with conventional milling (Leslie et al 1995, and Fig. 2). A problem encountered with such samples when prepared in the metallization stack is the absence of a reference zone axis to align the specimen accurately along the device normal. This can be overcome by the preparation of a second crater in the same specimen in the silicon substrate, which can then be used in the TEM for alignment.

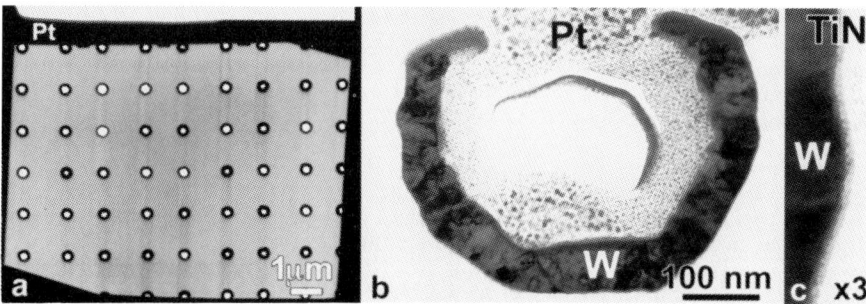

Fig. 2 : Plan-view FIB preparation through the middle of the W-plugs in an Al metallization. The thickness of the TiN barrier layer (c) can easily be measured whereas the presence of this layer is unclear on cross-section samples due to the geometrical blurring.

2.5 Marking techniques

For modern devices which have almost no topography, the site-specific capability of the FIB technique often requires a prior marking of the region of interest. Laser marking is frequently used for this purpose. Also the intermediate inspection with an optical microscope of the position of the rough FIB milled craters with respect to the feature of interest can be applied. Local in-situ deprocessing with gas-assisted etching in the FIB can be used to reveal buried structures of the metallization which can be used as reference points for design overlays. With the lift-out technique CAD-overlay methods (Burmer et al 1998) can be used more easily for localizing the device features, whereas this is less applicable in the narrow strips after sawing and polishing for the conventional milling procedure. Marking in the FIB by milling of small craters is reported prior to tripod wedge polishing preparation (Benedict et al 1998, Zhang 1998).

A combination of optical coordinates and FIB imaging of small etch pits is discussed for the investigation of low defect density studies on bare silicon wafers (Bender et al 1997). In that case marks are made by milling craters and deposition of Pt prior to the sawing of the strips for conventional FIB milling. For similar defect studies deposition of Pt blocks near the defects is applied for plan-view preparation with Ar ion milling (Bender et al 1999a).

3. SPECIMEN SLOPE AND DAMAGE LAYER

Since the emergence of FIB specimen preparation it was realized that the thinned slice has a sloped profile which is inherent to the beam profile (Kirk et al 1989, Park 1990) and therefore specimen tilting during the final preparation was introduced to compensate this slope (Yamaguchi et al 1993). Rather large tilt angles for both sides have been reported e.g. 4-5° for GaAs by Yamaguchi et al (1993) and for Si: 2° (Leslie et al 1995), 2.5° (Ishitani et al 1994), 3° (Pantel et al 1997), 0.5-2° (Young et al 1998). Based on modelling, an optimal tilt correction of 6° is found for W (Ishitani et al 1998). On the other hand it was shown experimentally that in open structures a slope of 1° can be obtained with a low beam current of 6 pA and of 4° for a high beam current of 7000 pA (Lipp et al 1996).

The damage introduced by the Ga^+ beam in the thinned foil has always been a major concern. Although beam induced lattice defects are not reported, the amorphized layer on the sidewalls reduces the quality of the high-resolution imaging and limits the minimal useful specimen thickness. A wide range of thicknesses as determined by various experimental techniques has been reported over the years for the damage layer on each sidewall induced by the 30 keV Ga^+ ion beam. That is, for Si: 20 nm (cross-section imaging, Kato et al 1998), 25 nm (EELS and wedge shaped specimens, Pantel 1996), 28 nm (cross-section imaging, Mardingly and Susnitzky 1998, Venables et al 1998). Also the damage depths calculated by Monte Carlo simulation for silicon samples vary strongly : 10 nm (Ishitani et al 1994), 15 nm (Ishitani et al 1998), and 30 nm (Mardingly and Susnitzky 1998). For W a damage depth of 8 nm is calculated (Ishitani 1995). For III-V materials a few data are reported as well : 4-5 nm on GaAs (diffraction analysis, Walker et al 1995), and 31 nm on InP (Yamaguchi et al 1993).

Various methods are proposed to reduce the damage depth: optimized final tilting (Leslie et al 1995), decreasing the energy of the Ga beam (Ishitani and Yaguchi 1996), short 5 keV sputtering on the tilted specimen (Young et al 1998), and Ar milling or wet etching after the FIB preparation (Kato et al 1998). The use of I_2 at the end of the thinning is shown to be unsuccessful for silicon (Kato et al 1998), but reduces the damage layer on InP to 2-5 nm (Yamaguchi et al 1993).

3.1 Experimental procedure

A set of dedicated specimens was prepared which allowed the simultaneous determination of the slope of the thinned slices and of the thickness of the damage layer. For this purpose a TEM specimen preparation is simulated by sputtering two large craters (8x4x4 μm^3) with a high beam current and subsequent further thinning of the intermediate material by milling of lines from both sides with decreasing beam currents as used under standard thinning conditions (Fig. 3a). In this way a TEM specimen is simulated with the thinned wall orthogonal to the actual TEM specimen. A similar technique has recently also been reported by Ishitani et al (1998) and Mardingly and Susnitzky (1998). The final beam current used is in the range 70-1000 pA, and the thinning is done with or without tilting during the last step of the simulated specimen. Two methods to reduce the damage depth are evaluated : 5 keV sputtering after 15° tilting, and exposure to XeF_2. After preparation of the simulated specimen, a Pt block is deposited over the thinned slice so that the craters are refilled (Fig. 3b) and subsequently standard thinning of the specimen is performed. During

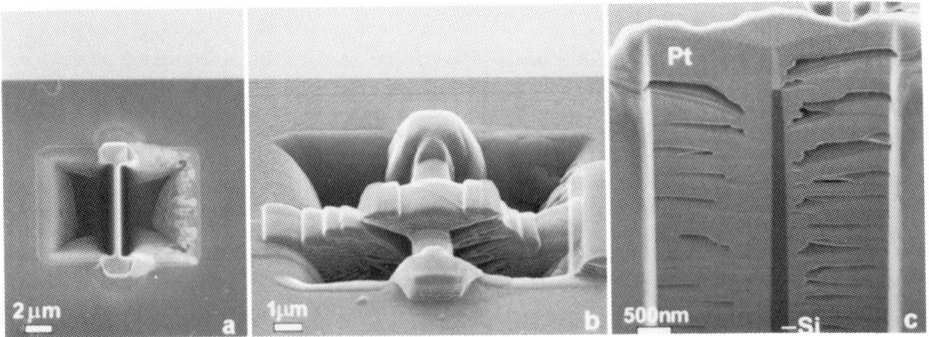

Fig. 3 : FIB preparation of the simulated specimens : a) simulated specimen prepared by milling craters and further standard line milling, b) deposition of a Pt block filling the craters near the simulated specimen (45° tilted view) and c) FIB cross-section image taken before the final thinning of the specimen (45° tilted view).

the preparation of the simulated specimen it is not intended to prepare the specimen ultimately thin, as this causes stability problems during the final preparation when only an obelisk of material remains supported by the weak Pt filling. For this reason also tilting is not used during the final thinning of the specimen.

3.2 Sidewall damage on silicon

Fig. 4a shows a TEM image of a specimen prepared by this method in a pure silicon sample. The top of the thinned slice is rounded which is due to the edges of the beam profile. This rounding is limited to the protective Pt layer, while the silicon shows a constant slope over a large depth (>6–8 μm for the used milling conditions). A summary of the slopes as obtained under different conditions is given in table 1. The slopes are measured on the TEM images as well as on FIB images (e.g. Fig. 3c) taken before the final thinning. The latter case avoids the curling of the specimen due to the bad support by the uneven filling of the Pt in the initial craters, as is sometimes the case for the TEM specimens. As expected, the slope increases for higher beam currents, but it saturates to 0.6–0.75° for the lower beam currents. In all cases the slope is smaller than previously reported. A nearly plan-parallel specimen can be obtained with a tilting of +/-0.4° during the final thinning (0.4° is the smallest step that the tilt angles can be measured on the FIB system).

The damage layer on the sidewalls has a thickness of 20-23 nm (Fig. 4c and 5, table 1) independent of the beam current and independent of the final tilt angle. It can indeed be reduced as proposed by Young et al (1998) by short sputtering of the tilted sidewall with a 5 keV Ga beam. In this way the amorphized layer is partially sputtered. The uniformity of the effect over the sidewall turns out to be difficult to control. Disadvantages of this method are also that the tilting and rotation of the specimen requires several intermediate images so that the risk for redeposition effects increases

Fig. 4 : TEM images of simulated specimens a) Si substrate, final milling with 150 pA beam, +/-0.4° tilt, b) W metallization structure, final milling with 150 pA beam, no tilt and with XeF$_2$ opened for 15 s after the milling, and c) Si substrate, final thinning with 70 pA beam, no tilt.

Table 1 : Sidewall slope and amorphous thicknesses on pure Si samples

| Beam current | Final thinning | | | Sidewall | | |
| | Tilt angle | Special condition | Slope (TEM) | Slope (FIB) | Amorphous thickness |
pA	°		°	°	nm
70	0	-	0.75	0.65	20-23
150	0	-	0.60	0.60	20-23
150	+/-0.4	-	0.25	0.08	20-23
150	+/-0.4	5 keV Ga	-	0	9 → 23
150	0	XeF$_2$ 60s 11pA line	-	<0	0
1000	0	-	1.15	1.1	20-23

and that the Ga implantation will be higher. In real structures with different types of materials, the differential sputter rates will also introduce uneven specimen faces.

A better method, which allows the total removal of the damage layer, is the exposure of the thinned slice to the XeF_2 gas after the final thinning, i.e. with the Ga beam blanked. It has been reported before that XeF_2 spontaneously etches silicon. The opening of the gas valve for 15 s with the needle inserted close to the specimen suffices for the full etching of the amorphized silicon. Opening the gas already during the final milling line of the Ga beam has a too strong etching effect leading to an uneven thinned slice.

3.3 Sidewall damage on Al metallization structures

Real specimens generally contain several layers of different materials, therefore similar experiments as for the pure silicon samples are performed on more representative layer structures. A first case considered consists of a stack $TiN/TiAl_3/Al/TiAl_3/SiO_2/Si$. The slopes (table 2) depend on the material: on the metal stack it is close to 1°, while in the silicon bulk a similar angle is found as in the pure silicon samples. The sidewall of the oxide is not flat but has a barrel shape : a strong increase just under the metal stack (slope ~ 6°), followed by a more steep slope in the middle of the layer and a negative slope just above the silicon. There is some indication that the barrel shape is at least accentuated by electron beam heating in the microscope.

The amorphized layer (table 2) on the silicon is similar to that on the silicon substrate samples. Also on the Al and $TiAl_3$ a thin amorphous layer is present, which is however absent on the TiN, which indicates that it is not due to redeposition. With a XeF_2 exposure for 30 s, the silicon below the oxide is some 20-30 nm thinner on each side compared to the oxide, and the amorphous layer is reduced to less than 3 nm on the silicon. Deep in the thinned crater a thin amorphous layer is still present, which is due to shadowing effects of the deep crater with respect to the gas. This shadowing effect also results in a slightly larger etching on the side directed to the gas nozzle. Hence the amorphized layer on the silicon can be fully etched away, while the oxide and the Al are not affected. Although there is the drawback that a thickness-step is created on the specimens, the XeF_2 etching can be applied for the HREM-study of defects in the silicon.

Table 2 : Sidewall slope and amorphous thicknesses on $TiN/TiAl_3/Al/TiAl_3/SiO_2/Si$ stacks

Final thinning			Sidewall		
Beam current	Tilt angle	Special condition	Slope (TEM)	Slope (FIB)	Amorphous thickness
pA	°		°	°	nm
150	0	-	Si : 0.8	Si : 0.83	Si : 20-23
150	0	XeF_2 30s, no beam	Al : 0.95 Si : 0.65	Si : 0.86	Al : 4.5 Si : 0 $\rightarrow$ 3

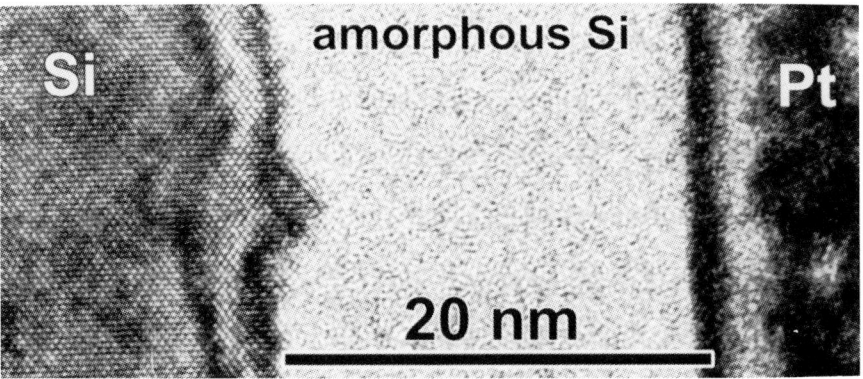

Fig. 5 : HREM image of the sidewall damage on a simulated specimen (final milling with 150 pA beam without tilting).

3.4 Sidewall damage on W metallization structures

Similar observations are also made on W/TiN/Ti/SiO₂/Si structures. The slope (table 3) of the sidewall in the W layer is larger, but in the silicon substrate the same steep slope as in bulk silicon is reached again. On the W no damage layer can be observed. XeF_2 allows the removal of the amorphous layer on the silicon (Fig. 4b) without damaging the other materials.

Table 3 : Sidewall slope and amorphous thicknesses on W/TiN/Ti/SiO₂/Si stacks

Final thinning			Sidewall		
Beam current	Tilt angle	Special condition	Slope (TEM)	Slope (FIB)	Amorphous thickness
pA	°		°	°	nm
150	0	-	W : 1.8 Si : 0.55	Si : 0.55	W : 0 Si : 19-23
150	0	XeF₂ 15s, no beam	W : 1.6 Si : 0.71	Si : 0.60	W : 0 Si : 0 → 4.5

3.5 Top surface damage and Pt/substrate interface layer

The damage caused by the ion beam in the top surface of the investigated structures during the localization of the region of interest and during the initial stage of the Pt deposition, is even more severe than the damage on the sidewalls. Electron beam navigation combined with in-situ electron beam induced Pt deposition in dual-beam FIB systems is the safest protection method to avoid damaging the top surface. With single beam FIB systems, the investigation of the top surface layer, as e.g. in partially processed or unpassivated devices, requires a protective layer to be deposited first (Bender et al 1998). It has been shown that for this purpose amorphous light element materials as e.g. poly- or amorphous silicon or CVD oxides, which can be deposited under standard device processing conditions are to be preferred. On the other hand, sputtered Au layers generally inadequately protect the bottom of structures with high aspect ratios. The use of plasma poly-merization layers (Miura et al 1996) and of resist layers (De Veirman and Weaver 1999) as surface protection have recently also been successfully applied.

Ishitani et al (1994) calculated by Monte Carlo simulation a damage depth of 30 nm for a 30 keV Ga ion beam orthogonally incident on silicon. From fig. 7 in the work of Mardingly and Susnitzky (1998) a damage depth of 60 nm can be derived. Lipp et al (1995), Walker and Broom (1997), Bender and Roussel (1997) and Bender et al (1998) have reported experimental thicknesses of the damage layer for different semiconductor materials. By comparison with an in-situ reference surface, it is shown that the dark layer in the top of the damage layer in the silicon has to be considered as an integral part of the interaction layer (Bender et al 1998). The experimental damage layer thicknesses in silicon for a normal incidence Ga beam are found to be : 30 keV : 60-65 nm, 10 keV : 35 nm and 5 keV : 21 nm. The thicknesses as obtained by cross-section TEM agree very well with measurements by means of Raman spectroscopy (Bender et al 1999b). These thicknesses are systematically larger than reported by Walker and Broom (1997).

For several materials, XTEM has shown that the interaction layer between the Pt and the substrate consists of a brighter and a dark layer (Bender et al 1998). For TiN and W also a darker layer is present above the bright layer. Auger depth profiling as well as PEELS/EDX line scans point out that the brighter layer is carbon rich, while the lower dark layer is a mixture of Pt, Ga and the substrate materials (Si or Ti-N). The upper dark layer is Pt enriched (Bender et al 1999b).

3.6 Discussion

The slopes of the thinned slice as found in this work are smaller than most data reported in the literature. Although, for multilayer samples, the slope slightly increases in harder materials (e.g. W), the overall slope is similar to that for pure silicon samples. This is also consistent with our finding that for thinning of 5-layer Al metallization structures, a tilt during the final thinning of only +/-4° is necessary to get a uniformly thinned slice. TEM observations on simulated specimens made through the via stack of such metallization structure indeed show that with that tilt angle a highly plan-parallel specimen can be obtained. As the slope is directly related to the Ga beam profile, the result will strongly depend on the ion column used and hence can be expected to be improved in newer FIB systems compared to older literature data.

On the other hand, the beam damage depth is related to the material and the energy of the ion beam. The present results for the damage layer on silicon sidewalls are in agreement with the data of Kato et al (1998) but are slightly lower than the data reported in other recent literature in which a similar preparation method is used (Mardingly and Susnitzky 1998, Venables et al 1998). As the protective Pt layer is deposited in-situ, artefacts due to e.g. oxidation of the surface before the deposition of the protective layer can be excluded. The amorphous thicknesses also agree with measurements with Raman spectroscopy (Bender et al 1999b).

4. HREM possibilities

The rather thick damage layer on both sides of the FIB specimens limits the possibilities for high quality structure imaging. Nevertheless high-resolution TEM results have been reported for several materials : SiO$_2$/Si (Park 1990, Ishitani et al 1994, Bender and Roussel 1997), TiSi$_2$/Si (Basile et al 1992), Al (Ishitani and Yaguchi 1996), C/SiC/Si (Tarutani et al 1992). For wide structures the preparation of the samples with the wedge technique guarantees the presence of a very thin area suitable for HREM (and EELS). Nevertheless also on conventionally milled FIB samples HREM imaging is possible on at least 50% of the specimens. The major limitation is actually not the presence of the damage layer, but the accurate control of the final thickness of the specimens. Promising techniques for in-situ monitoring of the specimen thickness in dual-beam systems have recently been discussed (Vieu et al 1993, Auvert 1998). Furthermore the possibility to etch the amorphous layer with the XeF$_2$ gas after the thinning, gives the possibility to increase the specimen quality for HREM imaging of the silicon severely. Some examples of HREM images obtained with the conventional milling approach are shown in Figs. 5 and 6.

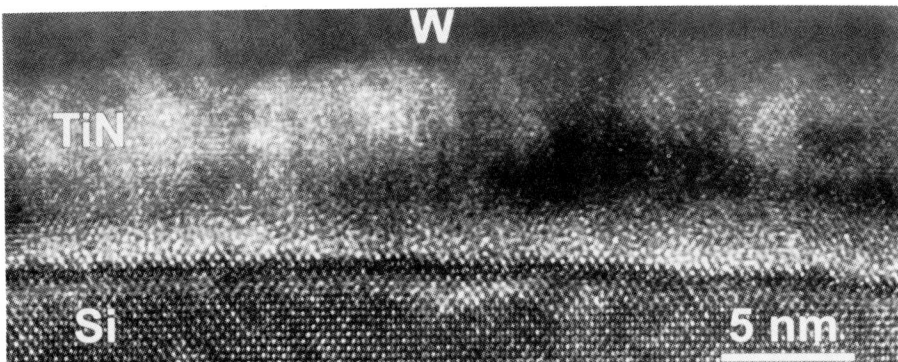

Fig. 6 : HREM image of the TiN/Ti barrier below a W contact plug on silicon.

5. Cu CORROSION IN FIB

Gas-assisted etching with a halogen gas (I$_2$ or Cl$_2$) is often used during the milling of the large trenches for TEM preparation. This has the advantage that the sputter rate is enhanced and that redeposition is minimized. For the preparation of TEM specimens with Cu metallization this turns out to be a disastrous procedure. Fig. 7 shows a TEM image of a structure with trenches in which a Cu-seed layer is deposited. After the FIB preparation, crystals are spread over the full area of the thinned region. By electron diffraction and EDS analysis these crystals are identified as CuI. It is shown that, in the FIB, all surfaces which have even been only slightly sputtered with the Ga beam strongly and quickly corrode even if only trace amounts of I$_2$ are present in the chamber (Bender et al 1999c). This corrosion is similar to that which has been reported after the exposure of Cu to Br$_2$ in FIB systems (Herschbein et al 1998). The corrosion on TEM specimens is still observable if I$_2$ has been used less than 24 h before the preparation of the specimen. This poses a strong limitation to the combined use of the FIB system for device modification during which halogen assisted etching is a standard procedure for cutting of Al lines and any FIB work on samples with Cu metallization.

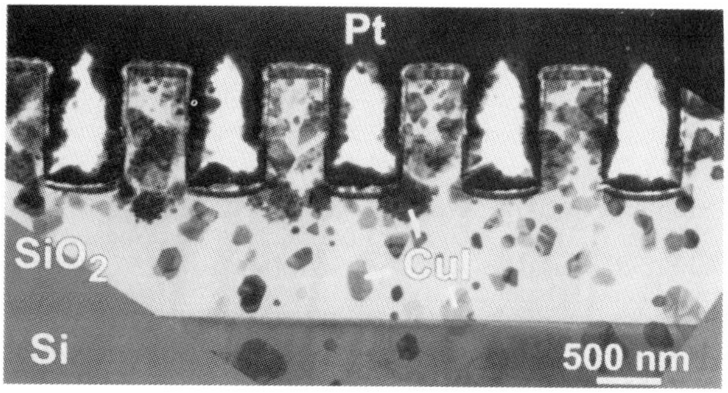

Fig. 7 : FIB prepared TEM specimen through trenches in which a thin Cu seed layer is deposited. Corrosion due to the I_2 background in the FIB chamber leads to the formation of CuI crystals on the specimen. I_2 was used in the FIB a few hour before the specimen loading and milling.

6. CONCLUSION AND OUTLOOK

Focused ion beam is a powerful technique for the site-specific preparation of TEM specimens of device materials. Highly plan-parallel specimens can be obtained with the present-day instruments by tilting the specimen during the final milling over only +/-0.4°. The sidewall damage in the thinned slice with a 30 keV Ga beam is for silicon on the order of 20-23 nm on each surface. For other device materials (Al, W, oxides, ...) it is much less. The damage layer on silicon can be etched in-situ with XeF_2 exposure, resulting however in a stepped surface in case of multilayer device materials.

A major improvement is still necessary in the possibilities to exactly control the thickness of the thinned specimen. Due to the rounding effect of the top of the thinned slice the real thickness is difficult to judge from a top view image and is nowadays still mainly based on experience. The demand for reproducibly very thin specimens is however large for applications that require high resolution structure imaging or electron energy loss spectroscopy. An even larger motivation for the preparation of ultimately thin specimens is the need to investigate barrier layers and plug filling in next generation devices with via nodes of only $0.18 \rightarrow 0.13$ µm. In such devices no topography except the top level metal will be visible in the top view image, so that an additional major problem becomes the positioning of the thinned slice through the middle of the plugs. As discussed by Hunt et al (1998) and Mardingly and Susnitzky (1998), the study of such plugs will require specimens thinner than 50-60 nm supposing that the positioning error is zero and specimens less than 30 nm for a positioning error of only 10 nm. As any intermediate imaging with the ion beam on the tilted TEM specimen must be avoided in the final thinning stage, the accurate preparation of such specimens will only be possible with dual-beam FIB systems so that the positioning can be controlled without tilting of the specimen with the SEM image. In such systems the thickness of the thinned slice can be monitored from the contrast of the SEM image as a function of the electron beam energy (Auvert 1998). The accurate milling of the thin and well-centered specimens requires a very fine ion beam and hence low beam currents will have to be used, of course with the drawback of increased preparation time. As these thickness and positioning requirements are so tough, plan-view FIB specimen preparation will become an important method for the study of barrier layers in next generation devices. As copper will replace the aluminum and tungsten in most of the metal layers, new gas chemistries will be necessary to avoid the in-situ copper corrosion. In conclusion, FIB preparation for next generation of devices will need dual-beam systems with still higher brightness ion beam and new gas chemistries.

ACKNOWLEDGEMENTS

It is the author's pleasure to acknowledge P. Roussel and S. Jin for their TEM contributions to this work and P. Van Marcke for most of the FIB preparation work. A. De Veirman and H. Peters (Philips Semiconductors, Nijmegen) are greatly thanked for the PEELS/EDS analysis. I. De Wolf and J. Chen are acknowledged for the Raman measurements. The TEM analyses were performed with the microscopes at the EMAT laboratory

602

(University of Antwerp, RUCA) which is acknowledged for the maintenance of the systems, the outstanding photography work and stimulating discussions. No analyses would have been possible without the materials inputs of the device processing groups of IMEC !

REFERENCES

Anderson R, Klepeis S J 1997 Mat. Res. Soc. Proc. Vol. **480**, 187
Auvert G 1998 2nd European FIB Users Group meeting, Copenhagen, unpublished
Basile D P, Boylan R, Baker B, Hayes K and Soza D 1992 Mat. Res. Soc. Proc. **254**, 23
Bender H and Roussel P 1997 Inst. Phys. Conf. Ser. **157**, 465
Bender H, Vanhellemont J and Schmolke R 1997 Jpn. J. Appl. Phys. **36**, L1217
Bender H, Van Marcke P, Drijbooms C and Roussel P 1998 AIP Conference Proc. **CP449**, 863
Bender H et al 1999a to be published
Bender H, Chen J, Dewolf I 1999b to be published
Bender H, Jin S, Vervoort I and Lantasov Y 1999c Proc. of the 25th International Symposium for Testing and Failure Analysis, ISTFA99, to be published
Benedict J P, Anderson R M, Klepeis S J 1998 Mat. Res. Soc. Symp. Proc. **523**, 19
Burmer C, Görlich S and Pauthner S 1998 Microelec. Rel. **38**, 987
De Veirman A and Weaver L 1999 Micron in press
Giannuzzi L A, Drown J L, Brown S R, Irwin R B and Stevie F A 1997 Mat. Res. Soc. Proc. **480**, 19
Herschbein S B, Fischer L S, Kane T L, Tenney M P and Shorein A D 1998 Proc. of the 24th International Symposium for Testing and Failure Analysis, ISTFA98, pp 127-130
Herlinger L R, Chevacharoenkul S, Erwin D C 1996 Proc. 22nd Int. Symp. Testing and Failure Analysis, ISTFA96, pp. 199-205
Hunt C A 1997 Proc. 23rd Int. Symp. Testing and Failure Analysis, ISTFA 97, pp. 97-101
Hunt C A, Zhang Y and Su D 1998 Mat. Res. Soc. Proc. **523**, 57
Ishitani T, Tsuboi H, Yaguchi T and Koike H 1994 J. Electron. Microsc. **43**, 322
Ishitani T 1995 Jpn. J. Appl. Phys. **34**, 3303
Ishitani T and Yaguchi T 1996 Microsc. Res. Techn. **35**, 320
Ishitani T, Koike H, Yaguchi T and Kamino T 1998 J. Vac. Sci. Technol. B **16**, 1907
Kato N I, Tsujimoto K, Miura N 1998 Mat. Res. Soc. Proc. Vol. **523**, 39
Kirk E C G, Williams D A and Ahmed H 1989 Inst. Phys. Conf. Ser. **100**, 501
Leslie A J, Pey K L, Sim K S, Beh M T F and Goh G P 1995 Proc. 21th Int. Symp. Testing and Failure Analysis, ISTFA95, pp 353-362
Lipp S, Frey L, Franz G, Demm E, Petersen S and Ryssel H 1995 Nucl. Inst. Methods Phys. B **106**, 630
Lipp S, Frey L, Lehrer C, Frank B, Demm E and Ryssel H 1996 J. Vac.Sci. Technol. B **14**, 3996
Mardingly J and Susnitzky D W1998 Mat. Res. Soc. Proc. **523**, 3
Miura N, Tsujimato K, Kanehara R, Tsutsui N and Tsuji S 1996 Proc. 22nd Int. Symp. Testing and Failure Analysis, ISTFA96, pp. 95-100
Overwijk M H F, van den Heuvel F C and Bulle-Lieuwma C W T 1993 J. Vac. Sci. Technol. B **11**, 2021
Pantel R 1996 Workshop notes "Focussed Ion Beam", ESREF96, Enschede, The Netherlands
Pantel R, Auvert G and Mascarin G 1997 Microelec. Engineering **37/38**, 49
Park K 1990 Mat. Res. Soc. Proc. **199**, 271
Shaapur F, Stark T, Woodward T and Graham R J 1997 Mat. Res. Soc. Proc. **480**, 173
Sheng T T, Goh G P, Tung C H and Wang L F 1997 J. Vac. Sci. Technol. B **15**, 610
Stevie F A, Irwin R B, Shofner T L, Brown S R, Drown J L and Giannuzzi L A 1998 AIP Conference Proc. **CP449**, 868
Su D H I, Shishido H T, Tsai F, Liang L and Mercado F C 1997 Mat. Res. Soc. Proc. **480**, 105
Tarutani M, Takai Y, Shimizu R 1992 J. Vac. Sci. Technol. B **31**, L1305
Yamaguchi A, Shibata M and Hashinaga T 1993 J. Vac. Sci. Technol. B **11**, 2016
Young R J, Kirk E C G, Williams D A and Ahmed H 1990 Mat. Res. Soc. Proc. **199**, 205
Young R J, Carleson PD, Da X, Hunt T and Walker J F 1998 Proc. 24th Int. Symp. Testing and Failure Analysis, ISTFA98, pp 329-336
Venables D, Susnitzky D W and Mardingly A J 1998 AIP Conference Proc. **CP449**, 667
Vieu C, Pepin A, Assayag G B, Gierak J, Ladan F R 1993 Inst. Phys. Conf. Ser. **134**, 385
Walker J F, Reiner J C, Solenthaler C 1995 Inst. Phys. Conf. Ser. **146**, 629
Walker J F, Broom R F 1997 Inst. Phys. Conf. Ser. **157**, 473
Zhang H 1998 Mat. Res. Soc. Proc. **523**, 45

Inst. Phys. Conf. Ser. No 164
Paper presented at Microsc. Semicond. Mater. Conf., Oxford, 22–25 March 1999
© *1999 IOP Publishing Ltd*

High sensitivity FIB-SIMS analysis of semiconductor devices

M Hughes, D S McPhail, R J Chater and J Walker[1]

Materials Dept., Imperial College of Science, Technology and Medicine, London SW7 2BP UK
[1] FEI Europe Ltd., Cottenham, Cambridge CB4 4PS UK

ABSTRACT: High energy gallium ion beams used in focused ion beam (FIB) instruments produce secondary ion fluxes from silicon that are comparable with sputtering by inert gas primary ion beams. Use of reactive primary ions such as oxygen and caesium lead to a large increase in the ionisation yield. This paper reports a systematic study of the ionisation yield enhancement by pre-treating the surface with implanted oxygen ions prior to compositional imaging by FIB-SIMS microscopy for common dopant and impurity elements. Yield enhancements of nearly three orders of magnitude are reported.

1. INTRODUCTION

Secondary ion mass spectrometry (SIMS) is a versatile analysis technique that can be used to investigate the spatial distribution of all elements and their isotopes with high sensitivity in vacuum compatible solids. As a result SIMS has become an indispensable tool in the semiconductor industry for the analysis of electronic and opto-electronic devices. With the shrinkage of semiconductor device sizes to the sub-micron scale and below, it has been necessary to restrict the lateral area of the analysis, placing great demands on the efficiency of analytical systems. The development of the liquid metal ion source (LMIS) has, however, allowed SIMS analyses to be conducted with a very high lateral resolution, in the order of tens of nanometres (Orloff 1993, Swanson 1994). The major difficulty over the more conventional oxygen or caesium ion source systems is in the reduction in the secondary ion yield. Reliable LMIS systems use typically either gallium or indium primary ions which have yields that are more comparable to inert gas beams in contrast to the yield enhancements obtained with sputtering by either reactive oxygen or caesium ions. Therefore at present LMIS SIMS systems have a high resolution but an undesirably low sensitivity (Schumacher et al 1991).

A limited amount of research into increasing the secondary ion yield of LMIS SIMS has been carried out (Crow et al 1995, Kinoshita et al 1993, Schumacher et al 1991). The findings of this work broadly agree with previous work that the introduction of oxygen locally into the sputtering system increased the secondary ion yield. It was the aim of this investigation to quantify the secondary ion yield enhancement that may be obtained by introducing oxygen into the system by ion beam implantation directly into silicon samples, a method that had not been previously examined.

We have determined how the secondary ion yield enhancement varied with energy and angle of the oxygen ion beam by measuring the depth distribution of the implanted oxygen using inert gas SIMS depth profiling and we have correlated this with the yield enhancement to the sub-surface oxygen distribution. The oxygen implantation was carried out using an ATOMIKA 6500 microprobe and the depth profiling and imaging of the samples using an FEI FIB 200 instrument equipped with a SIMS detection system (Dingle et al 1995).

2. EXPERIMENTAL METHOD

2.1 Oxygen implantation into Silicon

Oxygen was implanted into silicon using the ATOMIKA 6500 SIMS instrument to form thin crater base oxide layers at normal incidence with $^{16}O_2^+$ beam energies of between 3 and 15keV. The oxygen build up in the 250 x 250 μm² crater base of the sample was monitored using secondary ion detection for steady state sputtering conditions. A set of samples was similarly produced at incidence angles of between 5° and 35° at a fixed incident $^{16}O_2^+$ beam energy of 15keV.

2.2 Oxide layer characterisation with the FIB gallium primary beam

Depth profiles of the oxide layers in the crater base were obtained by raster scanning the gallium ion beam over an area of 50 x 50 μm² inside the original 250 x 250 μm² craters. The gallium primary beam parameters were fixed at 30keV and 5.7nA with a pixel width (beam spot diameter) of 150nm and pixel separation of 150nm. The profiling process was monitored by collecting Si^+ (28amu) and O^+ (16 amu) secondary ions collected from an area of 25 x 25 μm² (25% gate) at the centre of the crater. This electronic gating ensures that the recorded depth profile is free of crater edge effects. The depth profiling was continued through the oxide layer until the secondary ion signal indicated that the steady state sputtering conditions in the silicon substrate had been reached.

Two reference samples were used to interpret the secondary ion signals and calibrate the beam sputter-rate; unprocessed silicon and a 200μm thick thermal oxide layer. Analysis of the thermal oxide layer required a compensating electron gun. The depth of the craters formed after analysis were measured using an interferometric optical surface profiler (Zygo NewView200).

2.3 Oxygen irradiation and imaging of a silicon device

Craters were eroded on the surface of an Intel 486 microprocessor using five different oxygen primary ion beam energies between 3 and 15keV to form a surface oxide layer. These crater areas were imaged using an elemental mapping technique on the FEI FIB200. This technique involved scanning an area of 60 x 60 μm² inside one of the oxidised areas on the devices surface using the Ga^+ beam. The beam energy used was 30keV and the beam current 940pA. The dwell time of the ion beam on each pixel was automatically set by the FEI FIB200 so as to achieve the best contrast. The intensity of the Si^+ (28amu) secondary ions pixel by pixel produced by the Ga^+ ion beam bombardment gives a gray scale image of the area.

3. EXPERIMENTAL RESULTS AND DISCUSSION

3.1 Oxide layer characterisation with the Ga^+ LMIS primary beam

Figs. 1 and 2 show the calibrated secondary ion depth profiles of the oxide layer sputtered by Ga^+ beam bombardment. The relative sensitivity factor (RSF) demonstrates the increase in the

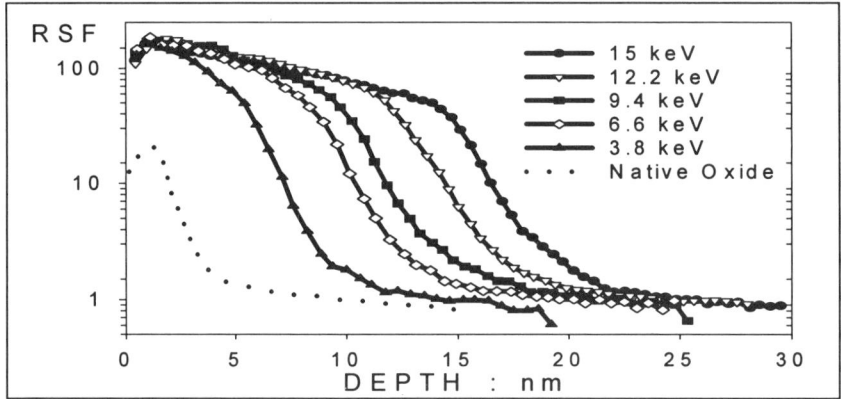

Figure 1.   Normalised $^{30}Si^+$ secondary ion depth profiles of the oxide layers implanted at normal incidence at five different $^{16}O_2^+$ beam energies and profiled by a 30keV Ga^+ beam.

sensitivity of the instrument to the detection of silicon positive ions with the presence of oxygen relative to the sensitivity without oxygen. It is clear from the figures that the presence of implanted oxygen significantly increases the secondary ion yield.

The results show that the RSF does not vary significantly with the oxygen beam energy but there is a clear decrease as the beam angle is increased above ~ 25°. This would support the oxide layer microstructure model proposed by Beyer (1994). It is also clear that oxide layers formed by oxygen implantation in silicon result in a higher secondary ion yield enhancement than the native oxide layer.

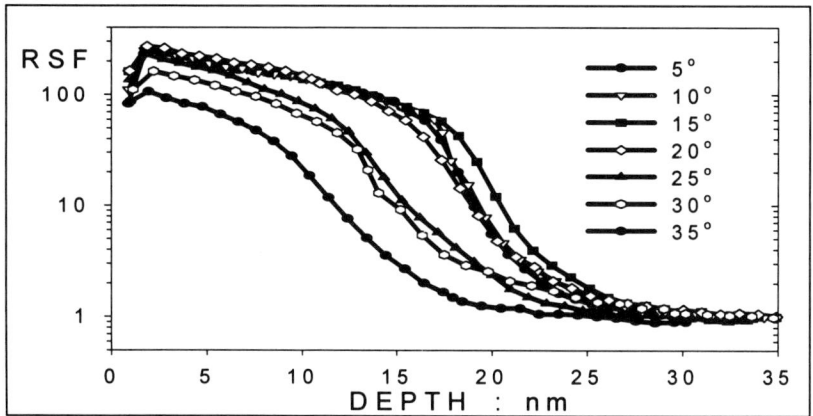

Figure 2.   Normalised $^{30}Si^+$ secondary ion depth profiles of the oxide layers implanted at various $^{16}O_2^+$ beam angles of incidence and profiled with a 30keV Ga^+ beam.

The gently sloping plateau region, which is a feature of all the depth profiles, is surprising since a stoichiometric layer should give a constant yield enhancement throughout its thickness. According to Beyer (1994), oxide layers formed during the implantation of O_2^+ ions in silicon vary in thickness from 11nm to 33nm with beam energies of 3 to 15keV. The projected range of gallium ions in silicon dioxide as calculated by TRIM95 is 21.1nm. Therefore the variation of the RSF in the plateau region could be due to gallium ion mixing of the oxide layer with the underlying silicon substrate, since the projected range of the gallium ions is comparable to the thickness of the oxide layers. As a consequence of this mixing effect, the concentration of oxygen at the sputtered surface would decrease below the stoichiometric concentration of silica as the depth profile progressed. Because the oxygen concentration is decreasing, there would be a corresponding decrease in the secondary ion yield enhancement. Thus the RSF profiles in Fig. 1 at the oxide layer/substrate interface show the same decay length.

In Fig. 2, this increase in sensitivity is approximately maintained for a wide range of incidence angles from normal to ~ 25° which suggests a useful independence from sample surface topography. The plateau RSF decreases with increasing implantation incidence angles. These secondary changes in sensitivity enhancement are believed to be due to a change in transmission coefficient of the SIMS instrument. By contrast the profiles in Fig. 2 show increasing decay lengths with increasing incidence angle of implantation. Thus these results suggest that the sensitivity enhancement is linked to the stoichiometry of the silicon dioxide layers that are produced by oxygen implantation in silicon.

3.2 Imaging a silicon device with the Ga^+ LMIS primary beam

SIMS images of the microprocessor are shown in Fig. 3. It can be seen from the images themselves that oxygen pre-irradiation affords a significant increase in contrast and brightness. Software analysis revealed that the image taken after oxygen irradiation was 23 times brighter than the image taken without oxygen irradiation.

No oxygen implantation. Maximum pixel dwell time : 2.0msec/pixel

15keV oxygen implantation. Pixel dwell time 0.68 ms/pixel

Figure 3.   LMIS-SIMS images of a 486 microprocessor obtained by monitoring the intensity of $^{30}Si^+$ ions produced by a 30 keV Ga^+ beam. Image dimensions, 60 x 60μm^2.

4. CONCLUSIONS

It was found that normal incidence pre-irradiation with O_2^+ ions of different energies, ranging from 3 to 15 keV, produced roughly the same increase in the sensitivity to the detection of Si^+ ions, that is an average increase of ~125 times. However, the depth of irradiated silicon that, once sputter removed, will produce a secondary ion yield enhancement was found to decrease with decreasing oxygen ion intensity. Pre-irradiation of silicon with 15keV O_2^+ ions at various angles of incidence, ranging from 5° to 35°, produced a roughly constant increase in the sensitivity of ~250 times for angles up to ~25°; thereafter the sensitivity enhancement decreased. These results suggest that the sensitivity enhancement is linked to the stoichiometry of the silicon dioxide layers that are produced by oxygen implantation in silicon.

Pre-analysis oxygen irradiation of silicon samples is a useful method of increasing the sensitivity and thus the accuracy of Ga^+ ion beam SIMS depth profiling and imaging. Whilst this has only been demonstrated for a silicon matrix ion, it is expected to be equally demonstrated for dopants and impurities in silicon

Pre-irradiation with oxygen ions in order to increase the sensitivity of Ga^+ ion beam SIMS imaging however has produced some very encouraging results. The pre-irradiation of a silicon micro-electronic device generated a clear increase in the quality of subsequent images in terms of both brightness and contrast. The RSF for the image was found to have increased by 23 times.

ACKNOWLEGEMENTS

One of the authors (RJC) thanks the Hilary Bauerman Trust for a grant to attend this conference. FEI UK Ltd is thanked for their generous assistance.

REFERENCES

Beyer G, 1994 PhD thesis, Imperial College of Science, Technology and Medicine, London
Crow G A, Christman L and Utlaut M 1995 J. Vac. Sci. Technol. B **13**, 2607
Dingle T, Bickford C U, Cook D B, Do D D, Gerlach R L, Li J-Z and Rumussen J 1995 Proc. 10th Int. Conf. on Secondary Ion Mass Spectrometry (SIMS X) pp 517-520
Kinoshita J, Shiozawa K and Yamasaki T 1993 Proc. 9th Int. Conf. on Secondary Ion Mass Spectrometry (SIMS IX) pp 569-572
Orloff J 1993 Rev. Sci. Instrum. **64**, 1105
Schumacher M, Migeon H N and Rasser B 1991 Proc 8th Int. Conf. On Secondary Ion Mass Spectrometry (SIMS VIII) pp 49-53
Swanson L W 1994 Appl. Surf. Sci. **76**, 80

Inst. Phys. Conf. Ser. No 164
Paper presented at Microsc. Semicond. Mater. Conf., Oxford, 22–25 March 1999
© *1999 IOP Publishing Ltd*

Focused ion beam control of sample cleaving for high resolution microscopy

R M Langford, C M Reeves, J F Findlay[1], C E Jeffree[1] and J G Goodall

Microelectronic Imaging and Analysis Centre, Department of Electronics and Electrical Engineering, The University of Edinburgh, Edinburgh, EH9 3JL, UK.
[1]Electron Microscopy Facility, Science Faculty, The University of Edinburgh, Edinburgh, EH9 3JL, UK.

ABSTRACT: The microelectronics industry places stringent demands on sample preparation for high resolution imaging. There is often a requirement for cross sectioning of samples to a positional accuracy of ± 500 nm or better. In this paper, we report on our latest work in which focused ion beam etching is used to control the location of the fracture plane during sample cleaving. This approach provides a method for fast preparation of site-specific cross sections for subsequent imaging by scanning electron microscopy, transmission electron microscopy and scanning probe microscopy. Details of the new technique are presented along with examples of its application in the field of microelectronics.

1. INTRODUCTION

As the dimensions of microelectronic devices continue to shrink cross sectional microscopy is becoming increasingly important for construction analysis of state of the art microelectronic devices. Examples of recent developments in this field include transmission electron microscopy (TEM) analysis of interconnects (Pantel et al 1997) and dielectrics (Cerva 1991) and scanning probe microscopy (SPM) analysis of dopant (Barrett et al 1996) and potential distributions (De Wolf et al 1996).

Cross sections for SPM and scanning electron microscopy (SEM) can be prepared by cleaving and polishing; cross sections for TEM can be prepared by tripod polishing, small angle cleaving or focused ion beam (FIB) milling. A problem using cleaving and polishing for the preparation of SPM and SEM cross sections is the difficulty in controlling the position of the cross section. Developed techniques, such as precision polishing for the preparation of cross sections, are suitable for arrays or device structures greater than 5 µm but become time consuming to implement and are less reliable for device structures at 0.5 µm dimensions and below (Madsen et al 1997). The standard FIB technique for the preparation of site specific cross sections for FIB and SEM imaging involves removing a wedge of the material and imaging at angles up to 45°. These cross sections are not readily compatible with SPM techniques which usually require open access to the region of interest.

In this paper, we report a new cleaving technique, which uses FIB milled grooves to control the position of the fracture plane during sample cleaving. The technique is useful for the fast preparation of site-specific cross sections for a range of high resolution microscopy techniques. The technique is described and factors which affect the positional precision of the cleave and the quality of the cross sections are discussed. Preliminary results are presented for SEM and SPM and a refinement of the technique is given for use with metallized devices.

2. THE FIB CLEAVING TECHNIQUE

The new technique is based on the concept of using FIB etched V-grooves to control the precise location of the cleave plane during cleave cross sectioning. The region of interest is identified and a surrounding rectangle 2-3 mm by 30 mm is diced out using either a diamond scribe or a dicing saw approximately centred, with the required cleave plane parallel with the shortest edge of the sample.

Using a 30 keV Ga$^+$ ion beam, V-shaped grooves are milled across the sample in the direction of the required cleave plane, leaving a deliberate gap at the precise region of interest to avoid ion damage or ion implantation at this location.

A simple approach is shown in figure 1(a) which guides the position of the cleave to ±1 µm while a more advanced pattern of grooves is shown in figure 1(b) which guides the position of the cleave to ±500 nm or better. Details of the length, width and depth of the V-shaped grooves are shown in table 1 for the more precise approach, along with details of the associated FIB parameters.

The grooves are designed to overlap by 5 µm and with their apexes aligned along the cleave plane. For the narrow grooves, to increase their etched depths, iodine is used to chemically assist the milling process. Figure 2 shows a FIB cross section through the three different types of V-groove, used

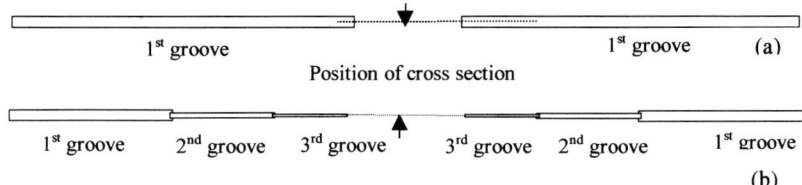

Figure 1 – Etch patterns used for (a) guiding the cleave to ±1 µm and (b) guiding the cleave to ± 500 nm or better.

Groove	Width (µm) (at surface of the Si)	Length (µm)	Depth (µm) (at surface of the Si)	Beam Current (pA)	I$_2$ used to chemically assist milling
1	1	500	4	12000	no
2	0.4	50	2	1000	yes
3	0.25	50	1.75	70	yes

Table 1 – Beam currents, pattern dimensions and associated FIB parameters used to etch grooves in figures 1(a) and 1(b).

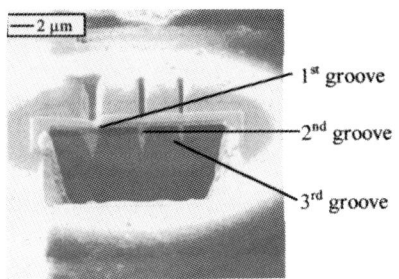

Figure 2 – FIB cross section of grooves milled using parameters in Table 1.

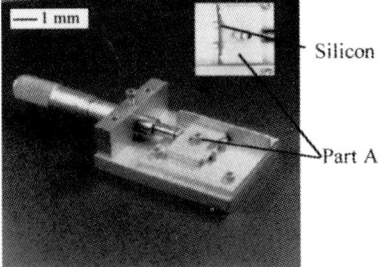

Figure 3 – The cleaving jig.

when guiding the cleave to within ±500 nm or better, milled using the parameters given in table 1.

The sample is then removed from the FIB workstation and is placed in a purpose built cleaving jig, shown in figure 3. The sample is cleaved by screwing Part A of the jig inwards. The insert in figure 3 shows the bowing of a sample just before it cleaves.

Figure 4 shows a FESEM secondary electron image of a coated silicon wafer cleaved using the new technique. The image clearly shows that the cleave plane has followed the V-groove producing a high quality cross section through the region of interest. In this example, the wafer has been prepared with a stack of thin films consisting of a thermal oxide layer, a polysilicon layer and an ECR (electron cyclotron resonance) deposited oxide layer.

Figure 5 shows a FIB secondary electron image of the transition region from the edge of the V-groove to the region of interest. In this image it can be seen that the cleave plane is coincident with the V-groove.

In this work 3 inch (100) silicon wafers were used, cleaving along the (110) plane. The V grooves were etched using a FEI FIB 200 system and the FESEM and AFM images obtained using a Hitachi S4500II FESEM and a Digital Instruments D5000 AFM respectively.

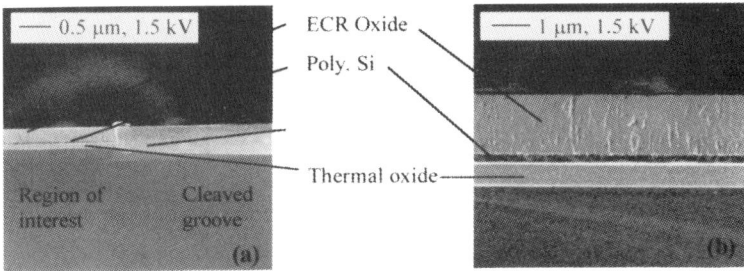

Figure 4 – FESEM secondary electron image of cleaved sample (a) a low magnification image showing the region of interest and the V-groove and (b) a higher magnification image of region of the interest.

3. APPLICATIONS

We believe the new technique will prove useful in many areas of high resolution microscopy especially TEM and SPM. As an example, we have explored compatibility with the SPM as shown in Figure 6. In this example a tapping mode atomic force microscopy (AFM) image has been obtained of the stack of thin films shown cross sectioned in Figure 4. In this image the thermal oxide layer is visible but, unlike in the FESEM image, the polysilicon cannot be differentiated from the ECR oxide.

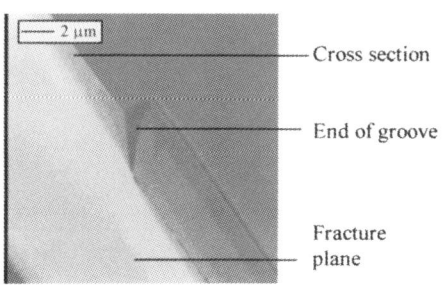

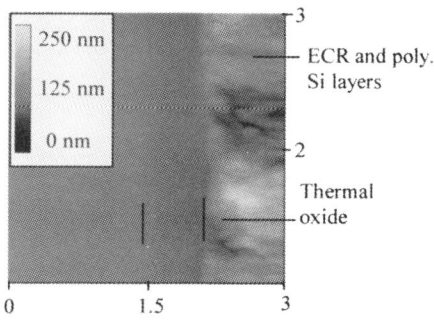

Figure 5 – FIB secondary electron image of a cleaved groove.

Figure 6 – AFM image of thermal oxide, polysilicon, ECR oxide layers.

4. METALLIZED SAMPLES COMPATIBILITY ISSUES

For metallized devices it was found that the metal layers could tear on cleaving. To overcome this problem an approach was chosen whereby the sample was reloaded in the FIB system to implement a short FIB 'clean up' step. Figure 7 (a) shows a FESEM secondary electron image of a FIB 'cleaned up' cross section through a MOSFET. It should be noted that the FIB 'clean up' introduces ion species at the region of interest. However, a further improvement in positional accuracy to ± 50 nm is also achieved. The FIB 'clean up' is further demonstrated in Figure 7(b) which shows a FESEM image of the FIB 'cleaned up' MOSFET having recessed the gate and field oxides by dipping the sample in HF(40%):H$_2$O (1:40) for 6 seconds.

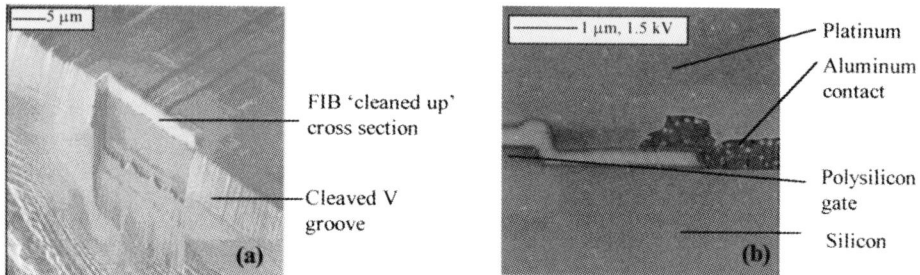

Figure 7 – (a) FIB secondary electron imaged of a cleaved cross section MOSFET after FIB 'clean up' and (b) FESEM secondary electron image of FIB 'cleaned up' MOSFET.

5. CONCLUSION

We have introduced a new technique for the preparation of site-specific cross sections for microscopy. This technique consists of FIB milling grooves in the substrate, to control the position of the fracture plane during cleaving. We have demonstrated good compatibility of this technique with FESEM, AFM and FIB imaging and we anticipate future applications to TEM analysis and other SPM approaches. For metallized microelectronic devices it was observed that the metal is prone to tearing. This effect is removed by a short FIB 'clean up' step.

REFERENCES

Barrett M, Dennis M, Tiffin D, Li Y and Shih C K 1996 J. Vac. Sci. Technol. **B 14(1)**, 447
Cerva H 1991 Inst. Phys. Conf. Ser. **No. 117** 155
De Wolf P, Trenkler T, Clarysse T, Caymax M , Vadervorst W, Snauwaert J J and Hellemans L
    1996, Scanning Microscopy **Vol. 10, No. 4**, 937
Pantel R, Auvert G and Mascarin G 1997 Microelectronic Engineering **37/38**, 49
Madsen L D, Weaver L and Jacobsen S N 1997 Microscopy Research and Technique **36**, 354

Inst. Phys. Conf. Ser. No 164
Paper presented at Microsc. Semicond. Mater. Conf., Oxford, 22–25 March 1999
© 1999 IOP Publishing Ltd

Interface-related artefacts in iodine-ion milling of ZnSe on GaAs grown by metal-organic vapour phase epitaxy

A C Wright

Advanced Materials Research Laboratory, NEWI Plas Coch, Mold Road, Wrexham LL11 2AW

ABSTRACT: Iodine-ion milling of TEM cross sections from ZnSe epilayers on GaAs substrates is shown to alter the structure and composition of the Zn_3As_2 interfacial layer formed by the initial zinc-exposure step prior to nucleation of the ZnSe. Here it is shown that iodine effectively leaches out the zinc component leaving an amorphous arsenic layer. Argon ion milling, in contrast, leaves the Zn_3As_2 layer intact. However, deliberate amorphisation of the Zn_3As_2 enables a clearer detection of its presence and allows one to distinguish better the coverage of very thin ZnSe epilayers.

1. INTRODUCTION

Iodine ion-milling of TEM specimens has been shown to greatly reduce ion damage in ionic II-VI compound semiconductors (Cullis et al 1985). Normal argon milling induces large numbers of stacking faults which are not present in the original material. The use of iodine has also proved useful for the elimination of indium droplets when thinning InP and its alloys (Chew and Cullis 1984).

However, the original work on ion-related damage of II-VI materials did not examine in detail the interface between a III-V substrate such as GaAs and a II-VI epilayer such as ZnSe. We have previously shown that the interface between ZnSe and GaAs can contain a thin layer of the II-V compound Zn_3As_2, depending on how the ZnSe epilayer was nucleated (Wright et al 1998). For growth by metal-organic vapour phase epitaxy (MOVPE), long (several minutes) exposure of the GaAs substrates yields the formation of Zn_3As_2 by a direct heterovalent exchange replacement mechanism. In molecular beam epitaxy (MBE) of ZnSe on GaAs, this reaction does not appear to occur: an initial zinc exposure is the usual step prior to 2D growth of low defect density ZnSe (Wu et al 1996).

The Zn_3As_2 layer is epitaxial with the GaAs but subsequent growth of ZnSe has very poor structural quality with a marked 3D growth mode, quite the opposite of that obtained when a similar procedure is used in MBE growth of ZnSe on GaAs.

Here, it is shown that iodine ion milling of such material for cross-sectional analysis results in amorphisation of this Zn_3As_2 layer by leaching out of the zinc leaving behind a layer rich in arsenic. Argon ion-milling, in contrast, dose not destroy the integrity of the Zn_3As_2 interlayer as shown by high resolution lattice imaging where the superstructure of this complex material can be directly observed. Even short exposures of this material to iodine ions results in loss of zinc and amorphisation, although this effect can result in enhanced visibility of the interfacial phase when diffraction contrast is used.

2. EXPERIMENTAL

All growth experiments were performed in a horizontal uncooled fused silica MOVPE reactor operated at atmospheric pressure. The details of the growth have been reported elsewhere (Wright et al 1998). All ion-milling was performed, after standard mechanical thinning, in an Atom-Tech 700 series mill fitted with both argon gas and a dual iodine reservoir for independent operation of the saddlefield ion guns. Samples were only exposed to iodine gas for a final five minutes after perforation had occurred. Only argon gas was used during the main length of milling time to reduce problems with the vacuum pumping system and with gun instability (aluminium cathodes) associated with the continuous use of iodine. The angle of incidence was 13° throughout and the specimen was uncooled in all cases. Gun voltage was 5 keV with termination at 2.5 keV. Gun currents were 40-50 microamperes with chamber pressure of 2×10^{-5} torr. TEM was performed in a Philips EM430T operated at 300kV using both diffraction contrast (g=(220)) and, for interfacial studies, lattice imaging along both the [110] and [1-10] poles. Cross-sectional specimens for TEM were prepared using standard techniques viz. bonding two wafers face to face using EPO-TEK 353ND epoxy and hand polishing to a 0.3 micron flat finish on both sides with no dimpling to a final thickness of 25 microns. These blanks were mounted onto Mo support grids (1mm hole).

3. RESULTS

Cross-sections prepared using iodine-ion milling clearly revealed that while an interfacial phase had occurred when GaAs substrates were exposed to 90 minutes supply of the zinc source, figure 1, diffraction contrast imaging indicated, from the lack of Bragg contours, that it was not of a crystalline nature.

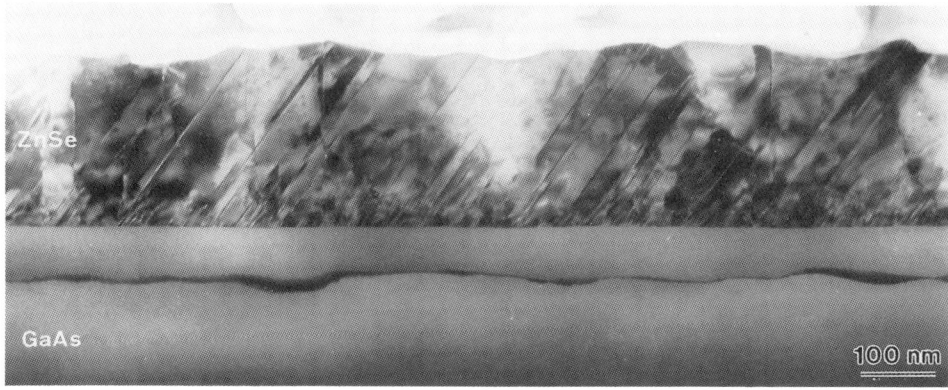

Figure 1. Iodine thinned cross-section of ZnSe on GaAs(100) with interfacial layer formed by zinc-exposure step prior to ZnSe growth.

This was confirmed by examination of a sample given a shorter (15 minutes) zinc exposure at high resolution, figure 2, where the interfacial material can clearly be seen to be amorphous in nature while the ZnSe layer is obviously crystalline, albeit highly defective with antiphase boundaries running vertically from the plane of the interface. Examination of many samples prepared for cross-section using iodine-ion milling showed no evidence for any crystallinity in this interfacial layer when prepared with a final iodine ion milling step.

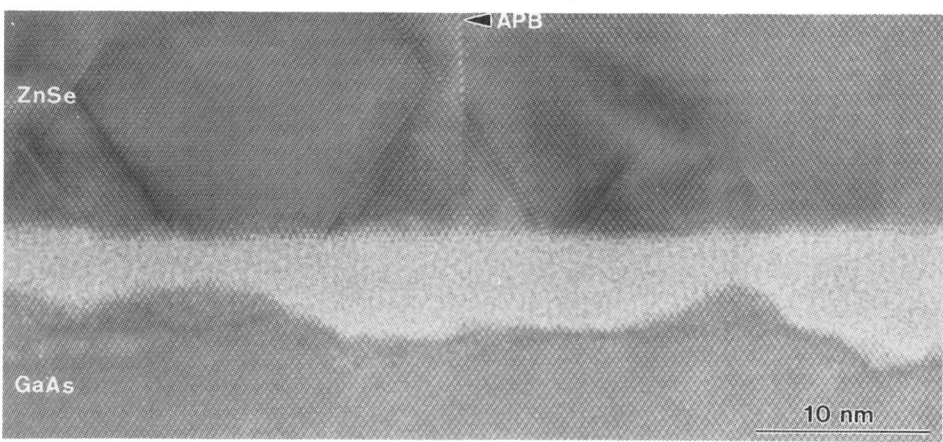

Figure 2. HRTEM cross-section showing interfacial amorphous layer between ZnSe and GaAs.

Microprobe analysis (nanoprobe, spot 5, LaB$_6$ source ~5 nm probe size) of this amorphous interfacial layer revealed that it was composed mostly of arsenic as shown in the spectrum of figure 3. No change to the 1:1 stoichiometry of the surrounding ZnSe and GaAs regions was observed. The small observed zinc level in the amorphous layer may be due to probe spreading into the adjacent ZnSe rather than a real signal from the layer.

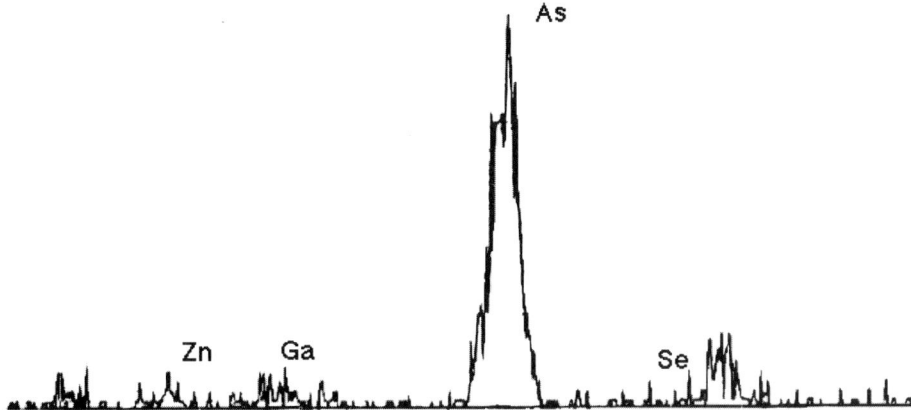

Figure 3. EDX microprobe spectrum from interfacial phase seen in figure 1 showing composition to be mainly composed of elemental arsenic.

By contrast, the same type of sample prepared using only argon gas throughout the entire milling procedure clearly showed a fully crystalline interfacial phase as seen in figure 4.

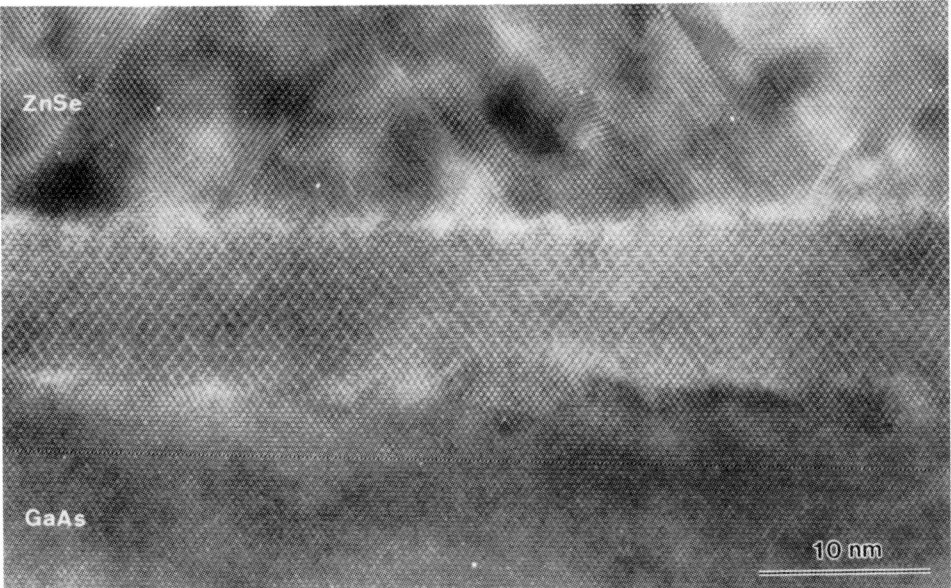

Figure 4. Argon-only thinned ZnSe on GaAs showing interfacial, fully crystalline, Zn$_3$As$_2$ layer.

The interfacial material is clearly epitaxial with both the GaAs substrate and the ZnSe layer and has a marked superstructure with a doubling of the {111} fringe periodicity. This phase has been identified by TEM in conjunction with EDX microanalysis, XPS and grazing incidence XRD as Zn$_3$As$_2$ and these structural studies are reported in full elsewhere (Wright et al 1999).

614

4. DISCUSSION

This work has clearly shown that the use of iodine gas as a milling agent for TEM cross-section preparation can give artefacts not present in the original material. For studies of II-VI layers on III-V substrates, this presents a choice. We can either use iodine and obtain the best possible quality specimen for the II-VI layer and ignore the interfacial region or use argon to obtain a chemically unchanged interfacial layer (if present) and suffer a high level of ion-damage created stacking faults in the II-VI layer. Of course, with careful specimen preparation, both outcomes can be achieved in two stages, first with argon and then with iodine. The use and effect of specimen cooling has not been examined and may have the effect of reducing a reaction between the iodine and the zinc within the Zn_3As_2 layer which clearly leaches out the zinc as a volatile iodide. This is a clear consequence of using a reactive milling gas instead of an inert gas such as argon.

Because of this reactivity, there may be other materials, which if grown as layers, would be corrupted by the use of reactive gases such as iodine. Other III_3-V_2 compound such as Zn_3P_2 have attracted interest as photovoltaic materials (Pawlikowski 1988) and may suffer similar zinc leaching effects as observed here. Clearly, cooling the sample should be investigated but its success would depend upon how well heat transfer from the sample can be effected.

However, for studies of the nucleation of ZnSe on these interfacial films, amorphisation can be used to enhance the visibility of the ZnSe layer when it is very thin as shown below in figure 5.

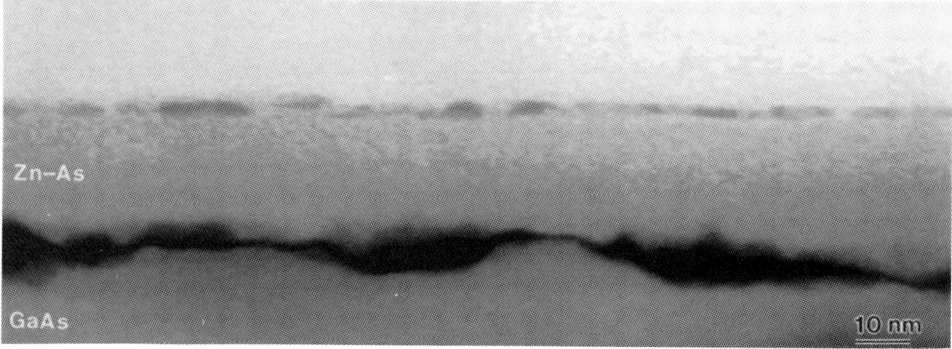

Figure 5. Thin ~1.5 nm ZnSe layer on amorphised Zn-As showing only partial coverage.

5. CONCLUSIONS

This work shows that the use of iodine as a reactive milling agent for TEM specimen preparation can yield artefacts, amorphisation and gross compositional changes, at the interfaces of II-VI films on III-V substrates particularly where interfacial phases such as the compound Zn_3As_2 have formed. Specimen preparation protocols for such materials must therefore be modified to take these effects into account. The work also indicated that other materials may be affected in terms of their structure and composition in a similar way and that care must be taken in the choice of milling gas for TEM specimen preparation.

ACKNOWLEDGEMENT

The author wishes to thank Dr. T L Ng and Dr. N Maung of the Advanced Materials Research Laboratory at NEWI for the MOVPE growth of the ZnSe/GaAs samples and to the Higher Education Funding Council for Wales (HEFCW) for support.

REFERENCES

Chew N G and Cullis A G 1984 Appl. Phys. Lett. **44**, 142
Cullis A G, Chew N G and Hutchison J L 1985 Ultramicroscopy **17**, 203
Pawlikowski J M 1988 Rev. Sol. State Sci. **2**, 581
Wright A C, Gnoth D, Ng T L and Maung N 1998 Appl. Surf. Sci. **123/124**, 555
Wright A C, Ng T L and Maung N 1999 Philos. Mag. **A70**, in press
Wu B J, Haugen G M, DePuydt J M and Salamanca-Riba L 1996 Appl. Phys. Lett. **68**, 2828

Inst. Phys. Conf. Ser. No 164
Paper presented at Microsc. Semicond. Mater. Conf., Oxford, 22–25 March 1999

Scanning capacitance microscopy imaging of state-of-the-art MOSFETs

R N Kleiman, M L O'Malley, F H Baumann, J P Garno and G L Timp

Bell Laboratories, Lucent Technologies, Murray Hill, NJ 07974

ABSTRACT: Using a scanning capacitance microscope (SCM) we have studied cross-sectioned n- and p-MOSFETs with gate lengths approaching 60 nm. In a homogeneous semiconductor, the SCM measures the depletion length, determining the dopant concentration. When imaging a real device there is an interaction between the probe tip and the built-in depletion of the p-n junction. Using a device simulator we can understand the relation between the SCM images and the position of the p-n junction, making the SCM a quantitative tool for junction delineation and direct measurement of the electrical channel length.

1. INTRODUCTION

With the aggressive scaling of MOSFETs approaching gate lengths of <100nm, the need for direct feedback to process simulation in becoming increasingly important. Scanning capacitance microscopy is a promising technique which convincingly provides quantitative information about the junction position in a MOSFET. Consequently junction depths and effective channel lengths can be measured. The interpretation of the SCM images is not completely straightforward, however with the assistance of process and device simulators we have been able to understand the imaging mechanism in more detail in the vicinity of the junctions, which is the crucial region where quantitative understanding is essential. This understanding allows us in principle to determine the entire dopant profile of the device, but it also allows us to develop a simple method for using the SCM to determine junction positions. While the entire dopant profile in unarguably valuable, a simple method for junction profiling addresses many of the crucial needs for device design.

At the same time, the aggressive scaling of MOSFETs dimensions places a high demand for steady improvements on the spatial resolution needed for characterization tools. We have attempted to understand the limits to the spatial resolution with SCM, theoretically with simulations, and experimentally by working with smaller probe tips. This has provided us with resolution suitable for sub-100nm devices.

2. PARALLEL PLATE MOS SIMULATIONS

There has been considerable effort to model the imaging with the SCM (Huang 1995 and Kopanski 1996). To introduce the technique we first consider the simplest possible geometry, that of the parallel plate capacitor (see Fig. 1). The results for the capacitance, C as a function of voltage, V, oxide thickness, t_{ox} and dopant concentration (assumed uniform), n are well known (Sze 1981) and will not be repeated here. For a variety of reasons it is more useful to measure dC/dV rather than just C. As a result one parameter is lost, but that is unavailable experimentally anyway due to the extremely low value of the measured capacitance relative to the ambient stray values. For the MOS capacitor there is a peak in dC/dV, which is positive for n-type materials and negative for p-type materials. The peak value of dC/dV is shown in Fig. 2 showing good sensitivity to dopant concentration in the relevant range for modern devices.

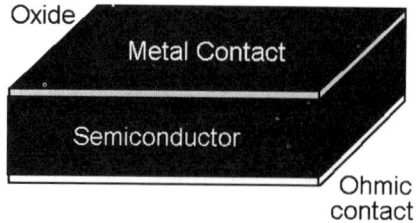

Figure 1: Device geometry for parallel plate capacitor MOS simulations.

3. MEASUREMENT TECHNIQUE

We have used a commercial (Digital Instruments) SCM to obtain the images presented in this investigation. The SCM consists of a contact-mode atomic force microscope (AFM) with a 915 MHz tuned circuit coupled to the tip-sample capacitance and is configured by a modulation technique to measure dC/dV. Modifications have been made such that both the magnitude and sign of the dC/dV signal are reflected in the data. This is essential to the ability to perceive the difference between n-type and p-type doped regions.

The semiconductor device samples have been cross-sectioned using standard techniques, finishing with a colloidal silica emulsion.

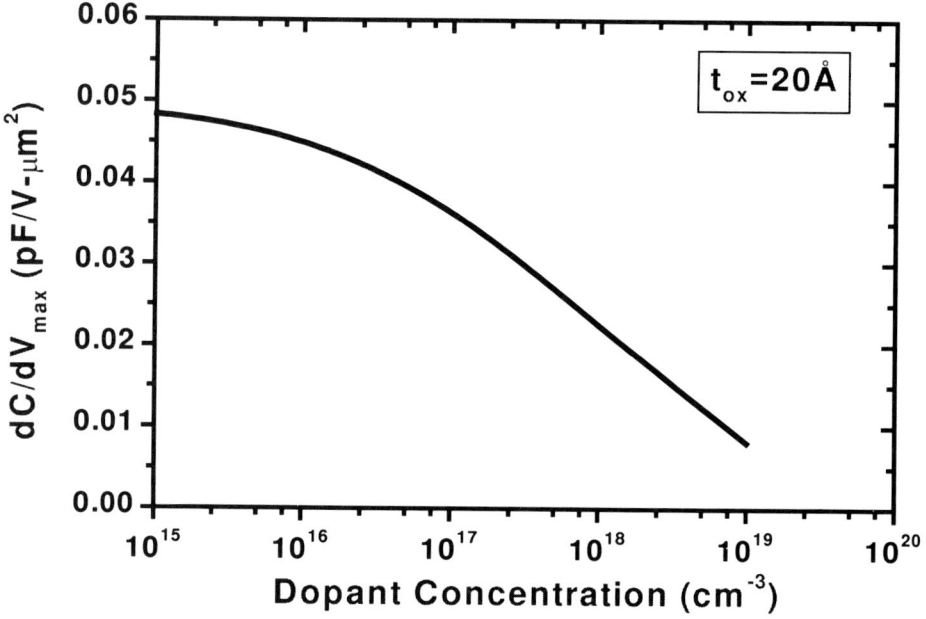

Figure 2: Calculated peak dC/dV value for a 20Å thick SiO_2 dielectric on a Silicon substrate of given concentration.

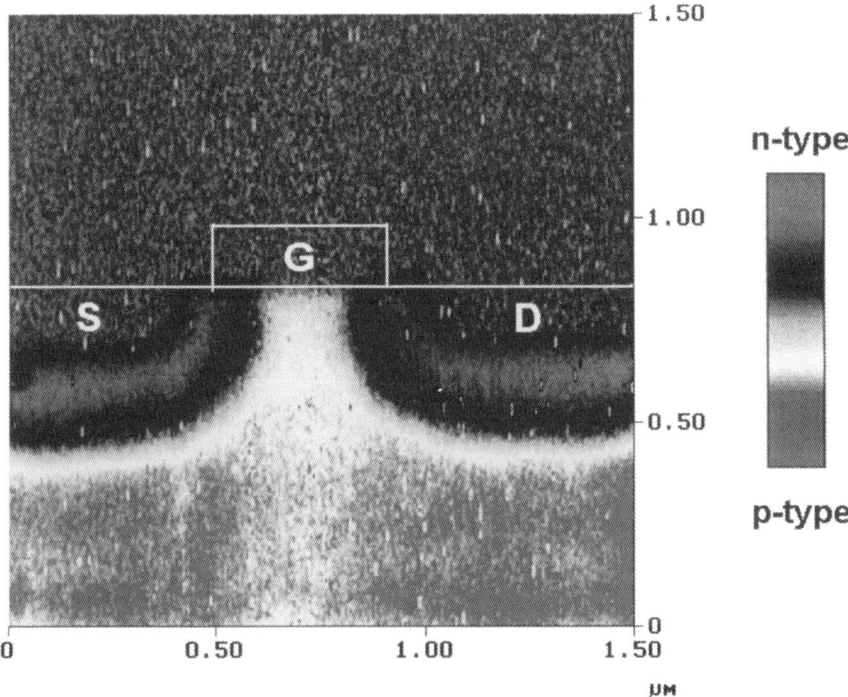

Figure 3: SCM Image of a 0.35 micron process n-MOSFET. The source, drain and gate are labeled. The red/yellow regions are p-type, the magenta/black regions are n-type, and the green region corresponds to dC/dV=0. The gate and the heavily-doped parts of the source and drain have dC/dV~0. The apparent junction position is the boundary where dC/dV=0.

4. SCM RESULTS ON 0.35μm PROCESS nMOS DEVICES

We began our investigations by studying a relatively large nMOSFET manufactured in 0.35 μm technology. A typical SCM image of this device is shown in Fig. 3. It clearly shows the source and drain regions, their heavily-doped parts with dC/dV~0, and their more lightly-doped parts a magenta color. The gate and the insulating region above it both have dC/dV~0, the former because it is heavily-doped and the latter because it has no carriers to deplete. The SCM image shows excellent contrast between the n- and p-type regions, clearly delineating the boundary between them, here shown in green and having dC/dV=0. The dimensions of the junction depths and of the effective channel length can be easily measured and are in good agreement with expectations.

However, when we change the voltage between the probe tip and the sample (all of the transistor leads are tied to each other and to the ohmic back contact) we see that the interpretation of the images is considerably more complicated than previously expected (see Fig. 4). The images change in such a way that the apparent pn junction (where dC/dV=0) moves with bias voltage, and consequently the sign of dC/dV changes in the active regions of the device. Clearly, for the relevant part of devices, SCM cannot be thought of as a simple dopant profiler, where the dopant concentration can be determined by a calibration such as shown in Figure 2 (modified for the realistic tip geometry). In order to use SCM as a quantitative tool we need to understand the change of the images with applied bias voltage.

618

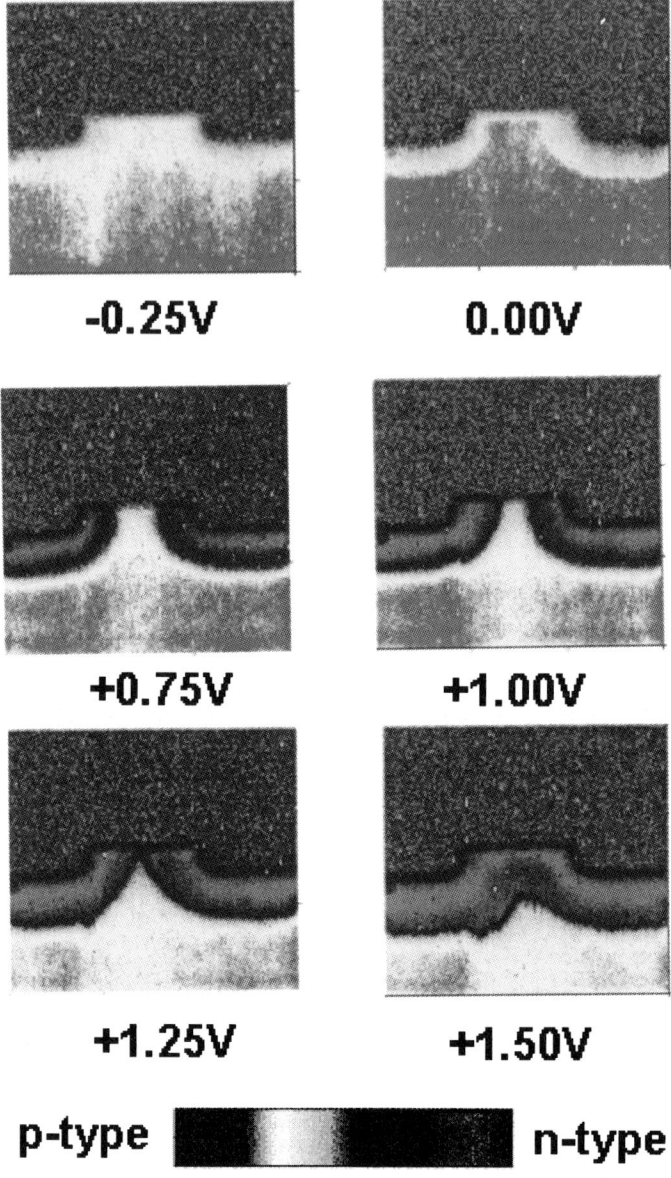

Figure 4: SCM images of a 0.35 micron process n-MOSFET with changing voltage V_B between the tip and the sample. The apparent junction position (dC/dV=0) moves significantly with V_B due to the interaction between the tip and the built-in depletion region of the device. The upper portions of each image have dC/dV=0, since that region is insulating. When V_B = +1.5 V the p-n junctions have moved to pinch off the channel. When V_B = −0.75V the junctions have moved to meet the heaviest parts of the source/drain implants. At intermediate voltages the junction sits beneath the gate.

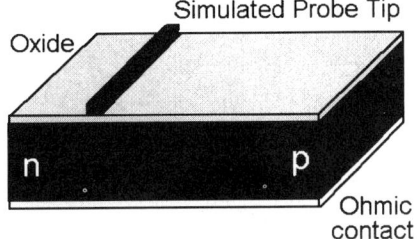

Figure 5: Device geometry for 1d simulations across a pn junction. The probe tip is chosen to be 100Å wide and infinitely long in our simulations. While this is far from a realistic tip shape, it is sufficient to introduce the new length scale of the tip diameter.

5. 1d MOS SIMULATIONS

The simulated dopant profiles were generated using the process simulator PROPHET which creates a 2-D doping profile with the subtleties of a fabricated pn junction. The electrostatic state of the system, including the carrier distributions, the potential distribution, as well as C and dC/dV were calculated using PADRE. PADRE simulates semiconductor device behavior, considering both minority and majority carriers and all relevant semiconductor physics while solving Poisson's equation, the continuity and energy balance equations for the relevant geometry.

While our simulations do not use a realistic tip shape, we believe that they capture the essential device physics to understand the SCM imaging. The geometry we have chosen (see Figs. 5 and 8) introduces the length scale of the tip radius, which must be small compared to typical device dimensions. A 2d simulator is only required for device regions with high radius of curvature.

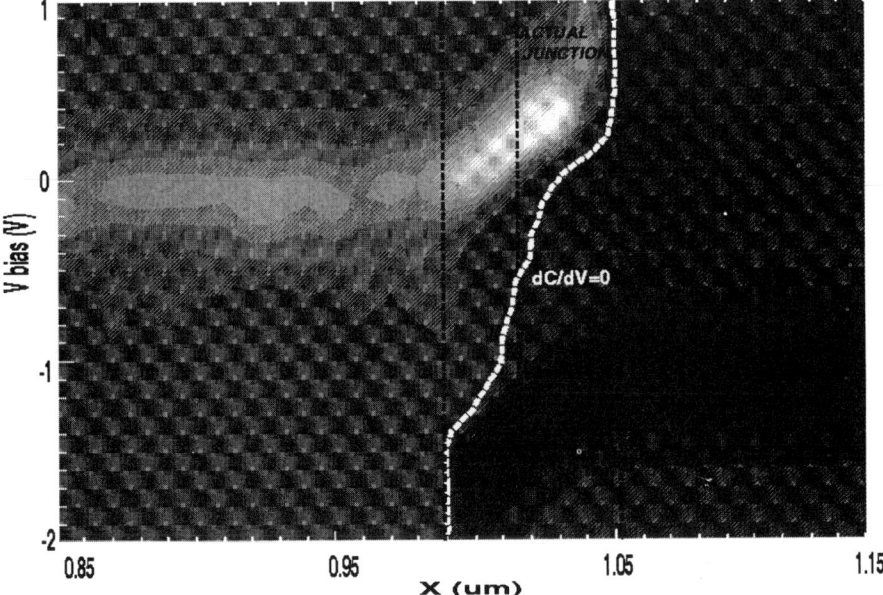

Figure 6: dC/dV plotted *vs* electrode position, x, and bias voltage, V_B. The physical p-n junction is at x = 1.015μm. The substrate is p-type with $n = 5 \times 10^{17} / cm^3$, while the implant side is n-type with $n = 3 \times 10^{18} / cm^3$. The light region corresponds to dC/dV > 0, while the dark region corresponds to dC/dV < 0. The apparent p-n junction where dC/dV = 0 is the nearly vertical contour separating the light and dark regions. The voltage scale is shifted from the experimental case, due to the work functions of the tip and the bottom electrode.

620

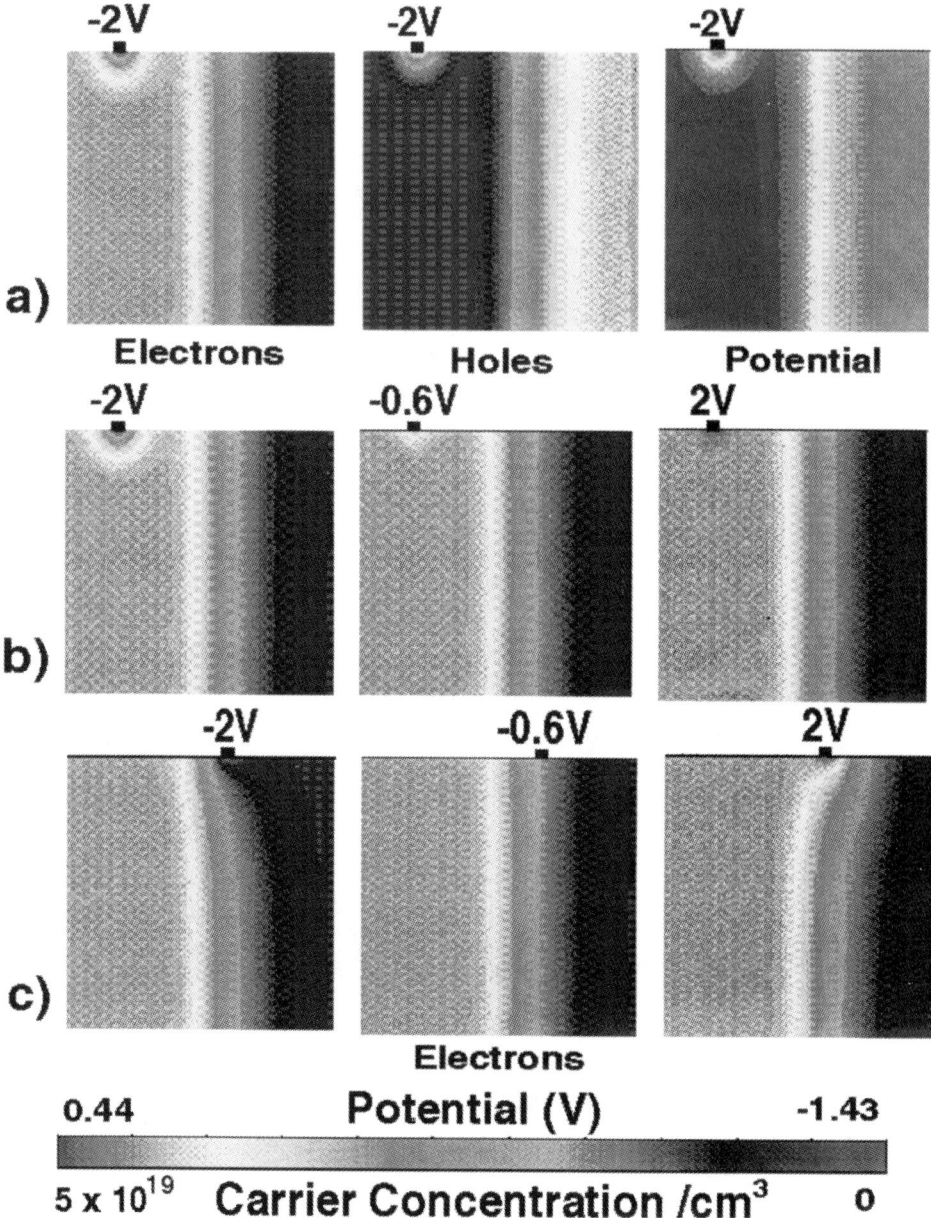

Figure 7. Simulation data for a pn junction. a) electrode placed over the n-type region: at −2V the electrons under the electrode are depleted, the holes are accumulated and there is a potential drop across the depletion region. b) electrode placed over the n-type region: the electron concentration changes as the electrode bias is changed from −2V to 2V. c) electrode placed over the depletion region: the real position of the pn junction moves as the electrode bias is changed from −2V to 2V.

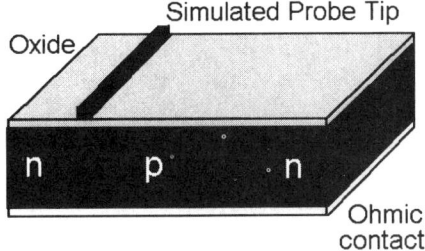

Figure 8: Device geometry for 1d simulations across an npn junction. This geometry is suitable to study the dC/dV response in the channel region. As the junctions get close to one another this simulation geometry is required to study the interaction between the junctions.

5.1 p-n junction results

Our simulation results are shown in Fig. 6 for a particular pn junction profile (Kleiman 1997). It clearly simulates the movement of the apparent pn junction with changing bias voltage. Outside of the junction regions we see simple peaks in dC/dV as expected from parallel plate results. In the junction region the sign of dC/dV changes with voltage as observed experimentally.

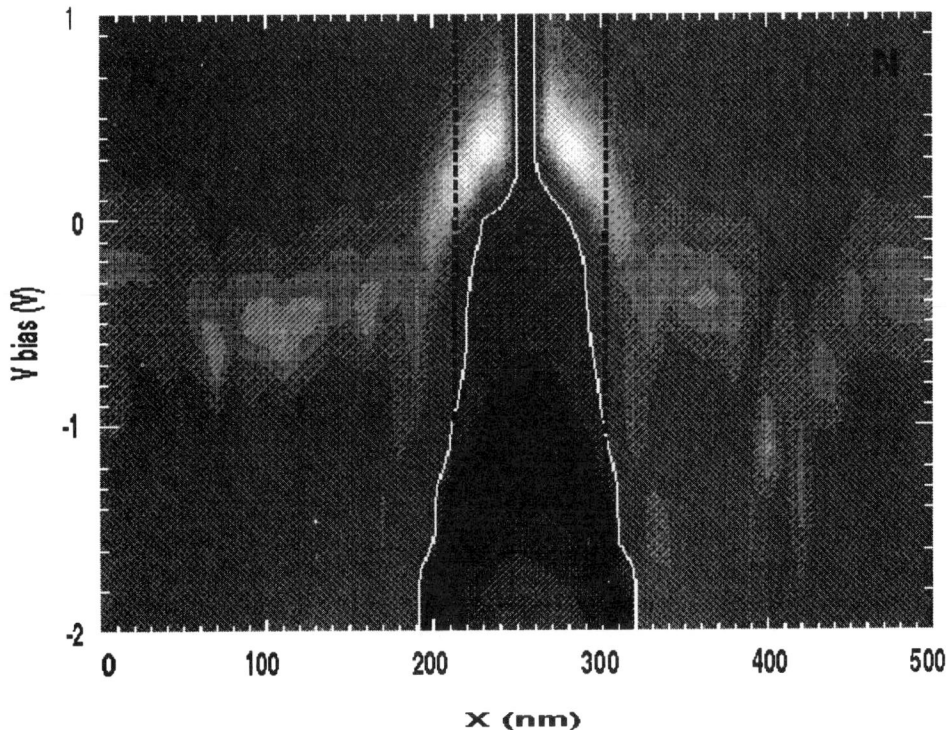

Figure 9: dC/dV *vs* electrode position, x, and bias voltage, V_B The substrate doping is $5 \times 10^{17}/$ cm^3 and the source/drain doping is $2 \times 10^{19}/cm^3$. The effective channel length, $L_0 \sim 90nm$. The p-n junctions are shown as dotted red lines. The light region corresponds to dC/dV > 0, while the dark region corresponds to dC/dV < 0. The apparent p-n junctions, where dC/dV = 0 are the white contours lines. The apparent channel length, L^*, is the spacing between those two lines and changes with V_B.

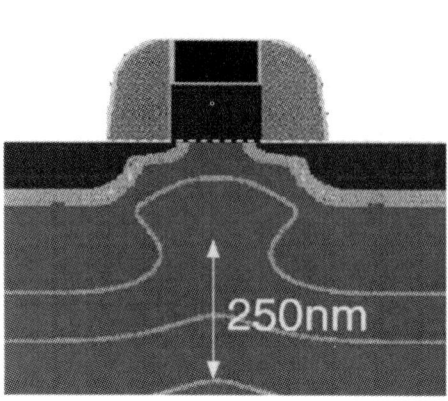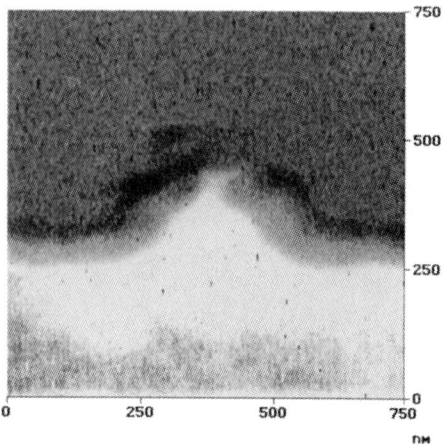

Figure 10: PROPHET process simulation of the intended doping profile of a 0.15 μm gate length nMOSFET and an SCM image of the same device at a representative bias voltage. The device is fabricated with an HDD and LLD, whose features are easily observed in the SCM image. The spacing between the junctions at the semiconductor surface is ≈0.12μm.

5.2 Physical significance of results

To better understand the physical significance of our results we investigated the carrier distributions at various points of Fig. 6 using our device simulator (O'Malley *et al* 1999). Typical results are shown in Fig. 7. Fig. 7 c) shows clearly that the junction position is actually moved by the tip, which can easily sweep electrons or holes into the depletion region of the device. However when the tip voltage is mid-band the junction is unperturbed. Note also that in Fig. 6 the dC/dV=0 boundary intersects the actual junction position (set in the simulation) when the voltage is roughly midway between the peaks in dC/dV outside of the junction region. This suggests a simple method for correct junction delineation with the SCM: set the voltage midway between the dC/dV peaks in the homogeneous regions (preferably regions with the same concentration), and the resulting dC/dV=0 boundary is a good approximation to the electrical junction position.

5.3 n-p-n junction results

Fig. 9 shows simulations of an npn structure. These simulations form the quantitative basis for using the SCM for the direct measurement of the effective channel length of a MOSFET. As with the pn junction the correct choice of bias voltage is crucial (Kleiman *et al* 1998).

6. SCM RESULTS ON 60nm PROCESS CMOS DEVICES

Figs. 10 and 11 show nMOS and pMOS transistors fabricated as part of an effort to understand the limits to conventional CMOS (Timp *et al* 1998). Figure 10 shows nice agreement with process simulation for an nMOSFET. We investigated a series of pMOSFETs and the results are shown in Fig. 12. The measurement of the minimum possible gate length (~53nm) shows the clear benefit of the direct measurement of the effective channel length using SCM.

7. ULTIMATE SPATIAL RESOLUTION WITH THE SCM

We have modeled the SCM response for npn structures with varying gate lengths (Kleiman et al 1998). We find that there is no intrinsic limit to resolving the channel in a realistic MOSFET structure. Fig. 13 shows that L_0 as measured by SCM is reliable until the two depletion boundaries meet. However sufficiently sharp probe tips must be used to achieve the ultimate resolution.

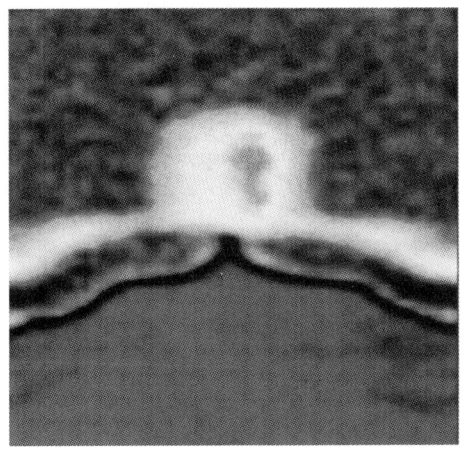

Figure 11: SCM image of a pMOSFET with a gate length of 80nm. The image is 400 X 400 nm. The shallow HDDs and ultra-shallow LDDs are easily observable and measurable from these images. The effective channel length is extremely narrow ~ 35 nm, demonstrating the high spatial resolution attainable.

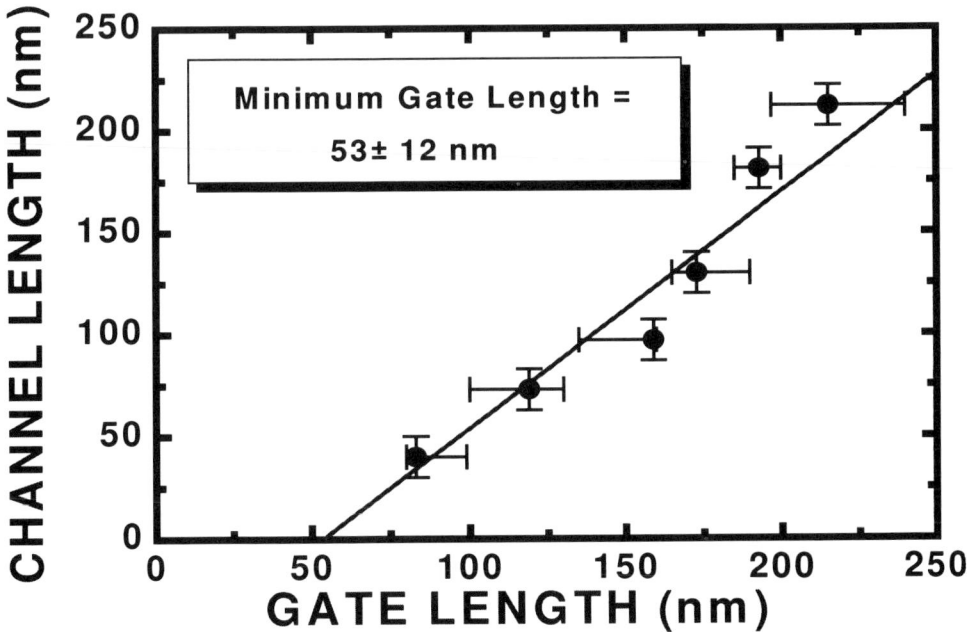

Figure 12: Effective channel length as determined from SCM imaging, plotted versus gate length as determined from SEM. Studying an array of different gate lengths improves the statistics in determining the minimum attainable gate length. Our measurements show that with the particular pMOS process used the minimum attainable gate length is ~53nm.

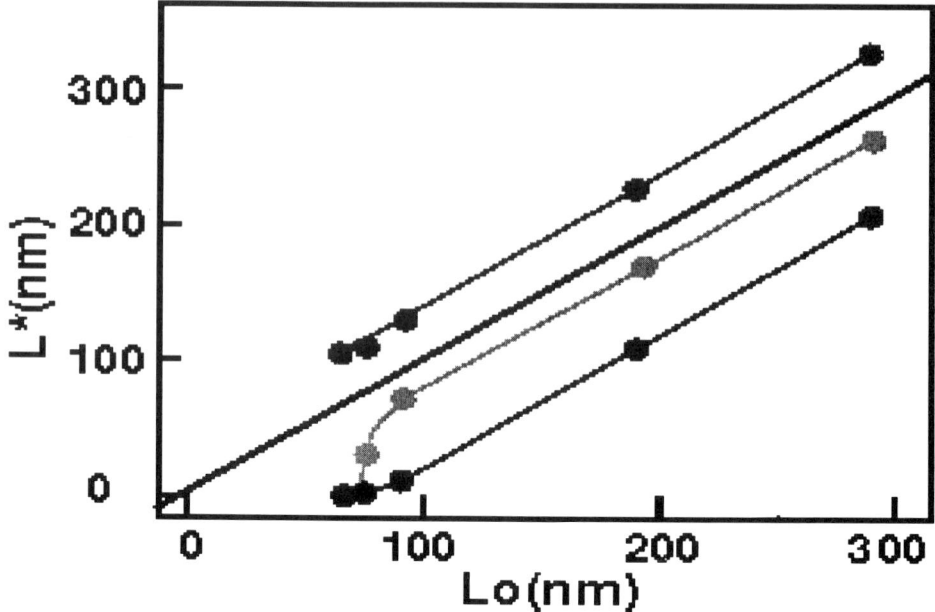

Figure 13: Apparent channel length, L*, *vs* effective channel length, L_0 for different choices of V_B. With V_B = 1.0V (lower), $L^* < L_0$ and the channel pinches off at 80nm. With V_B = -0.5V (middle), L* is slightly below L_0 due to tip broadening, but continues to track L_0 down to 80nm. With V_B = -2.0V (upper), $L^* > L_0$, but the channel structure is washed out.

8. CONCLUSIONS

We have shown experimentally that the SCM does not work as a simple dopant profiler in regions of a semiconductor device with pn junctions. Nevertheless our simulation work has given a clear understanding of the interaction between the probe tip and the built-in depletion of the junction region. This forms the quantitative basis for using the SCM to determine junction positions and to directly measure the effective channel length of a device. It has also led us to a simple method for implementing the SCM to measure junction positions.

We have used the SCM to study sub-100nm state-of-the-art MOSFETs, and have been able to achieve sufficient spatial resolution to study the smallest fabricated devices. We have found from simulations that there is no intrinsic resolution limit when studying MOSFETs, but experimentally this places a demand for suitable probe tips to achieve even higher resolution.

REFERENCES

Huang Y and Williams C C 1995 Appl. Phys. Lett. **66**, 344

Kleiman R N, O'Malley M L, Baumann F H, Garno J P and Timp G L 1997 IEDM Tech. Dig., 691

Kleiman R N, O'Malley M L, Baumann F H, Garno J P, Timp W G and Timp G L 1998 VLSI Tech. Symp. Dig., 138

Kopanski J J, Marchiando J F and Lowney J R 1996 J. Vac. Sci. Technol. **B14**, 242

O'Malley M L, Timp G L, Moccio S V, Baumann F H, Garno J P and Kleiman R N, 1999 Appl. Phys. Lett. **74**, 272

Sze S M 1981 Physics of Semiconductor Devices, (John Wiley and Sons: New York)

Timp G L, *et al* 1998 IEDM Tech. Dig., 615

Inst. Phys. Conf. Ser. No 164
Paper presented at Microsc. Semicond. Mater. Conf., Oxford, 22–25 March 1999
© *1999 IOP Publishing Ltd*

Scanning capacitance microscopy on cross section and bevelled samples

V Raineri, Cs S Daroczi*, S Lombardo and V Privitera

CNR - IMETEM, Stradale Primosole 50, 95121 Catania, Italy
* MTA - MFA, H-1525 Budapest-114, P.O. Box 49, Hungary

ABSTRACT: Scanning capacitance microscopy (SCM) was performed on p^+p, p^+n, pn, n^+n, n^+p, and np samples in cross section and bevelled configuration. The bevelling or sectioning sample preparation procedure was found to strongly influence the results of SCM. However, we demonstrate that a significant magnification can be obtained in determining the junction depth by using bevelled samples with respect to cross-sections.

1. INTRODUCTION

Scanning capacitance microscopy (SCM) is attracting an increasing interest as a 2D dopant profiling technique, because of the high sensitivity range (10^{16} - 10^{20} cm^{-3}), the good resolution (~100 nm) and the minimal sample preparation (Kopanski et al 1998). Measurements are taken in air and in a reasonable time (5 – 30 minutes). Plan-view measurements are non-destructive, while cross-sectional evaluations require sectioning and polishing (Erickson et al 1996). To date, the accuracy is sufficient to delineate 2D devices and to allow the comparison with numerical process simulation tools (Kim et al 1998). However, the accuracy should be improved to characterise minima features in submicron devices and subtle changes in junction position due to minimal diffusion, induced by transient or anomalous phenomena during device processing and induced by the device structure. However, resolution (>100 nm) is limited by the Debye length and the tip radius that can be reduced only with difficulties. In this paper we explore the possibility to enlarge the sample features by polishing at a bevelled angle.

2. EXPERIMENTAL DETAILS

2.1 Sample preparation

We adopted a method, described in the following, to ensure measurements reproducibility. They were repeated several times on different identical samples prepared using the same procedure. Only when measurements were reproducible we accepted them.

Cross-sectional sample preparation was performed in some steps. First, samples were cut on the region of interest and on the rear a thin ($\approx$10 nm) platinum layer was deposited. Afterwards, samples were glued together to obtain a sandwich. By a diamond saw the sandwich was cut in a parallelepiped of roughly 0.5×0.5×0.2 mm. On one of the two parallel sides with the specimen sections (the longer dimension) a platinum layer ($\approx$10 nm) was deposited. On the opposite side the sandwich was polished to obtain the flat and clean surface required by the scanning CV measurements. We used first 10 micron diamond lapping film to remove saw marks and flatten the sandwich side. We achieved a finer and finer grind by using decremental size lapping films (3, 1, 0.5, 0.1 µm). Lastly, 0.05 micron colloidal silica was applied on cloth for the final cleaning and polishing till a roughness value of 1 nm is reached.

Bevelled samples were prepared by using a support cut at 5°44' or 2° 52' with respect to the surface. These angle ensures a magnification of the depth dimension by a factor of 10 or 20. Samples were glued by wax on the support and polished on a glass plate by using 0.1 diamond suspension.

The rigid glass plate ensures a sharp angle. An oxide layer (≈500 nm) was deposited on the sample surface to avoid the small round between the surface and the lapped plane that could influence the angle measurement. Then, the sample was refined using colloidal silica on glass to eliminate the diamond scratches and reach a roughness of better than 5 nm.

Before measurements the samples were cleaned by a dip in HF (5%) and then re-oxidised by immersion in H_2SO_4 :H_2O_2 3:1 at 90 °C for 10 minutes. This process produces a uniform oxide thickness on the polished surface of ~2.0 nm.

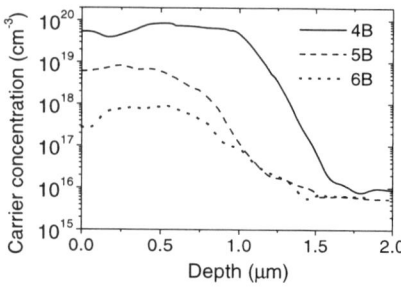

Fig. 1 *- Carrier concentration profiles on*
p⁺p samples obtained by SRP measurements

2.2 Capacitance measurements

Scanning capacitance measurements (SCM) were carried out by a Digital Instruments Dimension 3100 equipment. Capacitance derivative measurements were performed by optimising the DC voltage value. Samples with a junction were measured in capacitance feedback mode. Silicon etched and sharpened tips with radius < 30 nm covered by a cobalt layer were used. For a spherical metal-semiconductor junction with 30 nm radius of curvature, the depletion layer width is estimated to be less than 5 nm for doping concentration of 10^{18} cm^{-3} and about 1 nm for doping concentration of 10^{19} cm^{-3}. This value increases up to 100 nm for low doping concentration (<10^{16} cm^{-3}).

3. RESULTS AND DISCUSSION

In Fig.1 the carrier concentration depth profiles obtained by spreading resistance profiling (SRP) on multiple B$^+$ implanted (100) p-type 1-4 Ω·cm silicon samples are reported (samples 4B, 5B, and 6B). Multiple B ion implantation at several energies and doses was used to produce an almost flat carrier concentration profile. The same implants were performed on (100) n-type 1-4 Ω·cm silicon samples (samples 1B, 2B, and 3B corresponding to 4B, 5B, and 5B) to originate a junction. The SRP was carried out on p$^+$p samples to avoid spilling phenomena. These measurements are conventional to evaluate the carrier profile and the junction depth interpolating with the constant concentration of the n-type region. In general, the evaluation of the carrier concentration profile is critical when a junction is present. However, in all cases the difference between the p- and n-type region is at least two orders of magnitude so that the depletion region is extending in the n-type side. In these conditions the junction can be delineated by the depth extension of the p-type layer.

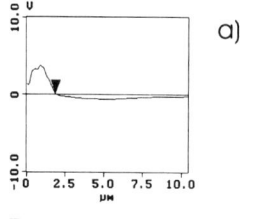

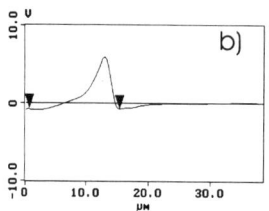

Fig.2 *- SCM measurements on*
a) a cross-section and b) a
bevelled region of sample 1B.

For each specimen cross-sectioned and bevelled samples were produced. In Fig.2a the typical average measured capacitance derivative (dC/dV) carried out on the sample section is shown. In sectioned sample the carrier concentration in the direction normal to the surface is constant. For this reason measurements are independent on the depletion layer due to capacitance measurements. The shown dC/dV profile reproduces the depth carrier concentration behaviour even if a conversion in a carrier profile is not immediate. Samples with uniform carrier concentrations can be used to calibrate SCM and convert SCM profiles in carrier concentration profiles for unknown samples (Clarysse et al 1998). This procedure can be adopted by using the samples with the carrier concentration profiles illustrated in Fig.1. We observed that for moderate doping concentration dC/dV increases more rapidly than carrier concentration.

The typical capacitance derivative profiles carried out on bevelled samples are quite different (see Fig.2 b). In this case the carrier concentration in a direction normal to the bevelled surface is not constant. It follows that the measured capacitance derivative is due to the interaction of the depleted region with the gradient of carrier concentration. Moreover, the depleted region itself is changing with the carrier concentration. For these reasons the extrapolation of a carrier concentration depth profile by these measurements can be quite intricate. Another important data that can be deduced by the measurements is the junction depth. It has been evaluated in bevelled and sectioned samples as indicated by arrows in Fig.2 and the values are reported in Table I. No difference both on the profile shape or in the junction depth has been measure by changing the bevelled angle. For comparison also the junction position according

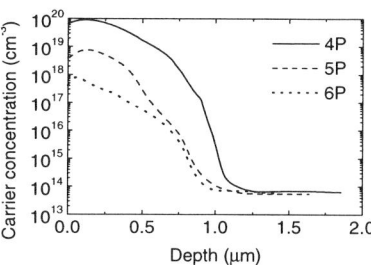

Fig.3-*Carrier concentration depth profiles obtained by SRP measurements on n⁺n samples.*

to the carrier concentration profiles obtained by SRP, and considering a substrate with a constant concentration of 5×10^{15} cm^{-3}, are reported. For this reason in the following the junction depth is also indicated as the depth profile concentration value at the substrate level (5×10^{15} cm^{-3}).

Sample	Junction depth by SRP (μm)	Junction depth on section (μm)	Junction depth on bevelled (μm)
1B (n-type substrate)	-----	1.68	1.65
2B (n-type substrate)	-----	1.52	1.50
3B (n-type substrate)	-----	1.52	1.48
4B (p-type substrate)	1.60	1.60	1.60
5B (p-type substrate)	1.50	1.50	1.51
6B (p-type substrate)	1.50	1.49	1.52

Tab. I - *Junction depth evaluation on bevelled and sectioned p⁺n samples.*

Sample 1B, 2B, 3B and respectively 4B, 5B, 6B have the same carrier concentration depth profile but different substrates. However, the obtained values are identical within an error of 5%. The valued carried on bevelled samples agree to the reference SRP data even better. This error can be attribute to the SCM resolution itself. By these data we can conclude that no influence on the junction depth determination is due to bevelling as expected in the used conditions.

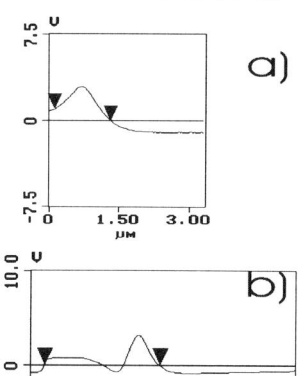

Fig.4- *SCM measurements on a) a cross-section and b) a bevelled region of samples n⁺n*

The depth carrier profiles, obtained by SRP measurements, of n⁺n samples are reported in Fig. 3 (samples 4P, 5P, and 6P). In this case (100) n-type 10-40 Ω·cm silicon samples were multiple implanted with P ions at several energies and doses. Also in this case the same implants were performed into p-type samples to produce n⁺p samples (1P, 2P, and 3P).

The typical derivative capacitance measurements obtained on cross-sections are reported in Fig.4a. Again the SCM profile agree with the carrier concentration profile and it can be used to obtain the carrier distribution depth profile by a deconvolution.

Once more the SCM measurement on bevelled samples has a profile that cannot be directly associated to the carrier distribution. The same behaviour already observed for p⁺n samples is determined. This is a confirm that the behaviour can be attributed to the bevelling, in particular to the interaction of the depletion region with a layer with a concentration gradient.

However, the junction depth can be estimated correctly. In fact, the values obtained in the different configurations are reported in Table II. Again samples 1P, 2P, 3P and respectively

628

4P, 5P, and 6P have the same carrier concentration depth profile. Within the SCM resolution the measured values are in agreement with the reference sample.

The SCM on bevelled sample procedure has been also applied to determine the junction depth in shallow junction into SiGe material. The SiGe layer has been obtained by high dose Ge implantation (130 keV 2×10^{16} cm^{-2}) in n-type 1-4 Ω·cm (100) silicon. In sequence a low energy BF$_2$ ion implant has been performed (40 keV 4×10^{13} cm^{-2}) in the amorphous region so that channelling tails are avoided. An annealing treatment has been performed at 1000 °C to recover the crystalline order. Finally, the samples were annealed at 1000 °C for 60 s and at 1100 °C for 60 s.

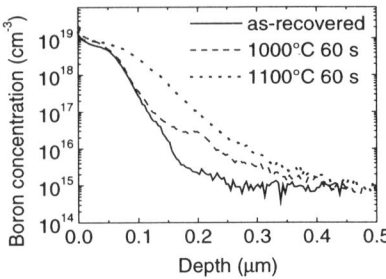

Fig.5 - SIMS depth profiles of B.

Secondary ion mass spectrometry (SIMS) has been performed to carry out the B concentration depth profile.

Sample	Junction depth by SRP (μm)	Junction depth on section (μm)	Junction depth on bevelled (μm)
1P (p-type substrate)	-----	1.17	1.22
2P (p-type substrate)	-----	1.13	1.19
3P (p-type substrate)	-----	0.95	1.00
4P (n-type substrate)	1.17	1.20	1.16
5P (n-type substrate)	1.00	1.05	1.13
6P (n-type substrate)	1.00	0.95	1.05

Tab. II - Junction depth evaluation on bevelled and sectioned n⁺ p samples.

The derivative measurements on cross-sections of the samples do not allow to determine the junction depth with a precision better than 10%. The SCM measured values on bevelled samples (respectively 0.16, 0.25 and 0.3 μm) are in full agreement with the values that can be extrapolated by SIMS profiles.

4. CONCLUSION

An expansion of the depth dimension can be reached by bevelling samples instead of sectioning. Sectioning or bevelling strongly influence the scanning capacitance measurements. However, an enlargement of the doped layers can be reached and a magnification of the junction depth delineation can be obtained. We believe that the adopted procedure can be used to evaluate the junction depth for shallow junction even in SiGe material. Its use can be particularly helpful for two-dimensional evaluation.

ACKNOWLEDGEMENT

The authors acknowledge MADESS II for financial support and the CNR-Hung.Acad. Sci. scientific co-operation for travel fund. We thank M. Furnari for technical assistance.

REFERENCES

Clarysse T, Caymax M, De Wolf P, Trenkler T, Vandervorst W, McMurray J S, Kim J, Williams C C, Clark J G and Williams C C 1998 J. Vac. Sci. Technol. B **16,** 394
Erickson A, Sadwick L, Neubauer G, Kopanski J, Adderton D and Rogers M 1996 Journal of Electronic Materials **25,** 301
Kim J, McMurray J S, Williams C C and Slinkman J 1998 J. Appl. Phys. **84,** 1305
Kopanski J J, Marchiando J F, Berning D W, Alvis R and Smith H E 1998 J. Vac. Sci. Technol. B **16,** 339

Inst. Phys. Conf. Ser. No 164
Paper presented at Microsc. Semicond. Mater. Conf., Oxford, 22–25 March 1999
© 1999 IOP Publishing Ltd

Investigations of p-i-n junctions with nanoscale corrugations

S Anand[*], D Söderström, S Lourdudoss, C F Carlström and G Landgren

Department of Electronics, Laboratory of Semiconductor Materials, Royal Institute of Technology, Electrum 229, Kista, Sweden
* anand@ele.kth.se

ABSTRACT: In this work, scanning capacitance microscopy (SCM) is applied to characterize p-i-n junctions with nanoscale corrugations. The samples are cross-sections of complex-coupled distributed-feed-back (CCDFB) InP/GaInAsP/InP lasers. The dry-etched gratings are overgrown either by InP or by an intermediate GaInAsP layer. We demonstrate high-resolution detection of the nanoscale modulation in the depletion region by SCM. The SCM results are shown to provide important information on the sample structure and doping variations. The sample topography is shown to be useful for interpreting the SCM data.

1. INTRODUCTION

Scanning capacitance microscopy (SCM) has emerged as a promising technique for high resolution two-dimensional doping profiling in semiconductor devices (Kopanski et al 1996). The ability of the SCM to image p-n junctions and its sensitivity to doping variations makes it an attractive tool to characterize complex device structures. Application of the SCM method for characterizing state of the art Si devices has been demonstrated (O'Malley et al 1999). Other applications include detection of buried nano-structures (Anand et al 1999), ordering in GaInP (Leong et al 1996) and orientation-dependent dopant-redistribution (Hammar et al 1998). Recently, it was shown that SCM images provide a complete two-dimensional map of buried-heterostructure (BH) lasers with p-n current blocking layers and gives important information on the current blocking properties (Bowallius et al 1999). Earlier, we had demonstrated high-resolution detection of nanoscale modulations in the p-n junction depletion width in CCDFB lasers (Anand et al 1999).

In this work, we present our recent SCM investigations of the DFB laser structures. Here, the fabricated gratings are regrown either by InP or by an intermediate GaInAsP layer that planarizes the grating. The latter procedure is used to reduce the index coupling (Schreiner et al 1999). In the present context, these devices can be viewed as periodically corrugated p-i-n junctions. The corrugation produces a corresponding modulation in the depletion region. The corrugation is geometric and is provided by the 240 nm period, 100 nm deep dry-etched grating. The gap between the grating teeth is about 100 nm. Since the formation of the junction requires epitaxial overgrowth of the gratings, regrown material quality, uniformity of doping and preservation of the gratings become important. Further, it is of interest to evaluate the effect of possible etch-induced damage. Imaging the cross-section of the lasers (after stain etching) by the scanning electron microscope (SEM) gives only the gross features and cannot provide electrical information. In contrast, the SCM technique can not only provide useful electrical information but also does not require any special sample preparation such as stain etching.

2. EXPERIMENTAL

The SCM measurements were performed with a Digital Instruments Dimension 3000 system operating in contact mode atomic force microscopy using commercial metal-coated tips. The capacitance sensor operating at 915 MHz is connected to the tip. The sample is biased with separately controlled ac and dc biases. The SCM data was acquired in dC/dV mode where the amplitude of the capacitance variations are recorded for a given ac modulation. For all measurements, the ac bias frequency was 50 kHz. The sample cross-section was prepared by cleaving and the measurements rely on the native oxide formed on the cleaved surface upon air-exposure. The p and n contacts of the device were electrically shorted.

A sketch of the CCDFB laser structure is shown schematically in Fig.1. The sample was grown by metal organic chemical vapour deposition (MOVPE) on n+ InP substrate ($4x10^{18}$ cm^{-3}). It consists of a 500 nm thick n-InP buffer layer ($1x10^{18}$ cm^{-3}), a 40 nm thick n-GaInAsP layer ($5x10^{17}$ cm^{-3}), a stack of 8 strain compensated GaInAsP multiple quantum wells and barriers, and, a 40 nm undoped GaInAsP layer. The 240 nm period and 120 nm deep grating was transferred onto the sample by electron beam lithography and low energy chemically assisted ion beam etching. The grating was overgrown by nominally undoped layer of GaInAsP, followed by p-InP ($1x10^{18}$ cm^{-3}). Samples in which the gratings were regrown directly by p-InP ($1x10^{18}$ cm^{-3}) are used for comparison.

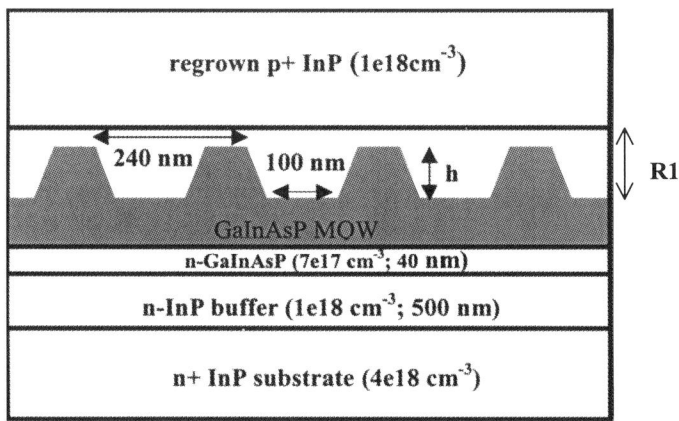

FIG.1. *Schematic sketch of the sample structure showing the essential layer sequence. The intermediate GaInAsP layer planarizing the gratings is labeled 'R1'. The height 'h' of the etched grating was typically 100 nm.*

2. RESULTS AND DISCUSSION

Fig.2(a) shows the typical dC/dV image of the device cross-section near the junction. The observed contrast variations in the image correspond to the doping differences in the different layers. The more heavily doped n-InP substrate appears darker than the lower doped layers, i.e., the n-InP buffer and n-GaInAsP layers. The p and n regions are separated by a dark band that corresponds to the depletion region. The detection of the modulation in the depletion region is remarkably clear; the bright region (n-layer) periodically weaves in and out of the dark region (depletion region). As expected, in regions of the device far away from the grating field, the depletion region did not exhibit any structure (modulation).

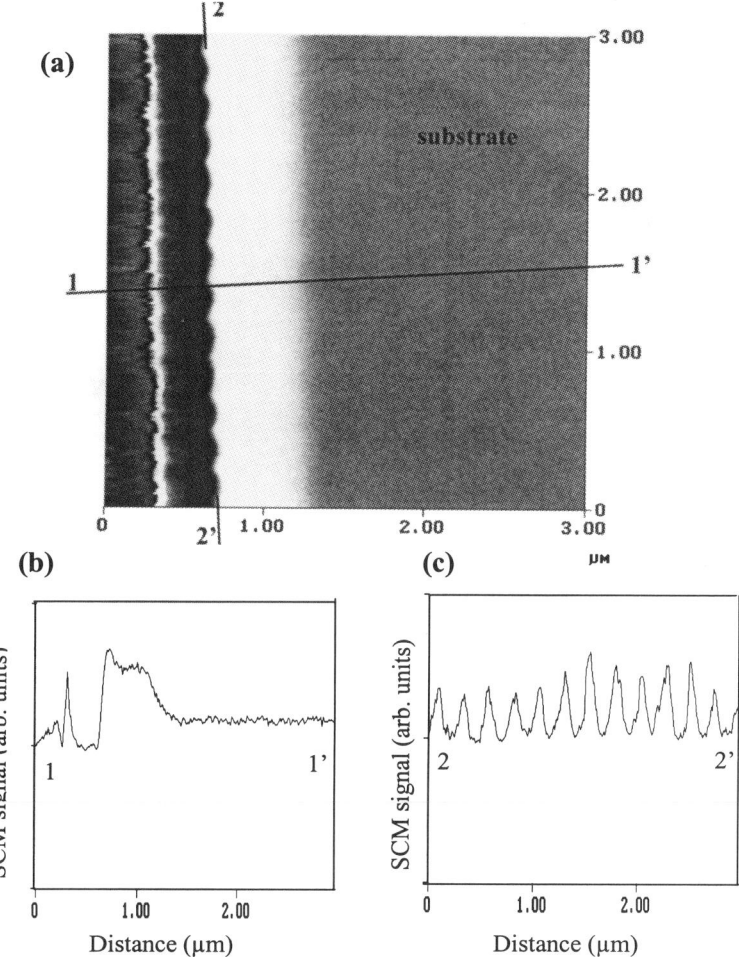

FIG. 2. *(a) Cross-sectional SCM (dC/dV) image of the CCDFB laser structure. The ac and dc biases were 1V and 0V, respectively. (b) Variation of the SCM signal along the line 1-1'. (c) Variation of the SCM signal along 2-2', showing the periodic oscillations corresponding to the modulation in the depletion region width.*

Variation of the SCM signal along the section (1-1') indicated on the image is shown in Fig.2(b). The observed variation of the signal for the n-InP buffer and n-GaInAsP layers relative to the substrate level is consistent with the doping levels of the layers. In the companion topography image (not shown), it was possible to distinguish the GaInAsP regions relative to InP due to differences in native oxide formation for the two materials. Interestingly, comparison of the topography and SCM data, show that the depletion edge on the n-side is within the doped GaInAsP layer. This is also consistent with the data of Fig.2(b).

Fig.2(c) shows the variation of the SCM signal along the section (2-2') cutting through the depletion edge on the n-side. The observed period of the oscillations is 240±10 nm and agrees very well with the grating period. This enables us to estimate the lateral resolution under the present experimental conditions and is clearly better than 100 nm. Further, by correlating the

SCM data with topography the observed modulation (Fig.2(c)) was found to be in phase with the grating. Similar modulation on the p-side is not seen (Fig.2(a)), indicating that the gratings are planarized by the GaInAsP layer. Importantly, if the background doping in the regrown GaInAsP layer and the MQW layers are similar, modulation in the depletion region will not be seen. In contrast, the SCM results (Fig.2 (a) and (c)) clearly show that this is not the case. The implication of this finding is that the doping in the regrown GaInAsP material in between the grating is indeed different from the nominal background doping level. The exact reason for this is not clear at present. Since the regrowth is performed on a surface that is processed *ex-situ*, increased doping due to contamination is a likely cause. However, possibility of diffusion of Zn during regrowth of the p-InP layer cannot be ruled out.

Samples where the gratings were partially planarized were also investigated. The SCM results obtained were consistent with those discussed above. However, in these samples, as expected, modulation in both p and n sides of the junction could be seen. It must be remarked that in the SCM data obtained for structures in which the gratings were regrown directly by p+ InP, the modulation in the depletion region width was observed only on the p-side. This is consistent with (a) the high doping level in the p-layer, i.e., the depletion edge closely follows the geometric shape of the grating, and, (b) the depletion edge on the n-side was inside the n-InP buffer layer (the n-GaInAsP layer is fully depleted). Correlating the SCM results with device performance will be the subject matter of our future investigations.

3. CONCLUSION

We have investigated CCDFB lasers by cross-sectional SCM. Detection of nano-scale modulation in the p-n junction depletion width was demonstrated and the lateral resolution was better than 100 nm. The SCM results were shown to give very important information on the doping in the layers and on the regrowth processes. Our experiments suggest that variation of the duty cycle of the grating can be a useful method to evaluate the lateral resolution of the SCM technique.

REFERENCES

Anand S, Carlström C F, Söderström D, Lourdudoss S and Landgren G 1999 Appl. Surf. Sci. (in press)
Bowallius O, Anand S, Hammar M, Nilsson S and Landgren G 1999 Appl. Surf. Sci.(in press)
Edwards H, McGlothin R, Martin R S, Gribelyuk E U M, Mahaffy R, Shih C K, List R S and Ukraintsev V A 1998 Appl. Phys. Lett. **72**, 698
Hammar M, Messmer E R, Luzuy M, Anand S, Lourdudoss S and Landgren G 1998 Appl. Phys. Lett. **72**, 815
Kopanski J J, Marchiando J F and Lowney J R 1996 J. Vac. Sci. Technol. B**14**, 242
Leong J K, Williams C C, Olson J M and Froyen S 1996 Appl. Phys. Lett. **69**, 4081
O'Malley M L, Timp G L, Moccio S V, Garno J P and Kleiman R N 1999 Appl. Phys. Lett. **74**, 272
Schreiner R, Wiedmann J, Coenning W, Porshe J, Gentner J L, Berroth M, Scholz F and Schweizer H 1999 Electronics Letters **35**, 146

Inst. Phys. Conf. Ser. No 164
Paper presented at Microsc. Semicond. Mater. Conf., Oxford, 22–25 March 1999
© 1999 IOP Publishing Ltd

Imaging sub-surface dopant and free electron distributions using scanning tunnelling microscopy

J Jacobs, B Hamilton, E Whittaker, U Bangert and M Missous[1]

Dept of Physics, UMIST, PO Box 88, Manchester, M60 1QD, UK.
[1]Dept. Electrical Engineering and Electronics, UMIST, Manchester M60 1QD, UK

ABSTRACT: The interaction between sub-surface charge and the physics of the tunnelling process has been studied for a number of doped semiconductor structures. The work was carried out using a UHV STM and measurements were made in two dimensions on (110) surfaces cleaved in UHV. It is clear that for such surfaces, tunnelling spectra, and hence the tip movement traditionally used to generate topology images, depend on the sub-surface depletion beneath the tip. The influence of the semiconductor screening properties and its impact on dopant related image-formation is discussed.

1. BACKGROUND TO 2-D DOPANT IMAGING

The trend for reduction in size of semiconductor devices has been a marked one over the last couple of decades. In silicon ULSI technology this process is driven by commercial forces viz. the competition to produce lightweight, low cost consumer equipment. This is an interesting time, in terms of controlling device properties, because the active regions within a device are soon set to approach nanometre dimensions. The electronic properties of the material will vary on a local scale in a manner which will differ from the properties of current devices. In order to understand and exploit the semiconductor there is a pressing need to measure and interpret properties of the solid on a nm scale. Of course, in the III-V device field, low dimensionality is already a key feature of many devices, especially optical devices in which quantum wells (and now quantum dots) play an active role.

For all devices, the positioning of electron sources and sinks locally is essential to the control of current flow. This means that dopant atoms must be located at precise planes and that dopant profiles must "survive" the thermal budget for any processing step following doping. The integrity of the dopant profile needs to be maintained in three dimensions, which (given conventional device symmetry), demands measurement on a 2-D scale (Hildner et al 1998, Chao et al 1997). This scale must be ~ nm to be of real use; a range highly suited to the STM provided that there is some inherent contrast within the dynamic range of tunnelling available. The optimum 2-D information set is on the (110) plane, assuming the usual <100> growth and processing directions for most semiconductor technologies. This is the plane investigated in this work, and was created in each case by cleaving in UHV just prior to measurement. We report here dopant/carrier related imaging for Boron diffused Si p-n junctions and for atomic plane doping in MBE grown GaAs. The key to generating image contrast lies, for all cases, in the role played by the semiconductor depletion which, for critical applied voltage regimes, forms part of a tip-gap-semiconductor dipole barrier system (Maboudian et al 1992). This is the dipole which sustains the potential drop across the tunnelling gap and which therefore controls the tunnelling spectrum. In analysing the images discussed here, we demonstrate that by understanding the nature of the barrier dipole, one can explain the image formation and begin to develop ideas for optimisation of this form of microscopy.

2. STM DOPANT RELATED IMAGES

2.1 Diffused p-n junctions: tunnelling spectrum derived images

The STM used in this work was an OMICRON UHV machine with facilities for in-situ cleaving and heating. The (110) plane is a natural cleavage plane for GaAs making the generation of atomically flat faces a relatively simple task. The bond strength for Si is weakest in the <111> direction and generating a flat (110) plane is extremely difficult. This has been achieved in this work by a combination of wafer thinning and the generation of stress concentration zones using anisotropic etching (Jacobs 1999).

The image shown in Fig. 1 was taken on a (110) cleaved face of boron diffused silicon. The scan region was near the edge of the SiO_2 mask used to define a p^+ stripe and is shown schematically in (a). The diffused stripes were oriented along the <011> direction. The diffusion source was a doped spin on glass and a maximum boron concentration of $\sim 10^{20}$ cm^{-3} is expected for the time/temperature cycle used.

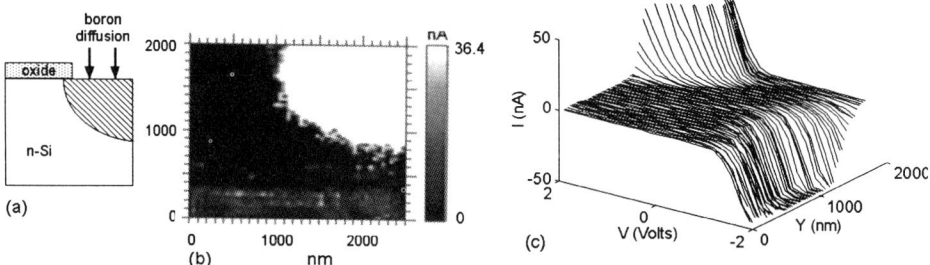

Figure 1. (a) The scan region used to acquire the image. (b) The image obtained by plotting the tunnelling current detected at a voltage of +4V at each pixel. Note that regardless of conductivity type, a +ve potential applied to the semiconductor will favour tunnelling of electrons from the tip into the conduction band DOS functions. (c) A line scan of the tunnelling I-V spectra obtained when moving from the n to p^+ regions.

The generation of such a "current" image rests entirely on the way that the tunnelling spectra are used; in this case by selecting for the pixel data the current at a fixed value of applied voltage. The tunnelling spectra are obtained by setting some fixing the tunnelling configuration (I and V) which then dictates a tip-sample gap. In all measurements the tip was grounded and the bias applied to the sample. Once the gap is chosen, the feedback to the system is turned off and the spectra are acquired at a constant gap for all pixels. Depending on the values I and V used, the gap may be relatively small (eg 3-4 Angstroms) or large (eg 100 Angstroms). For the dopant image shown in 1(b), it is clear that the contrast mechanism relates to the fact that the p^+ region permits a much larger current to flow for +ve sample bias than does the n-type substrate. We find that this is a common feature for dopant imaging and that the establishment of a small tunnelling gap is useful for enhancing the current asymmetry.

An important factor in determining the relationship between doping and image contrast is (to first order) the nature of the dipole associated with the tip-semiconductor system, (Stievenard et al 1998, Ness et al 1997). The physics of this dipole governs the form of the tunnelling spectrum since it describes the potential variation across the gap and the semiconductor. The semiconductor must always provide a charge Q_{sc} in order to support any potential across the gap. In some voltage regimes Q_{sc} may be provided by surface accumulation or inversion but when the externally applied voltage V_a is small, Q_{sc} is not a surface charge, but rather a distributed depletion charge. Now, a unique voltage V_{sc} must be produced across a sub-surface depletion width in association with any particular voltage dropped across the tunnelling gap, V_d. For high N_d or Na, only a small depletion depth is needed to supply the value of Q_{sc} required to sustain a given gap voltage. Since the depletion potential scales as the square of the depletion depth, V_{sc} is a relatively smaller fraction of V_a for high doping levels. In the metallic limit all the applied voltage drops across the gap. It is clear that for the p^+ material, when sensing the tunnelling current at $V_a = +4V$, a large tunnelling current is expected because the +ve polarity rapidly diminishes the extended, depletion component of Q_{sc}. Va is then used primarily in displacing the tip Fermi energy towards the conduction band DOS, resulting in a substantial tunnelling current. For the n material, this polarity of V_a increases Q_{sc} extending the depletion layer and this process will continue unless the surface of the semiconductor can invert. If the tunnelling barrier is thin, inversion may be difficult to sustain; quite high values of V_a may be required before the tip Fermi energy can access the conduction band DOS. This gives a qualitative view of why the p+ regions produce large contrast in the image shown.

2.2 Atomic plane doping: tip displacement derived images

The image shown in Fig. 2 was taken on a cleaved GaAs sample which contained ten silicon atomic plane (δ) doped layers; eight are imaged here. The samples were grown by MBE and the layers were separated by 100 nM, with the dopant flux adjusted to give a sheet density of 5×10^{12}cm^{-2}. The growth temperature was low (520°C), resulting in a very small dopant atom movement as gauged by SIMS. At this spacing there is no electronic interaction between the dopant planes (Ke et al 1993). A key experimental aspect of the image shown in Fig. 2 is that it was obtained by setting the tip-surface gap at a very small value, estimated to be 3.6 Angstroms. As the tip is scanned over the dopant planes, with feedback maintaining the tunnelling current at 0.3 nA, the image is formed because the tip retracts in response to the presence of the free electron gas associated with the delta doping.

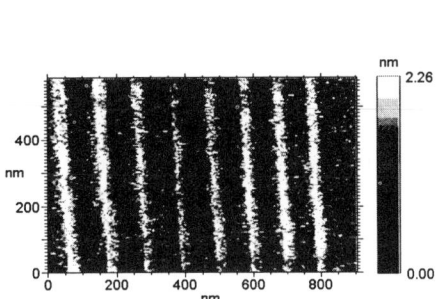

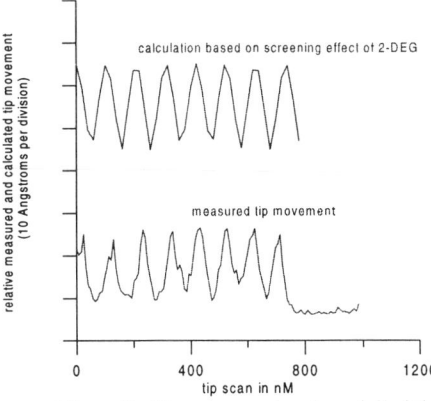

Figure 2. A 2-D image of atomically doped planes with sheet donor density of 5×10^{12} cm^{-2}. The layers were grown by MBE and the cross section obtained by cleaving the (110) face in UHV.

Figure 3. The measured and modelled tip displacement for the image of Fig. 2. The tip movement is in response to the screening properties of the 2-DEG.

As the tip-surface gap chosen for the scan was increased, the contrast rapidly diminished. With large tip spacing, the contrast disappeared. This proximity effect gives an important insight into the tunnelling processes responsible for image formation. The image reproduces the known spacing of the layers with good accuracy and if we associate tip retraction with a significant sub-surface electron density, then the implication is that the electron gas spreads to around 20 nm each side of the nuclear sheet charge on the δ plane. We have calculated the electronic structure of the multi-δ stack, in the effective mass approximation, by solving the Poisson and Schrodinger equations self consistently (Ke et al 1992). Screening from all sub-bands below the Fermi energy was included.

The dopant planes are characterised by three sub-bands below the Fermi energy, and it is the spread of the wave functions in the <001> direction (ie the growth direction) of electrons occupying these sub-bands which correspond to the spread in the electron gas imaged in Fig. 2.

The tip retraction as the tunnelling process "feels" the presence of the electron gas is again fundamentally related to the tip-gap-semiconductor dipole. Between the delta sheets the n-type background doping is modest. To drive the tunnelling current, voltage is dropped across both gap and semiconductor. The latter voltage is controlled by the depletion depth associated with the rather weak screening between the dopant planes. As the tip approaches the 2-DEG, the sub-surface voltage drop diminishes because the displacement of electron gas required to provide the dipole charge becomes very small. Now, the applied voltage switches mainly to the tunnelling gap and the tip retracts. Ideally, the as the tip scans laterally over the intersection of the δ plane and the surface it will probe the local electron gas density.

We have made an approximate model of the tip displacement for the structure measured here. The local electron density within a slab of width Δz was taken to be the integral $\sum \Psi^2(z)dz$, over the slab Δz where $z=0$ defines the δ plane and the summation runs over the normalised wave functions for all sub-bands. Poisson's equation was then solved for the tip-gap - electron gas system in one dimension and the relative displacement of d required to maintain the 0.3 nA tunnelling current obtained. The tunnelling current was assumed to be given by the WKB approximation, which is generally valid for the low voltage regime used in acquiring this image. The calculation produces surprisingly good agreement with the measured tip displacements for the structure modelled. This agreement rests in part on the choice of some minimum tip displacement which we took to be 3.6 Angstroms. This is a fairly sensitive fitting parameter for the amplitude of tip movement. However the low bias potential used ensures that the tip is in the strong tunnelling regime. The rather broad spread of the electron gas beyond the δ plane is a consequence of the spatial extent of the wave functions and the parity variation, which ensures that some wave functions are not centred on the plane.

3. CONCLUSIONS

The ability to image a variety of dopant structures in 2-D has been demonstrated and the physical basis for understanding and hence developing this form of microscopy has been discussed. Two basic techniques for producing image contrast were explored, viz tunnelling spectroscopy and z-imaging, though we point out that the fundamental physics is the same.

REFERENCES

Chao K J, Smith A R, McDonald J, Kwong D L, Streetman B G and Shih C K 1998 J. Vac. Sci. Technol. B **16**, 453
Hildner M L, Phaneuf R J, and Williams E D 1998 App. Phys. Lett. **72**, 3314
Jacobs J 1999 PhD thesis, UMIST pp 847.
Maboudian R, Pond K, Bressler-Hill V, Wassermeier M, Petroff P M, Briggs G A D and Weinberg W H 1992 Surf. Sci. **275**, L662
Ke M L, Rimmer J S, Hamilton B, Evans J, Missous M, Singer K E and Zalm P 1992 Phys. Rev. B **45**, 14114
Ke M L and Hamilton B 1993 Phys. Rev. B **47**, 4790
Ness H, Fisher A J and Briggs G A D 1997 Surf. Sci. **380**, L479
Stievenard D, Grandidier B, Nys J P, de la Broise X, Delarue C and Lannoo M 1998 App. Phys. Lett. **72**, 569

Inst. Phys. Conf. Ser. No 164
Paper presented at Microsc. Semicond. Mater. Conf., Oxford, 22–25 March 1999
© *1999 IOP Publishing Ltd*

An elevated temperature STM study of the Si(001) c(4×4) surface reconstruction

H Nörenberg and G A D Briggs

University of Oxford, Department of Materials, Parks Road, Oxford OX1 3PH

ABSTRACT: We have investigated structural changes of the Si(001) c(4x4) surface reconstruction in the temperature range between 600°C and 700°C in real time by elevated temperature STM. Shape changes of steps and trenches as well as smudges visible in STM images indicate that silicon is much more mobile on the (2x1) reconstructed than on the c(4x4) reconstructed areas at 600°C. Increasing the temperature to 700°C leads to the disappearance of the c(4x4) reconstruction. This happens by erosion from the circumference rather than by fragmentation of the c(4x4) reconstructed area.

1. INTRODUCTION

Si(001) is widely used in device fabrication and has been studied in great detail (Boland 1993). Recently, a Si(001) c(4×4) surface reconstruction has attracted much interest and it was argued that this surface reconstruction is caused by stress introduced by impurities such as carbon (Miki 1997). Starting from a (2×1) reconstructed surface, the c(4×4) surface reconstruction appears at around 500°C and is stable up to 750°C. Above 750°C, this reconstruction disappears and the original (2×1) reconstruction reappears. So far, this transition has been investigated by RHEED (Miki 1997), LEED, AES. The appearance of the c(4×4) reconstruction starting from a (2×1) reconstruction has been studied by elevated temperature STM (Lin and Wu 1998) The aim of this paper is to study the disappearance of the c(4×4) reconstruction in real time by elevated temperature STM.

2. EXPERIMENTAL

Samples of Sb doped silicon (001) with a (2×1) reconstructed surface were prepared by flashing in ultra-high vacuum (UHV). To obtain a c(4×4) surface reconstruction the samples were subsequently heated to between 500°C and 600°C in the low 10^{-10} mbar range for a few hours. STM investigations were carried at elevated temperatures with a JEOL STM 4500XT. PtIr-tips were used for the investigations. Samples have been monitored in-situ by reflection high-energy electron diffraction (RHEED) and were also investigated by low-energy electron diffraction (LEED) and Auger electron spectroscopy (AES).

3. RESULTS and DISCUSSION

Figure 1 shows a large scale STM image of a Si(001) surface taken at ~600°C. Large surface areas have already been transformed into the c(4x4) reconstruction. The remainder of the surface is covered by the original (2×1) reconstruction (if trench forming defects are considered, an appropriate description would be (2×n)-like (Lin 1998)). In our experiments the nucleation of a c(4×4) reconstructed patch always required a step, a trench or other defect.

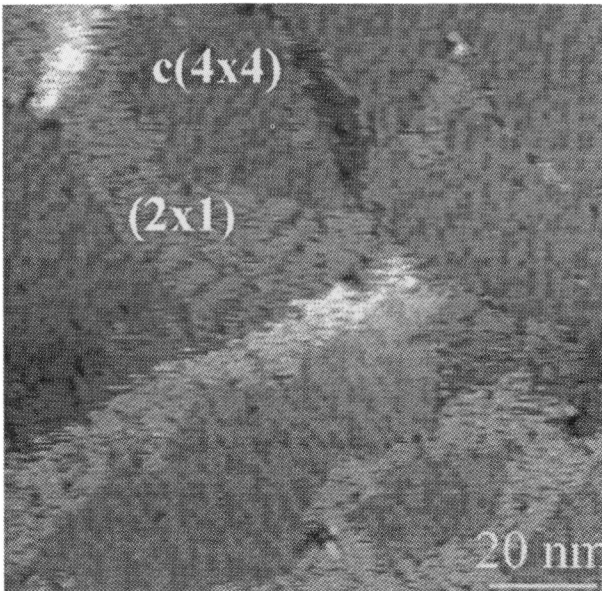

Figure 1: STM image of Si(001) surface taken at 600°C
V_{BIAS}=2 V I_T=0.08 nA

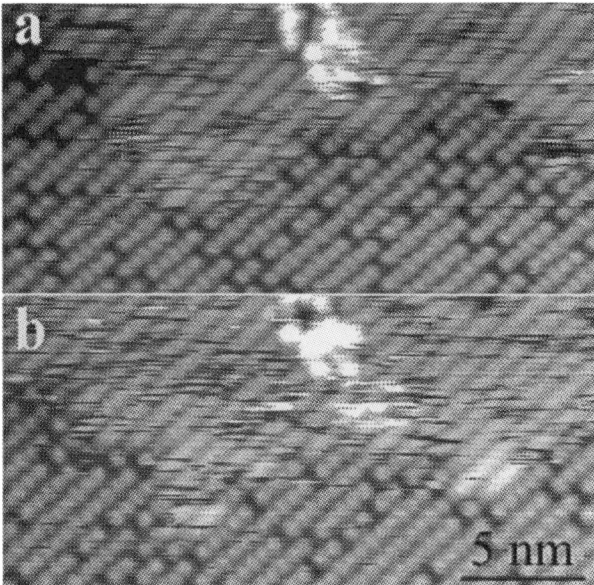

Figure 2: STM image of Si(001) surface taken at 600°C
V_{BIAS}=2 V I_T=0.08 nA, a) t=0   b) t=90 min

On the (2×1) reconstructed patches silicon is mobile in the temperature range investigated in this study. In Fig. 1 we observe a large number of smudges which are absent on the c(4×4) reconstructed areas. On the (2×1) reconstructed areas steps and trenches change their shape all the time. Si is likely to move as dimers on the surface. Because the formation energy of a dimer is 0.6 eV and its bond energy is 1.33 eV (Terakura et al 1997) it is believed that once the dimers are formed they are stable objects.

If impurity atoms such as carbon are supplied to the surface the appearance of the c(4×4) reconstruction proceeds quickly (Leifeld et al 1998). If the Si(001) sample is exposed to UHV background only, the net growth rate of the c(4×4) reconstruction was rather low and it took hours to see a noticeable increase in c(4×4) area.

The c(4×4) reconstructed areas show a number of defects where the characteristic bright spot is missing. These defects are mobile and if the time for one frame is longer than a few seconds (7 ms per line) the majority of these defects has moved. However, the rate of detachment and attachment is considerably lower than for Si-dimers on the (2×1) reconstructed areas.

As shown in Fig. 2 there are considerably more smudges (approximately one order of magnitude) on the (2×1) reconstructed areas than on the c(4×4) reconstructed areas.

Figure 3 illustrates how the c(4×4) reconstruction disappears after raising the temperature to about 700°C. The sequence of images was taken at the same location (the defect in the lower left corner serves as point of reference) at different times. In Fig. 3a approximately half of the area is covered by a patch of c(4×4) reconstruction. Three minutes later (Fig. 3b) this area has shrunk in size and a large number of silicon atoms have moved away leading to the formation of a new step. Another minute later (Fig. 3c) this new step reached the c(4×4) reconstructed patch which kept shrinking in size. After further 2 minutes (Fig. 3d) the patch of c(4×4) reconstruction has disappeared altogether. It shows that the disappearance of the c(4×4) reconstruction proceeds from the circumference of the patches. Fragmentation of patches has not been observed.

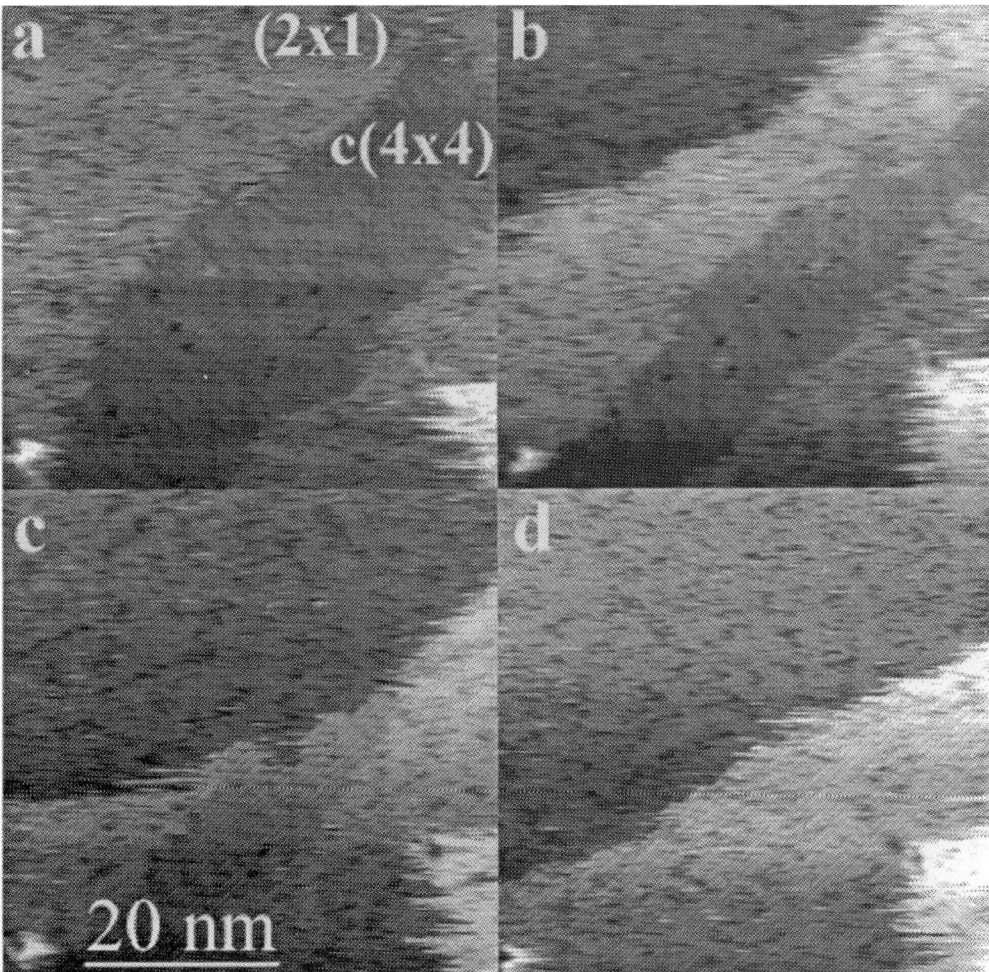

Figure 3: STM image of Si(001) taken from the same location on the surface at 700°C
a) t=0 b) t=3min c) t=4 min d) t=6 min V_{BIAS}=2 V I_T=0.08 nA

The sequence of images in Fig. 3 shows that the disappearance of the Si(001) c(4×4) surface reconstruction is related to silicon diffusion on the surface. This gives support to the concept of the c(4×4) reconstruction consisting of units of three silicon dimers (Wang et al 1987, Johnson et al 1993) with a 1-DV one silicon dimer vacancy (1-DV) (Owen et al 1995). The theoretical results on STM contrast by Owen et al (1995) are in qualitative agreement with STM observations at different bias voltages (Nörenberg and Briggs 1999). Impurities such as carbon located in subsurface locations that are supposed to be the cause of the c(4×4) reconstruction (Miki et al 1997) eventually form β-SiC islands which can be observed by RHEED and STM (Nörenberg and Briggs 1999).

4. CONCLUSIONS

STM at elevated temperatures is a suitable method to study transitions between surface reconstructions on Si(001) in real time. At 600°C c(4×4) reconstructed areas appear considerably more stable against detachment and attachment of silicon species compared to (2×1) reconstructed areas. At a substrate temperature of 700°C the c(4×4) reconstructed areas are shrinking from their edges and transformed back to the (2×1) reconstruction. These findings support surface models for the c(4×4) reconstruction where the top layer consists entirely of silicon, and impurities or adsorbates are in sub-surface locations.

REFERENCES

Boland J J 1993 Adv. Phys. **42**, 129
Johnson K E, Wu, P K, Sander M and Engel T 1993 Surf. Sci. **290**, 213
Leifeld O, Grützmacher D, Müller B and Kern K 1998 Mat. Res. Symp. Proc. **553**, 183
Lin D and Wu P 1998 Surf. Sci. **397**, L273
Miki K, Sakamoto K and Sakamoto S 1997 Appl. Phys. Lett. **71**, 3266
Nörenberg H and Briggs G A D 1999 Surf. Sci. in press
Owen J H G, Bowler D R, Goringe C M, Miki K and Briggs G A D 1995 Surf. Sci. **341**, L1042 Terakura K, Yamasaki T, Uda T and Stich I 1997 Surf. Sci. **386**, 207
Wang H, Lin R and Wang X 1987 Phys. Rev. B **36**, 7712

Inst. Phys. Conf. Ser. No 164
Paper presented at Microsc. Semicond. Mater. Conf., Oxford, 22–25 March 1999
© *1999 IOP Publishing Ltd*

A combined STM/AFM/TEM study of CdTe/CdS solar cell material grown on glass

P Buckle, C Bridge, J Jacobs, A J Harvey, U Bangert, P Dawson, E Z Luo[1], I H Wilson[1] and M E Özsan[2]

Dept. of Physics, UMIST, PO Box 88, Manchester, M60 1QD, UK
[1]Electronic Engineering Department and Materials Science and Technology Research Centre, The Chinese University of Hong Kong, Sha Tin N T, Hong Kong, China
[2]BP Solar, 12 Brooklands Close, Windmill Rd., Sunbury-on-Thames, Middx., TW16 7DX, UK

ABSTRACT : A microstructural investigation using TEM, X-ray microanalysis, STM and AFM was carried out on CdTe/CdS films deposited on tin oxide coated glass. Grain sizes and morphology of the CdTe and the CdS were studied prior to and after type conversion annealing. A correlation of surface morphology, conductivity and photo-current maps, obtained by STM and conducting AFM is attempted, revealing electrical behaviour on the nanometre scale.

1. INTRODUCTION

Thin film solar cells are of commercial interest because of their potential as a means of low cost power generation. CdTe, in particular, is a promising thin film device material because of its near ideal bandgap for solar energy conversion and high optical absorption coefficient. Device efficiencies in excess of 10% have been reported for a number of different growth techniques including electro-deposition (Woodcock et al 1991), closed space sublimation (CSS) (Tyan and Perez Albuerne 1988) and screen-printing (Ikegami 1988). However, irrespective of the growth method, most as-grown structures require some form of post-growth heat-treatment. The highest conversion efficiency, to date, of 16% (Aramoto et al 1997) was reported for a CdTe/CdS heterostructure grown by the CSS technique where CdTe acts as the absorber material and the wide band-gap CdS is the window layer. This figure remains some way short of theoretical predictions for this system which lie in the region of 30% (De Vos et al 1994). If significant further advances are to be expected, then an improved understanding of the material properties of the CdTe film and the effects of post-deposition processing must be achieved.

Cross sectional transmission electron microscopy (TEM) has been employed to compare the microstructural properties of as-grown and heat-treated polycrystalline CdTe/CdS heterojunction photovoltaic devices. Scanning tunnelling microscopy (STM) has led to complementary images of the CdTe surface whilst also providing novel nanometre resolution I(V) data. To our knowledge, the latter has not previously been reported in such structures. Furthermore conducting atomic force microscopy (AFM) has demonstrated that is possible to obtain spatially resolved photo-current images on the nanometre scale of the CdTe p-n junction.

2. EXPERIMENTAL

As-grown polycrystalline CdTe/CdS heterojunction devices were provided by BP Solar. The n-CdS film (~0.1μm) was chemical bath deposited on to a commercially available tin oxide coated (0.3-0.4μm) glass substrate (3mm). An n-CdTe:Cl film (1.6μm) was subsequently electrodeposited on to the CdS layer. The heat-treatment process (performed in-house) that is central to good device performance leads to type conversion of the CdTe layer. Other beneficial effects resulting from this process are thought to include grain growth and intermixing at the CdTe/CdS interface (Özsan et al 1994). Cross sectional specimens were prepared for inspection in a TEM (CM20) and for X-ray micro-analysis in a STEM (VG601UX). Surface studies were carried out in air in an STM and conducting AFM, whereby the sample was contacted on the tin-oxide after partial back etching of the CdTe/CdS film.

3. RESULTS AND DISCUSSION

TEM: The aim of the TEM work was to characterise the device layer structure in terms of composition and quality prior to, and following heat-treatment. TEM samples were prepared from as-grown and heat-treated structures. In both cases three distinct layers are observed in addition to the glass substrate: the tin oxide, CdS and CdTe films (see Fig.1). The measured thickness of each layer is in good agreement with the nominal figures provided by BP Solar.

In the as-grown sample the CdS grains appear to be spherical in shape and around 10nm across. The CdTe grains are quite different in appearance extending, in some cases, up to the full

layer thickness in length with a pyramidal, or faceted, form at the CdTe free surface. In contrast, they are only around 0.1μm thick. Additionally, the interiors of the CdTe grains show a high density of defects in the form of twin boundaries and stacking faults. The dark spots that are visible were identified as Te nucleation sites from scanning TEM (STEM) chemical X-ray analysis which also provided evidence for the accumulation of Cl and S at grain boundaries.

A number of differences are apparent in comparison with the images of the heat-treated sample. In the latter the CdS grain size is significantly larger, by around a factor of 10, whilst the layer itself also appears to have become porous. These effects suggest a restructuring of the CdTe/CdS interface. In contrast with the CdS film, however, the CdTe layer does not exhibit any grain growth. Nevertheless, some changes to the CdTe film are observed. The heat-treatment gives rise to a noticeably flatter surface morphology - the formation of a thin surface layer (~0.1μm), presumably an oxide, punctuated with large aggregates (0.5-1μm), is evident. There also appears to be a reduction in the dark regions associated with Te nucleation throughout the CdTe grain structure.

Fig.1 Cross-sectional view of the annealed sample (left) and the as-grown sample (right).

STM: STM data has provided topographic (z) images (typically 2μmx2μm) of the CdTe surface together with electrical data on a nanometre scale for as-grown and heat-treated structures. Surface height fluctuations were measured by maintaining a constant tunnel current between the STM tip and the sample surface. Regions of elevated topography can be correlated with CdTe grain size from the TEM view of the surface. In the as-grown material, the width of the elevated regions is consistent with the grain diameters of ~0.1μm observed in TEM images. The distinctly larger elevated regions in the annealed material are thought to be the surface aggregates seen in the TEM image since little grain definition is observed. Height differences measured by the STM correlate well with those observed in the TEM images of both samples (~0.1μm).

An altogether different STM representation of the structures was obtained by measuring I(V) curves at each z-image pixel. Displaying the current at a fixed voltage for each pixel generates high resolution electrical maps of the same scan area, thereby providing highly localised electrical information for the as-grown and heat-treated samples.

A comparison between the z-image and electrical map for the as-grown sample (Fig. 2) shows a clear correlation. Regions of enhanced conductivity broadly correspond to the low regions in the z-images, believed to be grain boundaries. The sharp change in tunnel current at the grain boundaries suggests that this is a real effect and that the current enhancement is not due to surface height fluctuations which would occur much more slowly with position.

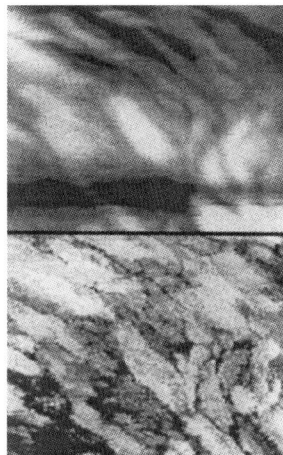

Fig.2: Topographic z-image (top) and electrical map (bottom) of as-grown material. Image 2 μm x 2

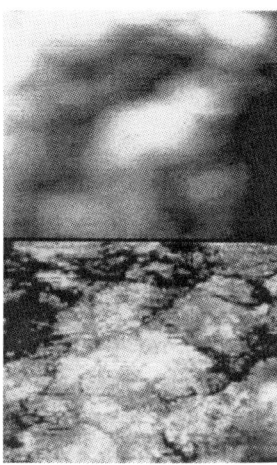

Fig. 3: Topographic z-image (top) and electrical map (bottom) of annealed material Image 3 μm x 3 μm

By contrast, no such correlation can be made between the z-image and electrical map of the heat-treated sample (Fig. 3). One possibility for this is that the electrical image, rather than being representative of the conductivity at the surface of the CdTe actually depicts the electrical behaviour from deeper within the structure. In the as-grown structure a proportion of the applied bias can be expected to be dropped across the Schottky barrier associated with the CdTe free surface as well as the sample-tip gap. By biasing the CdTe near-surface region it is reasonable to assume that the experiment is probing this region of the structure. By contrast, in the heat-treated material a p-n junction now exists at the CdTe/CdS interface. Consequently part of the applied bias must now be dropped across this part of the structure as well. If the voltage drop is greater across the interface region than the free surface, it is reasonable to suggest that the electrical image represents the interface region deep in the structure rather than the near-surface region. However, this interpretation must be approached cautiously since it is not clear what proportion of the applied bias is actually dropped across the device and what proportion is dropped across the sample-tip gap.

AFM: Conducting AFM was undertaken under forward and reverse bias of the as-grown and annealed samples whilst illuminating with 'white' light from a laboratory lamp. The contact geometry was similar to that for the STM studies. The as-grown sample showed no contrast for either biasing condition, and neither was there any current flow detected under reverse bias for the annealed sample. Forward bias revealed a locally non uniform current image in the annealed sample. The surface oxide accounts for the relatively large voltage of +3V that has to be applied to detect measurable current. In order to prove the existence of a photo current, the lamp was periodically switched on and off.

The dark stripes in Fig. 4b) correspond to 'light off', whilst the stripes with bright patches correspond to 'light on'. The fact that there is grain structure visible in the dark stripes is attributed to the AFM laser light shining on the cantilever. It is thus impossible to measure the true dark current. The apparent grain structure in the current image does not tie in with that of the topographic image. The grain diameters are, however, of similar magnitude as the grain diameters of the CdTe film near the p-n junction as deduced from cross sectional micrographs. This suggests that the current image originates from grains at the junction. The inhomogeneities are believed to be induced by the grain boundaries, which appear to prevent carriers from reaching the junction. This process might restrict the current overall efficiency of the cells.

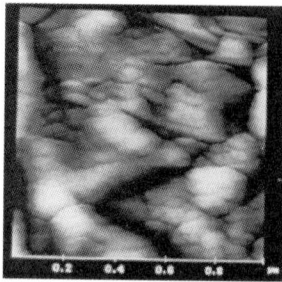

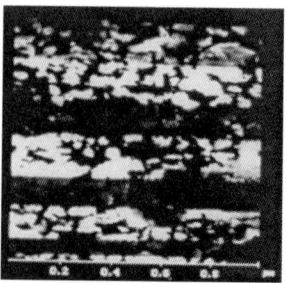

Fig. 4a: Topographic image
of annealed material

Fig 4b: Electrical with light
switched   on   and   off
periodically

4. CONCLUSIONS

In conclusion, we have demonstrated that there is evidence for regrowth of the CdS and for changes in the Te precipitates in the CdTe, but little change in CdTe morphology upon annealing. There is evidence for increased conductivity at the grain boundaries in as-grown samples. After annealing the newly formed junction appears to contribute to the electrical maps. Photo-current images reveal inhomogeneities of this junction on the nanometre scale.

REFERENCES

Aramoto T, Kumazawa S, Higuchi H, Arita T, Shibutani S, Nishio T, Nakajima J, Tsuji M, Hanafusa A, Hibino T, Omura K, Chijama H and Murozono M 1997 Jpn. J. Appl. Phys. **36,** 6304

De Vos A, Parrott J E, Baruch P and Landsberg P T 1994 12th European Photovoltaic Solar Energy Conf. p 1315

Ikegami S 1988 Solar Cells **23, 89**

Özsan M E, Johnson D R, Lane D W and Rogers K D 1994 12th European Photovoltaic Solar Energy Conf. p 1600

Tyan Y S and Perez Albuerne E A 1982 1991 Conf. 16th Photovoltaic Specialists' Conf. p 794

Woodcock J M, Turner A K, Özsan M E and Summers J G 1991 22nd Photovoltaic Specialists' Conf. p 842

Inst. Phys. Conf. Ser. No 164
Paper presented at Microsc. Semicond. Mater. Conf., Oxford, 22–25 March 1999

Contribution of cathodoluminescence and electron beam induced current to microscale evaluation of semiconducting heterostructures

B Sieber, J-L Farvacque*, J-L Lorriaux and A Fattorini**

* Laboratoire de Structure et Propriétés de l'Etat Solide, ESA 8008, Bât. C6, Université des Sciences et Technologies de Lille, 59655 Villeneuve d'Ascq Cédex, FRANCE
** IEMN, Université des Sciences et Technologies de Lille, 59655 Villeneuve d'Ascq Cédex, FRANCE

ABSTRACT: The microscale recombination properties of modulation-doped $(Al)GaAs/In_{0.15}Ga_{0.85}As/GaAs$ heterostructures are analysed at 87 K by means of plan-view electron beam induced current (EBIC) and cathodoluminescence (CL). Quantitative EBIC and CL measurements and imaging, associated with spectroscopic CL allow us to investigate a) the influence of the type of device (field effect transistor or conventional Schottky diode) on the collected signal b) the type of the dominating recombination at the quantum well.

1. INTRODUCTION

Since many years, beam injection techniques such as electron beam induced current (EBIC) and cathodoluminescence (CL) have been used in the scanning electron microscope (SEM) to characterize and image, at the microscale, the recombination properties of semiconducting material. Diffusion length measurements, surface and interface recombination velocities measurements require modeling of the EBIC and CL signals which have been developed for bulk as well as epilayers (De Meerschmann et al 1992, Cléton et al 1992). The spatial variation of point like and extended defects which act as recombination centers can also be detected by both techniques.

During the last ten years, CL has been demonstrated to be a very powerful method in the characterization of potential fluctuations in squared-shaped quantum wells (Bimberg and Christen 1993). Since such a determination is based on the radiative recombination of the exciton, it requires, for most of the semiconductors studied so far, a temperature as low as liquid helium. But, until now, no study has been devoted to the characterisation of quantum wells (QWs) when the exciton is dissociated, *i.e.* when the recombination occurs between free carriers. The experimental conditions which correspond to such a recombination can be encountered in i) squared-shaped quantum wells studied at a temperature such that the exciton is dissociated, ii) modulation-doped QWs where, for a sufficiently large channel density, the exciton is destroyed in free carriers. We have chosen to characterise modulation-doped quantum wells, in which the radiative recombination is dominated by band to band transitions. As a matter of fact, the electrons of the uppermost barrier are transferred to the QW where they form a 2D gas; since the QW channel is non intentionally doped (nid), this results in a drastic increase of the 2D electron mobility with respect to doped channels. Despite a current use of modulation-doped QWs as high electron mobility transistors (HEMTs), only very few EBIC (Kaufmann and Balk 1993, Holt et al 1993, Norman et al 1998) and even less CL (Fossaert et al 1997, Norman et al 1999) studies have been devoted to their local characterisation. These structures have been mainly studied by means of global optical techniques, such as photoluminescence, in order to derive the carrier sheet density in the well (Brierley 1993, Gilperez et al 1994) or to study many-body phenomena such as band-gap renormalization (Delalande et al 1987).

In this paper, we present the results of a combined plan view EBIC and CL analysis of the local recombination properties (Al)GaAs/InGaAs/GaAs modulation-doped heterostructures. The HEMTs which have been fabricated on these structures are field effect transistors (FETs). The electrical contacts, namely the drain, the source and the gate, are evaporated on the uppermost layer, the GaAs substrate being semi-insulating. This means that the ohmic contacts (the drain and the source) are directly connected to the quantum well InGaAs channel. In order to gain information about the effect of such a connection on the EBIC and CL signals, we have also studied a modulation-doped structure grown on a GaAs n^+ substrate on which an ohmic contact was deposited.

2. EXPERIMENTAL DETAILS

The specimens have been grown by molecular beam epitaxy at a temperature of 580°C for the GaAs buffer layers; it was lowered to 530°C for growing the InGaAs QWs.

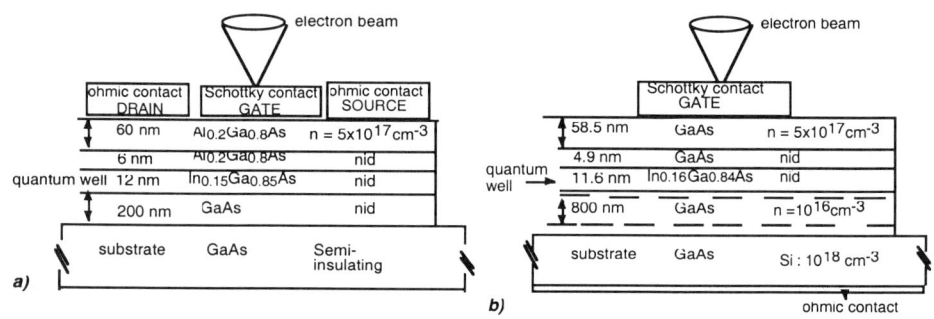

Fig. 1: Schematic representation of the two modulation-doped heterostructures.
a) the first specimen is an $Al_{0.2}GaAs/In_{0.15}GaAs/GaAs$ heterostructure evaporated on a GaAs semi-insulating substrate. The electron beam is injected through the Schottky contact and the electrical connections are made between the gate and the source of the FET. b) the second specimen is a $GaAs/In_{0.16}GaAs/GaAs$ heterostructure evaporated of a GaAs substrate silicon doped at a level close to 10^{18} cm^{-3}. The electron beam is injected through the Schottky diode.

The nid GaAs buffer layer was 200 nm thick in the first specimen (Figure 1 a: AlGaAs/In$_{0.15}$GaAs/GaAs) and close to 800 nm thick in the second one (Figure 1 b: GaAs/In$_{0.16}$GaAs/GaAs). The indium content of the nid InGaAs QW was estimated at 15 and 16 % in the first and second specimen respectively. The thickness of the well was close to 12 nm in both specimens. Growth of a 5 to 6 nm (Al)GaAs spacer layer on top of the QW allow to remote doping silicon impurities present in the (Al)GaAs uppermost layer at a level of $5x10^{17}$ cm^{-3}. The substrate of the first sample is semi-insulating GaAs (SI sample); that of the second sample is silicon doped at a level close to $1x10^{18}$ cm^{-3} (n^+ sample). FETs were fabricated on large AlGaAs/InGaAs/GaAs mesas of the SI sample(Figure 1 a), the mesas being isolated one from each other by the SI substrate. On each mesa, two ohmic contacts (drain D and source S) were fabricated first by evaporation of the Au/Ge eutectic followed by sputtering of nickel and annealing at 465°C for 90 seconds in a nitrogen-oxygen (90/10) atmosphere. Then, the Schottky diode – also called the gate G- was made by evaporating, in ultrahigh vacuum with an electron beam, titanium to a thickness of 10 nm, through a mask of 50 x 80 μm^2 dimensions, and by heating the specimen up to 280°C for 20 minutes in a nitrogen-oxygen (90/10) atmosphere. Only FETs were chosen for EBIC measurements which exhibited no reverse current in the (I-V)$_{GS}$ curve as well as good I$_{DS}$(V$_{DS}$) characteristics curves. The electrical connections for the EBIC measurements were made between the source and the gate of the FETs; this will be called the 'FET configuration' in the following. The n^+ sample was studied in the 'conventional configuration' (Figure 1 b), since the ohmic contact was deposited directly on the n^+ doped GaAs substrate. The specimens were mounted in a conventional scanning electron microscope (Cambridge 250) equipped with an Oxford CL collecting mirror. A homemade silicon photodiode and a homemade EBIC amplifier detected the CL and EBIC signals respectively. The specimens were cooled down to 87 K. An accelerating voltage of 15 kV is used and no reverse bias is applied to the structures, unless specified.

3. EBIC AND CL MEASUREMENTS

3.1. CL measurements of the modulation-doped quantum wells electronic charge

In order to check the presence of free carriers in the modulation-doped quantum wells, we have recorded their CL spectrum. As a matter of fact, in a thin InGaAs/GaAs QW, the binding energy of the ground state heavy-hole exciton is close to 8 meV (Ji et al 1987). Theoretical calculations give a similar value for a 12 nm InGaAs/GaAs QW (Iotti and Andreani 1995). The presence, in a quantum well, of a 2DEG density larger than $1-3\times10^{11}$ cm^{-2}, should lead to the dissociation of the exciton, whatever are the temperature and the electrical field (Schmitt-Rink et al 1989). The existence of free carriers induces a broadening of the QW luminescence linewidth (Skolnick et al 1987); then, recording of the CL spectrum allows to verify the presence of free carriers in the quantum well and to measure their density (Brierley 1993).

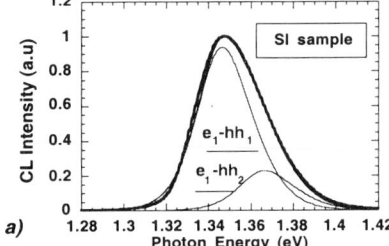 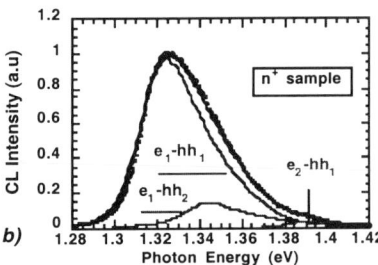

Fig.2: CL spectra recorded at 87 K on a) SI sample and b) n^+ sample. The accelerating voltage is 20 kV and the beam current 5 nA. The full width at half maximum is as large as 37 meV and 42 meV in SI and n^+ sample respectively. The sheet density is $6\times10^{11}cm^{-2}$ and 1×10^{12} cm^{-2} in the SI and n^+ sample respectively.

The very asymmetrical shape of typical CL spectra recorded on both modulation-doped InGaAs QWs (Figures 2), and their very large full-widths at half maximum confirm the presence of free carriers in the QWs. Thus, following Brierley (1993) we have measured the value of the electronic charge n_S of the well by fitting the luminescence spectra. Figures 2 show that the CL spectra can be deconvoluted in two optical transitions for the SI sample, and in three transitions for the n^+ sample. The low energy CL bands involve transition from the first electron subband to the first heavy-hole subband (e1-hh1). The second CL bands could be identified as resulting from the transition between the first electron subband and the first light-hole subband (lh1), but previous studies on InGaAs quantum wells have shown the lh1 subband to be resonant with the valence band (Marzin et al 1985, Lee et al 1997). Thus, the high energy luminescence band results from the optical transition between the first electron subband and the second heavy-hole subband (e1-hh2), in agreement with previous studies on modulation-doped QWs (Brierley 1993, Gilperez 1994). This parity-forbidden transition results from the presence of an electrical field in the modulation-doped QW which spatially separates electrons and holes, thus inducing a finite value of the envelope functions overlap and an enhancement of its squared optical matrix element. The fit of the CL spectra yields a value of the 2DEG density n_S close to $6 \times 10^{11} \text{cm}^{-2}$ and to 1×10^{12} cm^{-2} in the SI and n$^+$ samples respectively. The dark values of n_S should be a little different since neither the band gap renormalization (Zhang et al 1991, Delalande et al 1986, Das Sarma et al 1990) nor the filling up of the quantum well under illumination have been taken into account (Fafard 1993).

3.2. EBIC currents

In both FET and conventional configurations, the collection of holes by the Schottky diode gives rise to the EBIC current. By considering the energy band structure of both specimens in the dark (Figure 3) the EBIC current collected from the SI sample is expected to be much smaller than that collected from the n$^+$ sample. Experimental results displayed in Table 1 are in contradiction with such energy band structure considerations. As a matter of fact, the

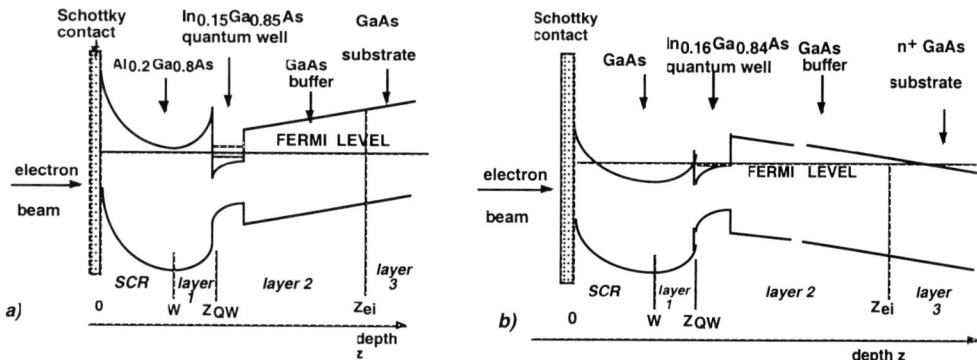

Fig. 3: Schematic representation of the energy band structure in the dark of the a) SI sample and b) n$^+$ sample.

EBIC current collected from the SI sample is about ten times larger than that collected from the n$^+$ sample.

Sample	Beam current	0.5 nA	2 nA
SI		0.785 µA	2.9 µA
n$^+$		0.07 µA	0.34 µA

Table 1: EBIC current collected at 87 K from both samples, at two beam current values.

A first explanation to this result is that the energy band diagram of both specimens is modified under beam injection. As a result of the semi-insulating type of the substrate of the SI sample, its band structure should tend towards flat band conditions. This would therefore lead to the collection of minority carriers from the buffer and substrate. Such conditions, have been taken as the starting point of an EBIC model we have developed in order to explain more precisely the striking differences observed in the EBIC currents collected from both samples.

3. EBIC MODEL

In a sample with a thick buffer, both carrier diffusion in the buffer layers and carrier capture by the quantum well will control the capture time. In the specimens we have studied, the capture will be diffusion-limited as a result of the large thickness of the buffer layer. In that case, it is possible to describe the diffusion-recombination phenomena by a diffusion of carriers in the barriers (Hillmer et al 1991). In the EBIC model, the thickness of the quantum well is neglected since it is small compared to that of the buffer layer. The QW influence on minority carrier transport is described by a recombination velocity V_{S1} that relates the carrier current densities to the carrier density at the quantum well plane. The EBIC current is the sum of the SCR current I_{SCR} and the diffusion current I_{BULK}. The first one is given by:

$$I_{SCR} = -e \int_0^w g_1(z)dz \tag{1}$$

where $g_1(z)$ describes generation of electron-hole pairs in the (Al)GaAs top layer (Akamatsu] et al 1989) and w is the SCR. The bulk current is:

$$I_{ccbulk} = -eD_1 \frac{\Delta p_1(z)}{dz}\bigg|_{z=w} \tag{2}$$

$\Delta p_1(z)$ is the minority carrier density in the first (top) layer. The minority carrier density in layer i obeys the one-dimensional diffusion equation:

$$D_i \frac{d^2 \Delta p_i(z)}{dz^2} - \frac{\Delta p_i(z)}{\tau_i} = -g_i(z) \tag{3}$$

where D_i is the minority carrier diffusion coefficient, $g_i(z)$ is the one-dimensional generation function, L_i is the minority carrier diffusion length.

The minority carrier density has the general form (de Meerschman et al 1992):

$$\Delta p_i(z) = C_i^+ \exp(z/L_i) + C_i^- \exp(-z/L_i) + \frac{L_i}{2D_i^*} \int_V \exp(-\frac{|z-z'|}{L_i}) g_i(z')dz' \tag{4}$$

For the configuration investigated in this study, the boundary conditions for the resolution of the diffusion equation are:

i) z = w: all the minority carriers are collected by the electrical field of the the SCR:

$$\Delta p_1(z = w) = 0 \tag{5}$$

ii) $z = z_{QW}$: a part of the minority carriers fluxes j_1 (in layer 1) and j_2 (in layer 2) are lost by recombination at the quantum well.

$$j_1(z_{QW}) - V_{S12}\Delta p_1(z_{QW}) = j_2(z_{QW}) \tag{6}$$

The recombination velocity V_{S12}, which describes such a process, is chosen identical on both sides of the QW: $V_{S12} = V_{S21} = V_{S1}$. Therefore, minority carrier densities are equal on both sides of the QW plane:

$$\Delta p_1 = \Delta p_2 \tag{7}$$

iii) $z = z_{ei}$: the same type of relations as in ii) hold for the buffer-substrate interface:

$$j_2(z_{ei}) - V_{S23}\Delta p_2(z_{ei}) = j_3(z_{ei}) \tag{8}$$

and

$$\Delta p_2 = \Delta p_3 \tag{9}$$

The last condition is issued from the equality of recombination velocities on both sides: $V_{S23} = V_{S32} = V_{S2}$.

iv) $z = \infty$

$$\Delta p_1(z = \infty) = 0 \tag{10}$$

The C_i^+ and C_i^- coefficients in eq. (4) are calculated numerically.

Numerical results are found in agreement with experimental ones by considering the EBIC current comes from i) both the SCR and the bulk region in the SI sample, ii) the SCR alone in the n^+ sample (Table 2). The flat band conditions assumed for the calculations of the EBIC current in the SI sample are therefore validated. But, beam injection dependence of energy band

Beam current	0.5 nA		2 nA	
Sample	*SI*	*n⁺*	*SI*	*n⁺*
I_{SCR} (µA)	0.096	**0.077**	0.38	**0.31**
I_{BULK} (µA)	0.690	0.375	2.7	1.5
I_{TOT} (µA)	**0.787**	0.450	**3.1**	1.8

Table 2: Calculated EBIC currents in both samples for two values of the beam current. No bias is applied to the structures. The parameters are as follows: SI sample: Barrier height = 1 V; V_{S1} = 2.5x10⁶ cm/s; V_{S1} = 1x10³ cm/s (Cléton et al 1992). n⁺ sample: Barrier height = 0.7 V; V_{S1} = V_{S2} = 5x10⁶ cm/s (these values are very large in order to decrease the bulk EBIC as much as possible). In bold are shown the calculated values which fit the measured ones (see Table 1).

structure does not allow to explain the difference in the EBIC currents measured on both samples, since the EBIC current collected from the n⁺ sample should be, at least, of the same order of magnitude as that collected from the SI sample (Table 2). Thus, another factor has to be considered: the way the electrical connections are made on both samples. The SI sample being studied in the FET configuration, this means that majority carriers of the QW channel can flow through the ohmic contact. This allows an easy collection of the minority carriers from the bulk, which have not been recombined in the QW or at its interfaces, with respect to a conventional configuration (n⁺ sample). But, this induces a decrease of the CL intensity from the QW compared to that emitted by the QW of the n⁺ sample in which the captured holes are forced to recombine with the electrons. This argument is supported by the fact that the QW CL intensity of the SI sample increases by a factor of 2.8 when the FETs connections are removed.

4. RECOMBINATION AT THE QUANTUM WELLS

It is well known that reverse bias leads to a carrier depletion of modulation-doped quantum wells, and, in the case of the n⁺ sample, to an increase of the GaAs buffer layer

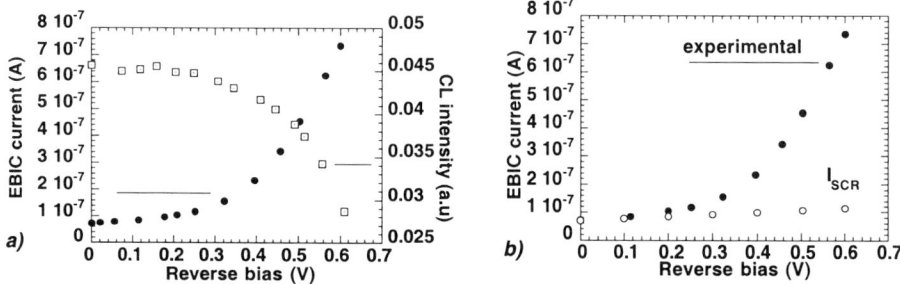

Fig. 4: Reverse bias dependence of the EBIC current and CL intensity in n⁺ sample. The beam current is 0.5 nA. a) experimental results. b) the measured EBIC current increases more rapidly than the EBIC current calculated in the space charge region.

electrical field. The decrease of CL intensity with reverse bias, as shown in Figure 4 a, is due to a decrease of the recombination efficiency of the quantum well. The very strong increase of the EBIC current with reverse bias (Figure 4 a) can be explained by the decrease of recombination efficiencies at the QW and in the buffer layer. As a matter of fact, the EBIC current increases by a factor 5, which is much larger than that calculated in the space charge region (Figure 4 b) and it reaches the same order of magnitude than that in the SI sample (Tables 1 and 2). Therefore, reverse bias of the n⁺ sample allows collecting minority carriers generated in the bulk, as in the case of the SI sample with no reverse bias. And also, the recombination velocity of the n⁺ sample should be very similar to that of the SI sample. If this parameter can be taken as a reliable measure of the recombination strength, as   is the case for planar interfaces, this would mean that the non-radiative recombination should be identical in both quantum wells. But, measurement of the minority carrier diffusion length gives quite different results. Usually, the value of the diffusion length allows the determination of the dominant recombination mechanism. In the case of a predominant radiative recombination, the minority carrier diffusion length is expected to be smaller in the n^+ QW than in the SI one, as a result of its larger electron density (Figure 2) (Arakawa et al 1985). We have measured it by the mask method (Zharem et al 1989), and found it is equal to 0.15 μm in the QW of the SI sample, and to 0.6 μm in that of the n^+ sample. Thus, the recombination is highly dominated by non-radiative mechanisms in the SI sample, as already shown by the large value of its QW recombination velocity. In the n^+ quantum well, the radiative recombination is thought to be more efficient, but non-radiative recombination cannot be ruled out completely. In the following we show that combined EBIC and CL plan view images of both quantum wells, allow to draw a more definite conclusion. These images, recorded at 15 kV, are shown in Figures 5 and 6. They all exhibit intensity fluctuations, but the EBIC-CL correlation seen in the SI sample differs from that seen in the n^+ sample.

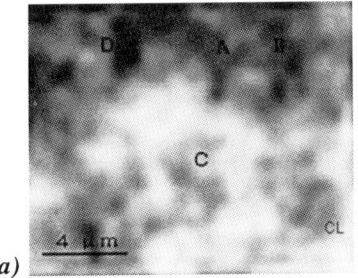
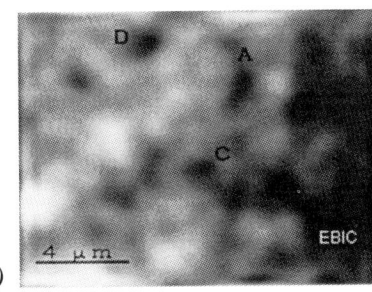

a)   b)

Fig. 5: Plan view CL and EBIC images recorded at 15 kV on the SI sample. No reverse bias is applied to the structure. Most of the dark areas in the CL image correspond to dark areas in the EBIC image.

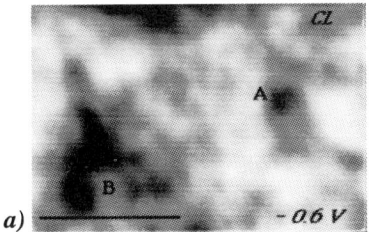
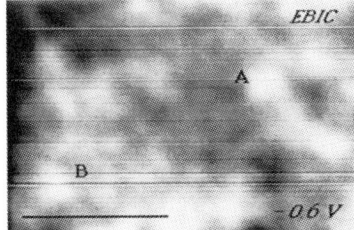

a)   b)

Fig. 6: Plan view EBIC and CL images recorded at 15 kV on the n^+ sample, for two values of the reverse bias. At a reverse bias of -0.6 V, dark areas in the CL image appear bright in the EBIC image, and vice-versa. The bar scale is equal to 4 μm.

A one-to-one correspondence between both images of the SI sample can be seen for many areas (Figure 5). On the basis of our previous analysis, this can be explained by the fact that the images display the spatial variations of the QW interface non-radiative recombination centers density. Lower CL and EBIC intensities correspond to a larger density of such centers. In the n^+ sample, no correspondence between EBIC and CL images is observed when the EBIC current is collected from the space charge region only. At − 0.6 V, when the participation of the minority carriers generated in the bulk to EBIC current is important, the EBIC and CL images exhibit inverse contrasts (Figure 6). A larger EBIC current ('bright' areas) and a smaller CL intensity ('dark' areas) should correspond to less radiative recombinations which is a result of fluctuations of the QW confinement potential. A smaller EBIC current ('dark' areas) and a larger CL intensity ('bright' areas) should mean that more radiative recombinations are present. In this case, this implies that the radiative constant experiences local fluctuations. Many mechanisms can be invoked to justify such fluctuations: local strains, electrical field variations, alloy fluctuations, well width fluctuations, etc...

5. CONCLUSION

The present EBIC and CL investigation has been undertaken in order to characterise the microscale recombination properties of modulation-doped quantum wells. We have shown that it is first necessary to use various configurations in order to be able to check the influence of beam excitation on the energy band bending of the structures. Then, we have demonstrated that numerical analysis and imaging are both important in the determination of the dominant recombination mechanism.

ACKNOWLEDGEMENTS

We should like to greatly acknowledge C. Vanmansart (LSPES) for his technical assistance in the EBIC and CL experiments, as well as for assembling the CL spectroscopic equipment, and for making the EBIC amplifier. We also thank D. Vandermoore for the electrical tests on the FETs.

REFERENCES

Akamatsu B, Henoc P and Martins RB 1989, J. Microsc. Spectrosc. Electron. **14**, 12a

Arakawa Y, Sakaki H, Nishioka M, Yoshino J and Kamiya T 1985 Appl. Phys. Lett. **46**, 519

Bimberg D and Christen J 1993, Inst. Phys. Conf. **134**, 629

Brierley SK 1993 J. Appl. Phys. **74**, 2760

Cléton F, Sieber B and Lorriaux JL 1992, Microsc. Microanal. Microstruct. **3**, 501

Das Sarma S, Jalabert R, and Yang SRE 1990, Phys. Rev. B, **41**, 8288

Delalande C, Orgonasi CJ, Meynadier MH, Brum JA, and Bastard G 1986, Solid State Communications **59**, 613

Delalande C, Bastard G, Orgonasi J, Brum JA, Liu HW, Voos M, Weimann G and Schlapp W 1987, Phys. Rev. Lett. **59**, 2690

De Meerschmann C, Sieber B, Farvacque JL and Druelle Y, Microsc. Microanal. Microstruct 1992, **3**, 505

Fafard S, Fortin E and Merz JL 1993, Phys. Rev. B **48**, 11062

Fossaert N, Dassonneville S, Sieber B, and Lorriaux JL 1997, Inst. Phys. Conf. **157**, 677

Gilperez M, Sanchez-Rojas JL, Munoz E, Calleja E, Davis JPR, Reddy M, Hill G, and Sanchez-Dehesa J 1994, J. Appl. Phys. **76**, 5931

Hillmer H, Forchel A, Kuhn T, Mahler G, and Meier HP 1991 Phys. Rev B, **43**, 13992

Holt DB, Napchan E, Wojcik A, Ammou M and Gibart P 1993 Inst. Phys. Conf. Ser **134**, 731

Iotti RC and Andreani LC 1995, Semicond. Sci. Technol. **10**, 1561

Ji G, Huang D, Reddy UK, Henderson TS, Houdré R and Morkoç H 1987, J.Appl. Phys. **62**, 3366

Kaufmann K and Balk LJ 1993, Inst. Phys. Conf. Ser **134**, 725

Lee KS, Kwon OK, Jeong BS, Han WS, Lee B, Lee EH, Kim Y, Lee CD and Noh SK 1997, Journal of the Korean Physical Society, **31**, 693

Marzin JY, Charasse MN and Sermage B 1985, Phys. Rev. **B 31**, 8298

Norman CE, Griffin N, Arnone DD, Paul DJ, Pepper M, Gallas B and Fernandez JM 1998, Solid State Phenomena **63-64**, 25

Norman CE et al 1999 this issue

Schmitt-Rink S, Chemla DS and Miller DAB 1989, Advances in Physics **38**, 89

Skolnick MS, Nash KJ, Saker MK, Bass SJ, Claxton PA and Roberts JS 1987, Appl. Phys. Lett. **50**, 1885

Zhang YH, Cingolani R, and Ploog K 1991, Phys. Rev. B, **44**, 5958

Zharem HA, Sercel PC, Lebens JA, Eng LE, Yariv A and Vahala KJ 1989, Appl. Phys. Lett. **55**, 1647

Inst. Phys. Conf. Ser. No 164
Paper presented at Microsc. Semicond. Mater. Conf., Oxford, 22–25 March 1999
© *1999 IOP Publishing Ltd*

Cathodoluminescence studies of individual GaN dots grown by MOCVD on SiC substrates

Anders Petersson, Anders Gustafsson, Lars Samuelson, Satoru Tanaka[1] and Yoshinobu Aoyagi[2]

Solid State Physics/nm Consortium, Lund University, Box 118, S-221 00 Lund, Sweden
[1]Research Institute for Electronic Science, Hokkaido University, Kita, 12-Nishi 6, Kita-ku, Sapporo 060-0812, Japan
[2]Institute of Physical and Chemical Research (RIKEN), Wako, Saitama 351-0198, Japan

ABSTRACT: We have investigated the structural and optical properties of free-standing GaN dots, grown on AlGaN barriers using metal-organic vapour deposition on SiC substrates. We have used high resolution scanning electron microscopy to study the size and position of the dots and cathodoluminescence imaging to study their optical properties. In our investigation, we have been able to study the emission of individual dots.

1. INTRODUCTION

There is currently a strong driving force for the fabrication of efficient light emitters in the blue and ultraviolet range of the spectrum. This has its origins in the large interest in flat screen displays, where efficient blue emitters are needed. Compared with today's technology, the shorter emission wavelength will increase the packing density of the information in CD storage. One of the most important material systems for this is quantum well (QW) structures of the III-nitrides (AlN, GaN and InN (Nakamura and Fasol 1997)). The major technological challenge in the fabrication of III-nitride based devices, is the lack of suitable substrates. The main contenders, sapphire (Al_2O_3) and SiC, both have a significant lattice mismatch and a substantial difference in thermal expansion coefficient (Strite and Morçok 1992). Both factors will inevitably introduce dislocations, generally associated with non-radiative recombination (Petroff et al 1980). It was considered that the dislocations were not as harmful in the III-nitrides as for conventional AlGaAs-based devices. Working III-nitride devices have been shown to have a threading dislocation (TD) density of $10^{10}cm^{-2}$, whereas the upper limit for a similar GaAs-based device is $\approx 10^4 cm^{-2}$ (Lester et al 1995). It has since been shown that dislocation in III-nitrides can be associated with non-radiative recombination (Sugahara et al 1998), as well as, deep level emission (yellow) (Christiansen et al 1996). As a confirmation of the negative effect of TDs, the latest generation of III-nitride based laser structure utilises a technique of epitaxial lateral overgrowth (Nakamura et al 1998) to reduce the density of TDs in the lower buffer layer. The effect is brighter emission and longer lifetime, where a life time of 10,000 hours has been reported (Nakamura et al 1997).

One way to try to avoid the effect of the threading dislocations (TDs) is to use quantum wires (QWRs, preferably in the *vertical* direction) or quantum dots (QDs) instead of QWs. If the density of QDs is higher than the TD density, some of the QDs will be unaffected. This, provided that the QDs are not preferentially formed at the dislocations (Keller et al 1998). Presently, there is a large interest in the fabrication of QDs by self-assembly in material systems with a significant lattice mismatch (>3%), where the QDs are formed through Stranski-Krastanow (S-K) growth. These are typically based on the conventional III-Vs, e.g., (Ga)InAs/(Al)GaAs and InP/GaInP (Seifert et al 1996). It is possible to form GaN QDs by S-K growth, but the required mismatch can only be achieved by growing on AlN barriers (Widmann et al 1998). Due to problems in obtaining suitable doping levels of AlN layers, a low Al-content AlGaN would be preferable from a device point of view. However, the low mismatch will not favour the S-K type growth, but a layer-by-layer growth. In order to grow self-assembled GaN QDs on AlGaN, a surface modifier is needed to induce the three-dimensional growth. By treating the AlGaN surface with Si, the normal step-flow growth can be inhibited, resulting in three-dimensional growth (Tanaka et al 1996).

2. GROWTH AND EXPERIMENTAL DETAILS

The QD structures in this study were grown by low-pressure metal-organic chemical vapour deposition (MOCVD) on 6H-SiC (0001) substrates, at a temperature of 1080°C. The QDs were grown on relaxed $Al_{0.12}Ga_{0.88}N$ barriers, generally with a thickness of 500 nm. To reduce the TD density, 1-2 nm of AlN was initially deposited on the SiC substrate (Tanaka et al 1997. The formation of the GaN dots was induced by treating the AlGaN surface with Si (tetraethyl-silane) during a growth interruption, resulting in a sub-monolayer coverage of Si. This changes the growth mode from step-flow growth to three-dimensional growth. Presently, the role of the Si is not fully understood. It is clear that a small amount of Si will inhibit the step-flow growth, therefore probably attaching itself to the step edges. However, it is not clear whether the Si will also act as a micro mask, leading to QD formation in the holes in the mask, or if the QDs are simply seeded on the terraces in between the steps.

To study luminescence properties of the GaN QDs, we have used, spatially resolved low-temperature cathodoluminescence (CL). This was performed in a conventional scanning electron microscope (SEM), equipped with a liquid-helium cold stage, with a temperature ranging from 20-300 K. The light was collected by a parabolic mirror and dispersed by a monochromator, using a spectral resolution of ~5 meV. Monochromatic images were recorded, using a photomultiplier tube in current mode, and spectra were recorded, using a liquid-nitrogen cooled CCD camera. The acceleration voltage used was in the range 2-5 kV, with a probe current of around 100 pA. All the CL work presented here was done in top view. The lateral shape of the QDs was studied by an SEM equipped with a field emission gun (FSEM), also in top view. Correlation between the images recorded in the CL-SEM and in the FSEM, was possible by using the slight build-up of contamination during the often, long CL investigations. The height of the QDs was studied using a comercial atomic force microscope (AFM).

3. RESULTS

We have studied a series of QD samples, produced under various growth conditions. In this report, we concentrate on one sample in the series. This particular sample is uncapped, in order to be able to directly compare the structural and optical properties. To ensure a strong CL intensity from the individual QDs, slightly more GaN than in the rest of the samples was deposited to produce large QDs. This also resulted in a broader size distribution in this sample, compared with the normal conditions. From FSEM investigations the lateral extension was found to range from a few tens of nm to about one hundred nm. AFM investigations revealed a height of the larger QDs of 40 nm. The size of the larger QDs means that the effect of the quantum confinement on the energy position of the emission peak is negligible, but we will for simplicity still refer to all dots as QDs.

Average CL spectra were recorded by scanning the beam of the SEM over a region of 50×50 μm^2. A typical spectrum is shown in figure 1, consisting of two peaks, at 3.71 eV and at 3.47 eV. This spectrum looks identical to spectra obtained with photoluminescence at the same temperature. The 3.71 eV peak corresponds to the energy of the near bandgap emission of the AlGaN barrier. The exact nature of the emission has not been investigated. The full width at half maximum (FWHM) and excitation density dependence indicates that it consists of several

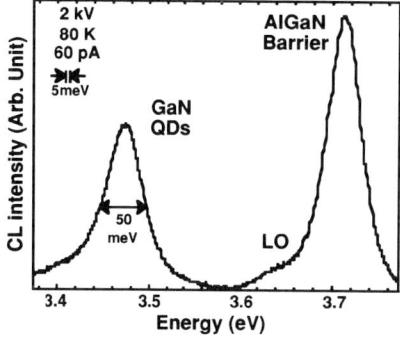

Figure 1. Spatially averaged CL spectrum.

emission peaks, including both free and bound excitons. In addition, on the low energy side of the peak, a shoulder appears. This is identified as a (longitudinal optical) phonon replica, separated by ~90 meV from the main peak. In samples without the GaN QDs, a series of replicas can be observed, with the 90 meV spacing. The spacing is independent of the layer thickness, excluding the possibility of these originating in Fabry-Perot interference in the AlGaN layer. The 3.47 eV is just above the bandgap of relaxed GaN, and can be attributed to emission from the QDs. The energy position of this peak increases with decreasing GaN deposition, i.e., QD size (Ramvall et al 1998).

Monochromatic CL images were recorded at various energy positions and at various temperatures. These images were compared with FSEM images in order to correlate the local luminescence intensity with the positions of individual QDs. A typical FSEM image is shown in figure 2(a) and the corresponding CL image at 3.47 eV is shown in figure 2(b). Figure 2(c) shows a composite image in order to better correlate the images. It was found that the 3.47 eV emission coincides with the large QDs and groups of QDs. An interesting feature is that the emission intensity goes down with the QD size. By varying the detection energy, we could conclude that this is not just a variation of emission energy, but a true decrease in the intensity. Furthermore, when recording images at 3.71 eV, weak intensity variations show complementary behaviour with the QD emission. This definitely identifies the 3.47 eV emission as the QD emission.

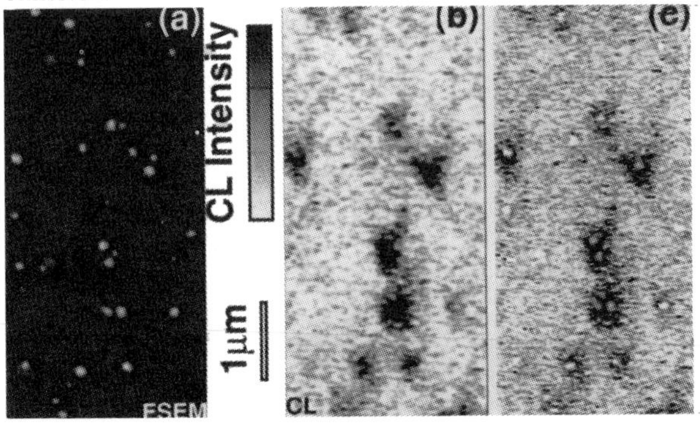

Figure 2. (a) FSEM image of an area of the sample, where the QDs appear white on a dark background. (b) monochromatic (3.47 eV) CL image of the same area (80 K, 2 kV, 60 pA). (c) composite image of (a) and (b), where a strongly luminescing QD appears white, surrounded by a dark halo and a weakly luminescing QD appears white on a white background.

To study the emission from individual QDs in more detail, we have made a series of linescans across the QDs, recording a spectrum in each point. One such linescan is presented in figure 3. Two important observations can be from this linescan: i) the FWHM of the emission from a single QD is about 50 meV. ii) The FWHM is not affected by excitation density, since varying the distance from the QD is effectively a variation in excitation density.

4. DISCUSSION

The first important observation to make is that we have demonstrated the possibility of performing spectroscopic investigations of individual QDs fabricated by S-K, or S-K like, methods. At the same time it is possible to study the structural properties of the same QD.

The decreasing intensity with QD size can have several origins. One important factor is the very short diffuison length in the III-nitrides, where a diffusion length of <100 nm has been reported in GaN at low temperatures (Chichibu et al 1997). This is consistent

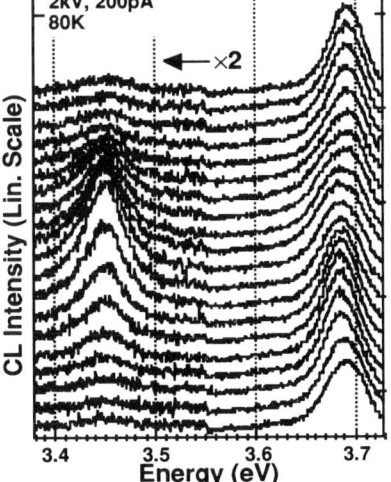

Figure 3. Series of spectra recorded during a linescan over an individual QD. The distance between consecutive spectra is 200 nm.

with the very high spatial resolution we have obtained in the CL images in the present sample. A very short diffusion length in combination with an extremely low acceleration voltage means that the main excitation in the QDs is the generation of electron-hole pairs directly in the QDs. The Grün range is ≤50 nm in GaN at 2 kV (c.f., the 40 nm height of the larger QDs). The number of electron-hole pairs generated in a QD depends on the height, which in turn means that the emission intensity increases with QD size. Another possibility is that the strain from the QDs can lower the bandgap of the barrier below the QDs, effectively creating a funnel for carriers in the barrier towards the QDs. It could be expected that a larger QD will create a larger funnel, thus capturing more carriers from the barrier. The fact that the QDs are uncapped can also affect the emission, where surface states can act as non-radiative centres and influence local band bending.

The FWHM in the spectra of individual QD is similar to the ensemble of QDs, and it is not affected by excitation density. This means that the broadening is intrinsic to the individual QDs and not an effect of the spread in size of the ensemble of QDs. It is unlikely that the broadening is caused by state filling induced by a high excitation density, this would have been observed in the linescan of figure 3. A possible explanation is the high concentration of Si in the QDs, which could lead to a significant spectral broadening. In bulk GaN, a FWHM of 50 meV corresponds to a Si concentration of $\sim 10^{18} cm^{-3}$ (Schubert et al 1997). The Si in the present sample series could result from the Si-treatment of the surface and from out-diffusion from the SiC substrate.

In summary, we have studied the structural and optical properties of individual free standing GaN QDs on an AlGaN barrier. At low temperature, there is a significant size dependence on the CL intensity. This is related to the very short diffusion length of carriers in the AlGaN barrier In combination with the low acceleration voltage used here, where the majority of carriers are generated inside the QDs, rather than in the barrier. We have also observed broad CL peak from the individual QDs, related to the high doping level (Si) within the QD.

Acknowledgements: This work was supported by NUTEK, NFR and TFR.

REFERENCES

Chichibu S, Wada K and Nakamura S 1997 Appl. Phys. Lett. **71**, 2346

Cristiansen S, Albrecht M, Dorsch W, Strunk H P, Zanotti-Fregonara C, Salviati G, Pelzmann A, Mayer M, Kamp M and Ebeling K J 1996 MRS Internet J Nitride Semicond. Res. **1** Article 19

Hessman D, Castrillo P, Pistol M-E, Pryor C and Samuelson L 1996 Appl. Phys. Lett. **69**, 479

Keller S, Keller B P, Minsky M S, Bowers J E, Mishra U K, DenBaars S P and Seifert W 1998 J. Cryst. Growth **189/190**, 29

Lester S D, Ponce F A, Craford M G and Steigerwald D A 1995 Appl. Phys. Lett. **66**, 1249

Nakamura S, Senoh M, Nagahama S-I, Iwasa N, Yamada T, Matsushita T, Kiyoku H, Sugimoto Y, Kozaki T, Umemoto H, Sano M and Chocho K 1998 Appl. Phys. Lett. **72** 211

Nakamura S, Senoh M, Nagahama S, Iwasa N, Yamada T, Matsushita T, Kiyoku H, Sugimoto Y, Kozaki T, Umemoto H, Sano M and Chocho K 1997 Jpn. J. Appl. Phys Part 2, **36**, L1568

Nakamura S and Fasol G 1997 The Blue Laser Diode (Springer, Heidelberg).

Petroff P M, Logan R A and Savage A 1980 Phys. Rev. Lett. **44**, 287

Ramvall P, Tanaka S, Nomura S, Riblet P and Aoyagi Y 1998 Appl. Phys. Lett. **73**, 1104

Schubert E F, Goepfert I D, Grieshaber W and Redwing J M 1997 Appl. Phys. Lett. **71**, 921

Seifert W, Carlsson N, Miller M, Pistol M-E, Samuelson L and Wallenberg L R 1996 Prog. Cryst. Growth Charact. **33**, 423

Strite S and Morçok H 1992 J. Vac. Sci. Technol. **B10**, 1237

Sugahara T, Sato H, Hao M, Naoi Y, Kurai S, Tottori S, Yamashita K, Nishino K, Romano L T and Sakai A 1998 Jpn. J. of Appl. Phys. **37** L398

Tanaka S, Iwai S and Aoyagi Y 1996 Appl. Phys. Lett. **69**, 4096

Tanaka S, Iwai S and Aoyagi Y 1997 J. Cryst. Growth. **170**, 329

Widmann F, Daudin B, Feuillet G, Samson Y, Rouvière J L and Pelekanos N 1998 Appl. Phys. Lett. **83**, 7618

Inst. Phys. Conf. Ser. No 164
Paper presented at Microsc. Semicond. Mater. Conf., Oxford, 22–25 March 1999
© 1999 IOP Publishing Ltd

Cathodoluminescence study of low-dimensional quantum structures grown on patterned GaAs(311)A substrates

U Jahn, R Nötzel, J Fricke, H-P Schönherr and K H Ploog

Paul-Drude-Institut für Festkörperelektronik, Hausvogteiplatz 5-7, 10117 Berlin, Germany

ABSTRACT: Molecular beam epitaxy on patterned GaAs(311)A substrates exhibits unique growth selectivity allowing the preparation of dense quantum wire arrays and coupled wire-dot structures. The temperature dependence of the cathodoluminescence contrast of a stacked quantum wire array with sub-micron periodicity reveals non-radiative recombination channels and a lateral as well as vertical exciton escape out of the wire regions as loss mechanisms in the range between 5 and 300K. Electronic coupling in wire-dot structures is established by spatially and spectrally resolved cathodoluminescence.

1. INTRODUCTION

In order to achieve two or even three-dimensional confinement of carriers in semiconductor hetero-structures, lateral potential variations on the nanometer scale are induced by growing quantum wells (Qwells) on patterned substrates. Most studies so far have been focused on the formation of GaAs/(Al,Ga)As quantum wires (Qwires) and quantum dots (Qdots) in V-grooves (Kapon et al 1989, Wang et al 1997) or on ridges (Walther et al 1993, Rajkumar et al 1993, Fujikura and Hasegawa 1995) on low-index GaAs(100) and (111) substrates. However, these structures are strongly non-planar implying difficulties for their integration in device structures. Moreover, the increase of the lateral Qwire density in V-groove wire arrays is limited by the self-ordering condition (Grundmann et al 1995). The latter demands, e. g., to grow a defined minimum thickness of (Al,Ga)As until the self-limited radius of the V-groove is reached, which in turn requires a certain etch depth, in particular if several Qwire arrays are stacked one upon another. The etch depth, however, limits the lateral distance of the Qwires. Recently, a new growth mechanism has been found in molecular beam epitaxy (MBE) of GaAs/(AlGa)As Qwell structures on stripe-patterned GaAs(311)A substrates (Nötzel et al 1996a and 1996b). Qwires are formed due to the preferential migration of Ga ad-atoms from the mesa top and bottom towards one of the sidewalls of mesa stripes oriented along [01-1]. For mesa stripes of a few 10nm height, this results in quasi-planar lateral Qwire structures. The selectivity of growth at this fast growing sidewall is even enhanced, if the lateral width and period of the mesa stripes are reduced below 1µm (Nötzel et al 1998).

In this work, we report on cathodoluminescence (CL) investigations of a vertically stacked array of sidewall Qwires with 0.5µm periodicity. The temperature dependence of the CL contrast as well as of the integrated CL intensities provide valuable information about the recombination dynamics and carrier escape behaviour in this system. Furthermore, the electronic coupling between sidewall Qwires and connected dot-like structures is investigated by spectrally and spatially resolved CL.

2. EXPERIMENTAL

For the formation of the Qwire array, a GaAs(311)A substrate was patterned with a 0.5µm-pitch periodic grating by holographic lithography. For single sidewall Qwires and wire-dot structures, standard optical lithography was used to define 10µm wide mesa stripes along the [01-1] direction and a zigzag pattern with 4.5µm long branches, alternatively misaligned by + or -30° with respect to the [01-1] direction, respectively. All samples were dry-etched to a depth of about 15nm. On the grating pattern, a 50nm thick GaAs buffer layer followed by 50nm $Al_{0.5}Ga_{0.5}As$, a stack of 3 Qwells, 50nm $Al_{0.5}Ga_{0.5}As$, and a 20nm thick GaAs cap layer were grown by MBE. The Qwells consist of 3nm thick GaAs layers separated by 10nm thick $Al_{0.5}Ga_{0.5}As$ barriers. The same structure, where the Qwell stack was replaced by a 3nm thick single Qwell, was grown on the single stripe and zigzag pattern. CL investigations were performed as a function of temperature in a scanning electron microscope equipped with an Oxford mono-CL and cooling stage system. The electron-beam energy amounted to 3–5keV and the current was 0.1–10nA. A grating monochromator and a cooled photomultiplier as well as a CCD photodetector were used to disperse and detect the CL signal.

3. RESULTS AND DISCUSSION

Fig. 1(a) shows CL spectra of the single Qwire (dashed line) and of the Qwire array averaged over an area of $20x20µm^2$ (solid line) at 5K. Both spectra consist of two CL lines, a low- and high-energy one originating from the Qwire and Qwell regions, respectively. The Qwell luminescence of the single Qwire is split into two lines, as discussed below. The structural arrangement of the Qwire array exhibits lens-shaped wire regions connected with thinner well regions in between (Nötzel et al 1998). The energy separation of the respective CL spectrum reflects the growth selectivity within the structure and corresponds to the lateral confinement potential of carriers captured in the Qwire regions. The comparison of the spectra in Fig. 1(a) illustrates that the selectivity of growth on the sub-µm-pitch grating is much higher compared to that on wide single mesa stripes. This behaviour is attributed to the small periodicity of the grating allowing Ga ad-atoms to migrate directly from the slow growing to the fast growing sidewall, which results in an increase of the Qwire thickness at the cost of the connecting Qwells. Consequently, we obtain a potential difference between both regions of 210 meV, which is more than three times larger compared with the single Qwire. The strong lateral confinement allows for a high luminescence efficiency of the Qwire array up to room temperature. The spectrum and image

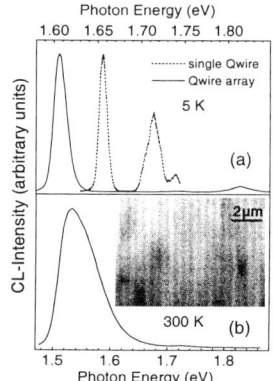

Fig. 1(a). CL spectrum of a single sidewall Qwire and of a stack of 3 sidewall Qwire arrays at 5K. (b) CL spectrum and image of the Qwire array at 300K. The detection energy was set to the peak position of the Qwire emission.

of the respective Qwire CL at 300K is depicted in Fig. 1(b). Since radiative recombination within the Qwell regions is hardly detectable in the spectrum, the surprisingly high contrast (C) in the image of the Qwire CL indicates non-radiative losses of carriers, in particular within the Qwell regions.

The temperature dependence of C and of the intensity ratio $R = I^R / I^W$ plotted in Fig. 2 provide information about the loss mechanisms within the considered system. I^R and I^W are the spectrally and laterally integrated intensities of the Qwire and Qwell regions, respectively. Starting at 5K, the value of C drops with increasing T, up to 100K. At the same time, R strongly increases. C increases again for T > 100K showing a drastic enhancement between 200 and 300K. Taking into account the temperature dependence of the exciton diffusivity D(T) and of the respective lifetime τ(T) for comparable Qwell widths (Hillmer 92 and Gurioli 91), the decrease of C and the increase of R between 5 and 100K can be attributed to an increase of both, D and τ. Since for T > 100K it is not expected that the diffusivity drops to values comparable to those at 5K, the contribution of D to the increase of C is probably small in this temperature range. Therefore, the re-appearance of the CL contrast above 100K is mainly governed by the T dependence of the exciton lifetime, in particular of its non-radiative component. Therefore, we conclude that besides exciton localization non-radiative recombination within the Qwell regions contribute to a reduction of the Qwire luminescence efficiency and to an enhanced contrast in respective CL images. We attribute the decrease of R for T >140K to a thermally activated lateral escape of excitons out of the Qwires into the Qwells. The resulting re-population of the Qwells

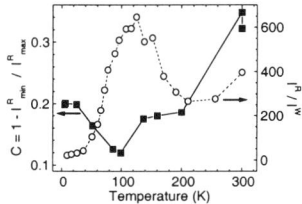

Fig. 2. Temperature dependence of the CL contrast C (squares) and the ratio of the integrated intensities of the Qwire and Qwell CL.

is negligible due to the large lateral confinement potential. The strongly enhanced contrast at 300K is attributed to a vertical escape of excitons into the (Al,Ga)As barriers, in particular out of the Qwell regions. Since the barrier material usually builds a reservoir of several kinds of non-radiative recombination centres, the vertical escape can be considered as a very efficient loss mechanism (Jahn 1999).

In the following, we will discuss the optical properties of a wire-dot structure formed during the growth on the zigzag-patterned GaAs(311)A substrate (Fricke et al 1999). The inset of Fig. 3(a) illustrates schematically the arrangement of 3 distinct regions differing remarkably in width. These are the Qwell regions on the top and bottom of the mesa, the sidewall Qwire branches of the zigzag pattern, and the corner, where two sidewall branches meet each other. The CL spectrum of Fig. 3(a) was recorded within a $3 \times 3 \mu m^2$ large scan field enclosing a corner of the structure. It consists of 3 main lines. The high-energy one, which again is split into two lines, originates from the surrounding Qwell regions. The component centred at 1.735eV represents thinner Qwell regions at the border of the sidewall and the top as well as the bottom mesa regions. The formation of these thinner regions is attributed to the migration of Ga ad-atoms towards the sidewalls during the growth process. They form potential barriers between the surrounding Qwell areas and the zigzag wire-dot regions as has been previously observed by near field optical microscopy at single sidewall Qwires on patterned GaAs(311)A substrates (Lienau et al 1998). The distance between the maximum position of these potential barriers and the Qwire amounts to less than 0.5μm.

With regard to the low-energy lines, the CL images in Fig. 3(b) and (c), for which the CL detection energy was set to 1.668 and 1.644eV, clearly reveal their spatial origin. While the CL line at 1.668eV represents the luminescence from the sidewall Qwire branches, the line at the lowest energy position stems from dot-like regions at the corners of the zigzag pattern [cf. arrows in Fig. 3(b) and (c)]. From the point of view of possible applications in device structures, it is important to establish, whether or not the Qwire branches are electronically coupled with the dot-like regions. For this purpose, CL spectra were recorded along two lines crossing a dot region. The results are depicted in Fig. 4. The first line follows one of the Qwire branches, crosses the dot, and ends on the top Qwell region [cf. inset of Fig. 4(b)]. The corresponding CL spectra are shown in Fig. 4(a). Far away from the corner (scan positions 1 to 10), 3 CL lines are observed. The line at 1.67eV corresponds to the Qwire luminescence. Since the diameter of the excitation volume (≤100nm) is larger than the lateral width of the Qwires (≈50nm), a weaker double line centred at 1.73eV appears additionally to the Qwire line due to the adjacent Qwell regions. Approaching the corner region, the Qwire CL line decreases, while at the

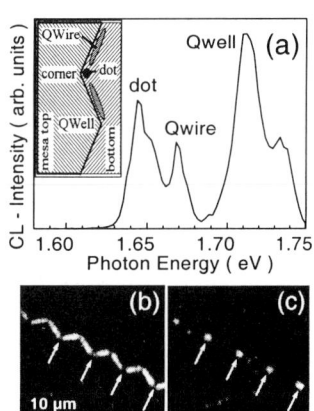

Fig. 3(a). CL spectrum from a $3 \times 3 \mu m^2$ scan field centred at the corner of the zigzag pattern at 5K (a). The inset marks the regions with different well widths. (b) and (c) are CL micrographs of the zigzag pattern at 5K. The CL detection energy was set to 1.668 and 1.644 eV, respectively.

same time the dot related CL line at about 1.65eV appears. When the dot region is reached, the CL intensity of the surrounding Qwell area increases remarkably. The integrated intensities of the Qwire and dot CL are plotted in Fig. 4(b) as a function of the scan position. The increase of the dot CL intensity is clearly correlated with the decrease of the Qwire CL intensity. Both, the decrease and increase, exhibit an exponential behaviour with comparable slopes (L^{-1}) indicating exciton transport from the Qwire towards the dot regions and, hence, electronic coupling between them. When the corner region is passed, the CL intensity of both, the Qwire and dot, decreases rapidly with a slope, which is much steeper than the one within the Qwire branch. Fig. 4(c) shows a reference line profile of the dot CL intensity, where the scan of the electron beam starts in the Qwell area at the bottom of the mesa, crosses the corner and arrives within the Qwell region on the mesa top. This profile is rather symmetric and exhibits exponential slopes similar to the profile in Fig. 4(b) after crossing the dot region. Comparing the CL line profiles in Fig. 4(c) and (b) the difference of the carrier

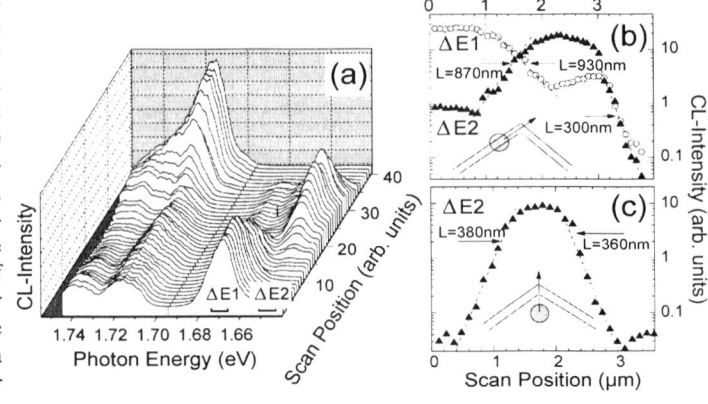

Fig. 4(a). CL spectra along a line following the Qwire branch towards the dot region of the wire-dot structure at 5K. (b) and (c) CL intensity as a function of the scan position integrated within the energy windows $\Delta E1$ (wire) and $\Delta E2$ (dot) for two different scan directions. L denotes the length, over which the CL intensity increases (decreases) by a factor of 10. The beam energy amounted to 3keV.

transport via the surrounding Qwell regions and via the Qwire branch towards the dot region becomes clearly visible. While the former is hindered by the adjacent barrier regions and/or by the low values of the exciton diffusivity within the thin Qwell regions, the latter is governed by a more efficient transport within the Qwires, which appear to be not influenced by additional energy barriers between the Qwire branches and dot regions.

REFERENCES

Fricke J, Nötzel R, Jahn U, Schönherr H-P, Däweritz L and Ploog K H 1999 J. Appl. Phys. **85**, 3576

Fujikura H and Hasegawa H 1995 J. Cryst. Growth **150**, 327

Grundmann M, Kapon E, Christen J and Bimberg D 1995 Int. J. Nonlinear Optical Phys. And Mat. **4**, 99

Gurioli M, Vinattieri A, Colocci M, Deparis C, Massies J, Neu G, Bosacchi A and Franchi S 1991 Phys. Rev. **B44**, 3115

Hillmer H, Forchel A, Tu C W and Sauer R 1992 Semicond. Sci. Technol. **7**, B235

Jahn U, Nötzel R, Ringling J, Schönherr H-P, Grahn H T, Ploog K H and Runge E 1999 Phys. Rev. **B** submitted

Kapon E, Hwang D M and Bhat R 1989 Phys. Rev. Lett. **63**, 430

Lienau Ch, Richter A, Behme G, Süptitz M, Heinrich D, Elsaesser T, Ramsteiner M, Nötzel R and Ploog K H 1998 Phys. Rev. **B58**, 2045

Nötzel R, Menniger J, Ramsteiner M, Ruiz A, Schönherr H-P and Ploog K H 1996a Appl. Phys Lett. **68**, 1132

Nötzel R, Ramsteiner M, Menniger J, Trampert A, Schönherr H-P, Däweritz L and Ploog K H 1996b J. Appl. Phys. **80**, 4108

Nötzel R, Jahn U, Niu Z, Trampert A, Fricke J, Schönherr H-P, Kurth Th, Heitmann D, Däweritz L and Ploog K H 1998 Appl. Phys. Lett. **72**, 2002

Rajkumar K C, Madhukar A, Rammohan K, Rich D H, Chen P and Chen L 1993 Appl. Phys. Lett. **63**, 2905

Walther M, Röhr T, Böhm G, Tränkle G and Weimann G 1993 J. Cryst. Growth **127**, 1045

Wang X L, Ogura M and Matsuhata H 1997 J. Cryst. Growth **171**, 341

Inst. Phys. Conf. Ser. No 164
Paper presented at *Microsc. Semicond. Mater. Conf., Oxford, 22–25 March 1999*
© 1999 IOP Publishing Ltd

Low-temperature spectral CL study of growth-induced defects in butt-coupled strain compensated InGaAs/ InGaAsP/ InP MQW amplifier-waveguide devices for semiconductor optical amplifiers

C Zanotti-Fregonara[+], L Lazzarini and G Salviati

CNR-MASPEC, Parco Area delle Scienze 37A, 43010, Fontanini - Parma, Italy
[+]also with Innovative Materials and Technologies Institute (MTI), Faculty of Engineering, University of Parma, Parco Area delle Scienze, 43100, Parma, Italy

ABSTRACT: Butt-coupled InGaAs/InGaAsP/InP multi-quantum-well (MQW) based amplifier-waveguide devices for semiconductor optical amplifiers (SOAs) were studied by spectral cathodoluminescence (SCL) at T=77K. The following growth-induced defects were observed: (1) a significant grading of the waveguide composition (up to ~ 160 meV), occurring up to ~ 200 μm from the amplifier-waveguide interface due to perturbations of the gas flow dynamics and surface kinetics of the regrowth process, caused by the Si based dielectric mask (2) etch residues of the MQW material inside the waveguide region due to poor etch selectivity and (3) multiple MQW emissions and variations of the MQW composition along the ridge due to temperature and/or flux instabilities during the growth.

1. INTRODUCTION

Opto-electronic device fabrication requires ever decreasing device size in order to increase the number of components which can be incorporated into the desired circuits. The superior spatial resolution of the cathodoluminescence (CL) technique is becoming extremely important in order to detect the presence of compositional variations and of non-radiative recombination centres in such small devices (Salviati et al. 1998). In addition, the technique is non-destructive.

The devices studied in this work are optical amplifiers used as gain-clamped amplifiers in semiconductor optical amplifier (SOA) circuits. In these devices, passive waveguides connect the amplifiers which amplify or switch the optical signals to various optical components. One of the key parameters in the production of these devices is the coupling efficiency between the amplifier and the waveguide. Butt-coupling of solid state devices is considered a high efficiency matching method that offers flexibility in the amplifier and passive waveguide design compared with other coupling methods like the evanescent (Wake et al. 1990) and the bundle integrated (Renaud et al. 1991) ones.

The presence of compositional grading and of non-radiative recombination impairs opto-electronic device performance, lowers the operating threshold, and reduces device lifetime (Ahn et al. 1996). The identification and elimination of these growth-induced defects is therefore necessary in order to increase device performance. The InGaAsP based quaternary waveguides used in these devices suffer from compositional grading (Mallard et al. 1995, Salviati et al. 1998, Zanotti-Fregonara 1998). Furthermore, it is difficult to remove the InGaAs MQW material in a uniform fashion due to reduced etch selectivity to certain compositions. This may leave behind unwanted material which can influence successive quaternary regrowth.

2. EXPERIMENTAL

All device structures were butt-coupled and had lattice-matched InGaAsP waveguide active-layers. The substrate was lattice-matched InGaAsP/InP for #qw1226, InP for the other specimens. #hrw827 and #hrw1000 had an InP:Zn-buried tensile/lattice-matched/tensile MQW having $In_{0.53}Ga_{0.47}As$ wells and tensile strained $In_{0.48}Ga_{0.52}As$ (-0.9 %) barriers. #qw1226 had an InP:Be-buried strain-compensated compressive/compressive/ tensile/compressive three-well MQW where the layer sequence was $In_{0.67}Ga_{0.33}As$ (+ 1% strain, 4 nm), InGaAsP (+0.5 % strain, 15nm), $In_{0.41}Ga_{0.59}As$ (-0.9% strain, 18nm), InGaAsP (+0.5% strain, 15nm). All structures had compressively strained $In_{0.53}Ga_{0.47}As$ cap layers which were removed before the CL study.

Device growth involved multiple step processes and included chemical beam epitaxy (CBE), photolithography for patterning, reactive ion etching (RIE), and metallo-organic chemical vapour deposition (MOCVD) for regrowths. A Si based dielectric mask was used in the patterning / etching steps. A detailed account of typical growth procedures may be found in (Salviati et al. 1998).

The specimens were studied by spectral cathodoluminescence (SCL) in a scanning electron microscope (SEM), in plan view, at T=77K and under low excitation conditions. A MonoCL2 spectrometer (Oxford Instruments) was used, equipped with a North Coast Ge diode detector (liquid nitrogen cooled) and a lock-in amplifier (optical chopper). The Cambridge S360 SEM was equipped with an Oxford Instruments liquid helium cryostat for cooling specimens down to T=5K.

3. RESULTS

3.1 #hrw827

Fig. 1a shows two spot-mode CL spectra taken at T=77K from #hrw827. Both spectra were taken from the waveguide stripe and show the extent of the compositional grading present in the device. The solid line was taken close to the amplifier-waveguide interface. The dotted line was taken far (~ 200 μm) from the interface. The peak shift is 43 meV (52 nm). This corresponds to a compositional variation of ~ 8% (Adachi 1982) which is in line with previous CL findings on similar device structures (Salviati et al. 1998). From the literature (Mallard et al. 1995, Salviati et al. 1998), the compositional variation of the waveguide layer regrowth has been ascribed to the influence of the Si based dielectric mask in perturbing the gas flow dynamics and the surface kinetics of the MOCVD regrowth process. This results in a modulation of the In and Ga concentration. The amount of compositional variation and the spatial extent of the modulation vary considerably between different growth apparatus (Salviati et al. 1998, Thompson et al. 1992). Secondary emission (SE) and monochromatic CL maps of the region of the waveguide far from the interface (~ 500 μm) are compared in Figs. 1b-c. The vertical stripe in Fig. 1b is the waveguide stripe. The surrounding region is dotted with etch residues (circular features). The monochromatic CL map in Fig. 1c was taken at 1168 nm. There is a strong variation of the CL intensity along the waveguide stripe region which suggests the presence of local variations of the optical confinement (since no local compositional variation was observed here). Electrical tests, carried out by the growers, showed that device threshold increased after growth of the waveguide. Device threshold also increased with waveguide length. This is consistent with the CL findings since the longer the waveguide is, the more optical losses are present in the device and therefore the threshold increases. Further CL studies also detected the presence of a strong modulation of the near-interface waveguide CL emission and of a larger than expected full-width at half height (FWHH= 44meV / 80 nm) from the MQW emission (which was centred at ~ 1500 nm). This suggests a possible involvement of plastic strain relaxation, which results from the unexpected composition and grading of the waveguide regrowth, and affects the MQW optical quality. This would also increase the device threshold (and reduce device lifetime). High resolution x-ray diffraction investigations revealed a partial strain relaxation inside the MQW after the regrowth, so fully confirming the CL results.

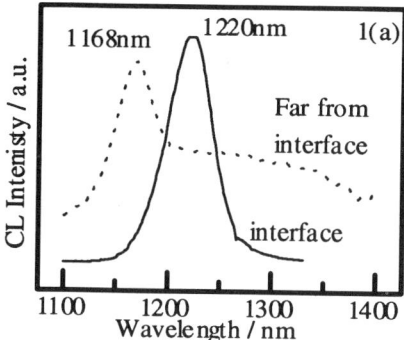

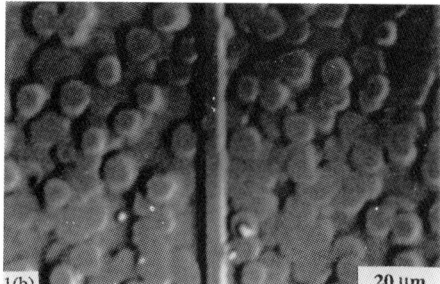

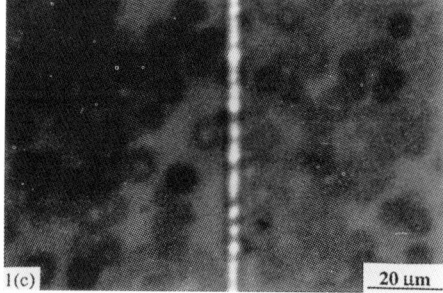

Fig. 1. #hrw827 at T=77K. (a) CL spectra taken in the waveguide close to (solid line) and at ~ 200 μm from (dotted line) the amplifier-waveguide interface. (b) Secondary electron (SE) and (c) monochromatic CL image of the waveguide, taken far from the interface (~ 500 μm).

3.2 #hrw1000

Fig. 2a is a CL spectrum of an area of #hrw1000 (shown in fig. 2b) containing a waveguide/amplifier junction. In Fig. 2b the waveguide-amplifier junction is seen in plan view. The junction is the oblique line that cuts the image in two: starting from the bottom left-hand corner and ending in the top right-hand corner. Anything to the left of this line is the amplifier, anything to the right of this line is the waveguide. The oblique line perpendicular to the junction corresponds to the gain-guiding stripe on the amplifier side and to the strip-load on the waveguide side. The CL emission from the amplifier (solid line) is broader than expected (~ 80 nm compared with ~ 20 nm expected). Deconvolution of this spectrum shows that it is made up of at least four CL emission bands (dotted lines) centred at ~ 1437 nm, 1470 nm, 1503 nm and 1524 nm. These emissions correspond to variations of the strained InGaAs (MQW) composition of up to 10% with respect to the expected ones. The presence of four distinct CL bands was confirmed by spot-mode CL spectra. Monochromatic CL maps (Figs. 2c-d) of the same area observed in Figs. 2a,b show that at least one of these emissions (1503 nm) is non-uniformly distributed within the material. From the literature (Salviati et al. 1998), this has been ascribed to compositional inhomogeneities induced by temperature and/or flux instabilities during growth. The waveguide emission spectrum was also broader than expected (FWHH ~ 137 meV / 180 nm compared with an expected 20 nm) and was centred at ~ 1240 nm. Fig. 2e is a monochromatic CL image taken at ~ 1360 nm which clearly shows strong CL

664

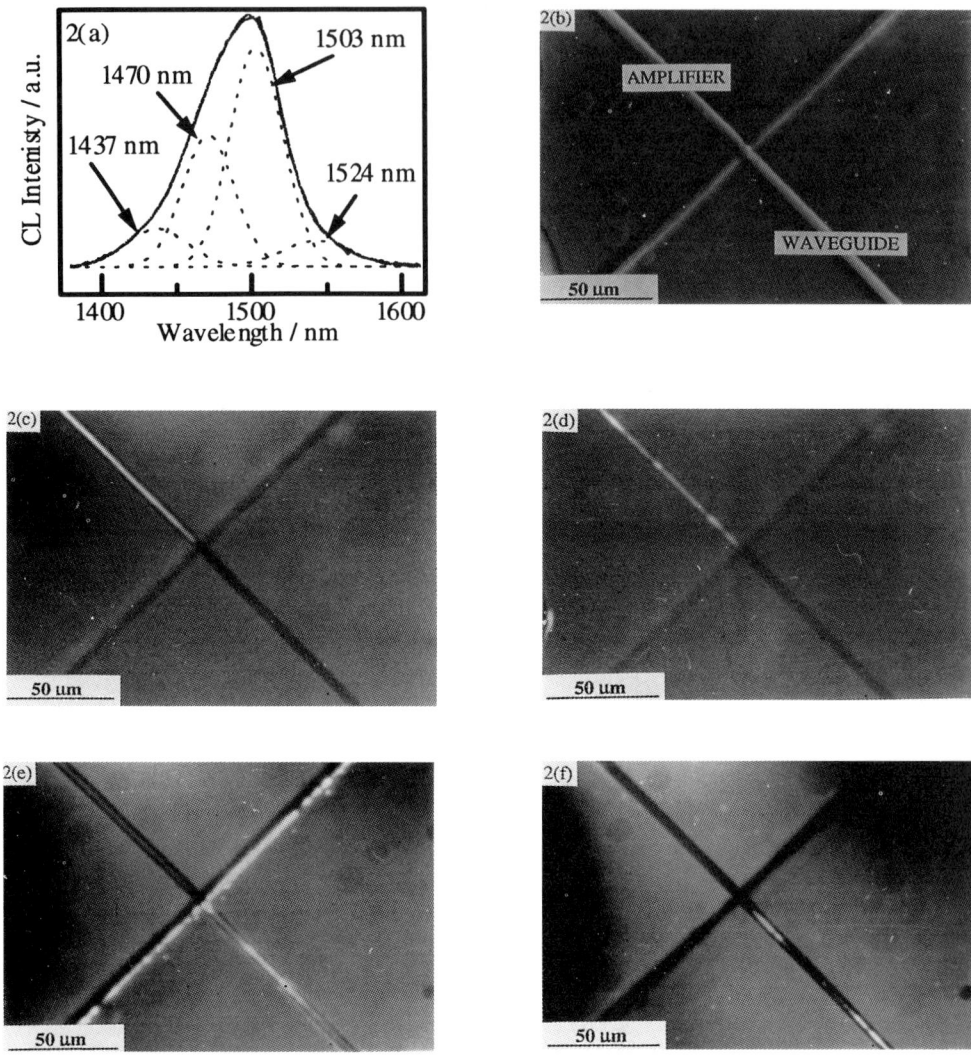

Fig. 2. #hrw1000 at T=77K. (a) CL spectrum (solid line) and deconvolutions (dotted lines) of the MQW emission from a region containing an amplifier/waveguide junction (b) secondary electron (SE) image of this region; (c-f) monochromatic CL micrographs of the same region, taken at (c) 1440 nm, (d) 1500 nm, (e) 1360 nm, and (f) 1215 nm.

intensity modulation at the amplifier/waveguide interface. Other monochromatic CL maps taken respectively at 1215 nm, 1180 nm and 1170 nm showed that there is a strong compositional variation of the waveguide layer in this structure close to the interface (up to ~ 30 µm), and non uniform CL emission far from the interface (Figs. 2e,f). The observed shift of ~ 160 meV (~ 200 nm) corresponds to a compositional variation of ~ 28% (Adachi 1982) which is higher (in magnitude) than those reported in some of the literature (Mallard et. al. 1995) for similar devices. The implication of these observations on device performance are: reduced device lifetimes and increased device thresholds.

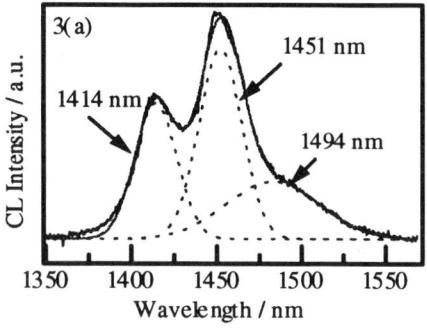
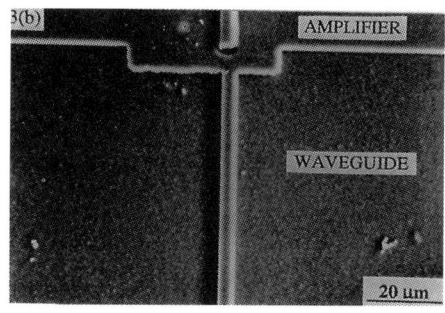
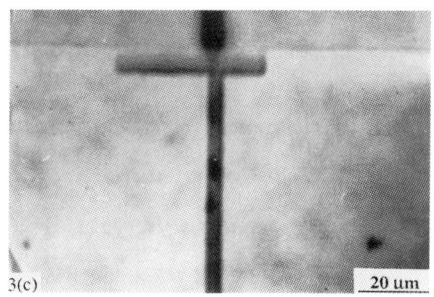
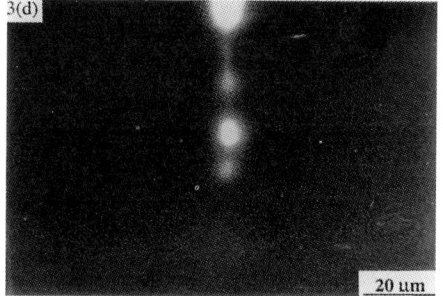

Fig. 3. #qw1226 at T=77K. (a) CL spectrum (solid line) and deconvolutions (dotted lines) of the MQW emission from a region containing an amplifier/waveguide junction (b) secondary electron (SE) image of this region; (c-d) monochromatic CL micrographs of the same region, taken at (c) 1114 nm, (d) 1416 nm.

3.3 #qw1226

Fig. 3a is a CL spectrum of #qw1226 taken from a region (shown in Fig. 3b) containing an amplifier/waveguide junction. Fig. 3b is similar to Fig. 2b except for the junction orientation. The the horizonal line crossing the image from left to right. The curve (solid line) shows that at least three distinct MQW emissions were present in the device. The deconvolutions (dotted lines) show that the emissions were centred at 1414 nm, 1451 nm and 1494 nm respectively. These emissions correspond to variations of the strained InGaAs (MQW) composition of up to 10% with respect to the expected ones.

Figs. 3c-d are monochromatic CL images taken respectively at 1114 nm and 1414 nm, corresponding to the waveguide emission and to one of the MQW emissions respectively. Comparison between these images clearly shows that the MQW emissions are present in the waveguide region and appear as 5 μm wide of circular areas, showing dark CL contrast in the 1114 nm image and bright CL contrast in the 1414 nm image. Additional monochromatic CL maps taken at 1451 nm and 1494 nm respectively and corresponding to the other two CL emissions from the MQW, were similar to Fig. 3d.

The presence of the MQW composition inside the quaternary waveguide region is due to incomplete etching during removal of the MQW material prior to waveguide regrowth. The variation of the refractive index in the waveguide, due to this unwanted material reduces the optical confinement, thereby increasing the operating threshold. In addition, Further CL spectra showed that at least two distinct waveguide CL emissions (1102 nm and 1114 nm) were present in the region studied. This suggests that in regions close to the MQW material, which was left-behind by the etch, the waveguide has grown with a different composition (from that expected), possibly due to the strain. Some of the other devices studied showed a reduction of the waveguide CL emission near the amplifier/waveguide interface. This suggests the presence of non-radiative recombination centres at the interface, as was found in previous work on similar structures (Salviati et al. 1998, Zanotti-Fregonara 1998) and may cause significant carrier leakage.

4. CONCLUSIONS

Growth induced defects in strain-compensated InGaAs/InGaAsP/InP multi-quantum-well (MQW) based butt-coupled amplifier-waveguide devices grown by CBE and MOCVD were studied by SCL at T=77K. The defects were found to increase device thresholds and to reduce device lifetimes.

In #hrw827 and #hrw1000 SCL revealed the presence of significant grading of the waveguide composition ($\sim$ 45 meV and 160 meV respectively), which occurred up to $\sim$ 200 μm from the amplifier/waveguide interface and corresponded to compositional variations of $\sim$ 8% and $\sim$ 28 % respectively. The compositional grading was attributed to perturbations of the gas flow dynamics and surface kinetics of the regrowth process, caused by the Si based dielectric mask. In #hrw1000 and #qw1226 multiple MQW emissions were found, corresponding to compositional variations of up to 10% of the strained InGaAs material with respect to the expected composition. In #qw1226 etch residues of the MQW material were found inside the waveguide region, corresponding to three distinct MQW compositions. These were leftovers from the etching process and were attributed to poor etch selectivity at specific atomic compositions. Further, more than one waveguide emission was observed in these regions, suggesting that growth on these MQW residues results in quaternary material of a different composition. Additionally, in #hrw1000, variations of the MQW composition were found along the ridge and attributed to temperature and/or flux instabilities during the growth. In some of the structures from growth run #hrw827 and far from the interface, a strong modulation of the waveguide CL emission intensity was observed, which suggested a modulation of the optical confinement. Electrical tests found that device threshold increased after regrowth of the waveguide and with increasing waveguide length. Therefore, a correlation seems to exist between the CL intensity profile and the operating threshold.

As a final comment, on the basis of the CL studies, the regrowth conditions have been optimised to give the best optical regrowth quality.

ACKNOWLEDGMENTS

Many thanks are due to Dr. D Campi, Dr. C Rigo, Dr. R Fornuto and Dr. R De Franceschi for supplying the specimens. This work is partially financed by the CNR-MADESS II project.

REFERENCES

Adachi S 1982 J. Appl. Phys. **53** 8775
Ahn J, Oh K R, Kim J S, Lee S W, Kim H M, Pyun K E and Park H M 1996 IEEE Photon. Technol. Lett., **8** 1041
Mallard R E, Puetz N, Miner C J, Zorzi J, Adams D, Cleroux M, Hillier G and Moore R 1995 Inst. Phys. Conf. Ser. 146 591 (Bristol: IOP)
Renaud M, Casivalles J A, Jarry Ph., Boucheres E and Erman M 1991 IEEE Photon. Technol. Lett. **3** 47
Salviati G, Zanotti-Fregonara C, Borgarino M, Lazzarini L, Cattani L, Cova P and Mazzer M 1998 J. Microelectronics Reliability **38** 1199
Thompson J, Wood A K, Carr N, Charles P M, Mosley A J, Pritchard R, Hamilton B, Chew A, Skyes D E and Seong T Y 1992 J Cryst. Growth **124** 227
Wake D, Judge S N, Spooner T P, Harlow M J, Duncan W G, Henning I D and O'Mahony M J 1990 Electron. Lett. **26** 1161
Zanotti-Fregonara C 1998 Ph.D. Thesis (London: University of London)

Inst. Phys. Conf. Ser. No 164
Paper presented at Microsc. Semicond. Mater. Conf., Oxford, 22–25 March 1999

Low temperature scanning cathodoluminescence studies of proton isolated GaAs / AlGaAs vertical cavity surface emitting lasers

G M Williams, P Parmiter, R A Wilson, J Heaton and G W Smith

Defence Evaluation Research Agency, St Andrews Road, Malvern, Worcs. WR14 3PS UK

ABSTRACT: Proton implantation into certain semiconductors results in regions of high resistivity non-radiative material. Implantation through appropriate masks has been used to isolate regions of GaAs / AlGaAs and produce arrays of discrete vertical cavity surface emitting lasers. Low temperature scanning cathodoluminescence was utilised to obtain spectra and images from devices. Information on the effectiveness and edge profile of the isolation is revealed and compared with predictions made using the "TRIM" ion implantation simulation software. The importance of the cathodoluminescence observations in relation to the laser performance is also discussed.

1. INTRODUCTION

Proton implantation into semiconductor materials results in the introduction of defect related mid band gap states, which act as traps for both electrons and holes. The fermi level of the implanted material is pinned at mid band gap and regions of high resistivity non-radiative material are created. With the use of appropriate masks this technique has been employed to isolate regions of semiconductor and produce arrays of discrete vertical cavity surface emitting laser (VCSEL) devices (see Jewell et al 1991, Sale 1995 and Tai et al 1989). This paper reports on a low temperature scanning cathodoluminescence (CL) study of p-n junction devices comprising of GaAs / AlGaAs / AlAs epitaxial multilayer structures grown on to n^+ GaAs substrates. The thicknesses, number of layers and composition of the various components of the laser (i.e. the top and bottom Bragg stacks of quarter wavelength dielectric layers and the multiple quantum well (MQW) cavity region between these mirrors) are designed to give the desired electrical and optical properties for visible wavelength laser emission. A number of devices were examined from two different basic structures.

Spectral information, line scans and monochromatic and panchromatic images will be presented. The cathodoluminescence data acquired from these structures was taken in plan view, at various electron beam energies to obtain depth information and in cross section. Information on the effectiveness and edge profile of the isolation is revealed and the observed depth of the isolation is compared with predictions made using the "TRIM" Monte Carlo simulation model.

2. EXPERIMENTAL METHODS

The scanning CL equipment used to study these devices is an Oxford Instruments MonoCL system with a GaAs PMT detector. The microscope on to which this equipment is built is a JEOL JSM6300 SEM. Samples were cooled to 4K and examined in plan view or cleaved carefully through individual devices for examination in cross section. The VCSEL device arrays were produced from material grown by molecular beam epitaxy or metal organic chemical vapour deposition and implantation and metalisation regions were defined using photolithography techniques. Proton implants with energies from 50keV up to 400keV were used to generate different depth isolations. Calculations using the "TRIM" model predicted this would damage from the semiconductor surface down to depths of 1.0μm to 4μm.

3. RESULTS AND DISCUSSION

3.1 Plan View CL Studies

The two plan view spectra shown in Fig. 1 below were acquired from a VCSEL device consisting of a 2.7μm top and bottom mirror each containing 81 layers of AlGaAs and AlAs. The top mirror is carbon doped p-type material and the bottom mirror is silicon-doped n-type. The centre cavity region is 0.5μm thick and contains three 6.0nm GaAs QWs. With a 20kV beam the top mirror and the MQW emission are observed (see Fig. 1, (a)). At 30kV, but keeping constant beam current, the relative intensity of the MQW luminescence increases (see Fig. 1 (b)), as would be expected from the increased sampling depth (Kanaya and Okayama 1972), but its wavelength is also seen to shift slightly. The observed shift from 794nm to 792nm represents a change in the thickness of the deeper QWs of approximately 0.5nm; constant e-beam power conditions ruled out thermal effects. Variations in the QWs thickness could clearly have effects on the wavelength of the laser emission and possibly the uniformity of emissions from an array of devices.

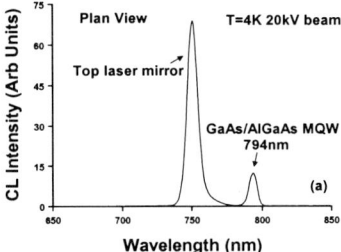 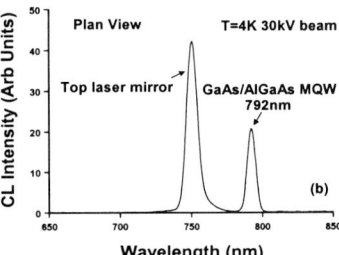

Fig. 1. Plan view spectra both from the same VCSEL device taken at (a) 20kV and (b) 30kV beam voltages with constant beam current.

Plan view monochromatic images acquired from a 50μm diameter VCSEL are shown in Fig. 2 (a) and (b). The 752nm luminescence is from the top mirror and the 794nm luminescence is from the MQW. As a result of the 4μm deep selected area proton isolation no luminescence is observed from the surrounding material. The spectrum from this structure was shown in Fig. 1. It is clear from images (a) and (b) that the 752nm mirror luminescence is much more uniform than the 794nm quantum well luminescence. The variations in the quantum well luminescence intensity could be associated with localised shifts in wavelength resulting from thickness variations as discussed above (see Stutzler et al 1988 and Oelgart et

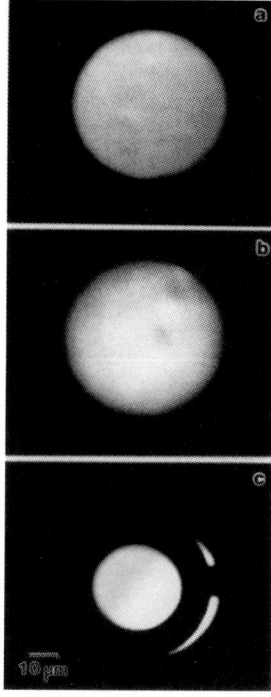

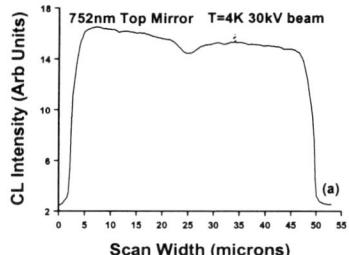

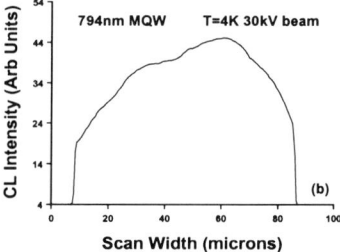

Fig. 2. Plan view monochromatic images of (a) 752nm top mirror and (b) 794nm MQW. Image (c) 656nm top mirror CL from a VCSEL in a different GaAs/ AlGaAs structure.

Fig. 3. CL plan view linescans across VCSELs (a) 752nm top mirror luminescence and (b) 794nm MQW luminescence.

al 1992). Image (c) is acquired at 656nm from a different VCSEL device structure. This luminescence is emitted by the 3.3μm thick AlAs / AlGaAs multilayer top mirror section of the device. The dark circle is the gold top contact and the image demonstrates how cathodoluminescence can reveal processing misalignments.

Images (a) and (b) in Fig. 2 revealed that the top mirror luminescence for the VCSEL structure examined was more uniform than the MQW luminescence for the same device. Careful examination of the images also revealed that the MQW luminescence around the edge of the circle was less intense than in the middle. Line scans were taken across several devices for both the MQW luminescence and the top mirror luminescence. On all devices examined the 752nm top mirror luminescence was uniform, as shown in the example in Fig. 3 (a), with the intensity dropping off sharply at the interface with the gold contact ring. The MQW luminescence at 794nm, however, has a different profile as is demonstrated by the example shown in Fig. 3 (b). The intensity is graded over a distance of some tens of microns and is then seen to drop off sharply when the scan reaches the contact ring.

It is possible that the difference in the line scans for the two components of the laser structure are due to varying degrees of lateral damage caused by the proton isolation process. Minority carrier diffusion lengths in GaAs are known to be significantly greater than in AlGaAs and this may partially explain the observation. Carriers generated in the quantum wells may be diffusing laterally to proton induced non-radiative centres. Although this effect is not yet fully explained if controlled it could be utilised to improve the modal structure of the laser. The observation is not believed to be due to absorption of the MQW luminescence in the top mirror, as previously reported by Cheng et al 1995 and Herrick et al 1995.

3.2 Cross Sectional CL Studies

Etching out a pillar to define the active region for a device is one method for producing VCSELs, however, this leads to poor thermal management and low output powers as the junctions are surrounded by air. Proton isolation enables the current path of the device to be defined without the removal of the surrounding semiconductor material and higher output powers of the order of milliwatts are achieved. To isolate material to a depth of several microns a technique has been developed which combines the effect of different energy proton implants. The depth profile for a series of different beam energies can be simulated using "TRIM" Monte-Carlo modelling software; this allows us to predict the expected depth and distribution of the damaged material. In the simulated data shown in Fig. 4 it can be seen how the combination of six different energies can result in material damaged to approximately 4.0μm.

The image in Fig. 5(a) shows the cross sectional spatial distribution of the 656nm luminescence from the top mirror of an individual 60μm diameter device; this is from the same set of devices as the example shown in Fig. 2(c). The layer around the VCSEL in Fig 5(a) is optically inactive to a depth of approximately 4μm as was predicted from the proton implantation simulation software. This region of material is the carbon doped 3×10^{18} cm^{-3} p-type AlAs / AlGaAs top mirror. Below this there is a stack of fourteen 3.5nm GaAs quantum wells which form the active part of the laser cavity and below this a bottom mirror of the same structure as the top mirror. Luminescence from the bottom mirror should extend across the image, as it has not been subjected to proton implantation. However, this material is n-type 3×10^{18} cm^{-3} silicon doped and exhibits weak luminescence with insufficient signal to collect an image. The edge profile of the luminescent region is seen, from the higher magnification image in Fig. 5(b), to be curved and this correlates with the increased lateral spread of proton damage as a function of increased energy. Figure 5(c) is a panchromatic image of a 7.0μm diameter device, which can be seen as the feature on the right of the image. The device shown in Fig. 5(c) is in the same material as (a) and (b) but this set of devices were only implanted to a depth of 2.0μm and so 2.0μm of luminescent top mirror material is still observed below the implantation area.

Figure 6 shows a monochromatic linescan taken in cross section at a wavelength of 656nm from the device shown in Figure 5 (a). The main feature to note is that the intensity builds up over a distance of some 5.0μm. This observation suggests that the damaged region of material is graded as well as being non-vertical as was suggested by Fig. 5 (b). The oscillations in intensity across the majority of the scan are unexplained but were observed for all of the cross sectional linescans and may be related to the complex 3.3μm thick structure containing 397 layers of AlGaAs / AlAs of varying composition. This feature was not observed in the image (Fig 5a) as the centre region intensity was saturated to more clearly reveal the edge profile of the isolation.

Spectra were taken at a number of positions from the cross sectional sample shown in Fig. 5 (a) and (b) and two of these are shown in Fig. 7 . The first cross sectional spectrum (Fig. 7 (a)) shows luminescence from the top mirror and the MQW region in the

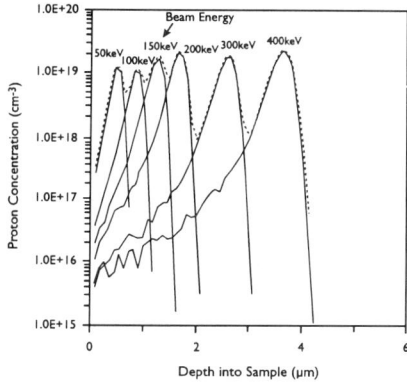

Fig. 4. "TRIM" simulation predicting the combined effect of six different energy proton implants into GaAs/AlGaAs

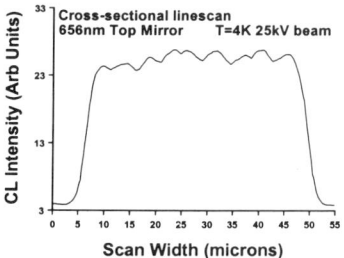

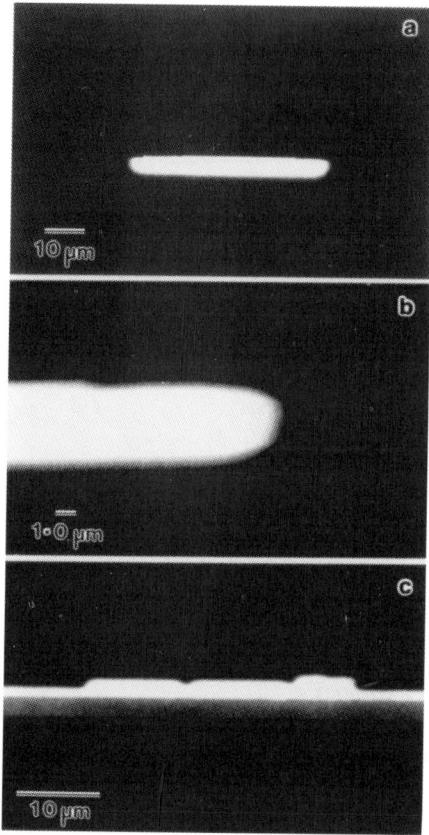

Fig. 6. Cross sectional linescan acquired using 656nm luminescence from device in image 5(a).

Fig. 5. (a) and (b) are cross-sectional images of the same device acquired using 656nm CL. Panchromatic image (c) shows a 2.0μm deep implanted sample.

laser cavity of the device. Spectrum Fig 7(b) is acquired from the bottom mirror and it is immediately obvious that the intensity is reduced and the peak broadened compared with the top mirror. This part of the structure is n-type doped ($\sim 3 \times 10^{18}$ cm^{-3} silicon) and it is believed that the high level of carriers results in the formation of a density of states in the conduction band which would cause broadening of the peak. The reduced peak intensity may result from Auger recombination or different radiative or non-radiative processes.

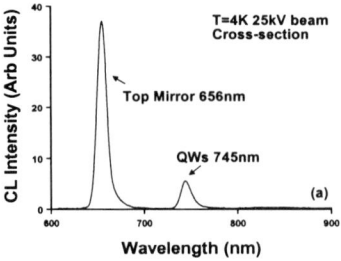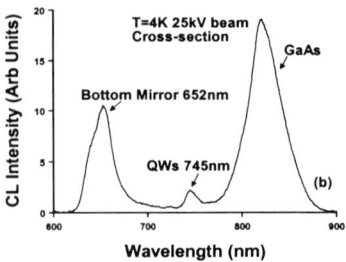

Fig. 7. Cross sectional spectra taken in (a) the region of the top mirror and the MQW cavity (Gain X1) and (b) the bottom mirror (Gain X10).

4. CONCLUSIONS

The application of CL to the characterisation of VCSEL devices has revealed a number of interesting effects associated with the proton damage profile. The observed preferential lateral damage of the MWQ region of the laser cavity can be utilised to achieve control over the laser output characteristics. A variation in QW thickness has also been detected and this demonstrates the power of CL for assessing the uniformity of the grown layers. It is planned to use CL and electrical measurements to characterise other proton isolated devices such as waveguide modulators.

REFERENCES

Cheng Y M, Herrick R W, Petroff P M, Hibbs-Brenner M and Morgan R A 1995 Appl. Phys. Lett. **67**, 1648
Herrick R W, Cheng M Y, Petroff P M, Hibbs-Brenner M and Morgan R A 1995 IEEE Photonics Tech. Letts. **Vol 7**, 1107
Jewell J L, Harbison J P, Scherer A, Lee Y H and Florez 1991 IEEE J. Of Quantum Elec. **27**, 1332
Kanaya K and Okayama S 1972 J. Phys. D: Appl. Phys. **5**, 43
Oelgart G, Lehmann L, Araujo D, Ganiere J D and Reinhart F K 1992 J. Appl. Phys. **71**, 1552
Sale T E 1995 Vertical Cavity Surface Emitting Lasers (Research Studies Press Ltd)
Stutzler F J, Fujieda S, Mizuta M and Ishida K 1988 Appl. Phys. Lett. **53**, 1923
Tai K, Fischer R J, Wang K W, Chu S N G and Cho A Y 1989 Electronic Letters **25**, 1645

Inst. Phys. Conf. Ser. No 164
Paper presented at Microsc. Semicond. Mater. Conf., Oxford, 22–25 March 1999
© 1999 IOP Publishing Ltd

Effect of proton irradiation on the spatial defect distribution of n+-p InP/Si solar cells

M J Romero, R J Walters*, D Araújo, S R Messenger[+], G P Summers* and R García

Departamento de Ciencia de los Materiales e I.M. y Q.I., Facultad de Ciencias, Universidad de Cádiz, Apdo. 40, E-11510, Puerto Real (Cádiz), Spain. E-mail: manueljesus.romero@uca.es
*Naval Research Laboratory, Code 6615, 4555 Overlook Ave., S W, Washington, DC 20375, USA
[+]SFA Inc., Largo, MD 20774, USA

ABSTRACT : For satellite communication systems the optimum orbits can be those in the 2,000 to 10,000 km range, known as medium earth orbits (MEOs). In these orbits, reliable satellite power generation is a critical need because the solar cells degrade their performance under the impact of high energy particle radiation (electrons and protons) in the space environment. Recently, n+-p InP/Si solar cells have been reported to be extremely radiation tolerant. We examine here the effect of proton irradiation on the spatial defect distribution of these cells making use of the cathodoluminescence (CL) technique. The CL data are discussed in terms of the displacement damage dose formulation of Summers et al (1995) and is then applicable to a wide energy spectrum of protons. The radiation-induced deactivation of Zn acceptors in the base region is shown by analysis of the CL spectra.

1. INTRODUCTION

Considering design parameters such as satellite coverage, transmission delays and launch costs, the optimum orbits for global coverage systems are often in or near the proton radiation belts, which extend from approximately 2,000 to 10,000 km above the earth's surface. Radiation effects are very severe in these orbits, and InP solar cells grown on Si substrates (InP/Si) offer one of the few possible power sources for such missions. The InP/Si technology combines the superior radiation resistance of InP with the strength and cost effectiveness of Si. These cells show essentially no degradation even after proton and electron irradiation up to fluences equivalent to ~ 1 x 10^{16} cm^{-2} 1 MeV electrons.

In this paper, the spatial distribution of proton irradiation-induced defects in InP/Si is analyzed using the cathodoluminescence (CL) technique. The nature of the radiation-induced defects is discussed using CL spectra. InP/Si solar cells that have been irradiated with protons with energies ranging from 100 keV up to 4.5 MeV are studied, and the measured data are correlated in terms of displacement damage dose, which is given by the product of the particle fluence and the non-ionizing energy loss (NIEL) (Summers et al 1995).

2. EXPERIMENTAL DETAILS

The cells were grown epitaxially by metalorganic chemical vapor deposition (MOCVD) on n-type Si wafers by Spire Corporation. The cell structure is described in Table 1. Several cells were proton irradiated up to a fluence of ~ 1 x 10^{12} cm^{-2} with proton energies between 100keV and 4.5 MeV. There were also cells irradiated with 3 MeV protons up to fluences of 1 x 10^{14} cm^{-2}, and we refer to them as heavily-proton irradiated cells (HPI). The irradiations were carried out at the Pelletron accelerator facility of the Naval Surface Warfare Centre in White Oak, MD, with a dosimetry accuracy of ~10%.

These cells were cleaved to carry out the CL measurements on {110} faces perpendicular to the epilayers. Micrographs were taken to monitor the CL on each (110) face and spectra were recorded from different locations of the InP buffer and base of the cell. The CL experiments were performed using a JEOL JSM820. A semi-parabolic mirror attached to an optical guide yields a highly efficient collection of the luminescence. A cryogenic CCD Photometrics SDS9000 is attached to an Oriel 77400 Spectrograph-

674

luminescence. A cryogenic CCD Photometrics SDS9000 is attached to an Oriel 77400 Spectrograph-Monocromator for the spectroscopic mode of the CL. A CTI-Cryogenics 22C/350C Helium closed-circuit cryostat attached to an anti-vibration system is used in low temperature CL experiments. The instrumentation and data acquisition are computer controlled.

3. RESULTS AND DISCUSSION

Figure 1 shows a set of CL micrographs recorded from cells prior to irradiation (a) and after irradiation with protons of energy 3 MeV (b), 1 MeV (c) and 100 keV (d) up to single fluence of $\sim 1 \times 10^{12}$ cm^{-2} and after heavy-proton irradiation (e). The displacement damage dose (D_d) increases from (b) 2×10^{10} MeV g^{-1} to (e) 3×10^{12} MeV g^{-1}. The level of excitation was E_b = 10 keV and I_b = 1 nA. As indicated at the left hand side of the Figure 1a, the CL emission is analyzed in the buffer region.

Prior to irradiation, a non-uniform spatial distribution is emitted throughout the buffer region, due to the activity as non-radiative recombination centers of dislocations threading into the active layers. The irradiation produces uniformly distributed defects that compete with the dislocations as non-radiative recombination channels. In extreme radiation conditions (Figure 1e), the radiation-induced defects govern the carrier dynamics in these cells so that the CL spatial distribution becomes uniform.

To quantify the increase in the CL spatial uniformity, we have analyzed the average lateral distribution of the CL, and estimated the standard deviation and maximum amplitude of these distributions in terms of D_d. The data are seen to correlate very well with D_d (Figure 2), and the difference between the maximum and the minimum of the lateral distribution, as well as the standard deviation, decrease with increasing D_d. This indicates that as the concentration of displacement defects in the epilayers increases, the lateral uniformity of the CL signal increases. Therefore, the dislocations do not getter the radiation-induced defects.

Figure 3 illustrates the effect of irradiation on the CL intensity emitted from the base of the cell. The radiation-induced degradation of the solar cell base minority carrier lifetime can be expressed as

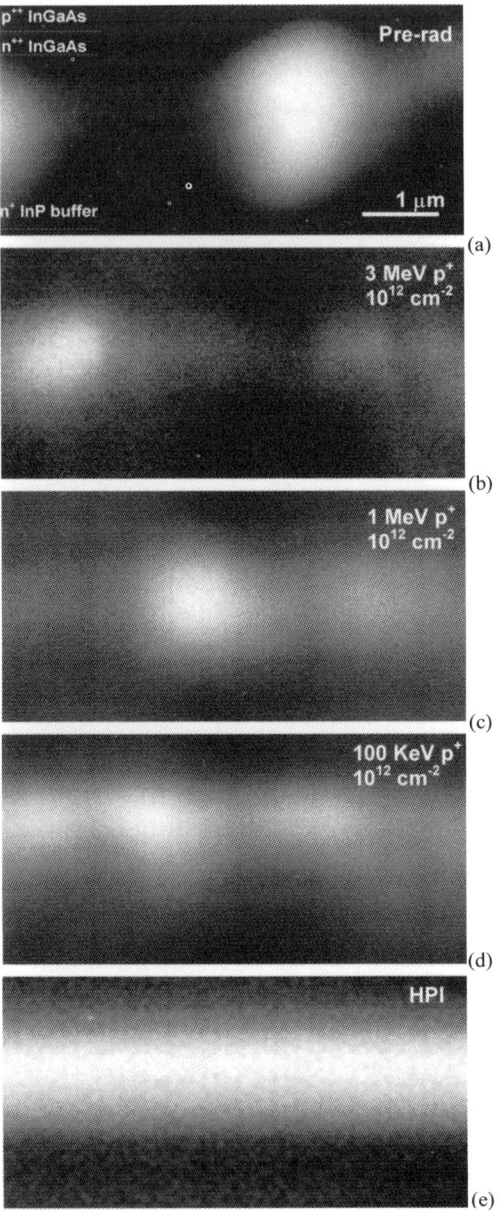

Figure 1. CL micrographs (T = 300 K) from the buffer layer of the n$^+$-p InP/Si cells prior to the irradiation (a) and after the irradiation (b-e) at different proton energies or fluences. The displacement damage dose increases from a to e. Before irradiation, a non-uniform spatial distribution of the emitted luminescence is recorded from the base region. The irradiation induces new recombination channels screening the prior non-uniformity. The level of excitation was E_b = 10 keV and I_b = 1 nA.

Material	Function	Thickness (μm)	Carrier density (cm^{-3})
			Electrochemical CV
n$^+$ InP	emitter	0.025	Si, 7×10^{18}
p InP	base	1.5	Zn, 4.5×10^{16}
p$^+$ InP	back-surface field (BSF)	1.5	Zn, 1×10^{18}
p^{++} InGaAs	tunnel junction	1	Zn, 2×10^{19}
n^{++} InGaAs	tunnel junction	0.5	Si, 7×10^{18}
n$^+$ InP	buffer	5	Si, $\sim 3 \times 10^{18}$
n$^+$ Si	substrate	400	Si, 2×10^{18}

Table 1. n$^+$-p InP/Si solar cell structure. The carrier concentrations from ECV measurements are also indicated.

where τ_0 is the minority carrier lifetime prior to irradiation (that includes the radiative and non-radiative recombination lifetimes, τ_r y τ_{nr}, respectively) and κ_τ (g s^{-1} MeV^{-1}) is the carrier lifetime damage coefficient. We substitute equation (1) in the expression of the luminescence intensity I, that is described under low injection by

$$ I \propto \frac{\Delta p}{\tau_r}\tau = \frac{\Delta p \tau_0}{\tau_r\left(1+\tau_0\kappa_\tau D_d\right)} $$

$$ I = \frac{I_0}{1+\tau_0\kappa_\tau D_d} \qquad (2) $$

where Δp and I_0 are the excess carrier density and the luminescence intensity prior to irradiation. A fit of the luminescence data to equation (2) is shown in Figure 3. The equation is seen to fit the data well, and the fit yielded an estimate of $\tau_0\kappa_\tau \sim 2 \times 10^{-14}$ g MeV^{-1}.

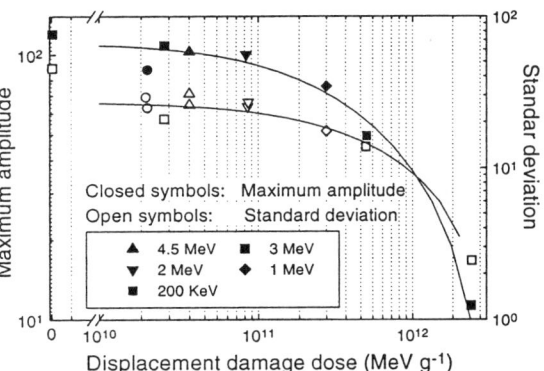

Closed symbols: Maximum amplitude
Open symbols: Standard deviation

▲ 4.5 MeV ■ 3 MeV
▼ 2 MeV ◆ 1 MeV
■ 200 KeV

Figure 2. D_d of the standard deviation and maximum amplitude estimated from the lateral distributions of the CL on the average for several n$^+$-p InP/Si solar cells irradiated with protons ranging from 100 KeV to 4.5 MeV

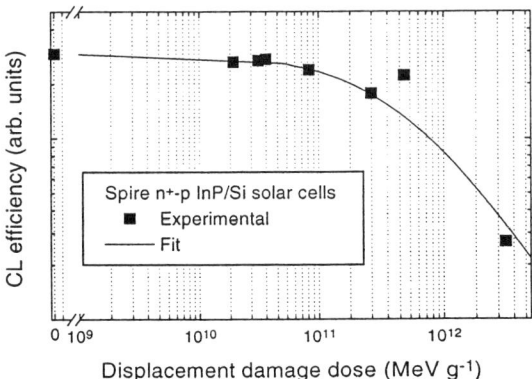

Spire n+-p InP/Si solar cells
■ Experimental
— Fit

Figure 3. Effect of irradiation on the luminescence intensity emitted from the base of the n$^+$-p InP/Si cells. The solid curve represents a fit of the data to equation (2).

Figure 4 shows the effect of irradiation on the CL spectrum emitted by the buffer of the InP/Si cells. Prior to irradiation, the CL spectra are characterized by the free-to-bound transition (h, Si^0), peaked at 1.414 eV, that involves Silicon donors that supply the n-type conductivity to the base. The luminescence band at $\sim$ 1.38 eV is attributed to the transition (e, Zn^0) associated with Zinc acceptors. The presence of Zn in the buffer layer is a result of the background level of Zn in the MOCVD reactor and the Zn diffusion from the tunnel junction during the growth (Romero et al. 1998).

676

The cells irradiated up to D_d levels below ~ 10^{11} cm^{-2} show essentially no change in the CL, with a CL intensity ratio, between the (h, Si^0) and (e, Zn^0) transitions, of ~ 1 :7. At higher levels of D_d, the relative contribution of the (e, Zn^0) emission decreases (Figure 4). Finally, at extremely high doses (3×10^{12} MeV g^{-1}), the 1.38 eV peak quenches and a broad band peaked at 1.35 eV emerges. This band is commonly observed in ion implanted InP (Bhattacharya et al. 1984, Rao 1986, Kim et al. 1987, Rao et al. 1988). From a previous analysis (Romero et al. 1998), this DAP transition is attributed to a complex involving the antisite defect In$_P$$^{2-}$ and Zn impurities.

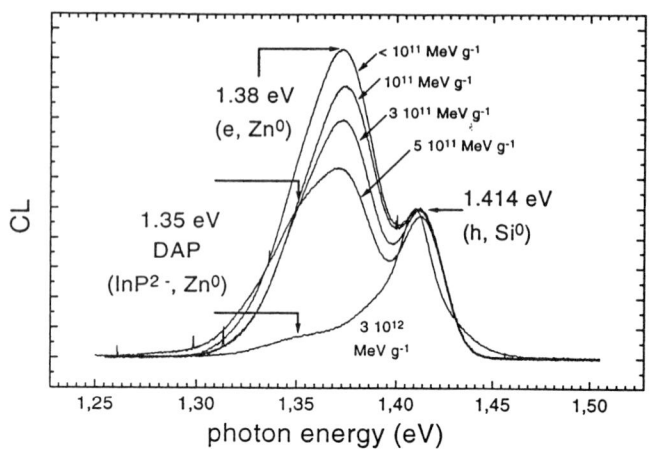

Figure 4. Effect of the displacement damage dose, D_d, on the spectral distribution of the CL (T = 65 K). The spectra are normalized to the 1.414 eV peak.

These results agree well with the ECV measurements of Walters et al. (1999) who showed that carrier removal effects in these cells, which cause a decrease of the hole concentration in p- type InP, become evident at D_d levels ~ 10^{11} MeV g^{-1}. The lowering of the relative contribution of the (e, Zn^0) transition with the dose can be correlated with the progressive deactivation of the acceptor character of Zn impurities, that in extreme irradiation conditions (~ 10^{12} MeV g^{-1}), type converts the base region. In this way, carrier removal essentially destroys the n$^+$-p InP junction so that the cell spectral response collapses.

4. CONCLUSIONS

A detailed analysis of the spatial defect distribution in proton irradiated n$^+$-p InP/Si solar cells has been presented. Prior to irradiation, we have found a non-uniform spatial distribution of the emitted luminescence throughout the cell active layers, due to the presence of dislocations acting as non-radiative recombination channels. The irradiation produces uniformly distributed displacement defects that compete with the threading dislocations so that the emitted luminescence gradually becomes spatially uniform. This study shows that the displacement damage dose model can be used to predict the degradation of this cell technology for the energy spectrum of protons found in the space environment.

ACKNOWLEDGEMENTS

This work was partially supported in Spain by the CICYT (Comisión Interministerial de Ciencia y Tecnología) under MAT94-0823-CO3-02 and by the Junta de Andalucía through group TEP-0120, and in the U.S. by the Office of Naval Research.

REFERENCES

Bhattacharya P K, Goodman W H, and Rao M V 1984 J. Appl. Phys. **55**, 509
Kim T S, Lester S D, and Streetman B G 1987 J. Appl. Phys. **62**, 1363
Summers G P, Burke E A, and Xapsos M A 1995 Radiation Measurements **24**, 1.
Rao M V 1986 Appl. Phys. Lett. **48**, 1522
Rao M V, Aina O A, Fathimulla A, and Thompson P E 1988 J. Appl. Phys. **64**, 2426
Romero M J, Walters R J, Araújo D, and García R 1998 Solid State Phenomena **63-64**, 497.
Walters R J, Romero M J, Araújo D, García R, Messenger S R, and Summers G P 1999 J. Appl. Phys. (in press)

Inst. Phys. Conf. Ser. No 164
Paper presented at Microsc. Semicond. Mater. Conf., Oxford, 22–25 March 1999
© *1999 IOP Publishing Ltd*

Novel applications of EBIC and CL to regrown, modulation-doped structures on patterned substrates

C E Norman[1], M L Leadbeater[1], T M Burke[2] and D A Ritchie[2]

[1] Toshiba Research Europe Ltd. Cambridge Research Labs. 260, Cambridge Science Park, Milton Road, Cambridge, CB4 0WE.
[2] Cavendish Laboratory, University of Cambridge, Madingley Road, Cambridge, CB3 0HE.

ABSTRACT: Modulation-doped structures regrown on patterned substrates have been characterised by EBIC and CL mode microscopy. EBIC reveals the delta-doped back gate used to form ultra-narrow undepleted channels in a regrown 2DEG with approximately 300nm resolution. It can also indicate the degree of Fermi level pinning at the regrowth interface, and the effectiveness of compensatory delta-doping. CL is used to map the 2DEG uniformity across the patterned structure, and can show variations in 2DEG carrier concentration on narrow facets.

1. INTRODUCTION

The field of regrowth on patterned substrates is of interest because it offers the possibility of extending the ultra-short length scales available in molecular beam epitaxial (MBE) growth to real device structures (Leadbeater *et al* 1999). As an example, if an MBE-grown layer only a few nm thick can be used "end-on" to modulate the electrical properties of a 2-dimensional electron gas (2DEG) grown in close proximity, ultra-narrow channels or barriers can be formed. This could be realised either by regrowth over in-situ cleaved edges or etched facets. One such example of this is the ultra-short gate length FET (Burke *et al* 1999) in which a delta-doped back gate intersecting a cleaved facet is used to deplete a very narrow region (only tens of nm wide) between two areas of regrown 2DEG to which source and drain contacts are made. Alternatively, as in the case of the structures reported here, such a narrow back gate can be used to prevent depletion of a narrow channel within a 2DEG, allowing 1-D transport through the channel to be studied.

Such devices depend on several factors for their successful operation, not least of which is the close proximity of the active layer (usually a 2DEG) to the regrowth interface (RI). Reducing this distance to a minimum not only offers advantages in terms of the length-scales achievable using regrowth, but also increased through-put in device growth by reducing the thickness of the buffer layers necessary to obtain high mobility structures. One key technological problem associated with regrowth of this type is the cleaning procedure to reduce contamination and resultant problems with regrowth layer quality and Fermi level pinning at the RI. We have previously demonstrated high (10^6 $cm^2V^{-1}s^{-1}$) mobility 2DEGs at distances of only 30nm from the RI (Burke *et al*, 1998), obtained using a specially developed hydrogen radical cleaning procedure.

This paper reports the application of cathodoluminescence (CL) and electron beam induced current (EBIC) mode scanning electron microscopy (SEM) to device structures regrown on patterned substrates.

2. EXPERIMENTAL

CL measurements were performed at 4.8K using an Oxford Instruments Ltd. MonoCL system and CF302TC liquid helium cryostat, attached to a Hitachi S-4500 cold cathode SEM upgraded with an oil-free vacuum system. At room temperature the chamber pressure is typically 6 x10^{-7} Torr, cryo-pumping to 6 x 10^{-8} Torr at helium temperature operation. A high sensitivity GaAs photocathode photomultiplier tube (PMT) was used to acquire images and spectra. EBIC measurements were performed using either a Stanford SR570 or a Matelect Ltd ISM-5 amplifier.

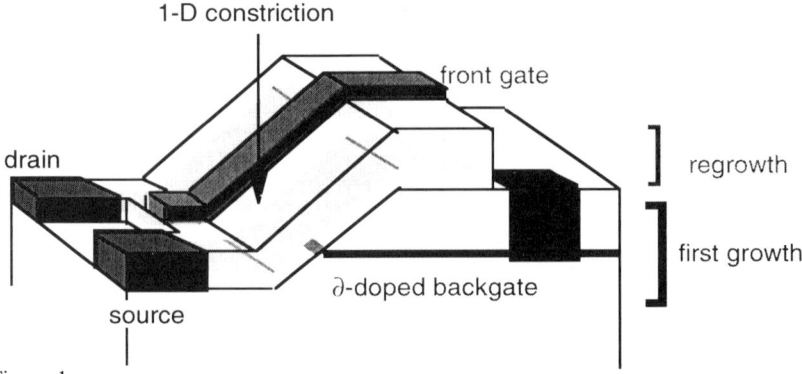

Figure 1.
A schematic representation of the device structure showing the ultra-narrow 1D channel formed under the front gate as a result of the delta-doped back gate.

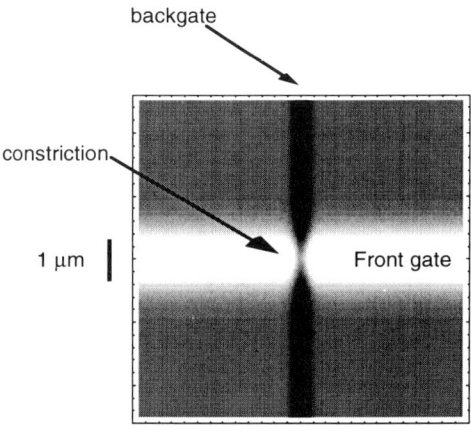

Figure 2.
A plan-view schematic representation of the electron potential expected for the structure shown in figure 1. Bright contrast indicates high potential.

The samples reported here are of the type shown in figure 1. An initial structure of 1.25μm of undoped GaAs containing a delta-doped layer (Si at 2 x 10^{12} cm^{-2}) was grown by MBE. The wafer was then removed from the MBE chamber, patterned by standard photolithography and wet etched in a HF/H$_2$O$_2$ mixture to reveal (411)B facets. The wafer was then replaced in the MBE machine to undergo an in-situ hydrogen radical cleaning procedure (Burke *et al*, 1998) and the subsequent regrowth of a quantum well containing a 2DEG. The structure is then re-processed to isolation-etch the 2DEG over the facet (see figure 1), and form Ohmic contacts to the back gate and 2DEG (source and drain). A narrow stripe of NiCr is deposited on the surface crossing the facet to form the front gate. The structure is designed to produce a narrow constricted region in the regrown 2DEG. Applying negative bias to the front gate depletes a stripe (the width of which depends on the front gate width) across the etched facet, except for an ultra-narrow line which is determined by the intersection of the delta-doped back gate with the etched facet.

Transport measurements have shown conductance plateaux in these structures, indicating ballistic transport through the 1-D channels. The form of the potential expected across the facet is shown schematically in figure 2. Bright contrast indicates high electron potential, and thence regions of the 2DEG which will be depleted.

The cleaning process is effective in removing many surface contaminants (in particular, O, OH, C$_2$H$_2$, O$_2$, Cl^{35}, Cl^{37}) but there is still significant pinning of the Fermi energy at the RI. Compensatory Si delta-doping is therefore also employed close to the RI in the regrown structure in an attempt to counteract this. The level of this delta-doping is important: if it is too low then the Fermi level pinning at the RI will have a significant effect on the 2DEG, if it is too high then the back gate may be screened, and will therefore less effective in forming the 1-D channel. Selecting the Si doping level which gives the highest mobility in the 2DEG (i.e. limiting the effect of the RI) has hitherto been the only way to assess the optimum doping level. If a technique such as EBIC can show both the effect of the RI pinning and the back gating, then it may prove to be a useful independent verification of the optimum compensatory doping level required.

The regrown structure is as follows (layers listed from substrate to surface, Al content): 2nm GaAs (580°C), 120s growth interrupt, 1nm GaAs, Si delta-doped layer (doping level varied from sample to sample), 5 period 1nm GaAs/1nm AlAs superlattice (all 490°C), 90s growth interrupt, 9 period 1nm AlAs/1nm GaAs superlattice (550°C), 14nm GaAs, 20nm AlGaAs, 40nm AlGaAs (Si doped 2 x 10^{18} cm^{-3}), 10nm GaAs cap (all 580°C).

3. RESULTS AND DISCUSSION

The band structure in the 1-D constriction device is shown schematically in figure 3. It is clear from the figure that there are two principal connection configurations for EBIC analysis. There will be significant band-bending between the front gate and the 2DEG (source and/or drain) which will allow the effectiveness of the front gate to be assessed. More interestingly perhaps, from the point of view of RI pinning and the compensatory delta-doping, there also will be band bending beneath the 2DEG, arising due to Fermi level pinning at the RI. Changes in the degree of Fermi level pinning at the RI, and hence the band bending, should be revealed by EBIC imaging.

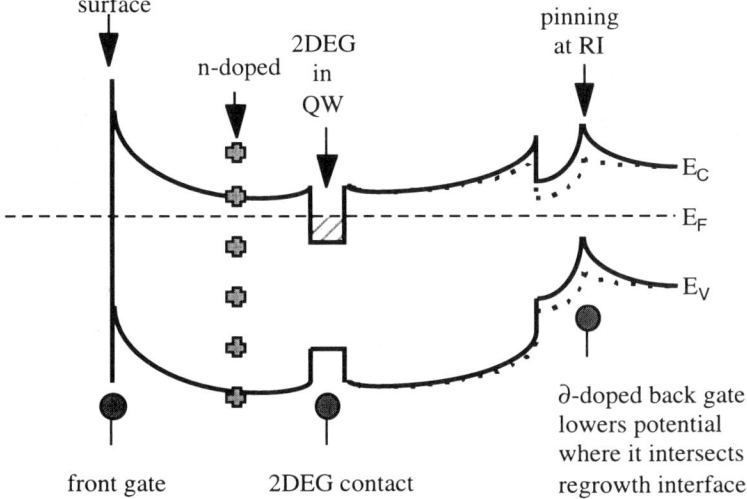

Figure 3.
A schematic representation of the band structure in the 1-D constriction devices. The dotted line represents the local change in potential at the intersection of the delta-doped back gate with the RI.

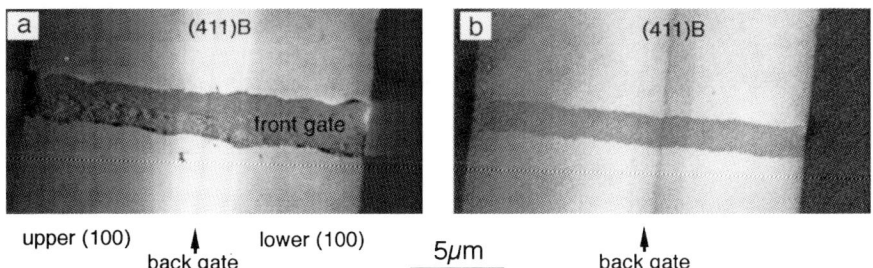

Figure 4.
Room temperature, 3keV EBIC images of two 1-D constriction devices. Specimen (a) has higher compensatory doping at the RI and exhibits better conduction on the (411)B facet in transport measurements than does specimen (b).

Figure 4(a) and (b) show room temperature EBIC images of two such devices. In both cases the front gates are floating and therefore appear dark as they partially block the beam entering the

680

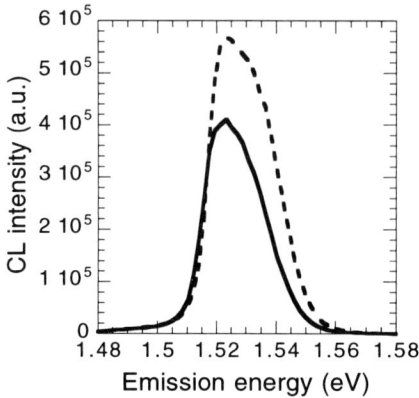

Figure 5.
4.8K CL point spectra from the centre of the (411)B facet (solid line) and the lower (100) plane of specimen (b) in figure 4. The larger FWHM of the (100) spectrum is indicative of a higher carrier concentration in the QW(2DEG).

specimens. Specimen (a) had the highest compensatory delta-doping at the RI, and showed better conduction on the facet than did specimen (b). Indeed, the EBIC images are qualitatively dissimilar in that figure 4(a) shows strongest EBIC from the (411)B facet region, whereas 4(b) shows stronger EBIC from the surrounding (100) planes. Both images show a dark line running approximately down the centre of the sidewall (411)B facet, which corresponds to the intersection of the delta-doped back gate with the sidewall. The n-type delta-doping is lowering the potential in this region, reducing the band-bending as shown schematically in figure 3, resulting in a lower EBIC collection. Quantitative EBIC linescans at 3keV show the half-width of this dark contrast line to be less than 300nm. The back gate is more clearly seen in sample (b) which has less compensatory delta-doping at the RI.

The quality of QW and 2DEG on the (100) and (411)B facets can also be monitored by CL. Figure 5 shows 5keV, 4.8K point spectra from the centre of the (411)B facet (solid line) and 2µm from the base of the sidewall on the (100) plane (dashed line) for sample (b). There is very little difference in the peak energy, but a marked difference in the peak intensity and FWHM. The stronger emission and larger FWHM from the QW(2DEG) on the (100) indicate inferior QW(2DEG) quality and lower carrier density on the (411)B facet in the case of sample (b), consistent with transport measurements. CL imaging also shows evidence of regions of slightly reduced (by around 3%) Al concentration in the AlGaAs barriers in 1µm-wide regions at the top and bottom of the (411)B sidewall facet, consistent with excess Ga diffusing off the (100) planes onto the upper and lower reaches of the (411)B sidewall during growth. Point spectra show no significant difference in carrier concentration across the (411)B facet, on the basis of a constant FWHM for the QW(2DEG) CL peaks.

4. CONCLUSIONS

EBIC and CL are very useful tools for the characterisation of modulation doped regrowth structures on patterned substrates. EBIC is able to reveal delta-doped layers with approximately 300nm resolution. EBIC imaging and linescans can also give direct, graphical evidence of the effectiveness of cleaning procedures and of doping intended to compensate for Fermi level pinning at the RI. CL is very effective as a contactless method for establishing the uniformity of QWs and 2DEGs grown over patterned substrates.

REFERENCES

Burke T M, Ritchie D A, Linfield E H, O'Sullivan M P, Burroughes J H, Leadbeater M L, Holmes S N, Norman C E, Shields A J and Pepper M 1998 Mat. Sci. Eng. **B51**, 202
Burke T M, Leadbeater M L, Linfield E H, Patel N K, Ritchie D A and Pepper M 1999 J. Crystal Growth (in press)
Leadbeater M L, Burke T M, Linfield E H, Patel N K, Ritchie D A and Pepper M 1999 Microelectron. J. **30**, 351

Inst. Phys. Conf. Ser. No 164
Paper presented at Microsc. Semicond. Mater. Conf., Oxford, 22–25 March 1999
© *1999 IOP Publishing Ltd*

EBIC and CL imaging of self-organised quantum dot clusters and wetting layers in modulation-doped structures

C E Norman[1], A J Shields[1], R A Hogg[1], M P O'Sullivan[1,2], I Farrer[2] and D A Ritchie[2]

[1] Toshiba Research Europe Ltd. Cambridge Research Labs. 260, Cambridge Science Park, Milton Road, Cambridge, CB4 0WE.

[2] Cavendish Laboratory, University of Cambridge, Madingley Road, Cambridge, CB3 0HE.

ABSTRACT: Modulation-doped structures containing single layers of quantum dots, intended for optical memory applications, have been studied by CL and EBIC in the SEM. Sub-μm resolution images of QD ground state emissions in the infra-red have been produced, and compared to images of excited state QD emissions, in addition to those of the wetting layers and neighbouring quantum wells containing 2-dimensional electron gases. EBIC measurements show weak contrast from features which appear to lie at a similar depth in the structure to the QD layers and which may correspond to potential perturbations in the structures arising due to clusters of charged QDs.

1. INTRODUCTION

There is currently great interest in the physics of quantum dot (QD) systems. Though such systems have yet to realise their commercial device potential, recent advances in the growth of QDs, most notably via the self-organised Stranski-Krastanow growth mode, has moved the technology to a stage in which prototype device structures are now being grown and characterised in the laboratory. The two principal areas in which QDs are foreseen as having the greatest likelihood of application are those of QD lasers (Arakawa and Sakaki, 1982) and memory devices. In the case of the latter, the resistance of modulation-doped field effect transistors containing a layer of self-organised QDs has been shown to display a bistability due to optically-induced changes in the channel density (Yusa and Sakaki 1997, Finley *et al* 1998) or mobility (Shields *et al* 1999).

The application of self-assembled QDs to memory devices is attractive due to their very high packing densities. Typical densities obtainable by self-organised growth are in the 10^9 to 10^{11} cm^{-2} range. If each dot could by some means be made to store one bit of information, this could represent a two hundred-fold increase in bit density over the best magnetic hard disk media currently available. This becomes even more attractive if one considers that the bit size of present day magnetic media based on the giant magneto-resistive (GMR) effect is approaching the super paramagnetic limit. The fact remains, however, that there are several major technological problems to be overcome regarding control of dot size and density, let alone in the addressing of individual dots, before QD memory devices would be a viable option. It is important at this stage to be able to reliably assess the electrical properties of QDs in real device-type structures, in order to gain a better understanding of the factors likely to limit QD applications, such as uniformity and overall structural integrity. Techniques such as scanning tunnelling, atomic force and near-field scanning optical microscopy can study individual dots, but only on or very near exposed surfaces, while transmission and scanning transmission electron microscopy can examine dots deeper in a structure, but not in a working device. For this reason, techniques such as cathodoluminescence (CL) and electron beam induced current (EBIC) can be of immense value in assessing working device structures. CL images of QD layers in the InAs/GaAs system have been appearing in the literature for some years, but to the best of our knowledge, all such images have been recorded at energies close to the GaAs band edge (Grundmann *et al* 1995), rather than at the lower ground state energies now commonly being observed for QDs in this system. As far as we are aware, no EBIC contrast associated with dot layers has yet been reported.

This paper reports initial results of a CL and EBIC study of QD layers in modulation-doped structures intended for optical memory applications. Such structures have been demonstrated (Shields *et al*, 1999) to show an optically induced bistability in the resistance of a Schottky-gated 2-dimensional electron gas (2DEG) which is grown in close proximity to the layer of QDs. The higher resistance state arises due to trapping of electrons in the QDs when a current is passed through the gate, producing local potential maxima in the 2DEG, resulting in a lowering of the 2DEG mobility. Illumination has been demonstrated to reduce the number of electrons trapped in the QDs, thereby reducing the potential fluctuations in the plane of the 2DEG, giving rise to the low resistance state.

2. EXPERIMENTAL

CL measurements were performed using an Oxford Instruments Ltd. MonoCL system and CF302TC liquid helium cryostat, attached to a Hitachi S-4500 cold cathode SEM upgraded with an oil-free vacuum system. At room temperature the chamber pressure is typically 6×10^{-7} Torr, cryopumping to 6×10^{-8} Torr at helium temperature operation. A high sensitivity GaAs photocathode photomultiplier tube (PMT) was used to acquire images and spectra at wavelengths below 900nm. Longer wavelengths were detected using a North Coast liquid nitrogen-cooled Ge detector. This type of detector is slow and inefficient compared to a GaAs PMT, and requires modulation of the electron beam. Nevertheless, by chopping at 70Hz, acceptable 512 x 416 pixel images could generally be acquired in around 6 minutes at the sort of signal levels obtained at low beam energies from our QD samples. EBIC measurements were performed using either a Stanford SR570 or a Matelect Ltd ISM-5 amplifier.

The samples reported here fit broadly into two categories: one with an InAs QD layer placed in the centre of a 20nm GaAs QW containing a 2DEG, and others with single InAs QD layers placed in close proximity to a 20nm QW containing a 2DEG. Where InAs layers were required to be grown outside the GaAs QW itself, (i.e. to be grown within one of the AlGaAs barriers), 2nm of GaAs was grown prior to InAs layer growth in order to ensure a smooth growth surface. In all cases the growth temperature of the InAs layer was 530°C, with self-organised dots appearing in samples with more than approximately 2 monolayers (ML) of InAs. Previous characterisation (Lian *et al*, 1998) of such layers grown in the same V80 MBE machine has established the importance of growing a relatively thick (several nm) GaAs cap layer at the same low temperature in order to prevent In desorption, and thus maintain the morphology of the dots. Growth interrupts under As stabilising conditions were employed whenever temperature changes were made during growth.

One important feature of the V80 MBE chamber design is that due to the position of the In K-cell, the In deposition is non-uniform across the wafer surface. The In inclusion is therefore greatest in the wafer centre and lowest at the wafer edge. This can lead to a variation in dot density across a wafer, and in the case of InAs layers at, or just above, the critical thickness for dot formation, can mean that the wafer centre contains dots whereas the wafer edge does not. This normally undesirable feature of the machine can, in this case, be advantageous - samples with different dot densities can be studied and compared, in the secure knowledge that they were grown under otherwise identical growth conditions.

The sample structures are given below, with layers listed in order of growth (i.e. from substrate to surface). The growth temperature is in brackets, all AlGaAs layers contained nominally 33% Al, all layers are undoped unless otherwise stated.

Wafer A is a double-sided modulation doped structure with a layer of QDs 10nm *above* a 20nm GaAs QW: 0.5μm GaAs, 0.25μm AlGaAs, 40nm AlGaAs (n-type 1×10^{18}), 40nm AlGaAs, 20nm GaAs (QW), 10nm AlGaAs, 2nm GaAs (all 590°C), InAs QD layer), 10nm AlGaAs (all 530°C), 50nm AlGaAs, 30nm AlGaAs (n-type 1×10^{18}), 10nm GaAs cap (all 590°C). This sample shows the most pronounced optically induced bistability of 2DEG mobility.

Wafer B is identical to sample A, but with a *thicker* (40nm) AlGaAs layer separating the QW and the layer of QDs.

Wafer C is an asymmetrically modulation doped structure with a layer of QDs incorporated *in the centre* of a 20nm GaAs QW: 0.6μm GaAs buffer, 50nm AlGaAs, 10nm GaAs (all 580°C), InAs QD layer, 5nm GaAs (all 530°C) 5nm GaAs, 40nm AlGaAs, 40nm AlGaAs (n-type 1×10^{18}), 17nm GaAs cap (all 580°C).

Wafer D is similar to sample C, but has the layer of QDs placed 10nm *below* the 20nm GaAs QW

3. RESULTS AND DISCUSSION

The QD density in the centre of wafer C has previously been determined by TEM (Lian et al, 1998) to be 3×10^9 cm^{-2}. The edge of this wafer does not contain any QDs, confirmed by the absence of dot luminescence in both PL and CL spectra. The QD density at the centre of wafers A, B and D is somewhat higher, probably around 2×10^{10} cm^{-2} cm for sample A, based on AFM measurements on an uncapped sample grown under identical conditions. All three wafers show QDs extending right up to their edges, which is also consistent with them having a higher overall dot density than wafer C.

Figure 1 shows 5keV, 4.7K spectra from two spots, approximately 1μm apart, on a central portion of wafer C. A strong wetting layer (WL) signal is observed at just below 1.44eV, together with up to six different transitions (labelled T_0 to T_5) associated with the QDs. These transitions have been identified as QD ground (T_0) and excited state transitions by varying the excitation conditions in both PL and CL experiments. The 1keV spot spectrum in figure 1 represents a much lower injection level than the two 5keV spectra, and shows only the ground state and four weak excited states. Interestingly, the wetting layer peak is greatly reduced in intensity, and is now comparatively much weaker than the dot emission. The energy of the ground state is slightly different in the two 5keV spectra, as are the apparent separations of the excited state emissions, suggesting the electron beam is sampling two different average QD sizes at the two different points.

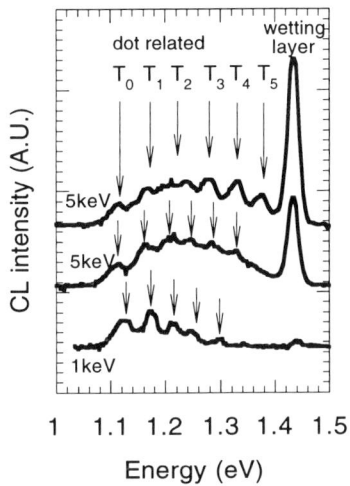

Figure 1.
Two 4.8K, 5keV point spectra of sample C, together with a lower injection level 1keV point spectrum of same. Several excited state transitions are seen at 5keV, relatively fewer at 1keV.

Regarding the spatial resolution achievable on these samples, since the QD layers are over 100nm below the sample surface, and the *average* dot spacing is of a similar magnitude, we will not be able to spatially resolve individual dots. A value of the beam energy which produces an acceptable trade-off between spatial resolution and signal level in both CL and EBIC is 5keV. Semi-empirical and Monte Carlo calculations both give the electron penetration range as approximately 150nm, and given that the band structure of such modulation doped structures should encourage carrier drift normal to the sample surface, we do not expect marked lateral diffusion leading to degradation of the lateral spatial resolution. We therefore reasonably expect a spatial resolution of no worse than 300nm. If sample C had a *uniform* QD density of 2×10^{10} cm^{-2}, such a probe size would sample approximately 14 QDs simultaneously. The clustering of the QDs shown by TEM, STM and AFM characterisation of similar dot layers, suggests that it may be possible to sample anything between approximately half and double this number, say, 7 to 30 QDs at any one position of the beam. In sample A, with a QD density of 3×10^9 cm^{-2} it may be possible to excite as few as two or three QDs.

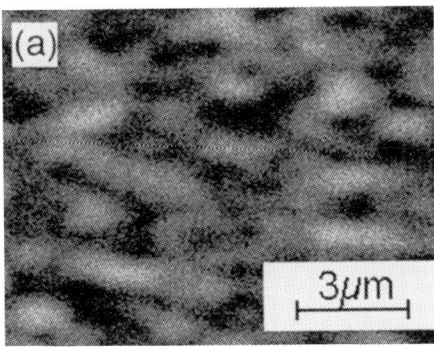

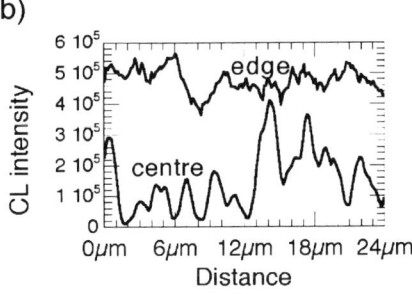

Figure 2.
(a) 1.433eV monochromatic image of the 4.8K WL emission from the centre of wafer C, together with (b) quantitative CL linescans at the WL peak energies for the centre and edge of the wafer. The centre contains QDs with a density of approximately 3×10^9 whereas the edge contains no QDs.

The wetting layer emission is different in the centre of wafer C and at the wafer edge, as shown in figure 2. The emission is seen to be strongly discontinuous (fig.2a). The contrast of the dark regions can be as high as 98%. Figure 2b shows quantitative CL linescans at the WL peak wavelength across this image and a similar image obtained at the edge of the wafer. The CL contrast is greatly (but not completely) reduced. The inference is that the presence of dots in the wafer centre is strongly modifying the wetting layer emission. One possible mechanism for the contrast is that the WL CL signal is weaker in regions where a higher QD density leads to more efficient carrier capture. On the other hand, the possibility that another mechanism associated with the higher indium dose, which results in the introduction of non-radiative recombination centres, cannot be ruled out.

The confinement of carriers in the GaAs QW surrounding the QD layer is likely to enhance the dot emission from samples such as wafer C, compared to sample A. It is noted that sample C gives measurable dot luminescence at room temperature, whereas sample A does not. This confinement may also act to reduce the lateral spatial resolution available in CL and EBIC by permitting lateral carrier diffusion in the QW prior to recombination in either the wetting layer or a dot. Better spatial resolution might therefore be expected on samples with the QD layer outside the GaAs QW.

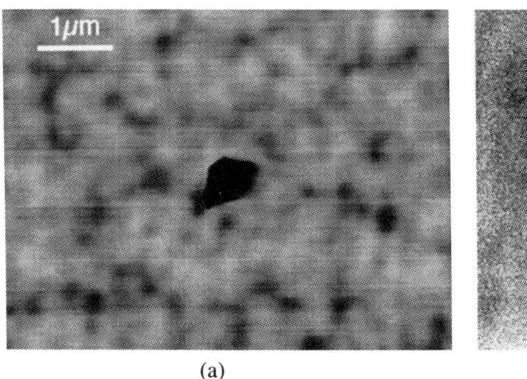
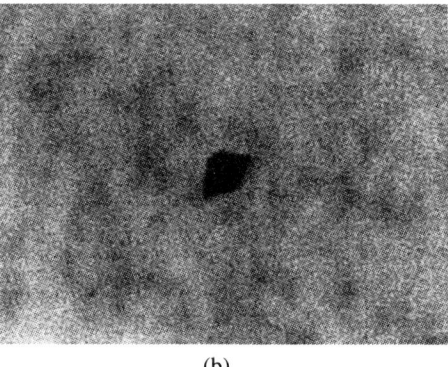

(a)   (b)

Figure 3.
(a) EBIC and (b) QW CL (1.519eV) 4.8K, 5keV images of a central region of wafer A. The strongest dark features in the EBIC image are also seen in the CL image.

Ideally we would wish to observe EBIC and CL from the same areas, in order to correlate the contrast mechanisms available in the structure. Unfortunately, even with a very thin (7nm) NiCr surface gate metallization, the QD luminescence intensity is greatly reduced compared to its non-gated intensity, making imaging at QD wavelengths very difficult. Figure 3 shows two images of the same area of sample A, recorded at 4.8K around a particle on the surface for fiducial reference. The dark spots seen in the QW luminescence correspond exactly to the darkest spots seen in the EBIC image, suggesting they arise due to the same non-radiative centres. The weaker EBIC contrast features, which appear as a fainter network between the points of strong contrast, are much less apparent in the CL image, and it is difficult to discern if there is indeed a one-to-one correspondence.

The weak and strong dark contrast features have rather different contrast behaviour as a function of beam voltage, as shown in figure 4. The strong features have typically 13% contrast at 5keV, and the contrast decreases linearly with increasing beam energy. Their strong contrast at short electron penetration ranges suggests they are either surface defects, or extended defects running from the surface down to the QW and possibly beyond. The weak features show a peak in their contrast at around 6keV, suggesting they lie deeper in the structure, probably around the depth of the QD layer and the QW, though it is impossible to be more

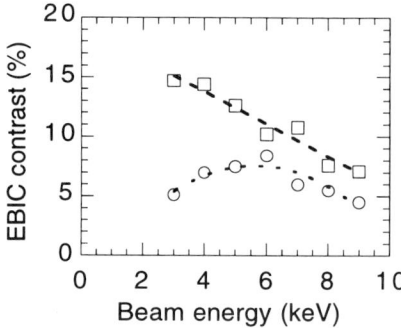

Figure 4.
The variation of EBIC contrast as a function of beam electron energy, showing two different behaviours for the "strong" and "weak" contrast features seen in figure 3(a).

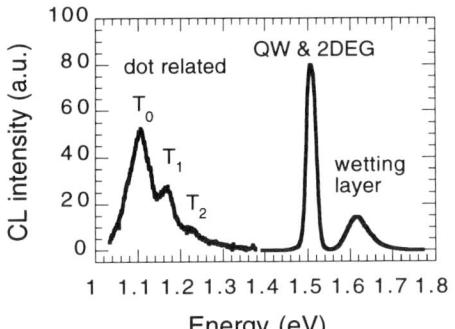

Figure 5.
4.8K CL spectra of sample A recorded with the Ge diode and GaAs PMT detectors, showing the full range of emission from the sample.

precise from the available data. The weak contrast seems to be relatively independent of temperature, and it occurs on a similar length scale ($\approx$ 200-300nm) over which TEM (Lian *et al*, 1998) shows the QDs clustering together. Furthermore, the network appears elongated in one of the [110] directions, consistent with the observation by TEM that the QDs align along step edges. These observations lead us to suggest that the weaker contrast may arise from fluctuations in the QD density. Since the dots are charged, due to the modulation doping, this produces spatial fluctuations in the lateral potential to which the EBIC signal is sensitive. Consistent with this contrast mechanism, we find the EBIC contrast measured from similar weak features observed in the samples with the QD layer *in* the QW (sample C) and *below* the QW (sample D) is much weaker than that observed when the QDs are above the QW (samples A and B), which is taken as further tentative evidence that the weak EBIC contrast is related to the plane of QDs, rather than the QW or the 2DEG therein.

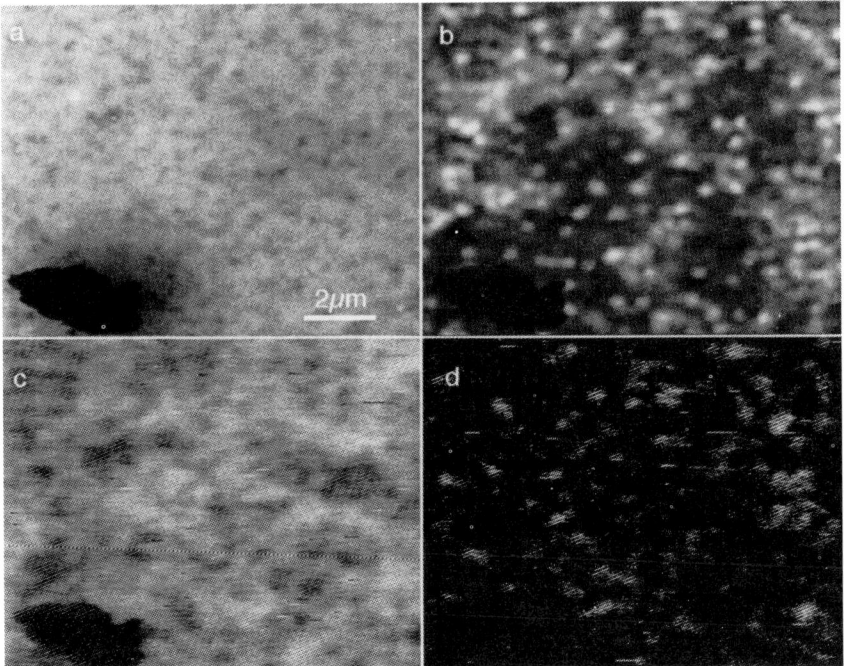

Figure 6.
4.8K images of sample A at wavelengths corresponding to (a) the QW peak (1.519 eV), (b) the WL peak (1.579 eV), (c) the QD ground state peak (1.097 eV) and (d) a point on the high energy side of the first excited state peak (1.194 eV). (N.B. Contrast and brightness levels have been adjusted so that each image is displayed over the full 0-256 grey scale, in order to permit effective reproduction.

Figure 5 shows spectra recorded from a $1\mu m^2$ area on sample A. The dot CL is detected with the Ge diode, and the QW and WL with the GaAs PMT, so the relative intensities are not comparable. Approximately 200-250 QDs will be excited, yielding a broad envelope of emission. Even so, the spectrum shows evidence of two excited states. Figure 6 shows images of the sample at four different energies: (a) the QW(2DEG) @ 1.519eV, (b) the wetting layer @ 1.579eV, (c) the centre of the dot ground state envelope @ 1.097eV, and (d) the high energy side of the first excited state @ 1.194eV. A surface particle in the bottom left of the images provides a fiducial reference. Many of the dark spots seen in the QW image can be seen as dark in both the WL image and the QD ground state image, suggesting that non-radiative centres are present and will therefore degrade regions of the QD layer.

The WL emission is strongly spatially inhomogeneous. Changing the imaging wavelength to monitor the emission in the high and low energy portions of the WL peak produces similar "spotty" images, but with different small regions emitting. The smallest spots visible are around 300nm in diameter, suggesting their apparent size is limited by the spatial resolution available at 5keV, and that the areas emitting at this wavelength may in fact be somewhat smaller. There is no clear complementary relationship between the WL image and the QD image, but to see any such relationship would entail imaging over the whole integrated range of discrete WL and QD wavelengths, which is not at all practicable. The spatially inhomogeneous nature of the dot ground state emission is apparent in figure 6(c). This arises from the fact that the beam is sampling, at most, tens of dots at any one time, and not all of those dots will emit within the 4meV bandpass of our experiment. Regions where there is a higher density of "detectable" QDs will appear brighter than regions containing fewer "detectable" QDs. Image (d) is very different from image (c): in this case, only regions with a low density of QDs emitting over the right energy range will be imaged. This is because the energy imaged in (d) is an excited state transition, in a region of the spectrum whose strength suggests there are not many dots emitting at this wavelength. For a given beam injection level, QDs in regions with lower QD density will be more likely to be stimulated to give higher order excited transitions. By selecting the appropriate energy, the CL images can thus become maps of dot density, for a given range of dot sizes.

4. CONCLUSIONS

We have studied a number of structures containing single layers of QDs and 2DEGs in QWs, designed to test the feasibility of using QDs in optical memory applications. Images of the WLs have been produced and shown to be highly spatially inhomogeneous, possibly on a length scale even smaller than the spatial resolution of the CL technique. Some as-yet unidentified non-radiative recombination centres are active in reducing emission from the QWs, the WL and the QD layers. A weaker EBIC contrast has been identified which may have its origins in the QD layers, as suggested by EBIC contrast levels as a function of beam energy, temperature, and the position of the QD layers relative to the QW(2DEG). A possible contrast mechanism is that clusters of negatively charged QDs are locally perturbing the electric field used to collect the EBIC. Further work is, however, required to confirm this. Cryogenic CL images of QD ground state emissions, showing sub-μm resolution in the InAs/GaAs and InAs/AlGaAs/GaAs systems have been demonstrated.

REFERENCES

Arakawa Y and Sakaki H 1982 Appl. Phys. Lett. **40**, 939
Finley JJ,Skalitz M, Arzberger M, Zrenner A, Bohm G and Abstreiter G 1998 Appl. Phys. Lett. **73**, 2618
Grundmann M, Christen J, Ledentsov NN, Bohrer J, Bimberg D, Ruvimov SS,Werner P, Richter U, Gosele U, Heydenreich J, Ustinov VM, Yu Egorov A, Zhukov AE, Kop'ev PS and Alferov Zh I 1995 Phys. Rev. Lett. **74**, 4043
Shields AJ, O'Sullivan MP, Farrer I, Ritchie DA, Cooper K, Foden CL and Pepper M 1999 Appl. Phys. Lett. **74**, 735
Yusa G and Sakaki H 1997 Appl. Phys. Lett. **70**, 345

Inst. Phys. Conf. Ser. No 164
Paper presented at Microsc. Semicond. Mater. Conf., Oxford, 22–25 March 1999

EBIC studies of heterojunction bipolar transistors

F Amin, D B Holt⁺, G A Hungerford*, E Napchan⁺ and A Rezazadeh

Department of Electronic Engineering, King's College London, Strand, London WC2R 2LS
⁺ Department of Materials, Imperial College of Science, Technology and Medicine, London SW7 2BP
* Present address: 6235 West Canyon Avenue, Littleton, Colorado 80123, USA

ABSTRACT: Measurements on GaAs homo- and heterojunction bipolar transistors followed the approach of Gonzales (1974) who considered the interaction of the two junctions in analysing such devices. The electron beam induced voltage produced across a charge collecting barrier without an electrical contact could be detected by a neighbouring junction. This can be useful for examining regions of integrated circuits to which no direct contact is made. Observations on the two forms of GaAs transistors are presented and discussed.

1. INTRODUCTION

Almost all published work on electron beam induced current (EBIC) deals with only one charge collection (CC) barrier i.e. with diodes. However, even the most basic of micoelectronic devices, transistors, contain more than one, so the final EBIC in devices is generally the result of two or more CC processes. The results of the interactions of these barriers in planar bipolar transistors were considered by Gonzales (1974). However, no measurements related to this analysis have been published to our knowledge. This paper reports such effects in experimental GaAs homo- and hetero-junction (or heterostructure) bipolar transistors (HBTs). EBIC studies, with no reference to the work of Gonzales, were reported on GaAs HBTs (e.g. Fitzgerald et al 1988) and MESFETs (metal Schottky field effect transistors) (Kaufmann and Balk 1993). The latter measured the gate EBIC current and used simulations to achieve quantitative interpretability.

Kroemer (1957) proposed the use of a second material with a wider energy band gap for the emitter to form heterostructure bipolar transistors. Steps in the band edges limit the undesirable base-to-emitter current and make large increases in the current gain possible. HBTs are of interest for use in monolithic microwave integrated circuits (MMICs) where they compete with MESFETs and HEMTs (high electron mobility transistors) (Hughes et al 1988 pp. 403 - 418).

Following Gonzales' approach, Hungerford (1988) studied GaAs bipolar homojunction transistors and some of those hitherto unpublished results are reported here together with the results of recent studies of GaAs-based HBTs.

2. EXPERIMENTAL METHODS

The initial observations were on an ion-implanted, large-area (100 μm diameter) circular geometry device kindly supplied by the GEC Hirst Research Centre. The (100) GaAs for the device was grown on semi-insulating Czochralski GaAs by molecular beam epitaxy (MBE). The device was produced by photo-masking and etching as shown in the schematic cross-section diagrams of Figures 1a and 1b. It was examined in a JEOL JSM-35 SEM with a Keithley 427 amplifier to detect the EBIC signals.

Emitter Cap	n+ GaAs 2E18 Si, 0.19 um	
Emitter	n GaAs 2E17 Si, 0.30 um	
		intrinsic GaAs, 0.01 um
Base	p+ GaAs 1E18 Be, 0.30 um	
		intrinsic GaAs, 0.01 um
Collector 1	n− GaAs 1E16 Si, 1.0 um	
Collector 2	n+ GaAs 1E18 Si, 0.50 um	
		intrinsic GaAs, 0.05 um
Super Lattice Buffer 0.195 um intrinsic		intrinsic GaAs, 0.055 um

Semi−Insulating Cz (100) GaAs substrate

(a)

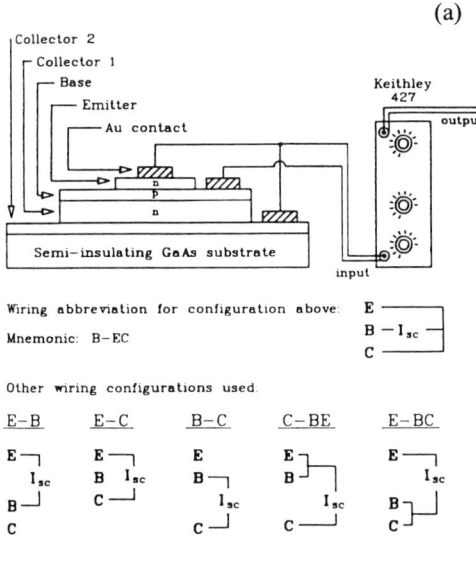

(b)

Figure 1. (a) The cross-sectional structure of the homojunction transistors. (b) The six possible ways of connecting the detecting amplifier (a Keithley 427) to a bipolar transistor: configuration B-EC is shown in the main Figure. The other connection configurations are only schematically represented (Hungerford 1988).

The later devices were produced on a single chip (Figure 2) at King's College London, with the layout and structure shown in Figure 3 and Table 1. Four devices were contacted. Of these, HBTS 1 and 3 were good and HBTs 2 and 4 failed. They were examined in a JEOL JSM 840A with either a Stanford Research Systems SR570 or a Matelect ISM-5 EBIC system, operated via TestPoint software, and images were processed by a Kontron processor.

Table 1. The layer structure of the KCL HBTs.

Layer No.	Description	Material and Grading	Doping (cm.$^{-3}$)	Thickness (nm)
9	Cap layer -2	n+ In$_{0.5}$Ga$_{0.5}$As	1 x 10^{18}	30
8	" " -1	n+ In$_x$Ga$_{1-x}$As (x = 0 $\Rightarrow$ 0.5)	1 x 10^{18}	30
7	" "	n+ GaAs	8 x 10^{18}	100
6	" "	n+ In$_{0.49}$Ga$_{0.51}$P	8 x 10^{18}	30
5	Emitter	n In$_{0.49}$Ga$_{0.51}$P	5 x 10^{17}	100
4	Base	p GaAs	2 x 10^{19}	80
3	Collector	n GaAs	3 x 10^{16}	500
1	Sub-collector	n+ GaAs	8 x 10^{18}	700
0	Substrate	Semi-insulating GaAs	Undoped	400 μm

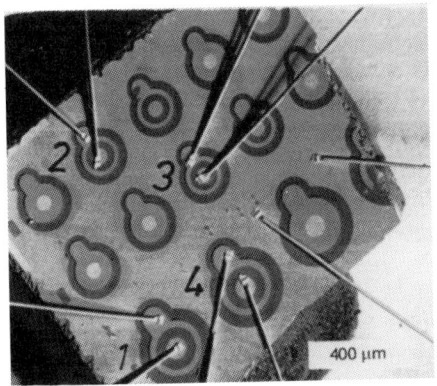

Figure 2. (a) Secondary electron and (b) EBIC images of the KCL (King's College, London) GaAs chip on a TO-18 header. HBTs 1 to 4 were connected to header pins. For (b) the EBIC system was connected to the HBT 4 emitter and a common collector contact. The other three transistors gave EBIC contrast but HBT 4, the failed device did not.

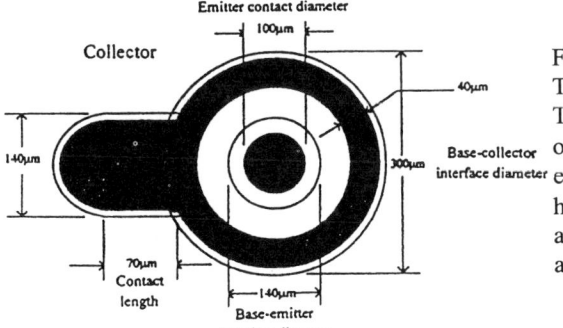

Figure 3. The layout of the KCL HBTs. The black areas are the metallization. There are two sizes of HBTs on the chip of Figure 3: the larger have 100 μm emitter contacts as indicated; the smaller have 70 μm e-contacts. The base-emitter and base-collector "interface diameters" are mesa edge positions.

2.1 EBIC ANALYSIS OF BIPOLAR TRANSISTORS

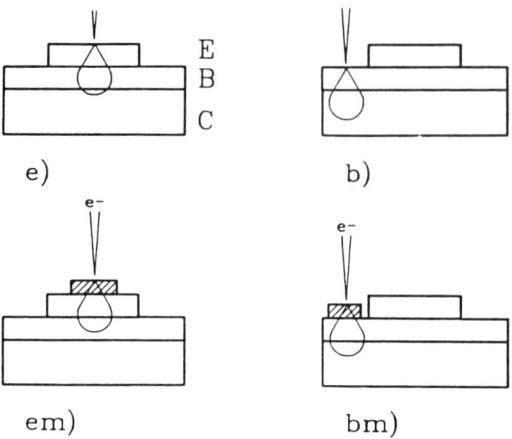

e)

E
B
C

b)

em)

bm)

Figure 4. The four types of electron beam excitation that are possible on an unpassivated transistor: (e) electron beam on the exposed emitter layer, (b) beam on the exposed base layer, (em) on the emitter metallization and (bm) on the base metallization. (Hungerford 1988).

Different EBIC results will be expected from the contact configurations for collecting the EBIC signal in Figure 1b. The notation is shown, where I_{sc} is the short circuit (EBIC) current. If the letter E, B or C has no wire (line) going to it then it floated electrically and the EBIC signal was taken across the other two contacts alone, as for configurations E-B, E-C and B-C. For C-BE and E-BC, the base/emitter and the base/collector respectively, were shorted together. The results also depend on the energy and point of incidence of the beam. That is, on which of the layers of the device are excited.

For E-C (Figure 1b) two junctions are in series and charge collecting in opposite directions so the EBIC signal is the difference. The dominant junction is determined by their diffusion potentials and the volume of minority carriers reaching the depletion regions (Hungerford 1988). At low beam voltages e-h (electron-hole) pairs are created in the emitter only and the direction of the EBIC is determined by the emitter-base junction. For the increasing beam voltage, e-h pairs are generated nearer to the base-collector junction.

3. RESULTS

3.1 Quantitative EBIC Analysis of the Transistors

EBIC linescans across the GaAs bipolar transistors were recorded for the six contact configurations of Figure 1b, for accelerating voltages from 2 to 25 kV. Each linescan contained a signal for each of the types of beam position (Figure 4). The charge collection efficiency, i.e. the I_{sc}/I_{max} , where I_{max} is the number of hole-electron pairs generated per second (after allowing for energy lost to the emitted electrons), was determined. Figure 5 reports for contacts E-C, the results for homojunction transistors as a function of beam accelerating voltage, V_b. With the beam incident on the emitter (e), initially, when the emitter-base junction alone is collecting charge, the current increases linearly with beam energy. From about 10 kV onwards, the lower base-collector junction begins contributing an opposing current which grows so the net current falls linearly with V_b. The maximum at $V_b = 12$ kV could be fitted to (Napchan's) Monte Carlo simulated curves to obtain the depth of the lower junction. The bm and b curves (Figure 4), for which only the base-collector junction lies below the beam impact point, show that the charge collection current did indeed begin to rise from zero, and in the opposite direction to that from the upper junction, at about 10 – 12 kV. At V_b = 23-24 kV, the e curve shows that the two currents are equal giving zero net EBIC signal, I_{sc}. The results for HBT1 were broadly similar.

Figure 2b is the EBIC image for connection E4-C, i.e. connection between the emitter of HBT4 and the n-type collector layer covering the whole area of the chip. All the HBTs connected to pins <u>except the one connected to the detecting amplifier</u> gave rise to EBIC signals. Clearly charge collection at HBTs 1 and 3 (good devices) and HBT 2 (failed - but not so completely as HBT4) was detected via the common collector. The same effect occurs when a good device (1 or 3) is contacted but the images are dominated by the very strong signal from the good, contacted device. In Figure 2b, the absence of signal from the contacted HBT4 meant that a high amplifier gain could be used.

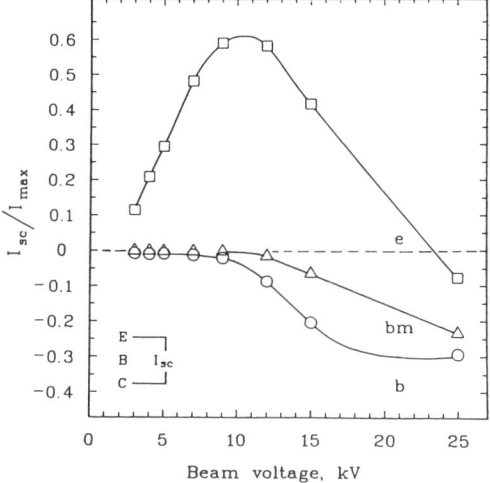

Figure 5. The variation of EBIC gain versus beam voltage for connection configuration E-C with the beam (e) on the emitter, (b) on the base and (bm) on the base contact metallization in the homojunction transistor (Hungerford 1988).

3.2 EBIC Defect Observations in the HBTs

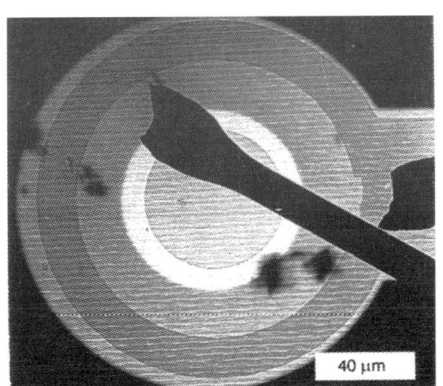

Figure 6. EBIC image of defects in the base material of one of the failed KCL HBTs (no. 2) with the signal extracted via the pins to the emitter and base of HBT 4, the other failed device.

Figure 6 shows dark EBIC defects in the base material of an HBT from KCL. These defects corresponded to visible surface damage apparently produced by the indenter used to bond the gold wire leads to the contact pads.

4. DISCUSSION

The importance of the different connection configurations (Figure 1b) and beam incidence positions (Figure 4) is clear. However, measurements showed many of the connection configurations produce similar results (Hungerford 1988). For connections E-C the two junctions work in opposition to produce the forms of beam accelerating voltage dependence shown in Figures 5. The form of Figure 5 could be simply understood as discussed above. However, the interpretation of the form of the similar curve for the heterojunction devices involves the greater complexity of the HBT energy band structure.

Figure 2b supports Gonzales' (1974) remark that an EBIC signal will be produced by a neighbouring junction that injects carriers into the charge collecting barrier so EBIC signals can be picked up from junctions and devices that are not themselves connected to the detecting amplifier. It has often been found in EBIC of integrated circuits, that junctions and devices, remote from the points of connection of the amplifier give EBIC contrast. Kolachina et al (1998) have treated such a case using an equivalent circuit approach. They point out that EBIC images of integrated circuits can even be obtained using only one contact.

Maxima in plots of EBIC gain versus V_b, i.e. beam penetration, like Figure 5, give information about the depths of the junctions, even in shallow devices.

REFERENCES

Fitzgerald E A, Ast D G, Ashizawa Y, Akbar S. and Eastman L F 1988 J. Appl. Phys. 64, 2473

Gonzales A J 1974 Scanning Electron Microscopy Vol. IV pp 941 – 948

Hughes W A, Rezazadeh A A and Wood C E C 1988 Chapter 7: Heterojunction Integrated Circuits; in GaAs Integrated Circuits (J. Mun, Editor) (BSP Professional Books: Oxford) pp 376 - 429

Hungerford G A 1988 Ph.D. Thesis, University of London

Kaufmann K and Balk L J 1993 in Microscopy of Semiconducting Materials. Inst. Phys. Conf. Ser. 134 (Inst. Phys.: Bristol) pp 725 – 730

Kolachina S, Phang J C H and Chan D S H 1998 Sol.-State Electron. 42, 957

Kroemer H 1957 Proc. IRE 45, 1535

Inst. Phys. Conf. Ser. No 164
Paper presented at Microsc. Semicond. Mater. Conf., Oxford, 22–25 March 1999
© 1999 IOP Publishing Ltd

Potential errors in quantitative EBIC studies caused by low resistance specimens

A G Wojcik

Faculty of Engineering, University College, Torrington Place, London, WC1E 7JE, UK

ABSTRACT: A source of serious error has been identified in quantitative EBIC measurements which results from the intrinsic operation and limitations of traditional current amplifier circuits. So called "dark currents", obtained without electron beam excitation and within the confines of the chamber of an SEM, may be substantially due to this error mechanism. This paper highlights the problem using a brief analysis of the operation of conventional current amplifiers and offers practical advice on reducing signal errors. Whilst zeroing an amplifier prior to taking a measurement can limit such errors, it is not always possible to completely eliminate them. A solution based on the use of a balanced pair of high value resistors can, in theory, be applied, and is here proposed.

1. INTRODUCTION

The electron beam induced current technique (EBIC) has been in use for many years as a valuable adjunct to SEM characterisation of the electrical activity of semiconductor materials and devices (Leamy 1982). The minute currents generated by electron bombardment require sensitive, low noise, DC amplification before they can provide qualitative images or quantitative linescan data of EBIC activity. Commercial current amplifiers are now highly sophisticated and offer the user additional facilities such as the ability to back-off standing currents and the inclusion of stable voltage sources for device biasing. Most modern current amplifiers rely upon a simple circuit configuration known commonly as a current to voltage converter. This, although a great improvement on earlier shunt circuit technology, suffers from a major drawback - namely an inability to handle specimens exhibiting a low internal resistance. The errors caused by this deficiency have largely been ignored by the research community principally because most EBIC measurements are used to generate qualitative images rather than quantitative data. A typical manifestation of the problem would be the generation of an apparent, and sometimes large, standing current even when a specimen is shielded within the confines of an SEM chamber and not subject to electron beam bombardment. This standing signal is often accorded the term "dark current" without too much concern for its origin.

In consideration of the renewed interest currently being shown in the EBIC technique this paper attempts to notify, clarify and explain the aforementioned problem through a simple theoretical examination of the amplification method used in current amplifiers. Practical advice on countering the problem is also given.

2. BACKGROUND THEORY

The simplest form of current to voltage converter is a resistor - an input current generates a measurable output voltage according to Ohm's Law. For the accurate measurement of small currents, the resistor has to be of a high value and this effectively means the input impedance of the circuit is high. In other words, the circuit loads the signal source.

Ideally a current to voltage converter should have zero input impedance and this is simply achieved by utilising an operational amplifier ("op amp"). An op-amp normally amplifies the difference between its two input terminals. Op-amps have very large amplification factors (Gain). They are normally operated by applying an element of negative feedback to the amplifier which allows the system gain to be accurately determined and controlled.

All op-amps suffer from deficiencies of design which generate non-ideal amplifier characteristics. Typical in this respect is the occurrence of an input bias current and input offset voltage. The former is a tiny current that is drawn by the input of the amplifier, whilst the latter is the voltage difference at the input terminals of the amplifier when no signal is applied.

Consider the amplifier in Fig. 1.

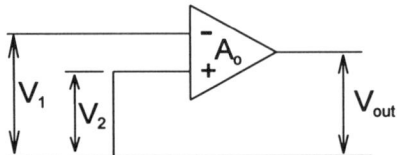

*Fig. 1. The generation of a virtual earth at the inputs
of an ideal operational amplifier.*

The output voltage V_o is given by $A_o(V_i)$ where V_i is the input voltage (i.e. the difference between V_1 and V_2) and A_o is the amplifier's Gain. Because A_o is very large (ca $\times 10^5$) and V_2 is zero (i.e. connected to Earth), V_i is small and therefore V_1 approaches zero (in the extreme case as A_o approaches infinity, V_1 reaches zero). The top terminal is therefore known as a Virtual Earth since it appears to the signal source as Earth potential. This makes the analysis of the current to voltage converter simple.

Adding our feedback resistor gives Fig. 2, which is the classic form of the current to voltage converter. If we assume the input bias current into the (-) terminal is negligible, then all the input current must flow through R_f, thus creating a voltage drop of $I_{in}R_f$ across the feedback resistor.

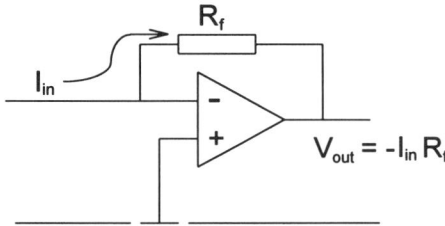

*Fig. 2. The generation of an output voltage proportional to
the input current.*

If the (-) terminal is considered to be a virtual earth, then the output voltage V_o must be $-I_{in}R_f$ for the current to flow as shown. Hence the output voltage is proportional to the input current. In order to obtain a 1 volt output for a 1 nA input, R_f must be 1 GOhm. In this analysis we have assumed that no input offset voltage exists. In reality this is never the case and a few millivolts difference may exist at the input terminals.

3. THE EBIC MEASUREMENT CONFIGURATION

A typical EBIC measurement set up is shown in Fig. 3a, below. This is similar to the diagram that applies to the current to voltage converter above but an additional resistance (that of the specimen) has been added. In order to analyse the effect of an input offset voltage on the EBIC measurements, we need to re-draw Fig. 3a to that shown below in Fig. 3b, and ignore the current to voltage converter analysis given previously.

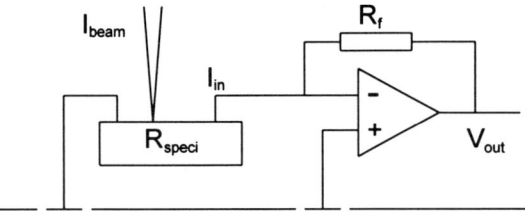

Fig. 3a. Schematic of the current to voltage converter in use to measure EBIC signals.

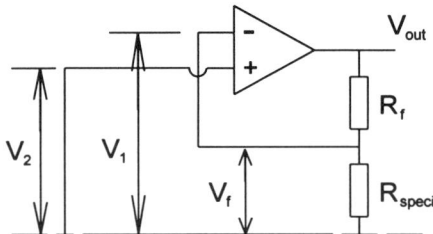

Fig. 3b. Schematic representation of Fig. 3a, as a traditional non-inverting op-amp circuit.

This diagram now effectively represents a classic non-inverting operational amplifier with the input tied to ground potential. The output voltage equation for this amplifier is easily defined as follows: The feedback voltage V_f is determined by treating the resistor chain formed by the feedback resistor R_f and the specimen resistance R_{speci} as a simple voltage divider thus:

$$V_f = V_o(R_f/(R_f + R_{speci})), \text{ and rearranging gives:} \qquad V_o = (1 + R_f/R_{speci})V_f \qquad (1)$$

An ideal op-amp's output will move so that the difference between the input terminals becomes zero, thus V_1 will equal V_2 and V_f (all of which move to zero since V_2 is connected to zero). The non-ideal amplifier will also try to achieve this but its output will eventually settle at one which results in the difference between the input voltages settling at the input offset voltage ΔV.

With V_2 at zero, V_1 must be at $(V_2 + \Delta V)$, thus since V_f is equivalent to V_1:

$$V_f = (V_2 + \Delta V) \qquad \text{and from (1);} \qquad V_o = (1 + R_f/R_{speci})(V_2 + \Delta V)$$

In the EBIC case, V_2 is held at zero, therefore the output voltage V_o is given by:

$$V_o = (1 + R_f/R_{speci})\Delta V \qquad (2)$$

This equation clearly illustrates why the specimen resistance is so critical to correct EBIC measurement. If the input offset voltage is approximately 100 μV, then a feedback resistance of 1 GOhm combined with a specimen resistance of 1 MOhm, gives a voltage output of approximately 0.1 Volts. A 100 kOhm specimen would thus generate 1 Volt. Such voltages would be interpreted as large signal currents when employing typical current amplifiers. In extreme cases, the amplifier could see its output saturate completely and thereby swing to the power supply rails. To greatly reduce this error, therefore, the specimen resistance should at least be a factor of 10 *above* the feedback resistor value. It is important to note that these error outputs are for a no-beam condition.

The foregoing analysis suggests that a knowledge of the resistance of a semiconductor specimen, together with an understanding of the feedback resistors employed within the amplifier used to measure induced currents, is absolutely vital to the success of quantitative EBIC measurements. The vast majority of EBIC studies are, however, qualitative and EBIC is usually employed as an imaging tool rather than as a means of obtaining quantitative measurements. That said, a small proportion of workers perform quantitative EBIC studies and it is, therefore, important for this community to grasp the implications of the analysis given.

4. IMPLICATIONS AND PRECAUTIONS

The error voltage produced at the output of the current amplifier will normally manifest itself as a constant offset imposed upon the displayed current value. Many commercial current amplifiers incorporate offset facilities where initial readings can be zeroed out before commencement of testing. It should therefore be possible to compensate for the error by using such offset facilities. It must be noted, however, that offset facilities are finite and it is thus not possible to counter every occurrence of a low resistance specimen. Furthermore, switching between current measurement ranges, on instruments that provide this facility, switches in different current measuring resistors (R_f) thus necessitating strict zeroing of the error signal each time a range is changed. Additionally, the input offset voltage from which the error originates is not a constant factor. Temperature has a particular influence on the offset voltage and this could result in significant drift of the error voltage over the period during which measurements are taken - hence requiring regular adjustment of the "zero" point with the electron beam blanked.

A possible solution to the offset error problem may reside in the deliberate use of additional resistive elements in the input lines to the current amplifier. Figure 4 schematically illustrates a suggested realisation of this concept.

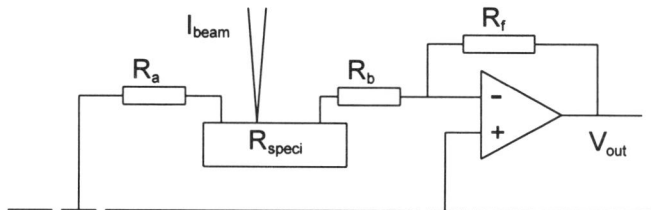

Fig. 4. Schematic of a modified EBIC measurement circuit.

It can be seen that the total resistance presented to the amplifier is now $R_a + R_b + R_{speci}$ and the values of R_a and R_b can be chosen to lie well within the factor of 10 requirement. It is important to note that R_a and R_b should be precisely selected/adjusted so that their values are the same otherwise an imbalance will be set up in the circuit whereby more of the signal current will flow in the lower resistance arm. Additionally, the high value of R_a and R_b will necessitate careful screening to avoid the pick-up of unwanted noise components. The resistors are best placed, therefore, within the SEM chamber and close to the specimen. This method can be shown to work from a purely theoretical viewpoint if it assumed that the electron beam constitutes a constant current source (i.e. a source of electrons with infinite voltage compliance). This is a reasonable assumption given the very large electrical potential that is theoretically possible to be built up on the surface of a specimen. A constant current source will supply current into the amplifier input that is independent upon total specimen resistance and hence unaffected by the presence of R_a and R_b.

5. CONCLUSION

A source of serious error has been identified in quantitative EBIC measurements which results from the intrinsic operation and limitations of traditional current amplifier circuits. So called "dark currents", obtained without electron beam excitation and within the confines of the chamber of an SEM may be substantially due to this error mechanism. Whilst zeroing an amplifier prior to taking a measurement can limit such errors, it is not always possible to completely eliminate them. A solution based on the use of a balanced pair of high value resistors can, in theory, be applied, however the limits of its applicability remain to be experimentally determined.

REFERENCE

Leamy H J 1982 J. Appl. Phys. **53**, R51

Inst. Phys. Conf. Ser. No 164
Paper presented at Microsc. Semicond. Mater. Conf., Oxford, 22–25 March 1999
© *1999 IOP Publishing Ltd*

Growth spirals and morphology in vapour deposited ZnS

Y Brada, D B Holt* and E Napchan*

Racah Institute of Physics, The Hebrew University, Jerusalem, Israel
*Materials Department, Imperial College of Science, Technology and Medicine,
London SW7 2BP, UK

ABSTRACT: Samples of ZnS from one supplier, when sublimed in a flow of H_2S, and deposited in a particular temperature zone of the furnace grew as flat platelets. On these platelets were numerous stepped hexagonal pyramids often with flat tops. In some cases spiral growth steps arising from screw dislocations were seen on flat areas on the platelets or on the flat tops of pyramids. In adjacent areas, platelets and pyramid tops were free of growth spirals and showed evidence of two-dimensional nucleation. Monochromatic CL images at the peak wavelengths of each of the three emission bands showed up the growth steps in two cases but not the third.

1. INTRODUCTION

The anomalous photovoltaic effect (APE) i.e. the generation of much larger than band gap photovoltages in striated platelets of vapour grown ZnS was first reported by Merz (1958). The structure of these platelets was found to consist of bands of a multitude of structures, ranging from the hexagonal wurtzite via a multitude of polytypes, to the cubic sphalerite. These structures and their formation mechanism were studied in detail over many years. The "anomalous" photovoltages and short-circuit currents will also sometimes reverse sign as the wavelength of the illumination changes from the visible to the near uv. This is due to the opposing direction of the electrical fields in the predominantly cubic, versus the more hexagonal, strips of the same single crystal (Shachar et al 1970, Yacobi and Brada 1974, Holt and Brada 1997). These strips also have different absorption edges. So, at the longer wavelength, the more cubic structures with their field direction are decisive for the direction of the short-circuit currents. At the shorter uv wavelengths, when all the structures present are strongly absorbing, the opposing internal fields attempt to drive current in opposite directions. If the larger part of the illuminated surface is hexagonal, the short-circuit current will reverse. These and other properties of these platelets were also systematically studied mainly in Jerusalem. For a review see Steinberger (1983).

Due to these interesting properties, a great deal of work on the vapour phase growth of ZnS and other II-VI compounds and alloys was done. It was found that many other morphologies than the striated platelets occurred and both the morphology and properties differed depending on which supplier's starting material was used. Recently we have examined a number of samples from the many crystals grown at the Racah Institute. Samples of striated platelets, some of which exhibit the APE, were studied using the scanning electron microscope (SEM), quantitative electron beam induced current (EBIC) (Holt and Brada 1997) and spectroscopic cathodoluminescence (CL) (Brada et al 1997) techniques. In this paper we report the results of observations made by secondary electron imaging of topographic contrast and CL in the SEM, on one other particular form of ZnS crystal.

2. EXPERIMENTAL MATERIAL

ZnS obtained from one supplier (Koch-Light Ltd.) and sublimed at the Racah Institute in a flow of H_2S, deposited on the walls of the furnace in a certain temperature zone as thin, flat large area crystals. This contrasted with all the other materials which started as thin needles growing from the fused silica surface of the furnace tube. These crowded, intergrown platelets of hexagonal forms of varying sizes are the samples studied here. The c-axis, normal to the platelet surface, was the direction of growth. This was established by measuring optical absorption with the optical ray perpendicular to the surface of the platelet. No birefringence was observed. The only optical inhomogeneity was due to strains in the cooled-down crystals The most likely explanation for the differing morphologies of these platelets is the possibly different impurity admixture in the Koch Light source material from that in the materials from other suppliers. The growth conditions were always very similar.

Examination was carried out in a JEOL 840A SEM at Imperial College using the CL system developed in our laboratory and described previously (Napchan et al 1993).

3. RESULTS AND DISCUSSION

On the large flat crystalline areas were seen many striking, stepped, often flat-topped hexagonal pyramids reminiscent of the Mayan temples of Latin America (Figure 1a).

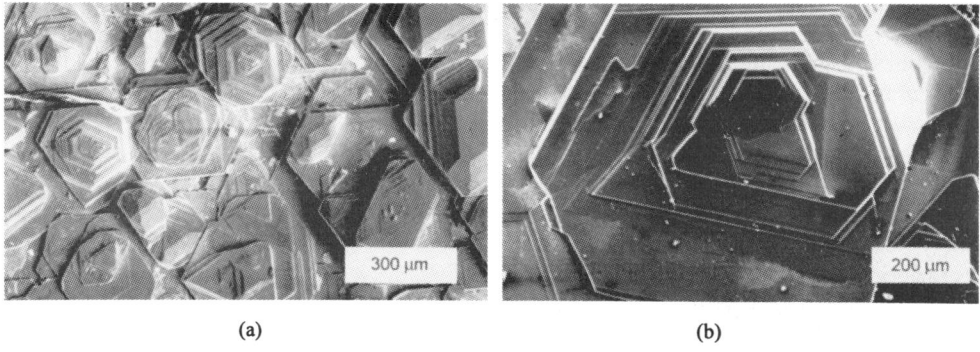

(a)   (b)

Figure 1. Scanning electron microscope secondary electron images (S.E.I.s) of (a) several flat-topped, stepped hexagonal pyramids and (b) one such pyramid (showing spiral growth steps at higher magnification in Figure 2) in samples of vapour deposited ZnS crystal grown from material supplied by Koch-Light Ltd.

A number of spiral growth steps, originating at the points of emergence of screw dislocations, were found in these samples. In a few cases they were found in the plateau atop stepped pyramids (Figures 1b and 2). Sometimes the growth spirals occurred in the midst of large flat pyramid-free areas. In other similar areas no spirals but numerous areas of the next growth layer occurred, apparently by independent two-dimensional nucleation (Figure 3).

Growth spirals were studied extensively at one time (see e.g. Verma 1953) as they were predicted to be necessary for rapid crystal growth from the vapour at low supersaturations by Burton, Cabrera and Frank (1949) and spiral growth steps were one of the

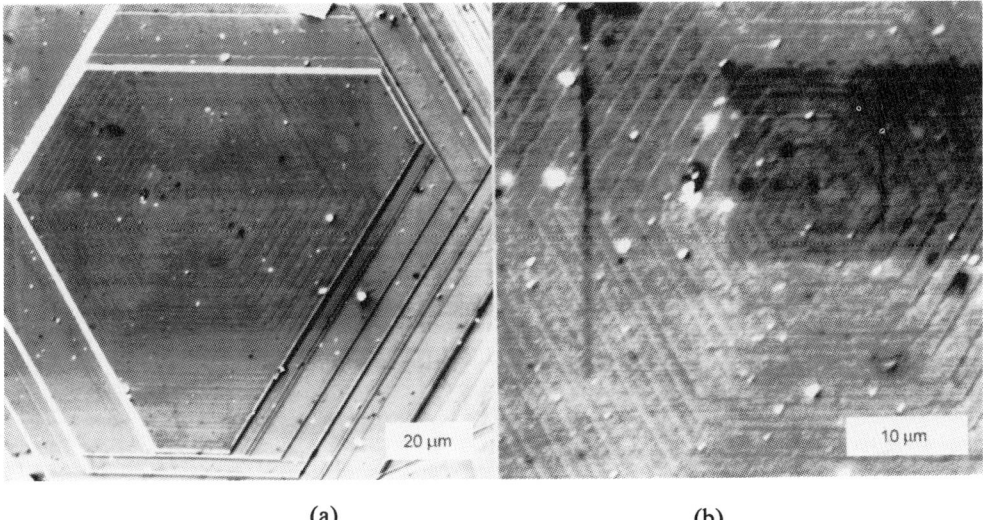

(a)   (b)

Figure 2. S.E.I.s showing (a) the spiral growth steps on the top of the pyramid in Figure 1b and (b) the point of emergence of the screw dislocation, at the centre of the spiral, from which the whole sample of Figure 1(b) grew. The dark vertical line at the left and the dark rectangle at the top right are are commonly-seen artifacts "written" by the beam during previous, higher-magnification scanning of this area. They are due to some form of "damage", perhaps involving charging of point defects to alter the secondary electron emission coefficient. Evidence for this type of mechanism for the darkening are that these artifacts appear too quickly to result from significant contaminant film deposition and fade and disappear in minutes to hours at room temperature.

Figure 3. The flat top of a large irregular pyramid on which a number of separately nucleated areas of the next growth layer (some marked by arrows) appear with a variety of shapes with crystallographically aligned edges.

earliest forms of evidence for the existence of dislocations in crystals. Both ordinary and giant screw dislocations were observed in the "needle-form" ZnS crystals by X-ray topography.

earliest forms of evidence for the existence of dislocations in crystals. Both ordinary and giant screw dislocations were observed in the "needle-form" ZnS crystals by X-ray topography. These crystals were mostly of hexagonal cross-sections and often developed a hollow core as they grew longer. These observations were made first by Mardix et al (1971) who also connected this fact with the periodic slip-processes which can also result in high-order polytypes. For the mechanism of such slip processes see Alexander et al (1970) and Steinberger et al (1973).

Figure 2 shows that growth in some of these samples was at low supersaturations so that deposition was nucleated predominantly if not solely at steps arising from the emergence points of screw dislocations as originally suggested by Burton et al (1949). Other similar samples presumably from adjacent positions in the furnace tube showed no evidence of the presence of the screw dislocation growth mechanism (Figure 3). This suggests that, although the samples of this morphology all grew in adjacent temperature zones of the continuous-flow furnace, the temperature and hence the supersaturation varied. Apparently this variation was sufficient so that, in some cases, the screw dislocation mechanism was required while, in others, two-dimensional nucleation could occur at the rate required. The latter condition resulted in the separate nucleation of new layers in nearby areas, as shown e.g. in Figure 3.

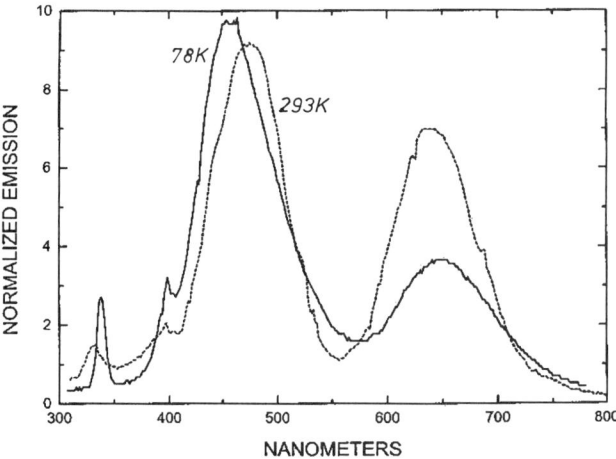

Figure 4. The emission spectrum of a crystal of Koch- Light ZnS. It consists of exciton emission at the shortest wave-length and two impurity activated bands at longer wavelengths.

SEM CL examination of another flat, hexagonal top of a pyramid like that shown in Figures 1(b) and 2 gave a three band emission spectrum (Figure 4) and showed growth-spiral-related contrast as can be seen in Figure 5. Comparison of the surface topography shown in Figure 5(a) with the contrast in the monochromatic CL micrographs recorded at the red, 650 nm, and blue, 455 nm, peak wavelengths (Figures 5b and c) shows that contrast occurs, related to the steps of the pyramid. Both these emission wavelengths are impurity activated, the red by Cd and Cu, as the band gap in ZnS is so large that the strong narrow exciton recombination emission band appears in the ultraviolet. The growth-step-related contrast in Figs. 5b and 5c is thus due to emission by impurity centres in the interior of the ZnS. Moreover, the contrast in the two visible CL Figures (5b and 5c) differs significantly. It cannot therefore be due entirely to spurious surface topography effects such as variation in the beam energy loss due to backscattering at the growth steps or greater emission through the two facets when the beam is incident near a step edge. That these spurious effects are negligible is confirmed by the fact that no step-related contrast appears in the exciton-radiation image, Figure 5(d).

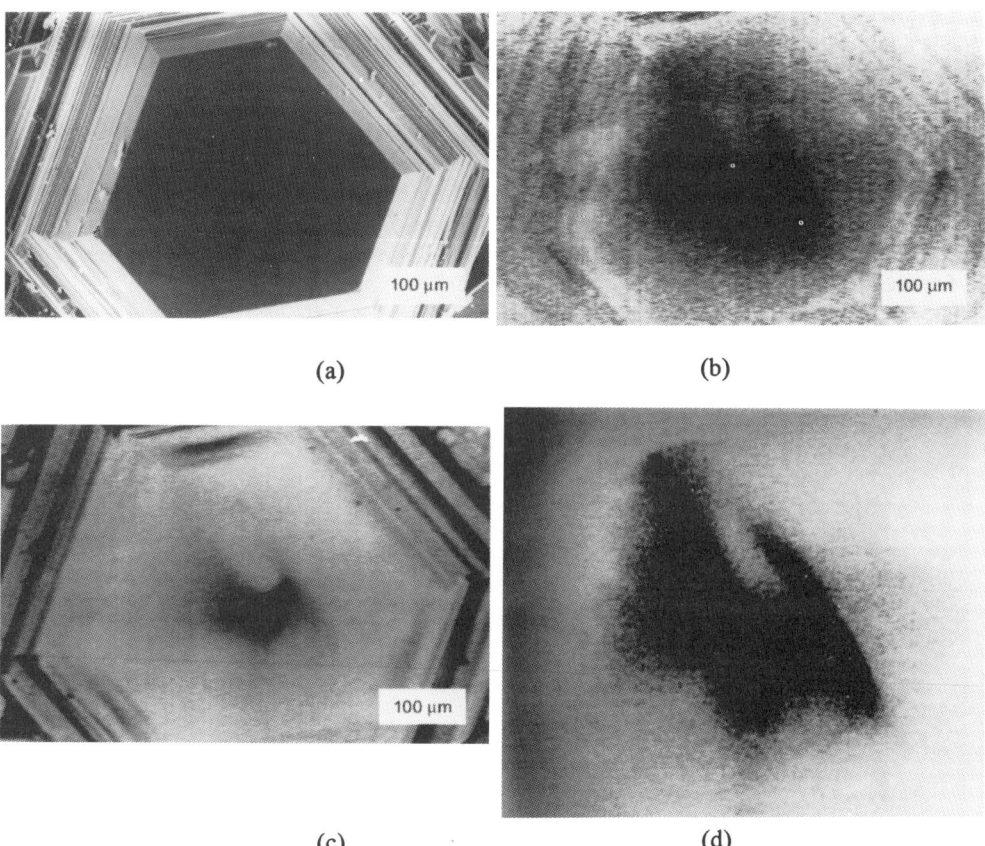

(a) (b)

(c) (d)

Figure 5. (a) S.E.I. of another hexagonal plateau like that in Figures 1 and 2, and monochromatic CL micrographs recorded at 80 K and (b) 650 nm (red), (c) 455 nm (blue) and (d) 330 nm (the UV, exciton peak wavelength).

The irregular dark central area in Figs. 5b, 5c and 5d is associated with the screw dislocation. This dislocation dark contrast may be due to either, or a combination, of two causes. The first is that the region may be denuded of the activating impurity atoms by segregation to the dislocation line where they precipitate out of solution and cease to be constitutents of the radiative recombination centers. The other is that the high central strain field near the dislocation increases the possibility of non-radiative recombination at the centres by creating additional energy levels of small spacing. It is noteworthy that the extent of the dislocation-related central dark contrast is quite different in Figures 5(a) and (b), which were recorded at the peak wavelengths of the blue and the red emission bands which were activated by different imputrity centres, and Figure 5(d), the exciton-CL image.

REFERENCES

Alexander E, Kalman Z H, Mardix S and Steinberger I T 1970 Phil. Mag. **21**, 1237

Brada Y, Holt D B and Mardix S 1997 in Microscopy of Semiconducting Materials 1997, Inst. Phys. Conf. Ser. **157** (Inst. Phys.: Bristol) pp 661 - 664

Burton W K, Cabrera N and Frank F C 1949 Nature **163**, 398

Holt D B and Brada Y 1997 in Microscopy of Semiconducting Materials 1997, Inst. Phys. Conf. Ser. **157** (Inst. Phys.: Bristol) pp 629 - 634

Mardix S, Lang A R and Blech I 1971 Phil. Mag. **24**, 683

Merz W J 1958 Helv. Phys. Acta. **31**, 625

Napchan E, O'Neill D and Zanotti-Fregonara C L M 1993 in Microscopy of Semiconducting Materials, Inst. Phys. Conf. Ser. **134** (Inst. Phys.: Bristol) pp 693 - 696

Shachar G, Brada Y, Alexander E and Yacobi Y 1970 J. Appl. Phys. **41**, 723

Steinberger I T 1983 Prog. in Cryst. Growth and Characterization **7**, 7

Steinberger I T, Kiflawi I, Kalman Z H and Mardix S 1973 Phil. Mag. **27**, 159

Verma A R 1953 Crystal Growth and Dislocations (Butterworth: London)

Yacobi B G and Brada Y 1974 Phys. Rev. B **10**, 665

Inst. Phys. Conf. Ser. No 164
Paper presented at Microsc. Semicond. Mater. Conf., Oxford, 22–25 March 1999

SEM EBIC characterization of SiC/Si solar cells

D B Holt, E Napchan, H Reehal* and S Toal*

Department of Materials, Imperial College of Science, Technology and Medicine, London SW7 2BP
*Department of Electrical, Electronic and Information Engineering, South Bank University, London SE1 OAA

ABSTRACT: A number of experimental solar cells with SiC emitters and Si bases, varying in photovoltaic efficiency from 0.1 to 5.8 %, were examined by EBIC. Defect particles were observed in the cells using EBIC. Evidence was found that suggests that the top contacts on the 0.1% cell were Schottky not ohmic in character. Quantitative EBIC linescans across these metal fingers were analysed to obtain values for L, the minority carrier diffusion length.

1. INTRODUCTION

The efficiency of Si solar cells can be increased by growing a top layer of a second semiconductor of forbidden gap energy $E_{g2} > E_{gSi}$. Incident light of photon energy $E_{ph} > E_{g2}$ will be absorbed in the layer. Then photons of energy $E_{ph} > E_{gSi}$ will be absorbed in the Si to form carrier pairs. In this way, more of the energy of the solar spectrum will be used to generate electrical power. One possible such wide gap "emitter" material is SiC.

A set of heterojunction solar cells consisting of nanocrystalline β-SiC p-type emitters on n-type Si (Figure 1) were grown and found to have photovoltaic efficiencies from 0.1 to 5.8% (Toal and Reehal 1998). This paper reports the results obtained in examining the cells in an SEM (scanning electron microscope) by the EBIC (electron beam induced current) method. By varying the beam energy and hence penetration range, measuring the EBIC gain and fitting these data to Monte Carlo simulation curves (Holt and Napchan 1994), in principle, it is possible to obtain the value of the minority carrier diffusion length in the base material and to determine its uniformity or variability (Hardingham and Holt 1995, Grunbaum et al 1995).

Monte Carlo electron trajectory simulation programs (Joy 1995) can be used to obtain data on the spatial distribution of hole-electron pairs generated by the beam and this can be applied to simulate EBIC data. EBIC curves were simulated by Napchan's well known MC-SET (Monte Carlo simulation of electron trajectories) suite of programs (Holt and Napchan 1994). Fitting experimental EBIC data to families of curves calculated in this way allows rapid evaluation of the material and device parameters involved.

2. EXPERIMENTAL METHODS

The solar cells were produced at South Bank University by electron cyclotron resonance plasma assisted chemical vapour deposition (ECR - PACVD) of the β-SiC. Their structure is shown in Figure 1.

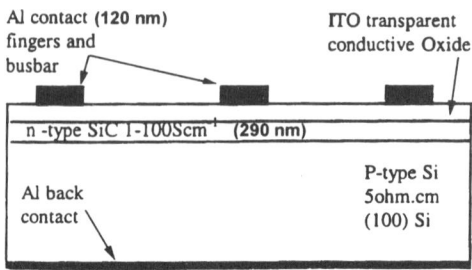

Figure 1. Schematic cross sectional diagram of the structure of the SiC/Si solar cells with ITO (indium tin oxide) layer.

The cells were characterized at Imperial College using a JEOL JSM 840A microscope plus either a Matelect EBIC system (mainly for imaging) or a Stanford Research Systems SR570 amplifier controlled and data-logged by TestPoint software (mainly for recording linescan profiles).

3. RESULTS

The lowest efficiency cell, "#116 with no ITO" appeared as shown in Figure 2. The black dots are suggestive of the incorporation of particles of foreign matter during growth. (Most of the dots in the EBIC image do not correspond to anything visible on the surface in the SEI.) The other cells all showed this same structure.

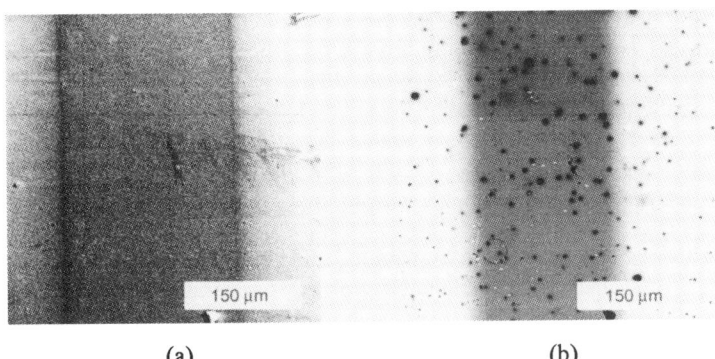

(a)   (b)

Figure 2. (a) SEI and (b) EBIC images of a typical area of the lowest (0.1%) efficiency cell (#116 - no ITO) sample. The vertical dark strip is a metal finger.

EBIC linescans were recorded across the parallel metallization lines over the cells as shown in Figure 3. The dips in the linescans as the beam crosses the metal fingers, are due to the absorption of the beam energy. The scan across the high-efficiency cell gives an approximately constant value of EBIC current (Figure 3a), but that across the low-efficiency one (Figure 3,b) falls exponentially away from the fingers. This is evidence that the contacts in the latter case were not ohmic.

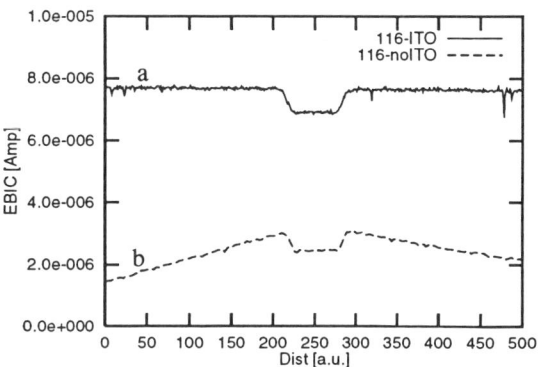

Figure 3. EBIC linescans recorded across one of the metal top contact 'fingers' of (b) the lowest efficiency (0.01 %) sample "#116 - no ITO" and (a) the highest (5.8%) efficiency cell (#116).

A series of such EBIC linescans at a series of increasing beam energies gave values for the collected current I_{EBIC}. They also give the (small) percentage variation in this quantity. This divided by the beam current, namely the gain = I_{EBIC}/I_b, was plotted against beam accelerating voltage and gave almost straight lines. Work is in progress on fitting this data to simulations calculated using the MC-SET suite of programs.

4. DISCUSSION

The reason for incorporating a layer of ITO, the well-known transparent conductor, is that the SiC layer is of high sheet resistance. Thus the ITO greatly improves current extraction. Problems with the shunt resistance were ascribed to the presence of pinholes in the film which Toal and Reehal (1998) ascribed to particulate formation in the plasma. It seems that the black spots in Figure 2(b) are direct evidence for such particulates.

Curve fitting the gain versus beam accelerating voltage involves the fitting parameters: L, the minority carrier diffusion length in the Si and the widths of the depletion regions in the Si and in the SiC. The process is a somewhat lengthy iterative one as there are a number of adjustable parameters. Initial indications are that the value of L is a few microns.

The fall of the collected current as the beam scans across the edges of the metal fingers in Figure 3,b is of exponential form. This can be explained as due to scanning across the edge of Schottky barriers so the signal falls as exp(-x/L) with distance from the edge of the metallization fingers. This is evidence of charge collection under the contact which is thus of Schottky not ohmic type. The flat EBIC response across the good device (Figure 3,a) is that to be expected for ohmic contacts so charge collection occurs only at the p-n heterojunction.

REFERENCES

Donolato C 1985 Sol. State Electron. **28**, 1143

Grunbaum E, Napchan E, Barkay Z, Barnham K, Nelson J, Foxon C T, Roberts J S and Holt D B 1995 Semicond. Sci. Technol. **10**, 627

Hardingham C and Holt D B 1995 in Microscopy of Semiconducting Materials, Inst. Phys. Conf. Ser. **146** (Inst. Phys.: Bristol) pp 621 - 624

Holt D B and Napchan E 1994 Scanning **16**, 78

Toal S J and Reehal H S 1998 Proc. 2nd World PV Conf., Vienna

Inst. Phys. Conf. Ser. No 164
Paper presented at Microsc. Semicond. Mater. Conf., Oxford, 22–25 March 1999
© 1999 IOP Publishing Ltd

A REBIC and EBSP study of a Σ = 13 grain boundary in silicon

D B Holt and E Napchan

Department of Materials, Imperial College of Science, Technology and Medicine, London
SW7 2BP

ABSTRACT: Remote electron beam induced current (REBIC) and electron back
scattering patterns (EBSP) were used in a preliminary study of a silicon Σ = 13 grain
boundary (GB) in a large bicrystal grown by Martinuzzi. The REBIC contrast indicated
that the boundary was of high resistivity at room temperature. EBSP mapping and REBIC
showed the boundary to contain a facet on a second crystallographic plane. The results are
discussed in relation to the literature.

1. INTRODUCTION

The fraction of coincidence sites for any possible orientation relation across a grain
boundary (GB) can only be $1/\Sigma$ where Σ is an odd integer. The smaller Σ is, the larger the
fraction of coincidence sites in the boundary and the greater the fraction of the area of the
boundary that is of good atomic fit. Small Σ GBs thus have simple structures, few broken or
strained bonds, low energies and are relatively electrically inactive so they have received
much attention. The $\Sigma = 13$ boundary in Si was studied by Bary et al (1987) and found to be
electrically active. Poullain et al (1987), however, found $\Sigma = 13$ boundaries to be always
EBIC inactive, like $\Sigma = 3$ twin boundaries. Ihlal and Nouet (1989) found EBIC GB contrast
due to Cu precipitates. Martinuzzi (1989) found $\Sigma = 13$ boundaries to give weak contrast
which increased steadily with both the time and temperature of annealing in a neutral gas. It
appears probable that the electrical activity of boundaries is dominated by impurities and that
the differences found by different workers are due to the differing types and amounts of
impurity segregation to the boundaries that they studied.

Kim et al (1992) studied the structure of a (510) oriented, [001] pure tilt $\Sigma = 13$
boundary in a Czochralski-grown Si bicrystal by high-resolution transmission electron
microscopy. They found this boundary to have a coincidence site lattice periodicity but an
aperiodic interface dislocation core structure. Discrete precipitate particles in the boundary
were shown by high-spatial-resolution EELS (electron energy loss spectroscopy) to be SiO_x
where $0 < x < 1$.

EBIC determines only the local level of recombination at a boundary, either
enhanced, giving dark contrast, or unaltered, giving no electrical activity. REBIC, however,
can also detect forms of contrast indicating charge on the boundary, the presence of high- or
low-resistivity boundary layers, etc. (Holt 1994). It is found (Hanoka et al 1980, Raza and
Holt 1995) that $\Sigma = 3$ twins in Si that are EBIC inactive at room temperature become active at
liquid nitrogen temperatures. This observation was explained as due to the Fermi level
movement with temperature resulting in filling the defect levels so the boundaries become
charged and, therefore, recombination active at the lower temperature.

708

This paper reports a REBIC observation of the contrast of a Σ = 13 boundary in Si together with EBSP (electron back scattering pattern) data to check its structure.

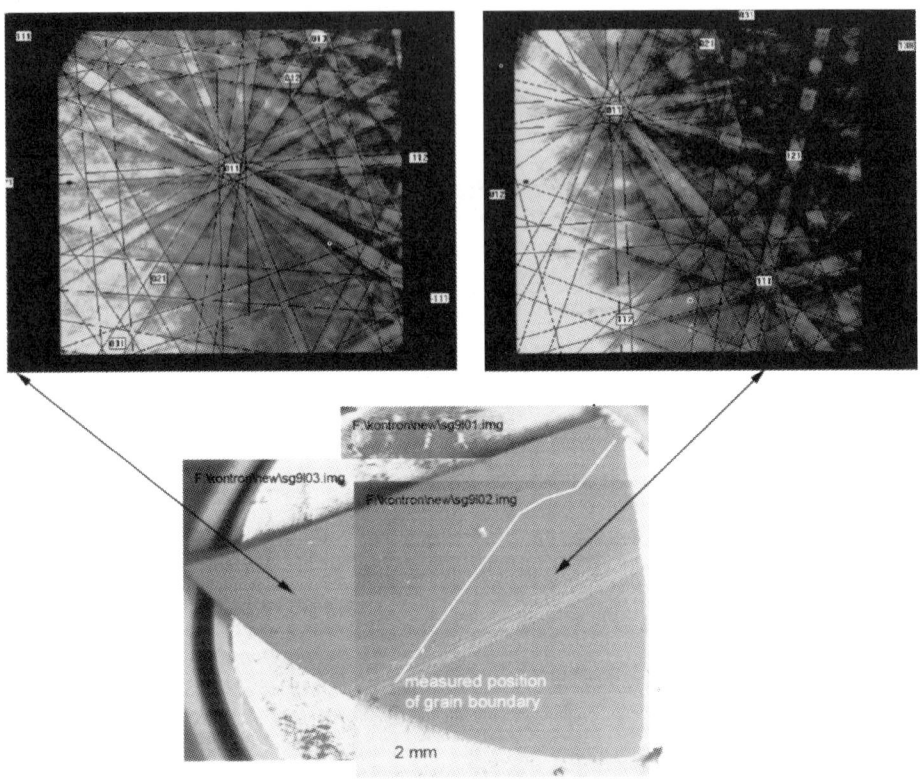

Figure 1. SEM secondary electron image of a specimen cut from a wafer. A scratch (along which the wafer did not crack!) can be seen. The position of the Σ = 13 boundary is marked. Also shown are the EBSPs of the grains on either side of the boundary.

2. EXPERIMENTAL METHODS

A cylindrical silicon bicrystal containing a Σ = 13 GB was kindly provided by Prof. Martinuzzi and Dr. Pasquinalli. The material of the bicrystal was p-type, boron doped to 1 ohm cm. They reported that the boundary in this material was only weakly electrically active in EBIC i.e. it produced slightly enhanced recombination. Long annealing at > 600 C was needed to make it strongly electrically active.

A number of slices were cut and polished by Dr. V. Higgs and specimens were cut from them. They were examined in a JEOL JSM 840A using a Matelect EBIC system or a Stanford Research Systems SR570 amplifier and computer running TestPoint software plus a Kontron image processing system. EBSP observations were carried out using equipment and Channel + software supplied by HKL Software.

3. RESULTS

Figure 1 shows a specimen cut from a circular wafer and containing the $\Sigma = 13$ boundary together with the EBSPs from the grains on either side. The boundary is marked on the SEI as found by noting the translation stage coordinates of the positions at which the EBSP changed from that for one grain to that for the other. It thus appeared that there was a step in the GB. EBSP confirmed the $\Sigma = 13$ nature of the boundary from the misorientation between the grains.

Figure 2. REBIC image showing the $\Sigma = 13$ boundary and the step therein.

Figure 2 is a REBIC image of part of the specimen of Figure 1 in which the boundary and the step are visible. The boundary also appeared as a line in secondary electron images due to low surface relief arising in polishing. The REBIC contrast is of the terrace type. This was confirmed by linescan contrast profiles which showed the characteristic step form.

4. DISCUSSION

The present work showed that this boundary, which was known to exhibit weak EBIC contrast (Martinuzzi private communication), also exhibited weak terrace contrast in REBIC in the as received condition. The occurrence of a small contrast step due to an interface layer of higher resistivity at $\Sigma = 13$ boundaries has not been known previously. The terrace contrast repeatedly reverses in Figure 2. That is the grain on one side is brighter, then further down it is the other grain and this reversal correlates with the step. The significance of this reproducible characteristic is not at present understood.

We intend to study the effects of both reducing the temperature and annealing at above 600 C to check the effect on the contrast, that is, on the boundary properties.

The boundary is normal to the (100) plane of the wafer and runs diametrically across it. In principle it will be possible, therefore, using the EBSP software to determine both the axis and angle of misorientation of the boundary and the crystallographic plane in which it lies. The occurrence of the step is interesting as it means that there is a component in this field of view that is on another plane. The step lies at 135° to the rest of the boundary, suggesting that they both lie in low-index, low energy orientations. The sections in the two orientations may differ in properties since they will differ in atomic structure.

It has thus been shown that a weakly EBIC-active grain boundary exhibits weak REBIC activity and the combination of REBIC and EBSP promises to be a powerful means for studying the structure and properties of grain boundaries in silicon.

ACKNOWLEDGEMENTS

We wish to thank Prof. S. Martinuzzi and Dr. M. Pasquinelli of the Laboratoire de Photoelectricite, Universite d'Aix Marseille III for the supply of the bicrystal and Dr. V. Higgs, then of Kings College London, who sliced and polished this material.

REFERENCES

Bary A, Mercey B, Poullain G, Chermant J L and Nouet G 1987 Rev. de Phys. Applique **22**, 597

Hanoka J I, Bell R O and Bathey B 1980 Symp. Electron. Opt. Prop. Polycryst. or Impure Semicon. Novel Cryst. Growth Techniques, eds K V Ravi and B O'Mara (Electrochem. Soc.: Princeton) pp 76 - 86

Holt D B 1994 in Polycrystalline Semiconductors III – Physics and Technology, Solid State Phenomena, eds H P Strunk, J H Werner, B Fortin and O Bonnaud (Scitec Publishers: Zurich) pp 171 - 182

Ihlal A and Nouet G 1989 in Polycrystalline Semiconductors, eds H J Moller, H P Strunk and J H Werner (Springer-Verlag: Berlin) Proc. in Phys. **35**, pp 77 - 82

Kim M J, Carpenter R W, Chen Y L and Schwuttke G H 1992 Ultramicroscopy **40**, 258

Martinuzzi S 1989 Polycrystalline Semiconductors, eds H J Moller, H P Strunk and J H Werner (Springer-Verlag: Berlin) Proc. in Phys. **35**, pp 148 - 157

Poullain G, Mercey B and Nouet G 1987 J. Appl. Phys. **61**, 1547

Raza B and Holt D B 1995 in Microscopy in Semiconducting Materials, Inst. Phys. Conf. Ser. **146** (Inst. Phys.: Bristol) pp 107 - 112

Inst. Phys. Conf. Ser. No 164
Paper presented at Microsc. Semicond. Mater. Conf., Oxford, 22–25 March 1999

711

Nanoanalysis of local electrical fields in silicon diodes by STEBIC

D Brouri, C Cabanel, J Y Laval, L Peymayeche and C Garrec

Laboratoire de Physique du Solide, CNRS-ESPCI, Paris, France

ABSTRACT: The scannning transmission beam-induced current (STEBIC) technique has been adapted to analyse Si diodes in cross-section. By working at 110K and with thin samples, the collection of carriers by surfaces is increased and the signal can be localised with a precision of 6 nm. Data were quantified by simulating STEBIC profiles from the the transport equation for excess carriers.

1. INTRODUCTION

The scanning transmission electron beam induced current technique (STEBIC) was developed by Sparrow and Valdrè (1977). It enables the study of local electrical activity in heterojunctions (Petroff et al 1980, Pennycook 1981, Brown and Humphreys 1995) or in polycrystalline silicon (Cabanel and Laval 1990) in plan view. This technique consists of *in situ* measurements of minority carriers induced by the electron beam of a STEM microscope. By synchronising measurements with each scan of the electron beam, the electrical activity can be viewed at a very local scale (5 to 100 nm). This method can be used in cross-section (X-STEBIC) to observe the recombination of carriers at defects in semiconductors (Cabanel et al 1999) and to analyse space charge zones related to junctions or heterojunctions.

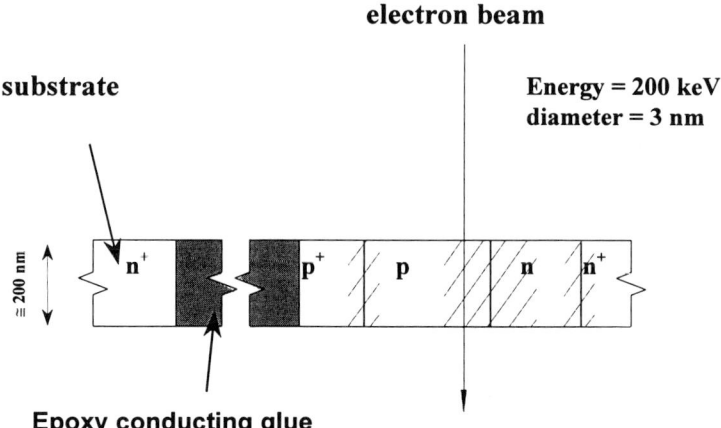

Fig. 1 : STEBIC experiment in cross-section.

2. EXPERIMENT DETAILS

This technique was applied to silicon IMPATT diodes (Impact Avalanche Transit Time) which correspond to a $p^+/p/n/n^+$ multijunction, and which are used as active elements for high power applications of oscillators (100 GHz) (Dalle and Rolland 1990). These diodes were prepared by chemical beam epitaxy at LCR-Thomson. The surface orientation was close to {111}. The dopant concentrations (As and B) were 10^{19} at/cm^3 for n$^+$ and p$^+$ layers and 10^{17} at/cm^3 for n and p layers. The experiment in cross-section (Fig. 1) sets the junction parallel to the electron beam. Ohmic contacts were produced by heating gold deposits for 1 min at 400°C in a flash furnace. The samples were thinned down to electron transparency by mechanical polishing (tripod method). We developed a specific STEBIC holder for a TEM-STEM, Jeol 2000FX. The signal was amplified by a Keithley 428 current-voltage amplifier. The 3 nm beam probes the space charge zone at each point. Electron-hole pairs are immediately separated in the electric field and then collected at the boundaries of the sample by the ohmic contacts. The lateral resolution of the signal is optimized in the field of the diode (≤ 10 nm).

3. RESULTS AND DISCUSSION

Fig. 2 displays the electrical activity of a diode, which cannot support an inverse current ≥ 10 mA. Sharp local variations of the signal (up to 12 %) occurred on the p/n junction whereas on a high performance diode, modulations in STEBIC intensity at the interface were limited to 0.5 %. This means that the poor properties of the diode of Fig. 2 are linked to the presence of recombination centres located in the vicinity of the p/n junction. This leads to strong local changes in the electric field, which can be detected by this method, under our experimental conditions, with a lateral resolution of 10 nm.

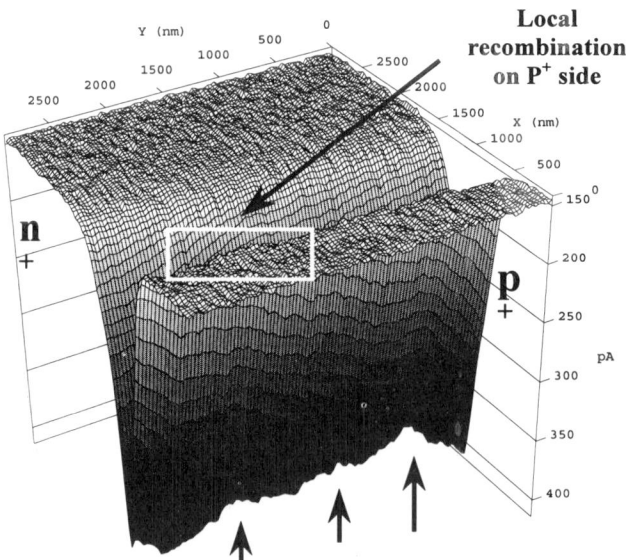

Fig. 2 : 2D STEBIC profile on an IMPATT diode with limited performance : *arrows indicate local variations in the electric field.*

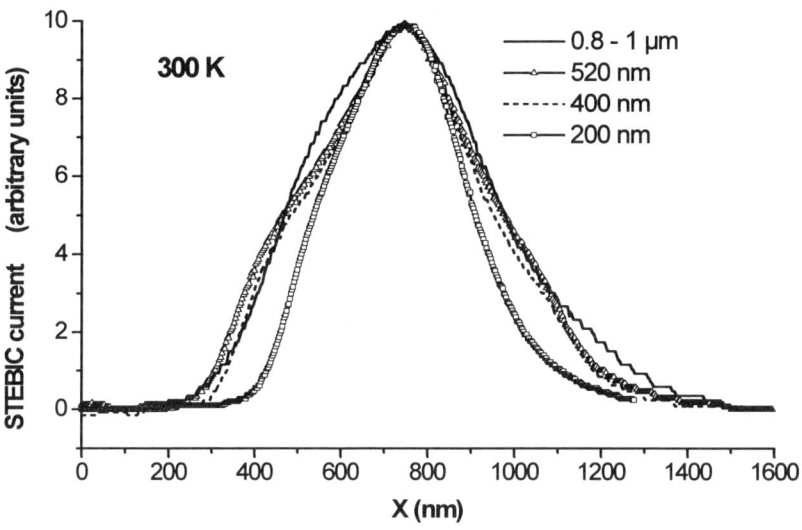

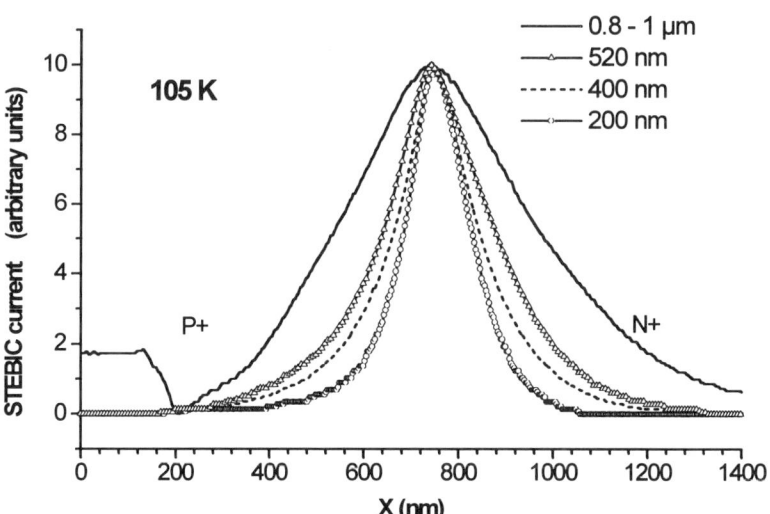

Fig. 3 : STEBIC profiles at 300 K and 105 K for different thicknesses measured on a bevelled sample.

714

In Fig. 3, STEBIC profiles carried out at room temperature and at 105 K are compared. Measurements were performed at varying thicknesses, using the same sample thinned as a bevel. Profiles recorded at 300 K display a widening which varies only slightly with the thickness of the sample. This widening is due primarily to the diffusion of minority carriers in the zones outside of the electric field. On the contrary, at low temperature, the width of the signal varies as a function of the thickness. For a thickness of 200 nm, we find that the width of the profile corresponds to the width of the depleted zone in the central field (300 nm). Since surface recombinations become more important at low temperature, the carriers are promptly collected by the surfaces and cannot diffuse in thin zones. The maximum of the signal can be localised with a precision of 6 nm compared with 15 nm at room temperature. From this quantitative data, it can be found that there is some interdiffusion of the dopants during the growth process.

In order to quantify these results, we have simulated the STEBIC-profiles by solving the transport equations for the excess carriers induced by the electron beam. The model developed by Marten and Hildebrand (1983) for EBIC, which takes into account diffusion and drift of carriers, was adapted for the conditions found in STEBIC. In this simulation the recombination rate of minority carriers Ro is the only adjustable parameter. For a thickness of 200 nm it was found that $Ro \cong 0.5$ cm^{-3}.s^{-1} at room temperature.

4. CONCLUSION

Cross-section STEBIC enables the imaging of electrical activity in Si multijunctions. Local variations of the electric field at junctions can be observed with a sensitivity of 5%. With this technique the electrical interface can be revealed with a precision ≤ 15 nm at room temperature. The performance of this method is significantly improved when working at low temperature and with thin samples.

ACKNOWLEDGEMENT

We are very grateful to C. Dua (Thomson-CSF, Corbeville, France) for providing us with the silicon diodes.

REFERENCES

Brown P D and Humphreys C J 1995, Inst. Phys. Conf. Ser. **147**, pp 285-288
Cabanel C, Maya H and Laval J Y 1999 Phil. Mag. Lett. **79**, 55
Cabanel C and Laval J Y 1990 J. Appl. Phys. **67**, 1425
Dalle C and Rolland P-A 1990 IEEE Trans. Microwave Theory Technique **38**, 366
Marten H W and Hildebrand O 1983 Scanning Electron Microscopy III, SEM Inc., p 1197
Pennycook S J 1981 Ultramicroscopy **7**, 99
Petroff P M, Logan R A and Savage A 1980 Phys. Rev. Lett. **44**, 287
Sparrow T G and Valdrè U 1977 Phil. Mag. **36**, 1517

Inst. Phys. Conf. Ser. No 164
Paper presented at Microsc. Semicond. Mater. Conf., Oxford, 22–25 March 1999

Optoelectronic characterisation of semiconducting materials in a scanning near-field optical microscope

R M Cramer, C Pagel and L J Balk

BUGH Wuppertal, Lehrstuhl für Elektronik, Fuhlrottstr. 10, 42097 Wuppertal, Germany

ABSTRACT: Scanning near-field optical microscopy (SNOM) allows to perform optical beam induced current analyses of semiconducting materials with a lateral resolution independent of the irradiation wavelength. In this paper, the influence of the SNOM probe on the obtained results is discussed with respect to both the presence of topographical artefacts and the influence of the probe potential on the induced current.

1. INTRODUCTION

Optical beam induced current analyses have long been carried out to analyse the optoelectronic properties of semiconducting materials and devices and to localise and characterise defects present in these specimens. The technique does not require vacuum, is in principle non-destructive and permits a localisation of device internal barriers and defects, a determination of diffusion lengths and a deeper insight into the physical properties of a sample due to its spectroscopic capabilities. Unfortunately, it suffers from a comparatively poor lateral resolution caused by the diffraction limit or Abbe-barrier, which prevents an optical beam to be focused to a spot smaller than approximately half the wavelength of the light used. This physical limit can be overcome by placing a sub-wavelength aperture in an otherwise opaque film in close proximity to the structure of interest, so that advantage can be taken of the lateral confinement of the electric field directly behind the aperture, the optical near-field (Betzig et al 1991). Over the past few years, optical near-field induced current (ONIC) techniques have been successfully applied to investigate relaxed GeSi films (Hsu et al 1994), quantum well lasers (Buratto et al 1994), silicon photodetectors (Cramer et al 1997) and Cu(In,Ga)Se$_2$ solar cells (McDaniel et al 1997).

2. THEORY

Although no consistent model including all effects related with the optical near-field could be developed, yet, in most cases the emission properties of a SNOM aperture can be discussed in terms of a radiating, Hertzian dipole. The emission of a such dipole can be divided into two separate components, the propagating component, which decays in intensity over distance with $1/r$, and the non-propagating or evanescent component, which decays with $1/r^6$. As only the non-propagating component is capable of probing physical properties of a sample with sub-wavelength resolution, the strong dependence of both signal intensity and resolution on the aperture-sample separation in scanning near-field optical microscopy becomes obvious. In particular, artefacts may be introduced in SNOM micrographs if either the topographical feedback is unstable or the position of the aperture is different from the most protruding part of the probe, which is responsible for distance control. Due to this so-called z-motion artefact (Hecht et al 1997), in many cases   contrast in near-field optical micrographs arises from the probe-sample distance control rather than from pure optical or optoelectronic properties. A good indication for the presence of z-motion artefacts is a spatial resolution in SNOM micrographs which is much better than expected from the size of the probe aperture or identical to the spatial resolution in the topographical representation of the sample. This situation is also valid for optical near-field induced current analyses, where the amount of photogenerated carriers is also strongly dependent on the optical power coupled into the sample from the near-field.

Another effect which has to date not been taken into consideration for ONIC investigations is the influence of the probe potential versus the sample surface, as in most cases the near-field probe potential is floating and charges could be easily influenced during scanning. As a result, local band bending underneath the probe may occur due to the nano-field effect (Koschinski et al 1995). This effect may on the one hand adversely influence the achieved results if neglected, on the other hand deliberately applying electrical potentials between probe and sample may allow to ionise trap states in the band gap to obtain information on these states. Also, lateral diffusion of minorities may be reduced due to the resulting band bending to increase the lateral resolution of the technique.

3. EXPERIMENTAL SET-UP

In our set-up, the near-field optical probe consists of a single-mode optical fibre heat-pulled to form a sharp tip and subsequently coated with aluminium to fabricate an aperture with a diameter of approximately 50 nm. The probe-sample separation is controlled by a tuning fork distance regulation scheme (Karraï and Grober 1995), where the fibre is glued to one of the prongs of a piezoelectric tuning fork commonly used in quartz resonators. As the tuning fork-fibre assembly is vibrated laterally above the sample surface at its resonant frequency with an additional piezo crystal, the amplitude and phase of the oscillation are represented by the output voltage of the tuning fork and measured by

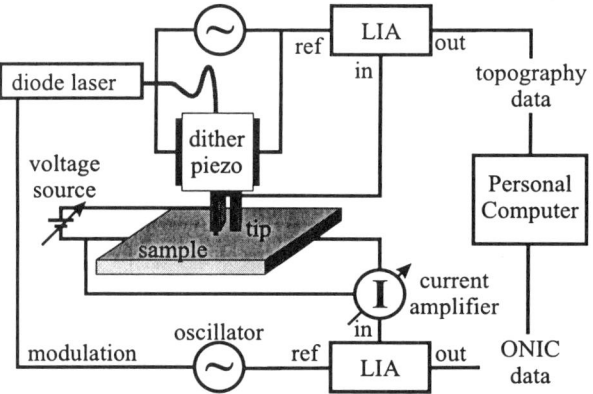

means of a lock-in amplifier. As the fibre approaches the surface, both amplitude and phase of the tuning fork signal change with distance, as the fibre penetrates a water absorption layer on the sample surface and is then subjected to inter-atomic forces between the probe and the sample. With a typical lateral oscillation amplitude in the range of 3 to 5 nm, the sample topography can be determined with a height resolution of about 0.1 nm. This scan mode is commonly referred to as constant gap width scanning.

Fig. 1 Optical near-field induced current analysis set-up.

Optical near-field induced currents are measured by a current pre-amplifier (Ithaco 1212) and further amplified by a second lock-in amplifier synchronised with the light output from the fibre-coupled diode laser ($\lambda = 635$ nm, P < 3 mW) used for optical excitation. This is necessary due to the attenuation of the light output at the aperture of about 10^{-5}-10^{-6}, so that routinely currents in the picoampere range must be determined. An electrical potential between the probe and the sample can be applied by connecting a ground-free voltage source to both the sample and the aluminium coating of the SNOM fibre. The complete set-up used in this work is illustrated in Fig. 1.

Scanning the probe in constant height over the sample tilt plane is accomplished by a home-made software package, where in a first, closed-loop scan the sample topography is recorded and the sample tilt is calculated based on a two-dimensional polynomial regression of the topographical data. Subsequently, the probe can be scanned over the determined tilt plane with an arbitrarily adjustable separation while optical near-field induced current data can be simultaneously acquired in both scanning procedures.

4. RESULTS

To investigate the influence of the z-motion artefact on the resolution and contrast in optical near-field induced current micrographs, a $CuInS_2$ absorber layer fabricated in a roll-to-roll process (Jacobs et al 1998) and coated with copper has been investigated with a near-field probe which has been damaged during a preceding scan, so that the aperture diameter has increased from about fifty to a few hundreds of nanometers. Fig. 2 shows the topography and optical near-field induced current micrographs obtained in constant gap width and constant height mode on this sample. In the to-pography micrograph, different grains can be clearly distinguished. The corresponding optical near-field induced current micrograph shows that variations of the collection efficiency of about 20 % are present in the investigated sample. The boundaries between the single grains can also be observed as distinguished features with a high clarity and contrast in the ONIC micrograph. However, the spatial resolution obtained in this image is about one order of magnitude better that could be expected from the probe aperture diameter. In a second scan, the same sample area has therefore been inves-tigated in the constant height scan mode, where the probe has been scanned at a height of about 5 nm over the highest elevation of the calculated sample tilt plane.

It becomes obvious from the data and the loss of resolution, that the sharp features observed at the grain boundaries have solely originated from the z-motion artefact, demonstrating the need for constant height scanning in SNOM techniques. Moreover, it appears that some ONIC con-trast, i.e. differences in the col-lection efficiencies of single CISCuT grains, has been con-cealed by the z-motion artefact.

Fig. 2: Topography (top left), constant gap width ONIC (top right), calculated tilt plane (bottom left) and constant height ONIC (bottom right) micrographs of CISCuT absorber.

For example, whereas the ring-like structure of grains showing a lower current in the lower left quadrants of the ONIC micrographs can be observed both in constant gap width and constant height scan modes, the enhanced collection efficiencies in the upper left part of the scan area can only be observed in the constant height scan ONIC micrograph.

The influence of the probe potential on ONIC analyses has been investigated on a silicon photo-detector destructively deprocessed to gain access to the active area. The results of the measurements are shown in Fig. 3, where the topographical image (top left) indicates the presence of surface contamination (bright structures). The calculated tilt plane of the sample required for constant height scanning shown below appears quite similar to the tilt plane of the CISCuT absorber. We believe that this is due to a slight tilt in the specimen holder itself, as this tilt is roughly equal for most samples.

From the optical near-field induced current micrograph obtained at zero potential between probe and sample (top right), a fairly homogeneous signal over the sample can be observed except for those areas which are covered with surface contamination. The grey scale in this micrograph correlates to a current variation from 1.8 to 2.1 nA. By applying a potential of -18 V to the probe, however, the contrast in the ONIC micrograph changes (bottom left). In addition to the decrease in current observed

718

at the surface contamination, a lower current can also be found in various areas distributed over the scan area. This may be the result of defects in the semiconductor introduced during sample de-processing and has been observed in many parts of the specimen.

One difficult issue in comparing scanning near-field optical microscopy micrographs, however, is the strong influence of the aperture shape on the light emission of the probe. As a consequence, the comparison of results achieved on different samples is still complicated, as the delicate aluminium coating is easily damaged during sample exchange and tip approach and some observed effects could be attributed to a varying probe rather than to differences in the investigated sample properties.

Fig. 3: Topography (top left), calculated tilt plane (bottom left) and constant height mode ONIC micrographs of photodetector measured at 0 V (top right) and −18 V probe potential.

5. SUMMARY

Although SNOM undoubtedly has the potential to improve the achievable resolution of OBIC measurements beyond the diffraction limit, great care must be applied in interpreting the acquired data. It has been demonstrated that high spatial resolution and distinguished sample features may be generated by a topographical artefact, so that scanning in constant height over the sample plane is mandatory. The influence of the probe potential on the contrast in ONIC micrographs shows that a defined probe potential is necessary to accurately determine the optoelectronic properties of a sample, but also allows to obtain information inaccessible in conventional OBIC measurements.

ACKNOWLEDGMENTS

The authors wish to thank O. Tober and M. Winkler, Institut für SolarTechnologien, Frankfurt(Oder), for providing the CISCuT absorber samples.

REFERENCES

Betzig E, Trautman J K, Harris T D, Weiner J S and Kostelak R L 1991 Science **251** (5000), 1468

Buratto S K, Hsu J W P, Trautman J K, Betzig E, Bylsma R B, Bahr C C and Cardillo M J 1994 J. Appl. Phys. **76** (12), 7720

Cramer R M, Heiderhoff R, Selbeck J and Balk L J 1997 Inst. Phys. Conf. Ser. **157**, 685

Hecht B, Bielefeldt H, Inouye Y, Pohl D W and Novotny L 1997 J. Appl. Phys. **81** (6), 2492

Hsu J W P, Fitzgerald E A, Xie Y H and Silverman P J 1994 Appl. Phys. Lett. **65** (3), 344

Jacobs K, Penndorf J, Röser D, Tober O and Winkler M 1998 Proc. 2nd World Conference on Photovoltaic Solar Energy Conversion, 409

Karraï K and Grober R D 1995 Appl. Phys. Lett. **66** (14), 1842

Koschinski P, Fiege G B M and Balk L J 1995 Inst. of Phys. Conf. Ser. **146**, 654

McDaniel A A, Hsu J W P and Gabor A M 1997 Appl. Phys. Lett. **70** (26), 3555

Inst. Phys. Conf. Ser. No 164
Paper presented at Microsc. Semicond. Mater. Conf., Oxford, 22–25 March 1999

Axial field measurements on a high resolution portable scanning electron microscope column

A Khursheed, Y Zhao *, N Karuppiah * and E J Chua

Electrical Engineering Department, National University of Singapore, 10 Kent Ridge Crescent, Singapore 119260.
*Institute of Materials Research and Engineering, c/o National University of Singapore

ABSTRACT: This paper presents Hall probe measurements on the recently proposed high resolution portable Scanning Electron Microscope (SEM) concept. A test column using permanent magnet lenses was constructed and has a height of 120 mm. Experimental axial flux density measurements were found to correlate well with simulation predictions. A method of varying the axial field strength by using magnetic shorting plates was investigated and found to be successful. In this way, the beam energy can be varied, and the portable permanent magnet column will be able to operate in a similar way to conventional SEMs.

1. INTRODUCTION

Recently, Khursheed et al (1998) proposed a high resolution portable Scanning Electron Microscope (SEM) column design which is based upon using permanent magnet technology. The overall height for such a column is typically under 120 mm. This column allows for two condenser lenses and a single pole magnetic objective lens. The demagnification of this miniature column is predicted to be comparable to conventional SEMs, so that it can operate with a variety of different electron gun sources.

The potential advantages of using portable SEM columns are, of course, many. They include the possibility of using different columns for a variety of different specimen chambers or of choosing the column that is optimised for the problem under investigation. In this proposal, the SEM chamber, vacuum system, gun assembly, driving and display electronics are fixed as normal, while an array of different columns are portable and can be fitted in as required. This cluster of portable SEM columns is predicted to be small enough to fit on a normal size shelf.

Grade 35 Neodymium-Iron-Boron (NbFeB) permanent magnets are used. They have a coercive force H_c of 0.9×10^6 A/m which translates into an effective coil excitation of 18,000 AT at the radial air/iron interfaces of a 2 cm long permanent magnet. Figure 1 shows a diagram of the magnetic circuit of the portable SEM. The permanent magnets are indicated by dark regions. The flux lines and axial field distribution were found via finite element programs written by Khursheed (1992). It should be noted that Figure 1 shows a free standing column. In practice, the specimen can be placed on a magnetic base plate that significantly increases the strength of the objective lens field. For thin specimens (less than 2 mm), the peak objective lens axial field can be increased to be over 1 Tesla.

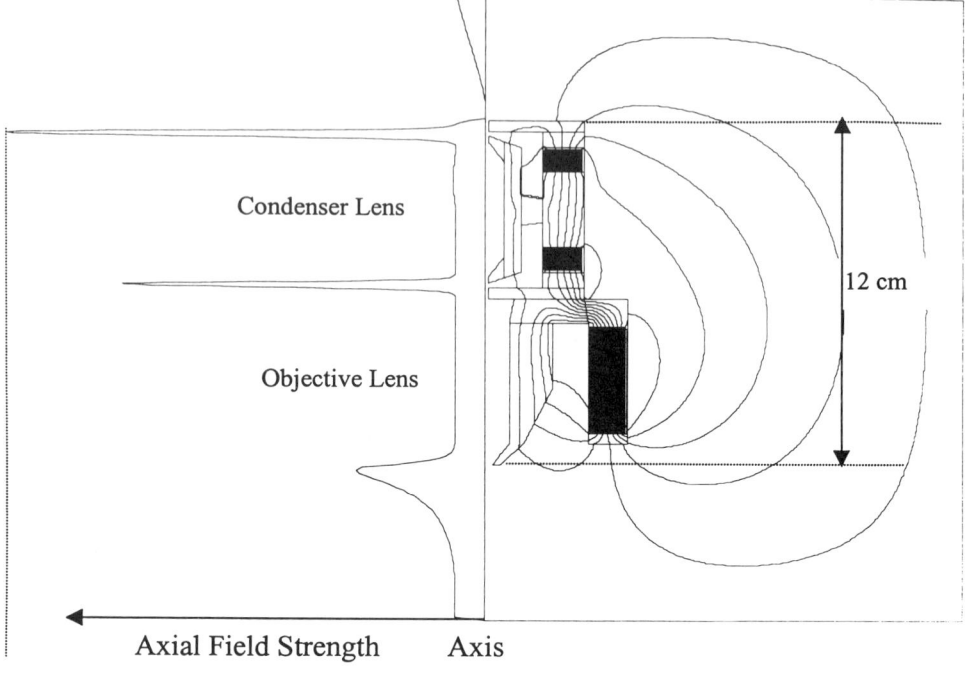

Condenser Lens

Objective Lens

12 cm

Axial Field Strength   Axis

1 Tesla

Figure 1. Simulated flux lines and axial field distribution for the portable SEM column, the permanent magnets are indicated by the dark regions

2. RESULTS

Experimental measurements of the axial field distribution were made by using an axial Hall probe (F.W.BELL series 9550 Gauss/Tesla meter with axial probe SAA99-0608). Figure 2 compares the experimental and simulated axial field distributions together on the same graph. The experimental values are close to the simulated ones, and confirm that permanent magnets can provide the necessary axial field strengths required to make an SEM column portable. The small difference between experiment and theory can be explained by several reasons. The B-H curve of the mild steel used for the magnetic circuit is not known with precision, and it is quite likely that the B-H curve used in the simulation is inaccurate. The magnetic properties of mild steel depend on many factors, including whether it has undergone thermal annealing treatment. The iron circuit used had not been treated in this way. Another reason that may explain the difference is due to the artificial closed boundaries assumed in the numerical field solving program. For a 30 keV primary beam energy, using the above experimentally measured axial field distribution, the demagnification of the condenser lens system is predicted to be around 325, which is comparable to conventional SEMs. For the free-standing objective lens axial field distribution operated at a 1 keV beam energy, the C_c and C_s are predicted to be 0.9 mm and 0.615 mm respectively, indicating that it is a high resolution lens.

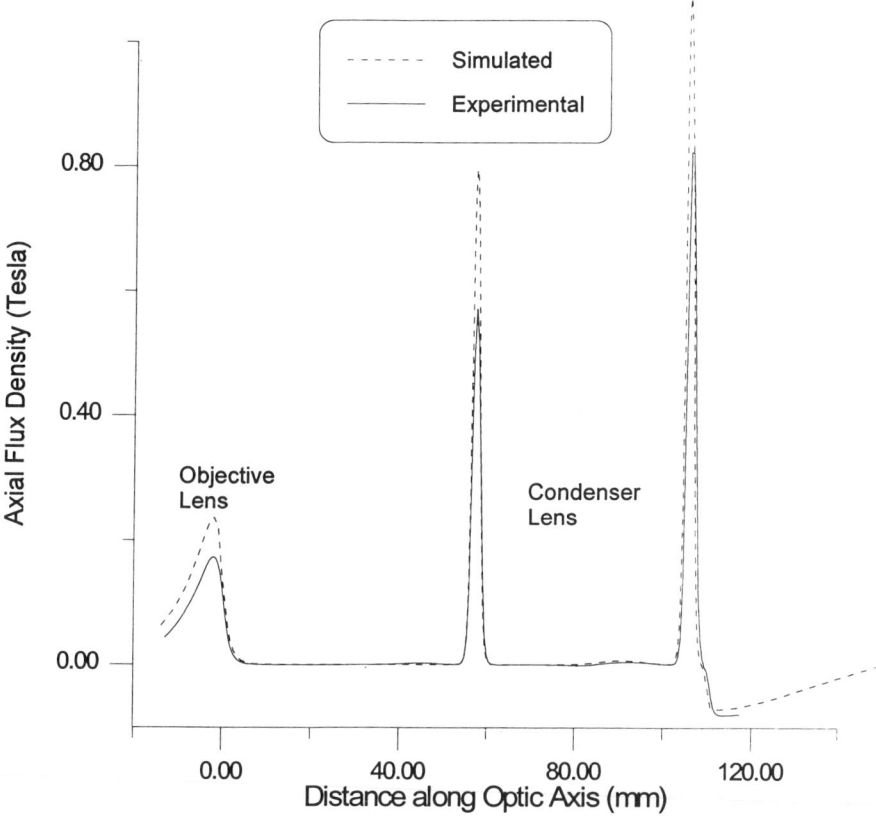

Figure 2. Comparison of Simulation Predictions with Hall Probe Measurements for the Axial Flux Density

3. MAGNETIC SHORTING PLATES

Magnetic shorting plates are used to vary the axial field strength so that the portable SEM column can operate in a similar way to conventional columns. By varying both the axial field strength of the condenser and objective lenses, a range of different beam energies can be used. Four magnetic shorting plates are placed around each magnet. Each plate covers a quadrant region. The plates are supported on non-magnetic structures so that they can be pushed towards or away from the magnet. In conventional SEMs, the beam energy typically ranges from 1 to 25 keV, which requires the peak axial field strengths of each gap to vary by a factor of 5. Figure 3 shows axial field measurements for the situation of no shorting plates and a full short. For the condenser lens gaps, the axial field strength is reduced by a factor of around 6.5 when the magnets are fully short. For the objective lens, it is not necessary to use shorting plates, since the axial field strength can be significantly changed by the use of a magnetic base plate. However, Figure 3 shows that for a free-standing column (with no base plate), the axial field can be reduced by a factor of two. These results demonstrate that the high resolution portable column should in principle be able to operate in a similar way to a conventional SEM, where the primary beam energy is typically varied from 1 to 30 keV.

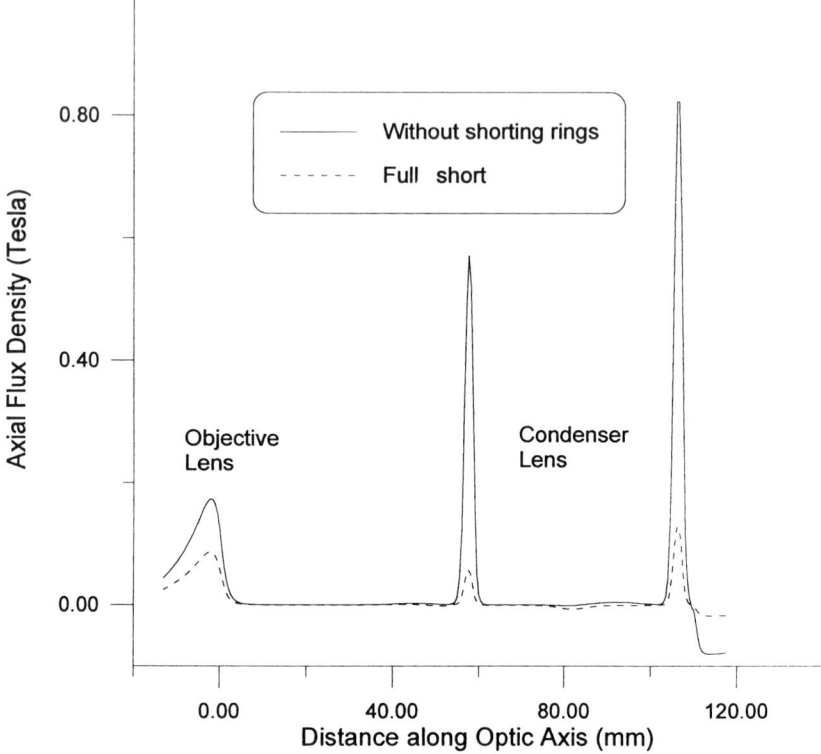

Figure 3. The effect of magnetic shorting plates on the axial field distribution

4. CONCLUSIONS

Experimental axial field measurements have been presented which confirm predictions made for the portable high resolution SEM design. A system of shorting plates has been designed which can be used to vary the axial magnetic field strength so that the column can be operated at different primary beam energies

REFERENCES

Kursheed A 1992 Kursheed Electron Optics Software, Department of Electrical Engineering, University of Edinburgh, Edinburgh, Scotland
Kursheed A, Phang J C and Thong J T L 1998 Scanning **20**, 87

Inst. Phys. Conf. Ser. No 164
Paper presented at Microsc. Semicond. Mater. Conf., Oxford, 22–25 March 1999
© 1999 IOP Publishing Ltd

SEM and TEM analysis of semiconducting SrTiO₃ ceramic devices

M Shiojiri, M Kawasaki[1], T Yoshioka[2], S Sato[3] and T Nomura[3]

Kyoto Institute of Technology, Kyoto 606-8585, Japan
[1]Electron Optics Division, JEOL Ltd., Tokyo 196-8558, Japan
[2]EO Applications Department, JEOL High-Tech Co Ltd., Tokyo 196-0022, Japan
[3]Materials Research Center, TDK Co., Narita 286-8588, Japan

ABSTRACT: The structure of boundary layer semiconducting SrTiO₃ ceramic condensers has been investigated using electron microscopy analytical techniques such as high-resolution transmission electron microscopy (HRTEM), field-emission SEM (FE-SEM), energy dispersive X-ray spectroscopy (EDS), and high angle annular dark field (HAADF) imaging in a FE-(scanning) transmission electron microscope (FE-(S)TEM). It has been found that the boundary layer consists of three layers; a thin grain boundary and a pair of so-called diffusion layers on both sides of the two facing grains. The analysis is focused on the distribution of Bi atoms that play an important role for the formation of these dielectric layers.

1. INTRODUCTION

SrTiO₃-based semiconducting ceramics are widely used in electric devices such as varistors and capacitors. The nano-structural analysis of these ceramics is very important to elucidate their characteristic electrical properties. We investigated SrTiO₃-based ceramic varistors by high-resolution transmission electron microscopy (HRTEM) (Koseki et al 1994) and cathodoluminescence scanning electron microscopy (CLSEM) (Kobayashi et al 1998; Shiojiri et al 1998). HRTEM revealed the structure of grain boundaries in the ceramics at the atomic scale. CL images, due to emission from the transition between the oxygen vacancy level and the valence band, revealed that each grain of the ceramics, particularly near varistor surface, has a highly conducting inner zone containing oxygen vacancies and an insulator boundary layer without oxygen vacancies. We have recently performed TEM and energy dispersive X-ray spectroscopy (EDS) analysis of very thin dielectric grain boundary layers and semiconducting grains inside a SrTiO₃ semiconducting ceramic condenser (Hitomi et al 1998).

In this paper, the boundary layer in a SrTiO₃ condenser is analysed using SEM and back scattering electron SEM (BE-SEM) in a field emission SEM instrument (FE-SEM) and EDS, and high angle annular dark field scanning transmission electron microscopy (HAADF-STEM) in an FE-(S)TEM instrument.

2. EXPERIMENTAL

Capacitors used in the present experiment are boundary layer (BL) semiconducting ceramic condensers of $(Sr_{0.94}Ba_{0.01}Ca_{0.05})_{0.99}TiO_3$, doped with Nb_2O_5 (~0.3 wt. %) and Y_2O_3 (~0.3 wt. %), annexed with SiO_2 (~0.2 wt. %). They were shaped into discs (9.0 mm diameter, 0.3 mm thick), after calcining (~1200 C, 2 h) a mixture of the raw oxide powders. Sintering was carried out in the reducing ambient (~1400 C, 2 h) of an H_2/N_2 atmosphere at an oxygen partial pressure of $P_{O_2}=10^{-8}$ Pa, followed by a reoxidizing process (~1100°C, 4 h) in the air. However, before this, the surfaces of the ceramics were painted with Bi, Pb and Cu compounds, which diffused into the ceramics during the reoxidizing process.

Specimens for TEM, EDS and STEM were prepared by the ion-thinning method, and SEM and BE-SEM observations were performed with a JEM-6330F scanning electron microscope. HRTEM, HAADF-STEM and EDS were carried out in a JEM-2010F FE-(S)TEM microscope equipped with a HAADF detector that collects scattered electrons in an angular range of 50-110 mrad (James et al 1998).

3. RESULTS AND DISCUSSION

The BL semiconducting condenser has the following electrical properties: capacitance of $C=7.6\times10^4$ pF, a loss tangent of $\tan\delta=2.5\times10^{-3}$, a relative dielectric constant of $\varepsilon_s=4.9\times10^4$, a dielectric breakdown voltage of $V_b=280$ V, and an insulation resistance of $R_i=6\times10^9\Omega$.

Figure 1 shows a TEM image of the condenser. Grains are seen to be surrounded by boundary layers as thick as ~100 nm. Each boundary layer consists of three layers that are a thin grain boundary and a pair of so-called diffusion layers on both sides of the two facing grains. It is often observed that the grain boundary is in the shape of a series of chained-triangles and that the boundary layers meander around the grain boundary. The diffusion layers exhibit dark contrast in the TEM image, which can be ascribed to Bi. Figure 2 shows an HAADF image of an area containing a grain boundary and EDS concentration profiles of Bi, Sr and Ti on the line indicated in the HAADF image. HAADF image contrast, or Z-contrast, should give intensity proportional to atomic number Z (Pennycook and Jesson 1991), so that the distribution of Bi in the ceramics is very explicitly indicated in Fig. 2, where Bi is enriched in the white contrast region. EDS analysis was carried out using a 0.5 nm

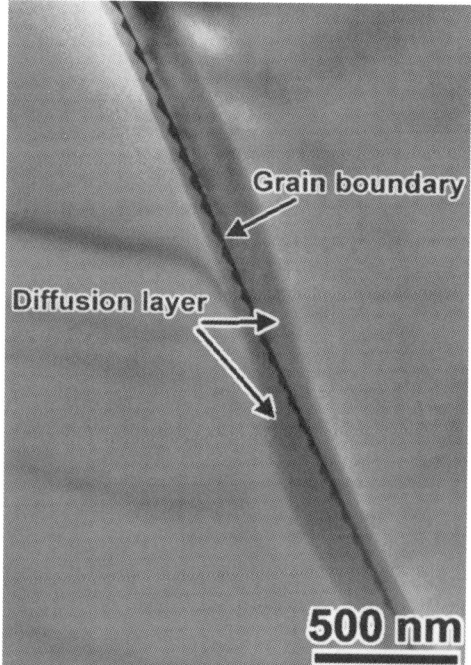

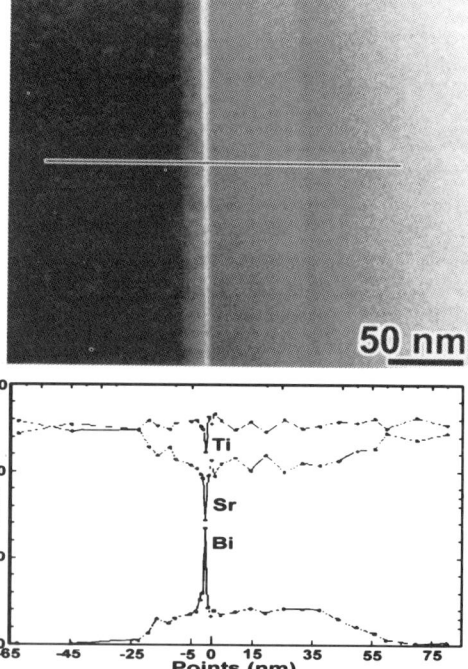

Fig. 1 Low magnification TEM image of SrTiO₃ BL semiconducting condenser.

Fig. 2 HAADF image of the SrTiO₃ condenser and EDS concentration profiles of Bi, Sr and Ti on a line in the HAADF image.

beam probe. For the concentration calculation, the *k*-factors of Bi-M and Sr-L to Ti-K were determined to be 0.5 and 0.9 using reference crystals with known stoichiometry. The standard deviations σ of the errors in these measurements are ± 0.7, ± 1.8, and ± 1.8 for Bi-M, Sr-L, and Ti-K, respectively. The EDS analysis confirmed that Bi exists at the grain boundary and in the diffusion layers. The Ti concentration is almost constant in the grains while Bi and Sr have inverse concentrations with respect to each other. This indicates that Bi atoms replace Sr atoms in the diffusion layers.

The ceramics sintered in the reducing ambient might contain O vacancies in the grains. During the reoxidizing process, the metal atoms, especially Bi, move from the ceramic surface by diffusion along the grain boundaries, together with O atoms. The metal atoms migrate deeply into the ceramics, but hardly penetrate the inside of the grains so that they replace Sr atoms in the close vicinity of the boundary to make the diffusion layers. The O atoms, which have an affinity for the Bi atoms, occupy exclusively O vacancy sites around these Bi atoms. This is a reason why a thin dielectric boundary layer (poor in O vacancies) is formed around each grain in the ceramics. From more EDS measurements, it was found that the other compound elements such as Pb and Cu were located at the triangles between the grains and at the triple points or the nodes of grain boundaries, but they could not be detected in the grains themselves (including the diffusion layers). Therefore, Bi is the only detected element to contribute to this diffusion process. The BL semiconducting ceramic condenser is integrated from many capacitor elements, which are constructed with a very thin (~100 nm) dielectric boundary layer and the semiconducting inner region (rich in O vacancies) as electrodes and, consequently, the device has a high capacitance.

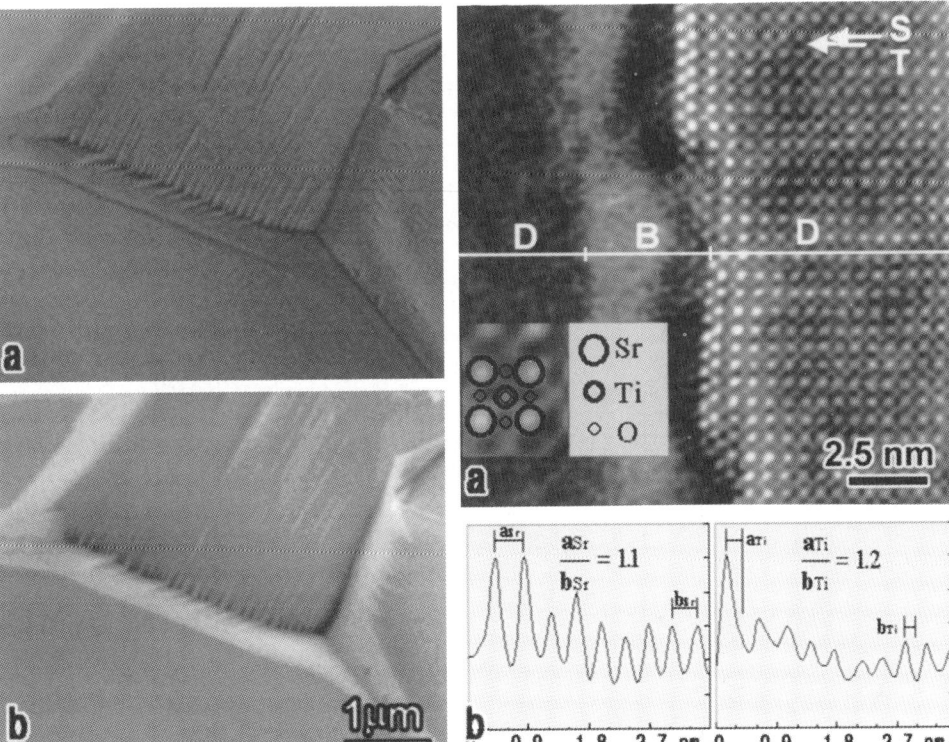

Fig. 3 (a) Secondary electron SEM image of a fractured surface of the condenser. (b) The corresponding back scattered electron image.

Fig. 4 (a) HAADF image of a grain boundary in the condenser. (b) Image contrast along arrows S and T in (a).

Figures 3a and 3b show a secondary electron SEM image and the corresponding BE SEM image of a fractured surface of the condenser, respectively. The characteristic strips appear on the surfaces of the grains and on the grain boundary layers. Also several triangles (strictly prisms) are seen in white contrast regions: these correspond to the Bi enriched grain boundaries which, in the BE image, give a higher intensity due to the presence of the high atomic number element. The triangles correspond to those seen in Fig. 1. During the reoxidizing process, the metallic (Bi, Cu and Pb) compounds coat and erode the surface of the sintered ceramics and then penetrate the inside through grain boundaries. The stripes and prisms seen in Fig. 3 can be regarded as etching lines and/or piles of the compounds along particular directions of the $SrTiO_3$ crystal. SEM observations thus revealed the morphological structure of the grain boundary in the ceramic condenser.

HAADF-STEM at the atomic level was used to observe a grain boundary area consisting of the diffusion layers D and the boundary layer B, shown in Fig. 4a. HAADF-STEM can give an image showing atom assignments by an intensity proportional to the atomic number (Pennycook and Jesson 1991). The right grain is orientated on [001] projection. Then, bright spots seen in the right diffusion layer should correspond to heavier atomic columns including Bi. The columns of the $SrTiO_3$ corresponding to the spots are indicated by the inset in Fig. 4a. The contrast of atomic columns along arrows S and T, which are on Sr-O and Ti/O-O planes parallel to the (200) plane, is shown in Fig. 4b. The distance was measured from the edge of the grain. The first two atomic columns at the edge are brighter than the other columns. This indicates that the outermost atomic layer is always composed of Ti columns and the second layer is composed of Sr columns, and that Bi occupies Ti and Sr sites in the first and second layers. These layers are considered to have a crystal structure similar to but not the same as that of $SrTiO_3$, having a certain concentration of Bi, Ti and Sr since the lattice distance becomes extended by ~10 % to the neighbouring lattice, as shown in Fig. 4b. In conclusion, during the reoxidizing process Bi atoms may first occupy most of Ti and Sr sites on the outermost atomic layer (erosion), and then they diffuse successively into the grain by replacing Sr atoms (diffusion). Ti atoms are not preferentially replaced inside the grains, as described above (in Fig. 2), as their charge is 4+.

REFERENCES

Hitomi A, Nomura T, Kawasaki M, Isshiki T, Nishio K and Shiojiri M 1998 J. Electron Micros. **47**, 603

James E M, Browning N D, Nicholls, A W, Kawasaki M, Xin Y and Stemmer S 1998 J. Electron Microsc. **47**, 561

Kobayashi Y, Sato S, Hitomi A, Isshiki T, Saijo H, Nomura T and Shiojiri M 1998 J. Electron Micros. **47**, 101

Koseki K, Nakano Y, Nomura T, Isshiki T, Nishio K, Saijo H and Shiojiri M 1994 Phys. Stat. Sol. (a) **143**, 245

Pennycook S J and Jesson D E 1991 Ultramicrosc. **37**, 14

Shiojiri M, Isshiki T, Saijo H, Sato S and Hitomi A 1998 Electron Technol. **31**, 170

Inst. Phys. Conf. Ser. No 164
Paper presented at Microsc. Semicond. Mater. Conf., Oxford, 22–25 March 1999

FEG-SEM imaging of semiconductor dopant contrast

S L Elliott, R F Broom, C J Humphreys, E J Thrush*, L Considine*, D B Thomson[+] and W B de Boer[#]

Department of Materials Science and Metallurgy, Cambridge University, Cambridge, CB2 3QZ, UK
* Thomas Swan and Co Ltd, Unit 1C, Button End, Harston, Cambridge, CB2 5NX, UK
+ Department of Materials Science and Engineering, NC State University, NC 27695-7907, USA
# Philips Research Laboratories, Eindhoven, The Netherlands

ABSTRACT: SEM SE contrast from doped GaN materials is reported in this work. Dopant contrast as a function of SEM working distance and tilt has also been investigated. Results given here show that a detailed understanding of the physical dopant contrast mechanism is far from being complete.

1. INTRODUCTION

The use of a field emission gun scanning electron microscope (FEG-SEM) as a dopant profiling tool is a relatively new technique, but one with tremendous potential. Doped semiconductors show secondary electron (SE) contrast, which is dependent on the dopant type and concentration. Requiring only minimal sample preparation (i.e. cleaving or polishing) and an in-lens SE detector, two-dimensional dopant profiles can be obtained quickly and easily with high spatial resolution. A range of Si, GaAs and InP materials and devices have been imaged (Sealy et al 1995, 1997a; Perovic et al 1995; Castell et al 1995; Turan et al 1996; Venables and Maher 1995, 1996; Venables et al 1998).

The dopant contrast mechanism has been explained in terms of surface band-bending and local electric fields, promoting or suppressing SE emission from p-type and n-type material respectively (Perovic et al 1995). However the development of this technique is still in its infancy, hampered by variable experimental data and possibly an incomplete model of the physical mechanism of this effect.

This paper reports dopant profiling in GaN for the first time. In addition, dopant contrast as a function of SEM working distance and tilt were investigated.

2. EXPERIMENTAL

A GaN device and a GaN test structure were imaged in this work. The device is a GaN/InGaN multiple quantum well (MQW) light emitting diode (LED) fabricated by Thomas Swan and Co. using metal-organic chemical vapour deposition (MOCVD). The structure has 0.2 μm of undoped GaN on a sapphire substrate, followed by 1.9 μm of n+ GaN, an InGaN/MQW undoped active region and 0.3 μm of p+ GaN. As determined by SIMS analysis, the p+ region is doped with Mg to a concentration of $3E19$ cm^{-3}; the n+ region is doped with Si to a concentration of $8E18$ cm^{-3}.

The n-type GaN test structure was fabricated by growth on a SiC wafer using MOCVD. The sample consists of an AlN buffer layer (100 nm), followed by 0.5 μm of undoped GaN and on top of this are five alternating layers (100 nm) of Si-doped GaN and undoped GaN. SIMS analysis shows the doping of the layers to be approximately $5E17$, $1E18$, $5E18$, $1E19$ and $3E19$ cm^{-3} (from the substrate).

The material used to study the dopant contrast dependency on SEM parameters was a B-doped Si test structure, supplied by Philips. The structure consists of five B-doped layers separated by undoped Si, on an n+ substrate (6-10 mΩcm Sb). The layer concentrations, as determined by SIMS analysis are $5E18$, $4E17$, $1E18$, $5E18$ and $3E19$ cm^{-3} (from the substrate). The structure was grown by

CVD in a commercially available epitaxy reactor: ASM's Epsilon One. The growth temperature was 850°C, and the Si growth rate was estimated at 0.15 μ/min.

The FEG-SEM used in this study was a JEOL 6340F, equipped with a semi-in-lens SE detector. It was operated at an accelerating voltage of 1.0 kV and a working distance between 4.0 mm and 7.0 mm. Materials were cleaved and immediately transferred to the SEM. Digital bitmap images were recorded, and line profiles produced using IDL.

3. SEM DOPANT PROFILING OF GaN

SEM dopant contrast has been observed in GaN. An SEM SE image of the GaN device is shown in Figure 1(a). The p+ region is the bright layer on the right side of the image; the n+ region is the central, wider layer which appears at a lower SE intensity. The dark area to the left is the sapphire substrate. The MQW region is not visible on the SEM image. The rough topographical features that can be seen on the device are a result of the cleave.

The p+ region is clearly brighter than the n+ region. A line profile taken across the device is shown in Figure 1(b). The origin of the extremely bright white line at the interface between the sapphire substrate and the GaN is not clear. TEM analysis on this device indicates a highly defective region along this interface. Possibly these are charged defects. (Tricker, unpublished)

An SE image of the GaN test structure is shown in Figure 2(a). The four higher doped layers are visible. A corresponding line profile is shown in Figure 2(b). Again rough topographical features are present.

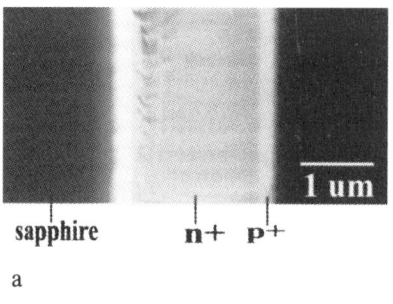
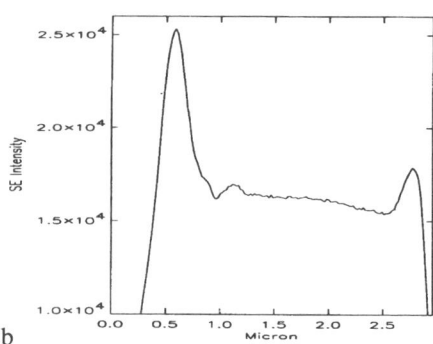

Figure 1 (a) SEM image of GaN device and (b) corresponding line profile.

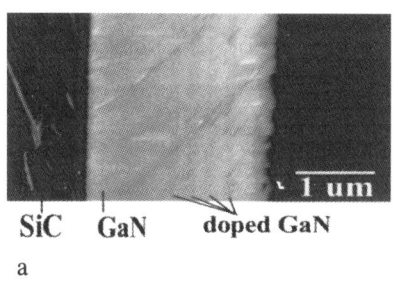
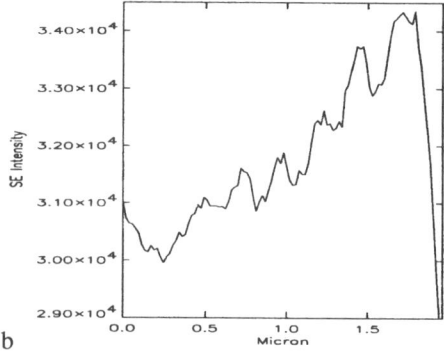

Figure 2 (a) SEM image of GaN test structure and (b) corresponding line profile.

4. DOPANT CONTRAST RELATIVE TO WORKING DISTANCE AND TILT

It has been reported that the effect of working distance on dopant contrast is substantial when an in-lens SE detector is used in comparison to a standard Everhart-Thornley detector (Sealy 1997b), and that dopant contrast is enhanced at specific working distances (Sealy 1997b) and tilt angles (Perovic et al 1995). As SEs provide the dopant contrast, it would be expected that maximum signal-to-noise is obtained when the SE yield is highest and the maximum number of electrons are collected. On the 6340F JEOL this was determined to be the smallest working distance possible, irrespective of tilt, as shown in Figures 3(a) and 3(b). For these measurements, a voltmeter was connected to the output of the SE detector and monitored as the working distance was varied between 2.0-10 mm and the tilt angle between −5.0 and 35 degrees. Au was sputtered onto the B-doped Si test structure to produce a homogeneous surface, with a SE yield that stays relatively constant for the tilt angles considered in this work (Reimer 1998).

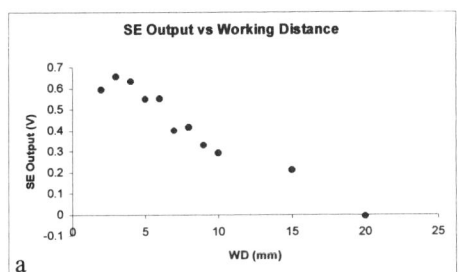 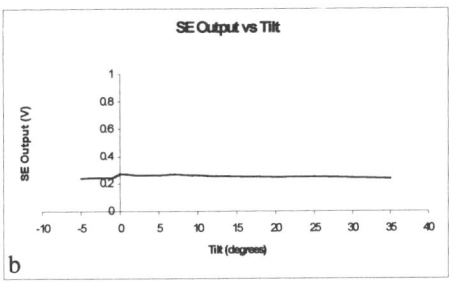

Figure 3 (a) graph of SE Output vs Working Distance and (b) SE Output vs Tilt measured on Au coated Si sample.

Doping contrast was determined according to the following relation

$$C = [(I_d - I_{ref}) - (I_{sub} - I_{ref})] / (I_{sub} - I_{ref}) \qquad (1)$$

where I_d is the intensity of the intentionally doped layer, I_{sub} is the intensity of the substrate and I_{ref} is the reference intensity obtained when the beam-blanking facility is switched on. On Si it was measured as a function of working distance and tilt. As the working distance was increased, dopant contrast scarcely changed, as shown in Figure 4(a).

As the sample was tilted, variable data were obtained. Two separate sets of measurements are shown in Figure 4(b). Series 1 and 2 were results from the same specimen collected on different days. Every effort was made to reproduce the same specimen orientation and tilt and, as can be seen, the results are very different. No trend is apparent in the data. Clearly dopant contrast is very sensitive to the tilt and orientation of the specimen.

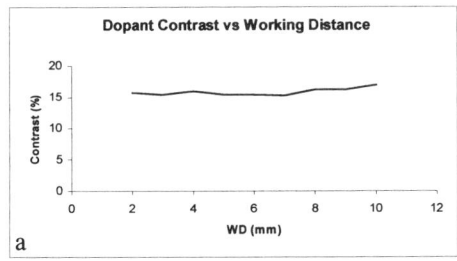 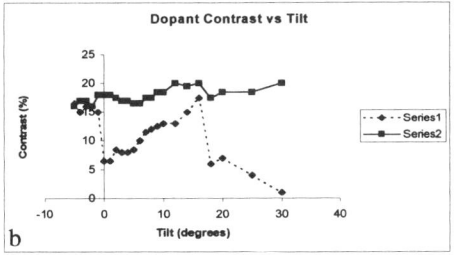

Figure 4 (a) graph of Dopant Contrast vs Working Distance and (b) Dopant Contrast vs Tilt measured on B-Doped Si sample.

5. DISCUSSION AND CONCLUSIONS

SEM dopant profiling of GaN is presented for the first time. p+, n+ and n-type materials have been successfully imaged with secondary electrons. Because of the substrate, it is difficult to obtain a smooth cleaved surface and the consequent roughness disturbs the SE contrast. Other means for obtaining a smooth clean surface should be investigated.

We describe preliminary results of the dependencies of dopant contrast on working distance and tilt. On a JEOL 6340F, it seems that dopant contrast is relatively constant as the working distance is increased. As the sample is tilted, dopant contrast changes irregularly, making quantitative dopant profiling difficult.

Previous studies involving working distance have shown that contrast increases when the working distance is increased. It has been suggested that this could indicate a directional dependency of dopant contrast (Sealy 1997b). Other studies have also indicated that dopant contrast is enhanced when the sample is tilted ~5.0 degrees from the normal orientation; however, no explanation for this is provided (Perovic et al 1995).

Although results in this study suggest an opposite trend in contrast with changing working distance, these results combined with those of tilt, do indicate a directional dependency of dopant contrast, resulting from electron channelling and other effects.

In conclusion, the results given here show that a detailed understanding is far from being complete. In particular, the angular behaviour suggests that the surface physics and crystallography play a crucial role.

REFERENCES

Castell M R, Perovic D D, Howie A, Ritchie D A, Lavoie C and Tiedje T 1995 Inst. Phys. Conf. Ser. **146**, 281

Perovic D D, Castell M R, Howie A, Lavoie C, Tiedje T and Cole J S W 1995 Ultramicroscopy **58** 104

Reimer L 1998 Scanning Electron Microscopy (Berlin: Springer) p157

Sealy C P, Castell M R, Wikinson A J and Wilshaw P R 1995 Inst. Phys. Conf. Ser. **146**, 609

Sealy C P, Castell M R and Reynolds C L 1997a Inst. Phys. Conf. Ser. **157**, 561

Sealy C P 1997b PhD Thesis, University of Oxford

Tricker D M unpublished

Turan R, Perovic D D and Houghton D C 1996 Appl. Phys. Lett. **69**, 1593

Venables D and Maher D M 1996 J. Vac. Sci. Technol. B **14**, 421

Venables D and Maher D M 1995 Inst. Phys. Conf. Ser. **146**, 605-608

Venables D, Jain H and Collins D C 1998 J. Vac. Sci. Technol. B **16**, 362

Inst. Phys. Conf. Ser. No 164
Paper presented at Microsc. Semicond. Mater. Conf., Oxford, 22–25 March 1999
© *1999 IOP Publishing Ltd*

Change of BSE spectrum on the edge of a surface step

L S Kokhanchik, E B Yakimov and S I Zaitsev

Institute of Microelectronics Technology, Russian Academy of Science, Chernogolovka, 142432 Russia

ABSTRACT: The Backscattering Electron (BSE) signal profile and BSE spectrum from a single surface step have been simulated by the Monte-Carlo method. The simulation has shown that the BSE spectrum is changed on the upper edge of a surface step. Results of BSE profile measurements on silicon samples with surface steps produced by reactive ion etching demonstrate a strong dependence on the detection mode used that confirms the results of the simulation. A qualitative explanation of BSE spectrum change on the upper edge of a surface step is proposed.

1. INTRODUCTION

The BSE mode seems to be very promising for the Scanning Electron Microscopy (SEM) metrology in the nanometric range (Postek 1994), and therefore the problem of correct simulation of BSE signal from a surface profile is of great importance. Also, a change of BSE coefficient due to a surface relief leads to a corresponding change in the absorption energy, which in turn should effect the signal formation in the Electron Beam Induced Current (EBIC), Cathodoluminescence (CL), X-Ray Microanalysis and other SEM modes. This effect could be rather complex, since a surface relief could change the BSE spectrum. The last point has not been discussed up to now although the effect of surface profile on the BSE energy spectrum has been already demonstrated (Dreomova et al, 1996, Zaitsev and Yakimov, 1997).

In the present paper the BSE signal profile and BSE spectrum from a surface with a single step was simulated by the Monte-Carlo method. The simulation revealed a change in the BSE spectrum when the e-beam was located on the upper edge of the step. A qualitative explanation of this result is proposed. BSE profiles measured on silicon with trenches produced by reactive ion etching were found to depend on the detection mode used that could be easily explained by the change in the BSE spectrum associated with the step. Thus these experimental results could be considered as an indirect confirmation of BSE spectrum modification associated with the surface relief.

2. BSE SIGNAL SIMULATION

The BSE signal profile for a single step on the silicon surface and the BSE energy spectrum in characteristic points of the surface with the step were simulated by the Monte-Carlo method. The program used was based on that developed by Reimer and Stelter (1986). To decrease the calculation time some modifications similar to that used by Firsova et al (1991) were made. As a result, it takes about 8 hours to simulate the BSE profile consisting of 50 points on a Pentium PC using 10^6 trajectories. The total energy range was divided into 20 boxes, the take-off angle range from 0 to $\pi/2$ was divided into 10 boxes, and the calculated BSE signal was accumulated taking into account the BSE energy and take-off angle. This allows the BSE energy and take-off angle dependence of BSE yield to be simulated at any point of surface relief.

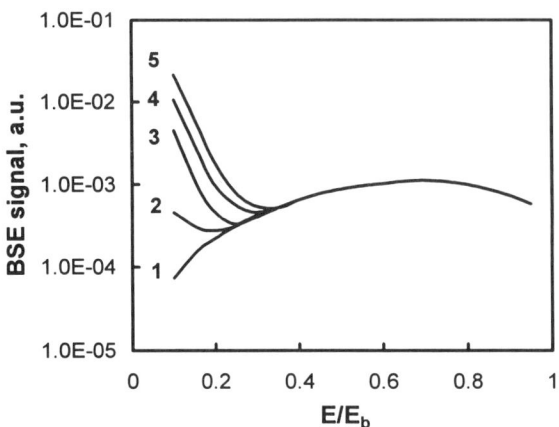

The simulated BSE profiles have a well-known form, increasing on the upper edge of the step and also decreasing on its bottom. In the next section they will be compared with experimental ones and will be discussed in more detail. The simulated BSE spectrum for the step bottom is practically the same as that obtained for the flat surface far from the step, i.e. has a form of wide peak with a maximum at about 0.7 E_b, where E_b is the beam energy (Fig.1, Curve 1). But on the upper edge of the step the spectrum changes. The main difference consists of an increase of BSE emission in the low energy part of the spectrum. This low energy electron emission was found to increase with the step height d for d smaller than about 3 R, where R is the electron range (Curves 2-5). For the higher step heights the spectrum is practically independent of d.

Fig. 1. BSE spectra simulated by the Monte-Carlo method for the upper edge of a surface step with a height of 0 (1), 0.05 (2), 0.36 (3), 0.5 (4) and 3.5 R (5).

3. EXPERIMENTAL RESULTS

Silicon samples with periodic trenches produced on the surface by reactive ion etching have been investigated in the BSE mode at e-beam energies E_b varied from 3 to 30 keV. The distance between trench walls in the structures studied was about 10 μm, which allowed us to consider them as individual steps. Steps with a height d varying from a few tenths to a several microns were investigated. As our measurements and simulations have shown, the BSE signal profile is mainly determined by the d/R relation. It is more convenient to change this relation by varying R (i.e. the primary electron energy) than to compare the samples with different step heights, and therefore the experimental results obtained on steps with a height d of 3.4 μm are presented. The investigations were carried out on a JSM-840 SEM. The BSE signal was detected by the standard Everhart-Thornley detector (ETD) with 0V or -300V on the collector, by the standard BSE semiconductor detector and, in addition, it was calculated as a difference between the beam and absorption currents. In the last case a voltage of about -50 V was applied between the sample and a grid located above it to suppress the Secondary Electron (SE) emission.

The BSE signal profiles measured on the same step using different detection mode at E_b= 25 keV normalized to the BSE signal far from the step are presented in Fig. 2. The upper curve shows schematically the corresponding step position. At other beam energies the relation between the profiles obtained using the different detection modes was similar but at small energies some features associated with a nonideality of step form were more

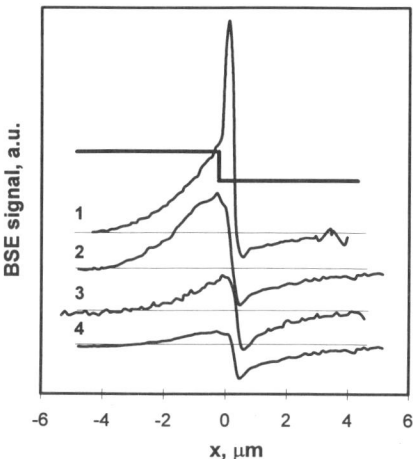

Fig. 2. BSE profiles for the surface step with a height of 3.4 μm obtained by the ETD with a collected voltage 0 (1) and -300 V (3), from the absorption current measurement (2) and by the semiconductor BSE detector (4). E_b= 25 keV.

pronounced. It is seen that a decrease in the BSE signal on the right side of the step associated with the step bottom is approximately the same for all profiles. But an increase of signal on the step upper edge depends essentially on the detection mode. When a voltage on the ETD collector was switched off (Curve 1) a narrow SE peak is superimposed on the BSE profile. A suppression of SE collection (Curve 3) leads not only to a pronounced decrease of this SE peak but also to a decrease of BSE signal maximum. It is seen that the BSE signal maximum-to-minimum ratio in accordance with the results of previous studies (Dreomova et al, 1996, Zaitsev and Yakimov, 1997) is different for all detection modes used. In Fig. 3 some of measured profiles together with the simulated one are presented. The profiles obtained by the absorption current measurements and that presented in Fig.1 (Curve 1) after subtraction of SE peak are very similar to the simulated one while on other profiles the maximum-to-minimum ratio of BSE signal is essentially smaller. This ratio is smallest in the case where the semiconductor BSE detector is used for the BSE profile measurements. Sometimes, especially for low E_b, a maximum in the BSE profile is practically absent in this detection mode.

4. DISCUSSION

First of all let us discuss a change in the simulated BSE spectrum due to a surface

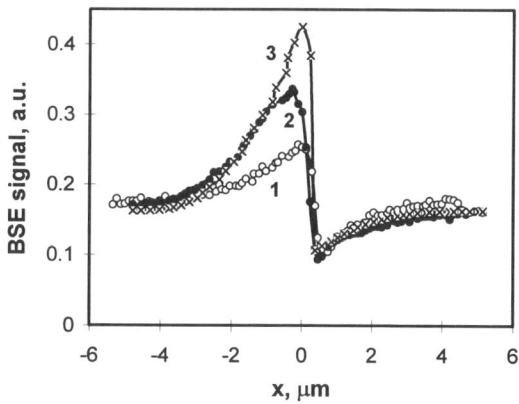

Fig. 3. BSE profiles for the surface step with a height of 3.4 μm obtained by the ETD with a collected voltage -300 V (1), from the absorption current measurement (2) and simulated by the Monte-Carlo method (3). E_b= 25 keV.

relief. As seen in Fig.4 a surface step could be described as removing the rectangular part of material OABC (Fig. 4). When e-beam is located on the left side of the step, the difference in the BSE signal between the flat surface and the step can be estimated as a number of electrons retarded in this region (marked by black circles). If a thickness of removed region (step height) is small, the energy of electrons stopped in the OABC region also should be small. From this consideration it is easy to understand why the increase in the BSE signal on the step upper edge is dominated by the low energy BSE. Their

734

number and energy should decrease with decreasing the step height in a close correlation with a simulated spectra presented in Fig. 1.

The change of the BSE spectrum obtained by the simulation allows to explain also the observed dependence of measured BSE profiles associated with the surface step on the detection mode. Indeed, only in the case of absorption current measurements was the real BSE coefficient measured. It should be noted that under these measurements all BSE with energies higher than 50 eV were detected while the simulation was carried out for the BSE with an energy higher than 1 keV. This could be a reason for a small difference between the simulated profile and that obtained from the absorption current measurements. Other types of detection modes are more or less energy sensitive.

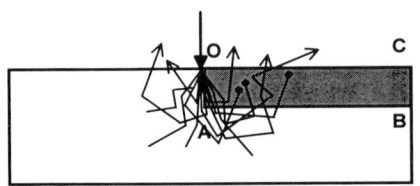

Fig. 4. A scheme illustrating a formation of BSE signal on the upper edge of a step.

When -300 V was applied to the collector of ETD, the BSE signal associated with the low energy electrons was suppressed due to a deflection of their trajectories that leads to a decrease of maximum-to-minimum ratio on the upper edge of the step (Fig. 3, Curve 1). When the semiconductor detector was used for the BSE detection the signal was proportional not to the BSE coefficient but to the backscattering energy. Thus this detector has a low sensitivity to low energy electrons and it is the main reason for the observed decrease in the maximum–to–minimum ratio in this detection mode. From the above considerations it follows that the difference in the BSE profiles for the surface step obtained using different detection modes correlates well with the simulated change in the BSE spectrum on the surface relief and could be considered as an indirect confirmation of such change.

Thus the surface profile can indeed essentially change the BSE spectrum. This change is determined by the relation of d/R, and increases with an increase of this relation with a saturation at large enough step heights. This effect should be taken into account under the simulation of not only the BSE signals but also other SEM signals if the surface of object under study is not flat. The results presented show that images obtained by the detection of BSE coefficient and of backscattering energy can differ significantly. To correct the surface relief effect on the images obtained in the EBIC or CL modes the detection of the backscattering energy, e.g. by the semiconductor BSE detector should be used.

ACKNOWLEDGMENTS

This work was partially supported by the Russian National Scientific and Technology Program "Physics of Solid State Nanostructures" (Grant 98-3007)

REFERENCES

Dreomova N N, Rau E I and Yakimov E B 1996 Izvestiya RAN SSSR, ser. Fiz. **60**, 101 (In Russian)

Zaitsev S I and Yakimov E B 1997 Izvestiya RAN SSSR, ser. Fiz. **61**, 1954 (In Russian)

Firsova A A, Reimer L, Ushakov N Gand Zaitsev S I 1991 Scanning **13**, 363

Postek M T 1994 J. Res. Nat. Inst. Stand. Technol. **99**, 641

Reimer L and Stelter D 1986 Scanning **8**, 257

Inst. Phys. Conf. Ser. No 164
Paper presented at Microsc. Semicond. Mater. Conf., Oxford, 22–25 March 1999
© 1999 IOP Publishing Ltd

Contactless characterization of semiconductor structures by the Surface Electron Beam Induced Voltage method

E I Rau, Zhu Shiqu* and E B Yakimov

Institute of Microelectronics Technology Russian Academy of Science, Chernogolovka, 142432 Russia
*Moscow State University, Vorob'evy gory, Moscow, 119899 Russia

ABSTRACT: An application of the Surface Electron Beam Induced Voltage (SEBIV) method for the contactless measurement of minority carrier diffusion length and lifetime in semiconductor crystals is discussed. The signal formation in the SEBIV mode is analyzed. Experimental results obtained on silicon samples are presented. The possibilities and limitations of this technique have been discussed in comparison with the conventional EBIC. It is shown that in many cases this technique allows to obtain data similar to those provided by EBIC but without any special sample preparation.

1. INTRODUCTION

A development of scanning electron microscopy (SEM) methods for studying a spatial distribution of electrical properties is very important for the characterization of semiconductor materials and structures. For these purposes the Electron Beam Induced Current (EBIC) mode is now widely used (Leamy 1982, Yakimov 1992), but for many applications it is very useful to develop contactless nondestructive methods. The surface electron beam induced voltage (SEBIV) method proposed by Gostev et al (1987a, 1987b) could be considered as one such method. It is the SEM method analogous to the well-known surface photovoltage (SPV) technique (Lagowski et al 1992) and it can be used for the nondestructive characterization of semiconductor structures and for revealing electrically active defects in semiconductor materials. Different applications of the SEBIV method for studying a distribution of electrical properties in semiconductors and semiconductor structures were demonstrated by Aristov et al (1990) and Rau et al (1997, 1998). It was shown that this method can reveal nondestructively p-n junctions and highly doped regions in semiconductors, carry out a contactless inspection of semiconductor structures, and reveal individual electrically active defects, such as misfit dislocations and dislocations introduced by plastic deformation. As shown by Rau et al (1997, 1998), the SEBIV image of dislocations is very similar to that observed by EBIC. But a question concerning a possibility of contactless measurements of minority carrier lifetime and diffusion length by this method is still open.

In the present paper an application of SEBIV for the contactless measurements of local electrical properties in semiconductor materials is discussed. The expressions describing the SEBIV signal formation are derived and analyzed. The results demonstrating the possibility of silicon parameter measurements are presented.

2. EXPERIMENTAL

The scheme of the setup used for the SEBIV investigations is presented in Fig. 1. In this method a grounded sample S is mounted on the stage of the SEM. The SEM is equipped with a beam blanking system controlled by a pulse generator G. A ring-like metal electrode D positioned above the sample close to the surface symmetrically to the electron beam (e-beam) is used as a signal detector. The signal induced on this capacitively coupled electrode due to a local change of surface potential induced by the e-beam excitation is transmitted to the charge sensitive preamplifier PA and then after the lock-in amplifier (LA) is displayed on the SEM

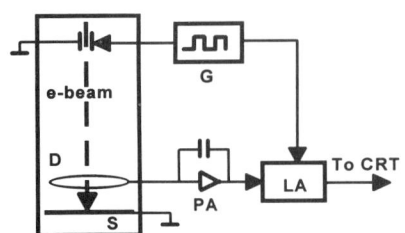

Fig. 1. Block diagram of setup for the SEBIV measurements.

CRT. It should be mentioned that the SEBIV image of inhomogeneous samples could be obtained without any blanking system. In this case the lock-in amplifier is not used.

3. SIGNAL FORMATION IN THE SEBIV MODE

To evaluate the charge induced on the capacitively coupled detector let us consider an n-type semiconductor with a flat surface. Similar to   EBIC the signal formation in the SEBIV mode includes the generation of excess carriers by a focused e-beam, their diffusion and recombination. A depletion-type surface barrier plays a role of a collector for the minority carriers generated by the beam. The change of surface potential under the e-beam excitation due to   capture of minority carriers (holes) can be described as

$$\Delta\varphi(\mathbf{r}) = \frac{kT}{e}\ln[\frac{J_c(\mathbf{r})}{J_{p0}}+1] \qquad (1)$$

where e is the electron charge, J_c and J_{p0} are the hole current densities with and without the excitation, respectively, $\mathbf{r}$ is the distance from the beam position. For simplicity, let us assume that the collected current is spread over the region with the characteristic radius σ and its spatial distribution can be described as

$$J_c(\mathbf{r}) = \frac{I_c}{2\pi\sigma^2}\exp(-\mathbf{r}^2/\sigma^2) \qquad (2)$$

where $I_c = \int\limits_W^\infty g(z)\exp[(W-z)/L]dz$ is the total collected current, z is the depth, $g(z)$ is the depth dependent generation function, W is the depletion region width, L is the minority carrier diffusion length. The charge induced on the ring electrode is proportional to the total change of the surface potential, i.e.

$$\Delta q = \frac{CkT}{e}\int\limits_0^\infty \ln(\frac{J_c}{J_{e0}}+1)2\pi\mathbf{r}d\mathbf{r} = \frac{CkT}{e}\int\limits_0^\infty \ln[\frac{I_c}{2\pi\sigma^2 J_{e0}}\exp(-\mathbf{r}^2/\sigma^2)+1]2\pi\mathbf{r}d \qquad (3)$$

where C is the coefficient proportional to the detector-sample capacitance.

As seen from (3) the SEBIV signal dependence on the diffusion length, leakage current, doping level and width of collected current spreading is rather complex. But under the assumption (2) Δq can be calculated as a function of I_c (Fig. 2). The procedures for I_c reconstruction using this calibration curve can be developed and then the methods similar to those proposed for the EBIC (see e.g., Yakimov 1992) can be applied to obtain the L value.

4. RESULTS AND DISCUSSION

As shown above, in a common case the procedure for the diffusion length reconstruction from the SEBIV data could be rather complex. Therefore in the present work possibilities of the SEBIV are demonstrated using some special structures which allow to simplify this procedure. Samples with small Schottky barriers or p-n junctions could be considered as an example of such structures. In this case $\Delta\varphi$ is constant over the area and could be made smaller than (kT/e). Under these conditions $\Delta\varphi$ is proportional to I_c and the diffusion length can be obtained from the SEBIV measurements by the methods similar to those used in the EBIC. The examples of such measurements are presented below.

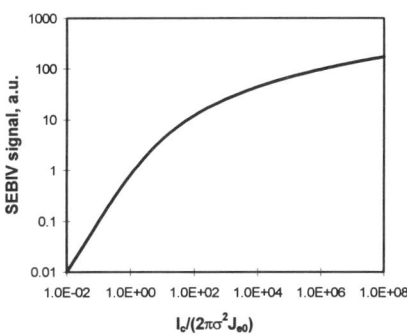

Fig.2. The dependence of Δq on I_c. calculated using the expression (3).

In Fig. 3 the Δq decay curves after the e-beam pulse excitation measured on two different Si samples with small p-n junctions are presented. It is seen that for both samples the relaxation is close to exponential with the characteristic times of 0.04 and 0.115 ms. Although it is not so straightforward to obtain the minority carrier lifetime from the I_c decay, it should be noted that the values obtained are close to those measured by the photocurrent decay in the same samples (the difference did not exceed 50%). Thus such method could be used at least for the contactless estimations of minority carrier lifetime.

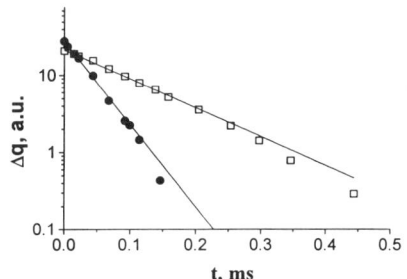

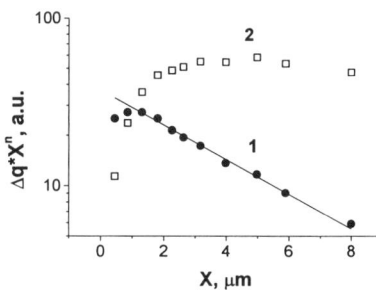

Fig. 3. Relaxation of Δq after the pulse excitation in two Si samples.

Fig. 4. Decay of $\Delta q*X^n$ as a function of $X,$ for $S \to 0$ (1) and $S \to \infty$ (2).

The decay of Δq as a function of the distance between the beam position and the Schottky barrier edge is presented in Fig. 4. In the EBIC the similar I_c decay is described as $I_c \sim \exp(-X/L) \cdot X^{-n}$, where $n = 1/2$ for the small surface recombination velocity ($S \to 0$) and $n = 3/2$ for $S \to \infty$ (Ioannou and Davidson 1979, Donolato 1985). The dependences $\Delta q*X^{1/2}$ and $\Delta q*X^{3/2}$ on X are presented in Fig.4 (Curve 1 and 2, respectively). It is seen that the assumption about the small S is more appropriate for the sample under study. Under such assumption the diffusion length of about 4 μm was obtained by fitting the Curve 1. It should be noted that   such structure measurements of **L** could be carried out also by studying the Δq dependence on the primary electron energy E_b similar to the well-known EBIC method. In this case the simplification of the problem of diffusion length (or I_c) reconstruction from the SEBIV data could be achieved also by defocusing the e-beam to a value of a few microns but the beam width should be the same for all beam energies used.

The results presented show that if the semiconductor structures under study contain collecting junctions or if a defocused e-beam can be used, the EBIC methods developed for the diffusion length measurements can be applied for the SEBIV practically without any modification. But contrary to the EBIC, the SEBIV does not need a contact preparation. When a focused e-beam is used the measurements of local semiconductor parameters are more complex, mainly due to the high excitation level. In this case the expression (3) should be used and unknown value of J_{p0} necessary for calculations could be determined using some additional measurements, e.g. by SEBIV measurements with different beam currents. In any case, first of all I_c should be obtained and then **L** could be calculated from the I_c dependence on E_b.

Thus the SEBIV can reveal the spatial variation of recombination properties in semiconductor structures, highly doped regions, p-n junctions and individual electrically active extended defects without any sample preparation. Moreover, this method allows contactless measurement of local semiconductor parameters, such as diffusion length. But as compared with the EBIC technique, usually this method needs a more complex calculation procedure to carry out such measurements.

ACKNOWLEDGEMENTS

This work was partially supported by the RFBR (Grant 98-02-16999) and by the Russian National Scientific and Technology Program "Physics of Solid State Nanostructures" (Grant 98-3007).

REFERENCES

Aristov V V, Kononchuk O V, Rau E I and Yakimov E B 1990 Microelectr. Engineer. **12**, 179

Donolato C 1985 Solid State Electr. **28**, 1143

Gostev A V, Kleinfeld Yu S, Rau E I and Surogina V A 1987a Poverkhnost No 5, 73 (In Russian)

Gostev A V, Kleinfeld Yu S, Rau E I and Spivak G V 1987b Mikroelectronika, **16**, 311 (In Russian)

Ioannou D E and Davidson S M 1979 J. Phys. D **12**, 1339

Lagowski J, Edelman P, Dexter M and Henley W 1992 Semicond. Sci. Technol. 7 A185

Leamy H J 1982 J. Appl. Phys. **53**, R51

Rau E I, Zhukov A N and Yakimov E B 1997 Inst. Phys. Conf. Ser. No 160, 75

Rau E I, Zhukov A N and Yakimov E B 1998 Solid State Phenomena **63-64** (Zurich: Scitec Publications Ltd) pp 327-332

Yakimov E 1992 Scanning Microsc **6**, 81

Author Index

Subject Index